Y. Chan
Location, Transport and Land-Use

Yupo Chan

Location, Transport and Land-Use

Modelling Spatial-Temporal Information

with 208 Figures and 113 Tables

 Springer

Professor Yupo Chan
University of Arkansas at Little Rock
Department Systems Engineering
Donaghey College
of Information Science
and Systems Engineering
South University 2801
Little Rock, 72204-1099
USA

E-mail: yxchan@ualr.edu

Shown on the cover is an image as acquired by the Spaceborne Imaging Radar-C/X-band Synthetic Aperture Radar (SIR-C/X-SAR) as part of NASA's Mission to Planet Earth. This radar image shows the massive urbanization of Los Angeles, California. Overlaid on top of the image is a graphical illustration of random or Poisson fields, which is a theoretical foundation governing spatial-temporal changes ranging from urban development to image processing.

Library of Congress Control Number: 2004111939

ISBN 3-540-21087-3 Springer Berlin Heidelberg New York

Springer is a part of Springer Science+Business Media GmbH
springeronline.com
© Springer-Verlag Berlin Heidelberg 2005
Printed in Germany

The use of general descriptive names, registered names, trademarks, etc. in this publication does not imply, even in the absence of a specific statement, that such names are exempt from the relevant protective laws and regulations and therefore free for general use.

Cover design: E. Kirchner, Heidelberg
Production: Almas Schimmel
Typesetting: Camera ready by Albert R. Anderson, Jr. and Susan Johnson Chan
Printing: Mercedes-Druck, Berlin
Binding: Stein + Lehmann, Berlin

Printed on acid-free paper   30/3141/as 5 4 3 2 1 0

To Susan,
for making this volume and
many things I treasure a reality...

PREFACE

1. Theme and focus

Few books are available to integrate the models for facilities siting, transportation, and land-use. Employing state-of-the-art quantitative-models and case-studies, this book would guide the siting of such facilities as transportation terminals, warehouses, nuclear power plants, military bases, landfills, emergency shelters, state parks, and industrial plants. The book also shows the use of statistical tools for forecasting and analyzing implications of land-use decisions. The idea is that land-use on a map is necessarily a consequence of individual, and often conflicting, siting decisions over time. Since facilities often develop to form a community, these decisions are interrelated spatially—i.e., they need to be accessible to one another via the transportation system. It is our thesis that a common methodological procedure exists to analyze all these spatial-temporal constructs.

While there are several monographs and texts on subjects related to this book's, this volume is unique in that it integrates existing practical and theoretical works on facility-location, transportation, and land-use. Instead of dealing with individual facility-location, transportation, or the resulting land-use pattern individually, it provides the underlying principles that are behind these types of models. Particularly of interest is the emphasis on counter-intuitive decisions that often escape our minds unless deliberate steps of analysis are taken. Oriented toward the fundamental principles of infrastructure management, the book transcends the traditional engineering and planning disciplines, where the main concerns are often exclusively either physical design, fiscal, socioeconomic or political considerations.

The analytical community has made significant progress in recent years in the basic building blocks of spatial analysis. Current models have captured accurately many of the bases of facility-siting decision-making—proximity to demand, competition among existing facilities, and the availability of utilities and other institutional supports. Throughout this text, accessibility (as afforded by transportation) and infrastructural support (as provided by utilities and sewers) are used as determinants of locational decisions. Equally covered are competitive, historical and other determinants that are not based on accessibility alone. In short, I try to survey the field and cover the subject comprehensively.

Simple examples and case studies are used throughout this volume. The examples serve to illustrate the fundamental principles and concepts while the case studies show how the principles can be applied in practice toward pluralistic decision-making. These case studies further assist in evaluating existing methodologies, showing successful and not so successful applications of these models. A unique feature of this book is that--unlike other treatments where case studies are separated out as appendages—the case studies are seamlessly blended into the methodological developments. The key to this approach is to be selective about which case studies to include.

Much of spatial analysis today is facilitated by the advances in remote-sensing imagery and geographic information systems. These include the use of electronic devices, which can make the collection and processing of data much more timely, organized and convenient. This in turn assists in the problem-solving process. I discuss how information can be stored so that it can be directly translated to a format for real-time decision-making. This means simple and transparent models that are data-base compatible, thus requiring minimal data manipulation in the solution process. This book gives the reader a comprehensive insight into the use of this "data mining" tool—identifying, assembling and utilizing the important information for problem solving. Perhaps most important, software and data-sets are available to support the analyses described here, including instructions on operating the software.

2. Anticipated audience

I have contacted two hundred authorities across the disciplines all over the world who have given me very useful suggestions--suggestions that help to shape the book to as large an audience as possible. As it is written to date, the book will be of great value to management-science analysts, transportation logisticians, civil and environmental engineers, economists, geographers, urban planners, developers, corporations, and government officials, who are involved in facility operation and planning. All of them wish to avoid the costly mistakes that can result from poor locational decisions. It also discusses facility consolidation, contraction, expansion and the traffic flow through these facilities.

More specifically, this book is tailored toward practicing professionals, graduate students and college seniors interested in locational analysis. Besides serving as a professional reference, it can be (and has been) used as an advanced textbook in curricula such as management science, operations research, economics, civil and environmental engineering, industrial engineering, geography, urban and regional planning, and policy sciences. To attract both the professional and academic markets, a more basic, companion book, Chan (2000)[1], has been prepared. This companion book is a primer that explains the first principles, assuming only basic scientific skills from the users. Mathematical backgrounds

---

[1]Chan, Y (2000) *Location Theory and Decision Analysis*, Thomson/South-Western, Cincinnati, Ohio.

specifically required for this current volume are reviewed in its four book-appendices on optimization, stochastic process, statistics, and systems theory respectively. Also of value are the extensive glossary of technical concepts, lists of symbols and acronyms used throughout this current volume.

This book purposefully accommodates the different technical backgrounds and career objectives of its readers. As mentioned, the background knowledge required to understand more advanced concepts is covered in a companion volume, providing a brush up on the required basic principles. For example, spatial-economics principles are introduced in this companion volume, allowing the non-economists to acquire the basic economic concepts that underlie many locational literatures. Most important, liberal numerical examples and graphics are used—both in the primer volume and the current volume—to "get the point across." My diverse background, which spans technical consulting firms, government, academia, and the defense community, enables me to communicate with different audiences in terms of a common language.

The current book should be of interest to the practice, research and development community as mentioned, since it represents a synthesis and extension of the state-of-the-art in this field. Rarely does one find a single volume that puts the latest developments in a didactic form. Rarer still is the fact that the volume is by a single author. This contrasts with an edited volume by dozens of authors who may not have achieved perfect coordination among themselves. Thus in the last two chapters —"Spatial temporal information" and "Retrospect and prospects"—I could relate seemingly diverse techniques to one another, with comments on their assets and liabilities.

3. Projected use and shelf life

As there is no `laboratory' contained within four walls to analyze locational decisions, the effective way to bring these concepts to life and to test their viability is through their capacity for problem solving. As mentioned, numerous case studies are blended into the body of the text. These case studies, drawn from around the world, address land-use planning, transportation, as well as the location of public and private facilities. They serve to validate the pertinence of the analysis procedures advocated in this book.

In view of the rapid progress in information technologies and to avoid obsolescence, the book is not specifically tied to a single generation of computational technology. Rather, the book is problem-oriented and provides a set of procedures and a set of data that can transcend the technological evolution. Hands-on experiences are discussed with respect to the basic models employed, rather than the particular software or hardware. This makes the book "computer independent." Thus, the shelf life of this book will extend well into the future until there are breakthroughs in *fundamental* concepts and procedures.

A novel feature of this book is the recognition that in today's service economy the traditional concepts of accessibility, a key factor in siting, need to be broadly interpreted. Prevailing thoughts suggest, for example, that half of the shopping will be executed by mail or through telecommunications at the turn of the century. Thus the definition of "a trip to the shopping center," and hence the conventional wisdom in siting a retail facility, will need to accommodate such a change. In the defense community, the idea of "global reach" suggests the need to redefine the concept of accessibility and distance. With the deployment of supersonic aircraft and command-and-control technology, half of the globe away is measured in terms of a few hours of flight time or fraction of a second of telecommunications time. Meanwhile, congested streets can make mere cross-town travel close to impossible-- thus encouraging tele-commuting. This again requires a re-definition of accessibility and hence the conventional wisdom in office site-selection. I am happy to report that at least a whole chapter of the book—the chapter on "Measuring spatial separation—is devoted exclusively toward this subject. The theme permeates throughout the volume, serving to unify many spatial-location models available to date. Available information suggests that housing and transportation consume consistently half of the household budget.[2] Such a situation is unlikely going to change, making a discussion on "Location, Transport, and Land-Use" cogent for a long time to come.

4. Organization of the book

As mentioned, a companion book contains a review of the fundamental principles. There I explain why facilities locate where they are, and why population and employment activities distribute on the map as they do. The chapters include not only the underlying determinants of facility-location and land-use, but also the techniques that are essential to analyze these locational decisions. In addition, these chapters discuss data-bases from remote sensing and geographic information systems, statistical tools for data analysis and forecasting, optimization procedures for choosing the desirable course of action, and multi-criteria decision-making techniques to tie the entire analysis procedure together. Key concepts in economics, a most important discipline in explaining the organization in space, are also reviewed as mentioned.

Let us now turn to the current volume. Observations regarding the way locational decisions have been made in the past are formalized in several accepted 'theories' and computational procedures. They include the classic *Weber problem, centers, medians, competitive locations, central place, economic-base, gravitational interaction, entropy and bifurcation.* These are described in terms of a set of 'regularities', or the standard procedures in which siting decisions are made. These regularities explain both individual facility-location decisions as well as the resulting land-use pattern. The book illustrates not the conventional single-criterion and single-stakeholder point of view, but a pluralistic decision-making environment in which these regularities typically occur.

---

[2]"Transportation Cost and the American Dream", a special report from the Surface Transportation Policy Project. July, 2003. (http://www.transact.org)

In subsequent chapters, individual (or *discrete*) facility-location is discussed in detail, including *deterministic* and *stochastic* formulations of classic *p-median, set-covering, maximal-coverage, quadratic assignment* and *obnoxious* facility problems. As part of stochastic facility-location (or location under uncertainty), the concept of accessibility is discussed in terms of *queuing* and *space-filling curves*. The relationship between facility location and routing--the counterpart of the land-use/accessibility couplet in population- and employment-activity allocation--is also examined. This is a subject that is covered inadequately in existing texts. We accomplish this through a companion chapter on "Measuring spatial separation," in which such concepts as l_p-metrics are discussed in detail.

The next section of the book delves into *continuous* land-use modelling on a map--the cumulative results of discrete facility-location. These land-use models deal with population and employment growth as well as activity allocation among regions in a study area. They cover *competitive* locational decisions, *Lowry-based* models, econometric models, *input-output* models, *spatial equilibrium* and dis-equilibrium. Included in this section are recent concepts of *catastrophe* and bifurcation, which explain sudden, precipitous land-developments much better than traditional models.

Perhaps the integrating element that ties many concepts together is *spatial statistics*. It is used "across the board " in remote-sensing/geographic-information-systems (GIS), facility-location models, gravitational interaction models, and *spatial time-series* models (of which econometric models are a subset). The key concepts in spatial statistics--spatial *homogeneity* and temporal stationarity--are fundamental properties for many analytic procedures to work properly. This includes the correct procedure to divide a map into regions, as illustrated by *Voronoi* diagrams. This section of the book concludes by suggesting that many concepts advanced in facility-location and land-use have extensive applications in satellite-imagery interpretation, and *spatial pattern recognition*--a "high tech " development today.

The future direction of the profession is then discussed, in terms of how analysis procedures may change over time. The book discusses the way equilibrium and dis-equilibrium theories may play an increasing role in future analysis. It suggests how spatial statistics may be a major player not only in land-use forecasting, but also in other high-tech applications, including "spatial-data mining. " Finally, stochastic facility-location and combined location-routing models illustrate a new paradigm for locating facilities amid a service-oriented economy and the age of advanced telecommunications. Unifying the entire volume, we show how the diverse number of descriptive and prescriptive models presented in this book can be explained in terms of *Random Field* theory. The book concludes on how management-science techniques and information technologies may support more imaginative locational decisions in the future.

5. Recommended usage

Depending on the background of the reader and the intended usage, the book can be read in different order. For a senior- or graduate-level course, the following sequence is suggested:

a. For a one-term course concentrating on facility-location, these three chapters are in order: facility location, measuring spatial separation, and location-routing models. Where there is time, the chapter on activity generation-and-distribution serves as an introduction to land-use models, viewed here as a generalization of a discrete facility-location model.

b. For a follow-on or stand-alone course specializing in land-use, start with the activity generation-and-distribution chapter and progress through Lowry-based models, chaos and bifurcation, equilibrium and disequilibrium, and finally econometric models.

c. Where there is room in the curriculum, a more advanced treatment will start with the state-of-the-art material from spatial econometrics, through spatial time-series, and then to spatial-temporal information, where facility-location and land-use are discussed mainly in terms of spatial statistics.

d. For professional reference or research use, the chapter outlines, the extensive index and references should provide the reader with a state-of-the-art, yet self-contained, source of information. I have taken the care to build elaborate cross referencing within the text itself. The same is true between the current book and the companion volume, particularly on where to find the prerequisites (within Chan [2000]) for a more comprehensive understanding of an advanced topic.

e. A set of homework and case studies is developed for both classroom and professional use. Rather than a regurgitation of mechanical calculations, it is designed to extend many concepts covered in the book. Although they are categorized by the major topics, these exercises play an important role in integrating the many diverse principles advanced in the text. In some quarters, they are given the name "mind expanding" exercises. One objective of these exercises is to challenge the readers in creatively using the data-sets and computer software that come with this book. In the exercises, the readers are often asked to perform their own case studies if desired and arrive at open-ended results. For the sake of synergy, all these exercises are placed side-by-side as an appendix to the book, rather than included separately at the end of individual chapters. To assist both the students, instructors and the professionals, answers to each and every one of these exercises are available. To facilitate classroom adoption, only half of these solutions are made available to students. The Solution Manual serves as a supplement to the text, where detailed explanations not needed for the development of subsequent topics are omitted for clarity and succinctness[3]. Instead, these explanations are provided in the Manual. For a copy of the Manual contact the author at his "E-mail address for life": ychan@alum.MIT.edu

Again, the strength of this book is that all the necessary information is contained in it and its companion volume. To the extent the field is getting increasingly interdisciplinary, it is advantageous,even imperative in the author's opinion, for the reader to be aware of related concepts. These related concepts are provided as I discuss a particular topic, all conveniently in one place. For example, a classic facility-location model is shown to solve a pattern-recognition problem in spatial statistics, in which the contours of pollutants are determined. We complement the discussions with an extensive bibliography, cutting across many disciplines. All referenced works are in the open literature, including military documents from the Defense Technical information Center.

6. Computer software

Computer software is supplied with this book. It is intended to provide the readers with sample software to supplement the discussions in the book. The platform is an IBM-compatible Personal Computer (PC), which at the time of this writing, is the most widely available computing-machinery. By adopting this platform, it is expected that the accompanying software will find its use among the largest number of readers.

Most software is an implementation of some facility-location, transportation and land-use models discussed in this book. While the book introduces the various analytical techniques pedagogically, more practical tools are provided in the software. The computer programs are therefore no longer purely for the classroom; they have real potentials among everyday, operational use.

These philosophies are followed in the preparation of the software:

a. To provide the widest dissemination possible, all files in software are ASCII-text files. Where possible, both source codes and executable codes are given--mainly for the ease of execution and modification by the users. Program documentation is included as README files. Highlights of Program documentation are also included in the Solution Manual for convenience.

b. We strive to provide stand-alone programs that do not require supporting software, including language compilers. While references are made to related software for extended use of some programs, all programs are self-contained . They are "tailor-made" in the sense that they have been developed or refined by the author and his associates.

c. Sample data-sets are provided to allow demonstration of the software. While `toy' problems are often used for introduction, most of the data are drawn from real-world case-studies discussed in the main body of this book.

---

[3] Availability of these detailed explanations are highlighted by a double asterisk (**) in the text.

The programs are distributed after extensively testing by the author and his associates. But with any programs, there is no guarantee that they are fully debugged. It is impossible for the author, with his limited resources, to provide any programming support for these programs. But the author is keenly interested in and would appreciate any feedback from the users regarding their experiences in using these programs (and this first-edition book in general). The objective is to allow distribution of an improved software package with a second-edition book at a future date.

To obtain a copy of the software and to provide your comments, simply contact Yupo Chan at his E-mail for life: ychan@alum.MIT.edu.

7. Acknowledgments

This volume is a result of over two-and-a-half decades of background work. During this time, I benefitted from friends and colleagues throughout the country and overseas, and most importantly my students. Many of these individuals are collaborators in various studies, such as a paper that won the 1991 Koopman Prize Award of the Operations Research Society of America (now the Institute for Operations Research and Management Sciences). I have also benefitted from my association with: the Massachusetts Institute of Technology, Kates, Peat, Marwick & Co., Pennsylvania State University at University Park, U.S. Congressional Office of Technology Assessment, State University of New York at Stony Brook, Washington State University, the Air Force Institute of Technology, and the University of Arkansas at Little Rock.

The following individuals by no means constitute the entire list of friends and colleagues who made this volume possible. They are simply names that I recall at the time of writing: S M Alexander, W Allison, K Ardaman, R L Argentieri, V Ashton, C S Bertuglia, B Boots, D Boyce, S E Brown, F Campanile, R Calico, O T Carroll, E Casetti, F S Chapin, J L Cohon, N Cressie, T Cronin, M Daskin, D S Dave, R deNeufville, A Denny, J W Dickey, D Ding, C B Doty, K J Dueker, E Erkut, R Esposito, J Farris, A R Fitzgerald, R L Francis, R J Gagnon, W Garrison, M Gilchrist, B L Golden, P Gray, P Guyer, L S Hager, S L Hakimi, W G Hansen, R G Haubner, D Hoover[4], C S Hsu, Z M Huang, A P Hurter, J J Jarvis, R Johnson, E J Kaiser, R Keeney, T J Kim, T S Kelso, R Klimberg, T Knowles, J Krarup, F Kudrna, G Laporte, D B Lee, G Leonardi, S M Levy, I S Lowry, R E Machol, D Marble, D H Marks, C McCormick, P B Mirchandani, S E Momper, E Morlok, J Morris, L Mugler, P Nijkamp, S Occelli, A Okabe, H Ohta, N Oppenheim, A Perez, M B Priestley, A B Pritsker, P Pruzan, P Purdue, S Putman, G A Rabino, D E Radwan, C ReVelle, B D Ripley, E Ruiter, J Sabatino, D Schilling, M H Schneider, T Sexton, R Silkman, J R Sitlington, H Slavin,

---

[4] Special thanks to Don Hoover, who is instrumental in editing the final draft of this huge volume and putting it in it final form.

B Smith, B Spear, R Steele, J Stein, R E Steuer, K Sugihara, J Szilagyi, R Tadei, J F Thisse, J Usher, R W Vickerman, G O Wesolowsky, S C Wheelwright, J A White, W B Widhelm, D Williams, A G Wilson, J R Wright, M Yeates, and M Zelany. All these people rendered invaluable guidance to this writing project.

The strength of this work is one of synthesis—cutting across disciplines, backgrounds and experiences—which is precisely where this field is heading. The diverse backgrounds of these friends and colleagues in private industry, government and academia make my job as synthesizer that much more streamlined. Often, I would paraphrase their findings (with due acknowledgment , any oversight is due to chance rather than by intention.) Thus I can rely on a management-science analyst to assist me in telling his/her story, while I inform him/her the related goings-on in planning and engineering in return. Likewise, I cover the traditional topics in geography and planning following the advice of the regional scientist, yet at the same time broaden his/her horizon to the latest in information technologies. In short, these people greatly help me in achieving the goal of informing the reader of the knowledge that falls not only within his/her discipline but also under other disciplines. This is written down in as readable, yet as precise and practical, a form as I can manage. I am happy to report that this goal has now been accomplished in the first draft of a single-authored volume, or should I say two volumes, based on these diverse experiences and observations.

Yupo Chan

TABLE OF CONTENTS

Chapter Seven
CHAOS, CATASTROPHE, BIFURCATION AND

Chapter Eleven
SPATIAL TEMPORAL INFORMATION:

*"The important thing in science is not as much
to obtain new facts as to discover new ways
of thinking about them."*
Sir William Lawrence Bragg

CHAPTER ONE

INTRODUCTION

We are in a stage of human development when information literally overwhelms our daily lives. The glut of information is precipitated to a large extent by fast computers and the Internet. While it may not be apparent, much of the information is spatial (or geographic) in nature. Let us take a familiar example (Engelhardt 2000). Selling goods and services on the Internet through the exploitation of information technology is most commonly called E-commerce. When one goes to make an on-line purchase, s/he is asked to enter his/her name, address, and credit card number. What the consumer may not know is that the computer is verifying his/her information by address matching his/her name to data on a separate postal-service list. At the same time, it is checking the credit-card company's records for a name and address match. The vendor's Internet site queries geo-demographic and consumer databases. It then stores this information in a Geographic Information System (GIS) program running simultaneously on that computer. The next time the consumer logs on to the Internet site, its serving computer might exploit the data embedded in his/her request to view the site, such as cookies, Internet Protocol, or E-mail address. This data allows the computer to postulate certain things about him/her. The serving computer could then compare this incoming data with its GIS, or query other Internet databases (such as census data) for additional information about this consumer. A separate program could then make assumptions about the consumer based on the resulting GIS queries to tailor a page specifically to him/her. For example, s/he would be presented with a selected basket of goods or service the serving computer has found will most likely appeal to him/her. Clicking on a particular item, the consumer could even be presented with a map and directions to the nearest mall that sells the item. If the traffic is not bad, s/he might just hop into his/her car and drive from home to the shop. This is most appealing since s/he has the chance to touch and feel the item s/he is purchasing. On the other hand, if traffic is intolerable, s/he might just decide to place an order directly on the web site. Here, both the industry and the public take advantage of spatial information, enlarging the market and the choice available. This is but one of many, many examples that illustrate the importance of spatial information in our daily lives. In this book (and the accompanying volume [Chan 200]), we present the *classic* and *modern* techniques to deal productively and intelligently with geographic data.

1.1   Aims

This book has three basic objectives. The first is *to identify the observed regularities in locational decisions*. This involves examining and answering questions such
as: Why do public and private facilities locate themselves the way they do? What
factors do real-estate developers consider when picking sites for development? Why
do people live in a certain location, and why do they often work in a location
different from where they live? Why are focal points such as shopping malls,
airports, terminals, and depots situated at certain 'nodes' in a network? Throughout
this book, we try to answer some of these questions, so that readers can judiciously
locate facilities and guide development toward desired goals (Shefer et al. 1997).

   We often take notice about why certain facilities are placed in certain areas. We
get as many explanations about such locational decisions as the number of 'experts'
we ask. Each seems to offer a plausible explanation. Such explanations can be any
combination of economic, technical, social, political, and behavioral reasons, not
to consider for the time being such oriental myths as "fung shui." Roughly
translated, Fung Shui means "location and orientation [of a facility] with respect
to the elements of nature" (Love et al. 1988). Are there really discernible patterns
about these locational decisions? Many of us have observed that—historically
speaking—ports and cities of the world are often located at major trade routes.
Usually, this means at the confluence of rivers, a convenient deep-sea harbor, or
where railroads come together. Scientists envision future habitats in the galaxies
being at 'Lagrangian' points. These points are locations that are stable enough that
a space station, when perturbed by slight impacts, will restore its position after
reasonable oscillations. Based on these examples, it stands to reason that there may
be some locational 'patterns' one can discern. These patterns, when observed to be
consistently recurring one after another, are called *regularities*. These regularities
are not anywhere as precise as scientific laws, nor can they often be explained in
terms of cause–effect relationships. One event does not necessarily occur because
of a previous event. As a result, we have to go by the observed *patterns* only and to
treat those recurring patterns merely as some generally agreed upon 'facts.' From
there on analytical models can be built to reflect these premises. The first objective
of this document then, is to understand systematically the regularity with which
different locational decisions are made. Thus systematic procedures can be defined
to anticipate similar situations that may arise in the future.

   We should quickly point out there is a difference between the systematic
analysis proposed here and comprehensive, or holistic planning—which goes under
different names such as morphology, 'concurrency,' or planning theory. That body
of knowledge, while extremely valuable, has been treated in excellent texts
elsewhere. These include those that are required readings in such professional
examinations as the American Institute of Certified Planners (AICP) and such
documents as Chapin & Kaiser (1979) and Davis (1976) on land-use. This book
aims at a different area, which is by definition more narrowly focused. We ask more
specific questions, such as "how does transportation affect locational decisions";
"how does infrastructural support (such as utilities and sewer) influence develop-

ment of a certain area?" or "how does transportation combined with infrastructural support affect facility location and land use?" In other words, we examine one factor at a time, one criterion at a time, and their cumulative effects. This is in lieu of the simultaneous effects of *all* factors across *all* criteria. Distinction is also made between the treatment here and an approach taken by two notable publications—one by the American Society of Civil Engineers (1987), the *Urban Planning Guide* and another by Brewer and Alter (1988), *The Complete Manual of Land Planning and Development*. The *Urban Planning Guide* is an excellent document regarding a whole host of planning topics, ranging from waste to energy planning, with a design flavor as an undertone, while Brewer and Alter's publication is a comprehensive description of site layout planning. As illustrated by the end-of-chapter examples, the focus here is on analytics, with quantitative model building as a key instrument. This way, this book serves as a useful companion to such documents as the *Guide*. Geared toward both the private and public sectors, it represents an area that has not been covered sufficiently. It is particularly targeted toward those who feel the need for state-of-the-art analytic tools to make capital-intensive locational decisions—the step before detailed design and implementation.

The second objective, *to review the operational analysis techniques that have been applied in the field*, follows that of the first. In this regard, we report on case studies that span most user groups—from public and private facility-location to land development. Thus we would look at the factors that go into the location of a nuclear power plant in seismically active areas in California, the consolidation of state parks in the greater New York metropolitan area, the choice of distribution centers for military logistics, the siting of satellite tracking stations in Canada, target location in search-and-rescue missions, the land development in several major North American cities, including a systematic study of bifurcation-development in a medium-size city, York, Pennsylvania. We also examine case studies around the world, including the economic impact of the Kansai International Airport outside Osaka, Japan. The common theme is on how locational regularities and spatial impacts can be quantified in a set of procedures or models.

The third objective, *to be able to stand back and critique some of these modelling experiences*, requires asking whether they have been successful, and how valid have these analyses been? In other words, what are the assets and liabilities of the various techniques that have been employed? Perhaps one can think of this book as a "consumers' guide" to locational analysis, transport and land-use models. A user can look up the price tag of using a particular model, and also the benefits, specifically regarding the problem being solved. The only time that a model or analysis procedure can help is when the user is fully aware not only of its strong points, but its shortcomings as well. Only under those circumstances can an analyst, an engineer, a planner, or a manager employ the most appropriate tool toward the problem at hand. S/he avoids overkills with `exotic' technology, below-par performances with an outdated tool, and misfits between problem and analysis tools.

What are the more 'visible' results and benefits from reading this text? For analysts, engineers, planners, and managers, the question is easy to answer. Since infrastructure represents a major capital expenditure and it supports economic well-being and quality of life, this book serves the important role of articulating investment in these infrastructural improvements. Such infrastructures may include tracking stations, depots, terminals, roads, factories, warehouses, hazardous facilities, office buildings, and housing, as alluded above. Both in public decisions and in corporate planning, the analytical skills discussed in this book can mean huge savings or benefits. Together with a companion volume (Chan 2000), this book serves as a useful compendium of spatial-analysis techniques under two covers. Between the companion volume and the current volume, it is a comprehensive (even encyclopedic) collection. The presentation style is pedagogic, starting with the basic building blocks in the companion volume to the more advanced concepts in the current volume. We point out the commonalities among models used to locate facility one-at-a-time and to forecast the development pattern in an entire area. In this regard, it is a unified volume on "spatial science"—defined here to mean the analytical techniques that explicitly recognize the spatial elements in a study. The term spatial science, when used in this context, encompasses the traditional disciplines of facility location, transportation, logistics, land use, regional science, quantitative geography and spatial economics. Hence, this book introduces to students and specialists in each of these disciplines the broader perspective as viewed from collective wisdom. This perspective is absolutely essential in the furtherance of the art of spatial science.

1.2   Locational decisions

As discussed, one goal is to uncover the observed regularities of locational decisions, i.e., the apparent underlying forces that shape development. We shall be examining four major determinants of location.

1.2.1   Technological determinants. The first determinant refers to physical principles that govern location and infrastructural supports such as highways, airports, railroads, power supply, sewers, and irrigation that make the functioning of the facility possible. Notice that these go beyond the *availability* of transportation and utilities. The example about building a space station at the galaxies drives this home. Only Lagrangian points in spatial mechanics will allow the location of a permanent habitat in deep space. This way, re-supply spaceships can dock conveniently with the assurance that it is a stable station that survives the impacts of objects—from as weak as a docking vehicle to as strong as a small meteorite. Similarly, satellite tracking stations must be where visibility is at its best to observe the desired orbits most of the year. It makes sense that a station too far north in the Northern Hemisphere will be unsuitable to track satellite orbits around the equator, not to mention that infrastructural support such as roadways and utilities will be scanty in these arctic regions. When the American West was developed, the railroad was the key instrument. Today, in the Midwest of the United States one can still trace the location of towns at regular intervals along the rail lines on the prairie. Apparently, they developed from water refilling stations required of the steam locomotives of the day. The separation represents the length during which all the water carried on a train evaporated--a technological factor in its truest sense.

1.2.2 Economic and geographic determinants. To illustrate this second determinant, consider a person lives at a location convenient for carrying out daily activities. These activities include both work and non-work. The location has to be commensurate with the ability and willingness-to-pay for the corresponding residential cost. For those who cannot afford the prime locations, housing a little bit further away is the only choice. A host of theories exists to explain this phenomenon, including Land Rent and Location Theories (Salomon and Mokhtarian 1998). Historically, cities have located on trade routes, perhaps due to accessibility to markets. To command a competitive edge in today's retail market, warehouses are more often than not situated at the midst of the demand. This way, they are most accessible to consumers through the retail outlets. The most graphic example may be in emergency planning. Thus quick, efficient medical evacuation of the wounded dictates a judicious placement of "hubs," to which the injured can be quickly transported and eventually delivered to hospitals for medical care. Even with today's Internet, locating a web site is often an economic decision. For example, economy of scale might have helped in the dramatic growth and expansion of certain sites that have started by selling specific merchandise.

1.2.3 Political determinants. An eminent example of the third determinant is zoning, which represents an institutionalized consensus among the community regarding the legitimate use of the land. Fiscal and jurisdictional considerations are also quite common. During the latter part of the 20*th* century, there have been free enterprise zones designated by the People's Republic of China. These zones manufacture export goods and conduct business with the free world. Some of these were located across the border from Hong Kong and Macau. These zones enjoy special jurisdictional and fiscal privileges—incentives for investment and workers. Finally there are eminent political decisions for location as well. The Dallas Fort-Worth Airport in the United States, for example, sprawls across two counties, apparently for political reasons. This in part explains its huge horizontal layout rather than a more vertically integrated structure. On a larger scale, many guidelines are enacted on a policy level as legislation. The location of airports, for example, is subject to a whole host of environmental regulations. Brewer and Alter (1988), and Chapin and Kaiser (1979), among others, have a good review of the various national, state and local legislation that govern land-use in general.

1.2.4 Social determinants. The fourth determinant includes dominance, gradient and segregation; centralization and decentralization; invasion and succession. Humans, a gregarious species, tend to congregate into communities. On the other hand, they tend to segregate themselves for certain other reasons, which results in the reservation of certain land accessible only by selected groups. Thus, there are segregated regions in a newly discovered land reserved for colonial citizens to the exclusion of natives of the land. Certain public facilities are segregated between women and men for privacy reasons. Between the phenomenon of 'togetherness' and 'separation,' all the shades exist in between, which explain to some extent the myriad of development patterns that we see through recorded history. These social and behavioral factors are varying depending on the values of the time and the context of the culture. They are difficult to quantify in a set of systematic procedures. One of the more feasible analysis procedures is gaming. In a game, interested parties participate in these decisions in a simulated environment.

1.3   The need for analysis

Some explanations of these perplexing issues can be found by the judicious employment of analysis techniques. Obviously, analysis of the problems posed above requires a set of very specialized skills. The techniques required of the analyst include *descriptive* and *prescriptive* tools. Descriptive tools pertain to the techniques that echo locational regularities that we observe around us. It is the representation of observed patterns by such methodologies as simulation and statistics, or more causal explanations such as regional economics. Through electronic computers, one can build a mathematical replica of the scenario and use it to test out alternate policies. This is akin to architects who build a scaled-down model of a building for study in a studio. Graphic display of information, afforded by today's geographic information systems, greatly eases such analysis (Thrall et al. 1995).

Prescriptive techniques, on the other hand, try to identify a course of action for decision makers. For example, to achieve the community goals and objectives, one specifies a set of policies and plans by goal-directed methodologies. A mathematical model can be formulated, from which one obtains a blueprint for future development and implementation. As with descriptive models, computers are often utilized to operationalize optimization models. These models include those that consider multiple criteria, echoing a pluralistic decision-making environment typical of locational decisions. Advances in computational techniques have made it practical to identify desired courses of action or facility locations that was impossible only a couple of decades ago (Churhill and Baetz 1998, Rodrigue 1997). While part of the advances has been due to the computational machinery, our understanding of prescriptive techniques has also made dramatic gains in the past decades.

Analysis can unveil *counter-intuitive* results that can easily be overlooked if such a set of rigorous thoughts is not carried out. This pertains obviously to complex situations where there are just too many factors to consider for the unaided mind to comprehend. What is more interesting is that they may arise in simple situations as well. We will demonstrate a couple of these immediately below, which hopefully make a strong case for the analysis procedures advanced in this text.

1.3.1 Airport example. Suppose a common airport is to be built to service New York City and New Haven, Connecticut—a distance of about 80 miles (128 km). Where is the best location considering the combined populations of the two cities—with, say, 14 million in New York and two million in New Haven? Notice the question asked here is a narrowly focused one. It is simply to lessen the travel requirement for all the 16 million residents of the area—in terms of total person-miles-of-travel. Most people who are asked the question responded by saying that the airport should be somewhere between the two cities. Some even pointed out that it should be closer to New York than to New Haven, since New York is a larger city, with more people living there. The more technically minded calculated that it should be 10 miles outside Manhattan and 70 miles away from New Haven on the major highway (Interstate 95) that connects the two cities.

The 'correct' answer here is that the airport, from a pure accessibility standpoint, should be as close to New York as possible. It is that location that will make the total person-miles-of-travel the least. To show this, just pick three possible locations:

a. Halfway between New York and New Haven, resulting in a travel requirement of $(40 \times 14 + 40 \times 2)$ or 640 million person-miles-of-travel (PMT) [1024 million person-km].

b. 10 miles outside New York and 70 miles (112 km) from New Haven, resulting in a PMT of $(10 \times 14 + 70 \times 2)$ or 280 million.

c. Located right at New York and a full length of 80 miles(128 km) from New Haven, resulting in $(0 \times 10 + 80 \times 2)$ or 160 million PMT (256 million person-km)!

When presented with this result, people quickly pointed out that it is impossible to locate a new airport at New York, since there is simply no land. Others pointed out that environmentally speaking, no one will accept an airport at New York City. But that was not the question. The question—which still appears in black and white above—simply focuses on one aspect: the total person-miles-of-travel!

We will come back to this in a case study later. There we will point out that those that having knowledge of Linear Programming—a prescriptive technique—will readily recognize an "extreme point" as the solution. This means that either New York or New Haven should be the airport site, not somewhere in between. We will then bring in other considerations, including the environment, and show how the location may change because of these *additional* factors. In other words, we answer the question for the accessibility factor, then the environmental factor and so on. We build up the complexity as we move along, rather than facing them simultaneously as in more holistic-planning methodologies.

1.3.2 Manufacturing-plant example. Another example equally illustrates the role of analysis. Suppose a major manufacturer opens an additional plant in Hometown, with a payroll of 1000 workers. What will the future population and employment increase be in Hometown? We further know that each household in Hometown has 2.5 people on the average, of which there is only one breadwinner

Time increment	Basic employ	Basic employ pop	Support service emp	Support service pop	Total employ	Total pop
1	1000	2500	500	1250	1500	3750
2			250	625	1750	4375
3			125	312.5	1875	4687.5
4			62.5	156.25	1937.5	4843.75
5			31.25	78.13	1968.75	4921.88
6			15.63	39.06	1984.38	4960.94
7						

Table 1-1   Economic forecast of Hometown

(i.e., one breadwinner and 1.5 dependents per household.) For every five additional people in the community, another support-service employee is required. In other words, there are multiplier effects on the economy, in which one dollar of payroll generates more that its value in the local economy. The manufacturing employees will require support services such as shopping, medical, recreation etc., involving new employees who also bring in their families who again require more services. According to the parameters given above, a moment's reflection will show that the 1000 new manufacturing jobs will bring into town 2500 people, including dependents. These 2500 people will require support services in Hometown, including medical, shopping, recreation etc. and generate 500 secondary jobs (since every five people require one secondary-service employee). These secondary-service jobs bring into town another 1250 population (500 ×2.5), including employees and family members. Now the total new employment in town is (1000+500) or 1500, and the total new population is 3750 (2500+1250). The process goes on as shown in **Table 1-1**, eventually stabilizing at about 2000 additional employees and 5000 additional people!

  Figure 1-1 depicts the growth profile of Hometown in terms of population and employment. The growth profile stabilizes in time-period 7. On the same Figure is shown the growth profile when household size is increased from 2.5 to 5. In this case, the growth will perpetuate forever, as shown by the straight line of **Figure 1-1**. When the support-service requirement is raised from 1 to 1.25-employees-for-every-5-people, totally uncontrolled growth will result—as shown again in **Figure 1-1**! Apparently, any slight increase beyond the 'watershed' points of 5 people in a household and 1-service-employee-for-every-5-people will fuel the fire-of-growth

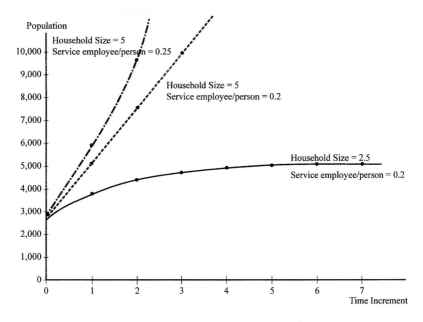

Figure 1-1 Bifurcation in population growth

to a fury. On the other hand, family size a tiny bit smaller than 5 or service requirement less than 1-employee-in-5 results in a stabilized growth. The `watershed' point is an important piece of information for all who are interested in the future of Hometown. A technical term for the dividing line between growth versus stagnation is `bifurcation.' Without *descriptive* analysis such as above, these bifurcation points are not obvious to simple, intuitive reasoning.

1.3.3 A composite example. Now, combining the above examples, if 1000 new jobs are added both to New Haven and to New York City, if the average family size is 2.5 people in New York and 5 people in New Haven, and if there is 1-service-employee-for-every-5-people in both places, New Haven will experience unlimited growth while New York will be stagnating. It does not take long for the labor force of New York to see job opportunities in New Haven and respond to them in terms of reverse commuting. Nor does it take long for the unlimited growth in New Haven to outgrow its physical or infrastructural capacity. Given the growth in New Haven must go somewhere, this will possibly mean the "spread of wealth" back to New York. **Figure 1-2** represents this interaction between the cities schematically. The time increment is on the vertical axis and the spatial interaction is on the horizontal axis. Different growth profiles, combined with the physical limitation to unlimited growth, result in a very interesting development pattern between the two cities.

With a changing demographic profile, the location of a regional airport must be reconsidered. We have already shown that from purely an accessibility

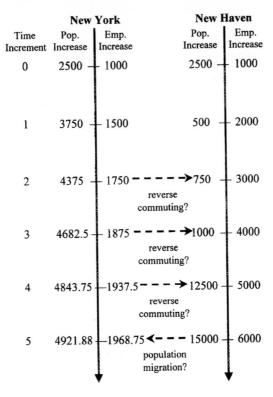

	New York		**New Haven**	
Time	Pop.	Emp.	Pop.	Emp.
Increment	Increase	Increase	Increase	Increase
0	2500	1000	2500	1000
1	3750	1500	500	2000
2	4375	1750	750	3000
3	4682.5	1875	1000	4000
4	4843.75	1937.5	12500	5000
5	4921.88	1968.75	15000	6000

Figure 1-2 Economic interaction between New York and New Haven over time

standpoint, the airport should be at the more populated of the two cities. Now with New Haven enjoying unlimited growth while New York is stabilizing at 14 million 5 thousand, it is only a matter of time before New Haven will surpass New York in terms of population (assuming the physical limitation to growth has yet to be reached.) The regional airport will eventually be at New Haven instead of New York. The interesting, somewhat counter-intuitive, fact is that the best location for the airport will switch abruptly the minute New Haven has one person more than New York—no sooner and no later. The moment that the New Haven population exceeds New York's is a bifurcation point, at which precipitous changes occur in the fundamental behavior of the development.

1.4 Analytical techniques

These examples drive home the point that analysis is an indispensable supplement to intuition in capital-intensive locational decisions. These examples are merely abstractions of case studies that will be presented in detail in later chapters. There the highly simplified situations used above are protracted into the multicriteria and pluralistic decision-making process common in locational debates (Antoine et al.

1997, Massam 1988). Suffice to say here that sophisticated analytical techniques have been developed in recent decades to perform these studies. These techniques are based largely on operations research, statistics, economic analysis, and systems science. The contribution of this book is not just the collection of these techniques, but more importantly the extension of them into the spatial context. Thus the well-known extremal-point optimality of linear programming (LP) is now extended from the Euclidean space of LP into the physical map of the Northeast, including the metropolis of New York and New Haven. It will be seen that when a triangle of three cities—say New York, New Haven, and Newark/New Jersey are involved, the complexity of locating a regional airport compounds many fold, resulting in the classic "brain teaser": the Steiner–Weber problem. The airport can now be located —again based on proximity—at either of the three cities or in the interior point of the triangle, as will be discussed in the "Facility Location" chapter. Any one who has worked with this problem can testify to the fact that the Steiner–Weber problem is not simply an extension of linear programming—it goes well beyond!

The same can be argued about the Hometown-development example. As seen above, the simple aspatial statement of the problem can quickly get complicated as we extend to two cities interacting with one another: say between New York and New Haven. It will be shown subsequently that the underlying theory is *input–output analysis*, a branch of knowledge economists since Leontief have developed. It explains trade between such economic sectors as manufacturing, service and housing. Including the spatial element into input-output analysis, however, compounds the model significantly, raising a whole host of conceptual and model-calibration problems as evidenced in the well-publicized Lowry–Garin-derivative models. When fully developed, several important factors have to be reckoned with here, including spatial competition such as in an oligopoly market consisting of several well-defined competitors. Now, other communities surrounding New York City—say Hartford and Newark—compete for employment and population surpluses from the City and vice versa. To compete, these surrounding communities may consider supply-side investment strategies to stimulate sub-areal and areal economic-growth. Intimately related is the fundamental assumption about "factor substitution"—for example, to what extent can labor be traded off against capital in the spatial production-process. In other words, can labor savings (because of local labor-shortage) be compensated by better local investment in equipment and production facilities? Simply put, the Lowry–Garin-derivative models are more than just a straightforward extension of aspatial input-output analysis. This will be shown in the "Activity-allocation," "Disaggregation and Bifurcation" and "Spatial Equilibrium" chapters.

1.4.1 Prescriptive Example. Given the complexity of including a spatial dimension, is there a fundamental basic-building-block of spatial interaction: the foundation that enables spatial generalization of most analysis techniques? Yes, there is a simple spatial law, credited to Tobler, which says the following. "Everything is related to everything else, but closer things are more related than distant things." The power lies in the beguiling simplicity of the statement, which finds its way into pervasive applications in facility location, transportation and land-use. It is also at the core of such analysis techniques as "spatial statistics." As it turns out, models of location, transport and land-use are nothing more than ways to process spatial-temporal information. For example, a traditional facility-location model places a facility proximal to the demands—since the best way to provide the best service is to be close to the customers. However, Tobler's `law' goes way beyond this simple example. It turns out that this facility-location paradigm can be used to analyze other spatial patterns, including a way to discern pollution plumes.

In discerning the pollution pattern from a point source, we use a digitized image consisting of *picture elements* (or pixels.) The concentration of pollutants is shown in terms of "grey values" of these pixels. An example of low-level radioactive-waste pollution is shown under **Figure 1-3**: "Example image of a pollution plume." The darkest grey value is 10 `notches'—with grey values ranging from 0 to 9—darker than the lightest. The `proximity' between the pixel with grey-value i and pixel with representative grey-value j is denoted by the `distance' $d_{ij}$, where $d_{ij} = |i\text{-}j|$. Here, proximity is interpreted as the `distinction' (or difference) in grey values between these two pixels. By minimizing the `distances' between pixels of grey-values i and j, we end up grouping pixels of similar grey values together in concentric rings, all centered at the point source. In other words we discern the contours of pollution concentrations from the point source. The number of such contours is determined by the "representative values" one wishes to specify, much like the separation between contour lines on a map is decided by the number of contours one uses to represent the topography. It can be shown that the following facility-location model will help to discern such a pollution pattern, in which there are three concentric rings.

$$\min \sum_i \sum_j f_i d_{ij} x_{ij}$$

$$\begin{aligned} s.t. \\ \sum_j x_{ij} = 1 \qquad & i = 0, \ldots, 9 \\ \sum_j y_j = 3 \\ x_{ij} \le y_j \qquad & i, j = 0, \ldots, 9 \\ y_i, x_{ij} \; binary. \end{aligned}$$

(1-1)

In this model, the objective function clusters all pixels of similar grey values together in a ring. Notice the objective function contains weights reflecting the `intensity' of a grey-value's occurrence. Thus, the most common and most similar grey-values are grouped together. Here f_i is the frequency of occurrence of grey-value i. The pixels in the data-set has the following distribution:

Grey value	0	1	2	3	4	5	6	7	8	9
Frequency f_i	109	40	32	28	20	14	6	4	2	1

Such $f_i d_{ij}$ values are displayed as the "[spatial-separation] `costs' between two grey values" Table in the "Facility Location" chapter. They are also shown as

0	0	0	0	0	1	1	1	1	0	0	0	0	0	0	0
0	0	0	0	0	1	2	2	2	1	1	0	0	0	0	0
0	0	0	1	1	2	3	3	3	2	2	1	0	0	0	0
0	0	0	1	2	3	4	4	3	3	3	2	1	0	0	0
0	1	1	2	3	4	4	5	4	4	3	2	1	1	0	0
0	1	2	3	4	5	5	6	5	4	3	3	2	1	0	0
0	1	2	3	4	5	7	8	6	5	4	3	2	1	0	0
1	2	3	4	5	6	8	9	7	6	5	4	3	2	1	0
0	1	2	3	4	5	6	7	7	5	4	3	2	2	1	0
0	1	2	3	3	4	5	6	5	5	4	3	2	1	0	0
0	0	1	2	3	3	4	5	4	4	3	3	2	1	0	0
0	0	1	1	2	2	3	4	3	3	2	2	1	0	0	0
0	0	0	1	1	2	2	2	2	2	1	1	0	0	0	0
0	0	0	0	0	1	1	1	1	1	0	0	0	0	0	0
0	0	0	0	0	0	0	0	0	0	0	0	0	0	0	0
0	0	0	0	0	0	0	0	0	0	0	0	0	0	0	0

Legend

Number in cells are grey values.

Figure 1-3 Example image of a pollution plume

coefficients of the objective function in model **(1-1)** immediately below. The first constraint in model **(1-1)** assigns each pixel to a `representative' value defining a contour ring. The second constraint establishes three contours or three rings of pollution from the outset. The last `precedence' constraint simply ties each pixel to a representative ring. The binary variable x_{ij} is zero if grey-scale i does not belong to the representative-ring j, and is unity otherwise. Similarly, $y_j = 0$ if grey-value scale i is not chosen as a representative grey-value of a ring, and $y_j = 1$ otherwise. Model **(1-1)** can now be written out in long hand:

min
$$+0x_{00} \quad +40x_{01} \quad +64x_{02} \quad +84x_{03} \quad +116x_{04} \quad +70x_{05} \quad +36x_{06} \quad +28x_{07} \quad +16x_{08} \quad +9x_{09}$$
$$+109x_{10} \quad +0x11 \quad +32x_{12} \quad +56x_{13} \quad +87x_{14} \quad +56x_{15} \quad +30x_{16} \quad +24x_{17} \quad +14x_{18} \quad +8x_{19}$$
$$+218x_{20} \quad +40x_{21} \quad +0x_{22} \quad +28x_{23} \quad +58x_{24} \quad +42x_{25} \quad +24x_{26} \quad +20x_{27} \quad +12x_{28} \quad +7x_{29}$$
$$+327x_{30} \quad +80x_{31} \quad +32x_{32} \quad +0x_{33} \quad +29x_{34} \quad +28x_{35} \quad +18x_{36} \quad +16x_{37} \quad +10x_{38} \quad +6x_{39}$$
$$+436x_{40} \quad +120x_{41} \quad +64x_{42} \quad +28x_{43} \quad +0x_{44} \quad +14x_{45} \quad +12x_{46} \quad +12x_{47} \quad +8x_{48} \quad +5x_{49}$$
$$+545x_{50} \quad +160x_{51} \quad +96x_{52} \quad +56x_{53} \quad +29x_{54} \quad +0x_{55} \quad +6x_{56} \quad +8x_{57} \quad +6x_{58} \quad +4x_{59}$$
$$+654x_{60} \quad +200x_{61} \quad +128x_{62} \quad +84x_{63} \quad +58x64 \quad +14x_{65} \quad +0x_{66} \quad +4x_{67} \quad +4x_{68} \quad +3x_{69}$$
$$+763x_{70} \quad +240x_{71} \quad +160x_{72} \quad +112x_{73} \quad +87x74 \quad +28x_{75} \quad +6x_{76} \quad +0x_{77} \quad +2x_{78} \quad +2x_{79}$$
$$+872x_{80} \quad +280x_{81} \quad +192x_{82} \quad +140x_{83} \quad +116x_{84} \quad +42x_{85} \quad +12x_{86} \quad +4x_{87} \quad +0x_{88} \quad +1x_{89}$$
$$+981x_{90} \quad +320x_{91} \quad +224x_{92} \quad +168x_{93} \quad +145x_{94} \quad +56x_{95} \quad +18x_{96} \quad +8x_{97} \quad +2x_{98} \quad +0x_{99}$$

$$+0y_0 \quad +0y_1 \quad +0y_2 \quad +0y_3 \quad +0y_4 \quad +0y_5 \quad +0y_6 \quad +0y_7 \quad +0y_8 \quad +0y_9;$$

subject to

$$x_{00}+x_{01}+x_{02}+x_{03}+x_{04}+x_{05}+x_{06}+x_{07}+x_{08}+x_{09} = 1$$
$$x_{11}+x_{11}+x_{12}+x_{13}+x_{14}+x_{15}+x_{16}+x_{17}+x_{18}+x_{19} = 1$$
$$\text{etc.;}$$
$$y_0+y_1+y_2+y_3+y_4+y_5+y_6+y_7+y_8+y_9 = 3;$$

$$x_{00} - y_0 \le 0 \qquad\qquad x_{01} - y_1 \le 0 \qquad\qquad x_{02} - y_2 \le 0$$
$$x_{10} - y_0 \le 0 \qquad\qquad x_{11} - y_1 \le 0 \qquad\qquad x_{12} - y_2 \le 0$$
$$\text{etc.;} \qquad\qquad\qquad\quad \text{etc.;} \qquad\qquad\qquad\quad \text{etc.;}$$

A linear program produced binary results for all decision variables, with representative grey-values 0, 1, and 4 chosen (i.e., $y_0 = y_1 = y_4 = 1$). For the grey values entered to the model as d_{ij}, the three-ring contours are illustrated in the **Figure 1-3** as mentioned. The outermost ring contains the pixels with grey values 0. The in-between ring has grey values of 1 and 2, and the innermost ring has grey-values between 3 and 9.

This example shows several important points. First, it illustrates Tobler's 'law' in an imaginative way, in this case through image-processing applications. The key is in defining *spatial separation*—a subject we will devote the entire chapter on later. It turns out that geographers, regional scientists, transportation planners, electro-optics researchers, and statisticians have all worried about this phenomenon for decades, if not centuries. In many ways, they have done the same thing without full knowledge of what the other discipline is doing. At the core is the concept of a 'neighbor', which is intimately tied to the definition of proximity or spatial-separation. In this example, pixel of grey-value i is classified as a neighbor of pixel of grey-value j according to Tobler's law. Separation between i and j here goes well beyond the Euclidean metric or 'bee line' distance-measure. It is defined—somewhat counter intuitively—as the difference between grey values of i and j. We will see many more examples of such counter-intuitive definition of 'distance' throughout this volume. This example shows a *prescriptive* application of Tobler's law in classifying or grouping neighbors together.

1.4.2 Descriptive example. In general, separation d_{ij} is best thought of as a spatial-price system that organizes locational decisions. This is similar to the familiar monetary-price that allocates scarce resources in microeconomics—a higher price discourages consumption while a lower price stimulates consumption. Proximity is the metric that establishes the level of interacton (correlation) among entities in space, such as the traffic between home and office. Thus accessibility in urban commuting takes on a very different light than proximity between two pixels in a satellite image. Yet in some ways, the fundamental principles governing both are remarkably similar in concept—namely Tobler's "first law" as outline above. Furthermore, both the urban planner and the remote-sensing analyst have the common goal of monitoring land-use or land cover. The "gravity model" relates spatial interaction as a direct function of activity-levels and inverse-function of spatial separation. It is a popular implementation of Tobler's first law. Thus more interaction (traffic) is found between high-density residential and employment centers that are close together than lower-density ones that are further apart. Similarly, satellites that monitor pollution will observe pollutants dissipating in inverse relationship to the point source. Calibration of the gravity model, however, is by no means simple. It often necessitates a fundamental re-examination of an entire array of basic statistical principles (Yun and O'Kelly 1997, Sen and Smith 1995). We shall give a *descriptive* application of the gravity model through the "weight matrix" below.

The ultimate objective of the use of a spatial weight-matrix is to relate a variable at one point in space to the observations for the same variable in other spatial units. This idea is prevalent in statistics. In time-series, for example, one relates the observation in time-period t to the observation to time-period $t-k$. This is achieved by using a *lag operator*, which shifts the variable by one or more periods in time. For example, $x_{t-k} = L^{(k)}x_t$ shows the variables x_t shifted k periods back from time t, as a result of the kth power of the lag-operator L. Thus a time series consisting of the following observations can be shifted by two periods as shown:

Time t	1	2	3	4	5	6	7	8
Observation x_t	8	12	15	9	30	19	13	17
Lagged series $L^{(2)}x_t$	-	-	8	12	15	9	30	19

Now the time-series x_t can be related to time-series $L^{(2)}x_t$ by regression if desired, for such purpose as forecasting the variable x over time.

While it is similar in concept, matters are not this straightforward in space due to the many directions in which the shift can take place. Subject pixel i, for example, can be related to pixel j directly `north', or pixel k due east, or pixel l due south and so on through all points of the compass. In most applied situations, there are no strong a priori motivations to guide the choice of the relevant form of spatial dependence. This problem is compounded when the spatial arrangement of observations is irregular, as in urban applications where population and employment are identified by census tracts, school districts or traffic zones instead of a square grid. In such spatial tessellations, many directional shifts become possible. Traditionally, this problem is resolved by considering a weighted sum of all neighboring values belonging to a given contiguity class, rather than taking each of them individually. This way, we get away from the open question of choosing a particular point on the compass. The terms of the sum are obtained by multiplying the observations in question by the associated weight from the spatial-weight matrix. Those familiar with the gravity model will recognize the model form in the following expression.

$$L^{(1)}x_i = \sum_{j \in N_i'} w_{ij}x_j \qquad \text{(1-2)}$$

Here $L^{(l)}$ is the spatial-lag operator associated with contiguity class l, and j is the index of the observations belonging to the contiguity class for i, N_i^l. In matrix form, this would be $L^{(l)}x_i = \mathbf{w}^{(l)}\mathbf{x}$. The spatial lag of observation x_i, $L^{(l)}x_i$ ($l = 1$, 2 or 3), would simply be the weighted sum of the four or eight neighbors in a lattice of observations. Here the subject observation is shown as a circle, the four immediate neighbors are black squares and the next set of neighbors are empty squares: $\begin{smallmatrix} \square & \blacksquare & \square \\ \blacksquare & \circ & \blacksquare \\ \square & \blacksquare & \square \end{smallmatrix}$.

Typically, the weights add up to unity, $\sum_j w_{ij} = 1$, as required in a gravity model.

Here is a typical application of the gravity model. Suppose we have 1000 commuting trips converging upon downtown zone-1 ($i = 1$) at the beginning of a work day from the closest-by suburbs (contiguity class $l = 1$). We wish to find out how many of these commuters come from locally downtown (zone $j = 1$), the immediate north (zone $j = 2$), east (zone $j = 3$), and south (zone $j = 4$) and west (zone $j = 5$) suburbs respectively. Through calibration, one obtains a weight matrix, which reflect the distance separations between each suburb and downtown. The first

column of such a spatial weight matrix $[w_{ij}]$ is as follows: $\begin{vmatrix} 0 & - & - & - \\ 0.401 & - & - & - \\ 0.298 & - & - & - \\ 0.199 & - & - & - \\ 0.102 & - & - & - \end{vmatrix}$, suggesting

that the northern-suburb zone-2 is the closest to downtown while western-suburb zone-5 is the furthest away. Here $w_{11}+w_{21}+w_{31}+w_{41}+w_{51} = 1$. The trivial computational result according to Equation (1-2) is that no commuters come from downtown, or $(0)(1000) = 0$; 401 or $(0.401)(1000)$ commuters come from the north suburb; 298 or $(0.298)(1000)$ come from the east; and 199 or $(0.199)(1000)$ come from the south, and 102 or $(0.102)(1000)$ from the west. It is clear that $0 + 401 + 298 + 199 + 102 = 1000$ as expected.

Equally, we can use the spatial-lag operator in spatial-filter applications, where filters are generally used to sharpen or blur an image. A filter 'mask' is often employed to process one part of the image at a time. For photographers taking pictures of the sky, they can think of a yellow filter used in front of a lens, which sharpens the object (the clouds). A 3×3 mask, for example, covers a nine-pixel object such as the grid shown above. In this grid, there are nine pixels under the mask—counting the subject pixel and the 8-connected neighbors, namely x_1, x_2, ..., x_9. The filter simply uses Equation (1-2) to produce a weighted sum for the subject pixel as $w_1 x_1 + w_2 x_2 + ... + w_9 x_9$. The grey value of the pixel at the center of the mask is then replaced by this weighted average. The mask is then moved to the next pixel location in the image and the process is repeated. This continues until all pixel locations have been covered. A "low pass spatial filter" for example, will simply average the pixels in the neighborhood, or set all mask coefficients w to 1, resulting in $x_1+x_2+ ... +x_9$. Not surprisingly, the effect is to blur the image, thus removing 'specks' of noise scattered around the image if desired. Clearly, the resulting notion of a spatially-lagged variable is not the same as in time-series analysis. It is similar to the concept of a distributed lag such as exponential smoothing in forecasting, where recent observations are weighted more heavily than those in the distant past according to a pre-determined set of weights. In this regard, it is important to note that the weights used in the construction of the lagged variables are taken as given, just as a particular time path can be imposed in estimating a distributed lag.

Besides the pre-determined weights used in filter masks, we have seen several general examples of this idea. A location study of competing Long-Island state-parks was performed in New York, where an additive 'utility-function' combining distance and travel-time was calibrated a priori: $w_1(distance)+w_2(time)+constant$. Here the weighted sum of distance from i and the time from i is computed, combining the features of both spatial- and time-correlated attributes centered at location i. Notice that this was performed exogenous to the model by valuating time using previous studies. In land-use studies, a "trip-distribution function" showing the percentage of trips of 5-minute, 10-minute, 15-minute durations, etc., was needed. It is used as part of an "accessibility function" for allocating population and employment around the study area (Civil Engineering 1998). Here, the trip-distribution function generates a set of weights $w_{ij}^{(l)}$ for the $(l) = 10$-minute, $(l) = 15$-minute etc. contiguity classes in the weight matrix $\mathbf{w}^{(l)}$. These two examples, one from state-park location and the other from residential land-use, represent an adequate illustration of generalized spatially-lagged variables and how they are determined. Following this latter land-use example, less-restrictive spatially-lagged-variable can be constructed from the notion of accessibility. As one recalls, accessibility is defined as the weighted trip-distribution function where the weights are population, employment or other activity variables x_j: $X_i = \sum_j f(d_{ij}, \mathbf{b})x_j$, where f is a trip-distribution function of distance d_{ij} and a vector of coefficients $\mathbf{b}$. One can think of $[f(d_{ij}, \mathbf{b})]$ as a weight matrix to the extent that $\mathbf{b}$ is a function of activity level $\mathbf{x} = (\leftarrow x_j \rightarrow)^T$ and distance d_{ij}. More will be said about this subject in the "Spatial Econometric Models" chapter.

One can see that Tobler's first law has a number of implications upon spatial-temporal information. While disciplines outside geography may not recognize this law, they have practiced it nevertheless. An example is the use of spatial weight matrix, which cuts across applications ranging from land-use to image processing. The weight matrix simply *allocates* economic activities among the study area in a land-use model. The same matrix can also serve as a filter to process an image in order to bring out the desirable features. As it turns out, when entries of the weight matrix are binary, it corresponds to an "all or nothing" assignment of demand

activities to the closest facility on the average. A matrix $\begin{bmatrix} 0 & 1 & 0 & 0 \\ 0 & 0 & 1 & 0 \\ 0 & 1 & 0 & 0 \\ 0 & 0 & 1 & 0 \\ 0 & 1 & 0 & 0 \end{bmatrix}$ assigns demands

at nodes 1, 3 and 5 exclusively to the facility at node 2, whereas the demands at nodes 2 and 4 are assigned to facility at node 4. This simply places a facility closest to the demands on the average. This can be viewed as the *locational* decision, instead of the *allocation* decision described above. The former is a prescriptive model while the latter is a descriptive model.

Tobler (1969) investigated the "inverse filter" as a way to correct for geographic data-collection errors. Recognizing the inconsistencies between collecting data on a county, census tract, or traffic zone basis, he suggested the inverse filtering concept. It provides a method of correcting for this geographic-aggregation effect. The important point here is that there exist some fundamental principles that govern location, transport and land-use (Louis Berger & Associates 1998). The same principles help to explain seemingly un-related applications. My intention is to uncover these principles to the best of my capability. Through the above examples, the reader should have by now a taste of the complexity of "spatial science" (ReVelle 1997). If nothing else, I hope that it arouses curiosity of the inquisitive, to whom we offer the following chapters of this book!

1.5   Summary

Making location, transport and land-use decisions is truly an art. But there is an information base that can or should explain, perhaps one factor at a time, these decisions (Transportation Research Board 1997). They range from technological, political, economic to social factors—just to name a few. Our purpose in this book is to trace the effects of these factors, not necessarily holistically, but "pulling one string at a time." Prescriptive and descriptive methodologies play a key role in clarifying these decisions. Some phenomena are "counter-intuitive." An analytical framework will extend our intuition a long way to seeing details that are not apparent to the unaided mind. The examples of airport location and the corridor development between New York and New Haven hopefully drive this home. There also exist fundamental principles underlying these prescriptive and descriptive techniques that govern the analysis of spatial-temporal information, such as Tobler's first law and the spatial weight matrix.

   Today's headlines are filled with competition among industries to locate in a certain locale, state or nation—perhaps for both economic and political gains. In fact, facility-location decisions have faced humankind throughout history. A familiar example can be found in the development of the steel industry in the United States. While iron ore was found in the convenient open pits of the Mesabi Range in Minnesota, coal was plentiful in Pennsylvania. Considering the amount of coal required, it was the more expensive of the two commodities to transport. Thus, we saw the historical development of steel mills in Pittsburgh, Pennsylvania. Whereas iron ore was shipped through Duluth, Minnesota, coal was collected at Pittsburgh via the Monongahela River. Perhaps this is another example of the linear-programming application, where the facility is located at one of the "extreme points," rather than somewhere in between.

   The tradeoff between fixed production costs and transportation cost is central to the spatial organization of an economy (Kilkenny and Thisse 1999). An economy characterized by large investments and low transport costs, although endowments may be spatially homogeneous, is likely to experience uneven spatial development with many activities concentrated in only a few places. Product differentiation fosters economic agglomeration to the extent that it relaxes price competition. This allows firms to select close-to-market locations that minimize transport costs. This seems to play a major role in our modern economies where differentiation in characteristics space substitutes for the more traditional form of geographic differentiation. As far as public policy, the more instruments firms have, including pricing policy, the more efficient they can be. Social welfare is often higher when firms are allowed to spatially price discriminate. Facility-location analysis would probably gain in relevance if pricing and strategic competition are integrated into the operational models. This difficult task could be achieved by combining operational location models with oligopoly models. The former tends to be discrete, while the latter continuous—with their different heritages.

In today's economy, globalization and technological innovation are dominant factors. It is critical to ask how locational conditions, compared with structure and strategies of the firm, play a part in the competitive market. For example, should we locate our manufacturing plants overseas where the labor is cheap? How many divisions or subsidiaries should we have in the company? When the innovation process is explained in terms of product-cycle and diffusion, relevant locational factors are stressed and a hierarchical pattern of innovation in space is arrived at. Thus, manufacturing, which is a routine operation, can be located where the factor inputs are the cheapest. On the other hand, evolutionary and network theories point to the relevance of historically evolved firm structures and strategies. Instead of deliberate planning, inertia comes into play. Empirical evidence seems to accommodate both schools of thought (Todtling 1992). In some firms there was a pronounced differentiation of innovation across space, such as concentration of research-and-development and product-innovation in the largest agglomerations as mentioned. However, strong innovation activities, corresponding more with the evolutionary model, were in addition identified in old industrial areas and newly industrialized rural areas simultaneously. It is required—more than ever—to discern the relevant factor that plays the pivoting role under these mixed development patterns, particularly when locational decision becomes paramount. With the advances in telecommunications, such as the Internet, accessibility (and hence location) takes on an expanded meaning (Gould and Golob 1998), particularly for information-based companies.

Today, mobile platforms allow full GIS functionality, perhaps with Global Positioning System as an option (Geo Info Systems 2000). Emerging remote sensing advances, the recent Shuttle Radar Topogrpahy Mission that mapped much of the globe, and the proposed National Spatial Data Infrasturcture all help to consolidate geographic information (Divis 2000). The development of integrated GIS systems and the availability of affordable differentially corrected GPS receivers have resulted in a paradigm shift in methods of infrastructure management (McNerney 2000). Practitioners are turning away from traditional forms toward map-based systems that have sophisticated analytical tools (Jha and Schonfeld 2000). Employing encryption technology, a computer can receive information in various open database compliant formats. This is accomplished by way of wireless links from mobile units. The data are then combined into a single database on the office server. This architecture allows users to conduct travel time studies at various level of resolution (Quiroga and Bullock 1999). Speeds and delays are readily computed along highway corridors for up-to-date accessibility measures. Obviously, these innovations come with its price tag. The greatest portions of costs, are for personnel and consultant services (Hall et al. 2000). There are general benefits associated with GIS. Efficiency benefits would result from the automation of previous manual efforts. Effectiveness benefits result from the traditionally intangible areas such as increased integration and accessibility of information for improved decision making.

Clearly the phrase "strategically located" will take on new meaning in real estate as the era of E-commerce comes into its own (Brennan 1999). Web use is doubling every 100 days. The Web has been adopted at a faster pace than even television or personal computers. Traditional retailers trying to make up for lost time by expanding quickly into E-commerce will create increased demand for bulk distribution, manufacturing and assembly buildings. On-line "book seller" Amazon.com, for example, recently selected a state-of-the-art cross-dock facility located near a major highway in Atlanta. This strategically located facility enables Amazon.com to provide easy access for small parcel trucks. These trucks deliver books and other products directly to its millions of customer in the Southeast United Sates. At the same time, the company benefits from state and county tax advantages. E-commerce is changing real estate's maxim from "location, location, location" to "location, price, service." The important point is that today's customers can order product on-line and choose to pick them up at the store or have them delivered. "Brick and mortar" properties have been converted into such use as mini-storage, telephone call centers even condominiums. The transformation is still in its infancy and we are only limited by our imagination about which direction we are heading.

Location, transport and land-use are highly capital-intensive and highly-valued decisions. In an emergency, a location decision often makes the difference between security and danger. Nowhere is it more apparent than the ongoing debate on hazardous-facility location, where fine line exists between perception and reality. Recent advances in multicriteria decision-making techniques can hopefully shed some light in the debate between the proponents of and the opponents of such facilities. In short, locational decisions have long-term effects on the regional and interregional economy, and profound implications on the quality of life. Modern analysis techniques can shed light on the matter and can have significant rewards in more informed choices (Karkazis and Thanassoulis 1998). Even more important is to be familiar enough with the analytical tools to guard against misuse of such techniques in arriving at half-baked 'truths'—which are often advanced by those possessing dangerous "half knowledge." While a single volume (or two including the companion volume) of this sort is certainly not a panacea, we hope it is a modest beginning in clarifying many ongoing debates.

"Familiar things happen, and mankind does not bother about them.
It requires a very unusual mind
to undertake the analysis of the obvious."
Alfred North Whitehead

CHAPTER TWO

FACILITY-LOCATION MODELS

While there are many variants to modelling location problems, it is the intent of this chapter to identify the basic building blocks of location models. We will present a taxonomy that is directed toward applications, in order to relate mathematical formulations to the standard types of problems encountered in location studies. These formulations are increasingly amenable to solution via integer programming (IP) or mixed-integer-programming (MIP) codes. Answers for decision-makers can be obtained in theory when we set aside algorithmic efficiency. Later part of this chapter will concentrate on a case study on locating satellite-tracking stations. This will show the applicability of location models. It will also illustrate an extension of the basic models to bypass some restrictive assumptions. Computational efficiency will then be discussed concerning the solution of this real-world application.

While there are whole books written on facility location (Daskin 1995, Drezner 1995, Francis et al. 1999, Love et al. 1988, Mirchandani and Francis 1990, Sule 1994), a distinguishing feature of this chapter is to relate facility-location models to the rest of this book. Thus we emphasize the profound implications of different spatial-separation metrics, making these models useful in situations well beyond traditional siting-decisions. Distance measures and facility-capacities are also shown to be intimately tied to stochastic location-and-routing, spatial competition, activity generation-and-allocation, and spatial-evolution decisions. The useful construct of multi–criteria decision–making is applied to practical site-selection problems (Malczewski and Ogryczak, 1995 & 1996). It is also exploited in the algorithmic employment of facility-location models to image classification. A final by-product of building these linkages is that efficient solution-algorithms for one type of models can now be transferable to variants of the model. A pre-condition is that the formal relationships are clearly delineated.

2.1 Private vs. public-sector models

By way of idealization, goods and services can be provided either by a commercial or governmental facility. While problems of locational decisions are common to both, the private and public sectors differ significantly in their operational contexts and basic practices. Ideally, we think the governmental sector has the public's convenience and well-being as its prime concern. The public agency, in carrying out its mission, is often subject to fiscal constraints as imposed by an openly scrutinized annual-allocation. The private sector, on the other hand, is operating under the main concern of conducting a successful business where the measure of success is often the `bottom line'. Such an objective is to be pursued within the context of market behavior and consumer preferences. While there is a definite budget constraint, the industry is often able to find additional capital and allocate its own resources freely without explicit public approval. Under these premises, we review some well publicized facility-location models.

2.1.1 *Simple plant-location problem.* The most basic location model is perhaps the *simple plant-location problem* (SPLP) or the *uncapacitated* facility location problem. This arises in the private sector where facilities are to be established with enough capacity to serve all demands. A lowest-cost objective is sought considering both facility and transportation costs:

$$minimize \ \sum_{j\in I} c_j x_{jj} + \sum_{i\in I}\sum_{j\in I} d_{ij} x_{ij} \tag{2-1}$$

$$subject \ to \qquad \sum_{j\in I} x_{ij} = 1 \qquad \forall i \in I \tag{2-2}$$

$$x_{ij} \geq 0 \qquad \forall i,j \in I \tag{2-3}$$

$$x_{jj} - x_{ij} \geq 0 \qquad \forall i,j \in I; i \neq j \tag{2-4}$$

$$x_{jj} \in \{0,1\} \qquad j \in I. \tag{2-5}$$

Here I is the set of locations for both demands for and supply of goods and services, the *location* variable x_{jj} is 1 if facility j is open and 0 otherwise, the *allocation* variable x_{ij} denotes that demand at location i is serviced from facility j if it is unitary valued, d_{ij} is the total of the variable capacity, production and distribution costs for servicing all demands at location i from facility j, and c_j is the fixed cost of establishing facility j. Here, we assume d_{ij} is proportional to the distance separation between i and j. Constraint **(2-2)** insures all demands are satisfied while constraint **(2-3)** requires a facility be *open* (available) to service a demand. More than one name has been given to this model. Some prefer to call it the "uncapacitated fixed-charge facility-location problem" (Daskin 1995, Nozick and Turnquist 1998).

Example

Figure 2-1 shows an SPLP, where there are two potential facility sites and three demand sites (Moore and Chan 1990). The fixed costs of siting the two facilities at the given

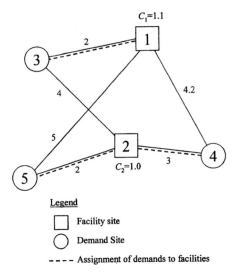

Figure 2-1 Simple plant location example

locations are displayed, so are the travel costs between any two nodes. The objective is to find the lowest cost to serve all the demands, considering both facility and transportation costs.

$$\min 1.1x_{11} + 1.0x_{22} + 2.0x_{31} + 4.2x_{41} + 3.0x_{42} + 2.0x_{52} + 5.0x_{51} + 4.0x_{32}$$

Demand for each location can be satisfied by either facility 1 or 2:

$$x_{31} + x_{32} = 1 \qquad x_{41} + x_{42} = 1 \qquad x_{51} + x_{52} = 1.$$

Demand cannot be supplied from facility 1 unless the facility is in place:

$$x_{11} - x_{31} \geq 0 \qquad x_{11} - x_{41} \geq 0 \qquad x_{11} - x_{51} \geq 0.$$

Similarly, demand cannot be supplied from facility 2 unless facility is in place:

$$x_{22} - x_{32} \geq 0 \qquad x_{22} - x_{42} \geq 0 \qquad x_{22} - x_{52} \geq 0.$$

The solution is graphically overlaid in **Figure 2-1**. It shows that facility 1 should serve demand 3 exclusively, while facility 2 serves demands 4 and 5. The total costs amount to 9.1 units. Sensitivity analysis shows that the cost will be increased by 0.9 with the elimination of facility 1, wherein all demands are now served by facility 2. An *uncapacitated* SPLP such as the one under discussion would allow this solution to specifically use the closest facility to serve a demand. There is no limitation on how much goods and services a facility can provide. This means a binary variable $x_{ij} \in \{0,1\}$, which does not allow a demand to be satisfied by more than one facility. □

2.1.2 Capacitated facilities, cost functions and candidate locations. The classic formulation of the *capacitated* location problem adds one additional constraint to Equations **(2-1)-(2-5)**. This constraint imposes a capacity on the production output of a plant to satisfy demands:

$$\sum_{i \in I} x_{ij} \leq \overline{P}_j x_{jj} \qquad \forall j \in I \tag{2-6}$$

where $\overline{P}_j$ is the production capacity of plant j. While this may be a more realistic formulation, it imposes computational intricacies in solution algorithms. Remember that solution to a location model is really composed of two steps: placement of the facility or plant j through the location variable $x_{jj} \in \{0,1\}$ and the assignment of demands to the selected facilities through the allocation variable $x_{ij} \geq 0$. In the uncapacitated model as formulated in the SPLP, the solution necessarily enforces the binary valuation $x_{ij} \in \{0,1\}$. The demands are allocated in an "all-or-nothing" fashion to the nearest open facility. The imposition of a production capacity destroys this property. x_{ij} can now logically assume a fractional value if $\overline{P}_j$ is non-integral.

There are a variety of formulations of the plant-location problem. The *fixed-charge transportation* problem, for example, has the transportation cost $d_{ij}(x_{ij})$, the second term of the objective function in Equation **(2-1)** as a non-linear function of the form $d_{ij}(x_{ij}) = C_i \delta_i + c_{ij} x_{ij}$. Here $\delta_i = 1$ if $x_{ij} > 0$, and 0 otherwise. c_{ij} is a constant unit-cost and the fixed cost C_i is to be activated if $x_{ij} > 0$. While this may introduce computational inconvenience, solution methods to the fixed-charge transportation model are well known[1]. The underlying cost-structure does not matter in the uncapacitated plant-location-problem. All that matters is the cost to transport demands from i to j. This is because flow will probably never be split in an uncapacitated problem; only one plant will supply each demand point's need (ReVelle and Laporte 1996). In the capacitated plant-location problem, though, the issue of the shape of the transport-cost function takes on importance in the calculation of the actual transport-cost. The shape of the cost function becomes important because, if the capacities are binding, flows to demand points can be less than the full values of demand. This results in some demand point receiving their needs from more than one source. If the transport-cost function were concave or a fixed-charge function, the actual cost would be higher than the linear estimate . This is because all the economies have not been captured with flows at split levels. Thus when plants are capacitated at levels that bind, a new and more difficult problem arises. If the transport cost has a fixed-charge component, the new problemmight be called the *capacitated-plant fixed-charge transport–location problem*. How this formulation might be exploited to yield a practical algorithm remains to be investigated. One possible approach would be to work along the lines

---

[1]See for example, Kennington and Unger (1976).

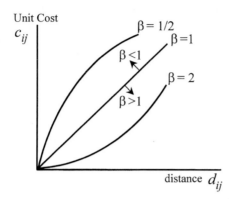

Figure 2-2 Cost functions of distance

suggested by Cabot and Erenguc (1984) for the fixed-charge transportation problem.

The more interesting problem is when c_{ij} is no longer a constant, but a convex or concave cost-function such as the form $c_{ij} = d_{ij}^\beta$. Here $\beta > 1$ means a convex cost-function and $\beta < 1$ means a concave function. A sketch of these cost functions is shown in **Figure 2-2**. In the first part of this chapter we will generally stay with linear distance-functions, where c_{ij} is a constant, or $\beta = 1$. This means the cost function is both concave and convex. We will also stay with formulations where a finite set of discrete locations I are the candidate sites for both facilities and demands. This will simplify a number of issues, particularly those associated with the shape of the cost function. Yet it is also a more practical formulation for application purposes, as we will explain in sequel.

2.1.3 *p-median problem.*

2.1.3 p-median problem. A fundamental location problem, the *p-median problem*, seeks to place p facilities strategically among the demands. The p-median problem is also referred to as a public-sector (rather than private-sector) model. It consists of finding the optimal location for exactly p facilities (corresponding to a fixed budget) to meet a specified demand at the lowest possible transportation cost. The p-median problem (Hakimi 1964) uses the average distance (or travel time) between service and demand points to determine the servicing costs of a given location. The fixed capital-cost for a facility c_j disappears from the objective function. This approach is particularly suited for the public sector. It avoids dollar comparisons between options and focuses on the average travel-distance or total weighted travel-distance (some measure of public convenience). The model assumes that each demand is being served by the nearest facility. Thus the goal is to minimize the average travel-distance between demand nodes and the p facilities in a network. Let I be a set of nodes, p be a specified number of facilities, f_i be the frequency of demand at node i (such as the surrogate measure of population), d_{ij} be the Euclidean distance between nodes i and j, and x_{ij} be a binary variable assuming

1 and 0 values depending on whether facility *j* serves *i*. The *p*-median problem can be stated as the binary program.

$$minimize \sum_{i \in I} \sum_{j \in I} f_i d_{ij} x_{ij}$$ (2-7)

subject to

$$\sum_{j \in I} x_{ij} \geq 1 \qquad \forall i \in I$$ (2-8)

$$\sum_{j \in I} x_{jj} = p$$ (2-9)

and Equations **(2-3)**, **(2-4)** and **(2-5)**. other words, we minimize the weighted distance between demand and supply points—sometimes called the *min-sum* objective. The constraints ensure that each demand point *i* is served by the closest facility *j*, a facility at point *j* servicing a demand point must be open, and exactly *p* facilities are selected.

A variation on the *p*-median problem is to impose a constraint to limit the distance that any user must travel to reach a service facility. This distance is called the maximal, desirable service distance. The use of the maximal service distance is discussed in depth in Toregas and ReVelle (1972): If *S* is the maximum, desirable distance between demand *i* and a facility *j*, let us define

$$N_i' = \{j \mid d_{ij} \leq S\} \qquad \forall i \in I.$$ (2-10)

Expression **(2-9)** in the original *p*-median problem is now modified to read

$$\sum_{j \in N_i'} x_{ij} \geq 1 \qquad \forall i \in I.$$ (2-11)

Limiting the distance that must be travelled to reach a service point also has the effect of reducing the average distance that must be travelled by any user. Assuming feasibility, it also results in no demand points relying on facilities further than the desirable distance *S* away. This model can be used to guarantee a worst possible performance of the system, which is important in siting critical facilities such as emergency shelters and hospitals. The *p*-median problem was first solved using classic LP and MIP methods. Different authors have attempted more efficient solutions to the p-median problem and are documented in such papers as Berlin et al. (1976), Swain (1974) and Mirchandani (1990). For the time being, we will concentrate on formulation. Solution efficiency will be discussed in the latter part of this chapter.

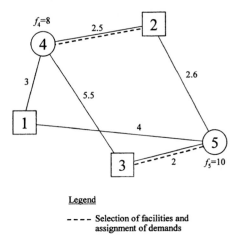

Legend

- - - - Selection of facilities and
       assignment of demands

Figure 2-3 *p*-median example

Example

The objective of a *p*-median problem is to find which facilities to use in order to serve the demand at the various locations. There is only enough budget to build two facilities in this example problem. **Figure 2-3** shows that demands 4 and 5 must be served using any two of the three facilities 1, 2, and 3 (Moore and Chan 1990).

$$\min\ (8)(3)x_{41} + (8)(2.5)x_{42} + (8)(5.5)x_{43} + (10)(4)x_{51} + (10)(2.6)x_{52} + (10)(2)x_{53}$$

Demands 4 and 5 must be served by one or more of the three facilities:

$$x_{41} + x_{42} + x_{43} \geq 1 \qquad x_{51} + x_{52} + x_{53} \geq 1.$$

Demand cannot be served by a facility unless the facility is in place:

$$x_{11} - x_{41} \geq 0 x_{11} - x_{51} \geq 0$$
$$x_{22} - x_{42} \geq 0 x_{22} - x_{52} \geq 0$$
$$x_{33} - x_{43} \geq 0 x_{33} - x_{53} \geq 0.$$

Two of the three facilities are to be selected:

$$x_{11} + x_{22} + x_{33} = 2.$$

The solution to this problem is again overlaid in **Figure 2-3**, which shows that facilities 2 and 3 were selected. Facility 2 serves locations 4 exclusively, while facility 3 serves location 5. The total costs were minimized at 40 units. Notice that no service distance restriction is placed on this example problem. □

2.2   Minimal requirement vs. maximal service

Services are either mandatory or discretionary. Again from a public- vs. private-sector standpoint, one can think of certain basic survival needs of the disadvantaged. They are requirements that must be fulfilled by a governmental facility, while certain luxury items are only sold by private companies as much as the market would bear. Having said that, one can also say that public well-being is best served if the services can reach as large a target-population as possible. Similarly, certain markets must be covered in order for a company to be competitive.

2.2.1   The set-covering problem.   Following the idea of maximal service-distance, Toregas (1971) presented a formulation of facility-location problems. His problem is to find a minimum number of p facilities $p^*(s)$ such that no user is further than a maximum service-distance away. Here p becomes a variable rather than being a constant. It is formulated as:

$$\textit{minimize} \sum_{j \in I} x_{jj} \qquad\qquad (2\text{-}12)$$

subject to constraints **(2-11)** and **(2-3)**. This results in a *set-covering* problem. This model formulation can be used for determining the feasibility of the maximal-service p-median problem. Suppose one compares the solution p^* of the p-median problem to the minimum number of facilities $p^*(s)$ obtained in the location set-covering problem .It can be found whether the p-median problem with a maximum-distance constraint is a feasible model, i.e., $p^* \geq p^*(s)$. The location set-covering problem is applicable when there is a no restriction on the number of facilities that can be constructed, that is, the resources and finances are there to implement the plan. Toregas et al. (1971) apply the location set-covering problem to the location of emergency service facilities, where the sheer urgency of the situation often overshadows other considerations. In the same spirit as the simple plant-location model, it is often used to model private-sector locational-decisions. In such industries as retail and service, the least number of stores are to be found in a master plan to cover the market.

Example

The objective of the set-covering problem is to find the minimum number of facilities required to ensure no demand is further than a maximum service-distance of S. **Figure 2-4** shows a five-node example, with a service-distance contour of 2 hours drawn from demands 3, 4, and 5 (Moore and Chan 1990): $\min x_{11} + x_{22}$. Demand must be served by a facility within the maximum service-distance S:

$$x_{31} \geq 1 \qquad x_{41} + x_{42} \geq 1 \qquad x_{51} + x_{52} \geq 1.$$

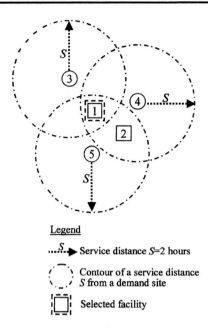

Legend

...*S*.➤ Service distance *S*=2 hours

(ˉ) Contour of a service distance
⌣ *S* from a demand site

▢ Selected facility

Figure 2-4 Set-covering example

Demand cannot be served unless the appropriate facility is in place:

$$x_{11} - x_{31} \geq 0 \qquad x_{22} - x_{32} \geq 0$$
$$x_{11} - x_{41} \geq 0 \qquad x_{22} - x_{42} \geq 0$$
$$x_{11} - x_{51} \geq 0 \qquad x_{22} - x_{52} \geq 0.$$

It turns out that only one facility is required for satisfying the demands. The solution is once again overlaid in **Figure 2-4**. This Figure shows that facility 1 was the only facility that could serve the demand within the 2-hour travel-time limit for each of the three locations. For this simple example, the solution is easily confirmed graphically. □

2.2.2   The maximal-coverage location problem.   In most public-sector locational decisions, there is a budget constraint for building facilities (as mentioned). If either the max-distance *p*-median or location set-covering method requires more budget than is available, modifications must be made to the model to take care of infeasibility. This results in a model that optimizes coverage for a fixed budget. To accommodate this, Church and ReVelle (1974) added a second distance *S'* greater than the desired service-distance *S*. This represents the *absolute* maximum distance any demand point can be away from a service facility. This assures that no user is outside the absolute distance *S'*, but allows *S* to represent a variable ranged between 0 and *S'*. This modification results in what is called the *maximal-coverage-location problem* with mandatory closeness constraints. The

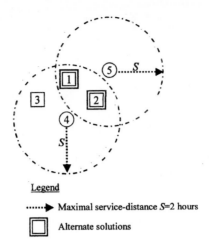

<u>Legend</u>

······▶ Maximal service-distance $S=2$ hours

◻ Alternate solutions

Figure 2-5 Maximal-coverage example

mathematical formulation of the maximal-coverage-location model is simple. It seeks to maximize the population served while meeting the constraints on the desired service-distance (Equation **(2-10)**) and the budget as represented by the fixed number of facilities p (Equation **(2-9)**). On top of these considerations, the complete formulation as developed by Church and ReVelle (1974) can be stated when the additional maximal-coverage objective and service-distance constraint are employed:

$$maximize \ \sum_{i \in I} f_i x_{ii} \tag{2-13}$$

$$subject \ to \ \sum_{j \in N_i'} x_{jj} \geq x_{ii} \quad \forall i \in I . \tag{2-14}$$

Suppose one relaxes the length of the desired service-distance S (up to S'). This type of model formulation is most valuable for trading off travel distance against the degree-of-coverage for various amounts of budget expenditure. The degree of coverage can be represented as a simple percent of all demand points. Imposing the absolute distance S', the various amounts of expenditure—ranging from the corresponding max-distance S to S'—can again be represented by the different number of facilities required. This is similar to the set-covering problem discussed previously.

Example

The objective of the maximal-coverage-location problem is to locate a fixed number of facilities while maximizing the service rendered to demands. **Figure 2-5** shows two demands at 4 and 5, with a maximal service distance of two hours. Three sites 1, 2, and 3 are also displayed, from which one facility is to be located: max $10x_{44} + 8x_{55}$. Demand can be served only if a facility is available within S hours:

$$x_{11} + x_{22} + x_{33} - x_{44} \geq 0 \qquad x_{11} + x_{22} - x_{55} \geq 0.$$

Only one facility is to be selected: $x_{11} + x_{22} + x_{33} = 1$. Solution shows that either facility 1 or 2 met each constraint. Either one could be used to serve the demands 4 and 5. The objective function was maximized at 18 units. Should the non-optimal facility 3 be used, the coverage metric would be compromised by 8 units. This is exactly demand 5, since facility 3 cannot reach demand 5. By relaxing the 2-hour service-distance requirement, however, the single facility 3 can presumably reach demand 5 as well. This can be shown by a larger circle drawn from demand 5. It is through such relaxation that maximal coverage can be traced as a function of S. □

2.3 Theoretical and computational links

If one reexamines the three location problems discussed above—p-median, set-covering and maximal-coverage—s/he will find the three are closely related. The set-covering objective **(2-13)** is obviously a special case of the p-median objective **(2-7)** if f_i is set at unity and i set equal to j (remembering d_{ij} is an arbitrary constant for all j and can be "factored out").

2.3.1 Computational links. At first examination, the maximization objective **(2-13)** for a maximal-coverage model seems different from the minimization objective **(2-7)** of a p-median model. But a more careful scrutiny would show that the two objectives are complementary and that the maximal-coverage location problem is again a special case of the general p-median problem. To see this let us define for mandatory closeness distance S' from j:

$$d'_{ij} = \begin{cases} 0 \ \ if \ d_{ij} \leq S' \\ 1 \ \ if \ d_{ij} > S' \,. \end{cases} \tag{2-15}$$

With this modified distance d'_{ij}, objective **(2-7)** becomes

$$minimize \sum_{i \in N_j'} \sum_{j \in I} f_i x_{ij}, \tag{2-16}$$

where N_j is the set of i *inside* the distance S' from facility j. The result is that the demand served *outside* S from a facility $\bar{N}_j$ is minimized. Now set j to be i and apply Expression **(2-11)** (which is satisfied at equality). With this notational restatement, we have

$$minimize \sum_{i \in \bar{N}_j} f_i x_{ii}. \tag{2-17}$$

Having the closeness constraint on **(2-13)**, one can rewrite the maximization objective as

$$maximize \sum_{i \in N_j'} f_i x_{ii} \ .$$

(2-18)

Expression **(2-17)** strives to minimize the demand served *outside* the distance S, while **(2-18)** maximizes the demand *within* the distance. They are effecting the same solution. Now the two problems—p-median and maximal coverage—are related to one another if one can see that constraint **(2-14)** and **(2-2)/(2-3)** both require that demand nodes be served. In the case study of a satellite-tracking station later, we will show that the graphs used to solve a maximal-coverage model and a set-covering model are similar. Both boil down to be variants of the p-median model. In Section 2.6 of this chapter, we will also show the relationship between *Center* and *Median* problems, where center is defined as a location that minimizes the furthest-away demand. Again the equivalence is established the two through a distance transformation along the line of Equation **(2-15)**.

Krarup and Pruzan (1990) point out that the p-median and SPLP models can be viewed as a location problem for which the underlying network is bipartite[2], with facilities represented as the left column of nodes and the demands as the right column of nodes. Two features distinguish SPLP from the p-median model: (a) a nonnegative fixed-cost is associated with each potential facility site and this fixed cost is incurred only if a facility is actually placed at that site, and (b) the number of facilities to be located is not pre-specified. A hybrid of the two models can be constructed: the *p-uncapacitated facility-location problem* (*p*-UFLP). *p*-UFLP can be viewed as an extension of the p-median problem in that fixed costs like those encountered in SPLP are associated with potential facilities. Alternatively, *p*-UFLP is similar to SPLP with the additional constraint that the number of facilities to be established is pre-specified as p. Objective function **(2-1)**, combined with constraints **(2-2)** through **(2-5)**, and **(2-9)** fully describe the model. In this formulation, f_i is assumed to be zero without loss of generality. If $c_j = 0$ for all j, *p*-UFLP reduces to the p-median problem. Similarly, SPLP is obtained if the requirement **(2-9)** that exactly p facilities be established is ignored. When $c_j = 0$ and the p-facility constraint is non-binding, both SPLP and p-median models yield the same solution.

2.3.2 Hierarchical generalizations. Recently there has been significant progress toward generalizing the p-median, location set-covering and maximal-coverage models. For example, hierarchical location models have been proposed (Moore and ReVelle 1982) to distinguish several levels-of-service facilities. Here a demand point is covered by a given level-of-service if some member of the facility hierarchy eligible to provide that service is present within an appropriate distance.

---

[2]For a definition of bipartite network, see the "Prescriptive Tools" chapter under the "Integer or mixed-integer-programming" section (Chan 2000).

Given a limited number of facilities, a maximal-coverage objective is adopted. Rado (1988) formulated a generalized multi-facility location problem in n dimensional space. Again hierarchical service-levels are defined. For example, a town is to be located to serve each region among a multi-region territory and a city is to serve the towns in turn. A more general formulation is given by Aykin (1988) in which partitioning of the territory into regions is not assumed. Perhaps the most general hierarchical facility-location model is proposed by Mirchandani (1987). The procedures for allocating demand types to facility types include all the three types: "successively inclusive," "successively exclusive," and "locally inclusive-successively exclusive." In a successively inclusive hierarchy, a type-k facility ($k = 1, 2, \ldots, K$) serves demands of types $1, 2, \ldots$, up to k. In a successively exclusive hierarchy, a type k facility serves only demands of type k. In the locally inclusive-successively exclusive case, a type k facility serves demand types 1 through $k-1$ locally, but serves only type k demands from outside its locality. It can be shown that the Mirchandani model incorporates the p-median model, uncapacitated plant-location model and others.

Most traditional facility-location models assume a one-to-one pairing between facility j and demands i, where i is served by j exclusively. As evidenced by the discussions above, there is a fair amount of recent interest in problems where service to i is rendered by more than one facility. Thus overlapping service is provided to demands i from facility $j_1, j_2 \ldots j_{|J|}$. While such an extension of classical facility-location models is fully justified by real-world applications, the problem becomes a good deal more complex to solve. Now consider another extension, in which only the appropriate facility of type k can be used to serve demand of type k (the successively exclusive hierarchy). This will no doubt increase the dimensionality of the problem, since the running index i-j becomes i-j-k. Imagine a facility can change from type k_1 to k_2 in order to cater for prevalent demand types, and several facilities of k_1 type can be co-located in j. Our job now becomes one of not only pairing i to j to k, but also in allocating facilities of type $k_1, k_2, \ldots, k_K$ among sites $j_1, j_2, \ldots, j_{|J|}$. As if life is not complicated enough, we further require that facilities $k_1, k_2, \ldots, k_K$ need to be assigned in bundles of m_j to site j. In other words, we assign either zero or m_j facilities to j; any number between 0 and m_j is not acceptable. To top all of these, we can have a nonlinear, probabilistic objective-function consisting of 0–1 and integer variables (Steppe et al. 1992). Variants of the above generalizations of the basic facility-location models will be illustrated later on in this chapter. They include satellite-tracking-station location, and satellite-imaging-station location. Reference is made here to the generalized search-and-rescue problem (Steppe et al. 1996) as an illustration of a hierarchical maximal coverage location-allocation problem. In our "Problems and Case Studies" web site, we construct this model and ask the readers to convert it to a solvable formulation. We will focus our attention not only on model formulations, but solution methods also as we go through these case studies.

2.3.3 Deterministic, probabilistic and stochastic methods. All of the location
problems discussed previously involve deterministic approaches to the model

formulation. That is to say distances between candidate sites and demands are assumed to be constant and known quantities. To say that any or all distances or demands are random variables would turn the problem into a probabilistic one. A probabilistic network was first applied to p-median problems by Mirchandani and Odoni (1979). A probability value P_k is defined for a link or a node if it is in state k, with as many states as K to model the full probability distribution. The formulation is then an extension of the p-median problem:

$$minimize \ \sum_{k=1}^{K} \sum_{i \in I} \sum_{j \in I} P_k f_{ik} d_{ijk} x_{ijk} \qquad (2-19)$$

$$subject \ to \quad \sum_{j \in I} x_{ijk} \geq 1 \qquad \forall i \in I \ and \ k = 1,2,...,K; \qquad (2-20)$$

$$x_{ijk} \leq x_{jj} \qquad \forall i,j \in I; \ i \neq j; \ and \ k = 1,2,...,K \qquad (2-21)$$

and constraint (2-9). The variable x_{ijk} simply takes on the interpretation of "a demand node i being assigned to facility j in state k." For example, the decision variable x_{ijk} may be used to reflect the different assignment of facilities to a demand point due to the variation in travel time between i and j at the peak and off-peak hours. Here K is 2 and d_{ij} has 2 corresponding values. Similarly, two sets of demands can be recorded for node i, f_{ik}, corresponding to the different periods of the day. Somewhat counter-intuitively the average travel-time in a probabilistic p-median problem is less than (or equal to) the average time of the equivalent p-median deterministic problem. One can perhaps 'explain' this result by remembering that both demand and separation are probabilistic in the 'stochastic' p-median problem. This provides a larger degree-of-freedom for location-and-routing decisions than the deterministic counterpart. Perhaps the most important result of this probabilistic extension is that at least one set of expected p-medians exists on the nodes in the network. Furthermore, *if the travel cost (or utility) is a concave and non-decreasing function of distance, then at least one set of expected optimal p-medians exists on the nodes.* This is a generalization of the result Hakimi (1964) found for deterministic networks.

Louveaux (1986) studied two classical location-models: the private-sector SPLP and the public-sector p-median problem. These were transformed into a two-stage stochastic-program with *recourse* when uncertainty on demands, variable costs *and* selling prices is introduced (to account for unmet demands). The first-stage decisions are the *location* and size of the facilities to be established and the second-stage decision is the *allocation* of the available service to the consumers. By recourse is meant that the second-stage decision-variables depend on the particular realization of the random events that govern revenue, cost and demand allocation. For the p-median problem, one way to handle the budget constraint (as represented by the fixed number of facilities) is through charging some 'price' for each unit of service. In this way, the 'revenue' offsets the expenditure. Note that a selling price allows tradeoffs to be made between the cost of increasing the capacity (as determined in the first stage of a stochastic program), the net 'profit' of selling goods and the probabilities of the various demand levels. Through this transformation, relationship between the stochastic versions of the SPLP and *p*-median is established. It was also found that for a given budget, the price to be charged in the stochastic *p*-median problem is larger than the deterministic case. This is a natural consequence of uncertainty. More important, *if the cost (or utility) function is linear, then the decisions taken in the public- and in the private-sector model will be exactly the same.* This extends the result from the deterministic case when the fixed cost $c_j = 0$ and the *p*-facility constraint is non-binding. Notice that the word 'stochastic' is used here typically in stochastic programming to mean probabilistic revenue, cost and demand-allocation rather than the strict definition of time-dependent probability.

In the general context, two types of uncertainty are found in *stochastic* (time-dependent) facility-location problems (Odoni 1987) in which services are rendered to geographically dispersed demands: (a) random travel-times along the arcs of the network, (b) queuing phenomena. The latter arises from a combination of finite capacity at the facilities and random location of demands on the network. It can also be a consequence of random arrival time of the demands and random service-times. Under uncertainties of type (a), problems can be treated analogous to the classical deterministic p-median problem, as we have shown above. With uncertainties of type (b), most research efforts have been focused upon single facility, due to the severe analytical difficulties associated with multi-facility problems. The two most commonly addressed problems are labelled *Loss Median* and the *Stochastic Queue Median*. The former refers to "balking" demands or demands that are "lost to the system" when all service facilities are busy. The latter allows a waiting line to be formed although no service can be rendered at the time. Both models seek locations that minimize a function of wait time (among other factors). Due to space limitations, however, these discussions are generally outside the immediate scope of this chapter. Interested readers are referred to the "Measuring Spatial Separation" chapter (besides Odoni's paper cited above). In this forthcoming chapter, the relationship between deterministic and stochastic facility location is discussed formally, including the conditions under which an optimal location occurs at a node.

Example of Probabilistic P-Median

Figure 2-6 gives an example of a probabilistic *p*-median problem (Moore and Chan 1990). The probabilistic states consist of off-peak (idol) and peak (busy) periods, with probabilities of 0.6 and 0.4 respectively. The objective is to minimize the cost of serving the demands while accounting for all possible states.

minimize

(off-peak) $(0.6)[(8)(3)x_{411} + (8)(2.5)x_{421} + (8)(5.5)x_{431} + (10)(4)x_{511} + (10)(2.6)x_{521} + (10)(2)x_{531}] +$
(peak) $(0.4)[(9)(3.5)x_{412} + (9)(3)x_{422} + (9)(6)x_{432} + (14)(5)x_{512} + (14)(3)x_{522} + (14)(2.5)x_{532}]$

Subject to

(off-peak)	$x_{411} + x_{421} + x_{431} \geq 1$	$x_{511} + x_{521} + x_{531} \geq 1$
	$x_{11} - x_{411} \geq 0$	$x_{11} - x_{511} \geq 0$
	$x_{22} - x_{421} \geq 0$	$x_{22} - x_{521} \geq 0$
	$x_{22} - x_{421} \geq 0$	$x_{22} - x_{521} \geq 0$
	$x_{33} - x_{431} \geq 0$	$x_{33} - x_{531} \geq 0$

(peak)	$x_{412} + x_{422} + x_{432} \geq 1$	$x_{512} + x_{522} + x_{532} \geq 1$
	$x_{11} - x_{412} \geq 0$	$x_{11} - x_{512} \geq 0$
	$x_{22} - x_{422} \geq 0$	$x_{22} - x_{522} \geq 0$
	$x_{22} - x_{422} \geq 0$	$x_{22} - x_{522} \geq 0.$
	$x_{33} - x_{432} \geq 0$	$x_{33} - x_{532} \geq 0.$

The solution is overlaid on **Figure 2-6**, in which facility 2 and 3 were selected to serve the demands. Facility 2 serves demand 4 exclusively, while facility 3 serves demand 5. The total costs were minimized at 48.8. Should one solve the two states—peak and off-peak—individually, we yield the same assignment. The off-peak solution was illustrated in the original discussion of the p-median problem already. The peak solution gives an objective function of $(9)(3)(1) + (14)(2.5)(1) = 62$. The 'average' objective-function value will be $(0.6)(40) + (0.4)(62) = 48.8$, which is identical to the probabilistic solution above. This is so since both peak and off-peak assignments are identical. Should the assignments be different, the probabilistic solution will be less than the 'average' value as mentioned previously. More will be said about this phenomenon in Chapter 3: "Measuring Spatial Separation." □

2.3.4 Location on a Plane and Network. In location problems, distances can be measured in three different ways. The classical method is to use the Euclidean metric, composed of straight-line distances between two points. This method is

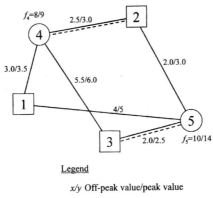

Legend

x/y Off-peak value/peak value

---- Selection of facilities and
assignment of demands

Figure 2-6 Probabilistic p-median example

most applicable to problems of air and ocean travel when the "great circle distance" can be approximated by a straight line in the limiting case. It can also be applied to problems such as siting an offshore natural-gas pipeline (Engberg et al. 1982) and power-plant siting (Cohon et al. 1980). In many location-and-routing methods the distances between nodes are measured using a *rectilinear* metric, often called the *Manhattan* metric. Travel is limited to lines parallel to the coordinate axis, in which the distance between two points (x^1, y^1) and (x^2, y^2) would be $|\mathbf{x}^1 - \mathbf{x}^2| + |\mathbf{y}^1 - \mathbf{y}^2|$. These types of problem formulations are most applicable to urban areas that have streets that are orthogonal. Travel is therefore limited to east-west and north-south directions (hence the word 'Manhattan' as in Manhattan, New York City). The types of distances mentioned above can also be called locational analyses on a plane and a semi-network, respectively. Finally, we have location in a full-fledged network that can assume any geometry. Locational analysis on a network is computationally more complex because of the need to determine the route. Example of such a route is the travel itinerary on a street grid consisting of east-west and north-south thoroughfares. However, network analyses address an important class of real-world problems, since very few distance-separations are Euclidean.

The "continuous p-median" of a network (Hansen and Labbe 1989) can be thought of as an example of a mixed planar/network facility-location problem. Let us define the distance between an arc and a point of a network as the maximum distance from that point to any point on the arc. A continuous median of the network is a point such that the sum of the distances between all arcs and the point is minimum. A continuous p-median is a set of p points such that the sum for arcs of the distance to the closest point of that set is minimum. It can be shown that *the set of nodes and mid-points of arcs always contain a continuous p-median*. Armed with this knowledge, powerful algorithms for the usual p-median problem can be used to solve the continuous version. In general, discrete facility-location problems on a network can be classified according to three characteristics (Evans and Minieka 1992):

a. The potential site of the facility to be located—either at a node/vertex or on an arc.
b. The location of demands—either at nodes/vertices or on an arc.
c. Objective function—either to minimize the total cost (or utility) to all demands or to minimize the maximum cost (or utility) to any demand. The former is often called *min-sum* (or equivalently minimum average-cost) and the latter *min-max* criterion.

The discussions thus far or Chan (2000) contain examples of these problems—the center, median and Steiner–Weber problems have been mentioned. Based on these previous discussions, the reader can probably relate to **Figure 2-7**. The Figure classifies the various types of discrete facility-location problems accordingly.

For the median problem we have been concentrating on so far, we can illustrate these terminologies in **Figure 2-7**:

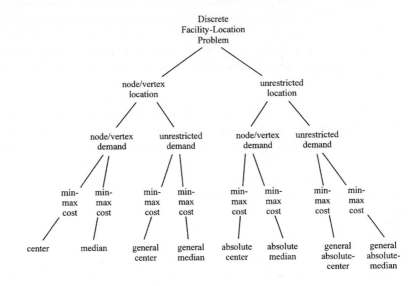

Figure 2-7 Types of discrete facility-location problems (Evans and Minieka 1992)

a. A median of a network or graph is any node or vertex that is closest to the demands on the average, where demands occur only at vertices.

b. A *general median* of a network or graph is any vertex that is closest to the demands on the average. Here the demands lie along links or edges (including nodes or vertices).

c. An *absolute median* is any point on an arc that is closest to the demands on the average, where demands occur only at nodes or vertices.

d. A *general absolute median* is any point that is closest to the demands on the average, where both facilities and demands are located anywhere on the arcs.

The *center* problems[3] turn out to be analogous to the median problems in terms of these definitions.

a. A *center* of a graph is any vertex whose furthest demand vertex is a close as possible. In this first case, both the facility and demand occur at vertices.

b. *A general center* of a graph is any vertex whose furthest demand point in the graph is as close as possible. Notice that the demand points can lie anywhere along arcs or vertices.

---

[3] A center is a location that guarantees proximity even for the most distant demand, as explained in the "Economics" chapter (Chan 2000).

c. An *absolute center* of a graph is any point whose furthest demand vertex is as close as possible, where the facility can be anywhere on an arc, including the vertices.

d. A *general absolute center* of a graph is any point whose furthest demand point is as close as possible. Here, both the facility and demand can be on an arc or vertex.

Obviously, there can be more than one facility in both the median and center problem. We call them the p-median and *p-center* problems respectively, of which the *p*-median model has been discussed in some length already earlier in this chapter.

2.4 Center Location Problems

Taking a leave from the medians, consider the problem of finding another centrally located node(s) (vertex(es)) in a network, the center. As previously mentioned, a center is any vertex *j* with the property that the most distant demand-node *i* from it is as close as possible. Let

$$D_i = \max_j(D_{ij}) \tag{2-22}$$

denote the maximum distance of any demand node *i* from facility node *j*, where D_{ij} is the length of the shortest path from *i* to *j*. A center j_c is $\min_i D_i$. If we replace the maximization operator in Equation **Figure 2-7** above with a summation operator, then we can convert back to the median problems discussed previously. We now write D_{ij} instead of d_{ij} for the center and median problems on a network, reflecting distance on a *route* from *i* to *j* rather than the rectilinear or Euclidean distance from *i* to *j*. Determining the center is then a straightforward min-max problem while finding the median is a min-sum problem. The p-center problem (Drezner 1989) is to find the locations for *p* facilities that minimize the maximal distance between demand points *i* and their closest facility *j*. In other words it picks *p* facilities based on the smallest of the metrics as defined by

$$D_j' = \max_{i \in I} \left\{ \min_{1 \leq j \leq p} D_{ij} \right\} \tag{2-23}$$

2.4.1 Center examples. A programming example of the *p*-center problem is shown in our now familiar network **Figure 2-8** (Moore and Chan 1990). The objective of the 1-center problem, for example, is to find the facility location that minimizes the maximum distance between it and the demands: min $(D_4 + D_5)$. A unique feature of the center problem is the constraints used to achieve maximum distance from both demands 4 and 5, D_4 and D_5 respectively: $D_4 - 3x_{41} \geq 0$, $D_4 - 2.5x_{42} \geq 0$, $D_4 - 5.5x_{43} \geq 0$, $D_5 - 4x_{51} \geq 0$, $D_5 - 2.6x_{52} \geq 0$, and $D_5 - 2x_{53} \geq 0$. Only one facility is to be selected: $x_{11} + x_{22} + x_{33} = 1$. The rest of the constraints are identical to the *p*-median problem, including specifying that each demand must

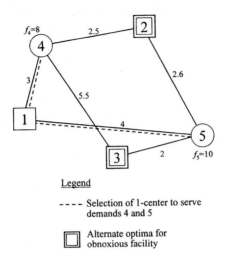

Figure 2-8 *p*-center example

be served, and demand cannot be served unless a facility is available. Notice the strict inequality is part of the model: $D_4 + D_5 > 0$.

The model solution is overlaid on **Figure 2-8**. Facility 1 was the facility of choice, serving demands 4 and 5. For a simple problem such as this, it is easy to check that this is the correct answer by inspecting the following Table.

(i) \ (j)	1	2	3	max
4	3	2.5	5.5	5.5
5	4*	2.6	2	4

The minimum of the maximum distances 5.5 and 4 is 4, which corresponds to choosing facility 1–same as the solution obtained previously. In locating two centers, Equation **(2-22)** is operationalized by this Table where the asterisk (*) identifies the best facility to serve a demand. Thus the best facility to serve demand 4 is 2, and the best facility to serve demand 5 is 3.

(i) \ (j)	1	2	3
4	3	2.5*	5.5
5	4*	2.6	2*

Here, if we pick facilities 1 and 2, the furthest demand proximal to a facility is 3 units away. If we pack facilities 1 and 3, the furthest demand is also 3 units away.

If we pick facilities 2 and 3, the furthest is 2.5 units. This results in selecting facilities 2 and 3.

In a sense, this is not a very good example since there really is no selection since facilities 2 and 3 dominate over 1. Perhaps a better example is when there is an additional demand at node 6, as shown below

(*j*) (*i*)	1	2	3
4	3	2.5*	5.5
5	4	2.6	2*
6	2.1*	5	6

Now out of the three viable facility-candidates, we choose two that minimizes the furthest distance from the proximal facility. Thus if we choose facilities 1 and 2, the furthest demand 5 can be covered in 2.6 units distance. If we choose 1 and 3, the furthest demand 4 is covered in 3 units distance. Finally, choosing 2 and 3 means a furthest demand of 5 units. The best choice is therefore to pick facilities 1 and 2.

2.4.2 *p-center.* To review and formalize the above examples, the problem of finding the one-node center is quite easy. The problem can be solved by inspection using the shortest-distance matrix $\bar{D}$. For each column of $\bar{D}$, we record the largest component. The node that corresponds to the column that gives the minimum among the numbers recorded is the one-node center. The *p*-center problem can be formulated in its complete form as follows in case of linear costs:

$$\min \left[\max_{i \in I; j \in J} f_i D_{ij} x_{ij} \right]$$

(2-24)

$$\sum_{j \in J} x_{ij} = 1 \qquad \forall i \in I$$

(2-25)

$$\sum_{j \in J} x_{jj} = p$$

(2-26)

$$\sum_{i \in I} x_{ij} \le (|I| - p + 1) \, x_{jj} \qquad \forall j \in J$$

(2-27)

$$x_{ij} \in \{0,1\} \qquad \forall i \in I, \, j \in J.$$

(2-28)

where *I* in this case is the set of demand nodes and *J* the set of candidate facility sites.

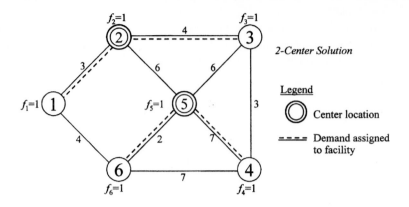

Figure 2-9 Network of Eastaboga, Alabama (Francis et al. 1999)

Example

Reference is made to a simplified network of Estaboga, Alabama shown as **Figure 2-9** . One, two and three centers are to be located to minimize the travel distance from the furthest demand to a center. For this example, each node has a demand of 1 ($f_i = 1$ for all nodes i) and the arc-distances are as shown. All six nodes are candidates for siting a center ($p = 1$). The corresponding model based on the above set of equations looks like:

$$\min \;[\max(3x_{21},7x_{321},6x_{561},\mathbf{10x_{4321}},4x_{61}),$$

(min-path tree to site 1)

$$\max(3x_{12},4x_{32},\mathbf{7x_{432}},6x_{52},7x_{622}),$$

(min-path tree to site 2)

$$\max(7x_{123},4x_{23},3x_{43},6x_{53},\mathbf{8x_{653}}),$$

(min-path tree to site 3)

$$\max(\mathbf{10x_{1234}},7x_{234},3x_{34},7x_{54},7x_{64}),$$

(min-path tree to site 4)

$$\max(6x_{165},6x_{25},6x_{35},7x_{45},2x_{65}),$$

(min-path tree to site 5)

$$\max(4x_{16},7x_{216},\mathbf{8x_{356}},7x_{46},2x_{56})\;]$$

(min-path tree to site 6)

where the shortest paths are explicitly spelled out in the x_{ij} variables. For pedagogic reasons, we highlighted the *absolute* maximum-distances to a demand with bold fonts, i.e., the furthest away possible from a demand. For example, the furthest demand is 10 units travel-time away from candidate site 1, and the furthest demands from site 2 is 7 units away and so on. Similar to the one-center example, this objective function can be simplified by introducing the constraints $D_1' \geq 3x_{21}$, $D_1' \geq 7x_{321}$, $D_1' \geq 6x_{561}$, $D_1' \geq 10x_{4321}$, $D_1' \geq 4x_{61}$, $D_2' \geq 3x_{12}$, ... , etc. The objective function will then be $\min(D_1',D_2', \ldots ,D_6')$ which can be further rewritten as *minimize* D_U', together with the additional constraints $D_U' = D_1'$, $D_U' = D_2'$, ... , $D_U' = D_6'$.

Thus the objective function is realized by first adding these constraints:

$D'_1 - 3x_{12} \geq 0$ $D'_3 - 7x_{321} \geq 0$ $D'_5 - 6x_{561} \geq 0$

$D'_1 - 7x_{123} \geq 0$ $D'_3 - 4x_{32} \geq 0$ $D'_5 - 6x_{52} \geq 0$

$D'_1 - 6x_{165} \geq 0$ $D'_3 - 3x_{34} \geq 0$ $D'_5 - 6x_{53} \geq 0$

$D'_1 - 10x_{1234} \geq 0$ $D'_3 - 6x_{35} \geq 0$ $D'_5 - 7x_{54} \geq 0$

$D'_1 - 4x_{16} \geq 0$ $D'_3 - 8x_{356} \geq 0$ $D'_5 - 2x_{56} \geq 0$

$\mathbf{D'_1 - 10x_{11} \geq 0}$ $\mathbf{D'_3 - 8x_{33} \geq 0}$ $\mathbf{D'_5 - 7x_{55} \geq 0}$

$D'_2 - 3x_{21} \geq 0$ $D'_4 - 10x_{4321} \geq 0$ $D'_6 - 4x_{61} \geq 0$

$D'_2 - 4x_{23} \geq 0$ $D'_4 - 7x_{432} \geq 0$ $D'_6 - 7x_{612} \geq 0$

$D'_2 - 7x_{234} \geq 0$ $D'_4 - 3x_{43} \geq 0$ $D'_6 - 8x_{653} \geq 0$

$D'_2 - 6x_{25} \geq 0$ $D'_4 - 7x_{45} \geq 0$ $D'_6 - 7x_{64} \geq 0$

$D'_2 - 7x_{216} \geq 0$ $D'_4 - 7x_{46} \geq 0$ $D'_6 - 2x_{65} \geq 0$

$\mathbf{D'_2 - 7x_{22} \geq 0}$ $\mathbf{D'_4 - 10x_{44} \geq 0}$ $\mathbf{D'_6 - 8x_{66} \geq 0.}$

Now we minimize the maximum distance D'_U, where D'_U is defined as

$$D' - D'_1 = 0 \qquad D' - D'_2 = 0 \qquad D' - D'_3 = 0$$
$$D' - D'_4 = 0 \qquad D' - D'_5 = 0 \qquad D' - D'_6 = 0.$$

This objective function then picks up $p \, x_{ij}$-binary-variables that are valuated at unity, placing p centers at the designated nodes (Thomas 1995). Notice that demand-weighted distances can be employed rather than strict distances. In this case, the set of 36 constraints that define the maximum distances is modified to read $D'_j - f_j D_{ij} x_{ij} \geq 0$.

$$x_{11}+x_{12}+x_{123}+x_{1234}+x_{165}+x_{16} = 1 \qquad \text{(demand node 1)}$$
$$x_{21}+x_{22}+x_{23}+x_{234}+x_{25}+x_{216} = 1 \qquad \text{(demand node 2)}$$
$$x_{321}+x_{32}+x_{33}+x_{34}+x_{35}+x_{356} = 1 \qquad \text{(demand node 3)}$$
$$x_{4321}+x_{432}+x_{43}+x_{44}+x_{45}+x_{46} = 1 \qquad \text{(demand node 4)}$$
$$x_{561}+x_{52}+x_{53}+x_{54}+x_{55}+x_{56} = 1 \qquad \text{(demand node 5)}$$
$$x_{61}+x_{612}+x_{653}+x_{64}+x_{65}+x_{66} = 1 \qquad \text{(demand node 6)}$$

There are p centers to be located: $x_{11}+x_{22}+x_{33}+x_{44}+x_{55}+x_{66} = p$. And the demand cannot be served unless the facility is in place:

$$x_{11}+x_{21}+x_{321}+x_{4321}+x_{561}+x_{61}-6x_{11} \leq 0 \qquad \text{(center at site 1)}$$
$$x_{12}+x_{22}+x_{32}+x_{432}+x_{612}+x_{612}-6x_{22} \leq 0 \qquad \text{(center at site 2)}$$
$$x_{123}+x_{23}+x_{33}+x_{43}+x_{53}+x_{653}-6x_{33} \leq 0 \qquad \text{(center at site 3)}$$
$$x_{1234}+x_{234}+x_{34}+x_{44}+x_{54}+x_{64}-6x_{44} \leq 0 \qquad \text{(center at site 4)}$$
$$x_{165}+x_{25}+x_{35}+x_{45}+x_{55}+x_{65}-6x_{55} \leq 0 \qquad \text{(center at site 5)}$$
$$x_{16}+x_{216}+x_{356}+x_{46}+x_{56}+x_{66}-6x_{66} \leq 0. \qquad \text{(center at site 6)}$$

For $p = 1$, the center is sited at node 2 or $x_{22} = 1$ (with an alternate optimum at node 5). This achieves a maximal travel-distance for demands 4 and 6 of 7 units. For $p = 2$, it is confirmed in the "Exercises and Problems" that solution of this binary program yields $x_{22} = 1$ and $x_{55} = 1$, which suggest placing the two centers at sites 2 and 5 respectively, with demands 1 and 3 served by site 2 and demands 4 and 6 served by site 5. The corresponding maximal travel-distance is for demand 4, at a cost of 4 units. For $p = 3$, the three centers are placed at 2, 6 and 4, for a maximal travel-distance of 3 for demands 2 and 4. These three models can be run sequentially by simply replacing a different p value in Equation **(2-26)** and the objective function each time. All other equations remain the same for all three models. The previous model fixes the maximal distance from a candidate site by exogenous computation. Now we rerun the model by omitting the bolded constraints above, which make these computational results. In other words, we let the model pick the furthest demand from each candidate site. For $p = 1$, node 2 (or 5) is still the optimal site as expected, and the objective function value remains at 27 units. For $p = 2$, the optimal sites are now 3 and 6 (at a reduced total cost of 13). For $p = 3$, the locations happen to be the same as before in the previous paragraph. ☐

Each optimal solution to the model as formulated above corresponds to a location in the nodes only. It represents a special version of the one-absolute-center general-case where the p points can also be chosen along the arcs (Averbakh and Berman 1997).

2.4.3 One-absolute center. The general problem of finding the one-absolute center is more difficult (Ahituv and Berman 1988). For a given network, let us consider any arc/link (a,b), any single site q in (a,b), and demand node i in the network. From **Figure 2-10**, the shortest distance between q and node i, D_{qi} can be expressed as

$$D_{qi} = \min \ [d_{qa} + D_{ai}, \ d_{qb} + D_{bi}] \tag{2-29}$$

which takes into account that the minimum path between q and i may be via node a or via node b. To simplify notation, let x stands for d_{qa}. The shortest path from q to i is via a if $x + D_{ai} \leq (d_{ab} - x) + D_{bi}$ or that $x \leq (D_{bi} + d_{ab} - D_{ai})/2 \equiv d_i$ where d_i can assume three different values depending on which is the shorter way to reach i: $d_i \leq 0$, $0 < d_i < d_{ab}$, and $d_i \geq d_{ab}$. Correspondingly, the minimum distance D_{qi} as a function of x is a one- or two-piece linear concave-function as shown in **Figure 2-**

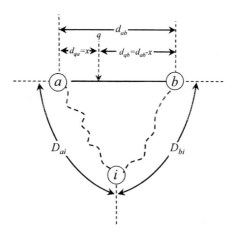

Figure 2-10 Determination of an absolute center
(Francis et al. 1999)

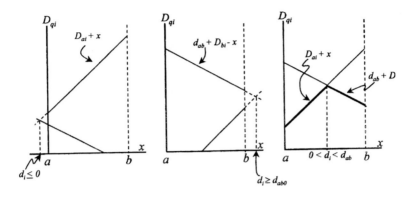

Figure 2-11 Absolute-center problem

11. In the figure, it is a linear function $D_{ai}+x$ as long as $d_i \geq d_{ab}$, a linear function $d_{ab}+D_{bi}-x$ as long as $d_i \leq 0$, and a two-piece linear concave function of x as long as $0 < d_i < d_{ab}$. If (a,b) is a directed arc, i.e., travel is allowed only from a to b, the first term in Equation 9 disappears and

$$D_{qi} = (d_{ab}-x)+D_{bi}. \tag{2-30}$$

Equations 9 and **Figure 2-11** are referred to as *point-vertex distances*.

Let us define the *local center* of a link, $q(a,b)$, as a point q in (a,b)—including nodes a and b. Here the maximum shortest-distance to a demand from q, $m'(q)$, has the inequality $D_{iq(a,b)} \leq m'(q)$ for all q on (a,b). The algorithm to find one-absolute center then involves these two iterative steps (Ahituv and Berman 1988):

Step 1 From any arc/link (a,b) find q.
Step 2 q^*'s picked as the absolute center by selecting the minimum $m'(q)$ from all (a,b) in the set of arc/links.∎

As shown in **Figure 2-11**, it can be seen that facility q is located in principle at the intersection of the linear functions $D_{ai}+x$ and $d_{ab}+D_{bi}-x$. The absolute-center location can be to the `left' of $(a,b)$, within $(a,b)$, to the `right' of (a,b), or at the vertices a or b. We can now conclude by stating this important observation: *There exists an optimal absolute center that is a vertex or an intersection point* (Hakimi 1964). Even11 through no weight f_i is involved in the above 1-absolute-center discussion, the results hold for weighted 1-absolute centers. Instead of a simple distance d or D, weighted distances $f_i(d_{qa}+D_{ai})$ or $f_i(d_{qb}+D_{bi})$ are used in their place.

2.4.4 One general center. A general center is any vertex q such that the most distant demand from it—including demand on an arc—is as close as possible (Evans and Minieka 1992). As discussed above concerning the one-absolute-center problem (and the associated **Figure 2-11**), if arc (a,b) is undirected, there are two ways to travel from demand point i on (a,b) to facility q: via vertex a or via vertex b. Observe that the two distances always sum to $[D_{qa} + d_{ai}] + [D_{qb} + (d_{ab} - d_{ai})] = D_{qa} + D_{qb} + d_{ab}$. In simplified notation where x replaces d_{ai}, $[D_{qa} + x] + [D_{qb} + (d_{ab} - x)] = D_{qa} + D_{qb} + d_{ab}$. Thus it follows that the maximum shortest-distance from vertex q to any demand point on arc (a,b), which occurs at the mid-point of arc (a,b), is denoted by

$$D_{(a,b),q} = \max_{i \in (a,b)} D_{i,q} = 1/2 (D_{qa} + D_{qb} + d_{ab}). \tag{2-31}$$

If on the other hand, arc (a,b) is directed, then a demand point on arc (a,b) can be reached only via vertex a. Consequently, the most distant points on (a,b) from any vertex q are the points closest to vertex b, that is the i points for which i approaches d_{ab}. In this case,

$$D_{(a,b),q} = D_{qa} + d_{ab}. \tag{2-32}$$

Now, number the arcs 1 through m in the graph. Let $\bar{D}'$ denote the $|I| \times m$ matrix whose q-kth element is the vertex-arc distance from vertex q to arc k. Observe that the vertex-arc distance matrix $\bar{D}'$ can be computed from the vertex-vertex distance-matrix $\bar{D}$ and the arc lengths by using Equations **(2-31)** and **(2-32)**. A general center can be found by identifying the row of $\bar{D}'$ with the smallest maximum-entry. The row corresponds to a vertex that is a general center. The reason is that the maximum vertex-arc distance equals the largest entry in the qth row of the vertex-arc distance-matrix $\bar{D}'$.

Example

For the graph shown in **Figure 2-12**, the arc lengths are shown (Evans and Minieka 1992). Using the vertex-to-vertex distance-matrix $\bar{D} = \begin{vmatrix} 0 & 2 & 3 & 3 \\ 4 & 0 & 2 & 1 \\ 6 & 2 & 0 & 3 \\ 3 & 5 & 4 & 0 \end{vmatrix}$, we can use Equations **(2-31)** and **(2-32)** to calculate $\bar{D}'$. For example, from Equation **(2-31)** $\bar{D}_{(3,4),\,q(1)} = \frac{1}{2}(\bar{D}_{13} + \bar{D}_{14} + d_{34}) = \frac{1}{2}(3 + 3 + 4) = 5$. From Equation **(2-32)**, $\bar{D}_{(2,4),\,q(1)} = \bar{D}_{12} + d_{24} = 2 + 1 = 3$. This results in

$$\bar{D}' = \begin{vmatrix} 2 & 3 & 3 & 3 & 3.5 & 5 \\ 6 & 7 & 4 & 1 & 2 & 3.5 \\ 8 & 9 & 6 & 3 & 2 & 3.5 \\ 5 & 6 & 3 & 6 & 5.5 & 4 \end{vmatrix} \tag{2-33}$$

Thus

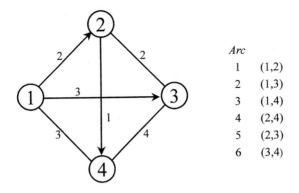

Figure 2-12 General center and median example

$$m'(1) = \max\{2,3,3,3,3.5,5\} = 5$$
$$m'(2) = \max\{6,7,4,1,2,3.5\} = 7$$
$$m'(3) = \max\{8,9,6,3,2,3.5\} = 9$$
$$m'(4) = \max\{5,6,3,6,5.5,4\} = 6,$$

and min$\{5,7,9,6\} = 5 = m'(1)$. Vertex 1 is a general center of the graph. The most distant demand-point from vertex 1 is 5 units away and lies on arc (3,4). □

2.4.5 One general absolute center. A general absolute-center is any point q such that the furthest demand i from facility q is as close as possible, where locations of both the facility and the demand need not be restricted to vertices. To find a general absolute-center we must find a point q on arc (a,b) such that the maximum point-arc distance is to be minimized: $\min_q m'(q(a,b))$. If arc (a,b) isundirected and different from the demand arc (i_1,i_2), the curve takes the same form as the point-vertex distances shown in , i.e., the "Absolute-center problem" figure.

$$D_{(i_1,i_2),\ q(a,b)} = \min\left\{\left[x + D_{(i_1,i_2),\ q(a)}\right],\ \left[(d_{ab}-x) + D_{(i_1,i_2),\ q(b)}\right]\right\} \qquad (2\text{-}34)$$

On the other hand, if the facility arc and the demand arc are the same, $(a,b) = (i_1,i_2)$, and if (a,b) is undirected, the maximum distance from q on (a,b) to any i point on (a,b) (where $d_{ai} < d_{aq}$) cannot exceed min $\{d_{aq},\ 1/2(d_{ab}+D_{ba})\}$ = min $\{x,\ 1/2(d_{ab}+D_{ba})\}$. The first term in this minimization accounts for routes from q to i. The second term in the minimization accounts for routes from q to i that traverse vertex b. Similarly, the maximum distance from q to i (where $d_{ai} > d_{aq}$) cannot exceed min $\{d_{ab} - x,\ \frac{1}{2}(d_{ab} + D_{ab})\}$. The first term here accounts for routes from q to i. The second term accounts for routes from q to i that traverse vertex a. Consequently, if arc (a,b) is undirected,

$$D_{(a,b),\ q(a,b)} = \max \left\{ \begin{array}{l} \min\left[x,\ \dfrac{1}{2}(d_{ab}+D_{ba})\right] \\[2ex] \min\left[(d_{ab}-x),\ \dfrac{1}{2}(d_{ab}+D_{ab})\right] \end{array} \right\}. \qquad (2\text{-}35)$$

When this distance, or more precisely the *point-arc distance*, function is plotted as a function of x, the curve takes the form shown in **Figure 2-13**. A moment's reflection would reveal that *no interior point of a directed arc can be a general absolute center*. Since all travel on a directed arc is in one direction, the terminal vertex of a directed arc is a better candidate for general absolute center than any interior point of this arc. Remember that the terminal vertex is closer to every arc in the graph. Consequently, we need only consider vertices and the interior points of undirected arcs in our search for a general absolute center. In summary, the technique for finding a general absolute center is the same as the technique for finding an absolute center except that point-vertex distances are replaced by the point-arc distances.

   2.4.6 p-absolute center. As a final topic in this section, we consider the *p*-absolute-center problem (Francis et al. 1999). Let $\Xi = \{q_1, q_2, \dots, q_p\}$ be the collection of centers, with each center on some arc, and $p < |I|$, where $|I|$ is the number of vertices with positive demands f_i. Now let $D_{i\Xi}$ be the distance between each demand i and a closest center in Ξ, so that we arE again if travel is between each demand and a closest center. We wish to find an absolute *p*-center Ξ^* that minimizes $M(\Xi) = \max \{f_1 D_{i\Xi}, \dots, f_{|I|} D_{|I|\Xi}\}$. Consider a situation with $|I| = 5$ vertices and $\Xi = \{q_1, q_2\}$, an absolute 2-center. Suppose that q_1 is closest to vertices 1, 2 and 3, while q_2 is closest to vertices 4 and 5. Let $m_1^i(q_1)$ denote the 1-center function (or, the maximal shortest-distance to a demand point) for vertices 1, 2 and 3, and $m_2^i(q_2)$ denote the 1-center function for vertices 4 and 5. It now follows that $M(\Xi) = \max \{m_1^i(q_1), m_2^i(q_2)\}$. If we solve the 1-center problems of minimizing m_1^i

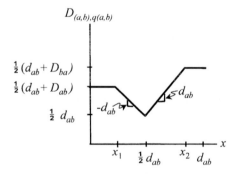

Figure 2-13 General absolute center problem
(Evans and Minieka 1992)

and minimizing m_2' individually, there are absolute-center locations q_1^* and q_2^*, each a vertex or an intersection point, for which $m_1'(q_1^*) \leq m_1'(q_1)$ and $m_2'(q_2^*) \leq m_2'(q_2)$, implying that max $\{m_1'(q_1^*), m_2'(q_2^*)\} \leq$ max $\{m_1'(q_1), m_2'(q_2)\} = M(\Xi)$. Now if we define the 2-center $\Xi^* = \{q_1^*, q_2^*\}$, the closest-center assumption implies that $M(\Xi^*)$ $\leq$ max $\{m_1'(q_1), m_2'(q_2)\}$. Hence we conclude that $M(\Xi^*) \leq M(\Xi)$. But since Ξ is an absolute 2-center and Ξ^* is a 2-center, we also know that $M(\Xi) \leq M(\Xi^*)$. Thus we conclude that $M(\Xi^*) = M(\Xi)$, so that Ξ^* is an absolute 2-center with each center being a vertex or an intersection point. The above establishes the finiteness of the search for p-absolute centers. *There exists at least one absolute p-center for which each center is at a vertex or an intersection point.* It suffices to consider only vertex and intersection point locations in order to find an absolute p-center. It is also to be noted that using a suitable reformulation, a p-center problem can be reduced to a p-median problem and *vice versa*. An algorithm that can solve one can also solve the other (Krarup and Pruzan 1981). We will discuss this in the section entitled *Equivalence Between Center and Median Problems*.

Example

Suppose one is to locate two fire stations in the network in **Figure 2-9** such that the maximum-response-time is to be minimized for everyone alike in Estaboga, Alabama (Francis et al. 1999). This reduces to finding a 2-absolute center for the network, in which the demand weights f_i are assumed to be unity and the arc travel-times are as shown. Because of the vertex-or-intersection property established above, the number of potential locations can be enumerated as shown in **Table 2-1** following the computation illustrated by the schematics of **Figure 2-10** and **Figure 2-11**. Notice that the intersection points, q_k, are given as a distance from a on the arc (a,b), and each has an associated bound value, B^k $= D_{iq}$. The bound value is simply the weighted distance from either one of the two demand vertices, i_L and i_R, used to define the intersection point. For example, the entry for $k = 9$ corresponds to an intersection point on the arc $(1,2)$. The intersection point is defined by demand vertices 5 or 6. It is located a distance of 2.5 from 1 and its weighted distance to either 5 or 6 is 6.5. The set S_k listed in **Table 2-1** is the set of demand vertices that would be covered by a center at q_k, with a weighted distance less than or equal to B^k. For example, a center at q_9 would cover every vertex except 4, with a maximum weighted distance of 6.5. The procedure for defining a p-center will employ a set $\Xi(z)$, defined as the collection of vertices and intersection points whose bound values are within z units. The last column in **Table 2-1** shows the intersection points that would be included in $\Xi(z)$ for $z = 3.5$. These include the qualified centers at 7, 10, 11, 12, etc. In other words, these centers can cover the respective demand sets within 3.5 units of weighted distance. □

2.4.7 *Relationship between p-absolute-center and covering*. For the above problem, the p-absolute-center problem now can be formulated and solved as

$$\min z_p \tag{2-36}$$

subject to

$$z_p \geq \min \left\{ f_i D_{iq} \mid q \in \Xi(X) \right\} \qquad i \in I \tag{2-37}$$

k	arc (a,b)	Demand i_L	Demand i_R	Intersection	Distance bound B^k	Demands S_k	Qualified set $\Xi(3.5)$
7		1	2	1.5	1.5	1,2	√
8	(1,2)	6	3	1.5	5.5	1,2,3,6	
9		6	5	2.5	6.5	1,2,3,5,6	
10		1	6	2	2	1,6	√
11		2	6	0.5	3.5	1,2,6	√
12	(1,6)	1	5	3	3	1,5,6	√
13		2	5	1.5	4.5	1,2,5,6	
14		3	4	2	9	1,2,3,4,5,	
15		2	3	2	2	2,3	√
16	(2,3)	1	3	0.5	3.5	1,2,3	√
17		2	4	3.5	3.5	2,3,4	√
18		1	4	2	5	1,2,3,4	
19		2	5	3	3	2,5	√
20		1	5	1.5	4.5	1,2,5	
21		2	6	4	4	2,5,6	
22	(2,5)	1	6	2.5	5.5	1,2,5,6	
23		1	4	5	8	1,2,3,4,5,	
24		3	5	1	5	1,2,3,5	
25		3	6	2	6	1,2,3,5,6	
26	(3,4)	3	4	1.5	1.5	3,4	√
27		1	6	1.5	8.5	1,2,3,4,5,	
28		3	5	3	3	3,5	√
29		2	5	1	5	2,3,4,5	
30	(3,5)	2	6	2	6	2,3,4,5,6	
31		3	6	4	4	3,5,6	
32		4	6	2.5	5.5	3,4,5,6	
33		4	5	3.5	3.5	4,5	√
34		4	1	6.5	6.5	1,2,3,4,5,	
35	(4,5)	4	2	6.5	6.5	1,2,3,4,5,	
36		4	6	4.5	4.5	4,5,6	
37		3	5	2	5	3,4,5	
38		3	1	5	8	1,2,3,4,5,	
39		4	6	3.5	3.5	4,6	√
40		4	1	5.5	5.5	1,4,5,6	
41	(4,6)	4	5	4.5	4.5	4,5,6	
42		3	6	2	5	3,4,6	
43		3	1	4	7	1,3,4,5,6	
44		3	5	3	6	3,4,5,6	
45	(5,6)	5	6	1	1	5,6	√

Table 2-1 Intersection points for centers (Francis et al. 1999)

$$\sum_{j=1}^{p} x_{jj} \leq p \tag{2-38}$$

$$x_{jj} \in \{0,1\} \qquad j = 1,\ldots,p \tag{2-39}$$

where $x_{jj} = 1$ if a center is at a vertex or an intersection point j and 0 otherwise (Francis et al. 1999). The set $\Xi(X)$ is the set of vertices and intersection points in the decision X that are used in the solution. Constraint (2-37) forces the objective function to be at least as large as the smallest weighted distance from a demand to any center location. This is a common way to take care of the maximum operator of a min-max objective-function, as we have seen previously. Constraint (2-39) ensures that at most p center-locations are selected. Taking the set of qualified candidate locations $\{q_7, q_{10}, q_{11}, q_{12}, q_{15}, q_{16}, q_{17}, q_{19}, q_{26}, q_{28}, q_{33}, q_{39}, q_{45}\}$ from **Table 2-1**, the two minimum values B^k that cover all nodal demands 1, 2, 3, 4, 5 and 6 are provided between q_{12} and q_{17}. Thus the absolute 2-median $\{q_{12}, q_{17}\}$, with a travel time not exceeding 3.5 units, can be picked out by enumeration and inspection in this case. In general, more efficient algorithms need to be used, one of which can be constructed out of the following relationship between the p-center and covering models.

Denote by (z^*, x^*) an optimal solution to the p-absolute center model above. Now consider the covering problem

$$\min p = \sum_{j=1}^{|J|} x_{jj} \tag{2-40}$$

subject to

$$\sum_{j=1}^{|J|} a_{ij} x_{jj} \geq 1 \qquad i \in I \tag{2-41}$$
$$x_{jj} \in \{0,1\} \qquad j \in J$$

where $a_{ij} = 1$ if $f_i D_{iq} \leq z$, and 0 otherwise. Suppose that z^1 and z^* are specific values for z, and the optimal solution to the covering problem is (p^1, x^1). The relationships between the p-absolute center and the covering models are:

a. If (z^*, x^*) solves the p-absolute-center model, then (n, x^*) solves the covering problem at z^*.
b. If (z^*, x^*) solves the p-absolute-center model and (n^1, x^1) solves the covering problem at z^1, then (a) if $p^1 > p$, $z^1 < z^*$, and (b) if $p^1 \leq p$, $z^1 \geq z^*$. Relationship 1 says that if we know z^*, the optimal solution value for the p-absolute-center model, we could use the covering model to find x^*, an optimal p-absolute center. This allows efficient computational schemes to be devised, since there exist several efficient methods to solve the covering problem. Relationship 2 says the following. If we guess a value for z^*, call it z^1, and solve the covering problem

solution and upper bound on z^*, case (b). A search algorithm and example are discussed in Francis et al. (1999).

2.4.8 *Obnoxious facilities*. The above discussions on centers have particular relevance to locating obnoxious facilities such as landfills, nuclear power plants, airports and the like. In locating obnoxious facilities, it is desirable to place them as far away as possible to the population. However, since most of these facilities are essential to providing a much needed service, it is usually desirable to place them proximal to the population as well. This constitutes a weighted *max-min* location problem:

$$\max_{i \in I} [\min_{j \in I} f_i d_{ij}] = \max_{i \in I} f_i' d_i' \tag{2-42}$$

2.4.8.1 AN INTRODUCTORY EXAMPLE. An example of the obnoxious facility-location problem is again shown in the familiar **Figure 2-8**. The objective for this problem is to find the facility location that maximizes the weighted minimum-distance between it and the demands 4 and 5, $f_4' d_4'$ and $f_5' d_5'$: max $8d_4' + 10d_5'$. Again the unique feature of this model is the constraints to achieve minimum distance: $d_4' - 2.5x_{42} \leq 0$, $d_4' - 3x_{41} \leq 0$, $d_4' - 5.5x_{43} \leq 0$, $d_5' - 2x_{53} \leq 0$, $d_5' - 2.6x_{52} \leq 0$, and $d_5' - 4x_{51} \leq 0$. The rest of the constraints are identical to the 1-center problem discussed above in connection with this figure, including the strict inequality $d_4' + d_5' > 0$. An alternate optima between facilities 2 and 3 were obtained. Such a solution was overlaid on **Figure 2-8**. For a simple problem such as this, the solution obtained above can again be checked by inspection. First the following table is constructed consisting of the weighted distances:

(*j*) (*i*)	1	2	3	min
4	24	20*	44	20
5	40	26	20*	20*

from which the row minima are recorded on the right hand side. In choosing the maximum of the two entries in the right-hand column, an alternate optima between facility 2 and 3 are obtained, as expected.

It can be seen that many parallels can be drawn between the min-max formulation for centers and the max-min formulation for obnoxious facilities. Much of our discussion on center location can therefore be carried over to obnoxious facility location, which is a very pertinent subject as long as environmental concerns are valued. Chaudhry and Moon (1987) classified the single obnoxious facility-location problem by first defining a minimum distance separating the customers from the 'obnoxiousness': $D_{ij} \geq S_i$ for all i. This constraint can be imposed on a median (min-sum) problem or a center (min-max) problem. Or it can be imposed on a *medi-center* or *cent-dian* problem, which combines both the min-sum and min-max criteria in a single formulation (Tamir et al. 1998). The opposite of each of these criteria constitutes an appropriate way for locating obnoxious facilities. The *max-sum* or *antimedian* model places the facility as far away from the average customer as possible. The max-min or *anticenter* model minimizes the displeasure of the worst disserviced customer by maximizing the minimum weighted-distance between all customers and the facility. Similarly, the *anti-medicenter* model maximizes the sum of the weighted-distance and minimizes the maximum weighted-distance at the same time. As we turn our attention to the problem of locating more than one obnoxious facility, we find it convenient—as in the regular center problem—to define the distance between customer i and the closest facility among all facilities. If we let J be the set of p unknown-locations for the facilities, the distance-of-interest is now $D_{iJ} = \min_k\{D_{ik}:k \in J\}$ for all i. Correspondingly, the distance constraint now can be written as or $D_{iJ} \geq S_i$.

2.4.8.2 A GENERAL FORMULATION. The max-min methodology finds the minimum distance between a proposed facility site and all of the demands and then selects the facility that provides the largest minimum distance. The inherent advantage in approaching the siting decision from this perspective is that the decision-maker is making a tradeoff between ease of access and distance from the community. The objective is no longer simply to maximize the distance from the obnoxious facility to each of the demand nodes, but to minimize the discomfort of the most impacted party. This class of location models is called the *p-anticenter* models. The model has the distinctly nonlinear objective: $\max \min_{i \in I, j \in J} f_i D_{ij} x_{ij}$ where f_i are the demand weights, D_{ij} is the distance between demand i and facility-site j and x_{ij} is the allocation decision-variable to assign demands to facilities. There are altogether $|J|$ facility-sites and $|I|$ demand nodes. We maximize the minimum of the shortest facility-to-demand distances. This objective function, because of its non-linearity, is inconvenient for standard computational procedures. Attempts have been made to reformulate the model so that it is convenient for mathematical-programming solution-codes (Daskin 1995, Church and Garfinkel 1978, Zhang and Melachrinoudis 1996). Here we present a formulation that is amenable to subsequent merging with a Data Envelopment Analysis model (Thomas 1995), to be described in Chapter 3 ("Measuring spatial separation"). Let us now define D_j' as the minimum distance between the facility-site j and all demand-nodes. The objective is to maximize D_j' for the case when one obnoxious facility j is to be sited. In other words, if a single site j is selected for an obnoxious facility, the goal becomes $\max D_j'$. For a multi-facility case where $|J|$ candidate-sites are involved, we have the following objective function, which maximizes sum of the distances from each demand to the closest selected facility: $\max \sum_{j=1}^{|J|} D_j'$.

At the center of an obnoxious-facility location-model is the way to reformulate the max-min objective. For the sake of clarity and without loss of generality, we will assume all demands to be equal. Thus we set $f_i = 1$ for all i's for the remainder of this section. The core of this approach is to generate constraints that force D_j' to assume the desired values. First, a set of constraints need to be included to ensure that the proper minimum-distance between a demand i and a facility j is chosen: $\sum_{i=1}^{|J|} D_{ij} x_{ij} \le D_{ij} x_{ij} + (1-x_{ij})M$ where M is any arbitrarily-selected large-number, and x_{ij} is the binary variable to signify whether a facility is open or closed. M has to be larger than D_{ij} in the minimum-distance matrix. This equation activates the appropriate x_{ij} corresponding to the shortest path between i and j. To understand the mechanics of this constraint set, assume $y_j = 1$ for a particular site j. In this case, the right-hand-side will simplify to $D_{ij} x_{ij}$ for each constraint in the set for the site j. Hence, the smallest D_{ij} will be the only one that satisfies the inequality, resulting in the corresponding x_{ij} to be activated. If site j is not activated, the constraint places only a ceiling M on the sum in the left-hand-side. Other constraints will intervene to force the x_{ij} to assume the proper values. In the one-facility case, for example, the constraints prevent an incorrect minimum-distance-assignment. If x_{11} is activated, indicating that facility one will be built, demand 2 can be served only with arc x_{21} or x_{22}. Other constraints in the program will make a final selection.

When the minimum distance D_j' are to be maximized, the following companion constraint determines the appropriate minimum-distance. For *each* proposed site j, the following constraint set will find the minimum-distance value to assign to D_j'.

$$D_j' + (M-D_{ij})x_{ij} \le M \qquad \forall\ i\text{-}j. \tag{2-43}$$

If a site is not open or activated (in which case the binary variable for site j is set to zero, or $x_{ij} = 0$), none of the x_{ij} is activated from the previous constraint. Now according to the current constraint, D_j' must be less than M. However, if a site is activated, or $x_{ij} = 1$, the constraints will force D_j' to be no larger than the smallest distance D_{ij}. In other words, if site j is selected ($x_{ij} = 1$), then logically at least one of eh corresponding xij will be activated. The utility of the arbitrary M is evident here. We place a `cap' on the value of D_j', which is under upward pressure from the objective function. This limiting value M must be larger than D_{ij} to enable the constraint to choose the correct minimum D_{ij}. We write this equation in long hand for the multi-facility case.

$$D_1' + (M-D_{i1})x_{i1} \le M \qquad \forall\ i$$
$$D_2' + (M-D_{i2})x_{i2} \le M \qquad \forall\ i$$
$$\dots \tag{2-44}$$
$$\dots$$
$$D_m' + (M-D_{im})x_{im} \le M \qquad \forall\ i.$$

These two constraints **28** and **(2-43)** are said to have taken care of the "criss cross" effects, helping to select the minimum distance to maximize. It is obvious that the objective function value in max $\sum_j D_j'$ is of little real utility. Only the D_j' values for the *selected* sites will be accurate, the key purpose being to force the proper choice of x_{ij} variables to activate.

The next constraint specifies the number of p facilities to be sited among $|J|$ candidate locations, $\sum_{j=1}^{|J|} x_{jj} = p$ where $p \leq |J|$. Another constraint that must be met is that demand must be served by a facility, as shown in Equation (2-25). Notice by an uncapacitated formulation, the demand at each node is met by only one facility, rather than being split up between several facilities. Demand at a node i (among $|I|$ demands) cannot be served by facility j unless the facility is built, as shown in Equation (2-27). If any xij are activated for a given facility j, this constraint will force the facility to be built at j ($x_{jj} = 1$). The second term in this inequality $(|I| - p + 1) x_{jj}$ ensures that the second facility chosen for construction (when $p = 2$ for example) serves at least a demand, while accounting for the case when a demand occurs at a facility site ($i = j$). Otherwise, the formulation could recommend the construction of a "white elephant" facility that serves none of the demands. It also compensates for the case where a demand node and a facility site are the same ($i = j$) via the +1 term.

Here, all the x_{jj} are 0–1 binary variables, which selects the appropriate facility, making this a mixed integer program. The allocation variable x_{ij}'s will be binary by the nature of this formulation although no integrality constraint is placed on the variable. The number of variables in this model is $|J||I| + 2|J|$ and the number of constraints is $9|I| + |J| + 1$. Thus while the number of constraints is linear with respect to the number of sites and demands, the number of decision variables goes up as the product of the number of sites and demands. To the extent that the combinatorial space is more critical in binary integer-programs, the number of decision variables governs the speed of the solution algorithm.

2.4.8.3 EXAMPLE OF THE GENERAL FORMULATION. The objective of this linear program is to maximize the minimum distance from the facility to the demand nodes for two facilities. We use the same data as the p-center example that appears earlier in this chapter:

$$\max \ D_1' + D_2' + D_3' + D_4' + D_5' + D_6'.$$

These constraints ensure that the maximum distance is chosen correctly. The constraints also account for the "criss-crossing effect." An arbitrary value of $M = 100$ is used. For site 1,

$$3x_{21} + 0x_{22} + 4x_{23} + 7x_{24} + 6x_{25} + 7x_{26} + 97x_{11} \leq 100$$
$$7x_{31} + 4x_{32} + 0x_{33} + 3x_{34} + 6x_{35} + 8x_{36} + 93x_{11} \leq 100$$
$$10x_{41} + 7x_{42} + 3x_{43} + 0x_{44} + 7x_{45} + 7x_{46} + 90x_{11} \leq 100$$
$$6x_{51} + 6x_{52} + 6x_{53} + 7x_{54} + 0x_{55} + 2x_{56} + 94x_{11} \leq 100$$
$$4x_{61} + 7x_{62} + 8x_{63} + 7x_{64} + 2x_{65} + 0x_{66} + 96x_{11} \leq 100.$$

For site 2,

$$3x_{21} + 0x_{11} + 7x_{31} + 10x_{41} + 6x_{15} + 4x_{16} + 97x_{22} \leq 100$$
$$4x_{32} + 7x_{31} + 0x_{33} + 3x_{34} + 6x_{35} + 8x_{36} + 96x_{22} \leq 100$$
$$7x_{42} + 10x_{41} + 3x_{43} + 0x_{44} + 7x_{45} + 7x_{46} + 93x_{22} \leq 100$$

$$6x_{52} + 6x_{51} + 6x_{53} + 7x_{54} + 0x_{55} + 2x_{56} + 94x_{22} \le 100$$
$$7x_{62} + 4x_{61} + 8x_{63} + 7x_{64} + 2x_{65} + 0x_{66} + 93x_{22} \le 100.$$

This is similar for sites 3 through 6.

The number of facilities to be built is two: $x_{11}+x_{22}+x_{33}+x_{44}+x_{55}+x_{66} = 2$. Demand must be met at each site by a facility:

$$x_{11} + x_{12} + x_{31} + x_{41} + x_{15} + x_{16} = 1 \qquad x_{41} + x_{42} + x_{43} + x_{44} + x_{45} + x_{46} = 1$$
$$x_{21} + x_{22} + x_{23} + x_{24} + x_{25} + x_{26} = 1 \qquad x_{51} + x_{52} + x_{53} + x_{54} + x_{55} + x_{56} = 1$$
$$x_{31} + x_{32} + x_{33} + x_{34} + x_{35} + x_{36} = 1 \qquad x_{61} + x_{62} + x_{63} + x_{64} + x_{65} + x_{66} = 1.$$

Demand cannot be served unless facility is in place:

$$x_{21} + x_{31} + x_{41} + x_{51} + x_{61} - 5x_{11} \le 0 \qquad x_{41} + x_{24} + x_{34} + x_{54} + x_{64} - 5x_{44} \le 0$$
$$x_{12} + x_{32} + x_{42} + x_{52} + x_{62} - 5x_{22} \le 0 \qquad x_{15} + x_{25} + x_{35} + x_{45} + x_{65} - 5x_{55} \le 0$$
$$x_{13} + x_{23} + x_{43} + x_{53} + x_{63} - 5x_{33} \le 0 \qquad x_{16} + x_{26} + x_{36} + x_{46} + x_{56} - 5x_{66} \le 0.$$

These equations determine the max-min distance. For site 1,

$$D_1' + 97x_{21} \le 100 \quad D_1' + 93x_{31} \le 100 \quad D_1' + 90x_{41} \le 100 \quad D_1' + 94x_{51} \le 100 \quad D_1' + 96x_{61} \le 100.$$

For site 2,

$$D_2' + 97x_{12} \le 100 \quad D_2' + 96x_{32} \le 100 \quad D_2' + 93x_{42} \le 100 \quad D_2' + 94x_{52} \le 100 \quad D_2' + 93x_{62} \le 100.$$

Similar constraints can be written for sites 3 through 6.

The optimal solution consists of facilities 3 and 4. Demands 1, 2 and 5 are allocated to facility 1, while demand 6 is allocated to facility 4.

2.5   Median location problems

Interesting as the center (and the related obnoxious-facility) problem may be, much of the facility-location and land-use literature is devoted to discussions of the median. Inherent in central-place theory is the assumption that consumers patronize the nearest facility on the average. Such spatial allocation can be modelled by minimizing the aggregated weighted distance (or transport cost), and it is called the median location-allocation problem. Here we will review the more advanced median problems similar to what we have done for the center, with one major exception: Computational solution to the p-median problem will be discussed more formally, since it is more developed. Its development is motivated by its relative computational ease in comparison to the center problems, and many applications.

2.5.1 One-general-median and one-absolute-median. A *general median* is any vertex in a network with the smallest total distance (or average distance) to every arc. Here the distance from a vertex to an arc is taken to be the maximum distance from the vertex to the demand points on the arc (Evans and Minieka

1992). Thus a general median is any vertex q such that the sum of vertex-arc distances $s(q)$ is minimized: $s(q) = \min_{i \in I} s(i)$. The sum of the entries in the ith row of the vertex-arc distance-matrix $\bar{D}'$ equals the sum of the distance from vertex i to all arcs. Hence, a general median corresponds to any row of $\bar{D}'$ with the smallest sum.

Example

For the network shown in **Figure 2-12**, we have already calculated the vertex-arc distance matrix as shown in Equation **22** [Evans and Minieka 1992]. Thus

$$s(1) = 2+3+3+3+3.5+5 = 19.5^* s(3) = 8+9+6+3+2+3.5 = 31.5$$
$$s(2) = 6+7+4+1+2+3.5 = 23.5 s(4) = 5+6+3+6+5.5+4 = 29.5.$$

Therefore, min $\{19.5, 23.5, 31.5, 29.5\} = 19.5$. Thus vertex 1 is the general median of this graph. The total distance from vertex 1 to all arcs is 19.5 units. □

An *absolute median* is a point in the network—which does not have to be a vertex—whose total distance to all demand vertices is as small as possible. *There is always a vertex that is an absolute median*. This is true not only for the linear or piecewise linear point-vertex distances defined by Equations **9** and **Figure 2-11**, but also *for any concave point-vertex distance function* of the facility q on an arc. Consequently, we need only consider the vertices in our search for an absolute median. Thus, *any median is also an absolute median*, and no new solution techniques are necessary. This computational ease can be credited to Hakimi (1964), who first pointed out these nodal-optimality conditions.

2.5.2 One general absolute median. A *general absolute median* is *any* point on the network that the total distance (or average distance) from it to all arcs is as small as possible (Evans and Minieka 1992). Again, the distance from a point to an arc is taken as the maximum distance from the point to all points on the arc. Thus a general absolute median is any point q such that the sum of point-arc distances $s'(q)$ is minimized

$$s'(q) = \min_{q(a,b) \in I \cup \underline{A}} s'(q(a,b)). \tag{2-45}$$

Here the facility q is picked among the union of both vertices I and arcs $\underline{A}$ that make up the graph.

As pointed out in the previous section, there is always a vertex that is an absolute median, should all the point-vertex distance functions be concave. If all the point-arc distance functions were also concave, analogous arguments could be made for the general absolute median. Unfortunately, this is not so, since the point-arc distance function $D_{(a,b),q(a,b)}$ is not concave, as shown in ?. It is possible, however, to eliminate from consideration the interior of some arcs. First, an interior point of a directed arc cannot be a general-absolute-median, because the terminal vertex of a directed arc is a better candidate for general-absolute-median than any interior point. Moreover, no interior point of undirected arc (a,b) is a general-absolute-median if

$$\left|s(a)-s(b)\right| > \tfrac{1}{2}(d_{ab}+D_{ba}). \tag{2-46}$$

This observation is useful because it can eliminate a number of edges from further consideration in a search algorithm. To check this condition, only the vertex-vertex distance matrix $\bar{D}$ and the vertex-arc distance matrix $\bar{D}'$ are needed. A third elimination rule is also available. This rule calls for checking

$$s'(q) \geq s(a)-\tfrac{1}{2}d_{ab} \quad and \quad s'(q) \geq s(b)-\tfrac{1}{2}D_{ba}. \tag{2-47}$$

This generates a lower bound on the total distance for every interior point on any edge that was not eliminated by the second rule. Each of these edge-lower-bounds can be compared to the least-total-distance-from-a-vertex, namely $s(q)$. If the lower-bound-for-an-edge is greater than the least-total-distance-from-a-vertex, this edge can be eliminated.

Each remaining, non-eliminated (or non-dominated) edge (a,b) must then be examined completely by evaluating $s'(q(a,b))$ for all q. Hopefully, the best candidate for general absolute median on the interior of the examined edge (a,b) will have a total distance that will be less than the lower bound of some non-examined edges. Then the non-examined edges can also be eliminated. Ultimately, all edges must be either eliminated or completely examined. A general absolute median is selected from the examined set of vertices and interior point candidates.

Example

For the network shown in **Figure 2-14** (Evans and Minieka 1992), the vertex-arc distance-matrix $\bar{D}'$ can be calculated as

$$\bar{D}' = \begin{bmatrix} 1 & 2 & 3 & 3 & 3 & 2 & | & 14 \\ 1 & 1 & 2 & 2 & 2 & 1 & | & 9 \\ 2 & 1 & 1 & 2 & 3 & 2 & | & 11 \\ 3 & 2 & 1 & 1 & 2 & 2 & | & 11 \\ 2 & 2 & 2 & 1 & 1 & 1 & | & 9 \\ 3 & 3 & 3 & 2 & 1 & 2 & | & 14 \end{bmatrix}.$$

From this matrix, row sums can be calculated as shown to the right of the matrix. The best vertex candidates for general-absolute-median are found to be vertices 2 and 5. Each has a total distance to all arcs of 9 units. We try next to eliminate the interiors of some edges by applying Equation **(2-45)**. Observe that in this graph all arcs are undirected, and consequently the right side of Equation **(2-45)** becomes d_{ab}. The following edges are examined and the pertinent ones eliminated by the calculations shown next to it:

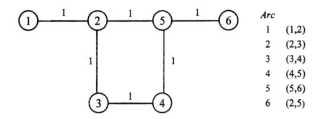

Figure 2-14 General absolute median example (Evans and Minieka 1992)

Edge	Equation (2-45) calculations
(1,2)	$\lvert s(1)-s(2)\rvert = \lvert 14-9\rvert = 5 > d_{12} = 1$
(2,3)	$\lvert s(2)-s(3)\rvert = \lvert 9-11\rvert = 2 > d_{23} = 1$
(3,4)*	$\lvert s(3)-s(4)\rvert = \lvert 11-11\rvert = 0 < d_{34} = 1$
(4,5)	$\lvert s(4)-s(5)\rvert = \lvert 11-9\rvert = 2 > d_{45} = 1$
(5,6)	$\lvert s(5)-s(6)\rvert = \lvert 9-14\rvert = 5 > d_{56} = 1$
(2,5)*	$\lvert s(2)-s(5)\rvert = \lvert 9-9\rvert = 0 < d_{25} = 1.$

Thus only edges (3,4) and (2,5) remain under consideration.

Next, let us apply Equations (2-47) to see if any further edge eliminations are possible: Edge (3,4) can be eliminated because according to Equation (2-47)

$$s'(q(3,4)) \geq s(3) - \tfrac{1}{2}d_{34} = 11 - 0.5 = 10.5,$$

which is greater than the 9 units achieved by selecting a vertex as general-absolute-median. Edge (2,5) cannot be eliminated by conditions (2-47) because

$$s'(q(2,5)) \geq s(2) - \tfrac{1}{2}d_{25} = 9 - 0.5 < 9 \quad \text{and} \quad s'(q(2,5)) \geq S(5) - \tfrac{1}{2}D_{52} = 9 - 0.5 < 9.$$

Only edge (2,5) remains under consideration. Equations (2-34) and (2-35) can be used to compute point-arc distances for edge (2,5):

$$D_{(1,2)}(q(2,5)) = 1+x \qquad D_{(4,5)}(q(2,5)) = 1+(1-x)$$
$$D_{(2,3)}(q(2,5)) = 1+x \qquad D_{(5,6)}(q(2,5)) = 1+(1-x)$$
$$D_{(3,4)}(q(2,5)) = 2 \qquad D_{(2,5)}(q(2,5)) = \max\{x, (1-x)\}.$$

Adding these point-arc distances yields $s'(q(2,5)) = 8 + \max\{x, (1-x)\}$. Since $\min_{0 \leq x \leq 1} \max\{x, (1-x)\}$ occurs at $x = 0.5$, the best candidate from general-absolute-median on edge (2,5) is the ½-point. Since 9 is the best possible total distance of a vertex location, we conclude that the ½-point of edge (2,5) is the general-absolute-median of the graph. Its total distance to all arcs is 8.5 units. □

2.5.3 *p-median and absolute-p-median*.

The *p*-median problem can be defined again in terms of $\min f(y) = \sum_{i=1}^{|I|} f_i \, D_i(y)$, where $\mathbf{y} = (y_1, \ldots, y_p)$ is a set of *p* medians on a network and $D_i(\mathbf{y})$ is the shortest distance between demand vertex *i* and its nearest median (Frances et al. 1999). Instead of writing x_{ij}, we have abbreviated the notation with binary variable y_j. We have already discussed at some length the *p*-median problem both in the "Prescriptive Tools" chapter in Chan (2000) and in the current chapter. To the extent that *there exists a least one absolute-p-median for which each median is at a vertex*, it suffices to consider only vertex locations to find an absolute-*p*-median. This also means that the discussions on *p*-medians carry over to the absolute-*p*-median. Here we will concentrate on the generalized *p*-median model and provide some solution algorithms other than regular mixed-integer-programming (MIP). The generalized *p*-median problem can be defined as

$$\min \; z = \sum_{j \in J} \left(g_j y_j + \sum_{i \in I} c_{ij} x_{ij} \right) \tag{2-48}$$

subject to

$$\sum_{j \in J} a_{ij} x_{ij} = 1 \qquad i \in I \tag{2-49}$$

$$-x_{ij} + y_j \geq 0 \qquad i \in I, j \in J \tag{2-50}$$

$$\sum_{j \in J} y_j \leq p \tag{2-51}$$

$$y_j \in \{0,1\} \qquad j \in J \tag{2-52}$$

$$x_{ij} \geq 0 \qquad i \in I, j \in J. \tag{2-53}$$

By appropriately specifying the parameters g_j, C_{ij}, and a_{ij}, and specialize on the inequalities, we obtain the simple *p*-median problem. It is represented by Equations (2-3) through (2-5) and (2-7) through (2-9), as the reader can verify.

The *partial-cover* problem and the *generalized-partial-cover* problem could be formulated as special cases of generalized *p*-median. The partial-cover problem given only p' (where $p' \leq p$) facilities can be stated as

$$\max \left(\sum_{i \in I} \max \{ a_{ij} y_j | j \in J \} \right) \tag{2-54}$$

subject to $\sum_{j \in J} y_j \leq p'$ and Equation (2-51). Notice that $a_{ij} y_j = 1$ only if vertex *i* is covered by a facility at location *j*, and the max operator ensures that at most one coverage is counted for each demand vertex. Thus the objective function measures the number of vertices covered. Notice the similarity between this model and the subject to $\sum_{j\in J} y_j \leq p'$ and Equation **(2-51)**. Notice that $a_{ij}y_j = 1$ only if vertex i is covered by a facility at location j, and the max operator ensures that at most one coverage is counted for each demand vertex. Thus the objective function measures the number of vertices covered. Notice the similarity between this model and the maximal-coverage model specified by Equations **(2-9)**, **(2-13)** and **(2-14)** or **(2-15)**. Equivalence was also established between the simple p-median and maximal-coverage models in the section entitled *Theoretical and Computational Links*. The generalized-partial-cover problem can be stated as: What is the minimum sum of 'distance' travelled between demands and their closest facilities if only p' facilities are available? Let C_{ij} be that total 'distance' travelled between demand i and facility j if the closest facility to i is at location j. The generalized-partial-cover problem can be formulated as

$$\min \left(\sum_{i\in I} \min\{C_{ij}|j\in\Xi(\mathbf{y})\} \right) \qquad (2\text{-}55)$$

subject to

$$1 \leq \sum_{j\in J} y_j \leq p' \qquad (2\text{-}56)$$

where $\Xi(\mathbf{y}) = \{j|y_j = 1\}$ and y_j is traditional binary variable for facility j.

Let us use the generalized median for solving partial-cover problems. Observe that maximizing the number of covered vertices is equivalent minimizing the number of vertices covered by an 'artificial' facility. Define $x_{i,|J|+1} = 1$ if demand i is assigned to an artificial facility and 0 otherwise; $a_{ij} = 1$ if $f_i d_{ij} \leq S_i^4$ or $j = |J|+1$, and 0 otherwise; $g_i = (|I|+1)$ if $= |J|+1$, and 0 otherwise; $c_{ij} = 1$ if $= |J|+1$, and otherwise; and finally $p = p'+1$. The optimal solution to this instance of the generalized-median-problem will always select facility $|J|+1$ for a median, whether or not it is 'used'. The remaining p' facilities will be selected to minimize the number-of-demands assigned to the artificial facility. A network-flow representation of this model is a convenient way to depict this, where the artificial facility will be a 'dummy' node where excess demand-flow assignments will be directed. An example sketch of this will be shown later in connection with the case study of set-covering and maximal-coverage models in **Figure 2-25**. This way solving the generalized-partial-cover-problem is even easier. Let us define $J = I$, $g_i = 0$ for $j\in J$, $c_{ij} = C_{ij}$ for $i\in I$ and $j\in J$, $a_{ij} = 1$ for $i\in I$ and $j\in J$, and finally $p = p'$. The generalized median problem, or more precisely the simple p-median version of it, is obtained.

Our point is that the generalized p-median model can be used to model many facility-location problems beyond simple median, set-covering, and maximal-coverage problems described initially in the section entitled *Minimal Requirement vs. Maximal Service*. Notice the generalized-median-problem makes no direct reference to an underlying network, although one usually is present in a situation

---

[4]Notice that a service distance is defined for each demand i.

under analysis. Because of the generality of this model, we introduced the subject of facility-location starting out with the *p*-median. Also, we have been referring to facilities and demands rather than vertices in a network. Having shown the versatility of the generalized-median-problem, let us now provide a specialized, yet analytical, solution algorithm that is more efficient than regular MIP.

2.5.4 Solving generalized medians using Lagrangian relaxation. Since the generalized median is an MIP, branch-and-bound (B&B) is a natural strategy to solve it. Recall from the chapter on "Prescriptive Tools" that a key element in this approach is the development of lower bounds for the candidate problems. We present here a *Lagrangian*, rather than a linear-programming, relaxation estimate of the lower bound (Frances et al. 1999). As described in the "Optimization" Appendix to Chan (2000), Lagrangian relaxation provides not only very good lower bounds, but also mechanism for constructing good feasible solutions. A relaxed problem is obtained from the generalized-median-problem (GMP) by assigning a multiplier or dual variable λ_i to each of the assignment constraints and transforming them into a "penalty function." The resulting objective function is

$$\min \sum_{j \in J} \left(g_j y_j + \sum_{i \in I} c_{ij} x_{ij} \right) + \sum_{i \in I} \lambda_i \left(1 - \sum_{j \in J} x_{ij} \right).$$

(2-57)

Rearranging terms gives us the relaxed problem:

$$\min \; z_{LR}(\lambda) = \sum_{j \in J} \left[g_j y_j + \sum_{i \in I} (c_{ij} - \lambda_i) x_{ij} \right] + \sum_{i \in I} \lambda_i$$

(2-58)

subject to Equations **(2-49)** through **(2-53)**.

The relaxed problem has two very important features. First, since it is a relaxation of GMP, it always provides the required lower bound to the original problem, i.e., $z_{LR}(\lambda) \le z^*$ for any choice of λ. Let us invoke the Karash-Kuhn-Tucker conditions [5]. If we are fortunate enough to find a dual vector for which the solution to the relaxed problem is feasible in GMP, this solution is also optimal. The second important feature of the relaxed problem is that it can be quite easily solved for a given multiplier vector, λ. For each possible facility location, define a subproblem

$$z_j'(\lambda) = z(\lambda, y_j) = g_j + \sum_{i \in I} \min(0, c_{ij} - \lambda_i)$$

(2-59)

---

[5] A discussion of Karash-Kuhn-Tucker conditions and an introduction to "Lagrangian relaxation" can be found in the Chapter on "Prescriptive Tools" (Chan 2000).

where we assume $a_{ij} = 1$ between all demands and facilities.[6] Now the relaxed problem of Equation **(2-57)** can be rewritten as

$$z_{LR}(\lambda) = \sum_{i \in I} \lambda_i + \min \sum_{j \in J} z_j'(\lambda) y_j \qquad (2\text{-}60)$$

subject to Equations **(2-51)** and **(2-52)**. Let us show that this is equivalent to the relaxed problem for a given λ. One can verify that when $y_j = 0$, there is no contribution to $z_{LR}(\lambda)$, while if $y_j = 1$, $z_{LR}(\lambda)$ includes g_j and all values $(c_{ij} - \lambda_i)$ that are negative. We shall now solve the relaxed problem. First, sort the $z_j'(\lambda)$ in increasing order, discard the non-negative values, and then select the p smallest remaining values or all values if there are fewer than p. The corresponding sites are selected for facility location. For a particular λ, it may happen that all $z_j'(\lambda) > 0$. Then it is legitimate to select the smallest $z_{ij}'(\lambda)$, which can be interpreted as adding the following redundant constraint to GMP: $\sum_{j \in J} y_j \geq 1$.

Example

The cotton farmers of Muddy Flat have compiled the following annual-cost estimates of operating a cotton ginning farm. Farms 1,2,3,4,5, and 6 have annual costs of 8000, 8000, 10000, 8000, 9000, and 8000 dollars respectively. It has been decided that each farm will require ginning for 1000 truckloads per year, and the average vehicle cost is $1 per mile ($0.63/km). What is the minimum total-cost, facility plus transportation, associated with operating two farms? (Francis et al. 1999)

Table 2-2 presents the result obtained by applying the algorithm to the problem with particular values for the dual variables. Notice that in this example, the solution to the relaxed problem $\bar{\lambda}$ is infeasible in GMP. The reason is that neither demand 2 nor 3 is assigned to a facility and both demands 5 and 6 are assigned twice. Also note, however, that using the set of facilities chosen, namely sites 5 and 6, it is trivial to construct a feasible solution. This is accomplished simply by assigning each demand to the cheapest facility in the set, as shown by the asterisks besides the c_{ij} in **Table 2-2**. For the example, the resulting feasible solution has a total cost of $(3+8)+(4+6+6+6+7+0+0) = 40$ according to Equation **(2-59)**. $\square$

Now let us discuss how one can obtain the dual variables λ. There are two strategies for obtaining such a vector. One is to use an *ad hoc* procedure, remembering that the dual variable has the interpretation of the marginal cost of serving a demand i. For example, the unit cost of servicing demand i can be taken as $\lambda_i = \sum_{j \in J} c_{ij} / |J|$. Such a procedure may work well for some problems, but in general is not robust. A second strategy is to solve the *Lagrangian dual* problem, $\max_{\lambda \geq 0} z_{LR}(\lambda)$ which requires searching in $|I|$-space for the optimal dual-vector λ^*. If we find λ^*, we are guaranteed to have the best lower-bound obtainable for the

---

[6] Notice this assumption is made without loss of generality, for one can prevent an assignment in an optimal solution by making the corresponding c_{ij} very large.

c_{ij}	$j=1$	2	3	4	5	6	$\bar{\lambda}_i$
$i=1$	0	3	7	10	6	4*	4
2	3	0	4	7	6*	7	4
3	7	4	0	3	6*	8	5
4	10	7	3	0	7*	7	8
5	6	6	6	7	0*	2	7
6	4	7	8	7	2	0*	7
g_i	8	8	10	8	9	8	
$z_j'(\lambda)$	-1	0	-1	-2	-4	-5	$\Sigma\bar{\lambda}_i$
z_5'					*		$=35$
z_6'						*	
$z_{LR}(\lambda)$				$35+(-9)=26$			

Table 2-2 Sample Lagrangian-relaxation solution

related problem. This is illustrated in another numerical example worked out in the "Prescriptive Tools" chapter of Chan (2000) under the "Decomposition" section. Referring to the appendix to this book entitled "Pertinent Optimization Schemes," *subgradient optimization* can be used to search for λ^*. The general idea is to increase λ_i if the corresponding demand is not assigned in the solution to the current relaxed problem. Similarly, we decrease if the corresponding demand is assigned to more than one facility. The resulting sequence of vectors can be shown to converge to if certain technical requirements are satisfied (Chhajed and Lowe 1998).

2.6   Equivalence between center and median problems

As hinted at by the numerical examples worked out in Section 2.4, the solution procedures for a discrete median problem on a network can be applied to a center problem and *vice versa* (Krarup and Pruzan 1981). For the unweighted case where all demands are assumed to have unitary weight, $f_i = 1$ $(i \in I)$. Let $X = \{(\mathbf{x}, \mathbf{y})\}$ be the feasible solution space for the p-median problem, the model can be written as min $\{\Sigma_i \Sigma_j \ c_{ij} x_{ij} | (\mathbf{x}, \mathbf{y}) \in X\}$. For a given feasible $\mathbf{y}$ we can find a minimum-cost transportation-plan simply by assigning each demand to the `cheapest' of the p open facilities. Consequently, only solutions with $x_{ij} \in \{0,1\}$ have to be considered. The corresponding p-center problem with cost matrix $\bar{D} = [D_{ij}]$ is

$$\min_{(x,y) \in X} \max_{i \in I, j \in J} \{D_{ij} x_{ij}\}. \tag{2-61}$$

Let z_c be the optimal value of the objective function in Equation **(2-61)**. z_c represents the largest demand-facility assignment-distance in an optimal solution. Suppose that the optimal value z_c is positive and known *a priori* and consider the following p-median problem for fixed $\beta \geq 1$:

$$\min \left\{ \Sigma_i \Sigma_j (D_{ij}/z_c)^\beta x_{ij} \,\middle|\, (x,y) \in X \right\}. \tag{2-62}$$

Obviously,

$$\lim \left(D_{ij}/z_c \right)^\beta = \begin{cases} \infty & \text{for } D_{ij} > z_c \\ 0 & \text{for } D < z \end{cases} \tag{2-63}$$

and $(D_{ij}/z_c)^\beta = 1$ for $D_{ij} = z_c$, all β. Thus as $\beta \to \infty$, $[(D_{ij}/z_c)^\beta]$ will approach a matrix of zeros, one and infinities. Since z_c is the optimal value of the objective function in the min-max problem **(2-61)**, there exists a solution to the min-sum problem **(2-62)**. The objective function of this solution is finite and such that the largest-assignment $\max_{i,j}\{(D_{ij}/z_c)x_{ij}\}$ will equal 1; this solution will therefore also solve the min-max problem **(2-61)**.

It has been established that a binary p-center problem with a min-max objective function, upon appropriate transformations of its cost coefficients, can be reduced to a min-sum p-median problem (Lin and Xue 1998). This has the same solution set such that an optimal solution to the latter will also solve the original center problem. This result can be generalized to weighted cases. Let **W** be an $|I| \times |I|$ diagonal matrix where the nonzero diagonal elements are the $|I|$ given demand weights $f_1, \dots, f_{|I|}$. Replace the $|I| \times |I|$ matrix $\bar{D}$ of shortest-path distance in Equation **(2-61)** by itself post-multiplied by **W**. Equation **(2-61)** is now interpreted as the problem of locating p facilities and assigning the $|I|$ demands to them. When the largest *weighted* distance between any demand-facility pair, $\max_{i,j}\{f_i D_{ij} x_{ij}\}$, becomes as small as possible, there must be a threshold level for $\beta = \beta^*$ such that $z_c(\mathbf{x}(\beta)) = z_c$ for all $\beta \geq \beta^*$. Notice that the weighted distances then will be transformed as $(f_i D_{ij}/z_c)^\beta$ when the min-max problem **(2-61)** is reduced to the corresponding min-sum problem. It should be emphasized that the results obtained will hold for 0–1 problems in general due to intrinsic properties of the min-max criterion. This result can be generalized. Chhajed and Lowe (1992), for example, have shown that a special version of the p-median problem can be solved as a flow graph. Furthermore, the graph can be used to solve the p-center problem. But the important result is that efficient solutions to the p-median problem can be employed to solve the center problem and vice versa. As mentioned more than once, the transformation of the distance measure D_{ij} is also of interest. In the next chapter on "Spatial Separation," we will show how the concept of spatial separation can be viewed from different angles. It will be shown that the l_p-metric bears resemblance to the transformation function employed above using Equation **(2-63)**.

Here we will illustrate the profound implications of distance-transformation by three *p*-median examples.

Example 1

Following the "Descriptive example" in the "Analytic techniques" Section of Chapter One, an adaptation of the *p*-median model yields the *p-medoid* method for forming *p* clusters in an image. This is particularly useful in defining pollution plumes from remote-sensing images (Wright and Chan 1994). Specifically, the *p*-medoid method takes on the form of Equation **(2-7)** through **(2-9)**, together with Equations **(2-3)** through **(2-5)**. Here f_i is the frequency with which the distance d_{ij} is encountered, for all *i*. This method partitions the pixels into clusters. The objective is to minimize the sum of the distance each pixel *i* is away from the pixel *j* representing the cluster to which it is assigned. In the original *p*-medoid model of Kaufman and Rousseuw (1990) f_i is simply unity. Instead of Euclidean distances, the distances will be represented by the differences between grey values from one pixel to the next. The result is a contour plot of grey values over the given area, which provides a graphic display of both the extent and concentration of the pollution plume. Instead of comparing each pixel to each other as called for by the original formulation, computational complexity is reduced by comparing each possible grey-value with each other. f_i is now no longer unity, but the number of pixels at the same 'distance' from *j*. Thus in a 256-shade grey scale associated with many satellite images,[7] there are only up to 256 nodes in each column of the bipartate graph representing as large an image as imaginable.

A 16×16-grid example-image is shown in the "Example image of a pollution plume" Figure of Chapter 1. The center area is highly polluted, and the pollution diffuses outward. The weighted cost matrix $[f_i d_{ij}]$ now looks like **Table 2-3**. For example the 'distances' between pixels of grey values 0 and 1, d_{10} have a uniform value of 1, and there are 40 pixels with grey-value of 1 in this image; hence $f_1 d_{10} = 40$. Similarly, the distances between pixels of grey-values 0 and 2 have a uniform distance of 2, and there are 32 of them. Hence $f_2 d_{20} = 64$, and so on. Based on these costs, a contour map of the pollution plume can now be delineated by solving the *p*-medoid model. It results in representative nodes 0, 2 and 5 here with $p = 3$. In other words, the representative node for the unpolluted area is associated with a grey-value of zero. The representative node for the moderately polluted area is 2. And the representative node of the lightly polluted area is associated with a grey value of 5. The plume-delineation result is overlaid in the "Example image of a pollution plume" Figure in Chapter 1. □

Example 2

Instead of a precise grey-value reading, most images have noise. To handle this, an important decision in image-classification is to strike a balance between spectral vs. spatial pattern-recognition—i.e., classification based on grey-values vs. contiguity respectively (Wright and Chan 1994). We have seen examples of this kind of classification technique in the "Remote sensing" chapter already. Here, we will show how a *p*-medoid method can be used to achieve

---

[7]This is the normal scale for a black-and-white image, as explained in the "Remote Sensing" chapter in Chan (2000).

Grey value	0	1	2	3	4	5	6	7	8	9
0	0	40	64	84	80	70	42	28	16	9
1	108	0	32	56	60	56	35	24	14	8
2	216	40	0	28	40	42	28	20	12	7
3	324	80	32	0	20	28	21	16	10	6
4	432	120	64	28	0	14	14	12	8	5
5	540	160	96	56	20	0	7	8	6	4
6	648	200	128	84	40	14	0	4	4	3
7	756	240	160	112	60	28	7	0	2	2
8	864	280	192	140	80	42	14	4	0	1
9	972	320	224	168	100	56	21	8	2	0

Table 2-3 'Costs' between two grey values

this function by a simple transformation of distance metrics. In other words, it is proposed that some weighted combination of contextual and non-contextual data could provide the best pollution contours, particularly in the presence of noise. If it can be assumed that contours generally form 'doughnut' shapes about some center point or area caused by the diffusion of pollutants, the difference in distances from that center-point to a pixel and to a potential representative-pixel can be used as a measure of distance. In essence, the contextual part of the formulation does the following. It minimizes the variance between the distance pixels are from the source-of-pollution, d_i and the distance their representative-pixels are from the source, d_j. The combination of spectral and spatial data can be accomplished through weights. These weights range from 0 to 1 for each data type, and the weights sum to unity. This is precisely the weighted-sum method explained in the "Multi-criteria Decision-making" chapter (Chan 2000). One criterion should be used in rating a weighing scheme: The weight applied to the spatial data should only be large enough to remove any noise from the data and no larger. In his way, the pertinent data, the spectral data, are allowed to "speak for themselves" as much as possible.

Consider the 6×6 grid shown in **Figure 2-15**(a). The very center of the physical grid—point (3.5,3.5)--is considered the source-point of the image, though the center-of-mass of the grey values—(3.5,3.525)—or the actual point-source of pollution could have been used. f_i now represents the grey-value of each of the pixels instead of the number of pixels with a grey-value of i. The cost of assigning a node i to representative pixel j is then $w|f_i-f_j|+(1-w)|d_i-d_j|$, where $0 \leq w \leq 1$. Such a cost matrix can be computed and displayed similar to **Table 2-3**, for each value of w. The first weighting-scheme attempted has a spectral weight of 1.0 and a spatial weight of 0.0. In other words, no contextuality is applied. The result is not expected to provide nice contours, particularly in the presence of noise. In fact, the result should be the same as example 1 above, since it is merely minimizing the

(a) Raw image

0	0	0	0	0	1
0	2	2	2	1	0
1	1	4	4	2	0
0	2	5	3	2	0
0	2	1	2	1	1
0	0	0	1	0	0

(b) Spectral pattern-recognition ($w=0$)

0	0	0	0	0	1
0	*2	2	2	1	0
1	1	2	2	2	0
0	2	2	2	2	0
0	2	1	2	1	1
0	0	0	1	0	0

(c) Spectral and spatial pattern-recognition ($0.5 \leq w \leq 1.0$)

0	0	0	0	0	0
0	1	*1	1	1	0
0	1	*2	2	1	0
0	1	2	2	1	0
0	1	1	1	1	0
0	0	0	0	0	0

Legend

*   Representative pixel

Figure 2-15 Contextual image-classification using p-methoid method

difference in grey-values between all pixels and their representative pixels. **Figure 2-15(b)** shows the result in which subregion 2 is the only one that is contiguous, but this happens only by chance. The noisy pixels on the edge are more typical results. Clearly little insight can be gained from this classification that examination of the raw data in (a) does not provide already. By contrast, the delineation of subregions 0, 1, and 2 for the spatial weight

$w \geq 0.5$ is shown in (c). It appears that 0.5 is sufficient in this example to factor out the noise found in some edge pixels. A 'correct' delineation should allow contiguity among members of each subregion to take place. This is shown here where perfect rings are obtained around the center point, suggesting a symmetric diffusion pattern. □

Example 3

Slight variation of the p-medoid method can be used to 'triangulate' the location of a 'target'—a common problem in defense operations (Fair et al. 1996). The U. S. Air Force, for example, needs a model to predict the location of electromagnetic radiation emitters. These emitters can be enemy search-radars or the radio-signal emission from a downed pilot. By taking advantage of the way in which terrain blocks signal energy, it should be possible to pinpoint the location of a radar or other signal emitters. Suppose the strength of the radar or signal energy could be measured near the ground at various geographic locations. By plotting the data on a grid, the pattern of signal energy strengths becomes apparent. A line could then be drawn from the edge of the area where the radar energy decreases to a known terrain feature. If several such lines were extended beyond the terrain feature, the point of their intersection would pinpoint the emitter's location, as illustrated in **Figure 2-16**. However, there is a problem with using this technique. Beyond blocking electromagnetic radiation, terrain features can interfere with a clear signal return by causing bouncing and scattering of the signal energy. This type of interference is common near the ground and can cause irregular or blurry results. Alternatively, the direction-finding signal from a survival radio could be jammed, or a depleted battery may render it weak in signals. Such blurring of the energy would make it very difficult to accurately determine the 'edge' of the area where the energy decreases. This would make it nearly impossible to pinpoint the source. The p-medoid method of pattern-recognition is one way to overcome this problem by 'smoothing' these noise signals[8].

We will now illustrate with a search-and-rescue example. A 20×20 data grid has been collected at Conchas Reservoir, New Mexico. A search-and-rescue helicopter has flown a grid pattern behind the high terrain on the south shore. The signal strength was recorded at regular intervals and rated from 1 to 9, 9 being the strongest signal. **Figure 2-16** shows the optimum delineation of the radio-signal strength. The white areas represent areas where the signal was strongest, and the dark represent areas of weak or no signal strength. This grid is affixed on the map in its true geographic location. It can be seen that there are two distinct shadows. The shadow on the left is from the Tenaja Mesa, and the right one comes from the 4020-foot (1206 m) peak east of Tenaja Mesa. We extend the lines from the edge of the signal shadows to the edge of the terrain feature causing the shadow and then beyond. These lines will fix the point where the signal is originating. These lines are drawn on **Figure 2-16**; the area around the intersection of the lines should contain the downed pilot. □

The above three examples show some highly unconventional ways to apply the p-median model. The resulting models can no longer be recognized under the median/center categorization (Carreras and Serra 1999). Aside from an innovative interpretation of 'distance', one interesting feature common to these three examples is the integrality of the solutions. Although the models have no restriction on

---

[8]For an introduction to noise removal, see the "Remote Sensing and Geographic Information Systems" chapter under the "Digital image processing" section (Chan 2000).

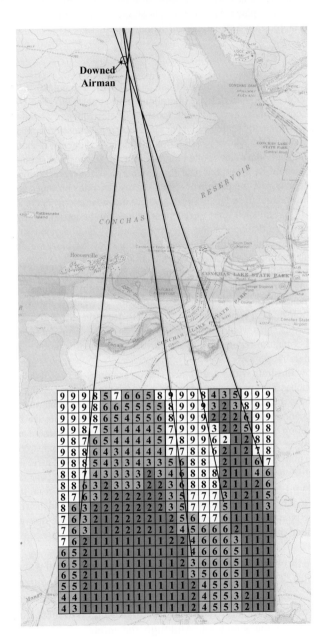

Figure 2-16 Using degradation of signal to pinpoint the signal source
(Fair et al. 1996)

maintaining integer solutions, all of the variables come out to be one's or zero's. It can be noticed that for the model formulation, all of the constraint coefficients in the tableau are 1, -1, or 0, rendering the matrix a totally unimodular[9] one. Since the right-hand-side consists of all ones, or the constraints require certain combinations of the variables sum to one, the decision variables are zero-one valued. It was shown also that the problem can be modelled by network flow[7], resulting in efficient solution to large-size problems (Wright and Chan 1994).

2.7   Planar location problems

In this chapter, we have been concentrating on facility-location on a *discrete* network. There is a parallel effort in locating facilities on a *continuous* plane—an effort that dates from the middle ages. We have referred to this as the Steiner–Weber problem in the "Prescriptive Tools" chapter (Chan 2000). It is appropriate at this juncture to devote some time formally to this important class of problems. **Figure 2-17** gives a geometric illustration of the type of problem that we consider in this section. Two facilities are to be located at two-dimensional (R^2) Cartesian points j_1 and j_2. Four demands are located at known Cartesian points $i_1,...,$ i_4, respectively.

2.7.1   Basic formulations. The lengths of the line segments $d(i,j)$ between a demand-and-facility pair represent the distance separation, such as the Euclidean

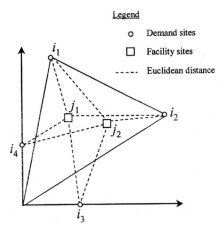

Figure 2-17 Illustrating a multi-facility location problem on a plane
(Francis et al. 1992)

---

[9]For a definition of total unimodularity and network-flow models, see the "Prescriptive Tools" chapter and the book appendix on "Review of Optimization Techniques" in Chan (2000).

l_2-measure shown in **Figure 2-17** (Colorni 1987).[10] Products or services are to be delivered between facility and demand, incurring a cost that is proportional to travel distance. Thus the service cost between a demand-facility pair is $c_{ij}d(\mathbf{i,j})$, where c_{ij} is the unit cost function. If each demand has a weight f_i, the multi-facility location problem on a plane can be defined as

$$\min\left(\sum_{i\in I}\sum_{j\in J}f_i c_{ij}d(\mathbf{i},\mathbf{j})x_{ij}\right)$$

(2-64)

where

$$x_{ij}=\begin{cases}0 & if\ j\neq l\\ 1 & if\ j=l\end{cases}\qquad i\in I.$$

(2-65)

l represents the facility that is closest to demand $\mathbf{i}$. Again, $d(\mathbf{i,j})$ is the l_p-metric as defined in the chapter on "Multi-criteria Decision-making" [Chan 2000] (and will be further discussed in the chapter on "Measuring Spatial Separation"). x_{ij} is the binary variable that assumes a value of 1 if demand i is served by facility j and 0 otherwise. Notice the notation $\mathbf{i}$ stands for the Cartesian coordinates of the point labelled i. This model is a continuous version of the median problem described previously, and is often called the *Weber problem* (short for the Steiner–Weber Problem). The *worst-case* version of the Weber problem corresponds to the use of the min-max criterion, or minimization of the maximum distance to be covered. Its formulation is therefore similar to the discrete center problem. $\min\left[\max_{i\in I,j\in J}f_i c_{ij}d(\mathbf{i,j})x_{ij}\right]$ In the case of obnoxious facilities, the figure-of-merit will be maximized instead of minimized and *vice versa* (Brimberg and Juel 1998). Thus in the Weber problem, the min-sum criterion of Equation **Figure 2-17** now reads: $\max\left(\sum_{i\in I}\sum_{j\in J}f_i c_{ij}d(\mathbf{i},\mathbf{j})x_{ij}\right)$ The worst-case problem now becomes $\max\left[\min_{i\in I,j\in J}f_i c_{ij}d(\mathbf{i,j})x_{ij}\right]$

If there is only one facility, the binary variable x_{ij} is a dummy variable. The reason is that the problem can be expressed adequately only in terms of the coordinates system as illustrated in **Figure 2-17**, which is part of the l_p-metric. The coordinates define the distances $d(\mathbf{i})$ ($i\in I$) from the single facility, say at $\mathbf{j}_1$, to the various demands $\mathbf{i}$, say at $\mathbf{i}_0$, $\mathbf{i}_1$ and $\mathbf{i}_2$. The worst-case versions parallel the discrete case introduced as Equations **(2-22)** and **(2-41)**, where the decision variable x_{ij} is missing. The presence of decision variables x_{ij} for the case of multi-facilities complicates things considerably. Nevertheless, the assumption of unlimited capacity, typical of this kind of problem, allows us to find the value of x_{ij}, as we have shown in the discrete case. Suppose that each facility can satisfy any demand, demand i will certainly make use of the nearest facility, as alluded to Equation **(2-64)** above.

---

[10]For a discussion of l_p-norms ($p = 1,2,...\infty$), refer to the "Multi-criteria Decision-making" chapter in Chan (2000).

The continuous case of median as described by Equations **Figure 2-17** through **(2-64)** dates back to Cavalieri and Fermat before being known as the classical formulation by Weber in 1909. Weber discussed the case of one facility and three demands. An early theoretical contribution concerned determination of the set of efficient solutions, i.e., points not dominated by other points, as a function of c_{ij} and f_i. More will be said about this shortly. Other results relate to the definition of the constrained case in which some arcs in the two-dimensional space are not feasible. In the example of **Figure 2-17**, this amounts to removing an arc such as (i_1, j_2) from consideration. Still other results include the determination of optimal solution for certain functional forms, of say c_{ij}. For the Weber problem, the main computational findings can be summarized as follows (Drezner and Goldman 1991).

2.7.2 Computational results. It is known that the planar Weber problem with l_p-metric has all its solutions in the convex hull of the demand points **i**. For example, $\mathbf{j}_1$ is to be found within the triangle $\mathbf{i}_0 \, \mathbf{i}_1 \, \mathbf{i}_2$ in **Figure 2-17**. For l_1 and l_∞, additional conditions are known which reduce the set of possible optimal points, or the efficient set. It is reduced to the intersection of that convex hull and the points defined by a certain grid. In **Figure 2-18**, for example, A, B, C, D, G, R make up the intersection points, or efficient set. F and T are not in the set and neither one can be an optimal point. Each point A, B, C, D is the only optimal solution when more than half the total weight f_i is placed on that demand point. R is the only optimal solution if $f_A = f_B = 3$, and $f_C = f_D = 2$. G is the only optimal solution if $f_A = f_B = 2$, and $f_C = f_D = 3$. To appreciate this result, we will give a numerical illustration.

Example

A way of locating an obnoxious-facility is described by Appa (1993), following the procedure of Drezner and Weslowsky (1983). Consider the problem of locating a single facility at (x,y) in a bounded and convex region. The objective is to maximize its rectilinear distance from the nearest of $|I|$ customers located in the region at (x^i, y^i) for $i = 1,\dots,|I|$. We construct a large rectangle in 2-dimensions that contain the region, and draw one horizontal and one vertical line through each customer, ending with $(|I|+1)^2$ rectangles as shown in **Figure 2-**

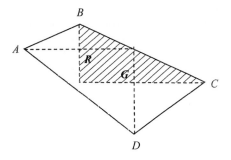

Figure 2-18 Example of Weber locations for l_1- and l_∞-metric

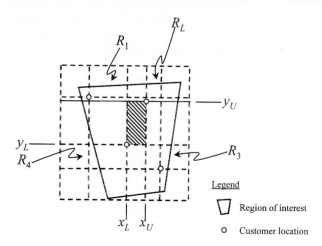

Figure 2-19 Locating an obnoxious facility via rectilinear distance

19, where $|I| = 4$. The optimal solution must lie inside the region in one of these rectangles. It can be obtained by picking the best of the solutions to a linear-programming (LP) problem defined over each rectangle—consisting of 25 LPs here. Let P_1 through P_4 be the nearest customers to the rectangle defined by $x_L \le x \le x_U$, $y_L \le y \le y_U$ in each of the four regions to the northwest, northeast, southeast, and southwest, as illustrated by R_1 to R_4 in **Figure 2-19** (Mehreq et al. 1986). Then the following LP finds the maximum distance D to the nearest of the four customers for the current rectangle. Closed-form formulas for computing the coordinates of local solutions in any rectangle were provided.

$$\max D$$
$$\text{s.t.} \qquad \begin{aligned} D - x + y &\ge -x_1 + y_1 \\ D + x + y &\ge x_2 + y_2 \\ D + x - y &\ge x_3 - y_3 \\ D - x - y &\ge -x_4 - y_4 \\ x_L &\le x \le x_U \\ y_L &\le y \le y_U. \quad \square \end{aligned}$$

It is known that for $1 < p < \infty$ a certain part of the convex hull is the smallest possible efficient set. But this set is different from the one for $p = 1$ and $p = \infty$. The airport-location problem discussed in the "Analytical solution technique" section of the "Prescriptive Tools" chapter (Chan 2000) illustrates the case when $p = 2$. Given demands f_i, we wish to find the location for the single median $\mathbf{j}^* = (x^*, y^*)$ in the Weber problem. In other words, we minimize the l_p-metric when $p = 2$: $z = \sum_{i=1}^{|I|} f_i \, r_i'(x,y;2)$ where r_i' is the Euclidean distance between demand i at (x^i, y^i) and the facility at (x^*, y^*). The necessary and sufficient conditions for optimality is

$$\frac{\partial z}{\partial x} = \sum_{i=1}^{|I|} \frac{f_i}{r_i'} (x - x^i) = 0 \qquad\qquad (2\text{-}66)$$

$$\frac{\partial z}{\partial y} = \sum_{i=1}^{|I|} \frac{f_i}{r_i'}(y - y^i) = 0. \tag{2-67}$$

The solution can be approximated to any degree of accuracy by the iteration equations

$$x(k+1) = \sum_{i=1}^{|I|} \frac{f_i x^i}{r_i'(k)} \bigg/ \sum_{i=1}^{|I|} \frac{f_i}{r_i'(k)} \tag{2-68}$$

$$y(k+1) = \sum_{i=1}^{|I|} \frac{f_i y^i}{r_i'(k)} \bigg/ \sum_{i=1}^{|I|} \frac{f_i}{r_i'(k)} \tag{2-69}$$

where k is the iteration number. While any point in the plane will do, the center-of-gravity is a good starting point for the iterations, yielding rapid convergence of the algorithm. Thus $x(0) = \sum_{i=1}^{|I|} f_i x^i / \sum_{i=1}^{|I|} x^i$ and $y(0) = \sum_{i=1}^{|I|} f_i y^i / \sum_{i=1}^{|I|} y^i$.

This algorithm is quite robust. There are reported problems, however, with a spatial separation function other than pure Euclidean. An example is *when the spatial-separation function d_i has a concave form* with respect to distance, such as

$$[r_i'(x,y;2)]^\beta \qquad (\beta < 1). \tag{2-70}$$

In this case, only local optimum (x', y') may be found. $(x', y') = (x^i, y^i)$ has been proven to be at least a locally minimal solution, and *multiple local minima exist at these demand points (x^i, y^i)*. There is no guarantee that a global solution can be obtained. Fortunately, the solution is either optimal or near optimal a large percent of the time. Referencing **Figure 2-2**, a concave travel-cost function is quite realistic in practice, reflecting economies of scale in transportation. Caution should therefore be exercised when the above algorithm is used in real-world applications. One can glean also from this experience and previous experiences with the airport-location problem (Chan 2000) that algorithms for the Weber problem are quite challenging (Brimberg et al. 1998). There are numerically intricacies when one goes beyond one single facility (Xue et al. 1996, Chen et al. 1998). Recently Sherali and Tuncbilek (1992) proposed a branch-and-bound procedure for locating p medians with known capacities, in which the squared Euclidean-metric is employed. Results within 2 percent of optimality were reported for problems where the median facilities range from 6 to 20, and the corresponding demands range from 20 to 60.

Example

Consider a triangle in which $i_0 = (0,0)$, $i_1 = (10,0)$, $i_2 = (0,10)$, and $f_0 = 3\,00$, $f_1 = 155$, $f_2 = 340$ (Hurter and Martinich 1989). The center of gravity is $(1.73, 3.80)$, which is used as the starting point. The following iterations are obtained

k	$(x(k),y(k))$	z
0	(1.73,3.80)	4851.40
1	(1.20,3.73)	4827.45
2	(1.11,3.65)	4825.69
3	(1.09,3.58)	4824.99
4	(1.09,3.52)	4824.54.

The algorithm stops within the error limit of $\varepsilon = 0.005$. Since there is no significant difference between $r_i'(3)$ and $r_i'(4)$, or $|r_i'(4)-r_i'(3)| \le \varepsilon$, the answer is taken to be acceptable. The optimal solution is actually $(1.04,3.13)$ which is not too different from the answer from the algorithm. $\square$

Application of this algorithm to the airport-location problem between Cincinnati, Columbus and Dayton Ohio—as first introduced in the "Prescriptive Tools" chapter- in Chan (2000)-yields the following optimal location:

Euclidean distance from Cincinnati	69.863 miles (111.78 km)
Euclidean distance from Columbus	0.141 (0.23 km)
Euclidean distance from Dayton	89.875 (143.80 km).

This essentially places the airport at Columbus, the largest of the three cities (Hurry et al. 1995). The same algorithm applied to the four-city case locates the airport at 25.172 miles (40.28 km) from Cincinnati, 73.981 miles (118.37 km) from Columbus, 34.848 miles (55.76 km) from Dayton, and 134.360 miles (214.98 km) from Indianapolis. In the three-city case, the result agrees with one of the alternate optima obtained from an LP model used previously to solve the problem. In the four-city case, however, the results from the two solution techniques differ, suggesting that improvements can be made to the LP model.

For the worst-case problem, an algorithm for one facility and $f_i = 1$ $(i \in I)$ was devised by Shamos and Hoey (as reported by Colorni (1987)) for Euclidean metrics. This algorithm divides the plane into a maximum of $|I|$ areas: each area R_i contains the points that are further from demand i than from any other demands; subsequently the algorithm finds the optimal solution by examining the set of the intersections of the straight lines dividing the region. **Figure 2-20** shows an example of $|I| = 5$. Boundaries of the 4 areas contain the points further from demand i than from any other are drawn as solid lines. The optimal center solution is point A. In locating obnoxious facilities, the partition of the plane is carried out so that each area R_i contains the points closer to demand i than any other. The optimal solution is found at the intersection of the straight lines dividing the region into the $|I|$ areas, or at the intersections of such lines with the boundary of the

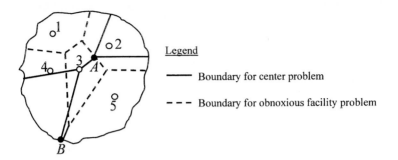

Figure 2-20 Planar one-center and obnoxious-facility example (Colorni 1987)

region itself. Here, boundaries of the 5 areas containing the points closer to demand *i* than to any other are drawn in **Figure 2-20**. Point *B* is found the optimum obnoxious-facility location. These procedures can best be explained in terms of *Voronoi diagrams*—a systematic effort to partition a plane into 'tiles', each of which defines an area according to proximity to a predetermined "focal point" (Ohsawa and Imai 1997). This subject is an exciting and promising research topic that will be treated formally later in a chapter entitled "Spatial-temporal Information."

2.7.3 Generalizations. Hurter and Martinich (1989) examine the Weber problem in terms of the theory of production. Consider a firm that produces one output using two spatially distinct inputs. The input sources are located on a plane at points i_1 and i_2, the output is shipped to a market at point i_0 (see **Figure 2-17**). Without loss of generality, we assume for the time being that the per-unit cost of transporting input *i* from i ($i = 1,2$) to plant location j is $c_i(j) = c_i d_i(j)$. Here d_i is the l_2 Euclidean distance between j and i, and $c_i > 0$ is the (constant) unit transport-cost. Let $c_0(j)$ be defined equivalently to represent the per unit-cost of shipping output from the plant to i_0. Let F' be the production function $z_0 = F'(z_1, z_2)$ where z_0 is the rate of output and z_1 and z_2 are the usage rates of the inputs. Let p'_0 be the selling price of the output, p_1 and p_2 be the purchase prices of the inputs. The firm's planar *single-facility production problem* can be written as

$$\max \left\{ \left[p_0' - c_0 d_0(j) \right] F'(z_1, z_2) - \sum_{i=1}^{2} \left[p_i + c_i d_i(j) \right] z_i \right\} \qquad (2\text{-}71)$$

subject to

$$z_i \geq 0 \qquad i = 1,2 \qquad (2\text{-}72)$$

$$j \in R^2. \qquad (2\text{-}73)$$

Notice the last constraint simply says the problem is a planar one in two-dimensional space.

When the output level is fixed at z_0' *a priori*, the above becomes the *design-location problem*

$$\min \left\{ c_0 d_0(\mathbf{j}) z_0' + \sum_{i=1}^{2} \left[p_i + c_i d_i(\mathbf{j}) \right] z_i \right\}$$ (2-74)

subject to

$$z_0' = F'(z_1, z_2)$$ (2-75)

and Equations **(2-72)** and **(2-73)**. In the special case when input factors are free, we have $p_i = 0$, $F'(z_1, z_2) = z_T - z_1 - z_2$ or $z_T = z_1 + z_2 + z_0$, where z_T and z_i' are constants. The problem reduces to the classic Weber problem, and in particular the airport-location problem first discussed in the "Prescriptive Tools" chapter in Chan (2000). The optimal facility location for the models from Equation **(2-71)** forward must be still in the convex hull of the set $\{\mathbf{i}_0, \mathbf{i}_1, \mathbf{i}_2\}$. *If the optimal solutions $z_i^* > 0$ (i = 1,2), it can be proved that an interior-edge location cannot be optimal* for the problems defined from Equation **(2-71)** forward (Hurter and Martinich 1989), where an interior-edge location is any point of the edge of the location triangle that is not a corner point. This can be confirmed by the airport-location results reported in the "Prescriptive Tools" chapter.[11] For the design-location problem defined by Equation **(2-74)** forward, the conditions for an interior location can be established. *Suppose there is some radial direction away from the output market that decreases total cost. As one moves away from input source I along edge (O,I) toward the output market, total cost decreases. Then the optimal facility location must be an interior point of the location triangle.* Again, the airport-location results in the "Prescriptive Tools" chapter confirm this, particularly when noise pollution is considered besides travel distance.

It should be noted that the two observations above, regarding the location of facility, do not rely on the transport cost functions being linear in distance. As long as the transport cost rates are increasing functions of Euclidean distance, i.e., $c_i(\mathbf{j})$ can be written as $c_i(d_i) d_i$, these observations are valid. Incidentally, this transport cost property will usually be true in the real world. Now define $I(i)$ as the set of input markets and $I(0)$ the set of output markets. The above results can be generalized to multiple inputs, where $|I(i)| > 2$, for the design-location problem. As stated earlier, it can be shown that the optimal facility-location for the $|I(i)|$-input design-location problem must be in the convex hull of $\{\mathbf{i}_0, \ldots, \mathbf{i}_{|I(i)|}\}$. *Suppose at least two inputs are used at positive levels and their source locations are not all collinear with the output market. Then no interior-edge-point can be an optimal location.* Again, this result can be confirmed with the computational experiences for the four-city airport-location problem described in the "Prescriptive Tools"

---

[11]For clarity of presentation, please refer to "Weber's industrial location model" in the "Economics" chapter (Chan 2000) and the airport-location figure in the "Prescriptive tools" chapter (Chan 2000) entitled "Solutions to the three-city configuration" within the "Analytical Solution Techniques" section.

chapter. Notice this observation implies that the optimal locations are either at extreme (corner) points or in the interior of the convex hull.

It so turns out that the same result can be extended to more general *multi-facility planar design-location model*:

$$\min_{z, \, \mathbf{j}^j} \left\{ \sum_{j \in J} \left[\sum_{i \in I(0)} c_0 \, d_0(\mathbf{j}^j) \, z_{0i}^j + \sum_{i \in I(i)} \left(p_i + c_i \, d_i(\mathbf{j}^j) \right) z_i^j \right] \right\} \tag{2-76}$$

subject to

$$z_0^j = \sum_{i \in I(0)} z_{0i}^j = F'(z_1^j, \dots, z_{|I(i)|}^j) \qquad j \in J \tag{2-77}$$

$$\sum_{j \in J} z_{0i}^j = z_i' \qquad i \in I(0) \tag{2-78}$$

$$z_i^j \ge 0 \qquad i \in I(i), j \in J \tag{2-79}$$

$$\mathbf{j}^j \in R^2 \qquad j \in J. \tag{2-80}$$

Here, J is the number of plants to be located, $\mathbf{j}^j$ is the location of the *j*th plant, $I(0)$ is the set of output markets, z_0^j is the amount of output produced at plant *j*, z_{0i}^j is the amount of output produced at plant *j* and sent to output market i ($i \in I(0)$), z_i^j is the amount of input i ($i \in I(i)$) used by plant *j*, z_i' is the amount of fixed demand at market i ($i \in I(0)$), and $I(0) + I(i) = I$ or a node is either an output market or input market. When restricted to one plant or $|J| = 1$, the model reduces to the multiple-output multiple-input model. When $|J| = 1$ and output market is restricted to one $|I(0)| = 1$, it further reduces to the multiple input model (Hansen et al. 1998). Finally, when $|J| = 1$, $|I(0)| = 1$ and we specify two input markets $|I(i)| = 2$, it is the two-factor, single-output, single-facility model we started this section with.

2.7.4 Discrete vs. planar facility-location modelling. In this chapter, we have considered a number of discrete and planar locations models, including the general location-production variety discussed immediately above. Let us now try to make some comparative remarks about them. In particular, we like to identify their respective roles in solving real-world problems. While planar location models are necessary for such applications as plant layout, Frances et al. (1999) observed that planar models are principally of value for the insight they provide, and the simplicity of their construction. Planar location models involve certain basic assumptions that limit their realism, however, including the approximation of spatial separation in terms of l_p distances rather than actual routes. This is a subject we will spend a lot more time on in subsequent chapters. Frances et al. further stated that construction of very detailed and involved planar models may be a self-defeating exercise. If a more detailed model is required, it may be appropriate to use a network or discrete model. In recent years, tremendous gains have been made in the development of discrete facility-location models, as evidenced by the lengthy bibliography included with this book. While there is active research activity in the planar, continuous models, the advances have no where been as dramatic. Over the remaining chapters, we will cover stochastic queuing location, quadratic assignment, combined location-routing and competitive location—spanning the broad spectrum of discrete-location literature available to date. In comparison, planar models can seldom go beyond single facility models as far as operational solution techniques are concerned (Krarup and Pruzan 1990).

Trading off between model tractability and model realism, however, one may want to have both a planar and a more sophisticated discrete model of the same problem. We then use the planar model to explain some behavior of the discrete model. There is in fact a fair amount of parallelism in both modelling approaches. This is quite evident from just a limited amount of discussions on planar models above, including the center and median problems. The insights gained may well provide food for thought in further development of discrete models as well. In the next chapter on "Spatial Separation," we will address the question of exactly how well Euclidean metrics approximate actual routing in the location of vehicle depots. We will spend the rest of this book on discrete facility-location models. The reason, once again, is that there is a wealth of solution algorithms available for discrete formulations. Fueled by explosive developments in discrete optimization techniques, the tools available for network facility-location seem endless. Although there really is no "one cure for all" approach to solving a discrete facility-location problem, there are often choices in operational algorithms that are computationally viable. In the case study below, we will illustrate some of these algorithms, ranging from enumeration, MIP, to network-flow procedures. This will be performed in both single- and multiple-criteria optimization contexts. Through the case study, we try to illustrate representative issues in computational procedures. While leaning toward analytical techniques, we will turn gradually to heuristics, starting with a humble beginning here and progressing toward more structural approaches in the next two chapters.

2.8   Long-run location–production–allocation problems

Thus far, we have been dealing with *static* facility-location formulations, where planning is performed for a single target date. Perhaps the more interesting and practical formulation of discrete facility-location problems is the long-run, or dynamic model. Here the infrastructural components such as facility capacity, production and demand levels can change over time. We will illustrate with two of such models here.

2.8.1 General network location–production–allocation problem. Hakimi and Kuo (1988) formulated a *general network location-production-allocation problem*, where up to p-facilities are to be located in a network to produce a product or render a service. Cost of production depends on the facility location at j, denoted by the binary variable y_j, and the level of production z_0^j, $c(x_j, z(y_j))$. A facility can be located along any arc in their general formulation. Benefit is measured as a function of the products or services sold at a node i, $T(z_i)$. The cost of provision is measured by the transportation cost along a shortest path from facility j to demand i, $d(y_j, z_i(y_j))$. The network location-production-allocation problem is to find the facility locations $\mathbf{x} = (x_1, \dots, x_p)$, the production level z_0^j and the allocation of product or service from supply to demand z_{0i} such that the benefit exceeds the cost as much as possible. Further generalization of this location–production–allocation model gives rise to the *spatial-price-equilibrium* model, which includes the nature of the market-place in its formulation—be it monopolistic, oligopolistic or pure-competition.[12] A special formulation of the location–production–allocation model, rather than the general model, is shown here:

$$
\max\left\{ \sum_{i \in I} f_i z_i - \sum_{i \in I} \sum_{j=1}^{p} d_{ij}(z_{0i}^j) - \sum_{j=1}^{p} c_j(z_0^j) \right\}
$$

$$
\begin{aligned}
s.t. \quad & \sum_{i \in I} z_{0i}^j = z_0^j && j = 1, \dots, p \\
& \sum_{j=1}^{p} z_{0i}^j = z_i && i \in I \\
& \sum_{j=1}^{p} z_0^j = \sum_{i \in I} z_i \\
& z_0^j \leq \overline{P}_j && j = 1, \dots, p.
\end{aligned}
$$

(2-81)

where $\overline{P}_j$ is the production capacity of facility j.

Example

An example of the general network location-production-allocation problem is shown in **Figure 2-21** (Moore and Chan 1990), in which the demands, costs and production capacities are given. The objective is to establish up to three facilities to serve the demands while maximizing net benefit, defined say as revenue minus cost.

Maximize

$$
\begin{aligned}
& 8z_4 + 10z_5 && \text{(benefit)} \\
& - (3z_{01}^4 + 2.5z_{02}^4 + 5.5z_{03}^4) && \text{(transport costs to demand 4)} \\
& - (4z_{01}^5 + 2.6z_{02}^5 + 2z_{03}^5) && \text{(transport costs to demand 5)} \\
& - (2z_{01} + 2.2z_{02} + 2.3z_{03}) && \text{(cost of production).}
\end{aligned}
$$

Production at each facility 1, 2, or 3 is to satisfy demands at 4 and 5:

$$
\begin{aligned}
z_{01}^4 + z_0^5 1 - z_{01} &= 0 \\
z_{02}^4 + z_0^5 2 - z_{02} &= 0 \\
z_{03}^4 + z_0^5 3 - z_{03} &= 0.
\end{aligned}
$$

---

[12]For a discussion of "spatial-price-equilibrium", see the forthcoming chapter on "Spatial Equilibrium and Disequilibrium".

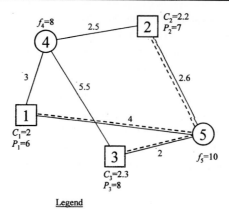

Legend

Selection of facility and allocation of demand

Figure 2-21 Example of general location–production–allocation

Individual demand at 5 and 4 is satisfied by supply from all three facilities:

$$z_{01}^4 + z_{02}^4 + z_{03}^4 - z_4 = 0$$
$$z_0^5 1 + z_0^5 2 + z_0^5 3 - z_5 = 0.$$

Total demands and supplies are balanced:

$$z_{01} + z_{02} + z_{03} - z_4 - z_5 = 0.$$

Levels of production at facilities 1, 2, and 3 are capacitated:

$$z_{01} \le 6 \qquad\qquad z_{02} \le 7 \qquad\qquad z_{03} \le 8.$$

The model solution is overlaid on **Figure 2-21**. Facility 1 should produce 6 units, all of which go to demand-point 5. Facility 2 should also produce 7 units to supply demand 5, and facility 3 produces 8 units to fulfil the remaining demand at 5. All together, demand at 5 is 21 units, while none is provided at demand-point 4. Even though this is a very specialized, linear version of the model, hopefully we have illustrated the generality of the original formulation. $\square$

2.8.2 Multiperiod capacitated location models. A facility-location problem can involve multiple time-periods. Consider the capacity-expansion plan of a single facility. **Figure 2-22** shows a schematic of how this can take place over time (Jacobsen 1990). Driven by the demands over time $f(t)$, $\bar{p}(t)$ capacity additions are built incrementally, resulting in a total capacity of $\bar{P}(t)$, where $t = 1, \dots, T$. The problem is one of supplying the given demand in each time-period at a minimal present-value of capacity-cost. The general multiperiod-capacitated-location model, given initial capacities at the $|J|$ plants, $\bar{P}_1(0), \bar{P}_2(0), \dots, \bar{P}_{|J|}(0)$, can be formulated as:

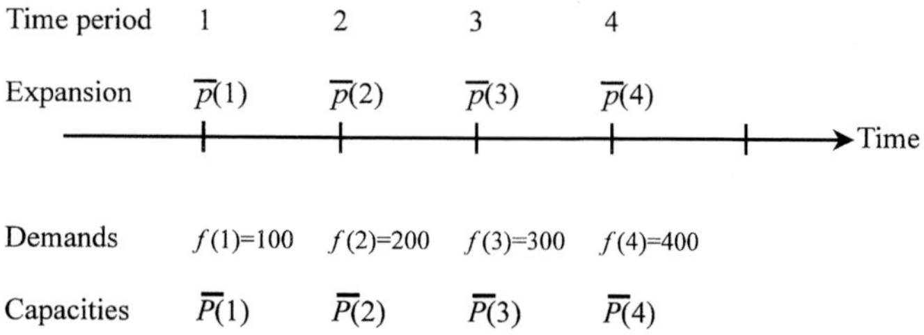

Time period	1	2	3	4
Expansion	$\overline{p}(1)$	$\overline{p}(2)$	$\overline{p}(3)$	$\overline{p}(4)$

Time

Demands	$f(1)=100$	$f(2)=200$	$f(3)=300$	$f(4)=400$
Capacities	$\overline{P}(1)$	$\overline{P}(2)$	$\overline{P}(3)$	$\overline{P}(4)$

Figure 2-22   Single-facility expansion-plan (Jacobsen 1990)

$$\min \sum_{t=1}^{T} \sum_{j\in J} \left[c_j(t, \overline{p}_j(t)) + \sum_{i\in I} d_{ji}(t) x_{ji}(t) \right]$$

$$\sum_{j\in J} x_{ji}(t) = f_i(t) \qquad i\in I, \ t=1, \dots, T$$

$$\sum_{i\in I} x_{ji}(t) \le \overline{P}_j(t) \qquad j\in J, \ t=1, \dots, T \qquad \qquad (2\text{-}82)$$

$$\overline{P}_j(t) = \overline{P}_j(t-1) + \overline{p}_j(t) \qquad j\in J, \ t=1, \dots, T; \ \overline{P}_j(0) \ given$$

$$\overline{p}_j(t) \ge 0, \ x_{ji}(t) \ge 0.$$

As introduced in the "Economics" chapter in Chan (2000), solution of dynamic facility-location problems of this type is most challenging conceptually, not to mention computationally. Two general approaches may be taken. First, we may look for possible system-growth paths through time under varying constraints and criteria-for-effectiveness, and try to identity stages at which such paths coincide with static-equilibrium solutions. Or we may set up static-equilibrium solutions and try to construct minimum-cost paths to connect them.

For a single-facility, the dynamic facility-expansion problem can be solved in a relatively straightforward manner. It can be formulated as follows:

$$\min \sum_{t=1}^{T} c(t, \overline{p}_t)$$

$$\overline{P}(t) = \overline{P}(t-1) + \overline{p}(t) \qquad t=1, \dots, T, \ \overline{P}(0) \ given$$

$$\overline{P}(t) \ge f(t) \qquad t=1, \dots, T \qquad \qquad (2\text{-}83)$$

$$\overline{p}(t) \ge 0 \qquad t=1, \dots, T.$$

Let $f'(t)$ be defined by the recursion

$$f'(t) = \max \left[f(t), f'(t-1) \right] \qquad t=1, 2, \dots, T; \ f'(0) = \overline{P}(0). \qquad (2\text{-}84)$$

It can be shown that when all capacity-cost functions $c(t,.)$, $t = 1, ... ,T$ are concave and increasing, then there exists an optimal solution to model **(2-83)** where $\bar{P}^*(t-1)=f'(t-1)$ whenever $\bar{p}^*(t)>0$. This is called the *regeneration-point theorem*. The main implication of the regeneration-point theorem is that the search for the optimal $\bar{P}(t)$ may be confined to the set $\{f'(\tau), \tau = t, t+1, ... ,T\}$ i.e., solution to Equation **(2-84)** from time period τ forward. The result in turn gives the values of capacity expansions $\bar{p}(t)$ by the first constraint of model **(2-83)**. The time period τ for which we set $\bar{P}^*(t-1)=f'(t-1)$ is called the *regeneration point*. At a regeneration point, the capacity equals demand before a possible expansion, and new expansions $\bar{p}(t)$ are built to serve additional demand. The solution procedure follows a dynamic-programming algorithm[13] that amounts to finding the shortest path through a network where the nodes correspond to the regeneration points. The arc-length w_{t_1,t_2} corresponds to the cost of providing the capacity $\bar{p}(t_1)=f'(t_2-1)-f'(t_1-1)$ at time t_1 to serve demand up to the next regeneration point t_2.

Example

This algorithm will now be illustrated by the following example (Jacobsen 1990). Consider a problem with a time-horizon of four years and assume that expansions may take place at the beginning of each year. The corresponding time-graph is shown in **Figure 2-23**, where each node corresponds to a regeneration point. Suppose the demands are $f(1) = 100, f(2) = 200, f(3) = 300, f(4) = 400$, and the initial capacity $\bar{P}(0)$ is zero. A path in the time-graph now suggests an expansion plan. For example, the path from node 1 via node 3 to node 5 corresponds to expansion of size 200 at nodes 1 and 3. Let the cost of an expansion of size $\bar{p} \geq 100$ at node t be

$$c(t,\bar{p}) = (98.5079+0.7585(\bar{p}-100))/(1.1)^{t-1}.$$

The arc lengths, w_{t_1,t_2}, representing the capacity-costs of expansion, can now be calculated in tabular form from this equation and are shown below:

	$t_2 = 2$	$t_2 = 3$	$t_2 = 4$	$t_2 = 5$
$t_1 = 1$	98.5079	174.3579	250.2079	326.0579
$t_1 = 2$		89.5526	158.5072	227.4617
$t_1 = 3$			81.4115	144.0974
$t_1 = 4$				74.0104

For example, the first row of elements ($t_1 = 1$) is computed by the formula with $t = 1$ and $\bar{p} = 100, 200, 300$ and 400 respectively. The next row ($t_1 = 2$) is computed by the formula with

---

[13]For a review of dynamic programming and a fully-worked-out numerical example similar to the current one, please consult the book appendix entitled "Markovian processes" under the "Markovian properties of dynamic programming" section.

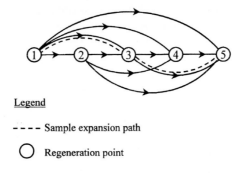

Legend

- - - - Sample expansion path

○   Regeneration point

Figure 2-23 Time graph for capacity expansion

t = 2 and $\bar{p}$ = 100, 200 and 300 respectively. In general, each succeeding row is computed by shifting the previous row one position to the right and dividing by the discount factor 1.1.

A path from node 1 to node 5 corresponds to an optimal expansion-plan satisfying the conditions of the regeneration-point theorem, and the length of the path is the cost of the expansion plan. Taking the path mentioned above from node 1 via node 3 to node 5, $w_{1,3}$ = 174.3579 corresponds to $\bar{p}(1)$ = 200, which is exactly sufficient to satisfy the demands before the expansion at node 3, that is $\bar{P}(2)$ = f(2) = 200. Similarly, $w_{3,5}$ = 144.0974 corresponds to $\bar{p}(3)=\bar{P}(4)$ = f(4) = 400, again satisfying the regeneration-point theorem.

The minimum-cost expansion plan is determined by finding the shortest path from node 1 to node 5. Any shortest-path algorithm will produce the matrix of shortest-path lengths shown below:

	$t_2 = 2$	$t_2 = 3$	$t_2 = 4$	$t_2 = 5$
$t_1=1$	98.5079	174.3579*	250.2079	326.0579
$t_1=2$		188.0605	257.0151	325.9696
$t_1=3$			255.7694	318.4553*
$t_1=4$				324.2183

The minimum cost of getting to node 5 (318.4553) is found in the column t_2 = 5 and row t_1 = 3, suggesting that node 5 is reached from node 3. Similarly, the minimum cost to reach node 3 is found in the column t_2 = 3 and row t_1 = 1 (174.3579). In aggregate the minimum-cost path is from node 1 via node 3 to node 5. The total cost of capacity-expansion is therefore 318.4553 (which is the sum of w_{13} = 174.3579 and w_{35} = 144.0974). Once the optimal path is known, the size of the expansions may be found by the regeneration-point property. The optimal state-variables are $\bar{P}^*(0)=0$, $\bar{P}^*(1)=\bar{P}^*(2)$ = f(2) = 200, $\bar{P}^*(3)=\bar{P}^*(4)$ = f(4) = 400, and consequently, $\bar{p}^*(1)=\bar{P}^*(2)-\bar{P}^*(0)=200$, $\bar{p}^*(3)=\bar{P}^*(4)-\bar{P}^*(2)=200$. □

The main result of this example is that an optimal-expansion program can be derived from the shortest path through the time- graph, which in turn is constructed by application of the regeneration-point theorem. Unfortunately, the regeneration-point theorem does not hold in the case of multiple locations, and the solution to the general model is complex. Because of the complexity, often heuristics are involved in the solution. The main problem here is, excepting the single-facility case, the discontinuity in transitioning from existing infrastructure commitments to possible new facilities. The key is to maintain an optimal spatial-distribution of services during expansion. In other words, oftentimes we may need to "clean the slate" and totally rearrange the service facilities in order to maintain optimality. We will come across a similar situation in land-use models when the ultimate capacity of a subarea is reached for growth, and additional population and employment need to find new homes elsewhere. This involves a discontinuity called *constraint catastrophe*—a phenomenon that is also difficult to model.

Shulman (1991) presents a solution to the dynamic capacitated-plant-location problem (DCPLP). He finds a time schedule and capacity expansions at plant locations to minimize the discounted cost of capital expenditures over the planning horizon. He considers a class of the DCPLP in which the available facilities have finite capacities (and the number of facility types is relatively small). In this case, the expansion sizes cannot be modelled by continuous variables. The DCPLP is formulated as a combinatorial-optimization problem that allows consideration of more than one facility type and finds the optimum mix of facilities in each location. Lagrangian relaxation is used to find a set of initial solutions. Heuristics are then used to covert infeasible solutions of the Lagrangian to a feasible solution of the DCPLP. Chardaire et al. (1996) formulated the classic dynamic, *uncapacitated*, facility-location problem as a quadratic program. Heuristic solutions are offered, bounded by Lagrangian-relaxation solutions. The duality gap, or the difference between the heuristic solution and the relaxation bound, was found to be very small. Daskin et al. (1992) recognize that on a limited scale, the regeneration-point theorem can be exploited in the multi-facility case. Specifically, the approximations that reduce the size as though the state-space could be generalized ,creating good heuristic procedures. They considered a dynamic facility-location model in which the objective is to find a planning horizon, t^*, and a first-period decision, x_1^*. Here x_1^* is a first-period decision for at least one optimal policy for all problems with planning horizons equal-to-or-longer-than t^*. In other words, a planning horizon t^* is found such that conditions after t^* do not influence the choice of the optimal initial-decision x_1^*. For the dynamic uncapacitated-fixed-charge-location problem, they showed that simple conditions exist such that the initial decision depends on the length of the planning horizon. A strictly optimal forecast-horizon and initial policy may not exist (Min et al. 1997).

With the dynamic facility-location formulation, we have finished our general survey of the various types of facility-location models. As promised, we will devote the rest of this chapter on more applications-oriented issues. This includes a couple of generalizations, including relaxation of the binary 0–1 allocation of a single-facility-to-a-single-demand, or the 'all-or-nothing' assumption. Also, the single-optimizing-criterion assumption will be relaxed, allowing us to optimize two or more criteria simultaneously. A case study of locating satellite-tracking stations will be introduced to drive home these concepts. Through solving this case study, essential   considerations   in   choosing   a   solution   algorithm   are demonstrated—although there really is no single, universal procedure for tackling all facility-location problems.

2.9   Another generalized version of the *p*-median problem

The *p*-median problem has been a subject of research and application for quite some time. Here, we extend the multi-product version of the model in two areas. First is to relax the single-assignment assumption in which a demand is served only by one nearest facility (the uncapacitated case). In the multi-product context, such relaxation is necessary so that an efficient utilization of the facilities can be obtained. Suppose the probability of being served by the closest facility from a given demand point is not 1, but depends on how `close' the other facilities are .We have the models described here. Under these circumstances, demands from a node are served by several facilities (Batta and Mannur 1990).

2.9.1   Multiple Coverage.   As explained previously, most of the familiarlocation models employ either a *min-sum* criterion or a *min-max* criterion. Neither of these criteria alone is sufficient in real-world applications. For example, the min-sum criterion often results in poor service rendered to certain demands, while the min-max criterion may result in too costly a solution. Different models have been proposed to examine these two seemingly opposite emphases including the hybrid, *Cent-dian* or *medi-center* models. Here, we will examine examples of both criteria: service coverage and equity among demands, and apply them toward two variants of the *p*-median problem: maximal coverage and set-covering. This results in a bicriteria maximal-coverage and a bicriteria set-covering model respectively. Solution techniques are offered to address the problems so formulated. Our emphasis here is on operational procedures, yet mindful of the generalization of our computational experience to other situations. Alternative solution algorithms were tried and several data sets were computed, including a case study of the location of Canadian satellite-tracking stations. It is our hope that such computational experiences will add toward the general body of knowledge for both practitioners as well as researchers in solving this generalized *p*-median problem.

In the classic *p*-median problem, a demand *i* is served by the least-cost facility *j* at a price of c_{ij}. Only one facility *j* is assigned to a demand *i* in an `all-or-nothing' manner, with x_{ij} assuming 0–1 values. Such a formulation has been generalized to include various states *k*, *k* = 1, 2, ..., *K*. The variable x_{ijk} simply takes on the interpretation of "a demand node *i* being assigned to facility *j* in state *k*." This is sometimes called the multi-product (or service) case of the *p*-median problem. Here x_{ijk} can assume a fractional value. The classic *p*-median formulation in Equations **(2-7)** through **(2-9)** can then be extended as

$$\min\, f_1(X) = \sum_{i=1}^{|I|} \sum_{j=1}^{|J|} \sum_{k=1}^{K} c_{ijk} x_{ijk}$$

$$s.t.\ \sum_{j=1}^{|J|} x_{ijk} \geq 1 \qquad \forall i,k$$

$$x_{ijk} - y_j \leq 0 \qquad \forall\ i,j,k;\ i \neq j \qquad \text{(2-85)}$$

$$x_{ijk} \geq 0$$

$$\sum_{j}^{|J|} y_j = p.$$

Several modifications are made in this extended model. First, we distinguish explicitly between the set of demands I and the facilities J. Second, the notation for decision variable x_{jj} is simplified to y_j. Now suppose that a demand i can be served by more than one facility j in state k. The binary assignment of demand to facility can consequently be generalized to

$$\sum_{j=1}^{|J|} w_{ijk} x_{ijk} \geq f_k \qquad \forall\ i,k \qquad \text{(2-86)}$$

where f_k is a minimum service-requirement placed upon the facilities and w_{ijk} is the contribution of facility j toward covering demand i in state k. This formulation considers redundancy-of-service satisfaction where needed—for cost or reliability reasons—as shown by the binary variable $x_{ijk} \in \{0,1\}$. It can be shown that the Hakimi *nodal-optimality condition holds for this generalized p-median problem* (Mirchandani 1990).

2.9.2 Two variants of the multi-product p-median problem. The classic *p*-median facility-location problem seeks to place p facilities among $|I|$ demand points. This type of problem seeks to minimize the average distance or time between facility and servicing locations. A maximal 'coverage' variety of the problem can be expressed as

$$\max \sum_{i=1}^{|I|} \sum_{j=1}^{|J|} \sum_{k=1}^{K} w_{ijk} x_{ijk} \qquad \text{(2-87)}$$

$$y_j - x_{ijk} \geq 0 \qquad \forall\ i,j,k \qquad \text{(2-88)}$$

$$\sum_{j=1}^{|J|} y_j = p \qquad \text{(2-89)}$$

$$\sum_{j=1}^{|J|} w_{ijk} x_{ijk} \geq f_i \qquad \forall\ i,k;\ f_i\ integer \qquad \text{(2-90)}$$

$$x_{ijk} \in \{0,1\}. \tag{2-91}$$

Here we seek to service "as much of the demands w_{ijk} as possible." Notice again this generalizes the classic formulation (Daskin 1995) to include not only demand i, but also the kth state of the system. Rather than an all-or-nothing assignment of facility to a demand, several facilities can be used to satisfy the demand. This is accomplished although we allow only one demand to be served discretely by one, two, three facilities etc. This is facilitated by a binary variable, rather than fractional coverage shared between facilities using a continuous assignment variable. There will be overlapping service regions among facilities, instead of service regions that are mutually exclusive of one another. In this maximal-coverage variant, the service-requirement constraint (2-90) is non-binding. We eliminate this redundant constraint and incorporate the precedence constraint (2-88) into the objective function .The original formulation is reduced to a *knapsack problem* of the form max $\{\sum_j U_j y_j | \sum_j y_j \leq p, y_j$ binary$\}$, where the inequality $\leq$ is satisfied as a strict equality, and where $U_j = \sum_i \sum_k w_{ijk}$.

Similarly, one can write a set-covering formulation of Equations (2-85) and (2-86) in the general format

$$\min \ p \tag{2-92}$$

subject to Equations (2-88), (2-89), (2-90) and (2-91). Again, the coverage constraint is generalized to model the situation where market or service areas overlap.

If one re-examines the three location problems discussed above—p-median, set-covering, and maximal-coverage—s/he will find the three are closely related. Mathematically speaking and setting aside physical interpretation, the set-covering objective in Equation (2-92) is a special case of the p-median objective Equation (2-87) . This can be seen if j is set equal to i and w_{iik} is `factored out' as an arbitrary constant. With p treated as a variable, the requirement that p facilities are to be located drops out and the following objective function, with Equations (2-88), (2-90) and (2-91), constitute the generalized set-covering problem. Notice the special assignment stipulations: $\sum_{i \in I} x_{iik} = p_k'$. Here, p_k' explicitly recognizes multi-products provided by dedicated kth-type facilities. We can write

$$\sum_{i=1}^{|I|} \sum_{k=1}^{K} x_{iik} = \sum_{k=1}^{K} p_k'. \tag{2-93}$$

In the case when every 'facility' can serve any 'demand' (i.e., a non-hierarchical facility-location problem), the objective function finally reduces to p, as shown in Equation (2-92). At first examination, the maximization objective (2-87) of the maximal-coverage model seems quite different from the minimization objective of the generalized p-median model as shown in Equation (2-85). But a more careful scrutiny would show that the constraints are identical and that the maximal-coverage problem is nothing more than a special maximization case of the

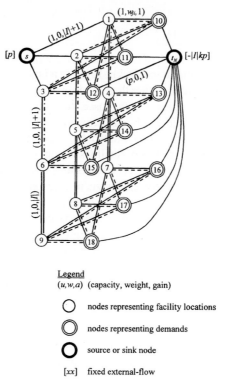

Legend
(u,w,a) (capacity, weight, gain)

○   nodes representing facility locations

◎   nodes representing demands

⬭   source or sink node

[xx]   fixed external-flow

Figure 2-24 Maximal-coverage network with gains

generalized p-median problem when $c_{ijk} = w_{ijk}$ (Church and ReVelle 1976, Martin 1999).

2.9.3   Network representation.   It has been shown above that—in their generalized versions—the maximal-coverage, set-covering facility-location models are again special cases of the classic p-median problem. It is well known by now that the classic p-median problem and other facility-location models can be represented by a bipartite graph (Krarup and Pruzan 1990, Dearing et al. 1992), where demands are being paired with facilities. However, the introduction of the w_{ijk} in the coverage constraint destroys such idealization since it invalidates any hope of total unimodularity in the constraint matrix. A more complex network will therefore be required for its representation, concomitant with a more complex solution procedure.

2.9.3.1 NETWORK WITH GAINS. A generalized network-with-gains[14] is proposed here to solve this multi-product p-median problem. An example is shown in **Figure 2-24**, where there are three candidate facilities ($|J| = 3$), three states ($K = 3$), and three demand points ($|I| = 3$). Nodes 1, 4, and 7 represent facility 1 in states 1, 2, and 3 respectively. Similarly, nodes 2, 5, and 8 represent facility 2, and nodes 3, 6, and 9 represent facility 3. Also, nodes 10, 13, and 16 correspond to demand point 1; nodes 11, 14, and 17 correspond to demand point 2; and nodes 12, 15, and 18 to demand point 3. The *fixed external flow* at the source is set equal to the number of facilities p to be selected. Arcs with capacity of 1, zero weight, and gain of $|I|+1$, connect the source with the facilities in state 1 only. Within each state, arcs link a given facility with all three demand points. The capacity of these arcs is 1 and the weights w_{ji} in states k are as shown in **Table 2-4**, **Table 2-5**, and **Table 2-6** corresponding to w_{ji1}, w_{ji2}, and w_{ji3} respectively. Interstate arcs, with capacity 1, weight 0, and gain $|I|+1$ (except the arcs between states $K-1$ and K, which have a gain of $|I|$) connect the nodes representing a given facility. Arcs connecting demand-point nodes to the sink are assigned a capacity of p and weight 0. Finally, the fixed external flow at the sink is set equal to $-|I|Kp$. The total number of nodes in the network for a problem with $|J|$ facility locations, K states, and $|I|$ demand points, is simply the source and sink nodes, plus the number of nodes in each state multiplied by the number of states: $2+K(|J|+|I|)$. The total number of arcs is the sum of the number of arcs out of the source, the number of arcs into the sink, the number of arcs connecting the facilities and demand points multiplied by the number of states, and the number of interstate arcs. This amounts to $K(|J|+|I|+|I||J|)$. Thus for the example problem with $|I| = |J| = K = 3$, there are 20 nodes and 45 arcs. The network representing any realistic problem, such as the one shown in the case stud y later, which include $|J| = 10$, $K = 12$, and $|I| = 10$, will therefore consist of 242 nodes and 1440 arcs. Notice that the demand (weight) specified at arcs incident upon the demands, (1,10), (1,11), (1,12), etc. are without the subscript k. The subscript k is not necessary since a sub-network is replicated for each state $k = 1,2,3$. Only one type of `commodity', corresponding to each state, can flow in a sub-network.

Again, the set-covering model can be represented as a network-with-gains. The resulting network is shown in **Figure 2-25**. The source has a positive *slack external flow*. The arcs leading out of the source have an upper bound of 1 and an arbitrarily large cost of M''. All other arcs have either a cost of zero, or, in the case of the arcs leading into the T_N node, a cost of -1. The gain of the arcs leading out of the source is $||I|+1)$ to allow for one unit of flow in all arcs leading out of a reached facility location. As with the maximal-coverage network, nodes 1, 4, and 7 represent facility location 1 in states 1, 2, and 3 respectively. Similarly, nodes 2, 5, and 8 represent location 2 and nodes 3, 6, and 9 represent location 3. Also, nodes 10, 13, and 16 represent demand point 1, nodes 11, 14, and 17 represent demand point 2,

---

[14]For an introduction to network-with-gains, see the "Prescriptive-tools" chapter under the "Integer or mixed-integer programming" section.

w_{ji}	$i = 10$	$i = 11$	$i = 12$	total
$j = 1$	8	5	4	17
$j = 2$	3	6	9	18
$j = 3$	2	7	5	14
total	13	18	18	49

Table 2-4 Weights for state 1

w_{ji}	$i = 13$	$i =$	$i = 15$	total
$j = 4$	6	2	7	15
$j = 5$	9	3	6	18
$j = 6$	4	6	5	15
total	19	11	18	48

Table 2-5 Weights for state 2

w_{ji}	$i = 16$	$i = 17$	$i = 18$	total
$j = 7$	4	8	9	21
$j = 8$	6	7	9	22
$j = 9$	3	9	7	19
total	13	24	25	62

Table 2-6 Weights for state 3

and nodes 12, 15, and 18 represent demand point 3. The arcs connecting facility locations and demand points have gains equal to the service capacity of that location-demand point pair. Since the flow through these arcs will either be 0 or 1 (as will be shown below), the flow into a given demand point represents the amount of service provided to that particular demand point. A lower bound for un-capacitated arcs leading from demand points to the t_N sink ensures that the demand-point requirement is met. The excess coverage is automatically routed to the un-capacitated arcs connecting the T_N sink since these have a cost set to -1. (Here we use M'', a large number, as the infinite arc-capacity.) Therefore the

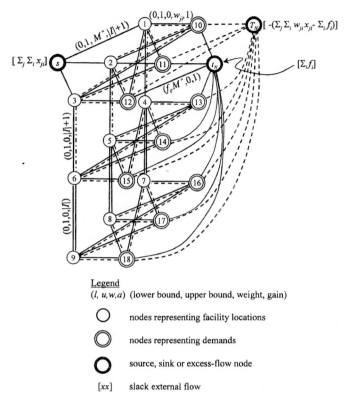

Legend
(l, u, w, a) (lower bound, upper bound, weight, gain)

○   nodes representing facility locations

◎   nodes representing demands

◯   source, sink or excess-flow node

[xx]   slack external flow

Figure 2-25 Set-covering network with gains

negative external-flow at sink t_N will equal $\Sigma_i f_i$, where $i = \{10, \dots, 18\}$. The external flow at sink T_N will equal $\Sigma_j \Sigma_i w_{ji} x_{ji} - \Sigma_i f_i$.

There is no requirement for the cost of all arcs leading into sink T_N to be set to -1. The relative costs of these arcs could for example represent the relative merit of obtaining excess coverage for one demand type i as compared to another. This allows a utility function to be constructed based on operational requirements. Perhaps the most practical way of assigning varying costs to the arcs leading into the sink would be to make the assignments such that for each state k, $-1 < w_{iT} < 0$ and $\Sigma_i w_{iT} = -1$. Notice again that the demands (weights) w_{ji} and w_{iT} are without the state subscript k. The same reasoning applies as before, namely that each sub-network is replicated for each state. Correspondingly, the summation is carried out only for subscripts i and j at the excess-flow sink T_N. Another way of thinking about it is that the source s generates only as much flow as $\Sigma\Sigma x_{ji}$. Granted that there is a gain of $|I|+1$ on arcs emanating from s, and again by w_{ji} at the subsequent arcs from the facility j. These gains, a total of $\Sigma_i f_i$, are however diverted to sink t, leaving as much excess flow to T_N as what starts out at the source.

The min-cost flow will minimize the slack external-flow into the source, while encouraging excess coverage, or maximizing flow into the T_N sink. The minimum requirements at each demand node, which are expressed by the lower bounds on the arcs into the t_N sink, determine how many arcs out of the source must be activated. In summary, the formulation will select the minimum number of sites required to fulfill demand requirements. This is accomplished by assigning a large cost M' to the selection of an arc leading out of the source. Discrimination between feasible solutions with equal number of sites is accomplished by rewarding total excess coverage. In this example, the amount of reward for excess coverage has been set constant across demands. However, a weighting scheme that would consider varying demands could be substituted.

Given $|J|$ locations, K states, and $|I|$ demand points, the total number of nodes in the network is 3 (source and two sinks) plus the number of nodes in each state multiplied by the number of states: $3+K(|J|+|I|)$. As calculated for the maximal-coverage network, the total number of arcs is the sum of the number of arcs out of the source, the number of arcs into the sink, the number of arcs connecting the facility-location and demand points multiplied by the number of states, and the number of interstate arcs. This amounts to $K(|J|+2|I|+|I||J|)$. Thus for the example problem, there are 21 nodes and 54 arcs. A real world problem, as will be discussed in the case study, contains 243 nodes and 1560 arcs.

With the network representation, the equivalence between the classic p-median problem and the two variations described above is clear. The core network remains the same for all three problems. While the classic p-median model seeks the lowest cost flow, the maximal-coverage variant finds the maximal cost solution. While the classic p-median and maximal-coverage models operate with fixed external flows, the set-covering variant has slack flows to find the least number of facilities required. The slack-flow formulation mandates the introduction of an extra sink T_N beyond the sink t_N that poses the coverage requirement. Obviously, the arc attributes—including costs and capacities—are specified quite differently between the two variants of the p-median model. This will effect the maximal-cost flow in the former and the min-cost flow in the latter. Let us note also one important difference between the network for maximal coverage and that for set covering. While there exists a distinct lower bound f_i for arcs incident upon the sink node t in the set-covering network, it is absent in the maximal-coverage model. The absence of a lower bound on arcs $(10, t)$, $(11, t)$, $(12, t)$ etc. confirms earlier observation that the constraints **(2-90)** are non-binding. But the significant point is that all three models have the same structural network.

2.9.3.2 SOLUTION PROCEDURES. Having represented the above problems as networks, the logical solution procedure is to employ a network-with-gains algorithm. The assumption is that network algorithms are computationally more efficient than regular integer-linear-programming techniques. Given the generalized-network formulation, the only question is whether the solution so obtained will be integral, inasmuch as the total-unimodularity property is now absent. A classic branch-and-bound (B&B) routine is proposed to integerize the solutions where necessary. These B&B procedures are applied to the flow variables at the arcs emanating from the source s. For the problems attempted, the solutions were invariably integerized simply by manipulating these selected flow-variables.

Due to the network structure apparent in the flow formulation, experiments were also performed using the node-arc-incidence-matrix way of representing the problem.[15] Regular LP codes were used for the solution of the above problems. An attempt was made to force an integer solution by adding a series of side constraints that impose conditions on the flow through the network. Three types of constraints were added to the network tableau for the maximal-coverage problem. First, the 'source-connector flow' constraints ensure that the upper bound on the flow through each of the 'source-connector' arcs is limited to 1:

$$x_{sj} \leq 1 \qquad j=1,2,\ldots,|J|. \tag{2-94}$$

Second, the "sink-connector flow" constraints ensure that the flow through each of the "sink-connector" arcs is set equal to p:

$$x_{it_N} - p = 0 \qquad i \in \{demand\ nodes\}. \tag{2-95}$$

Third, the "equi-distribution of flow" constraints ensure that the amount of flow through all arcs entering or leaving a facility-location node is equal:

$$x_{ji} - x_{kj} = 0 \qquad \forall\ i,k \in \{nodes\ adjacent\ to\ j\};\ i \neq k. \tag{2-96}$$

This last category of constraints requires four pairwise constraints at each facility-location node j as shown. Equation (2-96) requires one pairwise constraint for each facility-demand arc (including demands in the next state); in this case, four pairwise constraints for facility nodes 1, 2, 3, 4, 5 and 6, and three pairwise constraints for facility nodes 7, 8 and 9. 'k' in this equation prefers to any node preceding a facility node, such as a source node or another facility node from a preceding state. Similarly 'i' can be any node immediately after a facility node, such as a demand node or another facility node in the next immediate state. An example of this set of constraints for facility 1 in **Figure 2-24** looks like $x_{1,10} - x_{s,1} = 0$, $x_{1,11} - x_{s,1} = 0$, $x_{1,12} - x_{s,1} = 0$ *and* $x_{1,4} - x_{s,1} = 0$.

To ensure an integer solution to the set-covering model, however, this last set of constraints has to be applied with fixed integer-external-flow at the source, rather than the source-connector or the sink-connector flow-constraints. Obviously, this relapses back to the integerization philosophy adopted by the network-with-gains programming approach described earlier: implicit enumeration. The only difference is that the trial-and-error procedure replaces the branch-and-bound (B&B) tree. **Table 2-7** summarizes the integerization experience thus far. It can be seen that while the LP procedure is not as promising as the network-flow-programming

---

[15]For an illustration of the node-arc-incidence matrix, refer to the "Prescriptive Tools" chapter under the "Integer programming" section or the "Optimization Schemes" Appendix under "Network with side constraints", both in Chan (2000).

Test problems[1]	Network flow		Linear program	
	Maximal coverage	Set covering	Maximal coverage	Set covering
1	yes	yes	yes	yes
2	yes	yes	yes	no
3	-	yes	-	no

[1] Test problem 1 refers to $p = 1$, $f_1 = 2$. Test problem 2 refers to $p = 2$, $f_1 = 4$. Test problem 3 only refers to $f_1 = 6$.

Table 2-7 Success with integerization procedures

approach, it has some partial success nonetheless. An alternate solution-procedure besides network-related strategies is obviously mixed integer-programming (MIP). This will serve as a datum for comparison with the network-flow programming and network-related LP approaches, wherein the advantages and disadvantages of the network formulations can be entertained. However, MIP has its well-known shortcomings when the number of 0–1 variables becomes large—which can happen quickly when all the flow variables have to be integerized. It is therefore used only for validation purposes for small problems. Of the three solution procedures, the network-with-gains algorithm appears to be the most promising, as we will further illustrate.

2.9.3.3   BINARY NETWORK-WITH-GAINS ALGORITHMS. The network parameters, as shown in **Figure 2-24**, **Table 2-4**, **Table 2-5**, and **Table 2-6**, were input to a network-flow-programming code in MICROSOLVE (Jensen 1985) and an initial solution to the relaxed problem z_L^0 was obtained.[16] The $p = 1$ and $p = 2$ problems were solved using the standard 0–1 integerization branch-and-bound (B&B) routine, in which the arc flows leading out of the source node are integerized. All relaxed subproblems in the B&B tree were also solved using the same network solver. Using the MICROSOLVE software, the B&B algorithm for integerization was carried out. It was found that when 0–1 integer flow was imposed on all arcs emanating from the source, the flow through the remainder of

---

[16]For an introduction to the "Branch-and-bound" algorithm, see the "Prescriptive Tools" chapter (Chan 2000) under "Heuristic solution techniques".

the arcs in the network was automatically integer. To see this, refer to **Figure 2-24** and **Figure 2-25** again. It is clear from **Figure 2-24** that once the flows are integerized on arcs $(s,1)$, $(s,2)$, and $(s,3)$, one can follow the integral flows on three decomposed trees. The first tree has node 1 as its root. The second has node 2 as its root and the last has node 3. Except the individual interstate arcs, none of the arcs now has a gain parameter. Therefore the flow in each tree will likely follow a totally unimodular node-arc-incidence matrix, resulting in integer flow. **Figure 2-25** is more complicated. The trees rooted at 1, 2 and 3 will still have gains further downstream. It is possible, therefore, that additional integerization needs to be performed on additional arcs downstream. But we are encouraging maximal slack-flow out of node T_N. Given a fixed amount of external-flow depletion of $\sum_i f_i$, the x_{ji} are likely to be set at the maximal value of unity, rather than fractional. Again, the only flow variables that need integerization appear to be the ones emanating from the origin.

Due to the limited number of flow variables to be integerized, the B&B tree can be pruned rather quickly according to the following classic procedures [which is a *specialized version of the B&B procedure* described in the "Prescriptive Tools for Analysis" chapter in Chan (2000)].

Step 0   *Initialization.* Set z_U to infinity.

Step 1   *Branch.* Select an unfathomed node and partition it into two subsets, representing the two feasible values 0 and 1, for flow through an arc leading out of the source.

Step 2   *Bound.* For each new subset, find a lower bound z_L^i for the objective function of the feasible solutions in the subset, by solving a relaxed subproblem for the objective.

Step 3   *Fathom.* For each new subset i, exclude from further explicit enumeration if

(a) $z_L^i \geq z_U$;

(b) Subset i cannot have any feasible solution; or

(c) Subset i has a feasible solution. If $z_L^i < z_U$, set $z_U = z_L^i$ and store as the incumbent solution. Reapply test (a) to all unfathomed nodes.

Step 4   *Stopping rule.* If no unfathomed nodes remain, stop. Incumbent solution is optimal. Else, return to step 1. ∎

The B&B solution trees for the $p = 1$ maximal-coverage problem are shown in **Figure 2-26**. The node numbers shown in the solution trees indicate the order in which the subproblems were solved. Both the $p = 1$ and $p = 2$ problems only required that four subproblems be solved. The optimal solution obtained in both cases agreed with the solutions found using the MIP software. As shown in **Figure 2-26**, the initial solution to the relaxed problem was $z_L = -69.996$. This initial non-integer solution is a lower bound to the integer solution. The branching is first performed on the flow variable on arc $(s,1)$. The $x_{s1} = 1$ case produces an integer solution and thus a new upper bound to the solution, $z_U = -53$. Node 1 is therefore fathomed. The $x_{s1} = 0$ case does not produce an integer solution. Since the optimal

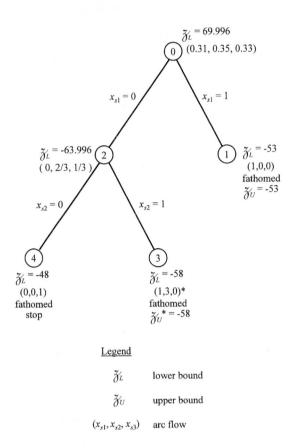

The figure shows a branch-and-bound tree:

Node 0: $\check{z}_L = 69.996$, (0.31, 0.35, 0.33)

Branch $x_{s1} = 0$ to node 2; branch $x_{s1} = 1$ to node 1.

Node 2: $\check{z}_L = -63.996$, (0, 2/3, 1/3)

Node 1: $\check{z}_L = -53$, (1,0,0), fathomed, $\check{z}_U = -53$

From node 2: branch $x_{s2} = 0$ to node 4; branch $x_{s2} = 1$ to node 3.

Node 4: $\check{z}_L = -48$, (0,0,1), fathomed, stop

Node 3: $\check{z}_L = -58$, (1,3,0)*, fathomed, $\check{z}_U^* = -58$

Legend

$\check{z}_L$	lower bound
$\check{z}_U$	upper bound
(x_{s1}, x_{s2}, x_{s3})	arc flow

Figure 2-26 Branch-and-bound tree for the maximal-coverage network problem

solution to this subproblem, $z_L = -63.996$, is less than the current upper bound $z_U = -53$, the node has not been fathomed. Node 2 is now partitioned into two further subproblems. The node 3 subproblem produces a new upper bound, $z_U = -58$. This stands as the optimal solution once node 4 is fathomed because the node-4 subproblem produces a feasible, but sub-optimal solution. The $p = 2$ problem is solved similarly using this integerization routine. It is important to note that although this is a maximal-coverage problem, the subproblems are solved as min-cost flow problems because the objective-function coefficients are negative. Thus the initial solution to the relaxed problem produces a lower bound (the largest negative) on the optimal integer-solution. Any integer solution will be greater (or less negative) than this relaxed solution and will provide an upper bound on the optimal solution. The optimal solution is therefore the integer solution that produces the biggest negative-value for the objective function. Using network-flow software instead of LP software to solve the subproblems obviously improves the computational efficiency of the B&B algorithm.

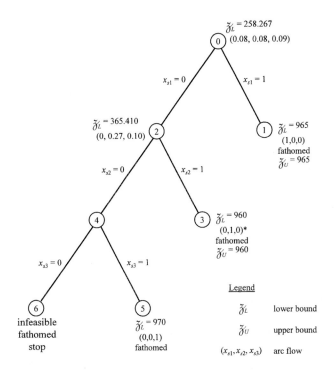

Figure 2-27 Branch-and-bound tree for the set-covering network problem

The above B&B algorithm was again used to find the integer solutions to the $f_i = 2$, 4, and 6 set-covering problems, including the use of a network solver for the subproblems. The initial relaxed solution to the $f_i = 2$ problem identified a lower bound of 258.267 on the integer solution. As shown in **Figure 2-27**, the node-1 subproblem was feasible and an upper bound of 965 was calculated by the network solver. Node 3 returned a new upper-bound of 960, which became the optimal solution once the remainder of the nodes was fathomed. These maximal-coverage and set-covering solutions agree with the MIP solutions, as expected. Not including the initial relaxed-problem, six subproblems had to be solved for the $f_i = 2$ problem, 10 for the $f_i = 4$ problem, and 8 for the $f_i = 6$ problem. On the average, eight subproblems needed to be solved, which is about twice the number of subproblem solutions that was required in the maximal-coverage formulation. The greater number of branching in the set-covering formulation may be because the source external-flow is slack instead of fixed. More experimentation with varying sets of data would be required to establish a definite correlation.

2.9.3.4 NETWORK WITH SIDE CONSTRAINTS: MULTI-COMMODITY FLOW. In a parallel case study of locating satellite-imaging stations (Schick 1992), the model for maximal-coverage was run again with a different solution technique. A multi-

Weights w_{ji} for	$i=1$	$i=2$	$i=3$	$i=4$
$j=1$	20/21/28	22/27/24	24/25/23	21/21/22
$j=2$	23/21/24	19/23/23	28/23/23	22/23/19
$j=3$	27/19/26	23/28/22	21/26/22	25/26/19

Table 2-8 Weights for states 1, 2 and 3 in imaging example

commodity flow network[17] was formulated and the network-with-side-constraints algorithm was used to solve the model. In this approach, the interstate arcs were removed, thus effectively making all but the first three facility-location nodes into source nodes. The network now resembles three identical sub-networks all connected to the same sink as shown in **Figure 2-28**, where four instead of three demands were also shown. The weights for the three states, corresponding to the three `commodities,' are shown in **Table 2-8**. Consonant with the multi-commodity formulation, side constraints were added to the facility-demand arcs to restrict the flows to like paths. For example, if arc $(1,1)$ in state 1 was taken, then the same arc must be taken in states 2 and 3 as well: $x_{1,1,1}-x_{1,1,2} = 0$, and $x_{1,1,1}-x_{1,1,3} = 0$. In reference to **Figure 2-28**, these equations use a different subscript notation than the one used in the figures drawn thus far. Here the first number denotes the jth facility and the second number the ith demand (instead of a continuous serialization [across all states] used previously).

Another set of side constraints was added to force only one facility selection per state for the $p = 1$ case. These constraints are such that the total flow along differing arcs emanating from different nodes within state 1, when multiplied by the number of satellites in that state,[18] is less than or equal to the total flow from t to s. Thus one such equation would be $4x_{1,1}+4x_{2,2}+4x_{3,3}-x_{ts}\leq0$ where the state subscript variable k is skipped for clarity. Because there are more demands than facilities per state (4 *vs.* 3), only three variables, other than x_{ts} would be included in that equation. Adding another variable such as $4x_{3,13}$ could force that variable and $x_{3,12}$ to be zero. For example, flow along x_{ts} was 4 from previous solutions using network-with-gains formulation. We knew that the optimum was achieved by using site 3. By setting $x_{3,12} = 1$ and $x_{3,13} = 1$ we would have $4x_{1,10} + 4x_{2,11} + 4x_{3,12} + 4x_{3,13}-4\leq0$ or $(4)(0) + (4)(0) + (4)(1) + (4)(1)-4\leq0$. This is not valid mathematically, and would force one or both of the variables from one to zero in

---

[17]For those who are unfamiliar with these terms, see the "Optimization" appendix (Chan 2000) for a background discussion of network-with-side-constraints and multi-commodity flow, including formulation and solution algorithm.

[18]The flow out is multiplied by the flow in so that when the equation is summed, the solution will be zero.

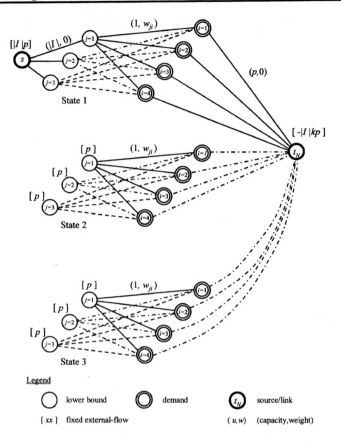

$[|I||p]$ $(|I|,0)$ $(1, w_{ji})$

$(p,0)$

$[-|I||kp]$

State 1

$[p]$ $(1, w_{ji})$

State 2

$[p]$ $(1, w_{ji})$

State 3

Legend

○ lower bound   ◎ demand   ⊙t_N source/link

$[xx]$ fixed external-flow (u,w) (capacity,weight)

Figure 2-28 Multi-commodity-flow network (Schick 1992)

order to keep the equation feasible. Thus site 3 would incorrectly be rejected. This led to a requirement to add more constraints by varying the variables used in this 'arc-restriction' arrangement.

In the case study of locating satellite-imaging stations, where $|J| = 12$, $|I| = 16$ and $K = 3$, the set of arc-restricting side-constraints was obviously not all inclusive, since it would have meant $|J|!/(|I|-|J|)!$ arc-restricting constraints. A minimal number of arc-restricting constraints, covering all the demands from at least one facility is sufficient. For the $p = 1$ problem, five arc-restricting side-constraints were included such that all of the satellites in state 1 were covered:

$$16(x_{1,1}+x_{2,2}+x_{3,3}+x_{4,4}+...+x_{11,11}+x_{12,12})-x_{ts}\leq 0$$
$$16(x_{1,1}+x_{2,3}+x_{3,4}+x_{4,5}+...+x_{11,12}+x_{12,13})-x_{ts}\leq 0$$
$$16(x_{1,1}+x_{2,4}+x_{3,5}+x_{4,6}+...+x_{11,13}+x_{12,14})-x_{ts}\leq 0$$
$$16(x_{1,1}+x_{2,5}+x_{3,6}+x_{4,7}+...+x_{11,14}+x_{12,15})-x_{ts}\leq 0$$
$$16(x_{1,1}+x_{2,6}+x_{3,7}+x_{4,8}+...+x_{11,15}+x_{12,16})-x_{ts}\leq 0.$$

The equi-distribution-of-flow constraints were added for the facility nodes to supplement the subset of these arc-restriction constraints that are actually included. These new equations for this case study amount to one pair-wise constraint for each facility-demand arc $(1,1)$ $(1,2)$ ... $(1,16)$, $(2,1)$ ... $(2,16)$, ... , $(12,1)$ ... $(12,16)$. Here, 16 pair-wise constraints for facility node j $(j = 1,2, ... ,12)$ are possible for the facility nodes in state 1. The equi-distribution-of-flow constraints were added for the facility nodes to supplement the subset of these arc-restriction constraints that are actually included. These new equations for this case study amount to one pair-wise constraint for each facility-demand arc $(1,1)$ $(1,2)$ ... $(1,16)$, $(2,1)$ ... $(2,16)$, ... , $(12,1)$ ... $(12,16)$. Here, 16 pair-wise constraints for facility node j $(j = 1,2, ... , 12)$ are possible for the facility nodes in state 1.

$$16x_{1,1} - x_{s,1} = 0 \qquad 16x_{2,1} - x_{s,2} = 0 \qquad\qquad 16x_{12,1} - x_{s,12} = 0$$
$$16x_{1,2} - x_{s,1} = 0$$

$$16x_{12,15} - x_{s,12} = 0$$
$$16x_{1,16} - x_{s,1} = 0 \qquad 16x_{2,16} - x_{s,2} = 0 \qquad 16x_{12,16} - x_{s,12} = 0.$$

These constraints force flow along all 16 arcs emanating from a facility, if there is flow into that facility. The factor of 16 is used in the constraints since any flow into the facility must be 16 to ensure a flow of one out of each of the 16 arcs leaving the facility.

For the one-facility maximal-coverage problem $(p = 1)$ under discussion, the 12 sets of equi-distribution constraints listed above were tried one at a time. This forces an integer solution for the corresponding `projected' facility, or `kicks' the facility out of the solution because of a fractional solution. None of the formulations resulted in a whole integer solution except facility 3. We concluded that facility 3 must be the optimal solution. The network-with-side-constraints algorithm was run again for $p = 2$ and $p = 3$ facilities to be chosen. The multi-commodity-flow formulations were identical to the $p = 1$ formulation. The only change is the limit of the capacity p for the 48 (3×16) demand-sink arcs between the three states. The capacity p corresponded to the number of facilities to be chosen.

2.10 Multi-criteria formulation

The network formulation presented above provides a good framework for addressing the single-criterion maximal-coverage and set-covering models. However, a second criterion is usually of interest in applying these models. It minimizes the variance in the coverage (Badri et al. 1998). To illustrate the implication of the mean vs. variance criterion function for facility location, we cite an example from Berman (1990). Consider a network with only two nodes and one link of one unit length. Suppose 99.9 percent of the demands reside on node 1 and 0.1 percent reside on node 2. The location that minimizes the variance criterion is the middle of the link whereas the median is at node 1—two drastically different

s_{jk}^2	$k =$	$k =$	$k =$	total
j	2.8	4.6	4.6	12.2
j	6	6	1.5	13.5
j	4.2	0.6	6.2	11.1

Table 2-9 Variances of weights

locations. Now let us turn back to the generalized *p*-median problem. From the maximal-coverage model with $|I|$ equally likely demand occurrences, the mean coverage of the demand by facility *j* in time period *k* is given by $U_{jk} = \sum_{i=1}^{|I|} w_{ijk}/|I|$. The variance is correspondingly $s_{jk}^2 = \sum_{i=1}^{|I|} (w_{ijk} - U_{jk})^2/|I|$. Such means for the example problem are computed from **Table 2-4**, **Table 2-5**, and **Table 2-6**, and the variance result is summarized in **Table 2-9**. In the Tables for the means, the row sums are $U_{jk}|I|$ and the average over these Tables are $U_j/|I|$. The criterion function to minimize the sum of the variances is then

$$\min f_2(X) = \sum_{j=1}^{|J|} \sum_{k=1}^{K} s_{jk}^2 y_j. \tag{2-97}$$

By including Equation **(2-90)** into this criterion function, it can be shown that it is equivalent to the second criterion in the set-covering formualation (Forgues et al. 1998). The set-covering model is also equivalent to the smallest f_2 value in Equation **(2-97)**, with the constraint $\sum_j y_j = p$ for all *p* in a knapsack formulation.

2.10.1 Mean-variance location problems.
Independent of our effort, Berman (1990) presented a $O(n^3)$ algorithm for the classic median-location problem when both the mean and variance criteria are considered. Three formulations of the problem were considered. First, he minimizes the mean measure $f_1(X)$ subject to a constraint on the variance measure $f_2(X)$. Second, he minimizes $f_2(X)$ subject to a constraint on the mean measure $f_1(X)$. Third, he minimizes the sum of the mean and a constant times the variance $f_1(X) + \lambda_2'/\lambda_1' f_2(X)$. The solutions so obtained in all three problems are nondominated solutions. In traditional multi-criteria optimization, we refer to the first two as ways to `slice' the *reduced feasible region* and the third as one instance of the *weighted-sum* solution procedure[19] (Steuer 1986). A polynomial algorithm based on shortest paths was presented to solve these three problems as mentioned. It was shown that the worst case complexity of $O(n^3)$ cannot be improved. However, a polynomial solution procedure could not be found for the

---

[19]These procedures are reviewed in the "Multi-criteria Decision-making" chapter (Chan 2000) under the "Exploring the efficient frontier" section.

general multi-criteria optimization problem when all slices of the reduced-feasible-region method and all instances of the weighted-sum approach were to be obtained. Again, the solution so obtained is only for the classic median problem, rather than the generalized models discussed here.

Here we use a modified simplex to solve each network-with-gains problem. It is well known that network-with-gains can be solved much more efficiently than a regular linear program. For example, any basis $\mathbf{A}_B$ of the node-arc incidence matrix $\bar{A}$ can be arranged into r diagonal blocks $\mathbf{B}_1, \dots ,\mathbf{B}_r$. Here, $\mathbf{B}_i$ represents the square (generalized) incidence matrix of a spanning subgraph. Put in this framework, the various simplex operations, including the computation of dual variables, the generation of an updated column, and the pivot and update operations can now be specialized (see Appendix on "Optimization procedures" in Chan (2000) and Bazaraa et al. (1990)). Such specialization exploits the network characterization and allows much of the operation to be performed on the graph itself. Active research activities are taking place in devising ever more efficient labelling algorithms for solving this generalized network problem (Chang et al. 1989). Polynomial algorithms based on interior-point methods are beginning to emerge (Goldberg et al. 1989). Perhaps the bottleneck of the computation is in the integerization part of the algorithm, where the B&B tree can grow large. This is in spite of a dramatic reduction in integerization variables by examining only the flows in arcs emanating from the source. The computational complexity is now determined by the tree search, rather than the network-with-gains algorithm. Suffice to say however that this represents a significant improvement over regular LP-relaxation techniques found in MIP codes.

2.10.2 Multi-criteria linear-programming solution. The node-arc-incidence matrix of the network remains the same as before in multi-criteria optimization. It is not totally unimodular and an integer solution is not guaranteed. However, imposing a binary integer-restriction on the variables for the arcs incident upon the sink is sufficient to guarantee an integer solution. Here, a series of side constraints that impose conditions on the flow through the arcs of the network is added to the node-arc incidence matrix. These were categorized as sink-connector-flow or equi-distribution-of-flow constraints respectively as previously mentioned. In the set-covering problem, merely imposing equi-distribution-of-flow constraint is sufficient. This computational experience is somewhat different from that of the previous single-criterion case. While the network-flow model is the more satisfactory solution method, the lack of operational software that combines branch-and-bound with network-flow-with-gains prevents its effective use. This is made even more complex in multi-criteria optimization, where the computational requirement is amplified manifold. The LP-with-side-constraints becomes a more feasible option purely because LP software—such as SAS/OR and CPLEX—is readily available and the inclusion of side constraints is straightforward.

Both the weighted-sum and reduced-feasible-region methods were used to solve the multi-criteria maximal-coverage integer LP (Steuer 1986). Results for $p = 1$ and $p = 2$ are presented in terms of the convex-weight-cones, the decision space X, as

w_1	$p=1$		$p=2$	
	X^*	Y^*	X^*	Y^*
0				
0.1	site 3	(48, –11.111)	sites 1,3	(101, –23.334)
0.2	site 2	(58, –13.556)		
0.3				
0.4				
0.5				
0.6			sites 1,2	(111, –25.779)
0.7	site 1	(53, –12.223)		
0.8				
0.9				
1.0				

[1] Sites 2,3 were revealed by the reduced-feasible-region method to be an additional non-dominated solution, with (106,-24.677) in the outcome space. It occurs between lambda weights (0.1,0.9) and (0.2,0.8).

Table 2-10 Multi-criteria maximum-coverage solutions

well as the corresponding outcome space Y' in **Table 2-10**. Here both criterion functions—max-coverage (mean) and set-covering (variance)—were transformed into maximization. For $p = 1$, three locations were identified depending on the weights used among the two criterion functions: sites 1, 2, and 3. However, the third nondominated-solution was missed by the weighted-sum method because of a nonconvex outcome space, only to be discovered by the reduced-feasible-region method. Three distinct set-covering problems were attempted by setting the demand requirements, f_i, equal to 2, 4, and 6 respectively. The results are summarized in **Table 2-11**, where the decision space X and the outcome space Y' are displayed, similar to the maximal-coverage problem. Here f_1 is shown in Equation **(2-87)** and f_2 in Equation **(2-97)**. Equivalently, the criterion $f_1 = \sum_j U_j x_j$ could have been used should we wish to adopt the knapsack model formulation.

It is a well-known fact that certain types of classic facility-location problems can be solved efficiently by network algorithms. A side benefit of such solution algorithms is—once again—that the solution can easily be integerized, since the constraints often form a nearly totally-unimodular matrix (TUM). The introduction of constraint **(2-86)** destroys such integrality. But *nodal- optimality is guaranteed*

$f_i = 2$		$f_i = 4$		$f_i = 6$	
X	Y'	X	Y'	X	Y'
site 2	(960,13.556)	sites 1,2	(1925,25.779)		
site 1	(965,12.223)	sites 2,3	(1930,24.667)	sites 1,3	(1935, 23.334)
site 3	(970,11.111)	sites 1,3	(1935,23.334)		

Table 2-11 Multi-criteria set-covering solutions

for the bi-criteria median problem (Forgues and et al. 1998). Kouvelis and Carlson (1992) have shown formally that TUM property of the constraint matrix in a bi-criteria optimization problem does not guarantee integer solutions. The only condition under which integer solution is obtained is when the criterion functions are concave for a maximization problem. This allows for a piecewise linear approximation of the criterion functions, where the intervals for such approximation are specified at the discrete points. Standard solution technique for multi-criteria integer program is still the time-honored LP-relaxation technique with its branch-and-bound integerization routines. Clearly, the bottleneck of such computation is in the integerization part of the algorithm, where the branch-and-bound tree can grow huge very quickly. Here we propose an interactive approach for discrete-alternative multiple-criteria-decision-making (MCDM) as a way to address the integrality issue (Koksalan and Sagala 1995). Let $\mathbf{y}^j = (y_1^j, y_2^j, \dots, y_H^j)^T$ denote site-alternative j that has performance-score $y_g^{\,i}$ on criterion g ($g = 1, \dots, H$), where each site alternative is defined by its scores in H criteria. Without loss of generality, assume that more is better in each criterion. Then site j is said to dominate j' if $y_g^j \geq y_g^j$, for all g and $y_g^j > y_g^j$ for at least one g. If there does not exist any alternative j satisfying the above conditions then j' is said to be non-dominated. We will use this simple property to design a solution algorithm below.

   2.10.3 Discrete-alternative MCDM algorithm. Selecting p facilities from a set of $|J|$ locations, the total number of possible solutions is given by $|J|!/(|J|-p)!p!$, as mentioned previously. The number of solutions is explosive with large $|J|$. But it is possible to enumerate all feasible solutions for problems with a limited number of candidate locations and a small p, if one uses the property observed above. Suppose we have two criteria- functions, $f_1(X)$ and $f_2(X)$. A feasible integer-solution, $\mathbf{x}^z$, is a nondominated solution $\mathbf{y}^z$ in the outcome space Y' if and only if ($y_g^z \geq y_g^h$) for all dimensions g and alternatives h (see the chapter on "Multi-criteria Decision-making" [Chan 2000] under the "Simple ordering" section). Based on this definition, a simple algorithm can be coded to generate the efficient frontier for relatively small, yet practical, combinatorial problems:

Step 1 Generate the alternative space X by testing each possible combinations of the decision variables against the constraints (if any).

Step 2 Map each point in the alternative space Y' to the outcome space using the criteria functions.

Step 3 Determine the set of nondominated solutions by testing each solution in the outcome space using the above definition. ∎

The algorithm is straightforward. First, it follows from the definition of dominance. Second, there is a finite alternative-space of the generalized p-median problem as suggested by the nodal-optimality property. Notice the mapping from X to Y' space is also straightforward. It bypasses any difficulty associated with a nonlinear criterion-function $f_g(X)$. For the maximal-coverage problem, $f_1(X)$ is defined by Equation **(2-87)** (or $\sum_j U_j x_j$) and $f_2(X)$ is defined by Equation **(2-97)**. Notice both criterion functions are easy to evaluate, as illustrated by the example problem above. The three-step algorithm developed for the maximal-coverage problem is modified slightly for the set-covering problem. The mapping from the discrete alternative-space X into the outcome space Y' is complicated. The reason is that the $f_1(X)$-criterion function (representing the arcs into the excess sink T_N) must be determined from each combination of sites. Let us set aside the limitation in problem size for the time being. When compared to the weighted-sum and reduced-feasible-region methods, the above algorithm has the advantage of ensuring that all non-dominated solutions be generated. It also bypasses the integerization routines required for the LP-relaxation and network-flow algorithms. Once the nondominated solutions are identified, the maximal-coverage and set-covering model can be readily solved by imposing a set of λ' weights.

2.10.4 Comparison of solution methods. By definition, both the maximal-coverage and set-covering problems are 0–1 integer programs. These types of problems are NP-complete. Depending on the number of variables involved, they may be difficult or impossible to solve using MIP software. The reason is that the number of feasible solutions that must be evaluated is explosive. The network-with-gains formulation in this research has resulted in approximately the same number of arcs as there were variables in the maximal-coverage and set-covering formulations. Thus, the graphical representation of these problems has not reduced the size of the problems. However, the network-flow model does provide an increased understanding of the dynamics of the problem. It allows the user to visualize units of flow progressing through the network and attach physical meaning to a mathematical solution process. One major advantage of the network model was to allow network-flow programming software to solve the subproblems in the B&B algorithm. Since the network-flow programming subproblems are polynomial, computational efficiency is better than solving the subproblems as standard LPs (which are NP-complete).

We consider a software package that interfaces a B&B algorithm and network-flow-with-gains software. This would be ideal for solving a realistic size problem ($|J| = 10$, $K = 12$, $|I| = 10$) which includes approximately 250 nodes and 1500 arcs. Software of this kind, unfortunately, is not readily available. Intuitively, once the arcs leading out of the source in both models have been integerized, the remainder of the arcs in the network must also carry integer flow. This characteristic of the network would greatly reduce the amount of branching required by the integerization algorithm. But the number of subproblems requiring solution could still reach $\sum_{j=1}^{|J|} 2^j$ in the worst case. Although B&B procedures greatly reduce this number in almost all situations, it could still amount to 2046 subproblems for $|J| = 10$, which is clearly prohibitive for manual B&B computations.

Again the network-with-gains formulation often reduces the number of 0–1 integer variables significantly in both the maximal-coverage and set-covering problems. Suppose imposing 0–1 integer restriction on the arcs leading out of the source does force integer flow in the remainder of the network. The total number of 0–1 variable would be equal to the number of locations $|I|$, and not the product $|I||J|K$ of the original mathematical-formulation. This means that the network-flow formulation decreases the number of 0–1 integer variables in the problem by a factor of $1/|J|K$. This savings will become particularly significant when multi-criteria optimization models are to be solved. We also have success in integerizing decision variables by placing side constraints on the node-arc-incidence matrix of LP's. Integer results are, however, not always guaranteed. More consistent results are obtained using network-with-gains algorithms. Oftentimes, the consistency issue becomes as important as—if not more important than—computational efficiency. Consequently we tend to favor the network-with-gains approach in principle, although no off-the-shelf software exists to combine B&B with network-with-gains algorithms.

Let us turn to the exhaustive-list solution algorithm. In the maximal-coverage problem, with $|I| = 3$ candidate locations, the number of alternatives is limited to 3 in both the $p = 1$ and $p = 2$ problems. In the case study with $|I| = 12$ locations, the alternative space includes 12 points in the $p = 1$ problem, $12!/2!10! = 66$ points in the $p = 2$ problem, and $12!3!9! = 220$ points in the $p = 3$ problem. These quantities are manageable and the alternative and outcome spaces can be readily generated using the discrete-alternative MCDM algorithm described above. For practical applications of this size, enumeration becomes a competitive procedure considering other complications such as integrality and 'hidden' alternatives on the efficient frontier.

2.11 Locating Satellite-Tracking Stations: A Case Study

Space surveillance is the detection, tracking, identification and cataloging of all man-made objects in space. Recently, attention is being focussed on an autonomous network of Ground-based-Electro-Optical-Deep-Space-Surveillance (GEODSS) sensors located in Canada. One-site, two-site and three-site options were considered for tracking satellites.

2.11.1 Problem Statement. The GEODSS sensor is an optical video-camera for tracking satellites. Such sensors require cloud-free-line-of-sight (CFLOS) for detection of their target. Beyond suitable environmental conditions, the satellite must remain in view for 5 minutes, be at least 15 degrees above the horizontal plane, and be illuminated when the sun is at least 6 degrees below the horizon. Given all of the above, the probability an observation time-block of five minutes being available is

$$P(ABCDEF) = P(A)P(B|A)P(C|AB)P(D|ABC)P(E|ABCD)P(F|ABCDE)$$

where A is the event that the sun is at least 6 degrees below the horizon, B is the event that wind speed is less than 25 knot, C is the event that the temperature is greater than 50 degree celsius, D is when the satellite is at least 15 degrees above the horizon, E is having CFLOS for five minutes, F is the satellite being illuminated by the sun, and AB is the intersection of events A and B etc.

The following assumptions of independence of events are made: Event D is independent of ABC, which implies that $P(D|ABC) = P(D)$. Event F is independent of BCE, which implies that $P(F|ABCDE) = P(F|AD)$. The above probabilities are obtained for each candidate location j, and for every month of the year k, to take into account seasonal variations of the climatological parameters. The probabilities, $P(A)$, $P(B|A)$, $P(C|AB)$, $P(D)$ and $P(E|ABCD)$ are based on a model from the Environmental Technical Applications Center (ETAC) at Scott Air Force Base, Illinois. It computes the probability of CFLOS. The probability, $P(F|AD)$, is derived using data from a software by Kelso (1990), which among other options, determines when events FAD and AD occur during a given time period, geographic location, and for a given satellite. The total number of minutes in a month when events FAD and AD occurred is divided by the total number of minutes in a month to obtain $P(FAD)$ and $P(AD)$ respectively. These probabilities are then used to calculate $P(F|AD) = P(FAD)/P(AD)$. The expected number of usable observation opportunities for each candidate location for every month of the year will be calculated using the equation $w_{ijk} = P_{ijk}B'_k$. Here w_{ijk} is the expected number of observation opportunities in month k at location j of satellite i; P_{ijk} is the probability that the five-minute block is observable; and B'_k is the total number of five-minute observation blocks in month k. Notice P_{ijk} is computed by the formula for $P(ABCDEF)$ at location j in month k for satellite i.

A maximal-coverage-facility-location model was set up to locate GEODSS sensors. Two criteria functions were used. The first is to maximize the total expected number of observation opportunities w_{ijk} for all sites, all year and all satellites: $\max f'_1 = \sum_i^{|I|} \sum_j^{|J|} \sum_k^K w_{ijk} x_{ijk}$. Again, this criterion function amounts to $\max \sum_j U_j x_j$. The second criterion function seeks to minimize the sum of the variances s_{ik}^2 in the number of observations collected from month-to-month over the entire year for site combination **y**: $\min f'_2 = \sum_{j=1}^{|J|} \sum_{k=1}^K s_{jk}^2(y)$. This variance criterion-function as expressed in **(2-97)** seeks to minimize the variance in the number of observation opportunities provided to individual satellites from location j in month k. For solutions where the number of facilities selected is greater than one ($p > 1$), another way of expressing the variance criterion-function is to minimize the

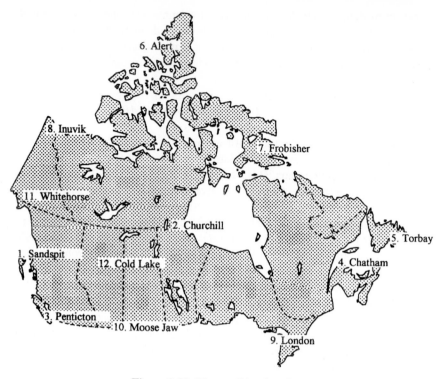

Figure 2-29 The candidate locations

variance in the number of observations provided by alternative $\mathbf{y}'$ in month k:
$s_{jk}^2 = \sum_{j=1}^{|J|} (w_{jk} - \bar{w}_k)^2/|J|$, where $\bar{w}_k = \sum_{j=1}^{|J|} w_{jk}/|J|$. The difference between the two is that in
the latter case, single site coverage is not penalized as long as a site in the
alternative set provides an `average' number of observations. In the case study
described here, the latter (or the alternative-based variance-criterion-function) is
used. The variance function seeks to ensure that satellites can be observed
throughout the year—rather than just one particular month of the year—for the
sensor configuration $\mathbf{y}'$.

The problem is formulated with the constraint: Equation **(2-90)**—a minimum
number of observations must be collected in any given month on a given satellite
to ensure that the constantly changing orbital parameters are adequately
maintained. f_i in Equation **(2-90)** represents the monthly `quotas' of observations
required on each satellite i to maintain the necessary update frequency of the orbital
parameters. Notice the other constraints, Equations **(2-89)** and **(2-90)**, are
implicitly considered in the generation of alternatives $\mathbf{y}'$ as explained previously.
They can be left out of the algorithm.

2.11.2 Solution of case study. Twelve sites are considered in the location of
p optical sensors ($p = 1,2,3$). These sites are each placed in a major Canadian
climatic zone, in which the meteorological parameters of interest do not vary

significantly. These sites are represented in **Figure 2-29**. First, w_{ijk} is computed from probability tables as discussed above. Besides the mean, feasible alternatives can then be identified by Pareto dominance through the variance equation **(2-90)**. Next, the values of both criterion functions are computed for all feasible alternatives providing a complete description of the outcome space for the one-site, two-site and three-site problems. Increasing the number of sites beyond these dimensions would mean a large increase in computer memory requirement and may negate the implicit-enumeration approach taken to rank order site locations.

The 12 sites in **Figure 2-29** have been carefully selected based on several factors: (a) historical weather record is available at the site, and (b) an evaluation of ambient lighting and particulate content of the atmosphere to ensure a minimum standard of optical transmisitivity. The satellite population consists of objects from 3125 miles (5000 km) altitude to geosynchronous altitudes in the western hemisphere and of sufficient brightness to be visible to the sensor. Six example satellites are considered: a semi-synchronous satellite at an inclination of 65 degrees, geostationary satellites at 50, 70, 100, 120 and 140 degrees west. In real-world applications, all synchronous satellites in inclined orbits should be included in the model. Elliptical orbits cannot be included at this point because the ETAC CFLOS model is limited to evaluate the observability of circular orbits only. The consequence of such omission should be assessed. By calculating the visibility limits of the geostationary belt from each candidate location, it is possible to see whether a 1-site, 2-site, or 3-site solution exists. The plot in **Figure 2-30**, for example, shows that it is impossible to cover the five geostationary satellites from a single site.

While the results show that there are no feasible $p = 1$ alternatives, the output shows that there are 12 feasible $p = 2$ alternatives, of which 10 are nondominated. The 10 are illustrated in **Figure 2-31**. It is noted that none of the northerly locations are included in the list of feasible alternatives. This is probably due in part to the fact that the long periods of daylight in the summer and the long winter nights introduce large variances between these seasons. Also, northerly locations afford very poor coverage of the geostationary belt and would rarely afford the necessary coverage in combination with only one southerly location. None of the feasible pairs are composed of adjacent sites either. Generally, each pair of sites includes a location in both eastern and western Canada. There are at least two reasons why this occurs. First, the fact that the geostationary belt was populated with eastern and western satellites forces a solution that includes east and west sites. Secondly, it is known that the P(CFLOS) increases with increasing elevation angles. Therefore, the optimal locations should also be the pair of sites that maximizes the elevation angle to the satellite populations. It is interesting to see that the nondominated set includes Chatham, New Brunswick, in four candidate solutions. This happens to be the location of the Canadian Baker-Nunn satellite-tracking-station that is currently operational as an optical (film) satellite-tacking system.

To locate three sensors, the program found that 90 of the 220 alternatives are feasible (see **Figure 2-32**). Of these, only 23 were nondominated. This means that only about 10 percent of the total number of alternatives would need to be

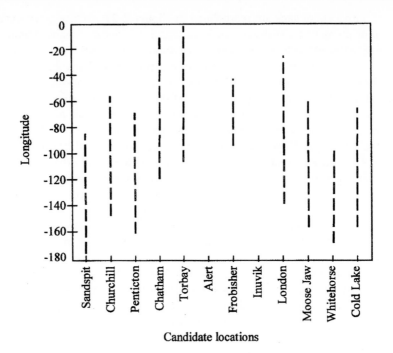

Figure 2-30 Limits of visibility

scrutinized to find which alternative is optimal. A marked difference between the two-sensor and three-sensor alternative sets is that the three-sensor alternatives favor the northerly locations while the two-sensor solutions do not. But it appears also that an alternative is never composed of two northerly sites. The popularity of northerly sites for the three-sensor problem may be explained. The three-sensor solutions can take advantage of the long winter observing period provided by northerly locations as long as a pair of southerly sites are available in the summer months to provide the necessary observations. A minimum of two southerly sites are required to provide coverage of the geostationary belt. An additional advantage of locating north is that in winter months, the northerly locations provide better coverage of the synchronous orbits. The chance of seeing a near-polar orbiting satellite simply increases as one increases the latitude (north or south).

Figure 2-32 shows some preferred locations should the two criterion- functions be equally weighted. The top seven nondominated solutions are identified. It is not surprising they all are congregated at the middle part of the efficient frontier, as one would expect. Another way to view these solutions is to minimize the deviation from the ideal. Here the ideal is the point (f_1^*, f_2^*) in the outcome space with $y_1^* = \max_i y_1^i$ and $y_2^* = \max_i y_2^i$. A Manhattan or rectilinear metric is used, or the l_p-metric with $p = 1$ and $w_i = 0.5$ for $i = 1,2$. The optimal solution to the set-covering problem is found simply by selecting the top-ranked alternative from the $p = 2$ and

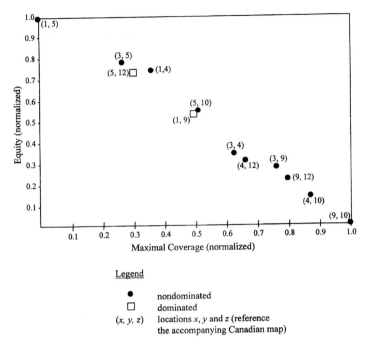

Figure 2-31 Two-sensor outcome-space

p = 3 solutions. For this case study, the two-sensor solution is the answer for the set-covering model.

2.11.3 Comments on solution methodologies and case study. Through the case study, let us make some concluding comments on various solution algorithms for facility-location models. The objective of the case study is to formulate and solve a more realistic version of the classic p-median problem. The classic model has been generalized to include overlapping service regions, multiple products or services, and the mean-variance criterion for multi-criteria optimization. The model was subsequently solved by a variety of methods, including LP relaxation, network-flow programming, network-with-side-constraints, heuristics and exhaustive enumeration. To the extent that the classic p-median model and extensions are special cases of this generalized model, the computational experienc e gained here has direct bearing upon many location models. According to Mirchandani (1990), the solution methods to the classic p-median problem can begrouped into enumeration, graph-theoretic, heuristic, primal-based mathematical-programming and dual-based mathematical-programming. This categorization applies to the extended p-median problems discussed here as well. The specific solution technique is driven by the nature of the problem. However, a common challenge belies all variants of the p-median problem—i.e., integrality of the solutions. Many solution techniques exploit the near total-unimodularity (TUM) of the constraint matrix.

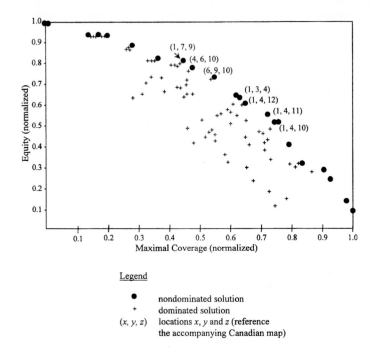

Figure 2-32 Three-sensor outcome-space

This approach is limited in multi-criteria optimization problems, where the TUM of the constraint set does not guarantee integer solution. Fortunately, in the generalized p-median problem formulated here, the solution has been shown to lie at the nodes. This property limits the solution space sufficiently for (implicitly) enumerating the dominated alternatives in the outcome space. Correspondingly, a discrete-alternative multi-criteria decision-making algorithm has been implemented that successfully exploits the basic definition of dominance. It has been shown that enumeration can be feasible in small problems and when the number of facilities is few. Under these circumstances, it can solve the problem quite expediently. The location of GEODSS sensors in Canada serves as a convincing case study that such an approach can solve realistic problems effectively, not just textbook examples.

The graph-theoretic approach exploits the inherent network-structure of the p-median problem. Where efficient labelling algorithms exist for the underlying network, this approach can be very rewarding. With the recent advances in generalized-network codes, polynomial algorithms are becoming feasible for the solution of network-flow-with-gains. This includes the work of Chang et al. (1989) and Goldberg et al. (1989). It will not be surprising that an even more efficient code will become available commercially soon beyond such packages as SAS/OR and CPLEX. The network-solution method can also be combined with LP-relaxation schemes to effect a viable binary-programming algorithm, as we have demonstrated above. Where integerization of the decision variables becomes a key issue, as it is in the network-with-gains model for the current problem, the network representation becomes a promising tool. This is due to its congruity with ad hoc integerization routines. The network approach also has bearing upon traditional-mathematical-programming approaches. It was found, for example, that an efficient network-with-side-constraints solution can often be obtained. Here side constraints fashioned after Bender's cuts are used to integerize decision variables and lead toward optimal solutions. The projection of the facility-location variable **y** on a multi-commodity formation appeared to solve the generalized p-median model very efficiently. Such a solution method adds to the emerging art of solving generalized p-median problems.

While not employed here in the case study, our previous discussion of Lagrangian relaxation shows that it is also a viable technique in solving a generalized median problem. Combined with its close 'cousin', Benders' decomposition[20], discrete facility location problems can often be handled expediently. As a conventional application, Lagrangian relaxation can replace LP relaxation to provide bounds within a branch-and-bound approach for solving location and other discrete optimization problems. Lagrangian relaxation can be viewed as a more general approach to network-with-side-constraints. In this approach, we take full advantage of any 'nice' constraints (not just the node-arc incidence matrices) while relaxing the complicating ones. We have already shown how this approach can be used to solve a generalized p-median model in Section 2.5 of this chapter, entitled "Solving generalized median using Lagrangian relaxation." The general role of decomposition algorithms in solving the classic p-median model was demonstrated in the "Prescriptive Tools" chapter (Chan 2000) under the "Decomposition methods in facility location" Section. In this previous chapter, both Benders' and Lagrangian relaxation methods were used to solve the same p-median example, and the two approaches were contrasted.

Obviously, the computational experiences reported here must be viewed only as a first step in a series of ongoing efforts to unify some very specialized tools for solving location problems. However, we have every reason to be optimistic. First, generalized p-median variants of the mean-variance location problem are formulated completely here as a multi-criteria optimization problem. A maximal-coverage and set-covering model have been modelled in a common underlying network, which also represents variants of the generalized p-median problem. Such a network was found to be amenable to a variety of efficient, albeit problem-specific, solution procedures. A feasible solution algorithm has been devised and illustrated. As part of this algorithm, integerizing only $|J|$ out of $|I||J|K$ binary variables was found sufficient in providing 0–1 solutions to the generalized network. This happens at a time when significant progress is being made day-by-day in generalized-network algorithms as well, as mentioned previously. When coupled with pre- and post-processor algebraic-language codes

---

[20]Benders' decomposition is explained in the "Optimization" appendix, where its relationship to Lagrangian relaxation is also drawn.

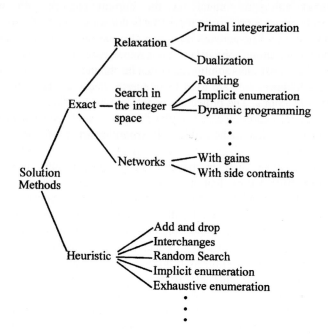

Figure 2-33 - Methods for solving location problems (Colorni 1987)

of integerization, the prototype network algorithm proposed here can truly become operational in solving generalized mean-variance location problems.

According to Colorni (1987), the main techniques for solving location problems consist of exact and heuristic methods, which are further broken down in **Figure 2-33**. Many exact methods make use of relaxation procedures, which reformulate the problem to obtain a more easily solved problem-. Here, care is being taken to ensure that the solution of the associated problem is also valid for the original problem. A first group of such algorithms relaxes integrality conditions for some integer variables, obtaining a linear (and sometimes nonlinear) programming problem. A classic procedure is branch-and-bound. The main advantage of these methods is that the relaxation of integrality conditions makes it possible to obtain problems for which very efficient algorithms exist (including network algorithms). A second group of relaxation algorithms is based on dualization using the Lagrangian function: i.e., the problem is reformulated by inserting some constraints, each multiplied by the relevant dual variable, in the objective function. There are two main advantages of dualization. One is the possibility of solving a series of simpler problems instead of a single complex one. The second is the possibility for separating the parts of the problem that interact using dualized constraints (Matta et al. 1999).

An alternate approach is to search in the space of integer solutions. This is possible because there are always some discrete variables in location models that by definition have only a finite number values. This means that the problem can be formulated in terms of combinatorial optimization. This results in studying the characteristics in the space of solutions (which are finite). There are only a few variants to this approach. One of them involves examining solutions through a ranking procedure that orders them according to an appropriate criterion. For example, one may rank by fixed plant-costs and generate solutions `adjacent' to those already examined in a search tree; a test based on over-estimate of the costs makes it possible to stop the algorithm, after examining only a limited number of the possible solutions. The main advantage of search methods in the integer-solution space is that we have an admissible solution to the problem any time (which does not occur with relaxation methods). The quality of the solution will depend on the depth of the search obviously. But very often, such methods give rise to simpler, repetitive heuristic-type procedures that are easy to solve.

Heuristic methods have been common in the solution of location problems since he beginning of the 1960s, and we will detail these procedures in the chapter on "location-routing models." By way of an introduction, *add-and-drop* methods begin with an initial feasible solution, i.e., from a set of open plants, and perturb this solution by adding or dropping plants. In practice we pass from a vector of 0 and 1 (corresponding to the closed-or-open state of the plants) to an adjacent one in binary-variable space, then to another and so on. This improves the objective function each time. When additional improvement is impossible, the algorithm stops and the last solution obtained is assumed final. Algorithms of this kind are strongly influenced by the starting values, as one may imagine. Some begin with the maximum number of plants open, others do the opposite (beginning with no open plants); and yet others start from the intermediate ground. They are mainly used for discrete problems with concave plant-costs or with fixed charges.

Another group of heuristic methods, which we shall call *interchange* methods, is based on iteration between the plant location and allocation of customers to the plant. First, an initial-feasible locational-decision is made; the next step is to optimally allocate customers to the open plants: this allocation step begins with a sub-division of the study area into zones, each assigned to one plant. From such sub-division, a new locational solution is calculated, followed by a new allocation-step and so on. The algorithm ends when the allocation-step does not modify the zone sub-division of the previous iteration. These methods also depend on starting values. They have been used in solution of problems with a fixed number of plants, both in continuous and in discrete space.

Recently, heuristic methods using random-search techniques have been proposed. The fundamental idea is to examine some solutions drawn at random from the set of admissible solutions, obtaining information helpful to the subsequent search. It is necessary, of course, to establish an upper limit to the number of draws and to use search methods that allow concentration on the most promising solutions. Algorithms of this kind are less dependent on starting values and may, theoretically, be used for any kind of location problem (continuous or discrete, capacitated or uncapacitated, with or without a fixed number of plants). In some cases, indicators of the closeness-of-the-solution-to-the-optimal exist.

Both exact and heuristic algorithms are, to a greater or lesser degree, conditioned by the data on which they are based. This concerns not only the quantity of data (i.e., the dimension of the problem) but also its structure. With problems of the same dimension certain algorithms work better if transportation costs prevail, others plant cost, some for uniform data-values, others for diversified data-values, and so on. We will take this point up further in the "Location-routing" chapter (Chapter 4) in sequel.

2.12 Summary

We have hopefully sketched in this chapter the state-of-the-art in facility-location models, inasmuch as one can do to the enormous literature within page limitations. For the interested reader, we have included a bibliography that documents a cross section of the current literature for further reading. We have taken an applicational standpoint in much of the discussions here. For that reason, we started with a given set of discrete locations for both demands and facilities—deemed the most practical with today's digitized world-. We prefer this over the axiomatic approach of distinguishing between planar vs. network-location problems. In discussing the network-location models, we have also skipped the interesting and often efficient results obtainable from a tree network. We judge that a more general (cyclic) network will cover the tree network as a special case—albeit at a compromise of computational efficiency and at the expense of mathematical elegance and mathematical insights. However, we have reported most the important analytical results as faithfully as we can. Parallels have been drawn between the planar case and discrete (network) case (Klamroth and Wiecek 1998). This is regarding nodal optimality and its relationship to the shape of travel-cost functions—whether they be concave or convex.

The chapter has been concentrating on the traditional median (min-sum) and center (min-max) problems, which work on the premise of proximity. They have been treated rather comprehensively in our opinion, particularly in network-location models where distinctions have been made between arc- *vs.* point-location of demand and facility. The discussions therefore encompass absolute, as well as general, medians and centers (including obnoxious facilities.) We have left the stochastic queue-median problems to the next chapter on "Spatial Separation," since the central issue of queue medians, in our judgment, is the response time, which cannot be adequately represented by the Euclidean or l_p-metrics used in traditional median and center problems (Taniguchi et al. 1999). The quadratic assignment problem, because of the spatial interaction between facilities and its relationship to routing, has been included in the "Spatial Separation" chapter as well. For similar reasons, we have deferred detailed discussions on *competitive* locations[21] to the

---

[21]In a competitive location problem, facilities are placed in order to command a fraction of the demand, commonly known as market share, against a competitor. Basic concepts on competitive facility-location were introduced in the "Economics" chapter (Chan 2000) under the "Spatial location of a facility" section.

chapters on "Evolution" and "Equilibrium," since it departs from the proximity locational-assumption. It also represents generalization of the median and center concepts to the extreme, when the demands and facilities are no longer given and fixed, but are generated within the system, as we allude to in the section on "Long-run location-production-allocation."

Within the limited number of pages, we attempted to cover the major solution strategies for these models. They include heuristics, linear-programming relaxation, Lagrangian relaxation, network-with-side-constraints and Benders' decomposition. There is no "sure fire" method of solving all facility-location problems. We show through a case study how different algorithms can be used according to the nature of the problem to achieve efficiency. Equally important are the relationships between apparently different models of median, center, maximal coverage, and set covering. Some of them can be equated either by transformation of the distance measure or represented in networks/graphs. By pointing out their similarities, it is hoped that efficient algorithms for one model can be transferred toward what appears on the surface as an unrelated model.

"From henceforth, space by itself, and time by itself,
have vanished into the merest shadows and
only a kind of blend of the two exists in its own right."
Herman Minkowski

CHAPTER THREE

MEASURING SPATIAL SEPARATION: DISTANCE, TIME, ROUTING, AND ACCESSIBILITY

Fundamental to location models is the measure of spatial separation. For the discussion here in this book, distance, time, impedance, and travel-cost are used interchangeably as units of spatial separation. Perhaps the most generic term in this context is distance, a term used for years by geographers to connote separation. Most introductory discussions on separation uses Euclidean ('bee-line') distance as the measure for distance mainly because it is convenient. A moment's reflection, however, will reveal that it is far from practical. Even the most common example of air distance or ocean travel is in fact measured in 'great-circle distance' which is far from Euclidean for any lengthy trip. Clearly, while Euclidean distance is an idealized, approximate measure for "fixing ideas", a more accurate measure is absolutely essential for any real applications. When travel is impeded by, say, road accidents or inclement weather, the routing and travel-time become highly uncertain. Now that locational decisions are dependent upon *time-varying* accessibility, and the quantification of distance then becomes extremely challenging. Under this *stochastic* situation, it is not at all clear the demand-for-service will stay fixed either. Should the customer cancels his/her demand for service, travel requirement goes away and spatial-separation then becomes a purely theoretical idea. The readers can see by now that the problem is a highly complex one, any light we can shed upon the problem will therefore be a great help.

3.1 Manhattan metric, Euclidean distance and generalization

Let us start with a simple routing example. In a city, the travel corridors are often well defined by a set of orthogonal streets. Distance measure between two points in this case is the total travel along east-west and north-south directions, sometimes called the *Manhattan metric*. More explicitly, such a measure is given by $l_1 = d_1 + d_2$ where d_1 is the absolute travel-distance along the east-west direction and d_2 the north-south direction. Example of such a measure is graphically illustrated in **Figure 3-1**. Euclidean distance between the same two points i–j is measured by $l_2 = (d_1^2 + d_2^2)^{1/2}$. Again, this is illustrated in **Figure 3-1**. We have already pointed out

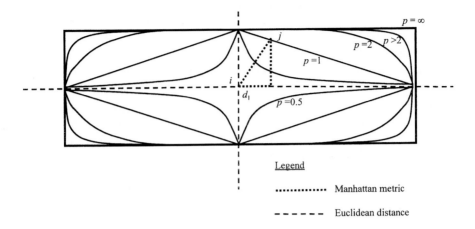

Figure 3-1 A family of separation measures

that such a distance is often too much of an idealization. In the United States, road distances between major cities are about 18 percent greater than the straight-line Euclidean distances (Love et al. 1988). The ratio between the idealized Manhattan metric and Euclidean is obviously greater than 1.18. The ratio is more like $l_1/l_2 = 1.42$. In other words, Manhattan metrics are 42 percent longer than Euclidean.

3.1.1 l_p-norms. Generalization of this concept leads us to the "l_3 measure," which is given by $l_3 = (|d_1|^3 + |d_2|^3)^{1/3}$. In general, we can have a family of measures called the l_p-norm. Such a concept has already been introduced in the chapter on "Multi-criteria Decision-making" (Chan 2000), where 2 we define the following norm-function in the absence of weights placed on each dimension

$$l_p = \left[\sum_i d_i^p \right]^{1/p} \quad (1 \le p \le \infty). \tag{3-1}$$

In this definition, we have generalized the distance measure to include not only the two-dimensional spatial map, but also the third and higher dimensions. For example, road distances are dependent upon the altitude besides the east-west and north-south separations. When measured in time, a trip on level ground will be a good deal shorter than through the mountains. Considering that travel through the mountain should be more time-consuming than on the plain, one can introduce a weight w_i to be applied toward the vertical dimension $i = 1$ (altitude) vis-a-vis the east-west ($i = 2$), north-south ($i = 3$) dimensions, with $w_1 > w_2$ and $w_1 > w_3$. Equation **(3-1)** is now generalized to read

$$l_p(w) = r'(d; p, w) = \left[\sum_i w_i^p d_i^p \right]^{1/p} \quad (1 \le p \le \infty) \tag{3-2}$$

as defined in the "Deviational measure" subsection of "Multi-criteria decision-making" chapter. For example, the altitude dimension may be weighted by 0.4 while the horizontal travel-distances by 0.3, then the above equation becomes $l_p(w) = (0.3d_1 + 0.3d_2 + 0.4d_3)^{1/p}$ where the three weights sum to unity.

In location models, the l_∞-norm becomes particularly important. The metric refers to the maximum of the distances, or $l_\infty = \max_i d_i$. In emergency facility-location, for example, a facility j is to be located so that the maximum distance to a demand i is to be minimized. Thus the figure-of-merit for locating such a facility becomes min-l_∞. There is no reason to suggest that p has to be greater than or equal to unity either. In circuitous routes, it could very well be such that the actual distance is quite a bit longer than 1.41 of the Euclidean distance (as in the Manhattan metric). Let us take the example of $p = 0.5$ in a two dimensional case, $l_{0.5}/l_2 = 2.84$, suggesting a much more circuitous path than the Manhattan metric between the two points i and j. **Figure 3-1** illustrates this case as well, along with $p \geq 1$. In general, **Figure 3-1** shows the contour of a constant distance away from demand i using the various l_p-norms. Thus for the same distance separation, a smaller p corresponds to a more circuitous routing than a larger p: the $l_{0.5}$-contour is `contained' within the l_1-norm contour, the l_1-norm contour is contained within the l_2-norm and so on. The contours all come together at the one dimensional degenerate case when $d_1 = 0$ or $d_2 = 0$. In other words, when single-dimensional distances are measured strictly in a single dimension (the east-west or north-south direction) only, there really is no such a concept as circuity. l_p-norms have a long history in mathematics, and are often called Minkowski (power) metrics in general.

3.1.2 Special Cases. The l_p-norm or Minkowski metric is an important measure in representing spatial separation. In the "Facility Location" chapter, we pointed out that spatial separation has profound impact upon locational deci-sions—as illustrated in the classic Steiner–Weber problem. In particular, there are totally different locational implications when the spatial cost-function $f(C)$ has constant-return-to-scale, $f(C) = C^1$; increasing-return-to-scale, $f(C) = C^{b_1}$ for $b_1 < 1$; and decreasing-return-to-scale, $f(C) = C^{b_2}$ for $b_2 > 1$. When spatial-cost C is measured in the Minkowski metric, $l_p(w)$, the cost function becomes $f(C) = C^\beta = [l_p(w)]^\beta = [(\Sigma_i w_i^p d_i^p)^{1/p}]^\beta$ where $\beta = 1$, b_1, b_2 above. In this case, $f(C)$ no longer satisfies the triangular inequality $f_{ik} \leq f_{ij} + f_{jk}$ when $\beta = b_2$ or when $\beta > 1$. To see this, consider the points $(-1,0)$, $(0,0)$ and $(1,0)$ on the horizontal axis. In this case, $l_p^\beta = 2^\beta = f_{ik}$ as measured from $(-1,0)$ to $(1,0)$, whereas $l_p^\beta = 1^\beta = f_{ij} = f_{jk}$ when measured from $(-1,0)$ to $(0,0)$, or from $(0,0)$ to $(1,0)$. Since $2^\beta > 1^\beta + 1^\beta$ for $\beta > 1$, the triangular inequality is violated. However, notice for $0 < \beta < 1$, the triangular inequality is preserved.

Now consider the metric $f' = \alpha[l_p(w = 1)]^\beta$ where the constant $\alpha > 1$, $p \geq 1$, $0 < \beta < 1$, and $d_i = x_i - x_i^0$. This means $f' = \alpha[(\Sigma_i\{x_i - x_i^0\}^p)^{1/p}]^\beta$, a metric that suggests increasing economy-of-scale that arises in many practical situations. Min-sum (median) and min-max (center) problems involving the Minkowski metric are normally convex minimization problems, where a local optimum is a global optimum. But f' is, strictly speaking, neither convex nor concave, and has multiple local minima at x_i^0, $i = 1, ..., n$. We have witnessed this in the "Facility Location" chapter under the "Planar location problem" section when the Steiner–Weber problem was first solved in the two-dimensional case using a gradient algorithm. There it was pointed out that—in spite of the theoretical difficulties—the solution is either optimal or near optimal a large percent of the time.

3.2   Measuring spatial price

Aside from a purely geometrical measure, spatial separation can be part of a "price system," in which the allocation of resources is coordinated through these spatial prices. Thus the components ($i = 1, 2, ...$) in Equation **(3-2)** can be interpreted to mean the time, distance, out-of-pocket cost, inconvenience etc. of making a trip. Together, they define the spatial price between a demand point and a facility. A common form of spatial price corresponding to $p = 1$ in Equation **(3-2)** is the linear combination of several 'costs':

$$C=f(\tau, d', c')=w_1\tau+w_2 d'+w_3 c'+r'' \qquad\qquad (3\text{-}3)$$

where τ is the travel time, d' is the distance, c' is the out-of-pocket cost and r' is the term that includes all other factors including inconvenience.

3.2.1   Accessibility.   After such a composite measure has been formed, it is often desirable to convert the measure into *accessibility* $f(C)$, which is built upon a fundamental function of spatial cost C. Accessibility is different from a spatial price (or spatial impedance) in that the two have an inverse relationship, when spatial price goes up, accessibility comes down and *vice versa*. A review of the literature shows that different formulations of $f(C)$ have been used in the past. Almost all of these, however, can be represented as a special case of the following function:

$$f(C)=\gamma C^{\alpha}\exp(-\beta C) \qquad\qquad (3\text{-}4)$$

where α, β and γ are calibration parameters, with the first unrestricted in sign, and the latter two being nonnegative. This general form accommodates the conflicting desires to choose a facility (or an activity zone) that is away from the demand (or zone of residence), but at the same time not too far to involve a huge travel effort. Oftentimes, the precise functional form has to be determined by trial-and-error process as a special case of Equation **(3-4)**.

A more practical way of calibrating such a function is to fit a curve through the trip-distribution[1] information. Such curves often assume a polynomial function of τ, and can include other travel 'costs'. In our notation, we write $C'(\tau)$ for work trips

---

[1]Trip distribution is a frequency plot of the percentage of trips of duration 0-5 minutes, 5–10 minutes, 10–15 minutes, etc.

and $C^k(\tau)$ for nonwork trips of purpose k where $C'(\tau) = 1/(\alpha\tau^\beta)$ and $C^k(\tau) = 1/(\alpha-\beta\tau+\tau^2)$, as we will see in the Lowry land-use model for Pittsburgh. A related function is the probability of residential location and retail location in zone j among n possible zones:

$$\Theta_{ij}(N_j,\tau_{ij}) = \frac{N_j C'(\tau_{ij})}{\sum_{l=1}^{n} N_l C'(\tau_{il})} = \frac{N_j C'(\tau_{ij})}{X_i(\tau_{ij})}$$

$$\Theta_{ij}(E_j^k,\tau_{ij}) = \frac{E_j^k C^k(\tau_{ij})}{\sum_{l=1}^{n} E_l C^k(\tau_{il})} = \frac{E_j^k C^k(\tau_{ij})}{X_i(\tau_{ij})}$$

(3-5)

where N_j is the population at destination zone j and E_j^k the kth-type employment. These expressions normalize the accessibility measure into percentages for convenience. While being more operational, such an approach carries with it a different assumption than Equation **(3-4)**. Aside from ignoring all travel costs except travel time, it also includes such explanatory variables as the attractiveness of the destination. Remember that trip-making is determined not only by travel time, but also by the socioeconomic value of the trip, including the activities executed at the destination. Conventional usage of the word accessibility usually includes socioeconomic variables besides travel time, which is different from the usage in Equation **(3-4)**. We will refer the readers to the chapter on the "Lowry Model" and the chapter on "Bifurcation and Disaggregation" as case studies of accessibility-function calibration.

While l_p-norms (Minkowski metrics) are used in the facility-location literature, accessibility measures are found largely in land-use applications. Both try to aggregate a number of distance attributes into a single measure. Minkowski metrics are purely geometrical in concept, whereas accessibility begins to introduce activities such as work, shop, recreation into distance separation, mainly as weights. There is no reason why Equation **(3-2)** cannot be used in land-use, nor Equation **(3-4)** in facility location. It is simply a matter of tradition, in my opinion, that the exchange has not taken place until recently.

3.2.2 Distance measure of efficiency. As an example of a hybrid between a geometric and spatial-price measure, the distance function has proved exceptionally useful as a theoretical description of efficiency. It is consistent with modern axiomatic production-theory in economics. In the section entitled "*Combined data-envelopment-analysis/location model*", we will marry the geometric and economic concepts formally. Grosskopf et al. (1993) proposed an extension of recent work on production-efficiency in terms of the Minkowski distance-function. Their *empirical* approach differs from previous work, including Grosskopf and Hayes (1993) and *data envelopment analysis*[2], in its capacity to allow for multiple outputs (from multiple input factors) and full measurement of statistical errors. It maintains the implicit-function form of the Minkowski distance-function, and minimizes deviations from its efficient-frontier contour of one, $r'(Z'') = 1$. Z'' in this case represents the deviation measures from the contour and r' stands for the contour value as illustrated in **Figure 3-1**. Denote the vector of inputs as $\mathbf{x} = (x_1, x_2, \dots, x_n)^T$ which is used to produce a vector of outputs $\mathbf{y'} = (y_1', y_2', \dots, y_m')^T$. The basic technological relationship is described by the mapping of a set of input factors $\mathbf{x}$ into output vector $\mathbf{y'}$, or in mathematical form $R'(\mathbf{y'}) = \{\mathbf{x}:\mathbf{x} \rightarrow \mathbf{y'}\}$.

We can interpret $R'(\mathbf{y'})$ as the set of input requirements $\mathbf{x}$, or $R'(\mathbf{y'}) = \{\mathbf{x}: \mathbf{x}$ can produce $\mathbf{y'}\}$. In words, the input distance-function seeks the greatest possible radial contraction of observed input-bundle $\mathbf{x}$ which still allows production of observed output-bundle $\mathbf{y'}$. The input distance-function provides a complete description of technology with minimal structure. It can easily model the most efficient multiple-output technology. The distance-function is now defined as $r'(\mathbf{y'},\mathbf{x}) = \sup\{\xi:\mathbf{x}/\xi \in R'(\mathbf{y'})\}$, or a set of measures generated as a result of fitting the inputs $\mathbf{x}$ statistically against the observed output $\mathbf{y'}$. In other words, we curve-fit the distance function as a pre-specified implicit function with a unit contour, $r'(Z_k';\boldsymbol{\beta}) = 1$, where $Z_k' = (y_k', x_k) = z_k''+\varepsilon_k$ and $\boldsymbol{\beta}$ is a set of calibration coefficients. Thus the kth pair of $x-y'$ values are mapped statistically into a z'' value, which in turn forms the distance function. Meanwhile, the error-terms for the random-variable Z'', $\varepsilon_k = (\varepsilon_k(x), \varepsilon_k(y'))$, is normally distributed with zero mean and a non-singular error-covariance matrix. Here $\varepsilon_k(y')$ denotes the measurement-error associated with the output-factors and $\varepsilon_k(x)$ is the measurement-error associated with the input-factors. ξ, which is to be maximized here, represents the proportion by which factor inputs have to be reduced at A to reach the efficient point on the isoquant at A'. This is illustrated in **Figure 3-2** for the two-dimensional case.

The distance function provides information on the technical efficiency of individual observations, i.e., whether an observation is on the frontier-of-technology or interior to the frontier. This measure of technical-efficiency captures deviation from the best-practice frontier, i.e., performance relative to the other observations in the sample. Again, technical efficiency is illustrated in **Figure 3-2**. Consider the observed input-bundle at $A = (x_i, x_j)$, which produces the same level of output(s) as point A'. The technical efficiency of observation A is measured as the ratio OA'/OA, i.e., the ratio of minimal to observed resource-use. Technical efficiency can be assessed by first taking the reciprocal of the distance function: $r'(\mathbf{y}^A, \mathbf{x}^A) = OA/OA'$. Subtracting OA'/OA from unity yields the percentage ξ by which inputs (and therefore cost) could be reduced and still produces the observed output. Note that $r'(\mathbf{y'},\mathbf{x}) \geq 1$ for all feasible input-bundles and $r'(\mathbf{y'},\mathbf{x}) = 1$ if and only if $\mathbf{x}$ is on the

---

[2]Data envelopment analysis, often shortened as DEA, is a study of the axiomatic underpinnings and technical characteristics of various measures of productive efficiency. It uses linear programming to compute technical efficiency, as outlined in the seminal paper by Charnes et al. (1978), which seeds a flurry of subsequent activities in this burgeoning field. DEA was introduced in the "Prescriptive Tools" chapter of Chan (2000) under the "Prescriptive analysis in facility location" section.

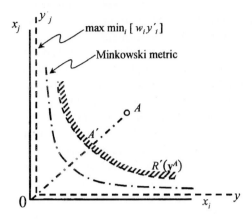

Figure 3-2 Illustrating technical efficiency

frontier, i.e., if and only if **x** is efficient. For example, if $OA = 15$ and $OA' = 5$, then $OA/OA' = 3$. The inverse of the distance function, and the efficiency of point A in relation to point A', would then be computed as 1/3. Hence the output bundle at A would only be 1/3 as efficient as the output bundle at A'. In short, the resulting distance function satisfies the following properties, $r_i(\mathbf{y}', \mathbf{x}) = 1$ if and only if $\mathbf{x} \in R'(\mathbf{y}')$, and $r_i(\mathbf{y}', \mathbf{x}) \geq 1$ if and only if $\mathbf{x} \in \text{Isoquant}\{R'(\mathbf{y}')\}$. If the economic system operates efficiently, the distance-function has a unit contour. Otherwise, the contour is likely to be outside unity. Other properties are as follows. The production function is homogeneous-of-degree-one[3], it has non-increasing inputs, and non-decreasing outputs. The distance-function is also quasi-concave in output and concave in inputs.

Grosskopf et al. (1993) parameterize the implicit distance-function as a generalized Leontief[4]:

---

[3]In economics, the transformation of input factors **x** to output factors **y'** is captured in a mathematical form called the *production function*. A production function is said to be homogeneous-to-the-first-degree if output increases in exact proportion to the increases in factor inputs. In other words, in the production function $y' = f(x_1, x_2)$, if $y'' = f(\gamma x_1, \gamma x_2)$ means $y'' = \gamma y'$, then $f(\cdot)$ is called a homogeneous-function-to-the-first-degree. For example, the Cobb-Douglas function $y' = a x_1^b x_2^{1-b}$ for $0 \leq b \leq 1$ is homogeneous-to-the-first-degree. It is also a well known fact from microeconomics that one can derive the cost-function $C = C(y)$ given the input-factor prices. In the spatial literature, this gives rise to the spatial-cost or accessibility function as shown in Equation **(3-5)**

[4]For a single-output production-function such as the Cobb–Douglas function $y' = a x_1^b x_2^{1-b}$ with $0 \leq b \leq 1$, the generalized Leontief-*cost*-function looks like $C = y' \sum_i \sum_j a_{ij} w_1^{1/2} w_2^{1/2}$ where w_1 and w_2 are input-factor prices. Rewriting the expression as $C/y' = (d_{12}' w_1^{1/2} w_2^{1/2} + d_{21}' w_2^{1/2} w_1'^{1/2}) = (d_1^{1/2} w_1^{1/2} + d_2^{1/2} w_2^{1/2})$, one can see the relationship of the generalized multiple-output Leontief-*production-function* contour to a Minkowski

$$r' = \sum_i \sum_j \alpha_{ij} x_i^{1/2} x_j^{1/2} + \sum_j \sum_k \beta_{jk} x_j \ln y_k' \tag{3-6}$$

where x_i is the input quantity and y_k' is the kth output. The index for each of the observation is omitted from the above equation for clarity. As alluded to already, $r' = 1$ and the frontier-of-technology is estimated. The generalized-Leontief form is chosen due to its flexibility, i.e., only minimal regularity-conditions are imposed on the distance function such as homogeneity-of-degree-one in input. Concavity assumptions can be easily examined. A third advantage is its ability to deal with zeros in the input and output vectors. A linear program can then be formulated to minimize the sum of the deviations of this production function for individual observations from the frontier of technology (where the distance function takes on the value of unity).

Example

In an aspatial example, Grosskopf et al. (1993) evaluated the relative efficiencies of public, not-for-profit and for-profit hospitals. They chose to use the distance-function to model the intermediate production of hospital services. These services include (a) acute patient-days, (b) intensive-care patient-days, (c) subacute patient-days, (d) number of surgeries, (e) outpatient visits, as well as (f) number of discharges. The inputs include (1) medical staff, (2) residents, (3) registered nurses, (4) licensed practical nurses, (5) other employees, and (6) number of beds. Preliminary results for the for-profit hospitals are given for illustration purposes in **Table 3-1** as a 6×6 matrix. Here only the β parameters for the variables x–y' in the implicit function $Z_k' = (y_k', x_k)$ [Equation **(3-6)**] are given. The reader can go to Grosskopf et al. (1993) for additional parameters, including the α parameters and the error covariance-matrix that correlates the error-terms in the function $\varepsilon_k = (\varepsilon_k(x), \varepsilon_k(y'))$. Technical details of the calibration procedure, following Fuller (1987), are somewhat irrelevant here. Notice the economic efficiency of the hospital has not been determined by this preliminary analysis, and no conclusion can be drawn from this study at this junction. □

The ideas in this simple example can be extended to general equilibrium in a spatial context, where every part of the economy is interdependent. In a subsequent chapter, entitled "Spatial equilibrium and disequilibrium," we will show how a production function for a hospital such as the one in the above example can be broadened to include many industries interacting spatially. There each industry employs the outputs of other industries as its raw materials. Transport in this case is one of the (exogenous) commodities, complete with a unit price, satisfying the same relation as the price of any other commodity. Given the name of *spatial input-output analysis*, empirical intransigence and the computational problems have forced on traditional input-output analysis several simplifying assumptions. Closely related to linear programming, it is assumed that in any productive process all

---

measure such as $\sum_i \left[w_i^{1/p} d_i^{1/p} \right]^{1/p}$ when $p = 1/2$. Meanwhile, max $r'(\mathbf{d}, \mathbf{0}, \mathbf{w}) = \max \min_i [w_i d_i]$, which is exactly the objective function of the obnoxious-facility location-problem.

Output Input	Acute patient days	Intensive care	Subacute	Surgeries	Outpatient visits	Discharges
Staff	-0.0161 (0.001)	-0.0331 (0.001)	0.0067 (0.000)	0.0504 (0.001)	0.0485 (0.001)	0.0992 (0.002)
Resi- dents	-0.00425 (0.001)	0.0136 (0.000)	-0.0008 (0.000)	-0.0568 (0.001)	-0.0157 (0.001)	-0.0246 (0.001)
Nurses	-0.6314 (0.025)	-1.3748 (0.026)	0.1246 (0.006)	-3.8117 (0.054)	1.3329 (0.032)	4.9505 (0.050)
Licensed nurses	0.0431 (0.001)	-0.0760 (0.001)	0.0379 (0.001)	0.0731 (0.002)	-0.0568 (0.002)	-0.1643 (0.003)
Other em- ployees	-0.1077 (0.002)	0.1349 (0.003)	0.0329 (0.004)	-0.0325 (0.004)	-0.2581 (0.003)	0.1175 (0.006)
Beds	0.0100 (0.001)	0.0221 (0.001)	-0.0283 (0.000)	-0.0368 (0.001)	0.0112 (0.000)	0.1028 (0.001)

Table 3-1 For-profit-hospital distance-function calibration

inputs are employed in rigidly fixed proportions and the use of these inputs expands in proportion with the level-of-output. This is a special case of constant-returns-to-scale. But the fixed-proportions assumption is far more restrictive than the linear homogeneous-production-function implied in constant-return-to-scale assumption used in the distance function above. A linear homogeneous-production-function requires only if the firm decides to triple the scale of either of the inputs, the result will be a tripling of output. A fixed-proportion premise, on the other hand, requires that a manufacturing process that is labor-intensive offers no option of a capital-intensive alternative. We mention the linkage here to put our discussion on distance function in broader perspective. In a spatial version of the input-output model, economic multipliers (or *technical coefficients*) are imbedded in the accessibility functions such as those shown in Equation **(3-5)**, which ultimately determine the efficiency of the spatial production-process. This can be thought of as a spatial generalization of the efficient-frontier concept employed in the simple hospital-production-function above.

3.2.3 Combined data-envelopment-analysis/location model. As explained above and in the "Prescriptive Tools" chapter of Chan (2000), data envelopment analysis (DEA) is a normative model to define a constant-return-to-scale efficient frontier, where the factor inputs are combined in the correct proportions to achieve the best efficiency. While the distance function and DEA are similar in the sense that both describe the efficient frontiers, DEA does not require the isoquant and its associated weights to be identified a priori. DEA determines the efficient frontier and optimal weighting scheme during the execution of the linear program. When DEA inputs and outputs are interpreted as costs and benefits, as we have done in the "Application to facility location" section of the "Prescriptive tools" chapter

(Chan 2000), it can be used for finding the 'best' facility location. In our presentations so far, location and DEA models have been solved separately. A closer examination of the DEA model used to site facilities also reveals that it does not really have a spatial dimension, since all spatial attributes are exogenously found and input to the model. No network representation is present. Similarly, classical facility-location models have only spatial attributes and typically lack the broad range of figures-of-merit typical of a siting decision. Take an obnoxious-facility model, for example, it is highly desirable to have the max-min objective as one of the several benefits of the DEA model, and have the cost-benefit analysis of DEA include explicitly the spatial dimension (Thomas 1995). In so doing, the distance function is now truly a composite of both physical distance-separation and economic benefit measures.

3.2.3.1 ONE-FACILITY CASE. The one-facility combined model is a simple extension of previous formulations. To develop this model, consider the DEA example in the "Prescriptive tools" chapter (Chan 2000) again, together with the max-min obnoxious-facility-location problem in the "Facility Location" chapter under the "p-center" and "Obnoxious facility" sections. Assume that all of the previously discussed inputs and outputs are the same. There is now additional information regarding each facility's proximity to population centers. If i is a population center (or demand point) and j is a facility, the distance-separation data are shown in **Table 3-2**. In order to solve the one-facility max-min DEA combined model, a new benefit-variable b_{Kji} must be introduced, which shows the relative importance of assigning the Kth benefit to the demand-facility pairing i-j. Specifically, this variable represents the locational benefit (number 4) provided by a facility j serving a demand at location i. The locational benefit for each facility can then be written as $4b_{441} + 6b_{451} + 8b_{461}$ for facility 1, $10b_{442} + 7b_{452} + 6b_{462}$ for facility 2, $7b_{443} + 5b_{453} + 9b_{463}$ for facility 3. Here the b are 0–1 ranged weights placed on the given benefits 4, 6, 8 etc. This new locational-benefit weight is now added as part of the overall cost/benefit structure that already exist in principle from the previous DEA formulation. The changes are reflected below. First, these constraints work in concert with the previous ones to bound the DEA score x to be between 0 and 100.

$$x_1 - 9b_{11} - 4b_{21} - 16b_{31} - 4b_{441} - 6b_{451} - 8b_{461} \geq 0$$
$$x_2 - 5b_{12} - 7b_{22} - 10b_{32} - 10b_{442} - 7b_{452} - 6b_{462} \geq 0$$
$$x_3 - 4b_{13} - 9b_{23} - 13b_{33} - 7b_{443} - 5b_{453} - 9b_{463} \geq 0$$
$$x_1 - 9b_{11} - 4b_{21} - 16b_{31} - 4b_{441} - 6b_{451} - 8b_{461} + 100y_1 \leq 100$$
$$x_2 - 5b_{12} - 7b_{22} - 10b_{32} - 10b_{442} - 7b_{452} - 6b_{462} + 100y_2 \leq 100$$
$$x_3 - 4b_{13} - 9b_{23} - 13b_{33} - 7b_{443} - 5b_{453} - 9b_{463} + 100y_3 \leq 100.$$

By limiting the sum of the costs c, the following constraints ensure that efficiency for the selected facilities cannot exceed the value of unity.

	Facility 1	Facility 2	Facility 3	Total	Minimum
Demand 4	4	10	7	18	4
Demand 5	6	7	6	23	6
Demand 6	8	6	9	21	5

Table 3-2 Distance-separation data for DEA model (Thomas 1995)

For facility 1

$$-9b_{11} - 4b_{21} - 16b_{31} - 4b_{441} - 6b_{451} - 8b_{461} + 5c_{11} + 14c_{21} \geq 0$$
$$-5b_{11} - 7b_{21} - 10b_{31} - 10b_{441} - 7b_{451} - 6b_{461} + 8c_{11} + 15c_{21} \geq 0$$
$$-4b_{11} - 9b_{21} - 13b_{31} - 7b_{441} - 5b_{451} - 9b_{461} + 7c_{11} + 12c_{21} \geq 0.$$

For facility 2

$$-9b_{12} - 4b_{22} - 16b_{32} - 4b_{442} - 6b_{452} - 8b_{462} + 5c_{12} + 14c_{22} \geq 0$$
$$-5b_{12} - 7b_{22} - 10b_{32} - 10b_{442} - 7b_{452} - 6b_{462} + 8c_{12} + 15c_{22} \geq 0$$
$$-4b_{12} - 9b_{22} - 13b_{32} - 7b_{442} - 5b_{452} - 9b_{462} + 7c_{12} + 12c_{22} \geq 0.$$

For facility 3

$$-9b_{13} - 4b_{23} - 16b_{33} - 4b_{443} - 6b_{453} - 8b_{463} + 5c_{13} + 14c_{23} \geq 0$$
$$-5b_{13} - 7b_{23} - 10b_{33} - 10b_{443} - 7b_{453} - 6b_{463} + 8c_{13} + 15c_{23} \geq 0$$
$$-4b_{13} - 9b_{23} - 13b_{33} - 7b_{443} - 5b_{453} - 9b_{463} + 7c_{13} + 12c_{23} \geq 0.$$

Another change needs to be made to the existing DEA portion of the combined formulation. This is to ensure that the locational benefit between facility j and demand i is not realized unless $i-j$ is the smallest arc in the network. This is in response to the max-min objective of the location model, and is accomplished by the unitary valuation of the binary variable $m'_{ij} = 1$. For example, if facility 3 is used, then the shortest distance from facility 3 to any of the three demands is $D_{53} = 5$. Since arc 5–1 is the shortest arc, m'_{51} would be set equal to 1. In the one-facility case, when a single facility serves all of the demands, this could be triggered by the y_j variable since the facility chosen will service all of the demands. However, to make the model more robust and the transition to the two-facility case clearer, the benefits will be realized as the arcs are turned on. The following either-or constraints accomplish this goal. With the following constraints, the reader can verify that if x_{41} is triggered to 1, then b_{441} can range from 0.0001 to 100. If x_{41} is not triggered, then b_{441} will be set equal to zero. The either-or constraints must be written for every arc in the network. But distance benefits will accrue only when the m'_{ij} variable is activated.

To implement these 'switches', we first define the 0–1 variable m'_{ij} in terms of x_{ij}. Thus for facility 1:

$$m'_{41} - x_{41} \le 0 \qquad m'_{51} - x_{51} \le 0 \qquad m'_{61} - x_{61} \le 0$$
$$-m'_{41} + x_{41} \ge 0 \qquad -m'_{51} + x_{51} \ge 0 \qquad -m'_{61} + x_{61} \ge 0.$$

For facility 2:

$$m'_{42} - x_{42} \le 0 \qquad m'_{52} - x_{52} \le 0 \qquad m'_{62} - x_{62} \le 0$$
$$-m'_{42} + x_{42} \ge 0 \qquad -m'_{52} + x_{52} \ge 0 \qquad -m'_{62} + x_{62} \ge 0.$$

For facility 3:

$$m'_{43} - x_{43} \le 0 \qquad m'_{53} - x_{53} \le 0 \qquad m'_{63} - x_{63} \le 0$$
$$-m'_{43} + x_{43} \ge 0 \qquad -m'_{53} + x_{53} \ge 0 \qquad -m'_{63} + x_{63} \ge 0.$$

To effect the triggering mechanism, we relate benefit valuation to the m'_{ij} variable for facility 1 to any of the three arcs 4–1, 5–1, and 6–1:

$$b_{441} - 100m'_{41} \le 0 \qquad b_{451} - 100m'_{51} \le 0 \qquad b_{461} - 100m'_{61} < 0$$
$$-b_{441} + 100m'_{41} \le 99.9999 \quad -b_{451} + 100m'_{51} \le 99.9999 \quad -b_{461} + 100m'_{61} \le 99.9999.$$

For facility 2, we write the following for arcs 4–2, 5–2, and 6–2:

$$b_{442} - 100m'_{42} \le 0 \qquad b_{452} - 100m'_{52} \le 0 \qquad b_{462} - 100m'_{62} \le 0$$
$$-b_{442} + 100m'_{42} \le 99.9999 \quad -b_{452} + 100m'_{52} \le 99.9999 \quad -b_{462} + 100m'_{62} \le 99.9999.$$

For facility 3 (arcs 4–3, 5–3, and 6–3):

$$b_{443} - 100m'_{43} \le 0 \qquad b_{453} - 100m'_{53} \le 0 \qquad b_{463} - 100m'_{63} \le 0$$
$$-b_{443} + 100m'_{43} \le 99.9999 \quad -b_{453} + 100m'_{53} \le 99.9999 \quad -b_{463} + 100m'_{63} \le 99.9999.$$

Now that the appropriate revisions have been made to the DEA sub-model, the locational half now needs to be addressed. On top of the classic anti-center problem, there are several new requirements posed by the current formulation. First, the constraints that allow the arcs to be activated must be included. These constraints ensure that the maximum distance is chosen correctly. The constraints also account for the "crossing effect" in that a demand can be served by any of the three facilities. Again an arbitrary maximum value of 100 is used for the score.

For site 1

$$4x_{41} + 10x_{42} + 7x_{43} + 96y_1 \le 100$$
$$6x_{51} + 7x_{52} + 5x_{53} + 94y_1 \le 100$$
$$8x_{61} + 6x_{62} + 9x_{63} + 92y_1 \le 100.$$

Here $96 = 100-4$, $94 = 100-6$, and $92 = 100-8$. For site 2

$$10x_{42} + 4x_{41} + 7x_{43} + 93y_2 \leq 100$$
$$7x_{52} + 6x_{51} + 5x_{53} + 95y_2 \leq 100$$
$$6x_{62} + 8x_{61} + 9x_{63} + 91y_2 \leq 100.$$

For site 3

$$7x_{43} + 4x_{41} + 10x_{42} + 90y_3 \leq 100$$
$$5x_{53} + 6x_{51} + 7x_{52} + 93y_3 \leq 100$$
$$9x_{63} + 8x_{61} + 6x_{62} + 94y_3 \leq 100.$$

Next, the model must ensure that demand is met at each site.

$$x_{41} + x_{42} + x_{43} = 1 \qquad x_{51} + x_{52} + x_{53} = 1 \qquad x_{61} + x_{62} + x_{63} = 1$$

Finally, demand cannot be met unless the facility is in place.

$$x_{41} + x_{51} + x_{61} - 3y_1 \leq 0 \qquad x_{42} + x_{52} + x_{62} - 3y_2 \leq 0 \qquad x_{43} + x_{53} + x_{63} - 3y_3 \leq 0.$$

Further revisions must be made in the locational sub-model. The model needs to identify the minimum distance in any portion of the network. The example below for facility 1 shows how this is accomplished

$$D_1' + 96x_{41} \leq 100 \qquad\qquad D_1' + 94x_{51} \leq 100 \qquad\qquad D_1' + 92x_{61} \leq 100.$$

As in previous obnoxious-facility formulations, D_j' is used here to identify the minimum distance. If facility 1 is built, then $x_{41} = x_{51} = x_{61} = 1$. This results in a D_1' value of 4, which indicates that 1–4 is the minimum-distance arc, or $x_{41} = 1$.

In previous max-min formulations, it was not necessary to identify which arc supplied the minimum distance. The overall minimum-distance was the goal. However, in the combined formulation, explicit identification of the minimum-distance arc is of great concern—as alluded to earlier. This is where the m_{ij} variables come in to play. Once the minimum distance has been identified, it is possible to find which arc supplies this distance by applying the following. The first three constraints identify the minimum-distance arc. The final two constraints ensure that if a facility is selected, exactly one of the m_{ij}' variables will be activated.

$$4m_{41}' - 100y_1 - D_1' \leq -100m_{41}' + m_{51}' + m_{61}' - y_1 \leq 0$$
$$6m_{51}' - 100y_1 - D_1' \leq -100m_{41}' + m_{51}' + m_{61}' - y_1 \geq 0.$$
$$8m_{61}' - 100y_1 - D_1' \leq -100$$

Recall that $D_1' = 4$. Since y_1 must be activated for any x_{ij} arc to be turned on, the first three inequalities reduce to $4m_{41}' \leq 4$, $6m_{51}' \leq 4$, $8m_{61}' \leq 4$. The only m_{ij}' arc that can be activated without breaking the three constraints above is m_{41}', which is the minimum-distance arc.

As indicated previously, once the proper m_{ij}' variable is identified the appropriate locational-benefit is activated, and the efficiency score can subsequently be computed. The locational part of the efficiency score can be interpreted as the ability of a minimum-distance arc to provide separation. This is in comparison to the other arcs that can serve that demand node. The scores from each facility are now 1.000 for facility 1. This is compared with 1.000 for both facilities 1 and 3 in the standard DEA model as shown in the "Prescriptive Tools" chapter of Chan (2000). Note the inclusion of locational benefit into the DEA formulation has 'improved' the first facility's DEA score vis-a-vis facility 3. This makes sense, since site 1 was optimal from a location point of view.

3.2.3.2 TWO-FACILITY CASE. Very few changes need to be made to implement the two-facility combined model. Obviously, the first change that must be made is to allow a pair of facilities—instead of one facility—to be chosen. This is done by the following constraint: $y_1 + y_2 + y_3 = 2$. In addition, three of the location constraints must be revised. Since this is a two-facility case, the second facility chosen must serve at least one of the demands. This is accomplished by changing the value of $3y_j$ to $2y_j$ for each of the following constraints. The constraints specify that demand i cannot be served unless facility j is in place:

$$x_{41} + x_{51} + x_{61} - 2y_1 \leq 0 \qquad x_{42} + x_{52} + x_{62} - 2y_2 \leq 0 \qquad x_{43} + x_{53} + x_{63} - 2y_3 \leq 0.$$

All other variables and constraints in the formulation remain the same as the individual DEA and location models. The only other changes that must be made to develop the two-facility max-min model is to allow two facilities to be considered and to ensure that one of the facilities cannot serve all of the demand. These alterations have already been discussed extensively in the "Facility Location" chapter and will not be discussed further.

3.2.3.3 MATHEMATICAL FORMULATION. Define M'' as the maximum score (taken to be 100 in the examples above) and ε as the minimum (taken to be 0.0001), the max-min one-facility combined-model looks like:

$$\max \sum_{j=1}^{|J|} x_j \tag{3-7}$$

$$s.t.$$

$$x_j \leq M'' \qquad \forall j$$

$$\sum_{j \in J} y_j = 1$$

$$x_j - \sum_{k=1}^{K-1} a_{kj} b_{kj} - \sum_{i \in I} D_{ij} b_{Kij} \geq 0 \qquad \forall i$$

$$x_j - \sum_{k=1}^{K-1} a_{kj} b_{kj} - \sum_{i \in I} D_{ij} b_{Kij} + M'' y_j \leq M'' \qquad \forall j \tag{3-8}$$

$$- \sum_{k=1}^{K-1} a_{kl} b_{kj} - \sum_{i \in I} D_{il} b_{Kil} + \sum_{h=1}^{H'} g_{hl} c_{hj} \geq 0 \qquad \forall j, l.$$

Here a and g are benefit and cost measures (for each facility j) in the DEA model. The weight-forcing constraints for non-distance factors appear as:

$$\sum_{h=1}^{H'} g_{hj} c_{hj} - y_j = 0 \qquad \forall j \tag{3-9}$$

which scale the denominator.

Here are the next two groups of constraints for the triggering mechanism:

$$
\begin{aligned}
b_{kj} - M'' y_j &\le 0 && \forall j;\ k=1,\ldots,K-1 \\
c_{hj} - M'' y_j &\le 0 && \forall j;\ h=1,\ldots,H' \\
\varepsilon - b_{kj} &\le M''(1-y_j) && \forall j;\ k=1,\ldots,K-1 \\
\varepsilon - c_{hj} &\le M''(1-y_j) && \forall j;\ h=1,\ldots,H'
\end{aligned}
\tag{3-10}
$$

They constitute the weight-forcing constraints for non-distance factors.

Introducing the binary 'marker' variable $m'_{ij} = \{0,1\}$ to 'register' the locational benefit of alternative j in serving demand i (if any), we have:

$$
\begin{aligned}
m'_{ij} - x_{ij} &\le 0 && \forall i,j \\
-m'_{ij} + x_{ij} &\ge 0 && \forall i,j \\
b_{Kij} - M'' m'_{ij} &\le 0 && \forall i,j \\
-b_{Kij} + M'' m'_{ij} &\le (1-\varepsilon) && \forall i,j \\
\sum_{j=1}^{|J|} x_{ij} &= 1 && \forall i \\
m'_{ij} \in \{0,1\},\ x_{ij} &\in \{0,1\}\ \ and\ \ y_j \in \{0,1\}
\end{aligned}
\tag{3-11}
$$

Demand cannot be served unless facility is in place

$$\sum_{i \in I} x_{ij} - |I| y_j \le 0 \qquad \forall j \tag{3-12}$$

The remaining obnoxious-facility location-constraints ensure the minimum distance is maximized:

$$
\begin{aligned}
\sum_{j=1}^{P} D_{ij} x_{ji} &\le D_{ij} y_j + (1-y_j)M'' && \forall i,j \\
D'_j + (M'-D_{ij}) &\le M'' && \forall i,j \\
D_{ij} m'_{ij} - M'' y_j - D'_j &\le -M'' && \forall i,j \\
\sum_{i \in I} m'_{ij} - y_j &\le 0 && \forall j \\
\sum_{i \in I} m'_{ij} - y_j &\ge 0 && \forall j.
\end{aligned}
\tag{3-13}
$$

For a p-facility model we replace the appropriate equations with

$$\sum_{j\in J} y_j = p$$

$$\sum_{i\in I} x_{ij} - (|I| - |J| + 1) y_j \le 0 \qquad \forall j. \tag{3-14}$$

The number of variables required to build the model can be determined by $2|J||I|$ $+ 3|J| + |J|(K-1) + |J| + |J|H'$. The number of constraints is $5p + 2|J|H' + 2|J|(K-1) + 5|J||I| + 1$.

3.2.4 Discussion. It is interesting to note that the binary variables y_j and m'_{ij} are here to effect selection between a DEA and max-min location-submodel, with y_j serving as the variable that links the two submodels. Once these binary variables are set, the two submodels are effectively unbundled. The resulting relaxed DEA LP at optimality assumes the form of the multi-alternative DEA model as discussed in the "Prescriptive tools" chapter of Chan (2000). The 'floating' variable x used to account for the multiple alternatives also becomes dis-functional. The resulting unbundled models are traditional DEA models for each decision-making unit (DMU). Furthermore, it carries with it the duality conditions of the DEA model. Since the y_j variables are now fixed, the remaining DEA model simply scales the cost and benefit weights b and c in the fashion of the traditional DEA model, with the models for each of the three DMUs laid side-by-side. Here is the primal and dual model for the DMUs (or facility) respectively for a targeted facility (j) (Charnes et al. 1978):

$$\max \sum_{k=1}^{K+|I|-1} b_k h_{k(j)}$$

s.t.

$$\sum_{h=1}^{H'} c_h g_{h(j)} = 1 \tag{3-15}$$

$$\sum_{k=1}^{K+|I|-1} b_k a'_{kj} - \sum_{h=1}^{H'} c_h g_{h(j)} \le 0 \qquad j = 1, \ldots, |J|$$

$$b_k \ge \varepsilon \qquad k = 1, \ldots, K+|I|-1$$

$$c_h \ge \varepsilon \qquad h = 1, \ldots, H'$$

and

$$\min \quad z(j) \; - \; \varepsilon \sum_{k=1}^{K+|I|-1} u_k - \varepsilon \sum_{h=1}^{H'} v_h$$

s.t.

$$\sum_{j=1}^{|J|} a'_{kj}\lambda_j - u_k = h_{k(j)} \qquad k = 1, \ldots, K+|I|-1 \tag{3-16}$$

$$z(j) g_{h(j)} - v_h - \sum_{j=1}^{|J|} g_{hj}\lambda_j = 0 \qquad h = 1, \ldots, H$$

$$\lambda_j, u_h, v_k \ge 0 \qquad \forall \; j, h \text{ and } k$$

where a'_{kj} incorporate both non-spatial and spatial output (benefit) measures a_{kj} and D_{ji} in Equation (3-8).

In matrix notation, the Charnes/Cooper/Rhodes model looks like:

$$\begin{array}{ll} \max_{b,c} \ b^T h & \min_{\theta,\lambda,u,v} \ z(j) - \varepsilon^T(u+v) \\[4pt] c^T g = 1 & A\lambda - u = h \\[4pt] b^T \overline{A} - c^T G' \le 0^T & z(j)g - G'\lambda - v = 0^T \\[4pt] b \ge \varepsilon, \ c \ge \varepsilon & \lambda \ge 0, u \ge 0, v \ge 0 \end{array} \qquad (3\text{-}17)$$

Here, the $K+|I|-1$ by $|J|$ matrix of output measures (including the locational benefit of locating at a site) is denoted by $\overline{A}$. The H' by $|J|$ matrix of input measures is denoted by $\mathbf{G'}$. Further, $g_{h(j)}$ denotes the amount consumed of the hth input measure and $h_{k(j)}$ denotes the amount produced of the kth input measure by the jth DMU. Finally, $a(j)$ and $g(j)$ denote, respectively, the vectors of output and input measures for the jth DMU, and $\mathbf{a} = [a(1), ... , a(|J|)]^T$ and $\mathbf{g} = [g(1), ... , g(|J|)]^T$. ε is the vector consisting of the same entry of an arbitrarily small number, and $\mathbf{u}, \mathbf{v}$ are surplus variables.

The dual variables in vector λ are of particular interest. It represents the dual price or weight placed on a DMU to form weighted average of two alternatives on the efficient frontier. θ is a scalar in the dual model, allowing us to see that the conditions for a DMU to be Pareto efficient are $z^*(j) = 1$, and $\mathbf{u} = \mathbf{v} = 0$. If a DMU is not efficient, $z(j) < 1$ and/or the surplus variables are positive. The constraints in the dual model imply that by increasing $h_{k(j)}$ by u_k and decreasing $g_{h(j)}$ by $(1-z^*(j))g_{h(j)} + v_h^*$, the associated DMU becomes efficient. Viewed in this light, the full power of duality interpretation applies in spite of the additional locational binary variable m'_{ij} beyond the selection variables y_j identified in the multi-alternative DEA model.

In the example above, the nonzero dual-variables are found in the constraints that deal with scaling and ranging the DEA scores. These occur in the first part of the model, namely the third through sixth group of constraints. Facility 1 is the chosen facility in this example. This result can be verified by the corresponding tight constraints or nonzero dual variables in the third, fourth, and sixth groups. If we pick out the output and input for these facilities using the absolute value of the dual price as the weight, we can see why facility 1 is most efficient. Thus taking the nonzero dual-prices for the two efficient facilities 1 and 2, we perform these computations to move along the production possibility frontier of the isoquant.

Averaged output vector:
$1.00000 \ (9, 4, 16, 4, 6, 8)^T + 1.00000 \ (5, 7, 10, 10, 7, 6)^T = (14, 11, 26, 14, 13, 14)^T$

Averaged input vector:
$1.00000 \ (5, 14)^T + 1.00000 \ (8, 15)^T = (13, 29)^T.$

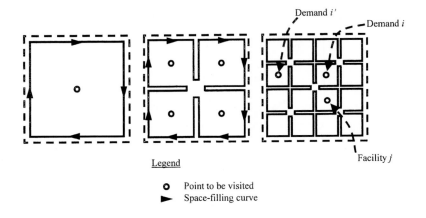

Figure 3-3 Space-filling curve 'visits' (encircles) all points in unit square

These are compared with the output vector of $(5, 7, 10, 10, 7, 6)^T$ and input vector of $(8, 15)^T$ for facility 3. One can see that facility 3 is clearly dominated by 1 and 2. This simple example shows how 'distance' from an efficient frontier can be considered along side of spatial separation in site evaluation.

3.3 Space-filling curves

The measurement of spatial separation invariably involves the transformation of two or more dimensions of spatial and non-spatial attributes into a composite metric. Thus in the l_p-norm, *"the x-y-z axes"* are collapsed into a contour of certain magnitude. In the case of land use, travel time, distance, cost and the like are aggregated into a single spatial price. Along this line of thinking, we would like to introduce an aggregation scheme that is particularly suited for facility-location and routing computations: The Space Filling Curve (SFC) is more than just another method to transform multidimensional spatial data into a single dimension ofseparation. It has the distinct advantage of being a location-and-routing solution technique of its own right. To the extent that we are discussing normally the spatial separation between a facility and a demand point in this chapter, SFC is a cogent technique.

   3.3.1 Algorithm. The SFC dates back as far as 1890. It is a part of the family of fractal curves (Bartholdi and Platzman 1988). A SFC is capable of mapping any number of dimensions to any other. For our discussion, only two- and three-dimensional mapping will be covered, particularly in their transformation to one-dimension. As **Figure 3-3** shows, the SFC is simply a line joining all the points

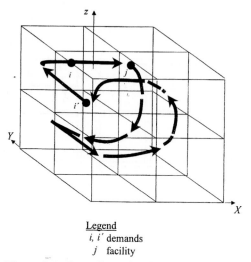

Legend
i, i' demands
j   facility

Figure 3-4 Three-dimensional space-filling-curve

within a space. The two-dimensional case here illustrates that the square is divided into four quarters, each a square in itself. As the curve passes through each quarter it assigns a 'psi' value Ψ, sometimes called the Sierpinski-curve function $f(d_1, d_2)$, to all demand points within the quarter. In order to distinguish among points within the quarter, each quarter is broken to quarters again, allowing the curve to assign up to 16 different psi values. This process of individual squares breaking down into four more squares continues until each demand point is assigned its own unique value of Ψ. It follows that points close to each other in the curve, such as between i and j, should be close on the plane as one can verify visually. Conversely, if two points i'-j are close in the plane, then they are only *likely* to be close on the curve. Again, this is shown in **Figure 3-3** The title SFC now seems logical since the curve continues to fill the square until every demand point—no matter where it is in the nit square—is visited.

The three-dimensional curve would operate much the same way except it would have eight separate cubes to travel through. Once again, each of the eight cubes is divided into eight more, all of which are visited. This process, illustrated in **Figure 3-4**, continues until each point 'owns' its own cube. Notice the unit square or unit cube is adopted for convenience. In practice, any shape in two dimensions or three dimensions and higher can be `molded' into a normalized square, cube and so on, as long as the appropriate bookkeeping scheme is used to identify where the actual points are within the square or cube. In the same spirit, a unit interval for Ψ is also adopted for convenience only. In actuality, the point Ψ =0 "wraps around" to become Ψ = 1 again in a continuous fashion. In this regard, the Sierpinski curve is 'circular'—it has no beginning or end. The generic SFC heuristic works as follows:

i	Hospital	X_i Latitude	Y_i Longitude	Z_i Patients	Ψ
1	Charlotte	35.21	80.44	0	0.03125
2	Ft Gordon	33.37	81.97	39	0.8125
3	Ft Bragg	35.17	79.02	234	0.8594
4	Ft Jackson	33.94	81.12	44	0.9063
5	Charleston SC	32.90	80.04	29	0.9531

Figure 3-5 Charlotte hub of the medical evacuation problem

Step 1 Transform the problem in the unit square or cube, via an SFC, to a problem on the unit interval. In other words, given the $|I|$ coordinates (X_i, Y_i, Z_i) of the nodes (demand points), compute the Sierpinski number $\Psi(X_i, Y_i, Z_i)$ for each node.

Step 2 Solve the easier location or/and routing problem on the unit interval $0 \le \Psi \le 1$. ∎

3.3.2 Applications. SFC is particularly adept in measuring spatial separation. It is common to apply the SFC to routing problems, such as the *travelling-salesman problem* (TSP), defined as the tour that visits all demand points from a depot and returns in the shortest length. Such a tour can be approximated by the Sierpinski curve by sorting the numbers Ψ in ascending order and visit the nodes in the same order producing a *tour*. An example is shown in **Figure 3-5** for a medical evacuation problem from the U.S. Department of Defense, where the wounded soldiers are brought back to the continental United States for medical care in hospitals around Charlotte, North Carolina. The three dimensions here are the latitude, longitude and 'demands' (which correspond to the number of patients delivered to the hospital under consideration). The cluster of points on the Sierpinski curve, in this case Ft. Gordon, Ft. Bragg, Ft. Jackson and Charleston, forms the tour. The cluster approximates the spatial separations between the Charlotte 'depot' and the four 'demand' points. In other words, the tour as computed by SFC consists of Charlotte–Ft. Gordon–Ft. Bragg–Ft. Jackson–Charleston –Charlotte. Notice that the Sierpinski curve "wraps around" from $\Psi = 0$ to $\Psi = 1$. The points on the tour are closer to one another than other points, including the depot, thus forming a logical cluster. In closing, the three-dimensional SFC solves the *weighted* TSP, where the distance to a demand point is weighted by the number of patients to be delivered to the point. An example executable code (or the BASIC program) by Carter (1990) is posted at my web site, including the data-sets. The software is also distributed in Chan (2000) as a CD.

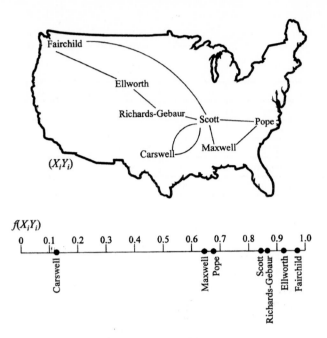

Figure 3-6 The multiple-space-filling-curve heuristic

   If two clusters of points are discerned instead of one, two salesman-tours are
suggested. Here, the demands are to be covered between the two salespersons, as we
will demonstrate below. In this case, the spatial separation between depot and
demands will be shorter than the one-vehicle (one-tour) case on the average, since
each of the two tours will be shorter than a single tour to cover the same demand
points. For the *multiple-travelling-salesmen problem* (MTSP), it is possible to
solvethese problems by partitioning routing tasks among the salespersons. Demand
 nodes are assigned to a salesperson, who in turn performs a TSP on his/her
assigned nodes. An example was mentioned (Batholdi and Platzman 1988) in
which demand nodes are uniformly distributed among a square. By partitioning the
TSP heuristic into segments consisting of equal number of nodes, a solution is
obtained corresponding to segments of roughly equal lengths. Extending the SFC
heuristic to handle the multiple-salesmen problem through the preservation-of-
nearness of data points for each salesperson, we obtained a solution to the MTSP.
Thus in **Figure 3-6**, we discern three clusters of demand points on the space-filling
curve, Carswell, Maxwell/Pope, and Richard-Gebaur/Ellsworth/Fairchild. A flight-
inspection aircraft fleet based in Scott Air Force Base (AFB) will send one aircraft
to cover the requests from Carswell AFB, a second aircraft to cover Maxwell and
Pope, and a third aircraft to cover Richard-Gebauer, Ellsworth, and Fairchild.
   In general, a set of $|I|$ nodes projected into a square, cube or hypercube of area
or volume A can be mapped onto an SFC. In a square, it will always result in a

curve no longer than 25 percent than an optimal Euclidean tour in the expected case among the same points, as the number of randomly distributed points becomes large. Furthermore, the tour length will be no greater than $2\sqrt{|I|A}$, which is also the limit for the *travelling-salesman tour* (TST). For a unitary square this reduces to $2\sqrt{|I|}$. The ratio of heuristic tour-length to theoretical is $O(\log|I|)$. The algorithm is extremely fast, consisting of essentially sorting. Its computational complexity is $O(|I|\log|I|)$ in the worst case, and $O(|I|)$ in the expected case.[5] In the terms of Bartholdi and Platzman (1988), the heuristic is *abstemious* in its data requirements. Only the $O(|I|)$ coordinates of the points to be visited are necessary, while the $O(|I|^2)$ distances between points are ignored. In other words, information on the separation metric and the distribution from which the points are drawn is not required. The algorithm is *agile* in that demand points may be inserted into or deleted from the tour and the solution can be updated within $O(\log|I|)$ steps. Finally, the computer coding is trivial as shown in a sample program as mentioned. The technique is also ideally suited for a geographic information system, which hasall the data required to generate such routings very quickly. We will come back to these points in a case study later.

3.4 Routing analysis

The most precise measure of spatial distance is determined through routing models. A routing model constructs the path from i to j in accordance with the specification desired. For example, routing analysis considers routing of populations to and from a service facility, as with visits to a store, library or school; routing of services to populations, as in the routing of emergency medical and fire services (Berlin et al. 1976, Toregas et al. 1972); or routing between a series of pick-up and drop-off points, as with the routing of garbage trucks and airline travel (Phillips et al. 1982, Beltrami et al. 1971, and Chan 1979). Golden (1984) and Bodin (1990) gave excellent reviews in vehicle routing. The basic routing problem can be stated as follows: Given a set of nodes and/or arcs, construct a feasible set of routes to optimize some objective, such as minimizing the cost of travelling from origins m to destinations n via a number of intermediate nodes, allowing for the case where m and n are the same node. A feasible route may be required, for example, to visit a sequence of locations to provide a specific service at these locations.

3.4.1 Shortest route. Finding the shortest route from origin m ($i = m$) to destination n ($j = n$) in the shortest distance (with $m \neq n$) is the celebrated shortest-path problem

$$\min \sum_{i \in I} \sum_{j \in I} d_{ij} x_{ij}. \tag{3-18}$$

---

[5]Computational complexity measures the efficiency of an algorithm. See an explanation in the Appendix on "Optimization procedures."

The convention throughout all routing discussion below is that $i \neq j$. To specify that the origin is m and destination is n, we write

$$\sum_{j \varepsilon M_n} x_{mj} = 1 \quad and \quad \sum_{i \varepsilon M_m} x_{in} = 1,$$

(3-19)

where M_j is the set of nodes that have arcs terminating at node j, and $\bar{M}_j$ is the set of nodes that have arcs originating from node j. The model then seeks a continuous path made up of a subset of the remaining nodes:

$$\sum_{i \varepsilon M_j} x_{ij} - \sum_{i \varepsilon \bar{M}_j} x_{ji} = 0 \quad \forall j \varepsilon I, \ j \neq m,n$$

(3-20)

$$x_{ij} = \begin{cases} 1 & if \ arc \ (i,j) \ is \ in \ the \ path \\ 0 & otherwise \ . \end{cases}$$

The solution of this binary 0-1 program for decision-variable x_{ij} would yield the desired set of arcs that link m to n.

It is often of interest to compute the kth shortest paths (Dreyfus 1969), where k can denote the *second* shortest path ($k = 2$), the *third* shortest path ($k = 3$) and so on. Finding the kth shortest time/distance paths is an important part of routing. Another variant of the shortest-path problem is where lengths of the arcs are time dependent. If t is the time of departure from m for n, $d^{mn}(t)$ represents the path length. Such a formulation is motivated by such applications as peak-hour travel where it takes longer than off-peak travel. Often, we are also interested in finding paths that include some specified intermediate nodes. For example, one may want to find the shortest path between nodes 1 and n that passes through $k-1$ 'specified' nodes 2, 3, ... , $k \leq n-1$. It can be seen that if $k = n-1$ and the last node n is the same as node 1, it becomes a travelling-salesman problem.

Example

An example of the shortest-route problem is shown in **Figure 3-7**, with the link travel times as given (Moore and Chan 1990). The objective is to construct the minimum path from origin m to destination n:

$$min \ 2x_{m1} + 2.6x_{m2} + 2.1x_{13} + 3x_{14} + 2.2x_{24} + 2.4x_{3n} + 2.1x_{4n}$$

subject to

$x_{m1} + x_{m2} = 1.0$	(specifies arcs leaving origin)
$x_{4n} + x_{3n} = 1.0$	(specifies arcs entering destination)
$x_{m1} - x_{13} - x_{14} = 0$	(node 1: arc in and arcs out)
$x_{m2} - x_{24} = 0$	(node 2: arc in and arc out)
$x_{13} - x_{3n} = 0$	(node 3: arc in and arc out)
$x_{24} + x_{14} - x_{4n} = 0$	(node 4: arcs in and arc out)
$x_{ij} \in \{0, 1\}.$	

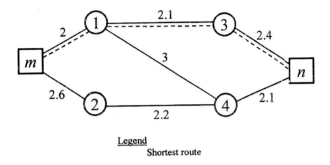

Legend

Shortest route

Figure 3-7 Shortest-route example

The solution is overlaid on **Figure 3-7**, where arcs x_{m1}, x_{13}, and x_{3n} make up the shortest route from origin m to destination n. The objective function was minimized at 6.5. Notice this way of finding the shortest route is for illustrative and pedagogic purposes only, it is not performed in actual applications. The same comments will apply throughout the numerical examples used for the routing models discussed in the current section. Starting with the section entitled "*Heuristic solutions for the travelling-salesman problem*," we will introduce more practical algorithms. □

3.4.2 Travelling-salesman problem. If the origin is the same as the destination ($m = n = 1$), the path is called a tour with the origin/destination as the home base or the depot. Such a tour is often referred to as a *Hamiltonian circuit* or cycle. Generally, one wants to find the shortest tour that visits all the nodes. This is formally called the TSP in the literature. The formulation includes Expressions **(3-18)** through **(3-20)** (if one reads $m = n = 1$), and the addition of a 'subtour-breaking' constraint

$$\sum_{i \in J} \sum_{j \in I-J} x_{ij} \geq 1 \qquad (3\text{-}21)$$

for every non-empty proper subset J of I. This constraint ensures that every node is visited in a depot-based continuous tour.

Although Equation **(3-21)** is easy to state, there are more efficient subtour-breaking constraints:

$$\sum_{i \in L} \sum_{j \in L} x_{ij} \leq |L| - 1 \qquad \forall \text{ nonempty subset } L \text{ of } I\text{-}1 \qquad (3\text{-}22)$$

or

$$\delta_i - \delta_j + |I| x_{ij} \leq |I| - 1 \qquad for \ 2 \leq i \neq j \leq |I| \qquad (3\text{-}23)$$

where L is the set of demand nodes, and δ_i are real numbers that record the number of 'legs' of the trip. In subtour-breaking constraint such as Equation **(3-23)**, we break up Equation **(3-20)** into two parts:

$$\sum_{i \in M_j} x_{ij} = 1 \qquad \sum_{i \in \overline{M}_j} x_{ji} = 1 \qquad \forall j \varepsilon I,\ j \neq m, n. \tag{3-24}$$

If the distance matrix $[d_{ij}]$ is symmetrical, the TST does not depend on the direction of travel and every node has two arcs incident to it. For asymmetrical distance matrix, however, there is one arc entering a node and another out of a node specifically. The former is called a *symmetrical* TSP, and the latter an *asymmetrical* TSP. Computationally speaking, symmetrical TSPs are much easier to solve.

The multiple travelling-salesman problem (MTSP), which we sometimes shortened as travelling-sales*men* problem, is a generalization of the travelling-salesman problem where more than one tour is used to cover all the nodes, all based at the depot. For the case of G tours, we modify Equation **(3-20)** as

$$\sum_{j \in I} x_{ij} = \begin{cases} G & \text{if } i = 1 \\ 1 & \text{if } i = 2, 3, \ldots, |I| \end{cases} \qquad i \neq j\ ; \tag{3-25}$$

$$\sum_{i \in I} x_{ij} = \begin{cases} G & \text{if } j = 1 \\ 1 & \text{if } j = 2, 3, \ldots, |I| \end{cases} \qquad i \neq j \tag{3-26}$$

which specify that each node is visited only once except for the depot, where G tours are based. The constraints **(3-21)** through **(3-23)** remain applicable to break subtours that do not include the depot node 1.

Single-Salesman example

An example of the TSP is shown in **Figure 3-8**, where node 1 is the home depot (Moore and Chan 1990). The objective of the TSP is to find the shortest tour that visits all the nodes:

$$\min\ 3x_{12} + 3.5x_{13} + 2x_{14} + 3x_{21} + 1.8x_{23} + 3.1x_{24} +$$
$$3.5x_{31} + 1.8x_{32} + 2.9x_{34} + 2x_{41} + 3.1x_{42} + 2.9x_{43}$$

subject to:

$$
\begin{array}{ll}
x_{12} + x_{13} + x_{14} = 1 & \text{(arcs out of depot)} \\
x_{21} + x_{31} + x_{41} = 1 & \text{(arcs going into depot)} \\
x_{12} + x_{32} + x_{42} - x_{23} - x_{24} - x_{21} = 0 & \text{(node 2: arcs in/out)} \\
x_{23} + x_{13} + x_{43} - x_{31} - x_{32} - x_{34} = 0 & \text{(node 3: arcs in/out)} \\
x_{14} + x_{24} + x_{34} - x_{41} - x_{42} - x_{43} = 0 & \text{(node 4: arcs in/out).}
\end{array}
$$

Now comes the most intricate part of the TSP: the subtour-breaking constraints. We will illustrate all the common ones, starting from the most straightforward to the most compact:

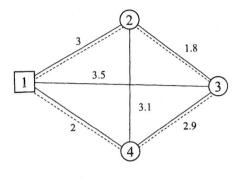

Legend
------- Optimal tour

Figure 3-8 Travelling-salesman-problem example

Equation **(3-21)**:

$$x_{12} + x_{13} + x_{14} \geq 1 \qquad\qquad J = 1$$
$$x_{21} + x_{31} + x_{41} \geq 1 \qquad\qquad J = 2, 3, 4$$
$$x_{12} + x_{32} + x_{42} \geq 1 \qquad\qquad J = 1, 3, 4$$
$$x_{21} + x_{23} + x_{24} \geq 1 \qquad\qquad J = 2$$
$$x_{13} + x_{23} + x_{43} \geq 1 \qquad\qquad J = 1, 2, 4$$
$$x_{31} + x_{32} + x_{34} \geq 1 \qquad\qquad J = 3$$
$$x_{14} + x_{24} + x_{34} \geq 1 \qquad\qquad J = 1, 2, 3$$
$$x_{41} + x_{42} + x_{43} \geq 1 \qquad\qquad J = 4$$
$$x_{13} + x_{14} + x_{23} + x_{24} \geq 1 \qquad\qquad J = 1, 2$$
$$x_{31} + x_{32} + x_{41} + x_{42} \geq 1 \qquad\qquad J = 3, 4$$
$$x_{12} + x_{13} + x_{42} + x_{43} \geq 1 \qquad\qquad J = 1, 4$$
$$x_{21} + x_{31} + x_{24} + x_{34} \geq 1 \qquad\qquad J = 2, 3$$
$$x_{12} + x_{32} + x_{14} + x_{34} \geq 1 \qquad\qquad J = 1, 3$$
$$x_{21} + x_{23} + x_{41} + x_{43} \geq 1 \qquad\qquad J = 2, 4.$$

Equation **(3-22)**:

$$x_{23} + x_{32} \leq 2 - 1 = 1 \qquad\qquad L = 2, 3$$
$$x_{24} + x_{42} \leq 2 - 1 = 1 \qquad\qquad L = 2, 4$$
$$x_{34} + x_{43} \leq 2 - 1 = 1 \qquad\qquad L = 3, 4$$
$$x_{23} + x_{24} + x_{32} + x_{34} + x_{42} + x_{43} \leq 3 - 1 = 2 \qquad L = 2, 3, 4$$

Equation **(3-23)**:

$$\delta_2 - \delta_3 + 4x_{23} \leq 3 \qquad \delta_3 - \delta_2 + 4x_{32} \leq 3 \qquad \delta_4 - \delta_2 + 4x_{42} \leq 3$$
$$\delta_2 - \delta_4 + 4x_{24} \leq 3 \qquad \delta_3 - \delta_4 + 4x_{34} \leq 3 \qquad \delta_4 - \delta_3 + 4x_{43} \leq 3.$$

With the last two subtour-breaking constraints, modifications must be made to the other constraints as follows. Constraint **(3-19)** remains the same and **(3-20)** is eliminated, but modified constraints are used to develop equations for the demand locations:

$$x_{12} + x_{32} + x_{42} = 1 \text{ (arcs into 2)} \qquad x_{21} + x_{23} + x_{24} = 1 \text{ (arcs out of 2)}$$
$$x_{13} + x_{23} + x_{43} = 1 \text{ (arcs into 3)} \qquad x_{31} + x_{32} + x_{34} = 1 \text{ (arcs out of 3)}$$
$$x_{14} + x_{24} + x_{34} = 1 \text{ (arcs into 4)} \qquad x_{41} + x_{42} + x_{43} = 1 \text{ (arcs out of 4)}.$$

All three subtour-breaking constraints yield the same solution, as expected. The TST starts at the depot (node 1), visits nodes 2, 3, and 4 sequentially before returning home. The tour, at total cost 9.7, is overlaid on **Figure 3-8**. Notice this is a symmetrical TSP, even though the formulation above treats it as an asymmetrical model. The symmetry can be exploited to reduce the number of equations written. We will illustrate this in the "Locating-routing" chapter under the "Facility/tour/allocation" section. □

Multiple-Salesmen example

The multiple-TSP uses more than one tour to cover all the nodes. All the tours are based at a depot. In the example of **Figure 3-9**, the objective is to find the minimum cost for two tours based out of depot node 1 (Moore and Chan 1990). The only changes from the classic TSP are the constraints that govern the depot *degrees* or the number of traversed arcs incident upon the depot:

$$x_{21} + x_{31} + x_{41} = 2 \text{ (arcs going into depot)}$$
$$x_{12} + x_{13} + x_{14} = 2 \text{ (arcs going out of depot)}$$

All the compact subtour-breaking constraints **(3-23)** apply as in the previous case. The minimum cost for two tours was 12.3 and consisted of one tour 1–2–3–1 and the other tour 1–4–1. Both are overlaid in **Figure 3-9**. The computer output shows a solution with a tour 1–3–4–1 and a tour 1–2–1 would increase the objective function to 16.5 (increase of 4.2). A solution consisting of 1–2–4–1 and 1–3–1 would increase the objective function to 16 (increase of 3.7). □

3.4.3 *The Vehicle-routing problem.*

If delivery requirements are placed upon the various demand points of a TSP, one ends up with a "vehicle-routing problem" (VRP). The multi-vehicle-type version can be stated as having a set of lowest-cost tours:

$$\min \sum_{i \in I} \sum_{j \in I} \sum_{h \in H''} d_{ij} x_{ij}^h \tag{3-27}$$

where H is the set of vehicle types ranging from $h = 1, 2, \dots |H''|$. The first two constraints ensure that each demand point is served by only one vehicle and that all vehicles are used:

$$\sum_{i \in I} \sum_{h \in H''} x_{ij}^h = \begin{cases} |H''| & \text{if } j = 1 \\ 1 & \text{if } j = 2, \dots, |I| \end{cases} \tag{3-28}$$

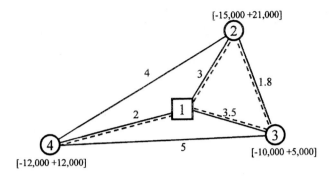

Figure 3-9 Example for the multiple-traveling-salesman problem

$$\sum_{j\in I}\sum_{h\in H''} x_{ij}^h = \begin{cases} |H''| & \text{if } i=1 \\ 1 & \text{if } i=2,\dots,|I|. \end{cases} \tag{3-29}$$

Route continuity is maintained for all vehicle types:

$$\sum_{i\in M_p} x_{ip}^h - \sum_{j\in M_p} x_{pj}^h = 0 \qquad \forall h, \ \forall p\varepsilon I. \tag{3-30}$$

Vehicle-capacity constraints are enforced for each vehicle type as it picks up traffic (or makes deliveries) along the tour:

$$\sum_{j\in I} f_j \sum_{i\in I} x_{ij}^h \le V(h) \qquad \forall h, \tag{3-31}$$

where V_h is the vehicle capacity. Constraints are placed upon the maximum 'time' $U(h)$ a vehicle h spends "on the road" (or sometime called the vehicle 'range' in the aeronautical profession):

$$\sum_{i\in I} t_i^h \sum_{j\in I} x_{ij}^h + \sum_{i\in I}\sum_{j\in I} d_{ij}^h x_{ij}^h \le U(h) \qquad \forall h \tag{3-32}$$

where t_i^h is the amount of time vehicle type h spends at demand point i and d_{ij}^h is now interpreted as the "link time" from i to j. The next two constraints guarantee that vehicle availability is not exceeded at depot 1 for vehicle h:

$$\sum_{j\in M_1} x_{1j}^h \le 1 \quad \forall h, \ \ and \ \ \sum_{i\in M_1} x_{i1}^h \le 1 \quad \forall h. \tag{3-33}$$

Connections (regardless of vehicle type) to constitute a tour are ensured by the subtour-breaking constraints **(3-21)** and **(3-23)**, with minor modifications such as the following to account for multiple vehicles

$$x_{ij} \rightarrow \sum_{h\in H''} x_{ij}^h .$$

(3-34)

It is interesting to observe that when vehicle-capacity constraints **(3-31)** and route-time constraint **(3-32)** are non-binding (and can be ignored), this model reduces to a multiple-travelling-salesman problem. This can be seen by eliminating constraints **(3-31)** and **(3-32)** and back-substituting Equation **(3-34)** in the objective functions and the remaining constraints. The model can easily be extended to include both pickup and delivery (Bauer et al. 1992). Let us say that the pickup at the first stop are required at the set of nodes $D'' = \{i\}$. Denote g_i' as the pickup requirement at node i. Then the vehicle-capacity constraint will mandate that the load picked-up, on top of the load yet to be delivered, is less than capacity.

$$\sum_{j\in I} g_i' x_{ij}^h + \sum_{j\in I-i} f_j \sum_{k\in D} x_{kj}^h \le V(h) \qquad \forall h, i \in D''.$$

(3-35)

The associated constraints to be written after picking up at the first two stops are

$$\sum_{k\in D-i} \sum_{j\in I} g_k' x_{kj}^h + \sum_{k\in D} f_i x_{ki}^h \le V(h) \qquad \forall h, i \in D''$$

(3-36)

and so on. Finally, a total pickup constraint, which is symmetrical with the delivery constraint, is added:

$$\sum_{i\in I} g_i' \sum_{j\in I} x_{ij}^h \le V(h) \qquad \forall h.$$

(3-37)

Minor alterations in the formulation of the basic vehicle-routing problem yield the multiple-depot model. Letting nodes 1, 2, ..., K denote the depots and $K+1$, $K+2$, ..., $|I|$ the demand nodes, we obtain the model by changing the index in constraints **(3-28)** and **(3-29)** to $j = K+1, ... , |I|$ and $i = K+1, ... , |I|$ for demand nodes, respectively, and by changing constraints **(3-33)** to

$$\sum_{i=1}^{K} \sum_{j=K+1}^{|I|} x_{ij}^h \le 1 \qquad \forall h$$

(3-38)

$$\sum_{p=1}^{K} \sum_{i=K+1}^{|I|} x_{ip}^h \le 1 \qquad \forall h .$$

(3-39)

Finally, we refine our subtour-breaking constraints such as **(3-21)** through **(3-23)** to read

$$\sum_{i \in J} \sum_{j \notin J} \sum_{h \in H^{\prime\prime}} x_{ij}^{h} \geq 1 \qquad \forall \ proper \ subset \ J \ of \ I$$

$$containing \ nodes \ \{1, ..., K\} \tag{3-40}$$

or

$$\sum_{i \in L} \sum_{j \in L} x_{ij}^{h} \leq |L| - 1 \qquad \forall \ nonempty \ subset \ L \ of$$

$$\{K+1, \ K+2, \ ...|I|\} \ ; \forall h \tag{3-41}$$

or

$$\delta_{i} - \delta_{j} + |I| \ x_{ij}^{h} \leq |I| - 1 \qquad for \ K+1 \leq i \neq j \leq |I| \ for$$

$$some \ real \ number \ \delta_{i}; \ \forall h. \tag{3-42}$$

Viewing the multiple-depot vehicle-routing problem (MDVRP) as the most general of routing problems, Kulkarni and Bhave (1985) gave a compact formulation for a single type of vehicle as follows. Expressions **(3-18)** through **(3-19)** and **(3-24)** remain the same. Then we supplement with these equations that guarantee all vehicles are used:

$$\sum_{i=1}^{K} \sum_{j \in I} x_{ij} = |H^{\prime\prime}| \qquad \sum_{j=1}^{K} \sum_{i \in I} x_{ij} = |H^{\prime\prime}| \ . \tag{3-43}$$

Three equations are written for node, load-capacity and route-time or vehicle-range constraints, respectively. The node constraint remains the same as **(3-42)**. The load-capacity $V(1)$ and subtour-breaking constraints are combined:

$$\pi_{i} - \pi_{j} + V(1)x_{ij} \leq V(1) - f_{i} \qquad \forall \ K+1 \leq i \neq j \leq |I| \tag{3-44}$$

where π_{i} are real numbers that can be thought of as the load carried on board. The vehicle-range and subtour-breaking constraint are also combined as

$$\sigma_{i} - \sigma_{j} + U(1)x_{ij} \leq U(1) - d_{ij} \qquad \forall \ K+1 \leq i \neq j \leq |I| \tag{3-45}$$

where σ_{i} are real numbers that can be thought of as 'odometer' readings. The principal advantage of this formulation is the reduction of the number of variables by a factor of $|H^{\prime\prime}|$. However, the number of non-assignment constraints increase. Kulkarni's method ignores distance to and from the depot. This problem was noted earlier by Brodie and Waters (1988) and Achutan and Caccetta (1991), but virtually no solution was offered thus far to redress the omission.[6] Clearly the σ associated with the first demand point on a tour should equal the distance to the home depot. Additionally, the distance from the last node prior to returning to the depot cannot

---

[6]Working independently, Desrochers and Laporte (1991) also corrected Kulkarni and Bhave's (1985) mistakes and constraints.

exceed the remaining range of the vehicle. The following equations, when used with Kulkarni's, sufficiently restrict range and complete the proposal originated by Kulkarni (Baker and Chan 1991):

$$d_{ij}x_{ij} \leq \sigma_j \qquad 1 \leq i \leq K, \ K+1 \leq j \leq |I| \tag{3-46}$$

Multiple frequency can be introduced, allowing a node to be visited more than once (Baker and Chan 1996).

$$d_{ji}x_{ji} + \sigma_j \leq U(1) \qquad 1 \leq i \leq K, \ K+1 \leq j \leq |I|. \tag{3-47}$$

A way to introduce multiple service-frequency is to split a node into a group of co-located nodes, between which no arcs exist. Thus a node i would become $i_1, ..., i_s$, where s is the prescribed frequency of visit. Such a scheme retains the binary valuation of variables x_{ij}. The reason is that each node i in the group would be visited only within the group, to a node outside the group, and back to a node within. Because these visits all occur on the same tour, however, such a scenario does not conform with the intent of forcing multiple visits by different tours (conceivably at different times.) To force all visits within a node group to occur via different tours, all nodes of a group are assigned only one 'odometer' variable σ. This prohibits a tour from serving any group of nodes more than once, effectively forcing as many tours as there are nodes into a given group. Thus Kulkarni's constraint now is modified sufficiently to introduce multiple service-frequency, as well as subtour-breaking and range-limitations. We will give a numerical example on multiple service-frequency in the "Location-routing" chapter under the "Depot location in commodity distribution" section.

Vehicle-routing example

Let us say there are two vehicles (C-130 Hercules transport planes) with a cargo capacity of 30 000 pounds (15 000 kg) each (Moore and Chan 1990). The crew-duty-day is 10 hours and the time spent loading at each station is 0.5 hour. We use the same network as the one used in multiple-TSP (**Figure 3-9**). Except for the extra superscript, the integer program looks similar to the TSP's:

minimize

$$3x_{12}^1 + 3.5x_{13}^1 + 2x_{14}^1 + 3x_{21}^1 + 1.8x_{23}^1 + 4x_{24}^1 + 3.5x_{31}^1 + 1.8x_{32}^1 + 5x_{34}^1 + 2x_{41}^1 + 4x_{42}^1 + 5x_{43}^1 +$$
$$3x_{12}^2 + 3.5x_{13}^2 + 2x_{14}^2 + 3x_{21}^2 + 1.8x_{23}^2 + 4x_{24}^2 + 3.5x_{31}^2 + 1.8x_{32}^2 + 5x_{34}^2 + 2x_{41}^2 + 4x_{42}^2 + 5x_{43}^2$$

subject to

$$x_{21}^1 + x_{31}^1 + x_{41}^1 + x_{21}^2 + x_{31}^2 + x_{41}^2 = 2$$
$$x_{12}^1 + x_{32}^1 + x_{42}^1 + x_{12}^2 + x_{32}^2 + x_{42}^2 = 1$$
$$x_{13}^1 + x_{23}^1 + x_{43}^1 + x_{13}^2 + x_{23}^2 + x_{43}^2 = 1$$
$$x_{14}^1 + x_{24}^1 + x_{34}^1 + x_{14}^2 + x_{24}^2 + x_{34}^2 = 1$$
$$x_{12}^1 + x_{13}^1 + x_{14}^1 + x_{12}^2 + x_{13}^2 + x_{14}^2 = 2$$
$$x_{21}^1 + x_{23}^1 + x_{24}^1 + x_{21}^2 + x_{23}^2 + x_{24}^2 = 1$$
$$x_{31}^1 + x_{32}^1 + x_{34}^1 + x_{31}^2 + x_{32}^2 + x_{34}^2 = 1$$

$$x_{12}^1 + x_{32}^1 + x_{42}^1 - x_{21}^1 - x_{23}^1 - x_{24}^1 = 0$$
$$x_{12}^2 + x_{32}^2 + x_{42}^2 - x_{21}^2 - x_{23}^2 - x_{24}^2 = 0$$
$$x_{13}^1 + x_{23}^1 + x_{43}^1 - x_{31}^1 - x_{32}^1 - x_{34}^1 = 0$$
$$x_{13}^2 + x_{23}^2 + x_{43}^2 - x_{31}^2 - x_{32}^2 - x_{34}^2 = 0$$
$$x_{14}^1 + x_{24}^1 + x_{34}^1 - x_{41}^1 - x_{42}^1 - x_{43}^1 = 0$$
$$x_{14}^2 + x_{24}^2 + x_{34}^2 - x_{41}^2 - x_{42}^2 - x_{43}^2 = 0.$$

$$x_{41}^1 + x_{42}^1 + x_{43}^1 + x_{41}^2 + x_{42}^2 + x_{43}^2 = 1.$$

Given a delivery requirement of 15 thousand at node 2, 10 thousand at node 3, and 12 at 4, the vehicle-capacity constraint now looks like

$$15{,}000(x_{12}^1 + x_{32}^1 + x_{42}^1) + 10{,}000(x_{13}^1 + x_{23}^1 + x_{43}^1) + 12{,}000(x_{14}^1 + x_{24}^1 + x_{34}^1) \le 30{,}000$$
(for vehicle 1)

$$15{,}000(x_{12}^2 + x_{32}^2 + x_{42}^2) + 10{,}000(x_{13}^2 + x_{23}^2 + x_{43}^2) + 12{,}000(x_{14}^2 + x_{24}^2 + x_{34}^2) \le 30{,}000$$
(for vehicle 2).

Notice that with these equations we model the delivery of cargo from one depot (node 1) to three bases. Here no consideration is made for carrying cargo back to the depot. The 'range' equation now looks like

$$0.5(x_{12}^1 + x_{13}^1 + x_{14}^1) + (3x_{12}^1 + 3.5x_{13}^1 + 2x_{14}^1) + 0.5(x_{21}^1 + x_{23}^1 + x_{24}^1) + (3x_{21}^1 + 1.8x_{23}^1 + 4x_{24}^1) +$$
$$0.5(x_{31}^1 + x_{32}^1 + x_{34}^1) + (3.5x_{31}^1 + 1.8x_{32}^1 + 5x_{34}^1) + 0.5(x_{41}^1 + x_{42}^1 + x_{43}^1) + (2x_{41}^1 + 4x_{42}^1 + 5x_{43}^1) \le 10$$
(for vehicle 1)

$$0.5(x_{12}^2 + x_{13}^2 + x_{14}^2) + (3x_{12}^2 + 3.5x_{13}^2 + 2x_{14}^2) + 0.5(x_{21}^2 + x_{23}^2 + x_{24}^2) + (3x_{21}^2 + 1.8x_{23}^2 + 4x_{24}^2) +$$
$$0.5(x_{31}^2 + x_{32}^2 + x_{34}^2) + (3.5x_{31}^2 + 1.8x_{32}^2 + 5x_{34}^2) + 0.5(x_{41}^2 + x_{42}^2 + x_{43}^2) + (2x_{41}^2 + 4x_{42}^2 + 5x_{43}^2) \le 10$$
(for vehicle 2).

Vehicle-availability constraints are

$$
\begin{aligned}
x_{12}^1 + x_{13}^1 + x_{14}^1 &\le 1 \qquad &\text{(vehicle 1)}\\
x_{12}^2 + x_{13}^2 + x_{14}^2 &\le 1 \qquad &\text{(vehicle 2)}\\
x_{21}^1 + x_{31}^1 + x_{41}^1 &\le 1 \qquad &\text{(vehicle 1)}\\
x_{21}^2 + x_{31}^2 + x_{41}^2 &\le 1 \qquad &\text{(vehicle 2)}.
\end{aligned}
$$

For the example here, a strict equality can be used instead of inequalities, given the requirement to use both aircraft in the fleet. The basic subtour-breaking constraint **(3-21)** now looks like

$$
\begin{aligned}
x_{12}^1 + x_{13}^1 + x_{14}^1 + x_{12}^2 + x_{13}^2 + x_{14}^2 &\ge 1 \qquad & J = 1\\
x_{21}^1 + x_{31}^1 + x_{41}^1 + x_{21}^2 + x_{31}^2 + x_{41}^2 &\ge 1 \qquad & J = 2, 3, 4\\
x_{12}^1 + x_{32}^1 + x_{42}^1 + x_{12}^2 + x_{32}^2 + x_{42}^2 &\ge 1 \qquad & J = 1, 3, 4\\
x_{21}^1 + x_{23}^1 + x_{24}^1 + x_{21}^2 + x_{23}^2 + x_{24}^2 &\ge 1 \qquad & J = 2\\
x_{13}^1 + x_{23}^1 + x_{43}^1 + x_{13}^2 + x_{23}^2 + x_{43}^2 &\ge 1 \qquad & J = 1, 2, 4\\
x_{31}^1 + x_{32}^1 + x_{34}^1 + x_{31}^2 + x_{32}^2 + x_{34}^2 &\ge 1 \qquad & J = 3\\
x_{14}^1 + x_{24}^1 + x_{34}^1 + x_{14}^2 + x_{24}^2 + x_{34}^2 &\ge 1 \qquad & J = 1, 2, 3\\
x_{41}^1 + x_{42}^1 + x_{43}^1 + x_{41}^2 + x_{42}^2 + x_{43}^2 &\ge 1 \qquad & J = 4\\
x_{13}^1 + x_{14}^1 + x_{23}^1 + x_{24}^1 + x_{13}^2 + x_{14}^2 + x_{23}^2 + x_{24}^2 &\ge 1 \qquad & J = 1, 2\\
x_{31}^1 + x_{32}^1 + x_{41}^1 + x_{42}^1 + x_{31}^2 + x_{32}^2 + x_{41}^2 + x_{42}^2 &\ge 1 \qquad & J = 3, 4\\
x_{12}^1 + x_{13}^1 + x_{42}^1 + x_{43}^1 + x_{12}^2 + x_{13}^2 + x_{42}^2 + x_{43}^2 &\ge 1 \qquad & J = 1, 4\\
x_{21}^1 + x_{31}^1 + x_{24}^1 + x_{34}^1 + x_{21}^2 + x_{31}^2 + x_{24}^2 + x_{34}^2 &\ge 1 \qquad & J = 2, 3\\
x_{12}^1 + x_{32}^1 + x_{14}^1 + x_{34}^1 + x_{12}^2 + x_{32}^2 + x_{14}^2 + x_{34}^2 &\ge 1 \qquad & J = 1, 3\\
x_{21}^1 + x_{23}^1 + x_{41}^1 + x_{43}^1 + x_{21}^2 + x_{23}^2 + x_{41}^2 + x_{43}^2 &\ge 1 \qquad & J = 2, 4.
\end{aligned}
$$

The solution is similar to the multiple-TSP shown in **Figure 3-9**. The minimum cost for two tours was 12.3 and consisted of one tour for vehicle 1 (1–2–3–1) and another tour for vehicle 2 (1–4–1). There is excess capacity for vehicle 1 of 5000 pounds (2500 kg) and for vehicle

2 of 18 000 (9000 kg). Vehicle 1 used most of the crew-day with 0.2 hour left while vehicle 2 had 5 hours to spare.

Bauer et al. (1992) included both pickup and delivery requirements to allow for re-supply of the depot. Pick-ups of 21,000(10,500), 5000(2500) and 12,000(6000) lbs(kg) are required at nodes 2, 3 and 4 respectively. This involves replacing the above vehicle-capacity constraints with these pickup constraints after the first stop:

$$21000(x_{21}^1+x_{23}^1+x_{24}^1)+10000(x_{23}^1+x_{43}^1)+12000(x_{24}^1+x_{34}^1) \leq 30000$$
(vehicle 1: first stop pick-up at node 2, deliveries at 3 and 4)

$$5000(x_{31}^1+x_{32}^1+x_{34}^1)+15000(x_{32}^1+x_{42}^1)+12000(x_{24}^1+x_{34}^1) \leq 30000$$
(vehicle 1: first stop pick-up at node 3, deliveries at 2 and 4)

$$12000(x_{41}^1+x_{42}^1+x_{43}^1)+15000(x_{32}^1+x_{42}^1)+10000(x_{23}^1+x_{43}^1) \leq 30000$$
(vehicle 1: first stop pick-up at node 4, deliveries at 2 and 3)

$$21000(x_{12}^2+x_{23}^2+x_{24}^2)+10000(x_{23}^2+x_{43}^2)+12000(x_{24}^2+x_{34}^2) \leq 30000$$
(vehicle 2: first pick-up stop at node 2, deliveries at 3 and 4)

$$5000(x_{31}^2+x_{32}^2+x_{34}^2)+15000(x_{32}^2+x_{42}^2)+12000(x_{24}^2+x_{34}^2) \leq 30000$$
(vehicle 2: first pick-up stop at node 3, deliveries at 2 and 4)

$$12\,000(x_{41}^2+x_{42}^2+x_{43}^2)+15000(x_{32}^2+x_{42}^2)+10000(x_{23}^2+x_{43}^2) \leq 30000$$
(vehicle 2: first pick-up stop at node 4, deliveries at 2 and 3).

The pickup constraints after the second stop are:

$$21000(x_{21}^1+x_{23}^1+x_{24}^1)+5000(x_{31}^1+x_{32}^1+x_{34}^1)+12000(x_{24}^1+x_{34}^1) \leq 30000$$
(vehicle 1: pickups at nodes 2 and 3, delivery at node 4)

$$5000(x_{31}^1+x_{32}^1+x_{34}^1)+12000(x_{41}^1+x_{42}^1+x_{43}^1)+15\,000(x_{32}^1+x_{42}^1) \leq 30000$$
(vehicle 1: pickups at nodes 3 and 4, delivery at node 2)

$$12000(x_{41}^1+x_{42}^1+x_{43}^1)+12000(x_{21}^1+x_{23}^1+x_{24}^1)+10000(x_{23}^1+x_{43}^1) \leq 30000$$
(vehicle 1: pickups at nodes 4 and 2, delivery at node 3)

$$21000(x_{21}^2+x_{23}^2+x_{24}^2)+5000(x_{31}^2+x_{32}^2+x_{34}^2)+12000(x_{24}^2+x_{34}^2) \leq 30000$$
(vehicle 2: pickups at nodes 2 and 3, delivery at node 4)

$$5000(x_{31}^2+x_{32}^2+x_{34}^2)+12000(x_{41}^2+x_{42}^2+x_{43}^2)+15000(x_{32}^2+x_{42}^2) \leq 30000$$
(vehicle 2: pickups at nodes 3 and 4, delivery at node 2)

$$12000(x_{41}^2+x_{42}^2+x_{43}^2)+12000(x_{21}^2+x_{23}^2+x_{24}^2)+10000(x_{23}^2+x_{43}^2) \leq 30000$$
(vehicle 2: pickups at nodes 4 and 2, delivery at node 3)

The total pickup constraints are:

$$21000(x_{21}^1+x_{23}^1+x_{24}^1)+5000(x_{31}^1+x_{32}^1+x_{34}^1)+12000(x_{41}^1+x_{42}^1+x_{43}^1) \leq 30000$$
(for vehicle 1)

$$21000(x_{21}^2+x_{23}^2+x_{24}^2)+5000(x_{31}^2+x_{32}^2+x_{34}^2)+12000(x_{41}^2+x_{42}^2+x_{43}^2) \leq 30000$$
(for vehicle 2).

The optimal tours are 1–3–2–1 for vehicle 1, and 1–4–1 for vehicle 2—the same as before.
□

Multiple-depot example

An example of the multiple-depot problem is shown in **Figure 3-10**. Depots are shown as nodes 1 and 2, while demands are shown as 3, 4, and 5 (Moore and Chan 1990). This problem has two aircraft: a C-130 with a cargo capacity of 30,000 pounds (15,000 kg) and a C-141 with a capacity of 60 000 pounds (30,000 kg). Again, the crew-duty-day is 12 hours and the time spent loading at each station is 0.5 hour. Instead of showing all the glorious details, we choose only to highlight the modifications to the basic formulation to accommodate multiple depots.

(in-degrees at depots and demand nodes)

$$x_{21}^1 + x_{31}^1 + x_{41}^1 + x_{51}^1 + x_{21}^2 + x_{31}^2 + x_{41}^2 + x_{51}^2 +$$
$$x_{12}^1 + x_{32}^1 + x_{42}^1 + x_{52}^1 + x_{12}^2 + x_{32}^2 + x_{42}^2 + x_{52}^2 = 2$$
$$x_{13}^1 + x_{23}^1 + x_{43}^1 + x_{53}^1 + x_{13}^2 + x_{23}^2 + x_{43}^2 + x_{53}^2 = 1$$
$$x_{14}^1 + x_{24}^1 + x_{34}^1 + x_{54}^1 + x_{14}^2 + x_{24}^2 + x_{34}^2 + x_{54}^2 = 1$$
$$x_{15}^1 + x_{25}^1 + x_{35}^1 + x_{45}^1 + x_{15}^2 + x_{25}^2 + x_{35}^2 + x_{45}^2 = 1.$$

(out-degrees at depots and demand nodes)

$$x_{12}^1 + x_{13}^1 + x_{14}^1 + x_{15}^1 + x_{12}^2 + x_{13}^2 + x_{14}^2 + x_{15}^2 +$$
$$x_{21}^1 + x_{23}^1 + x_{24}^1 + x_{25}^1 + x_{21}^2 + x_{23}^2 + x_{24}^2 + x_{25}^2 = 2$$
$$x_{31}^1 + x_{32}^1 + x_{34}^1 + x_{35}^1 + x_{31}^2 + x_{32}^2 + x_{34}^2 + x_{35}^2 = 1$$
$$x_{41}^1 + x_{42}^1 + x_{43}^1 + x_{45}^1 + x_{41}^2 + x_{42}^2 + x_{43}^2 + x_{45}^2 = 1$$
$$x_{51}^1 + x_{52}^1 + x_{53}^1 + x_{54}^1 + x_{51}^2 + x_{52}^2 + x_{53}^2 + x_{54}^2 = 1.$$

(tracking each vehicle based at depot 1 or 2 on its way out)

$$x_{13}^1 + x_{14}^1 + x_{15}^1 + x_{23}^1 + x_{24}^1 + x_{25}^1 \le 1$$
$$x_{13}^2 + x_{14}^2 + x_{15}^2 + x_{23}^2 + x_{24}^2 + x_{25}^2 \le 1.$$

(tracking each vehicle based at depot 1 or 2 on its way in)

$$x_{31}^1 + x_{41}^1 + x_{51}^1 + x_{32}^1 + x_{42}^1 + x_{52}^1 \le 1$$
$$x_{31}^2 + x_{41}^2 + x_{51}^2 + x_{32}^2 + x_{42}^2 + x_{52}^2 \le 1.$$

It is to be noted that when writing some of these constraints, redundant equations may be generated. Analysts may need to guard against this for efficiency. For a multiple-depot case such as this, the basic subtour-breaking constraint such as Equation **(3-21)** becomes explosive in dimension. It is much more compact to use constraint **(3-41)**, as shown below:

$$x_{34}^1 + x_{35}^1 + x_{43}^1 + x_{53}^1 + x_{45}^1 + x_{54}^1 \le 2 \quad \text{vehicle 1} \quad L = 3, 4, 5$$
$$x_{34}^2 + x_{35}^2 + x_{43}^2 + x_{53}^2 + x_{45}^2 + x_{54}^2 \le 2 \quad \text{vehicle 2} \quad L = 3, 4, 5$$

(For this problem, only one constraint is required for each vehicle since there are only 3 demand locations.) The solution to this multiple-depot problem is overlaid in **Figure 3-10**. The tour from depot one is 1–3–4–1, and the tour from depot 2 is 2–5–2. ☐

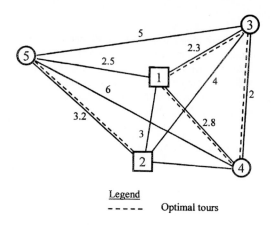

Legend

- - - - -   Optimal tours

Figure 3-10 Multiple-depot example

Extended Kulkarni-Bhave subtour-breaking example

We illustrate the Kulkarni-and-Bhave model using **Figure 3-10**. In the spirit of saving space, only the key equations are shown.

(all vehicles are used)

$$x_{12} + x_{13} + x_{14} + x_{15} + x_{21} + x_{23} + x_{24} + x_{25} = 2$$
$$x_{21} + x_{31} + x_{41} + x_{51} + x_{12} + x_{32} + x_{42} + x_{52} = 2.$$

For demands of 10 thousand pounds (5000 kg) at node 3, 13 thousand (6500 kg) at node 4 and 17 thousand (8500 kg) at 5, a combined load-and-subtour-breaking constraint is written

$$\pi_3 - \pi_4 + 30000x_{34} \le 20000 \qquad \pi_3 - \pi_5 + 30000x_{35} \le 20000$$
$$\pi_4 - \pi_3 + 30000x_{43} \le 17000 \qquad \pi_4 - \pi_5 + 30000x_{45} \le 17000$$
$$\pi_5 - \pi_3 + 30000x_{53} \le 13000 \qquad \pi_5 - \pi_4 + 30000x_{54} \le 13000.$$

A combined vehicle-range-and-subtour-breaking constraint is written for a crew-duty-day of 12 hours

$$\sigma_3 - \sigma_4 + 12x_{34} \le 10 \qquad \sigma_3 - \sigma_5 + 12x_{35} \le 7$$
$$\sigma_4 - \sigma_3 + 12x_{43} \le 10 \qquad \sigma_4 - \sigma_5 + 12x_{45} \le 6$$
$$\sigma_5 - \sigma_3 + 12x_{53} \le 7 \qquad \sigma_5 - \sigma_4 + 12x_{54} \le 6.$$

The Baker-Chan equations now look like

$$2.3x_{13} \le \sigma_3 \qquad\qquad 4x_{23} \le \sigma_3$$
$$2.8x_{14} \le \sigma_4 \qquad\qquad 3x_{24} \le \sigma_4$$
$$2.5x_{15} \le \sigma_5 \qquad\qquad 3.2x_{25} \le \sigma_5$$
$$2.3x_{31} + \sigma_3 \le 12 \qquad 4x_{32} + \sigma_3 \le 12$$
$$2.8x_{41} + \sigma_4 \le 12 \qquad 3x_{42} + \sigma_4 \le 12$$
$$2.5x_{51} + \sigma_5 \le 12 \qquad 3.2x_{52} + \sigma_5 \le 12.$$

The solution replicates that obtained previously in **Figure 3-10**, as expected. □

3.4.4 Scheduling Restrictions. The Time-Constrained Travelling-Salesman Problem (TCTSP) is a variation of the familiar TSP that includes time-window constraints on a time a particular stop(s) may be visited. While there are several formulations of the time-constrained travelling-salesman problem (TCTSP), perhaps the most elegant is the dual model. The dual of TCTSP (Baker 1983) yields a 'disjunctive-graph' model commonly used in scheduling theory. The model uses a directed graph to represent precedence constraints between 'tasks' to be scheduled. If d^i is the time that the salesperson visits node i, then $l^i \leq d^i \leq u^i$, where l^i and u^i are lower and upper bounds of a specified time-window. Additionally, the model allows the salesperson to wait at a city, if necessary, for a time-window to open. The TCTSP is to find the tour that visits each node within an open time-window and minimizes the total length of the tour:

$$minimize \ d^{|I|+1} - d^1 \tag{3-48}$$

subject to

$$d^i - d^1 \geq d^{1i} \qquad i = 2, 3, \ldots, |I| \tag{3-49}$$

$$|d^i - d^j| \geq d^{ij} \qquad i = 3, 4, \ldots, |I|; \ 2 \leq j < i \tag{3-50}$$

$$d^{|I|+1} - d^i \geq d^{i \, |I|+1} \qquad i = 2, 3, \ldots, |I| \tag{3-51}$$

$$l^i \leq d^i \leq u^i \qquad i = 2, 3, \ldots, |I|. \tag{3-52}$$

$$d^i \geq 0 \qquad i = 1, 2, \ldots, |I| + 1 \tag{3-53}$$

Here d^{ij} is the time between arrival at node i to arrival at node j. Notice this includes the wait time at node i.

To complete the dual formulation, we will present below the primal version of the time-constrained vehicle-routing-problem (TCVRP). If we interpret d^i as the arrival time at node i, then restrictions on delivery deadlines u_i and earliest delivery-times l_i can be represented by the nonlinear constraints

$$d^i = \sum_{h \in H} \sum_{j \in I} (d^j + t_j^h + d_{ji}^h) x_{ji}^h \qquad i = 1, \ldots, |I|$$
$$d^1 = 0 \tag{3-54}$$

together with constraint **(3-52)**. Alternatively the above can be replaced by the following linear constraints:

$$d^i \geq (d^j + t_j^h + d_{ji}^h) - (1 - x_{ji}^h) U'$$
$$d^i \leq (d^j + t_j^h + d_{ji}^h) + (1 - x_{ji}^h) U' \tag{3-55}$$

where $U' = \max_{h \in H} U(h)$. When $x_{ji}^h = 0$, these two constraints are redundant. Suppose $x_{ji}^h = 1$. The constraints determine d^i in terms of the arrival time d^j at node j preceding node i on a tour, the delivery or dwell time t_j^h at node j, and the travel time d_{ji}^h between nodes j and i. Although not identical, the readers can no doubt see the similarity between the dual variable d^i and the "odometer" variable σ_i used in subtour-breaking constraints such as Equation **(3-45)**.

Example

The model as formulated really consists of two parts: a TSP, or more correctly VRP, and an additional set of linear constraints that form the time-window portion (Bergevin et al. 1992). The VRP portion is the same as before except the following assumptions are made to make the example more compact: Nodes 1 and 2 are the depot nodes as before; aircraft 1 departs and returns to node 1; and aircraft 2 departs and returns to node 2. Aircraft 1 never goes to node 2; and aircraft 2 never goes to node 1. Other than the above modifications, the rest of the constraints and the figure that accompanies this example are the same as the multiple-depot example and adequately represented by **Figure 3-10**. In the time-window portion, these assumptions are made: (a) Aircraft 1 and 2 both start the day at time 0 on the ground at their respective depots. (b) It is acceptable that an aircraft will delay at any node, including the depot (except the other's depot) in order to wait for a time-window to open.

These are the specific constraints that effect the time windows: First, the following ground-time terms were added to the TSP objective-function: $t_1^1 + t_2^2 + t_3^1 + t_3^2 + t_4^1 + t_4^2 + t_5^1 + t_5^2$. Since loading/unloading requires 0.5 hour and is not performed at the depots, the dwell times at each node are constrained as follows: $t_1^1 \geq 0$; $t_2^2 \geq 0$; $t_i^1 \geq 0.5$ ($i = 3, 4, 5$); $t_i^2 \geq 0.5$ ($i = 3, 4, 5$). Equation **(3-55)** now reads: $d^3 \geq d^1 + t_1^1 + 2.3 - (1 - x_{13}^1)(12)$ and $d^3 \leq d^1 + t_1^1 + 2.3 + (1 - x_{13}^1)(12)$, where d^3 is the arrival time of an aircraft at node 3 given that it takes off from node 1 after a time delay of t_1^1, d_{13}^1 is the travel time for aircraft 1 from node 1 to node 3, which is 2.3 hours in this case. The same set of equations can be written for all node pairs. We show the complete set of equations below, after the equations such as the above two are simplified and terms organized between the left-hand-side and right-hand-side of the equation. For example, the above two equations for Equation **(3-55)** appear as the first and second groups below:

(arrival time at node 3 for aircraft 1)
$$12x_{13}^1 + t_1^1 + d^1 - d^3 \leq 9.7 \qquad 12x_{43}^1 + t_4^1 - d^3 + d^4 \leq 10 \qquad 12x_{53}^1 + t_5^1 - d^3 + d^5 \leq 7$$

(delivery time at node 3 for aircraft 1)
$$-12x_{13}^1 + t_1^1 + d^1 - d^3 \geq -14.3 \qquad -12x_{43}^1 + t_4^1 - d^3 - d^4 \geq -14 \qquad 12x_{53}^1 + t_5^1 - d^3 - d^5 \geq -17$$

(arrival time at node 3 for aircraft 2)
$$12x_{23}^2 + t_2^2 + d^2 - d^3 \leq 8 \qquad 12x_{43}^2 + t_4^2 - d^3 + d^4 \leq 10 \qquad 12x_{53}^2 + t_5^2 - d^3 + d^5 \leq 7$$

(delivery time at node 3 for aircraft 2)
$$-12x_{23}^2 + t_2^2 + d^2 - d^3 \geq -16 \qquad -12x_{43}^2 + t_4^2 - d^3 + d^4 \geq -14 \qquad -12x_{53}^2 + t_5^2 - d^3 + d^5 \geq -17$$

(arrival time at node 4 for aircraft 1)
$$12x_{14}^1 + t_1^1 + d^1 - d^4 \leq 9.2 \qquad 12x_{34}^1 + t_3^1 + d^3 - d^4 \leq 1 \qquad 12x_{54}^1 + t_5^1 - d^4 + d^5 \leq 6$$

(delivery time at node 4 for aircraft 1)
$$-12x_{14}^1 + t_1^1 + d^1 - d^4 \geq -14.8 \qquad -12x_{34}^1 + t_3^1 + d^3 - d^4 \geq -14 \qquad -12x_{54}^1 + t_5^1 - d^4 + d^5 \geq -18$$

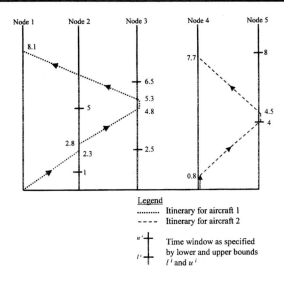

Figure 3-11 Time constrained vehicle-routing-problem example

(arrival time at node 4 for aircraft 2)
$$12x_{24}^2+t_2^2+d^2-d^4\leq9 \qquad 12x_{34}^2+t_3^2+d^3-d^4\leq10 \qquad 12x_{54}^2+t_5^2-d^4+d^5\leq6$$

(delivery time at node 4 for aircraft 2)
$$-12x_{24}^2+t_2^2+d^2-d^4\geq-15 \quad -12x_{34}^2+t_3^2+d^3-d^4\geq-14 \qquad -12x_{54}^2+t_5^2-d^4+d^5\geq-18$$

(arrival time at node 5 for aircraft 1)
$$12x_{15}^1+t_1^1+d^1-d^5\leq9.5 \qquad 12x_{45}^1+t_4^1+d^4-d^5\leq6 \qquad 12x_{35}^1+t_3^1+d^3-d^5\leq7$$

(delivery time at node 5 for aircraft 1)
$$-12x_{15}^1+t_1^1+d^1-d^5\geq-14.5 \quad -12x_{45}^1+t_4^1+d^4-d^5\geq-18 \qquad -12x_{35}^1+t_3^1+d^3-d^5\geq-17$$

(arrival time at node 5 for aircraft 2)
$$12x_{25}^2+t_2^2+d^2-d^5\leq8.8 \qquad 12x_{45}^2+t_4^2+d^4-d^5\leq6 \qquad 12x_{35}^2+t_3^2+d^3-d^5\leq7$$

(delivery time at node 5 for aircraft 2)
$$-12x_{25}^2+t_2^2+d^2-d^5\geq-15.2 \quad -12x_{45}^2+t_4^2+d^4-d^5\geq-18 \qquad -12x_{35}^2+t_3^2+d^3-d^5\geq-17$$

Equation **(3-52)** becomes $1\leq d^3\leq5; 2.5\leq d^4 \leq 6.5; 4 \leq d^5\leq8$. Bergevin et al. (1992) added an extra set of constraints to prevent the penalty of off-loading at a node which is not to be visited:

(node 3)
$$x_{13}^1+x_{43}^1+x_{53}^1-2t_3^1\leq0$$
$$x_{23}^2+x_{43}^2+x_{53}^2-2t_3^2\leq0$$

(node 4)
$$x_{14}^1+x_{34}^1+x_{54}^1-2t_4^1\leq0$$
$$x_{24}^2+x_{34}^2+x_{54}^2-2t_4^2\leq0$$

(node 5)
$$x_{15}^1+x_{35}^1+x_{45}^1-2t_5^1\leq0$$
$$x_{25}^2+x_{35}^2+x_{45}^2-2t_5^2\leq0.$$

The TCVRP as formulated above gives an answer that is readily verifiable. These routes are flown: 1–3–4–1 for aircraft 1, and 2–5–2 for aircraft 2. The various delays at each node, including at the depots, are best illustrated in **Figure 3-11**. Having delayed aircraft 2 for 0.8 hour on the ground at the beginning, the readers can see that both aircraft make their time-windows and crew-duty-days. □

Solution by Benders' decomposition

This version of the numerical problem has been modified to explicitly set the delivery times for the two origin nodes to zero ($d^1 = 0$, $d^2 = 0$). We also assume that each vehicle has to be used. We then follow the Benders' algorithm outlined in the "Optimization" appendix to Chan (2000) for integer solutions.[7] Twenty iterations are required for convergence. The computer output for the restricted master-problem at the 20*th* and final iteration is shown below (Wiley and Chan 1995):

```
MIN     Z
SUBJECT TO
      - 3 X112 - 2.3 X113 - 2.8 X114 - 2.5 X115 - 3 X121 - 2.3 X131
      - 2.8 X141 - 2.5 X151 - 4 X123 - 3 X124 - 3.2 X125 - 4 X132 - 3 X142
      - 3.2 X152 - 2 X134 - 5 X135 - 2 X143 - 5 X153 - 6 X145 - 6 X154
      - 3 X212 - 2.3 X213 - 2.8 X214 - 2.5 X215 - 3 X221 - 4 X223
      - 2.8 X241 - 2.3 X231 - 2.5 X251 - 3 X224 - 3.2 X225 - 4 X232
      - 3 X242 - 3.2 X252 - 2 X234 - 5 X235 - 2 X243 - 5 X253 - 6 X245
      - 6 X254 + Z >=    0
      - X114 - X124 - X134 - X154 - X214 - X224 - X234 - X254 <= - 1
      - X115 - X125 - X135 - X145 - X215 - X225 - X235 - X245 <= - 1
      - X113 + X131 - X123 + X132 + X134 + X135 - X143 - X153 <=    0
      - X121 - X131 - X141 - X151 <= - 1
      - X221 - X223 - X224 - X225 <= - 1
      - X115 + X151 - X125 - X135 - X145 <=    0
      - X113 - X123 - X143 - X153 - X213 - X223 - X243 - X253 <= - 1
      - X213 - X223 + X231 + X232 + X234 + X235 - X243 - X253 <=    0
      - X212 - X232 - X242 - X252 <= - 1
      - X114 + X141 - X124 + X142 - X134 + X143 + X145 - X154 <=    0
      - X214 + X241 - X224 + X242 - X234 + X243 + X245 - X254 <=    0
        12 X243 <=    10
        12 X224 <=     9
        X224 + X234 + X254 <=    0
      - 3 X112 - 2.3 X113 - 2.8 X114 - 3 X115 - 3 X121 - 2.3 X131
      - 2.8 X141 - 2.5 X151 - 4 X123 - 3 X124 - 3.2 X125 - 4 X132 - 3 X142
      - 3.2 X152 - 2 X134 - 5.5 X135 - 2 X143 - 5 X153 - 6.5 X145 - 6 X154
      - 3 X212 - 2.3 X213 - 2.8 X214 - 2.5 X215 - 3 X221 - 4.5 X223
      - 2.8 X241 - 2.3 X231 - 2.5 X251 - 3 X224 - 3.2 X225 - 4 X232
      - 3 X242 - 3.2 X252 - 2 X234 - 5 X235 - 2.5 X243 - 5.5 X253 - 6 X245
      - 6 X254 + Z >=    0
      - X112 - X113 - X114 - X115 <= - 1
      - 3 X112 - 2.8 X113 - 3.3 X114 - 2.5 X115 - 3 X121 - 2.3 X131
      - 2.8 X141 - 2.5 X151 - 4 X123 - 3 X124 - 3.2 X125 - 4 X132 - 3 X142
      - 3.2 X152 - 2.5 X134 - 5 X135 - 2.5 X143 - 5.5 X153 - 6 X145
      - 6.5 X154 - 3 X212 - 2.3 X213 - 2.8 X214 - 2.5 X215 - 3 X221
      - 4 X223 - 2.8 X241 - 2.3 X231 - 2.5 X251 - 3 X224 - 3.7 X225
      - 4 X232 - 3 X242 - 3.2 X252 - 2 X234 - 5.5 X235 - 2 X243 - 5 X253
      - 6.5 X245 - 6 X254 + Z >=    0
      - X215 + X251 - X225 + X252 - X235 + X253 - X245 + X254 <=    0
        X214 - X241 + X224 - X242 + X234 - X243 - X245 + X254 <=    0
        X215 - X251 + X225 - X252 + X253 - X245 - X254 <=    0
        X114 - X141 + X124 - X142 + X134 - X143 - X145 + X154 <=    0
      - 3 X112 - 2.8 X113 - 2.8 X114 - 2.5 X115 - 3 X121 - 2.3 X131
      - 2.8 X141 - 2.5 X151 - 4 X123 - 3 X124 - 3.2 X125 - 4 X132 - 3 X142
      - 3.2 X152 - 2 X134 - 5 X135 - 2.5 X143 - 5.5 X153 - 6 X145 - 6 X154
      - 3 X212 - 2.3 X213 - 2.8 X214 - 2.5 X215 - 3 X221 - 4 X223
      - 2.8 X241 - 2.3 X231 - 2.5 X251 - 3 X224 - 3.7 X225 - 4 X232
      - 3 X242 - 3.2 X252 - 2 X234 - 5.5 X235 - 2 X243 - 5 X253 - 1.5 X245
      - 6 X254 + Z >=    0
        12 X143 <=    10
        12 X114 <=   9.2
      - X114 - X134 - X154 <=    0
      - 3 X112 - 2.8 X113 - 3.3 X114 - 2.5 X115 - 3 X121 - 2.3 X131
      - 2.8 X141 - 2.5 X151 - 4 X123 - 3 X124 - 3.2 X125 - 4 X132 - 3 X142
      - 3.2 X152 - 2.5 X134 - 5 X135 - 2.5 X143 - 5.5 X153 - 6 X145
```

---

[7]An introductory numerical example for the Benders' discretization procedure can be found in the "Prescriptive tools" chapter of Chan (2000) under the "Decomposition Methods" section.

```
 - 6.5 X154 - 3 X212 - 2.3 X213 - 2.8 X214 - 2.5 X215 - 3 X221
 - 4 X223 - 2.8 X241 - 2.3 X231 - 2.5 X251 - 3 X224 - 15.7 X225
 - 4 X232 - 3 X242 - 3.2 X252 - 2 X234 - 5.5 X235 - 2 X243 - 5 X253
 - 6.5 X245 - 6 X254 + Z >= - 11.2
   X213 + X223 - X231 - X232 - X234 - X235 + X243 + X253 <=   0
 - 3 X112 - 2.3 X113 - 2.8 X114 - 3 X115 - 3 X121 - 2.3 X131
 - 2.8 X141 - 2.5 X151 - 4 X123 - 3 X124 - 3.2 X125 - 4 X132 - 3 X142
 - 3.2 X152 - 2 X134 - 5.5 X135 - 2 X143 - 5 X153 - 6.5 X145 - 6 X154
 - 3 X212 - 2.3 X213 - 0.5 X223 - 2.5 X251 - 3 X224 - 3.2 X225
 - 4 X232 - 3 X242 - 3.2 X252 - 2 X234 - 5 X235 - 2.5 X243 - 5.5 X253
 - 6 X245 - 6 X254 + Z >=   0


   OBJECTIVE FUNCTION VALUE

      15.80000

   VARIABLE           VALUE          REDUCED COST
   X113             1.000000          2.800000
   X141             1.000000          2.800000
   X134             1.000000          2.500000
   X225             1.000000         15.700000
   X252             1.000000          3.200000
      Z             15.800000          .000000
```

Now solve the dual subproblem with $x_{13}^1 = x_{41}^1 = x_{34}^1 = x_{25}^2 = x_{52}^2 = 1$ and all others variables set equal to zero:

```
MAX    2.3 λ3 - 10 λ4 - 7 λ5 - 2.3 λ6 - 14 λ7 - 17 λ8 - 8 λ9 - 10 λ10
   - 7 λ11 - 16 λ12 - 14 λ13 - 17 λ14 - 9.2 λ15 + 2 λ16 - 6 λ17
   - 14.8 λ18 - 2 λ19 - 18 λ20 - 9 λ21 - 10 λ22 - 6 λ23 - 15 λ24
   - 14 λ25 - 18 λ26 - 9.5 λ27 - 6 λ28 - 7 λ29 - 14.5 λ30 - 18 λ31
   - 17 λ32 + 3.2 λ33 - 6 λ34 - 7 λ35 - 3.2 λ36 - 18 λ37 - 17 λ38 + λ39
   - 5 λ40 + 2.5 λ41 - 6.5 λ42 + 4 λ43 - 8 λ44 - λ60 - λ61 - λ63 - λ64
   - λ65 - λ66 - 2 λ67 - 2 λ68 - λ69 - 2 λ70 - 2 λ71 - 2 λ72 - 2 λ73
   - 2 λ74 - 3 λ75 - 5000 λ85 - 18000 λ86 + λ47 + λ49 + λ52 - λ79

SUBJECT TO
   - λ3 + λ6 - λ15 + λ18 - λ27 + λ30 + λ1 <=   1
   - λ9 + λ12 - λ21 + λ24 - λ33 + λ36 + λ2 <=   1
   - λ16 + λ19 - λ29 + λ32 + 2 λ47 <=   1
   - λ4 + λ7 - λ28 + λ31 + 2 λ49 <=   1
   - λ5 + λ8 - λ17 + λ20 + 2 λ51 <=   1
   - λ22 + λ25 - λ35 + λ38 + 2 λ48 <=   1
   - λ10 + λ13 - λ34 + λ37 + 2 λ50 <=   1
   - λ11 + λ14 - λ23 + λ26 + 2 λ52 <=   1
   - λ3 + λ6 - λ15 + λ18 - λ27 + λ30 + λ45 <=   0
   - λ9 + λ12 - λ21 + λ24 - λ33 + λ36 + λ46 <=   0
    λ3 + λ4 + λ5 - λ6 - λ7 - λ8 + λ9 + λ10 + λ11 - λ12 - λ13 - λ14
   - λ16 + λ19 - λ22 + λ25 - λ29 + λ32 - λ35 + λ38 + λ39 - λ40 <=   0
   - λ4 - λ7 - λ10 + λ13 + λ15 + λ16 + λ17 - λ18 - λ19 - λ20 + λ21
   + λ22 + λ23 - λ24 - λ25 - λ26 - λ28 + λ31 - λ34 + λ37 + λ41 - λ42 <=   0
    λ5 - λ8 - λ11 + λ14 - λ17 + λ20 - λ23 + λ26 + λ27 + λ28 + λ29
   - λ30 - λ31 - λ32 + λ33 + λ34 + λ35 - λ36 - λ37 - λ38 + λ43 - λ44 <=   0


   OBJECTIVE FUNCTION VALUE

      2.300000

   VARIABLE           VALUE          REDUCED COST
   λ36              1.000000           .000000
   λ43              1.000000           .000000
   λ47               .500000           .000000
   λ49               .500000           .000000
   λ52               .500000           .000000
   λ46             -1.000000           .000000
```

It can be seen that the dual subproblem yields 2.3 in its objective function. The optimal value of the dual subproblem is equal to the value of the master-problem optimal-value minus the value of the fixed integer-vector: 15.8 – (2.3)(1) – (2.8)(1) –(2)(1)– (3.2)(1) –

(3.2)(1) = 15.8 − 13.5 = 2.3. Optimality conditions have been therefore satisfied according to the algorithm. The optimal set of tours uses vehicle 1 to make deliveries at nodes 3 and 4 and then return to node 1. Vehicle 2 waits at node 2 for 0.8 hour and then makes a delivery to node 5 and return. The minimal cost in time units for this set of tours is 15.8 hours of travel and wait time. It is noticed that the solution does not provide any sensitivity analysis on the impact of time-window constraints. In other words, should the time windows change slightly, the set of optimal tours does not change. A vehicle simply "waits it out" upstream to comply with the time-window constraints downstream. □

3.5 Heuristic solutions for the travelling-salesman problem

While the mathematical-programming formulations outlined above are useful and becoming increasingly practical (Haouari et al. 1990, Desrochers et al., 1992), most TSPs—because of their huge dimensionality—are solved by heuristics. Here we will describe two basic heuristic procedures, from which many others have sprouted: the *spanning tree* heuristic and the *Clarke-Wright* procedure. The two also serve to illustrate the fundamental concepts in heuristic solutions to the TSP.

3.5.1 Double spanning-tree heuristic. Consider a TSP with m nodes, where all link/arc costs, c_{ij} satisfy the triangle inequality (i.e., $c_{ij}+c_{jk} \geq c_{ik}$). The double-spanning-tree heuristic operates in the following manner. Determine a minimum-cost spanning tree in the graph (to be explained below). Let this be T' and we duplicate every arc of it. The resulting *multi-graph*, say T'', will be such that it must be possible to find (starting at any node) a closed path that uses each arc in T'' exactly once. Taking this path to be given by an ordered list of nodes, we begin with an initial node and proceed through the list jumping over nodes that have been previously encountered in the list. The result is a TST.

Example

Given the network as shown in **Figure 3-12**, a spanning tree can be constructed (Parker and Rardin 1988). We first list all the arcs according to increasing magnitude of arc cost:

  BF* CH* DH* EG* AF* CD FG* GH* AB EF AE AG BG EH BH DE BC

  2.8   2.9   2.9   2.9   3.0   3.0   3.1   3.2   4.0 4.5   5.0   5.1   5.1   6.0   7.0   8.0   10.0

Then we select arcs, beginning from the top of the list, until we have a tree for the graph. This means we may skip an arc in the list to avoid loops or cycles. The sum of the selected arcs' length is the minimum length. This results in a minimum spanning-tree. In this example, the arcs selected are marked by asterisks and the resulting tree is highlighted in solid lines in **Figure 3-12**(a). With a total length of 20.8. we call this T'. We now duplicate every arc of T' in **Figure 3-12**(b), resulting in T''. From T'', we find a closed path that uses each arc in T'' exactly once, starting from node A. Consider the ordered list of nodes as represented by the letters $A, B, C, D, …$ in alphabetical order. We jump over nodes already encountered. This results in the TST as shown in **Figure 3-12**(c), marked by Arabic numerals 1–2–3–4–5–6–7–8–1, indicating the sequence in which the nodes are included in the tour. The total length is 38.6. Notice that depending on the way one defines the ordered

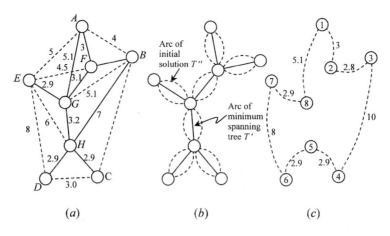

Figure 3-12 Network for double spanning tree heuristic

list in T'', a different tour may result. Another trial, for instance, yields the tour $A–B–C–D–H–E–G–F–A$, with a shorter total length of 34.8. ☐

Let $z^* > 0$ be the value of an optimal solution to a TSP and z be the value produced by a heuristic algorithm .It is conventional to measure accuracy in terms of the ratio z/z^*. A 'performance guarantee' is an upper bound or worst-case limit of the ratio for any problem instance. The simple procedure of jumping over duplicated nodes in the list described is nothing more than a scheme of shortcutting from one node to another on the spanning tree, T'. Since the minimum spanning tree is an $O(m \log m)$ algorithm, and the subsequent steps to construct T'' and the TSTs are also $O(m \log m)$, the entire procedure is $O(m \log m)$. Let the total cost of T' be $l(T')$ and of T'', $l(T'')$. However, if z^* is the length of an optimal tour, $l(T') \le z^*$ or $2l(T') \le 2z^*$ because no tour can cost less than the minimum-cost spanning-tree. Hence, $l(T'') \le 2z^*$. But the shortcutting process in the heuristic cannot add length due to the triangle inequality. That is, the heuristic solution $z \le l(T'')$, and so $z \le 2z^*$. We have a performance guarantee of 2 for the stated heuristic. Checking the two instances of TSTs, their lengths of 38.6 and 34.8 are well within the worst case of $(2)(20.8) = 41.6$. We choose to discuss the double spanning-tree heuristic because it is a prime example of a greedy algorithm. It introduces other greedy heuristics such as *nearest insertion* and *k interchange*, both of which will be discussed in sequel.

3.5.2 Spanning-tree/perfect-matching heuristic. The factor-of-two guarantee can be improved. This is afforded by the concept of an *Eulerian graph* (Lawler et al. 1985). An Eulerian graph is a connected graph in which each vertex has *even degree*, or that there are an even number of edges incident upon each vertex. An Eulerian graph contains an *Eulerian tour*, i.e., a cycle that passes through very edge exactly once, and such a tour can be found in $O(m)$ time. Suppose we had an

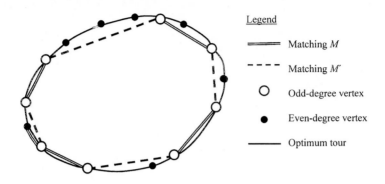

Figure 3-13 Travelling-salesman tour with matchings (Lawler et al. 1985)

Eulerian graph with the cities of a TSP-instance as its vertices. Then we could use the Eulerian tour for this graph to obtain a TST, merely by using the 'shortcut' technique described previously. Thus if we want to find better TSTs, all we need is a better way of generating an Eulerian graph connecting the cities. Christodfides suggested an elegant way of generating such an Eulerian graph, involving a *minimum-weight matching* procedure. Given a set containing an even number of cities, a *matching* is a collection of edges *M* such that each city is the endpoint of exactly one edge in *M*. A minimum-weight matching is one for which the total-length of the edges is minimum, and such matching can be found in time $O(m^3)$.

Let us now look once again at a minimum spanning-tree T' for a TSP instance. Certain vertices in T' already have even degree and hence do not have to receive more edges if we wish to turn the tree into an Eulerian graph. The only vertices we need to worry about are the ones with odd degree. Note further that there must be an even number of these odd-degree vertices, since the sum of all vertex-degrees must be even (it counts each edge exactly twice). Thus one way to construct an Eulerian graph that includes T' would be simply to add a matching for the odd-degree vertices. (See **Figure 3-14**(b) of the numerical example below, where the odd-degree vertices are marked by circles). This would increase the degree of each odd-degree vertex by one, while leaving the even-degree vertices alone. It is not difficult to show the following. If we add to T' a *minimum-weight* matching for its odd-degree vertices, we will obtain an Eulerian graph that has minimum length among those that contain T'. **Figure 3-13** shows a TST, with the cities that correspond to odd-degree vertices in T' emphasized. The tour determines two matchings $\tilde{M}$ and $\tilde{M}'$. Let T'' denote the given TSP instance, and let $l(T')$, $l(\tilde{M})$, $l(\tilde{M}')$ and $l^*(T'')$ denote the sums of the edge lengths for T', $\tilde{M}$, $\tilde{M}'$, and T'' respectively. By triangle inequality, we must have $l(\tilde{M})+l(\tilde{M}') \le l^*(T'')$. So one of M and M' must have length less than or equal to $l^*(T'')/2$. Thus the length of a minimum-weight matching for the odd-degree vertices of T' must also be at most $l^*(T'')/2$. Since $l(T')$ is less than $l^*(T'')$, as argued in the previous subsection, we conclude that the length

To From	2	3	4	5	6	7	8	9	10
1	96	105	50	41	86	46	29	56	70
2		78	49	94	21	64	63	41	37
3			60	84	61	54	86	76	51
4				45	35	20	26	17	18
5					80	36	55	59	64
6						46	50	28	8
7							45	37	30
8								21	45
9									25
10									

Table 3-3 Edge lengths for Christofides' algorithm

of the Eulerian graph constructed is at most $(3/2)l^*(T'')$. This provides an error-bound of 50 percent.

Let us now formalize an algorithm for constructing a TST with this tighter 50-percent error-bound:

Step 1 Construct a minimum spanning tree T' on the set of all cities in T'', with a computational complexity of $O(m^2)$.

Step 2 Construct a minimum-matching $\tilde{M}^*$ for the set of all odd-degree vertices in T', at a complexity of $O(m^3)$.

Step 3 Find an Eulerian tour for the Eulerian graph that is the union of T' and $\tilde{M}^*$, and convert it to a tour using shortcuts, at a complexity $O(m)$. ∎

The running time for the algorithm is dominated by the time for finding the matching in Step 2, or an algorithm complexity of $O(m^3)$. The Christofides algorithm illustrates how error-bounds for heuristics can be tightened by exploiting some basic concepts in graph theory. It represents a major emphasis of today's computational algorithms—posing theoretical underpinning on top of heuristics.

Example

Edge lengths for this example are given in **Table 3-3** (Nemhauser and Wolsey 1988). A minimum-weight spanning-tree and a minimum-weight perfect matching on the vertices of odd-degree in the tree are shown in **Figure 3-14**. An Eulerian cycle obtained from the double spanning-tree heuristic is

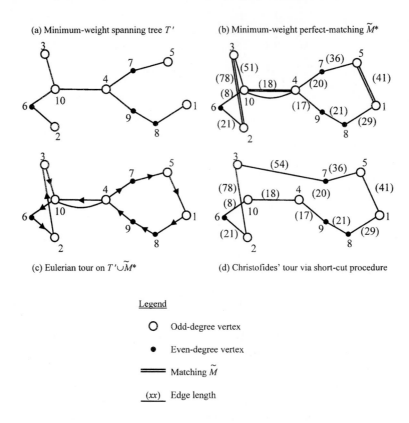

(a) Minimum-weight spanning tree T' (b) Minimum-weight perfect-matching $\tilde{M}^*$

(c) Eulerian tour on $T' \cup \tilde{M}^*$ (d) Christofides' tour via short-cut procedure

Legend

○   Odd-degree vertex

●   Even-degree vertex

═══ Matching $\tilde{M}$

(xx)   Edge length

Figure 3-14 Illustrating the Christofides algorithm

$$1\text{--}8\text{--}9\text{--}4\text{--}10\text{--}6\text{--}2\text{--}6\text{--}10\text{--}3\text{--}10\text{--}4\text{--}7\text{--}5\text{--}7\text{--}4\text{--}9\text{--}8\text{--}1,$$

yielding the tour $1\text{--}8\text{--}9\text{--}4\text{--}10\text{--}6\text{--}2\text{--}3\text{--}7\text{--}5\text{--}1$ of distance 323. An Eulerian cycle obtained from the spanning-tree/perfect-matching heuristic is $1\text{--}8\text{--}9\text{--}4\text{--}10\text{--}6\text{--}2\text{--}3\text{--}10\text{--}4\text{--}7\text{--}5\text{--}1$, which yields the same tour via the shortcut procedure. As mentioned in the previous subsection, each of these heuristics could have produced several other tours, depending on the Eulerian cycle chosen. □

3.5.3 Clarke-Wright Procedure. A classic heuristic for the VRP is the Clarke-Wright (CW) procedure, which in essence is a 'savings' approach. Suppose the 'out-and-back' routes to demands i and j in **Figure 3-15**(a) are combined into one route as shown **Figure 3-15**(b), the savings δ_{ij} incurred is

$$\delta_{ij} = D_{ik} + D_{jk} - D_{ij} \tag{3-56}$$

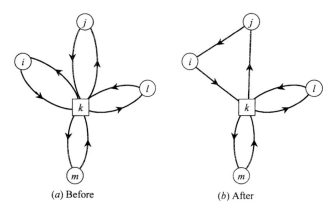

(a) Before   (b) After

Figure 3-15 Initial and improved solution for the Clarke-Wright procedure

where k is the depot, and D_{pq} is the travel cost of servicing between points p and q. If the savings is positive, it is worthwhile to combine the routes. Specifically, the CW algorithm consists of these steps:

Step 1 Compute the savings δ_{ij} for all pairs of demands i-j according to Equation (3-56).

Step 2 Choose the pair of demands with the largest savings and determine if it is feasible to link them together.[8] If so, construct a new depot-based route k-·-i-j-·-k by joining them. If not, choose the pair with the next largest savings.

Step 3 Continue with step 2 as long as the savings is positive. When all positive savings have been considered, stop. ■

Example

The demands at each node and the travel costs between nodes are given in **Table 3-4** (Evans and Minieka 1992). The vehicle capacity is 80. We begin with the initial solution consisting of seven 'out-and-back' routes from depot 1 to the demands 2, 3, ... , and 8: 1-2-1, 1-3-1, 1-4-1, 1-5-1, 1-6-1, 1-7-1, and 1-8-1. If we link demands 2 and 3 together, for example, the savings is $\delta_{23} = D_{21}+D_{31}-D_{23} = 20+57-51 = 26$. The savings associated with each pair of demands is computed in **Table 3-5**. First, select the largest savings of 130 between demands 5 and 8. Joining demands 5 and 8 results in a feasible route because the demand is 30+30 = 60, which is less than the vehicle's capacity of 80. The new set of routes is 1-2-1, 1-3-1, 1-4-1, 1-5-8-1, 1-6-1, and 1-7-1. Next, try to join demands 4 and 8, since this represents the next-largest savings of 103. However, the demand on this route (60+24 = 84) would exceed the vehicle capacity of 80, so this possibility is ignored. Moving down the list of savings, combining demands 3 and 8, 8 and 7, 3 and 5, and 3 and 7 all violate the capacity constraint. The next largest savings of 61, corresponding to demands 4 and 2,

---

[8]A number of constraints can be considered, including vehicle capacity and vehicle range.

From /to	Demand	Travel costs							
		1	2	3	4	5	6	7	8
1	-		20	57	51	50	10	15	90
2	46			51	10	55	25	30	53
3	55				50	20	30	10	47
4	33					50	11	60	38
5	30						50	60	10
6	24							20	90
7	75								12
8	30								

Table 3-4 Demands and travel costs for Clarke-Wright example

From /to	Travel cost savings						
	2	3	4	5	6	7	8
2		26	61	15	5	5	57
3			58	87	37	62	100
4				51	50	6	103
5					10	5	130
6						5	10
7							93
8							

Table 3-5 Savings from linking each pair of demands

results in a feasible route. Continuing in this manner, we obtain the final set of routes: 1–5–8–1, 1–4–2–1, 1–3–6–1, and 1–7–1. □

A number of heuristics are variations of the CW procedure, including the *nearest-insertion* heuristic for TSP (Rosenkrantz et al. 1974), Gillett and Miller's 'sweeping' algorithm (Gillett and Miller 1974), and the Fisher-Jaikumar algorithm for VRP (Fisher and Jaikumar 1981). Burnes (1990) extended the heuristic to allow more than one vehicle to visit a demand and in so doing, minimized the number of vehicles used. The computational complexity of the procedure remains at $O(|I|^3)$. Baker (1991) introduced the vehicle range in a multiple depot location-routing model through the CW heuristic and an extension of the minimum spanning tree (called the *minimum spanning forest* [MSF]). In the chapter on combined location-routing models, we will illustrate some of these extensions. A code SPANFRST is posted on my web site and distributed on a CD in Chan (2000). It implements boththe CW and MSF procedures.

3.5.4 Variation on the Theme. Routing models can assume either a single-criterion or multiple-criteria objective function. Martins (1984), for example, formulated multi-criteria shortest-path problems:

$$minimize \sum_{i,j \in I} d_{ij}^k x_{ij} \qquad k=1,\ldots,K \qquad \text{(3-57)}$$

subject to the same constraints as shown in Equations **(3-19)** through **(3-20)**. Here arc lengths are state dependent, suggesting different travel time during rush-hour vis-a-vis off-peak hours. Altman et al. (1971) used a multi-criteria routing model to simulate the routing of garbage trucks, where environmental concerns are balanced against cost efficiency. For multi-criteria analysis in network routing, the reader is referred to Current et al. (1987).

We will now concentrate on a couple of path-building problems here. The first among them is the shortest covering problem, which finds the shortest path from a starting node to a terminus node, covering all demand nodes in between. A node is *covered* if it is visited directly or passed within a distance S away. The introduction of S results in two criteria functions, one seeking to minimize the total path length while a conflicting criterion seeks to minimize S. The covering salesman problem (Current and Schilling 1989) is a variation of the TSP except that demand nodes need only to be *covered*, but not actually *visited*. Again, multi-criteria result, with a path length minimized on the one hand and S minimized on the other.

A *patrol-routing* problem (Ahituv and Berman 1988) refers to a server (such a vehicle) travelling from an origin to a destination under non-emergency circumstances. While the server is on the move, a call-for-service may arrive, and consequently the moving server is assigned to that call. Had there been no calls during that period, the travel would have terminated at the destination node. The problem is how to select an optimal path or route for the server so that the expected response time to stochastic calls of this nature is minimized. Note that the origin and destination nodes are not required to be two distinct nodes; they are often the same node, representing the depot at which a vehicle is based. Also, the call can be issued either while the server is in motion or at the destination node (the depot). **Figure 3-16** shows an illustration of a sample service network. It consists of five nodes and six links. The figures near the links designate travel time and the figures near the nodes stand for the average-percentage-of-demand originating at each

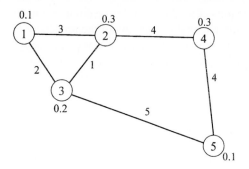

Figure 3-16 Illustration of a patrol-routing problem (Ahituv and Berman 1988)

node. For the time being, we will assume there are no stationary servers positioned in the network. A service unit is to travel from node 1 (origin node) to node 5(destination node) under non-emergency circumstances. Obviously, the shortest path is via node 3, with a total travel time of 7. However, most of the demand is generated at nodes 2 and 4, comprising of 0.3 each. Intuitively, one suspects that if the server selects the path from node 1 to node 5 via nodes 2 and 4, it will be closer to the places where "the action is." Consequently, if a call is received while the server is on the way, the response time is likely to be smaller than if the server had chosen the shortest path already. Ahituv and Berman (1988) show that there is an optimal route from node 1 to node 5 via nodes 3, 2, and 4—covering all the nodes in this case.

3.6   Travelling salesman and quadratic assignment-problems

So far we have been discussing the TSP as a routing model, which measures spatial separation precisely. We can show how the TSP may be formulated as a special case of the quadratic assignment-problem (QAP), a location model based on spatial interaction and has been defined as such in the "Prescriptive Tools" chapter of Chan (2000) and elsewhere. Define the discrete variable $x_{ki} = 1$ or 0, depending on whether workstation k is assigned to location i. Similarly, x_{lj} denotes whether workstation l is assigned to location j. Consider a piece of material moves from workstation k to workstation l, the total material movement (in say lb-ft or kg-m) per unit-time is to be minimized. For four workstations to be placed in four locations, we have:

$$\min \sum_{k=1}^{4} \sum_{l=1}^{4} \sum_{i=1}^{4} \sum_{j=1}^{4} c_{kl} d_{ij} x_{ki} x_{lj}$$

$$\sum_{i=1}^{4} x_{ki} = 1 \qquad k=1,2,3,4$$

$$\sum_{k=1}^{4} x_{ki} = 1 \qquad i=1,2,3,4 \qquad \text{(3-58)}$$

$$x_{ki} \in \{1,0\} \qquad k=1,2,3,4; \ i=1,2,3,4.$$

The connection between the two models, namely between location and routing analysis, may not be suspected based solely on intuition. But more in-depth examination would reveal that they are built upon the same fundamentals. Aside from a theoretical linkage, the discussion below has major computational importance (Burkhard 1990, Francis et al. 1992). In solving discrete location problems, we have been working directly with the location decision variables x_{ki} or x_{lj}, assigning facility k to location i and facility l to location j. In the "Prescriptive Tools" chapter of Chan (2000), we discover that QAP is not just another facility-location problem, since two decision variables x_{ki} and x_{lj} are multiplied together. In addition, it involves both spatial cost d_{ij} between locations i and j and transaction cost c_{kl} between workstations k and l. The latter represents the material flow between them in the assembly line. For the QAP, it is more convenient to represent a solution as a permutation vector π. Let $\pi(i)$ represent the location to which workstation i ($i = 1, 2, ..., 6$) has been assigned. A solution $\pi = [\pi(1), \pi(2), \pi(3), \pi(4), \pi(5), \pi(6)] = (2, 4, 5, 3, 1, 6)$ simply says that workstation 1 is now placed at location 2, workstation 2 is placed at location 4, workstation 3 at 5 and so on. With these locations fixed, or the x_{ki} and x_{lj} variables assigned, the transaction and spatial costs can then be easily computed and summed up. An optimal solution is simply the permutation that yields the lowest total cost.

3.6.1 *Quadratic assignment-problem revisited.*
Let $v = \{1, 2, ..., n\}$ be a collection of integer numbers. The permutation $\pi: v \rightarrow v$, or a re-arrangement of the sequence of four numbers in the set v, defined by $\pi(1) = 4$, $\pi(2) = 2$, $\pi(3) = 3$, $\pi(4) = 1$, corresponds to the permutation matrix $[x_{ij}] = \begin{bmatrix} 0 & 0 & 0 & 1 \\ 0 & 1 & 0 & 0 \\ 0 & 0 & 1 & 0 \\ 1 & 0 & 0 & 0 \end{bmatrix}$. In general, in an $n \times n$ permutation matrix $x_{i,\pi(i)} = 1$ for $i = 1, 2, ..., n$ and all other elements of $[x_{ij}]$ are zero. Thus a permutation matrix has in every row and column exactly one nonzero entry, namely 1; all other entries are zero. Using the concept of permutation matrices, QAP can be reformulated in the following manner. Suppose two $n \times n$ matrices $[c_{kl}]$ and $[d_{ij}]$ are given, where c_{kl} is the interaction between facilities k and l, and d_{ij} is the distance between sites i and j. A feasible placement of a facility at a site, as indicated by the decision variables $\mathbf{x}$ in the traditional formulation of QAP (See Chan [2000] for example), can be conveniently expressed in terms of a

permutation π of the n integers in the set v. That is, if facility i is assigned to site $\pi(i)$, for all i, the QAP can be expressed as

$$\min_{\pi} \sum_{i \in v} \sum_{j \in v} d_{ij}\, c_{\pi(i)\pi(j)}. \tag{3-59}$$

A permutation $\pi{:}\,v{\rightarrow}v$ is called *cyclic* if for all $k = 1, 2, \ldots\ n{-}1$, $\pi^k(1){\neq}1$, where $\pi^k(1)$ is defined recursively by $\pi^k(1) = \pi(\pi^{k-1}(1))$ for $k = 2, \ldots, n{-}1$. In other words, the permutation is performed repeatedly $n{-}1$ times on each entry of the set. For example, the permutation π defined by $\pi(1) = 2$, $\pi(2) = 3$, $\pi(3) = 4$, $\pi(4) = 1$ is a cyclic permutation. However, the permutation π' defined by $\pi'(1) = 2$, $\pi'(2) = 1$, $\pi'(3) = 4$, $\pi'(4) = 3$ is not cyclic, since $\pi^2(1) = 1$ and the exponent $2 = k < n{-}1 = 3$ — constituting a violation of $\pi^k(1){\neq}1$. Expressed in terms of permutation matrices, the cyclic permutations above correspond to

$$[x_{ij}] = \begin{bmatrix} 0 & 1 & 0 & 0 \\ 0 & 0 & 1 & 0 \\ 0 & 0 & 0 & 1 \\ 1 & 0 & 0 & 0 \end{bmatrix} \tag{3-60}$$

and

$$[x_{ij}]' = \begin{bmatrix} 0 & 1 & 0 & 0 \\ 1 & 0 & 0 & 0 \\ 0 & 0 & 0 & 1 \\ 0 & 0 & 1 & 0 \end{bmatrix} \tag{3-61}$$

where it is also clear that $\pi^2(1) = 1$ in $[x_{ij}]'$. For those that are familiar with 'list processing' in computer science, the notion of permutation and cyclic permutation is very clear. If the 'pointers' are systematically chained together in a sequential way among all the members of the integer set v, as in **Figure 3-17**(a), the permutation is cyclic. On the other hand, if they are not sequential among all members, as illustrated in **Figure 3-17**(b), then it is simply a permutation, but not a cyclic one. The difference lies in whether a *complete* cycle is obtained.

3.6.2 The travelling-salesman problem. Cyclic permutations correspond in a unique way to Hamiltonian circuits in complete graphs. If the vertices of the graph correspond to the cities that the salesperson in a TSP has to visit, the permutation $\pi(i) = j$ expresses that the salesperson travels from city i to city j. Let us compose an arbitrary permutation π of v, a fixed cyclic permutation δ' of v, and the inverse permutation of π, π^{-1}. We again obtain a cyclic permutation δ'' via $\pi{\rightarrow}\pi{\circ}\delta'{\circ}\pi^{-1}{\equiv}\delta''$. *All* cyclic permutations of v can be generated in this way with any fixed cyclic permutation δ'. In the case of TSP, this generates all feasible Hamiltonian cycles or tours. For example, let $v = \{1, 2, 3, 4\}$ and consider the cyclic permutation as shown before $\delta' = \begin{bmatrix} 1 & 2 & 3 & 4 \\ 2 & 3 & 4 & 1 \end{bmatrix}$ which forms a feasible TSP

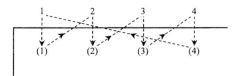

(a) Cyclic permutation

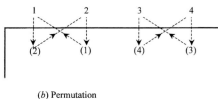

(b) Permutation

Figure 3-17 Illustration of cyclic permutation vs. permutation

tour 1–2–3–4–1. If we take any other permutation π (which may not be cyclic), say $\pi = \begin{bmatrix} 1 & 2 & 3 & 4 \\ 2 & 1 & 4 & 3 \end{bmatrix}$, then we obtain by $\pi \circ \delta' \circ \pi^{-1}$ another cyclic permutation δ''. In our example π equals π^{-1}—or the inverse permutation is the same as the original permutation—and is therefore $\delta'' = \pi \circ \delta' \circ \pi^{-1} = \begin{bmatrix} 1 & 2 & 3 & 4 \\ 4 & 1 & 2 & 3 \end{bmatrix}$ which yields the cycle 1–4–3–2–1, which is another TSP tour. Again, for those that are familiar with list processing, an inverse permutation is defined as retracing the pointers backward. Thus in **Figure 3-18**, the permutation π and its inverse π^{-1} are laid side-by-side for comparison. It can be seen that tracing in both directions, representing π and π^{-1}, yields the same result, namely 2–1–4–3. The permutation operation $\pi \circ \delta' \circ \pi^{-1}$ can again be illustrated step-by-step in **Figure 3-19**. The end result is δ'' as before.

Let the (fixed) cyclic permutation δ' correspond to the matrix $\mathbf{B}'' = [b'_{ij}]$, that is $b'_{i,\delta(i)} = 1$ and all other elements $b'_{ij} = 0$ (as illustrated in Equation **(3-60)** when x is now replaced by b). Then the matrix corresponding to $\pi \circ \delta' \circ \pi^{-1}$ is obtained by permuting simultaneously the rows and columns of $\mathbf{B}''$ according to π. The row and column permutation—in this case yielding the same matrix—has been illustrated in Equation **(3-60)**. Therefore $\pi \circ \delta' \circ \pi^{-1}$ corresponds to $\tilde{B}'' = \left[b'_{\pi(i),\pi(j)} \right]$, which is effectively the matrix multiplication $[b'_{ij}]'[b'_{ij}][b'_{ij}]' = \begin{bmatrix} 0 & 1 & 0 & 0 \\ 1 & 0 & 0 & 0 \\ 0 & 0 & 0 & 1 \\ 0 & 0 & 1 & 0 \end{bmatrix} \begin{bmatrix} 0 & 1 & 0 & 0 \\ 0 & 0 & 1 & 0 \\ 0 & 0 & 0 & 1 \\ 1 & 0 & 0 & 0 \end{bmatrix} \begin{bmatrix} 0 & 1 & 0 & 0 \\ 1 & 0 & 0 & 0 \\ 0 & 0 & 0 & 1 \\ 0 & 0 & 1 & 0 \end{bmatrix} = \begin{bmatrix} 0 & 0 & 0 & 1 \\ 1 & 0 & 0 & 0 \\ 0 & 1 & 0 & 0 \\ 0 & 0 & 1 & 0 \end{bmatrix}$.

Thus the TSP can be written in the form

$$\min_{\pi} \sum_{i=1}^{n} \sum_{j=1}^{n} d_{ij} \, b'_{\pi(i),\pi(j)} \tag{3-62}$$

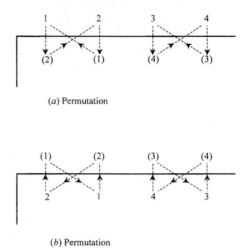

(a) Permutation

(b) Permutation

Figure 3-18 Illustrating an inverse permutation

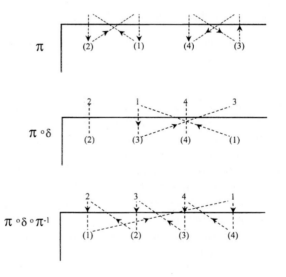

Figure 3-19 Illustrating the permutation operation for
the travelling-salesman problem

where $[d_{ij}]$ is the distance matrix between n cities and $[b'_{ij}]$ is a fixed cyclic-permutation-matrix. For example

$$B'' = [b'_{ij}] = \begin{bmatrix} 0 & 1 & 0 & 0 & \ldots & 0 & 0 & 0 \\ 0 & 0 & 1 & 0 & \ldots & 0 & 0 & 0 \\ 0 & 0 & 0 & 1 & \ldots & 0 & 0 & 0 \\ \cdot & \cdot & \cdot & \cdot & & \cdot & \cdot & \cdot \\ \cdot & \cdot & \cdot & \cdot & & \cdot & \cdot & \cdot \\ \cdot & \cdot & \cdot & \cdot & & \cdot & \cdot & \cdot \\ \cdot & \cdot & \cdot & \cdot & & 0 & 1 & 0 \\ 0 & 0 & 0 & 0 & \ldots & 0 & 0 & 1 \\ 1 & 0 & 0 & 0 & \ldots & 0 & 0 & 0 \end{bmatrix} \qquad (3\text{-}63)$$

Comparing with the QAP as discussed in the "Prescriptive tools" chapter in Chan (2000), this shows that TSPs are special quadratic assignment-problems with special coefficients of interactions b'_{ij}. In other words, a feasible QAP placement is conveniently expressed in terms of permutation π of the n vertices in the set $v = \{1, 2, \ldots, n\}$. That is, let a tour be constructed by assigning 'unit' i to location $\pi(i)$, for all i; the QAP can be expressed as Equation **(3-62)**, which is the TSP.

3.6.3 Computational considerations. Burkhard (1990) reported that even with today's computing machinery, it is time consuming to solve QAPs optimally for problems with more than 20 locations. Heuristics are therefore indispensable in practical applications. The above discussion on permutations allows efficient heuristics to be designed.

    3.6.3.1 SOLUTION METHODS. The *construction heuristic* is a prominent solution method. It begins with basic problem data and builds up a solution, or permutation, in an iterative fashion. Initially, none of the 'new' or candidate facilities will have been assigned to a location, which we indicate by the convention $\pi(k) = 0$. At any intermediate point in the construction process, we will have a set, F, of new facilities that have been assigned to a location. Given a partial solution, ... we can compute an associated total cost, by ignoring all the new facilities not yet assigned locations. If we now augment the partial solution by assigning new facility k to location i, the increment in total cost is $\sum_{l \in F} c_{kl} d_{i,\pi(l)}$. Note that to compute an *increment* in total cost, we must have at least one new facility already assigned. Thus we must specify an initial assignment for the construction heuristic. Let $[d_{ij}]$

$$= \begin{bmatrix} 2 & 3 & 4 & 0 \\ 3 & 0 & 5 & 7 \\ 8 & 2 & 1 & 7 \\ 0 & 0 & 4 & 2 \end{bmatrix} \text{ and } [c_{kl}] = \begin{bmatrix} 1 & 8 & 3 & 1 \\ 0 & 4 & 2 & 7 \\ 0 & 0 & 3 & 7 \\ 2 & 1 & 0 & 7 \end{bmatrix}.$$ Let us fix the indices in the order they appear, that is,

for $j = 1, 2, \ldots, n$. In the first step, we have to compute four values, namely $d_{11}c_{kk}$ for $k = 1, 2, 3, 4$. The minimum is attained for $k = 1$. Therefore we obtain $\pi(1) = 1$. Now we have to compare the three values $d_{11}c_{11} + d_{12}c_{1k} + d_{21}c_{k1} + d_{22}c_{kk}$ for $k = 2$, 3, 4. The minimum is attained for $k = 3$ and $k = 4$ and has the value 11. Since the sums $d_{11}c_{kk} + d_{12}c_{k1} + d_{21}c_{1k} + d_{22}c_{11}$ for $k = 2, 3, 4$ all have a value greater than 11, we obtain $\pi(1) = 1$ and $\pi(2) = 3$.

    In the third step, we have to evaluate six sums, namely

$(d_{11}c_{11}+d_{22}c_{33}+d_{12}c_{13}+d_{21}c_{31}) + (d_{13}c_{1k}+d_{23}c_{3k}+d_{31}c_{k1}+d_{32}c_{k3}+d_{33}c_{kk})$ for $k = 2, 4$

and

$(d_{11}c_{kk}+d_{22}c_{33}+d_{12}c_{k3}+d_{21}c_{3k}) + (d_{13}c_{k1}+d_{23}c_{31}+d_{31}c_{1k}+d_{32}c_{13}+d_{33}c_{11})$ for $k = 2, 4$

and

$(d_{11}c_{11}+d_{22}c_{kk}+d_{12}c_{1k}+d_{21}c_{k1}) + (d_{13}c_{13}+d_{23}c_{k3}+d_{31}c_{31}+d_{32}c_{3k}+d_{33}c_{22})$ for $k = 2, 4$.

The minimal value is 48 and is attained by the last sum; therefore we let $\pi(1) = 1$, $\pi(2) = 4$, and $\pi(3) = 3$.

In the last step we have to evaluate four permutations, namely

$$\pi(1) = 1, \pi(2) = 4, \pi(3) = 3, \pi(4) = 2;$$
$$\pi(1) = 2, \pi(2) = 4, \pi(3) = 3, \pi(4) = 1;$$
$$\pi(1) = 1, \pi(2) = 2, \pi(3) = 3, \pi(4) = 4;$$
$$\pi(1) =1, \pi(2) = 4, \pi(3) = 2, \pi(4) = 3.$$

The first permutation gives the initial value $z = 63$. This is optimal.

Construction methods are easy to implement, have short running times and can be run on a personal computer. However, the quality of their solutions, in particular for large sized problems, may not be very good. Let us now turn to analytic solutions. In the "Prescriptive Tools" chapter of Chan (2000) or elsewhere, we linearized the example QAP into the following form:

$$\min \sum_{k=1}^{4}\sum_{l=1}^{4}\sum_{i=1}^{4}\sum_{j=1}^{4} r_{klij} y_{klij} \tag{3-64}$$

s.t.

$$x_{ki}+x_{lj}-2y_{klij}\geq 0 \qquad k, l, i, j = 1,\dots,4 \tag{3-65}$$

$$\sum_{i=1}^{4} x_{ki}=1 \qquad k=1,\dots,4$$
$$\sum_{k=1}^{4} x_{ki}=1 \qquad i=1,\dots,4 \tag{3-66}$$

$$y_{klij}\in\{1,0\} \qquad k,l,i,j=1,\dots,4. \tag{3-67}$$

It is clear that the QAP, similar to TSP, is a regular assignment problem compli cated by Equation (3-65). Equation (3-65) ensures that y_{klij} equals 1 only if both the corresponding x_{ki} and x_{lj} both equal 1; in other words, y_{klij} equals 1 only if station k is at location i and station l is at location j.

To gain further insights into the similarity between QAP and TSP, we formulate the Lagrangian relaxed problem[9] as follows:

$$\min \sum_{k=1}^{4} \sum_{l=1}^{4} \sum_{i=1}^{4} \sum_{j=1}^{4} r_{klij} y_{klij} + \lambda_{klij}(x_{ki} + x_{lj} - 2y_{klij}) \tag{3-68}$$

with Equations **(3-66)** and **(3-67)** as the remaining constraints (Combs et al. 1996). In the formulation, we move the third set of constraints into the objective function and introduce the product term with Lagrangian multipliers. We solve the problem using the Subgradient Lagrangian Search as outlined in the "Optimization" book appendix of Chan (2000). We initially set $\lambda = 0$, solve the Lagrangian-relaxation problem, and immediately found the optimal solution $x_{14} = x_{23} = x_{32} = x_{41} = 1$. The zero dual variables suggest that all the constraints for Equation **(3-65)** are nonbinding, confirming the solution of all y equal zero in the numerical example worked out in the linear formulation section of the "Prescriptive Analysis" chapter in Chan (2000). This is similar to a small-size TSP where the subtour-breaking constraints happen to be nonbinding (but still necessary as part of the overall formulation). The interesting point is that we are now dealing with a regular linear assignment problem consisting of constraints **(3-66)** only, which automatically yields 0–1 decision variables due to total unimodularity.

From an application standpoint, QAPs are regarded as facility-location problems while TSPs are routing problems. The above discussion shows that the fundamental algebra underpinning the two formulations is the same. They are all related to the assignment problem. Viewed in this light, the artificial distinction between location and routing can be broken down. This paves the way for the chapter on "Simultaneous location-and-routing models," where such combined problems are systematically formulated and solved in an integrated fashion, particularly through heuristics that exploit the list-processing structure as demonstrated above. To the extent that the assignment problem is a special case of spatial-interaction models, the linkage between location-routing and land-use models can also be postulated, since the former is an assignment model and the latter a spatial-interaction model.

3.6.3.2 EQUILIBRIUM AND DISEQUILIBRIUM. While we are on the subject, we like to close by making some observations on *spatial equilibrium*, or the stability of locational decisions in general (Beckmann 1987)—a subject we will come back to in the chapter under that name. Consider four workstations in an assembly line with the following relationship: Workstation I supplies workstation II, workstation II supplies workstation III, workstation III supplies workstation IV, and workstation IV supplies workstation I. Let the four locations consist of two pairs of adjacent locations. Initially, as a result of quadratic assignment the workstations may be

---

[9]See the "Optimization" book appendix for an introduction to Lagrangian Relaxation.

located like this: $\begin{smallmatrix} I & II \\ & \\ III & IV \end{smallmatrix}$ which is a feasible solution but not in equilibrium. Now workstations II and IV can both gain by exchanging locations. This would make workstation II closer to its customer station III and work station IV closer to its customer station I. As a result of the exchange, the second quadratic-assignment solution may be: $\begin{smallmatrix} I & IV \\ & \\ III & II \end{smallmatrix}$ which is still not in equilibrium. Now it is workstations III and I's turn to have an incentive to exchange locations. After that, the reader may show that IV and II will want to exchange locations again, and there after I and II once more, whereupon the original location-assignment has been reached once more (i.e., a cyclic permutation). In this way, work-stations will be induced to rotate and no equilibrium exists. The point is clear: under the quadratic assignment-problem when supply relationships among workstations create 'externalities', there exist situations where lowest-cost assignment fails to achieve and maintain spatial equilibrium in the location of workstations in an assembly line. We will come back to this in subsequent chapters on "Chaos" and "Spatial Equilibrium." Suffice here to point out one physical interpretation of a cyclic permutation that equivalenced QAP and TSP.

3.7   Relationship between approximate and exact measures of distance

In this chapter thus far, we have proposed four different ways of measuring spatial separation: Minkowski's metric, accessibility, SFC and routing models. We made a distinction between approximate methods (the first three) and exact method (the last one). Here, we would like to add another to the repertoire of exact methods, namely measuring spatial-temporal separation in terms of stochastic queuing delays. This type of problems arises in the location of fire stations, where the response time to a call is very important. Unfortunately, it is subject to delay because the fire trucks may be busy somewhere else or the streets through which the fire trucks travel are crowded. The same problems are encountered in the defense community in the United States. The prevailing thinking calls for the remaining military bases to be located so that the stochastic response time to an incident anywhere around the globe can be minimized.

3.7.1 Stochastic facility location. Berman et al. (1990) describe a general model in which p mobile service- units are stationed at up to p different home-depot locations throughout a service region (p = 1, 2, ...). Usually the service region is modelled as a transportation network with deterministic travel times. Demands, assumed to be located only at nodes of the transportation network, are called by telephone into a designated switchboard, resulting in on-scene service by a service unit. In response to such a service request, one of three things happens:

a. If there is at least one available service unit, it is immediately dispatched to the scene.

b. If all service units are simultaneously busy, and if it is the operating policy of the system not to enter service requests into a queue, the demand is referred to a backup system—such as the fire station next town—for the desired service. With this operating policy, customers who call when all units are busy are 'lost' to the system under analysis.

c. If all service units are simultaneously busy and if system operating policy allows queuing, then the demand is entered into a queue with other waiting demands; the queue, having potentially unlimited capacity, is depleted in a first-in, first out (FIFO) manner.

d. Only one demand is serviced per trip. There are no 'back-to-back' assignments by which the server is directly dispatched to a waiting customer from the scene of a previous service request.

The system is complicated by three major probabilistic features of systems operation:

a. the unpredictability of calls for service;
b. the stochastic origin of a call;
c. the variability of service times, due both to differing travel times to different nodal demand-locations and to differing on-scene service-requirements.

The service-time random-variable associated with a demand from node i is characterized by $\tilde{\tau}_{j|i} = \tilde{t}_i + \beta d_{ji}/v''$. Here d_{ji} is the minimal travel-distance from depot j to demand i, v'' is the unit's response speed, $\tilde{t}$ is the non-travel component of service-time random-variable (i.e., on-scene service). The term $(\beta/v)d_{ji}$ accounts for the round-trip travel time between the home location of j and the service request at node i. For example, $\beta = 2$ implies that the unit requires as much time travelling from the facility at j to node i as it does in the return trip. For modellingpurposes, we require that $\beta \geq 1$. **Figure 3-20** shows the temporal sequence associated with a service response for a demand from node i, a sequence that occurs with probability f_i each time a random-demand arrives.

3.7.2 Loss system model. In this version of the model, there is no queuing capacity. In other words, demands that arrive when all p service units are busy are lost (Berman et al. 1990). The objective function is to minimize the weighted sum of mean-response-time and the cost-of-rejecting-the-demand. The respective weights are the probabilities of immediate response and rejection respectively. Our goal is to find a single home location at which a facility would be constructed to house all p mobile units. The location is chosen to minimize the weighted total cost. Berman et al. call this the p-server-single-facility-loss-median, or p_1-median. It can be shown that p_1-median coincides with the deterministic 1-median location (sometimes called the Hakimi median). The provisions are:

a. The locations of successive requests are statistically independent.
b. All p service units are indistinguishable in terms of their service-time-distribution.

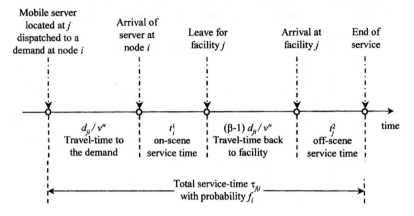

Figure 3-20 Spatial separation in terms of response time to a service call
(Berman et al. 1990)

c. The objective function is

$$[1-P_j(p)]\bar{\tau}_j+P_j(p)Q'' \tag{3-69}$$

Here $P_j(p)$ is the steady-state probability of saturation of all stationed at facility j service units p. $\bar{\tau}_j$ is the expected one-way travel-time to a random demand. Q'' is the cost per rejected demand, $Q'' \geq 0$. For a Poisson input, general independent service-time, p servers, and zero queuing capacity, we use the Erlang Loss Formula that is well documented in queuing texts[10]: $P_j(p)=\rho_j^p/p!\left/\sum_{k=1}^{p}\rho_j^k/k!\right.$ where ρ_j is the utilization rate of a service unit stationed at j. As a direct carry over from queuing literature $\rho_j=\lambda''\left(\bar{t}_j+\beta d_j/v''\right)$ where λ'' is the arrival rate of demands for service, and $\bar{t}_j=t_i^1+t_j^2$ in reference to **Figure 3-20**. Thus the major feature of the stochastic-median model here is the spatial-separation measure. Instead of a deterministic quantity, it is now a stochastic quantity $\bar{\tau}$.

Given the above assumptions regarding a loss system, the p_1-median coincides with the traditional Hakimian 1-median location if $Q'' \geq 0$ for all j in the network. *The optimal home location for a facility is to station p service units operating in a loss system at the 1-median location.* Therefore, congestion of the type observed in a loss system does not affect optimal facility location. One could use the traditional 1-median model and its simple spatial-separation measure to find the optimal location for a stochastic p_1-median.

---

[10]A review of the Erlang Loss Formula is documented in the "Markovian Processes" appendix to Chan (2000).

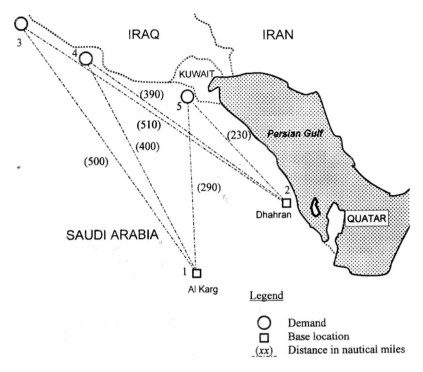

Figure 3-21 Gulf-War air-defense network

Example

The second-layer of a U. S. air-defense system in the theater is called the alert-mission. It fills the gap when the dedicated intercept-aircraft in any sector becomes unable to meet its commitment (see the map in **Figure 3-21**). This mission is often called "putting out fires," because of its similarities to the operation of a fire department. If all aircraft in this second-layer defense-system are busy, attackers are simply passed to the ground defenses, making this a 'loss system'. Judicious choice of a base-location-of-aircraft for this alert-mission is critical to an agile response-time (Sosobee and Chan 1993). In this example, all travel-times are in hours, all distances in nautical-miles (n mi) and speed in knots (n mi/hour). Given the arrival rate of attacking aircraft at 10 per hour (or $\lambda_i'' = 10$ for $i = 3, 4$, and 5), the probability of a demand at node i is $f_i = \lambda_i'' / \sum_{k=1}^{3} \lambda_k'' = \lambda_i'' / \lambda'' = 0.333$ for $i = 3,4,5$. Referring to **Figure 3-21**, the weighted distance to all the demands from base-location j is simply $d_1 = \sum_{i=1}^{3} f_i d_{1i}$ = 396.667, and $d_2 = 376.667$. Given an aircraft-speed of 550 knots ($v'' = 550$), the expected one-way response-time is $\tau_1 = d_1/v'' = 0.721$ hour and $\tau_2 = 0.685$ hour. Given an off-scene service-time of 0.9 hour at both base 1 and 2 ($\bar{t}_1 = \bar{t}_2 = 0.9$) and zero on-scene service-time ($\bar{t}_3 = \bar{t}_4 = \bar{t}_5 = 0$), the utilization-rate for aircraft based at j is therefore $\rho_1 = \lambda''(\bar{t}_1 + \beta d_1/v'') = 16.521$, $\rho_2 = 15.96$. Here the time for the return-trip is 1.57 times the response time, or $\beta = 2.57$. With 10 mobile-unit elements (or $p = 10$), the probability of system saturations is $P_1(10) = p_1^{10}/8! \Big/ \sum_{k=1}^{10} p_1^k/k! = 0.455$, and $P_2(10) = 0.439$. The expected-cost objective-function Z_j from Equation **Figure 3-20** is then calculated in hours for three different assumed costs-per-rejected-demand Q'':

Q''	0	2	10
Z_1	0.393	1.303	4.946
Z_2	0.384	1.263	4.778

With a lower objective-function Z_j, the base-location 2 (Dhahran) is chosen always, regardless of the cost assigned to a missed intercept. This answer checks out with the 1-median model when the weighted distances $f_i d_{ij}$ are directly used, as expected. □

3.7.3 Infinite-capacity queue.
Instead of a loss system, consider servicing every demand. This may result in an infinite queue of service requests (Berman et al. 1990). Assume there is only a single service-unit and our location problem is to find that home location for the unit that minimizes its average response-time to a random demand. As before, service demands occur exclusively at the nodes, with each node i generating an independent Poisson stream of demand rate $\lambda'' f_i$ where $\Sigma_i f_i = 1$. The queuing discipline is $M/G/1/\infty$ (or random arrival, general independent-service-times, a single server, and infinite queuing-capacity). The expected response-time is the sum of the mean-queuing-delay and the mean-travel-time:

$$\bar{r}_j(\lambda'') = \bar{q}_j(\lambda'') + \bar{\tau}_j. \tag{3-70}$$

The mean-travel-time $\bar{\tau}_j$ is the usual 1-median function $\left(\sum_{i \in I} f_i d_{ji}\right)/v''$. The mean queuing-delay for an $M/G/1/\infty$ is

$$\bar{q}_j(\lambda'') = \begin{cases} \dfrac{\lambda'' \sigma_j^2}{2(1 - \lambda'' \bar{\tau}_j)} & \text{for } \lambda'' \bar{\tau}_j < 1 \\ \infty & \text{otherwise} \end{cases} \tag{3-71}$$

where the service time $\bar{\tau}_j = \bar{t}_j + \beta d_j/v''$ with $\bar{t} = \sum_i f_i \bar{t}_i$. σ_2 is the second moment of service time.

The location problem becomes $\min_{j \in I} \bar{r}_j(\lambda'')$. The solution to this problem $j^*(\lambda'')$ yields a *stochastic queue-median*. A *stochastic queue-median coincides with a regular median when λ'' is sufficiently small as one would expect intuitively. A stochastic queue-median coincides with a regular Hakimian median having minimum second-moment-of-service-time when λ'' is sufficiently large but less than $\lambda''_{\max}$.* The second part of the finding needs a bit more explanation, and the best way is through an example.

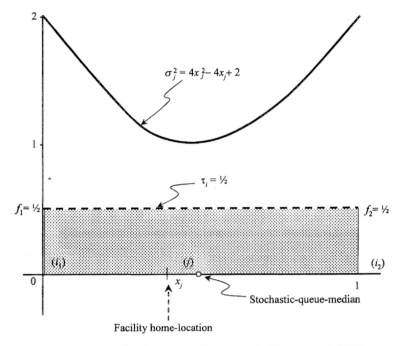

$\sigma_j^2 = 4x_j^2 - 4x_j + 2$

$\tau_i = \frac{1}{2}$

$f_1 = \frac{1}{2}$ $f_2 = \frac{1}{2}$

(i_1) (j) (i_2)

x_j 1

Stochastic-queue-median

Facility home-location

Figure 3-22 Stochastic-queue-median example (Berman et al. 1990)

3.7.3.1 A STOCHASTIC-QUEUE-MEDIAN EXAMPLE. To better understand the stochastic queue-median, consider the simple symmetric 2-node, 1-arc system shown in **Figure 3-22**. Assume $f_1 = f_2 = \frac{1}{2}$, $v'' = 1$, $\beta = 2$, $\bar{t} = 0$. Then $\tau_j = 2\bar{d}_j = 1$ for all j in the interval [0,1]. This means all points of the 'network' are regular 1-medians (see a similar airport-location problem in the "Prescriptive Tools" chapter in Chan (2000).) Similarly, $\sigma_j^2 = 4x_j^2 - 4x_j + 2$ defines the second moment of service-time at each location within this 0–1 interval. According to Equation **(3-71)** the mean queuing delay is $\bar{q}_j(\lambda'') = \lambda''(4x_j^2 - 4x_j + 2)/2(1 - \lambda'')$ for $0 \leq \lambda'' < \lambda_{max} = 1$. This quadratic function is minimized when σ_j^2 is minimized, or when j is at $x_j = 1/2$. Since $\sigma_j^2(x_j = 1/2) = 1$, $\bar{q}_{x_j=1/2}(\lambda'') = \min_{j \in [0,1]} \bar{q}_j(\lambda'') = \lambda''/2(1 - \lambda'')$. σ_j^2 is maximized at $x = 0$ and $x = 1$, at which points $\sigma_j^2(x_j = 0) = \sigma_j^2(x_j = 1) = 2$. Thus $\bar{q}_{x_j=0}(\lambda'') = \bar{q}_{x_j=1}(\lambda'') = \max_{j \in [0,1]} \bar{q}_j(\lambda'') = \lambda''/(1 - \lambda'')$. This represents a doubling in mean queuing-delay in comparison to the minimum possible. While traditional p-median location models would be indifferent to all positions within 0 and 1, the queuing result reveals a factor of two variation in the queuing component ($\bar{q}_j(\lambda'')$) of the objective functio n over [0,1]. The entire response-time objective function is $\bar{r}_j(\lambda'') = \bar{q}_j(\lambda'') + \bar{\tau}_j = \lambda''(4x_j^2 - 4x_j + 2)/[2(1 - \lambda'')] + 1/2$. This function is plotted in **Figure 3-23** over λ'' parametric on $x_j = 0$ and $x_j = 1/2$. As observed from the figure, the queuing component of the objective function becomes an arbitrarily large fraction of the total

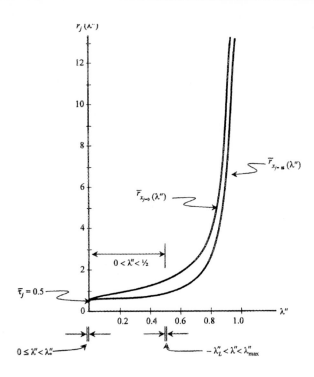

Figure 3-23 Average response-time for queue median (Berman et al. 1990)

as λ'' increases. Also, the queuing component is significantly larger at $x_j = 0$ than at $x_j = \frac{1}{2}$. This shows the limited value of traditional p-median models for the general case of stochastic systems, since such deterministic models focus on only one component of the objective function that may be dominated by the stochastic component.

$\lambda'' = 0$ is the only value of λ'' for which any location j within the $0-1$ interval is a stochastic queue-median. **Figure 3-23** shows the two curves for $x_j = 0$ and $x_j = \frac{1}{2}$ come together only at $\lambda'' = 0$. In most networks, however, the traditional median is at a single node. (Again, see the airport-location problem in "Prescriptive Tools" chapter of Chan (2000).) In these cases, there exists a strictly positive λ'' value, say λ_0'', such that the regular median is the stochastic queue-median for all λ'' in the range $0 \le \lambda'' < \lambda_0''$. Since point $x_j = 1/2$ is a regular median with minimum second-moment-of-the-service-time, this point is the stochastic queue-median at λ'' values 'near' $\lambda_{max} = 1$. In the above example, it turns out that $x_j = 1/2$ is the stochastic-queue-median for all positive permissible λ'' values, $0 < \lambda'' < 1$. In most examples containing a unique deterministic median, there exists a large λ'', say $\lambda_L'' < \lambda_{max}$, such that the deterministic median is the stochastic queue-median for all λ'', $\lambda_L'' < \lambda'' < \lambda_{max}$.

For intermediate values, that is $\lambda_0'' \leq \lambda'' \leq \lambda_L''$, the stochastic-queue-median tends to be pulled away from the deterministic median toward points having smaller values of σ_j^2.

Examining the objective-function Equation **(3-69)**, we can understand intuitively the placement of the optimal location for three different ranges of λ'' values. For small λ'', queuing rarely occurs and thus the travel-time component of the objective function dominates, yielding the optimal location at the deterministic median. For large λ'', one is concerned that the queue remains stable, implying finite mean-queuing-delay. Queue stability requires avoiding the singularity in the denominator of Equation **(3-71)**. This implies adherence to the inequality $\lambda'' \bar{\tau}_j <$ 1, a condition that is eventually violated from any x_j other than the deterministic median as λ'' increases to $\lambda_{\max}$. Therefore, to guarantee queue stability λ'' at near $\lambda_{\max}$, the stochastic-queue-median must return to the deterministic median. For intermediate λ'' values, the numerator of Equation **(3-71)** plays an important role. It pulls the stochastic queue-median away from the deterministic median for those cases in which the minimum value of the second-moment-of-service-time is not obtained at the deterministic median.

3.7.3.2 A STOCHASTIC QUEUE-MEDIAN HEURISTIC. The above discussion suggests that for an arrival rate within an interval between 'break' points, the stochastic queue-median can be approximated by a regular median, with its simple measure of spatial separation. This is particularly true for the lower and upper ranges of λ''. For the intermediate range, we have gained an insight into where the stochastic queue-median may be compared with the regular median. In general, Equations **(3-69)** through **(3-71)** suggest that in optimizing $\bar{r}$ there is a 'tug-of-war' between minimizing σ^2 and $\bar{r}$. A traditional Hakimian median does not minimize σ^2 over a network. Thus, starting at a traditional median, with $\lambda_L'' < \lambda_{\max}$, one seeks to direct the stochastic-queue-median search toward decreasing values of σ^2. This is the intuition behind the following heuristic that builds upon a sub-network and then locates the stochastic-queue-median over this sub-network. Here we assume uniqueness of the Hakimian median at a node to simplify exposition.

Step 1 Find the traditional 1-median of a network. Calculate the second-moment-of-the service-time at the median. Set s^2 to be equal to that second moment and label the node where the median exists.

Step 2 For all unlabelled nodes j connected directly by an arc to a labelled node, compute σ_j^2. If $\sigma_j^2 > s^2$ for all j, go to step 3. Otherwise, determine node j^* having the smallest second moment $\sigma_{j^*}^2$ of all eligible unlabelled nodes. Set $s^2 = \sigma_{j^*}^2$, label node j^*, and repeat step 2.

Step 3 Call the last labelled node j^*. Define the set I_N to be all the unlabelled nodes j connected directly by an arc to node j^*. Focus the search for stochastic

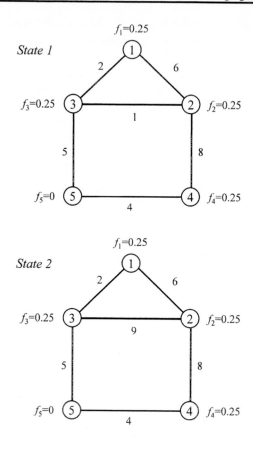

Figure 3-24 Sample stochastic-network (Berman et al. 1990)

queue-median only on the sub-network defined by all the labelled nodes and the set I_N. ∎

In approximately 10 test problems of networks with up to 30 nodes, the above heuristic has successfully located the stochastic queue-median.

3.7.4 Networks with random travel times. This section introduces the concept of a stochastic network with random arc lengths. We show how to locate service units on such a network based on spatial separations that are now dependent upon the state of the system. Thus the travel time between two points will be very different during "rush hours" and 'off-peak' hours. Because of the change in travel time, it may be necessary to reposition a facility between these two time-periods of the day (Ahituv and Berman 1988). Consider a two-state network as shown in **Figure 3-24**, with the demands f_i as shown. We assume, for example, that the arc connecting nodes 2 and 3 may have two travel-times, say $c_{23}(1) = 1$ and $c_{23}(2) =$

i	j										f_i
	1,2	1,3	1,4	1,5	2,3	2,4	2,5	3,4	3,5	4,5	
1	0/0¹	0/0	0/0	0/0	2/2	3/	3/	2/2	2/2	7/7	0.25/0.5
2	0/0	1/6	3/6	3/6	0/0	0/	0/	1/8	1/8	6/8	0.25
3	1/2	0/0	2/2	2/2	0/0	1/	1/	0/0	0/0	5/5	0.25
4	8/8	9/9	0/0	4/4	8/8	0/	4/	0/0	4/4	0/0	0.25
5	6/7	5/5	4/4	0/0	5/5	4/	0/	4/4	0/0	0/0	0
$\bar{\tau}'_j(k)$	2.25/ 2.5	2.5/ 3.75	1.25/ 2²	2.25/ 3	2.5/ 2.5	1/ 3.5	2/ 4.5	0.75³/ 2.5	1.75/ 3.5	4.5/ 5	
$P'_1\bar{\tau}'_j(1)+$ $P'_2\bar{\tau}'_j(2)$	2.4	3.25	1.7⁴	2.7	2.5	2.5	3.5	1.8	2.8	4.8	

¹ The first number in the entry refers to state 1 while the second number refers to state 2.
² Optimal expected response-time in state 2 $\bar{\tau}'_j = (2)$.
³ Optimal expected response-time in state 1, $\bar{\tau}'_j = (1)$.
⁴ Optimal overall expected response-time for the stochastic network $\min_{j \in I} \left(0.4\bar{\tau}'_j(1)+0.6\bar{\tau}'_j(2)\right)$.

Table 3-6 Expected response-times for all combinations of two home-nodes

9 units with probabilities of $P_1 = 0.4$ and $P_2 = 0.6$ being in states 1 and 2 respectively. This is similar for other arc travel-times as shown. Naturally, the spatial separation between two points i and j, measured along the shortest path, $d^{ij}(k)$, will correspondingly change depending on state k. We wish to find *permanent* locations for placing two service-units, considering the stochastic nature of the network where changes in travel times do occur. Later, we will consider repositioning the facilities according to the state of the system.

In state k, we wish to find a subset of two nodes where the two servers will be placed such that the expected response-time (here the same as travel time) to all calls will be minimized:

$$\sum_{i=1}^{5} f_i d^{j^* i}(k) \le \sum_{i=1}^{5} f_i d^{ji}(k) = \bar{\tau}'_j(k) \qquad \forall j \in I \qquad (3\text{-}72)$$

where j^* stands for two nodes that in combination will serve the demands in a non-overlapping way. Notice the similarity between Equation **Figure 3-24** and the denominator of Equation **(3-5)**. **Table 3-6** displays the shortest distances for all possible combinations of two 'home' nodes, under two different states. For state 1, the optimal home-locations would be in nodes 3 and 4, with an expected response-time $(\tau)'_j.(1)$ of 0.75 unit. For state 2, the home locations are 1 and 4 at an expected response-time $(\tau)'_j.(2)$ of 2 units. For permanent locations in considerations of a probability of being in state 1 $P'_1 = 0.4$ and the probability of being in state 2 $P'_2 = 0.6$, the minimum weighted-average is computed

$$\min_{j \in I} \left(P'_1 \, \overline{\tau}'_j(1) + P'_2 \, \overline{\tau}'_j(2) \right). \tag{3-73}$$

As shown in the last line of **Table 3-6**, the pair (1, 4) yields the least overall expected-response-time for the stochastic network, 1.7 units.

The above enumeration assumes that the optimal locations are still at nodes, rather than on links. This assumption was true for the deterministic network (as shown in the "Economics" chapter in Chan (2000) and "Facility Location" chapter in this volume.) But it is not obvious that it still holds for the stochastic case. It is possible to prove that under a reasonable set of assumptions that *there exists at least one set of optimal locations that fully reside at the nodes of the network.* The solution procedure for this problem was by enumeration. For large problems, it is much more practical to solve by mathematical programming or heuristic. The reader may recall that such an integer program was formulated and solved in the "Facility Location" chapter for demonstration.

Example

Refer to the example shown in **Figure 3-24**. Our objective here is simply to show how **Table 3-6** is computed to help locating the optimal two facilities (Horton and Chan 1993). To perform the calculations in Equation **Figure 3-24**, we define all the relevant entries one by one. For example, the arc-distance d_{12} is given as 6 in both states 1 and 2. On the other hand, the demand at node 1 changes from 0.25 in state 1 to 0.5 to state 2. The shortest path between a facility-site j and a demand i for each state k is computed. Thus, the shortest path between 1 and 5 in state 1 is $d^{15}(1) = \min(7,12,18,15) = 7$. With the shortest paths identified

as $[d^{ji}(1)] = \begin{bmatrix} 0 & 3 & 2 & 11 & 7 \\ 3 & 0 & 1 & 8 & 6 \\ 2 & 1 & 0 & 9 & 5 \\ 11 & 8 & 9 & 0 & 4 \end{bmatrix}$ and given demands $(\leftarrow f_i(1) \rightarrow) = (0.25, 0.25, 0.25, 0.25, 0)$, it is easy

to compute the average distance $\tau_j(1)$ in state 1 for each of the facility-sites j. This is similar for state 2. Now for all two-base combinations—(1, 2), (1, 3), (1, 4), (1, 5), (2, 3), (2, 4), (2, 5), (3, 4), (3, 5) and (4, 5)—the minimum base-dependent distances can likewise be computed for both states. The resulting response-times in state 1 are respectively 2.25, 2.5, 1.25, 2.25, 2.5, 1, 2, 0.75, 1.75, and 4.5 for each of these basing strategies. And this is similar for state 2. Then according to Equation **(3-73)**, the average response-times can be obtained as shown in **Table 3-6**, from which the best basing-strategy can be identified as (1, 4), at the minimum average-response-time of 1.7. □

3.7.5 Repositioning in a stochastic network. Suppose the dispatching center is fully informed on the state of the network and the location of the units. Whenever the network changes from state 1 to state 2 or *vice versa*, the dispatcher can react immediately and alter the location of home nodes if so desired (Ahituv and Berman 1988). These assumptions are made for this *repositioning* problem:

1. Time intervals between changes of states are much longer than relocation travel-times on the network so that the effect of repositioning can be felt.

2. *The state of the network is the sole determinant of the location of the home nodes. It is not important what the history of the network had been before it reached a certain state.*[11] Once a certain state is known to have been reached, the same locations are used. In the above example, the fire trucks are regularly at nodes 3 and 4 during period 1 to be close to the fire hazards. They are at nodes 1 and 4 during period 2 for similar reasons. These are kept as optimal locations for period 1 and period 2 respectively. This is irrespective of the fact that all fire trucks might have been employed in the part of town furthest away from home nodes recently, suggesting a departure from the norm.

3. There is a cost function associated with the relocation of idle service units. The cost function can be either linear to the distance (travel time) or a non-decreasing concave function. The latter implies `economies of scale.' *This assumption is required for proving that the optimal solution lies on nodes.*

Let us examine again the figures in **Table 3-6**. The optimal location under state 1 is (3, 4), yielding an expected response time of 0.75; the optimal location under state 2 is (1, 4), yielding an expected response-time of 2. The optimal location for permanent home nodes without distinguishing between time periods is also (1, 4), yielding an expected-response-time of 1.7. Suppose now the dispatcher can relocate service units. An intuitive guess would be to relocate the unit from node 3 to node 1 whenever the state of the network changes from 1 to 2, and reverse the relocation decision when a counter-transition occurs. Had the relocation cost been zero, the expected response-time could be calculated as (0.4)(0.75)+(0.6)(2) = 1.5. This weighted expected-response-time of 1.5 is better than 1.7 obtained in the previous section where relocation was not allowed. It is not necessarily true to assume the following: When repositioning is considered, the optimal home-nodes under each possible state conform with the optimal home-nodes associated with that particular state had this state been the only state of the network. For example, it might be possible that, because of cheaper cost, repositioning can take place not between the pairs (3, 4) and (1, 4), but between other pairs that might not be optimal under each time-period separately. They could be optimal when repositioning costs and expected response-time are examined simultaneously over the history of the network. This in fact relaxes assumption 2 above. An example will be offered to illustrate this situation later.

The above line of argument shows the need for a comprehensive model over the periods that takes into account both expected response time and cost. To do that, we have to define an explicit cost-function for relocation; and the transition

---

[11]This is referred to as Markovian property. Consult the "Markovian Process" appendix in Chan (2000) for a review of this concept.

probabilities between the various states of the network. An example Markovian matrix for our ongoing example is shown below

$$\Pi = \begin{bmatrix} \pi_{11} & \pi_{12} \\ \pi_{21} & \pi_{22} \end{bmatrix} = \begin{bmatrix} 0.25 & 0.75 \\ 0.5 & 0.5 \end{bmatrix}. \qquad \textbf{(3-74)}$$

This matrix says that if it is known that the network is now at state 1, it has a 25% chance of being in state 1 in the next time interval, and a 75% chance of switching to state 2. When the network is in state 2, it has 50-50 chance of being in either of the possible states during the next time interval. Ignoring the relocation costs for the time being, the expected response-time can be expressed as $\bar{r}(\tau') = P_1'\bar{\tau}_j(1) + P_2'\bar{\tau}_j(2)$ where P'$_1$ and P'$_2$ are the steady-state probabilities of being in states 1 and 2 respectively. These probabilities can be computed from the transition matrix by the system of equations (of which one is redundant):[12]

$$\begin{pmatrix} P_1' & P_2' \end{pmatrix} \Pi = \begin{pmatrix} P_1' & P_2' \end{pmatrix}$$
$$P_1' + P_2' = 1. \qquad \textbf{(3-75)}$$

For our contrived example, these equations are to be solved

$$0.25 P_1' + 0.5 P_2' = P_1'$$
$$P_1' + \quad P_2' = 1$$

yielding $P_1' = 0.4$ and $P_2' = 0.6$ as given before.

The objective function is to minimize the sum of the expected response-time and the expected relocation-costs, $\min\ [w\,\bar{r}\,(\bar{\tau}'(d)) + g(d)]$ where w is the number of dispatches per one transition epoch and $g(d)$ is the expected relocation cost (in time units). When $j^*(1) = j^*(2)$ or when no depot-relocation takes place, the term $g(d)$ becomes zero. Let us define by a relocation strategy any pair $\{j^*(1), j^*(2)\}$ such as $\{(1, 3), (2, 4)\}$, showing the two home-nodes or depots in states 1 and 2 respectively. Observe that there are 100 strategies for our problem representing pairing between all combinations of two home-nodes taken out of five, viz. $\binom{5}{2}^2 = 100$.

Define a *simple strategy* to be a no-relocation strategy where $j^*(1) = j^*(2)$. Among the simple strategies $\{(1, 4), (1, 4)\}$ is the best based on our calculation thus far as indicated in the last row of **Table 3-6**. It can serve as an *incumbent*[13] solution of cost 1.7 in our search for the best strategy.

---

[12]For a review of this concept, refer to the "Markovian Process" appendix in Chan (2000).

[13]For a formal definition of incumbency and how it assists in a search for best solution, see the "Prescriptive tools" chapter in Chan (2000) under branch-and-bound.

Let us assume that $w = 1$ and that the relocation cost-function is $g(d) = (0.1)(d^{ij})^{1/2}$. The partial cost $wr(\tau(d))$ of the non-simple strategies can be obtained easily from **Table 3-6** by simply summing relevant terms in the second-to-last row. For example, strategy $\{(1, 3), (2, 4)\}$ has a partial cost of $(0.4)(2.5) + (0.6)(3.5) = 3.1$. All those strategies that are dominated by the incumbent can be eliminated. A close examination of **Table 3-6** will suggest that all the non-simple strategies except $\{(2, 4), (1, 4)\}$ and $\{(3, 4), (1, 4)\}$ can be discarded. These represent partial costs of 1.6 and 1.5 respectively. Now let us compute the relocation costs. Suppose we start with strategy $\{(3, 4), (1, 4)\}$. When the network makes a transition from state 1 to state 2, there are two relocation possibilities:

a. From 3 to 1 and "stay put" at 4, resulting in a relocation cost of $(0.1)(d^{31}(2))^{1/2} + (0.1)(d^{44}(2))^{1/2} = 0.141$;
b. From 3 to 4 and from 4 to 1, resulting in a cost of $(0.1)(d^{34}(2))^{1/2} + (0.1)(d^{41}(2))^{1/2} = 0.631$.

The first strategy is preferred, with a lower repositioning cost of 0.141. Also, we have to compute the relocation cost for strategy $\{(3, 4), (1, 4)\}$ when a transition takes place from state 2 back to state 1. Again there are two possibilities.

a. From 1 to 3 and "stay put" at 4, with a relocation cost of $(0.1)(d^{13}(1))^{1/2} + (0.1)(d^{44}(1))^{1/2} = 0.141$;
b. From 1 to 4 and from 4 to 3, with the cost of 0.631.

The first possibility, being the lower cost of the two, is again chosen.

The total cost of strategy $\{(3, 1), (1, 4)\}$ is made up of the partial cost of 1.5 and the relocation cost of $(0.6)(0.5)(0.141)+(0.4)(0.75)(0.141) = 0.0846$, making a total of 1.5846. In the relocation cost, we consider with probability 0.6 that the system is in state 2, and with probability 0.5 that the network switches from state 2 to state 1. This is similar for the other way around from state 2 to state 1. Since $1.5846 < 1.7$, strategy $\{(3, 4), (1, 4)\}$ is better than the incumbent $\{(1, 4), (1, 4)\}$. The partial cost of the second candidate-strategy $\{(2, 4), (1, 4)\}$ is 1.6. This is higher than 1.5846. This strategy can be eliminated and $\{(3, 4), (1, 4)\}$ is the optimal strategy. The conclusion is that the optimal solution is to locate the facilities at nodes $(3, 4)$ when the network is in state 1 and at nodes $(1, 4)$ when the network is in state 2. When the system changes from state 1 to state 2, the service units on node 4 will remain in its place and the ones at node 3 will move to node 1. The reverse will take place as the state changes from 2 to 1. *It can be shown (Berman et al. 1990) that for this repositioning model, at least one set of optimal locations exists on the nodes of the network.*

The entire problem can be formulated as an integer program. Let $y_{ijk} = 1$ when node i is served by facility j when the network is in state k, and 0 otherwise; $y_{jk} = 1$ if the facility is at node j when the network is in state k, and 0 otherwise; $y_{u(k),v(l)} = 1$ if the facility u in state k is relocated to v when the network changes to state l, and 0 otherwise. The integer linear-program seeks to

$$\min \left\{ \sum_{k=1}^{K} P_{k}' \sum_{i \in I} \sum_{j \in I} f_{ik} d^{ij} y_{ijk} + \sum_{k=1}^{K} P_{k}' \sum_{l=1}^{K} \pi_{kl} \sum_{u \in I} \sum_{v \in I} y_{u(k),v(l)} g[d^{uv}(k)] \right\}. \qquad (3\text{-}76)$$

In this objective function, it is interesting to note the following. There is similarity between the "relocation term" notation and the objective functions documented in Equations **(3-59)** and **(3-62)**, particularly between "state transition" and "cyclic permutation."

The first constraint of this integer program assures that every node is served by a facility

$$y_{ik} + \sum_{j \in I(j \neq i)} y_{ijk} = 1 \qquad i \in I; \ k = 1, \dots, K. \qquad (3\text{-}77)$$

The second constraint limits the dispatching of services only from nodes where facilities are located

$$y_{jk} \geq y_{ijk} \qquad i \in I; \ i \neq j; \ k = 1, \dots, K. \qquad (3\text{-}78)$$

The third and fourth constraints limit the relocations only to nodes that are also home locations

$$\sum_{v \in I} y_{u(k),v(l)} = y_{uk} \qquad u \in I; \ k, l = 1, \dots, K \qquad (3\text{-}79)$$

$$\sum_{u \in I} y_{u(k),v(l)} = y_{vl} \qquad v \in I; \ k, l = 1, \dots, K. \qquad (3\text{-}80)$$

The last set restricts the number of facilities to p

$$\sum_{j \in I} y_{jk} = p \qquad k = 1, \dots, K. \qquad (3\text{-}81)$$

Using this integer program to solve the sample problem again yields the same optimal strategy $\{(3, 4), (1, 4)\}$ as expected (Berman et al. 1990). In fact this solution makes intuitive sense. The server at node 3 moves to node 1 in order to maintain some proximity to node 2 whenever the network undergoes a transition from state 1 to state 2. It turns out that this is the optimal strategy as long as the relocation cost function is $g(d) = \alpha d^{1/2}$ with $0 \leq \alpha \leq 0.2357$. When $\alpha > 0.2357$, the simple strategy $\{(1, 4), (1, 4)\}$ becomes optimal. This is not surprising since the relocation cost is high when α is large. The integer program becomes extremely large as $|I|$, p, and K grow. Heuristic is the only way to solve the problem under these circumstances (Ahituv and Berman 1988).

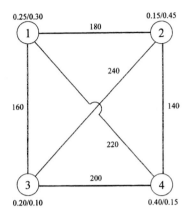

Figure 3-25 Facility-relocation problem

Example

The demand-generation rate is given as 3.6 services per day, f_{ik} is correspondingly the product of 3.6 and the probability of demand from each node (as shown in **Figure 3-25**.) For clarity, this simplified example will only place one home-node or facility in **Figure 3-25**. The cost of relocation in minutes between nodes u and v, defined as $P'_k \pi_{kil} g[d^{in}(k)]$, are the same in both states and are given below:

Relocation between	Relocation time (min–both ways)
1–1, 2–2, 3–3, 4–4	6
1–2 / 2–1	42
1–3 / 3–1	36
1–4 / 4–1	54
2–3 / 3–2	60
2–4 / 4–2	28
3–4 / 4–3	48

Here the relocation times for both directions are separated by a 'slash'. Using the same state-transitions as Equation **(3-74)** with the corresponding $P'_1 = 0.4$ and $P'_2 = 0.6$, we formulate the binary integer-program as follows. Here we simplify the notation $y_{u(k),v(l)}$ as $y_{u,v}$, inasmuch as the context will make it clear regarding the proper state transitions:

min $[64.8y_{121}+57.6y_{131}+79.2y_{141}+97.2y_{211}+129.6y_{231}+75.6y_{241}+$
$19.2y_{311}+28.8y_{321}+24y_{341}+39.6y_{411}+25.2y_{421}+36y_{431}+$
$64.8y_{122}+57.6y_{132}+79.2y_{142}+48.6y_{212}+64.8y_{232}+37.8y_{242}+$
$86.4y_{312}+129.6y_{322}+108y_{342}+138.6y_{412}+88.2y_{422}+126y_{432}+$
$6y_{1,1}+42y_{1,2}+36y_{1,3}+54y_{1,4}+42y_{2,1}+6y_{2,2}+60y_{2,3}+28y_{2,4}+$
$36y_{3,1}+60y_{3,2}+6y_{3,3}+48y_{3,4}+54y_{4,1}+28y_{4,2}+48y_{4,3}+6y_{4,4}]$

subject to

$y_{11}+y_{121}+y_{131}+y_{141} = 1$ (node 1 is served by a facility in state 1)
$y_{21}+y_{211}+y_{231}+y_{241} = 1$ etc.
$y_{31}+y_{311}+y_{321}+y_{341} = 1$
$y_{41}+y_{411}+y_{421}+y_{431} = 1$
$y_{12}+y_{122}+y_{132}+y_{142} = 1$ (node 1 is served by a facility in state 2)
$y_{22}+y_{212}+y_{232}+y_{242} = 1$ etc.
$y_{32}+y_{312}+y_{322}+y_{342} = 1$
$y_{42}+y_{412}+y_{422}+y_{432} = 1$
$-y_{21}+y_{121} \le 0$ (in state 1, location at node 2 precedes service)
$-y_{31}+y_{131} \le 0$ etc.
$-y_{41}+y_{141} \le 0$
$-y_{11}+y_{211} \le 0$
$-y_{31}+y_{231} \le 0$
$-y_{41}+y_{241} \le 0$
$-y_{11}+y_{311} \le 0$
$-y_{21}+y_{321} \le 0$
$-y_{41}+y_{341} \le 0$
$-y_{11}+y_{411} \le 0$
$-y_{21}+y_{421} \le 0$
$-y_{31}+y_{431} \le 0$
$-y_{22}+y_{122} \le 0$ (in state 2, location precedes service at node 2)
$-y_{32}+y_{132} \le 0$ etc.
$-y_{42}+y_{142} \le 0$
$-y_{11}+y_{212} \le 0$
$-y_{32}+y_{232} \le 0$
$-y_{42}+y_{242} \le 0$
$-y_{12}+y_{312} \le 0$
$-y_{22}+y_{322} \le 0$
$-y_{42}+y_{342} \le 0$
$-y_{12}+y_{412} \le 0$
$-y_{22}+y_{422} \le 0$
$-y_{32}+y_{432} \le 0$
$-y_{11}+y_{1,1}+y_{1,2}+y_{1,3}+y_{1,4} = 0$ (starting from node 1 & state 1, relocation to a home-node only)
$-y_{21}+y_{2,1}+y_{2,2}+y_{2,3}+y_{2,4} = 0$ etc.
$-y_{31}+y_{3,1}+y_{3,2}+y_{3,3}+y_{3,4} = 0$
$-y_{41}+y_{4,1}+y_{4,2}+y_{4,3}+y_{4,4} = 0$
$-y_{12}+y_{1,1}+y_{1,2}+y_{1,3}+y_{1,4} = 0$ (starting from node 1 & state 2, relocation to a facility only)
$-y_{22}+y_{2,1}+y_{2,2}+y_{2,3}+y_{2,4} = 0$ etc.
$-y_{32}+y_{3,1}+y_{3,2}+y_{3,3}+y_{3,4} = 0$
$-y_{42}+y_{4,1}+y_{4,2}+y_{4,3}+y_{4,4} = 0$
$-y_{11}+y_{1,1}+y_{2,1}+y_{3,1}+y_{4,1} = 0$ (relocating from a facility only to node 1 in state 1)
$-y_{21}+y_{1,2}+y_{2,2}+y_{3,2}+y_{4,2} = 0$ etc.
$-y_{31}+y_{1,3}+y_{2,3}+y_{3,3}+y_{4,3} = 0$
$-y_{41}+y_{1,4}+y_{2,4}+y_{3,4}+y_{4,4} = 0$
$-y_{12}+y_{1,1}+y_{2,1}+y_{3,1}+y_{4,1} = 0$ (relocating from a facility only to node 1 in state 2)
$-y_{22}+y_{1,2}+y_{2,2}+y_{3,2}+y_{4,2} = 0$ etc.
$-y_{32}+y_{1,3}+y_{2,3}+y_{3,3}+y_{4,3} = 0$

$-y_{42}+y_{1,4}+y_{2,4}+y_{3,4}+y_{4,4} = 0$

$y_{11}+y_{21}+y_{31}+y_{41} = 1$ (only one facility to be located in state 1)

$y_{12}+y_{22}+y_{32}+y_{42} = 1$ (only one facility to be located in state 2).

Solution suggests that the minimum objective-function-value of 399.8 minutes is achieved through a policy of operating the vehicles at a facility located at node 2 while the network is in state 1 and at node 4 while the network is in state 2 (Mohan and Chan 1993). ☐

3.7.6 Repositioning between non-state-specific home nodes.

When assumption 2 in the section entitled "*Repositioning in a stochastic network*" above is relaxed, the operator of the service does not have to use the same set of p locations for the p servers every time the network is in a given state k. Thus the set of p locations to be used when the system is in state k depends not only on the current state k, but also on the entire history of the system. For a Markovian network, this past history is of course summarized by the last state of the network. This means that the server assignment/location strategy of each state must be afunction of the last state and of the position of the servers in that last state, a total of $K\binom{|I|}{p}$ possible previous 'histories' and $\binom{|I|}{p}$ choices for server locations for each of the K states. In the ongoing example, this amounts to 200 possibilities. The integer-programming formulation is analogous to Equations **(3-76)** through **(3-81)**, so long as this higher dimensionality is recognized. For example, l can be the same as k in Equation **(3-76)** and Equations **(3-79)** and **(3-80)** no longer apply (see Berman and Rahnama [1985] for details). The optimal policy for location decisions in the stochastic network in **Figure 3-24** is identical to the one obtained when the assumption is made. In other words, the optimal strategy turns out to be independent of the network's Markovian `history'. Similar to the previous model when assumption 2 is enforced, *at least one set of optimal locations exists on the nodes of the network* (Berman et al. 1990).

Another example is offered to illustrate the case when relocation takes place between nodes that are not necessarily home nodes of the respective states. This example compares the optimal locations before and after one relaxes assumption 2 of the above relocation model. Consider the network of **Figure 3-26** with four nodes and six arcs whose travel times are denoted by $c_1, c_2, ... , c_6$ as shown. Let $f_i = 1/4$ for all i and let us locate optimally one server. This network is assumed to have six states. In state k, $c_k = 2$ and all other c are 1. All state transition probabilities π_{kl} are 1/6 ($k, l = 1, ..., 6$), leading to steady state probabilities $P_k' = 1/6$ ($k = 1, ..., 6$). We will assume that the cost of relocation $g(d)$ is very, very small in comparison to the expected response-time that it is negligible. There are $6\binom{4}{1} = 24$ states in this problem. For each state there are $4^6 = 4096$ possible strategies, for which enumeration would be impractical. It can be seen that in each state there are exactly two locations from which the server can optimally serve demand in that state. At least one optimal location exists on the nodes of the network. For example in state 1, placing the server at either node 2 or 3 is optimalfor that state alone. In

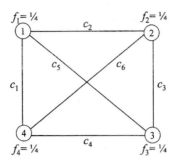

Figure 3-26 Example of repositioning between non-state-specific-home nodes
(Berman et al. 1990)

general, placing the server at either node that is not adjacent to arc k is optimal for
state k. Let us denote this set of two nodes by $j^*(k)$.

Now some thought will reveal the following if the relocation cost is sufficiently
small. The optimal strategy for operating this system under the problem scenario
here is "move the server to either node in $j^*(k)$ whenever the system enters state k,"
unless the server is already at a node in $j^*(k)$, in which case we do not move the
server. Clearly, in this optimal strategy the location of the server is allowed to vary
each time state k is entered, depending on the previous location of the server (which
obviously will depend on the state that the system leaves to enter current state k).
The optimal relocation-decisions, starting with state 1 is summarized in **Table 3-7**.
This Markovian relocation strategy, of course, is not permitted under assumption
2, which requires that the server be designated a single optimal-location in each
state $k = 1, \dots , 6$, or what we call state-specific home-nodes. It can be shown
without too much effort, that for assumption 2, there are several optimal strategies
in this example. One such strategy is $j^*(1) = 2$, $j^*(2) = 3, j^*(3) = 1, j^*(4) = 1, j^*(5)$
$= 2$, and $j^*(6) = 1$. All these strategies have an associated total-cost that exceeds
by an amount $g(d)/9$ the total cost achieved under the optimal strategy when
assumption 2 is relaxed.

3.7.7 Comments. Notice the mobile-server location problems presented here
are very general ones, in the sense that many known versions of facility-location
problems on networks can be viewed as special cases of the models above. While
we have pointed out some limiting cases where analytical properties are discernable, the general case can only be solved by simulation. The Hakimian p-median
model is the limiting case of the stochastic-queue-median model. Recall the
probabilistic-network problem discussed in the "Facility Location" chapter, in
which stationary facilities must be located on a network that undergoes probabilistic
transitions among K states. It is again a special case of the random-travel-time
model in the section entitled "*Networks with random travel times*" with a permanent
set of home nodes among all states. In others words, only simple strategies $j^*(1) =$

State k	State-specific home nodes[1]	Non-state-specific home nodes[2]					
		$k=1$	$k=2$	$k=3$	$k=4$	$k=5$	$k=6$
1	2*	2	3	1	2	2	1
	3	3	3	1	1	2	3
2	3*	3	3	1	1	2	3
	4	2	4	4	1	4	1
3	1*	2	3	1	1	2	1
	4	2	4	4	1	4	1
4	1*	2	3	1	1	2	1
	2	2	3	1	2	2	1
5	2*	2	3	1	2	2	1
	4	2	4	4	1	4	1
6	1*	2	3	1	1	2	1
	3	3	3	1	1	2	3

[1] Asterisks indicate the relocation pattern for state-specific home nodes starting from state 1 thorugh state 6.
[2] To read these columns, let us give an example. Suppose the home node is at 2 in state 1. Should the state change to 1, 4 or 5, the best relocation strategy is to stay at node 2. If state 2 is entered, relocate to node 3. Should the state be 3 or 6, then relocate to node 1.

Table 3-7 State-specific and non-state-specific relocations decisions
(Berman and Rahnama 1985)

$j^*(2) = \dots = j^*(K)$ are permitted. If in the probabilistic-network problem we further specify the number of states K to be reduced to 1, we obtain the standard p-median problem.

Measuring spatial separation can be as involved or as simple depending on what the application would dictate. Suffice to say that a modern-day-service-economy requires that we carefully review the idea of spatial separation that may not be as simple as geographic distances. With transportation and telecommunication advances in recent decades, sometimes it is easier to interact with distant contacts than the next office over. It may be attributable to congestion, random travel-time or a combination of both (setting aside sociological factors in our present discussion). In analyzing locational decisions today, it is important to build this phenomenon into the models we use. The step taken above is the beginning of an initiative that will continue, as long as the service economy and technology continue to develop. Of equal importance, however, is to outline the conditions under which sophisticated representation of spatial separation may not be necessary. Some of the nodal optimality conditions established above suggest that in limiting conditions, traditional approximations of spatial separation are entirely adequate. Today, a facility is no longer synonymous with a permanent location. Given time-dependent measures of separation, repositioning of facilities from one period to another becomes increasingly critical in responding to changing conditions on the delivery network. Here again, approximations are made to repositioning strategies, instead of truly time-dependent transitions. If nothing else, the above discussions hopefully demonstrate the repositioning idea lucidly through numerical examples.

3.8   Including random demands in accessibility measures

How well do the approximate methods estimate the actual distance in locational decisions? Chan and Merrill (1997) formulated a probabilistic-travelling-salesman-facility-location problem (PMTSFLP). Here demands f_i are random variables. The model locates a central depot that can respond to these sporadic demands with an *a priori* tour in the least expected-cost. The problem is formulated as a combination of a probabilistic 1-median problem and a multiple-TSP. The median problem looks exactly like the p-median problem in the "Facility Location" chapter when $p = 1$ and f_i is interpreted as the probability of demand.

   3.8.1 Routing as spatial separation. Previously, the separation between i and j, d_{ij} was modelled by Euclidean distance as typical with facility-location models. For the sake of comparison, a routing model will be used to compute the exact d^{ij}. Instead of the classic multiple-TSP formulation, these constraints are used for vehicle-fleet availability at the depot (node 1):

$$1 \le \sum_{j \in M_1} x_{1j} \le |H''| \quad and \quad 1 \le \sum_{i \in M_1} x_{i1} \le |H''| \tag{3-82}$$

where $|H''|$ was set at 4 in Merrill's experiments suggesting at most four vehicles being available. This allows the model to size the fleet based at a depot. Hence, PMTSFLP is characterized not only by uncertain demand, the number of 'salespersons' (airplanes in this case) sent on the road is also a random variable. In the Air Force, the actual number of airplanes used to solve the flight-inspection problem, for example, is determined by the commander based on timeliness-of-response and a 'reasonable' time for the flight crew to be "on the road."
   A trip must enter and leave a demand node via different ways except to and from the depot:

$$x_{ij} + x_{ji} \le 1 \qquad \forall j\,(j \ne 1), \forall i\,(i \ne 1) \tag{3-83}$$

In other words, these effectively keep vehicles from "doubling back" except to and

$$x_{i1} + x_{1i} \le 2 \qquad \forall\, i\,(i \ne 1). \tag{3-84}$$

from the depot, which means the minimum subtour would have to involve three nodes. Since Chan and Merrill's case study initially involves low nodes-visited-to-vehicle ratio, this effectively precludes subtours. Finally,

$$\sum_{i \varepsilon M_j} x_{ij} = 1 \quad and \quad \sum_{i \varepsilon M_j} x_{ji} = 1 \tag{3-85}$$

for all j except the depot (i not equal to j). Note that this constraint set satisfies the continuity requirement specified by Equation **(3-20)** above with $m = n =$ depot, but is more stringent. Furthermore, this constraint forces exactly one entry and one exit to and from every node other than the depot. This simplified multiple-TSP linear program works very efficiently up to seven nodes. To ensure that these simplified subtour-breaking constraints work, conventional subtour-breaking-constraint **(3-23)** is used to validate the resulting tours.

Crew-duty-day restrictions are enforced according to Equation **(3-45)** with $K = 1$. Here, $U(1)$ is the crew-duty-day limitation in hours. Subtour-breaking equation **(3-45)** is supplemented by Equations **(3-46)** and **(3-47)**. Thus the tours are designed, both manually by the commander-in-the-field as well as the analytical models here, to fit a mission (i.e., each salesman-tour) within the "time on the road."

In all cases, the simplified constraints and the crew-duty-day requirements are validated (Reynolds et al. 1990). While not a proof for the formulation, Equations **(3-82)–(3-85)** appear to work well to break subtours for small networks. The combined location-routing model is the combination of the 1-median and multiple-TSP. Here d_{ij} is replaced by d^{ij} to mean distance measured along a tour that connects depot j to demand i, rather than a Euclidean distance between i and j necessarily. Here the distance measure $d^{ij} = |\sigma_i - \sigma_j|$ is the difference in "odometer readings," where the 'odometer' records "time on the road."

3.8.2 Probabilistic Travelling-Salesman Facility-Location Problem. The Probabilistic Travelling-Salesman Facility-Location Problem (PTSFLP) can be defined as a location-routing problem in which the demands are random. Thus demand comes from node i with probability f_i. The demand list constitutes the basis for planning an optimal tour and locating the depot, so that the expected tour-length over all possible demand instances is at its minimum. Thus if demand at node i exists then the node must be visited; otherwise, a visit is not required. In flight inspection the size of the demand is not relevant; demand is therefore modelled as a random variable with realization zero and one. Obviously, one way to solve the PTSFLP is to solve all instances of the constituent Travelling-Salesman Facility-Location Problem (TSFLP), although this can be a laborious process. But there are some recent findings on the TSFLP that can be exploited to bypass this imposing (if not impossible) task.

Distinction is made here between *a priori* tours, where it is possible to minimize the expected tour-length by taking the expected-value of the random-demand vector. When operating *posterior* tours, one selects a minimum-length tour for each demand realization by complete enumeration. When the number of vehicles h is fixed in optimization, lengthy tours may result, which can be longer than the dispatcher likes to have the crew away from home base. *Post optimization* caters for these situations when *optimization* violates the "time on the road" constraint **(3-45)**. More vehicles are subsequently used to shorten the `range' of the vehicle. As h is increased to satisfy the "time on the road" constraint for each salesperson (aircraft crew), we pay for this through an increased total tour length among all salespersons (i.e., inducing feasibility for a "super-optimum"). The total tour-length is monotonically non-decreasing as we move toward a 1-median depot-solution (or when $h \rightarrow |I|$.) As we approach the 1-median solution, the total tour-length is at its longest while individual salesman-tour is at its shortest. Obviously, *post optimization* is of secondary importance here, since it is relatively straightforward. It is used to conduct control experiments, in which historical flight records are revisited to assess how performance measures against optimization.

Hakimi (1964) suggested that the medians of a network are found among the nodes. Berman and Simchi-Levi (1988) extended Hakimi's original finding to the TSFLP, suggesting that the optimal location have to occur at a node. The TSFLP then reduces to finding node j for all the instances of TSFLP that minimizes $v(j) = \sum_{S \subseteq I} P(L) \, \mathcal{L}(L \cup j)$. Here $P(L)$ is the probability that instance L occurs with nodal demands at a subset of nodes S in I. $\mathcal{L}(L \cup j)$ is the length of the optimal TST $T(L \cup j)$ for the instance L with the base located at j. And for independent f_i, $P(L) = \prod_{i \in L} f_i \prod_{i \in I-L} (1-f_i)$. Bertsimas (1989) showed that if $f_j = 1$, or demand exists at a facility with certainty, the optimal location for the facility (depot) is at j if the distance matrix satisfies the triangle inequality $d_{ij} \leq d_{ik} + d_{kj}$. Furthermore he extended a solution heuristic by Berman and Simchi-Levi, in which a relative worst case error of $(1-f_j)/2$ is achieved, when j is the optimal location. When $f_j = 1$ for all j, the problem reduces to a deterministic TSFLP, and the solution is expected to be exact.

As suggested above, it is cumbersome (if not impossible) to compute the optimal tour in every instance. In the Probabilistic Travelling-Salesman Problem (PTSP) literature such as that by Berman and Simchi-Levi (1988a) and Laporte et al. (1990), the idea of an *a priori* tour based at j is introduced, $T(j)$, visiting all potential demands. For a particular instance, one skips nodes without a demand. The problem then boils down to finding the location j and the 'master' tour T that minimizes $g(j,T) = \sum_{L \subseteq I} P(L) \, L_T(L \cup j)$, where $L_T(X)$ is the length of the master tour restricted to sites in X. This brings us face-to-face with the PTSFLP, where it has been shown that the optimal location occurs at a node also (Berman and Simchi-Levi 1988a). The PTSFLP can be reduced to the solution of n PTSPs. Given the vector of probabilities $(f_1, f_2, \dots, f_{|I|})$ and the optimal location at node j, the corresponding optimal tour $T(j)$ is the PTST with the vector of probabilities $\mathbf{P}_j = (f_1, \dots, f_{j-1}, 1, f_{j+1}, \dots, f_{|I|})$. Notice the depot carries with it a unitary probability. The problem then boils down to finding the $|I|$ optimal Probabilistic Travelling-

Salesman Tours (PTSPs) corresponding to the vectors of probabilities $\mathbf{P}_j$ ($j = 1, 2,$ ... ,$|I|$) and then select the one with the minimum expectation. The resulting j^* is the optimal location. It can be shown that this result can be generalized to the Probabilistic Multiple Travelling Salesmen Facility Location Problem (PMTSFLP).

Independent of one another, Bertsimas (1988) and Merrill (1989) arrived at a similar SFC heuristic for solving the PTSFLP and the PMTSFLP respectively. Here, we state an extended version of Merrill's solution, which is a more general algorithm for the multiple-salesmen application to be discussed in this chapter:

Step 1 Given the coordinates of the locations of all the demands in the set I, use SFC to find *a priori* master-tours T_k.

Step 2 Compute for every j ($j = 1, 2, ... ,|I|$), $G_n(j, T_k)$, where G_n is the generalization of the tour-length function g with k vectors of probabilities $\mathbf{P}_j(k)$ corresponding to k salespersons covering $|I|$ demand nodes without replication. Specifically $\mathbf{P}_j(k) = (f_1, ... ,f_{m_k} ; f_j = 1)$, where $m_k \leq |I|$, and with the set of nodes I partitioned into $I = I_1 \cup I_2 \cup ... \cup I_k$, where $I_i \cap I_j = \varnothing$ for $i \neq j$. The union notation simply denotes the partition of $|I|$ nodes on a space-filling curve into $m_1, m_2, ... , m_k$ number of nodes based on clustering, with $m_1 + m_2 + ... + m_k = |I|$. An extended SFC heuristic, the multiple-SFC is used to operationalize this.

Step 3 Select the point j^* ($j \in I$) that minimizes $G(j, T_k)$. Location j^* and the tours T_k^* constitute the solution to PMTSFLP, call this $G_n^*(j^*, T_k^*)$. If optimization is sought where k^* is given, stop; otherwise proceed.

Step 4 Perform steps 1 through 3 for all k where $k^* \leq k \leq |H''|$. The minimum tour-length $G_n^{**}(j^{**}, T_k^{**}) = \min_{k^* \leq k \leq |H''|} G_n^*(j^*, T_k^*)$ determines the optimal facility-location j^{**}, the optimal tours T_k^{**}, and the number of vehicles k^{**} to be employed to serve all *potential* demands. ■

Aside from the optimization vice post-optimization distinction, Step 4 is necessary in view of the "time on the road" constraint **(3-45)**, which could make T_k^* infeasible. Since SFC does not have a range constraint explicitly built in, Step 4 will enforce the "range constraint," assuming K is large enough to make it possible.

It can be shown that the algorithm has a complexity of $O(|I|^2)$ for $k < |I|$. The heuristic above yields a tour length at a factor of $O(\log |I|)$ from the optimal PMTSFLP-tour-length [Bartholdi and Platzman (1988, 1982)]. It is asymptotically optimal and equivalent to the Multiple-Travelling-Salesmen Facility-Location Problem (MTSFLP). Finally, for $|I|$ demand points that are randomly distributed, $G_n(j, T_k)/w(j^{**}) \leq G_n^*(j^*, T_k^*)/z(j^{**}) \leq G_n^{**}(j^{**}, T_k^{**})/w(j^{**}) = 0(\log |I|)$ where $z(j^{**})$ is the optimal tour-length from the analytic model. In the asymptotic case when $|I|$ is sufficiently large, $\lim_{|I| \to \infty} G_n(j, T_k)/z(j^{**}) = 1.25$.

Consider an instance L of the problem. Suppose the SFC heuristic produces a tour $T(L)$ if we run the heuristic on the instance L. Let $T(0)$ be the tour produced by the heuristic on the original instance. Then $T(L) = T(0)$, or the order of visit among the two tours is the same. The reason is that the SFC sorting preserves the order, which is exactly the property of the PTSP as well (Bertsimas 1989). This is akin to the observation by Hargrave and Nemhauser (1962) that the order of visitation around the convex hull circumscribing a network does not change in reaching the optimal solution. Suppose one can map points in Euclidean space onto a unit square and overlay a circuit generated by a SFC which visits all nodes within the square. Then the order that the nodes are visited as one travels along the curve will approximate the order the points would be visited by a travelling salesperson on an optimal tour. Closeness of orders here refers to tour length. Because the multiple-SFC is nothing more than the regular SFC with special partitioning, the asymptotic values of the regular SFC approximation are expected to hold. The reader is reminded that the order of visit in the master tour remains the same irrespective of the partitioning. Also, we should think about the correspondence between the ψ-space and x–y space in terms of proximity measures. Thus in the asymptotic case when there are ubiquitous demands over the x–y space (which mapped into ubiquitous distribution on the ψ-space), the partition among salespersons becomes unimportant. In general, the k intervals (tours in T_k) in the Sierpinski curve can be chosen in a number of ways (Bowerman and Calamai 1994). The first is the naive approach of breaking the curve up into intervals of $|I|/k$ points. The second is to break the curve up at the largest gaps. The third is to break the curve at k sites where the demands are all found within a small circle. We suggest breaking the curve up at the k largest gaps, for preservation of proximity and minimization of individual tour-lengths. These clusters will form the demands for which tours $T_k(j)$ will be constructed from a candidate home-base j.

To see the implications of this result, consider two special cases. Case 1 is when the $|I|$ demand points ($|I| \rightarrow \infty$) are randomly distributed on the Sierpinski curve. In this case, the clustering partition-scheme gives the same result as the scheme where the curve is equally divided into k parts. The end product is $r = 1.25$. Consider case 2 when the $|I|$ demand points are densely packed into k different clusters. In the limiting case, each cluster degenerates into a single point. Now the MSFC heuristic yields k "out-and-back" tours from the home depot. Then $r = 1$, or the clustering algorithm is perfect since no mis-classification is introduced due to distance distortions within a cluster during the SFC-transformation $\Psi(X_i, Y_i)$.

Notice that when k is determined this way, or by the LP-relaxation or mixed-integer-programming formulation of the section entitled "*Routing as spatial separation*," it is a random variable. Thus in the SFC heuristic, if all tours are selected *a priori*, and the partitioning into individual tours for each salesperson is performed *after* observing the demand realization, then the number of salespersons is a random variable. On the other hand, if all decisions are made based on the data in **Table 3-8** and following the LP-relaxation formulation in the section entitled "*Routing as spatial separation*," then the number of salespersons is given and no longer a random variable.

Mission Aircraft	No. of Bases Visited	Bases Visited					
1	2	Mood	Dobbins	MacDill	Tyndall	Hurlburt	
2	3	Patrick	MacDill	Kingsville	Ellsworth	Range1	
3	2	Williams	March	Moody			
4	3	Barksdale	Oklahoma City	Dallas	Ellington	Kingsville	
5	2	Williams	McClellan	McChord	Andrews	England	
6	2	Libby	Davis-Monthan	Reese	Columbus		
7	3	Keesler	Patrick	Volk	Richards-Bebaur		
8	3	Offutt	Little Rock	Robins	Myrtle Beach	Shaw	Moody
9	4	Homestead	Palmerola	Andrews	Kirtland	Offutt	Range1
10	3	Richards	Oklahoma City	Fort Smith	Barksdale	Redstone	
11	2	Cannon	Amarillo	Oklahoma City	Kirtland	Nellis	Range2
12	4	Little Rock	Wright-Patterson	Carswell	Nellis	Peterson	Langley
13	3	Pope	Richards-Gebaur	Fairchild	Carswell	Ellsworth	Maxwell

Table 3-8 Selected data-base of flight inspections from Scott

3.8.3 Case study of military-base location. The navigational-aid inspections by the U. S. Air Force to military facilities offer a unique opportunity to study the PMTSFLP. At the time of the original study (Merrill 1989), the Facilities Checking Squadron at Scott Air Force Base was scheduled to be operating six new C-29 aircraft to replace the then existing fleet. This created an environment for re-examining basing decisions. Operating cost and timeliness-in-response were only two among a myriad of considerations (which include political implications of base closing). But they will be the two main concerns of this study. This in turn leads to an implicit assumption: We consider only existing bases—rather than new bases—that can support such a fleet of inspection aircraft without additional capital investment. Four bases at the central part of the United States and two at the East and West coasts are candidates for close examination.

Historical data of flight-inspection missions over a one-year period were examined. The data were grouped by mission type and time frame. Only missions accomplished within ten duty-days of each other were grouped together considering crew-duty requirements. This is a reflection of timeliness-in-response as well as the common practice wherein a crew is out "on the road" for no more than ten days. Pure-inspection tours, operational-evaluation tours and the like were grouped separately to ensure a qualified crew could accomplish the tasking. **Table 3-8** shows a reduced data-base where each mission contains no more than seven nodes. These were extracted from the 1 September 1987 to 31 August 1988. This forms a comprehensive record for validation against the analytical LP model (which can take only up to seven nodes) as formulated in the section entitled "*Routing as spatial separation.*" The bases shown in **Table 3-8** are to be visited by one or more aircraft.

All together there are 13 instances (out of 28) consisting of subsets of bases requiring a visit by an inspection aircraft. Although not shown in the data-set in **Table 3-8**, it should be noted that Scott requires 15 visits during the year of study. In general, a zero frequency is assigned to the home base in the multiple-TSP analytical model (or its inspection needs are assumed to have been "taken care of"), whether it be Scott or other candidate home-base locations. However, the actual frequencies do play a role in locating median and center. While there are bases that have more frequent demands than others, the demand frequencies appear to be quite uniform. In other words, the f_i are fairly close to be equal to one another on the average, with only minor exceptions. An examination of their geographic locations also shows that the bases cover most of the continental United States. There is no particular region that has an overwhelming demand in comparison to others. Similar to others in the 98 total, each base in the continental United States is identified by its longitude and latitude. From this, great-circle distances between bases—reflecting the curvature of the globe—are readily computed. This paves the way for the implementation of SFC heuristic on a microcomputer.

While this constitutes a 'natural' case study for the PMTSFLP, there are some restrictions that prevent a direct implementation of the results reported here. Among the most prominent is the fact that these instances, or grouping of bases to be visited, are somewhat conditioned upon Scott being the base depot. In our study here, these instances are used as demand for all alternate basing-decisions. This may be a good assumption for depots close to Scott in the central United States. But the assumption may become indefensible should a coastal base become a reality. Another problem is that one year of data may not be sufficient to cover all instances to validate *a priori* optimization as discussed here. Because of all these factors, the re-optimization algorithm will give us the theoretical, but not necessarily the actual depot as defined by the evaluation function $G_n(j, T_k)$ in the section entitled "*Probabilistic Travelling-Salesman Facility-Location Problem.*"

3.8.4 Results of case study. As mentioned previously, space-filling-curve (SFC) heuristics were used to tackle the combined location-routing problem on potential demands that may exist in the 98 bases within the continental United

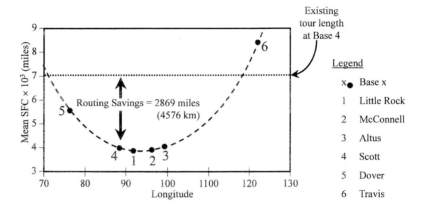

Figure 3-27 Comparison between existing basing location at Scott and alternate bases

States. We also extended the SFC heuristics to handle multiple-salesmen situations, implementing a mere suggestion by Bartholdi and Platzman (1988). In validating the multiple-SFC (MSFC) heuristics, results from the four candidate depots in the central part of the United States were compared with the optimal MTSP results based on the data in **Table 3-8**. In 8 of the 13 instances for McConnell AFB and Altus AFB, 9 of 13 instances for Scott AFB, and 10 of 13 for Little Rock AFB, the results were the same. For those cases where differences exist, the MSFC heuristics always provided answers within 17 percent of the optimal and averaged within 7.6 percent of the optimal among the four basing-decisions.

For the MTSP analytical solutions, we investigated only 13 out of the 24 missions (instances) in which six or fewer nodes are to be visited from the home base (see **Table 3-8**). The 11 missions not analyzed include anywhere from 7 demands (served by 3 aircraft) to 19 demands (served by 4 aircraft), with a majority in the 9- to 10-demand category. The data-base analyzed represents 54 percent of the entire set of missions during the year. Limiting the data-base to 6 demands or less does not seem to bias the data geographically toward any portion of the continental United States. These are the missions, as explained above, in which there are optimal MTSP solutions available, using the simplified set of subtour-breaking constraints contained in Equations **(3-82)–(3-85)**. Bases chosen for basing considerations include six locations: four at the center part of the continental United States (Little Rock, McConnell, Altus and Scott) and two at the coasts (Dover and Travis). The coastal locations are considered since in the larger data-base, occasional trips are made to Asia and Europe. The results of the MSFC heuristics on Scott, the existing base of operation, showed a distance savings of 2869 statue miles (4590 km) of en-route flying per mission over a year. Here we assumed the same number of missions, the same number of aircraft dispatched per mission, and

Base	Little Rock	Mc-Connell	Altus	Scott	Dover	Travis
Location-Routing (MSFC)[1]	1	2	3	4	5	6
Location-Routing (MTSP)[1]	1	2	4	3	5	6
Median[3]	1	3	2	4	5	6
Center[3]	1	2	4	3	5	6

1 The multiple-spacefilling-curve heuristic was run on the 98 bases within the continental United States.
2 This analytical model was run on the 13 instances.
3 The ranking was based on the 13-instance data-set.

Table 3-9 Base ranking using location vs. location-routing analyses

the same visiting locations. This translates to about three additional missions a year or seven additional flight hours per mission. While this, by itself, is not a significant saving, the effect of combined location-and-routing changes can be substantial.

Figure 3-27 shows the various basing decisions. The existing basing location at Scott is characterized by the horizontal straight line at an average of 6996 miles (11,194 km) per tour under existing operating conditions. This compares with the various location decisions at Little Rock, McConnell, Altus, Dover and Travis. It can be shown that the savings from improved routing at 2869 miles (4590 km) at Scott is significant enough to over shadow locational changes to the other bases unless the depot is at 6 (Travis) on the west coast. In reading **Figure 3-27**, one should keep in mind—once again—that the missions are grouped the same as in the mission data as shown in **Table 3-8**, irrespective of the basing decision. As suggested in **Table 3-8**, the number of aircraft used ranged from two to four. Should we allow the stops in the mission to interchange among missions, there would be substantially more savings involved, as alluded to above.

Table 3-9 compares the results of the combined location-routing analysis above with the median and center problems for the six candidate bases under consideration. Line two of the table shows the exhaustive enumeration of the 13 instances of the candidate basing-locations. This means running the LP MTSP 78 times (Moore et al. 1991, Stephens et al. 1991 and Shirley et al. 1992). The first line is the PMTSFLP as formulated above and solved by the MSFC heuristics on the same set of data. The third line is the median problem on the bases in the 13 instances. The fourth and last line is the center problem on the same large data-set. It is interesting that the ranking of the top four bases at the center of the United States is the same based both on the heuristic MSFC and optimal MTSP solutions. The only exception is the third and fourth, which are reversed between the two solution methods (Moore et al. 1991, Stephens et al. 1991). It is surprising to see the degree of similarity between the preference ranking among all four analyses. This is even more surprising considering that the 'valley' appears to be quite 'flat' in **Figure 3-27**, suggesting there is little difference between bases 1, 2, 3 and 4 at the central part of the United States. Perhaps the similarity between the median (for the 13 instances) and location-routing solutions can be explained in terms of an "evenly distributed" demand pattern. Likewise, the similarities between the center(for the 13 instances) and location-routing results can be attributed to a fairly symmetrical demand pattern about the candidate bases considered. Perhaps one can use the Euclidean median and center as the approximation for locational decisions in PMTSFLP when the demands are evenly distributed and are about equal. In the case of $f_i = f$ and $|I| \to \infty$. It can be shown that 1-center, 1-median and PMTSFLP all approach and are all tightly distributed near the same depot location at the center of a unit square (Bartholdi and Platzman 1982).

The data-base analyzed in **Table 3-8** represents 47 bases ($|I| = 47$) in the total of 98 included in the entire continental U.S. To the extent that the accuracy of the MSFC is $O(\log|I|)$, the sample data is expected to yield an estimate at $(\log 47)/(\log 98) = 84\%$ of the best we can do. Again, the amount of agreement between the locational ranking as obtained by MSFC and the remaining three analyses—MTSP, median and center—is striking. It appears that the only difference between MSFC, MTSP and the center is the rank reversal between Altus and Scott, two locations that are very close to one another in tour length according to **Figure 3-27**. The reversal between McConnell and Altus in the median ranking also reflects the minor difference between their tour lengths as shown again in **Figure 3-27**. Aside from the purpose of pre-positioning a home base, the MSFC analysis on the entire 98 bases can also be stored as in a 'Roledex' file *a priori*. This would be used for routing the inspection fleet on a day-to-day basis without any extra computation. This practice is totally compatible with the field environment under which the squadron level is operating, wherein a stop on the *a priori* tour is skipped when there is no demand.

The probabilistic aspect of the PMTSFLP model is so important that we like to re-emphasize the main thrust of this case study one more time. Line one of **Table 3-9** represents the *a priori* tours constructed for real-world considerations when deterministic demands are never available. In such state of ignorance, the decision-maker has no choice but to construct a priori tours out of the comprehensive set of bases that require flight checks—all 98 of them within continental United States. For validation purposes, we also show in line two *posterior* tours based on historical information on 13 instances of demands. It represents the tours selected *after* the demand realization shave been observed. The purpose is simply to show that the SFC heuristic can approximate an analytic location-routing model. It also serves to provide a necessary variant to justify the main thrust of this study, which again, is the construction of a priori tours in a world of uncertainty. Notice that line one is constructed from a random demand, but the number of aircraft (salespersons) is fixed. While the aircraft fleet is fixed in both line one and line two, it can be treated as a random variable as well. This is illustrated in the PMTSFLP algorithm, and in the mathematical programs.

While a comparison among all four lines in **Table 3-9** is worthwhile, the results need to be interpreted in terms of the data upon which the findings are based. It is understandable why a much larger data-base for line one does not deter from solution efficiency in that the MSFC, being of $O(|I|^2)$ complexity, is inherently a much simpler problem than the location-routing MTSP problem. The important point is that an extremely difficult problem can be approximated by a much easier one. A methodological finding is also worth discussing here. The simplified LP formulation for MTSP reported in the section entitled "*Routing as spatial separation*" yielded solutions satisfactorily for small networks. Granted that there are advances over the past decade in improving the compactness of the subtour-breaking constraint such as the one used in Equation **(3-45)**. Still, this formulation can be very practical for enumerative computation of all instances as performed in the MTSP solution above. It allows an integer solution to the MTSP using a regular linear programming code. Remember that the number of times we solve the NP-hard MTSP is in the order of $|I| |H''|$, where the number of instances $|H''|$ is a large number. Any savings effort will be necessary.

Again, there is a difference between the two ways of solving the location-routing model. The MTSP way of enumerating all 13 instances among the alternative basing-locations is the hard way, and the MSFC heuristic is the easy way. The former is NP-hard while the latter is $O(|I|^2)$. It is again gratifying to see that a difficult solution- algorithm can be approximated by an easy algorithm. The efficiency of MSFC, when compared with the formal MTSP, can be dramatically brought out by the computer execution-times. It requires several hours on the microcomputer to run the 13 instances via the LP model. In contrast, we are talking about 10 minutes using the MSFC heuristics to solve all 98 possible demand-points in the continental United States. Best of all, we need to do this only once, and the order of visit remains valid for all instances. It appears there is a definite role that SFC can play (Bertsimas 1989) in practical applications.

3.8.5   *Concluding comments on case study*.   The Probabilistic-Travelling-Salesman-Facility-Location Problem (PTSFLP) finds the whereabouts of a depot to minimize the tour among all nodes. We extended the PTSFLP to include multiple salespersons and offer a solution to the new model. We call this the Probabilistic-Multiple-Travelling-Salesmen Facility-Location Problem (PMTSFLP). With Space-Filling Curves (SFC), an efficient multiple-SFC heuristic is constructed to solve the PMTSFLP. The inherent robustness of SFCs, particularly their ability to preserve the order of visit among all nodes, allows for *a priori* optimization of various instances of the PMTSFLP, where instances refer to situations where demands are realized.

Notice the fleet size is intimately related to the time *each* crew spends "on the road." The larger the number of aircraft used, the shorter each *individual-tour* of a crew. Conversely, the smaller the fleet, the longer each crew is away from home base in his/her round-robin tours. A small enough fleet-size may result in an unduly lengthy-tour for each aircraft (crew). It may even violate the "range constraint" in a vehicle-routing formulation of the PMTSFLP. To introduce feasibility, one has to increase the number of dispatched aircraft from k to $k + 1$, $k + 2$, ... , up to the point when the time each crew spent on-the-road is brought down enough to fit the range constraint. Obviously, this is done at the expense of total system-mileage logged across the entire fleet. In a sentence, the individual-salesman tour is the shortest and the total mileage systemwide is the longest when $k = |H'|$, where $|H'|$ is the maximally available fleet-size.

The asymptotic properties of the PMTSFLP algorithm, deemed an important finding of this model, have been specified in terms of the performance bound $O(\log |I|)$, the accuracy bound 1.25, and the combined result of $r \, O(\log |I|)$. Here $|I|$ is a finite number of randomly-distributed demand-points and r ($1 \le r \le 1.25$) is problem-dependent constant. Another important contribution is that k, the actual number of salespersons deployed, is parametric in these performance bounds. An equivalent set of bounds can be written for a variable k (with $1 \le k \le |H'|$), parametric on the probabilistic-demand vector $\mathbf{f} = (f_1, f_2, ..., f_{|I|})$.

The flight-inspection mission of the Air Force Facilities Checking Squadron lends itself to be an extremely natural case study of the PMTSFLP, particularly when the basing decision is re-examined with the acquisition of a new fleet of inspection aircraft. The case study is facilitated by the availability of one-year of historical data, covering a variety of instances with varying demand patterns. A datum for comparison also exists, consisting of the existing practice of the Squadron at Scott Air Force Base. The one-year data and the way they are grouped does not allow coverage of all instances of demand to reach a definitive basing and routing decision. Nevertheless, it does allow us to illustrate with clarity the efficiency of the multiple-SFC heuristic. For comparison, we solved the equivalent MTSP linear program using a simplified formulation of the subtour-breaking constraints. Of the 52 instances solved, 35 instances yielded the same solution as the MSFC heuristic. Overall, the heuristic solution is within 7.6 percent of the optimal for each basing-location on the average.

Our analysis shows that the PMTSFLP yields a ranking of depot locations that are very similar to the medians and centers. We have attributed this in part to the even-distribution of demand over the study period and the ubiquitous geographic coverage of the nodes. Further analysis suggests there may in fact be some relationship between PMTSFLP and the Euclidean center and median problems, perhaps in an asymptotic sense. This could allow the PMTSFLP depot-location to be approximated very easily by solving the 1-center or 1-median problem (Bartholdi and Platzman 1982). This means that routing distances can be asymptotically approximated by Euclidean distances under these circumstances. We have verified the accuracy of the MSFC heuristic through solving 78 multiple-TSPs among six alternate basing decisions. From an application standpoint, the research offers a fast, yet accurate analysis tool for basing and routing decisions. Given sufficient data, definite basing recommendations can be forthcoming. For the Facilities Checking Squadron, the SFC heuristic is a convenient, yet rigorous, routing tool for the commander to respond to different demands on a daily basis. Best of all, this can be carried out with minimal computational requirement.

3.9 Summary

The art of measuring spatial separation has progressed significantly since the days of Euclidean geometry. To the extent that facilities and land development are coordinated by a set of spatial prices corresponding to separation, accurate yet operational determination of these spatial prices is critical. In this chapter, we have presented five different ways of representing separation, from l_p-norm (Minowski metric) to stochastic response-time. As always, each has its role to play in representing separation. While l_p-metric appears to be a convenient approximation, it was found that Space-filling curve (SFC) is not that much more difficult to compute, and has a good deal more to offer for routing-location applications, including probabilistic demands. Whereas l_p-metric ($1 \leq p \leq \infty$) is synonymous with triangular inequality, SFC is more robust and does not assume such inequality. In today's economy, advances in transportation and communication have made traditional concepts of distance obsolete. Interaction between activities takes place in spite of long distance-separation. Conversely, activities close by are prevented from interacting simply because of congestion that make contact difficult. In response to this, stochastic response-time has been used on top of other sophisticated measures in location models in recent years, in which queuing delays are explicitly considered as a factor of spatial separation. In spite of the increased complexity, it is gratifying to know that many stochastic location models can be approximated by traditional deterministic ones under a fair number of conditions. Of particular interest is the nodal-optimality properties that govern both stochastic and deterministic facility location. With this property, search for optimal locations in a network needs only be conducted at the nodes, in spite of the complexity brought forth by stochastic modelling.

The concept of accessibility is a cornerstone of land-use modelling. It has historically been developed independent of the other facility-location techniques: l_p-metric, SFC, routing and stochastic response time. In this chapter, we have tried in putting accessibility side-by-side the traditional measures used in facility location. In its role as weighted distance, it is no different from the other metrics such as a median. To the extent it bears an inverse relationship to distance, it is uniquely developed for its application context. Even then, it has some relationship to the Minkowski metric when $p < 1$. Accessibility has its measurement problems, not unlike l_p-metric or stochastic-queue median-models. The difficulty lies in its calibration, by which the spatial separation can be accurately represented. Because there are usually a number of calibration constants to calibrate, it is often not clear how to choose among several equally appealing ways to set these constants. Unfortunately, an accurate calibration is of critical importance in the operational use of many facility-location and land-use models, and to date there is no sure-fire way of overcoming this difficulty—a subject we will come back to later in the chapters on activity-allocation models. Suffice to say here that—similar to other modelling efforts—calibration lies at the heart of spatial organization. Accurate calibration of the general "distance function" allows one to approximate accessibility or spatial price. More important, it also helps to assess efficiency of the production function representing the spatial economy. This is particularly cogent in modelling a production function with multiple outputs from multiple input factors and a function in which the fixed-proportion assumption is relaxed to the regular linear-homogeneity premise and beyond.

Through our revisit to the quadratic-assignment problem, the artificial demarkation between a location and routing model has been somewhat demolished. This was demonstrated by the close mathematical relationship between a location model such as quadratic assignment and a routing model such as travelling salesman. Beyond their combinatorial-optimization connection, the linkage was highlighted through data-structure discussions, including the list structure. This paves the way for using data structure as a convenient way to model combined location-routing, as we will show in the chapter bearing the same name. Within this chapter, we have shown how traditional median and center location models serve as an excellent approximation for location-routing models such as the probabilistic-multiple-travelling-salesmen-facility-location-problem. This is found to be the case for ubiquitous demand sor when the cost function for facility location shows a "shallow valley" at the facility to be located. Under these circumstances, routing distances can be approximated asymptotically by Euclidean distances.

CHAPTER FOUR

SIMULTANEOUS
LOCATION-AND-ROUTING MODELS

It only takes a moment of reflection to realize that, realistically speaking, routing decisions cannot be separated from locational decisions, for a suitable location is one that is accessible. To the extent that accessibility is determined by routing, routing is an integral part of facility location. General methods for combining facility-location problems with network-routing problems were developed as early as 1974 by Chan (1974), whose solution method involved successive approximation. Current et al. (1983) reviewed other attempts to combine the two problems. They developed another more complete method (Current et al. 1987) for the solution of *median-path* problems. Some of the most recent formulations include Perl and Daskin (1983), Perl (1987), and Chan (1992). Laporte et al. (1986) performed pioneering work in an exact solution to a capacitated location-routing problem. A recent review article on integrated location-routing models was done by Balakrishnan et al. (1987) and Laporte (1988). Berger (1997) reviewed and extended current analytical results in solving such class of problems. Chan et al. (1998) reported advances on probalilistic inventory-routing models. The purpose of this chapter is to combine transportation-routing and facility-location problems in a single formulation. The combined models summarized herein have primarily been developed over the past decades with diverse applications to logistics, engineering, planning, emergency-service location, transportation networks and military-facility locations. The combined location-routing formulations will be presented, followed by solution algorithms and case studies.

4.1 Background

For the generalized formulations discussed in this chapter, a combined location-routing model has the following general formulation (Balakrishnan et al. 1987):

Maximize $[\lambda'$ (demand coverage) $-$ (depot costs + routing costs)$]$
  subject to
  (1) certain demand sites must be serviced
  (2) intermediate-stop requirements for vehicle routes   **(4-1)**
  (3) route restrictions
  (4) depot restrictions
  (5) forcing/linking constraints between location and routing decisions.

The decision variables include depot-location, routing and demand-allocation variables. Without loss of generality, all variables are integer-valued, with the first two assuming binary values of 0-1. In the objective function, the first term "demand coverage" refers to the service actually received by the consumer while the second term 'cost' refers to the service provided. The constant λ' simply puts the first term and second term in the same unit. It can also be viewed as a weight to trade off the two terms. The depot costs may consist of a fixed cost for establishing and operating it and variable costs that depend on the throughput of the depot. The routing costs may include the following components: route costs and variable transportation costs. The route costs are incurred when a vehicle operator makes one or more stops on a trip. From the vantage point of the traveller or shipper, the 'routing cost' is the time or charge incurred during transit. Generally speaking, both costs are directly proportional to the distance covered on the route. From both the operator and user points of view the variable transportation cost depends on the volume that travels on each arc of the network. It is important to remember that costs refer to both user cost and system-operator costs. Oftentimes, we cannot necessarily minimize both costs simultaneously. For example, an operator may wish to route a shipment in a circuitous way (to fill up a vehicle) in order to save costs. This works against a shipper who really wants to have his/her shipment reach the destination in the most direct way possible. This gives rise to multicriteria functions. Similar tradeoffs occur between service rendered and the cost of providing such service. On occasions, the demand-coverage criterion is incommensurate with the cost-of-providing-service; they have to be treated as separate criterion functions, since there is no simple way of equivalencing the incommensurate units-of-measurement. In this case, a multicriteria optimization model thus generates a set of *non-inferior* solutions rather than a single best answer.[1]

In the model as specified in Equation ?, the first constraint specifies the way demands are to be served. Selected demand sites must be visited—sometimes more than once—while others can be included as appropriate to optimize the figure(s)-of-merit in the objective function. In this general definition, the location decision does not only include the depots, but also the demand sites to be served. This is a generalization of the classic travelling-salesman or vehicle-routing models where all demands are to be served. Another way of viewing this generalization is that demand and depot sites are no longer separate and distinct sets. A site can be both a depot and a demand point (or none of the above). Thus the location decision now pertains to both demand and depot sites alike. A note on the demand for service is in order. Classic location-and-routing models typically have a fixed demand at site i. As long as delivery is made to that amount, the consumers are satisfied. In a generalized location-routing model, demands are specified by both the quantity and type (multicommodity). Thus, a certain amount of delivery is to be made specifically from node j. Also the quantity demanded can depend on the level-of-service

---

[1]For an explanation of non-inferior (or nondominated) solutions to a multicriteria optimization model, refer to the "Multicriteria decisionmaking" chapter in Chan (2000) for example.

in that the demand may disappear when service is tardy. Demands that can be delivered within a certain time-window are more 'popular' and correspondingly more numerous than demands delivered late.

The intermediate-stop requirements constitute the second constraint. It could specify a tour (as in the travelling salesman problem (TSP)) or a route with specified visitations where the starting depot need not be the termination depot. This constraint typically includes three types of restrictions. There are *degree* constraints on all nodes, defined as the number of 'active' arcs that are incident upon each node. For demand nodes, the in-degree must be equal to the out-degree, while for depots the degree would depend on the number of tours or routes originating and terminating at that depot. Physically, this means a salesperson or vehicle delivers a service or load at a demand node and then leaves. At a depot, the active sales force or fleet size determines how many salespersons or vehicles are sent out at the beginning of the period and report back at the end. Another type of constraint for the TSP is "subtour breaking," which eliminates illegal subtours that do not visit any depot. Finally, there are 'chain-barring' constraints for the TSP that prohibit two or more depots being covered by a single tour. The implicit assumption is that a salesperson or vehicle is based at only one single depot. We relax this assumption in the general formulation where the origin depot may be different from the termination depot.

In the third constraint, route restrictions refer to several items. An implicit upper bound is applied toward the routing variables if they are indexed by route. Alternatively, an explicit constraint on the total number of available salespersons or vehicles or tours may be specified, indicating the limited number of vehicles or tours that can originate from each depot. For the vehicle routing problem (VRP), there exist a vehicle capacity and an upper bound on the route distance, corresponding to the 'range' of the vehicle or the maximum amount of time the vehicle can be "on the road." For more general location-routing problems, an arc capacity may be imposed, limiting the number of vehicles that can traverse an arc during the same time period.

Next, depot restrictions might include upper and lower bound on the number of depots that can be established, and restrictions on the throughput or capacity of each depot. Notice that depot capacity sometimes has two separate meanings. The first is the amount of shipment or passengers it can handle. The second is the number of vehicles it can dock at anytime. What it amounts to are two types of flow that can go through a depot: vehicles and freight/passengers. The same applies to a demand node that is not a vehicle origination or termination point. When both vehicular and freight/passenger traffic are explicitly considered in a model, we call this a *nested* problem. Understandably, a nested problem introduces additional complexity. A vehicle is now assuming the additional role of facilities and there will be no passenger/freight traffic unless vehicle capacity is available, much like a vehicle cannot visit a depot when the depot capacity is exceeded. In the generalized location-routing model discussed here, multiple-frequency visits to a demand can be made by one or more vehicles.

Last but not least, the forcing/linking constraints relate the depot-location and/or demand-allocation variable to the routing variables. No routes are allowed to visit a depot facility or demand site unless a depot has been established there or a demand has been explicitly 'covered'. Simply put, both demand and depot have to be included in a vehicle route. While it sounds like common sense, its implementation is hardly trivial, since it represents the heart of an integrated location-routing model when both location and routing decisions are made simultaneously. The importance and complexity of the forcing/linking constraints are likened to the subtour-breaking constraints in a pure routing model such as TSP or VRP.

Besides model formulation, the survey here is intended to cover each of the viable solution techniques in solving a broadly defined class of location-routing problems characterized above. While there exist other recent surveys (see, for example, Srivastava and Benton (1990), Laporte (1988), Balakrishnan et al. (1987), Madsen (1983) and Berger (1997)), this survey differs from others in that we are as interested in model and solution diversity as well as commonality among models. This approach is perhaps justified considering the newness of combined location-routing problems and the vast differences possible between their formulations according to our generalized definition of location-routing above. We also distinguish ourselves in that self-contained algorithms are reported here, complete with numerical examples for each. The algorithms detailed here are judged to be either operational or pedagogically important. Aside from their ability to solve combined location-routing problems, it has also proven itself in solving either location or routing models individually—the two degenerate cases of a combined model. For that reason, the algorithms are useful for a better conceptual understanding of the structure of location models or routing models individually. In this regard, this chapter supplements the previous chapter on routing, particularly on computational efficiency of solution algorithms.

4.2   Single-facility/single-route/multi-criteria problem

A method for combined facility-location and routing analysis was developed by Current et al. (1987) for a single path from a starting node to a terminus node. It was formulated as an integration of the maximum-covering location problem and the shortest-path routing problem. The model formulation has two criterion functions. The first identifies the single shortest path from any given starting node (1) to a given terminus node (n), where $n > 1$. The second criterion is to maximize the demand that is satisfied, where a demand at node k can be satisfied only if the path enters node k directly or if it enters another node within a maximum-coverage-distance S from k. Because total path length must increase to expand the coverage the two criteria are in conflict with each other.

4.2.1 Model formulation. The model formulation as presented by Current et al. (1987) is as follows:

$$\max\ y_1' = \sum_{k=2}^{n-1} f_k y_k \tag{4-2}$$

$$\min\ y_2' = \sum_{i \in I} \sum_{j \in I} d_{ij} x_{ij} \tag{4-3}$$

subject to

$$\sum_{j \in M_1} x_{1j} = 1 \tag{4-4}$$

$$\sum_{i \in M_n} x_{in} = 1 \tag{4-5}$$

$$\sum_{i \in M_j} x_{ij} - \sum_{i \in M_j} x_{ji} = 0 \qquad \forall j \in I\ j \neq 1, n \tag{4-6}$$

$$\sum_{j \in M_i} \sum_{i \in N_k'} x_{ij} - y_k \geq 0 \qquad \forall k \in I,\ k \neq 1, n; \tag{4-7}$$

where

$$y_k = \begin{cases} 1 & \text{if demand node } k \text{ is covered} \\ 0 & \text{otherwise.} \end{cases} \qquad (k=2,\dots,n) \tag{4-8}$$

One will recognize that Equation **(4-3)** through **(4-6)** is nothing more than the eqivalent four equations in the "Routing analysis" section of the "Measuring Spatial Separation" chapter, with origin designated as 1 and destination n, assuming that there are n nodes in the network. Expression **(4-7)** simply says that k is covered only when a routing is within a distance S of the demand node k, as denoted by the set N_k. We call this an $s/s/m/-$ model, where the first s stands for single depot-pair (or single facility), the second s single-route, and m stands for multi-criteria. The hyphen "–" is simply a space occupier for the fourth attribute that is required to further specify this model. Unlike a traditional routing problem, TSP or VRP, there are two distinguishing features about this simple location-routing model. First, the route is not a tour, in that the starting depot is different from the terminating depot. Second, not all the demands are visited, only the subset that contribute toward an efficient or nondominated solution. Yet the model has all the features of a location-routing model, including the "forcing or linking" constraint **(4-7)** that establishes the precedence relationship between the nodal-coverage variable **y** and the arc-routing variable **x**. For pedagogic reasons, it is insightful to start with a simple-model such as this, where complicating subtour-breaking constraints are noticeably absent, making the model a good deal more transparent and simpler to solve.

Probabilistic formulation of the above problem is a simple extension:

$$\max \sum_{k=1}^{K} \sum_{l=2}^{n-1} f_{lk} P_k y_l \quad and \tag{4-9}$$

$$\min \sum_{i \in I} \sum_{j \in I} \sum_{k=1}^{K} P_k d_{ijk} x_{ijk}. \tag{4-10}$$

As before, the first criterion seeks to maximize demand coverage, with different nodal demands for each state k, while the second maps out the shortest route considering a full travel-time distribution. The constraint now can also be stated accordingly:

$$\sum_{j \in M_1} x_{1jk} = 1 \qquad \forall k \tag{4-11}$$

$$\sum_{i \in M_n} x_{ink} = 1 \qquad \forall k \tag{4-12}$$

$$\sum_{i \in M_n} x_{ijk} - \sum_{i \in Mj} x_{jik} = 0 \quad \forall j \in I, \; j \neq 1, n; \; \forall k; \tag{4-13}$$

$$\sum_{i \in N_l'} \sum_{j \in M_i} x_{ijk} - y_l \geq 0 \quad \forall l \in I, \forall k. \tag{4-14}$$

Thus it can be seen that the probabilistic formulation, as stated, simply increases the dimensionality of the problem without necessarily introducing too much additional insight to the problem.

4.2.2 Model solution. As shown above, the simplest case of a combined location-routing model consists of a single route between a specified origin and destination pair that selects visitation locations en route. The possible solutions to the problem are many because of the two disparate criteria of the model—one minimizing the operating cost while the other maximizing the demand coverage. It requires the development of a noninferior solution set, as opposed to a single optimal solution. The solution method suggested by Current et al. (1985) involves formulating a weighting method—sometimes called the *weighted-sum* technique—to be applied to both halves of the model formulation, allowing it to be solved as one formulation $\max \lambda' y_1' - (1-\lambda') y_2'$ where λ' assumes a value between zero and one. This weighted formulation is then solved as a linear program (LP) by relaxing the zero-one requirement on the routing and location variables. More often than not, an integer solution is obtained in spite of this relaxation. At times, a subtour is found in the solutions, consisting of circuits that pass through and return to the same intermediate node. The imposition of subtour-breaking constraint similar to $\sum_{i \in J} \sum_{j \in I-J} x_{ij} \geq 1$ in the "Measuring Spatial Separation" solves the problem, although fractional values for x_{ij} may result. To remedy this, a branch-and-bound (B&B) procedure is used to integerize the fractional variables.

Only noninferior solutions are considered in the final trade-off analysis. Notice the noninferior set is not convex. This nonconvexity is caused by the integer nature of the $s/s/m/-$ model. As a result, there exist noninferior-gap points. To identify these gap-point noninferior solutions by integer programming (IP), one must solve this model by a *reduced-feasible-region* method. This expresses one of the criterion functions as a constraint, thus specifically limiting the solution within a certain y_1' or y_2' range.[2] The generation of the entire noninferior set of a multiobjective problem is computationally intensive. Consequently, analysts are often content with an approximation to the noninferior set. For illustration purposes, a sample problem consisting of 15 nodes and 34 non-directed arcs was run (Current et al. 1987). Only thirteen noninferior solutions were generated using APEX III or a CDC Cyber 76. In all, 79 subproblems were solved to obtain the 13 noninferior solutions. Less than 34 central processing unit (CPU) seconds were required for this example. The case of an $s/s/m/-$ model illustrates a straightforward application of IP codes to the solution of a combined location-routing model. It also illustrates the solution of multicriteria functions via the use of a noninferior solution set. Finally, it represents a straightforward extension of a routing model, where the stops along a route from origin to destination are to be found. Here the location decisions do not apply toward the depots, but rather stops on the way, and the route is not a tour. It is interesting that subtours surface during this type of routing and subtour-breaking constraints similar to those used in TSP help to break such subtours.

Example

The example presented here consists of five nodes: depot pair 1-5, and three demand nodes 'en route' (nodes 2, 3, and 4) [Johnson and Chan 1990, Moore and Chan 1990]. The basic network under consideration is shown in **Figure 4-1**. A demand node is covered if the route visits any node within 2.5 miles of the node being considered. Note that a route entering node 2 will cover node 3 and a route entering node 3 will cover node 2. On the other hand, a path must actually enter node 4 to cover node 4. The first criterion is to maximize coverage of the demand nodes.

$$\max y_1' = f_1(\mathbf{y}) = 10y_2 + 8y_3 + 12y_4$$

The second criterion is to find the shortest route from node 1 to node 5.

$$\min y_2' = f_2(\mathbf{x}) = 3x_{12} + 3.5x_{13} + 4x_{14} + 6.5x_{15} + 2.5x_{23} + 3x_{24} + 4x_{25} + 2.5x_{32} + 3x_{42}$$
$$+ 3.5x_{34} + 6.5x_{35} + 3.5x_{43} + 2.9x_{45}$$

subject to

(degree at origin and destination depot)
$$x_{12} + x_{13} + x_{14} + x_{15} = 1 \qquad x_{15} + x_{25} + x_{35} + x_{45} = 1.$$

---

[2]For example, see "Multi-criteria decision-making" chapter in Chan (2000) for a more detailed discussion of the reduced-feasible-region method vis-a-vis the weighted-sum method.

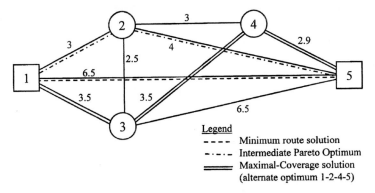

Figure 4-1 Example of a single-depot-pair/single-route/multicriteria model

(continuity equation for demand nodes)
$$x_{12} + x_{32} + x_{42} - x_{23} - x_{24} - x_{25} = 0$$
$$x_{13} + x_{23} + x_{43} - x_{32} - x_{34} - x_{35} = 0$$
$$x_{24} + x_{34} - x_{42} - x_{43} - x_{45} = 0.$$

(forcing/linking constraint between coverage and routing)
$$x_{23} + x_{24} + x_{25} + x_{32} + x_{34} + x_{35} - y_2 \geq 0$$
$$x_{23} + x_{24} + x_{25} + x_{32} + x_{34} + x_{35} - y_3 \geq 0$$
$$x_{42} + x_{43} + x_{45} - y_4 \geq 0.$$

Pareto-optimal solutions were generated by a weighting scheme between y_1' and y_2'.

λ'	y_1'	y_2'	route	demand coverage
0.0 – 0.05	0	6.5	1–5	
0.05 – 0.15	18.0	7.0	1–2–5	2, 3
0.15 – 1.0	30.0	8.9	1–2–4–5	2, 3, 4
			or	
			1–3–4–5	

The three solutions are overlaid in **Figure 4-1**. Evidently the first solution minimizes the route while the third maximizes demand coverage, with the intermediate Pareto optimum in between. □

4.3   Single-Facility/Multi-Tour/Allocation Problem

Federgruen and Zipkin (1984) presented a combined vehicle-routing and inventory-allocation model. A centralized depot at node 0 is to serve a number of demand points via a fleet of vehicles. First, the initial inventory for each site is reported to the depot. This information determines the allocation of the available product among locations (including zero allocation) z_i. The assignment of locations i to vehicles and the routes h are made, resulting in the valuation of route variables x_{ij}^h and allocation variables y_i^h.

4.3.1 Model formulation. The integer program seeks the lowest-cost solution, where cost consists of travel cost and inventory costs

$$\min \sum_{i\in I} \sum_{j\in I} \sum_{h\in H} d_{ij} x_{ij}^h + \sum_i q_i(z_i) \tag{4-15}$$

where x_{ij}^h is the 0–1 variable for arc coverage by vehicle h. A dummy route $h = 0$ is defined consisting of those locations to which nothing is to be shipped. The inventory-cost function is

$$q_i(z_i) = \int_{\beta_i+z_i}^{\infty} c_i(\xi - \beta_i - z_i)\,dF_i(\xi) +$$
$$\int_0^{\beta_i+z_i} C_i(\beta_i + z_i - \xi)\,dF_i(\xi) \qquad i\in I \tag{4-16}$$

in which β_i is the initial inventory at location i. z_i is a variable indicating the amount of delivery to i and $F_i(\cdot)$ is the cumulative distribution function of demand at location i, which is monotonically increasing. c_i is the unit shortage-cost in location i and C_i is the unit inventory-carrying-cost in location i.

The first constraint (which is nonlinear) ensures that the load assigned to each vehicle is within its capacity

$$\sum_{i\in I} z_i y_i^h \le V_h \qquad h\in H. \tag{4-17}$$

The second constraint guarantees that the total amount shipped is available at the depot:

$$\sum_{i\in I} z_i \le \overline{P}. \tag{4-18}$$

The remaining constraints are the same as in the vehicle-routing problem (VRP). Specifically $\sum_{i\in I} \sum_{h\in H''} x_{ij}^h = |H''|$ if $j = 1$ and 1 if $j = 2, \dots, |I|$; $\sum_{j\in I} \sum_{h\in H''} x_{ij}^h = |H''|$ if $i = 1$ and 1 if $i = 2, \dots, |I|$; and $\sum_{i\in M_p} x_{ip}^h - \sum_{j\in \overline{M}_p} x_{pj}^h = 0$ for all h, and $p\in I$ (as explained in the "Vehicle-routing problem" section of the "Measuring Spatial Separation" chapter). The subtour-breaking constraint looks like

$$\sum_{i\in L} \sum_{j\in L} x_{ij}^h \le |L| - 1 \qquad L\subseteq\{2,\dots,|I|\},\ 2\le|L|\le|I|-1;\ h\in H \tag{4-19}$$

and the `linking' constraint looks similar to Equation **(4-7)**:

$$\sum_{j\in M_i} x_{ij}^h - y_i^h \ge 0 \qquad \forall i, h\in H. \tag{4-20}$$

We label the entire model $s/m/-/a$, where the first s stands for single depot (or single facility), the second m stands for multiple salespersons (or multiple tours) and a stands for allocation of delivery among consumers.

Similar to the $s/s/m/-$ model, the current model selects a subset of the demands to visit. It only visits stops that need inventory replenishment and skips stops that have no inventory shortage. A node is only visited if the node is 'covered' by a route, and the linking/forcing constraint ensures that this is the case. Viewed in this light, the $s/m/-/a$ model is a generalization of a traditional median problem and a VRP. Unlike the $s/s/m/-$ model, however, the originating depot is the same as the terminating depot. Thus the model constructs tours rather than paths, and the subtour-breaking constraints **(4-19)** are required to obtain legitimate tours that are based at a depot.

4.3.2 Model solution. Being the logical extension of the $s/s/m/-$, the current location-routing model is to combine demand and routing decisions more formally. Federgruen and Zipkin (1984) used a generalized Benders' decomposition to solve the vehicle-routing and inventory-allocation problem, which we paraphrase below. A master problem in **y** variables, equivalent to the original, is derived by projection and dualization. A sequence of relaxed master problems is solved. Each such problem yields a tentative solution **y** that defines the subproblems. The subproblems are solved or determined to be infeasible. Dual solutions or extreme directions (rays) then define one or more constraints ('cuts') of the master problem. These cuts are appended to the previous relaxed master problem, and the process continues.[3]

4.3.2.1 MASTER PROBLEM. Recall that $\mathbf{x}^h = (x_{ij}^h)$ and $\mathbf{y}^h = (y_i)$ are vectors of decision variables for routing and coverage respectively. Let us rewrite the original integer program as

$$\min \sum_i \sum_j \sum_h d_{ij} x_{ij}^h + \sum_i q_i(z_i) \qquad \textbf{(4-21)}$$

subject to

$$\mathbf{x}^h \in X^h, \ \mathbf{y} \in Y, \ h \in H \qquad \textbf{(4-22)}$$

$$\varphi^h(\mathbf{x}^h, \mathbf{y}^h) \geq 0 \qquad \textbf{(4-23)}$$

and Equations **(4-17)** and **(4-18)**. In this formulation, the set Y'' represents all the possible assignment of demands i to vehicular routes h $\sum_{i \in I} y_i^h = 1$ for all h. As suggested by $\sum_{i \in I} \sum_{h \in H''} x_{ij}^h = |H''|$ and $\sum_{j \in I} \sum_{h \in H''} x_{ij}^h = |H''|$ above, a similar set **x** in X^h is defined for the routing variables $X^h = \{\mathbf{x}^h : 0 \leq x_{ij}^h \leq 1; i, j \in I\}$, and φ^h represents all the

---

[3]For an introduction to Benders' decomposition, one may wish to consult Chan (2000)'s appendix on "Optimization", his "Decomposition" section of the "Prescriptive Tools" Chapter, and the "Scheduling restriction" subsection of the "Spatial Separation" chapter in the current volume.

other linear inequalities defining the hth TSP polytope (or the set of constraints that define the "feasible region" for the hth TSP).

We now project this problem onto $\mathbf{y}$. The allocation problem and the TSPs are feasible for any $\mathbf{y} \in Y''$. For $\mathbf{y} \in Y''$, let $z'(\mathbf{y})$ be the minimal objective-value with $\mathbf{y}$ fixed. Then

$$z'(\mathbf{y}) = \sum_{h \in H} \sup_{\pi^h \geq 0} \min_{\mathbf{x}^h \in X^h} \left\{ \sum_{i \in I} \sum_{j \in I} d_{ij} x_{ij}^h - \boldsymbol{\sigma}^h \boldsymbol{\varphi}^h(\mathbf{x}^h, \mathbf{y}^h) \right\} +$$
$$\sup_{\Omega, \rho \leq 0} \min_{z \geq 0} \left\{ \sum_{i \in I} q_i(z_i) + \Omega(\bar{P} - \sum_{i \in I} z_i) + \sum_{h \in H} \rho^h(V_h - \sum_{i \in I} z_i y_i^h) \right\} \tag{4-24}$$

where 'sup' stands for the supremum function or the least upper bound (envelope) for the minima. Here $\boldsymbol{\sigma}^h$ is the vector of dual variables corresponding to the constraints defining the hth travelling-salesman polytope[4]. Ω is the dual variable corresponding to the terminal capacity constraint. ρ^h is the dual variable to account for the hth delivery-vehicle capacity. The following master problem is thus equivalent to the original model: $\min\{z \mid z \geq z'(\mathbf{y}), \mathbf{y} \in Y''\}$.

4.3.2.2   CUTS FOR THE RELAXED MASTER PROBLEM.   Cutting-plane methods produce a dual-optimal $\boldsymbol{\sigma}^h$ for each h. (Although $\boldsymbol{\sigma}^h$ is a vector of very large dimension, nearly all its components will usually be zero. Only the specified nodes visited by route h will record a nonzero "odometer reading" σ.) Solution of the allocation problem yields optimal multipliers (Ω, ρ). These values generate a cut, helping to approximate $z'(\mathbf{y})$ of the following form

$$z \geq \sum_{h \in H} \min_{\mathbf{x}^h \in X^h} \left\{ \sum_{i \in I} \sum_{j \in I} \left[d_{ij} x_{ij}^h - \boldsymbol{\sigma}^h \boldsymbol{\varphi}^h(\mathbf{x}^h, \mathbf{y}^h) \right] \right\} + \Omega \bar{P} + \sum_{h \in H} \rho^h V_h +$$
$$\min_{z \geq 0} \left\{ \sum_{i \in I} \left[q_i(z_i) - (\Omega + \sum_{h \in H} \rho^h y_i^h) z_i \right] \right\}. \tag{4-25}$$

The minimum over $\mathbf{x}^h$ is independent of $\mathbf{y}^h$, and reduces to the expression

$$\sum_{h \in H} \left(\kappa^h + \sum_{i \in I} \kappa_i^h y_i^h \right) = \sum_{h \in H} \left(\kappa^h + \kappa_0^h + \sum_{i \in I-0} \kappa_i^h y_i^h \right) \tag{4-26}$$

where all κ's represent constants. Here κ_i^h is the marginal cost of servicing node i with vehicle h, and κ^h is the basic operating cost of vehicle h. By marginal cost i, we mean the increment in "odometer reading" from the previous node to node i. By basic operating cost, we mean the total cost of executing the tour. The former is a result of the routing while the latter is given exogenously. In the minimum over $\mathbf{z}$ (the deliveries made), note that by constraint (4-17), exactly one y_i^h is 1 for each i. Thus the minimum equals

$$\sum_{h \in H} \sum_{i \in I} y_i^h \min_{z_i \geq 0} \left[q_i(z_i) - (\Omega + \rho^h) z_i \right]. \tag{4-27}$$

---

[4]Remember we have come across these dual variables before in the "Spatial Separation" chapter, in conjunction with subtour-breaking constraints.

For each i, h, the inner minimization is just a newsboy problem.[5] Call its cost φ_i^h. Thus the form of the cut is

$$z \geq \Omega\bar{P} + \sum_{h \in H}\left(\rho^h V_h + \kappa^h + \kappa_0^h\right) + \sum_{i \in I-0}\varphi_i^0 y_i^0 + \sum_{h \in H}\sum_{i \in I-0}\left(\kappa_i^h + \varphi_i^h\right)y_i^h. \tag{4-28}$$

The relaxed master problem is thus min z, subject to cuts of the form above, as well as constraints $\sum_{i \in I}\sum_{h \in H''}x_{ij}^h = |H''|$ and $\sum_{j \in I}\sum_{h \in H''}x_{ij}^h = |H''|$ above. In general, since Y'' is finite, only a finite number of subproblems can be generated, and the algorithm converges to the optimum in a finite number of iterations. Using the lower bound, we can terminate the procedure prior to optimality when any given error tolerance is achieved.

4.3.3 Numerical example. The $s/m/-/a$ problem is a combination of the shortest-routing problem for multiple vehicles and the allocation problem. The problem consists of a single fixed depot, and demand nodes that may or may not be visited. The objective is to minimize total costs where the total cost is composed of travel costs and inventory costs. Travel costs are directly related to distance travelled and inventory costs are a result of overstocking or understocking at the demand nodes. The problem is constrained by maximum inventory levels at each demand node, the capacity of the vehicles, the supply available at the depot, and the requirement that each node be visited by at most one vehicle. Johnson and Chan (1990) and Moore and Chan (1989) presented a four-node example: one supply depot (node 0) with 400 units of supply available and three demand nodes (nodes 1, 2, and 3). The basic network is shown in **Figure 4-2** complete with link costs. **Table 4-1** shows the inventory-allocation data. An examination of these data shows that a uniform probability-density-function (PDF), or a constant consumption-rate, is assumed. For certain shortage and surplus costs c_i and C_i, it can be assumed that the inventory-cost function **(4-16)** degenerates into linear functions $q_1(z_1) = 2600 - 8z_1$, $q_2(z_2) = 437.5 - 7.5z_2$, and $q_3(z_3) = 3012.5 - 6.5z_3$ where z_i is the amount delivered to node i.[6] These are all downsloping cost functions, meaning z_i are set to their upper limits. The problem is also constrained by the capacities of 150 and 200 respectively for vehicles 1 and 2. The problem can be formulated as a mixed-integer-programming problem. The problem contains three sets of decision variables: x_{ij}^h equals 1 if vehicle h travels directly from node i to node j, zero otherwise; y_i^h equals 1 if vehicle h travels to node i, zero otherwise; and z_i is a

---

[5] As defined in the "Descriptive Tools" chapter of Chan (2000), for example, the Newsboy problem finds the appropriate amount of newspapers to stock up in order to minimize the stockout cost and storage cost.

[6] *Example*: Suppose the shortage costs $c_1 = 15$ and the surplus cost $C_1 = 20$, $q_1(z_1) = \int_{100+z_1}^{\infty}(15)(0.002)d\xi + \int_0^{100+z_1}(20)(0.002)d\xi = 16 + 0.01z_1$.

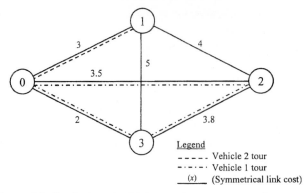

Figure 4-2 Single-depot/multiple-route/tour network with link costs

Node	1	2	3
Max inventory	500	100	500
Min inventory	0	0	0
Initial inventory	100	50	150
Demand PDF	0.002	0.010	0.002

Table 4-1 Inventory allocation data

continuous real-valued variable indicating the amount delivered to node i. The complete objective function is given by

$$\min \; 3x_{01}^1+3.5x_{02}^1+2x_{03}^1+3x_{10}^1+4x_{12}^1+5x_{13}^1+3.5x_{20}^1+4x_{21}^1+3.8x_{23}^1+2x_{30}^1+5x_{31}^1+3.8x$$
$$+3x_{01}^2+3.5x_{02}^2+2x_{03}^2+3x_{10}^2+4x_{12}^2+5x_{13}^2+3.5x_{20}^2+4x_{21}^2+3.8x_{23}^2+2x_{30}^2+5x_{31}^2+3.8$$
$$+6050-8z_1-7.5z_2-6.5z_3. \tag{4-29}$$

The only other unique feature of this model is constraints **(4-17)** and **(4-18)**. The first of these two constraints ensures that the load assigned to each vehicle does not exceed the capacity of the vehicle. Note that vehicle 0 is a 'dummy' vehicle with zero capacity:

$$z_1y_1^0+z_2y_2^0+z_3y_3^0 \le 0 \qquad z_1y_1^1+z_2y_2^1+z_3y_3^1 \le 150 \qquad z_1y_1^2+z_2y_2^2+z_3y_3^2 \le 200 \tag{4-30}$$

The second set of constraints ensures that the total amount delivered is available at the supply depot and that the amount delivered to a node will not result in an inventory greater than the maximum inventory level:

$$z_1+z_2+z_3 \le 400 \qquad z_1 \le 400 \qquad z_2 \le 50 \qquad z_3 \le 350. \tag{4-31}$$

The rest of the constraints follow the typical inequalities that define the TSP polytopes, as discussed in the "Routing analysis" Section of the "Measuring Spatial Separation" Chapter. Notice one intricacy in writing the in- and out-degree equations in view of the dummy vehicle. In the examples worked out here, we have assumed that both vehicles in the fleet will be used:

"In degree" at each node is unity, except the depot:
$$x_{10}^1 + x_{20}^1 + x_{30}^1 + x_{10}^2 + x_{20}^2 + x_{30}^2 = 2$$
(which states that 2 tours with 2 routes using real [rather than dummy] vehicle are needed into the depot).

"Out degree" at each node is unity, except the depot:
$$x_{01}^1 + x_{02}^1 + x_{03}^1 + x_{01}^2 + x_{02}^2 + x_{03}^2 = 2$$
(which states that 2 tours with 2 routes using real vehicles are needed out of the depot).

The problem can be solved as a mixed integer program (MIP) by fixing the y_i^h. These y_i^h are generated by the Benders' cut of Equation (**4-28**). In other words, each of the demand nodes is assigned to a specific vehicle (including the dummy vehicle) and the resulting routing/allocation problem is solved. The total number of possible assignments is finite; however the number can be quite large. Obviously, it is not practical to solve the problem for all possible assignments, therefore, the solutions from a subset of the feasible routings are generated here. As will be seen, Benders' decomposition is used instead of exhaustive enumeration, which solves subproblem as needed. It guarantees convergence toward the optimum. For this example, seven assignments are shown (out of $\binom{3}{1}^3 = 27$ possibilities). Based on solutions using the standard MIP package, the operating costs are much smaller than the inventory cost, therefore the solution is dominated by the optimal allocation. The results are summarized in **16** and the best solution is overlaid in **Figure 4-2**.

4.3.4 Generalized Benders' decomposition. While enumeration of the 27 solutions was performed (Loftus et al. 1992, Johnson and Chan 1990, Moore and Chan 1990), the approach is obviously not feasible for realistic size problems. We show here a few steps of Benders' decomposition in solving the problem, which promises to be more efficient (Weir et al. 1994). Recall that the solution is broken down into two subproblems: TSP and delivery-allocation. The TSPs are imbedded in the following equation: $\sum_{h \in H} \min_{x^h \in X^h} \sum_{i \in I} \sum_{j \in I} \left[d_{ij}^h - \sigma^h \varphi^h(x^h, y^h) \right]$. Recall also that a TSP involves the routing of each salesperson such that s/he visits all the locations required of him or her,[7] while minimizing the distance or costs incurred. Here the vehicles represent the salespersons, and the quantity to be minimized is the travel cost in delivering demands.

---

[7] One of the requirements is the linking constraint (**4-20**), which mandates routing be provided for the specified pairing between vehicle and demand nodes.

y_0^1	y_1^1	y_1^2	y_2^0	y_2^1	y_2^2	y_3^0	y_3^1	y_3^2	z_1	z_2	z_3	Operating cost	Inventory cost	Total cost
0	0	1	0	1	0	0	1	0	200	50	100	15.3	3425	3440.3
0	1	0	0	1	0	0	0	1	150	0	200	14.5	3550	3564.5
0	1	0	1	0	0	0	1	0	150	0	0	10	4850	4860
0	0	1	1	0	0	0	0	1	200	0	0	10	4450	4460
0	0	1	0	0	1	0	1	0	200	0	150	14.5	3475	3489.5
0	1	0	1	0	0	0	0	1	150	0	200	10	3550	3560
1	0	0	0	0	1	0	1	0	0	50	150	11	4700	4711

Table 4-2 Seven sample solutions for the routing/allocation model

4.3.4.1 TRAVELLING-SALESMEN SUBPROBLEM. The objective of the TSP subproblem is to derive the constants κ associated with the vehicles h. These κ are to be obtained by first solving the TSP for the σ^h multipliers, which corresponds to 'odometer' readings as explained in the "Spatial Separation" chapter. This amounts to solving a linear program (LP), with the σ showing up as the dual variables for the constraints that describe the hth TSP polytope. For an assignment of $y_1^0 = y_2^2 = y_3^1 = 1$ and the rest of the y set at zero, the TSP LP was solved for all three vehicles, 0, 1 and 2. Here the arc costs d_{ij} are the same between all three runs. The TSP subproblem that must be solved consists of all the constraints of the model except the nonlinear constraint (4-17) and the capacity constraints (4-18). The results are $\kappa^0 = 6, \kappa_0^0 = 3, \kappa_1^0 = 3$, $\kappa^1 = 4, \kappa_0^1 = 2, \kappa_3^1 = 2$, $\kappa^2 = 7, \kappa_0^2 = 3.5, \kappa_2^2 = 3.5$. Here κ^h is the total travel cost on tour h, and κ_i^h is the unit arc-cost (or the marginal cost) to reach node i. Thus $\kappa^0 = \kappa_0^0 + \kappa_1^0$, $\kappa^1 = \kappa_0^1 + \kappa_3^1$, and $\kappa^2 = \kappa_0^2 + \kappa_2^2$. We substitute the assignment vector $\mathbf{y}$ into the delivery-prohibition or linking constraints (4-20)

$$x_{10}^1 + x_{12}^1 + x_{13}^1 \geq 0 \qquad x_{20}^1 + x_{21}^1 + x_{23}^1 \geq 0 \qquad x_{30}^1 + x_{31}^1 + x_{32}^1 \geq 0$$
$$x_{01}^2 + x_{11}^2 + x_{31}^2 \geq 0 \qquad x_{20}^2 + x_{21}^2 + x_{23}^2 \geq 1 \qquad x_{30}^2 + x_{31}^2 + x_{32}^2 \geq 1$$
$$x_{10}^0 + x_{12}^0 + x_{13}^0 \geq 1 \qquad x_{20}^0 + x_{21}^0 + x_{23}^0 \geq 0 \qquad x_{30}^0 + x_{31}^0 + x_{32}^0 \geq 0$$

and solve the corresponding TSP. The end result is a routing pattern characterized by $x_{01}^0 = 1, x_{10}^0 = 1, x_{03}^1 = 1, x_{30}^1 = 1, x_{02}^2 = 1$, and $x_{20}^2 = 1$. Thus the optimal routing for the first TSP-polytope is for vehicle zero (the dummy vehicle) to go to node 1 and return to the depot, for vehicle 1 to go to node 3 and return, and for vehicle 2 to go to node 2 and return. Again, κ_i^h is the marginal cost of servicing node i with vehicle h, and κ^h is the basic cost of vehicle h. Assuming the basic cost κ^h is exactly equal to the sum of the marginal costs κ_i^h, vehicle operating cost would simply be twice the marginal costs of servicing a demand node here. For example, vehicle zero (the dummy vehicle) goes to node 1 and returns to the depot, the associated κ_i^h are $\kappa_1^0 = 3$

and $\kappa_0^0 = 3$ and corresponding basic cost κ^0 is 6, and total cost among all three vehicles is $6+4+7=17$. All other κ are zero.

4.3.4.2 DELIVERY-ALLOCATION SUBPROBLEM. Solution of the second subproblem, delivery-allocation, is represented by minimization over z, as shown in Equation (4-27). This equation is the result of the observation that each demand point will be visited by one and only one vehicle. Despite of the double summation over i and h, only $|I|=3$ terms will be generated by the expression. To see this, remember that for each i, only one y_i^h will be equal to one, the rest being 0. This annuls the corresponding terms in the minimization expression. For each i–h pair, we are left with a minimization taking the form of the classic Newsboy problem, which is to find the precise quantity of a perishable product such as newspaper that should be ordered to satisfy consumer demand. The dilemma of the news vendor is that if s/he orders too many papers from his/her distributor, whatever left unsold is wasted, incurring an over-stocking cost. On the other hand, if too few papers are ordered, s/he incurs an under-stocking cost of lost revenue. The news vendor thus seeks a balance by minimizing the sum of over- and under-stocking costs. To do this, s/he must know the probability that consumer demand will be less than (or greater than) the amount ordered, which of course is governed by the distribution function of the demand. In the case of the vehicle-routing/allocation problem, each consumer at a demand node can be thought of as a news vendor, since s/he will want to avoid both shortage and surplus allocations to the extent possible. The amount of product delivered to consumer i is the familiar z_i variable while $\varphi_i^h q_i(z_i) - (\Omega + \rho^h) z_i$ is the cost to be minimized according to Equation (4-27).

The overall cost φ_i is the weighted sum of the shortage unit cost c_i and the surplus unit cost C_i, where the weights are the probabilities:

$$\varphi_i = c_i P(z_i < d\xi) + C_i P(z_i \ge d\xi) \qquad i=1,2,3. \tag{4-32}$$

Given the probability density function (PDF) of $dF_1(\xi)/d\xi = 0.002$, $dF_2(\xi)/d\xi = 0.01$, and $dF_3(\xi)/d\xi = 0.002$, the overall cost thus becomes $q_i z_i - (\Omega + \rho^h) z_1 = \int_{\beta_i + z_i}^{M_i} c_i \, dF_i(\xi) + \int_0^{\beta_i + z_i} C_i \, dF_i(\xi)$. Here M_i is the maximum inventory carried at node i and the demand, represented by the random variable ξ, is distributed uniformly between the minimum 0 and maximum M_i inventory levels as shown by the PDFs above and in **Figure 4-3**. We solve the Newsboy problem parametric in terms of the dual variables, Ω and ρ, associated with the constraints of the allocation subproblem. We also make use of the newsboy property that the cost φ will be minimized by the smallest value of z_i satisfying $F(z_i) \ge c_i/(C_i + c_i)$ where $F(z_i)$ is the demand CDF evaluated at z_i.[8] These solutions are obtained for these inventory costs:

---

[8] For a discussion of this result, for example, see the "Descriptive Tools" chapter in Chan (2000) under "Inventory Control using marginal analysis".

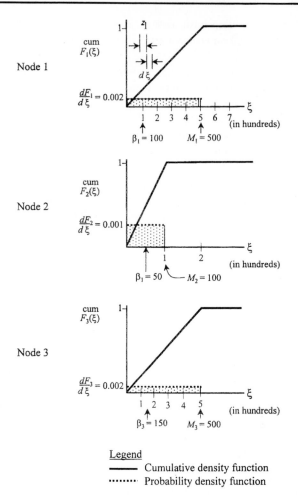

Figure 4-3 Demand functions at nodes 1, 2 and 3

$$\frac{d\varphi_1^h}{dz_1} = (5.6 - 0.6\Omega - 0.6\rho^h)z_1 - 100(\Omega + \rho^h) - 3400 = 0 \qquad \textbf{(4-33)}$$

$$\frac{d\varphi_2^h}{dz_2} = 8.75z_2 - 50(\Omega + \rho^h) - 812.5 = 0 \qquad \textbf{(4-34)}$$

$$\frac{d\varphi_3^h}{dz_3} = (9.45 - 0.4\Omega - 0.4\rho^h)z_3 - 150(\Omega + \rho^h) - 3987.5 = 0 \qquad \textbf{(4-35)}$$

The optimal orders of inventory for the newsboy problem are

$$z_1 = \frac{100(\Omega + \rho^h) + 3400}{5.6 - 0.6\Omega - 0.6\rho^h} \qquad h = 0, 1, 2 \qquad\qquad \textbf{(4-36)}$$

$$z_2 = \frac{50(\Omega + \rho^h) + 812.5}{8.75} \qquad h = 0, 1, 2 \qquad\qquad \textbf{(4-37)}$$

$$z_3 = \frac{150(\Omega + \rho^h) + 3987.5}{9.45 - 0.4\Omega - 0.4\rho^h} \qquad h = 0, 1, 2. \qquad\qquad \textbf{(4-38)}$$

Remember once again that each Newsboy problem is defined for an *i–h* pair. Obviously, the above expression poses a "chicken-and-egg" dilemma where ξ and the dual variables Ω and ρ are interdependent, reflecting the competition among the three demand locations for supplies. Furthermore, the upper bounds on ξ as imposed by Equation **(4-31)** poses additional complications, defying a traditional Newsboy solution.

The problem is best solved by taking out the z components of the original objective function, plus constraints **(4-31)** and **(4-30)**, with the assignment vector **y** as specified above. This constitutes the following delivery-allocation subproblem:

$$\min -8z_1 - 7.5z_2 - 6.5z_3$$

s.t.
$$z_1 + z_2 + z_3 \le 400$$
$$z_1 \le 400 \qquad z_1 \le 0$$
$$z_2 \le 50 \qquad z_2 \le 200$$
$$z_3 \le 350 \qquad z_3 \le 150$$

The usual nonnegativity constraints apply. Solution of this LP yields **z** as well as the dual variables Ω for the first constraint and ρ^1, ρ^2, ρ^3 for the last three constraints. Notice these dual variables are exactly the ones needed in the master problem.

4.3.4.3 MASTER PROBLEM.   Now we solve the master problem as defined by Equations **(4-21)** through **(4-23)**, plus **(4-17)** and **(4-18)**, again projecting on to the same **y** vector. The delivery-allocation subproblem determines the $\mathbf{z}^* = (0, 50, 150)$ vector, and more importantly the associated dual variables $\Omega = 0$, $\rho^0 = 8.0$, $\rho^1 = 6.5$, $\rho^2 = 0$. These dual values are now substituted into the general form of the cut according to Equation **(4-28)**:

$$z \ge 400\Omega + \sum_{h=0}^{2} (\rho^h V_h + \kappa^h + \kappa_0^h) +$$
$$\sum_{i=1}^{3} \varphi_i^0 y_i^0 + \sum_{h=0}^{2} \sum_{i=1}^{3} (\kappa_i^h + \varphi_i^h) y_i^h \qquad\qquad \textbf{(4-39)}$$

where all terms except y and z have been defined. Remember that the total costs φ have been defined in terms of Ω and ρ in the evaluation of Equation **(4-32)**. The above equation thus simplifies to

$$z \geq 1000.5 + 5203 y_1^0 - 675 y_2^0 + 1675 y_3^0 +$$
$$2600 y_1^1 - 262.5 y_2^1 + 1064.5 y_3^1 + 2600 y_1^2 + 66 y_2^2 + 2037.5 y_3^2 \tag{4-40}$$

which is now appended to the original set of equations $y \in Y$ (namely $y_1^0 + y_1^1 + y_1^2 = 1$, $y_2^0 + y_2^1 + y_2^2 = 1$ and $y_3^0 + y_3^1 + y_3^2 = 1$) plus Equation **(4-17)**, where the results obtained for the z vector are also applied:

$$50 y_2^0 + 150 y_3^0 \leq 0 \qquad 50 y_2^1 + 150 y_3^1 \leq 150 \qquad 50 y_2^2 + 150 y_3^2 \leq 200 \tag{4-41}$$

The above augmented equation set forms the master problem when the objective function min z is applied. Solution of this master problem suggests the next $\mathbf{y}$ vector $(y_3^1 = y_1^2 = y_2^2 = 1)$ upon which the projection is to be made and a corresponding z value (4731) found, and the process repeats itself from the start. In other words, it provides an improved assignment that we substitute into the TSP and the delivery-allocation subproblems. The starting point of the Benders' procedure, as represented by the initial $\mathbf{y}$ vector and the z value of 4731, is a bit far from the optimum of 3440.3. More binding cuts may be necessary to converge to the optimum. The next cut gets us to $y_2^0 = y_3^1 = y_1^2 = 1$ (with the remaining y values at zero). The corresponding upper bound is 3990. We continue in this manner for subsequent iterations. When the master problem returns the same solution as in the previous cycle we have found the optimum.

   Taking advantage of a linear objective function, possibly a better solution is to use the basic partitioning procedure of the Benders' decomposition method, rather than the generalized version described above (Weir et al. 1994). If we fix the y_i^h vector, the problem becomes a mixed integer program (MIP) that can be decomposed into a TSP problem and a Newsboy problem. Taking this fixed-$\mathbf{y}$ MIP, we can solve the dual of its relaxed LP. Although the dual will obtain a large number of variables, most of them will be zero (as mentioned). We then need to calculate the solution to the Newsboy problem given the y-vector used to create the MIP. With the solution to these two problems we can define the Benders' cut using the generalized formula $z \geq f(\mathbf{y}) + (\mathbf{b} - \mathbf{By})\lambda$. Here $f(\mathbf{y})$ is a function of the binary variables $\mathbf{y}$, and λ is the solution to the dual relaxation of the MIP, and $\mathbf{B}$ is a matrix that corresponds to the coefficient of the y variables in the original problem. We than append this cut to $y \in Y$ and the vehicle-capacity constraints, substituting in our solution of the Newsboy problem for the z_i values. Solving this integer program will give us our next y-vector, and we start over, solving the new MIP and Newsboy problems given this new y-vector. The Benders' cut generated by this new $\mathbf{y}$-vector will be appended to the Master problem and the vehicle-capacity constraints will be updated with the new Newsboy solution. Then the whole process starts over. The solution will provide a new upper bound on our original problem, and each solution to the relaxation of the Master problem provides a lower bound. The algorithm can terminate either at the optimal point by checking that the lower and upper bounds are equal or at some tolerance set by the user.

In the computational process, we realized that the master problem had multiple optimal points at some cuts (as is common with many integer programs.) To address this, we simply use branch-and-bound to explore multiple optima. Each optimal **y**-vector is used to create a new cut that is separately appended to the master problem. Each of these master problems can be solved and compared to the current best solution to decide if we should continue to evaluate that master problem. Starting from the same point as above, our first cut takes us to the point $y_3^1 = y_1^2 = y_2^2 = 1$ (with the remaining y at zero value). This gives us an upper bound of 4717. At the next step we are at the point $y_2^1 = y_3^1 = y_1^2 = 1$ (with the rest at 0). The upper bound is 3489.5. The next step remains at the same point and our upper bound and optimal solution is at the stabilized value of 3440.3. In comparison with generalized Benders, this procedure converges in fewer steps. Although this toy problem converges pretty fast, it is obvious that such a solution procedure is not practical for any practical size problem, even with the introduction of a Benders' cut. While Benders' decomposition may save computation in terms of the **x** and **z** space, the problem is in the sheer size of the **y** space that still remains.

4.3.5 Comments. In spite of this discussion, Federgruen and Zipkin (1984) used a modified interchange heuristic to solve this problem. The heuristic procedure consists of an 'inventory allocation' step, 'switching' step and 'sweep' procedure. For the sample problem, the authors conclude that "while the combined routing/allocation problem requires more effort than the vehicle-routing problem (as one would expect), the overall computational demands of the combined approach are reasonable for many applications." We choose to discuss this single-depot, multiple-vehicle routing-and-allocation (*s/m/–/a*) problem for several reasons. First, the projection and dualization approach described here represents a viable approach to solving this class of problems, particularly its kinship to Lagrangian-relaxation techniques[9] (Fisher and Jaikumar 1981). Second, it is pedagogically a halfway point between the simplest single-route/multi-facility problem (such as the *s/s/m/–* model) and the complex general location-routing problem to be discussed later. It further illustrates the class of location-routing problems in which the locational decisions apply toward a demand point rather than a depot. Only now we are talking about multi-vehicle tours rather than a route from origin to destination as with the *s/s/m/–* model. Finally, it brings out the role of heuristics in spite of an analytical formulation and solution procedure. Specifically, it stimulates further development of this type of heuristics in a more general location-routing formulation, as will be discussed in the model by Perl (1987), where the Federgruen-and-Zipkin type heuristic will be explained in its extended form.

---

[9]For an introduction to Lagrangian relaxation, see the "Optimization" appendix to Chan (2000) as well as the "Decomposition" section of the "Prescriptive Tools" chapter for example.

4.4   Single-Facility/Multiple-Tour/Probabilistic-Demand Model

The above routing/allocation model routes vehicles toward where inventory is running low. If deliveries are driven by random demands instead of stockout cost, a probabilistic model results. We call this the single-depot (or single-facility), multiple-tour, probabilistic-demand ($s/m/-/p$) model. Beyond visiting a subset of demands where the probability is nonzero ($I'' \subseteq I$), a location decision is to be made regarding the depot. This class of problem has been named probabilistic multiple-travelling-salesmen facility-location problem. The probabilistic multiple-travelling-salesman facility-location (PMTSFLP) problem is a combination of a facility-location problem and a probabilistic multiple-travelling-salesmen (PMTSP) problem. The problem consists of a network of demand nodes that may or may not have a particular demand instance for any particular tour. The objective is to base the travelling salespersons and find their tours so that the total expected travel cost is minimized. The 1-median facility-location problem is the p-median model as formulated in the "p-median problem" subsection of the "Facility Location" chapter when $p = 1$. We write y_j instead of x_{jj} for the location variable here for simplicity. An instance of the PMTSP is shown in the classic multiple-travelling-salesmen problem under the "Travelling-salesmen problem" subsection in the "Measuring Spatial Separation" chapter.

In the previous chapter on "Measuring Spatial Separation," we reported that Chan and Merrill formulated a probabilistic multiple-travelling-salesmen facility-location problem (PMTSFLP) as documented in the "Routing as spatial separation" subsection. For the forcing/linking constraint of this model, one writes (Loftus and Chan 1992)

$$\sum_{j \in M_i} x_{ij} \le |H''| y_i \qquad \forall i \in J \tag{4-42}$$

$$1 \le \sum_{i \in M_j} x_{ij} \le |H''| y_j \qquad \forall j \in J \tag{4-43}$$

$$x_{ij} + x_{ji} \le 1 + \sum_{k \in J} y_k \qquad \forall i \in J, \forall j \in J \tag{4-44}$$

$$x_{il} + x_{li} \le 1 + y_l \qquad \forall i \in I' - J, \forall l \in J \tag{4-45}$$

which established the precedence relationship between basing decisions **y** and routing decisions **x**. Equations (4-44) and (4-45) are simplified subtour-breaking constraints that work only for small networks with low sales-persons-to-nodes-visited ratio. Equations (4-42)–(4-45) cater for situations where there are no demands at a depot. Instead of an explicit linking/forcing constraint, however, Chan and Merrill (1997) enumerated all possible depots that can base a vehicle fleet and pick the one with the lowest cost. Their approach essentially uses a sequential approach to solve location-routing problems. It is justifiable since all depots are existing depots; little or no additional costs will be necessary to base a vehicle fleet there. Thus a location-routing model degenerates into several routing models based at candidate depot sites. The combined location-routing model is simply the combination of the median and TSP models in which the Euclidean distance d_{ij} is replaced by the tour distance d^{ij}.

More formally, since the optimal depot location must occur at a node it is possible to decompose the problem into p probabilistic multiple-travelling-salesmen problems, where p is the number of potential depots. Thus the number of depots located must be p:

$$\sum_{j\in J} y_j = p. \tag{4-46}$$

According to Berman and Simchi-Levi (1988), we solve a separate problem for each potential depot and find an *a priori* tour covering all of the demand nodes I'' while minimizing the expected travel cost. For any particular tour, nodes with no demand are skipped. This means the arc between node i and node j will only be traversed if there is demand at both nodes. The p solutions (one for each potential depot) are then compared to determine the optimal depot selection and the corresponding optimal *a prior* tour.

The objective of the probabilistic-travelling-salesman-facility-location problem is to find the location of a single depot that minimizes the least expected cost of satisfying the demand locations (see example in **Figure 4-4**). Since nodes with no demand are skipped, an arc between node i and node j will only be travelled ($x_{ij}=1$) if there is demand at both nodes, the expected travel costs for a tour that includes arc i–j can be computed by multiplication: the probability of demand for node i times the probability of demand at node j times the travel cost for the path. The probability of demand at the depot is always taken to be one—a tour must originate and terminate at the depot. The total expected travel cost is given by summing the expected travel costs across all tours. The particular numerical example included here has two possible depot locations (labelled 1 and 2) to choose from. There are two vehicles so there are two tours that will be used. Essentially, the problem breaks down into two multiple-travelling-salesmen problems through the use of location variable $y_j \in \{0,1\}$. The problem is solved by finding the two minimum-cost tours out of depot 1 while ignoring depot 2, and finding the two minimum-cost tours out of depot 2 while ignoring depot 1. Once the tours are found, the model picks the depot that has the minimum total-cost for the tours.

Example

Given an instance of the problem where $I'' = \{3,4,5\}$ nonzero demands are found among I nodes, solve this PMTSFLP as shown in **Figure 4-4**.

Here, $|I''|=3$ not counting the depots 1 and 2. In other words, no demands are placed at depot nodes. For this particular instance,

minimize $3x_{12}+3x_{13}+5x_{14}+5x_{15}+3x_{21}+3x_{31}+5x_{41}+5x_{51}+4x_{23}+5x_{24}+3x_{25}+$
$4x_{32}+5x_{42}+3x_{52}+4x_{34}+8x_{35}+4x_{43}+8x_{53}+7x_{45}+7x_{54}$

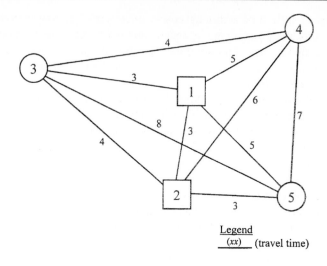

Figure 4-4 Example network of the PMTSFLP

subject to

$y_1+y_2=1$ (only one depot is to be located) $x_{13}+x_{14}+x_{15}= 2y_1$ (arcs going out of depot 1)
$x_{31}+x_{41}+x_{51}= 2y_1$ (arcs going into depot 1) $x_{23}+x_{24}+x_{25}= 2y_2$ (arcs going out of depot 2)
$x_{32}+x_{42}+x_{52}= 2y_2$ (arcs going into depot 2) $x_{31}+x_{32}+x_{34}+x_{35}= 1$ (arcs going out of node 3)
$x_{13}+x_{23}+x_{43}+x_{53}= 1$ (arcs going into node 3) $x_{41}+x_{42}+x_{43}+x_{45}= 1$ (arcs going out of node 4)
$x_{14}+x_{24}+x_{34}+x_{54}= 1$ (arcs going into node 4) $x_{51}+x_{52}+x_{53}+x_{54}= 1$ (arcs going out of node 5)
$x_{15}+x_{25}+x_{35}+x_{45}= 1$ (arcs going into node 5)

where the terms $2y_1, 2y_2$ need to be changed to $1-y_1$, and $1-y_2$ for the case where demand exists at a depot.

   (full subtour-breaking constraints)
   $\delta_3-\delta_4+4x_{34} \le 3$ $\delta_4-\delta_3+4x_{43} \le 3$ $\delta_5-\delta_3+4x_{53} \le 3$
   $\delta_3-\delta_5+4x_{35} \le 3$ $\delta_4-\delta_5+4x_{45} \le 3$ $\delta_5-\delta_4+4x_{54} \le 3.$

Here, the full set of subtour-breaking constraints are used instead of the simplified ones that work up to seven nodes:

   (simplified subtour-breaking for demand nodes)
   $x_{43}+x_{34} \le 1$ $x_{53}+x_{35} \le 1$ $x_{54}+x_{45} \le 1$

   (simplified subtour-breaking for depots)
   $x_{21}+x_{12} \le 1+y_1+y_2$
   $x_{31}+x_{13} \le 1+y_1$ $x_{32}+x_{23} \le 1+y_2$
   $x_{41}+x_{14} \le 1+y_1$ $x_{42}+x_{24} \le 1+y_2$
   $x_{51}+x_{15} \le 1+y_1$ $x_{52}+x_{25} \le 1+y_2$

When demands exist at the depots, the equation $x_{ij}+x_{ji} \leq 1+\sum_{k\in J} y_k$—as shown in Equation **(4-44)** and illustrated directly above as the first equation in the set of simplified subtour-breaking constraints for depots—is no longer required. (Shirley and Chan 1993) □

In theory, the equations listed above can be solved, albeit in an extremely costly way, inasmuch as all instances need to be solved. This is the reason why a space-filling curve represents the only feasible way of solving PMTSFLP, as described in the previous chapter on "Measuring Spatial Separation." The model described above is a variant of the *s/m/–/a* formulation incorporating probabilistic demands. Again, we call this the single-depot, multiple-tour, probabilistic-demand (*s/m/–/p*) model based on the least expected cost in view of uncertain demand. Rather than driven by inventory shortage, demands are explicitly treated as random variables. With the introduction of the forcing/linking constraint shown in Equations **(4-42)**–**(4-45)**, assignment of vehicles to demands and vehicle-basing decisions can be made simultaneously, rather than sequentially as illustrated above.

4.5   Multiple-Facility/Multiple-Tour/Allocation Model

It is natural to generalize the *s/m/–/p* model from a single depot to multiple depots. Laporte et al. (1986) address a symmetric location-routing problem which has the following properties: (a) The potential depots (set *J*) are distinguished among the demand sites (Set *I*), *J∈I*. (b) For compactness in bookkeeping, multiple passages through the same bi-directional symmetric arc are permissible provided distance savings are achieved. Thus $x_{ij} = 2$ means an "out-and-back" trip *i-j-i* or *j-i-j*. (c) A deterministic demand is assigned to each site and the total traffic picked up by the vehicle may not exceed capacity V. The problem consists of selecting depot sites, of determining how many vehicles are based at each selected depot, and of establishing vehicle tours in such a way that that the results in the following: (1) Each route starts and ends at the same depot in a tour. (2) All the pickup requirements at a node are met by only one vehicle. (3) Lower and upper limits are imposed upon a depot in terms of the number of vehicles that can be based there, labelled *h'* and *T* respectively. (4) The total cost is minimized.

4.5.1 Model formulation. The multiple-depot (or multiple-facility)/multiple-tour/fleet-allocation (*m/m/–/a*) formulation takes on the objective function of lowest travel, depot, and vehicle costs:

$$\min \sum_{i,j\in I} d_{ij}x_{ij} + \sum_{r\in J} (c_r y_r + b_r m_r) \qquad (4\text{-}47)$$

where m_r is the number of vehicles docked at depot *r*, and c_r, b_r are unit costs.

Sequencing an arc consistently by node labels in ascending order, the first constraint specifies that each demand site not used as a depot must be *serviced* exactly once by any given vehicle

$$\sum_{i<k} x_{ik} + \sum_{k<j} x_{kj} = 2 \qquad k\in I\text{-}J. \qquad (4\text{-}48)$$

The second constraint says that m_r vehicles must dock at depot r:

$$\sum_{i<r} x_{ir} + \sum_{r<j} x_{rj} = 2m_r \qquad r \in J.$$

(4-49)

The third constraint ensures that the solution does not contain any illegal subtours (subtours disjoint from J), or that the vehicle capacity be exceeded

$$\sum_{i \in L} \sum_{j \in L} x_{ij} \le |L| - \left\lceil \sum_{i \in L} f_i / V \right\rceil \qquad L \subseteq I - J, \ |L| \ge 3 .$$

(4-50)

Notice this is a generalization of the subtour-breaking constraint $\sum_{i \in L} \sum_{j \in L} x_{ij} \le |L| - 1$ used thus far. Introduced in the "Measuring Spatial Separation" chapter under the "Travelling-salesman problem" subsection, it incorporates the load constraint.

Because of a variable number of vehicles at a depot (rather than a fixed number), *chains* might be present in the solutions, where a chain is an itinerary consisting of more than one depot. The fourth and fifth are called "chain-barring constraints." They ensure that each route starts and ends at the same depot, rather than another depot. To prevent chains of four nodes, we write:

$$x_{i_1 i_2} + 3x_{i_2 i_3} + x_{i_3 i_4} \le 4 \qquad i_1, i_4 \in J; \ i_2, i_3 \in I - J$$

(4-51)

where i_1 and i_4 are depot nodes and i_2 and i_3 are demand nodes. Chains of five or more nodes are prohibited by constraints of the form

$$x_{i_1 i_2} + x_{i_{g-1} i_g} + 2 \sum_{\substack{i,j \in \\ \{i_2, \ldots, i_{g-1}\}}} x_{ij} \le 2g - 5 \qquad g \ge 5; \ i_1, i_g \in J; \ i_2, \ldots, i_{g-1} \in I - J.$$

(4-52)

The sixth constraint means that no vehicle can be based at a node that is not used as a depot. Moreover, if a node is used as a depot, it must have at least one vehicle assigned to it:

$$y_r \le m_r \le B y_r \qquad r \in J.$$

(4-53)

Here B can be an arbitrarily large number. The seventh constraint requires that the number of vehicles based at a depot must lie within specified bounds:

$$h' \le m_r \le G' \qquad r \in J$$

(4-54)

where G' has to be consistent with B by the relationship $B \ge G'$. Similarly, the eighth constraint limits the total number of nodes used as depots

$$p' \le \sum_{r \in J} y_r \le p .$$

(4-55)

Again, it is interesting to note that besides a 0–1 valuation of x_{ij}, x_{ij} can be as big as 2. The latter suggests an out-and-back trip through the same arc (i,j) from and to the depot (meaning a depot is at either node, i or j, remembering we are solving a symmetric TSP). Together with the x_{ij} notation where $i < j$, they form a distinguishing feature of Laporte's et al.'s symmetric formulation. We label this model $m/m/-/a$, suggesting its similarity to the Zipkin/Federgruen $s/m/-/a$ model and the Chan and Merrill $s/m/-/p$ model. The $s/m/-/a$ and $m/m/-/a$ models are similar in that a variable number of vehicle tours are assigned according to the demand load. While demand is no longer probabilistic, the $m/m/-/a$ model allocates the fleet of $|H''|$ vehicles among the depots much as the $s/m/-/p$ model does with a depot fleet of G'. Obviously, the major difference is that multiple depots are now employed.

4.5.2 Solution methodology. For this multi-depot location problem with vehicle-capacity constraints, Laporte et al. (1986) offer a variant of the iterative constraint- generation procedure. The solution algorithm is applied toward the m/m-/a symmetric location-routing problem formulated as Equations **(4-47)** through **(4-55)** above. We will paraphrase their procedures below. Instead of considering separate routing variables and capacity restrictions corresponding to each vehicle, vehicle capacities (assuming a single vehicle type) are directly incorporated in the subtour-elimination and chain-barring constraints iteratively. Integrality is enforced by embedding the iterative constraint-generation scheme in a branch-and-bound procedure.

Many constraints in the formulation are not binding and can be selectively eliminated. Generally, a constraint relaxation of the formulation is carried out iteratively. It begins by comparing a known feasible (but not optimal) solution to a subproblem solution that does not consider chain-barring or subtour-breaking. When such a subproblem offers solution improvement, integer, subtour, and chain-barring restrictions are added as necessary to induce subproblem feasibility. If that subproblem still offers improvement, it is stored as the best known incumbent and a new subproblem is considered. The process continues until all subproblems are implicitly considered. The solution algorithm (a) selects depot location, (b) finds how many vehicles should be based at each depot, and (c) determines the optimal routing for each vehicle.

Consider a relaxed problem obtained from the original formulation by removing chain-barring constraints **(4-51)** and **(4-52)**. The optimal solution to the relaxed problem may contain chains between two depots, i.e., sequence of nodes (i_1, i_2, ... , i_g) where

$$i_1, i_g \in J; \quad i_2, ..., i_{g-1} \in I-J; \qquad i_1 \neq i_g \qquad \text{(4-57)}$$

$$x_{i_t i_{t+1}} = 1 \qquad t=1,2,...,g-1. \qquad \text{(4-56)}$$

We wish to generate constraints that will eliminate such chains without cutting off any feasible solution to the original problem. It can be verified that the general conditions governing the cut generation reduces to Equation **(4-51)** for a 4-node chain, and Equation **(4-52)** for an over-5-node chain. Furthermore, we can show that chains involving less than four nodes can be ignored. Interested readers are referred to Laporte et al. (1986) for details if desired.

The algorithm developed to solve the original formulation can be summarized as follows:

Step 0 Obtain a first feasible-solution by means of an appropriate heuristic. Let z^* be the total system cost of that solution.

Step 1 Select a subproblem from the list generated from relaxing the original formulation. (The first subproblem will include constraints **(4-48)**, **(4-49)**, **(4-53)**, **(4-54)**, and **(4-55)** as well as the upper bound on the **x** variables (namely 2). The subtour-breaking and chain-barring constraints are deleted.)

Step 2 Solve the subproblem using the simplex or equivalent LP-solution method. Let z be the cost of the least-cost solution to the subproblem.

Step 3 If $z \geq z^*$, proceed to the backtracking step 9.

Step 4 The current solution contains: (i) sets of nodes $\{i_1, \dots, i_l\}$ ($l > 1$) corresponding to chains $(i_1, \dots, i_l)$ such that $\{i_2, \dots, i_{l-1}\} \cap J = \emptyset$ if $l > 2$ and for which all variables $x_{i_1 i_2}, x_{i_2 i_3}, \dots, x_{i_{l-1} i_l}$ have been previously fixed at 1, and (ii) nodes of $I - J$ not belonging to such chains (we define for each such node i a set of $\{i\}$). For convenience, we refer to these sets of nodes W_k (corresponding to chains or single nodes) as *components*. Each W_k has an associated weight $\sum_{i \in W_k} f_i$. Consider a component W_k. If it corresponds to a chain or a node sequence containing more than one depot, let p_k'' and s_k'' be the end nodes of that chain; if it corresponds to a node i, let $p_k'' = s_k'' = i$. In the first case where a chain is involved, consider two components W_u and W_v and let $i \in \{p_u'', q_u''\}$, $j \in \{p_v'', q_v''\}$. Then x_{ij} can be forced to zero if $i, j \in I - J$ and $\sum_{i \in W_u} f_i + \sum_{i \in W_v} f_i$. If step 4 has resulted in forcing any variable to zero, proceed to step 2. In other words, if the current solution contains chains whose node demands exceed vehicle capacity, they may be broken by forcing an arc to zero which splits the chains into pieces that do not exceed vehicle capacity. Should any variable be forced to zero, return to step 2.

Step 5 Check whether the current solution contains illegal subtours (i.e., subtours disconnected from J or having a total weight exceeding V). If there are no illegal subtours, proceed to step 6. Otherwise, generate a type **(4-50)** constraint for each illegal subtour and go to step 2.

Step 6 Check whether the current solution contains illegal chains between depots (in this context, a depot is a node r of J for which $y_r = 1$). If there are no illegal chains, proceed to step 7. Otherwise, generate for each chain a subtour-elimination constraint **(4-50)** or a chain-barring constraint **(4-51)** or **(4-52)** and go to step 2.

Step 7 If the solution is integer, store it and set $z^* = z$; proceed to step 9. Otherwise, execute the next step.

Step 8 Select a fractional variable to branch upon and create new branches in the search tree. Go to step 10.

Step 9 Backup the search tree. This consists of modifying the level of the search tree (from k' to k, where $k' \geq k$) by freeing all variables either forced to zero or fixed at some integer value at levels $k, k+1, \ldots, k'$. In other words, backtrack up the search tree by freeing variables that have been forced to zero or integer.

Step 10 Update the list of subproblems. If the list is empty, stop and print the best solution. Otherwise go to step 1. ∎

While the above represents the original detailed algorithm of Laporte et al., it is conceivable that the following simplification can be made (Johnson and Chan 1990). A review of the 10 steps above reveals that the methodology consists of relaxing the integer program (IP) by ignoring integrality, subtour-breaking, vehicle-capacity, and chain-barring constraints. The relaxed problem is solved using any LP software. Constraints are then added iteratively until a feasible solution is obtained. The approach is modified by including the integrality constraints into the relaxed problem and using available IP software to solve the resulting subproblems. Since the number of subtour-breaking/vehicle-capacity constraints and chain-barring constraints increase combinatorially with the number of demand nodes, this approach promises to be significantly more efficient than solving the completely-relaxed LP. It cuts out the branches for integerization, especially as the number of demand nodes becomes large:

Step 1 Construct the relaxed IP (the first subproblem) by deleting the subtour-breaking/vehicle-capacity and chain-barring constraints from the complete IP.

Step 2 Solve the subproblem using any IP software-package.

Step 3 If the solution contains an illegal subtour or violated vehicle-capacity constraints, add the appropriate constraint to form a new subproblem. Go to step 2.

Step 4 If the solution contains a tour starting at one depot and ending at a different depot, add the appropriate chain-barring constraint to form a new subproblem. Go to step 2.

Step 5 The current solution is feasible and optimal since it is nondominated. ∎

4.5.3 Example of Johnson-Chan's simplified algorithm. Johnson and Chan (1990) and Moore and Chan (1989) provided a numerical example of the symmetric m/m/-/a model. The problem consists of a five-node network of demand nodes and two potential depot locations (nodes 1 and 2, which may also be demand nodes). Costs consist of fixed depot-overhead-costs, fixed vehicle-overhead costs (per vehicle), as well as travel costs. Overhead costs for the depots are 5 and 6 units for depots 1 and 2, while those for vehicles stationed at a depot are 2.2 and 2.0 respectively for depots 1 and 2. The vehicle capacity is 30,000 pounds (15,000 kg). Demand at node 3 is 16,000 pounds (8,000 kg), demand at node 4 is 7,000 pounds (3,500 kg), and demand at node 5 is 14,000 pounds (7,000 kg). The link travel-costs are shown in **Figure 4-5**.

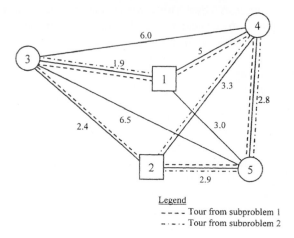

Legend
---- Tour from subproblem 1
--·--·· Tour from subproblem 2

Figure 4-5 Network for symmetric multiple-depot/multiple-tour model

The problem is formulated as an integer LP problem with three sets of decision variables. The first set of decision variables are x_{ij} indicating whether arc (i, j) is included in a tour. These variables can assume values zero, one, or two. In this symmetrical problem, x_{ij} is the same as x_{ji}; thus the number of these variables is half of what would be required (as mentioned). When an out-and-back tour is used, x_{ij} = 2 as mentioned. The second set are 0–1 decision variables y_r, where node r is a potential depot location. The third set of decision variables are m_r, which indicates the number of vehicles assigned to depot r. Although some expressions are similar to the asymmetric cases described earlier, we decide to write out each equation in long hand to illustrate the compact notations by Laporte et al.

The overall objective is to find the optimal depot locations, the optimal number of vehicles to be assigned to each depot, and the optimal routing for each vehicle in order to minimize total costs. For the example cited, we have

$$\min\ 1.9x_{13}+2.5x_{14}+3x_{15}+2.4x_{23}+3.3x_{24}+2.9x_{25}+$$
$$6x_{34}+6.5x_{35}+2.8x_{45}+5y_1+6y_2+2.2m_1+2m_2. \tag{4-58}$$

The first set of constraints ensures that each demand node is visited exactly once, remembering $i<j$ in our notation for x_{ij}.

$$
\begin{array}{ll}
x_{13}+x_{23}+x_{34}+x_{35}=2 & \text{(node 3)}\\
x_{14}+x_{24}+x_{34}+x_{45}=2 & \text{(node 4)}\\
x_{15}+x_{25}+x_{35}+x_{45}=2 & \text{(node 5)}.
\end{array}
\tag{4-59}
$$

The second set of constraints ensures that the number of tours entering and leaving a depot is the same as the number of vehicles assigned to the depot.

$$x_{13}+x_{14}+x_{15}-2m_1=0 \quad \text{(depot 1)}$$
$$x_{23}+x_{24}+x_{25}-2m_2=0 \quad \text{(depot 2)}. \tag{4-60}$$

The third set of constraints ensures that vehicles are only based at nodes that are selected as depots and that nodes designated as depots have at least one vehicle assigned. Additionally, upper and lower bounds of the number of vehicles assigned to each depot are established. For this example, the upper bound is two and lower bound is one at both depots.

$$1 \le m_1 \le 2y_1 \qquad 1 \le m_2 \le 2y_2. \tag{4-61}$$

A more general formulation may be $y_1 \le m_1 \le 2y_1$ and $y_2 \le m_2 \le 2y_2$.

Redundant as it may be here, the fourth set of constraints bound the total number of depots (for this example, the lower bound is one and the upper bound is two).

$$y_1+y_2 \ge 1 \qquad y_1+y_2 \le 2. \tag{4-62}$$

The fifth set of constraints combines subtour-breaking and vehicle-capacity limitations.

$$x_{34}+x_{35}+x_{45} \le 3-1.233 = 1.767. \tag{4-63}$$

The final set of constraints are "chain barring" constraints. Notice that these constraints are required for every possible combination of demand and depot nodes meeting the four- or five-chain criteria. Therefore the number of constraints increases combinatorially. For the four-chains

$$x_{13}+3x_{34}+x_{24} \le 4 \qquad x_{23}+3x_{34}+x_{14} \le 4$$
$$x_{13}+3x_{35}+x_{25} \le 4 \qquad x_{23}+3x_{35}+x_{15} \le 4$$
$$x_{14}+3x_{45}+x_{25} \le 4 \qquad x_{24}+3x_{45}+x_{15} \le 4. \tag{4-64}$$

For the five-chains

$$x_{13}+x_{24}+2(x_{34}+x_{35}+x_{45}) \le 5 \qquad x_{14}+x_{25}+2(x_{34}+x_{35}+x_{45}) \le 5$$
$$x_{13}+x_{25}+2(x_{34}+x_{35}+x_{45}) \le 5 \qquad x_{15}+x_{23}+2(x_{34}+x_{35}+x_{45}) \le 5$$
$$x_{14}+x_{23}+2(x_{34}+x_{35}+x_{45}) \le 5 \qquad x_{15}+x_{24}+2(x_{34}+x_{35}+x_{45}) \le 5. \tag{4-65}$$

For this small network, no chains longer than five nodes are possible.

The following represents the algorithmic steps for solving the problem. The first subproblem is defined by Equations (4-58), (4-59), (4-60), (4-61), and (4-62), with the subtour-breaking and chain-barring constraints removed (Equations (4-63), (4-64), and (4-65)). The solution for subproblem 1 is overlaid on **Figure 4-5**. The solution includes two illegal chains; the first one is 1–4–5–2 and the second 2–3–1. The chain-barring constraint $x_{14}+3x_{45}+x_{25} \le 4$ is now employed to form a new subproblem. The 3-node chain will be taken care of automatically without doing anything else, as suggested in the section entitled "Solution methodology".

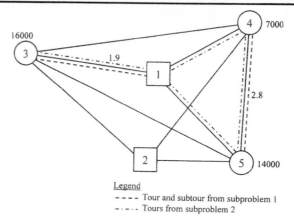

Legend
---- Tour and subtour from subproblem 1
-·-·- Tours from subproblem 2

Figure 4-6 Solutions to modified multi-depot/multi-tour problem

The solution for subproblem 2 is also overlaid in **Figure 4-5**. The combined solution is now free of illegal subtours, vehicle capacities are not exceeded, and there are no illegal chains. The solution is the optimal solution for the complete problem. One vehicle is assigned to each depot. The vehicle assigned to depot 1 is loaded with 16,000 pounds (8,000 kg) and travels from node 1 to node 3, where it drops off its load. It then returns to the depot. The vehicle assigned to depot 2 is loaded with 21,000 pounds (10,500 kg) and travels from node 2 to node 4, where it drops off 7,000 pounds (3,500 kg), then to node 5, where it drops off the remaining 14,000 pounds (7,000 kg), and returns to depot 2. The total cost for this solution is 28 units. Notice that this problem contains only 14 constraints as compared to 26 for the original problem. This illustrates the savings obtainable by this method. Since the number of subtour-breaking/vehicle-capacity and chain-barring constraints increases combinatorially with the number of nodes, the savings resulting from this solution methodology will become significant as the problem grows large.

The above example was varied by eliminating the requirement to base at least one vehicle at each depot. Thus Equation **(4-61)** is modified accordingly. The solutions for subproblem 1 of the modified problem is shown in **Figure 4-6**. All demand nodes are visited, but an illegal subtour 4-5-4 is included. Thus the subtour-breaking/vehicle-capacity constraint $x_{34}+x_{35}+x_{45} \leq 1.767$ is added to form a new subproblem. The solution network for subproblem 2 is shown in **Figure 4-6**. This solution is free of illegal subtours, vehicle capacities are not exceeded, and there are no illegal chains. The solution is optimal, suggesting depot location at 1, with two vehicles based there, and tours 1-3-1 and 1-4-5-1.

4.5.4 Example of an extended Laporte et al.'s algorithm. Another example is taken from the northwest United States of the Defense Courier Service Aerial network (Baker and Chan 1990, Green and Chan 1993). It is shown in **Figure 4-8** (Baker and Chan 1991). Nodes 1 and 5 are referring to the same geographic

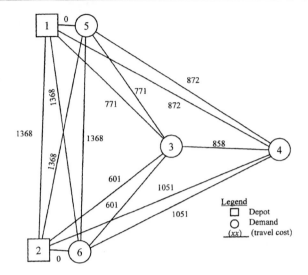

Figure 4-7 A network of the Defense Courier Service

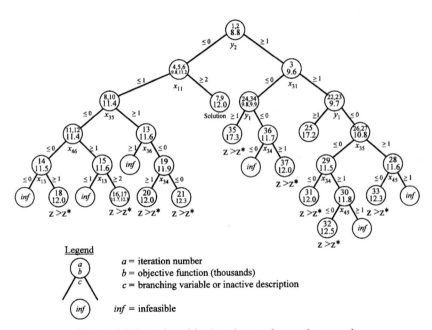

Figure 4-8 Branch-and-backtrack tree of example network

location, where two split nodes are created to model a demand that also occurs at a depot. The same for nodes 2 and 6. A demand of 25 is placed at each demand node, and vehicle capacity is restricted to 50. This effectively limits a vehicle tour to at most two demand nodes. Costs to be considered are the fixed depot overhead-costs and travel costs. While the travel costs are shown in **Figure 4-8**, the depot overhead-cost is 7538 dollars per depot.

First, subtour-breaking and chain-barring constraints are excluded from the model. A branch-and-backtrack tree is used to execute the LP-relaxation algorithm. Successive iterations are performed by modifying this original model with integrality, subtour-breaking, and chain-barring constraints as warranted. These iterations are documented in two ways: branch-and-backtrack tree and a constraint table. The results of the model are fully documented in **Figure 4-7**. Although the optimal solution appears at iteration 9, the algorithm requires 37 iterations to exhaust the set of subproblems. The only departure from the formal solution methodology, as outlined in the algorithm of the section entitled "Solution methodology," is found between iterations 8 and 9. Since iterations 7 and 8 offer a z of 12.0 and 11.4 respectively, the latter branch offers a potentially lower solution and should have been given precedence. Baker and Chan chose to complete the $x_{14} \geq 2$ branch before proceeding because it offered an imminent integer solution. Solution methodology is not compromised by this slight departure.

The constraints generated in each step of the branch-and-backtrack tree are documented below:

Iteration	Status	Constraint generated
1	subtour	$x_{34} \leq 1$
2	nonintegral	$y_2 \leq 0, y_2 \geq 1$
3	nonintegral	$x_{23} \leq 0, x_{23} \geq 1$
4	subtour	$x_{34}+x_{36}+x_{46} \leq 1.5$
5	subtour	x_{36}
6	nonintegral	$x_{14} \geq 2, x_{14} \leq 1$
7	capacity	$x_{35}+x_{36}+x_{56} \leq 1.5$
8	capacity	$x_{35}+x_{36}+x_{56} \leq 1.5$
9	solution	—
10	nonintegral	$x_{35} \leq 0, x_{35} \geq 1$
11	subtour	$x_{45} \leq 1$
12	nonintegral	$x_{46} \leq 0, x_{46} \geq 1$
13	nonintegral	$x_{36} \leq 0, x_{36} \geq 1$
14	nonintegral	$x_{13} \leq 0, x_{13} \geq 1$
15	nonintegral	$x_{13} \leq 1, x_{13} \geq 2$
16	subtour	$x_{46} \leq 1$
17	$z \geq z^*$	—
18	$z \geq z^*$	—
19	nonintegral	$x_{34} \geq 1, x_{34} \leq 0$
20	$z \geq z^*$	—
21	$z \geq z^*$	—
22	chain	$x_{15}+x_{23}+2(x_{34}+x_{45}) \leq 5$
23	nonintegral	$y_1 \geq 1, y_1 \leq 0$
24	subtour	$x_{34}+x_{35}+x_{45} \leq 1.5$
25	$z \geq z^*$	—
26	capacity	$x_{34}+x_{35}+x_{45} \leq 1.5$

27	nonintegral	$x_{35} \leq 0, x_{35} \geq 1$
28	nonintegral	$x_{45} \geq 1, x_{45} \leq 0$
29	nonintegral	$x_{34} \geq 1, x_{34} \leq 0$
30	nonintegral	$x_{45} \geq 1, x_{45} \leq 0$
31	$z \geq z^*$	—
32	$z \geq z^*$	—
33	$z \geq z^*$	—
34	nonintegral	$y_1 \geq 1, y_1 \leq 0$
35	$z \geq z^*$	—
36	nonintegral	$x_{34} \geq 1, x_{34} \leq 0$
37	$z \geq z^*$	—

It is interesting that producing the optimal solution requires five subtour-breaking or vehicle-capacity constraints at iteration 9, instead of the 37 used in the mixed integer solution generated from Equations (4-50), (4-51) and (4-52) in the original formulation. (The number 37 happens to coincide with the number of iterations in the branch-and-backtrack tree by sheer chance.) Even counting the entire branch-and-backtrack tree, there are altogether only nine subtour-breaking or vehicle-capacity constraints and one chain-barring constraint. It is clear from this comparison that Laporte et al.'s LP-relaxation offers considerable improvement in computational efficiency.

4.5.5 Comments. The original Laporte et al. algorithm was tested on several randomly generated problems with varying levels of capacity limitations and with or without fixed depot location and vehicle costs. The capacitated test problems ranged in size from 6 to 20 demand sites and 4 to 8 depot sites. The number of vehicles was not restricted; upper bounds of between 3 to 5 were imposed on the number of depot locations that were selected. The efficiency of the algorithm can be attributed to the initial relaxation of most of the problem constraints. It can be observed that the maximum number of effective constraints within the algorithm generally lies between $2|I|$ and $3|I|$, while the number of potential constraints is in the order of $2^{|I|}$. The difficulty of the problem is inversely related to the size of V, or the vehicle capacity. Imposing large depot costs tends to produce easier problems, but the same cannot be said about vehicle costs b_r, which do not seem to affect computation. Laporte et al.'s methodology, together with the Johnson-Chan simplification, represents a most exciting recent developments in exact solutions to location-routing models. Unlike the allied Federgruen and Zipkin model, it is an operational rather than theoretical, constraint-generation procedure. Similar to the network-with-side-constraint facility-location problem, the constraints are generated as needed and the augmented model is solved by an existing, "off-the-shelf" software.

Improvements to the solution algorithm can be made based on the special structure of this problem (Green and Chan [1993], Graham and Chan [1994]). Graham and Chan (1994), for example, exploited the network structure implicit in the formulation. In the problem above, a network of up to 10 nodes and 23 arcs can be recognized. Only four constraints remain as side constraints[10]. Computer codes such as CPLEX solves the network subproblem in 10 iterations and the side constraints in two LP iterations. Exploiting the embedded network structure allows a much more efficient integerization and chain-barring branch-and-bound tree to be constructed. In this case, the optimal solution can be obtained within two integerization branches and the addition of one chain-barring constraint.

4.6   Multi-Facility/Multi-Tour/Hierarchical Model

Perl and Daskin (1985) also formulated a generalized location-routing problem where the route is a tour, starting and terminating at the same depot. Their location-routing problem can be stated formally as follows: A company ships from one or more supply sources to a number of regional distribution centers (or depots for short). We call this the *trunk* traffic. The company then delivers goods from the depots to the demand points: the local-distribution traffic. The cost function for a facility j consists of a fixed charge b_j for facility construction and a linear variable-cost (as a function of traffic volume at unit cost c_j). Link costs are a linear function of traffic volume at unit costs of a_{ij} for trunk traffic z_{ij} or $g_h d_{ij}$ (without the fixed charge) for local-distribution traffic x_{ij}^h. Both the facility and the delivery vehicle have a limited capacity. Similar to the $m/m/-/a$ model, the model yields the *number* of the depot facilities and the tours beyond the regular outputs from a location-routing problem discussed previously. The Perl–Daskin model can be labeled as $m/m/-/h$. The h stands for "hierarchical," where two tiers of facilities are involved—the central warehouse and the regional distribution-centers.

   4.6.1   Formulation.   The objective function minimizes the sum of fixed facility-cost, trunking cost (defined as the cost to supply a depot), variable facility-cost and delivery-cost, as shown sequentially in the following expression:

$$minimize \ z(\mathbf{x}, \mathbf{y}, \mathbf{z}) = \sum_{j \in I} b_j y_j + \sum_{i \in I} \sum_{j \in I} a_{ij} z_{ij} +$$
$$\sum_{j \in I} c_j \sum_{i \in I} f_i x_{ij} + \sum_{h \in H} \sum_{i \in I} \sum_{j \in I} g_h d_{ij} x_{ij}^h. \tag{4-66}$$

Distinguished from previous formulations is the inclusion of trunking cost.
   Similar to the expression $\sum_{j \in I} x_{ij} \geq 1$ in the "*p*-median problem" subsection in the "Facility Location" chapter, the first constraint requires that each demand point be served:

$$\sum_{h \in H''} \sum_{j \in I} x_{ij}^h \geq 1 \qquad \forall i \in I. \tag{4-67}$$

---

[10]The approach utilizes a "network with side constraints" algorithm—an algorithm described in the Chan (2000) appendix entitled "Optimization procedures" (among other references).

The second constraint imposes the vehicle capacities on each tour exactly as the $\sum_{j \in I} f_j \sum_{i \in I} x_{ij}^h \le V(h)$ expression in the "Vehicle-routing problem" subsection of the "Spatial Separation" chapter. The third constraint limits the maximum length of a tour similar to the $\sum_{i \in I} t_i^h \sum_{j \in I} x_{ij}^h + \sum_{i \in I} \sum_{j \in I} d_{ij}^h x_{ij}^h \le U(h)$ expression in the same subsection, except without the stop-over time at a demand node. Furthermore, the same unit distance d_{ij} is used for all vehicles:

$$\sum_{i \in I} \sum_{j \in I} d_{ij} x_{ij}^h \le U(h) \qquad \forall h. \tag{4-68}$$

The fourth constraint requires that each delivery tour be connected to one of the K depots in a subtour-breaking constraint, following the $\sum_{i \in J} \sum_{j \in J} \sum_{h \in H''} x_{ij}^h \ge 1$ expression. The fifth constraint is merely a continuity constraint on the route identical to the $\sum_{i \in M_p} x_{ip}^h - \sum_{j \in \bar{M}_p} x_{pj}^h = 0$ expression. The sixth constraint places a similar continuity on the supply delivered to a depot j:

$$\sum_{i \in I} z_{ij} - \sum_{i \in I} f_i x_{ij} = 0 \qquad \forall j \in I. \tag{4-69}$$

The seventh constraint limits the trunk traffic supplied to a depot to its capacity $\bar{P}$:

$$\sum_{i \in I} z_{ij} - \bar{P}_j y_j \le 0 \qquad \forall j \in I. \tag{4-70}$$

The last constraint links the allocation and routing variables; it specifies that a demand node can be served only if a tour connects it to a depot:

$$\sum_{p \in M_i} x_{ip}^h + \sum_{p \in M_j} x_{pj}^h - x_{ij} \le 1 \qquad \forall i \in I;\ \forall j \in I;\ \forall h \in H''. \tag{4-71}$$

Notice the regular forcing/linking constraint is now split into three equations, **(4-69)**, **(4-70)** and **(4-71)**. This is a direct consequence of the hierarchical formulation. All variables are zero-one except z_{ij}, the amount of traffic supplied to facility j, thus making this a mixed integer program (MIP).

4.6.2 Solution algorithm. Perl's multi-facility/multi-tour/hierarchical problem $(m/m/-/h)$ includes multiple tours, supply sources, depot capacities, vehicle capacities and maximum tour length. As discussed previously, the problem is hierarchical in the sense that goods are shipped to two levels of facilities. First, goods are shipped from supply nodes to the depots, then the goods are shipped from the depots to the demand nodes. The problem from there on is a combination of a vehicle-routing and depot-location. The depots are capacitated by both the number and type of vehicles that may be assigned, as well as by the total amount of goods that may be shipped through the depot. The vehicles are capacitated by the amount of goods they may carry and the maximum tour length in distance or travel time. As mentioned, costs are given by the sum of the fixed facility costs, trunking costs from each depot, variable facility costs, and delivery costs. The problem solution includes the depot locations, the number of vehicles assigned to each depot, and the optimal routing for each vehicle. As usual, the overall objective is to minimize cost while satisfying demand. The sheer dimensionality requires decomposition and special solution algorithms.

Perl (1987) describes an optimization-based heuristic procedure for solving this complex multi-depot location-routing model. The algorithm consists of an initialization phase followed by an improvement phase made up of two iterative steps. The initialization phase assumes that all depots are open. A multi-depot vehicle-dispatch heuristic provides the initial set of minimum-cost tours satisfying the vehicle- and depot-capacity constraints but ignoring the depot fixed cost and throughput costs. The second phase focuses on selecting a subset of the potential depot-locations based on the fixed cost and throughput costs, while using only an approximation of the routing cost. The subproblem assumes that all demands served by each of the previously identified tours must be assigned to the same depot and served in the same order. Thus the only relevant routing cost is the 'stem' distance of each tour (i.e., the distance from each depot to the first and last facility visited in the tour), and the problem reduces to a depot-location/routing-allocation model. This problem is solved optimally using implicit enumeration. Having found the subset of depots that are open, the second step re-allocates each demand to an open depot and determines the best tours connecting each depot to its assigned demands. A multi-depot/routing-allocation heuristic is employed in this step. The above two steps in the improvement phase are iteratively applied until no further improvement in the total cost is obtained.

Specifically, the solution algorithm is made up of these two phases and steps:

Step 1 Initialization phase: Capacitated basic algorithm (CBA)
Step 2 Improvement phase: Single displacement (SD) step and Exchange (EX) step. ■

From this point forward, we paraphrase Perl's algorithm along these phases and steps.

4.6.2.1 CAPACITATED BASIC ALGORITHM (CBA). The CBA of initialization phase consists of assigning each demand to a depot and a tour, and then finding the shortest tour. This step builds upon the generalized savings measure (G'') for the single-depot problem

$$G_i^{''}(p,s) = v_1 d^{ji} + v_2 d^{ps} - d_{pi} - d_{is} \qquad\qquad (4\text{-}72)$$

where $G_i''(p,s)$ is the generalized savings from including demand i between points p and s; d^{ji} is the route distance between depot j and demand i; d^{ps} is the route distance between points p and s (a point may be either a depot or an actual demand); d_{pi} is the arc distance between point p and demand i; d_{is} is the arc distance between demand i and point s; and v_1, v_2 are route-shape parameters. The savings-equation is illustrated in **Figure 4-9**. The standard savings equation is obtained from the above Equation **(4-72)** by specifying $v_1 = 2$ and $v_2 = 1$, which implicitly

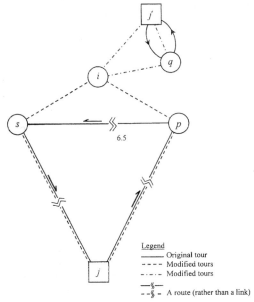

Figure 4-9 Generalized savings from a two depot case

assumes that the alternative to including i between p and s is serving it directly from the depot as a new tour. Equation **(4-72)** can be re-written as

$$G_i''(p,s) = v_1 d^{ji} - \gamma_i(p,s) \qquad \text{(4-73)}$$

where γ_i is the generalized 'strain' (triangular inequality) defined by

$$\gamma_i(p,s) = d_{pi} + d_{is} - v_2 d^{ps}. \qquad \text{(4-74)}$$

Modifying for the multi-depot case, we search over different depot-based tours for the inclusion of node i (assuming that we are currently considering serving i by epot j')

$$D^{j'i} = \min_{j \in J} d^{ji} - (d^{j'i} - \min_{j \in J} d^{ji}) = 2 \min_{j \in J} d^{ji} - d^{j'i} \qquad \text{(4-75)}$$

where $D^{j'i}$ is the modified distance between demand i and depot j'; and J is the set of all depot sites. Notice that if j' is the nearest depot to demand i, the modified distance is equal to the true distance d^{ji}, as we illustrated in **Figure 4-9**. Substituting Equation **(4-75)** into **(4-73)**, we obtain the modified generalized savings measure:

$$G_i'(p,s) = v_1 D^{j'i} - \gamma_i(p,s). \qquad \text{(4-76)}$$

The CBA considers two different cases in initiating *new* tours:

Case 1   There are no load-feasible demands, i.e., including any of the unassigned demands in an existing tour would violate either the vehicle-capacity constraint or depot-capacity constraint.

Case 2   The best G' value is smaller than a pre-specified lower bound while the number of tours initiated is less than the allowed number of tours.

The G' measure is an appropriate yardstick for selecting a demand to initiate a new tour under case 2.

The CBA starts by computing for each demand the ratio between the distances to the nearest and second-nearest load-feasible depots. During the execution of the algorithm this ratio is updated from the remaining unassigned demands if either the nearest or second-nearest depot becomes load infeasible. Notice the ratio indicates the ranking of demands in terms of their priority for inclusion in a tour. At each iteration, the algorithm determines if the emerging tour is expandable; i.e., if there is sufficient available vehicle capacity and depot capacity to include some unassigned demand. If the emerging tour is expandable, the unassigned load-feasible demand with maximum G' is identified. If the maximum G' is greater than a pre-specified minimum value, the demand is included in the emerging tour. The expansion process of the emerging tour continues until

a. the emerging tour is no longer expandable, or
b. the maximum G' for any unassigned demand is smaller than the pre-specified minimum value.

If the first condition occurs, a *new* tour is initiated by assigning the unassigned demands with lowest distance ratio to its nearest load-feasible depot (initiation under case 1). If the second condition occurs, a new tour is initiated by assigning the demand with the minimum G' to its nearest load-feasible depot (initiation under case 2). If the number of tours generated reaches the maximum allowed, existing tours are expanded based on maximum G', regardless of whether G' exceeds the pre-specified minimum value. The algorithm terminates when all the demands have been assigned. In summary, the CBA stage consists of identifying the demand nodes to be served by each tour and then determining the shortest path connecting the demand nodes and the depot for each tour.

4.6.2.2   SINGLE DISPLACEMENT (SD) STEP.   The first improvement stage is called the single displacement (SD) step. It consists of identifying demand nodes such that moving the demand node from one tour to another tour will improve the objective-function value. In other words, if moving a demand node from one tour to another tour will result in a decreasing overall-cost without violating any of the capacity or tour-length constraints, the demand is shifted. If there is more than one displacement meeting this criterion, the node is displaced to the tour that will result in the greatest decrease in overall cost.

The net change in travel distance from displacing demand i from its current position between i_L and i_R to a new position between p and s is given by

$$C_i(i_L, i_R, p, s) = d_{pi} + d_{is} + d_{i_L i_R} - d_{i_L i} - d_{i i_R} - d_{ps}. \tag{4-77}$$

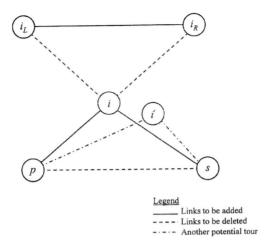

Figure 4-10 Reduction in travel distance from displacement

Figure 4-10 illustrates this situation. The C^* measure represents the new change in travel distance from displacing demand i to a tour h'' as given by

$$C_i^*(h'') = \min_{(p,s)\in l^{h''}} C_i(i_L, i_R, p, s) \qquad (4\text{-}78)$$

where (p,s), i_L and i_R are as previously defined, and $l^{h''}$ is the ordered set of points representing Candidate-tour h''. The G^* measure represents the net change in the objective-function value of the $m/m/-/h$ problem, from displacing demand i from tour h to h'', as given by

$$G_i^*(h'') = g C_i^*(h'') + (c_{h''} - c_h) f_i \qquad (4\text{-}79)$$

where g is the cost-per-mile (\$/km) or cost-per-hour of delivery vehicle, and c_h and $c_{h'}$ are the variable depot-costs at the depots which operate tours h and h'', respectively.

A *feasible* displacement of demand i is one that does not violate either vehicle or depot capacity. A *potential* tour for demand i is a tour that includes one or more demands that are closer to i than at least one of the points currently adjacent to i. As illustrated in **Figure 4-10**, the set of potential tours for demand i is defined by

$$H_i'' = \left\{ h'' \mid i' \in l^h \cap \left[(d_{ii'} < d_{i_L i}) \cup (d_{ii'} < d_{ii_R}) \right] \right\}. \qquad (4\text{-}80)$$

A potential displacement for demand i is a displacement to a potential tour. The displacement savings list for tour h includes all the potential displacements from that tour which result in improved solutions. It is straightforward to generalize the above equation to include different vehicle types. Johnson and Chan (1990) suggested a G^{**} measure to account for different operation costs from different fleets. The resulting G^{**} measure is given by

$$G_i^{**}(h'') = g_{h''}(d_{pi} + d_{is} - d_{ps}) + g_h(d_{i_L i_R} - d_{i_L i} - d_{i i_R}) + (c_{h''} - c_h)f_i \qquad \textbf{(4-81)}$$

where g_h is the vehicle cost-per-mile (\$/km) or cost-per-hour for tour h.

Now, for each tour, the SD step first identifies all the potential displacements from that tour. An iteration of the procedure is completed when all the tours have been examined. The potential displacement for any given tour is identified by noting for each demand on that tour the set of potential tours as defined by Equation **(4-80)**. If the best displacement of a given demand (based on the G^* value) would result in improved solution, it is recorded in the displacement savings list. When all the potential displacements for the tour under consideration have been evaluated, the procedure examines the displacement savings list. If the list is empty, the algorithm proceeds to consider the next tour. Otherwise the displacement savings list is scanned and the displacement with minimum G^* value is performed. The procedure terminates when a complete iteration does not result in a displacement.

 4.6.2.3 THE EXCHANGE (EX) STEP. The second improvement stage is called the Exchange Step. It consists of identifying pairs of demand nodes on different tours such that switching the nodes will cause an improvement in the objective-function value while maintaining feasibility. Suppose a demand i on tour h is closer to a demand node i' on another tour than it is to its closest adjacent node, and exchanging nodes i and i' will not violate any of the capacity or tour length limitation. Then node i' is identified as a potential exchange for node i.

The increase (decrease) in travel distance from including (removing) demand i between the adjacent points (p', q') is given by the 'triangular' relationship:

$$S_i(p', q') = d_{p'i} + d_{iq'} - d_{p'q'}. \qquad \textbf{(4-82)}$$

If demand i' is removed from tour h", the increase in travel distance from including demand i in h" (after i' has been removed) is given by

$$S_i(l_{h''/i'}) = \min_{(p',s') \in l_{h''/i'}} S_i(p', s'). \qquad \textbf{(4-83)}$$

Here $l_{h''/i'}$ is the ordered set of points in tour h' after removing demand i', (p', s') is a pair of adjacent points in $l_{h''/i'}$ and $S_i(p', s')$ is given by Equation **(4-82)**. If demand i_1 from tour h_2, and demand i_2 from tour h_1 are exchanged, the increase in travel distance from including each of the two demands in its new tour can be computed by Equation **(4-83)**. The reduction in travel distance from removing a demand from the current tour can be computed from Equation **(4-82)**. The net change in travel distance from the exchange is given by

$$E(i_1, i_2, h_1, h_2) = S_{i_1}(l_{h_2/i_2}) + S_{i_2}(l_{h_1/i_1}) - S_{i_1}(i_L, i_R) - S_{i_2}(k_L, k_R) \qquad \textbf{(4-84)}$$

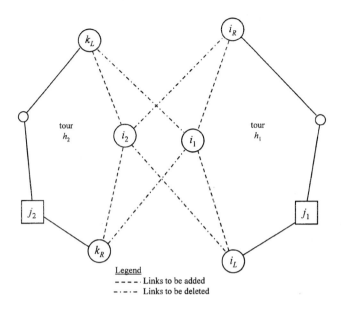

Figure 4-11 Change in travel from exchange

and is illustrated in **Figure 4-11**. The modified exchange-savings-measure (E') represents the net change in objective-function value (G') from exchanging demand i_1 from tour h_2 and demand i_2 from tour h_1, as given by

$$E'(i_1,i_2,h_1,h_2) = gE(i_1,i_2,h_1,h_2) + (c_{h_1} - c_{h_2})(f_{i_2} - f_{i_1}) \tag{4-85}$$

where f_{i_2} and f_{i_1} are the demands at i_2 and i_1 respectively.

As stated earlier, the EX step is not based on enumeration, but includes a component for identifying potential exchanges. A potential demand for tour h is any demand i', not in h, which is closer to some demand i in h than at least a point currently adjacent to i. The set of 'potential demands' for tour h, I'_h, is therefore defined by

$$I'_h = \left\{ i' \mid i' \in l_{h''}\{h'' \neq h\} \cap \{d_{ii'} < d_{i_Li}\} \cup \{d_{ii'} < d_{ii_R}\} \right\}. \tag{4-86}$$

A potential exchange for tour h is an exchange between a demand in tour h and a potential demand. Using Equation **(4-86)**, the EX step begins by identifying all the potential demands for a given tour. If the set of potential demands is empty, the algorithm proceeds to consider the next tour. An iteration of the procedure is completed when all the tours have been examined. For each potential demand the algorithm checks whether there is a feasible (does not violate either vehicle or depot capacity) exchange between that potential demand and any demand in the tour under consideration. If so, the associated E' measure is computed using Equation

(4-85). When all potential demands have been considered, the best exchange is identified. If the exchange results in an improved solution, it is performed. Otherwise, the algorithm proceeds to consider the next tour. The procedure terminates when a complete iteration does not result in an exchange.

4.6.2.4 STEP-BY-STEP ALGORITHM. Johnson and Chan (1990) modified the Perl algorithm to account for different operating costs, as indicated in **(4-81)**. They also summarized the entire algorithm in a step-by-step fashion.

Capacitated basic algorithm (CBA):

Step 1   For each demand node, determine
  ▸ nearest load-feasible depot,
  ▸ second-nearest load-feasible depot,
  ▸ the distance-ratio between the nearest and the second-nearest load-feasible depot.
Step 2   Initiate a tour by selecting the demand with lowest ratio and assigning it to the nearest load-feasible depot.
Step 3   Check to see if the tour can be expanded. In other words, can any demands be added to the tour such that
  ▸ vehicle/depot capacities are not exceeded,
  ▸ tour-length limitations are not exceeded,
  ▸ the maximum savings-measure G' is greater than a pre-specified minimum.
Step 4   Update ratios as necessary.
Step 5   Expand tour if possible and go to step 3. Otherwise, if all demands have not been assigned to a tour, go to step 2.
Step 6   Compute the shortest tour.

Single displacement (SD) step:

Step 1   Let $h=1$.
Step 2   Let k denote the kth demand on tour h. Let $k=1$.
Step 3   Let $h'=1$.
Step 4   If $h'=h$, $h' \leftarrow h'+1$.
Step 5   Let i be such that the kth demand on tour h is demand i.
Step 6   If displacing demand i to tour h' is not feasible, go to step 8.
Step 7   Compute cost-change G^{**}. If $G^{**} \leq 0$, add h' to the potential savings list.
Step 8 $h' \leftarrow h'+1$. If $h' \leq$ (number of tours), go to step 4.
Step 9 $k \leftarrow k+1$. If $k \leq$ (number of demands on tour h), go to step 3.
Step10   If the displacement list is not empty, select the smallest value on the displacement list, make the appropriate displacement and let DISPLACE-MENT=YES.
Step11 $h \leftarrow h+1$. If $h \leq$ (number of tours), go to step 2.
Step12   If DISPLACEMENT=YES, go to step 1.

Exchange (EX) step:

Step 1   Let $h=1$.
Step 2   Let k denote the kth demand on tour h. Let $k=1$.
Step 3   Let $h'=1$.
Step 4   If $h'=h$, $h' \leftarrow h'+1$.
Step 5   Let i be such that the kth demand on tour h is demand i.
Step 6   Determine potential exchanges as described in subsection entitled "The Exchange (EX) Step" If there are no potential exchanges, go to step 8.
Step 7   Compute E' for all the potential exchanges. If min $E' \leq 0$, select the smallest E', make the corresponding exchange and let EXCHANGE=YES.
Step 8 $h' \leftarrow h'+1$. If $h' \leq$ (number of tours), go to step 4.
Step 9 $k \leftarrow k+1$. If $k \leq$ (number of demands on tour h), go to step 3.
Step 10 $h \leftarrow h+1$. If $h \leq$ (number of tours), go to step 2.
Step 11 If EXCHANGE=YES, go to step 1. ■

4.6.3 Example. The example presented here consists of two depot locations and three demand nodes (Johnson and Chan 1990, Moore and Chan 1990). Depot 1 has one C-130 aircraft assigned with vehicle capacity of 30,000 pounds (15,000 kg), and depot 2 has one C-141 aircraft assigned with vehicle capacity of 60,000 pounds (30,000 kg). Additionally, depot capacity at both depots is limited to 50,000 pounds (25,000 kg). Costs associated with each depot are shown in **Table 4-3**. Operating costs for the vehicles are $1000 per hour for a C-130 and $2000 per hour for a C-141. Link travel times and demands are given in **Figure 4-12**. Each vehicle is further limited by a maximum 12 hour crew-duty-day. Pre-flight time at both nodes is 0.5 hour and stopover time at each node is 0.5 hour.

The following steps were followed to obtain the optimal solution for the example problem.

CBA:

Demand	Nearest Load-Feasible Depot	Second-Nearest Load-Feasible Depot	Ratio
3	1	2	0.5750
4	1	2	0.9333
5	1	2	0.7813

Steps 1 &2   Select the demand with the smallest ratio, demand 3, to initiate the first tour. Assign demand 3 to tour 1. Tour 1 uses vehicle 1 (C-130) from depot 1 (the nearest load-feasible (L-F) depot).
Step 3   Tour cannot be expanded since including either demand 4 or 5 would exceed vehicle capacity.
Step 4   Update ratios. In this case only one L-F depot remains, therefore the ratios are taken to be the actual distance from the depot to the demands.

Depot	Fixed	Trunking	Variable
1	100,000	10,000	1
2	120,000	20,000	1

Table 4-3 Depot costs for example problem

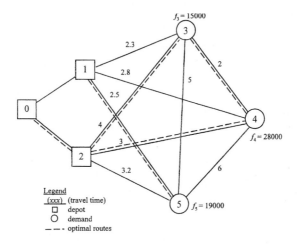

Figure 4-12 Hierarchical multiple-depot/multiple-tour problem

Demand	Load-Feasible depot	'Distance'
4	2	3.0
5	2	3.2

Step 2   Select shortest distance to initiate new tour. Assign demand 4 to tour 2. Tour 2 uses vehicle 2 (C-141) from depot 2.

Step 3   Tour can feasibly be expanded by including demand 5, therefore assign demand 5 to tour 2.

Steps 5-6   All demands have been included in a tour. The shortest tour length is computed to yield the CBA solution.

SD step:

Steps 1-6   Check tour 1: Check demand 3; tour 2 is not feasible for demand 3.

Steps 4-6   Check tour 2: Check demand 4; tour 1 is not feasible for demand 4. Check demand 5; tour 1 is not feasible for demand 5.

EX step:

Steps 1–6   Check tour 1: Check demand 3; $g_1 d_{13} = 23$, $g_2 d_{34} = 4000 \geq 23$, $g_2 d_{35} = 10000 \geq$ 23; no potential exchanges exist for demand 3.

Steps 4–7   Check tour 2: Check demand 4; $g_2 d_{24} = 6000$ and $g_2 d_{45} = 120$, $g_1 d_{34} = 20 \leq 60$; exchanging demand 4 and 3 is feasible. Therefore, demand 3 is a potential exchange for demand 4. The potential savings for this exchange is $E' = -129,999,000$.

Steps 4–6   Check demand 5: $g_2 d_{52} = 6400$ and $g_2 d_{45} = 12000$, $g_1 d_{35} = 5000 \leq 6400$, exchanging demand 5 and demand 3 is not feasible. The exchange would violate tour length limitations for tour 2.

Steps 4–8   Exchange demand 4 and demand 3.

Further iterations yield no exchanges. □

The final solution is overlaid on **Figure 4-12**. The solution contains no illegal subtours; depot capacities are not violated; vehicle capacities are not exceeded; tour length limitations are not violated; and there are no illegal chains. The problem was constrained such that one vehicle is assigned to each of the depots. The vehicle (C-130) assigned to the first depot is loaded with 19,000 pounds (9500 kg) and travels from the depot to demand-node 5 and returns to the depot. The vehicle assigned to node 2 (C-141) is loaded with 43,000 pounds (21,500 kg) and travels from node 2 to node 3, where it drops off 15,000 pounds (7500 kg), to node 4, where it drops off the remaining 28,000 pounds (14,000 kg), and returns to node 2. This means depot 1 is supplied with 19 000 pounds (9500 kg) of goods and depot 2 is supplied with 43,000 pounds (21,500 kg) of goods. The total cost for this solution is $1,050,307,800.

4.6.4 Comments. The above heuristic represents a more ambitious attempt to solve a simultaneous location-routing model, and is therefore unique in many ways. To make any comments on the heuristic, the algorithm had to be evaluated with respect to other methods for the multiple-depot *vehicle routing* problem (VRP) rather than the general multi-depot location-routing problem. This would constitute the closest comparison of a similar, validated heuristic. Thus the following comparison has to be taken in that light. Four test problems include 249 demands and differ in the number of depots. For the test problems here the number of tours generated are known, the Perl algorithm provided solutions with fewer tours. Perl concluded that if vehicle acquisition is considered (hence making the number of vehicles a variable), the solution by the present method may be superior to the standard VRP. Accounting for differences in hardware, Perl also concluded that the Central Processing Unit (CPU) times for this algorithm are significantly lower than the VRP.

Two additional problems were tested for sensitivity analysis, mainly to show the viability of the Perl algorithm. Specifically, they were to analyze the computational effect of changes in variable cost and depot capacity on the solution to the *m/m/–/h* problem. The first test problem includes 55 demand sites and 4 depots. The second problem includes 85 sites and three depots. Clearly, short-term costs are decreased as depot capacity increases. Also as expected, a decrease (increase) in variable cost at any given depot would usually increase (decrease) the demand allocated to that depot. A decreasing variable cost at a depot that already operates at or near capacity would result in only minor or no change in the allocation of demand in that route. Overall, depot capacity and variable cost may greatly affect the solution and the algorithm is shown to be useful in responding to these changes.

Notice that Perl's algorithm does not account for the different fixed costs for the depot. While this will probably not pose much difficulty when the costs are similar or when all depots must be used to satisfy demand, this deficiency must be considered when using Perl's algorithm. At about the same time Perl's work was originally performed (Perl and Daskin 1985), Madsen (1983) proposed other heuristics for solving two-tier hierarchical location-routing problems. First is the "tree-tour heuristic," which constructs a spanning tree from the supply source through the depots to the demands. This is to be performed satisfying timing and capacity constraints at minimum cost. The tree is constructed starting with the demands with the tightest time-constraints. The demands are then added one at a time, wherein cost minimization is indirectly taken into consideration. This method can be thought of as a cluster-routing heuristic.

Second, a three-stage heuristic is proposed. It begins with selecting a number of depots, with demands assigned to each depot. Secondary tours are formed at each depot, with attention paid to duration and delivery-time constraints. Primary tours (trunking routes) are formed with the central supply point as the 'depot' and the regular depots as 'demand' points. Attention is again paid to primary-vehicle capacity and delivery-time constraints. This heuristic is sometimes called the "alternate-location-allocation-savings" method.

Third, a variation of the three-stage heuristic approach is taken. Secondary tours are formed using the central-supply point as the depot, where feasibility checks include only the secondary tour-duration constraints. Depots are located at the first demand of each secondary tour. Feasibility checks are made on the delivery-time constraints assuming each transfer point is served by a primary tour (trunking route) coming directly from the supply point. Primary tours are constructed similar to the alternate-location-allocation savings method above. This is sometimes referred to as a "savings-drop" heuristic.

Computational experiences show that except for very tight time constraints, the tree-tour heuristic cannot be recommended. However, both the alternate-location-allocation-savings and savings-drop heuristics have promise in solving hierarchical location-routing problems. There appears to be a fair amount of flexibility in combining some common-sense basic-building-blocks into several good heuristics—as illustrated in both the Perl and Madsen algorithms. While there is no single heuristic that is clearly superior, Perl's algorithm shows how one can construct a set of heuristics step-by-step. We will continue this trend of thought in a Baker–Chan variation in the section entitled "Depot location in commodity distribution."

4.7   Multi-Facility/Multi-Route/Multi-Criteria/Nested Model

Perhaps one of the most general models of the location-routing problem was given by Chan (1974, 1992).   It includes multi-objective functions consisting of minimum-path chains (in addition to tours) and maximal coverage. Most importantly, it is a nested location-routing formulation in which both vehicle and commodity flows are analyzed. It is labelled the multi-depot (or multi-facility), multi-route/multi-criteria/nested (*m*/*m*/*m*/*n*) model. In the model, the term 'route' is used to include both chains and tours.

4.7.1   Formulation.   To formulate this as an integer-programming model, one has to break the problem into the distinctive phases of *synthesis* and *analysis*. In these two sequential phases, candidate routes are generated and then evaluated respectively. Chan (1979) presented a graph-theoretic method to synthesize vehicle routes. By way of a "contiguity matrix,"**A**", multi-stop routes can be obtained by raising the power of the matrix.

4.7.1.1   SYNTHESIS.   In a contiguity matrix **A**", a connection by a link between two nodes is recorded as a corresponding node pair in a square matrix. Thus this matrix carries the dimension of $|I|$, the total number of nodes in the network.   For unrestricted routing, the matrix **A**" will simply look like

$$A'' = [(i,j)] \qquad i \neq j \qquad\qquad \textbf{(4-87)}$$

where the diagonal elements (i,j) are zero by definition of a route. Here a non-stop connection is allowed between all node-pairs. When the contiguity matrix is raised to its second power, all the one-stop routes can be generated

$$A''^2 = [(i,k) \cdot (k,j)] \qquad i \neq k, \quad k \neq j. \qquad\qquad \textbf{(4-88)}$$

Again the diagonal elements are defined as zero, where all 'circuits' that start at i and end at i are not considered. If one reads '.' as 'and' and '+' as 'or', it can be seen that node pair (i,j) can be served by a one-stop route i–k–j in Equation **(4-88)**. The **A**" matrix contains $|I|^2$ non-stop routes, consisting of point-to-point arcs. The $\textbf{A}''^2$ matrix has $2|I|^2$ routes, as one builds alternate one-step routes. The $\textbf{A}''^3$ matrix expands to $4|I|^2$ routes etc., when two-stop routes are built. Let P be the set of vehicle routes in the network. Including self-circuits, the number of routes generated is $|P| = |I|^2(2^0 + 2^1 + 2^2 + 2^3 + ... + 2^\mu)$ which is an extremely large number for even a small μ, the number of stops in the route.   Some kind of rule is usually followed in picking a subset of *feasible* routes. Most of these rules will inevitably leave some routes out of consideration, providing no guarantee that the solution so obtained is optimal.

Generalization of this method can consider both intermediate-stop restrictions as well as tours (circuits). Interpreting '+' in the $\textbf{A}''^\mu$ matrix (with $\mu \geq 2$) as 'or', alternate route or tour can also be represented, such as the choice between a non-stop and a one-stop in the entry $[(i,j)+(i,k) \bullet (k,j)]$. Take another example, leaving

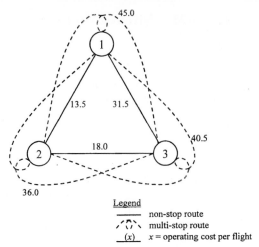

Figure 4-13 Template of all generated routes for a three-node airline network

an entry (*i,j*) as zero (*i≠j*) amounts to disallowing service between the origin-destination (O-D) pair concerned. On the other hand, a more complex connectivity requirement can be realized by introducing `dummy' nodes (Chan 1979). A point that bears commenting on at this juncture is the distinction between *itinerary r* and vehicle *route p*. The matrix method above generates vehicle routes. An itinerary, including connect itinerary, can be made up of any subsets of the nodes in routes *P*, I_p, where $I_p \in P$. Thus a one-stop route 1–2–3 can facilitate itineraries 1–2, 2–3 and 1–2–3. Combined with another non-stop route 3–4, a passenger or shipper has these additional itineraries available to him/ her: 1–2–3–4, 2–3–4, and 3–4. It is precisely these itineraries, rather than the routes, which provide the actual conductivity between O-Ds. Another way to look at it is that vehicular routes only serve the role of *facilitating* commodity flow from origin to destination. Consequently, vehicular routes of and by themselves have little value until commodity itineraries are executed via these routes. Itineraries can be viewed as chains, since they connect two or more 'depots' together. Here a depot can be both an origin and a destination. Instead of discouraging chains from being generated, we wish to encourage them since more chains means more coverage between origin-destination (O-D) pairs.

Example

For the three-node network shown in **Figure 4-13**, the **A"** matrix is 3×3 in dimension and looks like

$$
\mathbf{A}^{\prime\prime} = \begin{array}{c} \\ 1 \\ 2 \\ 3 \end{array} \begin{array}{ccc} 1 & 2 & 3 \\ \left[\begin{array}{ccc} 0 & (1,2) & (1,3) \\ (2,1) & 0 & (2,3) \\ (3,1) & (3,2) & 0 \end{array} \right] \end{array}
\tag{4-89}
$$

Here a non-stop connection is allowed between all node-pairs. When the contiguity matrix is raised to its second power, all the one-stop routes can be generated:

$$A''^2 = \begin{bmatrix} (1,2).(2,1)+(1,3).(3,1) & (1,3).(3,2) & (1,2).(2,3) \\ (2,3).(3,1) & (2,1).(1,2)+(2,3).(3,2) & (2,1).(1,3) \\ (3,2).(2,1) & (3,1).(1,2) & (3,1).(1,3)+(3,2).(2,3) \end{bmatrix} \qquad \textbf{(4-90)}$$

As mentioned, we interpret `.' as `and' and `+' as `or'. It can be seen that node pair 2-3 can be served by the one-stop route 1-3-2 (where 2-3 is the same as 3-2 if symmetry is assumed), and 1-3 by 1-2-3 and so on. Here the circuits in the diagonal elements form tours rather than routes. For example, the first diagonal element represents the "multiple-salesmen tours" 1-2-1 and 1-3-1. The two matrices A'' and A''^2 contain the nine different bi-directional routes possible within this simple network. They represent all the feasible routes in the search space, where `routes' here is a general term to include tours. In a larger network, raising such an A'' matrix to higher powers will generate all the two and longer multi-stop routes. But the combinatorial space is extremely huge even for the most modest networks. □

4.7.1.2 ANALYSIS. Once routes are generated, they are evaluated in terms of an integer program. In view of the importance of providing itineraries or routes among all O-D pairs, we proceed to establish the non-stop itinerary between m and n in the first constraint of the mathematical program. This constraint eventually selects the routes on which the itineraries are configured:

$$\sum_{p_0 \in P_0^{mn}} x_{p_0}^{mn} - y_0^{mn} \geq 0 \qquad m, n \in I. \qquad \textbf{(4-91)}$$

Here x stands for the non-stop itineraries between m and n, which in this case is precisely the segment m–n or non-stop route p_0 itself. As long as a single non-stop route (or itinerary) p_0 covers terminal pair m–n, connectivity can be rendered between the O-D pair. This results in a unitary valuation of zero-one variables x and possibly y. Conversely, connectivity is impossible between m-n (i.e., y=0) if the set of routes covering m–n via non-stop *itineraries*, P_0^{mn}, is empty. Notice the similarity of Equation **(4-91)** to other forcing/linking constraints such as **(4-3)**, **(4-7)**, **(4-20)**, **(4-42)** and **(4-70)**.

A similar connectivity prescription is given for one-stop *itinerary* given a route set P_1^{mn}:

$$\sum_{p_1 \in P^{mn}} x_{mip_1}^{mn} - y_1^{mn} \geq 0 \qquad i \in I_{p_1}; \; m, n \in I. \qquad \textbf{(4-92)}$$

Here one-stop itinerary between m and n is possible as long as there is a one-stop route p_1 within the itinerary set P_1^{mn} that connects, among other nodes, node pairs m–n via an intermediate stop. Thus the zero-one variable $x_{mip_1}^{mn}$ equals one when route p_1 is introduced. Notice the introduction of route p_1 is indicated by the usage of segment m–i within the route. Theoretically, usage of any segment within the route made up of nodes I_p is sufficient to show the introduction of route p_1, but the usage of $m-i$ makes it explicit. It should be emphasized again the difference between itinerary and route in mathematical terms: $P_0^{mn}+P_1^{mn} \subseteq I_{p_0}+I_{p_1}$ for all $m-n$.

Similarly, another prescription is written for two-stop itineraries:

$$\sum_{p_2 \varepsilon P_2^{mn}} x_{mip_2}^{mn} - y_2^{mn} \geq 0 \qquad i \in I_{p_2}; \ m, n \in I. \tag{4-93}$$

The zero-one variable $x_{mip_2}^{mn}$ assumes one in value if segment $m-i$ in route p_2 is used, establishing connectivity between $m-n$ via 2-stop itineraries. The latter is suggested by the possible unitary valuation of zero-one variable y_2^{mn}. The relationship between itineraries and routes now looks like: $P_0^{mn} + P_1^{mn} + P_2^{mn} \subseteq I_{p_0} + I_{p_1} + I_{p_2}$ for all m-n. In general, the r-stop itinerary coverage equation looks like:

$$\sum_{p_r \in P_r^{mn}} x_{mip_r}^{mn} - y_r^{mn} \geq 0 \qquad i \in I_{p_r}; \ m, n \in I. \tag{4-94}$$

Itineraries and routes are now tied together by: $P_0^{mn} + ... + P_k^{mn} \subseteq I_{p_0} + ... + I_{p_k}$ for all $m-n$.

Besides a direct service, an O-D pair can be served by a connect itinerary. Here a connect itinerary between m and n can be made when at least one of the connection points, say q in the set of I_c, is used. This is done after considering all routes that make up the connection P_c^{mn}.

$$\sum_{p_c \in P_c^{mn}} \sum_{q \in I_c} x_{iqp_c}^{mn} - y_c^{mn} \geq 0 \qquad i \in I_{p_c}; \ m, n \in I_c. \tag{4-95}$$

The coverage equations shown above represent a generalization of expressions **(4-69)** through **(4-71)**. Instead of three levels of hierarchy used in the $m/m/-/h$ model, we have generalized to rth-level connections (transshipments). Thus, the concept of a hierarchy is broadened to include not only a transfer through central supply to depot and to demand, but also all the transfers possible between any two levels—i.e., a one-stop connection is as likely as a two-stop and so on.

A connectivity requirement can be placed for each O-D pair if desired:

$$\sum_{t=0}^{r} y_t^{mn} \geq C^{mn}(r) \qquad \forall m, n; \forall r. \tag{4-96}$$

Irrespective of the way that connectivity is established between an O-D pair—via non-stop, multi-stop or connect itinerary—some assurance that a continuous path between O and D is helpful. Thus we add a continuity constraint to guarantee a continuous itinerary from origin to destination, regardless of route usage:

$$\sum_{p \in P} \sum_{i \in M_j} x_{ijp}^{mn} - \sum_{p \in P} \sum_{i \in \bar{M}_j} x_{jip}^{mn} = 0 \qquad j, m, n \in I; \ j \neq m, n. \tag{4-97}$$

In lay terms, the above expression simply says "flow in equals flow out" for all the intermediate nodes between m and n. This continuity equation is a generalization of **(4-6)** and $\sum_{i \in M_p} x_{ip}^h - \sum_{j \in \bar{M}_p} x_{pj}^h = 0$ in the "Vehicle-routing problem" subsection of the "Spatial Separation" chapter.

While establishing the vehicle routes (and hence the itineraries) is important, it is equally critical to specify the frequency on each route. It is frequency that allows a schedule to be eventually constructed. The following equation generalizes the TSP or VRP assumption that each demand is visited once only by a salesperson or a vehicle. Now a site can be visited several times:

$$\sum_{r=0}^{\mu}\sum_{m\in I}\sum_{n\in I} f_r^{mn} x_{ijp}^{mn} - V w_{ijp} \le 0 \qquad i,j\in I_p; \ p\in P. \qquad (4\text{-}98)$$

This equation says that the segment i–j in route p has to have enough frequency to cover the traffic it carries. Thus the first term shows the traffic bundled from all rth stop O-D demands f_r^{mn} on link (i,j) enroute route p. The second term is the total capacity on link (i,j) on route p as afforded by the frequencies w provided, considering there are say V seats on board each vehicle. Now the route frequency w_p has to accommodate the largest of all segment frequencies w_{ijp} on the route

$$w_p - w_{ijp} \ge 0 \qquad i,j\in I_p; \ p\in P. \qquad (4\text{-}99)$$

Notice Equations **(4-98)** and **(4-99)** are much more involved than the vehicle-capacity constraints in the "Vehicle-routing problem" subsection of the "Spatial Separation" chapter. There are two types of traffic flow in our model, one corresponding to passengers (or freight) while the other one vehicles. The word 'nested' is used to denote that both types of flows are involved. Previous models have concentrated on vehicular routes, with passenger or freight flow subsumed. Specifically, commodity flows along vehicular routes from a depot to demand points, except for prior transfer between a supply point and a depot. Now, commodities are identified by origin-destination m-n and they can be carried on any combination of vehicular routes to complete their itineraries. For compactness, Equations **(4-98)** and **(4-99)** can be combined to read (Mahaba et al. 1987)

$$\sum_{r=0}^{\mu}\sum_{m\in I}\sum_{n\in I} f_r^{mn} x_{ijp}^{mn} - V w_p \le 0 \qquad i,j\in I_p; \ p\in P. \qquad (4\text{-}100)$$

The next constraint has to do with the vehicle's 'range'. It follows $\sum_{i\in I} t_i^h \sum_{j\in I} x_{ij}^h + \sum_{i\in I}\sum_{j\in I} d_{ij}^h x_{ij}^h \le U(h)$ in the "Vehicle-routing problem" subsection of the "Spatial Separation" chapter and Equation **(4-68)**. The same equation can be used to enforce a "crew-duty-day" requirement, or the maximum allowable "time-on-the-road." A range equation is written for each route:

$$\sum_{i\in I}\sum_{j\in I} d_{ij} x_{ijp}^{mn} \le U_p \qquad \forall m,n,p. \qquad (4\text{-}101)$$

The last two constraints place a capacity limit on the terminal for passengers (or freight) at both the origin m and destination n:

$$\sum_{r=0}^{\mu}\sum_{n\in I} f_r^{mn} y_r^{mn} \le \bar{P}_m \qquad (4\text{-}102)$$

$$\sum_{r=0}^{\mu} \sum_{m\in I} f_r^{mn} y_r^{mn} \le \bar{P}_n.$$ (4-103)

A first criterion function of the location-routing problem is to maximize coverage as much as possible

$$\max \; y_1'(\mathbf{y}) = \sum_{r=0}^{\mu} \sum_{m\in I} \sum_{n\in I} f_r^{mn} y_r^{mn}.$$ (4-104)

Here the zero-one decision variable is y_r^{mn}, which assumes the numerical value 1 when O-D pair $m-n$ is served by the rth stop itinerary. This explicitly states that O-D demand varies depending on whether the itinerary type is a non-stop, 1-stop or 2-stop, etc. Thus the product is the total coverage among O-D pairs, summed up among a system of $|I|$ terminals and μ itinerary types. It represents a generalization of Equations **(4-2)** and $\max \sum_{i\in I} f_i y_i$ in the "Maximal-coverage location problem" subsection of the "Facility Location" chapter. The second criterion seeks to minimize the total cost of operation:

$$\min \; y_2'(\mathbf{x}, \mathbf{y}) = \sum_{p\in P} H_p w_p.$$ (4-105)

Here again the decision-variable w_p denotes the operating frequency of route p. The coefficient H_p stands for the unit cost of operating one dispatch on route p. Thus the total cost is obtained by summing up the costs from $|P|$ routes in the system. Notice that implicitly, this is a function of both the route variable $\mathbf{x}$ and the coverage variable $\mathbf{y}$. It is to be noted that the maximization criterion (Equation **(4-104)**) is measured in a different unit than the minimization objective (Equation **(4-105)**). For example, the former is in passengers while the latter in dollars. Also, the two objectives can be in conflict. Consequently, it is a multi-criteria optimization problem similar to the single depot/single route case documented in the section entitled "Single-facility/single-route/multi-criteria problem."

As mentioned previously, Chan (1974) and Chan (1997) outlined one of the most general combined location-routing models. The start and finish depots are identified for each route (notice that the two may be different as in a chain). The total amount of traffic that may be handled by any particular depot may be capacitated. It is a VRP since the amount routed is capacitated. In contrast to traditional routing problems, each node may be visited several times. This number can also be restricted. Furthermore, demands are only satisfied if it is profitable to do so. Associated with each possible route is an operational cost, and associated with each demand that is satisfied is some benefit. The objective is to maximize the benefit while minimizing costs. The problem solution includes the depots served by each route, the demands that are served, which are identified by their origin-destination and the itinerary. Instead of generating only one route for several demands/depots, the frequency of operating a vehicular-route is identified. This integer-programming model is labelled m/m/m/n, where the n stands for `nested' flow, including both vehicles and passenger/freight. With connect itineraries explicitly modelled, we can also handle `nested' vehicle fleets, say a transfer from a limousine to an airplane at an air terminal. The model generalizes on frequency of visitation, and it generalizes on the concept of hierarchy. Finally, it generalizes on the idea of fixed demand by modelling demand as a function of the itinerary.

Example

Again, let us refer to **Figure 4-13**, which shows an airline network consisting of three potential airports and a set of example routes between them. The travel demand function f_r^{mn} is given below in tabular form. Here in this table, we show the incremental demand as additional service is introduced. Starting with a nonstop, the airline can attract passenger patronage as shown in the first column of this table, namely 150, 80 and 20 passengers. Should a 1-stop service be introduced, an additional patronage of 75, 40 and 10 would result.

O-D	Routing Type (r)		
$(m-n)$	Non-stop $(r=0)$	1-stop $(r=1)$	Connect $(r=c)$
1-2	150	75	68
1-3	80	40	36
2-3	20	10	9

Notice demand is symmetric. It varies according to the travel itinerary--namely, demand is stimulated by better service as shown. The costs H_p of operating the six (symmetrical) routes in the set P are summarized below:

Route (p)	Cost/Flight (H_p)
1 (1-2)	458
2 (2-3)	917
3 (1-3)	940
4 (1-2-3)	1375
5 (1-3-2)	1857
6 (2-1-3)	1398

The average yield (revenue) per passenger, considered here as the sole measure of benefits, is also given:

O-D $(m-n)$	1-2	1-3	2-3
Yield	20	25	30

Without loss of generality, the demand-revenue and costs are assumed to be symmetric, i.e., the figures given apply to both directions of travel or the value from terminal A to B is the same as that from B to A. An effective seat-capacity V of 80 is used. An example IP-model formulation is shown in **Figure 4-14** for the case of equal weighting between y_1' and y_2'. The optimal solution carries an objective function of 3369. Frequency on nonstop route 1 is 2. Frequency on non-stop route 2 is 1, and frequency on one-stop route 5 is 1. □

In this IP, we have illustrated each of the model equations. The following table will establish the associations between the formal model and the computer listing in **Figure 4-14**. For example, the non-stop itinerary constraint as shown in Equation **(4-91)** of the IP will appear as lines 2 to 4 in the computer listing and so on.

	IP model Equation number	Lines in computer listing
Non-stop itineraries	**(4-91)**	(4)–(6)
One-stop itineraries	**(4-92)**	(7)–(9)
Connect coverage	**(4-95)**	(10)–(33)
Flow continuity	**(4-97)**	(34)–(36)
Frequency determination	**(4-98)**	(37)–(45)
Frequency bundling	**(4-99)**	(46)–(54)
Objective function	**(4-104)** and **(4-105)**	(11)–(12)

4.7.2 Solution algorithm. While the original integer program (IP) formulation was a good start to model the problem, it is not a practical way to solve problems that are more than three-node examples. Column generation is often used to handle synthesis of routes. Row generation takes care of resources or integrality. But regular decomposition based on column generation or row generation does not work here since it requires both row and column generation simultaneously. Rows are appended corresponding to additional demand coverage,[11] and columns are inserted for new routes (Chan 1972). Here we present a specialized decomposition procedure fashioned after recursive programming. We introduce the model by expressing the model in a functional form.

4.7.2.1 RECURSIVE PROGRAMMING. Recursive programming is a way to ameliorate the tautological "curse of dimensionality" between route-generation and route-selection as outlined above. By breaking up the generation and selection tasks into iterative increments, one can effectively reduce the problem size for feasible solution. Thus in one step, a handful of vehicle-route candidates are posed as alternatives. In a second step, these candidates are analyzed in terms of their desirability for inclusion in the route structure from both the operator and customer points of view. When these two steps are repeated, we have a recursive algorithm. The recursive format considers the relationship between a handful of alternatives and expected consequences and trace a sequence of decisions stagewise. This may or may not approach the optimality conditions according to a more rigorous

---

[11]Regular row-generation methods start with a relaxation of a constrained optimization problem and introduce cuts sequentially until feasibility is restored. It is the opposite situation here, where we start with an over-constrained system due to the limited numbers of routes, and relax the constraints as more itineraries become available to cover additional origin-destination demand.

```
MAX 3000Y120+1500Y121+1360Y12C+600Y230+300Y231+270Y23C+2000Y310+
1000Y311+900Y31C-458W1-917W2-940W3-1375W4-1857W5-1398W6
ST
X12121+X12124+X12126-Y120>=0
X23232+X23234+X23235-Y230>=0
X31313+X31315+X31316-Y310>=0
X12235-Y121>=0
X23316-Y231>=0
X31124-Y311>=0
X12232+X12313-Y12C>=0
X12235+X12313-Y12C>=0
X12234+X12313-Y12C>=0
X12232+X12315-Y12C>=0
X12234+X12315-Y12C>=0
X12235+X12316-Y12C>=0
X12234+X12316-Y12C>=0
X12232+X12316-Y12C>=0
X23121+X23313-Y23C>=0
X23124+X23313-Y23C>=0
X23126+X23313-Y23C>=0
X23121+X23315-Y23C>=0
X23124+X23315-Y23C>=0
X23126+X23315-Y23C>=0
X23121+X23316-Y23C>=0
X23124+X23316-Y23C>=0
X31121+X31235-Y31C>=0
X31124+X31232-Y31C>=0
X31121+X31234-Y31C>=0
X31121+X31232-Y31C>=0
X31126+X31232-Y31C>=0
X31126+X31235-Y31C>=0
X31124+X31235-Y31C>=0
X31126+X31234-Y31C>=0
X12313+X12315+X12316-X12232-X12234-X12235=0
X23313+X23315+X23316-X23121-X23124-X23126=0
X31121+X31124+X31126-X31232-X31234-X31235=0
150X12121+9X23121+36X31121-80W121<=0
150X12124+40X31124+9X23124-80W124<=0
150X12126+10X23126+36X31126-80W126<=0
20X23232+68X12232+36X31232-80W232<=0
20X23234+40X31234+68X12234-80W234<=0
20X23235+75X12235+36X31235-80W235<=0
80X31313+68X12313+9X23313-80W313<=0
80X31315+75X12315+9X23315-80W315<=0
80X31316+10X23316+68X12316-80W316<=0
W1-W121>=0
W2-W232>=0
W3-W313>=0
W4-W124>=0
W4-W234>=0
W5-W235>=0
W5-W315>=0
W6-W316>=0
W6-W126>=0
```

Figure 4-14 Integer program for 3-node network

stagewise optimization technique such as dynamic programming[12]. But recursive programming is more robust (Day 1973, Vickerman 1980).

Here, the maximand is a function of a series of decision vectors relating to the sequence for which decisions are taken. Thus we write the optimal payoff for cycle k in the algorithm as $I^*(k) = \max I[X(k)]$, where $\mathbf{X}$ is a collection of μ subvectors representing non-stop, one-stop and two-stop routes etc.: $\mathbf{X}(k) =)$, $\mathbf{x}^{k+1}(k), \ldots, \mathbf{x}^{k+\mu-}$, and the sub-vector $\mathbf{x}^{k+r}(k)$ refers to a known column vector of decision variables related to stage $(k+r)$ which are taken in cycle k $(0 \le r \le \mu-1)$. For the $(k+r)$th stage

---

[12]The readers may wish to consuilt Chan (2000)'s appendix on "Markovian Systems" for an introduction to dynamic programming and recursive programming.

in the sequence there exists a set of constraints governing the traffic flow, which is equivalent to Equation (4-96):

$$\Phi^{k+r}[\mathbf{x}^k(k), \mathbf{x}^{k+1}(k), \ldots, \mathbf{x}^{k+r}(k)] \geq \mathbf{c}^{k+r}(k) \qquad r=0,\ldots,\mu-1;\ k=1,\ldots,\infty. \qquad \textbf{(4-106)}$$

Here the 0–1 route-variables x are by definition nonnegative, $\mathbf{X}(k) \geq \mathbf{0}$. The traffic-flow variables Φ^{k+r} among all O–Ds are in the form of a vector of $\acute{\eta}_k$ constraint functions, which may vary over the stages but is fixed in cycle k. The $\acute{\eta}_k(r)$ functions form a delimited subsystem defining the local-flow pattern for the r-stop routings. Hence there are $\acute{\eta}_k \times \mu$ constraints in all, specifying non-stop, 1-stop and multi-stop O-D connectivity requirements. Each constraint is a function of the μm_k decision variables, where m_k is the number of decision variables to be determined for cycle k, $\mathbf{X}(k)$. This is a normal programming exercise for each cycle k, and as μ increases it becomes a growing sequence of such problems. It yields a solution in terms of an optimal decision-vector $\mathbf{X}^*(k) = [\mathbf{x}^{*k}(k), \mathbf{x}^{*k+1}(k), \ldots, \mathbf{x}^{*k+\mu-1}(k)]$, where the vectors $\mathbf{x}^{*k}(k)$ are a sequence of decision variables for the *states* $\{s\}$ at stages $k+0$, $k+1$, etc. for the kth cycle. In the location-routing problem discussed here, each O–D pair defines a state.

   In this algorithm, we would not normally expect the *ex ante* states for stage $k+r$ in cycle k, $\mathbf{x}^{*k+1}(k)$, to be equal to the *ex post* states for the same stage taken in cycle $k+1$, $\mathbf{x}^{*k+1+r}(k+1)$. Moreover, the realized vectors $\bar{\mathbf{x}}^k(k)$, $\bar{\mathbf{x}}^{k+1}(k+1)$, etc. at each stage of the cycle would not be equal to each cycle's decisions, $\mathbf{x}^{*k}(k)$, $\mathbf{x}^{*k+1}(k+1)$, because of changes in decisions from one stage to another and from one cycle to another. For each of the flow constraints Φ^{k+r} there exists a *dual* vector $\mathbf{U}^*(k) = [\mathbf{u}^{*k}(k), \ldots, \mathbf{u}^{*k+\mu-1}(k)]$, in which each of the $\acute{\eta}_k$ components of each sub-vector gives the marginal expected payoff $\mathbf{u}^{*k}$ from a relaxation of that constraint in the ensuing stage. The feature of the recursive program is that a relationship is introduced between the parameters of the objective and the constraint functions. Let the parameters of the payoff function[13] be

$$\Gamma(k) = [\gamma_1(k), \ldots, \gamma_q(k)] \qquad (q \leq \acute{\eta}_k\mu) \qquad\qquad \textbf{(4-107)}$$

where the number q is limited by the number of constraints $\acute{\eta}_k\mu$ in cycle k. Also let the parameters of each constraint function[14] be

$$\mathbf{Y}^{ij}(k) = [y_1^{ij}(k), \ldots, y_p^{ij}(k)] \qquad (p' \leq \acute{\eta}_k\mu) \qquad\qquad \textbf{(4-108)}$$

where $i = 1, \ldots, \acute{\eta}_k;\ j = k, \ldots, k+\mu-1$. The number of such parameters p' is also limited by the total number of constraints in a cycle.

---

[13]These parameters are reminiscent of the *return function* in dynamic programming, as explained in the "Markovian Processes" in Chan (2000), among other references.

[14]These parameters are equivalent to the *state variables* in dynamic programming.

We can now rewrite the problem as

$$I^*(k) = \max I(X(k), \Gamma(k)] \qquad \textbf{(4-109)}$$

subject to

$$\Phi^k[X(k), Y(k)] \geq C(k) \quad and \quad X(k) \geq 0 \qquad k=1,\ldots,\infty \qquad \textbf{(4-110)}$$

where $Y(k)$ is the matrix consisting of p'-entry vectors $\left[Y^{1,k}(k) \quad \ldots \quad Y^{\dot{n}_k,(k+\mu-1)}(k) \right]^T$ and $C(k)$ is a matrix of μ connectivity-requirement vectors $c(k)$.

The net revenue Γ in **(4-109)** can be written explicitly in terms of the difference between gross revenue and cost. The gross return (gross revenue in our location-routing example), being a function of the route decision-variables $X(k)$ and the dual variables $U(k)$ (or the relaxation of the flow constraints), is expressed in the form $\Pi(X(k), Y(k))$ in our location-routing problem, as we will see. The cost term, being a function of the decision variables and the flow pattern Φ^k and also parametric on the available vehicle capacity $V(k)$, is written as $T(X(k), \Phi^k; V(k))$. Thus we can now write $\Gamma(k) = [\Pi(X(k), Y(k)) - T(X(k), \Phi^k; V)]$ or the more generalized form $\Gamma(k) = [\lambda' \; \Pi(X(k), Y(k)) - (1-\lambda')T(X(k), \Phi^k; V)]$ where λ' is a 0-1 ranged weight.

The feedbacks are of three types: First,

$$\Gamma(k) = G[x^*(k-1), u^*(k-1), \ldots, x^*(k-h), u^*(k-h); V(k)] \qquad \textbf{(4-111)}$$

which established a connection of the payoff function to past decision variables in cycles $(k-1)$ through $(k-h)$. Exogenous influences also factor in, as shown by the vector $V(k)$. Such exogenous influences at cycle k include the `range' on route p, U_p, terminal capacity at origin-m $\bar{P}_m$, and the given vehicle-capacity V. Second, $Y(k) = Y[x(k-1), u^*(k-1), \ldots, x^*(k-h), u^*(k-h); V(k)]$ which sets up a linkage of the origin-destination connectivity-function to past decision-variables and exogenous influences. Third,

$$C(r) = C[x^*(k-1), u^*(k-1), \ldots, x^*(k-h), u^*(k-h); c(k)]. \qquad \textbf{(4-112)}$$

Here, the connectivity requirement between an O-D pair is precisely the right-hand side of the coverage constraints[15]. Hence the problem becomes

$$I^*(k) = \max I$$
$$\left\{ X(k); G[x^*(k-1), u^*(k-1), \ldots, x^*(k-h), u^*(k-h); V(k), c(k)] \right\} \qquad \textbf{(4-113)}$$

subject to

$$\Phi\left\{ X(k); Y[x^*(k-1), u^*(k-1), \ldots, x^*(k-h), u^*(k-h); V(k)] \right\}$$
$$\geq C[x^*(k-1), u^*(k-1), \ldots, x^*(k-h), u^*(k-h); c(k)] \qquad \textbf{(4-114)}$$
$$x(k) \geq 0.$$

---

[15]The second and third feedback relationship can be thought of as the *state transition function* of dynamic programming.

Again, the interesting feature of this is that the *ex post* decision vectors, i.e., $\mathbf{x}^{*k+r}(k)$, are not equal to the *ex ante* optimum for that stage, $\mathbf{x}^{*k+1+r}(k+1)$. But they will influence future decisions through the recursive structure. Hence the sequence of decisions is, in a successive approximation context, not the optimal sequence. In this way, while not necessarily producing a set of optimal decisions, the structure can be very useful in improving the current decisions. In dynamic programming, the Markovian property suggests that only the immediate past stage is of relevance. Recursive programming requires—strictly speaking—that we refer to all previous stages in the optimization procedure. The typical recursive-programming problem can be solved in that format presented above since it is essentially a partial one. The recursion between the air-carrier operator and the customer who travel or ship can be thought of as occurring in two different ways. Customers may respond directly to the operator's decisions through modifications to the payoff function $\Gamma(k)$. In this case we need to specify carefully the exogenous influences, $\mathbf{V}$, which could actually relate to the decision variables $\mathbf{x}_j = (k-i)$, $i=1, ..., h$; $j=1, 2$; where there are two decision makers: the operator and the customer. Secondly, they may respond indirectly through market signals. These signals particularly affect the matrix $\mathbf{Y}(k)$ relating to the constraint functions. The recursive-optimization procedure can either converge or diverge. The convergence of the algorithm obviously depends on the actual problem being solved.

While the framework outlined above for recursive programming is the most general, implementation—similar to dynamic programming—is problem specific. The key is to delimit the recursions sufficiently that computational savings is achieved to the extent possible. Another desirable goal is to show convergence of the algorithm in a finite number of steps. Finally, we wish to capture as many analytical properties of the problem as possible. Clearly, the integer linear program formulated for the analysis phase of the location-routing problem above has a clear recursive structure between non-stop, one-stop, and two-stop routes etc. This is evidenced from Equations **(4-91)** through **(4-95)**. Also, the route-synthesis phase of the problem—as described in Equations **(4-87)** through **(4-90)**—directly lends itself to dynamic programming, which is more analytic and much more efficient than recursive programming. We like to take advantage of these nice properties. A decomposition scheme is proposed below, in which we have adopted a lower-case variable-notation to replace the upper-case one used in the general recursive-program description above.

4.7.2.2 LOCATION-ROUTING MODEL AS A RECURSIVE PROGRAM. Here we show how a converging recursive program can be used to solve the terminal location-routing problem formulated above. First, we decompose the problem into two computational phases: *synthesis* and *analysis*. Due to the large combinatorial space, however, it is desirable to generate the routes *only as needed* to limit our search space to a controllable size. Within this limited space, one would search for partial solutions. Then we generate more search space and the process repeats itself until a satisfactory solution is reached, made up of a convolution of these partial solutions. With this premise in mind, a heuristic algorithm is proposed (Chan 1974, Chan 1997) to break down the synthesis/analysis problem into a series of synthesis/analysis couplets.

In this procedure, **x** is a proper subset of **X**. A screening rule such as the "most promising" or "least cost" routes help us to identify such a subset. Let us define a matrix operator such as

$$\left[l^{mn}(r) \right] = \left[\mathop{\Psi}_{k=1}^{|I|} \left(l^{mk}(r-1) + l^{kn}(0) \right) \right] \qquad \textbf{(4-115)}$$

where m,n are nodes within the network set I, and r is the number of intermediate stops in a route of length l. The operator Ψ is simply defined as $\Psi_{k=1}^{|I|}(l_k) = \min(l_1, l_2, \ldots, l_{|I|})$, or the familiar dynamic programming recursion for shortest paths (Chan 1979). Such an operator is to be applied toward the route candidates generated by Equation **(4-90)**. Thus Equation **(4-115)** defines $x \in X$ for us.

Example

Suppose we are given the inter-city non-stop 'distances' as shown in **Figure 4-13**. We put an ∞ between nodes that are not connected directly in an arc. The resulting adjacency matrix $[l^{mn}(0)]$ looks like $\begin{bmatrix} \infty & 13.5 & 31.5 \\ 13.5 & \infty & 18.0 \\ 31.5 & 18.0 & \infty \end{bmatrix}$. According to Equation **(4-115)**, the 1-stop-route distances are $[l^{mn}(1)]$ $= \Psi_{k=1}^{3}(l^{mk}(0) + l^{kn}(0))$ for $m-1, 2, 3$ and $n = 1, 2, 3$. For example, $l^{12}(1) = \Psi_{k=1}^{3}(l^{1k}(0) + l^{k2}(0))$ $= \min(l^{11}(0)+l^{12}(0), l^{12}(0)+l^{22}(0), l^{13}(0)+l^{32}(0)) = \min(\infty+13.5, 13.5+\infty, 31.5+18.0) = 49.5$. The minimum distance for 1-stop routes between nodes 1 and 2 is therefore 49.5. This corresponds to the route 1–3–2. □

Similarly, φ is a proper subset of Φ, showing a partial flow pattern among the subset of routes included in the decomposed search space under consideration at the time. These flow patterns are defined by the recursive equations **(4-94)** and **(4-95)**. The most critical step of the procedure has to do with the way the cost function $T(X,\Phi,V)$, benefit function $\pi(X,Y)$ and the objective function $I(X,Y,\Phi,V)$ are decomposed.

This is assisted by the recursive equations **(4-98)** through **(4-105)**. Both this task and the task of defining a decomposed search domain are greatly assisted by a properly defined *state-stage space*. Here the stage is labelled by index *r* and the state by an O-D pair *m–n*, shortened as *s*—an example of which is shown in **Figure 4-15**. Within this state-stage space, we compute the dual variables **u**, corresponding to 'benefits' by relaxing the flow constraints **(4-94)**[16]. We write

---

[16]In the presence of duality gap in integer programming, only *inference duality* applies. Hooker (1996) defined an inference dual as the problem of inferring from the constraints a best possible bound on the optimal value. The inference dual of a discrete optimization problem can be solved by examining the results of a primal branch-and-bound method, in particular an enumeration tree that is generated to solve the primal problem. He showed how an "optimality proof" can be recovered from an enumeration tree that solves the

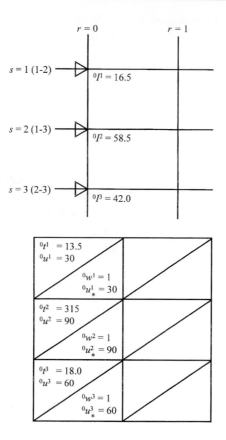

Figure 4-15 Tabular bookkeeping of state-stage space at stage 0

$^ru\,^s(x,y)=\,^r\pi^s(x,y)\otimes\,^rc^s(x)$ for each "grid point" $r.s$ of the state-stage space, correspond-ing to the sth state in stage r. Here π is the revenue vector and c denotes whether and how the relevant O–D's demands are served. y is the set of parameters for the constraint functions that govern the local traffic-flow pattern at grid points $r.s$ in cycle k, as defined by Equation (4-108). $\otimes$ is the relaxation operator to combine the payoff-function π with the coverage function c. This is the recursive-programming implementation of Equation (4-104) and (4-96) respectively. Compared with Equations (4-111) through (4-112), parameters π, y and c are defined considerably narrower in scope allowing for its evaluation readily.

---

that solves the primal problem. The most straightforward way is to branch on values of the variables until all feasible solutions have been found. The search backtracks whenever a feasible solution is found or some constraint is violated. The best feasible solution is optimal, from which the dual variables can be obtained. Inference duality thus permits a generalization of Benders' partitioning to any optimization problem.

While a relaxation of the service flow-pattern yields additional revenue, the other part of the payoff function is the cost function. Also, we decompose the cost function, reflecting how often a route is flown: $^r t^s(\mathbf{x}, \Phi, \mathbf{V}) = {}^r t^s(\mathbf{x}) \otimes {}^r w^s(\varphi, \mathbf{V})$. Again, this is the recursive-programming implementation of Equation **(4-105)**. Here, $[r.s]$ is defined as the set of shortest routes or itineraries serving a subset of the demands (customers) from origin m to destination n via an r-stop itinerary, $\{f_r^{mn}\}$. Recalling that in our notation s is a shorthand for the O-D pair m–n, one can begin to discern the equivalence between $[r.s]$ and $\{f_r^{mn}\}$. Both are precisely what was referred to as the `watershed' in the recursive-programming introduction-section. The watershed delimits the local traffic-flow pattern and is defined by $\acute{\eta}_k$ functions. The subscript k suggests that the "decomposed domain" definition may change depending on the cycle k through the algorithm. In essence, $r.s$ stands for a demand set $\mathbf{f}^*$ to O-D pair m–n and the partial route-structure $\mathbf{x}^*$. Here, $\mathbf{x}^*$ carries via its itineraries a broader set of O-D demands, collectively referred to as $\{f_r^{mn}\}^*$, which includes $\mathbf{f}^*$. Notice that such a partial-flow pattern is defined by Equations **(4-94)**, **(4-95)**, and **(4-98)**. These demands are labelled in the state-stage space, making a scanning among the decomposed space quite convenient in operationalizing the algorithm (see **Figure 4-15** and **Figure 4-16**). Now the partial passenger flow φ can be replaced by the given partial-demands $\mathbf{f}$ in this cost equation $^r t^s(\mathbf{x}, \mathbf{f}, \mathbf{V}) = {}^r t^s(\mathbf{x}) \otimes {}^r w^s(\mathbf{f}, \mathbf{V})$.

Finally and most important, the net-benefit function is decomposed cycle/loop k as

$$\max_{\mathbf{x} \in [r.s]} \left\{ {}^r I^s[\mathbf{x}, \mathbf{y}, \mathbf{f}, \mathbf{V}] = [\lambda^{\prime} \, {}^r u^s(\mathbf{x}, \mathbf{y}) - (1 - \lambda^{\prime}) \, {}^r t^s(\mathbf{x}, \mathbf{f}, \mathbf{V})] \right\} \tag{4-116}$$

subject to

$$\varphi[\mathbf{x}, \mathbf{y}, \mathbf{V})] \ge \mathbf{c}^r \tag{4-117}$$
$$\mathbf{x} \ge 0$$

where the only decision variable is the decomposed or partial set of routes and the associated itineraries $\mathbf{x}$, where the dual variable $^r u^s$ is obtainable from the relaxation of Equation **(4-117)**. In essence, Equation **(4-116)** is the recursive, decomposed re-formulation of the original IP as represented by Equations **(4-91)** through **(4-105)**. A search algorithm is constructed on the state-stage space using Equation **(4-116)** as the `gradient'. Notice that $^r I^s$ consists of two terms: the dual variable $^r u^s$ representing revenue and the cost term $^r t^s$. $^r I^s$ then represents the net revenue and constitutes the decomposed, recursive-programming adaptation of the convex combination of Equations **(4-104)** and **(4-105)**. A sequential scan is proposed to search one grid point at a time, starting with $r=0$ and progressing toward $r=\mu$. Likewise, within each stage, s starts with the numerical value of 1 and ends up with $|I|^2$ in value. The convolution of these hill-climbing steps is also taken care of by the state-stage space $\{r.s\}$ and the associated decomposed domains $[r.s]$. The stage-space is labelled with the domain definition for $[r.s]$ and the value of the gradient $^r F$ every time a grid point is examined. A table is used to do the associated bookkeeping for t, u and w (see **Figure 4-13** and **Figure 4-15** for the three-node network under discussion). The algorithm terminates when two passes through the state-stage space leave the labels unchanged. This means that $'I_*^s, 'u_*^s, 't_*^s$ from Equations **(4-116)** and **(4-117)** remain unchanged over two subsequent iterations. Without loss of generality we describe an algorithm below by ignoring range and capacity limitations as represented by constraints **(4-101)** through **(4-103)** in the original formulation.

Example

For the airline example shown in **Figure 4-13** and **Figure 4-15**, the first non-stop between nodes 1 and 2 (route 0.1) yields a revenue of $30,000. It is made up of $30,000 from coverage of non-stop demand for O-D pair 1-2. It carries an operating cost of $13,500 per flight for the frequency of one flight per day. The route yields a net profit of ($30,000 – $13,500) = $16,500 per day. Notice also that in this example, route 0.1 provides the shortest itinerary between nodes 1 and 2. Longer itineraries serving 1-2 are rejected. This is to isolate the non-inferior solutions corresponding to the sometimes conflicting objectives of user-optimization (least travel-stops) and system-optimization (maximum coverage *and* least travel-cost). □

RISE Algorithm

To realize these functional relationships, a set of computerized algorithmic steps have been devised (Chan 1974), carrying the name Route Improvement Synthesis and Evaluation (RISE). The steps can be grouped into two parts: (i) *intra* state-stage-space instructions, and (ii) *inter* state-stage-space instructions. The former pertains to the steps executed at the grid points of the state-stage space while the latter pertains to cycles or loops through the entire state-stage space. The intra state-stage-space instructions are again broken down into two parts: The first loop through the state-stage space requires the execution of the *synthesis*, *analysis* and *improvement* steps in their entirety. Subsequent loops require only the execution of analysis and improvement steps. The set of instructions can be formalized as follows (reference **Figure 4-15**):

Step 0   Initialize cycle counter to $k=1$.

Step 1   Synthesize a shortest route/itinerary $'x^s(k)$ at the grid point $r.s$ of the state-stage space $\{r.s\}$.

Step 2   Analyze (evaluate) the route/itinerary $'x^s(k)$ at the grid point $r.s$ by flowing demands $\{f_r^s\}$ over it from the decomposed domain $[r.s]$.

Step 3   Improve the existing route/itinerary $'x^s(k-1)$ and re-label if the marginal benefit from the new route/itinerary is larger than that from the existing; otherwise proceed.

Step 4   Now set the counter $k \leftarrow k+1$. If it is the first cycle or loop through the state-stage space, go to step 1; otherwise go to step 2. ∎

Let us now go through the entire algorithm from one state-stage-space iteration to another:

Step *i*: If no initial feasible-solution exists, impose the *penalty-cost method* to force mandatory service into the initial solution. Similarly, a prohibited service can be excluded from the solution base. This is carried out by assigning a $-\infty$ or $+\infty$ respectively to grid point $r.s$. If initial solution exists already, proceed.

Step *ii*: Execute intra state-stage space instructions as outlined by steps 0 through 4 above.

Step *iii*: If the labels in the state-stage-space remain unchanged over two successive loops, stop; a local optimum has been found; otherwise, go back to step *ii*. ∎

While the above algorithm operationalizes Equations **(4-87)** through **(4-105)**, however, a global optimum cannot be guaranteed. Clearly, the recursive-programming algorithm lacks separability and Markovian properties.[17] It is simply a hill-climbing procedure, as discussed in Chan (2000).

4.7.3 Examples of m/m/m/n model. Two numerical examples are worked out here. The first is particularly simple. No connect traffic is allowed, and the algorithm terminates within a single pass. The second example involves connect traffic and several coverage requirements for selected O-D pairs. Two passes or loops through the entire algorithm are required to ensure optimality. The former is simply to familiarize the reader with the notations in RISE. The latter, corresponding to the MIP model shown in **Figure 4-14**, is more illustrative of the RISE algorithm.

Example 1

Johnson and Chan (1990) worked out a simple airline example of the RISE algorithm. The airline example presented here consists of three depots/terminals with pairwise O-D travel-demands f_r^{mn} as indicated in **Table 4-4**. Notice that demand is associated with a terminalpair rather than just any node-pair; it is also expressed as a function of the service—i.e., a non-stop service induces more demand than a one-stop service. Operating costs for each route are shown in **Figure 4-13**, where a template of all generated routes are shown. The comprehensive set of routes are shown in **Figure 4-13** for illustration purposes only. In the execution of the algorithm, routes are generated only as needed. Vehicle seat-capacity V is 120 passengers (pax). Additionally, there is no demand for connect itineraries, or connect passengers are disallowed. The two criteria of maximal coverage (gross revenue) and least travel-cost are in dollars and weighted equally $\lambda'=0.5$, reducing the problem to a single objective: maximize profit.

The procedure begins with an unlabelled state-stage diagram $\{r.s\}$ and an empty table for tracking revenues and costs. Without loss of generality, the algorithm assumes a symmetric tabular format and routing pattern, much as Laporte et al. did. Thus 1–2 is treated the same as 2–1. Notice the state-stage diagram $\{r.s\}$ captures the combinatorial space used

---

[17]For a discussion of Markovian systems, one may wish to consult an appendix to Chan (2000) under the same name.

Terminal pair	Non-stop	One-stop	yield/pax
1–2	10	10	3
2–3	20	20	3
3–1	30	30	3

Table 4-4 Demand between terminal pairs

in the *synthesis* phase of RISE. The symmetrical origination-termination pairs of a route correspond to the states of the system. Thus a route that starts at node 1 and terminates at node 2 defines state 1. A route that starts at 1 and terminates at 3 defines state 2, and the route from 2 to 3 defines state 3. One can see that Equations **(4-87)** and **(4-88)**, used to generate non-stop and 1-stop routes, correspond to stage 0 ($r=0$) and stage 1 ($r=1$) of the diagram respectively. Unlike the contiguity matrix, the diagram is made extra compact by the pre-screening shortest-path-operator **(4-115)**. This operator picks the shortest distance between the origination and termination points of a route, commensurate with the intermediate-stop requirements.

In the first step of the algorithm, each non-stop route is considered (one at a time). If the route is profitable, the route is activated and an arrow is drawn at the appropriate grid point [$r.s$], along with the profit associated with that route. The bookkeeping table is updated as the routes are considered. In this case, all the non-stop routes are profitable. The state-stage diagram and bookkeeping table at the end of stage 0 is shown in ?.

The next step is to consider one-stop routes. In this example, the route 1–3–2 is not profitable since the operating cost is 40.5 and the revenue is only 30. This route will not be activated. Since connecting routing is not permitted, the demand for one-stop routes from 1 to 2 will not be satisfied. The route 1–2–3 is considered next, this route is profitable. Furthermore, the route can satisfy demand for non-stop routes 1–2 and 2–3 with no additional cost. The non-stop routes 1–2 and 2–3, along with their costs, are eliminated as a result. Here, $^1u_*^3(\mathbf{x},\mathbf{y}) = {}^1u^3(x_{121}^{13}, x_0^{12}, x_0^{23}; y_1^{13}, y_0^{12}, y_0^{23}) = {}^1\pi^3(x_{121}^{13}, x_0^{12}, x_0^{23}; y_1^{13}, y_0^{12}, y_0^{23}) \otimes {}^1c^3(x_{121}^{13}, x_0^{12}, x_0^{23}) = 150$. The companion cost is $^1t^3(\mathbf{x},\mathbf{f},\mathbf{V}) = {}^1t^3(x_{121}^{13}, x_0^{12}, x_0^{23}) \otimes {}^1w^3(f_1^{13}, f_0^{12}, f_0^{23}; 120) = 36$. The resulting net-benefit function is $^1t^3 = {}^1u_*^3 - {}^1t^3 = 150 - 36 = 114$. The decomposed domain [1.3], a consequence of the inference dual variable $^1u^3$, is defined by the **x**, **y**, and **f** vector above, and represented conveniently by the labelling procedure in the state-stage space. As mentioned, the decomposed domain changes from one cycle (loop) to another. The number of constraints that need to be relaxed in each cycle k to compute $^1u^3$, is bounded by $\acute{\eta}_k$, with $\acute{\eta}_k$ substantially smaller than the state space $|I|^2$. Here, $\acute{\eta}_k \leq 2$ and is expected to be much smaller in realistic-size networks.

Finally, route 3–1–2 is considered and found to be profitable. It can also satisfy the demand for non-stop routes 3–1 and 1–2, therefore routes 1–2 and 3–1 are also eliminated. The state-stage diagram and bookkeeping table at the end of stage 1 is shown in **Figure 4-16**.

At this point all possible routes and permutations have been considered (recall connect routings are not permitted) and no further improvement is possible. Thus the final solution is given by routes 1–2–3 and 3–1–2 (see **Figure 4-16**) and the total profits are 249. The solution happens to be identical to the analytical answer by mixed integer programming. While the algorithm yielded the optimal solution for this example, it should be noted that

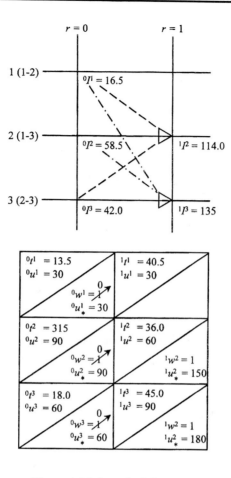

Figure 4-16 Stage 1 of algorithm

suboptimal (although good) solutions should generally be expected. Admittedly, this example is too simple to show the potency of the RISE algorithm, another more involved example is included below for reference. □

Example 2

This example follows the one used in the section entitled "Recursive programming." As an additional stipulation in our example, a non-stop route is mandatory between 1-2 due to a number of factors, such as competition and the like. Conversely a non-stop route between 1-3 by itself is to be avoided, possibly due to governmental regulation. Finally, O-D pair 2-3 is required to be served by a one-stop itinerary or better. The example problem will be solved using RISE, by which the various steps of the algorithm are illustrated in detail. Through this exercise, we hope to familiarize the readers with the RISE computer program that encodes the algorithm described here. Phase I of RISE is the first loop through the stages. It will provide an initial feasible solution, which will then be improved upon in subsequent loops of Phase II. These two execution steps are similar to the $m/m/-/h$ model, where the "capacitated basic algorithm" synthesizes the tours, to be improved subsequently by the "single displacement" and `exchange' steps.

Phase I

Stage 0   The algorithm starts out with an unlabelled state/stage diagram $\{r.s\}$ and an empty table (for keeping the basic unit costs/revenues and route frequency and fleet type). At grid point 0.1 (referring to the nonstop stage and the first state in this stage), the nonstop route 1–2 is synthesized.

Since the terminal pair 1–2 is required to be served by a non-stop itinerary an $-\infty$ is labelled on the grid point 0.1, by which it is `forced' by the penalty-cost method to enter the state-stage diagram. This way the coverage constraint on O-D pair 1–2 is immediately satisfied. Figure 2 documents the current state/stage diagram and the two corresponding tables. Notice that the entire demand quantity f_0^{12} is now carried by the non-stop route 0.1. To record this, the "demand arrow" is entered into the state/stage space at this grid point showing the loading of demand on this route.

Route 0.1 is flown twice a day, at a cost of $^0t^1_\bullet = 916$ dollars ($^0t^1_\bullet = {}^0t^1\ {}^0w^1$) and a gross revenue of $^0u^1_\bullet = 3000$ dollars, yielding a net profit of $^0I^1 = 2084$ dollars. The profit figure is labelled beside the circled grid point 0.1—showing acceptance of the non-stop route. At this time, the decision variable $^0x^1(0)=1$ and $y_0^{12}=1$. Here the itinerary and route are the same, being a non-stop, resulting in a scalar valuation of $^r\mathbf{x}^s(k)$.

Since no authorized non-stops can be flown between 1–3, the grid point 0.2 will be unlabelled—through assigning a cost of ∞ to the route. But the basic unit cost/revenue associated with this grid point is still entered into the unit cost/revenue table. This is shown in the corresponding cell in **Figure 4-17**, so that the information can be referenced at a later time.

At grid point 0.3, it is unprofitable to introduce non-stop route 2–3, since it will incur a loss. While the route is rejected, the basic unit costs/revenues for this grid point are again recorded in the table of **Figure 4-17** as explained previously. At this juncture Stage 0 computation is concluded.

Stage 1   At stage $r=1$, all the one-stop routes will be synthesized and evaluated. Around grid point 1.1, route 1–3–2 is being synthesized from segments 1–2 and 2–3. This is so indicated in the state/stage space (**Figure 4-18**). The basic unit-cost for this grid point is computed by simple addition. By assigning traffic from demands f_1, f_0 and f_0, route frequency can be computed, from which the operating cost and gross revenue figures are derived. Notice in here, demand for 2–3 is the same as 1–2. From these cost/revenue figures, it is decided that route 1–3–2 should be flown since it makes a profit. We label grid point 1.1 and do the appropriate bookkeeping in the table as shown in **Figure 4-18**. Here the decomposed domain [1.1] is the grid points 0.2, 0.3 and naturally 1.1, since the one-stop route 1–3–2 is made up of the itineraries 1–3, 2–3 and 1–3–2 corresponding to these three grid points. At this juncture $^1\mathbf{x}^1(0) = (x_0^{13}, x_0^{23}, x_{131}^{12}) = (1,1,1)$, and $y_1^{12}=1$.

At grid point 1.2, there are two ways to serve O-D pairs 1–3 via one-stop itineraries. The first is to institute a one-stop service 1–2–3. The second is to make a connect itinerary out of the non-stop 1–2 and segment 2–3 of the one-stop route 1–3–2. The additional revenues generated from both alternatives remain similar

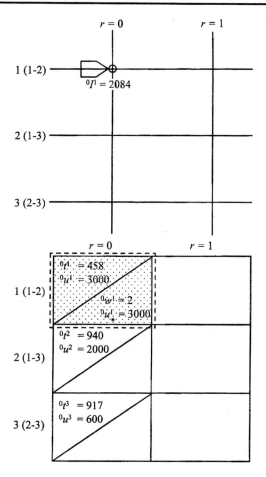

Figure 4-17 Stage 0: Grids 1, 2, and 3

[which is the product $^1u_c^2 = 900$; $(36)(25)$]. But the cost is drastically different. The new one-stop route costs $^1t_*^2$ is 1375 dollars from one flight. The connect itinerary means an additional cost of only $^0t_*^1 = 458$ dollars (which arises from adding one flight to the existing non-stop route 1–2). Clearly, the latter is preferred, resulting in a profit of $^1I^2 = 900 - 458 = 442$ dollars. Grid 1.2 in the state/stage space is labelled and marked with a 'triangle'. The necessary data are entered into the bookkeeping table. Notice the frequency of 2 is re-labelled 3 for non-stop route 1–2 to record the additional flight assigned to this leg of the connect itinerary.

Here the decomposed domain for grid point 1.2, [1.2], is composed of 0.1, 0.3 and 1.2, corresponding to non-stop itinerary 1–2, segment 2–3 of the one-stop route 1–3–2 and the one-stop connection 1–2–3. Notice that the only demand of interest is f_1^{13}. Hence the only additional 'arrow' introduced within [1.2] is at grid point 1.2, since the arrows for 0.1 and 0.3 have been inserted previously. The additional cost introduced with [1.2], on the other hand, is associated with grid 0.1 only and not for 0.3 or 1.2. Here $^1\mathbf{x}^2(0) = (x_{121}^{13}) = (1)$ and $y_c^{13} = 1$.

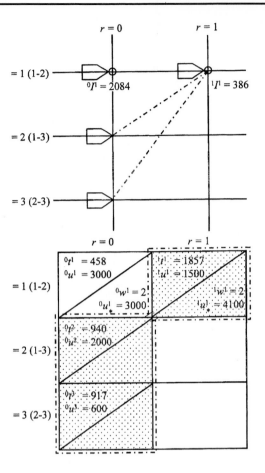

Figure 4-18 Stage 1: Grid 1

Since it is required that a one-stop or better itinerary be instituted for O-D pair 2–3, one-stop route candidate 2–1–3 is synthesized at grid 1.3. However, the do-nothing alternative with zero profit is found to be better than instituting the route (which incurs a loss). The decision is then to have the O-D demand unserved, noting that we have already satisfied the connectivity requirement $C^{23}(1)$ in Equation **(4-96)** by instituting non-stop service 2–3 previously.

At this point, we have come to the conclusion of phase I of the RISE algorithm, in which the initial feasible solution has been obtained. The necessary bookkeeping at the end of labelling grid point 1.3 is the same as the previous state/stage **(Figure 4-18)**. There is no additional entry except for the basic costs and revenue of route 2–1–3 (which we overlay in the table shown as part of **Figure 4-19**). We also label the synthesis trial for 1.3 in the state/stage space of **Figure 4-19** to define the decomposed domain [1.3], although it results in no new service (hence no arrow, circle or triangle drawn).

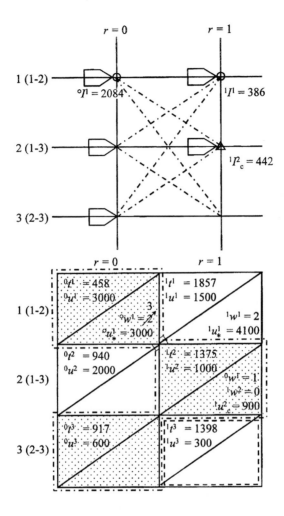

Figure 4-19 Stage 1: Grid 2

Phase II

Phase II begins with a second loop around the state/stage space. At each grid point, a "perturbation method" is used to examine whether there exists a comparable routing to serve the O-D demand under consideration that will result in a higher system profit. Notice this forms the heart and soul of a recursive program.

Stage 0 At grid point 0.1, the comparable routing that can carry demand from 1 to 2 can be readily picked out from the decomposed domain [0.1] in the state/stage diagram (**Figure 4-19**): Candidate one-stop route 2–1–3 can clearly cover O-D pair 1–2 on one of its non-stop segments, again assuming symmetry between 2–1 and 1–2, which enables us to write them interchangeably. This new itinerary would incur a cost of $^2w^3$ $^1t_c^3$ = (2)(1398) = 2796 dollars . It would yield a revenue of $^1u_c^3$ = $^1u^3+^0u^1$ = 300+3000 = 3300 dollars, resulting in a marginal profit of 504 dollars. But this is inferior to the existing itinerary that yielded $^0I^1$ = 2084 dollars. One retains the existing itinerary as a result.

We next consider whether the demand carried on the segment 1–2 of the connect routing 1–2–3 should be provided by the non-stop 1–2 (as is the current situation) or the one-stop 2–1–3. It is found that using 2–1–3 would cost more. Therefore the decision to retain the existing non-stop route 1–2 holds.

In this step we have illustrated graphically the basic workings of the state/stage diagram and the decomposition domain defined within this framework. Through the state/stage diagram, the decomposition domain for grid point 0.1 is clearly defined as 0.1, 1.3 and 1.2 corresponding to the demands carried by the non-stop route 1–2, segment 1–2 of the proposed route 2–1–3 and the connect itinerary 1–2–3. These are all the itineraries that can carry the traffic presently handled by the non-stop route 0.1. They are represented by the grid points reachable from 0.1 in **Figure 4-19** via the cross arcs from grid point 0.1 to 1.3 and 1.2. In lay terms [0.1] can be thought of as the "catchment area" (or watershed) of the non-stop route or segment 0.1.

Next, we find that no reexamination is necessary for grid point 0.2. The reason is that there is no route to consider in grid point 0.2 (as the grid point is not labelled). At grid point 0.3, no reexamination is necessary either for the same reason.

Stage 1   At grid point 1.1, the only alternative that will serve a subset of demand f_0^{13} and f_0^{23} and f_1^{12} is merely the proposed one-stop route 2–1–3. Unfortunately this other route can carry only f_0^{13}, yielding a profit of 106 dollars. In comparison to the existing profit of 386 dollars, it is wise to keep the existing route.

Under the given service restrictions, there is no alternate way of carrying f_1^{13} other than via the connect itinerary 1–2–3. With nothing changed in the state/stage space during Phase II thus far, the label on grid 1.3 also remains the same as before.

It is seen that the labelling in two consecutive loops remains identical. According to the method-of-successive approximation, an optimal solution has been found. This consists of non-stop route 1–2 operating at 3 flights a day and one-stop route 1–3–2 at 2 per day. The bookkeeping in **Figure 4-19** also shows a total profit of 2912 dollars $[^0I^1+^1I^1+^1I_c^2 = (^0u^1+^1u^1+^1u^2) - (^0w^1 \, ^0t^1+^1w^1 \, ^1t^1) = (3000+4100+900) - (3\times458+2\times1857).]$ □

The FORTRAN code of the RISE heuristic is distributed on my web site and at the back of Chan (2000). Also included are additional data sets for experimentation.

4.7.4 Validation of the RISE algorithm. Before deploying RISE in practical applications, it is useful to validate its performance. Toward this end, several validation runs were performed. Four experiments were conducted, in which the RISE solutions were compared with the IP solutions. The results are shown in **Table 4-5**, in which the first entry records the three-node example shown immediately above. Variations of this example were examined as experiments 2 and 3. Experiment 2 increases the aircraft capacity of the original example from 80 to

Experiment	IPSolution	RISE Solution	Percentage Difference
1	*2911	2912	0
2	3639	3639	0
3	4005	3639	-9.1
4	249	249	0
			AVERAGE -2.28

Table 4-5 Comparison between integer programming and heuristic solutions

90 seats and removes all route restrictions. Experiment 3 introduces an additional aircraft with a seat capacity of 150, at an operating cost 30 percent above the existing fleet. Again, no restriction is placed upon the routing. Finally, experiment 4 illustrates a case where travellers, while receptive to non-stop and one-stop itineraries equally, will not accept connections. It is the first numerical example worked out in the above section.

In experiment 1, the reason that the integer program yields a lower profit—as marked by the asterisk (*)— than the RISE heuristic can be explained. Due to the discretization of the x and y variables, the integer program assigned two nonstop flights to the O-D pair 1-2 to cater for the 150 passengers. Meanwhile, the 1-stop route 2-1-3 has only the leg 1-3 occupied, flying empty between 1-2 (or should I say 2-1). the underutilized leg is a product of the binary valuation of **x** and **y**. Examining the inequality $y_0^{12} \le x_{121}^{12} + x_{124}^{12} + x_{126}^{12} + x_{126}^{12}$, only x_{121}^{12} is "turned on" for cost savings (with $x_{126}^{12} = 0$). Here, x_{121}^{12} is set to unity in order to cater for the 150 passengers with two flights. None of the 150 passengers use the 2-1 leg of the 1-stop flight (route 6 that goes 2-1-3) as a result. The RISE heuristic, however, assigns passengers to available seats. When it sees the leg 2-1 of route 2-1-3 is not used, it fills up the seats, resulting in lower frequency for the non-stop route 1-2 and hence lower cost (and higher profit). [For those interested, consult the "Integer programming vs. heuristic" homework under the "Location-routing" section in the "Exercises and Problems" on my web site.]

These experiments cover a variety of situations ranging from different fleet specifications to demand patterns and form a fairly comprehensive set of tests for the RISE algorithm. All experiments are based on a three-node example although the current RISE software can run up to the size of 25 nodes (corresponding to the major markets of air carriers). It can easily be expanded beyond that. The three-node case is the only situation where the entire set of routes can be generated and included in the IP. It is the only controlled experiment that allows us to meaning-fully compare RISE solutions with IP solutions on an equal footing, without pre-screening of route candidates by Equation **(4-115)**. The computational bottleneck lies with the IP and the associated commercial software, not with the RISE algorithm. The latter can readily be extended to handle hundreds of airline terminals. As expected, RISE yields suboptimal solutions in two of four cases in **Table 4-5**. But nowhere is the solution worse than 9.1 percent of the IP solution--which occurs in experiment 3. In experiments 2 and 4, RISE and IP yield identical solutions. Considering all four experiments, the average difference between the two solutions is 2.28 percent, which is a reasonable margin of error in our judgment. The readers will recall that the validation IP for experiment 1 is listed in **Figure 4-14** for reference (Del Rosario and Chan 1992).

4.7.5 Remarks. In this m/m/m/n model, we have formulated and solved a class of location-routing problems that is much more general than the travelling-salesmen-type problems found in the literature. Instead of having a route start and terminate at the same depot as in a tour, the starting point can be different from the termination point. Furthermore, not all the nodes need to be visited as in the travelling-salesmen problem. Thus locational decisions are to be made among the candidate terminal pairs. In determining the subset of nodes to be visited en route, multiple coverage of a node by more than one route is allowed. In short, the model configures routes with frequency specification and locates terminals where service is to be rendered. This work distinguishes itself from others in that both vehicle and passengers are routed, rather than just the vehicle fleet. In this light, we combine the vehicle-routing/scheduling problem with the traffic-flow problem into a single formulation. The essential features of this class of problems are identified, together with the important characteristics of feasible solution-strategies. It was found, for example, that this general class of location-routing problems typically has two components that cannot be easily integrated, namely the route synthesis part and the analysis part. What it means is that an optimal solution is unattainable, practically speaking, mainly due to the huge dimensionality involved in both components (particularly the first). Notice the synthesis and analysis steps cannot be coordinated by a column-generation scheme as in classical decomposition problems, simply because `rows' are added (and existing rows changed) as new columns in the `tableau' are inserted.

　　For the first time, we formulate this general problem in a closed-form integer-linear-program (IP), which can be related to—but goes well beyond—other classical location-routing problems. While it consists of similar objective functions, demand specification, forcing/linking constraints, intermediate-stop requirements, route and depot restrictions, the IP location-routing model is also a multi-criteria optimization model. It can generate the entire set of non-dominated solutions for the first criterion of maximal-demand-coverage and the second criterion of least-operating-cost. It can be shown that the formulation reduces to the maximal-coverage location-routing problem when the second criterion is de-emphasized. Similarly, it reduces to the set-covering location-routing problem when the first is ignored.[18] The model degenerates into a routing problem when all demands are required to be satisfied. The model also shares the same shortcoming of other location-routing models in that an origin-destination demand is either carried by a route/vehicle in its entirety or not at all. Such "lumpiness" property will place solutions from location-routing models as suboptimal estimates of their real-world counterparts. In the real world, demands can be more evenly distributed—i.e., when the demand can be split among more than one vehicle. In fact, *lumpiness* distinguishes facility-location models from activity-allocation or land-use models, where the latter is a continuous, rather than discrete, modelling process.

Perhaps our biggest contribution is the structuring of such a general class of problems, for the first time, in the framework of recursive programming. Overcoming the *separability* and *additivity* restrictions of dynamic programming, recursive programs offer a very robust solution to variety of general location-routing problems. Most of all, the recursive program so formulated here is shown to converge in a finite number of iterations, although only a local optimum is guaranteed. Equally important, we can carry out the algorithm in a workable state/stage space where a state is defined for each O-D pair. Similarly, a stage is defined for non-stops, another for one-stop and so on. This combinatorial space offers several advantages:

a. Non-planar networks involving terminal location, vehicle route and traffic itineraries can be easily handled within this search space.

b. This space further allows a simple labelling procedure to be used, which decomposes the problem into an iterative series of synthesis/analysis couplets that effect a gradient search guided by a dual vector. Vehicle routes are synthesized as needed, similar to a column- or row-generation scheme.

c. The search space is compact. It is in the order of $|I|^2(\mu+1)$ where $|I|$ is the number of terminals in the network and μ is the number of intermediate stops in the longest route. This is much smaller than most integer programs, which spans a dimensionality of $|P||I|^4$. ($|P|$ is the number of routes that has a dimension in the order of $2^\mu|I|^2$: an astronomical number in any practical network).

d. From an application standpoint, the RISE algorithm proposed here yields a solution that is within 9.1 percent of the global optimum for all the controlled experiments we performed. This is a small price to pay in our judgment, considering the IP alternative cannot yield practical solutions to any realistic problem even of the modest size.

e. This state-stage space has been shown to accelerate search procedures used in general network-design problems, allowing a double-bounding scheme to be implemented (Chan 1986).

---

[18]For an explanation of the terms "maximal coverage" and "set-covering" consult the "Facility Location" chapter.

Another contribution of this recursive-programming model is the simplification of the way we view combined location-routing models. Through the multi-criteria optimization formulation, we are able to reduce a multi-decision-variable model involving at least routing and location decisions to simply a routing decision. Aside from computational advantages, this greatly enhances our conceptual understanding of this class of problems. Worthy of elaboration is the negative conclusion that IP plays an extremely limited role in this general class of combined location-routing problems. This is in view of the cumbersome formulation and the huge combinatorial space. These factors preclude the conventional approach of generating a feasible set of routes and then searching for an optimal solution among these candidates ($|P| \approx |I|^2 \cdot 2^\mu$). Even if the problem can be segregated into the distinct synthesis and analysis phases, a feasible set of routes have to be screened a priori by heuristics to avoid the curse of dimensionality. This automatically yields suboptimal solutions, thus defeating the purpose of an analytical solution procedure. With the efforts made in this study we hope that some insights have been gained in solving generalized location-routing problems. It is recommended that more computational experiments be conducted to gauge the performance of the recursive-programming approach, which will further generalize the findings reported here to an even broader class of location-routing problems.

4.8   Computational experiences

Table 4-6 shows the computational experiences recorded so far in solving combined location-routing problems. These computational requirements are referenced against different computers and different problem sizes and types. For these reasons, cross comparisons should be made only with extreme care and under special provisions. But it is of interest to note again that analytical techniques have only limited applications in solving this type of problem. For example, the only two problems that can be solved analytically appear to be the single-facility/single route/multi-criteria model ($s/s/m/-$) and the multiple-facility/multiple-tour/allocation model ($m/m/-/a$). While Benders' decomposition was devised to solve the single-facility/multiple-tour/allocation ($s/m/-/a$) problem, the problem was actually solved by heuristics. Location-routing problems are in general NP-hard. Only recently have exact procedures been proposed and most of these can only solve very special classes of problems. It is safe to say that most location-routing solutions are approximate, employing heuristic solution techniques. Recent emphasis is placed upon the performance bounds of these heuristics.

Perhaps the most exciting development in analytical solution-techniques is the relaxation techniques in polyhedron description. By relaxing the chain-barring constraints, Laporte et al. were able to design a set of deep cuts covering the combinatorial space for the routing variables. This, together with an innovative subtour-breaking- constraint incorporating vehicle capacity, allows solving up to 6–20 demands, 4–8 depots and 3–5 vehicles in the $s/s/m/-$ model. The analytical solution technique suggested that exact algorithms that iteratively add violated subtour and chain-barring constraints might perform well for medium-size location-

Model	Algorithm	Size of problems solved	Computer
s/s/m/–	noninferior solution of MIPs	15 demands	34 CPU sec CDC Cyber 76
s/m/–/a	Benders' decomposition	50-demands/ 6–9 vehicles	3–7 CPU sec
	also interchange heuristics	75-demands/ 3–4 vehicles	IBM 4341
	also Space Filling Curve for probabilistic demands	98 demands/ 3–4 vehicles	10 min on Apple Macintosh
m/m/–/a	composite, exact algorithm including B&B and heuristics	6-20 demands max of 3-5 vehicles 4-8 depots	< 300 sec Cyber 855
m/m/–/h	Sequential interchange heuristics	50-demands max of 80–160 veh 4 depots	< 5.5 CPU sec Univac 1100/80
	Alternate-location- allocation- savings heuristic	12 depots 12 primary trunking routes 213 secondary vehicle tours	2 min
	Savings-drop heuristic	13 depots 13 primary trunking routes 202 secondary vehicle tours	12 min Both on IBM 370/165 using FORTRAN-G
m/m/m/ n	State-space interchange heuristic	24 demands # veh as output 24 depots	51 system sec IBM 370/155

Table 4-6 Computational experiences

routing problems. In all the reported computational results, the number of subtour and chain-barring constraints that were required was significantly smaller than the total number of such constraints in the original formulation. Next in line in the set of analytical techniques is dualization, as illustrated by Benders' decomposition approach to solving the *s/m/–/a* problem: Although this may not be as well developed as the relaxation approach mentioned above, it has great promise as evidenced by recent developments in Lagrangian relaxation and decomposition techniques. Experience can also be drawn from the network-with-side-constraint models of relaxation as discussed in the "Facility Location" chapter and picked up again in the new trial-solution of Laporte's *m/m/–/a* model above. Magnanti and Wong (1990) made an assessment of the role decomposition methods play in facility-location problems. Their assessment has some bearing upon location-routing problems. Experiences show that Benders' decomposition is not competitive for solving large-scale p-median problems. Convergence problems were encountered, particularly with larger problems. Combining Dantzig-Wolfe[19] and Benders' decomposition techniques, on the other hand, has been more encouraging with the capacitated plant-location problem. The selection of an appropriate mixed integer programming model formulation is vital to the success of Benders' methods. Often, special side-constraints associated with the problem were exploited in an *ad hoc* fashion during decomposition to solve the problem efficiently. Such experiences, however, cannot be generalized to other problems.

Since the days of Clarke-and-Wright (1964), heuristics for routing and location have played a key role in solving realistic problems. While algorithmic designs differ in detail, they all share some fundamental characteristics. Not surprisingly, all algorithms start with an initial solution, followed by an improvement phase. The initial solution can be thought of as some kind of relaxation, say on vehicle capacity and depot capacity, which are re-introduced later. During this phase, some kind of synthesis heuristic is used to construct a set of routes/tours *ab initio*. This may involve insertion of additional nodes to existing routes/tours to configure a 'good' set of initial routes/tours. During the improvement phase, there is invariably some kind of *perturbation* consisting of interchanges of nodes that help to perform 'hill-climbing' incrementally. Each algorithm defines a solution space that can be the nodes and arcs of the original network or a transformed space. It is in this space that decomposition is carried out. A `sweep' operation is used to examine all potential candidates for improvement, often on a global level. Invariably, two successive sweeps that fail to reveal new interchanges signal a local optimum. Heuristics are quite powerful techniques. They are responsible for the solution of the largest size problems to date. The largest problem solved is reported to be 50 demands, 4 depots, and up to 80 to 160 vehicles. It refers to the multi-depot/multi-route/hierarchical ($m/m/-/h$) category (see **Table 4-6**).

Srivastava and Benton (1990) have classified heuristics into three categories. The *Savings-drop heuristic* uses a modified version of the travel-time-savings heuristic developed by Clarke-and-Wright for routing vehicles. This isextended to the multi-depot situation using Tillman's (1969) approach. This approach is most popular and generally yields good solutions. The solution procedure simultaneously considers dropping depots and assigning demands to routes developed from open depots. It iterates until the final solution is obtained. The *savings-add* heuristic uses a scheme similar to the savings-drop (S-D) approach. It assumes that all feasible sites are closed depots initially and will go on to open them one by one. In a distribution system where only a few depots are to be found from a large set of candidate sites, this approach will be computational easier and faster than the savings-drop (S-D) approaches. We can think of Perl's $m/m/-/h$ model as an

---

[19]The Dantzig-Wolfe decomposition procedure can be thought of as the primal version of Benders' decomposition.

example of S-D and S-A algorithms. The *cluster-routing* heuristic has been developed based on the assumption that if demands are distributed in clusters, a solution procedure that constructs routes among clusters should be more efficient. The cluster-routing (C-R) heuristic identifies the desired number of clusters and sets the number of depots equal to the number of clusters. The depot is located at the centroid of the cluster. The routing is achieved by first selecting the demands to be included in a single route based on their polar coordinates, i.e., sectors are identified. Then an optimizing TSP solution is applied to obtain the best route (Jacobs-Blecha et al. 1992). We can think of Merrill's *s/m/–/p* model as an example of C-R algorithm, although the procedure is executed in a transformed space using the Sierpinski curve.

The simultaneous location-routing model was compared with the sequential model in which locations were decided first and routing second. Different data were also input, including different location-to-routing cost ratios, spatial-distribution-levels in terms of clustering of demands, and the number of depot sites. The test problems have from eight to 12 depot sites, 40 demands that are grouped into one to five clusters. The overall performance of the S-D model was found to be 6.19 percent better than the sequential model as measured by total cost. The S-D procedure performs better when there are few depot sites and a uniform distribution of demands and sites. The cost structure accounts for the most variance, the S-D procedure clearly is best for scenarios with a low cost-ratio, i.e., when depot cost is small in comparison to routing cost. The general performance of the savings-add (S-A) heuristic was only 2.67 percent better than the sequential method. The S-A heuristic varies more for low cost-structure compared to the medium and high cost-structures. For a given number of depot sites and for a given cost structure, the heuristic performs poorer on all spatial distributions of demand and in general, when compared to the S-D heuristic. The C-R procedure performs 6.45 percent better than the sequential model on the average. The C-R model performs especially well for high cost-ratios, i.e., when depot cost is high in comparison to routing cost. It also did better for clustered distribution of demands. Statistically speaking, the three procedures score slightly better than the sequential model, except for low cost-structures, where only the S-D procedure is significantly better. Srivastava and Benton (1990) suggest using a combination of the S-D and C-R location-routing procedures over the sequential model under different environmental conditions such as demand distribution and cost ratios. The S-A should not be used since the sequential model is simpler and it outperforms this procedure.

However, for certain instances, there is not a strong justification for a combined location-routing model. For example, a combined location-routing model may not be necessary in the ranking of candidate sites. This is likely to happen when demands are ubiquitous and the routes can be statistically approximated by Euclidean distances. This was borne out by the case study on locating the home base for flight-inspection aircraft in the "Facility Location" chapter. The triangular inequality $C_{ij} \leq C_{ik} + C_{kj}$ often forms the underlying determinant of location. When Euclidean measures are valid metrics for spatial costs, location-routing models can boil down to simply location models. On the other hand, there are situations when the triangular inequality is no longer upheld. This is the case for special forms of spatial-separation measures (including stochastic location-routing models). Then routing may become important. The reader is referred back to the "Facility Location" and "Measuring Spatial Separation" chapters for a review of the underlying fundamental issues. Suffice to say here that different spatial-cost functions give rise to very different location implications. Notice this is independent of solution methods, since the triangular inequality arises both in heuristic procedures (such as the Clark-Wright algorithm) as well as analytical procedures (such as the network pricing-scheme used in column generation[20].)

Recent experiments to test variants of the deterministic fixed route problem were performed (Haughton and Rao 1995). These variants include modified-fixed-route-with-demand-unknown and modified-fixed-route-with-demand-known, a-priori-fixed-route, and re-optimization. Several measures-of-performance were used to evaluate each of these stochastic routing strategies, including transportation-cost, customer-service or the percentage-demand-covered. In conducting these experiments, three hypotheses were examined:

a. Service-level is an inverse function of the ratio-of-the-variance-for-each-demand, probability-of-a-nonzero-demand, and the tolerable-route-failure-probability;

b. The transportation-cost-increases-to-redress-route-failure is a function of the ratio-of-the-variance-for-each-demand, probability-of-a-nonzero-demand, and the tolerable-route-failure-probability.

c. The unavoidable-portion-of-transportation-cost of the modified-fixed-route-with-demand-unknown problem is a function of the probability-of-a-nonzero-demand.

It was found the within experimental errors, the ratio-of-the-variance-for-each-demand, the probability-of-a-nonzero demand, and the tolerable-route-failure-probability together greatly affect performance. While additional experiments are warranted, future extensions to include other exogenous factors beyond these three are definitely recommended. Meanwhile, it was shown that hypothesis 1 needs additional testing. Hypothesis 2 was adequately substantiated, and hypothesis 3 was supported also, but at a weaker level than the second one. All-in-all, routing appears to have a significant bearing upon location decisions, particularly when stochasticity is considered.

Except for limited verification, the 'optimality' of most locating-routing heuristics is not known accurately. The reason is that most analytical techniques are still too feeble to tackle any realistic size problem to produce an optimal solution for comparison with heuristics. Verification is likely to be carried out on small-size problems only, since they are often the only problems solvable by analytical techniques. Laporte et al. found out, for example, that when vehicle capacities are

---

[20]A good example of such a pricing scheme is found in the "Optimization" appendix to Chan (2000), under the "network with side constraints" section.

constraining and fixed depot-location costs are significant, their exact solution algorithm is difficult to solve optimally. Perhaps the most exciting development today are heuristics that have theoretical error bounds. The use of the space filling curve in the solution of the PMTSFLP is an example, where a 25-percent error bound has been established for an asymptotically uniform demand pattern.[21] Other asymptotic error-bounds have been uncovered by Chan and Simchi-Levi (1998) in multi-echelon inventory-routing problems. Here warehouses serve only as coordinator of the deliveries to the retailers but never holds inventory. In a related research by Chan et al. (1994), inventory cost is charged only at the retailers but not at the warehouse. This constitutes the antithesis of the former problem since now the incentive is to store at the warehouse. It can be shown, surprisingly, that this latter model becomes a bin-packing problem that can be readily formulated as a stylized location problem. Research in placing error bounds on location-routing heuristics is currently very active. It may very well be the answer to the heuristic-verification problem.

4.9   Case studies of location-routing models

In this section we assemble three case studies to illustrate the use of combined location-routing models. The three case studies are selected to cover various typologies of location-routing problems. The first case study examines an *obnoxious facility-location* problem, where we illustrate two major concerns facing the designers of nuclear power plants. The first is the seismic risks that could have catastrophic consequences on human lives and the environment. The second is the opposition by the public to siting such plants near their communities. Here, we present a model to configure a nuclear power system consisting of plants and the distribution network, taken into explicit consideration the two concerns cited above. In the model, we quantify seismic risk as comprehensively as possible in terms of ground acceleration, duration of shaking and probability of occurrence. We also model the risk faced by the entire power distribution system, rather than just the nuclear power plant. Perhaps the most interesting part of the model is that it balances the utility-company's valuation on reliability and cost against the public's concern for safety. In this light, the multi-criteria optimization model could provide valuable information in the political process of siting nuclear power plants. Similar location analysis can be performed for other obnoxious facilities, such as coal-burning power-plants and sewage-treatment systems, where the not-in-my-backyard (NIMBY) syndrome is ever present.

The second case-study concentrates on the siting of distribution centers such as warehouses. Timely distribution of supplies from manufacturer to consumers lies in the heart of an industrialized economy. An integral part of this distribution network is a number of depots or warehouses that serve to channel shipment to its final destinations. The distribution can either be performed by a dedicated fleet of

---

[21]See the chapter on "Facility Location" for details.

company-owned vehicles or by for-hire carriers. Recent interests in just-in-time deliveries have revolutionalized the idea of warehousing and inventory control. Equally innovative is the partnership between carriers and manufacturers in such concepts as "Parts Bank," a warehouse service offered by time-sensitive carriers such as Federal Express. Here a warehouse is located in the central hub of the carrier's operation at Memphis, where manufacturers actually have warehoused a large stock of commonly requested parts which the carrier can fly to almost any section of the country at short notice. This minimizes the "risk of stockout" for many receivers. Available information suggests that the hub-and-spoke route-structure is economical for serving thin-density markets while the point-to-point system is better for high traffic-volumes. Perhaps the middle ground, consisting of a mini-hub (or hub overlay) system, is a viable alternative for most industries. The problem now becomes one of locating the central hub and regional hubs by which materials can be shipped expeditiously from the central to regional facility to local demands. The question boils down to: How many regional hubs are required? Can some existing regional hubs be closed to save costs? What are the implications of Defense Courier Service, a public-sector counterpart to small-package carriers such as Federal Express and United Parcel Service.

The third case study locates transportation terminals. Specifically, it examines the market entry and exit practices of air carriers, in both times of prosperity and economic slowdown. This is particularly cogent as governmental regulation on carriers changes, governing the freedom with which a carrier can provide service to a terminal. The production process of an airline requires a combined location-routing approach. A route that goes from A to B to C carries not only the "local traffic" from A to B in segment $A–B$, but also "through traffic" from A to C (via B), and possibly connect traffic from cities other than A, B or C. In such a context, when a corporate planner considers whether to serve a market, s/he must include in the marginal revenue not only the yield from the passengers that travel among the three cities A, B and C, but also the connect traffic. The cost of service depends on the traffic density. Higher density and long-haul markets have the distinct advantage of economy-of-scale. But lower-density short-haul markets can add to the density of the trunk lines by contributing their share of traffic. There is this network effect inherent in the production process of an air carrier that a combined location-routing model is critical for the judicious location of transportation terminals. The major cities served by American Airlines will be studied to illustrate how techniques discussed in this chapter can be used in making corporate decisions.

4.10   A model of obnoxious facility location[22]

A major concern in the siting of any obnoxious facility is the performance of that facility during an emergency, when the undesirable effects are felt at its worst. The nuclear-power-plant system is an example of such an obnoxious facility that is

---

[22]This study is based on a paper presented under Chan (1990a).

highly susceptible to damage from an earthquake, resulting in immeasurable risk to people in the vicinity. Many empirical models have been developed in the past to analyze the seismic damage inflicted upon a particular component of the power system, such as the reactor. This case study brings together and extends these previously developed concepts in a model that analyzes the reliability of the nuclear-power *system* as a whole, from the reactor to the power distribution system that deliver electricity to the population. Through this model, we could examine the merits of an entire network-configuration rather than a component at a time. The model can trade off between the serious environmental impact from a reactor damage against the inconvenience (yet environmentally benign consequence) of the disruption of a power line. Thus while it is safer to place a reactor far away from the population, the lengthy power-lines that finally deliver the electricity to the users are that much more costly not only in its installation, but its maintenance. This study represents an extension of the single-facility-pair/single-route/ multicriteria (*s/s/m/-*) model discussed in this chapter. The idea is to have a reactor deliver power to a population via a route in consideration of both cost and safety.

4.10.1 A location-and-routing model. Once a seismic "hazard coefficient" is defined for a node or link, we can go on to construct a location-and-routing model. The optimization model has two objectives. The first is to minimize the seismic hazard the plants and transmission system are exposed to:

$$\min \; y_1' = \sum_{i \in I+J} \sum_{j \in I+J} H_{ij} d_{ij} x_{ij} + \sum_{i \in I} H_i' z_i' + \sum_{j \in J} H_j' y_j \;. \tag{4-118}$$

Here H_{ij} is the hazard on link (i,j); H_i' is the hazard on substation i (transformer); H_j' is the hazard on plant j; d_{ij} is the link distance; x_{ij} is the electric flow in link (ij); $y_i = 1$ if substation is located at node i, 0 otherwise; and $z_j'=1$ if a plant is located at node j, 0 otherwise. Hazardousness is computed as a function of (a) the magnitude of ground-motion acceleration, (b) duration of shaking and (c) probability of occurrence (Donovan 1973, Estera and Rosenblueth 1964, Matsuda 1982, Chan 1990a). Notice the possible locations for power plants are limited to the candidate set J, while the substations (transformers) are specified among the set of nodes I. This first criterion function also happens to minimize the construction cost. The reason is that the cost of the power system is strictly proportional to the length of its power line and the number of substations built, given a fixed number of power plants are to be located.

Also worthy of note is the fact that the three components of the criterion function have different units and orders of magnitude as function is now formulated. Specifically, the first term, representing the disruption potential for links, is of a much larger magnitude than the other two terms, since it includes the product of distance and kilowatts. To compensate for this, the H_{ij} are divided by the anticipated link-flow x_{ij} and the average link-length C_{ij} before being input to the model: $h_{ij}=H_{ij}/x_{ij}C_{ij}$. Besides equalizing the magnitude among the three terms, we also impart physical interpretation to the criterion function y_1'. For a power-system component of unit mass, the terms now show the unit of (force × time), or a seismic force acting through a certain time-duration.

A second criterion function of the model seeks to locate the plants away from population centers due to actual or perceived risks: $\max \ y_2' = \sum_{i \in I \cup J} \sum_{j \in J} d_{ij} f_i y_j$, where f_i denotes the frequency of demand at node i. This criterion function is considered in conflict with y_1'; it places the power plant away from the population while y_1' tends to place them close by. This is typical of a *max-min* model found in obnoxious facility-location. It also represents a generalization of the *p*-center model by virtue of its routing decisions.[23]

A measure of the correlation between the two criterion functions can be computed by the angle (Steuer 1986)[24]

$$\alpha' = \cos^{-1} \frac{(\mathbf{c}^i)^T (\mathbf{c}^j)}{\|\mathbf{c}^i\|_2 \|\mathbf{c}^j\|_2} \tag{4-119}$$

where $\|\mathbf{c}\|_2$ is the l_2-norm of the criterion-vector $\mathbf{c}$ in the criterion function.[25] The smaller the angle, the more correlated the two functions are. A 90-degree angle is expected for orthogonal (or independent) criterion functions.

The constraints to this two-criterion-function model consist, first of all, a statement on the desired number of power plants to be built, *p*:

$$\sum_{j \in J} z_j' = p. \tag{4-120}$$

This constraint fixes the order-of-magnitude of plant-construction cost. A given combination of power plants would cost roughly the same irrespective of where they are built, given the remote location of these sites. A similar constraint may be written for the number of substations to be located, *q*:

$$\sum_{i \in I} y_i \leq q. \tag{4-121}$$

Notice the inequality puts the number of substations as a bounded variable that would be a function of the mileage of power-lines in the distribution system. If the capital cost of plant construction is taken as fixed, then the total cost of the power system is roughly proportional to the length of the power-lines.

A set of equations can now be written for electrical flow. First among these states that the electric flow into and out of a substation is limited by the capacity T_i'' at a substation *i*:

---

[23]For an explanation of "*p*-center" and "obnoxious facility" models, please consult the "Facility location" chapter.

[24]The dependence among criterion functions is discussed—among other references—in the "Multicriteria Decisionmaking" chapter (Chan 2000) under the "Independence among criterion function" section.

[25] See a discussion on l_p-norm in Chan (2000)'s "Multicriteria Decision-making" chapter for example.

A set of equations can now be written for electrical flow. First among these states that the electric flow into and out of a substation is limited by the capacity T_i'' at a substation i:

$$\sum_{j\in I+J} x_{ji} \le T_i'' y_i \qquad \forall i\in I$$

$$\sum_{j\in I+J} x_{ij} \le T_i'' y_i \qquad \forall i\in I. \tag{4-122}$$

A similar inequality can be written for the power-generation output of a plant:

$$\sum_{i\in I+J} x_{ji} \le \bar{P}_j z_j' \qquad \forall j\in J. \tag{4-123}$$

Flow continuity among non-power-plant nodes is insured by the 'conservation' equation considering a depletion of f_i kilowatts at a demand node i:

$$\sum_{j\in I} x_{ji} - \sum_{j\in I} x_{ij} - f_i = 0 \qquad \forall i\in I. \tag{4-124}$$

The model formulation is a mixed integer program (MIP) in that the electric flow x_{ij} is a continuous variable while z_i' and y_j are binary variables. Existing MIP codes can be used to solve this multicriteria optimization problem. More astute observers, however, will note that there is a special structure to the mathematical program. There exists little forcing/linking constraint between the routing variable x and the location variables y and z'. The only place where the three variables come together is in Equations **(4-122)** and **(4-123)**. This allows more efficient decomposition algorithms to be deployed (Steppe and Chan 1990).

4.10.2 The case study. Central California—with the metropolitan San-Francisco-Bay-area being the focal point—is used as a case study. This is a seismically active area. The power-line system provided by Pacific Gas and Electric Company (PG&E) is also quite self-contained as a "closed system" (at least in this study). A schematic of the area is shown in **Figure 4-20**. Also shown in the figure are links representing possible rights-of-way for power lines, as well as nodes representing candidate plant and substation locations. The network consists of 14 nodes and 18 links. The demands are primarily at nodes A and D, where A is downtown San Francisco and D is the Southern Bay region of Palo Alto, San Jose and Santa Clara. The demands are placed at 2200 KW for each node. Two power plants are to be sited, each with a generating capacity of 2200 KW. In this case, the supply of power is exactly equal to demand, with no excess capacity. Finally, we limit the number of substations to be less than 8, and each with a capacity of 2200 KW.

Two possible sources for earthquakes are considered. The first is the epicenter of the San Andreas Fault, located just north of the city. Available information places the magnitude at 8.5, a focal depth—or depth below the ground surface—of 10 km (6.25 mi), a recurrence interval of 100 years (Wesson et al. 1975) and a duration of 12 seconds. A second fault is located near Humboldt Bay, with similar

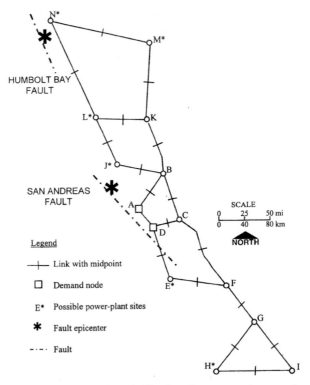

Figure 4-20 Obnoxious-facility-location case-study network

parameters as the San Francisco earthquake except the recurrence interval is 1000 years. An assumption is made here that the hazard indices are additive—i.e., a facility located between both the San Andreas and Humboldt Bay faults is exposed to the combined hazards of both faults. The slant distance from an epicenter to a facility such as a link or node is required for computing the hazard index. The slant distance to a link is measured to its mid point.

Based on these input coefficients, the angle α' of Equation **(4-119)** is calculated to be 76.4 degrees. Thus the two criterion functions are mostly independent of one another. As a "rule of thumb," an α' between 70 and 90 degrees suggests independence between the criterion functions. When α' is between 69 and 50 degrees, correlation could be a hindrance. An angle less than 50 degrees spells caution, and suspect results could be produced.

4.10.3 Solution. As we recall, the MIP model has two conflicting criterion functions. One seeks to minimize the distance between the demand and supply of power, while the other maximizes the distance. A single criterion function can be formulated to combine these two conflicting ones:

$$\min \ z = \lambda' y_1' - (1 - \lambda') y_2' \qquad\qquad \textbf{(4-125)}$$

Configuration	λ'	Plant Locations	Locations	Power Lines
I	1.0			
	0.9			
	0.7	M,E	K,B	MK,KB,BA,DE
	0.6			
	0.5			
	0.4			
II	0.35	M,H	K,B,E,F,G	MK,KB,BA, DE,EF,FG,GH
	3			
III	0.25			
	0.2	N,H	B,E,F,G,K, M	BA,DE,EF, FG,GH,KB, MK,MN
	0.1			
	0.01			
IV	0.0	N,H	L,J,B,C,F,G	NL,LJ,JB, BA,DCFC ,GF,HG

Table 4-7 Non-inferior solutions without redundant demand-coverage

where λ' is a positive weight between 0 and 1. The solution strategy involves finding a noninferior-solution set corresponding to a selection of λ' values ranging from 0 to 1, rather than a single optimal solution. (For more discussions of this methodology, see the chapter on "Multicriteria Decision-making" in Chan (2000) for example.) Solutions to this model were generated using the Mathematical Programming System (MPSX) on the AMDAHL 470. 5.5 CPU seconds are all that were required to obtain the solutions contained in **Table 4-7**. Further experimentation was conducted on LP/MIP83 (now XA), a mathematical-programming software for IBM-compatible personal-computers, GAMS (General Algebraic Modeling System) and ADBASE (a multi-criteria linear programming model by Steuer (1986)). An example input file (Steppe and Chan 1990) corresponding to Equations **(4-118)** through **(4-124)** is shown as **Figure 4-21**.

Runs with weights on the hazard/cost criterion-function ranging from 1.0 to 0.4 result in an identical power-system-configuration I as shown in **Table 4-7**. With little or no consideration for population proximity (y_2'), the selected plants, E and M, are not the nearest to the demand. The chosen locations reflect the seismic

Configuration	λ	Plant Locations	Substation Locations	Power Lines
	1.0			
	0.9			
	0.8			*DA(E)*,ED(E),*
I	0.7	E,H	C,D,F,G	*DA(H),CD(H),FC(H)*
	0.6			*GF(H),HG(H)*
	0.5			
	0.4			
	0.3			
	0.2			*DA(H),ED(H),FE(H)*
II	0.1	H,N	B,C,D,E	*GF(H),HG(H)*
	0.05		F,G,K,M	*BA(N),BC(N),CD(N)*
				MK(N),NM(N),KB(N)
				DA(H),ED(H),FE(H),
III		H,N	B,C,D,E	*GF(H),HG(H),*
	0.0		F,G,J,L	*BA(N),BC(N),JB(N)*
				LJ(N),NL(N),CD(N)

*The notation $DA(E)$ means electric flow on Link DA from power plant at E.

Table 4-8 Non-inferior solutions with redundant demand coverage

hazard at nodes J and L that are the closest sites to the population. The next cluster of runs with weights ranging from 0.35 to 0.3 for y_1', result in configuration II as shown in H is the plant location furthest away from the two faults and therefore most free from seismic risk. It is also far away from the population. The third configuration III comes from run cluster where the weights on the hazard/cost criterion-function range from 0.25 to 0.01. Power plants are located at H and N, the furthest sites away from the population. This is dictated by the large penalty placed on proximity—although this necessarily means N is now quite close to the Humboldt Bay fault. The reader will notice that links $N–L$ and $L–J$ were never used through all the runs thus far even though they are the 'shortcuts' between demand and supply. This is possibly due to the hazardous nature of these two power line corridors, which are close to the two faults. The fourth and last configuration IV,

```
Z1 PORTION OF OBJECTIVE FUNCTION*
PART 1 HAZARD AT LINKS*

&W (1.19 Xba + .75 Xbc + .75 Xcb + .59 Xad + .59 Xda +
.78 Xjb + .63 Xkb + .36 Xcd + .36 Xdc + .68 Xfc + .56 Xed
+ .40 Xef + .40 Xfe + .17 Xgf + .26 Xhg + .19 Xig + .22 Xhi
+ .68 Xjl + .68 Xlj + .44 Xlk + .44 Xkl + 1.02 Xnl + .51 Xmk
+ .61 Xnm +

PART 2 HAZARD AT SUBSTATIONS *
mixed integer restriction of double brackets retricts *
all variables to be zero/one variables.  i.e. only one *
plant or substations can be located at a given site. *

[[ 9.27 Sa + 4.01 Sb + 2.88 Sc + 3.72 Sd + 2.17 Se +
1.27Sf + .95 Sg + .80 Sh + .63 Si + 5.02 Sj + 2.23 Sk +
2.19Sl + 1.20 Sm + 1.29 Sn +

PART 3 HAZARD AT POWER PLANTS *

2.17 Pe + .8 Ph + 5.02 Pj + 2.19 Pl + 1.2 Pm + 1.29 Pn]])

Z2 PORTION OF OBJECTIVE FUNCTION *
mixed integer restriction of double brackets retricts *
all variables to be zero/one variables.  i.e. only one *
plant or substations can be located at a given site. *
SAFETY DISTANCE from population demand centers*

- &Z ([[1.5 Pe + 4.25 Ph + 1.49 Pj + 2.85 Pl + 4.42 Pm +
5.51 Pn]])
```

	Maximum number of plants is 2 *
MAXPLANT:	Pe + Ph + Pj + Pl + Pm + Pn = 2

	Maximum number of substations is 8 =
MAXSUBST:	Sa + Sb + Sc + Sd = Se + Sf + Sg + Sh + Si + Sj + Sk + Sl + Sm + Sn <= 8

Plants and substations cannot be colocated *

```
SITE E:    Se + Pe <=1
SITE H:    Sh + Ph <= 1
SITE J:    Sj + Pj <=1
SITE L:    Sl + Pl <=1
SITE M:    Sm + Pm <=1
SITE N:    Sn + Pn <=1
```

Flow out of a poser plant is limited by output of plant*
Flow out of a substation is limited by capacity of a * substation*

```
OUTFLO A:    Xad - 2.2 Sa = 0
OUTFLO B:    Xba + Xbc - 2.2 Sb = 0
OUTFLO C:    Xcb + Xcd - 2.2 Sc = 0
OUTFLO D:    Xda + Xdc - 2.2 Sd = 0
OUTFLO E:    Xed + Xef - 2.2 Se - 2.2 Pe = 0
OUTFLO F:    Xfc + Xfe - 2.2 Sf = 0
OUTFLO G:    Xgf - 2.2 Sg = 0
OUTFLO H:    Xhg + Xhi - 2.2 Sh - 2.2 Ph = 0
OUTFLO I:    Xig - 2.2 Si = 0
OUTFLO J:    Xjb + Xjl - 2.2 Sj - 2.2 Pj = 0
OUTFLO K:    Xkl + Xkb - 2.2 Sk = 0
OUTFLO L:    Xlj + Xlk - 2.2 Sl - 2.2 Pl = 0
OUTFLO M:    Xmk - 2.2 Sm - 2.2 Pm = 0
OUTFLO N:    Xnl + Xnm - 2.2 Sn - 2.2 Pn = 0
```

Flow into a substation is limited by capacity of substation*

```
INFLOW A:    Xba + Xda - 2.2 Sa = 2.2
INFLOW B:    Xcb + Xjb + Xkb - 2.2 Sb = 0
INFLOW C:    Xdc + Xfc + Xbc - 2.2 Sc = 0
INFLOW D:    Xad + Xcd + Xed - 2.2 Sd = 2.2
INFLOW E:    Xfe - 2.2 Se = 0
INFLOW F:    Xef + Xgf - 2.2 Sf = 0
INFLOW G:    Xhg + Xig - 2.2 Sg = 0
INFLOW I:    Xhi - 2.2 Si = 0
INFLOW J:    Xlj - 2.2 Sj = 0
INFLOW K:    Xlk + Xmk - 2.2 Sk = 0
INFLOW L:    Xjl + Xkl + Xnl - 2.2 Sl = 0
INFLOW M:    Xnm - 2.2 Sm = 0
```

Flow continuity / conservation of flow into and out of a node *

```
CONSER A:    Xba + Xda - Xad = 2.2
CONSER B:    Xcb + Xjb + Xkb - Xba = 0
CONSER C:    Xbc + Xdc + Xfc - Xcb - Xcd = 0
CONSER D:    Xad + Xcd + Xed - Xda - Xdc = 2.2
CONSER E:    Xfe - Xed - Xef + 2.2 Pe = 0
CONSER F:    Xef + Xgf - Xfc - Xfe = 0
CONSER G:    Xig + Xhg - Xgf = 0
CONSER H:    Xhi + Xhg - 2.2 Ph = 0
CONSER I:    Xig - Xhi = 0
CONSER J:    Xlj - Xjl - Xjb + 2.2 Pj = 0
CONSER K:    Xlk + Xmk - Xkb - Xkl = 0
CONSER L:    Xnl + Xkl + Xjl - Xlj - Xlk + 2.2 Pl = 0
CONSER M:    Xnm - Xmk + 2.2 Pm = 0
CONSER N:    Xnm + Xnl - 2.2 Pn = 0
```

Figure 4-21 Nuclear-power-plant siting model

generated from a weight of zero on y_1', considers only the proximity criterion-function without regard for the hazard/cost criterion-function. An examination of **Table 4-7** shows that plants are at H and N again, which are the furthest away from the population centers on the network.

Considering the runs as a whole, it can be seen that bifurcation occurs when the weight λ' takes up values of approximately 0.35, 0.25, and 0.005. The most sensitive political balance occurs at the weighting range of 0.3 to 0.35. This shows that nuclear-power system plans I and II can be acceptable to a fairly broad spectrum of political persuasions. Plan II is appealing only under a very special power-structure where the balance is approximately one-third "industrial" and, two-thirds "environmentalist." All these analyses consider the units used in the y_1' and y_2' (Hobbs 1980). One can say that the plans I and III are more 'robust' than plan II. They are more likely to "stand the test of time" as political affiliations drift over the years in the study area.

Compared with regular imp vector-optimization approaches, the *weighted sum* procedure above does not necessarily generate all the non-inferior extreme points. But it is a practical way to "map out" the general feature of the non-inferior frontier for an MIP. Through extensive computation, it was found that the non-inferior set is quite sensitive to the 'cost' vectors **c** in the criterion functions. Slight changes in the spatial-distance measures, for example, shift the non-inferior frontier. The results presented in **Table 4-7** represent the most representative set of non-inferior extreme points after extensive computational effort using various software packages.

4.10.4   Redundancy of power supply. In any power supply, a critical consideration is redundancy of coverage ,which ensures that backup electricity is available in case of a power failure. Toward this end, we have reformulated the model to include such redundancies. Notice this model is intended for planning rather than operational purposes. Toward this goal, we will ignore the transient effects typical of a power outage (Schiff 1981). These include operational policies such as reduced load, open branch or split node (opening a segmented bus). Here we assume redistribution of generation is the principal and normal procedure. Then we refer to the alternate power-plant to back up the demand left unserved by the disrupted power-supply from the primary generation-source. This assumption is considered valid under a scenario where two power-plants are theoretically available.

The model to include redundancy looks very similar to the basic model discussed above, however, the number of decision variables has been increased due to the *multicommodity-flow* formulation[26]. In other words, all link flows are identified by the supply sources, each of which is labelled a separate 'commodity.' An additional set of constraints specifies that a given power plant supplies each

---

[26]For an introduction to a multicommodity formulation, the reader can go to the "Optimization" appendix in Chan (2000).

demand with part of its power. The new model now looks like the following, plus Equations **(4-120)** and **(4-121)**:

$$\min \ y_1' = \sum_{k \in J} \sum_{i \in I+J} \sum_{j \in I+J} H_{ij} d_{ij} x_{ij}^k + \sum_{i \in I} H_i' z_i' + \sum_{j \in J} H_j' y_j \tag{4-126}$$

$$\max \ y_2' = \sum_{i \in I+J} \sum_{j \in J} d_{ij} f_i y_i \tag{4-127}$$

$$s.t. \quad \sum_{k \in J} \sum_{j \in I+J} x_{ji}^k \le T_i'' y_i \qquad \forall i \in I \tag{4-128}$$

$$\sum_{k \in I} \sum_{j \in I+J} x_{ij}^k \le T_i'' y_i \qquad \forall i \in I \tag{4-129}$$

$$\sum_{k \in J} \sum_{i \in I+J} x_{ji}^k \le \overline{P}_j z_j' \qquad \forall j \in J \tag{4-130}$$

$$\sum_{k \in J} \left[\sum_{j \in I} x_{ji}^k - \sum_{j \in I} x_{ij}^k \right] - f_i = 0 \qquad \forall i \in I \tag{4-131}$$

$$\sum_{j \in I+J} x_{ji}^k < f_i \qquad \forall i \in I, \ \forall k \in J. \tag{4-132}$$

The above constraint **(4-132)** forces each qualifying power-plant to supply part of its power to each demand at node i. In case i is cut off from the supply source k, it could be covered by k'. In other words, a power plant k or k' cannot be the sole source of supplying electricity to demand i. (Notice this constraint may give extraneous, yet benign, flow assignments when y_1' is given zero weight, but these can be screened out easily by inspection.) The solution of this model is shown in **Table 4-8**. Solution (I) represents the outcome of emphasizing the importance of reducing risk and cost to the utility rather than environmental concerns. Both plants are located in the southern network as a result, placing them not too far from the population. Solutions (II) and (III) reflect increasing environmental concerns, resulting in a more distant location of plants at H and N. Power routing changes from (II) to (III) while plant locations remain the same.

4.10.5 Comparison of results from the two models. Putting together the results from the models with and without redundancy, some very interesting observations can be made. Node H at Diablo Canyon appears to be a most pervasive location for siting one of the two power plants. It is favorable from both seismic-hazard, system-reliability, construction-cost and environmental standpoints. It is not surprising that PG&E is—at the time of the study—considering locating a plant at this site (as will be elaborated on in the next section). In the model with redundancy coverage, it is the all-time favorite irrespective of political persuasion as represented by the weights. On the other hand, site N or the Humboldt-Bay location only makes sense if proximity to population is of major concern (criterion function y_2'). It is not justifiable from the utility-company's point of view—considering both seismic hazard and overall costs (criterion function y_1'). It comes as no surprise that PG&E is closing the plant when negative seismic-hazard information surfaced. Location E is an acceptable choice for a second plant on top of Diablo Canyon. It represents a compromise location from both the utility's and the population's points-of-view—perhaps with a slant toward the utility's interest.

It should be noted that we have set aside power-system redundancy from the Western Systems Coordinating Council, which covers the adjoining states of California, Oregon, Washington, Idaho, Nevada, Arizona, Utah, Wyoming, Montana, Colorado and New Mexico. The models—both with and without redundancy—only reflect PG&E's framework. Thus the results from both models provide useful information for locating plants only when power supply from other states cannot be relied upon. Naturally, the model used above can be generalized to include redundancy from the Western Systems Coordinating Council if necessary.

4.10.6 Lessons learned. In the late 1970s, Pacific Gas and Electric (PG&E) closed the Humboldt-Bay nuclear-power-plant (located at node N). The plant was old and in need of major renovation. Geologic studies revealed that there were faults near the plant (beyond the major Humboldt-Bay fault). Considering the new seismic information that was not available at the time of original plant construction, PG&E decided the financial investment required to renovate the Humboldt-Bay plant was too much. Instead, PG&E examined a new plant at Diablo Canyon (node H). Naturally, questions were raised regarding its seismic safety in light of experiences with the Humboldt-Bay plant. A lot has happened since the original analysis was performed. A previous unknown fault was discovered which caused significant increase in cost. The model presented in this paper could be used to provide such locational analysis before construction and the concomitant capital investment. Obviously, in such applications, a complete list of faults—instead of just two major ones as shown in this study—has to be included in the model.

Another useful feature of the model is that it allows one to examine the cost, reliability and risk of a power-distribution network as a whole. This includes plants, substations and power lines, rather than one component at a time as in current practice. A re-examination of the process leading toward the results in **Table 4-7** and **Table 4-8** would show that the seismic risk of each network configuration (as represented by criterion y_1') is an important factor in the model. y_1' was defined as the product of ground-motion acceleration, duration of shaking and probability of an earthquake. It is a more comprehensive approach than using acceleration surrogates such as the Richter magnitude alone. Since the hazardousness model above considers a fuller spectrum of earthquake parameters and network redundancy in power supply, it is a useful extension of existing component-oriented measures of risk such as Operating Basic Earthquake (OBE) and Safe Shutdown Earthquake (SSE). This way, we are not only concerned with the reliability of a single plant alone, but a system of plants and substations serving the study area throughout the distribution network. Thus a disruption of power to the user can be traceable not only to the failure of a plant, but also to the substation or power line.

Perhaps the most interesting part of the model is the built-in tradeoff between the two criterion functions. The tradeoff is between cost-and-reliability to the utility company and the. risk to the population. Such an analysis helps to clarify a political decision, where a balance must be made between the focus of the utility company and the concerns of the public. Obviously, improvements can be made to the model in several areas. First, the model in its present form does not consider power loss that occurs in lines and transformers. Nor does it consider transient operational procedures during a power outage. Including such electrical-flow properties will transform the model to a nonlinear program (with integrality constraints), which may require more specialized solution techniques[27]. Second, other effects of an earthquake—those beyond acceleration and duration—have been neglected. These hazards, which may include floods, landslides, searches, ground rupture and liquefaction, are very much dependent upon the geotechnical conditions of the sites where the power plant, substations and power lines are located.

From an algorithmic standpoint, the model solution can obviously be further refined. A more systematic procedure of mapping out the non-inferior frontier is desirable. For example, interactive procedures with rigorous convergence properties such as the "interactive weight-sums/filtering" approach (Steuer 1986) may be desirable. The interactive Frank-Wolfe procedure, for example, can also assist in defining the underlying value function that represents the dynamics of pluralistic decision-making. It also helps in exploring the efficient frontier for consensus building among stakeholders. A Frank-Wolfe example is illustrated in the "Multicriteria Decision-making" chapter for airport location in which one decision maker is involved. But the concept can conceivably be generalized to two or more decision-makers. In short, experiences with the current solution procedure and this new interactive approach can be used to solve other obnoxious facility location problems. Here the tradeoff between stakeholders lies in the heart of the analyses.

4.11   Depot location in commodity distribution

In the distribution network of industry or government, a manufacturer or central store ships supplies to regional depots, from where they are in turn delivered to the consumers. Many problems of this kind involve the use of a trunk carrier to deliver

---

[27]For an introduction to nonlinear programming, refer to the chapter on "Prescriptive tools for analysis" in Chan (2000) for example.

supplies to regional depots, from where local distribution takes place via smaller vehicles. This is has been referred to as a *hierarchical* model earlier in this chapter. In this model, the first level consists of a central facility that ships commodities to regional depots, and the next level consists of regional depots delivering these commodities to consumers. Oftentimes, hundreds of nodes are involved. In this study we show how to solve this combined multi-facility/multi-tour/hierarchical $(m/m/-/h)$ and multi-facility/multi-tour/allocation $(m/m/-/a)^{28}$ problem in two practical steps. The regional depots are located first, from which fleets of smaller vehicles are routed to service the demands. Then the trunking costs by larger vehicles are added back to the system for the final evaluation of the entire distribution hierarchy, from central depot to the demands. Unlike more `elegant' schemes, this approach is shown to yield solutions well within the capabilities of today's regular computers. Here, we employ a case study of the huge Defense Courier Service—a counter part of the small-document express-service in industry.

4.11.1   Model formulation. Laporte et al. (1986) put forth a compact formulation of a multi-depot, multi-tour model. It describes the regional-depot location and local-distribution of the hierarchical problem, setting aside the central supply and trunking costs as described above. We will briefly review its discussion for convenience. The reader is referred to the section entitled "Multiple-Facility/Multiple-Tour/Allocation Model" for a more pedagogic development of the model, particularly Equations **(4-47)** through **(4-55)**.

4.11.1.1 VEHICLE RANGE EXTENSION. Aside from a huge dimensionality, there is another shortcoming of the Laporte et al. formulation. In the context of the Defense Courier Service (DCS), the "vehicle range" constraint, as specified by the range symbol U, is missing. To overcome this, a combined subtour-breaking and range constraint advocated by Kulkarni and Bhave (1985) is used to supplement the formulation:

$$Ux_{ij} + \sigma_i - \sigma_j \leq U - d_{ij} \qquad \forall\, |J| + 1 \leq i \neq j \leq |I| \,. \tag{4-133}$$

The set $\{|J|+1, |J|+2, \ldots, |I|\}$ represents the set of non-depot (demand) nodes, σ is an unrestricted variable associated with each node that indicates how much range capability remains in the associated tour. One can think of σ as some kind of 'odometer' reading on the vehicle, recording the mileage. The above range constraint simplifies to the following form when an arc (i,j) is used, or when $x_{ij}=1$: $\sigma_i + d_{ij} \leq \sigma_j$, which states that the odometer reading at a demand node must be incremented by at least the distance separating it from the prior node on the tour. If the arc (i,j) is not used, the equation's right-hand side becomes large relative to the left-hand side, and the constraint is non-binding.

---

[28]For an explanation of the $m/m/-/h$ and $m/m/-/a$ designations, please consult the section entitled "Multiple-Facility/Multiple-Tour/Allocation Model and the section entitled "Multi-Facility/Multi-Tour/Hierarchical Model" of this current chapter.

Kulkarni's range equation requires x_{ij} to be distinct from x_{ji}. Otherwise, the incrementing from i to j is nondirectional, and a running-total of mileage is not enforced according to Equation **(4-133)**. This complicates Laporte's previously symmetrical formulation by nearly doubling the number of variables, as well as forcing more rigorous conservation-of-flow equations. Specifically, Equations **(4-48)** and **(4-49)** now become

$$\sum_{i\in I} x_{ij} = \begin{cases} m_j & \text{for } j\in J \\ 1 & \text{for } j\in I-J \end{cases} \quad and \quad \sum_{j\in J} x_{ij} = \begin{cases} m_i & \text{for } i\in J \\ 1 & \text{for } i\in I-J. \end{cases} \tag{4-134}$$

Additionally, the chain-barring constraints must be altered to reflect the asymmetric formulation.

$$x_{i_1 i_2} + x_{i_2 i_1} + 3\,(x_{i_2 i_3} + x_{i_3 i_2}) + x_{i_3 i_4} + x_{i_4 i_3} \le 4 \qquad i_1, i_4 \in J; \ i_2, i_3 \in I-J$$

$$x_{1_1 i_2} + x_{i_2 i_1} + x_{i_{g-1} i_g} + x_{i_g i_{g-1}} + 2 \sum_{i,j \in i_2,\dots,i_{g-1}} x_{i,j} \le 2g-5 \qquad g\ge 5; \ i_1, i_g \in J; \ i_2,\dots,i_{g-1}\in I-J. \tag{4-135}$$

The resulting formulation is a binary mixed integer program, excluding the case when $x_{ij}=2$. If $i_1, \dots, i_h$ is a directed path from depot i_1 to depot i_n through demands $i_2, \dots, i_{n-1}$, then these constraints simplify to $\sum_{k=1}^{n-1} x_{i_k i_{k+1}} \le n-2$.

When we incorporated Kulkarni's range constraint into Laporte's model, the other problem we ran into is that Kulkarni's method ignores distance to and from the depot. This problem was discussed in the chapter on "Measuring Spatial Separation." $d_{ij} x_{ij} \le \sigma_j$ and $d_{ji} x_{ji} + \sigma_j \le U(1)$ in the "Vehicle-routing problem" subsection, when used with Kulkarni's, sufficiently restrict range and complete the proposal originated by Kulkarni.

4.11.1.2 SERVICING-FREQUENCY EXTENSION. The need to service a demand node more than once has to be addressed. The most obvious way to allow for different servicing-frequencies is to relax the binary mixed integer program into a simple mixed integer program. This would allow an arc flow greater than one, which permits a node to be served more than once. Because of a subtour-breaking constraint such as Equation **(4-50)** or its substitute Equation **(4-133)**, such repeat servicing would have to occur on separate tours. This necessitates reformulation of conservation-of-flow equations **(4-134)**, setting the right-hand-side equal to $\underline{s}_i$ and $\underline{s}_j$ for nodes i and j that are demand nodes:

$$\sum_{i\in I} x_{ij} = \underline{s}_j \quad j\in I-J; \qquad \sum_{j\in J} x_{ij} = \underline{s}_i \quad i\in I-J \tag{4-136}$$

where s is the specified frequency-of-visit. Closer examination of this prescription, however, reveals one significant shortfall: the range variable U in Equation **(4-133)** would no longer be multiplied by a *binary* variable x_{ij} and would no longer be enforcing the correct 'mileage' for the 'odometer' variables σ.

An alternate way to introduce multiple service-frequency—as mentioned in the chapter on "Measuring Spatial Separation"—is to split a node into a group of co-located nodes, between which no arcs exist. Thus a node i would become $i_1, i_2, \dots,$

i_s, where $\underline{s}$ is again the prescribed frequency-of-visit. To force all visits within a node group to occur via different tours, all nodes of a group are assigned only one 'odometer' variable σ (which can be distinguished by individual aircraft visit 1, 2, ... , $\underline{s}$). This prohibits a tour from serving any group of nodes more than once, effectively forcing as many tours as there are nodes in a given group. In other words, a tour cannot enter the group of split nodes, exit from the group, and then re-enter the group. Suppose demand node j is split into $\underline{s}$ nodes, $j_1, \ldots , j_{\underline{s}}$, representing $\underline{s}$ frequency-of-visit at the node. Equation **(4-133)** is now modified to read

$$Ux_{i,j_1} + \sigma_i - \sigma_j^1 \le U - d_{i,j_1}$$

$$Ux_{i,j_2} + \sigma_i - \sigma_j^2 \le U - d_{i,j_2}$$

$$\cdot$$
$$\cdot \qquad\qquad \forall |J| + 1 \le i \ne j \le |I| . \qquad\qquad \textbf{(4-137)}$$
$$\cdot$$

$$Ux_{i,j_{\underline{s}}} + \sigma_i - \sigma_j^{\underline{s}} \le U - d_{i,j_{\underline{s}}}$$

Similarly, if demand node i is split into $i_1, \ldots , i_{\underline{s}}$, it now reads as follows for the first and last leg of the tour:

$$d_{i_1,j} x_{i_1,j} + \sigma_i^1 \le U$$

$$\cdot$$
$$\cdot \qquad\qquad |J| + 1 \le i \le |I|, 1 \le j \le |J| . \qquad\qquad \textbf{(4-138)}$$
$$\cdot$$

$$d_{i_{\underline{s}},j} x_{i_{\underline{s}},j} + \sigma_i^{\underline{s}} \le U$$

Remember that the following equation, when used with Kulkarni's, sufficiently restrict range and complete the proposal originated by Kulkarni:

$$d_{ji} x_{ji} \le \sigma_i \qquad 1 \le j \le |J|, |J| + 1 \le i \le |I| \qquad\qquad \textbf{(4-139)}$$

Equation **(4-139)** is correspondingly modified to read

$$d_{j,i_l} x_{j,i_l} \le \sigma_i^l \qquad 1 \le j \le |J|; \; |J| + 1 \le i \le |I|; \; 1 \le l \le \underline{s} . \qquad\qquad \textbf{(4-140)}$$

As we notice, nodes have different service frequencies, meaning that all the routes flown in a planning period are not identical. While it might be optimal to serve a node early in a route for one service, it might not be optimal to retain that order for route with fewer (or more) stops. One possible way to redress this situation is by explicitly imposing nonnegative time-windows of visit for all $\underline{s}$-split demand nodes i: $\lambda_i^l \le \sigma_i^l \le u_i^l$ ($i = 2, 3, \ldots , |I|; \; 1 \le l \le \underline{s}$). Let σ_{ij}^l be the time between arrival at $\underline{s}$-split node i to arrival at node j, including the wait time at split node i, we impose the following constraints on the model (Baker 1983).

$$\sigma_i^l - \sigma_1 \geq \sigma_{i_l,1} \qquad i = 2, 3, \ldots, |I|; \ 1 \leq l \leq \underline{s} \tag{4-141}$$

$$|\sigma_i^l - \sigma_j| \geq \sigma_{i_l,j} \qquad i = 3, 4, \ldots, |I|; \ 2 \leq j < i; \ 1 \leq l \leq \underline{s} \tag{4-142}$$

$$\sigma_{|I|+1} - \sigma_i^l \geq \sigma_{i_l,|I|+1} \qquad i = 2, 3, \ldots, |I|; \ 1 \leq l \leq \underline{s}. \tag{4-143}$$

Now $d_{i_k,j}$ is replaced by $\sigma_{i_k,j}$ where there is a $\underline{s}$-split node i.

A numerical example of such formulation will be given in the "Validation" section under the subsection entitled "Experiments". Thus Kulkarni's constraint now is modified sufficiently to introduce multiple service-frequency, as well as subtour-breaking and range-limitations.

4.11.1.3 SUMMARY. Aside from the four above equations, the complete multiple-depot, multiple-travelling-salesmen facility-location model now consists of Equations **(4-47)**, **(4-134)**, **(4-137)**, **(4-138)**, **(4-53)**, **(4-54)**, **(4-55)**, **(4-136)** as well as the asymmetric modifications of **(4-135)**. The model can be solved by any mixed integer programming code, in which x_{ij} is binary, σ_i is continuous, and m_r and y_r are integers. It is worth repeating that this is an asymmetric formulation, rather than retaining the symmetry exploited by Laporte. Also, vehicle-capacity constraint is not part of the formulation, although its inclusion is straightforward. We call this complete model the Multiple-Depot Multiple-Travelling-Salesmen Facility-Location Problem (MDMTSFLP)

Admittedly, only a limited-size problem can be solved using this formulation. But it is a complete model that overcomes two key shortcomings—range and frequency—of the existing formulation of this problem, as mentioned above. In the author's opinion, such extensions are nontrivial in both conceptual and application efforts. For example, it is suitable for the validation of heuristic methods typically used to solve more realistic-size problems. For the first time, the model formulated above can measure the performance of a heuristic, albeit in a small problem. Only after such an assessment can a heuristic be employed in the field with some confidence level about its serviceability. An example of this validation model (Baker 1991) is shown in the software on my web site and the CD attached to Chan (2000). It is also illustrated in subsection entitled "Experiments."

4.11.2 Heuristic solution. Solving the travelling-salesman-problem using a spanning tree has been explained in the "Measuring Spatial Separation" chapter, where the linkage between minimum-spanning-tree (MST) and the travelling-salesman-problem (TSP) was pointed out. Based on this, we investigated the use of *Minimum Spanning Forest* (MSF) to extend the idea to a multiple-salesmen problem based at multiple depots. Prim's spanning-tree algorithm (Syslo 1983) successively finds the shortest arc that adds another node into the current collection of nodes and arcs. The spanning forests expand this idea to several unconnected node-arc collections. The $O(|I|^3)$ algorithm is a modified nearest-neighbor heuristic, and the procedure for forest selections is as follows:

Step 1 Select a combination of K depots for branching. Define this set as the nodes currently included in the forest.

Step 2 Admit the nearest node to the current forest unless the new branch connects that node to a depot that is greater than $U/2$ distance away. Here nearness is interpreted to mean weighted distance, where the demand serves as the weight.

Step 3 Repeat step 2 until all nodes are included in the forest. Save as the best solution if the weighted distance is the smallest yet found.

Step 4 Repeat steps 1–3 until all combinations of depots ($K \leq |J|$) have been explored, where $|J|$ is the maximum number of regional depots to be considered. ∎

This heuristic is similar to the ones adopted by Fisher (1994a), who presented a polynomial algorithm for the minimum, degree-constrained K-tree problem. His procedure is a generalization of the results of Glover and Klingman (1984) for finding degree-constrained minimum-cost spanning-trees. Consider a graph consisting of nodes I and edges $|\underline{A}|$. The algorithm first obtains a minimum-cost K-tree by generating a minimum-cost spanning-tree T and adding to it K least-cost edges from $|\underline{A}| - T$. If the resulting K-tree does not have degree K on node 1 (the home depot), he modifies it by a sequence of least-cost edge exchanges that increase or decrease the degree of node 1 as required. The key here is the greedy heuristic that considers the next least-cost edges.

In the above algorithm, depot selection will tend to favor proximity to multiple-frequency demand-points, unless the next 'hop' is further than the range U of the vehicle. As part of the algorithm, a demand node is weighted to reflect its required servicing-frequency. Thus despite the daunting size of the depot-selection process, the MSF coding is highly effective as it incorporates range and multiple-frequency weighting. Listing of a FORTRAN code is included in the software on my web site and the CD attached to Chan (2000). Once the optimal forest is chosen, the individual trees indicate depot-demand clusters. Although Held-and-Karp's MST-TSP transformation heuristic seems appropriate for use here, its application to a multiple-TSP with range restriction and multiple frequency is not straightforward. Since our focus has been on multiple tours and service frequencies, the Clarke-Wright (CW) heuristic appears to handle this requirement much more readily, as we will explain.

Araque (1989), as cited by Naddef (1994), suggested that the Vehicle Routing Problem (VRP) amounts to finding a partition of the nodes into node-disjoint paths of total minimum δ,[29] where δ represents the CW savings:[30]

$$\delta_{ij} = D_{ik} + D_{jk} - D_{ij} \qquad \forall k \in J \tag{4-144}$$

The multi-depot, multi-tour objective-function **(4-47)** now can be re-written as

$$\max \sum_{i \in I-J} \sum_{j \in I-J} \delta_{ij} x_{ij} + \sum_{r \in J} (c_r y_r + b_r). \tag{4-145}$$

Also, this constraint will replace Equation **(4-137)**:

$$\sigma_i' - \sigma_j' + (\delta_{ij} + U) x_{ij} \le U + \delta_{ij} - d_{ij} \qquad \forall \ |J| + 1 \le i \ne j \le |I| \tag{4-146}$$

where σ_i' and σ_j' are arbitrary real numbers (Naddef 1994). Note that this equation is derived simply by adding the variable δ_{ij} to both sides of the inequality **(4-137)**. When $x_{ij}=1$, the inequality reduces to Equation **(4-137)**. For $x_{ij}=0$, the inequality holds for $\delta_{ij} \ge 0$, or when the triangular inequality is guaranteed. The rest of the equations of the MDMTSFLP remain unchanged. The advantage of this new formulation is that it lends itself to the CW heuristic that effectively partitions the demand nodes *I-J* into node-disjoint paths for enumeration.

The original CW algorithm was designed as a multiple-tour, vehicle-capacity-constrained heuristic for solving single-depot VRPs as shown in the "Measuring Spatial Separation" chapter. Here, we relax the vehicle-capacity requirement, but include range and multiple-frequency servicing. After splitting a demand node into a group of co-located nodes corresponding to the required frequency, the CW algorithm is modified to keep track of which tour each node is on. The extended procedure follows these steps:

Step 1 Construct a savings matrix for one of the depot-demand subproblems (the depot within set *J*). Each saving between *i* and *j* is the difference between the combined cost of travel to the depot *k* from both nodes *i* and *j*, and the cost of travel between nodes *i* and *j* as defined in Equation **(4-144)**.

Step 2 Select the arc *(i,j)* offering the greatest savings unless the new tour-length would exceed range *U*, or the new tour involves revisiting a group already served along the route.

Step 3 Proceed until no further savings can be achieved due to the restrictions in step 2.

Step 4 Select the next depot-demand region and proceed with step 1. ■

Example results of this MSF/CW procedure is shown in **Table 4-9** and **Figure 4-22**. The two-stage MSF/MCW heuristic begins with a spanning forest (left), given

---

[29]A path may contain a unique node and therefore no arcs.

[30]If a path consists of a single node the δ-length is 0.

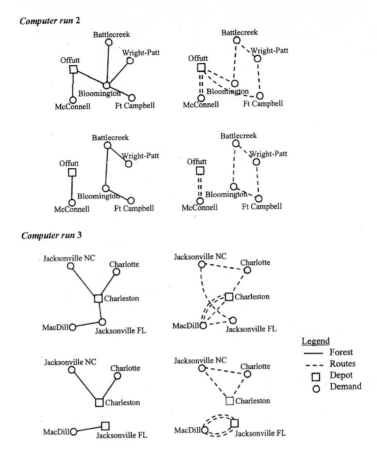

Figure 4-22 Examples of forests and routes

a number of depots. That number is varied between one and two in these examples. Once the depot-demand assignments are made, the Modified Clarke-Wright heuristic completes the routing (right). The combined process of MSF and CW takes up a considerable amount of computer time, with the MSF portion consuming most of the time. But it is an *operational* procedure to solve a huge problem, as we will demonstrate. Equally important is its ability to produce solutions close to the analytical model—a point we will come to immediately below.

4.11.3 Properties of the algorithm. Fisher's (1994a) nearest-neighbor algorithm yields the *minimum* (optimal) K-tree. The key is that there is a one-to-one pairing of the edges between the existing spanning tree T and the new tree T' during the least-cost edge-exchanges. Optimality is obtained when there are no admissible exchanges between edges. Now consider a spanning forest instead of a K-tree. **Figure 4-23** shows that one instance of the spanning forest is nothing but a K-tree with K additional 'stems'. Instead of the one-to-one exchange algorithm for a

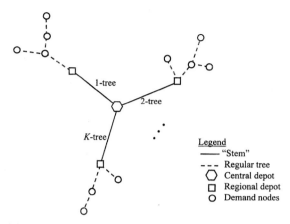

Figure 4-23 Relationship between minimum spanning forest and *K*-tree

specified *K* number of depots, we adopted an enumerative procedure to find the best *K* depots among $|J|$ candidate locations ($K \le |J|$). We are driven toward an enumerative procedure for two reasons. First, *K* is a variable in our case (rather than a constant). Our algorithm allows a fast evaluation of $\binom{|J|}{K}$ selections of depots. Second, we are interested in the sensitivity of base-closures. Through the MSF algorithm, the routing solutions turned out to be extremely stable from one base closure to another, which has great merits in the operational world. Another fringe benefit of an enumerative algorithm is that it is by definition optimal. The MSF algorithm is comparable in computation time to the *K*-tree procedure, but does a great deal more. The 8-depot/181-demand problem was solved, in which 8 depots are to be chosen from 27 candidate locations. It required just over 27 CPU hours on a Pentium 587 Personal Computer (PC). Consider the single instance of the hierarchical model of DCS, where the packages originate from a central depot and 'trunked' to *K* regional depots (from where they are delivered to their final destinations). This instance is identical to the *K*-tree. Now the additional 'stems' associated with DCS problem have a combined total cost. If this cost is "factored out" as explained in the introduction to this section, then we are left with the MSF problem. An instance of the MSF and the *K*-tree problem in this case are equivalent, as explained above.

From Fisher's experiences with six real problems with 25–134 demands, the *K*-tree approach to VRP has been shown to yield optimal solutions to five of the six real problems. The execution time ranges from 6 to 860 minutes (14.33 hours) on an Apollo Domain 3000 workstation. The remaining 134-demand problem was solved within 1.04 percent of optimality. Notice both the Fisher and the DCS problem employs a two-step procedure. The first step is the *K*-tree partitioning algorithm. The second is the conversion of a tree to routes. The difference between the two lies in the second step. Instead of a branch-and-bound procedure on the Lagrangian relaxation of the capacitated VRP, we use a modified CW heuristic for uncapacitated, but range-constrained, routing. Instead of a single vehicle to service each of the K trees, we have m_r vehicles stationed at each regional depot, where m_r is a variable. Because of the approximation inherent within a CW algorithm, solution to the DCS problem would have to be approximate rather than optimal. The error bound of our problem is correspondingly that of CW, which has been proven historically to be accurate. It sprang off several related heuristics, including nearest-insertion (Rosenkrantz et al. 1974), the sweeping algorithm (Gillet and Miller 1974), and the Fisher-Jaikumar algorithm (Fisher and Jaikumar 1981)—as mentioned in the "Measuring Spatial Separation" chapter. Burnes (1990) extended the heuristic to allow more than one vehicle to visit a demand and in so doing, minimize the number of vehicles used.[31] Baker (1991) further introduced the range of vehicle as a constraint. An important property of the CW-based heuristic is its computational efficiency.

4.11.4 Validation. The extended MDMTSFLP as formulated in the section entitled "Model formation" was solved using A Modeling Language for Mathematical Programming (AMPL) [Fourer et al., 1993], a PC mixed-integer-programming package. Run times on a Pentium class PC at 60 Mhz are all under three minutes. The validation process for each case consists of three steps: (a) exact solution of MDMTSFLP by mathematical programming using AMPL; (b) selection of the shortest weighted MSF by a FORTRAN code; and (c) tour selection by FORTRAN program using the CW algorithm, with the MSF results as input. Total system mileage is the measure-of-effectiveness used to evaluate the heuristic procedure. In this regard, all other costs such as station costs, have been converted to the equivalent 'mileage'. Each validation problem consists of two potential depot locations. A total of 16 validation runs are collected. The 16 runs are composed of eight geographic sets around the United States. The first run in each set does not restrict the number of depots, though a single depot is always chosen due to the high fixed cost of operation. The second run in each set forces two depots, which mimics the case-study solution process described later. Demand *locations* (as distinct from nodes, since the algorithm co-locates these nodes to model multiple-servicing frequency) were varied between 5 and 6.

4.11.4.1 EXPERIMENTS. The first validation-run consists of five nodes, all from the northwest of the United States. (See **Figure 4-26** for a rough orientation of the two trees where these five nodes are taken from.) These five nodes are split into eight by virtue of the frequency formulation. Thus in the distance-separation **Table 4-9**, nodes 7 and 8 are both Mountain Home Air Force Base (AFB), which requires two deliveries per week. The remaining nodes—McChord AFB (1 and 5), Travis AFB (2 and 6), Klamath Falls (3) and Boise (4)—all require one visit only. The double nodes at McChord and Travis permit both a depot and a demand to be

---

co-located at those places. To illustrate the idea of co-located nodes, let us write out in detail the range-and-subtour-breaking Equation **(4-137)**. Of particular interest are the equations written for demand nodes 7 and 8, both representing Mountain Home AFB, which requires two visits per week.

$$2600x_{73}+\sigma_7^7-\sigma_3\le2146 \quad 2600x_{74}+\sigma_7^7-\sigma_4\le2410 \quad 2600x_{75}+\sigma_7^7-\sigma_5\le2022 \quad 2600x_{76}+\sigma_7^7-\sigma_6\le1992$$
$$2600x_{37}+\sigma_3-\sigma_7^7\le2146 \quad 2600x_{47}+\sigma_4-\sigma_7^7\le2410 \quad 2600x_{57}+\sigma_5-\sigma_7^7\le2022 \quad 2600x_{67}+\sigma_6-\sigma_7^7\le1992$$
$$2600x_{83}+\sigma_7^8-\sigma_3\le2146 \quad 2600x_{84}+\sigma_7^8-\sigma_4\le2410 \quad 2600x_{85}+\sigma_7^8-\sigma_5\le2022 \quad 2600x_{86}+\sigma_7^8-\sigma_6\le1992$$
$$2600x_{38}+\sigma_3-\sigma_7^8\le2146 \quad 2600x_{48}+\sigma_4-\sigma_7^8\le2410 \quad 2600x_{58}+\sigma_5-\sigma_7^8\le2022 \quad 2600x_{68}+\sigma_6-\sigma_7^8\le1992.$$

Likewise, the range computation to depot (Equation **(4-138)**) can be written for demands 7 and 8 as

$$578x_{71}+\sigma_7^7 \le 2600 \qquad 578x_{81}+\sigma_7^8 \le 2600 \qquad 608x_{72}+\sigma_7^7 \le 2600 \qquad 608x_{82}+\sigma_7^8 \le 2600.$$

This formulation allows aircraft arrivals at different times—σ_7^7 and σ_7^8—to Mountain Home in making the two deliveries. While the above represents a general formulation, other constraints (such as time window constraints) need to be in place to prevent any tour from servicing any grouping of nodes more than once. An aircraft can be routed to node 7 or node 8 from any of the other nodes (depot nodes 1 and 2 for the outbound and the trip home, and demand nodes 3, 4, 5 and 6 when it is "on the road.")

Thus following the above example, it does not prohibit a single aircraft from revisiting Mt Home on the same tour. As an alternative to imposing time windows, we can invoke the following equation

$$\sum_{i,j\in l_n} x_{ij}\le |l_n'|-|I_n|+y_r|N_r|\,\mathrm{ind}\,(n\in NR) \qquad \forall l_n',\ \forall n\in N \tag{4-147}$$

Here, $\ell_n' \subset I$ is the range (U) limited subsets of I that include demand nodes within location n, with i, j: $I_n \subset \ell_n'$, $\sum_{i,j\in l_n} D_{ij} \le U$. $I_n \subset I$ are the demand nodes in location n. ind($\cdot$) *is the* indicator function, assuming unity if parenthetical argument is true. $NR \subset N$ is the subset of locations where demand nodes and candidate depots coexist.

This simple equation will preclude a Mt. Home revisit after servicing Klamath Falls: $x_{47}+ x_{74}+ x_{48}+ x_{84} \le3\ -2$. Since there are three total, and two co-located demands in this equation, $|\ell_n'| = 3$ and $|I_n| = 2$, suggesting that only one arc of the four possible can be used. This precludes visiting demands 4, 7, and 8 on the same tour. A similar constraint is written for each subset ℓ_n', and for each multiple-service location n.

4.11.4.2   JUSTIFICATION FOR MULTIPLE FREQUENCY.   In this example, range of the aircraft (U) is 2600 statue miles, Simple as these equations may seem when they are written out, they are certainly not intuitive (*a priori*)! The important result is that they work! We are the first to admit that the use of one 'odometer' variable per demand location appears to be a model restriction theoretically. It

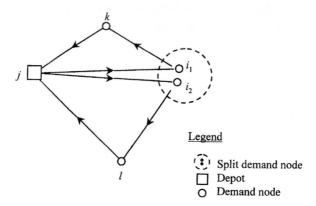

Figure 4-24 A split demand node

forces each co-located node $i_1,..,i_s$ to be visited at the same 'time' in each tour, as enforced by a common σ_i'. We recognize that it is not in the original requirements of the problem, and it appears to disallows a node to be visited early in one tour and again late in another tour, σ_i'' ($\sigma > \sigma_i'$). However, we have artificially accomplished the function of splitting the demand between two (or more) tours using co-located demand nodes $i_1,..,i_s$ (although not necessarily distinct tours as they are physically executed). In the case of a symmetric travel-distance matrix ($d_{ij}=d_{ji}$), and with the assumption that the order of visit is immaterial (i–k–j is as acceptable as j–k–i), we will show that this artifice provides no theoretical restriction at all. To the extent that the objective function **(4-145)** only accounts for the distance travel through the visitation-variables x_{ij} and arc-cost d_{ij} and not the `odometer' readings σ_i, we have accomplished our objective of travel-cost minimization. At the same time, we have introducedmultiple frequency of visit.

Perhaps an illustration will make this clear. **Figure 4-24** shows that demand-node i is artificially split into i_1 and i_2, which requires two visits by separate vehicles. Now an original vehicle-tour might consist of j–$i_{2\text{-}1}$–j, during which not all the demand at i is picked up. Later, a second vehicle picks up the rest of the demad at i during tour j–$i_{1\text{-}k}$–j. The second tour to go counter clockwise while the original goes clockwise. Since the order of visit is not important, we are only concerned that the two tours in aggregate satisfy the frequency specification $\underline{s}_i$ and the range constraint **(4-138)** at the lowest cost. We have explained how the lowest total-cost is accomplished by the objective function and its equivalent CW-formulation in Equation **(4-145)**. The demand is clearly satisfied by the exogenous specification of frequencies $\underline{s}$ at the demand nodes, as enforced by **(4-136)**.

The only remaining task is to show that in general, the CW-formulation **(4-145)** implements the range constraint in the multiple-frequency case. First, we will show that one can interpret the arc lengths in **Figure 4-24** as general distance-separations—D_{ij} from demand i to depot j and D_{ik} between two demand nodes. The optimization problem now becomes one of obtaining a pair of tours j–.–i_2–.–.–l–.–j and j–.–i_1–.–k–.–j to achieve the largest distance-savings. Now all we need to show

d_{ij}	\multicolumn{8}{c}{j}
	1	2	3	4	5	6	7	8
	McChord	Travis	Klamath Falls	Boise	McChord	Travis	\multicolumn{2}{c}{Mountain Home}
$i=1$		764	496	541	0	764	578	578
2	764		419	621	764	0	608	608
3	496	419		445	496	419	454	454
4	541	621	445		541	621	190	190
5	0	764	496	541		764	578	578
6	764	0	419	621	764		608	608
7	578	608	454	190	578	608		-
8	578	608	454	190	578	608	-	

Table 4-9 Distance between nodes in test network 1

is that this boils down to the same argument as the case of the arc-formulation in σ.

Let $\{p(1),\ p(2),\ \ldots,\ p(i),\ \ldots,\ p(m)\}$ be a path from which the tour $\{j,\ p(1),\ p(2),\ \ldots,\ p(m),\ j\}$ is constructed. We obtain from **(4-146)**: $\sigma'_{p(i+t)} - \sigma'_{p(i+t+1)} + \delta_{p(i+t),p(i+t+1)} + U \le U - d_{p(i+t),j} - d_{j,p(i+t+1)}$ where t is a non-negative integer. This is equivalent to the elemental building-block $\sigma'_{p(i+t)} - \sigma'_{p(i+t+1)} + d_{p(i+t),p(i+t+1)} \le 0$ by the definition of δ. Naddef further showed that in general this can be written as $\sigma'_i - \sigma'_j = D_{ij}$. Most importantly, the CW-formulation range-constraint **(4-146)** holds when i and j are on different paths, i.e., when they are picked up by different vehicles. In other words, there is a unique set of σ_i for the demand nodes I–J, with one σ'_i for each demand node i. This reduces to the original formulation of basic arc-triangular-inequality as illustrated in **Figure 4-24**. Similar to σ, σ'_i denotes the 'fastest' way to service node i to date, except that it allows for multiple frequencies of visitation through x_{ij}.

4.11.4.3 GENERAL OBSERVATIONS. For this case study, maximum fleet size at both potential depots (m_1 and m_2) is set at 4, and the fixed cost (in equivalent mileage) at each depot is 3382. For this case study, a fixed 'sortie' dispatch-cost is not considered, therefore, b_1 and b_2 are 0. The resulting location-routing decisions, after the model was solved, are shown in **Table 4-10**. This table contains the solutions by both the analytical model and heuristic, for the one and two depot cases. In **Table 4-11** the results of this first validation run show that in the 1-depot case, the analytical model yields an optimum of 3410 statue miles, compared with the MSF of 4303 and the CW of 3465. In the 2-depot case, the statistics are 2790,

	1-depot case	2-depot case
Analytical model *(all routes from depot(s))*	1. McChord-Boise-Mt Home-McChord 2. McChord-Mt Home-Travis-Klamath Falls-McChord	1. McChord-Boise-Mt Home-McChord 2. Travis-Mt Home-Klamath Falls-Travis
Minimum spanning forest *(all branches from depot)*	1. Travis-Klamath Falls-McChord 2. Travis-Klamath Falls-Boise-Mt Home	1. McChord 2. Travis-Klamath Falls-Boise-Mt Home
Clarke-Wright *(all routes from depot(s))*	1. Travis-Klamath Falls-McChord-Boise-Mt Home-Travis 2. Travis-Mt Home-Travis	1. McChord-McChord 2. Travis-Klamath Falls-Boise-Mt Home-Travis 3. Travis-Mt Home-Travis

Table 4-10 Validation run 1

3389 and 2874 respectively. In other words, the heuristic is sub-optimal by only 1.61 percent for the 1-depot case and just 3.01 percent for the 2-depot case.

Table 4-11 summarizes the results of all the validation runs. The heuristic solution averages 3.86 percent from optimal for the 16 problems considered. The range of error spans between 0 percent (9 occurrences) and 20.27 percent (1 occurrence). This latter case (6b) represents a controlled experiment where a special geographic coverage makes location-routing geometry difficult. Depot selection appears to be the weakest ability of the MSF/CW heuristic. In the MSF algorithm, all possible candidate depots must be tried exhaustively to reach a solution .Such solution may have little relationship to the optimal when routing is included. This problem is mitigated when the depots are preselected, as with most real-world problems. This reduces to the minimum *K*-tree. On the other hand, partitioning of nodes into service regions and routing within the region mimic the analytical model with reasonable accuracy. Another desirable feature of the MSF/CW heuristic is the consistency with which depots are selected for base closure, as one parametrically varies the number of depots. Such a phenomenon is observed in the 16 experiments conducted, where the 1- and 2-depot cases were run in parallel. This robustness in depot selection was independently confirmed by Ware (1990). This point will be examined further in the case study that follows.

4.11.5 A case study. As mentioned, DCS is responsible for the transportation of classified material between Department-of-Defense (DOD) installations. The current DCS distribution-network includes over 200 continental-United-States locations, most of which are serviced by aircraft. As suggested at the beginning of this study, the network is a multiple "hub and spoke" system. While each hub-and-spoke may be serviced by a different contract carrier, the trunking between hubs is provided by the United Parcel Service (UPS).

	(Total Demands = 6)				
Run Nbr	Nbr of Depots	Nbr of Demand Locations	Optimal Route Distance	Heuristic Route Distance	% Longer Than Optimal
1a	1	5	3410	3465	1.61
2a	1	6	2689	3099	15.25
3a	1	5	3163	3163	0.00
4a	1	5	2430	2494	2.63
5a	1	6	3435	3435	0.00
6a	1	5	6888	6888	0.00
7a	1	6	8410	8410	0.00
8a	1	5	4545	4545	0.00
1b	2	5	2790	2874	3.00
2b	2	6	2307	2307	0.00
3b	2	5	2374	2374	0.00
4b	2	5	2228	2412	8.26
5b	2	6	3125	3125	0.00
6b	2	5	4406	5299	20.27
7b	2	6	4280	4740	10.75
8b	2	5	3379	3379	0.00
				Average Error %	3.86

Table 4-11 Computational results of heuristic validation

4.11.5.1 THE PROBLEM. Over the years, the aerial portion of the DCS route-structure has grown incrementally to 12 regional-hubs serving 169 additional sites (Ackley 1990). Such a piecemeal growth-pattern has resulted in a less-than-optimal route-structure, though many current operating policies are sound. The DCS is satisfied with the current hub-and-spoke system of providing service at each regional-hub. The current security, frequency-of-service, and capacity of the system-components are adequate. This study concentrated on cost reduction by possibly reducing the number of depots, altering depot locations, and changing the routes flown—variables that can be manipulated within the current operating policy.

The costs associated with the aerial DCS-network can be broken into three categories: trunking cost paid to UPS, labor and overhead at regional-hubs, and contract cost to the regional carriers. In 1990, UPS charged the government a flat rate of 45 U.S. cents per pound (99.21 cents/kg) of freight carried. This flat rate means that we can simply concentrate on the MDMTSFLP, with the trunking part of the hierarchical route-structure subsumed. Clearly, adding more hubs will reduce the mileage flown by regional contract carriers, but may adversely affect the economies-of-scale associated with fewer but larger hub-operations. The regional contractors are paid by the miles flown. A separate contract is negotiated each year for each depot, and the 1990 agreements average a little over two U.S. dollars per mile ($1.24/km) [Thouvenot 1990]. Three of the hubs also serve as debarkation points for overseas cargo, effectively disqualifying them for potential relocation. Additionally, there are three priority-routes that are flown from the Baltimore-Washington hub despite the lower costs associated with providing service from the nearest depot.

Current analysis has been concentrated on restructuring the existing 12 depots, and considering station closures. Toward that end, the number of depots is varied between 11 and five in our analysis. Fewer depots than five will become infeasible due to inadequate aircraft range. In each run, the three unmovable depots are listed first, followed by the remaining 27 depot candidates and the 151 service-only locations.

4.11.5.2 SOLUTION. The MSF code is computationally intensive. In the 11-depot run, for example, the code must find the best of "27 choose eight" forests, which is more than 2.2 million iterations. Fortunately, the number of possible combinations drops quickly as the required number of depots is reduced. The 11-depot problem required just over 27-CPU hours on a Pentium 586 PC. Each subsequent reduction in the number of depots took less than half the CPU time of its antecedent. Each CW run took nominal CPU time and most were accomplished on a PC.

While the test networks are parts of the large network to be studied here, we have now the entire DCS network in the continental United States on which the MSF and CW algorithms are applied. In the 11-depot run, the MSF finds that nine of the 11 depots chosen are current depots. Only Griffiss AFB and Little Rock AFB are new to the system. The McChord AFB, Norfolk and Denver depots have been deleted from the original 12-depot system. In the 10- to 5-depot runs, each of the reduced depot models is similar to its predecessor model with regard to forest appearance. In all cases, the service stops of one depot are assimilated into a nearby depot, and the widowed depot is closed. This characteristic makes the new route-mileage very simple, since only one depot's routes are changed (due to the inclusion into its former neighbor's service locations). The only exception to the iterative depot-closure routine is between the six and five depot models. Two of the service locations (Des Moines and Minneapolis) of Wright-Patt are not within range of Baltimore, so the MSF attached those locations to the Offutt tree (see illustration in **Figure 4-26**.)

Table 4-12 shows a rank-ordered sequence of possible depot closings, where the 'next' closure is the one that incurs the minimal increase in cost. In other words, the mileage added is in an increasing order-of-magnitude in the table, starting out with the relatively 'painless' and progressing toward bigger penalties. To put it more technically, we minimize the marginal cost of closing another depot at each step. To assist in decision-making, the data in the above table is plotted in **Figure 4-25**. In the same figure is plotted a budget line, drawn from a slope of -5.681. This represents the national average-operating-cost of 2.229 dollars per mile ($1.39/km)

No of Depots in system	Depot closed	Depot with expanded service	Yearly mileage	Mileage added
10	Griffiss	Boston	1265368	13728
9	North Island	Travis	1289652	24284
8	Charleston	Baltimore	1317732	28080
7	Boston	Baltimore	1371864	54132
6	Little Rock	Jacksonville FL	1428284	56420
5	Wright-Patt	Baltimore	1635764	207480

Table 4-12 Cost-effective depot closure sequence

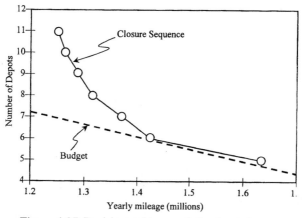

Figure 4-25 Decision making graph for depot closures

divided by the fixed cost of 0.392 million dollars per depot, as shown by the intercept of a regression of cost against the number of stations served (R^2=0.15, F=1.26, with $F(1,7)$=5.59). Obviously, such a result is only good for illustrative purposes due to its poor statistical fit. Used parametrically, the MDMTSFLP posed in this study is solved. For this example budget, a six-depot DCS-network that logs an annual routing-cost of 1.4 million miles (0.88 million km) is the answer.

4.11.5.3 A TOOL FOR DECISION MAKING. While the above is a good theoretical solution, political considerations tend to preclude such a drastic number of depot closures. If any depot can be closed, it may be Norfolk, mainly due to its proximity to Baltimore. Current operation places priority on routes from Baltimore to Wright-Patt, Langley and Huntsville, which should be preserved. Since these priority routes are flown at high frequency, there is considerable ability for aircraft flying these missions to also serve other locations along the route. **Figure 4-26** shows the group

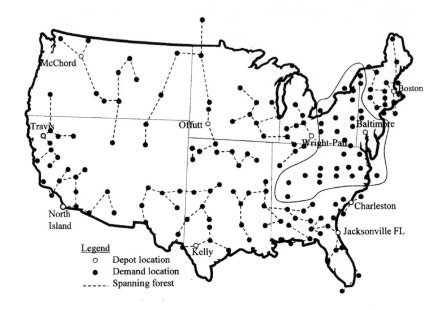

Figure 4-26 Near-term closure-solution produced by Multiple-Spanning-Forest

of stations that can logically be served this way. This "Baltimore cluster" is excluded from the MSF calculations.

An expedient near-term solution closes the Norfolk depot and preserves high-volume routes from Baltimore. Combining these demands with nearby demands forms the "Baltimore cluster," outlined above. A 10 depot model with these restrictions suggests that Denver be closed and its demands be served by McChord and Offutt. The increase in yearly system miles is nominal: 48,000 miles over a 1.4 million mile system total.

The MSF/CW heuristic determines that Denver is the most eligible depot for closure, based on route efficiency. At a flying cost of only 48,000 miles (76,800 km) per year above the 12-depot network, this solution provides the same level-of-service without the Denver or Norfolk depots. The point of the above 10-depot exercise and the construction of **Figure 4-25** is to show the versatility of the MSF/CW heuristic in combining judgment with computation, exploiting the interpolation that is possible by virtue of the stability of the MSF solution from one closure decision to another. Also exploited is the efficiency with which CW can restructure the set of routes in response to the partitioning of the nodes into groupings by MSF.

4.11.6 Concluding comments. In this study, the hierarchical location-routing model called Multiple-Depot Multiple-Travelling-Salesmen Facility-Location Problem (MDMTSFLP) was examined. We approached this problem from both a theoretical and practical standpoint. The Laporte formulation was combined with the Kulkarni/Bhave subtour-breaking technique and subsequently extended. The "range restriction" was included in Laporte's model, while the subtour-breaking concept suggested by Kulkarni/Bhave was made operational. Further, the subtour-breaking technique was extended to allow for multiple co-located demands. Such an improved model was used to validate a procedure made up of the Multiple Spanning Forest (MSF) and the Clarke-Wright (CW) procedure (with range-restriction imbedded). The MSF algorithm is an analytical generalization of the minimum K-tree. It executes in $O(|I|^3)$ time with comparable computation speed and yields optimal solutions. The CW heuristic facilitates the *analytic formulation* and implementation of multiple stops at a demand node with ease. This represents a departure from the traditional one-vehicular-stop restriction in vehicle-routing models for the symmetric travelling-salesman case. The 16 experiments conducted pointed toward the accuracy of the procedure in arriving at a solution that averages within 3.32 percent of the optimum. An additional strong point of the heuristic is the stability of the MSF solution as one explores base-closure options. While the MSF part of the algorithm is computationally intensive, the CW part is extremely efficient, in spite of the complications when range and multiple-frequency constraints have been included.

The validated algorithm was subsequently applied to a Defense Courier Service (DCS) network that has 12 depots and 181 demand locations in the continental United States. In particular, the decision on depot closure was explored. It was confirmed that the algorithm, although consuming hours of CPU time on a Pentium PC in the MSF portion, is very robust in applications where judgmental factors have to be included with computation. For example, the number-of-depots and the service-pattern can be determined *a priori*. These judgments can then be input to the heuristic without interfering with its computational procedure, as shown in the "Baltimore cluster" in the 10-depot case-study of **Figure 4-26**. Such flexibility is afforded, again, by the stability of the MSF solutions from depot closure to depot closure. It is also facilitated by the speed with which CW can re-optimize, even on a PC. The procedure can provide the framework in the determination of depot-routing configuration given a parametric budget level, as illustrated in the 6-depot DCS-case-study shown in **Figure 4-25**.

In spite of the pedagogic role it plays, the Kulkarni/Bhave Constraint Equation **(4-137)** and its CW representation **(4-146)** is a rather weak cutting-plane. It is desirable, where it is appropriate, to use it together with other constraints for computational efficiency. If we are dealing with a capacitated vehicle-routing problem (rather than the uncapacitated travelling-salesmen-problem here), for example, the inclusion of constraint Equation **(4-50)** is desirable. Laporte et al.'s analytic solution-procedure can also be improved. Johnson and Chan (1990) have simplified the Laporte algorithm. Instead of relaxing integrality, subtour-breaking, and chain-barring constraints, the improved procedure calls for including the integrality constraints into the relaxed problem. By using available integer-programming codes, one is able to solve the resulting subproblems. Since the number of subtour-breaking and chain-barring constraints increase combinatorially with the number of demand nodes, this approach promises to be significantly more efficient than solving the completely-relaxed linear-program. This is accomplished by cutting out the branches for integerization, especially as the number of demand nodes becomes large. Green and Chan (1993) have found that the branch-and-bound procedure can be further accelerated by an explicit introduction of the fleet requirement, namely that a minimum number of vehicles has to be used. If such a constraint is justifiable from the physical problem, the number of iterations in the branch-and-bound tree is again drastically reduced. Thus it is clear that although the analytic formulation can only solve problems conveniently up to six demand locations, advances in problem formulation and computing software/machinery can greatly expand the problem size that can be solved.

Admittedly, there are limitations on how large a problem of this sort can be solved by analytic means. Heuristics have to come in ultimately according to the current state-of-the-art. For this reason, the current research can be further refined by a more streamlined routing strategy starting with the output of the MSF. Given the efficiency of CW and with our focus on range and multiple-frequency extensions, we have not fully explored the natural linkage between a minimum-spanning-tree and a travelling-salesman-problem (TSP), which comes with *a priori* error bound and computational complexity. Carrying this result over to a MSF and a multiple-travelling-salesmen problem with range-restriction and frequency-implementation is not necessarily trivial, and constitutes a solid piece of research in the future[32]. Another area of research is to investigate the MSF as a formal partitioning scheme, particularly regarding the stability of its solution from one computer run to another. The work of Anily and Federgruen (1991) on "structured partitioning problems" may be useful in this regard.

4.12   Transportation terminal location

Traditional location-routing problems select depots and configure non-overlapping vehicle-routes that originate and terminate at the home depot. In the previous case study, we extended the model to account for multiple frequency-of-visit at a demand by overlapping routes. Here we present a general class of location-routing problems where the origin-termination depots are to be located as well as the stops en route. There is a special stipulation that the origin depot may be different from the termination depot (i.e., a 'chain' as referred to in the previous case study). Also a stop can be visited by more than one route. Finally, we present a way to combine the vehicle-routing problem with the *traffic-flow* problem, and this is performed on top of locating terminals. Although only vehicle routes are constructed in location-routing models traditionally, progresses have been made rapidly on the passenger or freight traffic carried on board the vehicles. Such traffic is now traced from an origin to a transfer point and finally to the destination. The resulting model

---

[32]The spanning-tree heuristic to solve a TSP is described in the "Measuring Spatial Separation" chapter, under the section "Heuristic solution to the TSP".

efficiently decides from where and how many vehicles are dispatched per time-period. These vehicles are routed around the network to serve as many origin-destination (O-D) depot-locations as desirable. Such a combined vehicle-routing and traffic-flow problem can be viewed as a general case of multi-facility/multi-route/multi-criteria/nested $(m/m/m/n)$[33] location-routing models. We pointed out the above problem of transportation terminal location is too huge to be solved by IP. A heuristic called RISE was proposed to solve realistic size problems, as explained in the section entitled "Solution algorithm."

4.12.1 A case study. RISE will be used to synthesize a route network in the American-Airlines system. A comparison will be made between this synthesized network and the existing network used by the carrier. Such a comparative study is a good way to bring to light exactly how serviceable the model is in transportation-terminal location, identifying the O-D markets the carrier wishes to serve. An analysis was carried out for the B707-320 fleet of American Airlines in the summer peak-season in early 1970's. The 707-320 fleet served 24 cities in the system. It was the longest-range aircraft with the largest seat-capacity at the time for American. The fleet mainly carried the east-west long-haul traffic. To carry out a comparative study, we attempted to follow the same "rules of the game" as the corporate planners in American in constructing our route network. The origin-destination (O-D) passenger demand carried by the B707-320 fleet is supplied directly from the carrier. Similarly, they have supplied the Civil-Aeronautics-Board (CAB)[34] route-authorities, the fleet characteristics (such as seat capacity, speed and range), intercity distances and block times[35], revenue function, etc. In summary, we would say that the route-network-configuration procedure of RISE was based upon the same premises as that used by the carrier.

In spite of the huge combinatorial degree-of-freedom that is allowed by the CAB route-authority, the terminals served by the RISE network bear remarkable similarity to the existing American system. An examination of **Figure 4-27** will bear this out. While there are less terminals served, it can be traceable to the fact that there are more 'excess' routes in the American system than RISE recommend as a whole. This is also due to the asymmetric American route-pattern, while RISE assumes symmetry in its route network[36]. For example, American may serve city-

---

[33]For an explanation of the $m/m/m/n$ designation, please refer to the section entitled "Multi-Facility/Multi-Route/Multi-Criteria/Nested Model" in the current chapter.

[34]The Civil Aeronautics Board was the regulatory agency in the United States that governed the market entry and exit of air carrier at the time of the study.

[35]Block times are defined as the amount of time an aircraft spends in covering a segment from A to B, starting from the moment the aircraft door is closed on terminal A to when the door is opened at terminal B.

[36]While symmetry is not essential, the current version of RISE assumes symmetry to save computation. This approach is analogous to the symmetry assumed by the Laporte model discussed previously.

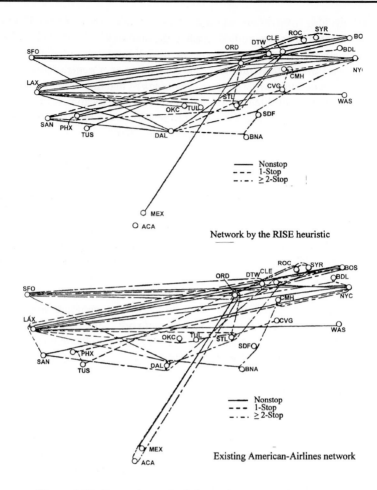

Figure 4-27 Comparison of existing and recommended networks

pair X–Y by a non-stop, but Y–X by a one-stop, resulting in two routes serving between the cities X and Y. RISE, with its symmetrical algorithm, would suggest serving X–Y in the same way as Y–X. There is only one route between cities X and Y. The asymmetry of the American network is rather puzzling to the author, since the CAB route-authorities are always symmetric and the O-D passenger demand of the American system is symmetric. The asymmetry could have been necessitated by fleet routing to meet maintenance requirements or most likely, the difference in eastbound vs. westbound schedules due to time zone changes.[37]

---

[37]In airline scheduling, it is desirable to have a convenient departure time and a convenient arrival time. This restricts the time-of-day an airline can schedule a flight, considering the three-hour time-zone change between the east coast and west coast of the United States.

On the whole, RISE suggested slightly more terminal *pairs* to be served than the existing American system—186 city-pairs rather than 137. This represents 33.7 instead of 24.8 percent of American's city-pair market. In a total of 46 city pairs where service is introduced by RISE, service neither existed in the eastbound nor the westbound direction before. These represent market entries for the best equipment of the time, the 707-320 fleet (Reference **Figure 4-27**). Although route competition has been incorporated into RISE only through the route generation phase, and the fact that we are analyzing only one of six fleet-types, 12 of the city-pair market-entries had been substantiated by the carrier as a sound decision. This is after considering all aircraft types and all other factors in schedule planning. There were additional city pairs, namely between Cleveland-Louisville (CVG–SDF) and reversely SDF–CVG, where American had been considering beginning service since passenger-demand potentials appear promising. While there are market entries recommended, there are also market exits as well. RISE suggests a number of city pairs where service in both directions should be discontinued altogether. **Figure 4-27** indicates that the carrier should consider stopping service in a total of 28 city pairs by the 707-320 fleet. An example of market exit is the service between Tulsa (TUL) and St. Louis (STL). In sum, RISE suggests 46 city-pair market-entries and 28 exits. In other words, for each market exit, RISE suggests 1.64 market entries. As explained above, 26.1 percent of the market-entries were further positively substantiated by the carrier's own analysis.

From a service standpoint, RISE provides more non-stop and one-stop connect itineraries than the existing American network. **Figure 4-27** shows that there are 86 city pairs served by non-stops, compared with 60 city pairs in the existing American system. In other words, RISE does not only enter a market, it also serves such O-D demands via improved, shorter itineraries. To be sure, there is service degradation in 53 city-pairs as well. But for every degraded city-pair, there are 4.82 upgraded city-pairs. Ten city pairs, or 18.9 percent of the city pairs with improved services, are positively endorsed by the carrier's own analysis. More capacity was provided to capture a higher market-share in the 707-320 system. With a 6.0 percent increase in available-seat-mile, passenger-traffic patronage would be increased by 8.0 percent. The expanded activities suggested by RISE are also justified by very favorable revenue figures. It is shown that the additional traffic would increase revenue from $555,000 to $613,000 per day—a 10.4-percent increase. This is accomplished by an 8.3 percent increase in direct operating cost. In other words, revenue increase more than offset cost increase; net revenue increases by approximately 2.1 percent. **Table 4-13** compares the RISE and existing American-system through the key statistics commonly used by air carriers.

In summary, it was observed that although the terminals served remain roughly similar between the existing American-system and that suggested by RISE, the level-of-service can be improved significantly by slightly expanding the city-pair markets. The system capacity expansion would also open up a number of services. While this would require a slightly higher fleet-requirement, the additional cost is more than justified by the increase in revenue. It is to be noted that such service expansion and the commensurate increase in net revenue are hardly achieved at the

	Existing American System	Suggested RISE network
City pairs served	137	186
Revenue pax miles (km)	11,263,000 (18,021,000)	12,166,000 (19,466,000)
Available pax miles (km)	17,579,000 (28,126,000)	18,796,000 (30,074,000)
System load factor	64.1 %	64.7 %
Fleet requirement (block hr/day)	282	305
Revenue ($)	555,000	613,000

Table 4-13 System summary statistics

expense of patronage on fleets other than the B707-320. Since the O-D-demand input is fixed at the potential patronage on the B707-320 fleet, all RISE is doing is re-distributing the fixed demand via more cost-effective routings. In the mean time, it finds ways to serve formerly uneconomical O-D markets. The shares between aircraft fleets would only come in when the full demand-function is used rather than some perfectly-inelastic-demand figures.

The RISE heuristic is surprisingly fast even when run on older-generation computers at the time (Amdahl 470V/8, IBM 370/155): 4 CPU seconds for 5-cities, 7 for 9-cities, 17 for 16, and 51 for 24. The 'modularized' 40-routine FORTRAN code is also very compact. The object or executable code takes up approximately 120 thousand bytes, including all the systems routines called and the COMMON data storage area used in FORTRAN codes. The array size required for a 25-city system is about 35 thousand bytes, and is projected to be about 353 thousand for an 80-city system. The core-storage requirement bears a roughly $O(|I|^2)$ relationship to the number of terminals. We have included a listing of the program on my web site and on the CD of Chan (2000), together with an executable code that runs on PCs. The complete data-input required for this case study is also included.

4.12.2 Extensions. In spite of the somewhat dated nature of this study, much of the model remains relevant to today's airline industry. RISE captures the key regulatory, competitive environment and the institutional factors of the industry. Even in today's deregulated market, the model serves as a laboratory to assess the consequences of managerial and regulatory decisions. It helps corporate planners to re-examine existing route-structure, fleet requirements and route frequency. In this regard, it holds promise to make the best out of the city-pair markets an airline has access to. By a careful channelization of the traffic-flow in a system of feeders and long-haul operations, formerly short-hop, thin-density, uneconomical segment could become profitable to serve. A traffic-flow pattern that results in high-route-density portions would bring about cost savings to the carrier. In short, systematic route planning minimizes the profit disadvantage of low-density and short-hop markets.

We can carry the application to the timely subject of mergers. Obviously, RISE cannot help merger decisions in its entirely (no single model can). It can, however, serve as a tool to predict the routing/scheduling implications when two sets of route structures are merged into a conglomerate network. Since route-structure complementarity is a prime consideration in mergers, the graph-theoretic way of encoding the combined city-pair markets expedites merger analysis. RISE may assist in answering questions like: "Could economies-of-scale in routing/ scheduling be expected from the merger?" and conversely "Would there be diminishing marginal-returns?"

Merger decisions obviously concern government agencies such as the U. S. Department of Transportation and its counter parts around the world. In this regard, such government agencies can conceivably find RISE useful. Many economists have used econometric models to assess the long-term health of the deregulated airline-industry. RISE can be used as a supplemental tool to predict the effect of different degrees-of-competition when mergers take place. This is readily done by parametrically varying the model inputs that represent competitive-market pressure exerted on an airline by the rest of the industry, particularly after a merger took place. When the total picture of all carriers is wanted, RISE can be applied in the following pairwise manner. We run RISE on airline *A* given the competition pressure from the rest of the industry. Then we run RISE on airline *B* with airline *A* now grouped into the rest of the industry. This repetitive application of RISE is a way to empirically trace out the trajectory of the partial equilibria for an oligopolistic industry. The critical point here is that the fast execution-speed of RISE makes this repetitive application possible. Market competition is a key feature in terminal-location decisions. In the following chapter, we will start modelling efforts to explicitly include competition in location-allocation decisions. The Cournot game, for example, assumes that a follower reacts to the decision of a "quantity setter," while the Bertrand game assumes reaction to a "price setter."[38]

4.13 Summary and Conclusion

In this chapter, methods for combining location and routing analysis were outlined. These combined methods typify a recent trend toward addressing the two problems simultaneously in a single model. These combined facility-location-and-routing models can be used for a variety of real-life applications. A major contribution of

---

[38]In this regard, we can view RISE as modelling a carrier's reactions to an overall price-and-quantity setter, in that both service-level and the amount-of-traffic are determined exogenously. An introduction to Cournot and Bertrand games was made in the "Game theory" subsection of the "Multi-criteria decision-making" chapter in Chan (2000).

LOCATION	ROUTING		
	Shortest Route	Vehicle Routing	Travelling Salesman
Simple-Plant-Location/*p*-median/*p*-center			
set-covering		multi-depot/multi-tour	
maximal coverage	multi-depot/multi-route		

Table 4-14 Taxonomy of combined location-routing models

this chapter is the stepwise development of combining facility-location and routing. Starting from the simplest formulation, we progress toward the many varieties one finds in logistics, engineering and planning applications. Of particular interest is how mathematical expressions and simple models can be put together to analyze problems of increasing complexity (see **Table 4-14**). If nothing else, we have demonstrated the versatility of this procedure.

One approach to integrating location and routing models is to select the depot sites, and then use a vehicle-routing model to design the service routes according to the selected sites. Facility-location models are often based on Euclidean distance, this assumes that a Euclidean-distance measure is a good approximation to routes or tours, which may not be valid. But if the site-development of a depot is capital-intensive, and much more costly than the actual cost of service-delivery, this sub-optimal solution may be totally acceptable in real-world applications. On the other hand, there are situations where sites already exist, and the cost of service delivery is highly labor-intensive. In this latter case, the fixed cost of location becomes minuscule compared to routing analysis. Combined location-routing models become correspondingly important, since a 'central' location will save routing costs. This is the reason for the discussions presented here, and represents the main focus of this chapter.

4.13.1 Formulation of a location-routing model. Due to the limited amount of effort in developing combined location-and-routing models, only a few observations can be made from the integration efforts as illustrated in **Table 4-14**. It appears that a 'cluster' of activities is found at the lower left-hand corner of the table. This shows a concentration of activities at merging "maximal coverage location" with 'routing' models, resulting in single-facility/ single-route/multi-criteria ($s/s/m/-$), multi-facility/multi-tour/hierarchical ($m/m/-/h$) and multi-facility/multi-route/multi-criteria/nested ($m/m/m/n$) models. Parallel efforts have been made in combining "set-covering location" and "routing/salesman-tour" models, as evidenced by the development of single-facility/multi-tour allocation (*s/m/–/a*) and single-facility/multi-tour/probabilistic (*s/m/–/p*) models. Little attempts have been made to combine the classic simple-plant-location model, *p*-median, or *p*-center models with routing models—perhaps a reflection of real-world needs.

We have tried to pose a typology among the various location-routing models in this paper. A four-letter code *a/b/c/d* is used to classify models according to

a. Single (*s*) or multiple (*m*) depots (or origin-destination pairs)
b. Single (*s*) or multiple (*m*) tours or routes employed
c. Whether the objective function is single (*s*) or multi-criteria (*m*)
d. Other special features of the model such as hierarchical facilities (*h*), nested flows (*n*), or allocation (*a*) of inventory/routes among demand sites.

The taxonomy is necessarily applications-oriented. Immediately below, we will examine these models differently from the vantage point of solution methodologies. But the typology here does accurately portray the complexity of the model from both application and methodological standpoints. For example, a single-depot model is usually less complex to solve than a multiple-depot problem and a single-tour model is again easier than a multiple-tour problem and so on.

While most of the models are deterministic, at least two location-routing models are presented in probabilistic formulations, where random variables are either the demand or the link distance. In the case of random distance, for example, it was shown that the average travel distance is shorter than in a deterministic case—a useful but somewhat counter-intuitive finding. Other exciting developments are observed in this chapter that prove to be invaluable in practical applications. For example, *a priori* tours can be constructed to cater for all instances in a probabilistic-multiple-travelling-salesmen-facility-location-problem. Such a configuration has a theoretical error bound of 25 percent for asymptotically ubiquitous demands.

Laporte et al. (1988) suggested another way of looking at a location-routing problem (LRP). Viewed as a graph of nodes and arcs, an LRP can be symmetrical or asymmetrical in the transportation costs d_{ij}. In this taxonomy, the set of nodes I are partitioned into J and I–J, where $J = \{1, ..., K\}$ represent a set of potential depot sites, and I–$J = \{K+1, ..., |I|\}$ are the non-depot demand sites. There can be at most $h = \{1, ..., |H''|\}$ types of vehicles in the fleet. Vehicles have a capacity of $V(h)$. There is a fixed cost of siting a depot at node j, b_j, and there is a variable unit-cost of dwell time at a node i by vehicle h, t^h_i. There are several "ground rules" for traditional LRP:

a. Each demand is satisfied once and exactly once by a vehicle;
b. Vehicle capacities are not to be exceeded;
c. The number of vehicles based at each depot j lies within pre- specified upper limit T;
d. The length of a vehicle route (including the dwell time) does not exceed a specified range, U_h;

e.   Depot sites or origin-destination pairs are to be selected from J ($J \subseteq I$) and vehicle routes satisfying a subset of constraints (i), (ii), (iii), (iv) are to be established simultaneously in order to minimize the sum of a subset of vehicle costs, depot operating-cost and routing-cost.

While there are separate reviews of location models and routing models such as that by Laporte et al., this chapter represents the first attempt—to the best of our knowledge—at synthesizing combined location-routing models that go beyond the traditional topology. For example, rule (i) is generalized to include multiple visits by vehicles. Commodity flows (passenger or freight) are explicitly modelled besides vehicle routes. Demand nodes need not be separate from depot nodes. A general hierarchy of facilities can be modelled (distinguishing, say, a supply from a distribution center). Probabilistic demands are included. Multiple criteria are considered. Revenue is considered in addition to cost. The formulation of such a combined location-routing model is no trivial task; it is much more general than the traditional framework.

We unify our models not only in terms of the $a/b/c/d$ topology described above, but also in terms of the forcing/linking constraint, the critical element in a location-routing model. In short, the forcing/linking constraint relates a location decision **y** to a routing decision **x**. In English, it says that both demand and facility have to be located on a route; a route with only facility or demand is meaningless. Expressed in this generalized statement, and in view of the linkage between a demand site and a facility site, routing decisions now include both the decision to visit a demand **x** and the decision to site a facility **y**. Introducing the additional decision-variable **z** poses a hierarchical structure to a location-routing model. The model can be used to study a factory/distribution centers/demand hierarchy. Instead of one set of forcing/linking constraints, additional sets need to be included for each pair of hierarchies, say between factory and distribution centers and between distribution centers and demands (Eren 2000). Thus we view the basic building block of a combined location-routing model as the forcing/linking constraint. The specific form of such a constraint characterizes the type of location-routing model, from the simple $s/s/m/-$ model to the complex $m/m/m/n$ model.

Unlike its constituent location and routing models (which are complex in and of themselves), solving the combined formulation is more of an art than a science. Nevertheless, simple examples of these formulations presented here can all be solved by standard MIP or IP codes, albeit in a rather clumsy way. (In the case of multi-criteria functions, the MIP and IP would have to be solved several times to generate the set of non-dominated solutions.) Thus we have in this chapter documented in a unifying set of nomenclature our ability to model generic classes of "real-world" problems. We presented formulations that are amenable to solution. An application-oriented analyst can run these formulations using off-the-shelf MIP or IP packages, at least to solve small-size problems. In the case studies, we further demonstrate that the formulations are amenable to shortcut solution algorithms. A good example of this is the "chain barring" cuts for eliminating more than one depot in a tour in the multi-depot/multi-tour model by Laporte et al. This lends itself to an efficient yet exact solution algorithm. Also, a variety of heuristic algorithms have been shown, including heuristics with theoretical error-bounds and polynomial computational complexity.

To bridge the discussions on model formulation and solution, it has been illustrated more than once that model formulation is an integral part of solution efficiency. As shown in the chapter on routing, different subtour-breaking constraints have profound implications upon the algorithms employed. Much of the progress made over the past few decades has been on the construction of an increasingly more compact and tight subtour-breaking constraint. In the case of a combined location-routing model, this relationship becomes even more pronounced as one combines two difficult problems into one. Astute formulation becomes ever more critical in the design of a practical solution-algorithm. An example is the general multi-depot/multi-route model (due to Chan [1972]) which has one formulation for regular integer-programming solution and yet another for heuristic solution. The readers may wish to review both to obtain complete information regarding the relationship between model formulation and solution.

4.13.2 Solution methodologies. Facility-location and vehicle-routing are two of the most pervasive topics in operations-research/management-science today. When the two problems are kept separate, model formulation and solution methods abound. For such combined location-routing models, research on both formulations and solutions has just begun. The literature reviewed here represents modest gains in the first phase of an ongoing research effort. In a combined location-routing model where demands and depots are co-located, where the starting point of a vehicle does not need to be the same as the termination point (i.e., chains), where more than one vehicle can visit a demand, where demand is random or actuated by inventory shortage, where multiple criteria are used for evaluation, where both vehicular and commodity flows are modelled, and where depot capacity and route length are constrained, the formulation becomes prohibitively complex. This chapter concentrates on the question: "What type of procedures are most suitable for solving these problems?"

As mentioned more than once, facility-location or vehicle-routing problems are difficult and challenging by themselves when treated separately. Thus it is to be expected that combined location-routing models will be *at least* as complex to solve. Since large-scale vehicle-routing problems are difficult to solve optimally, it is unlikely that optimal methods exist to solve location-routing problems in practice. As an alternative, a variety of heuristic approximation techniques appear unavoidable, and some of them perform quite satisfactorily for operational needs. Optimal solution methods such as the one suggested by Laporte et al. should be useful for evaluating the results of some heuristics—at least up to medium-size problems. Alternatively, carefully-designed heuristics today have error bounds already built in as part of the algorithm design, such as the space-filling-curve heuristic employed in the probabilistic multiple-travelling-salesmen-facility-location problem (PMTSFLP).

Due to problem complexity and the embryonic state-of-the-art, we have limited our review mainly to deterministic location-routing problems, although we have discussed in some depth the solution of the PMTSFLP, classified under $s/m/-/p$ in our topology. Recent development in probabilistic-vehicle-routing-location (PVRL) solution algorithms is extremely exciting. Similar to the PMTSFLP discussed, the PVRL problem calls for the location of a depot where $|H''|$ vehicle-routing tours are based ($|H''| > 1$). The vehicles must respond to a probabilistic demand list at most $|I|$ long. An $O(|I|^3)$ heuristic is suggested by Simchi-Levi (1989) to solve the problem, which has a worst-case error-bound of $3/2(1-V'(\cdot))$, where $V'(\cdot)$ is the normalized vehicle capacity. Thus when $V'(\cdot)$ is close to unity, error is less. When $V'(\cdot)$ is close to zero, more error is to be expected. In the worst case, $V'(\cdot)=0$ and the solution is 150 percent of the optimum. In the case where separations are measured in terms of Euclidean distances, the heuristic is proved to be asymptotically optimal. Thus in limited cases, one is able to put heuristic solutions in an analytical framework, including statements on computational complexity and error bounds—a point worthy of repeating more than once. This observation is echoed by Chan and Merrill (1997) in their attempt to solve the PMTSFLP, where an error bound limit of 25 percent is established, and a similar asymptotic optimality is found. Also, an $O(|I|^2)$ algorithm is obtained.

Given the importance of heuristics and the attractiveness of an analytical solution, the bridge between the two opens a brand new vista for designing innovative solution-methodologies. It promises many exciting possibilities that will give some well needed structure to the heuristics and computational efficiency to analytical models. The introduction of error-bounds and computational complexity to heuristics is only one way to bridge this gap. Relaxation techniques, including dualization (such as Benders') and the use of chain-barring constraints, are attempts in closing this gap starting from the analytical end. We have also seen how a Clarke-Wright heuristic can be embedded in an analytic formulation. Besides being a well-proven heuristic, it allows multiple-frequency of visit and range constraints to be implemented and solved within the framework of an analytic model. Similarly, we have witnessed how an analytic model can be transformed into a recursive program for effective solution. While there is much work to be done, the confluence from both ends represents a most promising research direction in location-routing analysis.

In this chapter, we have specifically surveyed a variety of location-routing models, where the definition of location-routing has been broadened to include not only depot location, but also demand locations as well. Routing is not limited to vehicles alone, but also commodities carried by the vehicles. In solution methods, we surveyed anywhere from dualization and relaxation methods to heuristics based on exchange and displacement. Instead of looking for commonalities alone, we celebrate diversity, in that we pointed out the very many different formulations and solutions that are possible. In the spirit of information dissemination, we have attempted to give self-contained description of each algorithm, complete with numerical examples. Although the algorithm is primarily designed for location-routing models, they have bearing upon individual location and routing models as well. The reason is that location and routing models are nothing but degenerate cases of a combined model. An algorithm that solves a combined model should therefore solve an individual, constituent model—albeit an 'overkill'. It is our hope that bringing these diverse techniques together in a chapter, together with an extensive bibliography, will stimulate further discussions and studies between practitioners and researchers alike.

4.13.3 Case studies. While the general methodology of model building and solution was discussed in detail above, some practical implications can only be spelled out in the case studies. First, let us point out—once again—that all the five types of simultaneous location-routing models have been illustrated by case studies. The only except ions arethe simple case of type 1 and type 2 (which is a special case of type 3):

a. Single-facility[-pair]/single-route/multi-criteria ($s/s/m/-$)
b. Single-facility/multi-tour/allocation ($s/m/-/a$)
c. Multi-facility/multi-tour/allocation ($m/m/-/a$)
d. Multi-facility/multi-tour/hierarchical ($m/m/-/h$)
e. Multi-facility/multi-route/multi-criteria/nested ($m/m/m/n$).

In this regard, the case studies represent a cross section of the model typology. In the siting of nuclear-power plant in California, electricity is to be supplied from a plant to population via a route in the least-cost and the most-risk-free manner possible. In locating depots to distribute time-sensitive packages, the Defense-Courier-Service plans the number of hubs and the trunking between these hubs in a two-level delivery system—between hubs and from a hub to the customers. In configuring an airline-network, the number of airports to be served and the routes between them are planned to achieve the highest profit for the carrier and the most convenience to the travelling public. The first and third case studies show that facility location involve both strategic and tactical decisions—that the longer-term siting decisions are ultimately related to shorter-term ones dealing with day-to-day deliveries. When siting cost is small compared with delivery cost, as in the second case-study, location-routing models still play a key role in decision-making although only tactical decisions are involved. This second case study illustrates the important decisions one faces in today's service-oriented economy.

The case studies above progress from the simple to the more sophisticated. In the first case-study of obnoxious facilities, for example, the routings have neither range nor vehicle-capacity constraints. There exists a single-level linking or forcing constraint that places a precedence relationship between location and routing decisions. We can conceptualize this as an un-capacitated location-model, representing the simplest of the three cases modelled here. A special feature of this model is the redundancy of service, in which a demand is served by more than one power plant. Most computational requirements of this model are exacted by multi-criteria optimization rather than the location-routing model itself. Largely, the computational requirement is handleable by off-the-shelf mixed-integer-programming software for even a realistic-size problem. Computational efficiency is further enhanced when the decision variables often become integer or discrete without any conscientious effort by the modeler. The challenge lies in completely capturing the efficient frontier representing all the non-inferior solutions.

In the second case study, a hierarchical location-routing model is solved, in which vehicle-routing considerations are an integral part, including vehicle-range and frequency-of-visit. Freight traffic is traced from a central supply through distribution centers to consumers. The inclusion of range and frequency places computational restrictions on the location problem that were absent in the first 'uncapacitated' case study. Determining the frequency-of-service poses another computational challenge that is not found in traditional location-routing models. Here, extended analytical-techniques such as binary-programming models are only good enough to validate the heuristic, which in turn has to be used to solve the real problem. A large depot-location problem was tackled by a skillful heuristic technique. Through this case study, we are able to place in perspective the role of an analytical vis-a-vis heuristic solution-procedure. In particular, we investigate the linkage between them—the fact that one can be transformed to the other. It also provides a computational benchmark on the size of problem that can be handled within the state-of-the-art.

In the third and final case-study, a most generalized location-routing model is applied toward transportation-terminal location. The model is characterized by intertwined relationships between terminal-location and the routing of vehicles *and* passengers, ending up with a 'nested' model. Here the forcing/linking equations are written not only for location of facilities, but also between vehicle-routes and passenger-itineraries[39]. Vehicle capacity is now explicitly introduced. The recursive-programming heuristics were shown to be the only viable solution-technique. While the heuristic procedure was validated by an integer-programming model, the validation was limited to a three-node network—a much more restrictive validation than in the second case-study. Once the heuristic was accepted, however, it executed in an extremely fast speed, in spite of the general nature of this model. In a broader context, this case study illustrates the transition from discrete modelling typically found in facility location toward continuous models of activity derivation-and-allocation where several the assumptions in facility location are relaxed. Thus the assumption regarding non-overlapping service regions each of which covered by a distinct facility is lifted, making way for overlapping service-regions in which activities such as freight or passenger traffic can be delivered by more than one vehicle. Certain discreteness or lumpiness is still found for sure, as in the assignment of frequencies to routes. Nevertheless, we provide the first of several bridges between discrete facility-location and continuous activity-allocation, where the latter is a key feature of land-use models. To the extent that demands are not perfectly inelastic, we have explicitly recognized the fact that traffic can be either stimulated or suppressed depending on the service provided.

---

[39]Again, these terms are defined in the section entitled "Multi-Facility/Multi-Route/Multi-Criteria/Nested Model" of the current chapter.

An intent of these case studies is to illustrate how the basic location-routing models can be modified and in turn be combined to fit a real-world problem at hand. This is shown in the first two case-studies, where extensions of the basic location-routing formulations and solution techniques are required for modeling and solving the problems at hand. Most important, we illustrate how these extensions allow problems of practical sizes to be solved. In all three case-studies, our discussions include modeling, computation and applications—i.e. the entire analysis process. As mentioned, the diversity of these three studies allows a representative cross-section of location-routing models to be included—from obnoxious-facility location to delivery logistics to transportation-carrier management. The first case-study shows how one can identify robust locations of obnoxious facilities that can survive changing political-persuasions. The second case-study shows how the sensitivity of base-closures can be mitigated by the range of options and gradualism, where gradualism is defined as providing explicitly the ease-of-transition. The third case-study suggests a prescriptive way to model market entries and exits that fully represent the detailed production-function of a carrier in terms of routing-and-scheduling. Most interestingly, the market is characterized here by oligopolistic competition with its complex characteristics.

CHAPTER FIVE

INCLUDING GENERATION, COMPETITION AND DISTRIBUTION IN LOCATION-ALLOCATION

An objective of this book is to marry the discrete facility-location literature with the continuous activity-allocation literature. For our purpose here, the former refers to the binary (all-or-nothing) decision in site selection. This is sometimes called un-capacitated assignment. Only when the facility capacity is reached before demand is assigned to a second (backup) facility. The latter deals with the equitable assignment of demand to alternate sites, in which a demand is assumed to be served by more that one facility at the outset. Extending this concept to a location-routing model, we may have competing vehicles (carriers) serving a demand and a facility. While we might have alluded to this idea, the discussions thus far have by-and-large fallen short of this objective. Starting with this chapter, it is our intention to address this integration issue in more detail. Obviously, the two types of models make both location and allocation decisions. But the emphasis differs, with discrete location-models more concerned primarily with site selection and continuous allocation focused on competitive demand assignment. This is understandable since they were developed to serve different needs, and they have performed their jobs adequately. It would be a false premise to claim that an integrated model that perfrom both the functions of a facility-location and activity-allocation model is preferable. Quite clearly, comprehensiveness tends to be gained at the expense of level of detail. The reader will find at the end of this chapter that, while we have extended the classic location-allocation models significantly, specificity such as routing and scheduling will be replaced simply by a more aggregate spatial-separation measure. In this chapter, we are looking for some fundamental principles that underline each type of models, and the theory that govern sboth types of models. It will help us understand such important ideas as competition, rather than offering a panacea that purports to do everything.

5.1 Inclusion of spatial-interaction in facility-location models

Most facility-location/routing models use proximity as a figure-of-merit in locational decisions. However, the assumption that demands patronize the nearest facility is an inaccurate model of observed behavior. Beaumont (1987) synthesized research performed in this area and extended the p-median formulation to a more general framework. We report this finding below with our interpretation and contribution. First, he relaxed the proximity or nearest-center assumption. A spatial-interaction term Θ_{ij} is introduced, in addition to V_i (the quantity demanded

$$\min \sum_{i=1}^{n'} \sum_{j=1}^{p} V_i \Theta_{ij} C_{ij} x_{ij} \tag{5-1}$$

at location i) in a classic p-median model
where C_{ij} is a distance measure between demand i ($i = 1, \dots, n'$) and facility j ($j = 1, \dots, p$), and x_{ij} is a 0–1 decision variable to pair i with j. A common form of spatial-interaction model looks like

$$\Theta_{ij}(W_j, C_{ij}) = \frac{W_j^a f(b, C_{ij})}{\sum_{j=1}^{n'} W_j^a f(b, C_{ij})} \tag{5-2}$$

where W_j is the attractiveness at location j, and a, b are nonnegative calibration constants. The trip-distribution function $f(b, C_{ij})$ shows the relative accessibility between i and j. It usually assumes the form of C_{ij}^{-b} exp$(-bC_{ij})$, as discussed in the "Economics" and "Descriptive Tools" chapters in Chan (2000).

5.1.1 Nearest center vs. probabilistic allocation. When demands travel to or are serviced by the nearest facility, and market areas do not overlap, the parameter b becomes infinite. Equation **(5-2)** degenerates to unity, and there is no spatial interaction; then regular location models, such as the p-median model, would apply when Θ_{ij} *disappears* in Equation **(5-1)** as it assumes the value of unity. Recall the constraints for the p-median model are:

$$\sum_{j \in J} x_{jj} = p \tag{5-3}$$

$$\sum_{j=1}^{p} x_{ij} \geq 1 \qquad \forall i \in I \tag{5-4}$$

$$x_{jj} - x_{ij} \geq 0 \qquad \forall i, j \in I; \ i \neq j \tag{5-5}$$

$$x_{ij} \geq 0 \qquad \forall i, j \in I \tag{5-6}$$

$$x_{ij} \in \{0, 1\} \qquad \forall i \in I, \forall j \in J. \tag{5-7}$$

For an uncapacitated p-median problem, x_{ij} are 0–1 valued under this formulation. On the other hand, when choices, tastes, and other non-transportation factors increase in importance, b tends toward zero, and the trip-distribution term, $f(b, C_{ij})$, in Equation **(5-2)** assumes unity and disappears. Hence

$$\Theta_j = \frac{W_j^a}{\sum_{j=1}^{p} W_j^a}. \tag{5-8}$$

Here, transport costs no longer determine a demand's choice of service-facility location. Demands base their decisions purely on the attractiveness of the destination facility.

Under spatial-interaction generalization, the p-median model would pair demand i with facility j so that the total spatial cost

$$\sum_{i=1}^{n'} \sum_{j=1}^{p} V_i \Theta_{ij} C_{ij} = \sum_{i=1}^{n'} \sum_{j=1}^{p} V_{ij} C_{ij} \tag{5-9}$$

is minimized. Notice the 0–1 (all-or-nothing) assignment of a demand i to a facility j is no longer mandated. In a capacitated p-median model, for example, it is therefore possible that a demand is served by more than one facility. It is for this reason that this generalization is sometimes referred to as probabilistic *activity distribution*. In such a distribution, demand i is now distributed among several facilities, rather than allocated toward a single facility. **Figure 5-1** illustrates the distinction between facility-location vs. activity-distribution models. Notice the overlap of service regions of facility in the illustration.

Example (de la Barra 1989)

Consider Equation **(5-2)** when $W = 1$ and $f(.)$ assumes the exponent form. As the value of b tends to infinity, Θ assumes the value of one—suggesting an "all-or-nothing" allocation in favor of destination j under consideration. By contrast, as b tends to zero, the percentage allocations among all destinations become equal, that is $\Theta_{ij} = 1/n'$, where n' is the number of destinations available. This corresponds to an *equal* allocation among all the destinations. To further illustrate this point, suppose there are two populations in the study area. The first group has a low sensitivity to cost and the second has a high sensitivity. These two groups are reflected, say, by the b values of 0.2 and 0.6 respectively. Suppose there are two destinations in the study area, with travel distances of $C_{i1} = 5$ and $C_{i2} = 8$, it can be shown that the zone-1 percentage-share is $\Theta_{i1} = 0.6457$ for the low-sensitivity population. This compares with a percentage of 0.8581 for the high-sensitivity population. It can be seen therefore that when a facility-location model includes spatial-interaction terms, the b parameter in spatial interaction plays an important role on activity allocation or distribution among destination facilities. □

When Equation **(5-9)** is written with the constraints on trip-productions and trip-attractions, it becomes the classic Hitchcock–Koopman model, which is a special case of the doubly-constrained gravity model, as shown in the "Descriptive" and "Prescriptive" tools chapters in Chan (2000):

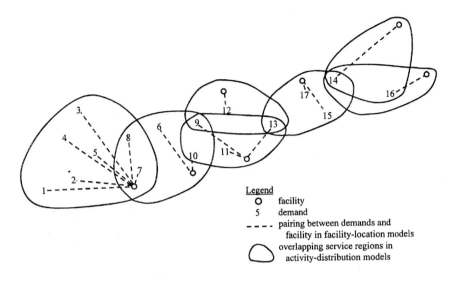

Figure 5-1 Facility-location vs. activity-distribution models

$$\sum_{j=1}^{p} V_{ij} = V_i \qquad \forall i \in I$$

$$\sum_{i=1}^{n'} V_{ij} = V_j \qquad \forall j \in J \tag{5-10}$$

$$V_{ij} \geq 0 \qquad \forall i \in I, \forall j \in J.$$

The "Descriptive" and "Prescriptive" tools chapters derived this relationship using entropy or information theory. It can be shown here simply that the objective function is a specialization of the basic building-block of the gravity model when $b = 1$, or $V_{ij} = z_{ij} F(C_{ij}) = z_{ij} C_{ij}^{-b} = z_{ij} C_{ij}^{-1}$ where z_{ij} and $F(C_{ij})$ are the non-distance and distance interactions between i and j respectively. A simple transposition of terms in this basic expression would yield $z_{ij} = C_{ij} V_{ij}$ and the objective-function $z = \sum_i \sum_j C_{ij} V_{ij}$ as shown in Equation **(5-9)**.

5.1.2 Optimal placement of activities. The quadratic assignment problem locates facilities in consideration of activity spatial interaction.[1] A parallel technique for The Optimal Placement Of Activities in Zones (TOPAZ) was

---

[1] The quadratic assignment problem was introduced in the "Prescriptive Tools" chapter in Chan (2000) and subsequently elaborated in the "Spatial Separation" chapter in the current volume.

proposed by Brotchie and Sharpe (1975), which considers the allocation of activities as well as the interaction between activities and traffic. We extend the latter model in our discussion here (Foot 1981). For a region divided into n' zones and considering the allocation of p types of land-use activities x_{kj} we maximize benefits and minimizes the travel-cost:

$$\min \sum_{k=1}^{p}\sum_{j=1}^{n'} -S'_{kj}x_{kj} + \sum_{i=1}^{n'}\sum_{j=1}^{n'} C_{ij}V_{ij}. \tag{5-11}$$

Here S'_{kj} is the benefit of developing activity-type k in zone j and x_{kj} is a 0–1 ranged continuous-variable denoting the fraction of type-k activities to be allocated in zone j. A singly- or doubly-constrained gravity-model is used to find the distribution of trips V_{ij}:

$$V_{ij} = V_i \frac{W_j \exp(-b\tau_{ij})}{\sum_j W_j \exp(-b\tau_{ij})} \tag{5-12}$$

where τ_{ij} is the travel time between zones i and j. Here the trip-productions V_i and trip-attractions V_j are generated from the activities assigned to the zones: $V_i = \sum_k x_{ki}K_k$ and $V_j = \sum_k x_{kj}K'_k$ where K_k is the trip-generation rate for a unit-of-activity k and similarly K'_k is the trip-attraction rate.

Equation (5-12) can now be re-written in the form in which V_{ij} is used in the overall objective-function (5-11):

$V_{ij} = (trip\ generation\ from\ i)(trip\ allocation\ among\ j)$

$$= \sum_k x'_{ki}K_k \frac{\sum_k x'_{kj}K'_k \exp(-b\tau_{ij})}{\sum_j \sum_k x'_{kj}K'_k \exp(-b\tau_{ij})} \tag{5-13}$$

where $x'_{ki} = x_{ki} + X^k_i$, with X^k_i denoting the existing level of activities. Notice here that W_j has been replaced by V_j. The cost-minimization objective takes place subject to certain constraints. The first is an equation that specifies the total amount of employment-type k required in the region

$$\sum_j x_{kj} = Z_k \qquad k = 1, 2, \ldots, p \tag{5-14}$$

where Z_k denotes the level-of-activity category-k, normalized by the regional-employment total. The second constraint is an equation that specifies the activity-

$$\sum_k x_{kj} = Z_j \qquad j = 1, 2, \ldots, n'. \tag{5-15}$$

level at zone j, again normalized by the regional total

Activity opportunities have to be placed among the zones in the region. In this case it also ensures that the sum of the parts equal the whole: $\sum_k Z_k = \sum_j Z_j = 1$. Together with the usual non-negativity constraint on the decision variables, a nonlinear-programming problem is now formulated.

Example (Dickey et al. 1973)

Consider a city in which 60 new residential-acres (24 ha) are to be built, and 20 industrial acres (8 ha) are to be placed in the forecast period. These two types of acreage are to be developed in two zones. Zone 1 has a development capacity of 30 acres (12 ha) and zone 2 50 (20 ha). The unit benefit of establishing an acre (ha) of activity-k in zone j (in millions of dollars) is denoted by S'_{kj}, where $S'_{11} = 0.18$, $S'_{12} = 0.14$, $S'_{21} = 0.70$ and $S'_{22} = 0.50$. The interaction-costs for trips between zones i and j C_{ij} are (in dollars per daily-trip over the planning horizon): $C_{11} = 400$, $C_{12} = 1600$, $C_{22} = 400$, and $C_{21} = 1600$. Trip production and attraction rates (in vehicles/acre (ha)/day) are $K_1 = K'_1 = 15$ and $K_2 = K'_2 = 13$. Suppose the existing amount of activities k in the zones i, X^k_i, are $X^1_1 = 10$ (4), $X^1_2 = 20$ (8), $X^2_1 = 3\ 0$ (12), $X^2_2 = 40$ (16) acres (ha).

To solve this nonlinear-optimization model, an initial feasible-solution to the model has to be estimated from which an iterative procedure follows, as explained in the "Prescriptive tools" chapter. Let us say that the initial trial solution x_{kj} is $x_{11} = 20$ (8), $x_{12} = 0$, $x_{21} = 10$, $x_{22} = 50$ (20) acres (ha), with the normalized fractions turned into acres (ha). This is done by multiplying the variables x_{kj} by the grand totals of 20 (8) and 60 (24), turning fractions into real numbers. Here, we can check, for instance, that $x_{11} + x_{12} = 20$ (8) and $x_{21} + x_{22} = 60$ (24). Instead of the exponential form, suppose a gravity model of the power-function form $(\tau_{kj})^{-2}$ is used, where the travel times between zones i and j τ_{11} are $\tau_{11} = 3$, $\tau_{12} = 10$, $\tau_{21} = 10$, and $\tau_{22} = 3$ minutes. The objective function **(5-11)** now looks like

$$
\begin{aligned}
&180{,}000(20) + 140{,}000(0) + 70{,}000(10) + 50{,}000(50) + \\
&\frac{400[15(20+10)+13(10+30)][15(20+10)+13(10+30)]1/3^2}{15(20+10)+13(10+30)]1/3^2 + [15(0+20)+13(50+40)]1/10^2} + \\
&\quad\text{other 3 travel-cost terms.}
\end{aligned} \tag{5-16}
$$

For the proposed solution, establishment costs total 6.8 million and travel costs total 1,146,540, for a grand total of 7,946,540. Now we substitute the solution x_{kj} for the proposed solution given above into the terms for trip attractions (but not for the productions) in the model. The objective function becomes

$$
\begin{aligned}
&180{,}000x_{11} + 140{,}000x_{12} + 70{,}000x_{21} + 50{,}000x_{22} + \\
&\frac{400[15(x_{11}+10)+13(x_{21}+20)][15(20+10)+13(10+20)]1/3^2}{[15(20+10)+13(10+20)]1/3^2 + [15(0+20)+13(5040)]1/10^2} + \\
&\quad\text{other 3 travel costs.}
\end{aligned} \tag{5-17}
$$

The objective function is now linear; for example, the expanded gravity-model-term in the above equation can be reduced to $5{,}185x_{11} + 4{,}493x_{21}$. Correspondingly, the original model becomes

$$
\begin{aligned}
\min\ &188\ 451x_{11} + 146\ 869x_{12} + 77\ 324x_{21} + 55\ 952x_{22} \\
&x_{11} + x_{21} = 30\quad x_{11} + x_{12} = 20 \\
&x_{12} + x_{22} = 50\quad x_{21} + x_{22} = 60
\end{aligned}
$$

with the usual non-negativity constraints. This linear model is in the form of the standard Hitchcock–Koopman transportation-problem[2] as shown in Equation **(5-15)** above and can be solved easily. The solution in this case would have $x_{11} = 0$, $x_{12} = 20$ (8), $x_{21} = 30$ (12), and $x_{22} = 30$ (12) acres (ha). It can be shown that this new solution improves the objective function from 7,946,540 to 7,544,950. Additional iterations show that the solution stabilizes at exactly the current solution. The current solution is therefore a local optimum and the solution algorithm terminates. □

5.1.3 Applicational considerations. As amply illustrated in the sections above, several issues arise when deterministic discrete-location and probabilistic activity-allocation models are placed side-by-side (O'Kelly 1987). The number of facilities or activities *p* may be decided exogenously by the decision-maker. For example, the value of *p* might be dictated by a target level of customer service or by a marketing strategy. If the value of *p* is itself a decision variable, then the problem becomes more complicated. Usually, *p* is a decision variable in models with explicit cost measures or coverage, such as the set-covering facility-location model (see the "Facility Location" chapter.) The discrete or deterministic problem with facility costs has highly efficient solution-methods. However, only very simple versions of the corresponding continuous or probabilistic problem have been solved before the TOPAZ model and the case study to be presented in this chapter. Two previous studies along these lines (Webber 1978, Leonardi 1983) point to services being provided at all feasible locations in a ubiquitous fashion. This is a direct contrast to all-or-nothing assignment to the nearest neighbor.

Regarding the performance of these models, Hodgson (1978) took the results of a doubly-constrained interaction-model with endogenous facility-location and compared them to a conventional *p*-median model. Ten possible locations were chosen at random from a square grid, and the Euclidean distances between the points were calculated. Intrazonal travel-times were defined to be one-half the distance from a zone to its nearest neighbor. Three facilities or centers from among the ten possible sites were chosen, and values of $b = 0.5$, 1.0, and 1.5 were used in the exponential form of the gravity model. In six out of the ten runs with $b = 1.5$, the activity-allocation or spatial-interaction model found the same location-solution as the deterministic or discrete p-median model. The results did not coincide in all cases because the minimum feasible-travel-times (and therefore the importance of *b*) vary between runs. Further research is needed to show how these solutions depend on distance-deterrence terms *relative* to this minimum feasible-travel-time.

Goodchild and Booth (1976) devised a probabilistic-interaction model to describe the patronage among 11 swimming pools in London, Ontario, Canada. Using this allocation model, two new pool-locations were chosen to minimize the total distance travelled by the patrons. The results of the analysis showed some discrepancy between the municipality's plans for a new site and the estimate of the

---

[2]The transportation problem allocates activities from supply depots to demand locations in the lowest cost. For an introduction and more explanation, see the "Prescriptive Tools" chapter in Chan (2000) or other mathematical-programming references

optimal location. Two important outcomes surfaced from the analysis. First, a realistic estimate of demand based on age profiles throughout the city was obtained. Second, despite the inclusion of a full-scale demand-model, the usage of the system seemed to be largely independent of the supply of facilities. The analysis showed that in only a few cases did the induced demands exceeds 10 percent of the current patronage level; and they concluded that "very few non-participants can be regarded as potential users of an increased supply of alternatives" for this recreational-activity. Webber et al. (1979) provided a more complex example of the interaction between facility location and residential demand for these facilities (such as shopping). In the model, both residential and facility locations are endogenous. Further, the model reflects the behavior of consumers through a large number of allowable trip-bundles; these bundles could contain 0, 1, or 2 visits to a facility. In the model application in Hamilton, Ontario, the level of demand proved to be relatively insensitive to facility-location patterns.

Both studies found that the highly-constrained nature of the interaction patterns made it impossible to induce variations in behavior by varying locations. Because they use parameters calibrated from existing travel patterns, there is really no way to judge from these results alone that demand is insensitive to facility locations. Nevertheless, two arguments support the assumption that perfectly-inelastic demand is a good initial approximation. First, many services have a fixed demand that must be filled regardless of the location of facilities. Second, once the initial estimate has been made, the conversion of the model to a full-scale-demand version is relatively straightforward. We visit many of these points in the following sections, ending with the state-parks study later in this chapter, which represents an integration of these ideas in a single model.

5.2 Private facility-location under spatial competition

We made a distinction between private and public facilities in the "Facility location" chapter. We suggested then that the success of a private facility is manifested in terms of its entrepreneurial viability such as profit and market share. A public facility, on the other hand, is often evaluated by its service to the population or demand, often called public convenience and necessity. Consider a demand pattern in which two homogeneous services or goods $k = 1,2$ are available. Each service is supplied by a single private facility (Thill 1992). Facilities are named after the service they supply (i.e., 1 or 2). Let p_k ($k = 1,2$) be the price charged for service i (exclusive of transportation cost). In addition, a composite service $k = 0$ is available at the same price throughout the market. It can be regarded as a mobile service "delivered to the door"; therefore transportation cost is nonexistent. Let p_0 be the given full price of this service.

5.2.1 Nash equilibrium. Consider an example in which the geographic space is a circumference of length l as shown in **Figure 5-2**. Circular markets are frequent substitutes for linear markets—whether bounded or unbounded—in models of spatial competition. Because there is no boundary, the effect of endpoints on optimal

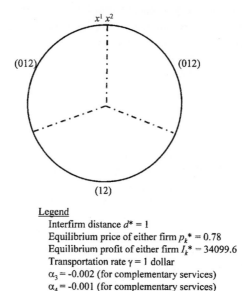

Legend
Interfirm distance $d^* = 1$
Equilibrium price of either firm $p_k^* = 0.78$
Equilibrium profit of either firm $I_k^* = 34099.6$
Transportation rate $\gamma = 1$ dollar
$\alpha_3 = -0.002$ (for complementary services)
$\alpha_4 = -0.001$ (for complementary services)
$(\cdot) = $ service baskets consumed

Figure 5-2 Example optimal pricing and location (Thill 1992)

strategies is nullified. An example of a circular market is the "space filling curve" discussed in the "Measuring Spatial Separation" chapter, in which a two or more dimensional space is transformed to a circle scaled between 0 and 1. Only the relative position of points on the circle is important, not the arbitrary beginning-point 0 and the end-point 1.

    5.2.1.1   SYMMETRICAL CASE OF TWO FACILITIES.   Without loss of generality, it is assumed that firms are located in the interval $[0, l/2]$ measured from the arbitrary origin 0 on the circumference. Let us denote their location by x^k ($k = 1, 2$). Therefore, the distance between facilities is given by $d = |x^1 - x^2|$. Customers or consumers are uniformly distributed at a uniform density around the circumference. When travelling to a facility, they follow the shortest path from their residence. The travel cost C is linear in the distance d: $C = \gamma d$, where γ is the transport rate. This cost is independent of the purchases (and carries no subscripts such as k). Therefore, it is possible for the demand to benefit from economies of scale on the quantity of each service that is purchased. Also, it benefits from *economies of scope*, or the number of services purchased at one time. The joint location of suppliers is not a necessary condition for demands to capture economies of scope. Depending on the relative location of facilities, it can be advantageous to visit two distinct locations at once instead of making a special-purpose trip to each. We call this a composite service.

Customers or consumers are perfectly informed and differ only in terms of their spatial location. They have the same fixed income I and the same preference structure defined by the following quadratic utility-function:

$$v(V_0, V_1, V_2) =$$
$$\alpha_0 V_0 + \alpha_1 (V_1 + V_2) - \alpha_2 (V_1^2 + V_2^2) - \alpha_3 V_1 V_2 - \alpha_4 V_0 (V_1 + V_2) \tag{5-18}$$

where $\alpha_0 > 0$, $\alpha_1 > 0$, $\alpha_2 > 0$, $4\alpha_2^2 \geq \alpha_3^2$, $4\alpha_2^2 \geq \alpha_4^2$, all of which are fixed parameters. Parameters α_3 and α_4 decide whether services are *substitutes* or *complements* to each other. If $\alpha_3 < 0$, services 1 and 2 are called complements; if $\alpha_3 > 0$, they are called substitutes. The two services are independent if $\alpha_3 = 0$. Similarly, the composite service 0 is a complement of services 1 and 2 if $\alpha_4 < 0$; these services are substitutes if $\alpha_4 > 0$, and independent if $\alpha_4 = 0$. As Equation (5-18) shows, services 1 and 2 are symmetric in consumption. Unit quantities of each provide the same satisfaction. In addition, they are identical regarding production (i.e., distribution). Both are produced at a constant marginal-cost c. If a demand at x purchases the basket (012) of three services, the problem is $\max_{V_0, V_1, V_2} v(V_0, V_1, V_2, x)$ subject to $I = b^U(x) + p_0 V_0 + p_1 V_1 + p_2 V_2$ where $b^U(x)$ in the budget constraint represents total travel costs to both facilities, whether they are visited on the same trip or on separate trips. Customers or consumers pay the full price—that is, the unit-price of the service p_k plus travel costs.

The demand-functions $V_k(x)$ ($k = 0, 1, 2$) are given by the first-order conditions. For example, **Figure 5-3** shows how the demand curve for a service V_1 may be constructed in principle from the indifference curve (an iso-utility function). The income line AB varies with each price change in V_1, as shown in AB^1, AB^2, AB^3 and AB^4. The points of tangency between the indifference curves and the expenditure lines define a price-consumption curve. The amounts of V_1 consumed $DD(1)$ may be projected on to the lower graph along with the corresponding prices to define the demand curve. The full specification of the customer's or consumer's problem finally requires that the conditions be determined for each basket of services to be selected. No closed-form expression has been obtained. Numerically determined, the resulting demand curves are generally discontinuous—each piece corresponding to the demand over a particular domain of prices at which a specific basket of services is purchased. The total sales D^k of facility k ($k = 1, 2$) are obtained by integrating the appropriate demands V_k over the circumference:

$$D^k(p_1, p_2, d) = \int_0^l V_k(p_1, p_2, x^1, x^2, x) \, dx \qquad k = 1, 2. \tag{5-19}$$

Notice that the interfirm distance d is used as an argument of the total sales-function instead of locations x^1 and x^2. A single locational argument is sufficient due to the absence of a boundary and the uniformity of our geographic space. Under the assumption of constant marginal-cost of production (as denoted by the unit production-cost c), 'profit' functions can be written as $I_k(p_1, p_2, d) = (p_k - c) D_k(p_1, p_2, d)$ for $k = 1, 2$.

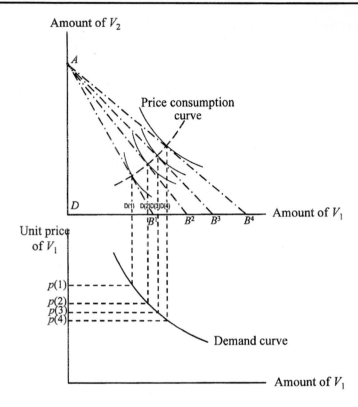

Figure 5-3 Derivation of a demand curve from a set of indifference curves

The profit-maximizing behavior of firms is conceptualized as follows. Each firm selects a best strategy (price or location) without any prior information from the competitor about its own action. In this *non-cooperative game* situation (or a game without collusion), the natural solution-concept to use is the *Nash equilibrium*. This equilibrium is a set of strategies such that no firm can improve its profit by a unilateral change of strategy. Specifically, the Nash criterion states the following. Suppose the status quo is (as a matter of convenience) evaluated at zero for both players in a game, and the players' payoffs are evaluated by the recipients at v_1 and v_2. Then a fair division is one that maximizes the product of these utilities, $v_1 v_2$. Nash derived this rather surprising arbitration formula under reasonable criteria for a fair division of the spoils—a no-regret criterion. It can be shown that the Nash equilibrium always ends up being a Pareto optimum or a non-inferior solution.[3] According to Baumol (1965), Nash has proved that *every* non-cooperative game in which each player has only a limited number of strategies open to him/her

---

[3]For a more detailed definition of Pareto optimality, or the win-win situation, see the chapter on "Multi-criteria Decision-making (MCDM)" in Chan (2000) or other MCDM references.

has at least one equilibrium point. In other words, there exists at least one combination of individual strategies such that, if they are employed by each player respectively, it will be unprofitable for any one of these players to switch to any other strategy. Thus there exist strategy combinations that have this self-policing feature: If all players but one follow this pattern the self-interest of this remaining player will also lead him/her to stick to the equilibrium pattern.

Example

We start with the utility function (Equation **(5-18)**) given by:

$$v = 1.0\,V_0 + 1.5(V_1 + V_2) - 0.001(V_1^2 + V_2^2) + \\ 0.002(V_1 V_2) + 0.001\,V_0(V_1 + V_2). \tag{5-20}$$

Choosing various values for the number of utiles v and plot V_1 against V_2, we have the family of iso-utility curves. Then we define expenditure curves for various ratios of products one and two. We know that the intersections of the iso-utility and expenditure curves can be joined to form the price-consumption curve. The price-consumption curves can then be used to derive a demand function. Since the sales make up the integral of the demand function (total demand) over the region (Equation **(5-19)**), the demand values were multiplied by 360 (uniform over the circle) to obtain sales values D_1. These values were paired against the unit price of V_1, p_1, which is obtainable from the slopes of the expenditure curves. The resulting graphs have these values when referenced against **Figure 5-3** (Burton and Chan 1993). □

V_1	D_1	p_1
185.69	$D(1) = 66{,}849.36$	$p(1) = 1.46$
374.63	$D(2) = 134{,}868.02$	$p(2) = 0.81$
557.91	$D(3) = 200{,}845.66$	$p(3) = 0.51$
738.41	$D(4) = 265{,}826.18$	$p(4) = 0.33$

5.2.1.2 ANALYTICAL STRUCTURE. Decisions regarding location and price are assumed to be sequential: first, location strategies are determined, then price strategies. The rationale for this two-stage process is that pricing is typically a short-term strategy while location is decided over the long term due to the considerable investments involved. To analyze the location game, the solutions of the price sub-game must be found first. Therefore, the solution of the location-price game goes backward. First, each firm anticipates, for any possible locational configuration of firms d, the price that maximizes profit once the price of the competitor is taken as given. The equilibrium pair $[p_1^*(d), p_2^*(d)]$ must be such that the following two conditions are met simultaneously:

$$I_1(p_1^*, p_2^*, d) = \max_{p_1} I_1(p_1, p_2^*, d) \tag{5-21}$$

$$I_2(p_1^*, p_2^*, d) = \max_{p_2} I_2(p_1^*, p_2, d).$$ (5-22)

The conditional-equilibrium prices are then substituted into the profit functions and the equilibrium distance d^* is determined by solving

$$I_k[d^*, p_1^*(d^*), p_2^*(d^*)] =$$
$$\max_d I_k[d, p_1^*(d), p_2^*(d)] \quad k = 1, 2.$$ (5-23)

Firms' planning relies on their knowledge of the consequence of any strategic move in geographic locations and in prices. The impact of any firm's—say firm k—changing its price and location on its own market-demand and on the demand addressed to the competitor can be decomposed into three effects: an aggregate substitution-effect, an aggregate income-effect, and *change-of-basket* effect. The first two effects are those identified for individual demands by neo-classical demand theory. They are negative on firm 1's own sales in general. While the aggregate substitution-cross-effects are positive if goods are complements and negative if goods are substitutes, the aggregate income-cross-effect is always negative. The cross change-of-basket effect corresponds to a shift in some marginal customers' consumption. The consumption shifts from a basket containing the service whose full price has been raised—say basket (12)—to another basket: for instance (012), (02), (2) or (0). This full-price increase is brought up by an increase of the unit-service price. Alternatively, it can be caused by the relocation of the firm supplying this service away from the location minimizing the visitation-tour's length—firm 1 in our example. The aggregate change-of-basket effect is always negative. The corresponding cross-effect on the firms' sales may be either positive or negative.

5.2.1.3 ALGORITHM. Once the demand curves have been derived as shown in Section 5.2.1.1, we could now start our search for the equilibrium prices (Burton and Chan 1993, Forsythe and Chan 1993). Each search is for a specific interfirm distance d. Given the problem symmetry, the travel-cost function is $2\gamma \sum_{i=0}^{180} |x(i) - d/2|$. The remaining computational search-procedures can now be summarized as follows:

Step 1. For a given price of the competing product 2, search along the range of demand for product 1, yielding a price ratio and hence price p_1^* that maximizes the profit of firm 1. Remember the profit function is given by $I_k = (p_k - c)D_k$, where k identifies the firm and c denotes the constant marginal-cost of production that is given in this problem.

Step 2. Having found the price of product 1 at this point, we hold it constant. Now search along the demand curve of product 2 for p_2^* to maximize profit for firm 2.

Step 3. This process continues until profits are maximized simultaneously for both firms. This constitutes the Nash price-equilibrium.

Step 4. Once we find the optimal prices for a specific inter-firm distance d, the search is repeated for all possible inter-firm distances. Determine the inter-firm distance d^* that maximizes overall profit. This constitutes the Nash distance-equilibrium. ∎

5.2.2 Some preliminary results. Most of the literature on competition in space is implicitly based on the premise that markets for physically different goods and services are perfectly independent. Yet this postulate seems inappropriate in most market situations. The Nash-equilibrium paradigm above fully accounts for inter-dependencies on the demand side, and where goods and services supplied are not independent in consumption. In addition, consumers can capture economies of scope in travel by purchasing several goods or services on one trip. Unfortunately, the model is far from analytical for closed-form solutions. An enumerative search algorithm was used to solve Equations **(5-21)** through **(5-23)**. An initial price is solved for given firm-locations as a start. The search for optimal prices is then repeated for other inter-firm distances. It is assumed that firms have 360 possible locations that are regularly spaced around the circumference ($l = 360$ units). The individual income I amounts to 600 dollars, the unit price of the composite service p_0 is 0.60 dollar, the marginal production-cost c of symmetric services 1 and 2 is 0.50 dollar, and $\alpha_0 = 1.0$, $\alpha_1 = 1.5$, $\alpha_2 = 0.001$ as mentioned above. To represent the inter-dependence between markets, these input ranges are specified: transportation rate γ ranges from 0.75 dollar to 1.25 dollars, α_3 ranges from -0.002 (for complementary services) to 0.002 (for substitute services), and α_4 ranges from -0.001 (for complementary services) to 0.001 (for substitute services). A higher transportation-rate implies large potential-savings in bundling purchases on the same trip and, other things equal, a strong mutual dependence among markets. On the contrary, a low transportation-rate suggests loosely dependent markets.

Examples of two particular simulations are shown in **Figure 5-2** and **Figure 5-4**. The former figure shows strong complementary between the two markets. The latter curve shows the typically U-shape and local maxima for distances occurring near 1 or 180. Competition is more intense when firms are neither too close nor too far apart. The complementary between the composite service 0 and each of the other two services makes consumption baskets (012), (01), and (02) quite desirable for consumers and suppliers. With a decrease in this complementary, an increasing number of customers will stop consuming the composite service. They will substitute basket (12) for (012), basket (1) for (01), and basket (2) for (02). In addition, due to change-of-basket effects, it becomes optimal for firms to locate at the antipodes of the circumference. By the same token, overlapping market areas for services 1 and 2 give way to mutually exclusive areas. These factors explain the service baskets in the example of **Figure 5-2**. They also explain the antipodal positions of x^1 and x^2 in **Figure 5-4** as well.

5.2.3 General observations. Thill (1992) performed comprehensive parametric simulation of the simple demand-system outlined in the Nash-equilibrium discussion above. The results below pertain to only two firms in

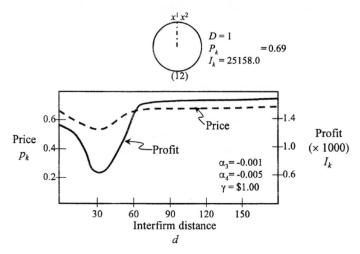

Figure 5-4 Example price and profit as a function of interfirm distance (Thill 1992)

symmetrical situations—i.e., their utility functions are symmetrical. The main results of the numerical analysis of optimal-price-location strategies are:

a. Depending on the parametric specification of the model, two alternate locational configurations of firms happen to be optimal: one of minimum differentiation, the other of maximum differentiation.

b. It is optimal for firms to seek joint locations if the goods and services they supply sufficiently complement each other. On the other hand, price-location competition results in perfect geographic-dispersion of firms if goods and services are either remote complements or substitutes. With these results, the simulation results rule against the conclusion reached earlier by Stahl (1982) and by Fujita et al. (1988) that the complementary among goods and services is a sufficient condition for spatial agglomeration. According to Thill (1992), the property needs to be restated accordingly: Complementary appears as an agglomerating factor whose impact can be thwarted by other mechanisms as enumerated below (Anderson and Neven 1991).

c. Other things being equal, it is optimal for firms to locate jointly if the goods and services they supply complement the composite good sufficiently; otherwise, they locate at antipodal sites.

d. Other things being equal, rising transportation-costs imply a spatial dispersion of retail firms. Ingene and Ghosh (1990) reached the same conclusions with a model where the potential for economies-of-scope in travel creates market interdependencies.

e. Optimal prices set by firms at their optimal spatial configuration exhibit complex patterns of variation. We have variation in parameters of complementary /substitutability among goods and services, with transportation costs, and with the geographic dispersion of the game solution.

f. If pricing is designated as the decision-variable of firms at fixed-locations, there is no systematic relationship between firms' optimal prices and their locations.

The numerical simulations have proved valuable. It provides comparative statistics of the optimal price-location solutions for the problem of inter-market competition. One can now evaluate linkages between markets in shaping the firms' competitive strategies. The symmetry in the equilibrium properties no longer holds when the symmetry of the utility function is relaxed. In such a case, one good or service would be in higher demand and its supplier would be granted a clear competitive advantage over its rival. The presence of a third competitor is also likely to disturb the symmetry property.

Locating facilities in the face of competition makes up the competitive-location problem. Suppose demands or customers at each node of a network choose from available facilities to minimize the distance travelled. The problem is to find a set of facilities that is stable—in the sense that each facility is adequately covering demands and no competitor can successfully open any additional facilities. Viewed in a different light, the focus is on minimizing the demands who would prefer some alternate location to the chosen one. A *condorcet point* is any point of a network such that there is no other point closer to a strict majority of perfectly inelastic demands. Stated differently, it is a location for which no more than half the customers would prefer any other point on the network. In fact condorcet points and medians coincide on a tree. For some networks, however, there are no condorcet points. This suggests looking for points such that the maximum number of consumers closer to another point is minimum. Such points are called S*impson points*. Thus a Simpson location is least objectionable in that no other location has fewer customers wanting to move to another point. For a treatment of these developing ideas on competitive location, the reader is referred to the "Spatial Equilibrium" chapter, Hansen et al. (1990) and Hakimi (1990). We will supplement these idealized studies with a more thorough study below. It outlines the tradeoffs between aggregate substitution effects, aggregate income effects, and change-of-basket effects in response to the firms' strategic choices of prices and locations.

5.3 Public facility-location under discretionary demand

Most of the facility-location models described in this book thus far have demands that are fixed, or perfectly inelastic. Such an assumption is valid for such requirements as observing a particular satellite as discussed in the "Facility Location" chapter. In many other facility-location decisions, we are faced with *demands* rather than requirements, and these demands can be discretionary. This happens when there is more than one type of satellites, each of which can be covered by one or more tracking stations. These stations offer different services at different months of the years. A better example is the location of recreational facilities, where the demand is certainly not fixed. One can always forego a trip to a state park, or replace it with a fishing trip on the river. In these discretionary demands, or demands that are sensitive to the service-level rendered, most classic p-median literature fails to address this class of problems.

5.3.1 Model formulation. Discretionary demands arise with overlapping service regions where customers or consumers have a choice between more than one provider of service, and that they can substitute one type of service with another. Under these conditions, least-cost objective-function is an inadequate evaluation-measure of performance. Particularly in public facilities, the maximization of "social welfare" minus cost is more appropriate. Under some common assumptions about income distribution, consumers' surplus is widely accepted as a measure of user benefit.[4] Perl and Ho (1990) pointed out an important previous finding. *If the consumers' surplus (CS) function is convex with respect to distance, optimal locations can be found among nodes of the network under CS maximization.* Most common demand-functions satisfy this property. We will cite below the results of Perl's investigation into this subject without the derivation details.

Figure 5-5 shows an example of public facility-location under discretionary demand. Node A has the larger demand of the two nodes, where a node has a certain *intra*nodal area instead of being a discrete, dimensionless point. The location of the facility can be expected to depend on the internodal distance d_{AB}; on the relative demands at A and B, as defined by a_i'—the demand-generating potential of node i; d_{ii}, the intranodal distance at node i; d_{if}, and the maximum travel-distance for node i. Three separate demand-functions were discussed, including a linear function, a modified constant-elasticity function, and a modified exponential-function. An example of the linear demand-function looks like $V_{ij} = 1\ 0(1-d_{ij}/d_{if})$ where i is node A and j is node B in **Figure 5-5**. We further define these ratios:

$$s' = \frac{a_B'}{a_A'}, \quad u = \frac{d_{Bf}}{d_{Af}}, \quad \bar{t}' = \frac{d_{BB}}{d_{AA}} . \tag{5-24}$$

Based on the 'demand' as characterized by the ratios defined by the above equation, the location of a facility is found. Nine combinations of the three ratios s', $\bar{t}'$ and u are possible as shown in **Table 5-1**, characterizing the different demand types.

For the case of a linear demand-function, Perl and Ho pointed out that under types V, VI, VII, and IX demand characteristics, or when $s' < 1$, $u \le 1$ and $\bar{t}' \ge 1$, the optimal location of the facility is always at node A. This result is the same for the classic p-median problem with least-cost objective. Under types I, II, III, IV and VII, the optimal location of the facility depends on the actual relative values of the demand characteristics as well as on internodal distance. When generalized to a network, a Maximum Consumers' Surplus Location Problem (MCSLP) can be defined. This refers to the case when there exists no facility cost, in which the CS

---

[4]For a more detailed discussion of this statement, see the chapter on "Economics."

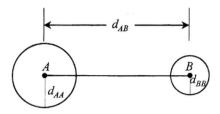

Figure 5-5 Example of facility - location under discretionary demand

$s'<1$	$u''>1$	$u''=1$	$u''<1$
$\bar{t}'<1$	I	II	III
$\bar{t}'=1$	IV	V	VI
$\bar{t}'>1$	VII	VIII	IX

Table 5-1 Characterizing demand function for facility location

is maximized in a binary program as follows: $\max \sum_{i \in I} \sum_{j \in I} S'_{ij} x_{ij}$ subject to Equation
(5-4) with equality replacing the ≥ plus Equations **(5-6)**, and **(5-3)**. In the objective
function, S'_{ij} stands for the consumers' surplus associated with the demand at node
i when i is assigned to facility at node j. An alternate objective-function is the net-
benefit measure, defined as the difference between CS and the cost of providing the
service, with λ' as a weight to put CS and cost in the same unit:

$$\max \lambda' \sum_{i \in I} \sum_{j \in I} S'_{ij} x_{ij} - \sum_{j \in I} c_j x_{jj}. \tag{5-25}$$

Perl and Ho named this the Maximum Net Benefit Location Problem (MNBLP). In
the above objective-functions, the values for S'_{ij} are determined exogenously by
analytical functions that depend on the demand characteristics as discussed above:
$S'_{ij} = f(d_{ij}, d_{if}, d_{ii}, a'_i)$.

5.3.2   *Illustrating solution methodologies*.   Solution of these models
invariably involves a large integer-programming (IP) model. This may present
computational difficulties particularly when the traditional LP-relaxation leads to
a non-integer termination. We have illustrated this phenomenon in our previous

discussion on network solutions. Several early studies proposed reducing the number of constraints of the form of $\sum_{j \in J} x_{ij} \geq 1$, which often leads toward an integer-solution termination. Other solution methods include dual-based branch-and-bound-methods as discussed in the "Optimization" appendix in Chan (2000). Perl concentrated on the reduction of the number of assignment variables x_{ij} in the primal formulation without affecting the likelihood of obtaining an integer solution from LP relaxation.

Experiments were conducted in a network from Austin, Texas, consisting of 55 traffic zones. The variable-reduction methods trimmed down the number of variables and constraints by 55 to 77 percent depending on the demand function used. The reduction methods are based on the observation that many potential assignments are not likely to be part of any optimal solution. For each demand node i, only a fraction of the assignments that correspond to a subset of the nearest nodes to i need to be represented explicitly. The reduction method for the problems with no facility-cost establishes a dummy node. This is to represent the assignment of any node i to a node not in the subset of nearest nodes. If the optimal solution to the reduced problem does not include any assignment to the dummy node, it is the optimal solution to the original problem. Suppose the solution to the reduced problem includes an assignment of any node i to the dummy node. Then the subset of assignments of node i that are represented explicitly is expanded. The reduction of the MNBLP model is carried out in two stages. In the first stage, any assignment of node i to node j that would generate no CS is replaced by an assignment to a dummy node. In the second stage, the assignment of node i to another node j is eliminated under the following condition: If establishing a facility at node i would increase the monetary value of the CS generated from i—compared with that generated from an assignment to j—by a value that exceeds the facility cost at node i.

5.3.3 Concluding comments. These are the main findings from the above experiments:

a.   Locational behavior becomes very disparate under different demand-functions as the size of the network increases.
b.   As demand decreases more rapidly with distance, more facilities are located in nodes with high demand-density.
c.   Locational behavior differs more between discretionary and perfectly-inelastic demands as the size of the network increases.
d.   As demand decreases more rapidly with distance, the locational behavior differs more between discretionary and perfectly-inelastic demands.
e.   The maximization of net benefits under MNBLP is likely to lead toward fewer facilities than the minimization of total cost under the classic p-median problem.

These observations are very interesting findings. In the following sections, we will pick up this subject on discretionary demand again. Instead of a discrete facility-location model, we will consider a combined discrete-continuous model of location-allocation. Such a discussion will provide further insight into public-facility location under discretionary demand. Aside from a disparate formulation, solution techniques tend to be quite different also. They are often tailor-made to the problem at hand, as illustrated amply by the heuristic schemes discussed throughout the chapter on discrete facility-location problems.

The above results, particularly the expression in Equation **(5-25)**, can be put into more general terms based on Bertuglia and Leonardi (1982). As reported by Colorni (1988), Bertuglia and Leonardi put consumers' surplus maximization within the framework of an entropic model[5], which takes on the following form:

$$\max \left[\lambda' V(x_{ij}) - \sum_{i \in I} \sum_{j \in I} d_{ij} x_{ij} - \sum_{j \in I} c_j y_j \right] \tag{5-26}$$

$$\sum_{j \in I} x_{ij} = z'_i \quad \forall i \in I \tag{5-27}$$

$$\sum_{i \in I} x_{ij} \le \bar{P}_j y_j \quad \forall j \in I \tag{5-28}$$

$$x_{ij} \ge 0 \quad \forall \; i, j \in I (i \ne j) \tag{5-29}$$

$$y_j \in \{0,1\} \quad \forall j \in I. \tag{5-30}$$

Here, $V(x_{ij})$ represents the 'scattering' of a fixed amount of flow among different demands i from origin j, according to the spatial price d_{ij}. $\bar{P}_j$ is the capacity of the supply source at facility j. An entropy function for V is assumed. Taking the usual logarithm and Stirling approximation, the first term in Equation **(5-26)** can be expressed as

$$\frac{1}{b} \ln V(x_{ij}) = -\frac{1}{b} \sum_{i \in I} \sum_{j \in I} x_{ij} (\ln x_{ij} - 1) + K \tag{5-31}$$

where b is the exponent of the exponential function measuring the accessibility between i and j: $f(b, d_{ij}) = \exp(-b d_{ij})$. Equation **(5-26)** now becomes

$$\min \left[\frac{1}{b} \sum_{i \in I} \sum_{j \in I} x_{ij} (\ln x_{ij} - 1) + \sum_{i \in I} \sum_{j \in I} d_{ij} x_{ij} + \sum_{j \in I} c_j y_j \right]. \tag{5-32}$$

As b becomes large, the overall model degenerates into the traditional cost-minimization facility-location model. Thus as the spatial price becomes unbearable, demands patronize the nearest facilities to avoid travel. This result agrees with Perl and Ho's observation that as demand decreases more rapidly with distance, more facilities are located at nodes with high demand-density. Between this more general formulation and the Perl–Ho approach, this formulation is conceptually more

---

[5]For a discussion of entropic model, see the chapter on "Prescriptive Tools" in Chan (2000).

complete, with its full recognition of non-unitary assignments from supply source to demand. However, this comes with it a computational price tag. The Perl–Ho case study above illustrates how a linear model of this sort can be solved by network-flow techniques. The nonlinear case proposed by Bertuglia and Leonardi becomes more problematic—particularly when both continuous and discrete variables are involved. We will have a solution for this dilemma in the following sections of this chapter.

Example

O'Kelly (1987) conducted an experiment on Hamilton, Ontario, Canada using a model similar to the one above. Two b values were used, 0.025 and 0.20. The results on activity distribution are very different. The smaller impedance-parameter b results in a clustered set of facility locations, and the large parameter shows a more dispersed set of locations. The major contrast is that the land-use pattern with $b = 0.025$ exhibits a very uniform distribution of facilities, with little or no preference for the nearest facility. On the other hand, the pattern with a larger impedance-parameter ($b = 0.20$) displays greater dispersion on facility locations as well as more marked use of the nearest facility. This results in shorter average-travel-times in general.[6] □

5.4 Generation, competition, and distribution considerations

It can be seen that there are common factors between private and public facility-location. The figures-of-merits may be different between these two types of problems. The underlying principles are, however, quite similar. Through the spatial-interaction parameter b, for example, one model can be shown to be a special case of the other. Having said that, what are the fundamental factors underlying discrete facility-location and continuous activity-allocation then? Based on the discussions in the previous sections of this chapter, we identify three. The first is the measure of benefits and costs (or the figures-of-merit as mentioned,) which drives the rest of the other two considerations. The second is the complementary or substitution between the goods-and-services rendered, which have profound implications on the customer-demand and relative locations of the providing facilities. Once these two issues have been resolved, the last consideration is the appropriate number of facilities to build and the size of each facility. This is to provide the best response to the customers or public.

5.4.1 Locational-surplus maximization. As soon as we include spatial interaction in the classic location- models, the total spatial-cost as represented by Equation **(5-9)** becomes suspect. Minimizing transportation cost alone is considered

---

[6]Notice this result is consistent with that found in the numerical example of the "Economic Methods" chapter of Chan (2000), under "Economic basis for the gravity model" section.

inadequate as a figure-of-merit when choice is embedded in the model.[7] As mentioned, an alternate welfare function, generally referred to as consumers' surplus, is more appropriate. It is only when there is no differentiation between facilities that demands or customers patronize the nearest facility, and transportation cost becomes the degenerate measure of performance. Suppose there is a choice of facilities. Then the customers' benefit-increase, when a system is improved from spatial cost $C^0 = [C^0_{ij}]$ to $C' = [C'_{ij}]$, is measured by[8]

$$\frac{1}{b} \sum_{i=1}^{n'} V_i \ln \frac{\sum_{j=1}^{p} W_j^{a'} \exp(-bC'_{ij})}{\sum_{j=1}^{p} W_j^{a'} \exp(-bC^0_{ij})} . \qquad (5\text{-}33)$$

where $f(b, C_{ij})$ takes on the exponential format $\exp(-bC_{ij})$. This expression (Williams 1976) can be maximized in the decision to locate p un-capacitated facilities at n' given locations. It should be noted that in this definition of locational surplus, only transportation costs are allowed to vary; the *in situ* attractiveness of facilities, W_j, is held constant. It is possible to include both types of changes simultaneously when both travel and attractiveness contribute toward the utility of tripmaking.

It can be shown, once again, that the consumers'-surplus expression degenerates into total transportation-cost when the parameter b goes to infinity. The consumers'-surplus expression in Equation (5-33) simply vanishes and the Θ_{ij} term likewise disappears from Equation (5-1). The classic p-median model takes over with its travel-distance minimization and non-overlapping markets. Thus the degree of market-area overlap depends on the value of the b parameter. A small value of b means large overlapping market areas through longer trips, and a large value of b means little market overlap with least travel—as mentioned previously.

Given a distribution pattern as described by the value of b, an optimization model can be constructed to prescribe the size of the facilities. Specifically, this finds the value of the *nonnegative* W_j, by maximizing the surplus expression of Equation (5-33), $\max_{W_j} \sum_{i=1}^{n} V_i \ln \sum_{j=1}^{p} W_j \exp(-bC_{ij})$ where the denominator is dropped since it is a constant. Following the doubly-constrained format of a gravity model, the sizing of facility W_j involves the allocation of $\Sigma_j W_j = W$ among up to p facilities ($p \le n'$). Let us form the Lagrangian L and carry out the necessary first-order conditions for maximization as shown in the chapter on "Prescriptive Tools" of Chan (2000):

---

[7]Again, this statement was explained in the "Economic Methods" chapter of Chan (2000), complete with an extensive numerical example under the "Economic basis of the gravity model" section. Alternatively, consult Cochrane (1975).

[8]For a derivation of this expression, the reader may wish to consult the "Doubly-constrained gravity model" section in the "Economic Methods" chapter in Chan (2000).

$$L = \sum_{i=1}^{n} V_i \ln \sum_{j=1}^{p} W_j \exp(-bC_{ij}) + c\left(W - \sum_{j=1}^{p} W_j\right) + \sum_{j=1}^{p} \gamma_j W_j . \tag{5-34}$$

$$\frac{\partial L}{\partial W_j} = \frac{1}{W_j} \sum_{i=1}^{n'} \frac{V_i W_j \exp(-bC_{ij})}{\sum_{j=1}^{p} W_j \exp(-bC_{ij})} - c = 0. \tag{5-35}$$

Here c and γ, are dual variables or Lagrange multipliers. Using V_{ij} to denote the trips from i to j as distributed by the gravity model, the above expression can be simplified to $\left(\sum_{i=1}^{p} V_{ij}\right)/W_j - c = 0$ or $\sum_{i=1}^{n'} V_{ij} = c W_j$. Thus the facility is sized according to the level-of-demand $\Sigma_i V_{ij}$ and following the "unit cost" c at each facility to cater for this demand (see also the forthcoming chapter on the "Activity-allocation/ derivation Models").

If the aggregate demand V of the study area is increased, proportional increase in facility size W_j would result. On the other hand, should spatial costs (including the fixed cost at facility j) vary among i–j pairs, a selective placement of facilities will occur. This is similar to the exogenous specification of the p number of facilities to be located. Expansion of the central-place system would, in this case, size the facilities according to not only V, but also the relative spatial-cost advantages.

5.4.2 Elasticity of demand for service. Most facility-location models assume a perfectly-inelastic demand V_i. Here we broaden the concept to include cases where demand is a function of the spatial-price system. In other words, potential interaction between i and j, V_{ij}, will be less when demand i, V_i, is far away from facility j: $V_{ij} = V_i \exp(-bC_{ij})$. V_i is now interpreted as the potential level-of-demand for a customer or consumer located at i if a 'center' is also co-located at i. Thus $V_{ii} = V_i$ under this definition. Such a relationship is captured by the parameter b. When b is zero, V_{ij} degenerates to V_i and demand is perfectly inelastic; and a non spatial allocation is obtained as shown in Equation **(5-8)**. When $b > 0$, demand is responsive to the spatial price, resulting in the more general case reported here.

It can be argued that consumer demand is not only affected by spatial separation C_{ij}, but also by the in-situ attractiveness of facility j, W_j. A spatial-interaction model that considers this more complete picture can be constructed:

$$V_{ij} = V_i \exp\left[b'\left(\ln \sum_{j=1}^{p} W_j \exp(-bC_{ij}) \right) \right] \Theta_{ij}(W_j, b, C_{ij}) \tag{5-36}$$

where Θ_{ij} is defined in Equation **(5-2)** when $f(\cdot)$ assumes the exponential form. Now the parameters b and b' will be shown below to determine the elasticities. When b' is zero, the "trip-generation" term $\exp[b'(\ln\sum_{j=1}^{p} W_j \exp(-bC_{ij}))]$ vanishes and the above model degenerates into $V_{ij} = V_i \Theta_{ij}(W_j, b, C_{ij})$ showing again the demand V_i distributed among facilities according to the dispersion parameter b. Once a demand model such as the above is set up, neoclassical economic-concepts of *income* and *substitution* effects can be readily discerned. For example, if the distance to all the facilities increased in identical proportions for customers or consumers in one zone, the pattern of trip distribution would remain the same. But the level of effective demand would decline, as clearly shown by the trip-generation term in Equation **(5-36)**. This is referred to as the income effect. On the other hand, if a large new facility was co-located in zone i, consumers in that zone would now tend to patronize that facility, rather than more distant facilities in other zones j ($j \neq i$) as they did in the past. This is called the substitution effect.

One can derive an expression for the consumers' surplus for the demand model of Equation **(5-36)**, from which the following objective function can be derived for consumers'-surplus maximization (Neuberger 1971): $\sum_{i=1}^{n'} V_i \exp(b' \ln \sum_{j=1}^{p} W_j \exp(-bC_{ij}))$.

The above equation shows that the aggregate consumers' surplus for the study area is made up of individual components for origin i and facility j. The distribution effects among these origins and facilities become a central issue in this type of modelling, among which are the equity concerns involved in the placement of many facilities. An explicit formulation of this, for example, can be found in the *center* problems in facility location, where the weighted distance to the furthest consumer is to be minimized.

5.4.3 Competition and market share. An important property in analyzing choice among service facilities is the "independence of irrelevant alternatives" (IIA) property. This property suggests that the ratio of choice probabilities of any two alternatives is unchanged no matter which other alternatives are in the choice set. In other words, if a third alternative k is introduced into a choice set when two alternatives m and n are already present, then k will draw its share proportionately from m and n in the same ratio as their current shares. IIA can be explained by examining the relative odds of choosing one destination i over another destination j as:

$$P(i)/P(j) = W_i f(b, C_{hi})/W_j(b, C_{hj}) \qquad i=m,n,k; \; j=m, \; n,k \; (i \neq j). \qquad \textbf{(5-37)}$$

Here the relative odds between reaching two different destinations from origin h are only a function of the attributes of these two and are independent of any other alternatives that may be available. *IIA allows activity-allocation models to consider new service facilities, while traditional facility-location models can only explain choice among a pre-determined number of service facilities.*

Most choice models work at the aggregate or group level, which assume homogeneity among the individual decision-makers. In the gravity model or logit model, this means that the valuation of time or the tradeoffs between attributes for individuals in a group is the same, as explained in the "Descriptive Tools" chapter in Chan (2000) under "Aggregate vs. disaggregate modelling." Also shown in the "Descriptive Tools" chapter is the fact that disaggregate models developed at the individual level can be applied at the aggregate level only after exercising due caution (and vice versa). Correspondingly, the IIA property needs to be interpreted carefully in a disaggregate vs. aggregate model (Gensch and Ghose 1997). Similarly, it is just not sufficient to evaluate IIA for a singe pair of alternatives, whether $m-n$, $m-k$, or $n-k$, as suggested in Equation **(5-37)**. It is important to take a "full-choice set" perspective. In other words, possible IIA violations should be examined for all pairs of alternatives in the choice set. We argue that for competition and distribution models at the aggregate level, IIA should be considered for the full choice-set rather than for an independent single-pair of alternatives. In this section, we will show that unless there is perfect homogeneity of choice, it is virtually impossible to meet the IIA assumption at the aggregate level. In doing this, we try to make clear some of the differences between individual- and aggregate-level IIA-violations, and under what conditions knowledge of one implies knowledge of the other. Let us introduce these ideas via a couple of examples.

Example 1

In this first example, we will show that the IIA property is observed both at the disaggregate and aggregate levels for a pair of alternatives m and n, but not necessarily for the full choice set. Suppose there were only these two destinations, and a new destination k becomes available. This new destination draws proportionately from both m and n. This will ensure that there is no resultant IIA violation with respect to m-n. But there would be IIA violations with respect to, say, the k-m pair and the k-n pair.

Thus according to **Table 5-2**, $P[m,(m,n)]/P[n,(m,n)] = \{215/300\}/\{85/300\} = 2.53$ and $P[m,(m,n,k)]/P[n,(m,n,k)] = \{150.5/300\}/\{59.5/300\} = 2.53$ at the aggregate level. Again for each individual in a disaggregate level, $P[m,(m,n)]/P[n,(m,n)] = \{60/100\}/\{40/100\} = 1.50$ and $P[m,(m,n,k)]/P[n,(m,n,k)] = \{42/100\}/\{28/100\} = 1.50$ for individual 1, and similarly for individuals 2 and 3. It can be shown that IIA holds in that $P(m,(m,n))/P[n,(m,n)] = P(m,(m,n,k))/P(n,(m,n,k))$. However for the complete choice set, $P[m,(m,n,k)]/P[k,(m,n,k)] = \{42/100\}/\{30/100\} = 1.40$ for individual 1, $P[m,(m,n,k)]/P[k,(m,n,k)] = \{56/100\}/\{30/100\} = 1.87$ for individual 2, and $P[m,(m,n,k)]/P[k,(m,n,k)] = \{52.5/100\}/\{30/100\} = 1.75$ for individual 3. These three ratios, applied toward alternatives m and k here, are certainly different. □

Example 2

It is possible to have pairwise preference-heterogeneity. This would violate the aggregate IIA property. Here the disaggregate ratio as shown in Equation **(5-37)** may be the same for m and n (whether k is present or absent). But the aggregate ratios will be different. To illustrate this, consider the data in **Table 5-3**. For the first individual, $P[m,(m,n)]/P[n,(m,n)] = \{60/100\}/\{40/100\} = 1.5$, and $P(m,(m,n,k))/P[n,(m,n,k)] = \{30/100\}/\{20/100\} = 1.5$. For the second individual, $P[m,(m,n)]/P[n,(m,n)] = \{80/100\}/\{20/100\} = 4$, $P(m,(m,n,k))/P[n,(m,n,k)] = \{72/100\}/\{18/100\} = 4$. For the third and final individual, $P[m,(m,n)]/P[n,(m,n)] = \{75/100\}/\{25/100\} = 3$, $P(m,(m,n,k))/P[n,(m,n,k)] = \{60/100\}/\{20/100\} = 3$. This shows that IIA holds at the disaggregate level for individual 1, 2, and 3. However, at the aggregate level, $P[m,(m,n)]/P[n,(m,n)] = \{215/300\}/\{85/300\} = 2.53$ and $P(m,(m,n,k))/P[n,(m,n,k)] = \{162/300\}/\{58/300\} = 2.80$. Since $2.53 \neq 2.80$, there

Individual	Pairwise choice		Full-choice set		
	m	*n*	*m*	*n*	*k*
1	60	40	42	28	30
2	80	20	56	14	30
3	75	25	52.6	17.5	30
sum	215	85	150.5	59.5	90
Average	73.67	28.33	50.17	19.83	30

Table 5-2 Choice between an existing pair at the aggregate and disaggregate levels (Gensch & Ghose 1997)

Individual	Pairwise		Full-choice set		
	m	*n*	*m*	*n*	*k*
1	60	40	30	20	50
2	80	20	72	18	10
3	75	25	60	20	20
sum	215	85	162	58	80
Average	71.67	28.33	54	19.33	26.67

Table 5-3 Heterogeneous individuals example (Gensch & Ghose 1997)

is IIA violation at the aggregate level. Notice that instead of 300, the average among the three individuals, or 100, can be used. □

The conclusion is clear. If one considers the full set of alternatives, it is not possible to satisfy IIA at both individual and aggregate levels unless there exists complete preference homogeneity. Even when the IIA assumption is valid for each individual, IIA is always violated at the aggregate level unless all individuals have identical choice patterns. This has serious implications for the wide range of empirical works that use statistical techniques that depend on IIA when using real world samples of individuals. The *nested logit-model* relaxes the IIA property of the multinomial logit-model by grouping alternatives based on their *degree of substitution*. Alternatives in a `nest' exhibit an identical degree of increased sensitivity compared with the alternatives not in the set. An assumption maintained in the nested logit is that the degree of sensitivity among nested alternatives is invariant across individuals making the choice. However, this assumption might be untenable in many practical situations. Bhat (1997) proposed an extension of the nested logit model to allow heterogeneity (across individuals) in the covariance among nested alternatives based on observed individual characteristics. He formulated and estimated a nested logit-model that accommodates differential levels of sensitivity among nested alternatives across individuals. The resulting nested logit-model with covariance heterogeneity generalizes the extant, standard nested logit-model which imposes an equal-correlation restriction among utilities of nested alternatives across all individuals. He used a full-information maximum-likelihood-method for estimating his model, showing that the existing nested logit-model by contrast has an inferior data-fit and biased estimates.

5.4.4 Size and number of facilities. An individual facility often tries to locate in order to capture the lion share of the patronage among p competing facilities:

$$\max \ V_j = \sum_{i=1}^{n'} V_{ij} \qquad j=1,\dots,p. \tag{5-38}$$

With perfectly-inelastic demands, the result of this maximization replicates the well-known pattern of spatial clustering. When demand elasticity is introduced, spatial clustering is an exception rather than a rule. Compare the hub-and spoke assignment of demands to facilities vs. overlapping service zones in **Figure 5-1** as examples where clustering is present or absent respectively.

In a patronage-maximization model on a network, a 'watershed' area defined by a maximum distance of S' is defined, beyond which no customers or consumers can be captured:

$$V_{ij} = \begin{cases} V_i - \theta C_{ij} & C_{ij} < S' \\ 0 & C_{ij} \geq S' \end{cases} \tag{5-39}$$

where θ is a parameter reflecting the decline in demand per unit-of-spatial-separation between i and j. Following the p-median model described in the "Facility Location" chapter, the "maximal coverage" now looks like $\max \sum_{i=1}^{n'} \sum_{j=1}^{n'} V_{ij} x_{ij}$. If there are enough facilities in the system to ensure that the nearest facility to each demand node i is within a distance of S', the optimal solution to this model is the same as that by the conventional p-median formulation.

Instead of a given number of facilities, it is desirable to compute the number that is required. A formulation may seek to minimize the number of facilities required to completely satisfy the demand of a set of customers or consumers. An example is found in the set-covering problem discussed previously in the facility-location chapter. This would provide the absolute minimum number of facilities irrespective of the distance that a demand needs to travel to access a facility. To insert an element of equity into the problem, a maximal service-distance S' is again introduced. This restricts the service range of a customer from a facility. This alternate model, location set-covering, would determine a more realistic number of facilities required. Finally, the maximal-coverage location-problem seeks to trace the service coverage by a fixed number of *p* facilities as one relaxes the maximal service distance S. This would parametrically trace the demand served as a function of the service rendered, and represents a more useful set of information where tradeoff is required.

All these models are limited in that they made simplifying assumptions. First, there is often no capacity placed on the facility. Procedures that would redress this problem exist; but they are often heuristic in nature. Second, the cost functions used are by-and-large quite simplistic—either a unit cost is assumed or at most an additional fixed-cost associated with the facility is included. They do not necessarily reflect an adequate utility function for travel; neither do they adequately model the investment decisions often associated with improving a facility. In order to perform the latter function satisfactorily, it is required to represent the attributes that characterize the facility of interest. Finally, there exists no procedure to capture the demand function at the origin *i*, V_i. Rather than a fixed number, to be allocated according to accessibility, V_i needs to be dependent upon the socioeconomic variables at the origin. This is in addition to the attraction at the facilities.

5.5 A generalized location-allocation model

To address some of these shortcomings, consider the following model-form for ubiquitous demand and discrete facility-location (Chan and Carroll 1985):

trips from demand-origin *i* to facility *j*
    = function of (socioeconomic factors at origin *i*, facility at *j*,
      accessibility between *i* and *j*).

In mathematical terms, the above can be written as $V_{ij} = f(W_i, W_j, C_{ij})$ where C_{ij} is the generalized travel-cost between *i* and *j*. One possible form of the model, for example, is multiplicative

$$V_{ij} = K W_i^\alpha W_j^\beta C(C_{ij}) \tag{5-40}$$

where $C(\cdot)$ is the trip-distribution function from *i* to *j*, and K, α, β are calibration constants. For both conceptual as well as mathematical reasons, it is convenient to separate Equation **(5-40)** into the trip generation and distribution parts:

$$V_{ij} = \{K_0 W_i^\alpha [X_i(C_{ij})]^b\} \{K_i W_j^\beta C(C_{ij})\}. \tag{5-41}$$

In the above equation, the first term is the trip activity-derivation model at the origin *i* (V_i). It is modelled as a function of socioeconomic variables at origin *i*, W_i, and total accessibility from origin *i* to all facilities, $X_i(C_{ij})$. The second term is the activity-allocation model (Θ_{ij}), represented as a function of relative accessibility to each facility $[C(C_{ij})]$ and the attractiveness of the facility W_j. These two terms are combined in a single multiplicative model as $V_{ij} = V_i \Theta_{ij}$.

To proceed further, one needs to identify the functions C_{ij}, $C(C_{ij})$, $X_i(C_{ij})$ and the K_i term. We recall that C_{ij} is a composite spatial-cost function including travel time, travel expenditure and other 'costs' associated with tripmaking. $C(C_{ij})$ is a trip-distribution function that shows the relative accessibility between origin i and facility j. It has to have an inverse relationship with the user cost between i and j. One common form of $C(\cdot)$ is either the power function with a negative exponent or the negative exponential function as mentioned previously. The second function to be specified is $X_i(C_{ij})$: the accessibility from origin i to all the facilities. Once the relative accessibility between i and j is defined, a logical way to write $X_i(\cdot)$ is $X_i(C_{ij}) = \sum_{j=1}^{p} W_j^\beta C(C_{ij})$ for $p \le n'$ which is a weighted sum of individual accessibility $C(C_{ij})$. The weights W_j^β are used to sum among the p facilities involved since a more attractive facility should be counted more than a less attractive one. The remaining term K_i must be defined so that the allocation model would have the property that the sum of all the trip allocations among p facilities would be equal to the total. In other words, $\sum_j V_{ij} = V_i$. To ensure this, it can be shown that $K_i = 1/X_i(C_{ij})$. Now our model is completely specified:

$$V_{ij} = \left\{ K_0 W_i^\alpha \left[\sum_{k=1}^{p} W_k^\beta C(C_{ik}) \right]^{b'} \right\} \left\{ \frac{W_j^\beta C(C_{ij})}{\sum_{k=1}^{p} W_j^\beta C(C_{ik})} \right\}. \tag{5-42}$$

Notice the similarity between Equations **(5-42)** and **(5-36)**, excepting for the logarithmic transformation of the trip-generation term. There are several other ways to derive the above equation. It can be derived, for example, from a first principle that says that individuals execute a trip from i to j if there is a positive consumer surplus (Cochrane 1975). While these other derivations have their merits, the above is probably the most straightforward for our purpose. Readers interested in these other derivations may consult the "Economic Methods" chapter in Chan (2000) under the "Unconstrained gravity model" subsection.

5.5.1 Use of the model. Such a generalized location-allocation model is both a private-sector and public-sector formulation and can be useful in three different ways:

(a) The trip pattern from the origins i to facility j, V_{ij}, is readily discerned, identifying the association of demands with a facility. The total demand served by location j, V_j, is of particular interest since it shows the level of activity at facility j.

(b) The direct expenditure by the provider and users of the system—including facility improvement, user fee, travel cost, travel time, and other out-of-pocket expenditures—is directly discernable from the model.

(c) For the public sector, the consumers' surplus can be easily calculated, which measures the direct social benefits of the facilities.

Of particular interest is the viability of the facility in terms of the demands served by a facility. We are also interested in where the users come from. The model gives these figures readily as a direct output of demand V_{ij}. Obviously, these service patterns will change in the future as facility-management policies and socioeconomic conditions change. For example, the user charge at the facility may change, or the facility may be improved. The model will help forecast the corresponding service pattern as a result of these policy changes. From the forecast demands served by the facility, we can easily compute the patronage or revenue to these facilities by multiplying the activities by the appropriate user fee. Similarly, the user non-fee expenditure—at least the transportation part of the expenditure—can also be computed. Note that included in the non-fee expenditures are such items as travel cost, travel time, etc. Note also that facility patronage or revenues and direct user-expenditures are responsive to policy and socioeconomic changes in the same way as the V_{ij}.

A measure of direct socioeconomic benefit is consumers' surplus, which shows the welfare to users based on their willingness-to-pay for services rendered by the facility. Such benefit measures can be computed for the whole study area, for each facility, or for each demand origin.[9] A figure for the whole area, for instance, would show the total direct benefit of all facilities in the area. Benefits by each facility *j* will allow for a welfare ranking among the facilities of interest. Finally, the benefit measures by each origin *i* would show the benefit distribution and incidence among the demand-points served. Mathematically, we can find the consumers' surplus for the residents of origin *i* provided by facility *j* by integrating the function $V_{ij}(C_{ij})$ as C_{ij} varies from the prevailing cost to infinity. We can then determine the consumers' surplus that accrues to the population of origin *i* by adding all *j*'s holding *i* fixed. Similarly, the consumers' surplus generated by facility *j* is obtained by adding all *i* holding *j* fixed. Finally, the total consumers' surplus generated by all facilities and accruing to all population of the region is the sum of all *i* and *j*. While there are several analytical ways to compute these surpluses, the most straightforward way, given the availability of the computer, is to integrate numerically using Simpson's rule.

5.5.2 Model calibration. Before one can calibrate the model, several details need to be discussed. For example, what is the minimum number of parameters that must be calibrated within the model (see Equation **(5-42)**) The consideration is based on several factors, including the tradeoff between policy sensitivity, data requirement, ease of calibration, and available resources. Ideally, the cost of making a trip C_{ij} should encompass all the expenditures that go into tripmaking, such as

---

[9]Consumers' surplus was originally introduced in the "Economic Methods" chapter in Chan (2000) and subsequently extended, including under Section 5.4.1 of the current chapter.

time, out-of-pocket expenditures, comfort, convenience and the like. However, in practical situations, many psychological factors cannot be quantified easily and are often omitted (Brog 1982). In this discussion, we include only time and out-of-pocket expenditures. Among the out-of-pocket expenditures in tripmaking are such items as the perceived cost of operating a vehicle and other user charges. Suppose the cost of operating a vehicle is c dollars a mile (km), direct user-charge is r_{ij} dollars, and the monetary value of an hour to a traveler from origin i is q_i dollars. Then

$$C_{ij} = c\,d_{ij} + q_i \tau_{ij} + r_{ij} \qquad (5\text{-}43)$$

where d_{ij} is the trip distance in miles (km), and τ_{ij} is the trip time in hours. Rigorously speaking, c and q_i are model coefficients to be calibrated. However, past experience in calibrating both the allocation part of the model (Stopher and Ergun 1982) and the combined derivation-allocation formulation (Cesario 1975) underscores the computational difficulty of nonlinear-optimization techniques. A simpler approach based on decomposition is proposed here. As a first step of the decomposition, we set exogenously for each trip the individual unit-cost of travel c and q_i. Such exogenous specification saves a great deal of effort in calibration. At the same time, it maintains the policy sensitivity by retaining the individual policy variables such as trip-time τ_{ij} and user-charge r_{ij}.

Another key policy-variable is the attractiveness-index W_j, which is used to characterize the facility j. From the policy-sensitivity point of view, W_j should be the variable directly related to facility improvements (or degradation). The exponent of W_j, denoted by ß, is then to be calibrated and included in the model. Similar to using decomposition to save computational resources, W_j^β is calibrated exogenously. After much experimentation, the attractiveness indices are found to be best calibrated using regression[10]

$$W_j = a_0 + a_1 S_{j1} + a_2 S_{j2} + \ldots + a_k S_{jk} \qquad (5\text{-}44)$$

where the S_j represent site-specific attributes of facility j. Such exogenous definition for W_j effectively bypass the requirement to calibrate the coefficient β. In other words, we have set β equal to unity and calibrate attractiveness outside the model. In spite of such exogenous procedure, we have retained policy sensitivity for the attractiveness variable by defining it in terms of attributes of the facility.

5.5.3 Decomposition procedure. In summary, the generalized facility-location-allocation model assumes the following format:

---

[10]For a review of linear regression, the reader may wish to consult the appendix to Chan (2000) entitled "Statistical Tools."

$$V_{ij} = b_0 W_i^{b_1} \left(\sum_{k=1}^{n'} W_k C_{ik}^{b_3} \right)^{b_2} \frac{W_j C_{ij}^{b_3}}{\sum_{k=1}^{n'} W_j C_{ik}^{b_3}} . \tag{5-45}$$

Here the trip-distribution function assumes the form $C_{ij}^{b_3}$ just for illustration purposes. The parameters b_0, b_1, b_2, and b_3 are the calibrated values for K_0, α, β, and γ respectively. Coefficient b_3 is expected to be negative in calibration value. As pointed out earlier, there are two parts to the model: the derivation part and the allocation part. It turned out that the two parts of the model can be calibrated individually and then combined through iterations. This forms the core of the decomposition technique:

First, we calibrate the allocation component of the model:

$$\Theta_{ij} = \frac{W_j C_{ij}^{b_3}}{\sum_{k=1}^{n'} W_k C_{ik}^{b_3}} . \tag{5-46}$$

Rather than following the conventional approach of pooling all the data to estimate the parameter b_3 (say, using least-squares regression), each origin-destination pair is treated individually in a 'disaggregate' fashion.[11] If θ_{ij} stands for the observed fraction of the trips from origin i that terminate in facility j, a best fit for Equation **(5-46)** is obtained for parameter b_3 when this function is minimized

$$g_{ij}(b_3) = (\Theta_{ij} - \theta_{ij})^2 \tag{5-47}$$

for all origin-destination pairs i–j. To carry out this minimization, the Fibonacci search technique[12] is carried out through the interval from -10 to 0, the plausible range for the exponent b_3. As described earlier, an initial estimate on the attractiveness index W_j is obtained from Equation **(5-44)** to operationalize the search technique. Once the best b_3 have been obtained for each i–j that minimizes the corresponding $g_{ij}(b_3)$, the median of all b_3, say $\hat{b}_3$, is adopted for the allocation model Θ in Equation **(5-46)**. This technique can be considered a hybrid of a disaggregate and aggregate calibration. The Fibonacci search performed for each i–j pair is disaggregate, in that it caters toward each data point. Taking the median of all b_3, however, is an aggregation. Compared with taking the average, this

---

[11]For a definition of "disaggregate calibration," see "Descriptive Tools" chapter and also the "Competition and Market Share" subsection of the current chapter.

[12]See the "Heuristics Solutions Techniques" section in the "Prescriptive Techniques" chapter of Chan (2000) for a description of the Fibonacci search.

approach mitigates the undesirable effects of outlier values of b_3. It tends to 'homogenize' the travel-cost sensitivities among the diverse origin-destination pairs.

The second step is to calibrate the derivation model based upon the estimated value for b_3, $\hat{b}_3$:

$$V_i(b_0, b_1, b_2, b_3) = b_0 W_i^{b_1} \left(\sum_{k=1}^{n} W_k C_{ik}^{\hat{b}_3} \right)^{b_2}. \tag{5-48}$$

The best values for b_0, b_1, and b_2—$\hat{b}_0$, $\hat{b}_1$, and $\hat{b}_2$—are obtained with regular log-linear regression technique. Finally, we adjust $\hat{b}_0$ so that the regional total of all estimated trips is the same as that observed after the logarithm is transformed back to the original data for model application.

The derivation and allocation parts, Equations **(5-46)** and **(5-48)**, are now put together, yielding new estimates on the trips from i to j:

$$\hat{V}_{ij} = V_i(\hat{b}_0, \hat{b}_1, \hat{b}_2) \Theta_{ij}(\hat{b}_3). \tag{5-49}$$

New values on Θ_{ij} are then obtained $\hat{\Theta}_{ij} = \hat{V}_{ij} / \sum_{j=1}^{n'} \hat{V}_{ij}$. Such $\hat{\Theta}_{ij}$ are subsequently used for Θ_{ij} in Equation **(5-47)** to arrive at new estimates on b_3. The procedure then goes through another iteration until two successive estimates of b_3 are consistent. Note that the cycle between Equations **(5-46)**, **(5-47)**, **(5-48)**, and **(5-49)** converges very fast, usually within two iterations. After such convergence is obtained, the attractiveness index W_j are usually adjusted to ensure that the estimated visitations to a facility j are compatible with the observed: $W_j' = W_j [V_j / \hat{V}_j]$

The goodness-of-fit parameters for the model include the coefficient-of-multiple-determination R^2, F-ratio and t-statistics, which are used to measure the 'fit' of the derivation component of the location-allocation model. On the other hand, the sum of the squared deviations (sum of squares) between the observed v_{ij} and the estimated trips V_{ij} is an indicator of the fit of the allocation component. Finally, the convergence of the combined derivation-allocation model can be measured by the overall variance-reduction, defined as the percentage reduction in the sum of squares throughout the iteration sequence: $1 - (V_{ij} - v_{ij})^2 / (v_{ij} - \bar{v}_{ij})^2$, where $\bar{v}_{ij}$ is the average value of all observed trips between i and j.

5.5.4 Sensitivity analysis. An important part of modelling is sensitivity analysis, for it is only through such a procedure that various alternatives can be tested. The two key policy-variables in the model are the user-cost C_{ij} and the attractiveness-index W_j. Three sensitivity analyses can be made to these two variables (as well as the socioeconomic variables W_i): (a) by using demand elasticities, (b) by varying user charges, and (c) by changing investment policies on the facilities. Demand elasticities $\eta(V_{ij}, Y)$ show the percentage change in demand V_{ij} corresponding to one-percent change in the policy or socioeconomic variables Y : $(dV_{ij}/V_{ij})/(dY/Y) = (dV_{ij}/dY) \cdot (Y/V_{ij})$. These direct demand-elasticities can be derived from the model: elasticity-of-cost $\eta(V_{ij},C_{ij}) = b_3[1+(b_2-1)\Theta_{ij}]$, and elasticity-of-attractiveness $\eta(V_{ij},W_j) = 1+(b_2-1)\Theta_{ij}$.

Of equal interest are the cross-elasticities between alternate-facility j and the candidate-facility l, where cross-elasticity of cost is

$$\eta(V_{ij},C_{il})=b_3(b_2-1)\Theta_{il} \tag{5-50}$$

and cross-elasticity of attractiveness is

$$\eta(V_{ij},W_l)=(b_2-1)\Theta_{il}. \tag{5-51}$$

A close examination of Equations **(5-50)** and **(5-51)** shows that the sign and magnitude of b_2 have very important policy implications. If facilities are *substitutes* among themselves, the following condition has to apply, given a negative value for b_3: $b_2 < 1$. On the other hand, a *complementary* effect is discerned if $b_2 > 1$. This latter condition says effectively that the *derivation* effect is stronger than the *allocation* effect. (It is possible that one can interpret this to mean that 'income' effect is stronger than substitution effects.[13]) Interestingly enough, an inherent property of our model is that the cross elasticity (Equations **(5-50)** and **(5-51)**) is the same among the alternate facilities j given a change in the cost to a candidate-facility l or the attractiveness W_l. It is a direct result of the IIA property as explained in the "Competition and market share" subsection of the current chapter. To the extent that this model allows us to assess the effect of introducing a new facility, it is a desirable feature (compared with traditional discrete facility-location models that require a full alternative-set to be in place before modelling.) This property is, at the same time, somewhat counter-intuitive. It may represent an area for improvement in model specification (Fotheringham 1983). Finally, it can be shown that the elasticity of the socioeconomic variables is: $\eta(V_{ij},W_i)=b_1$.

The elasticity concept is rather limited in our analysis, since most changes in C_{ij} and W_j are much bigger than the 'differential' change suggested by elasticities. Consequently, a supplemental sensitivity-analysis calls for rerunning the model with different C_{ij} and W_j as inputs. For example, if the user charge r_{ij} at a facility is uniformly raised from 3 to 4 dollars, the accessibility to a facility j, $X_j(C_{ij})$, will be significantly different. There will also be a significant change in the demand V_{ij}, consumer expenditure $C_{ij}V_{ij}$ and consumers' surpluses. The change in consumers' surplus at facility j, for example, can be approximated by comparing a 'before' and 'after' computation using the 'trapezoid' method: $\frac{1}{2}$ $[C_j(\text{after})-C_j(\text{before})]$ $[V_j(\text{after})-V_j(\text{before})]$, where C_j stands for the cost of accessing facility j. The same incremental analysis can be used to assess the effect of facility investment or closing. Therefore, the third and final sensitivity analysis involves rerunning the model with different attractiveness indices (W_j). Normally, one facility is tested at a time. The attractiveness index here can be varied from zero to a large magnitude—representing any policy from closing a facility to expanding it significantly.

---

[13]For a definition of "income effect," see the "Economic Methods" chapter in Chan (2000). The income effect is interpreted in Subsection 5.4.2 of the current chapter in terms of the model under consideration.

Please note an additional property of the model regarding the introduction of new facilities or the closing of facilities. The percentage change (Ω) of visitation at facility k is the same as another facility j before and after an 'irrelevant' facility l is introduced. In other words

$$\frac{\Omega_{ik}(k|S)}{\Omega_{ij}(j|S)} = \frac{\Omega_{ik}(k|S'')}{\Omega_{ij}(j|S'')} = 1 \tag{5-52}$$

where S stands for the set of n' facilities and S'' stands for the set of $n+1$ facilities. This property is a result of Luce's Theorem of irrelevant alternatives. This inherent property follows directly from the constant cross-elasticity property of Equation (5-51), representing both a desirable and an unsatisfying property of the model as mentioned.

5.6 A case study of state parks

Decisions often have to be made on the provision of recreational facilities such as state parks. The basis for making such decisions often includes: (a) How does park-use change with park improvements and user-charge increases? (b) what population groups in the regions are served by these state parks? and (3) aside from recreation, what are the economic values of state parks? Central to analyzing these issues is a location-allocation model such as the one above that can explain the travel pattern among the various state parks. Such a demand model would then be used to forecast park visitation and economic measures corresponding to several policy scenarios. For this study, we have focused on the waterfront state park in the New York/Long Island area (see **Figure 5-6**). We report here a study that estimates the use and primary (rather than secondary) economic values of a specific state-park or system of state-parks. Built into the model explicitly are the substitution effects among various parks when the user cost or the facility of a competing park changes. At the same time, the model addresses the fact that improving the attractiveness of a single park may complement other parks. Collectively, they may induce additional outdoor recreation among all the facilities in the system. Included into such 'complementary' may be the income effect associated with reduced travel-costs to a closer, improved state-park.

Various models exist in the regional-science and transportation literature to handle the substitution and complementary effects among state parks. These have been reviewed in previous sections in this chapter (Cesario and Knetsch 1976, Cesario 1975). For example, a trip-distribution (or activity-allocation) model is often proposed to analyze the competing facilities in a region (Stopher and Ergun 1982, Fotheringham 1983). Such a model usually estimates the shares of a given total-demand among the candidate facilities based on accessibility costs. Another model, often called trip generation (or derivation) analysis, estimates the total demand for recreational travel based on the socioeconomic condition at the study area. Such a functional separation into distribution/allocation and genera-tion/derivation is justified in modelling non-discretionary activities such as work

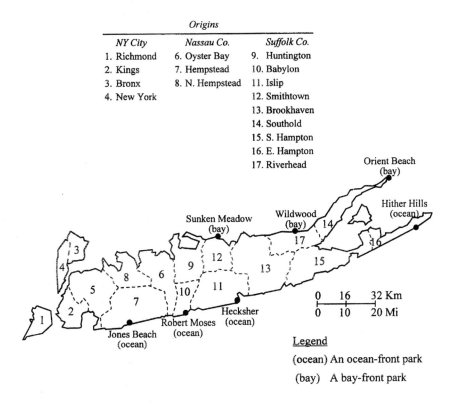

Figure 5-6 Map of origins and parks (Carroll et al. 1982)

trips that are highly inelastic. The functional separation also lends itself to more tractable models that are easier to calibrate. However, park visits are among those trips that are of a highly discretionary nature (Hughes and Lloyd 1977, Clawson and Knetsch 1966, Vickerman 1974). Thus a decision to visit a park is very sensitive to both the user cost to a park (or an alternate park) and the income and demographic profile of the study area as a whole. This calls for a combined derivation/ generation and allocation/distribution model in spite of the concomitant calibration and analytical complexities. Such a combined formulation—as outlined in the last section—has been referred to as an unconstrained distribution-model (Cochrane 1975).

5.6.1 Model specification. In the model specification, the socioeconomic variable term W_i in Equation **(5-40)** becomes two terms: population at origin i, P_i and income at origin i, I_i. As a result, Equation **(5-41)** now looks like

$$V_{ij} = K_0 P_i^\alpha I_i^\beta [X_i(C_{ij})]^b \{K_i W_j^\gamma C(C_{ij})\}. \qquad \textbf{(5-53)}$$

Correspondingly, the detailed model (Equation **(5-42)**) assumes the form

$$V_{ij} = \left\{ K_0 P_i^\alpha I_i^\beta \left[\sum_{k=1}^p W_k^\gamma C(C_{ik}) \right]^b \right\} \left\{ \frac{W_j^\gamma C(C_{ij})}{\sum_{k=1}^p W_k^\gamma C(C_{ik})} \right\}.$$
(5-54)

In the travel-cost function (Equation **(5-43)**), we set $c = \$0.10$/mile ($\$0.0625$/km) of perceived cost (Brog 1982), and $q_i =$ one-third the personal wage-rate at origin, in dollars (Cesario and Knetsch 1976). While this saves effort in calibration, there is one implication. The use of income to valuate trip time means that the income variable appears in two places in the model—once in derivation as per-capita income I_i in Equation **(5-53)** and another in C_{ij} as shown in Equation **(5-43)**.

From a policy-sensitivity point of view, W_j should be directly related to park improvements (or degradation). For example, if more parking spaces are contemplated for the state parks, the addition of parking spaces should be reflected in W_j. After much experimentation, the attractiveness-indices W_j calibrated for the model were found related to park size (in acres [ha]) most significantly. The relationship is, however, different for ocean-front vs. bay-front parks (see **Figure 5-6**). This equation is obtained to describe the relationship for the three ocean-front parks: $W_j = 0.0475 S_{j.} - 0.77$, where $S_{j.}$ is the park area in acres (ha) and the correlation coefficient (R) is 0.94. The equation for the four bay-front parks on the other hand is $W_j = 0.00198 S_{j.} + 12.5$, with a correlation coefficient of 0.47.

5.6.2 Calibration results. The model that we finally calibrate (Equation **(5-45)**) now looks like $V_{ij} = \beta_0 P_i^{\beta_1} I_i^{\beta_2} \left(\sum_{k=1}^{n'} W_k C_{ik}^{\beta_4} \right)^{\beta_3} \left[W_j C_{ij} \beta_4 / \sum_{k=1}^n W_k C_{ik}^{\beta_4} \right]$. We present the parameter estimates of the calibrated model in **Table 5-4**. While there are minor and not-so-minor discrepancies between estimated and observed values for particular origin-destination pairs, on balance the model is predicting visitation quite accurately. In fact, the model is successful in explaining 88.6 percent of the variance in the observed trip-table. As presented in **Table 5-5**, the model also predicts park attendance very well with the revised attractiveness W_j'. We view this as an excellent test of the analytical model and the decomposition procedure.

5.6.3 Analysis results. Of immediate interest to the parks are attendance figures. For example, the current day-use-travel from the 17 origins to the seven waterfront parks totals 14,130,000 a year, as shown in **Table 5-5**. Furthermore, we are interested in where the park users come from. The model gives these figures readily as a direct output of demand V_{ij}. From the forecast park-attendance-figures, the revenue to parks can be easily computed by multiplying the attendance by the appropriate entrance fee. **Table 5-6** allows for a ranking of the parks according to their socioeconomic benefits, where economic benefit is defined in terms of consumers' surplus. In this regard, Jones Beach stands out at the top while Orient Beach ranks last, perhaps due to the relative size of these two parks. Overall, the

Parameter	Estimate	Goodness-of-fit measures	
		t-statistic	Other parameters
β_0	5.20×10^{-8}	—	
β_1	0.93	7.48	$R^2 = 0.85$ Overall
β_2	2.79	2.81	$F\text{-ratio} = 24.6^1$ variance
β_3	1.81	6.44	reduction $= 0.886$
β_4	-2.10	—	$(V_{ij} - v_{ij})^2$ $= 20748$

Table 5-4 Parameter estimates and goodness-of-fit measures

Park	Attractiveness		Park attendance (in thousands of persons /yr)		Percent error
	W_j'	W_j	Estimated	Observed	
Orient Beach	13	13	102.0	106.0	-3.77
Jones Beach	121	114	7235.4	7300.0	-0.88
Robert Moses	53	47	3137.0	3136.0	0.03
Hecksher	14	16	1244.5	1194.0	4.23
Sunken Meadow	18	15	1418.4	1421.0	-1.83
Hither Hills	69	83	621.4	615.0	1.04
Wildwood	13	14	371.3	358.0	3.72
Total			14130.0	14130.0	0.00

Table 5-5 Calibrated attractiveness measures and park attendance
(Carroll et al. 1982)

model estimates an annual benefit of 44,600,000 dollars currently for all the waterfront parks per year in the metropolitan New York/Long Island area, considering day use alone.

With the calibration value of β_3 around 1.8 and the fractional value for Θ_{ij}, the cost elasticity is effectively β_4 in the expression

$$\eta(V_{ij}, W_j) = 1 + (\beta_3 - 1)\Theta_{ij}. \tag{5-55}$$

Park	Total value of all seven parks	Total value with park omitted	Economic value of park
Orient Beach		43,077	1,475
Jones Beach		12,569	31,983
Robert Moses		25,892	18,660
Hecksher	44,552	38,360	6,192
Sunken		37,019	7,533
Hither Hills		37,445	7,107
Wildwood		41,405	3,147
Total			76,097

Table 5-6 Consumers' surplus of each state park in thousands of dollars per year
(Carroll et al. 1982)

except for the heavily used parks. Thus when $\beta_4 = 0$, demand is perfectly inelastic with respect to travel cost. The trips are simply determined by the socioeconomic variables at the origin and the attractiveness of the parks. On the other hand, $\beta_4 = -\infty$ results in $V_{ij} = 0$ when recreational trips are so discretionary that none gets executed. When β_4 is somewhere in between, its exact value monitors the spatial allocation of recreational trips. The same reasoning about the fractional values for β_3 and Θ_{ij} leads us to infer a unitary elasticity for attractiveness:

$$\eta(V_{ij}, C_{ij}) = \beta_4 [1 + (\beta_3 - 1)\Theta_{ij}]$$ (5-56)

Thus visitation to a park goes up directly to its attractiveness on a one-to-one percentage basis. The situation in New York/Long Island, as shown by the calibration results in **Table 5-4**, shows that $\beta_3 > 1$, which says that the derivation effect is stronger than the allocation effect. Thus users value park recreation as a whole and are less discriminating about the specific park they visit. Cesario (1975) and Cesario and Knetsch (1976) found the same relationship in northeastern Pennsylvania. With a β_3 value of 1.81, the cost cross-elasticities are inelastic, as one can verify from

$$\eta(V_{ij}, C_{il}) = \beta_4 (\beta_3 - 1)\Theta_{il}.$$ (5-57)

This reinforces, once again, the complementary among state parks in providing recreational opportunities.

The income elasticity from the model takes on the form

$$\eta(V_{ij}, I_i) = \beta_2 + \beta_4 \left[\beta_3 - Q_i(\beta_3 - 1) - \frac{C_{ij} - q_i \tau_{ij}}{C_{ij}} \right]. \tag{5-58}$$

where $Q_i = \sum_k C_{ik}^{\beta_4 - 1} / X_i \leq 1$ Substituting the values of the calibration coefficients in the above equation, the income elasticity for the very poor, whose time is worth little, reduces to $1.7 Q_i + 1$. At the other extreme, the very wealthy assume the elasticity value of $1.7 Q_i - 1$. This shows that the income elasticity of the poor is bigger than the wealthy in general, reflecting the discretionary way the poor treats recreational opportunities (considering other pressing needs). In both cases, income elasticities are positive in value—that recreational travel increases as income gets higher—as expected.

5.6.4 Implications. Consider an across-the-board parking-fee-increase of $\Delta r_{ij} = \$1/\text{car}$—from three to four dollars for all users at all parks. Parking revenues within the seven parks would increase from 10,460,000 to 10,730,000 dollars; while the socioeconomic benefit of each park would fall due to the reduction in visitation (see **Table 5-7**). This corresponding loss in social benefit (consumers' surplus) for the seven-park system would be an estimated 6,510,000 dollars. Contrast this policy option with a more selective parking-fee increase of one dollar at Jones Beach only. This latter policy will yield systemwide revenue gains of roughly 3,000,000 dollars with only a 2,900,000-dollar reduction in the consumers' surplus of the park system. Since the cross-elasticities of other parks for Jones Beach is relatively price inelastic (according to Equation **(5-57)**), this selective increase in parking fees is a large net-revenue producer with only a small sacrifice in total park-visitation systemwide.

The across-the-board parking-fee increase is not only an inefficient way to raise revenues, it is also very inequitable. Its fiscal impact is not neutral with respect to the origins (see **Table 5-7**). Those origins closest to parks (hence frequent users of these parks) experience the greatest drops in visitation (according to Equation **(5-56)**) and thus the largest losses in social benefits. The relatively distant New-York-City residents, on the other hand, continue to use the parks at nearly the same rates. The reason is that the one-dollar increase in trip-costs is a relatively small percentage of total trip-costs to even the nearest of the parks. The increase in parking fees is not large enough to dissuade city residents from continued use of the parks. And since they continue to use the parks, they bear a large share of the increased parking fees. Before the fee increase, New-York-City residents pay an estimated 2,300,000 in parking fees or 22.5 percent of the total amount of fees collected. After the fee increase, they pay 2,700,000 or 25.4 percent of all fees collected.

By comparison, the selective parking-fee increase at Jones Beach would leave the distribution of parking-fee revenues throughout the park system virtually unchanged. The demands for Jones Beach from each of the 17 origins are such that

		Decrease in annual visits		Decrease in consumers' surplus	
		No of vehicles (thousands)	Percent	Thousands of dollars	Percent
	Richmond	5.4	9.6	31	6.8
	Kings Cty	99.0	13.4	397	9.7
	Bronx	37.0	12.8	157	9.4
	New York	39.0	8.7	246	6.3
	Queens	170.0	14.0	639	9.9
	Oyster Bay	350.0	22.9	767	16.3
	Hempstead	810.0	23.6	1608	16.8
o	N Hempstead	88.0	16.2	272	11.5
r i	Huntington	190.0	28.5	436	19.8
g i	Babylon	185.0	25.8	395	19.2
n	Islip	544.0	35.6	918	25.5
	Smithtown	149.0	34.0	279	23.5
	Brookhaven	81.0	21.9	242	16.0
	Southhold	2.4	19.5	9	13.4
	S Hampton	11.2	18.4	36	13.1
	E Hampton	51.1	36.0	58	23.5
	Riverhead	7.3	23.6	20	18.2
	Total	2810.0	23.0	6509	14.6
	Orient Beach	14.3	15.4	126	11.7
	Jones Beach	1340.0	21.0	2755	14.5
	Robert Moses	640.0	23.7	1613	15.5
P a	Hecksher	311.0	31.4	510	16.5
r k	Sunken Meadow	338.0	28.9	612	16.1
	Hither Hills	109.0	19.9	625	11.8
	Wildwood	69.0	21.2	268	14.0
	Total	2810.0	23.0	6509	14.6

Table 5-7 Effects of one-dollar across-the-board parking-fee increase

those origins facing the largest percentage increase in travel costs (the closer-by-origins) are those with the most elastic demand (according to Equation **(5-56)**). On the other hand, the effect of a selective fee-increase on demand for other parks is small due to an inelastic cross-elasticity (Equation **(5-57)**). Thus while there is a significant loss of revenue at Jones Beach, there is little effect on revenue collection from other parks. These two factors offset one another, culminating in an increase in park revenue systemwide. The distribution of these revenues remains pretty much as they were before the fee increase.

One very interesting and somewhat counter-intuitive result of these analyses, however, is that the sum of the economic values of the individual parks exceeds the total value of the seven-park system, as reported in **Table 5-6**. Again, this confirms the finding from the elasticities that the parks are complements to, rather than substitutes for, one another. Therefore, changes in any one park that affect visitation to that park will similarly affect visitation to the other parks in the system (via a uniform percentage as shown in Equation **(5-51)**). It appears that users view individual state parks a part of an integrated system. Correctly or incorrectly, they perceive conditions at any one state park to similarly exist at all state parks. The easy access that most residents of Long Island have to town, county, or private alternatives to these state parks—together with the discretionary nature of recreational trips—makes visitations to state parks highly volatile.

Similar to the case of the parking fee mentioned above, the closing of any one park will result in welfare losses for residents of the New York/Long Island region; however, these issues will tend to be most acute in those origins where demand is the highest (see Equation **(5-55)**). As a result, park policies here are not neutral with respect to origin. With the closing of a given park, the largest benefit losers tend to be those origins closest to the parks, that is, origins accounting for the largest percentage of visitations.

This case study shows the use of a simple analytical location-allocation model to determine the demand for and economic value of state park sites, including the effect of user-charge increases, facility improvements and park closing. We have achieved modest successes in this regard and in the process have reached these conclusions:

a. Recreational travel—unlike work trips—is highly discretionary. It is also sensitive to the location and attractiveness of a park. Making the facility attractive and setting a proper user-charge will not only have significant bearing upon the visitation of an individual park. It also affects the public's perception of the merits of the entire park system as a whole. A public park, therefore, besides being a substitute for other competing parks, represents the complementary set of parks in the family. This confirms findings from other studies, including one performed in Northeast Pennsylvania (Cesario 1975).

b. The analytical model permits sensitivity analysis of a variety of policy options affecting parks. The results of these analyses, in turn, provide valuable information, such as visitation, revenue, use expenditure and consumers' surplus, which allow decision-makers to judge the efficiency, effectiveness, and equity of policy options. A BASIC computer program is included on my web site for such state park studies, should the readers be interested. Besides New York/Long Island area, another data set from the State of Washington is included. Both programs are stored under the subdirectory STATEPRK, complete with run instructions in README files.

5.7 Concluding comments

In this chapter, traditional discrete facility-location models were extended beyond the quadratic assignment problem to include activity generation, overlapping allocation and spatial competition. In so doing a number of limiting concepts were relaxed. Distinction is again made between private-sector and public-sector models, in which profit motives are discussed side-by-side with public service objectives. A general model is proposed which allows extensive policy interpretation by way of the elasticity of demand. Aside from socioeconomic variables, the role of accessibility has been clearly analyzed. Many of the refinements achieved through the routing component of location-routing models are yet to be included in the accessibility definition. But the models presented here more than adequately illustrate satisfactory handling of locational surplus and demand elasticities that reflect substitution and complementary among facilities. As part of the discussion, we address such issues as policy sensitivities and computational rigor. In general, it moves us toward the next chapter: comprehensive modelling of activity derivation and allocation.

Through simple examples, we were able to define the gist of a competitive location-allocation problem, both in terms of an extension of the traditional median formulation as well as a more fundamental modelling of an oligopolistic economic system. We formulated the Nash equilibrium model for two firms in a circular market. The key concepts about substitutive and complementary goods and services are graphically illustrated. Market-baskets are also defined in terms of the combination of composite goods-and-services and individual goods-and-services between the two firms. Most important, we model the problem as a two-factor system consisting of unit-price for the goods-and-services and transportation-cost for the combined price-location decision. Besides profit-maximization for private firms, in a subsequent section locational surplus is included in the public-sector models explicitly because of discretionary-demand and competition. As we concentrate on the 'demand' side, however, service-delivery modelling—as represented by routing and scheduling—is compromised. Thus the logistics applications as described in the "Location-routing" chapter give way to loca-tion–allocation in socioeconomic planning. Unfortunately, this transition is not necessarily a continuum. The gap between location-routing models and general location-allocation models can possibly be bridged by extending the Perl–Ho formulation described in the section entitled *"Public facility-location under discretionary demand"* in the direction of the generalized location-allocation model in the section entitled *"Generation, competition, and distribution considerations"*.

We showed—through a case study—that it is possible to specify and obtain good estimates of generalized location-allocation models using a computationally-simple calibration-procedure. Such a procedure makes use of a conceptually-appealing decomposition of total-demand for facilities into its derivation and allocation components. These components correspond to the complementary and substitutive effects among facilities.

The location-allocation models discussed in this chapter are also called 'unconstrained' allocation-models, in that the originating demand is a function of the socioeconomic variables, rather than an exogenously-determined quantity. To the extent the model is really constrained at the destination end, such problems are typically much more difficult than the traditional models with a given number of originating activities. All allocation models of this kind face the problems of controls at the destination end, wherein the activities at the destination need to sum up to a given total. While the attractiveness measures do a credible job of destination-end control, room exists for further improvement in their measurement and estimation—both for calibration and policy-analysis purposes. Perhaps the calibration concepts of a doubly-constrained gravity model can be carried over to this model when the socioeconomic activities at i are treated as the 'origins'. In subsequent chapters on Lowry-derivative models, we will show how this can be achieved through adjustment of the accessibility function. Again, a decomposition scheme will be followed in the calibration process.

A unique feature of the activity-allocation generalization of a facility-location model is that it can consider new facilities, although the model was calibrated for a limited number of existing facilities. While the mathematical properties of the generation-allocation model are quite satisfactory, additional state-of-the-art improvements can be made in the model specification. A critical examination of the uniform cross-elasticity property (Equations **(5-50)** and **(5-51)**) is clearly worthwhile. This applies also toward the allied property of the model regarding the constant percentage-change-of-demand at other facilities as a new facility is built (Equation **(5-52)**). This is, in all honesty, not an easy problem to resolve in the author's opinion, since it is imbedded into the basic-building-block of all aggregate and disaggregate activity-allocation models: the independence-of-irrelevant-alternatives relationship in gravitational-interaction.

In this chapter, we have bridged part of the gap between discrete facility-location and continuous activity derivation-and-allocation. Our effort will not stop here. In future discussions, we will again attempt to compare and contrast the discrete and continuous versions of *dynamic* location-allocation models. This will be discussed in the context of the Garin–Lowry family of models and their general-equilibrium form: the spatial input-output model. This way, we include the temporal dimension into general location-allocation model, eventually leading toward locational conflicts and disequilibrium.

All things began in Order, so shall they end, and so shall they begin again,
according to the Ordainer of Order,
and the mystical mathematicks of the City of Heaven.
Sir Thomas Browne

CHAPTER SIX

ACTIVITY ALLOCATION AND DERIVATION: EVOLUTION OF THE LOWRY-BASED LAND-USE MODELS

An important class of location-allocation models deals with the forecast of population and employment activities over a map. To the extent these activities occupy land, such models can also be used to account for land use. There are two premises among these models: First, a study area develops from an economic base—process we call activity *derivation*. The spatial location of housing and employment is then determined by the interaction among these activities—a process called activity *allocation*. These interactions are postulated to be more intense when two locations are closer together and when the two locations have a concentration of activities. Conversely, the interaction is not as strong when spatial separation is huge and the activity levels are low. As pointed out previously, these two concepts bear the names of *economic-base theory* and *gravitational interaction* respectively. While they have a long history in geography and other disciplines, it was not until the arrival of the digital computer that these theories were made operational on a computational level. Lowry (1964) was attributed with the credit of implementing these ideas on a computer. His pioneering work stimulated a flurry of model developments—which has been referred to as the "Lowry heritage."

Lowry's model was a starting point for the development of a number of other models, including the Time-Oriented Metropolitan Model (TOMM-I) in 1964, the Bay Area Simulation Study (BASS-I) in 1965, the Community Land Use Game (CLUG) in 1966, TOMM-II in 1968, the Projective Land Use Model (PLUM) in 1968 and PLUM-IP (a modified version of PLUM using the Incremental Process) in 1971. While these developments in the 60's are by-and-large computational advancements, encouraging theoretical developments ensured from the late 60's to the present. These include the work of Wilson (1974) and Cesario (1975) which view spatial location as an entropy-maximizing formulation.[1] Cochrane (1975) provided an economic basis for the gravity model and related it to the intervening

---

[1]For an introduction to "entropy maximization" see the "Spatial interaction" section in the "Descriptive Tools" chapter and the "Prescriptive techniques in land use" section in the "Prescriptive Tools" chapter. Both chapters are in Chan (2000).

opportunity model and entropy models, as described in the "Economics Methods" chapter in Chan (2000). Recently, there are emerging theoretical interpretations of the development process in terms of catastrophe/bifurcation theories[2] (Wilson 1981, Varaiya and Wiseman 1981, Casetti 1981), and of an analytical decision-process (Beaumont 1982, Haag 1989). Perhaps representative of the new trend is the idea of self-organization and dissipative structures (Ulanowitz 1980, Allen et al. 1981), which interprets urban spatial modelling in a conceptually refreshing way.[3]

The development of each successive model in the 60's has enhanced the computational mechanism by Lowry, but the 'quality' of the resulting simulation has never been measured (Batty 1976, Putman 1983, Webber 1984, Putman et al. 2000). An experiment will be performed here to empirically test the refinement of the Lowry-based models on a common data base, allowing a comparative analysis of the model developments in the 1960's. Also included in this chapter is the review of the theoretical advancements in the 70's and 80's, which purports to depart from the traditional Lowry heritage. Similarities and differences are outlined between the 'old' and the 'new'. Experiments are also performed to interpret Lowry-based models in terms of disaggregation,[4] catastrophe, bifurcation and evolutionary concepts. This results in a disaggregate, bifurcation version of the Garin–Lowry model, as described in the next chapter. In the final part of the discussion, the model is extended to the more general input-output format,[5] rather than the traditional economic-base paradigm adopted thus far in our discussions (Golan and Vogel 1998). The disaggregation and bifurcation properties of this general model are also discussed. The general subject of spatial equilibrium, including competition among facilities, will conclude this part of the book with the "Spatial Equilibrium and Disequilibrium" chapter.

6.1 The Lowry model

Because of its pedagogic importance in the development of activity allocation, the Lowry model has been described many times and in many ways. This includes entire books written on the subject (see for example Webber [1984]). Our discussion here is not meant to duplicate other efforts. We review as briefly as possible the construct of the model in a didactic manner. We turn quickly to the matrix extension of the model by Garin; then focus on the calibration problems with the

---

[2]Refer to the "Control, Dynamics and System Stability" appendix in Chan (2000) for an introduction to catastrophe and bifurcation.

[3]See the "Discussion of Technical Concepts" appendix in Chan (2000) for a brief explanation of self-organization and dissipative structures.

[4]The concept of aggregate and disaggregate modelling was introduced first in the "Descriptive Tools" chapter inn Chan (2000) It was subsequently developed in the "Generation, Competition and Distribution" chapter in the current volume under the "Competition and market share" subsection.

[5]Input-output analysis was introduced in the "Economic Methods" chapter in Chan (2000).

model, including its doubly-constrained extension. The relationship to general input-output model and spatial equilibrium is discussed. Using the matrix notation, we introduce later the disaggregate formulation and the bifurcation version of the model, which represents some of the latest developments. The logical and mathematical structure of the Lowry model can be explained in three phases. First, one defines the basic mathematical nomenclature of the input and output variables. Second, the structural relationship between the three component sectors—basic[6], 'retail'[7] and household— is formulated in a set of mathematical equations. The three sectors form the backbone of the model. Finally, the working mechanism of these equations and the calibration of parameters is explained to make the model operational.

6.1.1 Input and output data. Let us, for the time being, treat the Lowry model as a "black box," which takes input information, processes it, and produces the desired forecast for the planning period. The following represent a list of the input data:

Activity: basic employment by zone*, E_j^B;
Land use: basic land per zone*, A_j^B; unusable land per zone*, A_j^U; total regional retail land, A^R;
Transportation network: interzonal separation between i and j, C_{ij};
Spatial-interaction function: $\Theta_{ij}(N_j, C_{ij})$;

Structural parameters: coefficients for specifying the spatial-interaction functions, b, c, w and g; number-of-households per basic-employee, f; number of nonbasic-(retail) employees necessary to serve one household for each trade-class k, d^k; land-consumption rate per retail employee of trade-class k, r^*;
Constraints: maximum allowable household-density per zone, D_j^H; minimum threshold of retail-employment by trade-class, d^k.

On the other hand, the data that are output from the model can be represented in the list below.

Activity: basic employment for each zone*, E_j^B; retail employment by trade class for each zone, E_j^k; number-of-households per zone, N_j;
Land use: basic land-area per zone*, A_j^B; retail land-area per zone, A_j^k; unusable land-area per zone*, A_j^U; residential land per zone, A_j^H.
Notice that certain information is both input and output. These are marked by an asterisk (*) in the list above. They represent information that is not changed from

---

[6]The basic sector in a Lowry model refers to that part of the economy that is providing goods and services for export consumption, rather than local consumption. For an introduction to "Economic-base theory", see the relevant section in the "Economic Methods" and "Descriptive Tools" chapters of Chan (2000).

[7]In the Lowry model, "retail" (or nonbasic) activity refers to the goods and services for the local population.

input to output. Basic activity and land-use are not changed, for example, because they refer to the forecast-period by definition, whether input or output. This point will become clear as we explain more about the model.

6.1.2 Theory. The model divides activities within the study area into three broad groups: the basic sector, the retail sector, and the household sector. The basic sector includes industries, business, and administrative establishments whose goods and services are generally exported. This is comparable to the basic activities in economic-base theory. The retail sector includes those businesses, administrative, and other establishments that deal directly with providing goods and services for the local residential population. Here the household sector consists of the resident population. The model is expressed in terms of nine simultaneous equations and three inequalities. First is the land-use equation, where the total area of each zone and the amount of land unusable for any activity is given in the input. The total amount of usable land is subsequently divided by the model into basic, retail and household as defined in:

$$A_j = A_j^U + A_j^B + A_j^R + A_j^H. \tag{6-1}$$

In more precise terms, this equation shows that the total area of zone j, A_j, is made up of the area of unusable-land in j plus the area used by basic establishments, retail establishments, and households.

The description of the basic sector is simple. For each zone j, the quantity of land used by basic establishments, A_j^B, and the employment opportunities provided by these establishments, E_j^B, are given exogenously for the forecast year. They are constant throughout the solution process, serving as the `seed' for economic development. In other words, the jobs created by outside investment in the forecast-year are input to fuel the local-economy. The retail sector is modelled by m groups of retail establishments. Each group has a characteristic production-function defined by the minimum efficient-size of the establishment, the number-of-clients required to support one employee, and the square footage or space required per employee. The employment in each retail group k is defined by

$$E^k = a^k N. \tag{6-2}$$

Thus retail employment is a function of the number of households in the area since the local consumer demand provides the market for each retail group. The term a^k can be viewed as a derivation coefficient applicable to the particular retail group. It is often called the *population-serving ratio*.

The distribution of the retail employment given in Equation **(6-2)** is dependent on the strength of the market in each zone. It is assumed that shopping trips originate either from homes or from workplaces. Therefore, the market potential of each zone is defined as the weighted sum of the number of households in the surrounding areas and the number of persons employed nearby—both of which are likely customers:

$$E_j^k = b^k \left[\sum_{i=1}^{n'} c^k N_i \, \Theta_{ij}(E_j^k, C_{ij}) + w^k E_j \right] \qquad \forall k.$$ (6-3)

The weighted term $w^k E_j$ gives the portion of market potential determined by the number of persons employed nearby. No distribution function, T^*, is needed for this term because the model assumes that only those employees within walking distance will patronize and contribute to the market. The weighted term $c^k N_i \Theta_{ij}(E_j^k, C_{ij})$ gives the portion of market potential determined by the number of households in the surrounding areas. Here $\Theta_{ij}(E_j^k, C_{ij})$ is positively related to the size of the service establishment (E_j^k) and negatively related to the length of the trip (C_{ij}). It is precisely the allocation function explained in the "Generation, Competition and Distribution" chapter, and is often determined by the trip-distribution curves that show the relationship between distance and the frequency of trips. The weights c^k and w^k represent the relative importance of home-based and workplace-based retail-trips for a particular purpose, k, whether it be shopping, laundry and the like. The coefficient b^k is a scale factor adjusting the retail employment in each loop to the areal total. Remember that such a total was determined in Equation (6-2).

The sum of zonal employment for each retail-trade group, k, is

$$E^k = \sum_{j=1}^{n'} E_j^k.$$ (6-4)

The total zonal employment is found by summing the employment for each retail group in the zone, plus the basic employment allocated to that zone: $E_j = E_j^B + \sum_{k=1}^K E_j^k$. The amount of land in each zone that is occupied by retail establishments is determined in

$$A_j^R = \sum_{k=1}^K r'^k E_j^k.$$ (6-5)

The term r^* is an exogenously-determined employment-density coefficient for each retail-group k, as mentioned previously. The total retail-land is found by summing the acreage for each retail-group k within zone j.

To describe the household sector, the area's total population is considered a function of total employment

$$N = f \sum_{j=1}^{n'} E_j.$$ (6-6)

The term f is a coefficient relating number-of-households to total employment. The reciprocal of f is usually called the *labor-force participation-rate* or the percentage of the population that is in the work force. The symbol N denotes both the number of households or the actual population in this book. The precise definition should be clear from the context. The number of households allocated in each zone j is

$$N_j = g \sum_{i=1}^{n'} E_i \Theta_{ij}(N_j, C_{ij}) \tag{6-7}$$

for a total of

$$N = \sum_{j=1}^{n'} N_j. \tag{6-8}$$

Equation **(6-7)** suggests that the number of households in zone j depends on the zone's accessibility to employment opportunity. The term g is a scale factor needed to make the sum of zonal population equal to the total areal-population.

6.1.3 Constraints. Three inequalities serve as the constraints on the system. The first limits the dispersion of retail employment by imposing a minimum size, d_j^k, for the activity to take place.

$$E_j^k \geq d_j^k, \ \ or \ else \ \ E_j^k = 0. \tag{6-9}$$

This, in effect, suggests that there needs to be a threshold activity before the business can be viable. The second constraint prevents the system from generating excessive population-densities in locations with high accessibility. This is done by imposing a maximum development-density, D_j^H, on a zone.

$$N_j \leq D_j^H A_j^H. \tag{6-10}$$

The superscript of H is used to show that while the maximum development-density is written specifically for housing, similar constraint can be written for other types of developments, including industrial. The third constraint insures that the amount of land set aside for retail establishments does not exceed the amount available.

$$A_j^R \leq A_j - A_j^H - A_j^B. \tag{6-11}$$

These constraints introduce attraction-constrained spatial-interaction into the Lowry model. In other words, there can only be so much activity at a target zone—a concept explained in quite some detail in the "Economic Methods" chapter in Chan (2000). Many solution sequences that do not explicitly recognize this structure at the outset have to resort to ad hoc procedures to enforce the above three relationships. We will show immediately below. In fact these three constraints represent a major computational burden.

6.2 Solution method

An iterative algorithm consisting of three general steps is used to solve the above set of equations. In *step 1*, Equations **(6-1)**, **(6-6)**, **(6-7)** and **(6-8)** are attended to. As a first approximation, the exogenously-determined value of E_j^B is assigned to the variable E_j and A_j^R is set to zero. Through an inner iteration, g, N_j, and N are estimated as shown in the flow chart of **Figure 6-1**. First, a value for A_j^H is obtained from Equation **(6-1)**: $A_j^H = A_j - A_j^U - A_j^B - A_j^R$. Then the value of N is found in Equation **(6-6)**. The next step is not always used, but it can in some contexts speed convergence. The final value of N can be anticipated by allowing for the labor-force requirements of retail establishments in Equation **(6-6)**:

$$N = fE^B \Big/ \left(1 - f \sum_{k=1}^{K} a^k\right).$$

(6-12)

This equation is the result of a geometric series as derived in the `Simulation' section of the "Descriptive Tools" chapter in Chan (2000). Alternatively, see the derivation discussion in the section entitled "Derivation."

Next, the population potentials for each zone are determined using Equation **(6-7)** in two steps. First, we compute:

$$N_j(1) = \sum_{i=1}^{n'} E_i \Theta_{ij}(N_j, C_{ij}).$$

(6-13)

The value of the scale-factor g is then determined by reference to the total population to be allocated: $g(1) = N \Big/ \sum_{j=1}^{n'} N_j(1)$. The scale-factor g is sequentially used to adjust the population potentials to a second approximation:

$$N_j(2) = g(1)N_j(1).$$

(6-14)

Using constraint **(6-10)**, N_j is tested against the maximum-density constraint D_j^H. For all cases that $N_j(2) \geq D_j^H A_j^H$, a third approximation is made by setting $N_j(3) = D_j^H A_j^H$. In zones where $N_j(2) > N_j(3)$, the difference $N_j(2) - N_j(3)$ is distributed among all other zones according to their population potentials by revising the scale-factor g. For all other cases, i.e., $N_j(3) = g(2)N_j(2)$, the inner iteration is finished by Equation **(6-8)**: $N = \sum_{j=1}^{n'} N_j(3)$.

In *step 2*, Equations **(6-2)**, **(6-3)** and **(6-4)** are partitioned and solved by repeated approximations. The solution of step 1 for N_j is fed into Equation **(6-3)**. The exogenously-determined value for E_j^B is used as a first approximation for E_j in Equation **(6-3)** and the employment potentials in retail-group k are found for each zone as follows

$$E_j^k(1) = \sum_{i=1}^{n'} c^k N_i \Theta_{ij}(E_j^k, C_{ij}) + w^k E_j.$$

(6-15)

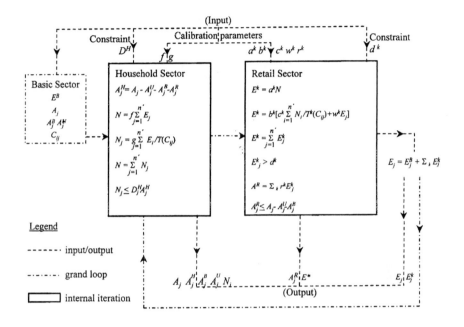

Figure 6-1 A framework of the Lowry model

Initial potential approximations are rescaled so they sum to the total employment that is determined from Equation **(6-2)**. Now a second approximation becomes

$$E_j^{\,k}(2) = b^{\,k}(1) E_j^{\,k}(1), \tag{6-16}$$

where $b^k(1)$ is found via Equation **(6-3)** as follows $b^{\,k}(1) = E^{\,k} \big/ \sum_{j=1}^{n'} E_j^{\,k}(1)$. This solution is tested against the minimum-size constraint, d^k. If such a constraint is violated, a search routine locates the smallest E_j^k, sets it equal to zero, and rescales employment in all other zones. The process is repeated until all $E_j^k(2)$ are greater than d^k and Equations **(6-4)** are satisfied.

In *step 3*, Equations **(6-4)** and **(6-5)** are solved by substitution for each trade class. The amount of retail land generated by Equation **(6-5)** is tested against constraint **(6-11)**. Where this constraint is not met, the model sets $A_j^R = A_j - A_j^U - A_j^B$. Overcrowding is allowed if insufficient land is available for retail employment at average densities. Retail land-uses have priority over residential land-uses. If residential population has been allocated to such a zone in step 1, it is removed by the residential-density constraint in the next grand loop of the iteration process. The values of E_j found from Equation **(6-4)** will be equal to or greater than the values of E_j that were used in the first trial of Equation **(6-7)**. The values of A_j^R are changed similarly in Equation **(6-5)**. The solution method slowly feeds in retail employment

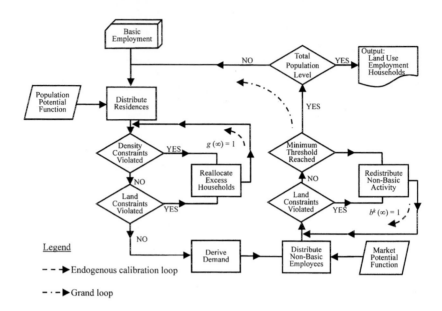

Figure 6-2 Lowry model flow-chart

and land-use as determinants of population distribution. The new values of A_j^R and E_j are used to begin a second iteration of the entire model beginning with step 1. Successive iterations are similarly performed, until the model converges on a stable set of values. A flow chart of the iterative scheme is shown in **Figure 6-2**.

6.3   A numerical example

We have gathered from above that the two basic computational steps of the Lowry model consist of an *allocation* technique using the gravity model and a *derivation* technique based on the Economic-Base Theory. Besides the diagrammatic representation in **Figure 6-1** and **Figure 6-2**, the best way to illustrate these two algorithmic steps is via an example (Reif 1973). A hypothetical area consisting of four zones has the spatial separation, say in dollars of travel cost, between each zone pair as $[C_{ij}] = \begin{bmatrix} 2 & 8 & 6 & 7 \\ 8 & 3 & 4 & 7 \\ 6 & 4 & 3 & 4 \\ 7 & 7 & 4 & 3 \end{bmatrix}$. Also shown in **Table 6-1** are the base-year employment and population figures for each of the four zones. It is to be noted that a straightforward way to calibrate the labor-force participation rate $1/f$ and the population-serving ratio a is by taking these ratios:

$$f = N/E = 3090/1200 = 2.575$$
$$a = E^R/N = 680/3090 = 0.220. \tag{6-17}$$

Zone	Basic emp E_j^B	Retail emp E_j^R	Total emp E_j	Population N_j
1	70	30	100	480
2	100	100	200	870
3	300	500	800	1020
4	50	50	100	720
Total	520	680	1200	3090

Table 6-1 Base-year activity data

6.3.1 Allocation. The purpose of this exercise is simply to illustrate the allocation and derivation relationships among the four zones. We will perform the detailed calculations for the first iteration only. Before the actual allocation calculations, two gravitational-interaction functions have to be defined, one for the choice of residential locations and other for the choice of retail- or service-employment locations. First, we define the distribution function for residential location as

$$\text{Prob (residential location)} = \text{prob (live in zone } j \mid \text{work at } i) =$$

$$\Theta_{ij}(N_j, C_{ij}) = \frac{N_j/C_{ij}^2}{\sum_l N_l/C_{il}^2}. \tag{6-18}$$

Notice it is assumed, for illustration purposes, that a square exponent is calibrated. In general, a trip-distribution function is defined $C = C(C_{ij})$. Here $C = C_{ij}^{-2}$ is merely a special case. The probability that an employee works and lives in zone 1 is therefore

$$\frac{N_1/C_{11}^2}{\sum_{l=1}^{4} N_l/C_{1l}^2} = (480/2^2)/(480/2^2 + 870/8^2 + 1020/6^2 + 720/7^2) = 0.68. \tag{6-19}$$

In a similar way, the probabilities of living in zone j for those working in zone i are computed by the following

$$[\Theta_{ij}(N_j, C_{ij})] = \begin{bmatrix} 0.68 & 0.08 & 0.16 & 0.08 \\ 0.04 & 0.53 & 0.35 & 0.08 \\ 0.06 & 0.24 & 0.50 & 0.20 \\ 0.06 & 0.10 & 0.37 & 0.47 \end{bmatrix}. \tag{6-20}$$

Notice the row sums are unity by definition.

The probability of retail location can be defined similarly as a gravity model. Thus the probability that a household that lives in zone i shops in zone j is

$$\Theta_{ij}(E_j^R, C_{ij}) = \frac{E_j^R / C_{ij}^2}{\sum_{l=1}^4 E_l^R / C_{il}^2}$$
(6-21)

where the trip-distribution function $C^R = C^R(C_{ij}) = C_{ij}^{-2}$ is exactly the same as the residential distribution-function above. Since there is only one retail trade-class, the following single matrix can be defined

$$[\Theta_{ij}(E_j^R, C_{ij})]^T = \begin{bmatrix} 0.31 & 0.01 & 0.01 & 0.02 \\ 0.07 & 0.25 & 0.10 & 0.05 \\ 0.58 & 0.71 & 0.84 & 0.79 \\ 0.04 & 0.03 & 0.05 & 0.14 \end{bmatrix}.$$
(6-22)

Here the column, rather than row, sums are unity, since the transpose of the matrix was taken. In summary, the allocation of the households and retail employment around the four zones is facilitated by the gravitational-interaction relationship $\Theta_{ij}(N_j, C_{ij})$ and $\Theta_{ij}(E_j^R, C_{ij})$ respectively. The process is identical to the allocation part of the "Location-allocation" model described in the previous chapter bearing the same name.

6.3.2 Derivation. The "activity derivation" part of the algorithm is akin to the "activity generation" portion of the 'unconstrained' gravity model, except that "time-series" projection of activities, rather than a "cross sectional" demand-function is used. To illustrate the iterative procedures, an explanation of the algorithm is in order. The derivation of secondary activities from basic employment is a continuous process until an equilibrium condition occurs—as manifested in terms of a stable total employment level, E, where $E = E^B + E^R$. Obviously, this is in reference to the similar process of Equation (6-12), where a converging geometric-series is also assumed. This is explained in the "Descriptive Tools" chapter of Chan (2000) and briefly reviewed below. Given basic-employment E^B, one can derive basic population in the first iteration $N(1)$ by multiplying it by the reciprocal of labor-force participation-rate:

$$N(1) = f E^B.$$
(6-23)

On the other hand, the amount of retail-employment $E^R(1)$ to serve the basic-population $N(1)$ is found by applying the population-serving ratio a:

$$E^R(1) = a N(1) = a f E^B.$$
(6-24)

For simplicity, the reader will recall that no categorization of the retail activity is considered in this example. For that reason, we write E^* simply as E^R. In the second iteration, the population $N(2)$ can again be derived from retail-employment $E^R(1)$ as follows: $N(2) = f E^R(1)$. The service employment required by the population is $E^R(2) = a N(2) = a^2 f^2 E^B$. Notice $E^R(2)$ also has its associated population $N(3)$ in the third iteration, and the process continues in this fashion. By recursion, the general expression for E^R can be obtained: $E^R(m) = a^m f^m E^B$.

As shown in the Chan (2000) chapters on "Descriptive Tools" and "Economic Methods," the total employment of the system is found by adding up all the increments of retail employment and basic employment:

$$E = E^B + E^R(1) + E^R(2) + \ldots + E^R(m) =$$
$$E^B(1 + af + a^2 f^2 + \ldots + a^m f^m). \tag{6-25}$$

Mathematically speaking, the summation results in $E = E^B(1 - af)^{-1}$ if $af < 1$ and m approaches infinity. Total population can then be calculated correspondingly by $N = f E = f E^B(1 - af)^{-1}$ (and hence the expression for Equation (6-12)). The system will only reach equilibrium when the employment and population reach these levels. Notice the assumption that $af < 1$ is reasonable because by definition of Equation (6-17) $af = (E^R/N)/(N/E) = E^R/E$, which is always less than unity. Applying the equation for total-employment calculation, one obtains $E = 520[1 - (0.22)(2.575)]^{-1} = 1200$, and the retail employment is $E^R = E - E^B = 1200 - 520 = 680$. These represent the equilibrium levels when the process finally converges. It is for this reason that Equation (6-12) may be helpful in speeding up this convergence of the algorithm.

6.3.3 An iterative process. The Lowry model, as extended by Garin (1966), forecasts future population and employment by executing the allocation and derivation steps iteratively. First, residential locations are assigned to total employment according to the distribution-function $\Theta_{ij}(N_j, C_{ij})$. Second, dependent population and retail employment are derived correspondingly. Third, the retail employees so generated are assigned work locations according to the distribution-function $\Theta_{ij}(E_j^R, C_{ij})$. Four, residential locations are assigned to the retail employment according to $\Theta_{ij}(N_j, C_{ij})$. Dependents are again generated, who in turn generate moreretail employment. Once more, residential locations are assigned to these retail employees via the function $\Theta_{ij}(E_j^R, C_{ij})$. This process simply repeats itself from thereon until convergence is obtained. Thus step 1 involves allocating employees to residences according to $\Theta_{ij}(N_j, C_{ij})$. The resulting allocation is shown in **Table 6-2**. Notice the $E_i(1)$ are merely basic employments since it is the first iteration. Also notice that the column sums are precisely the first approximations of the number of households, $L_j(1)$, in each zone j. This will eventually lead to an estimate on the total population, including dependents, as seen below.

Now one derives dependent population according to a zone-specific version of Equation (6-23): $N_j(1) = f L_j(1)$, where f has been given as 2.575, resulting in the following

Zone i	Zone j				Sum $E_i(1)$
	1	2	3	4	
1	48	5	11	6	70
2	4	53	35	8	100
3	18	72	150	60	300
4	3	5	19	23	50
Sum $L_j(1)$	73	135	215	97	520

Table 6-2 Residential location of retail employees

Zone	$L_j(1)$	$N_j(1)$
1	73	188
2	135	348
3	215	554
4	97	249

Retail employees can also be derived using a zone-specific version of Equations **(6-2)**, **(6-24)**: $M_j(1) = a\,N_j(1)$, where $a = 0.22$. The resulting first approximations of retail-employments $M_j(1)$ are

Zone	$N_j(1)$	$M_j(1)$
1	188	41
2	348	77
3	554	122
4	249	55

Retail employees so generated are assigned work locations according to a variation of Equation **(6-15)**. Here w^R is set to zero for simplicity, thus dismissing shopping made from the work location $E_i^R(1) = \sum_{j=1}^{4} M_j(1)\Theta_{ij}(E_j^R, C_{ij})$. Such calculations can be obtained in a most straightforward manner by referencing the $\Theta_{ij}(E_j^R, C_{ij})$ matrix as shown in Equation **(6-22)**. The table so generated is now shown in **Table 6-3**, where the end results are the row sums—the first approximations on retail employees.

The converging process is checked by comparing the total retail-employment derived thus far with the final retail-employment we expect from the summation performed previously in the section entitled "Derivation." Since $E(1) = 295$ is far

Zone *j*	Zone *i*				Sum $E_j^R(1)$
	1	2	3	4	
1	13	1	1	1	16
2	3	19	12	3	37
3	24	55	103	44	226
4	1	2	6	7	16
Sum $M_i(1)$	41	77	122	55	295

Table 6-3 Location of retail employment

short of $E = 680$, we know that additional iterations are necessary. For example, the next iteration is to assign residential locations to the retail employees generated above. Notice these calculations are performed for illustration purposes only. Many details are left out. For example, only economic activities such as population and employment are modelled, dropping the land-use derivatives. Other more notable omissions include the maximum and minimum constraints, which require extra iterations. As mentioned already, the retail-work-location probability $\Theta_{ij}(E_j^R, C_{ij})$ is simplified by omitting the influence of the other employment on retail locations. Finally, calibration constants such as g and b of the normalizing constraints **(6-14)** and **(6-16)** for locational-probabilities $\Theta_{ij}(N_j, C_{ij})$ and $\Theta_{ij}(E_j^R, C_{ij})$ are assumed to be unity. These simplifications allow us to show the essential features of the model without loss of generality—as we plan to address these simplifications in sequel. For those who wish to complete the six iterations required for convergence, we show the final results in **Table 6-4** and **Table 6-5**. It will be shown that a more straightforward matrix procedure exists to handle the same calculations.

6.4   A laboratory for experimentation

The city of York, Pennsylvania, was selected to illustrate the use of the Lowry model. York is located in the south-central portion of the State, about 25 miles (40.23 km) south of the state capital of Harrisburg and some 96 miles (154.46 km) west of Philadelphia. As such, York is not considered a "bedroom" suburban community to any of Pennsylvania's larger cities. It has its own identity and travel patterns reasonably isolated from other larger communities. Incidentally, York was reputed to be the first capital of the United States. This can be called a 'historic' case study, perhaps both due to York's place in history and the fact that the data base was obtained from 1963! The highway network included various smaller

Zone i	E_i^B	$E_i^R(1)$	$E_i^R(2)$	$E_i^R(3)$	$E_i^R(4)$	$E_i^R(5)$	$E_i^R(6)$	Total
1	70	16	6	3	2	1	0	98
2	100	37	21	11	7	4	2	182
3	300	226	130	76	41	23	14	810
4	50	16	9	4	3	2	1	85
Total	520	295	166	94	53	30	17	1175
Cumulative	520	815	981	1075	1128	1158	1175	

Table 6-4 Example employment iterations (Reif 1973)

Zone j	$N_j(1)$	$N_j(2)$	$N_j(3)$	$N_j(4)$	$N_j(5)$	$N_j(6)$	Total
1	188	67	36	18	8	5	322
2	348	198	111	64	36	23	780
3	554	348	196	113	62	31	1304
4	249	147	85	46	31	18	576
Total	1339	760	428	241	137	77	2982
Cumulative	1339	2099	2527	2768	2905	2982	

Table 6-5 Example population iterations (Reif 1973)

outlying communities, as shown in **Figure 6-3**. The York urban-area is most accurately described as having a core-concentrated structure with a well-defined c entral business-district and a radial system of streets and highways emanating from the core. The outlying communities of Dallstown and Red Lion constitute the only real departure from an otherwise dense core. These communities have a considerable amount of residential and commercial development, lying about six to seven miles (9.65–11.26 km) southeast of the city center.

York had a population of 125,300; labor force of 50,000; and a total employment of 58,500 in 1963. It was anticipated that the area would grow rapidly. In the twenty years from 1963, the land-development policy represented continued residential growth in the east-west and east-southeast corridors of York. While the regional role of downtown York was strengthened, new activity-centers were not located in relation to population-growth areas. Thus, employment and retail

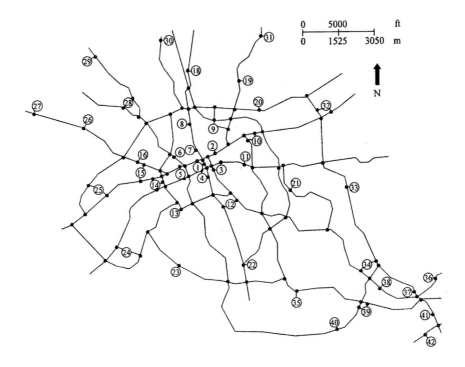

Figure 6-3 York, Pennsylvania, highway network and zone centroids

opportunities are not totally accessible to the population at large. The 1963 base-year activity-distribution is shown in the computer disk that comes with this book. The average trip-lengths remain relatively unchanged over ten years from 1963, with the average trip being seven to nine minutes depending on trip purpose. These trips are mainly executed by automobiles, with bus transit playing a very small role (1.5 percent of total vehicular trips).

A listing of the Lowry-model FORTRAN code is given in the computer software that comes with this book, with a brief description of the way the program can be run. The original code was obtained from the University of Pennsylvania, and subsequently modified at the Pennsylvania State University, State University of New York at Stony Brook, Washington State University, and the Air Force Institute of Technology. The model was used extensively in various case studies throughout this book. We have also included a listing of the input data-base in York, which parallels the input to the numerical example in the previous section. Such input data are supplemented by information shown above in other formats (such as in the York map of **Figure 6-3**). We hope that they, all together, form a self-contained package for experimentation.

6.5   Model calibration

A land-use model such as Lowry—like any other model—needs to be calibrated before it can be used. By calibration we mean the estimation of model parameters using statistical techniques to replicate the observed. Aprocedure to ensure that the calibration of these parameters has been performed satisfactorily is to run the calibrated model, using as input the base-year basic employment. The more the land-use and activity distribution replicate the existing condition, the better the quality of the calibration. The Lowry model employs eight parameters explicitly in its algorithm, as mentioned previously. These parameters are $a^k, b^k, c^k, w^k, r^k, f$ and g. It also employs a lower and upper bound on retail activities and residential density, d^k and D_j^H, as well as coefficients for work and nonwork trip-distribution function. The values of and calibration procedure for these parameters in York will be discussed one by one. Some parameters can be calibrated only if we analyze statistically the data gathered from the study area under consideration. For these parameters, we follow an *exogenous* procedure. On the other hand, other parameters are calibrated by an iterative algorithm built into the Lowry model. We call such calibration procedure *endogenous*.

The retail-employment coefficient, a^k, is the ratio of retail employment to population. It has a numerical value of 0.47. It is found in Equations **(6-2)** and **(6-24)**. The value is exogenously calibrated using a manual approach, where a ratio is taken between census data for areawide retail-employment and population as indicated in Equation **(6-17)**. Rather than an areawide average, the value could conceivably be more accurate if more than one trade-class is defined. The retail-employment scale-factor, b^k, is used in Equation **(6-3)** to ensure that the zonal retail-employment sadd up correctly to the areal total, as indicated in Equations **(6-4)**. The factor approaches a value of 1 in the final iteration to ensure a correct summation. The calibration is endogenous, i.e., it is part of the iterative algorithm. It is calibrated using the equation $b^k(l+1) = E^k / \sum_{j=1}^{n'} E_j^k(l)$ where l is the iteration number, as we have already seen in Equation **(6-16)**.

The two shopping-trip weight factors, c^k and w^k, give the relative importance of home-based shopping trips. They are used in Equation **(6-3)**. Their values are 0.9 and 0.1 respectively. They are exogenously calibrated using a constrained regression[8] or a maximum likelihood procedure with census and transportation data. While these are site-specific, they represent the relative importance of home- vs. work-based trips. The retail-employment density-ratio, r^k, represents the retail floor-space required per employee. It is found in Equation **(6-5)** and has a value 1/1.01. The calibration is exogenous and uses a regression procedure with census data.

The inverse of the labor-force-participation rate, f, represents the number of households needed to support one employee. It is found in Equation **(6-6)** and has

---

[8]An example of a constrained regression is given in the "Descriptive Tools" chapter (Chan 2000) under the section entitled "Gravity model revisited."

a value of 0.674. The calibration is exogenous and uses a manual approach with census data. A single coefficient is defined for the entire population and total employment areawide as suggested in Equation **(6-17)**. The population scale factor, g, is used in Equation **(6-7)** to ensure that in the final iteration Equation **(6-8)** is satisfied, which implies a final g value of 1.0. This calibration is endogenous, meaning that it is part of the iterative algorithm. The value for iteration number $l+1$ is $g(l+1) = N / \sum_{j=1}^{n'} N_j(l)$.

The minimum-retail-threshold constraint, d^k, limits the minimum-allowable-size of retail establishment. In Equation **(6-9)**, this threshold is set at 10. The calibration is exogenous and uses a manual approach with data from industry and census. The maximum residential-density-constraint, D_j^H, limits the maximum-allowable housing-development per zone. It is used in Equation **(6-10)**. It has a range of values, with the maximum density of 0.639 households per 1000 sq. ft. (90 m^2) at zone 5 and a minimum value of .034 at zone 31. The detailed data are listed in the software that comes with this book.

The coefficients of the retail-distribution function $C^k(C_{ij})$ are α, β and γ, carrying the values of 55.35, 14.60 and 0.665 respectively. This is for the special case for York, where only one retail class is defined, i.e., $k = R$. These coefficients are found in the function $dP^k/dC_{ij} = C^k(C_{ij}) = 1/(\alpha - \beta C_{ij} + \gamma C_{ij}^2)$ where P^k is the 'potential' and C_{ij} is the travel cost (or time) between zones i and j. The calibration is exogenous and uses a maximum-likelihood or regression procedure with data from travel studies. The coefficients of the work-trip distribution-function $C(C_{ij})$ are u and v', carrying the numerical values of 0.121 and 1.664 respectively. They are found in the equation $dP'/dC_{ij} = C(C_{ij}) = (v' C_{ij}^{v''})^{-1}$. Similar to the retail-distribution function, the calibration is exogenous and uses a maximum-likelihood or linear-regression procedure with data from travel studies.

While the basic ideas are amply described above, there are certain additional features and computational refinements that are only found in the computerized model. Such a model can carry out more computations on a complex set of data. Take the data-set for York as shown in the computer disk, for example. Convergence is shown after four grand loops in **Figure 6-2**, as suggested by the comparison between the two sets of numbers $E(3)$ and $E(4)$. However, the algorithm has not converged if the constraint equation **(6-10)** is imposed, or when N_j is compared to $D_j^H A_j^H$. In fact, the endogenous calibration process and the associated minimum and maximum constraints on activities represent a more complex part of the Lowry model. This complexity has not been adequately addressed until recently.

6.6 Matrix formulation and solution: an example

As the reader may have gathered by now, the unconstrained version of Lowry model is really the spatial version of the economic-base theory described at some length in the "Economic Methods" chapter of Chan (2000). Using matrix algebra, we will illustrate the additional spatial-dimension through a numerical example, using the same numbers as the four-zone illustration above. Aside from a more compact formulation using matrix notations, the maximum residential-density and minimum-threshold retail-employment will be explicitly considered for each zone.

6.6.1 Unconstrained version. Let the basic employment in the four zones be $\mathbf{E}^B = (70, 100, 300, 50)$. Thus in the initialization stage when $m = 0$, $\mathbf{E}(0) = \mathbf{E}^B$ by definition. It will be shown through matrix algebra that, with the maximum and minimum zonal-constraints enforced in sequel, the total employment in the mth grand loop will be $\mathbf{E}(m) = (100, 200, 800, 100)$, when m is a large number. Likewise, the number of households at the mth grand loop is $\mathbf{N}(m) = (480,870,1020,720)$. Differing from the conventional notation, these vectors are row instead of column vectors. Such convention is mainly adopted for convenience. It saves the extra step of writing out the vector transpose. Upper case letters are used to highlight this fact.

The inverse of the labor-force participation-rate, sometimes called the *activity rate*, can be expressed in the diagonal of a square matrix

$$[f_j] = \begin{bmatrix} f_1 & 0 & 0 & 0 \\ 0 & f_2 & 0 & 0 \\ 0 & 0 & f_3 & 0 \\ 0 & 0 & 0 & f_4 \end{bmatrix} = \begin{bmatrix} 2.575 & 0 & 0 & 0 \\ 0 & 2.575 & 0 & 0 \\ 0 & 0 & 2.575 & 0 \\ 0 & 0 & 0 & 2.575 \end{bmatrix}. \tag{6-26}$$

Likewise, the population-serving ratio can be expressed this way

$$[a_j] = \begin{bmatrix} a_1 & 0 & 0 & 0 \\ 0 & a_2 & 0 & 0 \\ 0 & 0 & a_3 & 0 \\ 0 & 0 & 0 & a_4 \end{bmatrix} = \begin{bmatrix} 0.22 & 0 & 0 & 0 \\ 0 & 0.22 & 0 & 0 \\ 0 & 0 & 0.22 & 0 \\ 0 & 0 & 0 & 0.22 \end{bmatrix}. \tag{6-27}$$

Work-to-home trip-probability-distribution can be expressed in matrix form as shown in Equation **(6-20)**. An equivalent home-to-work trip-probability distribution matrix can be written as Equation **(6-21)**.

6.6.1.1 INITIALIZATION. In the first iteration, we allocate employees to residences and derive dependent population:

$$\mathbf{N}(0) = \mathbf{E}(0)\,\mathbf{F} = \mathbf{E}(0)\left[\Theta_{ij}(N_j, C_{ij})\right]\left[f_j\right]. \tag{6-28}$$

where

$$\mathbf{F} = \begin{bmatrix} 0.68 & 0.08 & 0.16 & 0.08 \\ 0.04 & 0.53 & 0.35 & 0.08 \\ 0.06 & 0.24 & 0.50 & 0.20 \\ 0.06 & 0.10 & 0.37 & 0.47 \end{bmatrix} \begin{bmatrix} 2.575 & 0 & 0 & 0 \\ 0 & 2.575 & 0 & 0 \\ 0 & 0 & 2.575 & 0 \\ 0 & 0 & 0 & 2.575 \end{bmatrix} = \begin{bmatrix} 1.75 & 0.21 & 0.41 & 0.21 \\ 0.10 & 1.36 & 0.90 & 0.21 \\ 0.15 & 0.62 & 1.29 & 0.52 \\ 0.15 & 0.26 & 0.95 & 1.21 \end{bmatrix}. \tag{6-29}$$

The initial population-distribution can then be computed as $N(0) = (70, 100, 300, 50)$, $F = (188, 348, 554, 249)$.

6.6.1.2   GRAND LOOP 1.   Retail employees can be allocated to the various zones at the first loop through:

$$\mathbf{E}^R(1) = \mathbf{N}(0)\,\mathbf{A}' = \mathbf{N}(0)\left[\Theta_{ij}(E_j^R, C_{ij})\right]\left[a_j\right] \tag{6-30}$$

where

$$\mathbf{A} = \begin{bmatrix} 0.31 & 0.07 & 0.58 & 0.04 \\ 0.01 & 0.25 & 0.71 & 0.03 \\ 0.01 & 0.10 & 0.84 & 0.05 \\ 0.02 & 0.05 & 0.79 & 0.14 \end{bmatrix}\begin{bmatrix} 0.22 & 0 & 0 & 0 \\ 0 & 0.22 & 0 & 0 \\ 0 & 0 & 0.22 & 0 \\ 0 & 0 & 0 & 0.22 \end{bmatrix} = \begin{bmatrix} 0.068 & 0.015 & 0.128 & 0.009 \\ 0.002 & 0.055 & 0.156 & 0.007 \\ 0.002 & 0.022 & 0.185 & 0.011 \\ 0.004 & 0.011 & 0.174 & 0.031 \end{bmatrix}. \tag{6-31}$$

By substituting Equation **(6-28)** into Equation **(6-30)**, we have

$$\mathbf{E}^R(1) = \mathbf{E}(0)\mathbf{F}\mathbf{A}'. \tag{6-32}$$

Putting the numerical values into the equation above in a stagewise manner:

$$\mathbf{E}^R(1) = [\mathbf{E}(0)\,\mathbf{F}]\mathbf{A}' =$$
$$(188, 348, 554, 249)\mathbf{A}' = (16, 37, 226, 16). \tag{6-33}$$

Summing basic and retail employment together up to this juncture, and using the result from Equation **(6-32)**:

$$\mathbf{E}(1) = \mathbf{E}^B + \mathbf{E}^R(1) = \mathbf{E}(0) + \mathbf{E}^R(1) = \mathbf{E}(0)(\mathbf{I} + \mathbf{FA}') \tag{6-34}$$

where $\mathbf{I}$ is a 4×4 identity matrix. It follows that $\mathbf{N}(1) = \mathbf{E}^R(1)\mathbf{F} = \mathbf{E}(0)\mathbf{FA'F}$.

6.6.1.3 GRAND LOOP 2 AND BEYOND.   Similar to Equation **(6-32)**, retail employment can be derived from population by $\mathbf{E}^R(2) = \mathbf{E}(0)\mathbf{FA'FA'} = \mathbf{E}(0)(\mathbf{FA'})^2$. Likewise, population is derived from retail employment through the reciprocal relationship shown in Equation **(6-28)**: $\mathbf{N}(2) = \mathbf{E}(0)(\mathbf{FA'})^2\mathbf{F}$. This process is repeated iteratively through the mth iteration, resulting in

$$\mathbf{E}^R(m) = \mathbf{E}^R(1) + \mathbf{E}^R(2) + \ldots + \mathbf{E}^R(m) =$$
$$\mathbf{E}(0)\left[\mathbf{FA'} + (\mathbf{FA'})^2 + \ldots + (\mathbf{FA'})^m\right]. \tag{6-35}$$

In the general case, one can express total employment, which is the sum of basic and retail employment, as an infinite series $\mathbf{E} = \mathbf{E}(0)\left[\mathbf{I} + \mathbf{FA'} + (\mathbf{FA'})^2 + \ldots + (\mathbf{FA'})^m + \ldots\right]$. Likewise, total population can be expressed as an infinite series

$$N = N(0) + N(1) + N(2) + \ldots + N(m) + \ldots =$$
$$E(0)\left[I + FA' + (FA')^2 + \ldots + (FA')^m + \ldots\right]F.$$

(6-36)

If the sum of the elements on each row of **FA'** is less than unity,[9] then the series can be summed by the closed-form solutions

$$E = E^B(I - FA')^{-1} \quad and \quad N = E^B(I - FA')^{-1}F.$$

(6-37)

A check on the product **FA'** shows that it indeed satisfies the condition, as one can

see from below: $\mathbf{FA'} = \begin{bmatrix} 0.121 & 0.049 & 0.372 & 0.028 \\ 0.012 & 0.098 & 0.431 & 0.027 \\ 0.016 & 0.070 & 0.445 & 0.036 \\ 0.017 & 0.051 & 0.446 & 0.051 \end{bmatrix}$. Numerically, one can now compute the

closed-form solution for total employment in Equation **(6-37)**:

$$E = (70,100,300,50)(I - FA')^{-1}$$

$$= (70,100,300,50)\begin{bmatrix} 0.879 & -0.049 & -0.372 & -0.028 \\ -0.012 & 0.902 & -0.431 & -0.027 \\ -0.016 & -0.070 & 0.555 & -0.036 \\ -0.017 & -0.051 & -0.446 & 0.949 \end{bmatrix}^{-1}$$

$$= (98,182,810,85)$$

(6-38)

Similarly, the total population becomes $N = EF = (98,182,810,85)F = (322,780,1304,576)$.

6.6.2 *Adding zonal constraints.*
The maximum residential-density and minimum-retail-threshold constraints, as expressed in Equations **(6-10)** and **(6-9)** respectively, can be expressed in vector form. Each entry of the vector denoting the population and employment in a zone: $N \le N^c$ and $E^R \ge d^R$, where $N^c = D^H A^H$. Here

$\mathbf{D}^H = \begin{bmatrix} D_1^H & 0 & \ldots & 0 \\ 0 & D_2^H & \ldots & 0 \\ 0 & 0 & . & . \\ 0 & 0 & . & D_{n'}^H \end{bmatrix}$. Endogenous calibration procedures have been explained in the

section on calibration to enforce these constraints. In the example above, let us say $N^c = (480, 870, 1020, 576)$ and $N(m) = (322, 780, 1304, 576)$ as $m \to \infty$. The $N(m)$ vector has the population in zone 3 exceeding the maximum residential-density. Endogenous iterations are necessary to ensure that the zoning ordinances are followed. This consists of computing $N(1) = (125.1, 302.9, 506.4, 223.7)$ using Equation **(6-13)** and subsequent iteration-steps. Thus

$$g(1) = N(m) \Big/ \sum_j N_j(1) = \frac{322 + 780 + 1304 + 576}{125.1 + 302.9 + 506.4 + 223.7} = 2.573.$$

(6-39)

Now $N(2) = 2.573 N(1) = (322, 780, 1304, 576)$ as one would expect.

---

[9]Notice this is a generalization of the assumption $af < 1$ in Equation **(6-25)**.

For those zones with population exceeding the allowable, or $N_j > N_j^c$, we arbitrarily set the zonal population at the maximum density. Thus for zone 3, we set $N_3'(2) \equiv N_3^c = 1020$. Now for $N_j(2) > N_j'(2)$, the overage is redistributed to other zones. Take zone 3 again, the excess population $N_3(2) - N_3'(2) = 1302 - 1020 = 284$ is redistributed to zones 1, 2 and 4 according to their population potentials.

$$N_1(3) = g(2)N_1(2) = \frac{(2982)}{(322+780+1020+576)}(322) = 336.0. \qquad \textbf{(6-40)}$$

Similarly, $N_2(3) = g(2)N_2(2) = (1.043)(780) = 813.5$ and $N_4(3) = g(2)N_4(2) = 600.8$. Now $N(3) = \sum_j N_j(3) = 336.0 + 813.5 + 1020 + 600.8 = 2770.3$, which still falls short of the goal of 2982. More iterations are necessary; thus iteration 4 begins. The iterations end when $g(l) = N(m)/N(l) = N(m)/\sum_j N_j(l) = 1$.

An equivalent illustration can be given for the minimum-retail-threshold constraint. Again for demonstration purposes, let $\mathbf{d}^R = (40, 88, 500, 20)$ and $\mathbf{E}^R(m) = (28, 82, 510, 35)$. It can be seen that zone 1 (among two zones) falls short of the threshold activity by a hefty margin. Adjustments are needed to redress this violation. At iteration 1, $\mathbf{E}^R(1)$ is estimated using Equation **(6-15)** where w^R is set to zero to preclude shopping from the work location .Subsequently $b^R = c^R$; this reduces the two calibration constants to the single scaling constant b^R. The resulting $\mathbf{E}^R(1)$ vector is $(127.3, 372.7, 2318.0, 159.1)$. The first estimation of the scaling constant is $b^R(1) = E^R(m)/\sum_j E_j^R(1) = 0.22$. Hence $\mathbf{E}^R(2) = b^R \mathbf{E}^R(1) = (28, 82, 510, 35)$. Since the retail threshold constraint is violated in zone 1, or $E_1^R = 28 < d^R = 40$, set $E_1'^R(2) = 0$. The excess $E_1^R(2) - E_1'^R(2) = 28$ is redistributed by rescaling employment in all other zones. Thus

$$E_2^R(3) = b^R(2)E_2^R(2) = \frac{(28+82+510+35)}{(0+82+510+35)}(82) = 85.7. \qquad \textbf{(6-41)}$$

$E_3^R(3) = b^R(2)E_3^R(2) = (1.045)(510) = 532.8$, and $E_4^R(3) = b^R(2)E_4^R(2) = 36.56$. The iterations continue until $l = 3$, when $b^R(3) \approx 1$.

While the above procedures were explained in scaler notation, the results can be formalized in terms of matrix algebra. In general, when the maximum-density and minimum-threshold constraints are violated, new derivation-allocation matrices **FA'** need to be revised. Also, population/employment are recomputed according to $\mathbf{N} = \mathbf{E}\,\mathbf{F}(1)$ and $\mathbf{E}^R = \mathbf{N}\,\mathbf{A}'(1)$, where $\mathbf{F}(1)$ and $\mathbf{A}'(1)$ are revised matrices. The way to effect these revisions are many. We have described one approach above. $\mathbf{F}(1)$ and $\mathbf{A}'(1)$ can be defined in an alternate way, similar to attraction-balancing procedure of the gravity model.[10] For example, they can be defined with these entries to effect a distribution of activities that satisfies the zonal constraints

---

[10]For an explanation of gravity-model calibration, see the "Descriptive Tools" chapter in Chan (2000). The attraction-balancing procedure is one of several techniques to calibrate a doubly-constrained gravity model.

$$\Theta_{ij}(N_j, C_{ij}) = \frac{N_j'/C_{ij}^2}{\sum_j N_j'/C_{ij}^2} \qquad \forall j \tag{6-42}$$

where $N_j' = N_j^c\left(N_j^c/N_j\right)$. Similarly,

$$\Theta_{ij}(E_j^R, C_{ij}) = \frac{E_j'^R/C_{ij}^2}{\sum_j E_j'^R/C_{ij}^2} \qquad \forall j \tag{6-43}$$

where

$$E_j'^R = \begin{cases} d^R & for\ zones\ in\ which\ E_j^R \ge d^R \\ 0 & for\ zones\ in\ which\ E_j^R < d^R. \end{cases} \tag{6-44}$$

These perform similar functions as the endogenous-calibration loops described in Equations **(6-14)** and **(6-16)** *plus* the subsequent iterations in our initial description of the model constraints. This was illustrated by the numerical example above. In the "Chaos, Catastrophe, Bifurcation" chapter to follow, where we will generalize the results of the Lowry-derivative models, this idea will be developed. We will outline a simple, perhaps even elegant, way of handling the zonal constraints.

6.7   The Projective Land Use Model

A `descendent' of Lowry is the Projective Land Use Model (PLUM) by Rosenthal, Meredith and Goldner (1972). Having explained the Lowry model in detail, the basic ideas of PLUM can be illustrated with a numerical example (Williams and Tempio 1974), since PLUM is essentially a computational refinement of Lowry.

6.7.1 Input. Consider a three-zone study-area with interzonal travel-cost as defined by $[C_{ij}] = \begin{bmatrix} 5 & 10 & 10 \\ 10 & 5 & 20 \\ 10 & 20 & 5 \end{bmatrix}$. The base-year data are given by **Table 6-6**, consisting of population, employment, and land-use. This is supplemented by the distribution function for both work and nonwork trips, as characterized by the following matrix: $[\theta_{ij}(C_{ij})] = \begin{bmatrix} 0.3 & 0.4 & 0.4 \\ 0.4 & 0.3 & 0.1 \\ 0.4 & 0.1 & 0.3 \end{bmatrix}$. Thus the probability of making a trip or the percentage of trips between zones 1 and 2, which is the same as making a trip of 10 units in cost, is 0.4. Notice that while the trip-distribution function has the properties of a probability-density-function, the entries do not necessarily sum to unity row-wise or column-wise in this representation. The reason is that not every trip of every travel-cost is shown in the matrix. **Figure 6-4** shows the distribution--function $\Theta_{ij}(C_{ij})$, where intrazonal trips in zone 1 have a trip cost of 5 units and represent 30 percent of the trip etc.

Base-year data		Zone 1	Zone 2	Zone 3
Population	Dwelling units	100	200	200
	Residential population	300	400	600
	Employed residents	150	200	200
	Nonworking population	150	200	100
Employment (at place of work)	Basic emp	200	100	100
	Local-service emp	20	30	100
Land use	Total acreage	900	900	900
	unusable acreage	70	30	20
	Basic acreage	100	200	100
	Local-service acreage	28	28	56
	Residential acreage	100	100	200
	Other vacant acreage	250	250	200
	Street/highway acreage	130	70	80
	Vacant industrial acreage	200	200	200

Table 6-6 Base-year input-data for PLUM

The target-year data are shown in **Table 6-7**, consisting of both zonal and areawide figures. This is equivalent to the forecast-year data in the Lowry model, where basic employment—specified by zone—serves as the seed for future growth in the study area. But the idea is expanded to include the areawide total of local-service employment, nonworking population and housing units expected for the target year. The areawide figures serve as "control totals" for the forecasts.

6.7.2 Solution. *Step 1* of the algorithm calls for the distribution of basic employment. The incremental-distribution procedure is best explained in a tabular form, where the distribution function and the zonal allocations can be displayed side-by-side. For example, the growth increment from base to target year in zone 1 is (300–200) or 100. This amount of employment is allocated among the three zones in the study area according to the distribution probabilities 0.3, 0.4 and 0.4, entries of the matrix $[\Theta_{ij}(C_{ij})]$.

In *step 2*, basic population (pop) is derived from basic employment (emp) through a simple ratio, which corresponds to the average-population per dwelling-

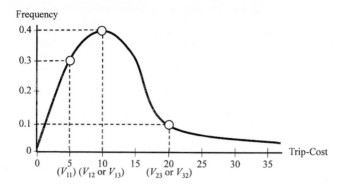

Figure 6-4 Work and nonwork trip-distribution-function

Target year data	Zone 1	Zone 2	Zone 3
Basic employment	300	150	150
Basic acreage	200	250	150
Total local-service emp		300	
Total nonworking pop		1120	
Total housing units		800	

Table 6-7 Target-year population and employment control figures

unit. In other words, (Δ basic emp at residence) (avg. pop /dwelling unit) = (Δ basic pop).

Zone	Calculation	Δ basic pop
1	(27+25+25)(300/100)	231
2	(37+19+6)(400/200)	124
3	(36+6+19)(600/200)	183
		538.

In *step 3*, a multiplier similar to the population-serving ratio in Lowry is calculated as follows:

(multiplier) = (Δ total local-service emp)/(Δ total basic pop) = (300−150)/538 = 0.28.

In *step 4*, the demand for local-service employment at basic-population residences is derived. This is simply the multiplier effect from basic-employment population at each zone.

Zone pair	Probability	Normalized probability	Total basic employment increment	Distributed basic employment
1–1	0.3	0.27	300–200=100	27
1–2	0.4 1.1	0.37	100	37
1–3	0.4	0.36	100	36
2–1	0.4	0.50	150–100=50	25
2–2	0.3 0.8	0.38	50	19
2–3	0.1	0.12	50	6
3–1	0.4	0.50	150–100=50	25
3–2	0.1 0.8	0.12	50	6
3–3	0.3	0.38	50	19

Table 6-8 - Result of basic-employment distribution

Zone	Calculation	Δ service emp
1	(231)(0.28)	64
2	(124)(0.28)	34
3	(183)(0.28)	51

In *step 5*, the demand for local-service employment at basic-employment locations is calculated.

Zone	Calculation	Δ service emp
1	(100)(0.28)	28
2	(50)(0.28)	14
3	(183)(0.28)	14

Step 6 calls for the increment to local-serving-employment allocation from resident demand—again a very similar concept as that in the Lowry model:

Zone-pair	Probability	Normalizing	Demand	Distribution
1–1	0.3	0.27	64	17
1–2	0.4	0.37	64	24
1–3	0.4	0.36	64	23
2–1	0.4	0.50	34	17
2–2	0.3	0.38	34	13
2–3	0.1	0.12	34	4
3–1	0.4	0.50	51	26
3–2	0.1	0.12	51	6
3–3	0.3	0.38	51	19

Here the normalization is carried out to make each row-sum and column-sum of allocation matrix $[\Theta_{ij}(C_{ij})]$ unity.

As one would expect, the other increment to local-service employment is from basic-employment demand. It is calculated in *step 7*. The allocation process follows the same format as above:

Zone-pair	Normalizing	Demand	Distribution
1–1	0.27	28	8
1–2	0.37	28	10
1–3	0.36	28	10
2–1	0.50	14	7
2–2	0.38	14	5
2–3	0.12	14	2
3–1	0.50	14	7
3–2	0.12	14	2
3–3	0.38	14	5

The allocations so far are summed for each zone in *step 8*, resulting in the total increment to local-service employment:

Zone	Calculation	Δ Service emp
1	17+17+26+8+7+7	82
2	24+13+6+10+5+2	60
3	23+4+19+10+2+5	63
Total		205

As differentiated from the Lowry concepts, the PLUM model has the checks built in through control totals. When the allocation sum above does not agree with the control total, a correction factor is applied to scale these allocation totals up or down, so that the correct total is obtained. Thus in *step 9*, we compute such a correction factor:

(input increment of local-service emp)/(calculated increment) = (150)/(205) = 0.73.

The corrected local-service-employment increment is correspondingly calculated in *step 10*:

Zone 1	(82)(0.73)	60
Zone 2	(60)(0.73)	43
Zone 3	(63)(0.73)	47
Total		150

The total increment to employment is the sum of local-service and basic increments. Thus in *step 11*, these totals are obtained:

Zone 1	60+100	160
Zone 2	43+50	93
Zone 3	47+50	97

Now is the time to find a home for the employment increment to be decided. In *step 12*, the total increment of employment is allocated spatially in the usual manner:

Zone-pair	Probability	Normalized	Increment	Distribution
1–1	0.3	0.27	160	43
1–2	0.4	0.37	160	59
1–3	0.4	0.36	160	58
2–1	0.4	0.50	93	47
2–2	0.3	0.38	93	36
2–3	0.1	0.12	93	11
3–1	0.4	0.50	97	48
3–2	0.1	0.12	97	12
3–3	0.3	0.38	97	37

The nonworking population is now incremented in *step 13*. It amounts to a fraction of the total increment of employment—in the same proportion as the base-year:

$$[\{(\text{base pop})/(\text{base employed pop})\} - 1](\text{total emp increment}).$$

Computationally, this amounts to

Zone 1	[(300/150)−1](43+47+48 or 138)	138
Zone 2	[(400/200)−1](59+36+12 or 107)	107
Zone 3	[(600/200)−1](58+11+37 or 106)	212
Total		457

Increment to the nonworking population is again corrected vis-a-vis the control total. This process is parallel to steps 9 and 10, where the local-service employment increment is adjusted. Thus we calculate the correction factor as follows in *step 14*:

[(input total nonworking)/(calculated total nonworking)] = [(1120−750)/457] = 0.81.

Corrected nonworking population is determined correspondingly as follows in *step 15*:

Zone 1	(138)(0.81)	112
Zone 2	(107)(0.81)	87
Zone 3	(212)(0.81)	171
Total		370

Now we take the sum of the residential-population increment between working and nonworking residents in *step 16*:

Zone 1	112+138	250
Zone 2	87+107	194
Zone 3	171+106	277
Total		721

A residential factor is defined to adjust the size of the family from base to forecast year in *step 17*:

$$\frac{[(\Delta forecast\text{-}yr\ pop)/(\Delta forecast\text{-}yr\ housing\text{-}unit)]}{[(base\text{-}yr\ pop)/(base\text{-}yr\ housing\text{-}unit)]} = [(721)/(800\text{-}500)]/[1300/500] = 0.92.$$

Population per housing-unit in forecast year can be found by adjusting the family size from base year. In *step 18*, we simply calculate on a zonal basis these figures:

Zone 1	(0.92)(300/100)	2.76
Zone 2	(0.92)(400/200)	1.84
Zone 3	(0.92)(600/200)	2.76

The increase in dwelling-units per zone is simply derived from population in *step 19*:

Zone 1	250 /2.76	94
Zone 2	194 /1.84	106
Zone 3	277 /2.76	100
Total		300

6.7.3 Land-use accounts and output. So far, we have been dealing with population and employment. At this juncture, the land-use consumption due to these activities is computed. In *step 20*, the land-absorption coefficient for local-service employment is calculated, showing the rate of land-consumption per local-service employee:

(base-yr local-service-acreage)/(base-yr local-service-emp) = 112/150 = 0.75.

In *step 21*, the increment of local-serving land is computed to be

Zone 1	(60)(0.75)	45
Zone 2	(43)(0.75)	32
Zone 3	(47)(0.75)	35
Total		112

In *step 22* industrial land is accounted for in the following table:

Zone	Vacant industrial	Other vacant increment	Basic + local-service	Remaining vacant
1	200	250	200+45=245	205
2	200	250	250+32=281	169
3	200	200	150+35=185	215

If the basic and local-service employment-increment exceeds available land, the excess activities are allocated among other zones as in the Lowry model (Goldner et al. 1972).

A parallel step is used to account for residential land-use. Thus in *step 25*, the land-absorption coefficient for residences is also computed by the ratio [(base residential-acreage)/(base dwelling-units)]:

Zone 1	100/100	1.0
Zone 2	100/200	0.5
Zone 3	200/200	1.0

The increment of residential land is now computed in *step 26:*

Zone 1	(94)(1.0)	94
Zone 2	(106)(0.5)	53
Zone 3	(100)(1.0)	100

All these are less than the available vacant land. Should they be greater, a reallocation procedure will be followed (Goldner et al. 1972).

The final results can be organized into a table as shown below. Notice all figures are expressed in terms of increments:

	Zone		
	1	2	3
Dwelling units	250	194	277
Residential population	160	93	97
Nonworking population	112	87	171
Basic employment at place of work	100	50	50
Local-service emp at place of work	60	43	47
Basic acreage	100	50	50
Residential acreage	94	53	100
Vacant	111	116	115
Local-service acreage	45	32	35.

6.7.4 Summary. We select the Projective Land Use Model (PLUM) to illustrate a step child of Lowry. It differs from Lowry in these respects. First, PLUM requires more time-series data, such as control totals of local-service employment, total nonworking-population and dwelling-units for the target year. Lowry, on the other hand, only requires target-year disaggregate (zonal)-basic-employment-data. Second, PLUM uses an incremental approach to activity allocation and derivation. This means a variation on model parameters such as population-serving ratio, residential factor, and land-absorption coefficients from one iteration to another. Third, both Lowry and PLUM are prototype models that have limited application experience. PLUM has been disseminated by the Federal Highway Administration of the United States Department of Transportation as a computational package. Meanwhile, Lowry has been a favorite of academics and researchers because of its methodological underpinning. Fourth, PLUM is more expensive to run on the computer, simply due to its incremental approach to computation. This is particularly the case when reallocation is required due to zonal holding-capacities, which is already cumbersome in Lowry. Finally, Lowry is more transparent in its model structure. While PLUM is more sophisticated, its extra burden on computation may not justify its additional features, as we will show in the section entitled "An experimental comparison."

6.8   Theoretical foundation

As discussed above, Lowry-based models provide spatial allocation of future population and employment based on a two-part theory: economic-base theory and spatial allocation. The classic economic-base (or export-service) theory attributes the growth of a study area to the economic opportunities generated between the basic and non-basic sectors. The operation of the economic-base theory rests chiefly on the observation that the 'exporting' goods and services of the basic sector has a multiplier effect on the non-basic local-economy. As such, the theory can be viewed as a special case of the input-output model (Webber 1984).

6.8.1   Economic-base theory.   Specifically, through the generation of population-serving employment, the incremental increase in employment is related to the previous increase in employment via an artificial time increment and a 'constant' γ, which is defined by the labor-force-participation-rate $1/f$ and population-serving-ratio a.

$$\Delta E(t) = (\gamma \Delta t)\, \Delta E(t-1). \tag{6-45}$$

A geometric series sums up these increments, where the total employment E in the study area is related to basic-employment E^B via the economic-base multiplier: $\gamma = \gamma \Delta t = af$

$$E = E^B(1 + \overline{\gamma} + \overline{\gamma}^2 + \overline{\gamma}^3 + \ldots + \overline{\gamma}^m) = E^B\,\Gamma. \tag{6-46}$$

Here the constant Γ is $1/(1-\gamma)$, or the gross economic multiplier, as defined in a "total stock" approach (Lowry 1964). Notice the multiplier geometric series converges (or reaches an equilibrium) because the ratio γ is normally less than unity and γ is assumed to be constant over the increments in Equation **(6-45)**. Throughout this process, the population increases correspondingly. While the above was developed for a scalar quantity initially, we have shown in the "Lowry-based model" discussions that the process can be carried over to matrix notations where population or employment in each zone is an entry in the vector.

6.8.2 Spatial allocation.   According to a set of spatial prices established by transportation costs, economic activities (such as population and employment) are allocated among the subregions or zones of a study area. Two fundamental concepts are employed in spatial allocation: 'accessibility' and 'opportunity'. The accessibility to a particular activity is a measure of the transportation cost to that activity. Opportunity, on the other hand, generally can be represented by the amount of land available for development (at a specified density). From these two parameters, three "relative accessibility" or 'distribution' functions $\Theta(C_{ij})$ are ideally defined: (a) the probability of locating a residence in site j given an employment location in site i, $\Theta_{ij}(N_j, C_{ij})$ (b) the probability of locating a shop in site j given a residential location in site i, $\Theta_{ij}(E_j^R, C_{ij})$ and (c) the probability of locating a business-serving establishment in site j given a work location i $\Theta_{ij}(E_j, C'_{ij})$, where the spatial cost C'_{ij} takes on a special combined form as illustrated in Equation **(6-15)**. A strong coupling exists, therefore, between residential and service-industry locations. Using these probability functions, there exist several mathematical techniques to allocate activities around the urban area, including the gravity model (as used in Lowry) and the intervening opportunity model[11] (as used in PLUM).

For example, population is distributed relative to work locations i by this function in the Lowry model: $N'_j = \sum_{i=1}^{n'} E_i \, \Theta(N_j, C_{ij})$ where N'_j is the total number of households in zone j as computed by the model, and E_i is the total employment in zone i. Lowry used the following definition originally for the distribution function $\Theta(N_j, C_{ij})$:

$$\frac{N_j / C(C_{ij})}{\sum_l N_l / C(C_{il})} . \tag{6-47}$$

$C(C_{ij})$ is an accessibility function. A related concept is a calibrated trip-distribution function or *potential surfaces* P' for destination choice. Thus a special form of potential surface may give rise to $dP'/dC_{ij} = (v'C_{ij}^u)^{-1}$. Following this line of thoughts,

$$\Theta(N_j, C_{ij}) = \frac{dP'}{dC_{ij}} N_j / R'_i = v^{l-1} N_j C_{ij}^{-u} / R'_i \tag{6-48}$$

where R'_i is the normalizing constant similar to the denominator of Equation **(6-47)**: $R'_i = \sum_{j=1}^{n'} N_j (dP'/dC_{ij})$. The extra term R'_i is introduced to correspond to the operation of the computer programs in the accompanying disk. In the Lowry model and the Yi–Chan model software, the number of zones reachable within x minutes of travel time from zone i is kept in a table. Depending on the origin zone i, the maximum travel-time to reach the furthest destination-zone in the study area is different. The total number of zones reachable within x minutes is also different.

In PLUM-IP, on the other hand, the equivalent distribution-function for residential location is

$$\frac{dP'}{dC_{ij}} = \beta \frac{W_j}{C_{ij}^3} \exp(\alpha - \beta/C_{ij}) \tag{6-49}$$

where W_j is the attractiveness at destination-zone j, and the following relationship is observed between calibration constants α and β: $\alpha = \beta /(\text{largest trip-cost})$. Given the largest trip-cost varies depending on the study area, the allocation distribution-function changes from study area to study area. To the best of the author's knowledge, such trip-distribution functions have to be calibrated exogenously for all activity-allocation models. It should be noted that all these various forms— $C(C_{ij})$, dP'/dC_{ij} and trip-distribution function—have a common characteristic. They

---

[11]For an introduction to the intervening opportunity model, see the "Economic Methods" chapter in Chan (2000).

all effect a 0–1 ranged value function.[12] For example, $C(C_{ij}) = \exp(-C_{ij})$ is a declining function of C_{ij} for a nonnegative C_{ij}. $C(C_{ij}) = C_{ij}^u$ declines from unity toward zero for C_{ij} ranged between one and infinity. dP'/dC_{ij} as shown in Equation (6-49) has an exponent that is always negative and a power function C_{ij}^u, thus effecting the characteristics of both functions mentioned above.

Lying in the heart of these seemingly diverse allocation formulations is an approximation of the economic concept of land rent.[13] Recent work by Cochrane, for example, put both the gravity and opportunity model on a multinomial logit formulation through a process of consumers' surplus maximization.[14] Bearing in mind Kain's (1961) rent model, however, the structure of the probability function should vary with respect to the location from the center city in addition to the spatial separation between two sites. Hence the specific functional form for the frequency-distribution dP'/dC_{ij} should vary. Quite clearly, there should also be incorporated into the distribution or probability function an increasing function expressing out-of-pocket expenditure as well as travel time, so that a generalized spatial-price C_{ij} can be obtained. As far as the definition of opportunity goes, the development opportunity of a site can be thought of as land availability, with some important details. First, opportunity is proportional to the developed-land area for the activity under consideration. Second, land area is the available land for such proposes as industrial, commercial and residential given the maximum development-density prescribed by the zoning ordinance. Although the implementation procedure is different in Lowry vs. PLUM, a holding capacity is imposed as an upper limit for residential activities in a zone: $N_j \leq N_j^c$. In addition, a threshold of minimum viable activity must be available in the Lowry model before retail development can take place: $E_j^k \geq d_j^k$. It is these 'truncations' that have profound implications on the way activities get allocated eventually—as we have illustrated. In short, the measure of opportunity represents an 'intrinsic' attractive-power of a site, generally expressing a measure independent of the proximity to other sites.

6.9   Lowry vs PLUM

We have seen that there are certain theoretical commonalities between Lowry and PLUM, which qualify them to be classified under a common school. There is, however, different extent to which these theories are incorporated into the models, as we have witnessed already. Since a decade has elapsed from Lowry's model to

---

[12]While the difference between value function and utility function is highlighted in the "Multi-criteria Decision-making" chapter of Chan (2000), we are using the terms interchangeably here.

[13]The concept of land rent was introduced in the "Utility theory" section of the "Economic Methods" chapter (Chan 2000).

[14]See the discussion in the "Economics Methods" chapter (Chan 2000) under "Economics basis of the gravity model" section..

PLUM-IP, the latter has incorporated some computational refinements not found in the original Lowry model.

6.9.1 Growth projection. The simulation of growth is accomplished by the economic-base theory in both models. However, the multiplier concept is incorporated into the structure of the model with important variations (Tiebout 1956, Sirkin 1959). Lowry's model merely assumes that basic vs. non-basic activities are defined exogenously to the model—possibly via a functional criteria established by the location quotient or minimum requirement method (Fitzroy 1978).[15] Goldner (1968), recognizing the importance of the problem and the difficulties in operationalizing the definition, advanced a locational-classification criteria.Non-basic activities are often located in reference to exogenously determined basic-activities (according to a location quotient). An operational way to group industries is to label any industry or trade whose location is dependent on the spatial distribution of households as "population-serving" industry. Otherwise, it is "locationally basic."

Table 6-9 classifies standard industrial categories into major groups by locational orientation and economic-base orientation. Again, economic-base theory makes a distinction between export industry and local-consumption industries. An export industry generates income from sources outside the metropolitan region while a local-consumption industry generates income from local population and firms. Locational classification distinguishes between basic and population-serving industries in geographic terms. A basic industry locates with respect to inter-regional transportation routes, resources and unique site-features, inter-industry linkages, and agglomeration economies. A population-serving industry, on the other hand, locates with respect to residential population and purchasing power, and daytime population concentrations. Obviously, the groups shown in the table have overlaps. Manufacturing in group I, for example, includes some non-durable manufacturing in food, apparel, printing and publishing for direct local-consumption that should be in quadrant III. Retail trade and services in IV include some trade and services generated by tourism that should be in quadrant II. Bank headquarters in I includes banks in the central business-district transferred from quadrant IV. Finally, major private universities in I include those transferred from quadrant IV.

The implied structure of Lowry's multiplier (Γ in Equation **(6-46)**) follows directly the functional classification of basic vs. non-basic activities inherent in the minimum-requirement method. However, elements of locational criteria have to be included in the multiplier definition also. It is the mixture of both the functional

---

[15]*Locational quotient* measures the amount of goods and services consumed locally vs. outside the local area. Basic activities are supposed to be exported and not sold to local residents. *Minimum requirement* measures the degree to which an activity requires other activities for support. A basic activity supports the development of other secondary activities, but most of it is for export consumption. For that reason, exports are estimated as the amount of a region's activity that exceeds the minimum amount required to meet local needs.

Economic-base orientation	Locational orientation	
	Basic	Population-serving
Exports	*I* Agriculture, mining, manufacturing, transportation, finance, government; Selected establishments: major private universities, bank headquarters	*II*
Local consumption	*III* Wholesale trade, retail trade, building materials, services	*IV* Construction, retail trade, transportation, communication, utilities, finance, services, local government

Table 6-9 Classification by economic-base vs. locational orientation
(Goldner 1968)

and locational criteria that lend the model to criticism. For example, a regional average non-basic employee per household (population-serving ratio) is used as one of the multiplier constants. Using such a constant for purposes of estimating non-basic activities at the zonal level can be criticized in that it assumes uniform regional family size, activity rates, economic status and life style. In addition, if the multiplier is formulated typically for the census period of 10 years, the result of such application could be biased. This is due to changes in local economic-structure that could occur within the decade. If only a regional aggregate for total employment was required, the "instant metropolis" phenomenon would probably be acceptable. However, since a spatial distribution of population and employment is required, estimation and allocation of total stock by zone can lead to serious bias. We will redress some of these issues in the "Disaggregation and Bifurcation in Garin–Lowry Model" chapter.

The definition of the multiplier has been reformulated in PLUM (Moody and Puffer 1970, Lane 1966); the multiplier is a function of the number of zonal residents assigned for the work sites. Recall the residential allocation of workers is dependent on the characteristics of the transport system in the region. The modified multiplier contrasts significantly with typical economic-base multipliers in its dependence on the spatial distribution of basic employees. Specifically, the multiplier γ depends on the average zonal-family-size, the number of non-working residents of each zone, the dependent population per employee resident, and the locational distribution-function. The multiplier, when so formulated, no longer remains constant over time or similar from zone to zone. Thus, the major improvement in PLUM is the treatment of the development process. Rather than using total-stock relationships, as does the Lowry model, increments between base-year levels for the entire inventory of regional economic-activity and land-use are computed. These increments produce a layered effect.[16] Beyond being more pleasing conceptually, the incremental approach may have a more detailed accounting of growth for short-period projections. It also allows for "urban inertia" to be modelled more effectively.

6.9.2 Activity allocations. There is a major difference between the two models in the mathematical form of the allocation function. Lowry's model uses a simple (singly-constrained) gravity-allocation function. Such gravity simulations tend to have biases that are strongly influenced by zone size. They also tend to be deficient in their allocations of employees making short trips. Using the intervening-opportunity formulation, the allocation procedure of PLUM is initially constrained by being normalized by the opportunity function, which is a measure of the holding capacity of each zone. The matrix of probability of household location is normalized by the holding capacity in addition to the time-distance function. Consequently, a conceptually better first approximation of worker locations can be determined than if a (singly-constrained) gravity simulation is used. Since the subsequent location of local-serving employees and their families depends heavily on the initial allocation of employees, the better the initial allocation, the less re-allocation is necessary for conformance to exogenous control totals. This is especially significant given the proportional re-allocation of excess employment and household required in both models (and in many other urban spatial models in general).

In terms of contribution to additional theories of residential location, the PLUM formulation adds little. Specifically, the entire area of land-rent theory and income substitution[17] is not dealt with at all by both models. In the terms of Webber (1984), it is based on consumer demand rather than the "supply side," as will be explained further in the conclusion to this chapter. As far as the service sector is concerned, PLUM does not deal with a regional hierarchy of non-basic activities as does Lowry's model (reference **Table 6-10**). Although both models are highly sensitive to accessibility measures, PLUM's original method of locating non-basic activities assigned too large a weight to the value of travel cost (see Equation **(6-49)**). The effect of the transport system can be over-emphasized in determining non-basic activity centers.

There is a promising aspect of PLUM's treatment of the service sector, however. With minor modification, the model could be made to reflect Greenhut's (1956) idea of the effect of arriving competitors on the market potential-function

---

[16]Refer to the latter part of the numerical example in the section entitled "The Projective Land Use Model" for the "layered effect."

[17]For an explanation of Land Rent Theory refer to the "Economic Methods" chapter in Chan (2000). For an explanation of Income Substitution Effect, consult both the "Economic Methods" chapter (Chan 2000) and "Generation, Competition and Distribution" chapter in the current volume.

		PLUM-IP	Lowry
Overall model structure	Incremental	√	
	Total stock		√
Allocation functions	Euclidean distance separation		√
	Network distance separation	√	
	Opportunity function	√	
Residential allocation	Gravity model		√
	Intervening opportunity model	√	
	Zonal activity considered		√
Residential reallocation	Proportional to initial allocation	√	√
	Density considered	√	√
	Supply constraint	√	
Non-basic activity allocation	Demand determined	√	√
	Inertia measure	√	
	Disaggregation by trade class		√
	Minimum activity threshold		√
	Reallocation necessary		√

Table 6-10 Functional comparison of Lowry model and PLUM-IP

P'. It would be possible to re-determine the opportunity component of the market potential of a zone. This would tend to decrease the opportunity for new non-basic industry to enter. But the effect of agglomeration could be reinforced by applying an attraction-modification factor a' (such as illustrated in the "Generation, Competition and Distribution" chapter and recapped in Equation **(6-50)**) to the service-employment opportunity-function W_j. Compared to the Lowry model, many elements in PLUM's equation structure, notably the allocation functions, tend to be refined by incorporating additional factors into the function. As yet another example, Lowry's model uses a variation of Hansen's (1959) accessibility measure to construct demand surfaces P' for allocating housing and non-basic activities. The principal elements used are spatial separation and zonal activity. Potential supply of the opportunity for satisfying the demand is left out. The measure, suggested by

Cordey-Hays et al. (1970), is incorporated in the PLUM formulation, but the concept of existing activity in a zone is dropped.

6.9.3 Further comments. It is interesting that there are some negative aspects to the increased sophistication in PLUM. For example, in the incremental structure of regional simulation evident in PLUM, the effects of past economic and social conditions affect future predictions of household and employment allocation. The feedback from these pre-existing conditions for development of the region in a subsequent time-increment may produce relationships between projected demand and constraints (such as the maximum-density zonal-constraints on residential activities) which are not temporarily valid. A way to overcome this is to make the arbitrary-constraint functions more flexible. In this way, the constraint functions are responsive both to endogenous (micro) factors which occur as incremental basic-employment in the region is added, and to exogenous (macro) re-specification of control parameters that reflect projected policies or socio-economic conditions. Such an approach, however, will introduce additional complexity in an already elaborate model.

6.10 An experimental comparison

From previous theoretical-discussion, the PLUM model conveys a more complex simulated structure of the systems and constraints affecting activity and land-use development than does Lowry's original model. One would suggest, therefore, that PLUM would provide a more reasonable set of forecasts than Lowry. However, there has been theoretical conjecture, but little empirical proof that supports the expectation of an improved model-output. An experiment was therefore conducted, in which a common data-base of York, Pennsylvania (as listed in the computer disks attached to this book) was input to both models. The two models were then evaluated in terms of their outputs and the degree of correlation between these outputs. The experiment simulates a field application, in which the two models are alternate candidates used in forecasting future land-use using 1963 as base year.

6.10.1 Input requirement. PLUM is much more intensive than Lowry in data requirements. For example, Lowry's model requires only one data-set for supplying base-year and target-year data. The only projections exogenous to the model are related to basic employment. In PLUM, both basic employment and basic-employment increments are required. These estimates are supplied as separate data-sets from the base-year data. The base-year data is also more extensive for PLUM. This is because zone-specific endogenous parameters are generated based on existing conditions; and through these internally-generated parameters zonal-allocations are controlled throughout successive iterations in the model. Lowry's model, on the other hand, relies on region-wide control (total stock) ratios passed to the model exogenously.

Another major difference is in the calibration formulation of employment and population allocation-functions. Exogenous pre-processing of both trip-information and spatial-separation data are required to specify the parameters in Lowry's model.

		Lowry	PLUM
Parameter	Employed residents per household	1.408	1.216 [1]
	Population per household	—	3.181[1]
	Population per employed resident	—	2.616[1]
	Non-basic employment per household	0.679	—
Trip distribution function	Work-home	2.196/0.272 [2]	28 [3]
	Home-shop	0.688/0.040 [4]	22 [2]
	Work-shop	0.100 [4]	14 [2]

[1] Endogenous parameters.
[2] Values representing v' and u in Equation **6-48** .
[3] These are ß values in Equation **6-49**.
[4] Values represent the weight placed on shopping trip from work; this implies the weight placed on home-based shopping trips is 0.900.

Table 6-11 Calibration parameters for Lowry model and PLUM

In PLUM, the shape of the distribution-curve, while dependent on calculations of the α and ß parameters (Equation **(6-49)**), is determined during specific re-allocation mechanisms. This is required for correcting the initial allocation of employment and households (as illustrated by the determination of the W_j parameter in the "Generation, Competition and Distribution" chapter and recapped in Equation **(6-49)**). The final difference in input requirements is reflected in exogenous parametric determination. **Table 6-11** summarizes the calibration parameters of both Lowry and PLUM. It can be seen in the Table that the two sets of input parameters can be equated after computational transformations since they are derived from a common set of input data in our current experiment.

6.10.2   Model outputs.   Both Lowry's model and PLUM produced projections based on nearly identical input-data; any resulting variation in output is therefore attributable to the internal structure of each model. A statistical correlation between the outputs from both models was attempted. First, a graphical comparison is made to help in the formulation of hypothesis about the correlation. Second, statistical tests were performed to either confirm or negate the correlation hypothesis. Together with the correlation analysis, a simple linear-regression equation was also produced, which shows the relationship between the outputs from the two models. As shown in **Figure 6-5**, Lowry's model and PLUM provide two

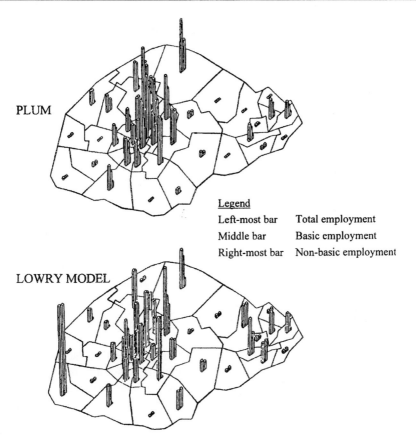

Figure 6-5 Zonal-employment estimates between Lowry and PLUM
(Chan and Fitzoy 1985)

very different employment-projection outcomes. Using Lowry's model, a dispersal
of employment from the central zones to several suburban areas is quite noticeable.
Many concentrations of non-basic employment occur in outlying zones in the
southern and western areas of the map. These graphic representations also display
a characteristic of the PLUM employment-allocations. PLUM tends to locate non-
basic employment closer to central positions of York. This is probably due mainly
to the allocation function that places a large weight on travel time (as previously
mentioned and shown in Equation **(6-49)**).

As presented in the upper part of **Table 6-12**, outputs from the two models are
compared, only household allocations provide a reasonably-significant correlation
($r = 0.78$). Nonbasic-employment correlation between the outputs from the two
models is discouragingly low $(r) = 0.11$). One possible circumstance that explains
this phenomenon can be conceived. It is suspected that zonal comparisons are too
detailed for the different activity-allocation-logic between the two models as

	Experiments (Blocks I and II)		No of households	Non-basic employment
Upper block I	Before aggregation (42 zones)	r(Lowry-forecast, PLUM-forecast)	0.783	0.111
		Adjusted-r	0.777	not applicable
	After aggregation (6 districts)	r(Lowry-forecast, PLUM-forecast)	0.970	0.857
		Adjusted-r	0.962	0.818
Lower block II	Accessibility parameters: r(1963-Lowry, 1963-observed)		0.810	0.544
	Basic employment parameter: r(before, after)		0.999	1.000

Table 6-12 - Correlation analysis of model outputs

summarized in **Table 6-10**. Due to the different allocation functions and re-allocation logic (e.g., total stock vs. incremental) adjacent zones in the same subregion were allocated quantities of employment or households that compensated for over-or-under allocation in any adjacent or nearby zone. To attenuate the spatial caprices of each model, the 42 zones were aggregated into six subregions or districts.

Results of the correlation developed using aggregated projections are also shown in **Table 6-12**. The correlation coefficients yield much higher values between the non-basic employment in the Lowry and PLUM models, where the correlation coefficient is upgraded dramatically from 0.11 to 0.86. This cannot be explained through the reduction in the degree-of-freedom alone, as shown by the adjusted-r values.[18] It is seen, therefore, that although population and non-basic employment was not allocated to the identical zones by the two models, it was allocated to the right general-subregion within York. Allocation of households, in particular, was only marginally affected by this lapse of allocation-ability between the two models. Thus household allocations correlated fairly well as long as non-basic employment was located within a reasonable distance from where it ought to be. Detailed examination of non-basic activities themselves, however, highlights the very different allocation-logic (see the last part of **Table 6-10**). Most notable of these is the frequent re-allocation necessary in Lowry due to the specification of minimum-activity threshold, while no such re-allocation is used in PLUM.

---

[18]For a definition of adjusted R^2, see the "Statistical Tools" appendix to Chan (2000).

Furthermore, the comparatively much larger weight (0.90) assigned to home-based shopping trips in Lowry's model—instead of equal weights between home-based and work-based (**Table 6-11**)—may also be responsible for the difference in the non-basic activity forecasts in specific, detailed locations.

The above experiments were performed in the original versions of Lowry and PLUM. Refinements to these original models were subsequently introduced in the 1970's and beyond by a number of researchers (including Batty, Brotchie, Echenique, Massere, Putman and many others). Conducting tests on the original model makes our experimental findings more significant. This is due to the degree of similarities that exists in spite of the accentuated, first-edition differences between the two original models—differences that were somewhat obliterated in subsequent refinements.

6.11 Theoretical thread

The empirical finding above on the effect of aggregation is really not mystifying if we review the fundamental theory behind both models. The theoretical underpinning is essentially the same except for computational refinements—as pointed out earlier. In fact, if we trace through the later developments in the seventies and eighties, the common theoretical thread continues, although new and refreshing conceptual-interpretations emerge. In the following sections, we synthesize the old and new efforts, hopefully laying the foundation for the next chapter, where a set of ideas on disaggregation and bifurcation is formulated.

6.11.1 Spatial location. It has been pointed out that the two basic-building-blocks of urban modelling are: (a) spatial location and (b) activity developments. The analysis of spatial location has centered around the multinomial-logit formulation. An example of the singly-constrained allocation-function (Equations **(6-48)** and **(6-49)**) is shown below as a generalized equation combining both accessibility and opportunity concepts:

$$V_{ij}(W_j, C_{ij}) = V_i \frac{W_j^{a'} f(b, C_{ij})}{\sum_{j=1}^{n'} W_j^{a'} f(b, C_{ij})} \tag{6-50}$$

where $W_j^{a'}$ is the 'opportunity' or intrinsic attraction of zone j, with a' as an exponent. The activities V_i and V_{ij} (such as population and employment) can be measured in various units, including their valuation in dollars (as will be illustrated in Equation **(6-52)**). Finally, the distribution-function $f(b, C_{ij})$ can take on a variety of forms depending on the values of b. Examples of these forms are shown in Equations **(6-48)** and **(6-49)**.

Much theoretical understanding has been obtained on this family of spatial-allocation models over the late sixties and seventies. The first to be recapped here is the derivation of Equation **(6-50)** using an optimization formulation, as explained in the "Economics" and "Prescriptive tools" chapters. The entropy-maximization approach to derive the equation and the doubly-constrained case has been advanced by Wilson (1974) and interpreted by Cesario (1975).[19] The explicit use of the doubly-constrained formulation represents a possible theoretical refinement on the computational procedures of the then existing urban spatial-models. Within the optimization framework, Cochrane (1975) derived the gravity-model format—the stochastic, disaggregate assumption that trip makers choose the trips that provide the greatest benefit (consumers' surplus) for them as individuals. Furthermore, he showed the intervening-opportunity model is a special case and a more limited version of the gravity model, as reviewed in the "Economic Methods" chapter in Chan (2000). This helps to accentuate, again, the common basis between Lowry's and PLUM's allocation functions.

Perhaps an exciting recent-development is the interpretations of spatial allocation in terms of catastrophe theory. Wilson (1981) interpreted the allocation-equation **(6-50)** as a fold catastrophe, where the optimal size of a shopping-facility W_j is to be found. Formulating the residential-allocation problem in such a "revenue-maximization" format, the 'cost' function $C_j(W_j)$ can be modelled as a straight line through the unit cost c_j ($= \Sigma_i c_{ij}$):

$$C_j = c_j W_j \tag{6-51}$$

while the 'revenue' function V_j, or more appropriately the 'benefit' function—is an S-shape curve (for $a' > 1$ in Equation **(6-50)**):

$$V_j = \sum_{i=1}^{n'} V_{ij}. \tag{6-52}$$

Depending on the unit cost c_j, the optimal size of the housing development in zone j, W_j^*, can be zero or a positive number when the criterion of net-benefit ($V_j - C_j$) maximization is adopted (see **Figure 6-6**(a) and (b)). Again V_j is the benefit (such as gross revenue at a shopping center) accrued at facility j. Mapping the size of W_j as a function of the unit-cost c_j, the classical fold-catastrophe curve can be plotted as **Figure 6-6**(c). In this Figure, the housing development appears or disappears as one passes through the critical c value in the fold. The maximum household-density $N_j \leq N_j^c$ and the minimum-threshold-activity constraint $E_j^k \geq d_j^k$ suggest the possible existence of constraint catastrophes in the Lowry-based models. Local maxima or minima may occur on the boundary imposed by these minimum-retail-threshold and maximum-residential-density constraints. Illustrated in **Figure 6-6**(c) is an example of such a local-maximum N_j^c corresponding to a reduced residential-holding-capacity constraint. Here N_j^c is less than the assigned amount of 2,420 households using the data-sets provided in -our distributed software—as will be explained in sequel.

---

[19]See the "Optimization Schemes" appendix in Chan (2000) for such a derivation.

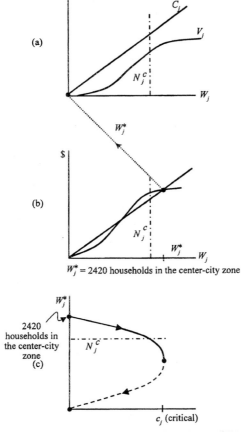

Figure 6-6 Activity allocation as a fold catastrophe (Wilson 1981)

6.11.2 Urban-activity development. While the multinomial-logit model is the common function for spatial allocation among the various models, there is less of an obvious commonality among the procedures for projecting the stock of population and employment in an urban area. But the fundamental Malthusian growth-equation seems to render an adequate starting-point for discussion:

$$\dot{Z} = \gamma Z \qquad (6\text{-}53)$$

where Z can be either total employment or population, $\dot{Z}$ is its derivative with respect to time, and γ is a constant defined by the labor-force participation rate and population-serving ratio. Thus the geometric series of Lowry/PLUM is simply a special case of Equation **(6-53)** if we write Equation **(6-45)** as $\Delta E(t)/\Delta t = \gamma \Delta E(t\text{-}1)$. The general solution to this class of equations is the familiar exponential-function

$$Z = Z_0 e^{\gamma t} \qquad (6\text{-}54)$$

where the constant Z_0 can be interpreted as the basic-employment level and Z as the total-employment in subsequent time-periods. Notice γ can assume different values in each computational-increment in PLUM. The important feature of Equation (6-54) is the existence of bifurcation behavior at $\gamma = 1$. If γ is bigger than or equal to unity, an exponential growth occurs. On the other hand, the system stabilizes to an asymptotic value when γ is less than one. This is the normal case for the urban systems modelled by Lowry/PLUM, as pointed out in the discussion of Equation (6-46) on several occasions.

One can recast Equation (6-53) as $\dot{Z}=G(Z,\gamma)+\varepsilon$ which introduces the random-element ε. Such 'randomness' can be the result of intravenous introduction of basic industries into the urban area. Instead of a closed system, the study area is now modelled as an open system. One can also interpret the randomness as a fluctuation of the economic-base multiplier γ (as simulated in PLUM). In this sense, complex evolution of the city can result depending on the frequency, magnitude, and type of 'noise' injections over time.

6.11.3 Activity development and allocation. A novel feature of the Lowry-based models consists of the repeated coupling of the development equation (6-53) and the gravitational-allocation equation (6-50). In matrix algebra, Garin's formulation (1966) is a generalization of Equation (6-46) and (6-54): $\mathbf{E} = \mathbf{E}^B(\mathbf{I}+\bar{\mathbf{Q}}+\bar{\mathbf{Q}}^2+\bar{\mathbf{Q}}^3+...+\bar{\mathbf{Q}}^m)$, where $\bar{\mathbf{Q}}$ is a matrix of γ incorporating both $\gamma\Delta t$ in Equation (6-45) and Equation (6-50), and $\mathbf{I}$ is an $n \times n$ identity matrix. The Garin geometric-series can now be written as a discrete solution to the general differential-equation

$$\dot{\mathbf{Z}}=\mathbf{H}(\mathbf{Z},\boldsymbol{\Gamma})+\varepsilon \qquad\qquad\qquad (6\text{-}55)$$

where $\mathbf{Z}$, $\boldsymbol{\Gamma}$ and ε are the corresponding matrix generalization of the scalars used before. The dimensions are simply in the order of n', the largest number of zones in the study area. For example, $\mathbf{Z}$ can be a vector of zonal population $\mathbf{N}$ or employment $\mathbf{E}$, with as many entries as the number of zones in the urban area. $\boldsymbol{\Gamma}$ is the set of parameters in $\bar{\mathbf{Q}}$ that include c_j of Equation (6-51) and γ of Equation (6-54), both of which are responsible for the bifurcation behavior discussed above. Finally, ε represents the set of perturbations on the zonal-basic-employments $\mathbf{E}^B$ or parameters γ in PLUM.

Thus in the general equation (6-55), we have tried to show the possibility of bifurcation in parameter $\boldsymbol{\Gamma}$, and *evolution* in parameter ε. Here evolution is defined as adaptation of the system to environmental changes in general. On a detailed level, one can illustrate the specific components of this dynamic equation. Wilson (1981), for example, suggested a way to embed the traditional static spatial-location models such as Equation (6-50) in a dynamic framework by this nonlinear differential equation

$$\dot{W}_j = h''(V_j-C_j) = h''(\sum_i V_{ij} - c_j W_j). \qquad\qquad (6\text{-}56)$$

Thus it can easily be seen that bifurcation can occur depending on the critical values of the set of parameters in the equation, including h'' and c_j. Due to the constraints (such as the minimum retail-threshold and maximum residential-density in Lowry) that govern the linkage between the retail and population sectors, however, the behavior of bifurcation is complicated by constraint catastrophes. Such behavior can best be observed through computer modelling. As a result, analytical properties that relate to bifurcation are often buried under the weight of constraint-catastrophe. Since the complexity lies in the constraint-catastrophe inequalities, the Garin formulation of the Lowry model for example—without the inequalities—offers a transparent demonstration of the evolutionary pattern. A full formulation will subsequently be offered in its glorious details, including constraint catastrophes—in the next chapter.

An alternative to the school of thought outlined above is to explore the feasibility of using the generalized Lotka-Volterra (predator-prey) equations. This is to model the competition between the population and employment sector for limited space (Bertuglia et al. 1987). The Lotka-Volterra model of interaction takes on the following form:

$$\dot{x}^l = x^l \left(a^l + \frac{1}{b^l} \sum_{k=1}^{m} a^{kl} x^k \right) \qquad l = 1, 2, \ldots, m \tag{6-57}$$

where m is the number of activities, such as population and various types of retail employment, etc. In this case, the population-growth rate is a linear function of the abundance of all competitive activities. The model was initially proposed as a non-spatial m-activity interaction model in mathematical ecology. Here a^l *is the "self-growth" rate, b^l is an "average weight" of the activities*, and the coefficients a^{kl} are the interaction parameters between the activities. Dynamic properties of this model have been shown to depend on the number of interacting activities. The model can produce complex behavior, even in a predator-prey (two activities) model (Dendrinos and Sonis 1990). In Equation (6-55), one may trace interesting bifurcations through the values of the coefficient set Γ, similar to the costs c_j in Equation (6-56). But the solution of the generalized Lotka-Volterra type equations in a spatial context is by no means simple. This is particularly so when one introduces the stochastic fluctuations—as suggested by Allen et al. (1981).

6.11.4   Synergetic models of spatial interaction.   Suppose the fish population in the Mediterranean Sea is divided into the prey-population $x(t)$ consisting of food fish, and the predator-population $y(t)$ consisting of selachians. Assume the food fish do not compete very intensively among themselves for their food supply. Hence, without the selachians, the food fish would grow according to the Malthusian law of population-growth as shown in Equation (6-53), for some positive-constant a. Next, the number-of-contacts per unit-time between predators and prey are bxy, for some positive-constant b. Therefore $\dot{x} = ax - bxy$ where the negative sign is to show a decrease in food-fish population because of their predictor, the selachians. Predators have a natural rate-of-decrease cy proportional to their present number (perhaps due to the limited food supply). They also increase at a rate $h''xy$ proportional to their present-number y and their food-supply x. Thus Equation **(6-57)** can be written as

$$\frac{dx}{dt} = ax - bxy = x(a-by)$$
$$\frac{dy}{dt} = -cy + h''xy = y(-c+h''x).$$
(6-58)

The system-of-equations above govern the interaction of the selachians and food fish in their natural habitat (Braun 1978, Nijkamp and Reggiani 1992).

Observe that Equation **(6-58)** has two equilibrium-solutions $x(t) = 0$, $y(t) = 0$ and $\bar{x}(t) = c/h''$, $\bar{y}(t) = a/b$. The first equilibrium-solution, of course, is of no interest to us. This system also has the family of solutions $x(t) = x_0 e^{at}$, $y(t) = 0$ and $x(t) = 0$, $y(t) = y_0 e^{-ct}$. Thus both the x- and y-axis are solutions of Equation **(6-58)**. This implies that every solution $x(t)$, $y(t)$ of Equation **(6-58)** that starts in the first quadrant $x > 0$, $y > 0$ at time $t = t_0$ will remain here for all future-time $t \geq t_0$. The solutions of Equation **(6-58)** for $x, y \neq 0$ are the solution-curves of the first-order equation $dy/dx = (-cy + h''xy)/(ax - bxy) = [y(-c+h''x)]/[x(a-by)]$. This equation is separable, since we can write it in the form $[(a-by)/y][dy/dx] = (-c+h''x)/x$. Consequently, a solution is $a \ln y - by + c \ln x - h''x = k_1$, for some constant k_1. Taking exponentials of both side of this equation gives

$$\frac{y^a}{e^{by}} \cdot \frac{x^c}{e^{h''x}} = K$$
(6-59)

for some constant K. Thus the solutions of Equation **(6-58)** are the family-of-curves defined by Equation **(6-59)**, and these curves are closed orbits as shown in **Figure 6-7**.[20]

Let us now include the effect of fishing in our model. Observe that fishing decreases the population of food fish at a rate $gx(t)$ and similarly decreases the population of selachians at $gy(t)$. The constant g reflects the intensity of fishing. Thus the true state-of-affairs is described by the modified system of differential-equations.

$$\frac{dx}{dt} = ax - bxy - gx = (a-g)x - bxy$$
$$\frac{dy}{dt} = -cy + h''xy - gy = -(c+g)y + h''xy.$$
(6-60)

This system is exactly the same as Equation **(6-58)** (for $a-g > 0$), with a replaced by $a-g$, and c replaced by $c+g$. Hence the average values of $x(t)$ and $y(t)$ are now $\bar{x} = (c+g)/h''$ and $\bar{y} = (a-g)/b$. Consequently, a moderate amount of fishing ($g < a$) actually *increases*

---

[20]For an introduction to the subject of "systems stability", please refer to the section bearing the same title in the "Control, Dynamics and System Stability" appendix to Chan (2000), where the different stable and unstable trajectories are discussed.

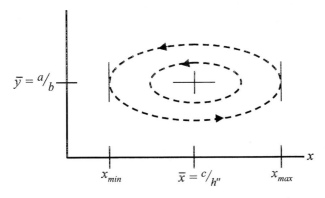

Figure 6-7 Orbit of two-species Lotka-Volterra equations (Braun 1978)

the number of food fish, on the average, and *decreases* the number of selachians! Conversely, a reduced level of fishing *increases* the number of selachians on the average, and *decreases* the number of food fish!

The models described above—predator-prey, symbiosis, and competing species—can be unified into the broad concept of so-called *synergetic models*. They refer in general to the cooperative interactions occurring between the various units of a system. In the specific context of *spatial*-interaction, we will present now a dynamic-spatial-synergetic model. It includes growth and decline cycles between the attractiveness of labor markets (measured in terms of available new jobs or workplaces W_i) in a certain place i and related inflow into i from other areas j, V_{ji}. The underlying hypothesis is that in a growing economy the number of new potential-jobs on a given labor-market i exhibits a growth pattern upon which inflows in i seek:

$$\dot{W}_i = (\alpha_i - \sum_j \beta_{ji} V_{ji}) W_i \qquad \alpha_i, \beta_{ji} > 0. \qquad \textbf{(6-61)}$$

It is clear the above dynamic equation typically represents the prey phenomenon described in Equation **(6-57)**. Here W_i is the prey population and V_i is the predictor population. It should be noted that α_i in Equation **(6-61)** indicates the growth rate of the vacant workplaces without the inflows. The coefficient β_{ji} represents the rate at which the jobs are filled by various inflows V_{ji}. Obviously the parameter β_{ji} suggests also that not all inflows can be accommodated by the labor market because of lack of matching caused by differences in qualifications, etc.

The second assumption is that the growth of inflows V_{ji} is encouraged by the number of vacant workplaces W_i, so that this equation is related to a predator phenomenon, as defined in Equation **(6-57)**:

$$\dot{V}_{ji} = (\gamma_{ji} W_i - \theta_i) V_{ji} \qquad \gamma_{ji}, \theta_i > 0. \qquad \textbf{(6-62)}$$

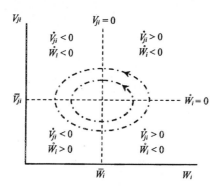

Figure 6-8 Spatial cycles between inflows and workplaces
(Nijkamp and Reggiani 1992)

In the above equation, θ_i represents the rate-of-decline in inflows V_{ji} as workplaces W_i are being filled, while γ_{ji} represents the rate at which job matching takes place (similar to β_{ji}). Obviously, this model assumes that no dissipative force exists (i.e., the absence of congestion externality). In other words, this model represents the orbital dynamic-equilibrium, as the effect of congestion disappears. **Figure 6-8** shows the various conditions under which certain oscillating behavior may occur. Similar to the fishing-policy-in-the-Mediterranean example, the precise equilibrium behavior may be the result of such decision as maximizing some social-benefits. This allows the system to reach the steady state over a pre-specified time horizon.

Now consider, once again, the *general*-predator-prey equations as outlined in Equation **(6-57)**. A two-species example ($m = 2$) would look like

$$\dot{x}^{(1)} = x^{(1)}(a^1 - a^{11}x^{(1)} - a^{12}x^{(2)})$$
$$\dot{x}^{(2)} = x^{(2)}(-a^2 + a^{21}x^{(1)}) \tag{6-63}$$

assuming a weight of unity, or $b^1 = b^2 = 1$. This model has two equilibrium solutions, a trivial one $x^{(1)} = x^{(2)} = 0$ and a more complicated one: $x^{(1)*} = a^2/a^{21}$ and $x^{(2)} = a^1a^{21} - (a^{11}/a^{12})x^{(1)*}$. Volterra-Lotka equations **(6-63)** cannot be solved in an analytical way due to their non-linearity,[21] although solutions can be found for the linearized approximations. If we restrict ourselves to the exploration of the optimal trajectories in a *phase diagram*[22], we can plot the integral curves of Equations **(6-63)** for one configuration of the coefficients. In **Figure 6-9**, the equilibrium is stable when $a^1/a^{11} > a^2/a^{21}$ (**Figure 6-9**(a)). But by changing the sign of the coefficient a^{11}

---

[21]A *linear* differential equation is characterized by the absence of products (or nonlinear function) of the dependent variable (unknown function) and its derivative.

[22]The concept of a phase diagram is explained in the "System stability" section of the "Control, dynamics and system stability" appendix to Chan (2000).

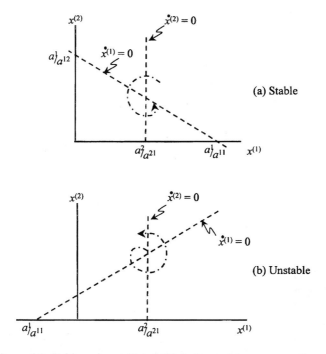

Figure 6-9 Stable and unstable equilibria for predator-prey equations
(Nijkamp and Reggiani 1992)

to negative, we obtain a structurally unstable equilibrium, as shown in **Figure 6-9(b)**.

From the above examples, clearly the coefficient $a^{11} = 0$ represents a critical value at which a bifurcation may emerge—from stability ($a^{11} > 0$) to instability ($a^{11} < 0$). It has been shown that when $a^{11} = 0$, the solutions to the system of equations are closed orbits. The stability conditions are represented by the Jacobian[23] of the matrix of the linearized system 1 (Equation **(6-57)**). If the Jacobian is evaluated at the non-trivial equilibrium $(x^*, y^*) = (c/h'', a/b)$, it assumes the values $\begin{bmatrix} 0 & -bc/h'' \\ h''a/b & 0 \end{bmatrix}$, leading to purely imaginary eigenvalues. This means that the solutions are closed orbits. The system is neutrally stable in that no unambiguous conclusion can be drawn on the global behavior from the analysis of the Jacobian. It is interesting that the sign of the *trace*[24] of the Jacobian plays a dominant role in deciding the kind of oscillating behavior of a two-dimensional dynamic-system. In particular, it is possible to identify the existence of 'dampening' when the trace is negative. We call

---

[23] Analysis of stability is discussed in the "System stability" section of the "Control, Dynamics and System Stability" appendix to Chan (2000), including the relationship to the stability-setting Jacobian-matrix.

[24] The trace of a symmetric matrix is the sum of the diagonal elements.

the system *dissipative*. A closed orbit emerges when the exploding and imploding forces both tend toward zero, i.e., when the trace vanishes.

6.11.5 Experiments.

To further explore catastrophe theory and bifurcation concepts in Lowry-based models, three experiments were performed using the data-sets shown in the accompanying software to the book. Using the York, Pennsylvania, data again (**Figure 6-3**), we test the sensitivity of the Lowry model to two sets of parameters: the relative accessibility-functions $\Theta(C_{ij})$ and the seeding of basic-employment E^B. Also, a dynamic evolutionary-pattern is introduced into the Lowry Model by randomizing the multiplier γ. In summary, we try to ascertain the existence and nature of fold catastrophes and bifurcation in Lowry-based models respectively through perturbing the parameter sets Γ and Z_0 to simulate ϵ.

In the first experiment, one compares the 'observed' and 'estimated' distribution of population and employment. The observed figures represent the then existing-condition in 1963, while the estimated are the output from the model using a poorly-calibrated accessibility-function $\Theta_{ij}(N_j, C_{ij})$ (Equation (**6-47**)). Essentially, we are placing a perturbation on a parameter in Γ, namely c_j, since the calibrated accessibility-functions are likely to be different from the 'actual' functions in the lack of calibration care. We then look for any constraint and fold catastrophes following Equations (**6-51**) and (**6-52**). In the second experiment, the seeding of basic employment among 20 of the 42 zones is altered between two runs. The basic employment of the inner city's 10 zones now becomes the basic employment of the geographic-ring of 10 zones immediately outside the center city. This represents a perturbation on the parameter Z_0. We try to detect any bifurcation that might result from this, as shown by Equations (**6-54**) and (**6-55**) when $Z = Z_0$. In other words, the 'seeding' of basic employment is repeated over time, rather than just at the beginning of the simulation. In the final experiment, we change the value of the multiplier γ between the grand loops of the Lowry run (see the section entitled "Matrix formulation and solution: an example" on the matrix formulation of the Lowry model for the definition of grand-loops). Thus in the first two grand-loops, we set $\gamma\Delta t = 0.276$: a 'normal' value for the multiplier. This guarantees a rapid convergence of the geometric series. In the subsequent grand loops, however, the $\gamma\Delta t$ is set at 0.974: a value close to the critical-value of unity. We are looking for possible bifurcation behavior and evolution consequently.

Results of the first two experiments are shown in the lower block of **Table 6-12**. The first experiment shows that the model is quite sensitive to the accuracy of the accessibility function. Since we purposefully perturbed the accessibility function, until costs c in Equation (**6-51**) is changed, and this results in a dramatic change in the estimated retail employment. The estimated population-distribution is also changed, but to a much lesser extent. What is interesting is that there is an exodus of population from the center city (2420 households in the inner-most zone and 495 in the next zone out). This is apparently triggered by the fold-catastrophe curve (?) as the unit-cost figure passes through the c_j(critical) point. Consequently, several zones in the center city lost all their population. While only two zones are affected in the population sector, 27 out of 42 zones lost all their retail activity because of the change in costs (although the amount of each 'loss' is less). Instead of having shopping centers scattered around the urban area, the retail activities are now all in and around the center city, leaving many zones (27 of them) in York with no shopping activities at all! The relative sensitivity of population and service activities to unit cost c_j—particularly service—is shown by the correlation coefficient of 0.81 and 0.54 respectively in **Table 6-12**.

The second experiment shows that due to the nature of the development process, perturbations on the basic-employment E^B in 20 zones do not affect the development and allocation results very much. This is shown by the correlation-coefficient of 0.99 and 1.00 for population and employment respectively in **Table 6-12**. This can be possibly explained by the proximity of the `before' and `after' locations of basic-employment and the rapid convergence of the geometric series, corresponding to a multiplier value that is far from the critical point for bifurcation to take place in any of the zones. The contrast between the results from this experiment and the previous one is critically important. The comparison highlights the model's sensitivity to accessibility parameters c_j in Γ and its robustness with respect to basic-employment $\mathbf{Z}_0$-distribution. The sensitivity to c_j in Γ may explain the detailed differences between the Lowry and PLUM output shown in the upper part of **Table 6-12**.

The final experiment shows there is distinct bifurcation-behavior correspond-ing to the value of the economic-base multiplier γ. As shown in **Figure 6-10**, the change of multiplier from a 'normal' value to a value close to the critical-value of unity causes a 'jump' in both population and retail employment in even the outer-most suburban zone. This occurs at the transition from the second to third grand loop. More important, the curves in **Figure 6-10** are 'flat' after the multiplier is changed to a value close to unity. Remember both curves were 'decaying' before, as one would expect from the mathematics of the geometric series (Equation **(6-46)**). While this is admittedly the result of externally introduced perturbations, it illustrates possible evolutionary behavior.

6.12 Dynamic, stochastic, and decision-theoretic formulation

The above deterministic-formulation and experiments of spatial-interactions can be viewed in a different light. Haag (1989) re-formulated the problem as a stochastic model[25] that focuses on the "supply side," concerning the developers, land owners, and retailers—particularly with respect to their investment. This contrasts with the demand-side models that mainly deal with the consumers who are viewed as the driving force behind the economy. In this re-formulation, we consider the customers or consumers in choosing a destination for shopping, entrepreneurs choosing a

---

[25]For a review of stochastic process, see the "Markovian Process" appendix to Chan (2000). Also reference "Compartmental models" in the "Control, Dynamics" appendix (Chan 2000) for general theory and specific examples of multivariate or vector process.

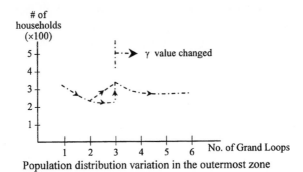

Population distribution variation in the outermost zone

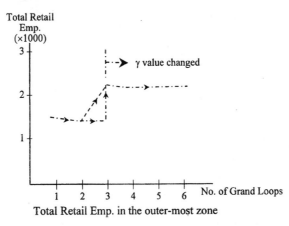

Total Retail Emp. in the outer-most zone

Figure 6-10   Illustrating bifurcation behavior

place to develop his/her facilities, retailers deciding on the price of goods and services they provide, and land owners setting the rent level. Their decisions are captured by the consumers' expenditure-flow $V_{ij}(t)$, developers' facility-stock $W_j(t)$, retailers' price-level $p_j(t)$, and landlords' rent-level $r_j(t)$ respectively.

6.12.1 The master equation.

At the heart of the model is monetary flow. The total revenue $V_j(t)$ attracted to each zone j is given by Equation **(6-52)** with the time dimension added to it. Similarly, the total-expenditure $V_i(t)$ of customers or consumers living in zone i is $V_i(t) = \sum_{j=1}^{n'} V_{ij}(t)$. Consistent with the supply-side school-of-thought, the utility-functions $v_{ij}(t)$ are assumed to depend on such variables as facility stock, price of goods and services, and land rent. The rest of the system is as defined in the deterministic case. Now how can the destination choice at time t, represented as $\{V_{ij}(t)\}$, the spatial and temporal pattern of development $\{W_j(t)\}$, the prices $\{p_j(t)\}$ and land rent, $\{r_j(t)\}$ be determined for given initial conditions $\{V_{ij}(0)\},\{W_j(0)\},\{p_j(0)\}$ and $\{r_j(0)\}$, respectively?

The system is characterized by the set of spatial-price or transportation-costs $\mathbf{C} = [C_{ij}]$. Since we are interested in the probability $P(\mathbf{Z},t)$ that a configuration $\mathbf{Z}$

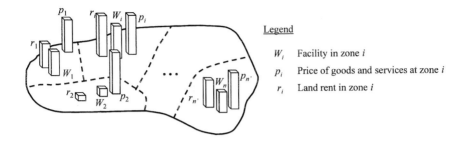

Figure 6-11 A probable configuration of the activities (Haag 1989)

$= \{\mathbf{V}, \mathbf{W}', \mathbf{p}, \mathbf{r}\}$ is realized at time t, a master equation[26] is defined for the probability distribution:

$$\dot{P}(\mathbf{Z}, t) = \sum_k \pi(\mathbf{Z}, \mathbf{Z} + \mathbf{k}) P(\mathbf{Z} + \mathbf{k}, t) - \sum_k \pi(\mathbf{Z} + \mathbf{k}, \mathbf{Z}) P(\mathbf{Z}, t) \qquad \text{(6-64)}$$

with $\sum_{\mathbf{Z}} P(\mathbf{Z}, t) = 1$. In the configuration space the maximum (or the maxima) of $P(\mathbf{Z}, t)$ represents the most probable expenditure-flow-pattern and the corresponding spatial-distribution of the facility stock, prices and rents. **Figure 6-11** shows a possible state $\mathbf{Z}$ consisting of facility-stock $\mathbf{W}'$, land-rent $\mathbf{r}$ and price-level $\mathbf{p}$ of goods and services in an n'-zone study-area. In the terminology of "Compartmental Models," $\mathbf{Z}$ denotes the number of units in each compartment (zone). Such a state occurs with probability $P(\mathbf{Z}, t)$. The row vector $\mathbf{k}$ consists of 0, +1, −1 entries marking an orthonormal base of the transition rate space,[27] where transition rates are defined immediately below. The master-equation **(6-64)** contains the full stochastic-information on the system. For that reason, we can determine not only the most probable decision-configuration or its mean values, but also the variances of the distribution-function.

6.12.2 *Total transition rates*. The total transition-rate $\pi(\mathbf{Z} + \mathbf{k}, \mathbf{Z})$ is obtained as the sum over contributions of different decision-processes:

---

[26]A "master equation" is defined in the "Compartmental models" section of the "Control, Dynamics" appendix to Chan (2000).

[27]For a definition of "orthonormal base", see the "Compartmental models" section of the "Control, Dynamics" appendix to Chan (2000).

$$\pi(\mathbf{Z}+\mathbf{k},\mathbf{Z}) = \sum_{i,j,k}^{n'} \pi_{ik,ij}^{(1)}(\mathbf{V}+\mathbf{k}^{(1)},\mathbf{W}',\mathbf{p},\mathbf{r};\mathbf{V},\mathbf{W}',\mathbf{p},\mathbf{r})$$
$$+ \sum_j^{n'}\left[\pi_{j,}^{(2)}(\mathbf{V},\mathbf{W}'+\mathbf{k}^{(2)},\mathbf{p},\mathbf{r};\mathbf{V},\mathbf{W}',\mathbf{p},\mathbf{r})+\pi_{j-}^{(2)}(\mathbf{V},\mathbf{W}'+\mathbf{k}^{(2)},\mathbf{p},\mathbf{r};\mathbf{V},\mathbf{W}',\mathbf{p},\mathbf{r})\right]$$
$$+ \sum_j^{n'}\left[\pi_{j,}^{(3)}(\mathbf{V},\mathbf{W}',\mathbf{p}+\mathbf{k}^{(3)},\mathbf{r};\mathbf{V},\mathbf{W}',\mathbf{p},\mathbf{r})+\pi_{j-}^{(3)}(\mathbf{V},\mathbf{W}',\mathbf{p}+\mathbf{k}^{(3)},\mathbf{r};\mathbf{V},\mathbf{W}',\mathbf{p},\mathbf{r})\right] \qquad \textbf{(6-65)}$$
$$+ \sum_j^{n'}\left[\pi_{j,}^{(4)}(\mathbf{V},\mathbf{W}',\mathbf{p},\mathbf{r}+\mathbf{k}^{(4)};\mathbf{V},\mathbf{W}',\mathbf{p},\mathbf{r})+\pi_{j-}^{(4)}(\mathbf{V},\mathbf{W}',\mathbf{p},\mathbf{r}+\mathbf{k}^{(4)};\mathbf{V},\mathbf{W}',\mathbf{p},\mathbf{r})\right].$$

The terms $\pi_{ik,ij}^{(1)}$ refer to changes in the expenditure-flow configuration $\mathbf{V}$, due to decisions of customers or consumers to change from a state of residing in i and using facilities in j to a state of still residing in i but now using facilities in k. Changes of residential location are not considered in the current formulation, since it is reasonable to assume that housing mobility will be considerably slower than shopping mobility. The $\pi_{j,}^{(l)}, \pi_{j-}^{(l)}$ describe decisions of entrepreneurs ($l = 2$), retailers ($l = 3$) and land owners ($l = 4$) to make available or to remove one unit of the facility-stock W_j, and to increase or decrease the prices p_j and the rents r_j, respectively. In Equation **(6-65)**, we have neglected contributions of simultaneous transitions to the total transition. This means that a transition to a neighboring state in the configuration space is assumed as a sequential process of single steps-of-changes in consumers' flows, facility stocks, prices and rents, instead of a simultaneous transition of all of them. This is according to typical practice in modelling stochastic process.

6.12.3 The demand side.

The next step is to specify the individual transition-rates. Let $N_{ij}(t)$ be the number of customers or consumers residing in i and 'shopping' in j. Then $N_i(t) = \sum_j N_{ij}(t)$ is the number of consumers living in i, and the total number of consumers is given by $N(t) = \sum_i N_i(t)$. The individual transition-rate $\pi_{ik,ij}(\mathbf{Z})$ at which a consumer living in i changes his/her shopping destination from j to k is assumed to be the result of certain expected utility-gain $(v_{ik}(t)-v_{ij}(t))$: $\pi_{ik,ij}(\mathbf{Z}) = h_1''(t)\exp[v_{ik}(t)-v_{ij}(t)]$, where $h_1''(t)$ is a time-scaling parameter and v_{ij} can be interpreted as the accessibility between zones i and j. Then the total number of changes of consumer trips from (i,j) to (i,k) is given by

$$\pi_{ik,ij}^{(1)} = N_{ij}(t)\,\pi_{ik,ij}(\mathbf{Z}). \qquad \textbf{(6-66)}$$

In many applications it is reasonable to assume that the expenditure flows $V_{ij}(t)$ can be linked to the consumer numbers $N_{ij}(t)$ via $V_{ij}(t) = g(t)N_{ij}(t)$, where $g(t)$ can be seen as the average of individual needs. By a comparison of the total stocks, $g(t)$ can be easily determined, similar to the way labor-force-participation rate and population-serving ratios are found. Thus we are able to derive from the consumer-dynamics-equation **(6-66)** transition-rates per unit-of-time for the expenditure-flows:

$$\pi_{ik,ij}^{(1)}(\mathbf{V}+\mathbf{k}^{(1)},\mathbf{W}',\mathbf{p},\mathbf{r};\mathbf{V},\mathbf{W}',\mathbf{p},\mathbf{r})=$$

$$\begin{cases} V_{ij}\,h_1''(t)\,\exp\!\left[v_{ik}-v_{ij}\right] & \mathbf{k}^{(1)}=(0,\dots,1_{ik},\dots,(-1)_{ip},\dots,0) \\ 0 & \text{for all other } \mathbf{k}^{(1)} \end{cases} \tag{6-67}$$

This equation can be inserted into the total-transition-rate equation **(6-65)** and subsequently into the master equation **(6-64)**. This gives the evolution of a certain expenditure-flow pattern. However, it must be remembered that the supply-side factors—facility stock, the prices and rents—interact with these flow patterns as well.

6.12.4 The supply side. The dynamics of the provision of facilities $\mathbf{W}'$, the prices $\mathbf{p}$, and rents $\mathbf{r}$ can be considered together because the same kind of methodcan be used. In particular, because there is no migration-of-flow involved,[28] they are each considered simply in a *birth/death process*[29], corresponding to the increase or decrease respectively of the corresponding quantities. Let $\mathbf{x}^{(l)} = (x_1^{(l)},x_2^{(l)},\dots,x_n^{(l)})$ be the *supply* quantities of facility-stock configuration ($l=2$, $x_j^{(2)} = W_j$), the price configuration ($l=3$, $x_j^{(3)} = p_j$), or the rent configuration ($l=4$, $x_j^{(4)} = r_j$), respectively. By decisions of the corresponding stakeholders the configuration $\mathbf{x}^{(l)}$ will change in time. As above, we will introduce individual transition-rates to these processes. Let $\pi_j^{(l)}(\mathbf{x}^{(l)}+\mathbf{k}^{(l)},\mathbf{x}^{(l)})$, $\pi_j^{(l)}(\mathbf{x}^{(l)}+\mathbf{k}^{(l)},\mathbf{x}^{(l)})$ be the birth-rate and death-rate per unit-time of the stock-variable $\mathbf{x}^{(l)}$. Then a rather general formulation for these rates can be obtained

$$\pi_{j\pm}^{(l)}(\mathbf{x}^{(l)}+\mathbf{k}^{(l)},\mathbf{x}^{(l)})=$$

$$\begin{cases} \dfrac{1}{2}\,f_j^{(l)}(\mathbf{x})\,\exp\!\left[\pm\zeta_j^{(l)}(\mathbf{Z})\right] & \mathbf{k}^{(l)}=(0,\dots,(\pm1)_j,\dots,0) \\ 0 & \text{for all other } \mathbf{k}^{(l)} \end{cases} \tag{6-68}$$

for $l=2,3,4$ and for rather general functions $f_j^{(l)}(\mathbf{x})>0$ and $\zeta_j^{(l)}$. In Equation **(6-68)** an in-migration term could be included which describes the possible settlement-of-facilities in initially-empty zones. The functions can also be selected to include saturation effects.

The factor $f_j^{(l)}(\mathbf{x})$ describes the speed of adjustment. Since the birth/death rates must not be negative, the condition $f_j^{(l)}(\mathbf{x})>0$ applies. Here we assume for simplicity that the speed-of-adjustment depends on the stock-level $\mathbf{x}^{(l)}$ in a linear way, with the time-scaling parameters $h_l''(t)$: $f_j^{(l)}(\mathbf{x}) = h_l''(t)x_j^{(l)}$. The function $\zeta_j^{(l)}(\mathbf{Z})$ takes into

---

[28]For an explanation of this concept, see the "Econometric modelling" section in the "Economic Methods" chapter 0f Chan (2000)

[29]An example of the birth-death process is illustrated in the "Markovian Process" appendix to Chan (2000), wherein a simple queuing-formula is derived.

account the difference between cost-of-supply and the revenue-attracted-to-facilities in a zone. If there is an economic-surplus $\zeta_j^{(l)}(\mathbf{Z}) > 0$ (or deficit $\zeta_j^{(l)} < 0$), it is likely that the facility stock is expanded, the prices and the land rents are increased (or vice versa). A reasonable assumption is that $\zeta_j^{(l)}(\mathbf{Z}) = \sigma_l(t)[V_j(t) - C_j(t)]$ Here, $\sigma_l(t) > 0$ describes the response-rate of a stakeholder to an economic-surplus $\zeta_j^{(l)} > 0$. The total revenue attracted to j is $V_j(t)$, and the total cost of supplying facilities of size $W_j(t)$ at zone j, where the unit land-rent is $r_j(t)$, is denoted by $C_j(t)$. If there is an economic surplus, namely if $V_j > C_j$, the probability of an expansion of the facility-stock exceeds the probability of a reduction due to profit motive.

   6.12.5 The resulting quasi-deterministic equations. By substitution of the transition rates **(6-65)**, **(6-67)**, and **(6-68)** into the master equation **(6-64)**, we get the evolution of the probability distribution $P(\mathbf{Z},t)$. The mean values of the expenditure-flows $\bar{V}_{ij}(t)$, the facility-stock $\bar{W}_j(t)$, the prices $\bar{p}_j(t)$, and the land-rent $\bar{r}_j(t)$ are obtained in the process:

$$\dot{\bar{V}}_{ij}(t) =$$

$$h_1''(t)\left\{\sum_{k=1}^{n'} \bar{V}_{ik}(t) \exp[v_{ij}(t) - v_{ik}(t)] - \sum_{k=1}^{n'} \bar{V}_{ij}(t) \exp[v_{ik}(t) - v_{ij}(t)]\right\} \qquad \textbf{(6-69)}$$

$$\dot{\bar{W}}_j(t) = h_2''(t)\bar{W}_j(t) \sinh\left\{\sigma_2\left[\sum_{j=1}^{n'} \bar{V}_{ij}(t) - (c_j + \bar{r}_j(t))\left(\bar{W}_j(t)/W_j(t)^\delta\right)\right]\right\} \qquad \textbf{(6-70)}$$

$$\dot{\bar{p}}_j(t) = h_3''(t)\bar{p}_j(t) \sinh\left\{\sigma_3\left[\sum_{i=1}^{n'} \bar{V}_{ij}(t) - (c_j + \bar{r}_j(t))\left(\bar{W}_j(t)/W_j(t)^\delta\right)\right]\right\} \qquad \textbf{(6-71)}$$

$$\dot{\bar{r}}_j(t) = h_4''(t)\bar{r}_j(t) \sinh\left\{\sigma_4\left[\sum_{i=1}^{n'} \bar{V}_{ij}(t) - (c_j + \bar{r}_j(t))\left(\bar{W}_j(t)/W_j(t)^\delta\right)\right]\right\} \qquad \textbf{(6-72)}$$

for $i, j = 1, 2, \ldots, n'$ and $\delta = 0$ or 1. Since the mean values—rather than the full distribution—of a stochastic process is estimated, we refer to this model as *quasi-deterministic*.

   For given initial conditions $\{\bar{V}_{ij}(0), \bar{W}_j(0), \bar{p}_j(0), \bar{r}_j(0)\}$, the trajectories for $t > 0$ can be computed. The hyperbolic sine function leads to an amplification of the reactions of the stakeholders in economic dis-equilibrium. On the other hand, Equation **(6-70)** boils down to Equation **(6-56)** for constant prices and rents and assuming consumers in equilibrium. By summing up Equation **(6-69)** over j, it can easily be seen that the total expenditure-stock $\bar{V}_i$ of zone i remains constant with time:

$$\dot{\bar{V}}(t) = 0 \qquad i = 1, \ldots, n'. \tag{6-73}$$

This arises from the fact that changes of residential location of individuals are not considered in the model.

6.12.6 Stationary solution.[30] The stationary solution to Equations **(6-69)** through **(6-72)** is a complicated nonlinear set of transcendental equations. They determine the expenditure-flows $\tilde{V}_j$, the scale-of-facilities $\tilde{W}_j$, the prices $\tilde{p}_j$ and rents $\tilde{r}_j$. It can be seen that the stationary version of Equation **(6-69)** yields for the expenditure flows

$$\tilde{V}_{ij} = \tilde{V}_i \frac{\exp(2\tilde{v}_{ij})}{\sum_{k=1}^{n'} \exp(2\tilde{v}_{ik})}. \tag{6-74}$$

This result can be verified easily by inspection. On the other hand, the stationary solution of Equations **(6-70)** through **(6-72)** is $\tilde{V}_j = \tilde{C}_j$ which must hold for all zones j. There exists a coupling between the stationary demand $\tilde{V}_j$ and the stationary values of the facility-stock $\tilde{W}_j$ and the rent-level $\tilde{r}_j$ for each zone since $\tilde{C}_j$ depends on $\tilde{W}_j$ and $\tilde{r}_j$. In other words, for each facility scale and rent level, revenue from the demands needs to cover the fully-compensated supply-cost of making the facility available.

Comparing Equation **(6-74)** with the special form of the flow pattern implied by Equation **(6-56)**, the following "utility functions" can be derived:

$$v_{ij}(t) = \frac{1}{2}\left[\alpha \ln W_j(t) - \gamma \ln p_j(t) - \beta C_{ij}(t)\right]. \tag{6-75}$$

In the utility function above, we have assumed that the expenditure flows are always in equilibrium. In other words, $h_1''(\infty) > h_2''(\infty)$, $h_3''(\infty)$—that the expenditure-flow component is more dominant than either the facility scale or the price system—can be justified. When the system is near equilibrium, $|\zeta_j(\mathbf{Z})| < 1$, convergence is observed as can be gleaned from Equations **(6-70)** through **(6-72)**. Assuming Equations **(6-74)** and **(6-75)** also, the familiar quasi-deterministic model as shown in Equation **(6-50)** is obtained, where the calibrated value of α is shown as a', where the trip-distribution function f(.) now assumes the exponential form, and where prices and rents are not considered. The corresponding general differential equation is $\dot{\bar{W}}_j(t) = h_2^*(t)\bar{W}_j(t)\left[\sum_{i=1}^{n'} \bar{V}_{ij}(t) - c_j\bar{W}_j(t)\right]$. Notice the parameter $h_2^*(t)$ is a scaled time-constant that is different from $h_2''(t)$, and Equation **(6-56)** is a special case of this differential equation when $h_2^*(t)\bar{W}_j(t) = h''$.

---

[30]For a definition of stationarity (or steady-state) see the "Markovian Process" appendix to Chan (2000).

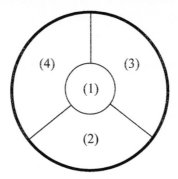

Figure 6-12 A four-zone urban area (Haag 1989)

The utility function given in **(6-75)** can assume other forms. Frankhauser (1987), for example, proposed that the shopping attitude of a consumer living in zone i will primarily depend on the offer of activities in a certain retail-area j and his/her travelling-costs C_{ij}:

where $X_j(t)$ is the unweighted accessibility to retail zone j:[31] $X_j(t) = \sum_{i=1}^{n'} \exp(-bC_{ij}(t))$.

Notice that prices and rents are again not considered. This utility function appears to be similar to the previous one when α is set to unity. However, in Equation **(6-75)** the attraction in zone j is only represented by the variable W_j, usually interpreted as an arbitrary quantitative-measure. Instead of W_j, the product $c_j W_j$ is used here in Equation **(6-76)**, representing the cost of providing the facility, C_j, which is a better defined economic variable.

6.13   An example of dynamic spatial-interaction

While the above model is quite comprehensive, its operational status is somewhat less impressive. For one thing, solving the equations analytically is difficult. Numerical techniques are often used in obtaining the solution. Using the utility

$$v_{ij}(t) = \frac{1}{2}\left[\ln(c_j W_j(t)) - \beta C_{ij}(t) - \ln X_j(t)\right] \qquad (6\text{-}76)$$

function as defined in Equation **(6-76)** and setting aside prices and rents for simplicity, a numerical simulation is performed using the equations of the dynamic model as formulated above (Haag 1989). We start our simulation by assuming that the urban area, as shown in **Figure 6-12**, is in its stationary state. The transportation costs C_{ij} between neighboring zones are assumed to be twice the cost within each zone. The facility-stock $\tilde{W}_j$ and its unit-cost c_j are three times higher in the central zone (1) than the suburban zones (2,3, and 4). We start our simulation by

---

[31]For an explanation of this accessibility function, see the "Generation, Competition and Distribution" chapter (Chan 2000) under the section entitled "A general location-allocation model."

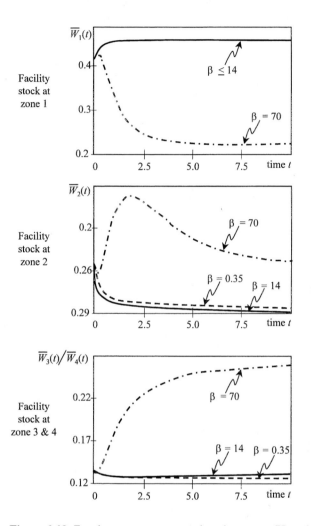

Figure 6-13 Zonal responses to a new shopping center (Haag 1989)

assuming an intermediate value $\beta = 14$ for the trip-distribution function. We shall now ask for the response to a sudden increase of the facility supply in a certain zone, such as the opening of a shopping center in zone 2, increasing the retail-floor space to twice the amount. Average zonal-responses for the β values of 0.35, 14 and 70, corresponding to low, medium and high transportation costs, are displayed in **Figure 6-13**. Because of the geometric symmetry, the responses at zones 3 and 4 are expected to be the same.

For a low transportation-cost, interzonal travel is easy. An increase of the number of shops, $\bar{w}_2(0)$, cannot create a sufficiently high attraction in zone

2—relative to the three other zones—to be self sustaining. In other words, a small β value of 0.35 leads to a decrease of facility stock in zone 2 eventually as shown in **Figure 6-13**. In contrast, the importance of zone 3 (or zone 4) remains almost constant here. However, it must be emphasized that the expansion of the facility stock in zone 2 has a positive impact on the demand level in the central urban-area, zone 1. It leads to a substantial increase of the supply of facilities, $\bar{W}_1$, in the long run. Thus the competition of service-sector zones due to low transportation-cost has some positive influence on the economic system as a whole, in response to an investment in zone 2. But the infrastructure in zone 2 cannot be self sustaining. Quite a different pattern is obtained when ß is increased to 70, or when transportation cost is high. After a short time of prosperity the facility stock of zone 1 decreases in a very dramatic way to nearly half its initial value . Meanwhile, the investment in zone 2 becomes stable at a rather high level after a small decrease, reflecting the localization of demand. The evolution in zone 3 (and 4) is surprising. The facility increase of $\bar{W}_2$ leads to an unexpected large-increase of the demand in zone 3 (and 4) from 8.3 percent to 20.3 percent of the total demand. The facility supply $\bar{W}_3$ (and $\bar{W}_4$) increases correspondingly, following the demand.

Sudden changes on the demand and/or supply side often lead to over-reactions and oscillations of the economic variables in the urban area—sometimes in unexpected ways. These effects are mainly caused by different speeds of adjustment of the supply and demand; they are an integral part of a dynamic system such as the one under discussion. In this particular case the initially doubled facility-supply in zone 2 results in more consumers attracted, but this large increase will diminish afterwards and stabilize at a *sustainable* level. Part of the value of a dynamic model is to observe the short-term and medium-term behavior, besides the long-term (steady-state) results.

6.14 Conclusion

In a personal letter from Lowry (1992), the originator of the model bearing his name states: "During the early 1960s, I did study the interactions of land-use and transportation ... I used the newly available resources of mainframe computers to solve (by iteration) a computationally tedious but mathematically-simple simultaneous-system. Because of its conceptual simplicity, and its promise of small-area solution-values, the model attracted considerable attention, and its mathematical foundation was elaborated by much more competent mathematicians [including Garin] than I can claim to be ..." The modest innovator continued: "The main warning I would offer you ... is that you should think carefully about point of view, otherwise known as 'objective function'. Most of the location decision and activity-allocation literature deals with efficiency and profitability at the level of the enterprise. Most of the location constraints are imposed by organizations that are not subject to that calculus. For example, in the location of airports, nuclear power plants, industrial parks etc., the permit-issuing authorities have quite different objectives—perhaps fatally ambiguous ones—than the entrepreneurs of the activity."

Apparently, Lowry recognizes clearly the pros and cons of his model and most importantly, concurs with the viewpoint of multiple criteria and pluralistic decision-making.

The role of comprehensive urban-spatial-modelling is an issue of active discussion since the beginning of computer-model development in 1960. The discussion is perhaps most intensified in the last three decades when there was an advocacy for simple planning procedures (Lee 1973) at first and a new school of dynamic modelling-structure (Wilson 1981) later. Recent debates (U S Department of Transportation 1996) center upon disaggregation, time-series information, integration with transportation models, environmental models and geographic information systems. Of equal interest is the relationship to interactive decision-making procedures. It was felt it is an opportune time for us to reexamine the debate via a set of experiments. This is a procedure rarely attempted by previous researchers—with the possible exception of Putman (1983). These experiments are supplemented by pertinent literature-review.

6.14.1 Findings. We suggest that the Lowry-based models (such as its computationally elaborate step-child PLUM) are founded on a common set of urban-economic theories that have remained essentially the same over the sixties. The economic-base theory is one of the few feasible means to project growth, while the general multinomial-logit-model is a key method to distribute population and employment within the study area. Advances in model building were made by-and-large in the computational procedures rather than in the basic concepts. For example, while the Lowry model did not address the problem of defining basic vs. non-basic activities, PLUM took a positive step to categorize activities, which resulted in an improved definition of multiplier. As far as the spatial-allocation of activities is concerned, the incremental process employed in PLUM represents an improvement over the total-stock approach. Correspondingly, the re-allocation of excess activities is handled more satisfactorily using a form closer to activity-constrained allocation-procedures.

These operational improvements in PLUM showed up in the results of our first experiment. While the forecasts in household activities are similar, the non-basic-employment allocations by the two models are quite different on a detailed zone-by-zone basis. This difference is understandable given the modifications on the multiplier and the re-allocation procedure. The differences, however, are quite minor when one examines the resolution on a more geographically aggregated scale. By collapsing zones into districts, the difference is dramatically mitigated. In spite of the computational refinements culminated during the 1960's, therefore, the forecast of future activities to the general subregions of the study area remains similar. The pendulum is swinging toward simpler methods and a new school of dynamic modelling theories. The elaborate details that the PLUM development tries to attain could possibly be replaced by more transparent (and analytic) models. This would help recasting existing models in terms of these new theories, which in turn sharpen our senses during verification and testing.

The basic theory underpinning the Lowry/PLUM models—to be distinguished from the computational procedures—continued to be used in model building through the seventies. The general multinomial-logit-model remains the cornerstone of activity allocation, although its specific formulation and interpretation vary (Putman 1983, Wilson 1981). For example, variants of the singly- or doubly-constrained gravity-model showed up as the intervening-opportunity model and entropy-maximizing model just to name two. Each of these variants can be interpreted in terms of different economic, physical, statistical, mathematical and behavioral theories. Many of these are quite refreshing and constitute some of the research advancements in the seventies.

As far as the theory of urban-activity development is concerned, the decade of the sixties witnessed a long-run equilibrium approach to forecasting. Conceptually this forecasting procedure can be viewed as a set of difference or differential equations. The solution of the difference equation in Lowry-based models gives rise to geometric-series increments of employment and population according to the economic-base theory. Recently, dynamic modelling procedures (Wilson 1981, Allen et al. 1981, Bertuglia et al. 1987) depart from the explicitly-stated equilibrium approach. They conceptualize the relationship between components of an urban system in terms of evolutionary and oscillatory behavior (although equilibrium state is not excluded). Strictly speaking, it is often a matter of semantics to distinguish between a development vs. evolution model. The reason is that--computationally speaking--evolution is often absent if it were not for the intravenous "seeding" of perturbations during model operations (as it happens in our experiments). Following this line of argument, it is also conceivable that evolutionary structure be introduced by recasting PLUM and Lowry in a dynamic framework with possible bifurcation points. (It may even be argued that the possible incremental seeding of basic-employment and the internal updating of labor-for-participation-rate in PLUM-IP as an example of evolution).

In a set of experiments, we showed the possible existence of bifurcation, fold and constraint catastrophes in the basic formulation of Lowry by recasting it in terms of dynamic equations. Particularly important is the fact that Lowry-based models appear to bifurcate more readily when the spatial cost C_{ij} and multiplier $\gamma \Delta t$ are perturbed—more so than the seeding of basic employment. In other words, the control parameters Γ are more critical than the initial conditions $\mathbf{Z}_0$ in Equation (6-55) as far as bifurcation is concerned. Webber (1984), Haag (1989) and others have labelled the traditional Lowry-based models as demand-oriented models, in which the customers or consumers compete for available facility stock. In contrast, these researchers advocated a supply-side model in which there are more explicit recognition of investment decisions, land rent, prices, and expenditure flow. Starting from a stochastic framework, a dynamic service-sector model has been derived from individual decision processes. By casting the model as a set of master equations and transition rates, the behavior of the system can be assessed—although numerical techniques, rather than analytical solution methods, are used. Through these equations, the short-, medium-, and long-term behavior of the system can be observed, including dis-equilibrium. Such a formulation also allows a more systematic parameter-calibration procedure to take place.

Conventional Lowry-based models can be shown to be special cases of this formulation. By assuming that consumers make their decisions independent of prices and land rent and for certain utility functions and time-scaling parameters, the conventional spatial interaction models result. An experiment with such a special case has confirmed that a decrease of transportation-cost leads to pronounced competition between the zones, and localization of the economy results from a higher transportation-cost. Even through the final stationary-state is independent of initial changes in the scale of provision, the long time to reach stationarity highlights the importance of these dynamic models in assessing the disparate zonal-responses in the interim. This is a distinguishing feature of these stochastic, dynamic models in comparison with the long-run-equilibrium approaches adopted in traditional Lowry-based models.

6.14.2   Supply-side models.   No doubt, the Lowry model is not a *comprehensive* planning model, as it is sometimes referred to. The model—and for that matter the location-allocation models surveyed thus far—adopted a dominant, neoclassical theory in economics. Several shortcomings, noted Webber (1984), arise from its focus on contingent rather than on determining relations. The required negative relationship between the price of capital (in relations to wages) and the capital-intensity of production is not reproduced by the theory. Also, the theory does not provide a means of interpreting the evolution of social systems from one form of production to another. The internal structure of the Lowry model is extremely simple. It is an idealized description of the appearance of a system, without a detailed explanation of why people behave the way they do. Lowry models a market—the matching of labor and jobs, and the equilibrium of the consumption and the supply of services—rather than production and consumption themselves. Thus the locational choice of manufacturing (or basic employment) is exogenous to the model.

These weaknesses can be exemplified by reference to the underlying principles: economic-base and spatial-interaction mechanisms. The equations are contingent relationships because they are based on given definitional-identities. In the economic-base model, the level of employment does equal the number of jobs and the existence of a given population does imply a particular activity-rate. But these facts ignore the processes by which jobs are created and workers offered employment. So the model cannot deal with realities of unemployment or of labor shortages. Equally, the number of journeys-to-work that begin in a zone equals the number of employed-workers in that zone, that the number-of-journeys terminating in a zone equals the number-of-jobs there, and that there is some observed average-expenditure on the journeys (usually called the "spatial price"). But spatial-interaction equations ignore the forces creating jobs and residences. In effect the spatial-interaction equations imply that all workers are instantaneously competing for all jobs and houses in a way that equilibrate so that there can be no shortages of workers or of jobs in particular zones.

A more appropriate approach would have been one of *capitalist production-theory*, contends Webber (1984). A capitalist holds money, uses that money to buy machinery, raw materials, and labor power. He uses these capital items to produce the commodities and sells the new commodities for money. The aim is to obtain more money at the end than at the beginning of the process. Now all production of commodities involves the use of plant and machinery that is long-lasting and immobile and capitalist decision-making is decentralized. Since the profitability of a decision depends upon later decisions made by other investors, capitalists are uncertain about the outcomes of their investment decisions. The first proposition is therefore, according to Webber, that capitalist society maximizes the amount-of-parts-made per unit-of-time and thus the rate of accumulation-of-capital—over the long run and based on anticipated changes in the economy.

The second proposition qualifies the first: at the level of the individual decision made by capitalist in a particular instance and at a specific geographic-location, profit-maximizing-behavior based on past performance is not necessarily apparent. These two optima--based on anticipation and past performance—may differ if the expected state-of-affairs does not come about. Long-run profit-maximization requires that the capitalist stays in business rather than simply maximizes the profit associated with each individual decision. The location, scale, and type of production of a single factory, for example, are not chosen solely with respect to the operation of that factory. Rather, the decision is made in the context of the profit-maximizing behavior of the exterior business-enterprise of the capitalist owner.

The third proposition of Webber directly challenges the central neoclassical tenet of consumer sovereignty. Consumer sovereignty is the postulate that all economic activity is directed to consumption that consumers have exogenously determined and they have insatiable wants. The contrasting view asserts that the economic system is dominated by the supply side, by producers. The conditions deduced by utility theory from the assumption that consumers choose commodities to maximize their welfare, subject to a budget constraint, have little effect upon relations in an economic system. Prices, including wages, are determined by technological conditions (as expressed in input-output coefficients[32]) and by the welfare level of workers. The marginal-utility conditions determine the quantities produced. But preferences are not independent: monopolistic or oligopolistic firms are free to set output levels and to advertise rather than to accept the production levels determined by existing tastes.

Such a change of view from the "individualistic neoclassical orthodoxy" requires a corresponding change in the central propositions of location theory in general. The orthodox view emphasizes consumption and interaction, particularly residential location and the journey to work and shop, but ignores production at locations of supply. Webber argues that models must change from an orientation toward demand to analysis of supply and of the concomitant social structure. The idea that communities are places of consumption and interaction is to be replaced

---

[32]For an explanation of input-output coefficients, see the "Economic Methods" chapter in Chan (2000).

by their role as places of production, of capital circulation, and of the accumulation of wealth. Whereas orthodox theory analyzes residential differentiation and population density in terms of distribution of choices and preferences, a supply-oriented theory examines building and transport technology at the time residences were built, the convenience of building homogeneous rather than diverse subdivisions, and the relative profitability of developing vacant land rather than redeveloping built-up land. Naturally this change in theoretical emphasis implies a fundamental revision of the form of operational models as we have discussed so far in this text. For example, Haag (1989) implemented the supply-oriented theory via several variables: the facility stock, land rent, and the spatial price. A birth-death process describes the status of the facility stock. Expansion of facility-stock takes place when the revenue attracted to these facilities exceeds the cost of supplying them, and vice versa. In subsequent chapters on "Spatial Equilibrium" and "Spatial Econometrics," we will tend to some of Webber's concerns.

6.14.3 *Operational models*.

In spite of these critiques, it must be realized that virtually all of the work in this area has been accomplished in the past four decades under very pragmatic conditions, argued Putman (1983). The work has not been very expensive in comparison with many other scientific endeavors. The complexities of problems involved are staggering. The work must be done with non-experimental techniques, as every urban area is in some way different from every other, and none are available for the researcher to manipulate at will. The available data range in quality from moderately good to un-usably bad. It is an era when most of publicly funded research can only be justified in terms of immediate applicability. Funds are not available merely for the pursuit of knowledge.

Location-allocation models, the Lowry land-use model, and subsequent developments such as PLUM share the same intellectual basis, namely neo classical economic-base and gravitational theories. They make no pretense of being the definitive description of the real world. Yet, here are a set of methods that have produced interesting, useful and substantive findings. This tradition has continued in forging even closer ties between land-use and transportation modelling (Shen 1997, Yen and Fricker 1997, Huang 1997). A recent count (U. S. Department of Transportation 1996) suggests at least 13 models are being used in 20 sites globally to further this effort. The methods can be used by planners and managers now to analyze policy issues in ways that were not possible before. The methods cannot answer all questions. No method can answer all questions. We may, however, expect that continued progress in this field will permit greater number of increasingly complex questions to be analyzed in more insightful and sensible ways.

6.14.4 *The next step*.

The greatest lesson to be learned from the above experimentations is not which model is correct or incorrect, which is a question of empirical verification—something yet to be performed satisfactorily. An obvious task is that we use post-implementation data to verify forecasts made in the fifties and sixties (Mihram 1972) in order to validate methodologies proposed at that time period, now that 30 or 40 years have elapsed, and the forecasts can either be correct or incorrect. It is the only means to avoid refining a methodology that was misdirected from the outset. However, most urban areas are not subject to controlled experiments of this kind, making the validation task close to impossible. To a limited extent, this has been performed by Putman (1983) in his "integrated model" project, with mixed results.

In carrying out further empirical work, it is also desirable that, to the extent possible, a more transparent/analytical model such as the Garin formulation of Lowry Model be used at the beginning. Such a formulation, because of its analytical property, will allow more reliable tracing of catastrophe, bifurcation and evolutionary behavior. Otherwise, such behavior may be buried under the weight of the massive data-processing effort—a lesson learned from our experiment with PLUM. Although the stochastic, dynamic models presented above appear analytic, their solutions are often simulation-based. This can make the model other than transparent, thus offsetting the thrust of these efforts.

The use of the Garin-type model in the first phase research effort means glossing over the examination of transient behavior and constraint catastrophe. A second phase will emphasize the use of attraction-constrained allocation models to explicitly address constraint catastrophes. In the linkage, a decomposition procedure such as the one used in the "Generation, Competition and Distribution" chapter will allow autonomy between the 'allocation' and 'development' component of the combined model, thus maintaining the transparency that is so critical in further modelling effort in this area. Although the dynamic and stochastic aspects are left out, we feel that it is still a worthwhile effort that results in tangible results for those in practice. The above form the guidelines for further examination of evolutionary behavior, including conditions for dis-equilibrium, to take place. It also makes a case for developing a disaggregate, bifurcation version of a Lowry-Garin based model, as we will do in the next chapter.

Time is a sort of river of passing events, and strong is its current;
no sooner is a thing brought to sight than it is swept by
and another takes its place, and this too shall be swept away.
Marcus Aurelius Antonius

CHAPTER SEVEN

CHAOS, CATASTROPHE, BIFURCATION AND DISAGGREGATION IN LOCATIONAL MODELS[1]

Recently, keen interest has been shown in such theories as chaos, bifurcation, catastrophe and disaggregation. This is in response to the observation that many locational decisions and urban/regional developments occur precipitously, rather than gradually as traditional theories would explain. Sometimes oscillatory behavior is observed. Additional interest has also been shown in disaggregate modelling, where finer levels of details about the *individual behavior* of the decision-makers and community (rather than the average behavior) are analyzed. Considering this interest, this chapter will revisit chaos, catastrophe theory and disaggregation. We will illustrate these concepts via a case study of urban spatial modelling along the Lowry–Garin tradition. The theoretical and practical results are shown to be a logical extension of some basic building-blocks of facility location and urban spatial models.

7.1 Chaos

A system is chaotic, if there exists an uncountable, invariant set A of initial conditions such that all trajectories starting in A meet the following requirements: (1) They never repeat themselves; (2) They neither attract nor are they attracted by other trajectories; and (3) They show a sensitive dependence on initial conditions. In practice, small uncertainties or small perturbations are amplified. In particular the errors or fluctuations may grow exponentially, and periodicities may become extremely irregular, so that—even if behavior is predictable in the short term—it may become unpredictable in the long term and it may lead to very different trajectories. Consequently, it is usually impossible to make accurate predictions of unstable behavior for other than very short time-periods. Thus the resulting aperiodic and interlaced trajectories can produce very complicated forms (Fischer et al. 1990).

---

[1]This chapter borrows heavily from Yi and Chan (1988).

7.1.1 A simple example from urban dynamics[2]. We will illustrate chaotic behavior with an example. The model here is assumed to reproduce the potential evolutionary pattern of a structurally declining area losing most of its population base. Three variables are assumed to play a key role: city size (population N), employment potential (share of population employed y), and urban attractiveness (share of immigrants amid total population). The model presented here is illustrative and extremely simple. It would have to be extended toward a more comprehensive model of urban dynamics in order to make it realistic.

In a structurally declining area, the growth-rate of population is negative, although this may be compensated by a rise in employment flows (or migration flows) from outside the urban system. Consequently, we may assume the following simple relationship: $\dot{N} = \gamma_1 E' - \gamma_2 N$, where N and E' represent population size and employment share (workers/nonworkers) respectively. γ_1 and γ_2 denote the related growth and decline rates respectively. Next we assume that for a structurally declining city the urban attractiveness has a negative growth-rate $(-\beta_1)$, while this negative trend maybe compensated by a rise in the employment rate:

$$\dot{G}' = -\beta_1 G' + \alpha E' \tag{7-1}$$

where G' represents the above mentioned immigration-share (immigrants/non-immigrants). The growth-rate of immigration related to employment (α) is assumed to be positively correlated (through the parameter β_2) with agglomeration economies emerging from city size, $\alpha = \beta_2 N$. We obtain for Equation **(7-1)** the following expression: $\dot{G}' = -\beta_1 G' + \beta_2 N E'$. Finally, the employment rate of a structurally declining city is assumed to have a negative growth rate $(-\delta_1)$. This may be reinforced by a change in immigration-rate $(-g)$ from outside, thus increasing the unemployment. This trend is reversed, however, by a growing city from activity-expansion rate δ_3:

$$\dot{E}' = -\delta_1 E' - g G' + \delta_3 N. \tag{7-2}$$

Next we assume that the change in immigration is affected by synergetic effects related to city size through the parameter δ_2, $g = \delta_2 N$. At the end we obtain the following expression for Equation **(7-2)**: $\dot{E} = -\delta_1 E' - \delta_2 N G' + \delta_3 N$. If $\gamma_1 = \gamma_2 = \gamma$, $\delta_1 = \delta_2 = 1$, $\delta_3 = r$, $\beta_1 = b$ and $\beta_2 = 1$, we have the Lorenz model (1963):

$$\dot{N} = \gamma(E' - N) \tag{7-3}$$

$$\dot{E}' = N G' + r N E' \tag{7-4}$$

$$\dot{G}' = N E' - b G'. \tag{7-5}$$

---

[2]For an introduction to urban dynamics, see the "Descriptive Tools" chapter in Chan (2000).

The generalized Lorenz equations would simply be $\dot{N} = -\gamma_2 N + \gamma_1 E'$, $\dot{E} = \delta_2 NG' + \delta_3 N - \delta_1 E$, and $\dot{G}' = \beta_2 NE' - \beta_1 G'$. The steady-state solution of this system can be analytically determined. For example, it is clear that the obvious steady-state solution $N = E' = G' = 0$ exists. When $\gamma_1 \delta_3 = \gamma_2 \delta_1$, we have a critical value (or bifurcation point). After this critical value, or when $\gamma_1 \delta_3 > \gamma_2 \delta_1$, motion begins.

A limit-cycle oscillation emerges from an equilibrium state as some system parameters are varied. This defines a *Hopf bifurcation*,[3] the most common bifurcation and stability conditions studied to date. It can be shown that for $\gamma_2 > \beta_1 + \delta_1$, a Hopf bifurcation occurs at $\dot{\delta}_3 = \gamma_2^2(\gamma_2 + \beta_1 + 3\delta_1)/\gamma_1(\gamma_2 - \beta_1 - \delta_1)$. Now it may be interesting to investigate the stability conditions and steady-state solutions of our model. A series of simulation experiments will be performed for varying parameter values. The first scenario is for modest urban decline. For this simulation the following parameter values will be assumed: $\gamma_1 = 0.1$, $\gamma_2 = 0.001$; $\delta_1 = 0.1$, $\delta_2 = 0.005$, $\delta_3 = 0.001$; $\beta_1 = 0.01$, and $\beta_2 = 0.0001$. The initial values are $N = 100$, $E' = 0.5$, and $G' = 0.1$. It should be noted that we have just chosen the critical values $\gamma_1 \delta_3 = \gamma_2 \delta_1$,[4] after which unstable motion begins. To show this, we summarize the results in **Figure 7-1** —a figure reminiscent of those obtained by the Limits to Growth model described in the "Descriptive Tools" chapter in Chan (2000).

Figure 7-1 shows that for parameter values and the initial values given above, the urban system concerned shows a gradual decline for total population, a significantly decreasing pattern for the employment share and a clear growth pattern for the immigration share (followed by a slight decline in later periods). The contrasting behavior between employment share and immigration share can be explained. People at the beginning do not respond directly to urban decline, so that the inflow increases until a certain value (corresponding to the minimum employment rate); after that a slight decline for immigration share appears. The net result for the urban system appears to be one of modest decline. Clearly, this result reflects transient motions that depend strongly on initial conditions, so that other initial conditions may lead to different results. Simulation 2 assumes the following parameter values according to the conditions of unstable motion shown in the steady-state solution, for which $\gamma_1 \delta_3 > \gamma_2 \delta_1$ and $\gamma_2 > \beta_1 + \delta_1$.[5] Specifically, $\gamma_1 = 0.1$, $\gamma_2 = 0.015$; $\delta_1 = 0.01$, $\delta_2 = 0.01$, $\delta_3 = 0.02$; $\beta_1 = 0.001$, and $\beta_2 = 0.001$. The initial values of the variable are again the same. The chaotic urban-decline results are shown in **Figure 7-2**, with employment and immigration shares assuming wildly fluctuating values. These results show an interesting chaotic-decrease of urban population,

---

[3]The Hopf bifurcation refers to critical parameter values of a differential equation set at which a stable equilibrium point becomes an unstable orbit or disappears. See the book appendix "Control, dynamics and system stability" in Chan (2000) for more details.

[4]These values are computed from steady state solution for a generalized Lorenz system, as described in Annex A, Chapter 3 of Fischer et al. (1990).

[5]These steady-state conditions are worked out in Fischer et al. (1990) as Annex A of Chapter 3.

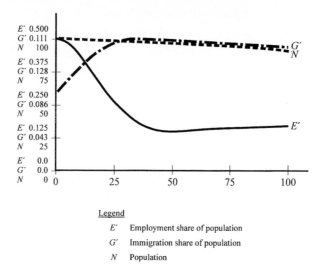

Legend

    E' Employment share of population

    G' Immigration share of population

    N Population

Figure 7-1 Modest urban decline (Fischer et al. 1990)

while the immigration share and employment share shown in the same period are oscillating in an irregular pattern. In this simulation we have opted for the parameter δ_3. This represents a magnitude below the critical value that is necessary to get a Hopf bifurcation,[6] which occurs when $\delta_3 = \gamma_2^2(\gamma_2+\beta_1+3\delta_1)/\gamma_1(\gamma_2-\beta_1-\delta_1) = 0.026$, since we are not interested in having a chaotic motion in the long run. The pattern shown in **Figure 7-2** will therefore, after initial chaotic motion, reaches a monotonically declining path.

    Besides simulation, the readers may be interested to know that Fischer et al. (1990) discussed an optimal-control model[7] of the general Lorenz equations. Such a model provides better insights into the critical values of the parameters for various system behavior. It was found, for instance, that depending on the values of the parameters, it is possible that optimal-control trajectories are oscillating. In other words, fluctuations that are usually considered as non-optimal may emerge as solutions of an optimization process related to a chaotic system of the Lorenz type. Consequently, decision-makers that are aware of this dynamical phenomenon can manipulate the parameters whose values are responsible for the oscillating behavior of the model. This will achieve the desired goal for the community.

---

   [6]For detailed derivation of the Hopf bifurcation condition, see the Annex to Chapter 3 of Fischer et al. (1990).

   [7]An introduction to control theory is found in the book appendix "Control, Dynamics and System Stability" of Chan (2000).

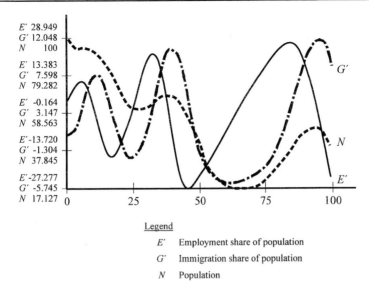

| E' 28.949 |
| G' 12.048 |
| N 100 |
| E' 13.383 |
| G' 7.598 |
| N 79.282 |
| E' -0.164 |
| G' 3.147 |
| N 58.563 |
| E'-13.720 |
| G' -1.304 |
| N 37.845 |
| E'-27.277 |
| G' -5.745 |
| N 17.127 |

Legend

E'	Employment share of population
G'	Immigration share of population
N	Population

Figure 7-2 - Chaotic urban decline (Fischer et al. 1990)

7.1.2   Dynamic spatial-interaction models. Here an optimal-control version of a dynamic entropy-model will be presented by assuming all flow-variables V_{ij} are time dependent (Nijkamp and Reggiani 1992). The choice of control variables and state variables in an optimal-control model is somewhat arbitrary, as it is not always sure whether a control variable means a direct-intervention measure or an indirect policy-instrument. In our model, the flow V_{ij} will be considered a control variable while the trip-origin variables V_i will be regarded as state variables. Their evolution is assumed to be dependent on the net push-out/pull-in effects of the flows at zone i:

$$\dot{V}_i = f_i V_i + \varphi_i \left(\sum_j V_{ji} - \sum_j V_{ij} \right). \tag{7-6}$$

This is a direct result of the well-known dynamic migration-model:[8]

$$\dot{N}_i = f_i N_i + \sum_j V_{ji} - \sum_j V_{ij}. \tag{7-7}$$

Here f_i is the natural growth of population N_i at the ith zone. Clearly, Equation (**7-6**) is derived from Equation (**7-7**) by assuming that the trip-ends V_i are linearly dependent—through the parameter φ_i—on population N_i: $V_i = \varphi_i N_i$. It follows $\dot{V}_i = \varphi_i \dot{N}_i$. Substitution of these last two equations into Equation (**7-7**) leads to Equation (**7-6**).

---

[8]For an introduction to population-migration models, see "Interregional growth and distribution" and "Interregional components of change" sections in the "Economic Methods" chapter (Chan 2000).

The objective function of the present spatial-interaction system is a multi-period cumulative entropy-function, representing a general utility (or social welfare) of the system during the time-horizon T: $\max \int_0^T -\sum_i \sum_j V_{ij}(\ln V_{ij} - 1)dt$, subject to—besides Equation **(7-6)**—the by now familiar trip-end (origin and terminating trips) and travel-cost constraints:

$$\sum_j V_{ij} = V_i \qquad \forall i$$
$$\sum_i V_{ij} = V_j \qquad \forall j \qquad\qquad (7\text{-}8)$$
$$\sum_i \sum_j V_{ij} C_{ij} = C.$$

Quite interestingly, the solution is again the generalized spatial-interaction model: $V_{ij} = k_i l_j V_i V_j \exp(\beta C_{ij})$, which is identical to the static-model result. It is not surprising therefore that the generalized dynamic multinomial logit-model follows: $V_{ij} = [l_j V_j \exp(-\beta C_{ij})] / [\sum_k l_k V_k \exp(-\beta C_{ik})]$. Please note, again, that here all variables are a function of time. Furthermore, the balancing factor l_j is dynamic, since it depends on the costate variable[9], which might be interpreted as dynamic accessibility to zones j.

7.1.3   Chaos in spatial-interaction models.   Nijkamp and Reggiani (1992) showed that a dynamic multinomial-logit model (MNL) emerges as a solution to an optimal-control problem whose objective function is a cumulative entropy function[10]. The spatial interaction assumes the form of a dynamic logit model[11]:

$$\frac{d\Theta_j}{dt} = \dot{\Theta}_j = \frac{d}{dt} \frac{\exp(v_j)}{\sum_i \exp(v_i)}, \qquad\qquad (7\text{-}9)$$

where Θ_j is the subareal-share of activities at zone j. It can be shown that this expression leads toward

$$\dot{\Theta}_j = \dot{v}_j \Theta_j (1 - \Theta_j) - \Theta_j \sum_{i \neq j} \dot{v}_i \Theta_i, \qquad\qquad (7\text{-}10)$$

where $\dot{v}_j = dv_j/dt$ represents the time rate-of-change of the utility function v_j. The latter term at the right-hand-side of Equation **(7-10)** represents essentially interaction effect. Now impose the condition that the sum of the subareal shares add

---

[9]As explained in the "Control, Dynamics and System Stability" appendix to Chan (2000), the costate variable in control theory can be interpreted as the marginal value of the stock, if the state variable is interpreted as inventory stock in an inventory-control application.

[10]The relationship between entropy maximization and the logit form of the gravity model has been developed in both the "Descriptive Tools" and "Prescriptive Tools" chapters of Chan (2000)—albeit in a static, rather than dynamic, form.

[11]See the "Descriptive Tools" chapter in Chan (2000) for an introduction to the Logit model.

up $(\sum_j \Theta_j = 1)$ in Equation **(7-10)**. We can derive from Equation **(7-10)** that the related dynamic condition $\sum_j \dot\Theta_j = 0$ is automatically satisfied.

Now it is interesting to examine in more detail possible trajectories of $\dot v_j$ and its effect on system stability. By assuming, for instance, that the utility of choosing zone j increases linearly with time through a fixed parameter a_j, we would have $\dot v_j$ = constant = a_j. Then the evolution of an MNL model becomes:

$$\dot\Theta_j = a_j \Theta_j (1 - \Theta_j) - \Theta_j \sum_{i \neq j} a_i \Theta_i. \qquad (7\text{-}11)$$

The reader can see it is a system of predator-prey equations[12] with a limited prey-share Θ_j; Θ_i can be interpreted as a predator whose influence will be the reduction of population Θ_j through parameters a_j. It is interesting that the first term in Equation **(7-11)** represents the continuous-time logistic-equation depicting logistic growth. The second term represents interaction effects, which may clearly affect the dynamic trajectory of the system.

We shall now approximate Equation **(7-11)** in discrete time by considering a unit time-period:

$$\Theta_{j,t+1} - \Theta_{j,t} \approx a_j \Theta_{j,t} (1 - \Theta_{j,t}) - \Theta_{j,t} \sum_{i \neq j} a_i \Theta_{i,t}, \qquad (7\text{-}12)$$

where a_j is now approximated by $v_{j,t+1} - v_{j,t}$, so that we have the following final expression in discrete terms:

$$\Theta_{j,t+1} = (a_j + 1) \Theta_{j,t} - a_j \Theta_{j,t}^2 - \Theta_{j,t} \sum_{i \neq j} a_i \Theta_{i,t}. \qquad (7\text{-}13)$$

An interesting observation is that the first two terms at the right-hand-side represent a difference version of the standard discrete logistic-growth model for a biological population

$$x_{t+1} = b x_t (1 - x_t) \qquad (7\text{-}14)$$

where $x < 1$, and b is a parameter reflecting the growth rate ($0 < b < 4$). By deleting for the time being the last term of Equation **(7-13)** (i.e., the interaction term), by assuming a constant utility-increase a_j, and by putting $b_j = a_j + 1$, we can rewrite Equation **(7-13)** as follows: $\Theta_{j,t+1} = b_j \Theta_{j,t} [1 - (b_j - 1) \Theta_{j,t} / b_j]$. If we make the following transformation $x_{j,t} = \Theta_{j,t}(b_j - 1)/b_j$, this equation can be written in the canonical form **(7-14)** as originally advanced by May (1976): $x_{j,t+1} = b_j x_{j,t} (1 - x_{j,t})$. According to the findings of May, our simple dynamic MNL model may in principle embody unstable or chaotic behavior when the model is specified in a difference-equation form, at least in its degenerated version. It is interesting to see that this happens for

---

[12] The predator-prey equations govern an ecological growth or decline of species. It was introduced in the "Activity Allocation and Derivation" chapter under the "Theoretical thread" section.

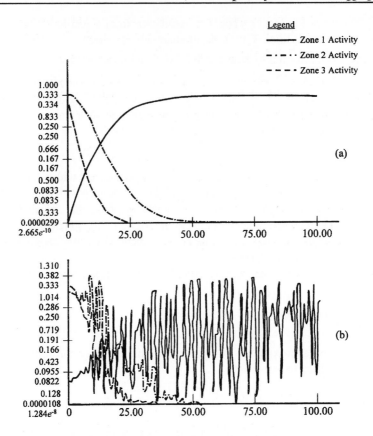

Figure 7-3 Stability of a spatial-interaction model

the values $3 < b_j < 4$ (or $2 < a_j < 3$, where $a_j = b_j - 1$). In other words, only when the slope a_j of the utility-function v_j with respect to time is less than 2 we have stable solutions in choosing the subarea j. High values of the marginal utility-function can lead to unpredictable movements in spatial patterns.

Let us now return to the full model of Equation **(7-13)**. We consider the equivalent formulation $\Theta_{ij} = W_j \exp(\beta C_{ij}) / \sum_i W_i \exp(-\beta C_{ij})$ where $W_j = V_j$, or the attractiveness of the destination-j is represented by the trips terminating in zone j. Let us say the additivity conditions $\sum_j \Theta_j = 1$ do not need to be satisfied in each point in time. In other words, we have a spatial-interaction model now the additivity constraint has been relaxed. Let $a_1 = 1.2$, $a_2 = 1.1$, and $a_3 = 1$, with the initial conditions: $\Theta_1 = \Theta_2 = 0.333$, and $\Theta_3 = 0.334$. Similar to the case in **Figure 7-1**, a stable behavior emerges in the long run, as illustrated in **Figure 7-3**(a). On the other hand, when $a_1 = 2.9$, $a_2 = 2.85$, and $a_3 = 2.8$, and $\Theta_1 = \Theta_2 = \Theta_3 = 0.333$, chaotic behavior emerges, as shown in **Figure 7-3**(b). The oscillating behavior results mainly from the fact that the values of the parameters no longer satisfy the stability conditions, as will be shown below. We can also notice from this pattern the influence of the utility-increase a_j in the amplitude of the trajectories. Evidently, activities Θ_2 and Θ_3 are the first variables reaching an equilibrium in the long run, since their related parameters a_2 and a_3 have the lowest values.

7.1.4 Stability conditions of a dynamic spatial-interaction model.

To explain the stability and instability behavior observed in the above example, we can reexamine Equation **(7-12)**. In long-hand format, this set of equations for the dynamic spatial-interaction model is written for the three-zone study area[13]:

$$\Theta_{1,\,t+1} = a_1\Theta_{1,\,t}(1-\Theta_{1,\,t}) - \Theta_{1,\,t}(a_2\Theta_{2,\,t}+a_3\Theta_{3,\,t}) + \Theta_{1,\,t}$$
$$\Theta_{2,\,t+1} = a_2\Theta_{2,\,t}(1-\Theta_{2,\,t}) - \Theta_{2,\,t}(a_1\Theta_{1,\,t}+a_3\Theta_{3,\,t}) + \Theta_{2,\,t} \qquad \textbf{(7-15)}$$
$$\Theta_{3,\,t+1} = a_3\Theta_{3,\,t}(1-\Theta_{1,\,t}) - \Theta_{3,\,t}(a_1\Theta_{1,\,t}+a_2\Theta_{2,\,t}) + \Theta_{3,\,t}\ .$$

It is evident that the trivial fixed points[14] are $(0,0,0)$, $(1,0,0)$, $(0,1,0)$ and $(0,0,1)$. For this difference-equation set, the Jacobian matrix[15] of $\Theta_{t+1} = (\Theta_{1,t+1}, \Theta_{2,t+1}, \Theta_{3,t+1})^T$ with respect to $\Theta_t = (\Theta_{1,t}, \Theta_{2,t}, \Theta_{3,t})^T$ is

$$\nabla_{\Theta_t}\Theta_{t+1} =$$

$$\begin{bmatrix} \dfrac{\partial\Theta_{1,t+1}}{\partial\Theta_{1,t}} & \dfrac{\partial\Theta_{1,t+1}}{\partial\Theta_{2,t}} & \dfrac{\partial\Theta_{1,t+1}}{\partial\Theta_{3,t}} \\[2ex] \dfrac{\partial\Theta_{2,t+1}}{\partial\Theta_{1,t}} & \dfrac{\partial\Theta_{2,t+1}}{\partial\Theta_{2,t}} & \dfrac{\partial\Theta_{2,t+1}}{\partial\Theta_{3,t}} \\[2ex] \dfrac{\partial\Theta_{3,t+1}}{\partial\Theta_{1,t}} & \dfrac{\partial\Theta_{3,t+1}}{\partial\Theta_{2,t}} & \dfrac{\partial\Theta_{3,t+1}}{\partial\Theta_{3,t}} \end{bmatrix} = \qquad \textbf{(7-16)}$$

$$\begin{bmatrix} a_1-2a_1\Theta_1-(a_2\Theta_2+a_3\Theta_3)+1 & -a_2\Theta_1 & -a_3\Theta_1 \\[1ex] -a_1\Theta_2 & a_2-2a_2\Theta_2-(a_1\Theta_1+a_3\Theta_3)+1 & -a_3\Theta_2 \\[1ex] -a_1\Theta_3 & -a_2\Theta_3 & a_3-2a_3\Theta_3-(a_1\Theta_1+a_2\Theta_2)+1 \end{bmatrix}.$$

---

[13] Variations of this equation set will be given in the forthcoming "Spatial Econometric" and "Spatial Time Series" chapters.

[14] A fixed point refers to an equilibrium point in systems stability, such as a `sink' or `focus'. For an introduction to these concepts, see the book appendix "Control, Dynamics and System Stability" in Chan (2000).

[15] Let $\mathbf{x} = (\cdots x_i \cdots)^T$ be a vector of decision variables and $\mathbf{F}(\mathbf{x}) = (\cdots F_i(\mathbf{x}) \cdots)^T$ be a vector of functions for $i = 1, \dots, n$. These functions are characterized by asymmetric interactions if $\partial F_i'(\mathbf{x})/\partial x_j \neq \partial F_j'(\mathbf{x})/\partial x_i$ $(i \neq j)$.

Suppose that $\nabla\mathbf{F}(\mathbf{x}) = \dot{F}'(\mathbf{x})$ is continuously differentiable. A Jacobian matrix is the generalization of the gradient for asymmetric interactions. It is defined as $F'(x) = \begin{bmatrix} \dfrac{\partial F_1'}{\partial x_1} & \cdots & \dfrac{\partial F_1'}{\partial x_n} \\ \vdots & & \vdots \\ \dfrac{\partial F_n'}{\partial x_1} & \cdots & \dfrac{\partial F_n'}{\partial x_n} \end{bmatrix}$. For more detailed explanation, see the "Control, Dynamics" appendix in Chan (2000) under the "Variational inequality" section.

Therefore, the Jacobian related to the fixed point $(1,0,0)$ is $\begin{bmatrix} -a_1+1 & -a_2 & -a_3 \\ 0 & a_2-a_1+1 & 0 \\ 0 & 0 & a_3-a_1+1 \end{bmatrix}$,

which is a triangular matrix. The related eigenvalues are $q_1' = -a_1 +1$, $q_2' = a_2-a_1 + 1$, $q_3' = a_3-a_1 +1$. The stability conditions for each fixed point then become $|q_1'| < 1$, $|q_2'| < 1$, and $|q_3'| < 1$.[16] For $|q_1'| < 1$, the following conditions must be met for a_1: $0 < a_1 < 1$ or $1 < a_1 < 2$. From the second condition on q_2', we obtain the following values for a_2: $a_1- 1 < a_2 < a_1-2$. In an analogous way, we obtain the following stability conditions for a_3: $a_1-1 < a_3 < a_1-2$. Let us now consider the values a_1, a_2, a_3 assumed in our simulation experiments: $a_1 = 1.2$, $a_2 = 1.1$, $a_3 = 1$ for case (a) and $a_1 = 2.9$, $a_2 = 2.85$, $a_3 = 2.8$ for case (b). While the first case satisfies the stability conditions laid out above the second case does not. Hence, we have stability in the former and instability in the latter as demonstrated in simulation experiments (Nijkamp and Reggiani 1992).

7.2   Spatial dynamics: competition

Most of the efforts in dynamic, competitive location-theory are concerned with spatial oligopoly processes, in which several decision-makers are siting their facilities simultaneously. This process is modelled with lagged reactive and conjectural variation-strategies. The former refers to time delays in response to an initiating action, and the latter to the basic contexts of the problem—whether it be a leader-follower situation or a simultaneous-gaming situation. Two questions are relevant in the analysis: (a) the locational structure of the steady-state equilibria if they exist, and (b) the convergence or non-convergence of the reactive process to such steady states. *Catastrophe theory* is a way of examining such dis-equilibria (Andersson and Kuenne 1986).

7.2.1 Basic premises. Most work centers around the first of these questions, namely the steady-state spatial- equilibria. Because of the complexity of solving for steady states even in relatively simple models or establishing their existence, the second question on reactive process has become important. Often, simulation analysis is used lacking any analytical solution—as evidenced by the examples worked out starting with the "Generation, Competition and Distribution" chapter. In the analytical solutions available to date, the economies of interest are described by the following simplifying assumptions: First, the spatial configuration for the market is either a finite line segment, a circle, or a disc (circle and its interior). Second, demands are distributed over the line segment and the circle either uniformly or according to a density function that is integrable and at least once differentiable. In the case of the disc, the demand is uniformly distributed. Third,

---

[16]The stability conditions for differential equations are explained in the book appendix "Control, Dynamics, and System Stability" of Chan (2000). The equivalent stability-conditions for difference equations are discussed in Lorenz (1993).

each firm chooses a location strategy based on either a *Cournot* conjecture that other firms will not respond to its move; or a game-theoretic conjecture that some rivals will react by causing the largest possible loss of the initiating firms's market. Finally, the steady state is identified with a *Nash* equilibrium—result of a fair division of the spoils.[17]

7.2.2 Cournot equilibrium. The classic Cournot solution assumes an oligopolistic naive behavior in which there is no collusion among the competitors[18]. For a duopoly, an individual firm perceives of the selling price of its products or services as constant. S/he then decides how much s/he is willing to supply at that price. The naive behavior refers to the assumption that the other seller holds constant his/her output and price. Yet each time such an assumption is shown to be falsified by the actual response of the competitor. In short, the Cournot conjecture says that each duopolist determines a quantity to be sold. Both together contribute to a combined output that determines the price, and each competitor tries to maximize his/her own profit assuming the other to hold his/her quantity constant.

The Eaton-Lipsey necessary and sufficient condition for a steady-state solution to a linear or circular model with Cournot conjectures are the following: (a) No firm's total sales are less than some firm's long-side sales, where a firm's long-side sales are the sales to customers in the longer of its market segments left and right of its site. (b) Peripheral firms, where existent, are located next to (are paired with) other firms, so that the short-sides of the paired firms are (nearly) zero. (c) Let $V(x)$ be the patronage at point x. Then, for every unpaired firm $V(B_L) = V(B_R)$, where B_L and B_R are the left- and right-hand boundaries of the firm's market area. In other words, the patronage at the left boundary is the same as the patronage at the right boundary. (d) On the other hand, for each paired firm, $V(B_s) \geq V(B_l)$, where B_s and B_l are the boundaries of the short- and long-sides of the firm's market areas. Put it in plain English, when the patronage from the short-side market is different from the long-side market, firms agglomerate.

For circular market areas, peripheral firms do not exist, so consideration (b) can be eliminated. For uniformly distributed demand, conditions (c) and (d) are met trivially and can be omitted. The above two to four conditions give rise to interesting locational decisions. In general, an oligopoly market allows much greater strategic freedom to the participants than would be possible under the iron discipline of competitive markets. For the Hotelling economy—linear market with uniform demand distribution—the famous case for $n = 2$ firms featured a solution with pairing at the center of the line, which follows from conditions (a) and (b). It is well known that under duopoly and when demand is inelastic, the two firms will

---

[17]For a definition of the Nash equilibrium, see "Game theory" subsection under the "Multi-criteria Decision-making" chapter in Chan (2000). Subsequent discussions can be found in the "Generation, Distribution and Competition" chapter under the section "Private facility location under spatial competition".

[18]The Cournot and Bertrand/Edgeworth equilibria were introduced in the "Economic Methods" chapter (Chan 2000) under the "Imperfect information" subsection.

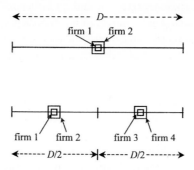

Figure 7-4 Hotelling equilibrium points

gravitate to the median—the point that is closest to the demands on the average.[19] In one-dimension, equilibrium solution of the Hotelling type exists for even numbers of firms: Clusters of two are formed at the centers of segments of lengths $2D/n$, when D is the length of the line-market segment. **Figure 7-4** illustrates the case of duopoly ($n = 2$), showing the locations of the firms. For $n = 3$, condition (b) cannot be met and the steady state is nonexistent. When $n = 4$, condition (b) requires pairing and condition (a) dictates pair locations at the first and third quartiles. **Figure 7-4** again illustrates the case of quadropoly ($n = 4$). For $n = 5$, pairs are located at $x = 1/6$ and $x = 5/6$. and the unpaired firm is at the center. For $n > 5$, an infinite number of steady-state solutions exist.

When Hotelling's economy is bent into circular form, condition (a) alone dictates location. For $n = 2$ all configurations are steady state, and when $n > 2$ an infinite number of solutions exist.[20] For the linear market with non-uniform demand distribution, and when the market has no rectangular demand distribution segments, conditions (c) and (d) imply the following: *A necessary condition for the existence of a steady-state solution is that n be no more than twice the number of modes in the population distribution.* Therefore, on a unimodal distribution, no equilibrium exists for $n > 3$, and for $n = 2$, conditions (a), (b) and (d) dictate a solution at the median of the distribution. Indeed, the latter proposition for $n = 2$ holds true whatever the form of the distribution of population (including the uniform). When the circular market with non-uniform distribution is considered, condition (b) is eliminated. But since it does not enter into the implication of the theorem, the latter continues to hold. The results, therefore, are not changed from those of the linear model. Generalization to a two-dimensional space is discussed in the "Spatial-temporal Information" chapter using Voronoi diagrams.

---

[19]For a definition of the median, see the "Economic Methods" chapter of Chan (2000) and "Facility Location" chapter in the current volume.

[20]A circular market was discussed in the "Generation, Competition and Distribution" chapter.

7.2.3 *Game-theoretic solution*.

For the linear model with uniform-demand distribution, and when firms have game-theoretic conjectures, the firms are led to a min-max solution. That is, they will maximize the short-side of their market areas by locating at their centers[21]. For all $n \neq 2$, this implies that peripheral firms will locate one-third of the distance from the end of the line to their neighbor and interior firms midway between their neighbors. When $n = 2$ firms will pair at the midpoint of the line. Each firm will have a market area of $1/n^2$.

Since peripheral firms are not paired when the economy is bent into circular form, fewer changes in conditions occur than under Cournot conjectures. Specifically, the solutions of $n = 1$ and 2 are no long determinate. For both linear and circular market areas with non-uniform distributions under game-theoretic conjectures, the existence or nonexistence of a steady state depends upon the form of the demand distribution. It is possible, for example, for each firm to be a local equilibrium but not a global, so that the firm may better its position by a quantum leap. On the other hand, some forms of distribution will yield a global steady state.

One can conclude that there are many interesting questions that remain to be investigated about the above games and its equilibrium. For a two-person game, a general question is: can we find the class of networks for which the game reaches an equilibrium for certain starting choices of the first player? This problem seems difficult. However, Hakimi (1990) pointed out that for a tree network, if the second player can respond immediately, the best choice for the first player is a centroid (center on a tree) of the network. Of course the second player will subsequently pick a medianoid (median on a tree). No further move by either player can lead to an advantage. We have given examples of this observation in the "Economic Methods" chapter in Chan (2000). Suppose the game is played on a network with discretionary demands, or when the game is played on a plane. Proving the existence of an equilibrium state, or its nonexistence, is a fertile ground for further study.

As in all realistic oligopoly analysis, the strength of game theory ironically coincides with its disappointing results. It succeeds in highlighting the richness of realistic processes through the methodology's indeterminateness. It serves to reinforce the conclusion of our earlier presentation: spatial oligopoly presents dynamic adjustment potentialities that in the general case are essentially indeterminate in outcome. Hence, such problems must be accepted as of their own kind. Work is most profitably confined to specific cases via simulation, and the theoretical ambitions of the analyst should be constrained to the limited applicability to the solutions.[22]

---

[21]For a classic definition of center, the point that minimizes the maximum distance away, see the "Economic Methods" chapter of Chan (2000) and the "Facility Location" chapter in the current volume.

[22]See the chapter on "Generation, Competition and Distribution," particularly under the "Public-facility location" section.

7.2.4 *Chaotic evolution in spatial competition.*

To bring the concepts home, we conclude this section on competition with the case of two competing regions (Nijkamp and Reggiani 1992). The following is a slightly varied version of the predator-prey model.[23]

$$\dot{x} = x(\alpha - bx)$$
$$\dot{y} = y(d - ex - \zeta y). \qquad (7\text{-}17)$$

Its discrete counterpart is the following difference-equation set:

$$x_{t+1} = L''x_t(1 - x_t)$$
$$y_{t+1} = y_t(m' - ex_t - \zeta y_t) \qquad (7\text{-}18)$$

where $L'' = \alpha + 1$ and $m' = d + 1$. To interpret this equation, it is best to re-write it in terms of the multi-region model $\dot{x}_i = \delta_i x_i (\bar{P}_i - x_i - \sum_j h_{ij} x_j) - \gamma_i x_i$ where x_i is the production at location i, δ_i is the expansion rate of i, γ_i is the contraction rate of i, $\bar{P}_i$ is the production capacity, and h_{ij} is the competition rate between i and j. Here, a positive h_{ij} reflects competition between different types of production, while a negative h_{ij} stands for the stimulation in region i by the growth of other regions. Thus this model may describe a wide variety of episodes, such as competition for the same resource, competition for the same market, or synergetic reinforcement. Specialization of this model in two dimensions (with $x_1 = x$ and $x_2 = y$), leads after some mathematical derivations to the two-region model

$$\dot{x} = x(\alpha - bx - cy)$$
$$\dot{y} = y(d - ex - \zeta y). \qquad (7\text{-}19)$$

If we assume that y does not drain the resources used by x, the competition coefficient c is equal to 0 and Equation (7-17) results.

The equilibrium analysis related to system (7-18) then shows that two fixed points exist: a trivial one (0, 0) and a non-trivial one $[(L''-1)/L'', (m'L'' - eL'' + e - L'')/L''\zeta]$. The stability analysis for the first point shows unstable behavior for $L'' > 1$ and $m' > 1$. Examine the stability of the second point, we get the critical value

$$m^* = (eL''^2 + 2L''^2 - 3L'' - 3eL'' + 2e)/(L''^2 - 2L'') \qquad (7\text{-}20)$$

at which a Hopf bifurcation (i.e., a bifurcation of a fixed point into a closed orbit) occurs. This implies that for $m' > m^*$ the second fixed-point becomes unstable, with the possibility of oscillations. This result shows that if one equation in the general system

$$x_{t+1} = x_t p(x_t, y_t)$$
$$y_{t+1} = y_t q(x_t, y_t) \qquad (7\text{-}21)$$

---

[23]As mentioned, the predator-prey ecological model of the populations of two species was introduced in the "Activity Allocation and Derivation" chapter under the "Theoretical thread" section.

is reduced to an equation of a May type (leading to chaos), we get oscillating behavior based on Hopf bifurcations, and therefore unpredictable movements. This result is indeed interesting since it shows that a 'chaotic' evolution can introduce oscillations in a system that in itself is non-oscillatory as the one defined in Equation **(7-21)**.

At an e-value of 0.8, Equation **(7-20)** shows that when the (m^*, L'') pair assumes the values of (3.45, 3.1) or (3.20, 3.6), irregular behavior results. At the values of (3.12, 3.9) and (3.10, 3.99), chaotic behavior ensures in the May equation. Further, we like to point out the relevance of the competition-coefficient e, since it is directly proportional to the critical value m^*. In particular, by decreasing the competition effect captured by the parameter e, the instability of system **(7-18)** occurs at a lower value of m^*. For example, at $L'' = 3.9$, these matching pairs of e and m^* values are observed:

e	0.80	0.69	0.50
m^*	3.12	3.04	2.90

Supplemented by simulation analysis, Nijkamp and Reggiani (1992) emphasized the relevance of a chaotic pattern, and also the growth rates of competing regions. The chaotic behavior in a region emerges only in certain limited range of its growth rate. It influences (in terms of irregularity) the competing region only if the related intrinsic growth-rate (and the interregional-competition coefficient) exceeds a critical value at which a Hopf bifurcation occurs. We will revisit this point in terms of the derivation and allocation components of a Lowry-based model.

7.3 Spatial dynamics: discontinuities

Consider the decision-making of a seller of goods or services who find it most advantageous to locate at the point of maximum population concentration. This is illustrated by the radial highway that links suburban commuters to the central city in **Figure 7-5** (Andersson and Kuenne 1986). At time $t = 0$, suppose that the distribution of population, $N = f(x; I', T)$, is as depicted in **Figure 7-6**, where I' is the aggregate income, T' is an exogenous variable such as institutional factors, technological factors etc., and that the firm's location x_0 is optimal. We now assume a time trajectory for control points (I'_t, T_t) and reproduce five population distributions on the assumed time path in the Figure. Finally, we assume that data on population distribution at any time t is fragmentary, so that the firm must relocate based on local information. Its instantaneous adjustment is assumed to follow the gradient from $\dot{x} = \partial x/\partial t = \partial N/\partial x$. In other words, the firm moves to the direction of increasing population-gradient.

7.3.1 Catastrophe theory. In $t = 1$, income I' rises while other factors T remains constant, dispersing the population and leading in to a local maximum near B. As time progresses the firm moves to x_1 continuously. By $t = 2$, the rising I' continues to move commuters farther from the central city, and the global

Figure 7-5 A linear transportation route (Andersson and Kuenne 1986)

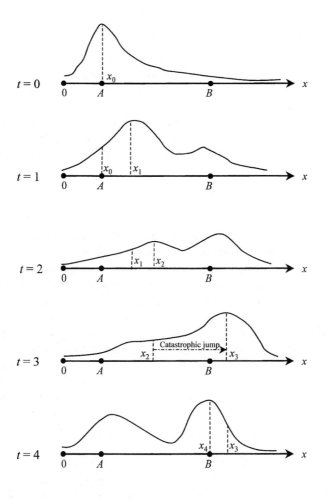

Figure 7-6 Population distributions at $t=0,1,2,3,4$ (Andersson and Kuenne 1986)

maximum moves to the outermost concentration. However, following its myopic reaction pattern, the firm moves continuously to x_2, the location of the local maximum. In $t = 3$, however, the local maximum and the minimum of $t = 2$ have converged as I' continues to rise, leaving the firm at an unstable equilibrium. Following its gradient rule, the firm now 'jumps' discontinuously to x_3. At this point a sharp or 'catastrophic' break in the smooth adjustments of the firm occurs. In $t = 4$, the firm resumes its continuous adjustments—stabilizing at x_4. The process is akin to wave propagation on a beach.

Catastrophe theory is designed to examine the *qualitative* characteristics of sudden, discontinuous behavioral changes. It is also used to generalize the nature of functions that move abruptly from one smooth path of adjustment to another fundamentally or 'catastrophically' different smooth path. These are in response to control trajectories. Two characteristics of the theory must be stressed: the theory is not designed to predict the path of the endogenous variable x_t, but merely to describe the nature of such paths for specified exogenous variable (control point) sequences; and the insights it yields into these paths are qualitative and not quantitative. Catastrophe theory answers the question: in what manners can regimes of smooth adjustment in state variable evolve into regimes of abrupt change, and what governs those rules of evolution?

7.3.2 A numerical example. To further illustrate these points with our example, consider the manner in which the firm's policy relates to the control variables. For any given control point (I',T) in the control space, the firm chooses its location from among those points where $\partial N/\partial x = 0$. Therefore, we are interested in the function that relates these potential equilibrium-states to the control variables. Moreover, because we are interested in qualitative results only, we may deal with any function that is *qualitatively equivalent* to the specific function of the problem. Catastrophe theorists have shown that the types of catastrophic changes that can occur are decided by the number of control variables in a problem. Since these types of catastrophes can be obtained qualitatively by simple polynomials that are qualitative equivalents of the original behavioral functions, we may deal with these simple *canonical models*.

For example, the canonical model for population distribution $N(x;I',T)$ as described in our hypothesis about location behavior is $0.25x^4 - 0.50\,Tx^2 - I'x = 0$, whose equilibrium surface $\partial N/\partial x$ is

$$x^3 - Tx - I' = 0. \tag{7-22}$$

This surface is graphed in **Figure 7-7**, but is lifted above the indicated origin at the cusp on the control space for convenience. The surface consists of two single-sheeted sections linked by a double-sheeted (shaded) portion. The fold curve is the boundary of the single and double-sheeted portions, and is the locus of points tangent to vertical lines, that is, where the derivative of the surface **(7-22)** is zero, or

$$3x^2 - T = 0, \quad \text{or} \quad (1/3\,T)^{1/2} = x. \tag{7-23}$$

Its projection onto the control space yields a bifurcation set, B', with a cusp, from which the specific form of the catastrophe takes its name. Thus from Equations **(7-23)** and **(7-22)**, elimination of x yields the bifurcation set $27I'^2 = 4T^3$ when the origin is defined at the cusp. All control points within B' possess multiple maxima (as in the $t=1$ and $t=4$ cases of **Figure 7-6**.) The folded sheet contains the minimum points and the firm therefore will never be on it.

Suppose the trajectory of control points is that shown by the dot-dash curve on the control set. The firm's location remains on the lower sheet (closer to A on the $t=1$ graph in **Figure 7-6**) even as the control point enters B' on **Figure 7-7**. On the surface, it travels under the folded sheet until it reaches the fold curve at J' in the bifurcation set. At that critical point it jumps discontinuously to the upper sheet at J''. Hence, we can explain qualitatively how a continuous, dynamic model can yield sudden large jumps in the location of the firm. Moreover, catastrophe theory assures us that the only manner in which a two-peaked population distribution can evolve from a one-peaked distribution is via the cusp catastrophe graphed in **Figure 7-7**, and the catastrophe requires two control variables. In similar fashion, a three-peaked distribution can evolve from a one-peaked distribution only by a different type of catastrophe, the *butterfly* that requires four control variables.[24] It is in cataloging the qualitative forms of continuous surfaces that give rise to discontinuous jumps in endogenous variables, and in the specification of the number of control variables needed to produce them that catastrophe theory yields insights into dynamic behavior.

7.3.3 Catastrophe in spatial-interaction models.

In the "Theoretical thread" section of "Activity Allocation and Derivation" chapter, we have developed the relationship between revenue from an activity $(\sum_i V_{ij})$ and cost to provide that activity $(c_j W_j)$ at zone j. Setting revenue to cost at the break-even point, we have $\sum_i V_{ij} = c_j W_j$ for all zones j. Correspondingly, this equilibrium condition arises from the differential equation

$$\frac{dW_j}{dt} = \dot{W}_j = \varphi\left(\sum_i V_{ij} - c_j W_j\right) \tag{7-24}$$

where φ is the response rate of the system. This equation suggests that W_j will increase if revenue exceeds cost and decrease if revenue falls short of cost. Taking stochasticity into account, Equation **(7-24)** takes on the form:

$$dW_j = \left[\varphi\left(\sum_i V_{ij} - c_j W_j\right)\right]dt + W_j \theta_j \, dA_j = g_i'' dt + W_j \theta_j \, dA_j \tag{7-25}$$

---

[24] The butterfly catastrophe will be discussed in the chapter on "Spatial Equilibrium and Disequilibrium." It is also catalogued in a table of elementary catastropes in the "Control, Dynamics and System Stability" appendix to Chan (2000).

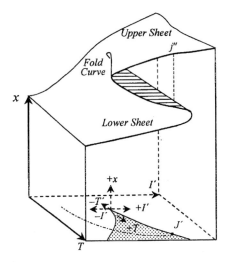

Figure 7-7 Illustrating the cusp catastrophe (Andersson and Kuenne 1986)

where g''_i represents the spatial 'drift' of activities and where the last term is the stochastic perturbation-force (proportional to the number of shops). Thus θ_j is the diffusion component and dA_j is the incremental change (white noise) in a stochastic-process A_j satisfying a Brownian process.[25]

Let us formulate now the following stochastic optimal-control model (Nijkamp and Reggiani 1992):

$$\max \; E\int_0^T \left[-\frac{1}{\beta}\sum_i \sum_j V_{ij}(\ln V_{ij}-1)+\sum_i \sum_j V_{ij}\left(\frac{a'}{\beta}\ln W_j - C_{ij}\right) \right] dt. \qquad (7\text{-}26)$$

This objective function maximizes the expectation of the weighted sum of total entropy and net benefits of transacting activities between i to j. The optimization is subject to the singly-constrained origin-trip-ends $\sum_j V_{ij}= V_i$ and Equation (7-25). It can be shown that the solution has the familiar form $V_{ij}= MV_i W_j^a \exp(\beta\varphi\Phi_j - \beta C_{ij})$ where Φ_j is the stochastic costate (dual) variable. Thus, owing to the term Φ_j, V_{ij} is a stochastic expression, incorporating a social cost-benefit measure; the term $\beta\varphi\Phi_j$ reflects the (stochastic) shadow price of employment growth in j, while the term βC_{ij} reflects the distance-friction costs.

It is interesting to observe that if we impose the optimal-path boundary-condition $\Phi_j(t_{\max}) = 0$, we find for $t = t_{\max}$ the very familiar solution $V_{ij}=V_i\left\{\left[W_j^{a'}\exp(-\beta C_{ij})\right]\middle/\left[\sum_k W_k^{a'}\exp(-\beta C_{ik})\right]\right\}$ Here, when the deterministic part of Equation (7-

---

[25]One way to think of this is through "linear regression" as explained in the "Statistical Tools" appendix to Chan (2000). The dependent variable (dW_j in this case) can be explained by the independent variables ($g_i dt$), as well as the residuals ($W_j\theta dA_j$). A better analogy is the autoregressive moving-average models covered in the forthcoming "Spatial Time Series" chapter.

(a) Inflow and equilibria

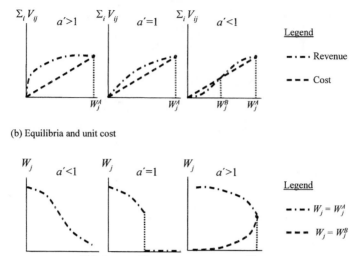

(b) Equilibria and unit cost

Figure 7-8 Stable and unstable points (Nijkamp and Reggiani 1992)

$$W_j = W_j^o \exp(-\theta_j^2 t/2 + \theta_j A_j) \qquad (7\text{-}27)$$

25) is in equilibrium, we get $\Sigma_i V_{ij} = c_j W_j$ and consequently $dW_j = W_j \theta_j dA_j$. The solution of this differential equation is which represents the optimal trajectory of the state-variable W_j under deterministic equilibrium. In the plane $(\Sigma_i V_{ij}, W_j)$, we have clearly the possibility of both stable and unstable points. In the plane (W_j, c_j) we have the possibility for catastrophe behavior, as can be shown in **Figure 7-8**. In this Figure, phase diagrams are constructed by eliminating t and A. **Figure 7-8(a)** illustrates the functional form of the inflow $\Sigma_i V_{ij}$ toward the work-places W_j for different values of a'. In **Figure 7-8(b)**, we consider the effect of varying c_j, obtaining a set of equilibrium points linked to the fold catastrophe.

Clearly, there are values of c_j (c_j') beyond which W_j will jump from W_j^A to zero (for $a' \geq 1$). This means that for the case $a' \geq 1$ there is a critical value c_j' in the accessibility term beyond which the number of work-places jumps to zero. This could be explained by acute congestion that chokes commerce to a standstill. By contrast, from Equation **(7-27)** we can see that the diffusion components θ_j do not influence this dramatic change, since the only way to obtain the condition $W_j = 0$ is a sudden decrease to zero of the initial condition W_j^o. The conclusion is that at time $t = t_{max}$, and under the condition that the deterministic part of Equation **(7-25)** is in equilibrium, certain smooth parameter changes may lead to discrete changes in the state variable W_j. It is observed therefore that the inclusion of stochastic perturbation in this singly-constrained dynamic model affects the optimal trajectories, so that we get again stochastic movement. However, if we have conditions for the appearance of catastrophe behavior, the stochastic disturbances do not necessarily account for these drastic changes. This interesting result shows the characteristic of structural stability for dynamic models that *their qualitative properties stay intact when they are subject to small perturbations.*

7.4   Discrete networks over time

The location-allocation or multi-source Weber problem can be dynamicized.[26] Necessary conditions for its solution can be obtained using a modification of conventional *optimal control theory*[27], which we will explain below. The static problem is to locate p facility *sources* from where goods and services are supplied among n fixed nodes. Each of these nodes can be a *sink*, or consumption point with a given demand. We also wish to allocate sinks to sources in such a manner as to minimize total transportation costs. The dynamic problem requires treatment of product flows as rates over time $V_{ij}(t)$, instead of a fixed amount (Tapiero, as cited by Andersson and Kuenne [1986]).

7.4.1 Model formulation. In this model, there are certain given quantities, the exogenous variables[28], which are the: planning period from $t=0$ to t_{max}, $(0, t_{max})$; nodes representing all possible sink locations in Euclidean two-dimensional space (x^j, y^j), $j = 1, 2, \dots, n$; demands at these sinks z_j; source capacity limitations $\bar{P}_i$, $i = 1, 2, \dots, p$; transportation-cost/unit-distance from source i to sink j, c_{ij}; unit storage-costs at source i, c_i; and unit storage-costs at sink j, c_j. There are certain quantities that are determined within the model, the endogenous variables, which are the: source locations, (x^i, y^i); shipment from source to sink, $V_{ij}(t)$; goods in storage at sources, $\dot{z}_i$; and goods in demand at sinks $\dot{z}_j$. Thus the locations of suppliers (sources) are theoretically determined as part of the model.

The functional (objective function) to be minimized, when transportation costs are proportional to units shipped and Euclidean distance, is

$$\min \int_0^{t_{max}} \sum_{i,j} \left\{ V_{ij}(t) c_{ij} \left[(x^j - x^i)^2 + (y^j - y^i)^2 \right]^{0.5} + c_i \dot{z}_i(t) + c_j \dot{z}_j(t) \right\} dt \qquad \textbf{(7-28)}$$

---

[26]For a discussion of the location-allocation problem and Weber problem, see the "Economic Methods" and "Prescriptive Tools" chapters in Chan (2000) and the "Facility Location" chapter in the current volume.

[27]The theory is reviewed in the book appendix "Control, Dynamics and System Stability" to Chan (2000). An example of optimal control theory is also illustrated in the dynamic programming discussion in the "Markovian Process" appendix.

[28]Exogenous variables are the 'givens', while endogenous variables are determined within the process itself. Both terms are defined in the "Descriptive Tools" chapter in Chan (2000).

subject to the demands at j

$$\int_0^{T_{max}} \dot{z}_j(t)\, dt = z_j, \qquad j=1,2,\dots,n; \tag{7-29}$$

the availability of supplies at i

$$\int_0^{t_{max}} \dot{z}_i(t)\, dt = \overline{P}_i \qquad i=1,2,\dots,p; \tag{7-30}$$

the depletion rate at sources including the initial and final inventory

$$\dot{z}_i(t) = -\sum_j V_{ij}(t) \qquad \dot{z}_i(0)=\overline{P}_i \qquad \dot{z}_i(t_{max})=0 \qquad i=1,2,\dots,p; \tag{7-31}$$

the rate of consumption at demand points commensurate with the total demand z_j

$$\dot{z}_j(t) = \sum_i V_{ij}(t) \qquad \dot{z}_j(0)=0 \qquad \dot{z}_j(t_{max})=z_j \qquad j=1,2,\dots,n; \tag{7-32}$$

and, for the time being, assuming permanent supply-facilities (rather than relocatable facilities)

$$\dot{x}^i(t)=0 \qquad i=1,2,\dots,p \tag{7-33}$$

$$\dot{y}^i(t)=0 \qquad i=1,2,\dots,p. \tag{7-34}$$

In this formulation, V_{ij}, x^i, and y^i are treated as control variables, or the dynamic version of allocation-and-location decision-variables, and $\dot{z}_i$ and $\dot{z}_j$ are state variables. The choice of time paths for the control variables dictates, via the equation of motion, movement of the state variables. In other words, the depletion of inventory at the supply sources and the consumption rate at the demand points are found by the location of the supply sources and the shipment over time from sources to sinks. The objective functional, total costs, is minimized by optimal choices of the time paths for the control and state variables. The constraints **(7-33)** and **(7-34)** are somewhat restrictive in that the sources are to be located, once and for all, at the first instant of time and are not to be changed over time. Hence their paths cannot be altered, or they cannot be relocated, to control the state variables. We will come back to this assumption later. The readers may notice this model is a time-dependent counterpart of the general network location-production-allocation problem discussed in the "Facility Location" chapter—for which a numerical example has been worked out.

7.4.2 Model solution. Let us solve the above dynamic location-allocation model and discern the properties of the solution. Given X and U are vectors of state and control variables respectively, the objective functional **(7-28)** can be written as $Z^* = \min_U \int_0^{t_{max}} Z(X,U,t)\, dt$ subject to constraints **(7-29)** through **(7-34)**. The model again can be put into the following form: $\dot{X} = F(X,U,t)$ where the boundary or end-point conditions $\mathbf{X}(0)$, and $\mathbf{X}(t_{max})$ are given. Typically, we form the augmented functional $Z^* = \int_0^{t_{max}} \left[Z(\mathbf{X},\mathbf{U},t) + \lambda_1(\dot{\mathbf{X}} - F(\mathbf{X},\mathbf{U},t)) \right] dt = H dt$ where λ_1 is a vector of co-state variables, or Lagrange multipliers. First-order conditions for a minimum of Z derived from the Hamiltonian H are

$$\dot{\lambda}_1 = -H_\mathbf{X}$$
$$\dot{\mathbf{X}} = H_{\lambda_1} \qquad \mathbf{X}(0), \mathbf{X}(t_{max}) \text{ } fixed \qquad (7\text{-}35)$$
$$H_\mathbf{U} = \mathbf{0}$$

where H_ξ is the derivative of the Hamiltonian with respect to ξ. When a subset of the control variables $\mathbf{u}$ is fixed in time, they are converted to state variables with constraints $\dot{u} = 0$, which in turn are introduced into Z^* and H with co-state variables λ_2. The resulting first-order conditions then become

$$\dot{\mathbf{X}} = H_{\lambda_1} \qquad \mathbf{X}(0), \mathbf{X}(t_{max}) \text{ } fixed$$
$$\dot{\lambda}_1 = -H_\mathbf{X}$$
$$\dot{\lambda}_2 = -H_\mathbf{u} \qquad \lambda_2(0) = \lambda_2(t_{max}) = \mathbf{0} \qquad (7\text{-}36)$$
$$H_{\bar{\mathbf{U}}} = \mathbf{0}$$

where $\bar{\mathbf{U}}$ is the subset of control variable from which $\dot{\bar{\mathbf{U}}} \neq \mathbf{0}$.

When these general conditions are applied to the model as represented by Equations (7-28) through (7-34), we have $4p + 2n + pn$ simultaneous differential-equations to be solved for the source coordinates and shipment flows. Although in general the equations must be solved by simulation, an analytic solution for quadratic transport-cost- functions has been obtained, similar to the network location-production-allocation problems mentioned above. Because of the restrictive non-state-specific assumption expressed in Equations (7-33) and (7-34), these results are not as interesting for their analytical content as for their broader methodological contribution. We have succeeded in confronting a peculiarly spatial-analytic problem using control theory. We fix the locations through time, when these locations are specified as control variables, and deriving the necessary conditions for solving it. We hypothesize that the derivation will permit the solution of other more involved dynamic optimization problems in space. Thus this exercise illustrates the initial step of a potentially promising direction for further research.

7.5 Continuous networks over space

Consider a simple economy consisting of region R in Euclidean 2-space with a boundary that is a closed curve. Consistent with Economic-base theory, all fuel is imported from abroad over the boundary and is paid for from exports of the single produced good. The price of fuel, p_f, is given parametrically within R, as is the price of the good, p_g. Both commodities must be transported within R, and we may define iso-price contours for fuel by adding transport costs to p_f, and of goods by subtracting them from p_g (Puu as reported in Andersson and Kuenne 1986).

7.5.1 Assumptions. The good is producible at every location in R under identical Cobb-Douglas technology with labor, fuel, capital and land, and is consumed at each location: production $= K(\text{labor})^a(\text{fuel})^b(\text{capital})^c(\text{land})^d$, where K, a, b, c and d are calibration constants. The utility obtained from consumption is additive, and the objective function is obtained from the surface integral of the local utilities over R. Land is immobile and given at each location, but capital and labor are free to move costlessly among locations. At each point transport services needed to move units of goods and fuel are produced from the same four factors by a Leontief input-output technology[29] whose production coefficients may vary among locations. The flows of goods and fuel at a given point (x,y) will be characterized by (a) direction, (b) magnitude, and (c) net addition and subtraction at location (x,y)—namely the *source* and *sink* respectively. The trajectories through space traced out by goods and fuel flows will connect points in patterns derived from constrained optimization, when such flow lines are completely un-channeled. This contrasts with channeled flows along the links in a discrete networks. What, then, will be the defining characteristics of such flow lines, and what can be said *qualitatively* of the stability of such spatial structures?

The solution is obtained by maximizing the integral of the utility functional subject to the constraints that:

a.   The Cobb-Douglas production-technology rules at each point;
b.   The integrals of capital and labor demands at each point equal supplies of these resources;
c.   Demand equals supply of land at each point;
d.   Net additions or subtractions to or from product flow at each source or sink equals (positive or negative) excess supply at that point;
e.   Subtraction of fuel from the flow at each point equals demand for it at that point; and
f.   The value of product exports equals the value of fuel imports for the region.

Calculus of variations[30] yields first-order Euler-Lagrange conditions that require (a) the marginal utility of a dollar's worth of product must be equal at each location; (b) the marginal value products of land, capital, labor, and fuel must equal their prices at each location; and (c) most importantly, the flows of goods and fuel at each point must be orthogonal to the relevant p_g and p_f iso-price contours at that point, representing the direction of steepest ascent. The two relationships containing (c) are differential equations with respect to *space* whose solutions yield

---

[29]See the "Economic Methods" chapter of Chan (2000) for an explanation of input-output analysis.

[30]Calculus of variation is a special case of the control problem and is explained in the "Control, Dynamics and System Stability" appendix to Chan (2000).

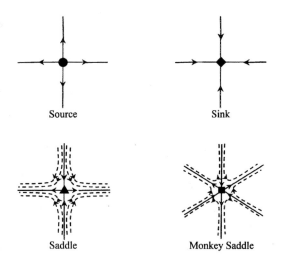

Figure 7-9 Types of singular points (Andersson and Kuenne 1986)

the patterns of flow lines for goods and fuel in R. Most of the points on such flow lines will be 'regular,' i.e., with only one flow path entering and leaving. However, 'singular' points on more than one flow path can arise. These will be (a) source nodes from which all flow lines in a neighborhood originate; (b) sink nodes into which all flow lines in the neighborhood terminate; (c) simple saddles, in which two pairs of in-going and outgoing flow-lines serve as *separatrices* dividing flow lines into four sectors with hyperbolic flow-lines in each sector; and (d) various unstable forms, most particularly monkey saddles into which three pairs of flow lines are incident dividing the neighborhood into six sectors with hyperbolic flow-lines. Each of these flow patterns is illustrated in **Figure 7-9**. Further comments on these "fixed points" can be found in the "Control, Dynamics and System Stability" appendix to Chan (2000).

7.5.2 Results. Analogous to temporal-stability analysis, spatial stability implies a structure of flow lines and points whose form will remain invariant to perturbations. Unstable patterns are assumed to disappear quickly in the face of continuous perturbations. Stable flows, on the other hand, are characterized by three qualities: (a) unstable singular-points like monkey saddles cannot exist; (b) all points except a finite number of isolated stable singular- points will be regular; and (c) no flow lines can join saddle points. It has been shown that the most basic spatial structure consistent with these stability conditions is a quadratic structure of the type illustrated in **Figure 7-10**. Only those flow lines incident to saddles are included to avoid clutter. Note that each sink and source is surrounded by four saddles, and each saddle by two sinks and two sources. *In triangular or hexagonal tessellations of space where flow lines are the sides of triangles or diagonals of hexagons, it is conjectured that their non-conformance to the stable quadratic pattern as shown may make them nonexistent in real economies.*

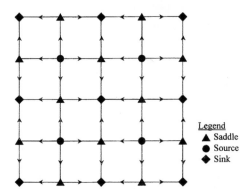

Figure 7-10 Basic quadratic stable-spatial-structure (Andersson and Kuenne 1986)

Catastrophe theory was applied to study the transition from unstable to stable configurations as control variables are changed. With three control variables for a two-dimensional state space, two forms of the bifurcations set are possible. These are the elliptic umbilic[31], with canonical equation $x^3 - 3xy^2 + a(x^2 + y^2) + bx + cy$, and the hyperbolic umbilic, with canonical equation $x^3 + y^3 + axy + bx + cy$. Parallel to temporal dynamics, the above discussion outlines the essential features of alternative optimal-spatial-structures under idealized conditions, and their survivability under parametric displacements. Again, the value lies in gaining insights through the process, rather than assessing its operational status. For this reason, we have dispensed with the detailed mathematical-model formulations here. Suffice to say that the structural equations are of the type discussed in the "System stability" section of the "Control, Dynamics and System Stability" book appendix to Chan (2000). The curious reader can find a similar example in the "Multiregion neoclassical growth model" subsection in the "Spatial Equilibrium and Disequilibrium" chapter.

7.6   Aggregation vs disaggregation

Thus far, we have been concentrating on chaos, catastrophe, bifurcation, and temporal-spatial stability in general. We now turn to the definition of a geographic decision-making-unit. The question is: what should be the basic building block in spatial-temporal analysis: a region, a zone, a household or some other units (Brown 1992, Manheim 1984)? We have already drawn a distinction between spatial vs. non-spatial analysis earlier as examples of disaggregate vs. aggregate modelling respectively. Here, finer stratifications are introduced, including zonal vs. area-wide

---

[31]Both the elliptic and hyperbolic umblics are catalogued among the table of elementary catastrophes in the "Control, Dynamics and System Stability" appendix to Chan (2000).

	Dwelling unit 1	Dwelling unit 2
Cost	4	5
Time	2.5 hours	2 hours
Amenities	1.0	0.3

Table 7-1 Characteristics of two competing dwelling–units (Manheim 1984)

characterization of such parameters as population-serving ratio and labor-force participation rate—the economic multipliers in a Lowry-based model.

7.6.1 Utility. As a start, we will concentrate on the spatial allocation part of the process, rather than the activity-derivation part. Although widely employed in practice, the traditional gravity model and its theoretical underpinning have been subject to substantial scholarly stricture. Among these are its erroneous assumptions concerning the nature of consumer demand. Much of it has to do with the behavioral basis of such demand, which in turn is determined by the assumption about definition of the decision-maker—should it be such disaggregate units as a household or a zone, or an aggregate unit such as a region? The difference lies in the homogeneity of growth and the spatial interaction among subregions.

Consider a household has a choice between two dwelling units. We assume that the only attributes that affect its choice are monthly housing-cost (including mortgage, maintenance and utilities) c, travel-time to work round-trip τ, and other amenities x, which incorporate such factors as site characteristics, and attractiveness of the neighborhood. The present characteristics of the two dwelling units are shown in **Table 7-1**. Let us use a linear additive-utility-function.[32] We predict the choice made by this consumer by estimating the relative weights it—as an individual household—places on cost, time, and amenities (w_c, w_τ, and w_x respectively). With these weights the three attributes can be collapsed into a single measure of utility:

$$v = w_c c + w_\tau \tau + w_x x. \tag{7-37}$$

Note that all the weights take negative values, since an increase in any of the attributes reduces the utility of the dwelling unit. Using our model, we then assume that the consumer will select the dwelling unit that has the highest value of this utility, that is, the least value of a negative number. We assume that the weights are -2 thousand dollars per month for each hour of commuting time (w_τ) and -4

---

[32]For a discussion of utility functions, see the "Multi-criteria Decision-making" chapter in Chan (2000).

House-hold	Relative			Utilities		Choice (Dwelling unit type)
	w_c	w_τ	w_x	Dwelling unit 1	Dwelling unit 2	
A	-1	-2	-4	-13	-10.2	2
B	-1	-3	-8	-19.5	-13.4	2
C	-4	-2	-3	-24	-24.9	1
D	-8	-2	-3	-40	-44.9	1

Table 7-2 Example residential-choice base on a simple utility model

thousand dollars per month for each unit of amenities (w_x), relative to a weight of −1 (thousand) for the housing-cost w_c.

The utilities, in thousands of dollars, for the two competing dwelling-units are as follows:

$$v_1 = (-1)(4) + (-2)(2.5) + (-4)(1.0) = -13$$
$$v_2 = (-1)(5) + (-2)(2.0) + (-4)(0.3) = -10.20.$$

Here, we can predict that the consumer will select dwelling unit 2 as the preference because it has a 'higher' utility. This model is very useful for understanding the effects for a potential or actual change in the factors governing residential choice for a household, a fundamental unit in locational decisions. Any such change can be expressed as a change in one or more attributes for one or more of the alternatives. Through the weights they place on the attributes, this would cause a change in the utility this consumer places on the affected alternatives.

7.6.2 Different consumers. A different consumer faced with exactly the same choice would use different weights on the attributes to evaluate dwelling units. **Table 7-2** shows the behavior of consumers with different preferences. The resulting choice is shown for each consumer. Some choose dwelling-unit type 1 and others choose dwelling-unit type 2 because of the difference in preferences, reflected in the differences in weights and in turn the utility functions. Consider household A in **Table 7-2**. While under present conditions this consumer would choose dwelling-unit 2, it is possible that changes in attributes could alter this. For example, if the cost of dwelling-unit 2 is increased by 4 thousand dollars, the utility of dwelling-unit 2 becomes −14.2 thousand, while the utility of dwelling-unit 1 stays at −13 thousand. The household now shifts to dwelling-unit 2. The effects of such variations are shown graphically in **Figure 7-11**. Here residential choice is shown as a function of housing cost. For cost below 7.8 thousand dollars, the household chooses dwelling unit 2; above 7.8 thousand, it chooses dwelling unit 1. The *change* in housing cost is shown in the same figure.

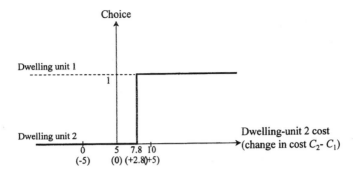

Figure 7-11 Housing cost as a determinant of residential choice (Manheim 1984)

So far we have looked at each of our four households as an individual consumer. We now examine the behavior of the group as a whole. Referring to **Figure 7-12(a)**, each household represents 25 percent of the group of four households. At the prevailing dwelling-unit 2 cost of 5-thousand dollars, 50 percent of the group (two households) choose dwelling-unit 1 and 50 percent choose dwelling-unit 2. If we increase the dwelling-unit 2 cost, we change the percentage of dwelling-units chosen: household A will shift to dwelling-unit 1 at a dwelling-unit-2 cost of 7.8 thousand, thus increasing the dwelling-unit-1 share of the market to 75 percent. If we increase the dwelling-unit-2 cost further to 11.1 thousand, household B will also shift to dwelling unit 1, resulting in a 100 percent dwelling-unit 1 chosen. Conversely, if we decrease dwelling-unit-2 cost to 4.78 thousand, we will attract household C to dwelling-unit 2, for a 25 percent dwelling-unit 1 choice; and a decrease to 4.39 thousand will yield a 0 percent dwelling-unit-1 choice. We can thus derive the aggregate behavior of the group from the behavior of each individual in the group. This relationship is shown in **Figure 7-12(b)**. Usually a group will be composed of many more than four consumers. In that case, the relationship between the housing attribute and the fraction of the group making a particular choice will be much smoother. This can be approximated by a continuous function, as shown in **Figure 7-12(c)**.

Of course, the residential choice of a particular group can be affected by changing one or more of the housing attributes, including the other amenities that go into residential choice (variable x in Equation **(7-37)**). To represent this we might show on the horizontal axis the differences in utilities, leading to a function like the one shown in **Figure 7-12(c)**. The intercept on the vertical axis, a_0, represents the percentage of dwelling-unit 1 chosen even if both dwelling-unit types had the same utilities (that is, zero difference in utilities). Notice this intercept may or may not be at the 50-percent mark. It depends on the socioeconomic condition or activity system under which the residential choice is made, often represented as a calibration constant w_1^2 in $\Delta_1^2 = (v_2 - v_1) + w_1^2$, where $\Delta_1^2 = (v_2 + w_2) - (v_1 + w_1)$ and $w_1^2 = w_2 - w_1$. In general, for a choice between alternative i and j

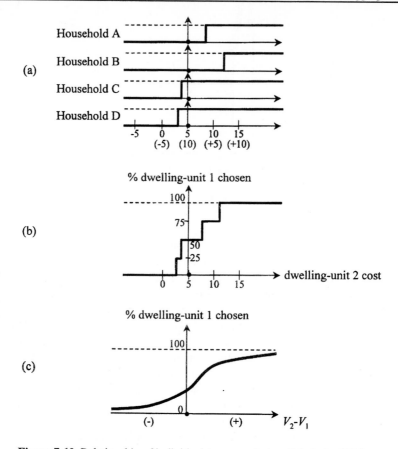

Figure 7-12 Relationship of individual to group choices (Manheim 1984)

$$\Delta_i^j = (v_j + w_j) - (v_i + w_i) = (v_j - v_i) + (w_j - w_i) = (v_j - v_i) + w_i^j. \tag{7-38}$$

Now for reference dwelling-unit 2 with activity constant w_2, this curve would represent the demand function

$$\Theta_1^2 = V_2 / (V_1 + V_2) = V[(v_2 - v_1), w_1^2] \tag{7-39}$$

where Θ_1^2 is the share or fraction choosing dwelling unit 2, V_2 and V_1 are the number of households choosing dwelling unit 2 and 1 respectively. For the general case, the fraction of unit-k chosen is

$$\Theta_k = V_k / \sum_l V_l = V(v_k, w.) \tag{7-40}$$

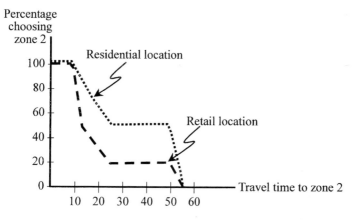

Figure 7-13 Demand functions for residential and retail location

for a given set of candidate houses {·} and their associated utilities and activity constants.

Assuming an activity generation-rate of f_i at work location i, the number of houses chosen in area j can be estimated by the 0–1 ranged utility functions $v_j = \exp(v_{ij}+w_i)$ and $v_k = \exp(v_{ik})$:

$$\Theta_{ij}\, f_i = \frac{V_{ij}}{\sum_k V_{ik}} f_i = \frac{\exp(v_{ij}+w_i)}{\sum_k \exp(v_{ik})}, \tag{7-41}$$

where $f_i{=}\exp(w_i)$. Likewise, for an activity-generation-rate a_j at residential-location j, the number of service establishments located at k to serve the population is determined by

$$\Theta'_{jk}\, a_j = \frac{V'_k}{\sum_l V'_{jl}} a_j = \frac{\exp(v_k+w_j)}{\sum_l \exp(v_l)}, \tag{7-42}$$

where V'_{jk} is the number of nonwork trips, and $a_j{=}\exp(w_j)$. The reader may recognize that Equations **(7-41)** and **(7-42)** include both "generation and distribution" as described in the chapter under a similar name. Activity rates f and a are examples of activity-specific exogenous constants w_2 in Equation **(7-39)** and w. in Equation **(7-40)**. The behavior of a group, considering both generation and distribution, can be derived by aggregating the individual decisions represented by Equations **(7-41)** and **(7-42)**. A group of workers finding residences will have a different demand function than a group of shoppers looking for a place to shop. Examples of the two demand functions are shown in **Figure 7-13** for a two-zone case. These represent generalization of the sketches in **Figure 7-12**, when the horizontal axis is now replaced by travel time, and the vertical axis by percentage locating in zone 2. This effectively reverse the zonal share, making the curve down-sloping rather than upward-sweeping.

7.6.3   *Calibration considerations*.   To operationalize the above demand functions, a disaggregate calibration procedure is needed. In this procedure, the weights for the utility function need to be estimated considering the individuality of each worker or household. Recall that in the "Descriptive Tools" chapter of Chan (2000), we made a distinction between aggregate vs. disaggregate calibration. For an aggregate model, a uniform set of weights are calibrated for the entire study area, say $\mathbf{w}$. For a disaggregate model, a zone-specific set of weights is calibrated, each of which is different, $\mathbf{w}_k$. It was also shown that aggregate vs. disaggregate models are based on distinctly different behavioral assumptions and care should be exercised in applying the correct model.

Irrespective of the exact calibration procedure—regression, maximum likelihood, and the like—model specification is the key. Equations **(7-41)** and **(7-42)** can be explicitly written as $V_{ij} = \left\{ \left[N_j C'(C_{ij}) \right] \middle/ \sum_l N_l C'(C_{il}) \right\} f_i$ and $V'_{jk} = \left\{ \left[E_k^R C^R(C_{jk}) \right] \middle/ \sum_l E_l^R C^R(C_{jl}) \right\} a_j$. Now the travel-propensity functions (or trip-distribution curves) $C'(\cdot)$ and $C^R(\cdot)$ can contain a zone-specific set of calibration constants or an area-wide set, depending on whether a disaggregate or aggregate model is wanted. Examples for zone-specific or disaggregate travel-propensity functions are: $C'(C_{ij}) = \exp(w_j C_{ij})$ and $C^R(C_{ij}) = \exp(w'_j C_{ij})$ where w_j and w'_j are negative weights to be calibrated for each zone. Likewise, examples of areawide or aggregate travel-propensity functions are: $C'(C_{ij}) = \exp(w C_{ij})$ and $C^R(C_{ij}) = \exp(w' C_{ij})$ where w is a negative weight to be calibrated for the entire study area.

Combining generation and distribution, we have a choice among several combined functions, depending on whether a zone-specific or an area wide activity-generation parameter is used: z_i or z, where z stands for both f or a. Using $C(\cdot)$ to stand for either $C'(\cdot)$ or $C^R(\cdot)$, we can have these example choices of demand functions:

$$V_j(C_{ij},w_j) = \Theta_{ij}(C_{ij},w_j) \qquad z\, V_j(C_{ij},w) = \Theta_{ij}(C_{ij},w)\, z$$
$$V_j(C_{ij},w_j) = \Theta_{ij}(C_{ij},w_j) \qquad z_i V_j(C_{ij},w) = \Theta_{ij}(C_{ij},w)\, z_i. \qquad \textbf{(7-43)}$$

The above examples encompass all possible combinations of zone-specific propensity functions and activity-generation rates. Later on in this chapter, we will illustrate the last two forms of the demand function in Equation **(7-43)**, namely an areawide set of calibration parameters for the propensity functions, with a distinction made between an areawide activity-generation rate and a zone-specific generation rate. This will round out previous discussions on a zone-specific set of weights w_j vis-a-vis an areawide set of weights w for the propensity functions.

Now we have two sets of parameters to calibrate, the weights for the propensity functions and the activity-rates. It will be shown that the propensity-function weights will be calibrated by Fibonacci technique while the activity rates by explicit equations. The calibration of individual activity rates is similar to the input-output model explained in the "Economic Methods" chapter (Chan 2000) and later in the

"Spatial Equilibrium" chapter. As an example of disaggregation, consider the calibration multipliers in economic-base theory. The highest level of aggregation calls for the ratio of region-wide nonbasic (or service) employment to population (population-serving ratio) at one point in time. The next level calls for the disaggregation of the ratio into several sub-regional ones. At its finest level, data sets are collected from two periods in time—rather than from a single period—to estimate the sub-regional ratios (Chan and Yi 1987). Each level of aggregation has profound effects upon the predictive accuracy of such models.

7.6.4 Calibration procedures. On the surface, the gravity model or the more general multinomial logit format can be converted to a log-linear form, toward which linear-regression calibration procedure can be applied. For example, we recall that the unconstrained gravity-model from the "Economic Methods" chapter (Chan 2000) and "Generation, Competition and Distribution" chapter looks like $V_{ij}=V_i\Theta_{ij}$, or $V_{ij}=KW_i^\alpha W_j^\beta C(C_{ij})$. These functions can be linearized and subsequently fitted by regression—at least in theory any way. However, ordinary least squares (OLS) presuppose continuous variables with a normal error structure, whereas data in the field have integer-valued origin-destination tables, and many entries will be small. It can be argued that a poisson error-structure would be more appropriate in this case (Upton and Fingleton 1989).

A second problem is the *map-pattern effect*. This problem occurs when larger areas of populations figure among the closer pairs of locations. Correspondingly, there will be very considerable migration back and forth between these locations, which overshadow other commutes. Under these conditions, it becomes very difficult to estimate the parameters such as α, β and those in $C(C_{ij})$. The reason is that the effect of activity level W and short inter-locality distance C_{ij} becomes confounded. This confounding results in linear collinearity of the regression equations and hence in a wide range of parameter estimates. Unfortunately, the problem is not solved by dis-aggregating large centers of activities into smaller subareas.

For this type of models, some practical and theoretical problems inherent in the maximum-likelihood estimation of their parameters are also raised. It was found that the observed frequencies in a contingency table of migration counts will not be independent of one another. This is clearly the case. Not only do people move in groups, but new houses are often created an estate at a time; old houses are cleared *en masse*, and similarly collective movements occur in other natural populations. It may be more practical to maximize the so-called quasi-likelihood to avoid making the full distributional assumptions required by maximum likelihood. This involves exploring the autocorrelation structure of the origin-destination matrix of residuals using a variant of the *joint-count statistics*.[33]

---

[33]The joint count statistic is a parallel statistic to the correlation coefficient r. It measures whether neighboring locations are more likely to assume similar or opposite values. We will discuss this statistic in more detail starting with the "Spatial Econometrics" chapter.

The practical difficulties associated with many origin-destination tables can readily be imagined if one visualizes the 1000 by 1000 flow-matrices typical in many urban transportation studies. A table like this would challenge any "off-the shelf" statistical software to the limit. The scarcity of available data and pressing schedules in the field have sometimes rendered statistical theory irrelevant and have fostered ingenuous numerical solution schemes. We have already witnessed some of these in the "Generation, Competition and Distribution" chapter. There the unconstrained model $V_{ij}=V_i\Theta_{ij}$ is decomposed into two stages—calibration of the allocation part Θ_{ij} and then generation-allocation combined-model $V_i\Theta_{ij}$. Also, both endogenous and exogenous calibrations take place in the Lowry model. We will further illustrate these points in the remaining sections of this chapter.

7.7   Applications to Garin-Lowry models

In spite of the many initiatives taken (Clarke and Wilson 1983), researchers are still putting together a general theory of bifurcation for urban spatial-models. As a senior relative of catastrophe theory, bifurcation models describe systems characterized by multiple equilibria, in which shifts from one equilibrium to another may involve discontinuous jumps (Wilson 1981). According to Wilson, there are three general types of discontinuities: (a) a sudden jump, (b) a reverse path to some point different from the original, and (c) a small difference in approach toward a critical point leads to a very different state of the system. We are particularly interested in the third type, dealing with critical parameter values at which a stable equilibrium point becomes unstable.

7.7.1   Basic theories.   As described in some detail in two or more previouschapters,[34] geometric series are used in the Lowry-Garin family of models to forecast the total employment vector, $\mathbf{E}$, and the population vector, $\mathbf{N}$:

$$\mathbf{E}=\mathbf{E}^B\left[\mathbf{I}+\mathbf{FA}'+(\mathbf{FA}')^2+\ldots+(\mathbf{FA}')^m\right], \qquad m\to\infty \qquad (7\text{-}44)$$

$$\mathbf{N}=\mathbf{E}^B\left[\mathbf{I}+\mathbf{FA}'+(\mathbf{FA}')^2+\ldots+(\mathbf{FA}')^m\right]\mathbf{F}, \qquad m\to\infty \qquad (7\text{-}45)$$

where $\mathbf{E}^B$ is the basic-employment vector for n' zones in the study area, $\mathbf{F}$ is an $n'\times n'$ matrix for the derivation and allocation of population, and $\mathbf{A}'$ is another $n'\times n'$ matrix for the derivation and allocation of local-service employment.

We wish to make some remarks about the convergence of the series. Specifically, under what conditions will the $\Theta_{ij}(N_j,C_{ij})$ se series converge, diverge, and bifurcate? Can these conditions be expressed in terms of the elements of $\mathbf{F}$ and $\mathbf{A}'$ in the series? Recall the matrices $\mathbf{F}$ and $\mathbf{A}'$ are given by

$$\mathbf{F}=\left[\Theta_{ij}(N_j,C_{ij})\right]\left[f_j\right]=\left[t_{ij}\right]\left[f_j\right], \qquad (7\text{-}46)$$

---

[34]For example, see chapter on "Generation, Distribution and Competition".

$$\mathbf{A'} = \left[\Theta_{ij}(E_j^R, C_{ij})\right][a_j] = [u_{ij}][a_j];\qquad(7\text{-}47)$$

where short-hand notations t_{ij} and u_{ij} are used for work and nonwork accessibility-functions instead of and $\Theta_{ij}(E_j^R, C_{ij})$ respectively. Quantitatively, they are defined as

$$N_j C(C_{ij}) \Big/ \sum_{k=1}^{n} N_k C(C_{ik})\qquad(7\text{-}48)$$

and

$$E_j^R C^R(C_{ij}) \Big/ \sum_{k=1}^{n} E_k^R C^R(C_{ik})\qquad(7\text{-}49)$$

respectively. In the above two equations, N_j and E_j^R are the population and service employment at zone j; C and C^R are work and nonwork trip-distribution functions respectively. f_j is the inverse of the labor-force participation rate, or the number of residents per employee; and a_j is the population-serving ratio, defined again as the number of service jobs generated from one resident. $[f_j]$ and $[a_j]$ are $n' \times n'$ matrices with diagonal elements f_j and a_j.

Furthermore, can we say something about the estimation of zonal parameters such as f_j and a_j, particularly the effect of aggregation on convergence? To answer this question, we recall, first of all, the two series in Equations **(7-44)** and **(7-45)** can be written as

$$\mathbf{E} = \mathbf{E}^B \left(\mathbf{I} - \mathbf{FA'}\right)^{-1}\qquad(7\text{-}50)$$

$$\mathbf{N} = \mathbf{E}^B \left(\mathbf{I} - \mathbf{FA'}\right)^{-1}\mathbf{F}\qquad(7\text{-}51)$$

if and only if the biggest row-sum of **FA'** is less than unity (Noble 1969, Webber 1984).

7.7.2 Uniform areawide multipliers. In the case of a uniform service-employment and population multiplier f and a for all zones areawide, the row sums of **F** and **A** are simply: $F_i = \sum_{j=1}^{n'} t_{ij} f_j = f \sum_{j=1}^{n'} t_{ij}$ and $A_i' = \sum_{j=1}^{n'} u_{ij} a_j = a \sum_{j=1}^{n'} u_{ij}$. It follows that an entry in the **FA'** matrix is $(FA)_{ij} = af \sum_{k=1}^{n'} t_{ik} u_{kj}$ and the row sum of **FA'** is

$$(FA)_i = \sum_{j=1}^{n'} (FA)_{ij} = \sum_{j=1}^{n'} \left(af \sum_{k=1}^{n'} t_{ik} u_{kj} \right) = af \sum_{k=1}^{n'} \left(t_{ik} \sum_{j=1}^{n'} u_{kj} \right) = af \sum_{k=1}^{n'} t_{ik} = af\qquad(7\text{-}52)$$

where by definition $\Sigma_j u_{kj} = 1$ and $\Sigma_k t_{ik} = 1$. It is clear from Equation **(7-52)** above that if the product of f and a is less than unity, then the row sum of matrix **FA'** will also be less than one, and the series in Equations **(7-44)** and **(7-45)** will converge.

Notice we calculated a and f as a constant value for all zones in a study area. This can only be done if development is stable enough for their estimation:

$$f = \bar{f} = \sum_{j=1}^{n'} N_j \left(\sum_{j=1}^{n'} E_j \right)^{-1}, \qquad a = \bar{a} = \sum_{j=1}^{n'} E_j^R \left(\sum_{j=1}^{n'} N_j \right)^{-1};\qquad(7\text{-}53)$$

where stability means a more-or-less constant-rate for f and a over the forecast period. One can argue that we can use zonal weights of population (w_j^N) and employment (w_j^E) to compute f^w and a^w:

$$f^w = \sum_{j=1}^{n'} w_j^E f_j = \sum_{j=1}^{n'} f_j \left(E_j \bigg/ \sum_{j=1}^{n'} E_j \right) = \sum_{j=1}^{n'} \left(\frac{\Delta N(j)}{\Delta E_j} \right) \left(E_j \bigg/ \sum_{j=1}^{n'} E_j \right). \qquad (7\text{-}54)$$

Here f_j is now defined as the incremental increase in population, $N(j)$, as a result of an increase in employment at zone j, ΔE_j. (Notice the population increment, $\Delta N(j)$, does not have to be located in zone j.) Similarly,

$$a^w = \sum_{j=1}^{n'} w_j^N a_j = \sum_{j=1}^{n'} a_j \left(N_j \bigg/ \sum_{j=1}^{n'} N_j \right) = \sum_{j=1}^{n'} \left(\frac{\Delta E^R(j)}{\Delta N_j} \right) \left(N_j \bigg/ \sum_{j=1}^{n'} N_j \right). \qquad (7\text{-}55)$$

Here a_j is the incremental increase in local-service employment, $\Delta E^R(j)$, owing to population increase, ΔN_j. (Again, $\Delta E^R(j)$ does not have to be located in zone j.)

Should an areawide uniform multiplier be used following Equations (7-53), the row sum of **FA'** will be less than unity:

$$(FA)_i = \overline{af} = \sum_{j=1}^{n'} E_j^R \left(\sum_{j=1}^{n'} E_j \right)^{-1} < 1 \qquad (7\text{-}56)$$

since $\Sigma_j E_j^R < \Sigma_j E_j$ by definition. On the other hand, if the weighted average definitions of f and a are used,

$$f^w = \sum_{j=1}^{n'} f_j w_j^E \le f^{\max} \sum_{j=1}^{n'} w_j^E = f^{\max} \qquad (7\text{-}57)$$

$$a^w = \sum_{j=1}^{n'} a_j w_j^N \le a^{\max} \sum_{j=1}^{n'} w_j^N = a^{\max}. \qquad (7\text{-}58)$$

As long as $f^{\max}$ and $a^{\max}$ are less than one individually, the row sum is again less than unity. This gives a necessary condition for the convergence of the series in Equations (7-44) and (7-45). According to the development through Equation (7-52), it is also the sufficient condition.

7.7.3 Different zonal multipliers.
What if we explicitly use different a and f for each zone? Will the condition for bifurcation change? This is actually a very cogent question as disparity in zonal socioeconomic-status often dictates different a_j and f_j, instead of the areawide averages, $\overline{a}$ and $\overline{f}$, or a^w and f^w. An interesting example has been worked out in **Table 7-3** under Group 1, where for all three

Item Group	Zone j	E_j	N_j	f_j	a_j	$\bar{f}$	$\bar{a}$	w_j^N	w_j^E	f^w	a^w	$(FA)_i$
Group 1												
$f_3a_3=1.100$	1	40	200	5.0	0.12			0.17	0.13			1.08
$\bar{f}a=0.919$	2	60	150	2.5	0.32	3.83	0.24	0.13	0.20	3.83	0.25	
$f^wa^w=0.977$	3	200	800	4.0	0.27			0.69	0.67			
Group 2												
$f_1a_1=1.100$	1	40	200	5.0	0.22			0.22	0.13			0.78
$\bar{f}a=0.735$	2	60	180	3.0	0.20	3.50	0.21	0.20	0.20	2.93	0.20	0.57
$f^wa^w=0.601$	3	200	500	2.5	0.20			0.56	0.67			0.53
Group 3												
$f_ja_j<1.000$	1	40	200	5.0	0.12			0.20	0.13			0.90
$\bar{f}a=0.792$	2	60	198	3.3	0.25	3.77	0.21	0.19	0.20	3.33	0.22	0.78
$f^wa^w=0.746$	3	200	600	3.0	0.25			0.60	0.67			0.75

Table 7-3 Sample calculations

groups, the interzonal spatial-cost matrix is $[C_{ij}] = \begin{bmatrix} 2 & 6 & 4.5 \\ 6 & 2 & 4 \\ 4.5 & 4 & 2 \end{bmatrix}$. In the Group-1 example, a and f in each zone are different. In zone 3, for example, we have $a_3f_3 = 1.1 > 1$. One of the row sums of matrix **FA'** is also bigger than one. Although the areawide averages for af remain less than one (that is, $\bar{a}f < 1$ and $a^wf^w < 1$), they no longer guarantee convergence.

Considering this example, the question is whether the stringent condition of

$$a_jf_j<1 \qquad \forall j \qquad (7\text{-}59)$$

can be treated as the criterion for convergence. This is obviously motivated from the Group 1 calculations in **Table 7-3**, where divergence might have resulted from having some $a_jf_j \geq 1$. Our rationale is that for the zone with a large multiplier a_jf_j, the zone will always be in need of more local-service employment E_j^R, and these service employment will generate more population and so on. This may lead toward exponential growth in the entire study area. Inequality **(7-59)** is a sure means of preventing this from happening. But we recall that Lowry-based models are built upon two basic premises: spatial-allocation theory and the economic-base theory. It is possible that even if some zones have $a_jf_j > 1$, the other zones can `absorb' the activities from those `exploding' zones because of the allocation process. This dampens the exponential growth. An example is shown in **Table 7-3**, Group 2, in which convergence is illustrated for the case of $a_1 f_1 > 1$.

Now we are ready to analyze the general criteria for convergence. Once again let $\mathbf{F} = [t_{ij}][f_j]$, then $F_{ij} = t_{ij} f_j$. In the same way, $\mathbf{A'} = [u_{ij}][a_j]$ and $A'_{ij} = u_{ij} a_j$. Now $(FA)_{ij} = \sum_{l=1}^{n'} F_{il} A'_{lj} = \sum_{l=1}^{n'} [(t_{il} f_l)(u_{lj} a_j)]$ and the row sum of matrix $\mathbf{FA'}$ is given by

$$(FA)_i = \sum_{i=1}^{n'} (FA)_{ij} = \sum_{j=1}^{n'} \sum_{l=1}^{n'} [(t_{il} f_l)(u_{lj} a_j)] = \sum_{l=1}^{n'} (t_{il} f_l) \sum_{j=1}^{n'} (u_{lj} a_j). \qquad (7\text{-}60)$$

If $a^{\min}$ and $f^{\min}$ are the smallest values of a_j and f_l, let us replace a_j and f_l with these minimum values in Equation **(7-60)**. We can now write

$$\sum_{l=1}^{n'} (t_{il} f_l) \sum_{j=1}^{n'} (u_{lj} a_j) \geq f^{\min} \sum_{l=1}^{n'} \left(t_{il} a^{\min} \sum_{j=1}^{n'} u_{lj} \right) = a^{\min} f^{\min} \sum_{l=1}^{n'} \left(t_{il} \sum_{j=1}^{n'} u_{lj} \right) = a^{\min} f^{\min}. \qquad (7\text{-}61)$$

In much the same way, replacing a_j and f_l with the maximum a_j and f_l, respectively, yields

$$\sum_{l=1}^{n'} (t_{il} f_l) \sum_{j=1}^{n'} (u_{lj} a_j) \leq f^{\max} \sum_{l=1}^{n'} \left(t_{il} a^{\max} \sum_{j=1}^{n'} u_{lj} \right) = a^{\max} f^{\max} \sum_{l=1}^{n'} \left(t_{il} \sum_{j=1}^{n'} u_{lj} \right) = a^{\max} f^{\max}. \qquad (7\text{-}62)$$

Combining Equations **(7-61)** and **(7-62)** one can write

$$a^{\min} f^{\min} \leq \sum_{j=1}^{n'} (FA)_{ij} \leq a^{\max} f^{\max}. \qquad (7\text{-}63)$$

This gives the feasible range for the row sums of matrix **FA'**; the equality sign applies only if $a^{\min} = a^{\max} = a$, and $f^{\min} = f^{\max} = f$, or if the labor-force participation rate and population-serving ratio are uniform areawide. From Equation **(7-63)**, it is also obvious that the development in the area will explode when $a^{\min} f^{\min} > 1$.

7.7.4 General convergence criteria. Now let us look at the row sum of **FA'** in more detail,

$$(FA)_i = \sum_{j=1}^{n'} (FA)_{ij} = \sum_{l=1}^{n'} (t_{il} f_l) \sum_{j=1}^{n'} (u_{lj} a_j). \qquad (7\text{-}64)$$

The two summations on the right-hand-side may be written as convex combinations of a_j and f_l, where the weights are u_{lj} and t_{il}, respectively. Calling these convex combinations a''_l and f'_i, Equation **(7-64)** can now be written as

$$(FA)_i = \sum_{l=1}^{n'} t_{il} f_l a''_l \leq \sum_{l=1}^{n'} t_{il} f_l a^{\max} = a^{\max} \sum_{l=1}^{n'} t_{il} f_l = a^{\max} f'_i. \qquad (7\text{-}65)$$

Expressions **(7-64)** and **(7-65)** show that individual $a_j f_j$ can be bigger than unity and convergence can still be guaranteed if the convex combination is less than one:

$$a^{\max} f_i^t < 1 \qquad \forall i. \qquad (7\text{-}66)$$

Physically this means that there may in fact be some sort of 'dampening' effect on a popular residential zone, such that the explosive derivation effect can be absorbed during the allocation process. This will happen when the contribution of the terms is small, even though may be greater than unity. This will happen if the corresponding t_{il} is small. This means that 'explosive' zone l is either far away from zone i, or that it is a relatively small fraction of the regional population (according to the definition of t_{il} in Equation **(7-48)**). On the other hand, the system diverges when a number of explosive zones are close to zone i. These zones reinforce each other in the deviation and allocation process, making the row sum in Equation **(7-65)** bigger than or equal to one.

The row sum of matrix **FA'** will be less than unity, or would never reach the bifurcation point, under two conditions:

$$a_l^u \le a^{\max} \qquad and \qquad f_i^t \le \left(1 / a^{\max}\right). \qquad (7\text{-}67)$$

For a two-zone case, one can write

$$u_{11} a_1 + u_{12} a_2 \le a^{\max}, \qquad u_{21} a_1 + u_{22} a_2 \le a^{\max} \qquad (7\text{-}68)$$

and

$$t_{11} f_1 + t_{12} f_2 \le \frac{1}{a^{\max}}, \qquad t_{21} f_1 + t_{22} f_2 \le \frac{1}{a^{\max}}. \qquad (7\text{-}69)$$

If the spatial-allocation process is perfectly symmetrical, or

$$u_{11} = u_{22} = t_{11} = t_{22} = X, \qquad u_{12} = u_{21} = t_{12} = t_{21} = Y; \qquad (7\text{-}70)$$

the above set of equations reduces to

$$a_1 Y + a_2 X \le a^{\max}, \ f_1 X + f_2 Y \le \frac{1}{a^{\max}}, \ a_1 X + a_2 Y \le a^{\max} \ and \ f_1 Y + f_2 X \le \frac{1}{a^{\max}}. \quad (7\text{-}71)$$

Figure 7-14 illustrates the relationship between a_1, a_2, f_1, f_2, X and Y when $a_1 = a^{\max}$ and $a_1 f_1 \ge 1$. In the sub-figures (a) and (b), remember that $X + Y = 1$ by definition of the accessibility index. It defines the possible values of u_{ij} and t_{ij}. In **Figure 7-14**(a), we show that convergence is not possible because there is no feasible region obtainable as defined by the intersection of $X + Y = 1$ and Equations **(7-70)** through **(7-71)**. In **Figure 7-14**(b), on the other hand, X and Y can assume any value to satisfy the equation $X + Y = 1$, and convergence is guaranteed. Physically, we are saying that, within bounds, there is a dampening effect among the two zones to moderate the explosiveness at zone 1.[35]

7.7.5 Summary. These basic theoretical discussions simply illustrate the complexity of bifurcation and aggregation issues in urban/regional spatial-systems. Although we have defined the general condition for bifurcation and interpreted its physical properties, there are virtually infinite combinations of the zonal labor-force participation rate, population-serving ratio, and accessibility indices that can give rise to bifurcation. In comparison with aggregated or averaged labor-force participation rates and population-serving ratios, the disaggregated case—the more realistic of the two—is much more complex, giving rise to interesting behavior.

7.8   Constrained version of the Lowry–Garin model

Thus far, we have been concentrating on the unconstrained version of the Lowry–Garin model. Here we impose maximum residential-density and minimum threshold for retail activities:

$$N_j \le N_j^c \qquad \forall j, \tag{7-72}$$

$$E_j^R \ge d^R \quad \text{or otherwise} \quad E_j^R = 0 \qquad \forall j. \tag{7-73}$$

The traditional way of dealing with these constraints is to re-allocate the excess activities—should capacity be exceeded or threshold not reached—to other zones according to some heuristic in an ex post facto manner. This approach has its flaws, most notably in the rationale of the reassignment procedure. Here we introduce a more behaviorally consistent and hopefully also a more elegant approach. The approach can be thought of a hybrid of the singly-constrained and doubly-constrained gravity models.

7.8.1 Modified accessibility indices. In terms of computation, the procedure used here simply eliminates from the zone concerned those households that exceed the zoning-density limit and those retail employees that do not reach the threshold during iteration m. The so modified zonal population and service employment are referred to as $N_j(m)$ and $E_j^R(m)$. The excess attractions are then recycled through the *existing* accessibility expressions as in the unconstrained case:

$$t_{ij}(m) = h_{ij}^b N_j(m) \left[\sum_{k=1}^{n'} h_{ik}^b N_k(m) \right]^{-1} \tag{7-74}$$

and

---

[35]Based on the "spectral radius" of a matrix and eigenvalues, a more formal proof of the convergence criteria is discussed in the book appendix entitled "Control, Dynamics and Stability" in Chan (2000). It provides a better linkage to the stability conditions established previously for spatial-interaction difference-equations.

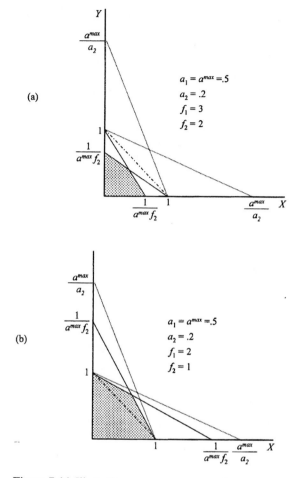

Figure 7-14 Illustrating nonconvergence and convergence

$$u_{ij}(m) = g_{ij}^{c} E_{j}^{R}(m) \left[\sum_{k=1}^{n'} g_{ik}^{c} E_{k}^{R}(m) \right]^{-1} \qquad (7\text{-}75)$$

where in the spirit of simplified notation h_{ij} is used to stand for $C(C_{ij})$ and g_{ij} for $C^{R}(C_{ij})$. Notice that b and c are negative calibration parameters that are introduced to guide the re-allocation procedure by modifying the shape of the trip-distribution curves. Below, we will say more about these 'elasticity' parameters, particularly concerning the reasoning behind the computational procedure.

The modified t_{ij} and u_{ij} [Equations (7-74) and (7-75)] are used to forecast again. Households that exceed the zonal capacity are again removed (as are the activities in which the number of retail employees is inadequate). This procedure is repeated until the final run when not even a single constraint is violated. The reasoning behind this method is to let the system of equations redistribute activities following the zoning and the minimum-threshold constraints *within the existing framework of accessibility functions*. Allocation adjustments according to the zoning and threshold constraints are summarized by the endogenous parameters b and c, which are embedded within the accessibility functions. For all practical purposes, the set of equations for our urban spatial-model remains the same as in the unconstrained case outlined under the "Basic theory" subsection (Section 7.7.1). The constraint equations **(7-72)** and **(7-73)** are taken care of endogenously through parameters b and c. Therefore, the bifurcation criteria given in Equation **(7-67)** remain valid for the constrained model.

It can be seen that b and c have very important physical interpretations. If one examines the travel-cost elasticity of modified accessibility:

$$\eta(t,h)=\frac{dt_{ij}}{dh_{ij}}\frac{h_{ij}}{t_{ij}}=b(1-t_{ij}) \qquad \eta(u,g)=\frac{du_{ij}}{dg_{ij}}\frac{g_{ij}}{u_{ij}}=c(1-u_{ij}) \qquad (7\text{-}76)$$

A large b or c (in absolute value) suggests that travel-impedance improvement has a large influence on accessibility and vice versa. Remember b and c are negative, and t_{ij} and u_{ij} are less than unity. Equation **(7-76)** tends to discourage additional assignment of activities to an already 'popular' zone (as characterized by a large t_{ij} and u_{ij}) in comparison to a less popular one. This latter statement is significant in that this re-allocation procedure has the inherent property of redistributing activities from overcrowded zones to zones that have not yet reached their capacities. This is exactly the objective we are trying to achieve.

A sensitivity analysis was also performed for the bifurcation conditions in Equation **(7-67)** with respect to the propensity measures h_{ij} and g_{ij}:

$$\frac{df_i^t}{dh_{ij}}=b\sum_{j=1}^{n'}F_{ij}\frac{1-t_{ij}}{h_{ij}} \qquad \frac{da_i^u}{dg_{ij}}=c\sum_{j=1}^{n'}A_{ij}'\frac{1-u_{ij}}{g_{ij}}. \qquad (7\text{-}77)$$

These equations show that as propensity-function h_{ij} or g_{ij} goes up the row sum for the bifurcation condition gets smaller, since b and c are negative in value. More important, the bigger b and c are in absolute value, the faster the row sums go down. Thus for given values of the 'modified' row sums of F_{ij} and A_{ij}', large values for b and c (in absolute value) will show *bigger* 'dampening' effects for exponential growth (remembering again that b and c are negative). In other words, all else being equal, f_i' and a_i' tend to increase more rapidly (thus reaching the condition for bifurcation) for larger values of b and c (including the negative sign) than for smaller values. Subsequent computational experiences associate a smaller b or c as zones reach their capacities. This reinforces the useful role these two parameters play in redistributing activities from overcrowded zones to those that are less crowded. In other words, there is an inherent mechanism for dampening explosive growth at certain zones by re-allocating activities to other zones. More will be said about this in the case-study section of this chapter.

7.8.2 Calibration of elasticities. The Fibonacci-search technique[36] (Bazarraa and Shetty 1979) is used to calibrate the parameters b and c. The procedure seeks to minimize the deviation between *each* revised accessibility-function, $t_{ij}(m)$ and $u_{ij}(m)$, and their original values, $t_{ij}(0)$ and $u_{ij}(0)$:

$$\min[t_{ij}(m)-t_{ij}(0)]^2 \qquad \forall\, i,j \tag{7-78}$$

$$\min[u_{ij}(m)-u_{ij}(0)]^2 \qquad \forall\, i,j. \tag{7-79}$$

For each calibration, *median* of the b and c values are obtained, which become the current values of $b(m)$ and $c(m)$ in Equations **(7-74)** and **(7-75)**. The corresponding $t_{ij}(m)$ and $u_{ij}(m)$ are used again to forecast according to Equations **(7-50)** and **(7-51)**. Thus the calibration is performed endogenous to the forecast process. Fibonacci search is carried out instead of regular regression or other curve-fitting procedures to be consistent with the 'disaggregate' formulation advocated in this chapter.

7.8.3 Calibration of multipliers. To operationalize fully the derivation and allocation of Equations **(7-44)** and **(7-45)**, one has to use the observed data E and N to calibrate the disaggregate multipliers f_j and a_j. To achieve this, we can rewrite Equation pair **(7-50)** and **(7-51)** as

$$\mathbf{E}(\mathbf{I}-\mathbf{FA}')=\mathbf{E}^B, \qquad \mathbf{NF}^{-1}=\mathbf{E}^B(\mathbf{I}-\mathbf{FA}')^{-1}; \tag{7-80}$$

or

$$\mathbf{E}-\mathbf{E}^B=\mathbf{EFA}', \qquad \mathbf{NF}^{-1}-\mathbf{E}^B=\mathbf{NA}'; \tag{7-81}$$

where

$$\mathbf{F}=[t_{ij}][f_j]=\mathbf{t}\varphi \qquad \mathbf{A}'=[u_{ij}][a_j]=\mathbf{u}'\sigma. \tag{7-82}$$

Combining Equations **(7-81)**–**(7-82)** we get

$$\mathbf{E}-\mathbf{E}^B=\mathbf{E}\,\mathbf{t}\varphi\,\mathbf{u}'\sigma, \qquad \mathbf{N}\varphi^{-1}\mathbf{t}^{-1}-\mathbf{E}^B=\mathbf{N}\mathbf{u}'\sigma; \tag{7-83}$$

or combining with Equations **(7-81)** and **(7-82)**

$$\mathbf{E}\,\mathbf{t}\varphi=(\mathbf{E}-\mathbf{E}^B)\sigma^{-1}\mathbf{u}^{-1}=\mathbf{E}\,\mathbf{F}=\mathbf{N} \qquad \mathbf{N}\mathbf{u}\sigma=\mathbf{N}\varphi^{-1}\mathbf{t}^{-1}-\mathbf{E}^B=\mathbf{E}-\mathbf{E}^B \tag{7-84}$$

Equations **(7-84)** form a pair of equation set which enables us to solve for a_j and f_j explicitly.

Experimentation with numerical-solution techniques, however, reveals that owing to the nonlinearity of the equation sets, questions of solution uniqueness and algorithmic convergence arise. We recognize this difficulty, and the fact that in the equation sets thus far we have yet to recognize zonal constraints explicitly via the calibration data. The base-year data **E** and **N** is now supplemented with a "calibration year" data set. In this way, the effects of zoning constraints are

---

[36]The Fibonacci search technique is reviewed in the "Prescriptive Tools" chapter of Chan (2000).

explicitly considered—via the two sets of real-world data *over time*. Also, the extra data may help settle the problem of solution uniqueness. This supplemental calibration-procedure has the distinct advantage of establishing a "trend line" over two data sets from two different time-periods. It ensures temporal (on top of spatial) replication and thus paves the way for forecasting applications. The results of the calibration between "calibration year" to base-year become the input for the forecast from base year to "target year." Conceptually, one can think of this as solving Equations **(7-84)** with each variable replaced by its increment from calibration year to base year (excepting and—the unknowns to be solved). Results of this calibration approach are extremely satisfactory (Yi 1986, Smith 1996). This calibration procedure is included as an integral part of the attached software under the YiChan subdirectory.

7.8.4 A numerical example. We illustrate the theoretical discussion thus far with a numerical example consisting of four zones. Here are the computational steps:

a. The calibration-year data are given as follows. Similar to population and employment figures, trip-distribution curves for work and nonwork trips are taken from 'textbook' sources for simple illustrative purposes only: $h_{ij} = C(C_{ij})$ $= 345 + C_{ij}^{1.8} - 15C_{ij}$, $g_{ij} = C^R(C_{ij}) = 6.4 + 120C_{ij}^{-0.8}$; basic employment: $\mathbf{E}^B =$ (150,180,130,100), service employment: $\mathbf{E}^R = (30,25,20,15)$, population: $\mathbf{N} =$ (400,350,200,220).

b. The base-year data are given as follows: basic employment: (the same as in the calibration year), total employment: $\mathbf{E} = (200,205,160,120)$, population: $\mathbf{N} =$ (410,345,230,240).

c. The spatial-separation (or travel-cost) matrix is $[C_{ij}] = .$

d. Using the definition for t_{ij} and u_{ij} in Equations **(7-48)** and **(7-49)**, and with the help of the calibration-year data, we obtained the basic accessibility-functions as follows:

$$[t_{ij}(0)] = \begin{bmatrix} 0.3717 & 0.2877 & 0.1618 & 0.1788 \\ 0.3354 & 0.3185 & 0.1668 & 0.1793 \\ 0.3298 & 0.2997 & 0.1922 & 0.1783 \\ 0.3391 & 0.2928 & 0.1675 & 0.2006 \end{bmatrix}$$

and

$$[u_{ij}(0)] = \begin{bmatrix} 0.4778 & 0.2421 & 0.1624 & 0.1177 \\ 0.3307 & 0.3797 & 0.1586 & 0.1311 \\ 0.3019 & 0.1814 & 0.3852 & 0.1315 \\ 0.2873 & 0.2660 & 0.2049 & 0.2417 \end{bmatrix}.$$

e. Using base-year $t_{ij}(0)$ and $u_{ij}(0)$ as t and u in Equations **(7-84)**, and with the base-year data **(E,N)**, we calibrated the disaggregate multipliers as:

$$\varphi = \begin{bmatrix} 1.770 & 0 & 0 & 0 \\ 0 & 1.779 & 0 & 0 \\ 0 & 0 & 1.783 & 0 \\ 0 & 0 & 0 & 1.840 \end{bmatrix} \qquad \sigma = \begin{bmatrix} 0.094 & 0 & 0 & 0 \\ 0 & 0.103 & 0 & 0 \\ 0 & 0 & 0.108 & 0 \\ 0 & 0 & 0 & 0.115 \end{bmatrix}.$$

f. By Equations **(7-46)** and **(7-47)**, the initial matrices **F** and **A'** are determined as

$$= \begin{bmatrix} 0.6578 & 0.5118 & 0.2885 & 0.3291 \\ 0.5935 & 0.5666 & 0.2974 & 0.3300 \\ 0.5837 & 0.5331 & 0.3427 & 0.3282 \\ 0.6002 & 0.5208 & 0.2986 & 0.3692 \end{bmatrix} \qquad A' = \begin{bmatrix} 0.0449 & 0.0250 & 0.0176 & 0.0135 \\ 0.0310 & 0.0392 & 0.0172 & 0.0151 \\ 0.0283 & 0.0187 & 0.0418 & 0.0151 \\ 0.0270 & 0.0275 & 0.0222 & 0.0278 \end{bmatrix}$$

Here the $[t_{ij}]$ and $[u_{ij}]$ matrices are from the calibration year and $[f_j]$ and $[a_j]$ from the base year.

g. The forecast-year (or target-year) basic-employment is given as $\mathbf{E}^B =$ (180,180,150,130). The zoning constraint for residential housing is $\mathbf{N}^c =$ (150,480,170,700). For the sake of clarity and owing to the symmetry of the minimum-threshold constraint, however, we do not include $\mathbf{N}^c$ in this illustrative example.

We use Equations **(7-44)** and **(7-45)** as forecast models; Equations **(7-74)** and **(7-75)** for $t_{ij}(m)$ and $u_{ij}(m)$, and the Fibonacci search as an endogenous calibration-technique for parameters b and c. Then we obtain target-year forecast within six iterations:

First iteration:
N(1) = (469,409,234,258), modified N: N(1)* = (150,409,170,258); $b = -1.52$, $c = -1.19$.
Second iteration:
N(2) = (207,566,235,355), modified N: N(2)* = (150,480,170,355); $b = -1.86$, $c = -0.81$.
Third iteration:
N(3) = (183,574,204,431), modified N: N(3)* = (150,480,170,431); $b = -2.11$, $c = -1.25$.
Fourth iteration:
N(4) = (163,520,183,457), modified N: N(4)* = (150,480,170,457); $b = -0.92$, $c = -0.80$.
Fifth iteration:
N(5) = (178,537,193,522), modified N: N(5)* = (150,480,170,522); $b = -1.43$, $c = -1.13$.
Sixth and final iteration:
N(6) = (149,477,168,501); $b = -1.43$, $c = -1.13$.

Here the b and c parameters are applied to both the calibration-year and base-year t and u, thus generating revised $[f_j]$, $[a_j]$, $[t_{ij}]$ and $[u_{ij}]$ in Equations **(7-46)** and **(7-**

47). As none of the zonal constraints are violated, the forecast is now complete and the calibrated values b and c are adopted. The numerical example presented here not only illustrates the algorithmic steps, but also highlights the fast convergence-rate of this algorithm. For illustration purposes, this numerical example is part of the software data sets under the YiChan subdirectory.

Further details by way of a case study will be presented in the "Spatial-temporal Information" chapter. A complete set of source and executable codes of the 'dynamicized' Garin–Lowry model is included in a computer disk attached to the back of this book. These include the data-sets of the numerical example as well as the York case-study. As with our other computer programs, we have included detailed operation instructions in a README file for those who wish to experiment with the software.

7.9   Summary and conclusion

In this chapter, we have reviewed recent developments in chaos, as well as spatial dynamics involving competition and discontinuities. Couched in a control-theoretic framework, we have illustrated these ideas both in terms of differential and difference equations, discrete and continuous networks, and over time and space. We also discussed the implication of aggregation and disaggregation on model calibration and application. Most important, we showed that these are not just theoretical concepts. They can be implemented in a spatial-interaction model, as shown in the York case-study. These discussions pave the way for general spatial equilibrium and disequilibrium: the subject of a following chapter.

7.9.1   Precipitous happenings in location and land use. The subject of mathematical chaos has certainly become newsworthy over recent years. Many popular magazines have carried articles on these new studies of chaotic dynamics. These are in part motivated by the discovery of a seemingly underlying order with the potential of predicting certain properties of noisy behavior. While its roots are in physical systems, chaos theory has found its way to social science, as illustrated by chaotic models of urban change in this chapter. We applied the by now well-known techniques, including Lorenz equations, to investigate the stability characteristics of this structure.

Non-regular, periodic movements have been observed in the growth and decline of industries, cities, and national economies for quite some time. However, by far the most frequently observed phenomena in socioeconomic systems are turbulence and chaos. A number of novel and interesting dynamic features are exposed in this chapter. They include dynamic bifurcations that are discrete analogues of the Hopf bifurcation found in continuous dynamics. Other dynamic events include stable and unstable competitive exclusion. The models defined here are intrinsically dynamic because the state of the system is the consequence of the various past transitions. The results derived describe the transient and stationary regimes of the models.

A controversy about catastrophe theory is: does the theory say enough to justify the conclusions that are drawn? Variables are often defined to describe some systems of interest, including an appropriate number of control variables. These variables can then be used to determine the nature of the worst kind of degenerate singularity that can occur. The canonical form of the response surface is then used to draw a number of conclusions about the possible, and often asserted-as-likely behavior of the system. While considerable insight into possibilities of bifurcation can be gained from a *qualitative* analysis, our ambition here is to carry this through to building a *quantitative* model. While it is valuable to know the possible existence of critical points, it is often even more valuable to quantify them.

The customary use of models in planning has been of conditional forecasting: a set of values is assumed for some controllable variables and the models are used to make an assessment of impact. This method will continue to be useful, especially for short and medium-term forecasting. However, dynamical system theory offers an additional and important focus: a concern with criticality and stability. Suppose a major change can take place in a system at some critical parameter value. It is obviously valuable to a planner to know both that this possibility exists and also the actual value of the critical parameter. The knowledge may be important for two kinds of reason. First, the planner may wish to prevent the change from taking place; second, s/he may wish to encourage it. In the first case s/he would be attempting to maintain the system in a stable state and not let conditions change. In the second case, s/he may be trying to get the system into a new stable-state by changing a parameter that may not obviously be connected with such transitions.

Simply put, a dynamic model is one whose structural equations contain non-trivial temporal forms of the endogenous variables. Such temporal forms include inter-temporally subscript variables, time derivatives, and integrals of sums over time. By 'non-trivial' means the exclusion of such practices as simply defining investment to be the derivative of capital stock with respect to time. This definition excludes models of inter-temporal comparative statics which move their endogenous variables through time wholly by parametric ranges administered in a pre-determined time pattern. Truly dynamic models internalize their laws of change with their structural definition, giving rise to an endogenous time process.

The goals sought, problems encountered, and method employed in regional dynamics are parallel to but distinct from those of its static counterparts. The aims are generally to determine specific values for the endogenous variables in a numerically specified model; or, less completely, to gain qualitative insights into the structure of the solution. There are several methods for deriving these structural equations from the set of assumptions and parameters. They include outright assumptions of form, the use of first- and second-order optimization conditions, or the use of dynamic forms of spatial-interaction theory.

In recent years, the level of aggregation has been shifting from macro to micro-based analyses. This new trend has fostered a closer orientation toward the following: (a) behavioral approaches, (b) a more precise description of actual spatial interaction, (c) a better possibility for analyzing choice processes on a longitudinal, event-history or dynamic basis, and (d) a greater flexibility in specifying choice processes. The price to be paid, however, is the complexity of model design and calibration effort. In the present chapter, we provided the linkages between spatial-interaction models and two particular classes of disaggregate models, namely, the conditional utility maximization models and the discrete choice models. The former is based on deterministic utility theory while the latter on random utility theory.

7.9.2 Application to spatial-interaction models. The Lowry–Garin derivative
models are pedagogical cornerstones of urban spatial-analysis. In this chapter we reexamined this classic school of thought in light of bifurcation theory, disaggregation, and emphasis on more transparent modelling and computational structure. A main thrust of this chapter is the reformulation of the matrix Lowry–Garin model to include zonal constraints. Both the maximum-residential-density and the minimal-service-employment constraints are included in the forecast sequence within the equations of the unconstrained model. This is accomplished with an introduction of simply two endogenous parameters b and c. The parameters carry with them the physical interpretation of travel-cost elasticity. Such parameters are policy-sensitive in that the development implication of accessibility improvement is shown analytically by the magnitude of these parameters. This relieves the sheer reliance on traditional simulation approaches to answer 'what-if' questions.

Also included in this chapter are discussions on the level of aggregation for multipliers such as the population-serving ratio and the labor-force participation rate. A disaggregate calibration procedure is recommended to estimate parameters b and c endogenously. It is also used to estimate the population and employment multipliers exogenously by use of time-series datasets drawn from two points in time. This ensures better replication of development pattern temporally as well as spatially, as a result of the explicit recognition of the effect of zonal constraints.

Perhaps the most significant contributions are the general conditions laid down for bifurcation. Expressed in terms of accessibility, labor-force participation-rate, and population-serving ratio, the various conditions for convergence and divergence of the spatial systems are delineated. These conditions include the influence of travel costs, multipliers, and zonal constraints. They encompass and illustrate the myriad of combinations under which bifurcation can take place. Of particular interest is the discussion on the 'dampening' effects of spatial allocation on the 'explosive' growth at selective zones. The disaggregate analysis allows a better understanding of bifurcation in terms of the economic-base and the spatial-interacting components of Lowry–Garin derivative-models.

The methodological findings above were tested on a medium-size city of a-quarter-million population. In comparison with aggregate analysis, disaggregate modelling was found to yield better reproduction of the existing development pattern--as reported elsewhere (Webber 1984). The bifurcation behavior was then interpreted in terms of the theoretical conditions laid out earlier, showing the serviceability of such conditions in explaining an admittedly complex urban-development pattern. The role of endogenous parameters b and c in the forecasting sequence is again reconfirmed. Unlike other algorithms, the calibration and solution procedure presented here converges very fast for those cases where bifurcation does not occur. The disaggregate calibration-procedure for zonal multipliers is also quite serviceable. The procedure merges within the forecasting sequence--namely, the calibration result from the calibration year to the base year serves directly as input to the forecast from the base year to the target year.

The findings here can be generalized to input-output modelling of a regional economy, since the economic-base paradigm can easily be extended to this more comprehensive framework. In a following chapter, we will develop these general equilibrium conditions in more detail.

7.9.3 Comments. In an era of reconstruction and rehabilitation of public infrastructure, the analyses here have potentials, in our judgment, to guide the judicious investment of public and private funds in infrastructural improvements. We are interested in investments that will result in the most desirable economic developments in the shortest possible time by capitalizing on precipitous happenings. Such analysis has the distinct advantage of relying on analytical properties that are by definition more transparent than many simulation approaches in the past.

As always, this work can be further refined by more case studies. This will allow even more accurate verification of the analytical bifurcation conditions in terms of field data. Better still, before-and-after studies of a community should be conducted, as they provide the only way to validate the predictive power of the model. A side benefit of these empirical exercises is to stimulate further basic research. This ranges from a 'tighter' set of bifurcation conditions to an even more satisfactory formulation and a more efficient calibration and forecasting procedure.

We used to think that if we knew one, we knew two,
because one and one are two.
We are finding that we must learn a great deal more about "and."
Sir Arthur Eddington

CHAPTER EIGHT

SPATIAL EQUILIBRIUM AND DISEQUILIBRIUM: LOCATIONAL CONFLICT AND THE INPUT–OUTPUT MODEL

The concept of spatial-price equilibrium underlies most studies of spatial movement and competition. However, this concept is based upon a notion of perfect competition that is often unrealistic for many markets and ignores the existence of local spatial-monopolies. After reformulating the classical spatial-price-equilibrium model, we state and analyze a Cournot–Nash model and discuss solution algorithms. We present alterative economic models of spatial competition and show how all these models can be solved with essentially the same solution algorithm. We then generalize these models and algorithm to networks with transshipment nodes. The economic-base and spatial-allocation concepts described in most of the Garin–Lowry based models can also be generalized. We wish to develop a procedure to describe spatial economic-equilibrium in an input–output model. It will be found that such developments as bifurcation and disaggregation can again be carried over to this general context, defining a framework of disequilibrium for spatial economics.

8.1 Conflicts in spatial choice

The ideas of market equilibrium in economics can be transferred to the spatial context in such discussions as spatial games and competition (Vickerman 1980, Kanafani 1983). We make a distinction between individual and system optimum: what is good for a participant in the group may not be beneficial for the community as a whole. To see this, we consider a simple multi-region example. Here the excess supply in each region is an increasing function of its price, or that goods or services do not sell when the price is too high. Commodity flow will occur—i.e., goods will be traded—between any pair of regions if the price at the destination less the price at the origin equals the transportation cost. Notice that this represents a steady-state equilibrium condition. A higher price at the destination, for example, will not result in a sustainable trade between the two regions. In other words, the flows between regions have to satisfy the conservation property that the total net export from any

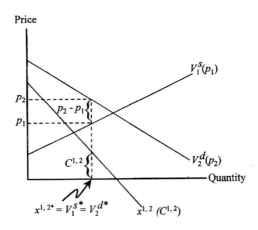

Figure 8-1 Spatial equilibrium in a two-region example

region be limited by the excess supply in that region and that the total net import be limited by the excess demand. Both excess supply and demand are limited quantities. The problem can be simply solved for the two-region case, but it will require a formal equilibrium model for the multi-region case. If the decision process is one of maximizing net-benefit and if the net exports in each region are fixed, then the problem becomes a linear-programming problem.

8.1.1 Spatial equilibrium. A simple two-region case was illustrated by Chiang et al. (1980) using a graphical solution. In a two-region system we let V_1^s be the excess supply in the first region and V_2^d be the excess demand in the second. These values are related to the prices p_1 and p_2 in the two regions, as shown in **Figure 8-1**, where demand V_2^d is a decreasing function of p_2 and supply V_1^s an increasing function of p_1. Flow from region 1 to region 2 will occur if the cost of transportation $C^{1,2}$ is equal to the price difference (p_2-p_1). Equilibrium requires that the quantity shipped between the two regions x^{12} be equal to V_1^s and to V_2^d at any given transportation cost. In other words, if the demand curve for transportation is given by $x^{1,2}(C^{1,2})$, then the following equality must hold: $x^{1,2*}(C^{1,2}) = x^{1,2*}(p_2-p_1) = V_1^{s*}(p_1)$ $= V_2^{d*}(p_2)$. Graphically this is shown by the transportation-demand curve $x^{1,2}$ in **Figure 8-1**. Here for each value of transportation-cost $C^{1,2}$ the amount shipped is such that $V_1^{s*} = V_2^{d*}$ and $(p_2-p_1) = C^{1,2}$. Optimization and microeconomic analysis combine when the approach used—as illustrated here—is to incorporate a spatial element into the firm's production-cost function.

Now spatial competition comes out of the old issue of market areas and the extent to which a spatial economy tends to be monopolistic. The classic treatment is Hotelling's now famous example for the beach ice-cream salespersons.[1] Optimal

---

[1]For an introduction to the Hotelling problem, see the chapter on "Chaos, Catastrophe, Bifurcation and Disaggregation."

positions for two ice-cream salespersons on a linear beach with uniformly distributed customers is for each to be at a distance one quarter of the total length of the beach from each end. This location minimizes total travelling costs for the public and is therefore Pareto optimal. The ice-cream salespersons are, however, motivated by the desire to increase their market share. Each is induced to move toward the center to capture some of the other's market. This process of trying to capture a larger share of the total can only reach an equilibrium distribution when each is at the center for the beach. While they are still splitting the market equally, the travel costs for the customers are now markedly increased. From this result we can deduce that equilibrium locations are inclined to be Pareto-inefficient.[2] When the number of salespersons is increased, the problem changes. It is interesting to note that, in this case, spatial equilibrium is less likely to be achieved in a linear market. The reason is that there is usually an incentive to 'leapfrog' over one's competitors. But the resultant positioning more nearly approximates the optimal.

The existence of space markets does tend to produce local monopolies. A firm in a particular location is protected from the competition of his next nearest neighbors by the transport costs necessary to compete. A more important question, however, is whether that monopoly power can be eroded. The traditional view tends to be that with free entry the monopoly power of firms is limited: by locating on the market area boundary for two existing firms a new entrant captures part of each market and reduces the existing firms' profits. Logically new entry could continue until all profits above normal were eliminated. The fact that this does not happen in the real world can be ascribed to imperfections in the market mechanism. The monopoly power exercised by firms in a spatial economy is seen as largely due to imperfect information.

This view has only been seriously challenged by Eaton and Lipsey (1977, 1978), who have worked through the standard neoclassical theory of the firm with the addition of space. They explicitly recognize that any spatial concentration of production, whatever the spatial distribution of demand, depends on the existence of increasing returns-to-scale over the relevant range of output. Suppose increasing returns-to-scale exists. Then there must be a critical minimum size of plant that a new entrant must be able to attain before entry can be profitable. The extent to which the new entrant can undercut his/her rivals even when s/he has a location advantage is thus limited. Most important, however, is the limit to new entry. In the traditional model this is logical, but unrealistic, when there are an infinite number of infinitely small producers evenly distributed spatially. In the revised model the minimum size considered to ensure the existence of spatial-monopoly power is at a level of price above normal. This is not because of imperfect information in the spatial economy, but because of the nature of the production function itself.

One further point emerging from this view of the spatial organization of markets is that it also suggests that the relevant responses are not smooth and

---

[2]The concept of Pareto optimality or non dominated solutions was introduced in the "Multi-criteria Decision-making" chapter in Chan (2000).

continuous. Because of the minimum constraints on new entrants, existing firms may have considerable freedom over pricing and output decisions. This results in a wide range of potential profit levels, without invoking retaliatory action from new entrants. Their strategy may thus be more concerned with preserving market areas vis-a-vis existing firms. Since existing firms have power relative to potential entrant much will depend on the number of firms in the industry. In practice a wide range of solutions may be possible, each of which is stable, with the production conditions only determining the minimum size for profitable operation. It is in this indeterminate world that we can see the need for a game-theory approach. We can also see the difficulty in predicting both what will occur and what should occur in an optimal situation.

Our general conclusion from these various considerations of the structure of spatial markets, both markets for normal goods and services in a spatial economy and those for specially spatial activities such as land use and transport, is as this. Equilibrium is unlikely to be achieved and that some latent conflicts will exist within the economy. Our next task is, therefore, to consider those conflicts that are to be resolved by the market. This includes their effect on the structure of the economy and how they can be resolved in the interest of removing any implicit losses of welfare. Here the term welfare is used in the context of welfare economics. It covers the broad questions of the optimal allocation of inputs among industries and the optimal distribution of goods and services among consumers.

8.1.2 Disequilibrium. To illustrate non-existence of spatial equilibrium consider the following four firms and locations (Beckmann 1987). In this quadratic-assignment problem[3], part of the output of each firm is used as input by another firm. In particular suppose that firm 1 supplies firm 2, firm 2 supplies firm 3, firm 3 supplies firm 4, and firm 4 supplies firm 1. Let the four locations consist of two pairs of adjacent locations as shown in **Figure 8-2**. Initially the firms are located as shown in part (a) of the figure. But firms 2 and 4 can both gain by exchanging locations, since firm 2 is closer to its customer 3 and firm 4 is closer to its customer 1. We assume for the time being that the producer pays all transportation costs. The result of the exchange is then shown in **Figure 8-2**(b). Now firms 3 and 1 have an incentive to exchange locations for similar reasons. The reader may show that 4 and 2 will want to exchange locations again and after that 1 and 3 once more, whereupon the original location assignment has been revisited. In this way firms will be induced to rotate and no equilibrium exists. More complicated examples may be constructed to avoid obvious objections. The point remains: under the quadratic-assignment problem, trade relationships among firms create externalities, or cost/benefits not accounted for by the price system.[4] In this case there exist

---

[3]For a discussion of the *quadratic assignment problem*, see the "Prescriptive Tools" chapter in Chan (2000). For its relationship to the travelling-salesman problem, see the chapter on "Measuring Spatial Separation" in the current volume.

[4]For a discussion of *externality*, see the "Economic Methods" chapter in Chan (2000).

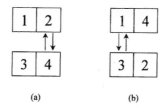

Figure 8-2 - Non-equilibrium example from quadratic assignment

situations where land rents fail to achieve and maintain spatial equilibrium in the location of firms.[5]

Reference back to our abstract model of conflict and equilibrium in game theory tells us that there are two basic reasons for a non-stable solution: a conflict in optimal strategies in a simple constant-sum game or the existence of a non-constant sum to the game.[6] The former is more easily dealt with since it only implies that the payoff matrix as currently constructed for the pure strategies gives us no solution. This is the situation that would exist if there were excess demand for a particular item and therefore re-consideration of the various strategies in the game is necessary. In these circumstances the solution can be found by mixing strategies as discussed in the "Economic Methods" and "Multi-criteria Decision-making" chapters of Chan (2000). The key factor is the time taken to search for the optimal mix of strategies that determines the game. Conflicts are thus short-term phenomenons that will be eliminated when an appropriate period for identifying the correct mix of strategies has elapsed.

Non-constant-sum games are rather different propositions. These cannot usually be solved by a simple search for appropriate strategies and may instead settle at an apparent equilibrium that is not optimal. Optimally here involves looking not just at the individual behavior and payoffs of each participant in the game in question. It involves looking at the total payoffs to all participants, which may be termed the social payoff. Individual economic agents can in isolation only take decisions with reference to private payoff. This is all that is perceived without the extra information, which would come either from collusion with other interested parties or from additional information provided by a 'social' organization such as a government. This extra information, which may often take the form of penalties such as a user-charge, causes the individual to perceive the full social payoff as if it were a private decision. S/he now takes decisions that are both individually and socially optimal.

---

[5]For a discussion of land rent, see the "Economic Methods" chapter in Chan (2000) also.

[6]In a noncooperative game, the goal of each player is to achieve the largest possible individual gain irrespective of the other player(s). Many such games lead to a subdivision of a certain fixed (constant) amount of payoffs among the players. We call this a constant sum game. See game theory discussion in the "Economic Methods" and "Multi-criteria Decision-making" chapters in Chan (2000).

There is a basic case of divergence of social from private payoffs that is relevant here: externalities. This pertains to those cases where there is some interference of one person's action on the well-being of others. This may be detrimental in that the individual's private payoff is largely at the expense of payoffs to others, or it may be beneficial. In the latter case, the private payoff to the individual decision-maker fails to reflect the additional benefit his/her decision may confer on others. In both cases the nature of the interaction is such that the spill-over effect cannot be traded, since if it were possible to so trade the externality would soon disappear. An equilibrium strategy involving all participants would be found.

8.1.3 Braess' paradox with perfectly inelastic demand. This situation can be demonstrated by a Cournot–Nash spatial-equilibrium model, which characterizes the non-cooperative equilibrium of several players, each minimizing his/her own cost (Catoni and Pallottino 1991). Externalities can very well result in such an oligopolistic market. If the number of Cournot–Nash players is one, the Cournot–Nash equilibrium model is reduced to the system-optimizing or prescriptive model, as player cost and total cost are the same. One can also think of this as a monopolistic market, where there is centralized decision-making. Increasing the number of players, however, the Cournot–Nash equilibrium model becomes a *user-optimizing* or descriptive model. Here 'user' refers, for example, to each shipper that uses the transportation system to trade. It transitions from a monopoly to an oligopoly and eventually to a purely competitive market when the number of players in the game becomes large. The user and system optimum can be related to Wardrop's equilibrium conditions respectively, which were initially developed for road traffic:

Wardrop's first principle: The journey times of all routes actually used are equal and less than those which would be experienced by a single vehicle on any unused route.

Wardrop's second principle: The average journey-time is minimal.

Building on Wardrop's principles, the most graphic illustration of the stability and sensitivity of market equilibrium is perhaps the celebrated Braess' paradox. Braess illustrated this through an example in which six units produced at node-1 V_1^s are to be supplied to the demand at node-4 V_4^d, and there are two routes to ship the commodity. This situation is shown in **Figure 8-3**, in which the example from Subsection 8.1.1 is now expanded to a network consisting of four nodes. Here the quantity shipped between the two regions $x^{1,4}$ is equal to V_1^s and to V_2^d. The shipment demand here is perfectly inelastic: $x^{1,4}(C^{1,4}) = 6$. Let x^1 be the equilibrium flow on path 1–2–4 and x^2 be the flow on path 1–3–4. As a first step, the following equations are developed to represent the transportation costs along each path for the given network:

$$\text{Path 1: } C^1 = (60+x^1)+(10+10x^1) = 70+11x^1$$
$$\text{Path 2: } C^2 = (60+x^2)+(10+10x^2) = 70+11x^2.$$

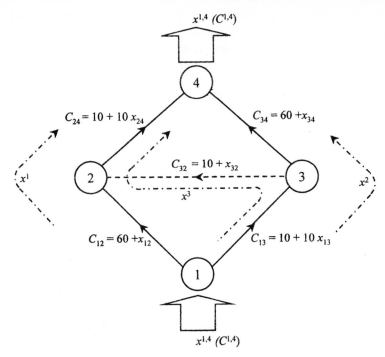

Figure 8-3 - Braess'-paradox example with a demand function

Before another 'cross' link (2,3) is added to improve the transportation network, the path-cost functions are really 'link' cost functions, since we can add the constituent link costs on path 1 and path 2 separately, as we have done above. There are really no transshipment nodes or 'network interaction' effects in the complete sense of the term.

Now user equilibrium can be obtained by considering the first unit of shipment. It can be shipped via either path 1 or path 2, since both have the same transportation cost. Let us say we take path 1. Based on the Cournot assumption, the shipper, in dispatching the second shipment, will take the current market condition as given and optimize his/her shipment strategy accordingly. The second unit shipped will take path 2 since it has a lower transportation-cost of 70 dollars instead of 81 dollars (70+11(1)) on path 1. Given the current market condition again, the third unit will be shipped along path 1 once more to minimize cost. In this alternating fashion, a Nash equilibrium will be obtained finally. Three units are shipped via path 1 and the other three via path 2, i.e., $x^1 = 3$ and $x^2 = 3$. The transportation cost on both paths will be 70+11(3) = 103 dollars, i.e., $C^1 = C^2 = 103$, at equilibrium. Similarly, we can compute the system equilibrium, in which the total cost in unit-dollars is to be minimized for the entire network. By trial and error, one can also arrive at $x^1 = x^2 = 3$, at a total cost of (3)(103)+(3)(103) = 618. One can show that this is indeed the optimum by switching one unit of flow between the two paths, which will yield a sub-optimal (higher-cost) solution of $(2)(70+11(2))$ + $(4)(70+11(4)) = 640$.

The heuristic procedure can be formalized in terms of nonlinear programming. The user-optimizing model looks like

$$\min \; f(\mathbf{x}) = \int_0^{x^1}(70+11x)\,dx + \int_0^{x^2}(70+11x)\,dx$$
$$s.t. \qquad x^1 + x^2 = 6 \tag{8-1}$$
$$x^1, x^2 \geq 0$$

yielding $x^1 = x^2 = 3$ and the dual variable $\lambda = -103$ dollars/unit, suggesting that another unit of shipment will *increase* the total transportation cost by exactly 103. In other words, the marginal cost to a seventh unit shipped will be 103. The corresponding system-optimizing model looks like $\min \; F(\mathbf{x}) = (70+11x^1)x^1 + (70+11x^2)x^2$ subject to the same constraints. The solution of $\mathbf{x} = (3,3)$ is again confirmed, with the extra information on the dual variable of $\lambda = -136$. In other words, the marginal cost to total system transportation cost is 136 unit-dollars for an additional unit shipped—higher than the user-optimum case.

Upon the addition of a 'cross' link that connects nodes 3 and 2, network interaction effects will now take place. Here a link carries more than one path flow, and transshipment truly occurs at nodes 2 and 3. For example, **Figure 8-3** shows that link (2,4) carries path-flows x^1 and x^3 and link (1,3) x^2 and x^3. The nonlinear-programming model for user optimization now looks like

$$\min \; f(\mathbf{x}) = \int_0^{x^1}(60+x)\,dx + \int_0^{x^1+x^3}(10+10x)\,dx$$
$$+ \int_0^{x^2+x^3}(10+10x)\,dx + \int_0^{x^2}(60+x)\,dx + \int_0^{x^3}(10+x)\,dx \tag{8-2}$$
$$s.t. \qquad x^1 + x^2 + x^3 = 6$$
$$x^1, x^2, x^3 \geq 0.$$

The solution to this model is $x^1 = x^2 = x^3 = 2$ and $\lambda = 112$. The transportation costs on the three paths are

Path 1–2–4: $C^1 = (60+x^1) + (10+10(x^1+x^3)) = 70+11x^1+10x^3$
Path 1–3–4: $C^2 = (60+x^2) + (10+10(x^2+x^3)) = 70+11x^2+10x^3$
Path 1–3–2–4: $C^3 = (10+10(x^1+x^3)) + (10+x^3) + (10+10(x^2+x^3)) = 30+10x^1+10x^2+21x^3$,

which correspond to $C^1 = C^2 = C^3 = 112$. This says that improvement in the transportation network results in an *increase* in transportation cost from 103 dollars to 112 dollars! This is clearly a paradox since one would expect the opposite to be true—that an improvement of the transportation network will lower shipping cost for everyone concerned.

One can write the Karash-Kuhn-Tucker (KKT) conditions for this spatial equilibrium. The one that is of particular interest is dual complementary-slackness:[7]

$$x^1(70+11x^1+10x^3+\lambda) = 0$$
$$x^2(70+11x^2+10x^3+\lambda) = 0$$
$$x^3(30+10x^1+10x^2+21x^3+\lambda) = 0.$$

These conditions stipulate that if a shipper could receive at the destination a price λ that equals the transportation cost, or $(C^i+\lambda) = 0$, s/he will continue to ship ($x^i >$ 0) the good until the sale price falls below transportation cost .Remember λ is negative in value and that there is no production cost in this example. On the other hand, if the delivery cost is greater than the sale price, or $(C^i+\lambda) > 0$, the shipper will lose money in trading. Hence s/he will not trade ($x^i = 0$).

The system-optimizing model after network improvement will assume the following objective function:

$$\min F(\mathbf{x}) = x^1(70+11x^1+10x^3) + x^2(70+11x^2+10x^3) + x^3(30+10x^1+10x^2+21x^3)$$

subject to the same constraints as the user-optimizing model. Solution of this model yields $x^1 = x^2 = 3$ and $x^3 = 0$. Notice the new link that purports to improve the transportation network is never used—a "white elephant'" in its truest sense! Travel pattern and transportation costs, including the shadow price λ, remain the same as before–another manifestation of the paradox! The irony of the situation is as follows. To prevent a user equilibrium from taking place after network improvement, with its concomitant higher transportation cost for every shipment, a centralized decision-maker, such as a regulatory agency, would have to bar shippers from using the cross link (2,3). This example clearly shows the externality incurred with a Cournot–Nash market. Also illustrated is the instability and sensitivity of market equilibrium.

8.1.4 User vs. system optimization. As mentioned, user optimization in the above example can be thought of as effecting a Cournot–Nash equilibrium while system optimization as a monopolistic-market equilibrium. The transition from the former to the latter can be traced out by a multicriteria-optimization model, in which a weighted sum is formed between the user-optimizing and system-optimizing criteria (Barondas and Chan 1991). In this model an increasing weight is placed upon the system-optimizing criterion as λ' decreases from unity to zero:[8] $\lambda' f(\mathbf{x}) + (1-\lambda')F(\mathbf{x})$ where $0 \le \lambda' \le 1$. The constraints of the three-path optimization model remain the same.

For a fixed demand of six units, the following summarizes the solutions for different values of λ'.

---

[7]For an explanation of KKT conditions and complementary slackness, see the "Prescriptive Tools" chapter in Chan (2000).

[8]This approach is explained in the "Multi-criteria Decision-making" chapter (Chan 2000) as the weighted-sum method of multicriteria optimization.

λ' weight	User-opt. criterion f (unit-dollars)	System-opt. criterion F (unit-dollars)
1.0	506	672
0.9	507	652
0.8	509	638
0.7	513	628
0.6	516	621
0.5	519	618
.	.	.
.	.	.
.	.	.
0.0	519	618

Setting aside the paradoxical results between user and system optimization, these columns clearly show the results of decreasing competition. As the amount of competition decreases (or as λ' decreases), user cost increases while system cost decreases. This says that the lack of competition results in higher user costs. The 'monopolist' achieves his/her objective in cutting down system-wide operating cost (and increasing 'revenue'). The users, meanwhile, are not as well served since they are paying more in aggregate as a group. At λ' = 0.52, three-path flow changes to two-path flow. More will be said about this in a more general model.

8.1.5 Braess' Paradox with a demand function. Suppose there is a trade flow between production site 1 and the consumers at site 4 as expressed by this function $x^{1,4}(C^{1,4}) = 118 - C^{1,4}$ where $C^{1,4}$ is the shipment or transportation cost from 1 to 4, as shown in **Figure 8-3**. This transportation-demand function and the average-cost function $C(\cdot)$ for path 1–2–4 and path 1–3–4 can be graphed in **Figure 8-4**. Also shown in the graph are the marginal-cost function $S(\cdot)$; so are the user- and system-equilibrium flows x and the associated costs C. The user equilibrium is obtained by the intersection of the average-cost function of travel from 1 to 4 $C^{1,4}$ and the transportation-demand function $x^{1,4}$. The system equilibrium, on the other hand, is the intersection of the marginal-cost function $S^{1,4}$ (which happens to be the same as path-cost functions C^1 (or C^2) here) and the transportation-demand function. Equilibria are shown for both the entire network (throughput x and travel cost C) and by path (flow x^1, x^2 and travel costs C^1 and C^2 respectively).

Algebraically, the cost function between origin (O) 1 and destination (D) 4 is the sum of these two path functions along the horizonal axis, since the total flow consists of flows on both path 1 and 2: $C^{1,4}(x^1+x^2) = 70 + 11/2(x^1+x^2)$ for O–D pair 1–4. The marginal-cost functions—the incremental cost of producing an additional unit of output—are simply the first derivative of the total-cost function $C^i(x^i) = (70+11x^i)x^i$ for $i = 1,2$ with respect to x. This means $S^i(x^i) = 70 + 22x^i$ for $i = 1,2$. The marginal-cost function for the entire network is again the aggregation of path functions along the horizontal axis: $S^{1,4}(x^1+x^2) = 70 + 11(x^1+x^2)$. These algebraic

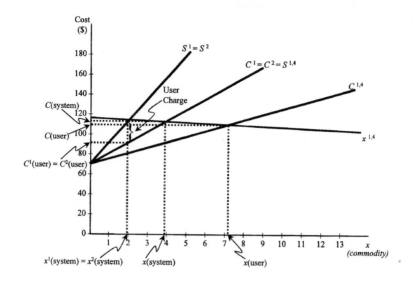

Figure 8-4 - Graphical solution to Braess'-paradox example

derivations confirm the graphical plot presented earlier in **Figure 8-4**, as the reader can verify.

A quadratic program can now be formulated for user optimization by minimizing the transportation cost for each participant in the game:

$$\min \left[\int_0^{x^1}(70+11x)\,dx + \int_0^{x^2}(70+11x)\,dx \right] \tag{8-3}$$

such that $x^1 + x^2 = x^{1,4} = 118 - C^i(x^i)$ and $x^i \geq 0$ for $i = 1,2$. Similarly, a quadratic program can be written for system optimization, in which the total cost of shipping is minimized by a centralized decision-maker: $\min [x^1(70+11x^1) + x^2(70+11x^2)]$ such that $x^1 + x^2 = x^{1,4} = 118 - S^{1,4}(x^1 + x^2)$ and $x^i \geq 0$ for $i = 1,2$.

User optimum can be found by first forming the Lagrangian

$$\varphi(\mathbf{x}, \boldsymbol{\lambda}) = \int_0^{x^1}(70+11x^1)\,dx + \int_0^{x^2}(70+11x^2)\,dx$$
$$+ \lambda_1(x^1 + x^2 - 118 + (70+11x^1)) + \lambda_2(x^1 + x^2 - 118 + (70+11x^2)) \tag{8-4}$$

Taking partial derivatives yield four equations and four unknowns:

$$\frac{\partial \varphi}{\partial x^1} = 11x^1 + 12\lambda_1 + \lambda_2 + 70 = 0 \qquad \frac{\partial \varphi}{\partial \lambda_1} = 12x^1 + x^2 - 48 = 0$$

$$\frac{\partial \varphi}{\partial x^2} = 11x^2 + \lambda_1 + 12\lambda_2 + 70 = 0 \qquad \frac{\partial \varphi}{\partial \lambda_2} = 12x^2 + x^1 - 48 = 0 \tag{8-5}$$

where λ's are unrestricted in sign. Solution of this set of simultaneous equations yield $x^1 = x^2 = 3.69$ units shipped, $\lambda_1 = \lambda_2 = -0.51$ dollars/unit. The shipping cost on either paths is $C^1 = C^2 = 70 + 11(3.69) = 110.59$ dollars. The negative sign in front of the Lagrange multiplier or dual variable suggests that additional demand for shipping will *increase* the travel cost.

System optimum can be obtained similarly by forming the corresponding Lagrangian:

$$\Phi(\mathbf{x},\lambda) = x^1(70+11x^1) + x^2(70+11x^2) + \lambda[x^1+x^2-118+(70-11(x^1+x^2))]. \qquad \textbf{(8-6)}$$

Taking partial derivatives yield three equations and three unknowns here:

$$\frac{\partial\Phi}{\partial x^1} = 22x^1 + 12\lambda + 70 = 0$$

$$\frac{\partial\Phi}{\partial x^2} = 22x^2 + 12\lambda + 70 = 0 \qquad \textbf{(8-7)}$$

$$\frac{\partial\Phi}{\partial\lambda} = 12x^1 + 12x^2 - 48 = 0.$$

Equilibrium solution consists of $x^1 = x^2 = 2.00$ units shipped and $\lambda = -2.17$ dollars / unit. The corresponding transportation cost is $S^{1,4} = 70 + 11(4.00) = 114.00$ dollars.

An interesting KKT condition from this system-optimization model is $(70 + 22x^1 + 12\lambda) x^{1,4}$ and $(70 + 22x^2 + 12\lambda) x^{1,4}$. This states that if there is trade flow between 1 and 4, the marginal economic-cost of transportation $70 + 22x^i$ has to be equal to the revenue 12λ. Remember again that λ is negative in value and there is no production cost in this example. If the total marginal cost $70+22x^i$ exceeds the revenue 12λ, then nothing would be shipped.

If desired, a user-charge can be used to effect system optimation, wherein the instability and sensitivity of a user-optimum can be avoided. By levying a user-charge of $(114.00 - [70 + 11(2.00)]) = 22.00$ dollars, the participants in the shipping game will be coerced to behave in a way they perceive to be for their best interests (or user-optimal). In reality, however, their behavior will result in system optimum. This is best illustrated in **Figure 8-4**, in which the user charge is shown as the difference between the system-optimum transportation-cost and the user-optimum cost on each path. We have seen through this example that whatever the nature of the externality, it is possible to provide a theoretical means of internalizing the effect such that a stable equilibrium can be achieved. Internalization and hence elimination of the external effect involves ensuring that the social cost of production of the external effect appears in the appropriate constraint of the causer(s), in the form of an effective shadow price. As noted above, this can be done through imposition of an optimal user-charge.

8.2 The diagonalization network-equilibrium algorithm

Most of the examples solved thus far assume that link-performance functions are independent of each other. In other words, transportation cost on a link depends on the flow through that link and not on the flow through any other link. This assumption is not always valid, given that market equilibrium is often achieved through a strategic game between participants. The shipping decision of one often affects the other, since they use the same transportation network to reach their markets. We will therefore have to find the equilibrium when the interactions of flows and transportation costs across links are recognized explicitly (Sheffi 1985).

8.2.1 Symmetric vs. asymmetric interaction. Flow interactions can be eithersymmetric or asymmetric. When the interactions are symmetric, the marginal effect of one link flow, say x_a on the transportation cost on any other link, say C_b, is equal to the marginal effect of x_b on C_a. In other words,

$$\frac{\partial C_a(x_a, x_b)}{\partial x_b} = \frac{\partial C_b(x_b, x_a)}{\partial x_a}.$$ (8-8)

In this case, the equilibrium-flow pattern can be found by applying the equivalent minimization approach used in the previous examples. This symmetry condition can be applied toward path flows as well. In the Braess' paradox where there is a cross link, for example, marginal cost on path-1 depends on the path-3 flow x^3, as much as marginal cost on path-3 depends on path-1 flow x^1, and similarly for each pair of paths:

$$\frac{\partial C^1}{\partial x^2} = \frac{\partial C^2}{\partial x^1} = 0 \qquad \frac{\partial C^1}{\partial x^3} = \frac{\partial C^3}{\partial x^1} = 10 \qquad \frac{\partial C^2}{\partial x^3} = \frac{\partial C^3}{\partial x^2} = 10.$$ (8-9)

When the link or path interactions are asymmetric, there is no known equivalent minimization program that can be used to find the equilibrium-flow pattern. In other words, traditional optimization-procedure breaks down when

$$\frac{\partial C[a](x)}{\partial x[b]} \neq \frac{\partial C[b](x)}{\partial x[a]} \qquad \forall a \neq b$$ (8-10)

or when the following *Jacobian* matrix is asymmetric

$$\nabla_{\mathbf{x}}\mathbf{C}^T = \begin{bmatrix} \dfrac{\partial C[1](\mathbf{x})}{\partial x[1]} & \dfrac{\partial C[2](\mathbf{x})}{\partial x[1]} & \cdots & \dfrac{\partial C[i](\mathbf{x})}{\partial x[1]} & \cdots \\[2ex] \dfrac{\partial C[1](\mathbf{x})}{\partial x[2]} & \dfrac{\partial C[2](\mathbf{x})}{\partial x[2]} & \cdots & \dfrac{\partial C[i](\mathbf{x})}{\partial x[2]} & \cdots \\[2ex] \cdot & \cdot & \cdot & & \\ \cdot & \cdot & \cdot & & \\ \cdot & \cdot & \cdot & & \\[1ex] \dfrac{\partial C[1](\mathbf{x})}{\partial x[i]} & \dfrac{\partial C[2](\mathbf{x})}{\partial x[i]} & & \dfrac{\partial C[i](\mathbf{x})}{\partial x[i]} & \\[2ex] \cdot & \cdot & & & \cdot \\ \cdot & \cdot & & & \cdot \\ \cdot & \cdot & & & \cdot \end{bmatrix} . \qquad \textbf{(8-11)}$$

Here, the notation [·] denotes either a link or a path. For convenience, we will carry out the discussion in terms of link (arc) flow below. The result can be carried over to path flow in a straight-forward way.

It can be found that all the above equilibria, including asymmetric-flow interactions, can be obtained by a general network-equilibrium solution-algorithm. The algorithm presented here is based on solving a series of standard user-equilibrium mathematical programs. Each iteration of this procedure requires the solution of one such program. Assume that, at the kth iteration of the procedure, the values of all the link-flow variables $(x_1(k),\ldots, x_{|A|}(k))$ are known .Here $|A|$ denotes the maximum number of arcs in a network made up of arc-set $\underline{A}$. The following mathematical program can be formulated:

$$\min \sum_a \int_0^{x_a} C_a(x_1(k),\ldots,x_{a-1}(k),\omega,x_{a+1}(k),\ldots,x_{|A|}(k))\,d\omega \qquad \textbf{(8-12)}$$

subject to

$$\sum_p x_p^{i,j} = V^{i,j} \qquad \forall\, p,i,j \qquad \textbf{(8-13)}$$

$$x_p^{i,j} \geq 0 \qquad \forall\, p,i,j. \qquad \textbf{(8-14)}$$

This program will be referred to later as the `subproblem'. It is a standard user-equilibrium minimization. The objective function includes only simple link-performance-function in one argument each (i.e., C_a is a function of x_a only); all other flows that may affect the link are fixed at their values during the kth iteration. This formulation does *not* fix the flows themselves, only their cross-link effects. Accordingly, the subproblem is also known as the 'diagonalized' problem–i.e., the Hessian of Equation **(8-12)**

$$\nabla^2 f(\mathbf{x}) = \left[\frac{\partial^2 f(\mathbf{x})}{\partial x_i \partial x_j}\right] = \begin{bmatrix} \frac{\partial^2 f}{\partial x_1^2} & 0 & \cdots & 0 \\ 0 & \frac{\partial^2 f}{\partial x_2^2} & \cdots & 0 \\ & & \cdots & \\ 0 & \cdots & & \frac{\partial^2 f}{\partial x_{|A|}^2} \end{bmatrix}$$

is diagonal since all cross-link effects have been fixed, or all off-diagonal elements $\frac{\partial^2 f}{\partial x_i \partial x_j}$ $= 0$ $(i \neq j)$.

8.2.2 The solution algorithm. The general process of finding the equilibrium flows requires the following steps:

Step 0: *Initialization.* Find a feasible link-flow vector $\mathbf{x}(k)$. Set $k = 0$.

Step 1: *Diagonalization.* Solve subproblem as defined by Equations **(8-12)** through **(8-14)**. This yields a link-flow vector $\mathbf{x}(k+1)$.

Step 2: *Convergence Test.* If $\mathbf{x}(k) \approx \mathbf{x}(k+1)$, stop. If not, set $k \leftarrow k+1$, and go to step 1. ∎

The convergence test required in step 2 of the algorithm can be based on the maximum difference in link flow between successive solutions or on any other criterion measuring the similarity between $\mathbf{x}(k)$ and $\mathbf{x}(k+1)$. For example, the algorithm can be stopped if

$$\max_a \left\{\frac{|x_a(k+1) - x_a(k)|}{x_a(k+1)}\right\} \leq \varepsilon \tag{8-15}$$

where ε is a dimensionless pre-determined threshold.

8.2.3 An example. Consider the simple two-link network shown in **Figure 8-5**, with the following 'cross-over' performance functions: $C_1 = 2+4x_1+x_2$ and $C_2 = 4+3x_2+2x_1$. The origin–destination demand is 5 units: $x_1+x_2 = 5$. The equilibrium solution of this system can be found analytically, as done in previous examples. At equilibrium, Wardrop's first-principle applies: $C_1(x_1, x_2) = C_2(x_1, x_2)$. The resulting equilibrium solution is then $x_1^* = 3$ and $x_2^* = 2$. Due to the cross-over effects in the link-performance functions, however, the equivalent mathematical-programming approach to this problem cannot be used since the symmetry condition does not hold. The Jacobian of the link-travel-cost function is the following asymmetric (though positive definite) matrix $\begin{bmatrix} \frac{\partial C_1}{\partial x_1} & \frac{\partial C_1}{\partial x_2} \\ \frac{\partial C_2}{\partial x_1} & \frac{\partial C_2}{\partial x_2} \end{bmatrix} = \begin{bmatrix} 4 & 1 \\ 2 & 3 \end{bmatrix}$.

The problem can be solved by the diagonalization algorithm. To apply this algorithm, an initial feasible solution is needed. Take $x_1(0) = 0$, $x_2(0) = 5$, which can be easily checked to be feasible. Furthermore, assume that the convergence

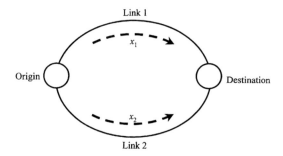

Figure 8-5 - Network illustrating the diagonalization algorithm

criterion is based on the relative change in the flow being less than 1.5 percent, as specified in Equation **(8-15)**. The algorithmic iterations can be described then as follows:

FIRST ITERATION:

Step 1: *Diagonalization.* Using traditional programming-techniques, solve the subproblem:

$$\min\ f(x_1, x_2) = \int_0^{x_1} (2 + 4\omega + 5)\,d\omega + \int_0^{x_2} (4 + 3\omega)\,d\omega \qquad (8\text{-}16)$$

subject to $x_1 + x_2 = 5$ and $x_1, x_2 \geq 0$. We have $x_1(1) = 3.286$, $x_2(1) = 1.714$.

Step 2: *Convergence test.* Compare $(\mathbf{x}(0))^T = (0, 5)$ with $(\mathbf{x}(1))^T = (3.286, 1.714)$ or $\max_a\{\,|x_a(1) - x_a(0)|\ /\ x_a(0)\,\} = |3.786 - 0|\ /\ 0 = \infty$. Clearly the convergence criterion is not met.

SECOND ITERATION:

Step 1: *Diagonalization.* Solve the new subproblem

$$\min\ f(x_1, x_2) = \int_0^{x_1} (2 + 4\omega + 1.714)\,d\omega + \int_0^{x_2} (4 + 3\omega + 7.572)\,d\omega \qquad (8\text{-}17)$$

subject to $x_1 + x_2 = 5$ and $x_1, x_2 \geq 0$. The solution is $x_1(2) = 3.265$, $x_2(2) = 1.735$.

Step 2: *Convergence test.* Compare $(\mathbf{x}(1))^T = (3.286, 1.714)$ with $(\mathbf{x}(2))^T = (3.265, 1.735)$ or $\max_a\{\,|x_a(2) - x_a(1)|\ /\ x_a(1)\,\} = |1.735 - 1.714|\ /\ 1.714 = 0.012$. The convergence criterion is now met and more iterations are not necessary.

8.2.4 *Comments.* In spatial competitive-equilibrium, clearly there is a fair amount of interaction between the participants in the game. Except for the case of a monopoly, such interaction has to be explicitly addressed. The interaction is made more complex by the spatial dimension through which each participant ships his/her goods to market .within a common network. Flow interaction in the network assumes an extremely interesting pattern. The diagonalization algorithm has been presented here to solve for the general network equilibria. Such an algorithm is robust enough to address many applications. Its computational efficiency, however, is not necessarily as impressive. Several improvements over the original diagonalization algorithm have been reported, including those by Sheffi (1985), Theise and Jones (1988), Schneider and Zenios (1990), Nagurney (1993)., and Miller et al. (1996).

8.3 Alternative models of spatial competition

The above discussions on "Conflicts in spatial choice" can be generalized to include both nonlinear cost-functions and nonlinear demand- and supply-functions. Instead of a simple four-node example, a general network can be solved for pure competition, oligopoly and monopoly alike. Even more satisfying is that all these network models can be solved by a unifying diagonalization algorithm, including transshipment networks (Harker 1986). Later in this chapter, we will show how the general algorithm described in this section can be further improved for computational efficiency.

8.3.1 *Classical spatial-price-equilibrium model*. Consider a network $N(I,\underline{A})$ composed of a set of nodes I and a set of arcs $\underline{A}$. Each node in I represents a centroid of a region. Each arc in $\underline{A}$ represents an origin–destination (O–D) pair connecting two regional centroids. Let us define, for each region l in the set of nodes, the variable V_l^s as the supply quantity, V_l^d the demand quantity, and V_{ij} the flow between O–D pair (i,j). The conservation of flow in every region can be written as

$$V_l^s - V_l^d + \sum_{i \in I} \sum_{(i,l) \in \underline{A}} V_{il} - \sum_{j \in I} \sum_{(l,j) \in \underline{A}} V_{lj} = 0 \qquad \forall l \in I. \tag{8-18}$$

The market-clearing condition requires that total demand equal total supply, or that

$$\sum_{l \in I} V_l^d - \sum_{l \in I} V_l^s = 0. \tag{8-19}$$

This condition is redundant, since summing **(8-18)** over all $l \in I$ will yield the same result. However, Equation **(8-19)** will play an important role in the network-flow-programming solution-algorithms.

For each O–D pair, let us define a function $C_{ij}(V_{ij})$ that represents the *average* cost of transporting goods between O–D pair (i,j). Let us also define $S_i(V_i^s)$ as the inverse supply function and $D_i(V_i^d)$ as the inverse demand function for each region l, representing the price charged at supply quantity V_l^s and the price paid at demand quantity V_l^d respectively. The spatial-equilibrium conditions can then be written as

(a) if $V_{ij} > 0$, then $S_i(V_i^s) + C_{ij}(V_{ij}) = D_j(V_j^d)$ for all (i,j);

(b) if $S_i(V_i^s) + C_{ij}(V_{ij}) > D_j(V_j^d)$, then $V_{ij} = 0$ for all (i,j).

These conditions specify the conditions under which a trip will be executed or a shipper will purchase the good in a perfectly competitive market at node i for price $S_i(V_i^s)$ and will incur an economic price of transporting $C_{ij}(V_{ij})$ for the move from i to j. S/he takes this economic price of transportation as given. If the price $D_j(V_j^d)$ s/he could receive in the perfectly competitive market at region j exceeds the sum of the purchase price and the transportation charges, then s/he will continue to ship the good until the sales price and the delivered price are equal (condition (a)). On the other hand, if the delivered price is greater than the sale price, the shipper will lose money in shipping from i to j and hence will not ship (condition (b)). Thus the classical spatial-price-equilibrium (SPE) model assumes that each shipper acts between perfectly competitive markets and that s/he treats transportation as a perfectly competitive factor of production.

Example

To a bipartite network with two supply nodes and two demand nodes, we include the artificial source s and sink d, framing a full network structure (Nagurney 1993). The inverse supply-functions are given as $S_1(V^s) = S_1(V_1^s) + S_2(V_2^s) = 5V_1^s + V_2^s + 2$ and $S_2(V^s) = S_1(V_1^s) + S_2(V_2^s) = V_1^s + 2V_2^s + 3$. The average-cost functions are $C_{11}(V_1) = C_{11}(V_{11}) + C_{12}(V_{12}) = V_{11} + 0.5V_{12} + 1$, $C_{12}(V_2) = C_{12}(V_{12}) + C_{22}(V_{22}) = 2V_{12} + V_{22} + 1.5$, $C_{21}(V_1) = C_{21}(V_{21}) + C_{11}(V_{11}) = 3V_{21} + 2V_{11} + 15$, $C_{22}(V_2) = C_{22}(V_{22}) + C_{12}(V_{12}) = 2V_{22} + V_{12} + 10$. The inverse demand-functions are $D_1(V^d) = -D_1(V_1^d) - D_2(V_2^d) = -2V_1^d - V_2^d + 28.75$ and $D_2(V^d) = -D_1(V_1^d) - D_2(V_2^d) = -4V_2^d - V_1^d + 41$.

    Following conditions (a) and (b) above, the equilibrium production, shipment and $= 1.5$, $V_{12} = 1.5$, $V_{21} = 0$, $V_{22} = 2$; $V_1^d = 1.5$ and $V_2^d = 3.5$. The equilibrium supply prices are $S_1(V^s) = 19$, $S_2(V^s) = 10$. The costs are $C_{11}(V_1) = 3.25$, $C_{12}(V_2) = 6.5$, $C_{21}(V_1) = 18$, $C_{22}(V_2) = 15.5$. The demand prices are $D_1(V^d) = 22.25$ and $D_2(V^d) = 25.5$. Note that supply-market 2 does not ship to demand-market 1. This is due, in part, to the high fixed-cost associated with trading between this market pair. In essence, this model says that there is a single most-efficient route-of-trading between a supply market i and a demand market j, as represented by link (i,j). □

Conditions (a), (b) and Equations (8-18), (8-19) can be formulated as a mathematical program:

$$\min \sum_{l \in I} \int_0^{V_l^s} S_l(x)\,dx - \sum_{l \in I} \int_0^{V_l^d} D_l(x)\,dx + \sum_{(i,j) \in A} \int_0^{V_{ij}} C_{ij}(x)\,dx \qquad (8\text{-}20)$$

subject to Equations (8-18) and (8-19)

$$\begin{aligned} V_l^d, V_l^s \ge 0 &\qquad \forall l \in I \\ V_{ij} \ge 0 &\qquad \forall (i,j) \in A. \end{aligned} \qquad (8\text{-}21)$$

Here the objective function simply states that the revenue received by the supplier equilibrates with the expenditure of the consumer, accounting for the transportation

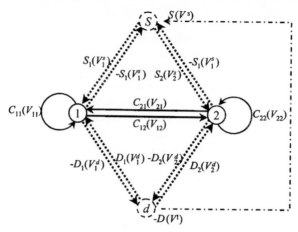

Figure 8-6 - Full network structure for a two-region example

costs. Equations **(8-20)** through **(8-21)** form the classical spatial-price-equilibrium problem (CSPE).

In the case the supply function is not given explicitly, we can write the following expression. We write the minimum cost of producing V_i^s units of output as one at which price equals the marginal cost of production in that region:

$$C_l(V_l^s) = \int_0^{V_i^s} S_l(x)\, dx + constant.$$

(8-22)

Such a relationship can be graphed in **Figure 8-7**, which shows a set of typical cost-functions. Let us substitute Equation **(8-22)** into **(8-20)**, drop the constant term, and place upper and lower bounds on the supplies, demands and O–D flows that rise from production capacities, minimum consumption levels, and other problem features. Now we can rewrite CSPE as

$$\max \sum_{l \in I} \int_0^{V_i^d} D_l(x)\, dx - \sum_{l \in I} C_l(V_l^s) - \sum_{(i,j) \in \underline{A}} \int_0^{V_{ij}} C_{ij}(x)\, dx$$

(8-23)

subject to Equations **(8-18)** and **(8-19)** and these bounds

$$\tilde{P}_l \le V_l^s \le \overline{P}_l \qquad \forall l \in I$$

(8-24)

$$\tilde{Q}_l \le V_l^d \le \overline{Q}_l \qquad \forall l \in I$$

(8-25)

$$\tilde{V}_{ij} \le V_{ij} \le \overline{V}_{ij} \qquad \forall (i,j) \in \underline{A}.$$

(8-26)

Here the objective function maximizes profit, and we call this reformulated model *pure-competition*.

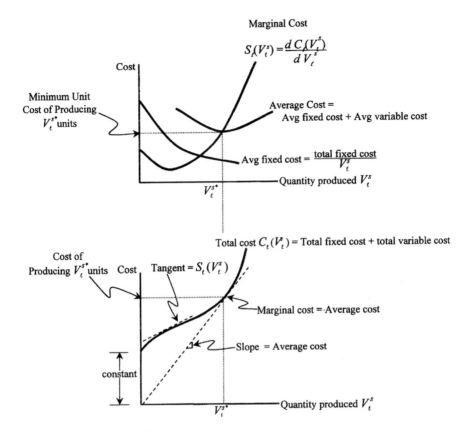

Figure 8-7 - Marginal, fixed, average and total-cost functions

8.3.2 Monopoly models. Let us assume that there exists only one firm that produces a single homogeneous-commodity, and that this firm has prefect information concerning the demand behavior in each region and fully controls the transportation system. That is, the monopolist is either a monopsonist in the transportation market or owns its fleet of vehicles. Now, the firm's profit-maximization problem can be written as

$$\max \sum_{l \in I} D_l(V_l^d) V_l^d - \sum_{l \in I} C_l(V_l^s) - \sum_{(i,j) \in \underline{A}} C_{ij}(V_{ij}) V_{ij} \tag{8-27}$$

subject to

$$V_l^s - V_l^d + \sum_{i \in L} \sum_{(i,l) \in \underline{A}} V_{il} - \sum_{j \in L} \sum_{(l,j) \in \underline{A}} V_{lj} = 0 \qquad \forall l \in I \tag{8-28}$$

$$V_l^d, V_l^s \geq 0 \qquad \forall\, l \in I. \tag{8-29}$$

$$V_{ij} \geq 0 \qquad \forall\, (i,j) \in \underline{A}. \tag{8-30}$$

Upper- and lower-limit constraints **(8-24)** through **(8-26)** and the redundant constraint **(8-19)** has been temporally ignored for ease of presentation.

Suppose the revenue $D_l(V_l^d)V_l^d$ is concave for all $l \in I$, price $D_l(V_l^d)$ is strictly decreasing and continuously differentiable, total-cost of production $C_l(V_l^s)$ is convex and continuously differentiable, and total transportation-cost $C_{ij}(V_{ij})V_{ij}$ is convex and the cost of transportation $C_{ij}(V_{ij})$ is strictly increasing and continuously differentiable for all (i,j). Then the Karash–Kuhn–Tucker (KKT) conditions[9] of this problem are necessary and sufficient for a solution. Letting π_l denote the dual variable of constraint **(8-18)**, we can write the KKT conditions as

$$\begin{aligned}
(D_l + V_l^d \dot{D}_l - \pi_l)V_l^d &= 0 \qquad \forall\, l \in I \\
D_l + V_l^d \dot{D}_l - \pi_l &\leq 0, \quad V_l^d \geq 0;
\end{aligned} \tag{8-31}$$

$$\begin{aligned}
(-\dot{C}_l + \pi_l)V_l^s &= 0 \qquad \forall\, l \in I \\
-\dot{C}_l + \pi_l &\leq 0, \quad V_l^s \geq 0;
\end{aligned} \tag{8-32}$$

$$\begin{aligned}
(-C_{ij} - V_{ij}\dot{C}_{ij} + \pi_j - \pi_i)V_{ij} &= 0 \qquad \forall\, (i,j) \in \underline{A} \\
-C_{ij} - V_{ij}\dot{C}_{ij} + \pi_j - \pi_i &\leq 0, \quad V_{ij} \geq 0;
\end{aligned} \tag{8-33}$$

where the $\bullet$ denotes differentiation.[10]

According to "Theory of the firm," condition **(8-31)** says that if there is demand in region l, the shadow price π_l will equal the marginal revenue in region l, $D_l + V_l^d \dot{D}_l$. Similarly, if there is supply in region l, condition **(8-32)** states that the shadow price π_l equals the marginal cost of production $\dot{C}_l$. The boundary conditions (V_l^s or $V_l^d = 0$) reflect the fact that no production will occur if marginal costs are greater than marginal revenue at the supply and a similar condition exists for demand. Condition **(8-33)** says the following. Suppose` there is flow between regions i and j, then the marginal economic-cost of transportation (MTC), $C_{ij} + V_{ij}\dot{C}_{ij}$, plus the

---

[9]For an introduction to the KKT condition, see the chapter on "Prescriptive Tools for Analysis" in Chan (2000).

[10]The "Optimization" book appendix in Chan (2000) reviews KKT conditions as applied toward networks.

marginal production-costs $\pi_i = \dot{C}_i$ equals marginal revenue (MR) $= \pi_j$. In other words,

$$\text{(c) if } V_{ij} > 0, \text{ then } \dot{C}_i + \text{MTC}_{ij} = \text{MR}_j.$$

If the total marginal-costs exceeds the marginal revenue, then nothing should be shipped from i to j:

$$\text{(d) if } \dot{C}_i + \text{MTC}_{ij} > \text{MR}_j, \text{ then } V_{ij} = 0.$$

The equilibrium conditions (c) and (d) are very similar to the SPE conditions (a) and (b) except that the average transportation-costs and average revenue are replaced by their marginal values.

8.3.3 Oligopoly model.

Between pure competition and monopoly is a model consisting of a few firms operating in spatially separated markets, which is often the more realistic case for discrete facility-location. Let Q' denote the set of firms operating in the market and let J_q denote the set of production sites or regions under firm q's control, where $q \in Q'$. We assume that at most one firm operates in each region, and that each firm has knowledge of the demand behavior in each region and is neither a monopsonist nor controls the transportation system as in the monopoly model. Instead, it takes the economic price of transportation-service as given. This results in the *average* economic price of transportation being used rather than the marginal values as in the monopoly model. Finally, let us assume that the producing firms behave in a Cournot–Nash manner, in which each firm takes the other firms' production decisions as fixed when deciding upon its own supply/distribution strategy.

Given these assumptions, let us define V_{lq}^d as the amount supplied by firm q to region l (or the amount demanded by the consumers in region l from firm q), and $\tilde{V}_{lq}^d$ as the amount supplied by all other firms to region l: $\tilde{V}_{lq}^d = \sum_{j \in Q', j \neq q} V_{lj}^d$. The total amount demanded, which is the same as the total amount supplied, in region l, is given by $V_l^d = \sum_{j \in Q'} V_{lj}^d$. Firm q now decides on its optimal strategy vector $x_q^T = [(V_l^s | l \in J_q), (V_{lq}^d | l \in I), (V_{ij} | i \in J_q, j \in I, (i,j) \in A)]$, consisting of the total-amount supplied locally to region $l \in J_q$, the amount supplied by firm q to consumers elsewhere in the network, and the specific shipment between production site i and consumer site j. The set of constraints $\Omega_q = \{x_q\}$ it faces are:

$$V_l^s - V_{lq}^d + \sum_{i \in J_q} \sum_{(i,l) \in A} V_{il} - \sum_{j \in I} \sum_{(l,j) \in A} V_{lj} = 0 \qquad \forall l \in J_q \tag{8-34}$$

$$-V_{lq} + \sum_{i \in J_q} \sum_{(i,l) \in A} V_{il} = 0 \qquad \forall l \in I, l \notin J_q \tag{8-35}$$

$$\tilde{P}_l \leq V_l^s \leq \overline{P}_l \qquad \forall l \in J_q \tag{8-36}$$

$$\tilde{Q}_l \leq V_l^d \leq \bar{Q}_l \qquad \forall l \in I \tag{8-37}$$

$$\tilde{V}_{ij} \leq V_{ij} \leq \bar{V}_{ij} \qquad \forall i \in J_q, (i,j) \in \underline{A}. \tag{8-38}$$

Constraints **(8-34)** and **(8-35)** state that there must be conservation of firm q's flow in those regions in which firm q does and does not produce, respectively. The rest are bounds placed on supplies, demands, and shipments, similar to the purely competitive model.

Firm q's profit-maximization problem now can be written as

$$\max \sum_{l \in I} D_l(V_{lq}^d + \tilde{V}_{lq}^d) V_{lq}^d - \sum_{l \in J_q} C_l(V_l^s) - \sum_{l \in J_q} \sum_{(i,j) \in \underline{A}} C_{ij}(V_{ij}) V_{ij} \tag{8-39}$$

subject to $\mathbf{x}_q \in \Omega_q$. Notice again that firm q takes $\tilde{V}_{lq}^d$ and C_{ij} as fixed, according to a Cournot–Nash assumption and is a price-taker in the transportation market. Let us assume that total revenue $D_l(V_l^d) V_{lq}^d$ is a strictly concave function, market price $D_l(V_l^d)$ is a strictly decreasing function, total-cost of production $C_l(V_l^s)$ is a strictly convex function, and that the feasible region Ω_q is nonempty. Then problem **(8-39)** is completely equivalent to the following *variational inequality*[11] problem: Find the optimal production-vector for firm q

$$(\mathbf{x}_q^*)^T = \left[(V_l^{s*} | l \in J_q), (V_{lq}^{d^*} | l \in I), (V_{ij}^* | i \in J_q), j \in I, (i,j) \in \underline{A} \right]$$

such that

$$\dot{\mathbf{F}}_q^T(\mathbf{x}_q)(\mathbf{y}_q - \mathbf{x}_q) =$$

$$\left[\sum_{l \in J_q} \dot{C}_l(V_l^s)(V_l^s - V_l^{s*}) - \sum_{l \in I} MR_{lq}(\mathbf{v}^{d*})(V_{lq}^d - V_{lq}^{d^*}) + \sum_{l \in J_q} \sum_{(ij) \in \underline{A}} C_{ij}(V_{ij}^*)(V_{ij} - V_{ij}^*) \right] \geq 0 \tag{8-40}$$

for all $\mathbf{y}_q \in \Omega_q$, where $\dot{\mathbf{F}}$ is the gradient of the function being maximized, MR_{lq} denotes the marginal revenue in region l for firm q: $MR_{lq}(\mathbf{v}^d) = D_l(\tilde{V}_{lq}^d + V_{lq}^d) + V_{lq}^d + \partial D_l(\tilde{V}_{lq}^d + V_{lq}^d)/\partial V_{lq}^d$ and C_{ij} and $\tilde{V}_{lq}^d$ were taken as fixed when calculating the gradient of Equation **(8-39)**.

It can be shown that *a unique equilibrium exists in this model when these conditions are satisfied:*

a. $C_l(V_l^s)$ *is a strictly convex, continuously differentiable function for all l,*

---

[11]Variational inequality is a technique to provide unified treatment of equilibrium and optimization problems. For a formal discussion of variational inequality, see the "Control, Dynamics" appendix to Chan (2000).

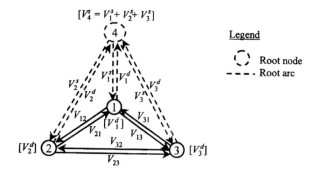

Figure 8-8 - Network structure for a three-region example

b. $C_{ij}(V_{ij})$ *is a monotone (nondecreasing), continuous function for all (i,j), and*
c. $-MR(V^d) = (\dots , -MR_{lq}(V^d), \dots)^T$ *is a strictly monotone, continuous function.*

These prerequisites for existence and uniqueness can be shown to be specialization of the more general conditions for a concave *n*-person games[12]. From the experience in the "Generation, Competition and Distribution" chapter, however, the last condition may be restrictive, particularly in discrete facility-location.

8.3.4 Solution algorithms. In designing a solution algorithm, the first fact one should notice is this. The constraint set in pure-competition and the monopoly models is a network-flow problem due to the addition of constraint **(8-19)**. To see this result, the reader need only note that each variable will appear with a +1 in one row and a -1 in one row. It makes up a pure network-flow problem.[13] **Figure 8-8** illustrates the type of network structure that underlies these problems. The only difference between the various models is the arc-cost definition. Therefore, one general nonlinear-network-algorithm can be used for all these models. It so turns out it through a variational-inequality formulation, the oligopoly model can also be solved by this network-algorithm. The variational-inequality formulation looks like:

$$\max \sum_{l \in I} D_l(\tilde{V}_{lq}^d(k-1) + V_{lq}^d(k)) V_{lq}^d(k) - \sum_{l \in J_q} C_l(V_l^s(k))$$
$$- \sum_{l \in J_q} \sum_{(i,j) \in A} \int_0^{V_{ij}(k)} C_{ij}(s)\,ds \qquad \textbf{(8-41)}$$

---

[12]For a description of gaming, see the "Economic Methods" and "Multi-criteria Decision-making" chapters in Chan (2000).

[13]For a discussion of network-flow programming, please see the "Prescriptive Tools" chapter and the "Optimization procedures" appendix to Chan (2000).

Region l	α_l	β_l	σ_l	δ_l
1	1.0	0.5	19.0	0.20
2	2.0	0.4	27.0	0.01
3	1.5	0.3	30.0	0.30

O–D Pair (i,j)	φ_{ij}	μ_{ij}
(1,2)	1.0	0.1
(1,3)	2.0	0.4
(2,1)	1.0	0.2
(2,3)	3.0	0.3
(3,1)	1.0	0.1
(3,2)	4.0	0.4

Table 8-1 Coefficients for example 1 (Harker 1986)

subject to $\mathbf{x}_q \in \Omega_q$ *and*

$$\sum_{l \in I} V_{lq}^d - \sum_{l \in J_q} V_l^s = 0. \tag{8-42}$$

The *diagonalization* network-algorithm will be described below for the most challenging of the three models, the oligopoly case:

Step 0. Choose an initial $\mathbf{x}(0) = (\mathbf{x}_q(0)) \in \Omega = \bigcup_{q \in Q} \Omega_q$, set $k = 0$.
Step 1. Solve Equation **(8-41)** for $\mathbf{x}_q(k+1)$ for each $q \in Q$.

Step 2. If

$$|V_l^s(k+1) - V_l^s(k)| \le \varepsilon_1 \qquad \forall \; q \in Q, \, l \in J_q \tag{8-43}$$

$$|V_{lq}^d(k+1) - V_{lq}^d(k)| \le \varepsilon_2 \qquad \forall \; q \in Q, \, l \in I \tag{8-44}$$

$$|V_{ij}(k+1) - V_{ij}(k)| \le \varepsilon_3 \qquad \forall \; (i,j) \in \underline{A}, \tag{8-45}$$

where the ε are preset tolerances; stop, the current solution is a Cournot–Nash equilibrium. Else, set $k \leftarrow k+1$ and return to step 1. ∎

This diagonalization algorithm will converge for any $\mathbf{x}(0) \in \Omega$ if

$$n \left| -\dot{D}_l(V_l^d) - V_{lq}^d \ddot{D}_l(V_l^d) \right| \leq w\alpha \qquad 0 < w < 1 \qquad \text{(8-46)}$$

for all $l \in I$, $q \in Q$, and $(V^s, V^d, V) \in \Omega$, where

$$= \min \left\{ \ddot{C}_l(V_l^s) \ \forall l \in I; \ 2\dot{D}_l(V_l^d) - V_{lq}^d \ddot{D}_l(V_l^d) \ \forall l \in I, q \in Q; \ \dot{C}_{ij}(V_{ij}) \ \forall (i,j) \in A \right\} \qquad \text{(8-47)}$$

and $n = |I| \, |Q'|$.

8.3.5 *Example 1*. Consider the network in **Figure 8-8**. The regional cost-functions, inverse demand- functions and origin-destination (O–D) transportation-cost functions are given by $C_l(V_l^s) = \alpha_l V_l^s + \beta_l \{V_l^s\}^2$, $D_l(V_l^d) = \sigma_l - \delta_l V_l^d$, and $C_{ij}(V_{ij}) = \varphi_{ij} + \mu_{ij} V_{ij}^2$ respectively. **Table 8-1** lists the coefficients used for this example; all lower bounds and upper bounds are taken to be zero and infinity, respectively. The nonlinear network-code by Dembo (1983) was used in the solution of all three models. **Table 8-2** summarizes the results of the various models. The diagonalization algorithm took eight iterations to converge when $\varepsilon_1 = \varepsilon_2 = \varepsilon_3 = 0.002$. By using these equilibrium values for supplies, demands and flows, the reader can easily check to see that the desired equilibrium conditions are achieved in all three cases. Also shown in **Table 8-2** is a comparison among the three models, ranging from the most competitive to least competitive. As one can see, the total profit achieved in the market increases with decreasing competition, as expected. Also, the increasing lack of competition tends to restrict output and raise prices. Thus this small example illustrates the ease with which a wide range of spatial-economic behavior can be computed.

8.3.6 *Models on a general transportation network*. Consider a general transportation-network that contains nodes that are neither origins nor destinations, but represent transfer or transshipment facilities such airports, rail yards or ports. The introduction of these transshipment nodes and their associated arcs creates a path-variable formulation. This general network is illustrated in **Figure 8-9**, where nodes 1 through 3 represent origins and destinations, while nodes 4 through 6 represent transshipment nodes. The solution to any model of spatial competition must now include a description of the *path* that the passengers or goods follow from origin to destination.[14]

To formalize this path dependency, let us define P as the set of paths in the network, P^{ij} as the set of paths between O–D pair i–j, x^p as the flow on path $p \in P$, C^p as the cost associated with traversing path p, x_a as the flow on arc a, $C_a(x_a)$ as

---

[14]This extension is similar to the extension from "multi-facility/multi-tour/hierarchical model" to "multi-facility/multi-route/multi-criteria/nested model" in the "Simultaneous Location-routing" chapter, where the discrete-facility analogue of these models is discussed.

Variable	Pure-competition	Oligopoly	Monopoly
V_1^d	6.36	5.10	8.79
V_{11}^d		4.25	
V_{13}^d		0.85	
V_2^d	41.79	43.28	38.20
V_{21}^d		9.14	
V_{22}^d		30.33	
V_{23}^d		3.81	
V_3^d	31.20	23.48	22.65
V_{31}^d		2.74	
V_{33}^d		20.74	
V_1^s	16.73	16.13	14.19
V_2^s	30.73	30.33	30.30
V_3^s	31.90	25.39	24.85
V_{12}	8.86	9.14	5.70
V_{13}	1.51	2.74	
V_{31}		0.85	
V_{32}	2.20	3.81	2.20
Total profit	822.64	837.40	887.00
Total supply	79.36	71.85	69.64
D_1	17.73	17.98	17.24
D_2	26.58	26.57	26.62
D_3	20.64	22.96	23.21

Table 8-2 Nonzero equilibrium results for example 1 (Harker 1986)

the transportation-cost function for traversing arc a, and $a_{ap} = 1$ if path p traverses arc k and 0 otherwise. Using the above notations, we can define the following relationships:

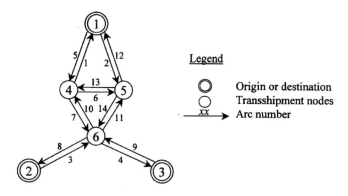

Figure 8-9 - General network with transshipment nodes

$$V_{ij} = \sum_{p \in P^{ij}} x^p \qquad \forall i,j \tag{8-48}$$

$$x_k = \sum_{p \in P} a_{kp} x^p \qquad \forall k \in \underline{A} \tag{8-49}$$

$$C^p = \sum_{k \in \underline{A}} a_{kp} C_k(x_k) \qquad \forall p \in P. \tag{8-50}$$

Equation **(8-48)** says O–D flow is the sum of all path flows between an O–D pair. Equation **(8-49)** states that path flows bundle up in an arc, while Equation **(8-50)** states that the sum of the relevant arc costs is the path cost.

This path-variable formulation lead to the following additional equilibrium-conditions in the purely competitive spatial-price-equilibrium model:

(e) if $x^p > 0$, then $C^p = C^{ij}$ for all i,j and $p \in P^{ij}$
(f) if $C^p > C^{ij}$, then $x^p = 0$ for all i,j and $p \in P^{ij}$

where costs C^{ij}, with superscript ij to highlight an origin-destination pair rather than an arc, is now defined as $\min_{p \in P^{ij}} \{C^p\}$. That is, a Wardropian (1952) path-cost equalization occurs between every O–D pair at equilibrium, wherein travel costs on all paths that serve the same O–D pair are all at a minimum. The following mathematical program yields the desirable equilibrium conditions (a), (b), (e) and (f):

$$\max \sum_{l \in I} \int_0^{V_l^d} D_l(s)\, ds - \sum_{l \in I} C_l(V_l^s) - \sum_{a \in \underline{A}} \int_0^{x_a} C_a(s)\, ds \tag{8-51}$$

subject to Equations **(8-18)**, **(8-19)**, **(8-24)** through **(8-26)**, **(8-48)** and **(8-49)**.

Similarly, the arc-cost functions and pertinent members of constraints **(8-48)** through **(8-50)** can be added to the monopoly and oligopoly models to yield

$$\max \sum_{l\in I} D_l(V_l^d) V_l^d - \sum_{l\in I} C_l(V_l^s) - \sum_{a\in\underline{A}} C_a(x_a) x_a \tag{8-52}$$

subject to Equations **(8-18)**, **(8-19)**, **(8-24)** through **(8-26)**, **(8-49)** and **(8-50)**. For the oligopoly model, we find $\mathbf{x}^*\in\Omega$ such that the game expressed by Equation **(8-39)** holds for all $\mathbf{x}\in\Omega$, where $\mathbf{x}^T = (x_q^T|q\in Q) = [(V_l^s|q\in Q, l\in J_q),\ (V_{lq}^d|q\in Q, l\in I),\ (x_a|a\in\underline{A})]$. $\Omega = \{\mathbf{x}|$ **(8-34)–(8-38)**, **(8-49)** and **(8-50)**$\}$, and the variational-inequality problem

$$\dot{\mathbf{F}}^T(\mathbf{x}^*)(\mathbf{x}-\mathbf{x}^*) =$$
$$\sum_{q\in Q}\left[\sum_{l\in J_q}\dot{C}_l(V_l^{s*})(V_l^s-V_l^{s*}) - \sum_{l\in I}\mathrm{MR}_{lq}(\mathbf{V}^{d*})(V_{lq}^d-V_{lq}^{d*})\right] +$$
$$\sum_{a\in\underline{A}} C_a(x_a^*)(x_a-x_a^*) \geq 0. \tag{8-53}$$

In the monopoly model, the marginal economic-prices on all used paths between any O–D pair will be equilibrated. In the pure competition and oligopoly models, average cost will be equilibrated. The two cases are often called Wardropian system and user equilibrium respectively. Existence and uniqueness properties can be deduced from Equation **(8-53)** correspondingly.

8.3.7 Example 2. Consider the network of **Figure 8-9** and let the cost and inverse demand-functions for origin-destination nodes 1 to 3 be the same as in Example 1 of Section 8.3.5. For each arc a, let us define the following cost function: $C_a(x_a) = K_a + K_a' x_a^2$. **Table 8-3** lists the coefficients of this function for the example network, while **Table 8-4** summarizes the results from the three models. It can be checked that the O–D equilibrium conditions are achieved. The same goes for the equilibrium of marginal path-cost in the monopoly model and the equilibration of average path-costs in the oligopoly and pure-competition models. Although commodity supplies and demands and arc flows can be shown to be uniquely determined, path flows x will not be unique—any path flows adding up to the x_a in **Table 8-4** are a solution.

8.3.8 Extensions to multiple commodities. All models considered thus far deal with only one commodity. In other words, there is no distinction between where the goods or services come from, viewing from the angle of the consumer. Also to the supplier, they do not care where their goods and services go—as long as demands are satisfied and suppliers are duly compensated. In reality, however, there is a distinction between commodities. An example is the state-park problem addressed in the "Generation, Distribution, and Competition" chapter. There the historical clientele for each state park is distinct, and state parks are not necessarily

Arc a	K_a	K_a'
1	1.0	0.5
2	2.0	0.2
3	3.0	0.3
4	1.0	0.4
5	2.0	0.3
6	1.0	0.1
7	1.0	0.1
8	2.0	0.25
9	2.0	0.2
10	1.0	0.9
11	3.0	0.8
12	3.0	0.5
13	2.0	0.2
14	1.0	1.0

Table 8-3 - Link-cost functions for example 2 (Harker 1986)

interchangeable from the eyes of the visitors. For these reasons, it is desirable to extend these models to handle multiple commodities. Here a *multi-copy network* is used and each element of the network is characterized by both its physical location *and* a particular commodity.[15] Again in the state-park case study, we specify the visitations at a state park by the points of origin of the visitors. We also specify the breakdown of the population from each region that patronize a particular state park.

In the multi-copy network, a region in which there are four commodities is represented as four nodes in the multi-copy network. In forming this new network, however, we destroy the separability of the functions in the problem. To show this recall the diagonalization or nonlinear Jacobi algorithm used to solve the variational-inequality problem:

$$\dot{\mathbf{F}}^T(\mathbf{x})(\mathbf{y}-\mathbf{x}) \geq 0 \qquad \forall \mathbf{y} \in X \qquad (8\text{-}54)$$

---

[15]The concept of a multi-commodity flow network is discussed in the "Optimization Schemes" appendix to Chan (2000).

Variable	Pure	Oligopoly	Monopoly
V_1^d	12.10	10.35	11.66
V_{11}^d		10.35	
V_2^d	34.37	34.82	32.20
V_{21}^d		1.85	
V_{22}^d		30.43	
V_{23}^d		2.53	
V_3^d	31.62	23.73	23.20
V_{31}^d		1.66	
V_{33}^d		22.07	
V_1^s	15.58	13.86	13.34
V_2^s	30.82	30.43	30.42
V_3^s	31.69	24.60	24.30
V_{12}	3.48	1.85	1.68
V_{13}		1.66	
V_{32}	0.07	2.53	1.10
x_1	2.21	2.23	1.11
x_2	1.27	1.28	0.58
x_4	0.07	2.53	1.10
x_7	2.21	2.23	1.11
x_8	3.55	4.38	2.78
x_9		1.66	
x_{14}	1.27	1.28	0.58
Total	804.54	825.73	849.91
Total	78.09	68.90	68.06
D_1	16.58	16.93	16.67
D_2	26.66	26.65	26.67
D_3	20.51	22.88	23.04

Table 8-4 - Spatial-equilibrium results for example 2 (Harker 1986)

where X is a closed convex subset of the n-dimensional Euclidean space. A smooth function $\mathbf{G}(\mathbf{x,y})$ is defined such that $\mathbf{G}(\mathbf{x,x}) = \dot{\mathbf{F}}(\mathbf{x})$. By *separability* is meant that for the function $\mathbf{G}(\mathbf{x,y})$, this equality holds for each firm, path or arc i in successive iterations $k-1$, k, $k+1$ etc.: $G_i(\mathbf{x,y}) = G_i(\mathbf{x}_i(k), \mathbf{x}_{-i}(k-1))$.[16] Here in the n-dimensional Euclidean space $\mathbf{x}_{-i} = (x_1, \dots, x_{i-1}, x_{i+1}, \dots, x_n)^T$, with the subscript i referring to each of n possible firms, paths or arcs.

For the monopoly model, this approach causes no problems other than the need to switch to a nonlinear network-algorithm that can handle non-separable arc-cost functions. However, for the remaining models, the objective-function integral will no longer be path-independent; since we cannot assume in general that the integrands, or the expressions to be integrated, will be functions with a symmetric Jacobian. That is to say, for every fixed $\mathbf{x}$, $\mathbf{y} \in X$, the n by n matrix $\mathbf{G}_x(\mathbf{x,y}) = [\partial G_i(\mathbf{x,y})/\partial x_j]$ is *symmetric positive-definite* or $\partial G_i(\mathbf{x,y})/\partial x_j = \partial G_j(\mathbf{x,y})/\partial x_i$ and $\mathbf{x}^T \nabla^2 G_x(\mathbf{x,y})\,\mathbf{x} = \mathbf{x}^T[\partial^2 G_x/\partial x_i \partial x_j]\mathbf{x} > 0$.

While there is an integrability issue, we can still cast all the models in their variational inequality form, and the diagonalization algorithm can be successfully applied to this problem. The point to be made here is this. The single-commodity models of the previous section can, with judicious computational schemes, be extended to the multiple-commodity case. Also, the earlier solutions using the diagonalization algorithm can be found by solving the same sub-problems as in the single-commodity case. As a last comment, it should be noted that the spatial models that are currently available assume the arc-cost functions $C_{ij}(V_{ij})$ is a constant or increasing function. In reality, freight-rate discounts for large shipments make this a decreasing function of V_{ij}, which leads to non-convex optimization problems.

This and other complexities re-state an earlier comment: The general oligopolistic market problem is to be solved on a case-by-case basis in spite of the significant progress reported above. Examples of these case-by-case solutions have been included in the "Generation, Distribution and Competition" chapter. Instead of the diagonalization algorithm, Ke (1997) used the *Expanding Algorithm* (Theise and Jones 1988) to solve the multi-commodity problems with nonlinear (increasing) shipping-cost functions. He solved a series of linearized subproblems, exploiting network-data structure and regional dominance inherent in such problems. The prototype algorithm was demonstrated by a two-commodity version of Example 1 above. Although the problem was solved via a different algorithm, we recap Ke's numerical example below. For the interested readers, they may wish to repeat the results using the algorithm presented here in Subsection 8.3.4.

---

[16]The concept of separability is illustrated in dynamic programming. See the "Markovian Systems" appendix to Chan (2000) under the "Markovian properties of dynamic programming" section.

Region	α_l^1	β_l^1	α_l^2	β_l^2	σ_l^1	δ_l^1	σ_l^2	δ_l^2
1	1.0	0.5	2.0	0.3	19.0	0.20	27.0	0.30
2	2.0	0.4	1.5	0.5	27.0	0.01	30.0	0.20
3	1.5	0.3	1.0	0.4	30.0	0.30	19.0	0.01

O-D	φ_{ij}^1	μ_{ij}^1	φ_{ij}^2	μ_{ij}^2	ω_{ij}^1	ω_{ij}^2
(1,2)	1.0	0.10	2.0	0.4	0.02	0.02
(1,3)	2.0	0.40	1.0	0.1	0.03	0.03
(2,1)	1.0	0.20	3.0	0.3	0.01	0.01
(2,3)	3.0	0.30	1.0	0.2	0.04	0.04
(3,1)	1.0	0.10	4.0	0.4	0.03	0.03
(3,2)	4.0	0.41	10.0	0.1	0.02	0.02

Table 8-5 - Coefficients for multi-commodity example (Ke 1997)

Example 3

The notation follows example 1, except that the regional production-cost-function is generalized to $C_l^r(V_l^{sr}) = \alpha_l V_l^{sr} + \beta_l \{V_l^{sr}\}^2$, the inverse demand-function becomes $D_l(V_l^{dr}) = \sigma_l^r - \delta_l^r V_l^{dr}$, and O–D transportation-cost function now becomes $C_{ij}^r(V_{ij}^r) = \varphi_{ij}^r + \mu_{ij}^r \{V_{ij}^r\}^2 + \sum_{k \neq r} \omega_{ij}^k V_{ij}^k$. Here r is the index for each commodity. Notice the transportation-cost function explicitly recognize the interaction between the various commodities. It suggests that the more commodities compete for the limited network resources, the more costly (congested) the traffic flow is. Instead of using node-arc incidence to model the flow interaction, a more explicit form is taken. The coefficients are displayed in **Table 8-5**, and the equilibrium solutions in **Table 8-6** Of all three market structures, oligopoly is most sensitive to the interaction between commodities (Ke 1997). □

The spatial-equilibrium models presented thus far in this chapter is a generalization of the Lowry-tradition and discrete location-routing models. For example, the diagonalization algorithm allows us to relax the "independence of irrelevant alternatives" assumption inherent in gravitational interaction. A different approach, harnessing the power of variational inequality, is employed to assess equilibrium and disequilibrium conditions. Miller et al. (1996) and Nagurney (1993) used the spatial-equilibrium framework to generalize location-routing models. The spatial-equilibrium models are in general not in the same computational status as the Lowry-heritage models (and for that matter location-routing

Variable	Pure-competition	Oligopoly	Monopoly
V^1_{11}	6.373	4.285	8.788
V^1_{12}	8.85	9.124	5.697
V^1_{13}	1.502	2.707	
V^1_{21}			
V^1_{22}	30.729	30.331	30.295
V^d_{22}			
V^d_{23}		0.851	
V^1_{32}	2.139	3.769	2.196
V^1_{33}	29.74	20.763	22.652
V^2_{11}	26.101	17.394	18.801
V^2_{12}	2.515	3.522	1.594
V^2_{13}		3.356	2.472
V^2_{21}			
V^2_{22}	22.289	19.017	19.324
V^2_{23}			
V^2_{31}			
V^2_{32}	6.25	5.858	2.023
V^2_{33}	16.049	16.196	19.917

Table 8-6 - Equilibrium flow for multi-commodity example (Ke 1997)

models).[17] But it opens up alternate tools for examining spatial-activity allocation and competition. In parallel, we will now re-visit the Lowry-based models and generalize from that school of thought via an input–output viewpoint. It will be seen that this latter school, in turn, has its unique strength in activity *derivation* and *dynamics* not found in the former.

---

[17]Significant progress has been made in spatial-equilibrium solution algorithms recently, as evidenced from Portugal and Judice (1996).

The "Spatial competition models" presented in this section can be further extended to include activity derivation, as alluded to already. Furthermore, we can extend the *generation*-distribution model described in the "Activity Generation, Competition and Distribution" chapter to include multi-commodity network-flow (or traffic assignment). Several formulations and solution procedures have been offered in the literature to effect this integration either in part or in whole. These include Chu (1999), Oppenheim (1995), Putman (1991), Boyce and Lundquist (1987), Evans (1976) and Florian (1975). Even though a combined model could be formulated, it is quite another matter to produce efficient numerical solutions. It is only by decomposing the problem that a solution is possible, as demonstrated in the "Activity Generation and Distribution" chapter. More interestingly, questions exist regarding the existence of unique equilibrium.

8.4 Growth of regional economic activities

In the previous sections, the problem of regional interactions as a market-clearing phenomenon is discussed without specific mention of the general economic-theory of *growth*. Here, it is our intention to formally introduce the theory of growth (Andersson and Kuenne 1986). After a short review of Keynesian growth theory for one closed region, the focus will be shifted to a pair of inter-connected regions and then generalized to a finite number of interacting regions. The distinguishing feature of the discussion here is the inclusion of *investment* as a key factor in economic growth—a factor that is absent in earlier discussions in this chapter. The development follows the same pattern as that for the Lowry–Garin models, in which a market-oriented model was put forth first, followed by a generalization to an investment-based model.

8.4.1 Single-region growth models. Keynesian theory assumes that a growth equilibrium requires market clearing at the macro, rather than the micro, level, i.e., the *total* supply, $V^s(t)$, equals total demand, $V^d(t)$, plus investment demand, $W(t)$:

$$V^s(t) = V^d(t) + W(t). \tag{8-55}$$

Furthermore, consumption demand $V^d(t)$ is assumed to be determined by total supply $V^s(t)$ only, because of the assumed constancy of relative prices:

$$V^d(t) = \alpha V^s(t) \qquad (0<\alpha<1). \tag{8-56}$$

Investment in new capacity, $W(t)$, which acts as the energizer of the system, is assumed to be proportional to the growth of total-supply $\dot{V}^s(t)$.

$$W(t) = \sigma' \dot{V}^s(t) \equiv \sigma'\left[dV^s(t)/dt\right]. \tag{8-57}$$

The simplest growth models can then be completed by introducing the initial condition: $V^s(0) = V^s_0$.

By substituting Equations **(8-56)** and **(8-57)** into Equation **(8-55)**, we arrive at the first-order differential-equation

$$V^s(t) = \alpha V^s(t) + \sigma' \dot{V}^s(t) \qquad (V^s(0) = V_0^s).$$ (8-58)

This equation has the economic-growth solution:

$$V^s(t) = V^s(0) \exp\{[(1-\alpha)/\sigma']t\}.$$ (8-59)

The equation shows that increasing the propensity to consume α will decrease the rate of growth. An increased σ', equivalent to a decreased capital productivity, would also decrease the rate-of-growth of the economy. The whole analysis is valid only in a situation of *stock* equilibrium at the outset where the economy starts from a position when capital K at the beginning $K(0) = \sigma' V^s(0)$. From this it follows that $K(t) = \sigma' V^s(t) = \sigma' V^s(0) \exp\{[(1-\alpha)/\sigma']t\}$ or that $\dot{K}/K = \dot{V}/V$, i.e., the rate-of-growth of capital equals the rate of economic-growth. It is interesting to compare this single-region growth-equation **(8-59)** with the economic-base theory[18] wherein the effect of capital investment, in addition to consumption, is now *explicitly* included. It has been argued that the assumption of constant coefficients α and σ' in the growth process is too restrictive, because of the need to incorporate substitution phenomena. This is however a quite legitimate assumption as long as the analysis is performed under the assumption of a constant relative-price of capital. We will relax this assumption at a stage when it is meaningful, i.e., when we are considering a multi-sectoral economy.

8.4.2 Two-region equilibrium growth. We now model the trade between two regions We assume that the supply of commodities in region i is decided by production in the region, R_i, and by imports assumed to be proportional to production, $\rho_i R_i$. The import of one regions is of course constrained to be the export of the other region. The system can now be written in the following form. The first equation describes total supply in region i: $V_i^s = R_i + \rho_i R_i$. The consumption demand at region i is $V_i^d = \alpha_i R_i$. Finally the investment demand in region i is $W_i(t) = \sigma_i' \dot{R}_i$. Naturally, all these equations are written for both regions ($i = 1, 2.$)

For a two-region economy, and accounting for the fact that the imports to region 1 have to be equal to the exports from region 2 and vice versa in an interregional-equilibrium equation-set results:

$$R_1 + \rho_1 R_1 = \alpha_1 R_1 + \sigma'_1 \dot{R}_1 + \rho_2 R_2$$
$$R_2 + \rho_2 R_2 = \alpha_2 R_2 + \sigma'_2 \dot{R}_2 + \rho_1 R_1.$$ (8-60)

Simplifying, we have

$$\dot{R}_1 = \rho_{11} R_1 - \rho_{12} R_2$$
$$\dot{R}_2 = \rho_{21} R_1 - \rho_{22} R_2,$$ (8-61)

---

[18]For a discussion of economic-base theory, consult the "Economics Methods" chapter and the "Descriptive Tools" chapter of Chan (2000).

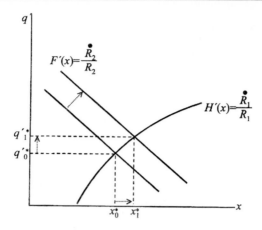

Figure 8-10 - Balanced growth and income share in a two-region economy

where the trade coefficients are defined as $\rho_{ii} = \left[(1+\rho_i-\alpha_i)/\sigma_i'\right]$ and $\rho_{ij} = \left(\rho_i/\sigma_j'\right)$. This coupled system will now be studied from the following point of view: Does there exist any balanced-growth solution at growth-rate q' in the sense that the two regions will remain in constant relative economic-positions indefinitely?[19] In other words, does there exist a balanced-growth solution such that $\dot{R}_i = q'R_i$ for $i = 1,2$; or using Equation **(8-61)** to express R_i in matrix form,

$$\left[\rho - Iq'\right]R = 0 \qquad\qquad (8\text{-}62)$$

where I is the identity matrix and $\rho = \begin{bmatrix} \rho_{11} & \rho_{12} \\ \rho_{21} & \rho_{22} \end{bmatrix}$. Notice that this problem has a non-trivial solution (i.e., with $R_i \neq 0$ for at least one of the regions) only under the condition of Equation **(8-62)**.

The growth-rate q' is an unknown to be found by the roots of the characteristic equation: $q'^2 - (\rho_{11}+\rho_{22})q' + (\rho_{11}\rho_{22}-\rho_{12}\rho_{21}) = 0$ or $q' = 1/2\{\rho_{11}+\rho_{22} \pm \left[(\rho_{11}+\rho_{22})^2 - 4(\rho_{11}\rho_{22}-\rho_{21}\rho_{12})\right]^{1/2}\}$. The solution will give exponential growth without oscillation if and only if $(\rho_{11}+\rho_{22})^2 > 4(\rho_{11}\rho_{22}-\rho_{21}\rho_{12})$, or when there are no imaginary roots. Assuming symmetry of behavior and size of the regions, exponential growth occurs if and only if $\rho_1/\sigma_2' > 0$, which is evidently always true. Furthermore, a partially increasing (decreasing) propensity to consume for a capital-output ratio decreases (increases) the *common* rate of growth.

The determination of the rate-of-growth common to both regions can be illustrated with **Figure 8-10**. We introduce a transformation: $R_i \geq 0$, $R_1 + R_2 = 1$ and $R_1/(1-R_1) \equiv x$.

---

[19]The "Control, Dynamics" appendix to Chan (2000) discusses solution to a simultaneous differential-equation set.

	Households	Industry 1	Industry 2	Exports	Row total
Households	0	yz^{e_1}	yz^{e_2}	0	yz^e
Industry 1	H^1	G^{11}	G^{12}	O^1	R^1
Industry 2	H^2	G^{21}	G^{22}	O^2	R^2
Imports	H	G^1	G^2	0	R

Table 8-7 - Aspatial input–output table (Webber 1984)

$$H'(x) = \dot{R}_1/R_1 = \rho_{11} - \rho_{12} x^{-1}$$
$$F'(x) = \dot{R}_2/R_2 = \rho_{22} - \rho_{21} x \qquad\qquad (8\text{-}63)$$
$$q'^* = H'(x) = F'(x).$$

Assume that α_2 decreases, with the consequence that ρ_{22} and $F(x)$ increases from one position to another as shown in the Figure. The decreased propensity-to-consume in region 2 implies an increased *common* rate-of-growth but also *increased share of region 1* in total production. Only if both regions increase their respective propensities-to-save (invest) could the growth rate increase with an unchanged relative production in the two regions.

8.4.3 Input–output model. It is well known that Leontief's input–output model can be used to describe the spatial economic-activities of a study area. We have alluded to this point already in the "Economics" chapter. Webber (1984) introduced this concept using an example, which we will summarize below. In his aspatial example, there is one type of household and *two* industrial sectors. Both economic sectors export their goods, and imports to the study area are modelled explicitly. There are all together two geographic entities in the area: the local economy and the external world. **Table 8-7** shows the transactions between the household sector, the two industrial sectors, imports and exports. The level of export is assumed to be *fixed* exogenously.

In **Table 8-7**, each entry represents a sale (in, say, dollars) made by a row sector to a column sector. For example, the first row shows the sale of labor by households to the two industrial sectors, which amounts to the total wage paid to the workers: yz^{e_1} and yz^{e_2}, where y is the wage rate and the z represent the two types of industrialemployment, designated by e_1 and e_2 respectively. It is assumed that households do not sell labor to other households or export their labor directly. The total of this row becomes $y(z^{e_1} + z^{e_2}) = yz^e$. The industrial sectors 1 and 2 sell goods and services to households for H^1 and H^2. They also sell to each other in the tune of G^{11}, G^{12}, G^{21}, and G^{22}. Finally, they export in the amount of O^1 and O^2, where $H^1 + G^{11} + G^{12} + O^1 = R^1$ for industry 1. A similar row sum can be written for industry

2. The fourth row in the table records imports: H is the direct shopping from outside the area by the household sector, the G are imports by industries, and the total import is R.

Now let us define these *technical coefficients*: $\rho^{hh} = 0$, $\rho^{h1} = yz^{e_1}/R^1$, $\rho^{h2} = yz^{e_2}/R^2$, $\rho^{ph} = H^p/yz^e$ ($p = 1,2$), and $\rho^{pq} = G^{pq}/R^q$ ($p = 1,2$; $q = 1,2$). With these definitions, the above two equations can be rewritten as

$$yz^e = \rho^{h1}R^1 + \rho^{h2}R^2 \tag{8-64}$$

$$R^p = \rho^{ph}yz^e + \rho^{p1}R^1 + \rho^{p2}R^2 + O^p \qquad p=1,2. \tag{8-65}$$

Further discussion of the coefficients ρ is in order. The coefficients ρ^{hp}, for example, represents the dollar value of labor sold to sector p per unit-value-of-output in sector p. In other words, for every dollar worth of output, industry p must buy ρ^{hp} dollars worth of labor from households. Conversely, ρ^{ph} measures the value of goods bought by households from sector p per dollar-of-labor supplied by households. Now ρ^{pq} becomes the value of inputs bought by sector q from sector p per unit-of-output in sector q, i.e., every dollar of output of sector q needs ρ^{pq} dollars of input from sector p.

It is assumed that these coefficients are given and fixed, no matter what the level of output of each industry. Thus irrespective of the level of R^q, $\rho^{pq}R^q$ is the amount of output that sector q must buy from sector p. The matrix

$$\rho = \begin{bmatrix} \rho^{hh} & \rho^{h1} & \rho^{h2} \\ \rho^{1h} & \rho^{11} & \rho^{12} \\ \rho^{2h} & \rho^{21} & \rho^{22} \end{bmatrix}$$

is the matrix of input–output coefficients. The exogenous variables in Equations (8-64) and (8-65) are exports, O^p, which are assumed to be fixed and given, in the same manner that Garin-Lowry assumes that basic employment was given. The vector of exports contains no exports from the household sector: $\mathbf{O} = (0 \; O^1 \; O^2)^T$. The unknown variables are the total wages earned by labor, yz^e, and the value of output of each industrial sector (R^1 and R^2), which form a vector of $\mathbf{R} = (yz^e \; R^1 \; R^2)^T$. The total value of import is to be found at a later stage.

Now Equations (8-64) and (8-65) can be rewritten in vector notation:

$$\mathbf{R} = \rho\mathbf{R} + \mathbf{O} \tag{8-66}$$

or after rearrangement

$$\mathbf{R} = (\mathbf{I} - \rho)^{-1}\mathbf{O} \tag{8-67}$$

where $\mathbf{I}$ is the identity matrix, and $\mathbf{R}$ is a vector of transaction between the household sector, industrial-sector 1, and industrial-sector 2. The above is the solution to Equations (8-64) and (8-65), expressing the wage income and value of output in terms of the input–output coefficients and the export levels. The equation set can be solved either as a set of simultaneous equations or iteratively in terms of multipliers ρ in a geometric series, as was done in economic-base theory. A solution exists, or there is an equilibrium, if certain conditions are imposed upon ρ. Specifically, convergence of the series occurs when each row sum of the matrix ρ is less than unity. This is similar to the conditions laid out for ρ in the "Two-region equilibrium growth" section above (Subsection 8.4.2), where we show exponential growth will result if and only if $\rho_i/\sigma_j' > 0$ $(i = 1,2; j = 1,2; i \neq j)$. The difference is that exponential growth is part and parcel of the previous growth model with investment while this is not necessarily true for the input–output model. This is traceable to the fact that investment is explicitly built into the former model but not the latter.

8.4.4 *Example of aspatial input–output model.*

As a numerical example, let us assume that there are two industries: food and clothing (Oppenheim 1980). The level of monthly basic (export) production has been set at $O^1 = 50$ and $O^2 = 75$ thousand dollars of food and clothing respectively. 30% of the total output of each industry is in payroll. 26% of output of the farmers goes back to the farming sector for food, and 1% to the clothing sector for clothes. Similarly, 21% of output of the tailors goes for food and 6% clothes. This mean that $\rho^{h1} = 0.3$, $\rho^{h2} = 0.3$, $\rho^{11} = 0.26$, $\rho^{21} = 0.01$, $\rho^{12} = 0.21$, and $\rho^{22} = 0.06$. In their private lives, the household sector spends 30% of their income on food and 30% on clothing ($\rho^{1h} = \rho^{2h} = 0.3$). In so doing, the labor or household sector consumes every dollar of income on the local economy. Equation **(8-65)** can now be written as

$$R^1 = 0.3x + 0.26R^1 + 0.21R^2 + 50$$
$$R^2 = 0.3x + 0.01R^1 + 0.06R^2 + 75, \qquad \textbf{(8-68)}$$

where x is the shorthand for yz^e, the total wages for the household sector that supply labor to industries. Equation **(8-64)** can correspondingly be written as $x = 0.3R^1 + 0.3R^2$. The solution can be obtained by solving the three equations with three unknowns according to Equation **(8-67)**:

$$\begin{pmatrix} x \\ R^1 \\ R^2 \end{pmatrix} = \left(\begin{bmatrix} 1 & 0 & 0 \\ 0 & 1 & 0 \\ 0 & 0 & 1 \end{bmatrix} - \begin{bmatrix} 0 & 0.3 & 0.3 \\ 0.3 & 0.26 & 0.21 \\ 0.3 & 0.01 & 0.06 \end{bmatrix} \right)^{-1} \begin{pmatrix} 0 \\ O^1 \\ O^2 \end{pmatrix} \qquad \textbf{(8-69)}$$

The solution to this simultaneous equation set is $(226.75\ \ 124.25\ \ 102.5)^T$. Alternatively, the solution is obtainable through a geometric series as defined by Equation **(8-67)** with the economic multipliers contained in the 3×3 matrix $[\rho^{ij}]$ as shown in Equation **(8-69)**. Notice each row sum of matrix ρ in Equation **(8-69)** is less than unity, ensuring convergence of the series or that an equilibrium exists.

8.5 Generalized multi-regional growth equilibria

The Keynesian-growth model described in the "Growth of regional economic activities" section (Section 8.4) easily generalizes to any finite number of regions:

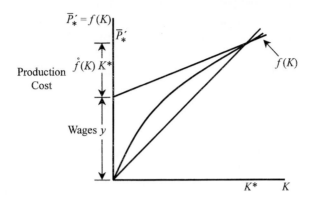

Figure 8-11 - Steady-state solution of a neoclassical growth model

$$\dot{R}_i = \rho_{ii} R_i - \sum_{j \neq i} \rho_{ij} R_j \qquad i = 1, \ldots, n'. \tag{8-70}$$

where n' replaces 2 in Equation **(8-61)**. This ordinary-linear-differential-equation system can be re-written in matrix form: $\dot{R} + \rho R - \hat{\rho} R = 0$ where $\rho = [\rho_{ij}]$ and $\hat{\rho} = [\rho_{ii}] \, I$, or $\hat{\rho}$ is a diagonal matrix. More compactly $\dot{R} = Q' R$ where $Q' = (\hat{\rho} - \rho)$. Keynesian linear models of growth such as the above, with many rigid relations built into them, are perhaps too simplistic to capture the observed adaptability of real-world spatial economies. It is possible to substitute capital for labor because of changing relative wage rates. This would secure a smooth, adaptable development path for the macro economy (Andersson and Kuenne 1986).

8.5.1 Multiregion neoclassical growth model. In **Figure 8-11**, let $\bar{P}' = f(K)$ be the aggregate production function with capital K as input, i.e., the overall relationship between maximum production and capital investment is modelled by $\bar{P}' = f(K)$. As can be seen, production $\bar{P}'$ increases in direct proportion to capital investment K. Further, let us define y as wages. As a steady-state-equilibrium investment-level remains constant, or $\dot{K} = 0$, zero profit results. Here production income pays for total cost of production and total wages, i.e., $\bar{P}'_* = \dot{f}(K^*)K^* + y$. Now suppose the macro production-function $f(K)$ is sufficiently smooth and linearly homogeneous, or that output increases in exact proportion to increase in factor input, resulting in a linear production schedule. This is illustrated by a production-function of capital and labor in **Figure 8-12** It is then possible to have a real rate-of-interest matching the marginal productivity of capital. Together with wages, and when used to compensate each unit of capital, such interest rate will just exhaust product $\bar{P}'_*$.

For two (or n') regions the problem becomes less trivial. For such a system we observe the basic macro-equilibrium-conditions to be that savings equal investment

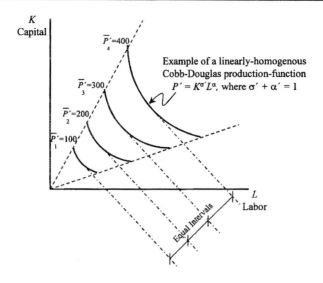

Figure 8-12 - Illustrating a linearly-homogeneous production-function

(or capital accumulation), i.e., an equilibrium will result with such an equality. Suppose no depreciation and savings to come out of total internally-produced and net-imported supply. We have $\dot{K}_r = k_r(\bar{P}'_r - O_r + H_r^G)$ where $\dot{K}_r$ is the instantaneous rate-of-capital-accumulation in region r, k_r is the *propensity-to-save (invest)*[20] in region r, $\bar{P}'_r$ is the gross product of region r, O_r is exports from region r, and H_r^G is imports to region r. Further, assuming a certain export-share o_r for each region, we have the following two-region application of this model:

$$\dot{K}_1 = (k_1 - o_1)\bar{P}'_1 + o_2\bar{P}'_2$$
$$\dot{K}_2 = o_1\bar{P}'_1 + (k_2 - o_2)\bar{P}'_2,$$

(8-71)

with the assumption that $k_r > o_r > 0$, or the propensity-to-save is greater than the export-share.

Now assume a regional linearly-homogeneous production-function $\bar{P}'_r = K_r^{\sigma'_r}L_r^{\alpha_r}$ and $L_r = L_r(0)\exp(a_r^f t)$, where L_r is the regional labor-input-factor to the production-function, and a_r^f is the growth-rate-of-labor over time t. A possible two-region neoclassical-growth-model would then be:

$$\dot{K_1} = (k_1 - o_1) K_1^{\sigma_1'} L_1(0)^{\alpha_1} \exp(\alpha_1 a_1^f t) + o_2 K_2^{\sigma_2'} L_2(0)^{\alpha_2} \exp(\alpha_2 a_2^f t)$$

$$\dot{K_2} = o_1 K_1^{\sigma_1'} L_1(0)^{\alpha_1} \exp(\sigma_1' a_1^f t) + (k_2 - o_2) K_2^{\sigma_2} L_2(0)^{\alpha_2'} \exp(\sigma_2' a_2^f t).$$

(8-72)

A difference-equation variant of this model can easily be simulated for a set of positive parameters k_r and o_r, when there is no growth in labor-input $a_r^f = 0$, and for given initial input-values of labor $L_r(0)$ and capital $K_r(0)$. Such computer simulations show that the two regional capital-stocks develop at *converging* growth rates. We thus can conjecture that a neoclassical multiregional-growth-model has some stable steady-state properties, notably balanced growth of regional-capital-stocks, a fact that can be established formally. Further comment on the *stability* of the interregional neoclassical-growth-model is now in order. If started at some suitable random point, simulation iterations indicate that the discrete version of the model approaches an equilibrium structure at $x_i/\Sigma x_i$. Also, a steady-state growth-rate would result. These all point toward stability of the solution.

8.5.2 Spatial formulation of the input–output model. The models of growth of preceding sections are based on a value-added representation of the spatial economy according to Keynes. More complete inter-regional analysis—especially for applications—requires a full representation of the industrial structure of each region. A convenient starting point for such an analysis is the static aspatial input–output analysis, which we described at some length in Section 8.4.3. It was pointed out that intermediate-good interactions between sectors, each one producing one commodity with one linear activity, are regulated by input–output coefficients $\rho^{pq} = x^{pq}/x^q$. Here x^{pq} is the flow of commodity from sector p to sector q, and x^q is the gross output of sector q. This basic description of the economy can be extended to a system of n' zones. Each transaction listed in **Table 8-7** is now detailed in terms of the origin and destination zones. The resulting set of transactions for a two-zone system is given in **Table 8-8**. The former transactions, yz^{e_1} for example, are now decomposed into four transactions: the supply-of-labor by household in zones 1 and 2 to sector 1 firms located in zones 1 and 2, or $yz_{ij}^{e_1}$ for $i = 1,2$ and $j = 1,2$. Again, the level of exports from each sector in each zone is given.

The row-total constraints corresponding to Equations **(8-64)** and **(8-65)** are simply

$$yz_i^e = y \sum_j (z_{ij}^{e_1} + z_{ij}^{e_2}) \qquad \forall i$$

(8-73)

The input–output-coefficient-matrix ρ now becomes

$$R_i^p = \sum_j \left(H_{ij}^p + G_{ij}^{p^1} + G_{ij}^{p^2} \right) + O_i^p \qquad \forall p, i.$$

(8-74)

		Households		Industry 1		Industry 2		Export	Row total
		zone 1	zone 2	zone 1	zone 2	zone 1	zone 2		
House-holds	zone 1	0	0	$yz^{e_1}_{11}$	$yz^{e_1}_{12}$	$yz^{e_2}_{11}$	$yz^{e_2}_{12}$	0	yz^e_1
	zone 2	0	0	$yz^{e_1}_{21}$	$yz^{e_1}_{22}$	$yz^{e_2}_{21}$	$yz^{e_2}_{22}$	0	yz^e_2
Industry 1	zone 1	H^1_{11}	H^1_{12}	G^{11}_{11}	G^{11}_{12}	G^{12}_{11}	G^{12}_{12}	O^1_1	R^1_1
	zone 2	H^1_{21}	H^1_{22}	G^{11}_{21}	G^{11}_{22}	G^{12}_{21}	G^{12}_{22}	O^1_2	R^1_2
Industry 2	zone 1	H^2_{11}	H^2_{12}	G^{21}_{11}	G^{21}_{12}	G^{22}_{11}	G^{22}_{12}	O^2_1	R^2_1
	zone 2	H^2_{21}	H^2_{22}	G^{21}_{21}	G^{21}_{22}	G^{22}_{21}	G^{22}_{22}	O^2_2	R^2_2
Imports		H_1	H_2	G^1_1	G^1_2	G^2_1	G^2_2	0	R

Table 8-8 - Two-zone input–output transactions (Webber 1984)

$$\rho = \begin{bmatrix} 0 & 0 & \rho^{h1}_{11} & \rho^{h1}_{12} & \rho^{h2}_{11} & \rho^{h2}_{12} \\ 0 & 0 & \rho^{h1}_{21} & \rho^{h1}_{22} & \rho^{h2}_{21} & \rho^{h2}_{22} \\ \rho^{1h}_{11} & \rho^{1h}_{12} & \rho^{11}_{11} & \rho^{11}_{12} & \rho^{12}_{11} & \rho^{12}_{12} \\ \rho^{1h}_{21} & \rho^{1h}_{22} & \rho^{11}_{21} & \rho^{11}_{22} & \rho^{12}_{21} & \rho^{12}_{22} \\ \rho^{2h}_{11} & \rho^{2h}_{12} & \rho^{21}_{11} & \rho^{21}_{12} & \rho^{22}_{11} & \rho^{22}_{12} \\ \rho^{2h}_{21} & \rho^{2h}_{22} & \rho^{21}_{21} & \rho^{21}_{22} & \rho^{22}_{21} & \rho^{22}_{22} \end{bmatrix}. \tag{8-75}$$

Each of these coefficients now becomes $\rho^{hp}_{ij} = yz^{e_p}_{ij}/R^p_j$, $\rho^{ph}_{ij} = H^p_{ij}/yz^e_j$, and

$$\rho^{pq}_{ij} = G^{pq}_{ij}/R^q_j. \tag{8-76}$$

As in the aspatial case, these coefficients—now specified on an inter-zonal or interregional level—are assumed to be fixed. ρ^{hp}_{ij} is the value-of-labor supplied by residents of zone-i per industry-p unit-of-output in zone-j; ρ^{ph}_{ij} is the sale of sector p in zone-i to the residents of zone-j per value-of-labor unit supplied by those residents; and ρ^{pq}_{ij} represents the sales of sector-p firms in zone-i per unit-of-output of sector-q firms in zone-j. The vectors **O** and **R** are similarly expanded to account for spatial details:

$$\mathbf{O} = \begin{pmatrix} 0 & 0 & O^1_1 & O^1_2 & O^2_1 & O^2_2 \end{pmatrix}^T \tag{8-77}$$

$$\mathbf{R} = \begin{pmatrix} yz_1^e & yz_2^e & R_1^1 & R_2^1 & R_1^2 & R_2^2 \end{pmatrix}^T \tag{8-78}$$

where **O** is again given exogenously. With these expanded definitions, Equations
(8-66) and **(8-67)** still hold for the spatial version of the input–output model.

   8.5.3 Example of spatial input–output model. This numerical example is on
an economic system of two regions and two industrial sectors, with the household
sector left out (Kanafani 1983). The zonal or regional inter-sectoral trade-flows are
given in the following tables, one for each zone or region. Notice that we have disaggregated **Table 8-8** into its submatrices here, for ease of presentation and for
reasons that will become clear. Also we equate export with exogenous *demand*, and
we equate imports with *"value added."* Let us first show the table for zone-1
(region-1):

Sector	Industry 1	Industry 2	Demand (export)	Total con-sumption
Industry 1	5	4	10	19
Industry 2	3	6	7	16
Value added (imports)	4	10		
Total production	12	20		

Similarly the data for zone-2 (region-2) are:

Sector	Industry 1	Industry 2	Demand (export)	Total con sumption
Industry 1	2	2	8	12
Industry 2	3	2	5	10
Value added (imports)	14	2		
Total production	19	6		

The inter-zonal or inter-regional trade-flows by commodity are also given in the
following tables. For the commodity manufactured by industry-1, we have:

	Destination 1 (subarea 1)	Destination 2 (subarea 2)	Total production
Origin 1 (subarea 1)	10	2	12
Origin 2 (subarea 2)	9	10	19
Total consumption	19	12	

where 'subarea' refers to both a zone or region. Similarly for the commodity of industry-2, we write:

	Destination 1 (subarea 1)	Destination 2 (subarea 2)	Total production
Origin 1 (subarea 1)	12	8	20
Origin 2 (subarea 2)	4	2	6
Total consumption	16	10	

When disaggregated into these submatrices, the subareal input-output tables—namely the first pair of tables above—show the total consumption and production of each commodity (by industry) as the row and column sums for each subarea. Notice that the following relationships hold for these data. The totals of each commodity received (consumption) by each subarea are

$$G_{11}^{11} + G_{11}^{12} + O_1^1 = G_{11}^{11} + G_{21}^{11} = 19$$
$$G_{22}^{11} + G_{22}^{12} + O_2^1 = G_{12}^{11} + G_{22}^{11} = 12$$
$$G_{11}^{21} + G_{11}^{22} + O_1^2 = G_{11}^{22} + G_{21}^{22} = 16$$
$$G_{22}^{21} + G_{22}^{22} + O_2^2 = G_{12}^{22} + G_{22}^{22} = 10$$

and commodities shipped (productions) are

$$G_{11}^{11} + G_{11}^{21} + G_1^1 = G_{11}^{11} + G_{12}^{11} = 12$$
$$G_{22}^{11} + G_{22}^{21} + G_2^1 = G_{21}^{11} + G_{22}^{11} = 19$$
$$G_{12}^{11} + G_{11}^{22} + G_1^2 = G_{11}^{22} + G_{12}^{22} = 20$$
$$G_{22}^{12} + G_{22}^{22} + G_2^2 = G_{21}^{22} + G_{22}^{22} = 6.$$

A shorthand notation for these equations is

$$G_1^1 + O_1^1 = G_1^1 = 19 \qquad\qquad G_1^{\cdot 1} + G_1^1 = G_{1\cdot}^1 = 12$$
$$G_2^1 + O_2^1 = G_2^1 = 12 \qquad\qquad G_2^{\cdot 1} + G_2^1 = G_{2\cdot}^1 = 19$$
$$G_1^2 + O_1^2 = G_1^2 = 16 \qquad\qquad G_1^{\cdot 2} + G_1^2 = G_{1\cdot}^2 = 20$$
$$G_2^2 + O_2^2 = G_2^2 = 10, \qquad\qquad G_2^{\cdot 2} + G_2^2 = G_{2\cdot}^2 = 6$$

respectively.

Using the shorthand notation, the subareal technical-coefficients ρ_{ij}^{pq} can be computed for each subarea and pair of commodities by dividing flow G_j^{pq} by the row sums $G_j^{\cdot q}$, following Equation **(8-76)**. For example, $G_1^{11}/G_{1\cdot}^1 = 5/12 = 0.417$. They are arranged in a multi-subareal matrix P_s as follows:

$$\begin{array}{cc} & \begin{array}{cc} \textit{Industry 1} & \textit{Industry 2} \end{array} \end{array}$$

$$\begin{array}{c} \textit{Industry 1} \\ \\ \textit{Industry 2} \end{array} \begin{bmatrix} 0.417 & 0 & | & 0.200 & 0 \\ 0 & 0.105 & | & 0 & 0.330 \\ \cdots & \cdots & \cdots & \cdots & \cdots \\ 0.250 & 0 & | & 0.300 & 0 \\ 0 & 0.158 & | & 0 & 0.330 \end{bmatrix} \qquad \textbf{(8-79)}$$

where the 2×2 submatrices describe the flow between subareas 1 and 2. Next, the column input–output coefficients are calculated for each commodity and pair of subareas. We divide the entries of the trade table by the column sums: $G_{ij}^q/G_{.j}^q$, again following Equation **(8-76)**. For example, $G_{11}^1/G_{.1}^1 = 10/19 = 0.53$. The values are obtained in one table for each commodity (or sector):

		Industry 1		Industry 2
Origin	Destination		Destination	
	1	2	1	2
1	0.530	0.167	0.750	0.800
2	0.470	0.833	0.250	0.200

These are arranged in the multi-subareal matrix ρ_T as follows:

$$
\begin{array}{c}
\\
\text{Industry 1} \\
\\
\text{Industry 2}
\end{array}
\overset{\begin{array}{cc}\textit{Industry 1} & \textit{Industry 2}\end{array}}{
\begin{bmatrix}
0.530 & 0.167 & 0 & 0 \\
0.470 & 0.833 & 0 & 0 \\
0 & 0 & 0.750 & 0.800 \\
0 & 0 & 0.250 & 0.200
\end{bmatrix}}.
\qquad \textbf{(8-80)}
$$

Again, each 2×2 submatrix for an industry pair contains entries between subareas 1 and 2. Note that both ρ_S and ρ_T have the same dimension and format. Also note that the column sums of ρ_T equal unity.

With these matrices prepared, it is possible now to apply the multi-subareal input–output model of Equation **(8-67)** to compute the equilibrium flows. To do this we obtain the matrix

$$
\mathbf{I} - \rho = \mathbf{I} - \rho_S \rho_T =
\begin{bmatrix}
0.4579 & 0.6968 & -0.1500 & -0.1600 \\
-0.0493 & 0.9126 & -0.0826 & -0.0660 \\
-0.1325 & -0.0418 & -1.9750 & -0.2400 \\
-0.0074 & -0.1316 & -0.0825 & 0.9340
\end{bmatrix}.
$$

Notice here that we have finally combined the two matrices ρ_S and ρ_T into the composite matrix ρ as shown in Equation **(8-67)**. Applying Equation **(8-67)**, we now verify with $\mathbf{O} = (10\ 8\ 7\ 5)^T$ and $\mathbf{R} = (12\ 19\ 20\ 6)^T$ that

$$
\mathbf{R} = [\mathbf{I} - \rho]^{-1}\mathbf{O} =
\begin{bmatrix}
0.7502 & 0.1959 & 0.2778 & 0.1985 \\
0.3374 & 1.1925 & 0.2590 & 1.2391 \\
0.7418 & 0.2812 & 1.0215 & 0.2715 \\
0.1024 & 0.3860 & 0.0739 & 0.3361
\end{bmatrix}
\begin{pmatrix}
10 \\ 7 \\ 8 \\ 5
\end{pmatrix}
=
\begin{pmatrix}
12 \\ 20 \\ 19 \\ 6
\end{pmatrix}.
\qquad \textbf{(8-81)}
$$

The effects of a change in the final demand (export) $\mathbf{R}$ can now be traced back to the output vector $\mathbf{O}$ and the flows of commodities between subareas $(\mathbf{I}-\rho)$.

8.5.4 *Estimation of input–output coefficients*. Equation **(8-67)** is a standard interregional input–output model, where the input–output-coefficient matrix ρ is assumed to be known. In practice, these coefficients are difficult to estimate due to data availability. This problem prevents widespread use of the model. While gravitational-interaction models have been proposed to overcome this problem, it was not until recently that the problem was solved. For example, the Lowry–Garin model, with its singly-constrained gravity-model, was somewhat at odds with the more satisfactory doubly-constrained model. The latter is used to estimate input–output coefficients, where each row sum and column sum is enforced.

To see how these coefficients can be estimated by a gravitational-interaction model, re-definitions and extensions of the Lowry–Garin model is necessary. We recall that the supply of labor to industry is described by the first two rows of **Table 8-8**. The number of persons who live in zone-i and work in industry-q in zone j, $z_{ij}^{e_q}$, is the same as the journey-to-work variable

$$K_j^{\prime q} z_j^{e} W_i^{h} C^{\prime q}(C_{ij}), \tag{8-82}$$

Here z_j^{e} is the total-employment by industry-q in zone-j, the calibration-constant K' ensures that the distribution is constrained at the destination end $\sum_i z_{ij}^{e_q} = z_j^{e_q}$, W_i^{h} is the attractiveness of zone-i as a residence, and $C^{\prime q}(C_{ij})$ is the industry-specific property function of commuting-cost between zones. Since $z_j^{e_q} = (\rho^{hq}/y)R_j^{q}$ and $\rho_{ij}^{hq} = yz_{ij}^{e_q}/R_j^{q}$, one can write the input–output coefficient as

$$\rho_{ij}^{hq} = y K_j^{q}\left(\frac{\rho^{hq}}{y}R_j^{q}\right)\frac{W_i^{h}C^{\prime q}(C_{ij})}{R_j^{q}} = K_j^{q}\rho^{hq}W_i^{h}C^{\prime q}(C_{ij}) \tag{8-83}$$

where ρ^{hq} is the value-of-labor required per dollar-of-sales in industry-q.

The total value-of-labor supplied by zone-i is $yz_i^{e} = y\sum_j\sum_q z_{ij}^{e_q} = \sum_j\sum_q \rho_{ij}^{hq}R_j^{q}$. Its population is found from $z_i^{h} = f_i z_i^{e}$, where f_i is the number-of-residents per employed-worker (or the inverse of the labor-force-participation rate). The first two columns of **Table 8-8** describe the sale of commodities by industries to households. H_{ij}^{q} stands for the sales by sector-q firms in zone-i to residents of zone-j. The income of the residents in zone-j is yz_j^{e}. If y^q/y of income is spent on goods from industry-q, the expenditure by zone-j residents on industry-q is $y^q z_j^{e}$. Let W_i^{q} denote the attractiveness of zone-i as a place to purchase industry-q's goods. Then the typical spatial-interaction model of the expenditure-flows of the residents of each zone is given by $H_{ij}^{q} = K_j^{\prime q} z_j^{e} W_i^{q} C^{q}(C_{ij})$. Here K_j^{q} is a balancing factor that ensures that $\sum_i H_{ij}^{q} = y^q z_j^{e}$.

Now the input–output coefficients can be expressed as

$$\rho_{ij}^{qh} = \frac{H_{ij}^{q}}{yz_j^{e}} = K_j^{q}\left(\frac{y^q}{y}\right)W_i^{q}C^{q}(C_{ij}). \tag{8-84}$$

The remaining elements of the matrix G_{ij}^{pq} denote the interzonal-intersectoral flows, which are normally ignored by the Lowry–Garin model. However, a spatial-interaction equation in the spirit of Lowry–Garin is $G_{ij}^{pq} = \overline{K}_j^p \overline{R}_j^p \overline{W}_i^q C^q(C_{ij})$. Here $\overline{R}_j^p$ is the total value of non-labor inputs to sector-p in zone-j, $\overline{K}_j^p$ is the scaling constant to ensure $\sum_i \sum_q G_{ij}^{pq} = \overline{R}_j^p$, $\overline{W}_i^q$ is the attractiveness of zone-i as a location for industry-q, and $C^q(C_{ij})$ is a function that describes the effects-of-distance on trading by firms in industry-q. But the total non-labor input to sector-q in zone-j are $\left(\sum_q \rho^{qp}\right)R_j^p$ and $\rho_{ij}^{qp} = G_{ij}^{qp}/R_j^p$, hence

$$\rho_{ij}^{qp} = \overline{K}_j^p \left(\sum_q \rho^{pq}\right)\overline{W}_i^q C^q(C_{ij}). \tag{8-85}$$

Equations **(8-83)**, **(8-84)**, and **(8-85)** allow the input–output-coefficient matrix to be estimated and thus allow the spatial version of Equation **(8-67)** to be solved. The data required by Equations **(8-83)**, **(8-84)** and **(8-85)** are the labor-input coefficient ρ^{hq}, zonal-residence attractiveness W_i^h, and work-trip-impedance functions $C'^q(C_{ij})$; the relative expenditure by households on different industries y^q/y, zonal-shopping attractiveness W_i^q, and the shopping-impedance functions $C^q(C_{ij})$; the inter-industry input-coefficients ρ^{qp}, the zonal-attractiveness for industry $\overline{W}_i^q$, and the trading impedance-functions $C^q(C_{ij})$. In addition, the model needs to know the exports of each sector for each zone R_i^q.

8.5.5 *Relationship to Lowry–Garin model*. The above is a derivation of the 'unconstrained' version of the input–output model, or the version of the model that does not necessarily have the rows and columns of **Table 8-8** balanced. Translated into the context of activity derivation-allocation models such as Garin–Lowry, this means that the formulation does not necessarily have a land-use capacity constraint on each zone. The extension to a doubly-constrained model involves no additional theoretical extension if one considers the iterative procedure to ensure land-use capacity as described in the "Chaos, Catastrophe and Bifurcation" chapter. Our job now is to establish the formal relationship between input–output models and the Lowry–Garin model in the first place.

Suppose industry-sector 1 exports, or it is the basic employment in the study area. Sector 2, on the other hand, is nonbasic employment. All inter-industry flows are zero. Sector 2, being nonbasic industry, sells to households, while sector 1 does not. Then the matrix of input–output coefficients, Equation **(8-75)**, degenerates into

$$\rho = \begin{bmatrix} 0 & 0 & \rho_{11}^{h1} & \rho_{12}^{h1} & \rho_{11}^{h2} & \rho_{12}^{h2} \\ 0 & 0 & \rho_{21}^{h1} & \rho_{22}^{h1} & \rho_{21}^{h2} & \rho_{22}^{h2} \\ 0 & 0 & 0 & 0 & 0 & 0 \\ 0 & 0 & 0 & 0 & 0 & 0 \\ \rho_{11}^{2h} & \rho_{12}^{2h} & 0 & 0 & 0 & 0 \\ \rho_{21}^{2h} & \rho_{22}^{2h} & 0 & 0 & 0 & 0 \end{bmatrix}. \tag{8-86}$$

The vector of basic-employment, $\mathbf{O}$, as shown in Equation **(8-77)** becomes $\mathbf{O} = (0\ 0\ O_1^1\ O_2^1\ O\ O)^T$ while the total-employment vector as represented in Equation **(8-78)** remains the same. With these new definitions, the spatial version of Equation **(8-67)** is exactly the same as the Lowry–Garin model.

The above generalization of the Garin–Lowry model is much more satisfying than the economic-base paradigm. What prevents it from its widespread use may be the non-availability of essential data such as the inter-sectoral input–output coefficients ρ^{pq}. These coefficients are required to estimate the technical coefficients in Equation **(8-83)** for the Garin–Lowry model and both Equations **(8-83)** and **(8-85)** for the input–output model.

8.5.6 Equivalence between coefficients and multipliers.

To fully equivalence the Garin–Lowry and input–output model, maybe a comparison between the technical-coefficients ρ and the labor-force-participation-rate $1/f_j$ and the population-serving-ratio a_j is helpful. The matrix ρ is equivalent to the matrix $\mathbf{FA}$ in the Garin–Lowry model, where $\mathbf{F}$ is the derivation-allocation matrix associated with work trips and $\mathbf{A}$ is the corresponding matrix for nonwork trips (see the chapter on "Chaos, Catastrophe and Bifurcation"). To this extent, $(FA)_{ij} = \rho_{ij}$ or

$$
\begin{aligned}
\rho_{ij} &= \sum_{l=1}^{n}(t_{il}f_l)(u_{lj}a_j) \\
&= t_{i1}f_1u_{1j}a_j + t_{i2}f_2u_{2j}a_j + \ \ldots \ + t_{in}f_nu_{nj}a_j \\
&= (t_{i1}u_{1j}f_1 + t_{i2}u_{2j}f_2 + \ \ldots \ + t_{in}u_{nj}f_n)a_j
\end{aligned}
\tag{8-87}
$$

where $t_{il} = N_l / \sum_k N_k C'(C_{ik})$ and $u_{lj} = E_j / \sum_k E_k^R C^R(C_{lk})$

In the special case when a uniform multiplier applies toward all zones, or $f_j = f$ and $a_j = a$ for all j, Equation **(8-87)** becomes $\rho_{ij} = (t_{i1}u_{1j} + t_{i2}u_{2j} + \ldots + t_{in}u_{nj})fa = \mathbf{t}_i\mathbf{u}_j fa$. Notice that the input–output coefficient is defined between the household sector and the retail sector, . In other words, a more precise way of writing the coefficient is ρ_{ij}^{hp} or ρ_{ij}^{2h}, where work trips pertain to both basic (sector 1) and nonbasic (sector 2) employment while shopping trips are only to nonbasic (sector 2) employment locations. We now write out the expressions for the coefficient according to Equation **(8-83)**, and lay them side-by-side with the work and nonwork derivation-allocation matrix-elements in Garin–Lowry: $K_j^q W_i^h C'(C_{ij})\rho^{hq} = t_{ij}f_j$ and $K_j^q W_i^q C^R(C_{ij})(y^q/y) = u_{ij}a_j$. This reveals respectively that $\rho_{ij}^{hq} = t_{ij}f_j$ and $\rho_{ij}^{qh} = u_{ij}a_j$. Put it in another way, the spatial labor-force participation rate is equivalent to the value-of-labor required per dollar-of-sales in industry and the spatial population ratio is the fraction of family-income spent on goods and services from retail industry, depending on whether we are talking about work trips or nonwork trips. It is worthy to note once more that these coefficients do not cover inter-sectorial transactions between industries. For that reason inter-sectorial multipliers need to be obtained exogenously, either from a survey or from secondary data.

Example

Kim (1989) estimated the input–output technical coefficients for the Chicago metropolitan area from the 1977 U.S. national input–output tables using a *non-survey* technique. Instead of carrying out Equation **(8-85)**, a simple *location quotient* is used for constructing an aspatial input–output-coefficient matrix for the Chicago area from the national input–output table. The U.S. table consists of two parts: the "Make Table" ρ_T and the "Use Table" ρ_S. As defined in Subsection 8.5.3, the former is the production of commodities by industries and the latter are commodities consumed. The following steps are used to construct a regional input–output-coefficient matrix and to aggregate 88 sectors into four commodity groups:

a.   Update the 1977 national commodities-to-industries used-table (ρ_S matrix) to reflect the process level of the study period, say 1980. Prices were adjusted using the *Consumer and Producer Prices Indexes.*

b.   Remove imports (the last row of **Table 8-8**), because imports to the nation as a whole would have considerably less impact on the regional trade pattern in a subregion of the nation. The Use table (ρ_S) was subsequently adjusted by the factor $1+a/(a-b)$, where a is the import amount for sector p, and b is the total commodity output for sector p.

c.   Convert cells of use (ρ_S) and make (ρ_T) matrices from transactional volumes to coefficients, following Equation **(8-76)**.

d.   Adjust the input–output coefficients for regional trade patterns. The *simple location quotient* is used. Here the quotient is defined as $[E(R)_p/E(R)]/[E(N)_p/E(N)]$, where $E(R)$ is regional employment in Chicago and $E(N)$ is the national employment level. Apply this quotient to the Use matrix (ρ_S) for each sector p.

e.   Aggregate all sectors into four commodity groups: manufacturing, trade, services, and households. Weights are used to compute weighted averages among sectors within each group.

f.   Estimate the direct requirements table by multiplying the Use matrix by the Make matrix, following the same idea as $\rho_S \rho_T = \rho$ in Subsection 8.5.3. □

8.6 Generalization of the input–output model

Although the input–output model is a better model than Lowry–Garin, it has its limitations as well—some of which are carry-overs from the Lowry–Garin model. First, the model is linear: an x percent increase in output requires x percent more labor and x percent more of each input (including imports.) Economies of scale are thus disallowed, as are input-substitution and the existence of production bottlenecks. Equally, the spatial-interaction representations of the input–output coefficients in Equations **(8-83)**, **(8-84)** and **(8-85)** presume the production technology of each sector does not vary spatially. Also, the consumption patterns do not vary over space. Steps are taken incrementally below to relax these assumptions.

8.6.1 Inclusion of capital investment.   One can define σ_{ij}^{pq} as the subareal investment-coefficient or the marginal capital-output ratio, which is analogous to the traditional input-output technical-coefficients ρ_{ij}^{pq}. More precisely, $\sigma_{ij}^{pq} = W_{ij}^{pq} \big/ \dot{x}_j^{\,q}$, which quantifies the multiplier effects of investment. Let us now express the growth at subarea-i as a function of the common rate-of-growth q':

$$\dot{x}_i^P = q' x_i^P.$$

(8-88)

The equilibrium economic-activity at each subarea-i and sector-p, x_i^p, is then the result of both trade and investment:

$$x_i^P = \sum_q \sum_j \rho_{ij}^{pq} x_j^q + q' \sum_q \sum_j \sigma_{ij}^{pq} x_j^q \qquad \forall\, i,p$$

(8-89)

where q' is un-determined at this point. Equation **(8-89)**, or its matrix form x = $\rho x + q' \sigma' x$, must hold at equilibrium (Andersson and Kuenne 1986).

We can now ask: what is the highest rate-of-capacity-utilization γ' if the sectors were expected to grow at a constant-rate q'? In this case, the equilibrium condition would be characterized by Equation **(8-88)**, showing a constant growth-rate for each subarea and each economic sector. Correspondingly, we could formulate the entire problem in the following way: $\max_x \gamma'(x)$, subject to $\gamma' x = \rho x + q' \sigma' x$, $x \geq 0$ and $\gamma' > 0$. Expressing this optimization problem in terms of its Lagrangian, we have

$$\max_{x,\gamma',\lambda} \; H = \gamma' - \sum_{i,p} \lambda_i^p \left(\gamma' x_i^P - \sum_j \sum_q \rho_{ij}^{pq} x_j^q - q' \sum_j \sum_q \sigma_{ij}^{pq} x_j^q \right).$$

(8-90)

The necessary (and here the sufficient) conditions of a maximum are:

$$\frac{\partial H}{\partial \gamma'} = 1 - \sum_i \sum_p \lambda_i^P x_i^P = 0$$

(8-91)

$$\frac{\partial H}{\partial x_i^P} = \lambda_i^P \gamma' - \sum_j \sum_q \lambda_j^q \rho_{ij}^{pq} - \sum_j \sum_q \lambda_j^q \sigma_{ij}^{pq} q' = 0 \qquad \forall\, i,p$$

(8-92)

$$\frac{\partial H}{\partial \lambda_i^P} = \gamma' x_i^P - \sum_j \sum_q \rho_{ij}^{pq} x_j^q - \sum_j \sum_q \sigma_{ij}^{pq} x_j^q = 0 \qquad \forall\, i,p.$$

(8-93)

From this follows that:

$$q' = \frac{1 - \sum_j \sum_q (\lambda_j^q / \lambda_i^p) \rho_{ij}^{pq}}{\sum_j \sum_q (\lambda_j^q / \lambda_i^p) \sigma_{ij}^{pq}} \qquad \forall\, i,p$$

(8-94)

if γ' is required to be at the full capacity level, i.e., equal to unity. Thus the relative shadow-value of each sector-product must be adjusted until the profit ratio, relative to the marginal capital-output-ratio equals the common balanced-growth-rate q'.

This result is patently similar to the condition of equilibrium of a one-sector-one-region economy. The first requirement in Equation **(8-91)** states furthermore that all shadow values should be orthogonal, so that the sum of products of shadow values and economic production quantities x_i^p are "on the unit circle." This implies that if the production of commodities goes toward infinity, shadow values go toward zero. In other words, a high level of economic activity for a commodity proportion-ately diminishes its intrinsic value.

8.6.2 Dynamic version. In the above discussion, the coefficients ρ_{ij}^{pq} and σ_{ij}^{pq} are assumed constant throughout. Obviously, this is viable only in the short run. As long as a longer time-frame is used, production capacities can and do change, and thus, non-linearities become a necessary part of interzonal economic analysis. Here we introduce explicitly the projection of employment and population into the spatial input-output model described above. Suppose each zone j has an industrial-sector p, and the corresponding employment-level at time 0 is $E_j^p(0)$. The total employ-ment in sector-p in all zones is then $\Sigma_j E_j^p(0) = E^p(0)$. At the beginning of the next time-period the employment level in sector-p in zone-j is now $E_j^p(1)$. This is the result of growth at the rate of a_j^p, where $a_j^p = E_j^p(1)/E_j^p(0)$. The overall rate-of-growth in the study area is correspondingly $a^p = E^p(1)/E^p(0)$. If the zone grows at the overall rate, the expected level-of-employment will be $\tilde{E}_j = \sum_{p=1}^{K} a^p E_j^p(0)$ representing the sum over all sectoral increases.

The difference between the expected and the observed level-of-employment can be expressed in terms of (a) the change in total employment and the distribution among economic sectors, and (b) the relative share of each zone in the production:

$$(E_j(1) - \tilde{E}_j(1)) = E_j(1) - \sum_{p=1}^{K} a^p E_j^p(0)$$
$$= E_j(1) - \sum_{p=1}^{K} a_j^p E_j^p(0) + \sum_{p=1}^{K} a_j^p E_j^p(0) - \sum_{p=1}^{K} a^p E_j^p(0) \qquad \textbf{(8-95)}$$
$$= \left(E_j(1) - a_j E_j(0)\right) + \sum_{p=1}^{K} (a_j^p - a^p) E_j^p(0).$$

In other words

$$E_j^P(1) - \tilde{E}_j^P(1) = (E_j^P(1) - a_j^P E_j^P(0)) + (a_j^P - a^P) E_j^P(0). \qquad \textbf{(8-96)}$$

In the last line of the above expression, the first term represents the change in importance of industrial-sector p in zone j over the time-period. The second term represents the increase (or decrease) in employment due to the relative competitive-ness of zone j vis-a-vis other zones. They are called the *shift* and *share* components of employment–change.[21] If the first term is negative, it means the sector's share of the market in the zone has decreased. If the second term is positive, it means that the industrial-sector p in zone j has grown faster than the overall areal-rate and so on. The above represents a projection of zonal employment (and population), possibly by trade class, expressed in terms of its explanatory factors: growth and migration. While it is easy to state, such projection is difficult to carry out. It will be shown below how the empirical relationship in Equations **(8-95)** and **(8-96)** can be estimated via input–output modelling, thus actually implementing the projection defined above.

---

[21]*Shift-share analysis* was introduced in the "Economic Methods" chapter in Chan (2000).

8.6.3 Inclusion of production function into input–output models. The first way of implementation observes that the substitution between products and competition between subareas is often modelled with a production function that—for convenience—is assumed to be concave and at least twice differentiable (Andersson and Kuenne 1986). One extremely simple example, illustrated in **Figure 8-12** is the Cobb–Douglas function $x_i^p = a_{ip}(x_{ii}^{pp})^{\alpha_{ii}^{pp}} (x_{ii}^{qp})^{\alpha_{ii}^{qp}} (x_{ji}^{pp})^{\alpha_{ji}^{pp}} (x_{ji}^{qp})^{\alpha_{ji}^{qp}}$ where $\alpha_{ii}^{pp} + \alpha_{ii}^{qp} + \alpha_{ji}^{pp} + \alpha_{ji}^{qp} = 1$. Here x_i^p is the output of commodity-p in subarea-i, x_{ji}^{qp} is the input of commodity-q from subarea-j in producing commodity-p in subarea-i. α_i^p and α_{ji}^{qp} are examples of technological parameters (similar to population-serving-ratio[22] and input–output technical-coefficients). In general, one can write the following Cobb–Douglas production-function:

$$\ln x_i^P = \ln a_i^P + \sum_q \sum_j \alpha_{ji}^{qp} \ln x_{ji}^{qp} \qquad \left(\sum_q \sum_j \alpha_{ji}^{qp} = 1 \right). \tag{8-97}$$

A production function can be regarded as a technological *constraint* on a decision-maker operating a set of plants.

Supposed some projected sales of product-p are available, say according to Equation **(8-89)**. Based on these projected sales, $\tilde{x}_i^p$, a reasonable objective is to minimize the costs of producing the planned-amount $\tilde{x}_i^p$. Remember that factor-of-production input-prices $p_j^{\prime q}$ include unit transport-costs C_{ji}^q. Formally, we construct this mathematical program: $\min_x \sum_q \sum_j \left(p_j^{\prime q} + C_{ji}^q \right) x_{ji}^{qp}$, subject to

$$\ln a_i^q + \sum_q \sum_j \alpha_{ji}^{qp} \ln x_{ji}^{qp} = \ln \tilde{x}_i^q \tag{8-98}$$

with $x_{ji}^{qp} \geq 0$. The following equivalent optimization problem involving a Lagrangian can then be formed with $\sum_q \sum_j \alpha_{ji}^{qp} = 1$.

$$\min_x L_i^P = \sum_q \sum_j \left(p_j^{\prime q} + C_{ji}^q \right) x_{ji}^{qp} + \lambda_i^P \left(\sum_q \sum_j \alpha_{ji}^{qp} \ln x_{ji}^{qp} - \ln \tilde{x}_i^P \right) \tag{8-99}$$

A typical optimum-condition can be written as

$$\frac{\partial L_i^P}{\partial x_{ji}^{qp}} = \left(p_j^{\prime q} + C_{ji}^q \right) - \lambda_i^P \alpha_{ji}^{qp} \frac{\tilde{x}_i^P}{x_{ji}^{qp}} = 0 \qquad \forall q, j \tag{8-100}$$

or $\left(p_j^{\prime q} + C_{ji}^q \right) x_{ji}^{qp} / \alpha_{ji}^{qp} = \left(p_r^{\prime 1} + C_{ri}^1 \right) x_{ri}^{1p} / \alpha_{ri}^{1p} = \dots$ and Equation **(8-98)**. This implies that all input demands are decided by all prices and transportation costs: $x_{ji}^{qp} = \ln K_{ji}^{qp} + \ln \tilde{x}_i^P + \sum_r \sum_j \alpha_{ji}^{qp} \ln(p_j^{\prime r} + C_{ji}^r) / (p_j^{\prime q} + C_{ji}^q)$. This derivation implies that $a_{ji}^{qp} = K_{ji}^{qp} \prod_r \prod_j \left[(p_j^{\prime r} + C_{ji}^r) / (p_j^{\prime q} + C_{ji}^q) \right]$. Although the above derivation is based on the

---

[22] Population-serving-ratio is defined as the number of retail employees to support one household. It is defined in detail in the chapters on "Activity Derivation and Allocation" and "Chaos, Catastrophe and Bifurcation."

Cobb–Douglas production function, there is no need to specify the technological constraints of interzonal trading by production functions. It is enough if cost functions can be postulated. *Knowing these cost functions allows the determination of optimal factor demand and input–output functions with input prices and transport costs as arguments.*

8.6.4 Inclusion of migration into input–output models.

A second way of setting up a dynamic input–output model is through labor migration. Suppose we are given an employment–migration matrix of the form $M = [m_{ij}]$, where $i = 1, \ldots, n'$ and $j = 1, \ldots, n'$. m_{ij} stands for the proportion of jobs in the production sector that were previously located in zone i and have, over the time period, moved to zone j (zones i and j are within the same study area). The term m_{jj} in the main diagonal represents the net rates of change, after considering the gain and loss of jobs from and to the outside.[23] The employment vector in time period 1 can now be estimated from period 0 by $E^p(1) = M\, E^p(0)$. In terms of the input–output lingo of production, the above equation can be rewritten as $R(1) = M\, R(0) = M\, O(0)$.

Referring back to the geometric-series expansion of the basic input–output equation **(8-67)**, and interpret each term of the series as a time period, we have at time-1: $R(1) = \rho\, O(0)$, and if we include migration by replacing ρ by ρM: $R(1) = (I + \rho M)\, O(0)$. After 2 time periods, $R(2) = (I + \rho M + \rho^2 M^2)\, O(0)$. The above equation can be extended to the total output levels at time m as functions of the given basic-production levels at time 0:

$$R(m) = (I + \rho M + \rho^2 M^2 + \ldots + \rho^m M^m)\, O(0). \tag{8-101}$$

If the migration changes from one period to another, then G is no longer a constant. Equation **(8-101)** can now be written as

$$R(m) = (I + \rho M_1 + \rho^2 M_1 M_2 + \ldots + \rho^m M_1 M_2 \ldots M_m)\, O(0). \tag{8-102}$$

Thus far, no procedure was offered to estimate the M matrix. Available information points toward a tautological impasse since information for the future time-period 1 is not available to implement the definition of Equations **(8-95)** and **(8-96)**. It is possible at this juncture to refer back to the chapter on "Chaos, Catastrophe, Bifurcation and Disaggregation." There we used the procedure for estimating the zonal labor-force participation-rate and population-serving ratio from a calibration year (past) to a base year (at the present) and project forward one period each time. This breaks the impasse. A further advantage is that each term ρM_m can be treated as a composite term similar to the product of the work-derivation–allocation matrix F and the nonwork-derivation–allocation matrix A in the Garin–Lowry series, where the inherent migration between zones is already built in. Thus:

$$R(m) = (I + \overline{Q} + \overline{Q}^2 + \overline{Q}^3 + \ldots + \overline{Q}^m)\, O(0) \tag{8-103}$$

---

[23]For a discussion of migration modelling, see the "Interregional components of change" section of the "Economic Methods" chapter (Chan 2000). Simply replace population with employment.

where $\bar{Q}$ is the composite disaggregate matrix where migration, derivation, and allocation are all included. Should the technical coefficient be estimated by gravitational interaction, the migration component of employment–population projection is then explicitly recognized.

8.6.5 Disequilibrium and multiplier calibration. As alluded to earlier, perhaps the biggest advantage of a dynamic model of the type shown in Equation **(8-103)** is the analytical properties. Clearly the general bifurcation-condition for such a series has been well established. In the chapter on "Chaos, Catastrophe, Bifurcation and Disaggregation," it was pointed out that each row sum of $\bar{Q}$, or $\bar{Q}_i$, has to be less than unity in order to avoid bifurcation. Depending on the makeup of $\bar{Q}$, further details can be specified regarding bifurcation.

Through an elasticity parameter on the trip-distribution function C^p, endogenous calibration techniques can be employed to ensure both constraints (inputs and outputs) are complied with. As part of the process, the individual labor-force participation rate $(1/f_j)$ and population-serving ratio (a_j) are updated for each time-period by solving a set of simultaneous, albeit nonlinear, equations in the Garin–Lowry models. Since the multipliers or technological coefficients ρ are estimated by a gravitational-interaction model, the form of the series **(8-103)** is expected to bear resemblance to the generalized Garin–Lowry series. Many of the findings on dynamic formulation and disequilibrium can be directly transferred to the input–output model—at least in the case of a uniform multiplier areawide.

Perhaps it bears further commenting on the calibration of multipliers in the input-output model. The way that disaggregate f_j and a_j are updated cannot be carried over to the spatial version of the input-output model. The reason is that there was no equivalence between these multipliers and the technical coefficients ρ_{ij}^{pq}. Also, the gravitational-interaction formulation of the input-output model, which links it directly to the Garin–Lowry model, bogs down in the spatial formulation, as discussed previously. To the best of the author's knowledge, econometric estimation techniques, rather than analytical modelling, have to come in at this point. In the "Spatial Econometric Models" chapter, we will discuss how this is performed.

8.6.6 Structural change of regional labor supply. Another place where analytical modelling has faced its limitation can be described (Mees as reported in Andersson and Kuenne 1986). The sudden and unexpected collapse or explosion of regional populations leading to "ghost towns" or "boom towns" have traditionally been explained by indivisibilities and similar large-scale causalities[24]. These

---

[24]Examples of indivisibility include fixed-costs-of-provision or capacity constraints of facilities. These factors lead to combinatorial 0–1 allocation problems: how to locate p service facilities, where $p = 1,2,\ldots$ etc. Another example is the existence of increasing-return-to-scale in production, which can lead toward discontinuities.

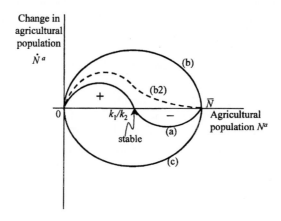

Figure 8-13 - From a stable employment-mix
to alternative specialized-labor-allocations (Andersson and Kuenne 1986)

explanations have been typically ad hoc in nature. Recent advances make it possible
to apply catastrophe theory to the problem of structural change in population and
labor force, lending a much more satisfactory explanation of such precipitous
events. This represents a starting point of a potentially exciting approach, as we
discussed in the "Chaos, Catastrophe and Bifurcation" chapter. In the context of
demographics, the concept is best illustrated in **Figure 8-13**. In a given region, let
N^a be the population working in agriculture, $\bar{N}$ be the total working-population,
and $Y(N^a)$ be the wage rate of agricultural workers as a function of agricultural
labor-supply, and C is the transportation cost. We can construct the following
model:

$$\dot{N}^a = k_0 Y(N^a) N^a (\bar{N} - N^a)$$
$$Y(N^a) = k_1 - k_2(C) N^a \tag{8-104}$$

or $\dot{N}^a / N^a = k_0 [k_1 - k_2(C) N^a][(\bar{N} - N^a)]$ where the k are calibration coefficients, $\dot{N}^a / N^a =$
0 if $N^a = k_1 / k_2$ or if $N^a = \bar{N}$.

Solution to this differential equation can be characterized by these three types:

a.   Interior solution at the stable equilibrium-point k_1 / k_2, yielding an integrated
     economy with part of the population in agriculture and the other part in urban
     employment;

b.   With the whole population in agriculture as indicated by the stable equilibrium
     point $\bar{N}$, exporting agricultural commodities in exchange for urban amenities;
     and

c.   With the whole population in urban-commodity production, as shown by the point $N^a = 0$.

A structural change from a type (a) to a type (b) or type (c) economy could occur if k_2, which is a function of transportation-cost C, declines below a certain critical value. It is hypothesized that because of declining cost no interior-solution exists, only the stable upper-corner solution occurs, where $N^a = \bar{N}$. In **Figure 8-13**, this transition is shown to have gone through the intermediate solution of type (b2) before reaching the type (b) pattern. A drop in k_2 also results in some other regions specializing at the equilibrium point $N^a = 0$, with their entire population engaging in the production of urban commodities only. Declining transportation costs are thus a necessary condition for specialization and interregional trade. A smooth gradual decline of transportation costs can lead to a sudden extinction of the stable interior equilibrium k_1/k_2 often referred to as a catastrophe.

To extend this concept further, a four-parameter version of this labor-allocation model can be constructed. Besides transportation-cost C and total-population N, the two other parameters are defined as follows. Let ρ^a be the productivity-in-agriculture per unit-of-labor, and ρ^s be the productivity-in-urban-occupations per unit-of-labor. After proper normalization, the canonical parameters are: ρ for the average productivity of an occupation, $\Delta\rho$ for the difference in productivity between occupations, and D' the population density (or a measure of crowding,) defined as population N divided by arable land. A four-parameter n'-region ordinary-differential-equation that acts to maximize a smooth potential (accessibility) function with a single canonical variable has a solution set called a *butterfly*[25], as shown in **Figure 8-14**.

Referring to **Figure 8-14**, let us characterize the solutions to the four-parameter model in terms of transport-cost C and congestion-level D'. Assume the starting point to be transport-cost level C_0 and congestion-level D'_0. Infrastructural investments such as road building gradually reduce transportation-cost to and from the region following the trajectory $(C_0, D'_0)-(C_T, D'_T)$ vertically. Then congestion sets in as the trajectory moves horizontally as shown. Up to the point marked as (1), there is *no* change in the mixed occupations. Not until the point marked by (2) is there an abrupt shift from an unspecialized mix to exclusive agricultural production (with imports of urban commodities.) Further reducing the transportation costs, combined with immigration of labor and increasing-congestion D', finally implies a drastic shift from complete specialization on agriculture to a complete specialization on city commodities. This occurs beyond the point marked by (3). It can be proven in general that a gradual improvement of transportation infrastructure with generally decreasing transport-costs leads to smooth or drastic increases in regional specialization and inter-regional trade. One can of course argue that the dynamics assumption is too special. This is a valid critique if completely general economic

---

[25]The butterfly catastrophe is one of seven elementary catastrophes introduced in the "Control, Dynamics" appendix to Chan (2000). It is a higher-order catastrophe in four dimensions. **Figure 8-14** represents a two-dimensional slice of the higher–order diagram.

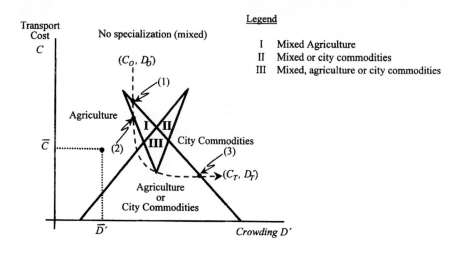

Figure 8-14 - Butterfly catastrophe (Andersson and Kuenne 1986)

systems are assumed. The Darwinian character of competitive systems (as well as the overall profit-maximizing character of Maxian or state capitalistic systems) does, however, lend credence to searching in the direction of total-cost minimization or profit-maximization for a large class of economic systems.

8.6.7 Remarks. Under Lowry's assumptions of a locationally and quantitatively given economic-base (especially manufacturing activity) and service, multipliers can be extended to a more general framework. Also, operational procedures for predicting bifurcation (and catastrophe) can be readily transferred from a previous chapter to the more general input–output formulation. Extending previous findings on the Garin–Lowry model, one can handle interzonal and interregional migration directly through multipliers. In the case of uniform areawide-multipliers, the calibration of these multipliers is easily handleable, in spite of the dynamic generalization to include labor migration in the input–output model.

While we have reviewed both descriptive and prescriptive extensions of the input-output model, Kim (1989) has shown independently that the input/output format has allowed computationally-feasible prescriptive models to be built. Experience with the Chicago-area studies is reported for both a linear- and nonlinear-programming model, with external calibration of the multipliers as illustrated in Subsection 8.5.6 above. He shows that the general-equilibrium modelling approach is desirable. The entropy-maximization approach combines land-use and transportation decisions regarding origins, destinations, routes and the amounts of goods to be produced at the optimal land-use density. The solution algorithm converges well, ensuring both system and user-optimizing equilibria. Kim also suggests multi-level programming as a potential way to address the multi-criteria, multi-stakeholder environment. In parallel, artificial intelligence and expert system are recognized as powerful tools for future analytical efforts in such interactive decision-making.

While the input–output model bypasses a fair amount of the difficulties associated with the artificial definition of basic activities, a weakest part of the modelling procedure still remains. Manufacturing jobs do depend locationally on the population, and the relations between them are cyclic. A simultaneous-equation regression-model will be offered in sequel to redress this shortcoming. Whereas Lowry-based and input–output based models assumed that the location of manufacturing was exogenous and that population follows jobs, the truth is that dependence is mutual. The structural simultaneous-equations explicitly recognize that the location patterns of populations and of jobs are dependent on each other. Furthermore, the simultaneous econometric-equations have the shift-share analysis built into the model naturally as part of the calibration procedure. We will point this out in a following chapter entitled "Spatial Econometric Models."

8.7 Optimization formulations

The models discussed here in this section can be viewed as optimization extensions of a spatially disaggregated input-output model. They all consist of an input–output multiplier linking the inputs to the outputs for each sector (Equation **(8-66)**), and a spatial-interaction component to describe the relative location of sectors as producers and users of inputs and outputs (Equations **(8-83)**, **(8-84)**, **(8-85)**). This is not at all unlike the derivation and allocation components of the Garin–Lowry-based models. Our objective is to establish linkages between prescriptive models and descriptive models. Here, prescriptive models use an optimization formulation and descriptive models are simply replications of the system—as pointed out in the beginning of this book. Here we will solve for spatial equilibrium again for both regular and dynamic conditions (Wilson et al. 1981). The solutions so obtained will then be compared with previous findings.

Let R_j^p be the total output of sector-p in zone-j, R_{ij}^{pq} be the output of p in i used by q in j (including the household sector), and O_j^p as the final demand for the output of sector-p in zone j. We are interested in the case in which the coefficients ρ_{ij}^{pq} can be represented by

$$\rho_{ij}^{pq} = \rho_j^{pq} C^P(C_{ij}) \Big/ \sum_i C^P(C_{ij}), \tag{8-105}$$

where ρ_j^{pq} is the technical coefficient specific to the receiving-sector zones, and the expression is nothing more than an attraction-constrained spatial-interaction model. The generalized Equation **(8-66)**, $R_i^p = \sum_j \sum_q \rho_{ij}^{pq} R_j^p + O_i^p$—including the household sector—now becomes

$$R_i^p = \sum_j \sum_q \frac{\rho_j^{pq} \exp(-\beta^p C_{ij}^p)}{\sum_i \exp(-\beta^p C_{ij}^p)} R_j^q + O_i^p = \sum_j \sum_q R_{ij}^{pq} + O_i^p \qquad \forall p, i \tag{8-106}$$

if the trip-distribution function C^p assumes an exponential form. Notice this is a more compact restatement of Equations **(8-82)** through **(8-85)**.

8.7.1 An entropy-maximization formulation. When expressed in the above form, it can be expected that a mathematical-programming formulation can be constructed from the same information. The solution to this program will also be consistent with the results from an input–output model. An entropy-maximizing model is used to reproduce the exponential trip-distribution functions:

$$\max_{R_{ij}^{pq},R_i^p} -\sum_{ij}\sum_{pq} R_{ij}^{pq} \ln R_{ij}^{pq} \qquad\qquad\qquad\qquad (8\text{-}107)$$

subject to

$$R_i^p = \sum_j\sum_q R_{ij}^{pq} + O_i^p \qquad\qquad \forall\, i,p \qquad\qquad\qquad (8\text{-}108)$$

$$\sum_i R_{ij}^{pq} = \rho_j^{pq} R_j^q \qquad\qquad \forall\, j,pq \qquad\qquad\qquad (8\text{-}109)$$

$$\sum_{ij}\sum_q R_{ij}^{pq} C_{ij}^p = X^p \qquad\qquad \forall\, p. \qquad\qquad\qquad (8\text{-}110)$$

The first constraint is a restatement of the generalized Equation **(8-66)**. The second constraint is based on the definition of technical coefficients in input–output models. Here the disaggregate multiplier ρ_j^{pq} is to be calibrated exogenously by a descriptive procedure. This procedure was outlined in Subsection 8.5.6. There, the ρ_{ij}^{pq} can be obtained from Equation **(8-105)**, albeit for only transactions between households and basic industry and between households and retail industries. The last constraint is the usual spatial-interaction travel-constraint placed on the control total of areawide-transportation-cost X^p. Our intent is to show the parallel between the solution to this mathematical program and the gravitational input–output model for estimating R_i^p and R_{ij}^{pq}, as outlined in Equation **(8-106)**.

Solution to the entropy-maximization model starts with the Lagrangian:

$$L = -\sum_{ij}\sum_{pq} R_{ij}^{pq} \ln R_{ij}^{pq} + \sum_i\sum_q \lambda_i^p (R_i^p - \sum_j\sum_q R_{ij}^{pq} - O_i^p) $$
$$+ \sum_j\sum_{pq} \gamma_j^{pq}\left(\sum_i R_{ij}^{pq} - \rho_j^{pq} R_j^q\right) + \sum_p \mu^p\left(\sum_{ij}\sum_q R_{ij}^{pq} C_{ij}^p - X^p\right). \qquad (8\text{-}111)$$

First-order conditions are

$$\frac{\partial L}{\partial R_{ij}^{pq}} = \ln R_{ij}^{pq} - \lambda_i^p + \gamma_j^{pq} + \mu^p C_{ij}^p = 0 \qquad\qquad \forall\, i,j,pq \qquad\qquad (8\text{-}112)$$

$$\frac{\partial L}{\partial R_i^p} = \lambda_i^p - \sum_q \gamma_i^{pq}\rho_i^{pq} = 0 \qquad\qquad \forall\, i,p. \qquad\qquad\qquad (8\text{-}113)$$

Together with the three original constraints **(8-108)**, **(8-109)** and **(8-110)**, the optimality conditions are defined. Equations **(8-112)** and **(8-113)** yield

$$R_{ij}^{pq} = \prod_r (B_i^{rp})^{\rho_i^{rp}} B_j^{pq} \exp(-\mu^p C_{ij}^p) \qquad \forall\, i,j,p,q \tag{8-114}$$

where $B_j^{pq} = \exp(-\gamma_j^{pq})$. μ^p can be evaluated from Equation **(8-110)**, B_j^{pq} from **(8-109)**, and R_i^p from **(8-108)**. Now we proceed with the result from Equation **(8-114)**: $R_{ij}^{pq} = b_i^p B_j^{pq} \exp(-\mu^p C_{ij}^p)$, where this means $b_i^p = \prod_r (B_i^{rp})^{\rho_i^{rp}}$.

Now using the solution to the entropy-maximization model, arrange the term B_j^{pq} in the above equation to be

$$B_j^{pq} = \frac{\rho_j^{pq} R_j^q}{\sum_i \exp(-\mu^p C_{ij}^p)}. \tag{8-115}$$

Neutralize the shadow-price b_i^p by setting

$$(W_i^p)^\alpha b_i^p = 1 \tag{8-116}$$

where the W term is the `attractiveness' used in gravity models. Recognizing β and μ are equivalent calibration constants, the R_{ij}^{pq} term in Equation **(8-106)** thus becomes $R_{ij}^{pq} = (W_i^p)^\alpha b_i^p B_j^{pq} \exp(-\mu^p C_{ij}^p)$. The programming model is shown to produce equivalent allocations R_{ij}^{pq} as the general input–output gravitational-model under the conditions specified in Equations **(8-115)** and **(8-116)**.

8.7.2 Dynamic models revisited. Dynamic models can be obtained by writing a differential equation for such equilibrium conditions as the generalized Equation **(8-66)**:

$$\frac{dR_i^p}{dt} = \tau'(R_i^p - \sum_j \sum_q R_{ij}^{pq} - O_i^p). \tag{8-117}$$

In other words, outputs will increase until they meet the demands placed on them. τ' is merely a constant in this equation. We can also derive dynamic input–output models from the first-order condition of the Lagrangian of the entropy model, Equation **(8-113)**: $dR_i^p/dt = \sum_q \gamma_i^{pq} \rho_i^{pq} - \lambda_i^p$. Substituting ρ_i^{pq} from Equation **(8-109)** and rearranging terms, the reader can check that $dR_i^p/dt = (\lambda_i^p R_i^p - \sum_j \sum_p \gamma_{ji}^{pq} R_{ji}^{qp}) R_i^p$.

Yet another dynamic embedding of the input–output equations is to write

$$\frac{dR_i^p}{dt} = R_i^p (R_i^p - F(R_i^1, R_i^2, \dots, R_i^K)) \qquad p=1,\dots,K;\ i=1,\dots,n' \tag{8-118}$$

where $F(R_i^1, R_i^2, \ldots, R_i^K)$ is defined by the right-hand-side of Equation **(8-106)**. This formulation has the special interpretation in the ecological literature as representing a number of interacting 'populations'(Hof 1998). This was explained in the "Activity Allocation" and "Chaos, Catastrophe" chapters. Perhaps most interesting is that the Lowry model, being a special case of the input–output model, can also be 'dynamicized' by Equation **(8-118)**. All these are continuous versions of the geometric series way of representing a dynamic model as discussed earlier in this chapter and in previous chapters on urban spatial-modelling and bifurcation.

8.8 Summary

We have developed a number of models in this chapter concerning the relationships between different decision-makers in spatial organization. This is beyond the classic facility-location and activity-allocation variety. Of particular interest are conflicts that can arise within groups of similar decision-makers, consumers and producers, and between such groups, including the supply of commodities from producers to consumers. The question is whether such interaction in spatial markets results in stable equilibria. Where markets do not converge readily on such an equilibrium, strategy becomes all-important. For this reason, we have illustrated classical cases of pure competition, oligopoly and monopoly, all in the context of gaming.

The conclusion is that frequently the introduction of spatial considerations does disturb achieving equilibrium and reduces the possibility of an easy resolution to conflict. It is not just the purely spatial sectors such as transport and housing which present problems of increasing returns-to-scale and indivisibilities. All production that concentrates on space must have increasing returns over the relevant range of output. Further analysis of this problem suggests that discontinuities may also be important, if not more important. Such discontinuities often give rise to multiple equilibria and sometimes precipitous and paradoxical results. Thus spatial equilibrium on discrete networks takes on extra levels of complexity, particularly in the face of market competition.

Spatial equilibrium involves the integration of the various sectoral equilibria (Vickerman 1980). The spatial version of the input–output model does a reasonable job of putting this notion together in a workable format. Export-base theories now are expanded into a more general concept involving trade among the industrial sectors--rather than just between the household, export, and service sectors. The gravitational-interaction version of the model allows technical coefficients to be calibrated. It also establishes the detailed linkages between the multipliers used in the export-base theories, most notably the Garin–Lowry variety, and the technical coefficients used in the more general spatial input–output models.

Since Leontiff's initial formulation, input–output analysis has been extended in many different ways. One area of steady improvement is the computational aspect of updating the input–output table. The fundamental procedure of row- and column-total consistency—a pervasive problem in location–allocation practice—has been made more efficient (see for example Nagurney (1993), Nagurney and

Eydeland (1992), Schneider and Zenios (1990)). Recent attempts were also launched to calibrate input–output tables for urban-regional applications [Jin et al. (1991), Kim (1989), Suzuki et al. (1988)]. This is distinguished from conventional applications on a larger scale in the order of state or national level. Although there is always room for improvement, these are encouraging signs for the use of input–output spatial equilibrium in day-to-day practices. Dynamic and disequilibrium formulations are accommodated within the spatial input–output models. Several versions of the model were presented, including the geometric series formulation and the differential-equation formulation. Bifurcation conditions are outlined for the geometric-series model, drawing the strong analogy with the Garin–Lowry models. It was pointed out that interzonal migration is an inherent property of this formulation. In special cases, the multipliers can be endogenously calibrated in the forecasting sequence. This bypasses a tautological impasse in previous attempts at solving the problem using migration matrices.

Alternate derivations of the same set of input–output solutions were attempted. It is apparent that optimization approaches yield consistent results with descriptive modelling, both under regular conditions and dynamic conditions. Thus entropy-maximizing models yield consonant solutions as input–output gravitational models. It was shown, however, that there are well-known limitations of the export-base concepts, even in its most generalized form in the input–output model. To the extent that population and manufacturing sectors are intertwined, rather than manufacture being the clear-cut seed for development, alternate approaches to modelling spatial equilibrium are clearly worthwhile. Econometric models using simultaneous equations may be a viable alternative to input–output modelling. An additional advantage of such an approach is the replacement of the multipliers by trend-line projections, a concept compatible with shift-share analysis of regional development.

In this chapter, we have also included the investment sector explicitly into the spatial input–output model. This broadens the application of input–output analysis in supply-oriented economies, with production functions built directly into the model. It was found, however, that there is no need to specify the technological constraints of interzonal trading by production functions. It is enough if cost functions can be postulated. Knowing these cost function automatically allows the determination of optimal factor demand and input–output functions.

Equilibrium means different things to different people today. We are often concerned with equilibrium paths for one zone or region. At the same time, we are equally concerned with the convergence or divergence of paths for other parts of the study area. A key issue is whether a failure to converge is due to misadjustments spatially or a more fundamental instability in the system. In several instances, we could make some fairly definitive comments on this issue throughout the last three chapters. Hopefully, this adds toward general understanding of these spatial equilibria.

CHAPTER NINE

SPATIAL ECONOMETRIC MODELS:
THE EMPIRICAL EXPERIENCE

Previous discussions on activity-allocation-and-derivation models have concentrated on gravity-interaction and economic-base concepts. Largely, these are well-structured modelling approaches founded on generally accepted theories of spatial development. It was observed that more interaction takes place more in closer proximity and the economy is fueled by the multiplier-effect of each dollar of investment. There is a parallel school of thought about the modelling process; however, in which empirical relationships are established between observed economic activities such as population and employment. Econometric or correlative equations are then constructed to explain the interaction between these activities. The EMPIRIC model is one of the most widely acclaimed urban models in the 1970s and 1980s. This chapter will review its development and application, and through such discussions, we can reexamine the different approach to comprehensive spatial modelling in general. It complement sthe more causal models presented in previous chapters.

The chapter starts out with a review of spatial econometrics. This special branch of statistics has been evolving rapidly in recent years, fueled mainly by applications ranging from image processing to renewed interests in geo-statistics. The discussion here can be viewed as part one of two parts. Part one concentrates on "spatial dependence" or the empirical relationship between adjacent geographic units; such analyses are based mainly on *cross-sectional* data. Part two delves into spatial-temporal models constructed upon *time-series* information. The idea of spatial dependence is not new, since we have seen many examples of spatial relationship already ranging from remote-sensing applications to gravitational interaction. The focus here is simply to formalize the information presented thus far, casting an *economic* framework within which diverse applications can find a home.

9.1 Spatial econometrics

In general, one can think of spatial econometrics as dealing with *spatial effects*, of which there are two kinds: *spatial dependence* and *spatial heterogeneity*. Spatial dependence is best manifested by Tobler's "first law of geography ."This law says that "everything is related to everything else, but near things are more related than distant things." Here the notion of 'nearness' goes well beyond Euclidean space, as we have discussed previously in the "Spatial separation" chapter. Spatial heterogeneity, on the other hand, takes exception to homogeneous dependence among all points in space. While it does not violate Tobler's "first law," it introduces additional complexity in modelling in that spatial relationships differ from one point to another (Anselin 1988).

9.1.1 Spatial dependence vs. heterogeneity.

9.1.1 Spatial dependence vs. heterogeneity. Perhaps the best known statistic to measure spatial dependence is *spatial autocorrelation*, defined as the relationship or correlation[1] between data points in specified orientation in space. It is analogous to classic Pearson correlation and particularly autocorrelation over time. But the multi-directional nature of spatial dependence—as contrasted with say a one-directional situation in time—necessitates the use of a different methodological framework.

The second type of spatial effect, spatial heterogeneity, suggests the lack of stability over space in the behavioral or other relationships under study. More precisely, this implies that functional forms and parameters vary with location and are not homogeneous throughout the data set. This is likely to occur in econometric models estimated on a cross-sectional data set—i.e., data taken at one point in time (a 'snapshot')—of dissimilar spatial units. For example, traffic zones are typically drastically different in size and shape at the city center compared to suburbs (with those at the center smaller than the suburban ones). Suppose a model is calibrated for a base-year. A separate model may have to be calibrated in the central city vice the suburbs, since the *homogeneity* assumption may be violated. In contrast to the spatial dependence case, the problems caused by spatial heterogeneity can for the most part be solved by means of standard econometric techniques. An example is the use of dichotomous 0–1 variables to 'switch' between a central city and suburban model. Specifically, methods that pertain to varying parameters, random coefficients and structural instability can easily be adopted to take into account such variation over space. We will illustrate how this can be done in sequel.

9.1.2 Connectivity in space.

9.1.2 Connectivity in space. The very notion of spatial dependence implies the need to decide which other units in space have an influence on the particular unit under consideration. Formally, this is expressed in the topological notions of neighborhood and nearest neighbor. This concept has been illustrated in the "Geographic Information Systems" chapter (Chan 2000). In raster files of digitized images, we pointed out that in such a grid system, a pixel has either four immediate neighbors each sharing a common side with the subject pixel, or eight adjoining neighbors, including the ones at the four corners. (Refer to the "Raster data structure example" figure in the chapter.) The four sharing a side each are closer

---

[1]For a formal definition of correlation, see the "Statistical Tools" appendix in Chan (2000). Autocorrelation simply extends the concept from the relationship between data points *i* and *j* to data points between period *t* and *t*+1, as will be explained more formally in sequel.

in Euclidean distance than the other four at the corners. We referred to the former as *first-order* neighbors and the latter as *second-order* neighbors. In analogy to the game of chess, we also call this the *rook* and *bishop* case respectively.

In general, consider a system of n' spatial units, labelled $i = 1, 2, ... ,n'$, and a vector measure $\mathbf{x}$ with a value x_i observed for each of these spatial units. The variable $\mathbf{x}$ may represent the grey values in satellite images for example. In such applications as image restoration in raster-files, a set of neighbors of a spatial unit i can be defined as the collection of those units j for which x_j is contained in the functional form of the probability of x_i, conditional upon $\mathbf{x}$ at all other locations. Formally, this would yield the set of neighbors for i as N'_i, for which $P(x_i|\mathbf{x}) = P(x_i| \mathbf{x}_{N'_i})$ where $\mathbf{x}_{N'_i}$ is the vector of x_j observations, for all j in N'_i, and $\mathbf{x}$ is the vector containing all x_k ($k \neq i$) values in the entire system. Alternatively, and less strictly, the set of neighbors j for i can be taken as $\{j\,|\,P(x_i) \neq P(x_i|x_j)\}$. These are the locations for which the conditional marginal probability for x_i is not equal to the unconditional marginal probability. As mentioned, application of this concept is graphically illustrated in the Bayesian method of image classification. This is discussed in the "Geographic Information Systems" chapter of Chan (2000), where the ith pixel is allocated based on an updated probability depending on the grey value of the neighboring pixels j. Note that, in general, neither of these neighborhood definitions includes information about the relative location of the two spatial units, but only the neighbor's influence via conditional probabilities.

In order to introduce a spatial aspect to these definitions of neighbors, which also makes the link with the notions of *spatial stochastic processes*, this working definition is suggested: $\{j\,|\,P(x_i) \neq P(x_i|x_j)$ *and* $d_{ij} < S_i\}$. Here d_{ij} is a measure of the distance between i and j in a properly structured space, and S_i is a critical cutoff distance for each spatial unit i, and possibly the same for all spatial units. The distance metric underlying d_{ij} is in the most general sense and can include a Euclidean-, Manhattan-, or general l_p-metric (sometimes referred to as the Minkowski metric.) We have seen the basic notion of cutoff distance illustrated in the discussion of deterministic facility-location models. There a maximal response time S_i is specified around a demand point i for such emergency services as fire and medical. In other words, a fire station should be able to respond to a fire in no more than say $S_i = 10$ minutes. This working definition introduces an additional structure in the spatial data set—by combining a notion of statistical dependence with a notion of space. Notice the definition does not preclude spatial units j that do not meet the distance criterion from exerting an influence on the conditional probability for x_i. But they are not considered nearest neighbors. For example, they can be included as higher-order neighbors, which implies that the influence of j on i works via other spatial units. The definition based on conditional probabilities only does not allow for this distinction between first-order and higher-order neighbors and is therefore more limiting.

9.1.3 Spatial weights. Through *weights*, the simple concept of contiguity was extended to include a general measure of the potential interaction between two spatial units. The larger the separation, the less the interaction is expected and hence the smaller the weight. This is expressed in a spatial weight matrix $\mathbf{W} = [w_{ij}]$ for all pairs of geographic units in the study area. The determination of the proper specification for the elements of this matrix w_{ij} is a more difficult and controversial methodological issue in spatial econometrics. For example, one can use a combination of distance measures and the *relative* length of the common border between two spatial units. The resulting weights will be asymmetric, unless both spatial units i and j have the same relative common boundary length, and the distance between the two units is the same both ways ($d_{ij} = d_{ji}$). Specifically,

$$w_{ij} = |s_{ij}|^a \Big/ |d_{ij}|^b \qquad (9\text{-}1)$$

where s_{ij} is the proportion of the interior boundary of unit i that is in contact with unit j, and a and b are calibration parameters. To the extent that $s_{ij} \neq s_{ji}$ or $d_{ij} \neq d_{ji}$ in general, one cannot expect $w_{ij} = w_{ji}$. In gravitational interaction, such a relationship is amply illustrated, with say population at i (N_i) and employment at j (E_j) as proxy for the geometric quantity s_{ij}. We can think of $s_{ij} = (N_i E_j)^a$ in this case.

In a similar way, weights can also take into account the relative area of the spatial units:

$$w_{ij} = \delta_{ij} \theta_i s_{ij} \qquad (9\text{-}2)$$

with δ_{ij} as a binary contiguity factor ($\delta_{ij} = 1$ if units i and j are neighbors and 0 otherwise,) and θ_i is the share of unit i in the total area of all spatial units in the system. A competitor of gravity model, the Intervening Opportunity Model, illustrates this weight definition. Here "annular rings" (or zones of influence) are sequenced in terms of distance away from the subject zone, each characterized by the number of 'opportunities' which are in effect proxies for 'area'. Each unit in a ring shares a fraction of the residential or employment 'opportunities' depending on the 'size' of the ring. The example in the "Descriptive Tools" chapter (Chan 2000) has an employment origin at downtown, and there are five destinations in the neighboring first ring, and the probability of residential location at any destination is 1/2. The probability of population living in any first-ring destination is simply $\theta = e^{-\frac{1}{2}(1)} - e^{-\frac{1}{2}(6)} = 0.556$. Here the probability of working at i and living in ring j can be interpreted as w_{ij}, $\delta_{ij} = 1$ between downtown and the first ring, and $s_{ij} = 1$ since the ring completely surrounds the downtown origin.

Both weight definitions, in their original design, are closely linked to the physical features of spatial units on a map, or its digitized image. As with the binary contiguity measures, they are less useful when the spatial units consist of discrete points or a lattice, since then the notions of boundary length and area are largely artificial. They are also less meaningful when the spatial-interaction phenomenon under consideration is determined by purely economic factors, which may have little to do with the spatial pattern of boundaries on a physical map. This is again illustrated by the gravitational-interaction example above, where economic activities such as population and employment have to be used as proxies for the boundary variable.

Consequently, some have argued for weights with a more direct relation to the particular phenomenon under study. For example, a general accessibility weight, calibrated between 0 and 1, combines in a logistic function the influence of one or more modal linkages between two regions—including roads, railways, and other means of transport. Formally,

$$w_{ij} = \sum_j k_j \frac{\alpha}{1 + \beta \exp(-c_j d_{ij})} \qquad (9\text{-}3)$$

in which k_j shows the relative importance of the transportation mode j. The sum is over all the modes, each of which separates the spatial units by a distance of d_{ij} at a different unit cost c_j. α and β are calibration constants. Similar to Equation (9-2), it should be noted that a smaller weight results from a longer 'distance' separation, which is really time separation in transportation modes.

An important problem results from the incorporation of calibration parameters in the weights. Typically, these weights are taken to be exogenous and the parameter values are determined prior to, or in steps separate from, the rest of the spatial analysis. This creates problems for estimating and interpreting the results. In particular, it could potentially lead to the inference of *spurious* relationships. Remember the validity of estimates is pre-conditioned by the extent to which the spatial structure is correctly reflected in the weights. More important, it could result in a circular reasoning, in that the spatial structure, which the analyst may wish to discover in the data, has to be assumed known before the data analysis is carried out.

When the weight matrix—determined say by Equations (9-1), (9-2) and (9-3)—is used in a hypothesis test, the null hypothesis[2] is one of spatial independence. Ideally, the weight matrix should be related to the relevant, alternative hypothesis of spatial dependence, in order to maximize the power of the test. However, even with an improperly specified weight matrix, a conservative interpretation of a rejection of the null will only imply a lack of independence, and not a particular type of dependence. This practice has its redeeming value however. Although the power of the test may be compromised, the potential for spurious conclusions is minimized. Unfortunately, application of such a hypothesis test is not simple in practice.

9.1.4 Spatial lag operators. The ultimate objective of a spatial weight matrix is to relate a variable at one point in space to the observations for that variable in other spatial units. In the more familiar time-series context, one relates the observation in time period t to the observation to period $t-k$. This is achieved by using a *lag operator*, which shifts the variable by one or more periods in time. For example, $x_{t-k} = L^{(k)}x_t$ shows the variables x_t shifted k periods back from time t, as a result of the kth power of the lag operator L. Now the time-series x_t can be related

---

[2]For an explanation of a null hypothesis, see the "Statistical Tools" appendix to Chan (2000).

(a) Shift to first-order neighbors

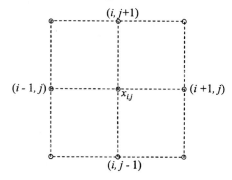

(b) Shift to second-order neighbors

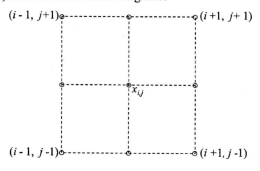

Figure 9-1 Illustrating spatial lags

to time-series $L^{(2)}x_t$ by regression if desired, for such purposes as forecasting. (The correlation coefficient between the two time series is the autocorrelation coefficient.)

While it is similar in concept, matters are not this straightforward in space due to the many directions in which the shift can take place. As an illustration, consider the discrete lattice structure in **Figure 9-1**. The variable x, observed at location (ij), can be shifted in separate ways using two simple contiguity criteria. If the first-order neighbors are the positions considered for the shift, the alternate positions are $x_{i-1,j}, x_{i,j-1}, x_{i+1,j}, x_{i,j+1}$. If the second-order neighbors are the shift positions, on the other hand, the spatial-lag positions are $x_{i-1,j-1}, x_{i+1,j-1}, x_{i+1,j+1}, x_{i-1,j+1}$. For an *eight-connected-neighbors*[3] shift, as defined in the "Geographic Information Systems" chapter (Chan 2000), we are talking about the combination of the four positions in **Figure 9-1**(a) and the four positions in **Figure 9-1**(b). Thus three different directions of spatial lag can be easily defined even for a very simple example such as

---

[3]The eight-connected neighbors are analogous to the eight more positions of a queen on a chess board.

this. Corresponding to each of these directions, correlation parameters such as autocorrelation coefficient need to be estimated respectively.

In most applied situations, there are no strong a priori motivations to guide the choice of the relevant form of spatial dependence. This problem is compounded when the spatial arrangement of observations is irregular, since then an infinite number of directional shifts becomes possible. Clearly in any statistical analysis, the number of parameters associated with all shifted positions quickly would become unwieldy and preclude any meaningful model to be constructed. Moreover, unless the data set is very large and structured in a regular way, the remaining degrees-of-freedom would be insufficient to allow an efficient estimation of these parameters. In the extreme case, this would be analogous to calibrating a two-variable linear regression line through two data points, where no degree of freedom is left.[4]

This problem is resolved by considering a weighted sum of all values belonging to a given contiguity class, rather than taking each of them individually. The terms of the sum are obtained by multiplying the observations in question by the associated weight from the spatial-weight matrix. Formally,

$$L^{(l)} x_i = \sum_{j \in N_i^l} w_{ij} x_j \qquad\qquad (9\text{-}4)$$

where $L^{(l)}$ is the spatial-lag operator associated with contiguity class l, j is the index of the observations belonging to the contiguity class for i, N_i^l. In matrix form, this would be $L^{(l)} x_i = \mathbf{w}^{(l)} \mathbf{x}$. Thus in the lattice of **Figure 9-1**, either the first-order neighbors are treated as a group, or the second-order neighbors, or the eight-connected-neighbors. Then a set of four weights would be defined for both the first and second order neighbors and a set of eight weights for the eight-connected neighbors. The spatial lag of observation x_i, $L^{(l)} x_i$ ($l = 1$, 2 or 3), would simply be the weighted sum of the four or eight neighbors.

We have seen examples of the spatial-lag operator in spatial-filter applications. In the "Remote Sensing" chapter (Chan 2000), we have seen that irrespective of the type of image filters used, the basic approach is to sum products between the mask coefficients and the intensities of the pixels under the mask. In a 3×3 mask, for example, there are nine grey levels of pixels under the mask counting the subject pixel and the 8-connected neighbors, namely $x_1, x_2, \dots, x_9$. The filter simply uses Equation **(9-4)** to produce a weighted sum for the subject pixel as $w_1 x_1 + w_2 x_2 + \dots + w_9 x_9$. The grey value of the pixel at the center of the mask is then replaced by this weighted average. The mask is then moved to the next pixel location in the image and the process is repeated. This continues until all pixel locations have been covered. A 'low-pass spatial-filter' for example, will simply average the pixels in the neighborhood, or set all mask coefficients w to 1, resulting in $x_1, x_2, \dots, x_9$. Not surprisingly, the effect is to blur the image, thus removing 'specks' of noise scattered around the image if desired. Clearly, the resulting notion of a spatially lagged variable is not the same as in time-series analysis. Instead, it is similar to a

---

[4]For more discussion of this problem, please consult the "Statistical Tools" appendix to Chan (2000).

distributed lag such as exponential smoothing in forecasting, where recent observations are weighted more heavily than those in the distant past according to a pre-determined set of weights. It is important to note that the weights used in the construction of the lagged variables are taken as given, just as a particular time path can be imposed in estimating a distributed lag. The joint determination of the weights and measures of statistical association, such as correlation or regression coefficients, becomes a nonlinear problem. However, by fixing the weights a priori, this is reduced to a more manageable linear problem—at the risk of imposing a potentially wrong structure as explained in the previous section.

Besides the pre-determined weights used in filter masks, we have seen several general examples of this practice previously. Recall that in the "Generation, Competition and Distribution" chapter, a case study of Long Island State Parks was performed in New York, where an additive utility function combining distance and travel time need to be calibrated a priori: $w_1(distance) + w_2(time) + constant$. Here the weighted sum of distance from i and the time from i is computed, combining—not unlike a filter mask—the features of both spatially correlated attributes centered at location i. Notice that this was performed exogenous to the model by valuating time using previous studies. Similarly car-operating-expenses per mile (km) were also found exogenously using established cost figures. In the chapter on "Chaos, Catastrophe, Bifurcation," a trip-distribution function showing the percentage of trips of 5-minute, 10-minute, 15-minute durations, etc., was obtained. It was needed as part of an accessibility function for allocating population and employment in a disaggregate Lowry–Garin Model. This trip-distribution function—an example of which appears as $constant + c_1(duration) + c_2(duration)^2$—was again calibrated outside the model. Here, the trip-distribution function generates a set of weights $w_{ij}^{(l)}$ for the $(l) = 10$-minute, $(l) = 15$-minute etc. contiguity classes in the weight matrix $\mathbf{w}^{(l)}$, as shown in the "Activity Allocation and Derivation" chapter. Additional examples can be cited. But these two, one from facility location and the other from land use, represent an adequate illustration of generalized spatially-lagged variables and how they are determined independent of measures of statistical association in a model.

Following this latter example, less-restrictive spatially-lagged-variable can be constructed from the notion of potential or accessibility. As one recalls, accessibility is the weighted trip-distribution function where the weights are population, employment or other activity variables x_j: $X_i = \sum_j f(d_{ij}, \mathbf{b}) x_j$ where f is a trip-distribution function of distance and a vector of coefficients $\mathbf{b}$. One can think of a weight matrix to the extent that $\mathbf{b}$ is a function of $\mathbf{x} = (\cdots x_j \cdots)^T$. The resulting expression is nonlinear. Estimation and hypothesis testing will correspondingly be more complex. To overcome this problem, a piecewise linear model can be constructed, in which Equation **(9-4)** is estimated for the current cross-sectional data (or 'snapshot'). Such an estimate is then used to compute the spatial lag. If the 'snapshots' are taken frequent enough, and the model is reasonably 'smooth', these piecewise linear models can be excellent approximations to the original model. We will illustrate this with a case study of the EMPIRIC model later in this chapter.

A related concept to spatial-lag operators is *cellular automata*. It is traditionally used to represent a dynamical system in terms of discrete economic-variables.
In this context, it describes the values assumed by a large number of identical cells
with local interactions among them. In this book, they can consist of a lattice of
sites, each with a finite set of possible values. A set of discrete rules is now defined
for the evolution of a spatial economy. The values assumed at each site evolve
synchronously in discrete time-steps according to identical rules. The value of a
particular site is decided by the previous values of a neighborhood of sites around
it. For example, if the system is in state *i* and a specific criterion is met, then the
state changes to *j*. Cellular automata may be divided into basic classes with
different behavior. It is through this taxonomy that it appeals to modelers. These
automata can occasionally generate chaotic trajectories over time. A major
drawback of these models, however, is a lack of rigorous connection between
physical laws in the continuum one is trying to model and the symbol rule that
generates the iterated coupled-cells. But it holds promise to explain the type of
spatial-temporal process discussed in this book, as demonstrated in examples of
spatial interaction and image processing above.

9.1.5 *Problems with modelling spatial effects*. A substantial part of spatial
econometric analysis is based on data collected for spatial units with irregular and
arbitrary boundaries, such as political subdivisions and traffic zones. Nonetheless,
the interpretation of the various models and their policy implications are often
made *spatially* in general. This implies that there is some unique and identifiable
spatial structure, with clear statistical properties, independent from the way in
which the data are organized in spatial units. Unfortunately, matters are not this
straightforward—the way subareal boundaries are drawn has a direct bearing upon
structure and the interpretation of a model. We have already alluded to this earlier
in this chapter, but it bears repeating.

The *modifiable-areal-unit* problem addresses the fact that statistical measures
for cross-sectional data are sensitive to the way in which the spatial units are
organized. Specifically, the level-of-aggregation and the way contiguous units are
combined affects the various measures of association, such as spatial-autocorrelation coefficients and parameters in a regression model. This problem is an old one,
and has been referred to as the micro-macro aggregation problem in econometrics
and the ecological fallacy problem in sociology. The modifiable-areal-unit problem
can best be explained as a combination of two familiar problems in econometrics:
aggregation and *identification*.

9.1.5.1 SPATIAL AGGREGATION. As is well known, aggregation is only
meaningful if the underlying phenomenon is homogeneous across the units of
observation. If this is not true the inherent heterogeneity and structural instability
should be accounted for in the various aggregation schemes (Chan 1991). In other
words, unless there is a homogeneous spatial process underlying the data, any
aggregation will tend to neutralize variation in the data and lead to misleading
results. Because a model's coefficients are found by explaining variations in

observed data, the less the variation to be explained in aggregated data, the less reliable the model will be. The reduced variability in aggregate data also results in a high level of collinearity between variables at the aggregate level, which does not exist in the disaggregate level. It would impart bias to estimated coefficients. For example, a coefficient of −0.4 may be estimated using disaggregate data, whereas a value of −0.3 may be obtained for the same parameter using aggregate data.

The concept of spatial aggregation goes well beyond the straightforward definition of areal units such as census tracts or traffic zones. When one employs Equation **(9-4)** to 'pre-process' the data, one also has performed spatial aggregation. For example, the membership in the contiguity class for areal unit i, N_i^I, can vary depending on whether a 4-connected neighbor is defined, or an 8-connected neighbor; there is no reason one cannot go beyond a 3×3 'mask' to a 4×4, 5×5 and so on. Depending on the definition adopted, the resulting data base would look very different. In short, spatial aggregation introduces a bias on the model-building process. To account for this bias, this aspect of the modifiable-areal-unit problem should be considered a specification issue, related to the form of spatial heterogeneity. It is not solely an issue determined by the spatial organization of the data. Also an integral part of model specification is *model identification*, which can be thought of as the ability to discern the underlying model structure.

9.1.5.2 IDENTIFICATION DEFINED. Consider a simultaneous-equation model as illustrated in the "Descriptive Tools" chapter (Chan 2000) under the "Econometrics approach" section. The identification problem arises when a system of more than one equation contains two-way causality[5], so that there is a sort of imbalance between the dependent (*endogenous*) and the independent (*exogenous*) variables. Since the exogenous variables are the source of information for estimating model parameters, too few of them might preclude parameter estimation of all the endogenous variables. Equations that do not contain the sufficient number of exogenous variables, and whose parameters cannot be estimated, are called *under-identified equations*.

The identification problem can be illustrated by a simple example (Kanafani 1983). Suppose a demand and supply model system is given by

$$V = \alpha_0 + \alpha_1 p'' + a_D$$
$$p'' = \beta_0 + \beta_1 V + a_S$$

(9-5)

where V is the inter-regional trade, p'' is the unit price, α_0, α_1, β_0 and β_1 are structural parameters to be estimated, and a_D and a_S are error or disturbance terms in each of these two equations. Both equations are said to be *under-identified* since it is not possible to estimate their parameter values from empirical observations.

---

[5]For a discussion of the confusion between causality and correlation, see the sub-section on "Arrow diagrams" in the "Descriptive Methods" chapter (Chan 2000). It is explained in the appendix that the direction of the arrow shows causality, while correlation between two factors exists irrespective of the arrow direction.

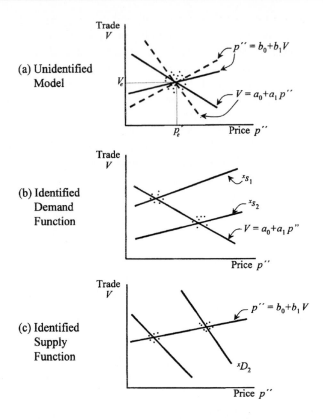

Figure 9-2 Under-identified and exactly-identified systems (Kanafani 1983)

There are simply not enough exogenous variable in both equations. Referring to **Figure 9-2**(a), it can be seen that the two equations are to be calibrated by a cluster of data at around only "one point" (V_e, p_e''). Hence, these observations on trade and price cannot be used to estimate parameters of either equation, since they will provide only "one point" on each curve. Observations thus obtained will result in an average of the location of the equilibrium point, and a rather poor average at that, as suggested by the possible scatter in the data illustrated in **Figure 9-2**(a). This scatter is reflected by the error terms a_D and a_S. In order to identify either the demand equation or the supply equation, it is necessary for this model to identify an additional exogenous variable. For the sake of fixing ideas, we will assume the error terms a_D and a_S to be negligible (or set at zero value) from this point forward until further notice.

As illustrated in **Figure 9-2**(b), an additional exogenous variable, say x_S that contributes toward locating the supply function but not that of the demand function, will allow making observations of trade and price at different levels of x_S. These observations made on different supply curves will permit estimating the demand curve. This is analogous to adding an exogenous variable in the model, which will change it to

$$V = \alpha_0 + \alpha_1 p'' + \alpha_2 x_S$$
$$p'' = \beta_0 + \beta_1 V + \beta_2 x_S. \tag{9-6}$$

When x_S does not influence the demand, i.e., when $\alpha_2 = 0$, it becomes possible to make observations of V and p' at different values of x_S. This way one obtains observations of the equilibrium points at different supply curves. Such observations will allow the identification of the demand function, as illustrated in **Figure 9-2(b)**. In other words, α_0 and α_1 can now be estimated.

Similarly, to estimate the supply function, it would be necessary to have an additional exogenous variable, say x_D, which affects the demand without affecting the supply. The model then becomes

$$V = \alpha_0 + \alpha_1 p'' + \alpha_2 x_S + \alpha_3 x_D$$
$$p'' = \beta_0 + \beta_1 V + \beta_2 x_S + \beta_3 x_D. \tag{9-7}$$

With both $\beta_2 = 0$ and $\beta_3 = 0$ or when exogenous variables x_S and x_D do not influence supply, it is now possible to obtain observations of the equilibrium point at different values of x_D. This will allow estimating β_0 and β_1 of the supply function, as illustrated in **Figure 9-2(c)**.

This example illustrates the fact that for an equation to be identifiable, it is necessary that a certain relation exists between the endogenous and the exogenous variables. In general, it is sufficient to have the vector of parameters for the model equations set all independent and orthogonal. In practice, it often suffices that the following relationship between the number of variables in an equation be met: Let E' be the number of exogenous variables in the whole model, E'' be the number of endogenous variables, e' be number of exogenous variables left in any model equation after estimation, and e'' be the number of endogenous variables left: then for any model equation to be identifiable, the following relationship must hold:

$$(E' - e') + (E'' - e'') \geq E'' - 1. \tag{9-8}$$

When it is strict equality, the equation is said to be *exactly identified*, and when a strict inequality holds, the equation is said to be *over-identified*. It is understood that an under-identified equation is one for which the inequality is reversed. It is possible to estimate parameters for exactly identified and over-identified equations, but not for equations that are under-identified. In other words, the total number of exogenous and endogenous estimated parameters, including the intercept, must be equal to or exceed the total number of exogenous variables. This is analogous to the degree-of-freedom in a regular regression equation, which is the number of 'useful' pieces of information after estimation, or [(total number of data) – (number of estimated parameters)]. A regression equation can be estimated only if the number of useful pieces of information equal to or exceed the number of estimated parameters, taking into account the 'intercept' parameter.

In the example used earlier in this section, the following relationships hold. For Equation **(9-5)**, $E' = 0$, $e' = 0$, $E'' = 2$, $e'' = 2$ for both equations. Applying Equation **(9-8)** yields $(0 - 0) + (2 - 2) < (2 - 1)$ which is the same for both equations and suggests that neither is identifiable. For Equation **(9-7)**, however, $E' = 2$, $e' = 1$, $E'' = 2$, $e'' = 2$ for both equations. Again, applying Equation **(9-8)** yields $(2 - 1) + (2 - 2) = (2 - 1)$, which shows that both equations are exactly identified. The same goes for Equation **(9-6)**, where $(1 - 0) + (2 - 2) = (2 - 1)$.

What should one do if a system of equations is under-identified? The natural course of action is to devise a reasonably realistic set of identifying assumptions (Kane 1968). Basically, there are only three ways by which under-identification can be removed. Either (a) we constrain the values of certain structural parameters, (b) we find enough exogenous variables that affect one equation but not the others, or (c) we constrain the relative behavior of the random-error terms $\sigma(a_S)$ or $\sigma(a_D)$. This last alternative effectively removes the error terms from the equations since the remaining constant terms can now be incorporated into the respective intercepts α_0 and β_0.

We have illustrated methods (a) and (b) adequately in our discussions above. Let us now give equal time to the error terms. Assuming the variance of the error terms $\sigma(a_D) = 0$, this would render the demand curve identifiable in Equation **(9-6)** when the 'error' term $\alpha_2 x_S$ can now the 'absorbed' as a constant in α_0. Similarly, setting $\sigma(a_S) = 0$ in Equation **(9-7)** would allow the supply curve to be identified. These assumptions would adequately portray the data having been generated by one curve shifting along a second roughly stationary one, as explained above. Because error terms arise as a result of a complex of unknown forces, however, assumptions regarding the properties of a are peculiarly hard to defend. At best, one's choice of additional exogenous variables is apt to be only slightly less arbitrary. Nor is it possible to guarantee in advance that the chosen exogenous variables will, in fact, have the hoped-for properties. Consequently, most econometricians prefer to constrain the structure of their model. Here at least, a priori reasoning can play a legitimate supporting role. Of the many restrictions one might impose on the structural coefficients, these represent the most frequently encountered varieties: (a) set a parameter to zero; (b) two coefficients are set equal; (c) coefficients are estimated outside the model.

9.1.5.3 SPATIAL IDENTIFICATION.

For proper identification of the structure of *spatial dependence*, an analysis of spatial association is typically carried out by relating a variable to its spatially lagged counterpart. The latter is constructed as a weighted sum (typically a linear combination) of the observations in the system, as alluded to above. The association is indicated by a correlation or regression coefficient. For example, a variable $\mathbf{x}$ would be related to $\mathbf{\Phi W}^{(l)}\mathbf{x}$, where $\mathbf{\Phi}$ is a spatial autoregressive coefficient matrix and $\mathbf{W}^{(l)}$ is the spatial weight matrix for the *l*th contiguity class. Clearly a different choice of $\mathbf{W}^{(l)}$ will result in a different $\mathbf{\Phi}$, and therefore the measure of spatial association is indeterminate. Drawing our experience from image processing, however, the model specification should look like

$$
\begin{bmatrix} x_1(l) \\ x_2(l) \\ \cdot \\ \cdot \\ \cdot \\ x_{n'}(l) \end{bmatrix} = \begin{bmatrix} \varphi_1 w_{11}^l & \varphi_2 w_{12}^l & \cdot & \cdot & \cdot & \varphi_l w_{1(l+1)}^l & 0 & \cdot & \cdot & 0 & \cdot \\ 0 & \varphi_1 w_{21}^l & \varphi_2 w_{22}^l & \cdot & \cdot & & \varphi_l w_{2(l+1)}^l & \cdot & 0 & 0 & \cdot \\ \cdot & \cdot & \cdot & \cdot & \cdot & \cdot & \cdot & \cdot & \cdot & \cdot \\ \cdot & \cdot & \cdot & \cdot & \cdot & \cdot & \cdot & \cdot & \cdot & \cdot \\ \cdot & \cdot & \cdot & \cdot & \cdot & \cdot & \cdot & \cdot & \cdot & \cdot \\ 0 & 0 & 0 & \cdot & \cdot & \varphi_1 w_{(l+1)1}^l & \cdot & \cdot & \varphi_l w_{(l+1)(l+1)}^l & 0 & \cdot \end{bmatrix} \begin{bmatrix} x_1 \\ x_2 \\ \cdot \\ \cdot \\ \cdot \\ x_{n'} \end{bmatrix}
$$

Here the $(l+1)$th entries in each row include the subject spatial-unit itself. Notice that there are only l calibrated autoregressive-coefficients, with the coefficient for the subject unit being unity by definition. Conceptually, we are expanding $\Phi W x$ in this expression. This is comparable to the autoregressive time-series:

$$
\begin{bmatrix} x(1+l) \\ x(2+l) \\ \cdot \\ \cdot \\ \cdot \\ x(n+l) \end{bmatrix} = \begin{bmatrix} \varphi_1 & \varphi_2 & \cdot & \cdot & \cdot & \varphi_l & 0 & \cdot & \cdot & 0 & \cdot \\ 0 & \varphi_1 & \varphi_2 & \cdot & \cdot & \cdot & \varphi_l & \cdot & 0 & 0 & \cdot \\ \cdot & \cdot & \cdot & \cdot & \cdot & \cdot & \cdot & \cdot & \cdot & \cdot \\ \cdot & \cdot & \cdot & \cdot & \cdot & \cdot & \cdot & \cdot & \cdot & \cdot \\ \cdot & \cdot & \cdot & \cdot & \cdot & \cdot & \cdot & \cdot & \cdot & \cdot \\ 0 & 0 & 0 & \cdot & \cdot & \varphi_1 & \cdot & \cdot & \varphi_l & 0 & \cdot \end{bmatrix} \begin{bmatrix} x(1) \\ x(2) \\ \cdot \\ \cdot \\ \cdot \\ x(n) \end{bmatrix}
$$

in which a forecast is made based on the l most immediate data points.

From an econometric standpoint, the problem can be viewed as an identification problem, since there is insufficient information in the data to allow for the full specification of the simultaneous interaction over space. In this sense, a formulation of linear spatial-association can be considered as a special case of a system of simultaneous linear-equations, with one observation for each equation:

$$ x_i = \sum_j a_{ij} x_j \qquad \forall i. \tag{9-9} $$

As mentioned previously, constraints need to be imposed for at least some parameters to render the model identifiable. The usual approach in spatial analysis is to introduce a spatially-lagged variable $\sum_j w_{ij} x_j$ and thereby to reduce the empirical problem to that of estimating one parameter φ, the l-entry vector in $x_i = \sum_j \varphi_i w_{ij} x_j$ We have seen examples of this in the spatial filters discussed in Section 9.1.4. Clearly, a choice of different weights w_{ij} is likely to result in a different estimate for φ, as mentioned previously. Interestingly enough, Anselin (1988) conjectured there may be some relationship between spatial weights and technical coefficients of an input-output model. Measures of interconnectedness that are based on such technical coefficients would be equally applicable to spatial weights. I share this stipulation in light of how we use gravitational interaction to estimate technical coefficients in the "Spatial Equilibrium" chapter.

The seeming indeterminacy of φ is mostly a problem in exploratory data analysis, since there is insufficient structure in the data as such to derive the proper spatial model. In the model-driven approach that is taken here, a priori (theoretical) reasons dictate the particular form for the identification constraints. This is similar to the approach taken for systems of simultaneous econometric equations, as we will show. Competing specifications can subsequently be compared by means of model specification tests and model selection procedures.

9.2   A taxonomy of spatial econometric models

In classifying spatial econometric models, a fundamental distinction is made between a *simultaneous* and a *conditional* spatial process. They have a significant impact on estimation and testing strategies, as we will see later. The conditional model, as the name implies, is based on a conditional probability specification, whereas the simultaneous model is expressed as a joint probability. Both models are most often presented in an *autoregressive* form, i.e., where the value of a variable at one point is related to its values in the rest of the spatial system (Anselin 1988).

9.2.1   Simultaneous vs. conditional model. The simultaneous model is constructed based on a stochastic process (or "random fields") on a regular lattice in the form of a stochastic difference equation. In the following equations $j, k = ..,$ $-1, 0, 1, ...$ are spatial indices, Z is the variable under consideration, α are parameters to be estimated, and a is the independent, normally-distributed random variables:

$$\sum_{l=0}^{m} \sum_{u=0}^{n} \alpha_{m-l,n-u} Z_{j+l,k+u} = a_{j,k} \qquad \forall\, j,k. \tag{9-10}$$

For $m = 1, n = 1, j = -1, 0; k = -1, 0$; for example, we can write:

$$
\begin{aligned}
\alpha_{1,1} Z_{-1,-1} + \alpha_{1,0} Z_{-1,0} + \alpha_{0,1} Z_{0,-1} + \alpha_{0,0} Z_{0,0} &= a_{-1,-1} & (j=-1, k=-1) \\
\alpha_{1,1} Z_{-1,0} + \alpha_{1,0} Z_{-1,1} + \alpha_{0,1} Z_{0,0} + \alpha_{0,0} Z_{0,1} &= a_{-1,0} & (j=-1, k=0) \\
\alpha_{1,1} Z_{0,-1} + \alpha_{1,0} Z_{0,0} + \alpha_{0,1} Z_{1,-1} + \alpha_{0,0} Z_{1,0} &= a_{0,-1} & (j=0, k=-1) \\
\alpha_{1,1} Z_{0,0} + \alpha_{1,0} Z_{0,1} + \alpha_{0,1} Z_{1,0} + \alpha_{0,0} Z_{1,1} &= a_{0,0} & (j=0, k=0).
\end{aligned}
\tag{9-11}
$$

Notice that Z is used from this point on as a more general notation than x to accentuate *spatial* data. Refer to **Figure 9-3**(a), where the equations are illustrated. Let us now re-label the four grid points as $i = 1, 2, 3, 4$ as shown in **Figure 9-3**(b) where $j = -1 / k = -1$ is equivalenced to $i = 1, j = -1 / k = 0$ to $i = 2, j = 0 / k = -1$ to $i = 3$, and $j = 0 / k = 0$ to $i = 4$. In matrix notation, this is equivalent to a first-order spatial autoregressive model: $\mathbf{Z} = \Phi \mathbf{WZ} + \mathbf{a}$ or in its more familiar form

$$Z_i = \varphi_i \sum_j w_{ij} Z_j + a_i, \tag{9-12}$$

(a) Spatial index *j, k*

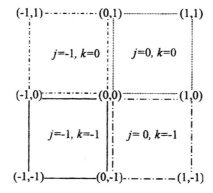

(a) Spatial index *i*

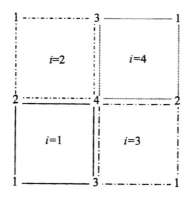

Figure 9-3 Simultaneous spatial econometric model on a lattice

where equivalence has been established between α_{ij} and the $\{\varphi_i, w_{ij}\}$. One can now write Equation **(9-12)** out in more detail:

$$
\begin{aligned}
Z_1 &= \varphi_1(w_{11}Z_1 + w_{12}Z_2 + w_{13}Z_3 + w_{14}Z_4) + a_1 \\
Z_2 &= \varphi_2(w_{21}Z_1 + w_{22}Z_2 + w_{23}Z_3 + w_{24}Z_4) + a_2 \\
Z_3 &= \varphi_3(w_{31}Z_1 + w_{32}Z_2 + w_{33}Z_3 + w_{34}Z_4) + a_3 \\
Z_4 &= \varphi_4(w_{41}Z_1 + w_{42}Z_2 + w_{43}Z_3 + w_{44}Z_4) + a_4
\end{aligned}
\tag{9-13}
$$

or

$$(1 - \varphi_1 w_{11}) Z_1 - \varphi_1 w_{12} Z_2 - \varphi_1 w_{13} Z_3 - \varphi_1 w_{14} Z_4 = a_1$$
$$\varphi_2 w_{21} Z_1 + (1 - \varphi_2 w_{22}) Z_2 - \varphi_2 w_{23} Z_3 - \varphi_2 w_{24} Z_4 = a_2$$
$$\varphi_3 w_{31} Z_1 - \varphi_3 w_{32} Z_2 + (1 - \varphi_3 w_{33}) Z_3 - \varphi_3 w_{34} Z_4 = a_3 \qquad \textbf{(9-14)}$$
$$\varphi_4 w_{41} Z_1 - \varphi_4 w_{42} Z_2 - \varphi_4 w_{43} Z_3 + (1 - \varphi_4 w_{44}) Z_4 = a_4 .$$

Clearly this is nothing but a re-labelling of the spatial indices.

In the matrix form of Equation **(9-12)**, Φ is a diagonal matrix of the form $[\varphi_i]$, $\mathbf{Z}$ and $\mathbf{a}$ are vectors of spatial variables and error terms respectively. Estimating this model involves the specification of a joint probability distribution (say the Gaussian) for $\mathbf{Z}$. It necessitates nonlinear optimization, from which the autoregressive coefficients Φ can be estimated (Cressie 1991). An example of such nonlinear optimization is the joint maximum-likelihood-function in disaggregate random-utility-choice-model of the "Descriptive Tools" chapter (Chan 2000).

The conditional model is constructed as an alternative to the highly nonlinear estimation-procedure above. In essence, it consists of a linear relationship between a conditional expectation of the dependent variable and its values in the rest of the system J:

$$E(Z_i | Z_j \; \forall j \in J, j \neq i) = \varphi^T \mathbf{W} \mathbf{Z} \qquad \forall i \in J. \qquad \textbf{(9-15)}$$

Here φ^T is a vector of coefficients $(\varphi_1, \varphi_2, \dots, \varphi_n)$, and the equation looks like: $Z_i = \varphi_1 \Sigma_j w_{1j} Z_j + \varphi_2 \Sigma_j w_{2j} Z_j + \dots + \varphi_n \Sigma_j w_{nj} Z_j = \beta_1 Z_1 + \beta_2 Z_2 + \dots + \beta_n Z_n$. The conditional structure of the model makes it possible for ordinary least-squares to be used as an estimation technique, provided that the data is coded in a particular manner. The coding essentially consists of eliminating observations so that the remaining dependent variables are independent (i.e., not contiguous). In a situation where few observations are available, this is clearly not a very practical approach. Also, the coding scheme is not unique and thus different estimates may be obtained for the same data set, depending on which observations are dropped.

An interesting feature of the conditional approach is that it has been suggested as a way to avoid some problems associated with the specification of the spatial weight-matrix. Nevertheless, most modelling situations in empirical regional science are easier to be expressed in a simultaneous form. This form is also closer to the standard econometric approach. Therefore, given the model-driven approach taken in this book and the focus of this chapter on econometric techniques, the emphasis here will be almost exclusively on the simultaneous model.

Analogous to the Box-Jenkins approach in time-series analysis (subject of the next chapter), spatial model specifications have been suggested that combine both autoregressive and moving-average processes. Here the variation of the error term can be captured in the model on top of variation in the data values themselves. We will defer the discussion on moving average to the next chapter entitled "Spatial Time Series." It turns out the moving-average model is completely analogous to the autoregressive model. Suffice to say here that when only cross-sectional data are available, an autoregressive model is the methodology to follow. When both cross-sectional and time-series information are available, however, spatial time-series may represent an attractive alternative to analyzing and forecasting land-use patterns.

9.2.2 Spatial linear regression models for cross-section data. The model specification here pertains to the situation where observations are available for a cross-section of spatial units at one point in time. Consonant with the thrust in this chapter, the general modelling framework is provided first, from which specific models can be obtained and illustrated. This is done by imposing certain constraints on the parameters of the general formulation. The general model consists of a set of simultaneous equations, including an equation on the error term:

$$z = \Phi W_1 z + Z\beta + a$$
$$a = \gamma W_2 a + \varepsilon. \tag{9-16}$$

In this specification, β is a $K \times 1$ vector of parameters associated with exogenous (i.e., not lagged-dependent) variables Z, which is represented in an $n \times K$ matrix. This is to be differentiated from the spatially-lagged dependent variable $W_1 z$. Φ is the coefficient matrix of the spatially-lagged dependent variable, and γ is the coefficient matrix in a spatial autoregressive structure for the disturbance a. The error vector ε is taken to be normally distributed with $\varepsilon \sim N(0, \Sigma)$, and the diagonal elements of the error-covariance matrix $\Sigma = [\text{cov}(\varepsilon_i \varepsilon_j)]$ as $\Sigma_{ii} = h_i(\alpha^T y)$ $(h_i > 0)$. It comes with a general diagonal covariance matrix Σ as mentioned. The diagonal elements allow for heteroscedasticity[6] as a function of $m'+1$ exogenous variables y, including a constant term. The m'-parameter α are associated with the non-constant terms such that for $\alpha = 0$, it follows that $h = \sigma^2$—the classic homoscedastic situation with a constant variance.

The two $n \times n$ matrices W_1 and W_2 can be standardized or unstandardized spatial-weight matrices, i.e., they can be normalized as fractions that sum to one or otherwise. They are associated respectively with a spatial autoregressive-process in the dependent variable and in the disturbance term. This allows for the two processes to be driven by a different spatial structure. For example, the regular variable may have spatial dependence on a spatial order defined for the 8-connected neighbors while the disturbance term for the 4-connected neighbors. In all, the model has $3n+K+m'$ unknown parameters, which are in vector form:

$$\Phi, \beta, \gamma, \sigma^2, \alpha^T. \tag{9-17}$$

---

[6] As mentioned in the "Statistical Tools" appendix to Chan (2000), residuals to a regression is homoscedastic if the error is normally distributed with a constant variance around the regression line. Heteroscedasticity is defined as the opposite phenomenon when this property is not found.

Several familiar spatial model structures result when subvectors of the parameter vector (9-17) are set to zero. Specifically, the following situations correspond to the four traditional spatial autoregressive-models discussed in the literature:

a. When $\Phi = 0$, $\gamma = 0$, $\alpha = 0$, we have $m'+2n$ parameters set to zero, the remaining equations make up the classical linear-regression model, with no spatial effects:

$$z = Z\beta + a = Z\beta + \varepsilon. \tag{9-18}$$

where $z = z$, $Z = (1, z_1, z_2 ..., z_{K-1})$, $\beta^T = (\beta_0, \beta_1 ... \beta_{K-1})$ and $a = a = \varepsilon$.

b. For $\gamma = 0$, $\alpha = 0$, $m'+1$ parameters are eliminated. We have the mixed-regressive-spatial-autoregressive model, which includes common-factor specifications (with WZ included in the explanatory variables) as a special case:

$$z = \Phi W_1 z + Z\beta + a = \Phi W_1 z + Z\beta + \varepsilon. \tag{9-19}$$

We will illustrate this case in detail later in this chapter, when a widely disseminated land-use model EMPIRIC is discussed.

c. The case of $\varphi = 0$, $\alpha = 0$, or fixing $m'+1$ parameters, results in the linear-regression model with a spatial autoregressive-disturbance:

$$z = Z\beta + (I - \gamma W_2)^{-1}\varepsilon. \tag{9-20}$$

d. For $\alpha = 0$ or fixing m' parameters, the mixed-regressive-spatial-autoregressive model with a spatial-autoregressive disturbance is obtained:

$$z = \Phi W_1 z + Z\beta + (I - \gamma W_2)^{-1}\varepsilon \tag{9-21}$$

which is the full model specification in Equation (9-15).

Four more specifications are obtained by allowing heteroscedasticity of a specific form (i.e., a specific h($\alpha^T y$) in the models (9-17) through (9-21).) However, our taxonomy above focuses primarily on the specification of spatial dependence and cross-sectional data. The introduction of the time dimension considerably increases the complexity of issues that can be taken into account in the specification of spatial econometric-models. We will defer to the "Spatial Time Series" chapter for handling these complexities. For the time being, we will simply illustrate through a case study of applying these cross-sectional models in real-world land-use forecasting. We will concentrate on the case of the mixed-regressive-spatial-autoregressive model of Equation (9-19), which has been implemented in a widely disseminated model: EMPIRIC. We will also show how one gets around the limitation of a cross-sectional model by a finite difference equation set, which introduces a small dose of time-series information when the difference is taken between two cross-sectional data sets.

9.3   The EMPIRIC model

EMPIRIC is a linear and simplified backup version of the original Polymetric Model that consists of a set of nonlinear finite-difference equations. The Polymetric Model, developed in the early 1960's by the Traffic Research Corporation (now part of Kates, Peat, Marwick & Co.), was discarded because it placed too much requirement on the user—not only in terms of resources but also his/her appreciation for technical intricacies. EMPIRIC became a widely used model in transportation planning partly because of its simplicity and partly because of its continual development and maintenance by Kates, Peat, Marwick & Co. (KPM & Co.). It was subsequently incorporated in the U. S. Federal Highway Administration's Urban Transportation Planning Package during the early 1970's and is therefore in the public domain.

The model is discussed for several reasons. First, it is a demonstrable example of an empirical, econometric model (rather than a structured, 'theory-based' model such as Lowry), using statistical correlations established among variables.[7] In its basic version, EMPIRIC is designed as a straightforward projection of future activities and land use, without any pretense of underlying theoretical construct that explains the development process. Second, it represents one of the most widely used land-use models, and as such, serves as an example of a 'successful' modelling effort. It is possibly the only model that had been applied across at least seven metropolitan areas in the United States and Canada. The question is: does its success have anything to do with its simple, empirically-based model structure or there other intangibles involved? Third, there is a pedagogic reason for discussing the EMPIRIC model. It illustrates how spatial econometrics play a role in the real world. The discussion here presents the model not only in terms of spatial statistics but also the general econometric literature.

9.3.1 An example of EMPIRIC model. Let us describe the EMPIRIC model in greater details through an example. Following the arrow diagram technique presented in the chapter on "Descriptive Tools" (Chan 2000), the structural relationship between the identified variables—population, white- vs. blue-collar employment, and accessibility is—shown in **Figure 9-4**. It is portrayed, for example, that the population in the forecast period is not only affected by the base-year population, but also employment in the forecast year. We call population the dependent (endogenous) variable and the rest independent variables, and an equation can be written to realize this relationship:

(forecast pop) = *a* (base-yr pop) + *b* (forecast white-collar emp) + *c* (forecast-yr access)

---

[7] See chapter on "Descriptive Techniques" in Chan (2000) for a comparison between theory-based and empirical models.

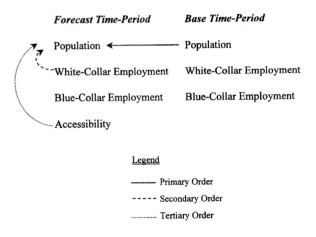

Figure 9-4 Arrow diagram for EMPIRIC

where a, b, and c are calibration coefficients. As noted in the chapter on "Descriptive Tools," the relationship and hence the equation is merely a postulation by the model builder. The validity of the postulation can only be established by statistical verification and validation using available data.

After the arrow diagram is completed in full, a final set of equations can be constructed as shown in Equation **(9-22)** below. Notice the intertwining relationship between the dependent and independent variables. A dependent variable such as forecast population (that appears in the left-hand-side in the first equation) serves as an independent variable in the second equation (on the right-hand-side), while accessibility remains an independent variable throughout. It is through such a simultaneous equation formulation that the interdependence of the various factors is modelled. This is an example of a simultaneous econometric model (rather than a conditional model.) We may also like to contrast this with our experience with the Lowry Model, where the interrelationship between the population (household) sector and the basic and service employment sectors is modelled in a sequential (rather than a simultaneous) fashion.

We show below the same three equations corresponding to the three dependent variables respectively: forecast population, white-collar, and blue-collar employment. Such equations remain the same for every subareal unit (such as a traffic zone) and each forecast period:

$$\Delta N = 0.3234\,\Delta W - 0.0064\,N + 1.9258\Delta t\,'$$
$$\Delta W = 0.4166\,\Delta N - 0.0061\,W + 0.9640\Delta u \qquad\qquad \textbf{(9-22)}$$
$$\Delta B\,'' = 0.1562\,\Delta N - 0.0130\,B\,'' + 0.9993\Delta u$$

where ΔN is the change-in-subareal-share-of-total-population, ΔW is the change in subareal-share-of-white-collar-employment, and $\Delta B''$ is the change-in-subareal-share-of-blue-collar-employment. N is the base-year-subareal-share-of-total-population, W is the base-year-subareal-share-of-white-collar-employment, and B'' is the base-year-subareal-share-of-blue-collar-employment. $\Delta t'$ is the change-in-subareal-share-of-transportation-accessibility-to-employment, and Δu is the change-in-subareal-share-of-accessibility-to-population. Notice this is in the form of a mixed-regressive-spatial-autoregressive model of Equation **(9-15)**, or Equation **(9-19)**.

To see this, one needs only to recognize these variable definitions: $\mathbf{z}^T = (\Delta N,$ $\Delta W, \Delta B'')$, $\mathbf{Z}_1 = \begin{bmatrix} N & 0 & 0 \\ 0 & W & 0 \\ 0 & 0 & B'' \end{bmatrix}$, and $\boldsymbol{\beta}_1 = (-0.0064, -0.0061, -0.0130)^T$ where $\mathbf{z}\boldsymbol{\beta} =$

$\mathbf{Z}_1\boldsymbol{\beta}_1 + \mathbf{Z}_2\boldsymbol{\beta}_2$. $\mathbf{W}_1 = \begin{bmatrix} 0 & 1 & 0 \\ 1 & 0 & 0 \\ 1 & 0 & 0 \end{bmatrix}$, $\boldsymbol{\Phi} = \begin{bmatrix} 0.3234 & 0 & 0 \\ 0 & 0.4166 & 0 \\ 0 & 0 & 0.1562 \end{bmatrix}$. As alluded to already, a common

factor is specified on top of other explanatory variables, with $\mathbf{Z}_2 = \begin{bmatrix} \Delta t' & 0 & 0 \\ 0 & \Delta u & 0 \\ 0 & 0 & \Delta u \end{bmatrix}$ and $\boldsymbol{\beta}_2$

$= (1.9258, 0.9640, 0.9993)^T$. We will specify in more detail the weight matrix $\mathbf{W}_1$ and the observation matrix $\mathbf{Z}$ in sequel. Meanwhile, let us define the concept of 'change-in-subareal-share'. Alternatively, one can view EMPIRIC as a special case of Equation **(9-21)** when $\boldsymbol{\beta}$ is a single vector of $(-0.0064, -0.0061, -0.0130)^T$ and

where $[\mathbf{I} - \gamma\mathbf{W}_2] = \begin{bmatrix} 1.9258 & 0 & 0 \\ 0 & 0.9640 & 0 \\ 0 & 0 & 0.9983 \end{bmatrix}$ and $\varepsilon = (\Delta t\, \Delta u\, \Delta u)^T$.

9.3.2 The concept of shares. Notice the variables in the equations do not represent the absolute values of the subareal population, employment and accessibility. Instead, they are shares of regional totals, where the 'shares' of each activity for each subarea is computed as the ratio of the activity-level for the subarea to the regional total. Thus if the calibration-year $(t - \Delta t)$ population in subarea i is 22,000 out of the regional total of 2,050,000, for example, the share of that subarea is 22,000/2,050,000 or 0.107. The difference between the shares so computed for two points in time (say between calibration year $t-\Delta t$ and base year t) is then taken as the change-in-share for the subarea of interest. Thus if subarea-i's share in population is 0.140 in the base year, the change over the two time-periods is 0.140 $-$ 0.107 = 0.033, which indicates that a higher percent of regional population is located in the subarea under discussion over the time-period Δt.

The policy measure such as accessibility is defined similarly as changes in share in the equations. Let us say the accessibility to employment from subarea i is $t_i(b)$ in the base year and $t_i(c)$ in the calibration year and the total regional accessibility is $t(b)$ and $t(c)$ respectively. Subarea-i's shares of accessibility are $t_i(b)/t(b)$ and $t_i(c)/t(c)$ correspondingly. Therefore,

$$\Delta t' = \frac{t_i(b)}{t(b)} - \frac{t_i(c)}{t(c)}. \tag{9-23}$$

There are several reasons why subareal shares, rather than absolute numbers, are used. First, they reflect the 'competitive' subareal component relative to the region, showing the relative concentrations of activities. Second, the numbers are bounded between zero and one, thus easing the calibration process, in that the coefficients are prevented from being too large or too small because of the values of the variables. In this dimensionless form, the calibrated coefficients can also be compared to the calibration experiences in another city more readily since a normalized form is used in both cities.

Changes in shares, instead of absolute numbers, are used again for good reason. In order to perform projection incrementally as in EMPIRIC, it is more statistically meaningful to relate variables such as population-growth to the improvement-in-accessibility, both of which are quantified in terms of changes. Thus if one forecasts from calibration-year $(t-\Delta t)$ to base-year (t) to forecast-year $(t+\Delta t)$ to future-year $(t+2\Delta t)$, these time periods are summarized into three Δt increments in the model: calibration-to-base years, base-to-forecast years, and forecast-to-future years.

In general, the concept of 'share' fits in well within the *allocation* function of an activity-allocation model, while the concept of incremental change is congruous to the *projection* function of land-use models. The future allocations are translated back into absolute population and employment levels by the following simple procedure. The subareal shares are multiplied by the exogenously specified control totals. The forecast increments of subareal growth or decline are then added to the observed initial values to obtain the final figure of interest. The EMPIRIC model is a good example of this *shift-share analysis* that was described previously in the chapter on "Economic Concepts" (Chan 2000).

Example

Suppose the following equations have been calibrated for a three-zone city: $\begin{array}{l}\Delta N = 0.25\Delta E + 1.93\Delta t\ ' \\ \Delta E = 0.50\Delta N + 0.96\Delta u\end{array}$

where E stands for employment. Part of the hypothetical data base used for this calibration is shown in the columns marked by "projection year" and "base year" in **Table 9-1**.

Using these given numerical values from **Table 9-1**, the two unknowns ΔN and ΔE can be solved by the two equations: $\begin{array}{l}\Delta N = 0.25\,\Delta E + 1.93\,(-0.0026) \\ \Delta E = 0.50\,\Delta N + 0.96\,(-0.0030)\end{array}$. The results are $\Delta N = -0.00617$ and $\Delta E = -0.00732$, which are entered in the 'forecast' columns of the table. (It will be a good exercise for the readers to fill in the missing entries of this table.)

To transform these shares into absolute activity levels, we use the following formula:

(forecast activity) = [(change in share) + (base-yr share)](forecast-yr control-total).

Thus

     (forecast pop) = [−0.00733 + 0.3872](116,230) = 44,148
     (forecast emp) = [−0.00617 + 0.3671](42 300) = 15,267.

Both numbers are included under the forecast columns in **Table 9-1**. □

Sub-area	Projection year				Access to pop		Access to emp		Base year		Forecast			
	Access to pop		Access to emp		Base yr Level	Chnge in share	Base yr Level	Chnge in share	Pop Level	Emp Level	Chnge in pop share	Chnge in emp share	Zonal pop Level	Zonal emp Level
	Level	Share	Level	Share										
1	24,016	0.3096	11,933	0.231	24,311	-0.0026	12,135	-0.0030	38,600	8,700	-0.0733	-0.00617	44,148	15,267
2	27,793		18,188		27,793		18,203		43,200	9,600				
3	25,771		20,213		25,771		20,213		17,900	5,400				
Regional total	77,580	1.00	50,334	1.00	77,875	0	50,551	0	99,700	23,700	0	0	116,230	42,300

Table 9-1 Partial data base for hypothetical city

9.4 Forecast sequence

We recall that the Lowry Model uses cross-sectional data for its calibration; and forecasting is performed only for a single time period into the future. EMPIRIC, on the other hand, is one of the few models that operates on cross-sectional data organized in two points in time. Consonant with census updates in the U.S., the EMPIRIC forecasting period Δt is in five- or ten-year increments. The model is applied typically for several forecasting increments—for example from the base-year t to forecast-year $t+\Delta t$, from forecast-year $t+\Delta t$ to future year $t+2\Delta t$ and so on. The output of EMPIRIC's first period forecast is input for the second etc. For the initial forecast increment, the model is invoked by inputting this information:

a. Calibration- and base-year activity and land-use by subarea (such as a traffic zone), corresponding to time periods $t-\Delta t$ and t;

b. Planning policy that represents changes in the freeway system, sewage facilities and zoning ordinances and the like during the time increment Δt;

c. Regional-control-totals of population and employment for the base-period t. Notice that the activity- and land-use data specified in (a) is generated by the model for each of the future time-increments once the information for the original period (between calibration-year and base-year) is assembled. The policy and regional control-total projections, however, have to be provided outside the model for each forecast increment. **Figure 9-5** illustrates the calibration and forecasting steps of EMPIRIC.

Referring to **Figure 9-5**, subareal activity- and land-use data are the first category of input into the EMPIRIC model. Activity data consist of demographic information such as the number of families/households, population for each subarea, and similar employment data. Land-use data, on the other hand, is usually

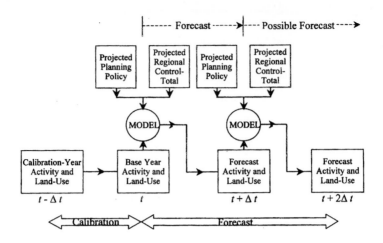

Figure 9-5 Block diagram of forecasting procedure

expressed in terms of acreage by type-of-use, again disaggregated by subarea. Output of the EMPIRIC model again includes subareal activity and land-use data, which are in turn used as input to the next forecast increment of the model.

EMPIRIC recognizes that the study area cannot be treated in isolation, the interaction with the neighboring areas has to be modelled. An external subarea is used to reflect the influence of the 'outside' (external world). The data used for input are labelled external household and employment. Regional activity is yet another category of input into EMPIRIC. The activity data consist of regional control totals of population and employment. Such regional-data input is required for each forecast period. They are allocated among the subareas by the model every time. Finally, policy-planning changes are input to the EMPIRIC model. Direct policy changes include infrastructure such as transportation and utilities (sewage, water etc). Indirect yet important policy changes such as density control and land availability are also used as inputs into the EMPIRIC model. Again, the policy variables have to be specified for each forecast period. **Table 9-2** summarizes the various inputs and outputs to the model. It is to be noted that activity/land-use data are needed for two periods of time—the base-year and calibration-year—for model calibration.

9.5   Model calibration

The technique used to calibrate the EMPIRIC model involves the determination of regression coefficient values. As mentioned previously, these values are calculated using data from two historic time-periods, simply because EMPIRIC is a difference-equation model. The regression coefficients so obtained are then verified by observing how well the calibrated model can reproduce the forecast-year data. Such

			Input	Output

Activity/ Land Use	Subareal	INTERNAL
		 ☐ *demographic*
		    No. of families
		    No. of households
		    Population
			✓
			 ✓
			 ✓
				✓
				 ✓
				 ✓

		☐ *employment* No. of employee	✓	✓
		☐ *land-use* Acreages by use	✓	✓
		EXTERNAL ☐ *total household* ☐ *total employment*	✓ ✓	✓ ✓
	Regional	☐ *demographic* ☐ *employment*	✓ ✓	
Policy		DIRECT ☐ *transportation* ☐ *utilities* (e.g. water, sewer)	✓ ✓	
		INDIRECT ☐ *density control* ☐ *land availability*	✓ ✓	

Table 9-2 EMPIRIC input-output information

a verification is facilitated by the FORCST block of the program, which 'forecasts' the base-year t activity distribution and land use from the calibration-year $t-\Delta t$. The forecasted data are subsequently verified by the "reliability testing" block.

An examination of the regression package used in the model calibration process reveals that it consists basically of five programs. The programs are: (i) Graphing block, (ii) Data correlation block, (iii) Factor analysis block, (iv) Regression block, and (v) Reliability testing block. While the functions of these programs may be quite apparent from their names, some discussion on the regression block is necessary. The regression block is the program that estimates the model coefficients. Included in the estimation procedure are these components:

a. Ordinary least squares: a standard least-squares regression package, used in the initial stages of calibration and in the construction of independent sub-models operating on the output of the simultaneous-equations model. (More will be said about these sub-models later.)

b. Two-stage least-squares: a standard, `simultaneous'-equations regression-package, used to estimate the coefficients in the basic EMPIRIC model using the sequential, iterative method described in the chapter on "Descriptive Tools for Analysis."

c. Maximum likelihood: same as for above, but used in unbiased estimators, again as described in the "Descriptive analysis" chapter for an introduction to this technique.

d. Simultaneous least squares: same as for (b) and (c) above, but based on the method of indirect least squares for special 'triangular' form of the simultaneous equation set. (See "Descriptive Tools" chapter in Chan (2000).)

Item (a) is used in variable definition, as input to calibration and in sub-model development. One of the other sub-programs (b) through (d) is used to calibrate the coefficients of the simultaneous-equation model. The output of each of these programs is a set of coefficients, together with associated goodness-of-fit statistics such as the coefficient-of-multiple-determination R^2, the t and F statistics. Together all the five programs (i) through (v) formalize the variables, estimate the coefficients, and re-estimate coefficients using input from the factor-analysis-program block and the initial stages of calibration. In the calibration process, two-stage least squares is the main tool employed, rather than indirect least squares, as the reader can imagine.

As an example, take the three-equation model presented earlier:

$$\Delta N = a_1 \Delta W + b_1 N + c_1 \Delta t\,'$$
$$\Delta W = a_2 \Delta N + b_2 W + c_2 \Delta u \qquad\qquad\qquad \textbf{(9-24)}$$
$$\Delta B\,'' = a_3 \Delta N + b_3 B\,'' + c_3 \Delta u.$$

Using the calibration and base-year data sets, the coefficients of a, b, and c are estimated through regression calibration. It follows the procedure outlined in the"Descriptive Tools" chapter[8] and the "Statistical Tools" appendix in Chan (2000). The estimated coefficients (shown above in Equation **(9-22)**) are then statistically evaluated and tested for their significance. These coefficients can show either positive or negative relationships among the variables, which either agree or disagree with intuition. Further evaluation of the coefficients is done by forecasting the distribution of the three dependent variables for the base-year t based on the

---

[8]For a numerical example showing how two-stage least squares is performed step-by-step, see the "Calibration" section of the "Descriptive Tools" chapter (Chan 2000).

calibration-year data $t-\Delta t$. The committed transportation system for the base year is used as policy input. The results of the 'forecast' are then statistically compared with the observed data for the base-year t. Through an iterative process such as two-stage least-squares, the coefficients are fine tuned until they replicate base-year data satisfactorily.

It should also be noted that, strictly speaking, the calibration of the model has to be performed for each study area that is being applied. However, there are new research efforts to transfer calibration results from one city to another. The calibration process is a major part of the modelling effort and requires a large quantity of data to be used. Being able to transfer calibration results from a similar city will save a great deal of time and money. Putman (1979) tabulated the regression coefficients of EMPIRIC studies performed in six cities in the United disagree with intuition. Further evaluation of the coefficients is done by forecasting the distribution of the three dependent variables for the base-year t based on the calibration-year data $t-\Delta t$. The committed transportation system for the base year is used as policy input. The results of the 'forecast' are then statistically compared with the observed data for the base-year t. Through an iterative process such as two-stage least-squares, the coefficients are fine tuned until they replicate base-year data satisfactorily.

It should also be noted that, strictly speaking, the calibration of the model has to be performed for each study area that is being applied. However, there are new research efforts to transfer calibration results from one city to another. The calibration process is a major part of the modelling effort and requires a large quantity of data to be used. Being able to transfer calibration results from a similar city will save a great deal of time and money. Putman (1979) tabulated the regression coefficients of EMPIRIC studies performed in six cities in the United States. A fair amount of consistency is found among the population equations in **Table 9-3**. Although there are occasional exceptions both in signs and magnitude, some interesting patterns emerge. The change-in-share of a region's total population found in each subarea moves with the change-in-share of the adjacent population class. For example, lower income moves with change-in-share of lower income, and lower middle income moves with change-in-shares of lower income and upper middle income and so on. On the other hand, change-in-share moves in opposition to its own concentration in the base year and moves with concentrations of the next higher income group. In other words, changes in subareal shares of each income group move with changes in shares of the next higher and next lower income group, and away from concentrations of their own income group toward concentrations of the next higher income group.

Putman pointed out that the patterns found in these coefficients are quite consistent with hypotheses regarding peoples' desires for increased socioeconomic status. They are also consistent with hypotheses regarding peoples' unwillingness to live among groups very different from their current economic status. The coefficients of other variables in the population equations as well as those of the employment equations do not exhibit any where the same degree of uniformity. They point to the site-specific nature of the calibration process. Counterintuitive

Dependent variable	Population by income (independent variable)							
	Change in share				Base year share			
	Low	Lower middle	Upper middle	Upper	Low	Lower middle	Upper middle	Upper
Change in share low income population	-0.119 (1)[2]	+0.129 to +0.637 (6)	-0.295 to -0.367 (2)	-0.281 to -0.39 (2)	-0.199 to +0.133 (5)	+0.294 to +0.36 (3)	-0.109 to +0.258 (2)	N.A.
Change in share low-middle income population	+0.194 to +0.53 (5)	N.A.[3]	+0.307 to +0.781 (6)	N.A.	N.A.	-0.054 to -0.353 (4)	-0.334 to +0.10 (3)	N.A.
Change in share upper-middle income population	-0.125 to -0.14 (2)	+0.434 to +0.658 (6)	N.A.	-0.16 (1)	-0.16 (1)	N.A.	-0.219 to -0.27 (3)	-0.155 to +0.113 (2)
Change in share upper income population	-0.42 to -0.507 (2)	-0.282 to +0.685 (3)	+0.504 to +0.83 (4)	N.A.	N.A.	N.A.	+0.219 (1)	-0.278 to -0.481 (4)

[1] The six calibrations include Atlanta, Denver, Washington DC, Twin Cities Puget Sound and Boston.
[2] The number of data points are included in parenthesis. For example, (3) means three coefficients are included in the range of numbers citied.
[3] N.A. stands for not applicable, meaning no coefficients were specified in the structural equation to be calibrated.

Table 9-3 Population coefficients in six EMPIRIC calibrations[1]

signs of the coefficients may even suggest spurious correlations that may work in short-term forecasts, but are not suitable for long-run applications.

9.6   Definition of accessibility and developability

A really key item in each land-use model is the way accessibility and developability measures are quantified, since they really make up the pivotal concepts in activity allocation and forecasting. Throughout this book, accessibility has been recognized as one of the determinants of facility location and land development. It lies at the heart of spatial price theory. Developability, on the other hand, reflects the development potential of a subarea from such considerations as zoning ordinances and the economic and non-economic opportunities that exist. Much of regional-science research has been focusing on the capacity of a subarea in taking on new activities.

9.6.1 *Accessibility.* Instead of presenting the equation for accessibility directly, let us try to derive it. First, we recall that accessibility is a meaningful measure only when applied toward a particular type of activities and with respect to a particular location in the study area. For example, the accessibility to employment opportunities is defined for a specific residential zone, and likewise the accessibility to shopping can also be defined for the same zone. This reflects that locations accessible to work may be different from those to shops, and the accessibility to shops may be different from home versus the office.

Accessibility is related to travel time. In general shorter travel times mean better accessibility or proximity, and vice versa. Finally, accessibility for a subarea has to reflect travel not only to one work or shop location, but also other possible locations. In this way it summarizes the proximity to job and shop opportunities of the region as an aggregate. This is normally taken into account by taking a weighted average of all possible proximities to these opportunities. Employment or shopping activities in the target subareas are often used as weights. Since employment or shopping activities are often manifested in terms of work or shop trips, trip volumes are often used as weights in the place of actual number of jobs or retail floor space. Such a concept of accessibility has been introduced earlier in this book including in the "Generation, Competition and Distribution" chapter, where state parks are located vis-a-vis accessibility to the local population.

An equation that takes all the above factors into account is given below for a study area consisting of n' subareas, tracts, zones or districts:

$$t_k = \sum_{g=1}^{n'} E_g \exp(-\alpha_t \tau_{gk})$$
$$u_k = \sum_{g=1}^{n'} N_g \exp(-\alpha_u \tau_{gk}).$$

(9-25)

Here τ is the travel time, and the interaction propensity α is corresponding to the *L*-value of the accessibility-opportunity model, representing the probability of location (see chapters on "Economic Methods" and "Descriptive Tools" in Chan (2000)). The exponential-function formulation considers both the attractiveness of destination and its proximity, since it discounts the interaction with subareas far away from *k*. It should be noted that the above accessibility definitions are somewhat different from that used in the Lowry model. Instead of distributing population and employment, the econometric model here relates the location of population and employment to a number of factors, including proximity to jobs and proximity to the labor market respectively. The difference relates to the fundamental conceptual distinction between the two modelling approaches. We highlight this difference by the index *g–k* instead of *k–g*, although the two are equivalent so long as travel time and cost between *g* and *k* are symmetrical.

For example, in the hypothetical case study to be discussed later in this chapter, the accessibility to employment for the calibration period is computed for all the 10 districts. The numerical value ranges from 154 807 for district 3 to 31 039 for district 6, showing that the proximity to work locations in the Boston area is best if one lives in district 3 and worst in district 8. Notice the numerical value for accessibility is purely an index and best thought of as simply a scale. (For this reason, we normalize accessibility eventually between 0 and 1 as described earlier.)

The accessibility definition adopted in Equation **(9-25)** can be used to compute the spatial weights w_{ij} in the mixed-regressive-spatial-autoregressive model of Equation **(9-15)**. One can readily see the connection between accessibility and the spatially-lagged variable—alias for spatial weights—in Equation **(9-4)**. In brief, $w_{ij} = f(\tau_{ij}, \alpha) = \exp(-\alpha\tau_{ij})$, and the spatially-lagged variable is $L^{(l)}x_i = \Sigma_j w_{ij}x_j = X_i = \Sigma_j x_j \exp(-\alpha\tau_{ij})$. The change-in-accessibility matrix $\begin{bmatrix} \Delta t' & 0 & 0 \\ 0 & \Delta u & 0 \\ 0 & 0 & \Delta u \end{bmatrix}$ as defined by Equation

(9-23) is simply the common-factor specification Z_2. Notice only a single class of spatial contiguity ($l = 1$) is involved here, and the spatially-lagged variable is exactly the definition of accessibility, with the activity variable x being employment E or population N, depending on whether accessibility-to-work (t') or accessibility-to-shop (u) is considered. Remember these accessibilities are used to locate residence and white/blue collar employment respectively in the EMPIRIC Model of Equation **(9-22)**. To be noted also is that the spatial-lag is normalized to be 0–1 ranged, as alluded to already. In sum, the normalized common-factor-specification term $Z_2\beta_2$ is simply

$$\begin{bmatrix} \dfrac{t_i(b)}{t(b)} - \dfrac{t_i(c)}{t(c)} & 0 & 0 \\[2ex] 0 & \dfrac{u_i(b)}{u(b)} - \dfrac{u_i(c)}{u(c)} & 0 \\[2ex] 0 & 0 & \dfrac{u_i(b)}{u(b)} - \dfrac{u_i(c)}{u(c)} \end{bmatrix} \begin{bmatrix} 1.9258 \\ 0.9640 \\ 0.9993 \end{bmatrix} \tag{9-26}$$

where Z is now a matrix consisting of the combination of three weights corresponding to activity variables E, N, N.

9.6.2 Developability. The other key concept—the development opportunity for a particular activity type—is defined in a rather ad-hoc fashion. It essentially assumes that a subarea's development potential (called developability) is related to the amount of developable land and the fraction of the total gross area devoted to the activity of interest. This is accomplished in the following manner:

$$\left(\hat{A}_j^H / A_j\right)\bar{A}_j, \quad \left(\hat{A}_j^R / A_j\right)\bar{A}_j; \tag{9-27}$$

where $\hat{A}_j^H$, $\hat{A}_j^R$ represents the net area for housing and service-industry development in subarea j respectively, A_j is the gross area, and $\bar{A}_j = A_j - A_j^U - A_j^B - A_j^R - A_j^H$ is the developable area after accounting for the undevelopable and used areas. Here the used area includes all the functional purposes such as residential (H), manufacturing (B) and service (R). EMPIRIC normally assumes that undevelopable areas have been accounted for exogenously by redefining gross area as A'_j, which already excludes the undevelopable area (i.e., $A'_j = A_j - A^U_j$).

In the next section, the reader will see how the capacity or land developability index for population is computed and accounted for by the first of the two in Equation **(9-27)**, while the index for service industry by the second of the two. Notice there is one serious inadequacy of the land developability definition in general—that an area is not developable unless prior developments have been located there.

9.7 Supportive analysis

The EMPIRIC model consists of 16 subroutines divided into four groups. The first group, Data Assembly Programs, prepares basic data and develops historical records. The second group, Calibration Programs, computes the various coefficients in the regression programs, and calibrates the simultaneous equations in general (as described previously). The third group, Forecasting, substitutes the base-year and policy data into the calibrated equations for projecting subregional or subareal activities. The fourth group, Land Consumption, is an accounting routine used to determine the amount of acreage by use from the projected activity levels.

The Data Assembly and Land Consumption programs can be thought of as supportive programs. Under the Data Assembly Program, one has a variety of utility programs:

a. *Data stack block* – expand data column-wise for inclusion of new variables,
b. *Data modification block* – for insertion of a new value in a table,
c. *Subregion stacking block* – expand data row-wise for inclusion of a new subregion or subarea,
d. *Town aggregation block* – used to combine data row-wise from two or more subareas into a single area,
e. *Data combination block* – combine or transform two or more of the variables column-wise into a new variable(s),
f. *Data difference block* – computes changes in subregional or sub-areal activities,
g. *Percent and normalization block* – converts variable columns to subregional or subareal shares,
h. *Accessibility block* – computes subregional or subareal accessibilities to an activity,
i. *Report generation block* – prints selected information from input or output data.

As expected, the Calibration and Forecasting programs form the core of the model. They were discussed in some detail in sections 9.5 and 9.4 respectively and will not be elaborated on further here. The initial forecasts of subregional or subareal population and employment are translated into equivalent changes in classified small-area land use by means of the fourth, Land Consumption, program.

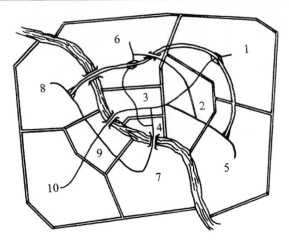

Figure 9-6 - Base-year network

This is again carried out using data assembled for the same two points in time. It accepts as input the output of the simultaneous equation module, together with developability in each small area, including the range of permissible development densities for each activity. Its output is a detailed updated accounting of land use, broken down by type, within each small area for each forecast year. Where a subregion or subarea is capacitated, the acreage assigned for new development is scaled down. This way, the total acreage would be reduced to exactly the number at the base year of the forecast interval plus that becoming available during the interval due to declines of activities. It can be seen, therefore, that the EMPIRIC model, besides the basic set of simultaneous equations, is supported by a number of routines. It can be thought of as a collection of programs and is really a package of econometric and special purpose software adapted for land-use forecasting. Again, the model is available from the U.S. Federal Highway Administration under their urban transportation planning computer package.

9.8   A hypothetical study

A hypothetical city is divided into a lattice structure of 10 subregions as illustrated in **Figure 9-7** (Thompson 1972, Kates, Peat, Marwick & Co. 1971). The area is traversed by a river in the middle. Subregions 3 and 4 represent downtown, while subregions 2, 7 and 10 represent relatively new and rapidly growing suburban areas. The assumed base-year highway network at time *t* shown in **Figure 9-6** consists of a predominantly radial arterial highway system, supplemented by one cross-town link and a partially completed inner beltway at the northern part of town. The network includes only two river bridges. Also shown is **Figure 9-7**, the calibration-year network at time *t*–Δ*t*. The only difference between the calibration- and base-year networks is that the partially completed innerbelt is built during the elapsed time increment Δ*t*. The study area is reminiscent of Boston, Massachusetts.

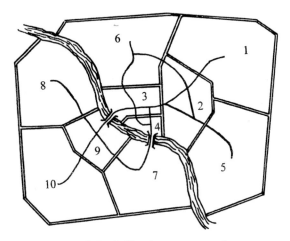

Figure 9-7 - Calibration year network

To simplify the example, the following assumptions are made:

a. No transit system exists or is proposed for the future;
b. Urban growth during the period between calibration- and base-year was influenced by concurrent improvement in the highway system (i.e., the partially completed beltway);
c. The committed network in the forecast year consists of an inner beltway around the city, and including two additional river bridges at where the innerbelt crosses the river.

A simple three-equation model is to be developed involving three dependent and five independent variables:

Dependent endogenons variables
Calibration-to-base-year-change in subregional-share-of-total-population
Calibration-to-base-year-change in subregional-share-of-white-collar-employment
Calibration-to-base-year-change in subregional-share-of-blue-collar-employment

Independent exogennons variables
Calibration-year subregional-share-of-total-population
Calibration-year subregional-share-of-white-collar-employment
Calibration-year subregional-share-of-blue-collar-employment
Calibration-to-base-year-change in subregional-share-of-accessibility-to-total-employment
Calibration-to-base-year-change in subregional-share-of-accessibility-to-total-population.

Table 9-4 summarizes the raw data for the ten subregions, with separate accessibilities for the calibration and base years, where accessibility is defined by

Sub-region	Population		White Collar Employment		Blue Collar Employment		Accessibility to Population		Accessibility to Employment	
	Calibration year	Base year	Calibration year	Base year	Calibration year	Base year	Calibration year	Base year	Calibration year	Base year
1	220,000	386,000	22,000	47,000	21,000	35,000	139,963	171,661	69,683	82,748
2	415,000	432,000	55,000	56,000	110,000	128,000	228,669	228,669	130,501	138,542
3	205,000	179,000	184,000	240,000	40,000	44,000	214,657	214,657	154,807	154,807
4	190,000	186,000	91,000	115,000	24,000	22,000	196,850	196,850	137,335	137,371
5	140,000	228,000	9,000	30,000	38,000	58,000	121,945	166,945	61,288	71,892
6	290,000	390,000	830,000	133,000	7,000	14,000	172,577	201,490	97,898	103,817
7	230,000	321,000	22,000	28,000	36,000	43,000	184,564	196,260	110,007	118,818
8	80,000	196,000	5,000	26,000	14,000	25,000	75,996	111,321	31,039	42,627
9	185,000	256,000	35,000	51,000	18,000	24,000	165,163	179,105	87,012	94,967
10	95,000	176,000	4,000	21,000	5,000	10,000	88,379	131,966	36,010	52,297
Total	2,050,000	2,750,000	510,000	747,000	313,000	403,000	1,588,763	1,798,924	923,760	997,887

Table 9-4 Raw data for calibration

Equation **(9-25)**. Notice it is through this separation that non-linearity is taken out of the model. As discussed in Section 9.1.4. this represents a piecewise linear approximation of the accessibility term X_i (or the spatially-lagged variable $L^{(l)}x_i$). The shares and change-in-shares of each activity and policy variable for each subregion is computed using a similar formula as shown for accessibility in Equation **(9-23)**. **Table 9-5** summarizes the changes-in-shares together with the calibration-year activity shares.

The three-equation model relates the changes in shares of three activities—total population, blue-collar employment, and white-collar employment—to one another and to changes in highway accessibility. It is calibrated and shown earlier as Equation **(9-22)**. The first equation of the set postulates that changes-in-a-subregional-share-of-population can be expressed as a linear function of its change-in-share-of-white-collar-employment, its calibration-year-level-of-population and its change-in-share-of-accessibility-to-employment. Similarly, the changes-in-share-of-white-collar-employment and blue-collar-employment for a given subregion are related to the parallel changes in the same subregion's share-change-of-population, the subregion's calibration-year share of the appropriate employment category and its change-in-share-of-accessibility-to-population.

The calibration result appears to be reasonable. Changes-in-shares-of-activity are positively related to the changes-in-shares-of-accessibility. The negative

	Changes in Shares		Calibration Year Shares			Changes in Shares		
Sub-region	Population	White Collar Employment	Blue Collar Employment	Population	White Collar Employment	Blue Collar Employment	Accessibility to Population	Accessibility to Employment
1	+0.03305	+0.01978	+0.01976	0.10732	0.04314	0.06709	+0.00733	+0.00729
2	−0.04535	−0.03288	−0.03382	0.20244	0.10784	0.35144	−0.01682	−0.01110
3	−0.03491	−0.03950	−0.01861	0.10000	0.36078	0.12780	−0.01579	−0.01245
4	−0.02505	−0.02448	−0.02209	0.09268	0.17843	0.07668	−0.01447	−0.01101
5	−0.01462	−0.02251	+0.02251	0.06829	0.01765	0.12141	+0.01605	+0.00570
6	+0.00035	+0.01530	+0.01238	0.14146	0.16275	0.02236	+0.00338	−0.00194
7	+0.00453	−0.00565	−0.00832	0.11220	0.04314	0.11502	−0.00707	+0.00002
8	+0.03225	+0.02500	+0.01730	0.03902	0.00980	0.04473	+0.01405	+0.00912
9	+0.00285	−0.00035	+0.00205	0.09024	0.06863	0.05751	−0.00440	+0.00097
10	+0.01766	+0.02027	+0.00884	0.04634	0.00784	0.01597	+0.01773	+0.01343

Table 9-5 Input data for Calibration

relationship with the base-year level is also understandable; crowded subregions with high base-year values do not grow as rapidly as suburban subregions with low base-year values. No statistical test on the region's equations is presented here due to the small size of the data set (10 subregions), and the consequent reduction in statistical degrees-of-freedom. Such tests would, however, form the basis for a detailed evaluation of a larger model.

The calibrated model is then used to forecast. For the forecast year, alternate transportation networks were generated. Scenario 1 calls for the completion of the innerbelt as shown in **Figure 9-8**. Scenario 2 calls for a reversal in highway-department policy between base and forecast year. Instead of completing the innerbelt, a series of radially oriented freeways might be constructed instead, as shown in **Figure 9-9**. Here the accessibility for the downtown subregions has been radically altered, while the southern-most outlying subregions remain relatively isolated.

Changes-in-accessibility are computed and changes-in-shares are forecasted corresponding to these two alternate scenarios. The results are shown in **Table 9-6** and **Table 9-7**. The situation under the radial alternative is somewhat more complex. Although accessibility to population for the central-city subregions increases markedly, their accessibility-to-employment declines even more than under the first accessibility term is a function not only of travel time, but also the activity level. Also, the 'share' concept comes in, where each subregion is in competition with others in a "zero-sum-game"—one subregion gains at the expense

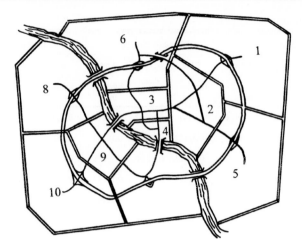

Figure 9-8 Scenario 1 forecast: completed innerbelt

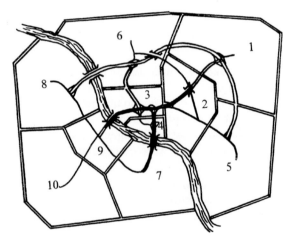

Figure 9-9 Scenario 2 forecast: radial freeways

of another. Subregions 3 and 4 are basically employment areas, with the population employed there dwelling in the outer areas. Thus provision of a set of radial freeways increases the accessibility of the central city to part of the population, but decreases the alternative. This apparent contradiction may be explained by recalling that the accessibility of the other subregions (including central-city subregions) to the employment in subregions 3 and 4. This is evidenced by the high positive changes noted in **Table 9-6** and in the shares-of-accessibility-to-employment for

Sub-re-gion	Projection Year Accessibility to Population Value	Accessibility to Employment Value	Forecast #1 (Completed Beltway) Accessibility to Population		Accessibility to Employment		Forecast #2 (Radial System) Accessibility to Population		Accessibility to Employment	
			Value	Change in Share	Value	Change in Share	Value	Change in Share	Value	Change in Share
1	240,166	119,339	243,114	−0.00503	121,355	−0.00217	317,695	+0.00510	163,260	+0.00668
2	277,935	181,888	277,935	−0.00726	182,031	−0.00543	340,626	−0.00368	207,719	−0.01419
3	257,716	202,134	257,716	−0.00673	202,134	−0.00615	360,632	+0.01246	244,710	−0.00755
4	235,825	178,711	235,825	−0.00616	178,808	−0.00537	308,393	+0.00375	216,424	−0.00663
5	202,209	98,142	228,757	+0.00591	104,423	+0.00156	211,701	−0.01548	104,363	−0.01223
6	253,098	143,941	267,371	−0.00059	146,278	−0.00269	301,058	−0.00658	174,899	−0.00500
7	227,677	143,494	249,405	+0.00321	160,043	+0.00762	320,498	+0.01169	211,346	+0.01695
8	165,512	67,387	177,575	+0.00076	70,613	+0.00029	178,116	−0.01096	90,344	+0.01529
9	239,382	131,225	241,561	−0.00533	134,200	−0.00184	316,654	+0.00508	192,923	+0.01529
10	134,381	58,281	193,009	+0.02121	80,286	+0.01417	165,798	−0.00139	80,978	+0.00400

Table 9-6 Changes in accessibility for forecasts

such subregions as 7 and 9, which have been made much more accessible to the jobs in the central city than other subregions.

The following regional totals of activity are projected for the forecast year

population	3,530,000
white-collar employment	985,000
blue-collar employment	465,000

The calibrated model produces estimates of future subregional activities under the two alternate networks as shown in **Table 9-7**. Examination of these two forecasts, typically through graphic displays, shows some interesting contrasts. Population appears to be more evenly distributed by the circumferential innerbelt around the city under scenario 1, as one would expect. By contrast, subregions 7 and 9 enjoy more population growth under scenario 2, with its radial freeways making jobs more accessible. As far as employment, subregions 3 and 4 again experience white-collar employment growth more under scenario 2, with their increased access to the labor force. The radial highway's stimulation on blue-collar employment is less pronounced, partly because of the relatively small number of blue-collar jobs in 3 and 4.

Sub-region	Population			White Collar Employment			Blue Collar Employment		
	Base year	Forecast year	Forecast year	Base year	Forecast year	Forecast year	Base year	Forecast year	Forecast year
1	386 000	475 300	548 100	47 000	55 100	72 100	35 000	39 500	45 400
2	432 000	509 300	452 600	56 000	62 200	58 600	128 000	150 100	150 400
3	179 000	177 700	189 600	240 000	305 700	323 100	44 000	49 000	57 500
4	186 000	194 000	196 400	115 000	141 600	150 200	22 000	22 800	27 200
5	228 000	310 500	190 800	30 000	46 400	14 600	58 000	72 800	61 000
6	390 000	481 100	458 300	133 000	172 700	165 100	14 000	16 100	13 000
7	32 100	469 100	543 200	28 000	46 000	61 700	43 000	54 300	59 600
8	196 000	254 700	258 100	26 000	35 200	25 700	25 000	30 500	25 500
9	256 000	310 100	440 500	51 000	60 500	84 400	24 000	26 200	33 400
10	176 000	348 200	252 400	21 000	59 600	29 500	10 000	23 700	12 000
Total	2 750 000	3 530 000	3 530 000	747 000	985 000	985 000	403 000	485 000	485 000

Table 9-7 Forecast activity levels

9.9   Concluding remarks on EMPIRIC

EMPIRIC is a most flexible and versatile land-use model. It lacks a unique theoretical structure. It is essentially an empirically-based, spatial-econometric--model and does not build on any theory of urban change. It operates on a series of very general hypotheses concerning urban growth, by which the major factors influencing past development are extrapolated into the future. In many ways, it is less of a model than a parcel of techniques for model building. This structure is both the main strength and the main weakness of the model. In one sense it provides the analyst with a powerful and extremely flexible device for assessing established patterns of development over a particular time period. As such, it may be used effectively as a research tool. It can both test specific theories of development and assess the impact which specific past developmental policies have exerted on urban growth. More important, it has been used to project the future growth pattern in response to a variety of alternate policy considerations. With this flexibility, the model builder may take full advantage of a rich data set or a relatively complex theory of development, or s/he may rely only upon relatively simple data sources and a very crude conceptualization of the development process.

On the other hand, the lack of a strong, theoretical foundation such as that embodied in the Lowry-based models places a heavy burden on the ability of the analyst to develop his/her own conceptual structure and to translate this into a set of meaningful, valid statistical relationships. The effective calibration and application of EMPIRIC require a significant degree of familiarity both with the development process itself and also the fundamental principles of spatial-econometric techniques. This is not to argue that the model can only be used by the specialist. Rather it is to emphasize that professional maturity is required in the course of model calibration, and developing the model for continuing application within the work program of a conventional planning agency. This can often be accompanied by a significant degree of staff training.

The model has a history of some significant operational successes. It has been applied in Boston, Washington, D.C., Denver, Winnipeg, Toronto, Seattle and Atlanta with proven operational status. It has been used to deal with issues relating to both highway and transit system development, evaluation of alternate fair-share housing policies, development controls, open-space policies and moratoria on additional sewer-system construction. It has not, however, stimulated any amount of independent research activities as have flowed, for example, from the Lowry Model. In part, the widespread use of the model derives from its inherent flexibility and proven operational record. It was also supported by a major consulting organization, and had been continuously updated over a decade.

Its data requirements are extensive—particularly on gathering two consistent data sets over time, although this is not significantly more so than other land-use models. They may, however, be readily integrated within the continuing work program of a typical transportation or regional planning agency that is at least minimally committed to analytically-oriented activities. As with most of the other models, much of the data assembled for calibration and forecasting purposes have widespread application in other related areas of planning.

We choose to discuss EMPIRIC in this book for a pedagogic reason. It is an excellent example of a spatial econometric model. As pointed out earlier, it is a special case of a spatial time-series called mixed-regressive-spatial-autoregressive model. The most interesting part is that the model has captured both `causal' relationship and spatial-temporal correlations. Thus an a priori relationship is hypothesized between population and employment, including the one-lag momentum between subareal population at t and $t+\Delta t$ (and similarly for employment.) At the same time, spatial weights are used to model the interaction between subareal employment and population, again couched within a one-lag time frame. It can be said that it has "the best of both worlds" in this regard. One thing worth mentioning again is that the weight matrix appears two or three times. The first weight matrix $\mathbf{\Phi W}_1$ is to quantify the 'causation' between population and employment. The second one $\mathbf{Z}_1\mathbf{\beta}_1$ quantifies the correlative relationship between population at t and $t+\Delta t$ (and again employment between t and $t+\Delta t$.) The third one $\mathbf{Z}_2\mathbf{\beta}_2$—which is truly a spatial weight matrix—models the *spatial* correlation between subareal population activities and employment activities. In this regard, we view EMPIRIC as an introduction to the forthcoming chapter on "Spatial Time Series."

9.10 Integrating macro and micro models

Land-use/facility-location models are disparate in their scale and function. For our purposes, let us call the large-scale general-purpose land-use models *macro* models and the small/specialized ones *micro* models. For example, the EMPIRIC model described above is a macro model, while a location-allocation model that sites industrial plants may be classified as a micro model. These two types of models are often regarded as mutually exclusive and have been presented in separate parts of this book. At best micro models are often employed in a preliminary or a later step—before or after use of the macro model—to obtain more specialized or detailed estimation of activities. Macro models have been developed to cope with urban analysis comprehensively, but their aggregation level and their descriptive generalization make them good estimators of activity allocation only at a regional scale. On the other hand, micro models go into a level of specialization that preclude consideration of interaction among activities in the study area as a whole.

A number of urban planning theories have been hypothesized over the years, most notable of which are the economic-base theory and its input-output generalization. Other methodologies extrapolate existing development patterns, often utilizing statistical or econometric techniques, such as the EMPIRIC experience described above. The former has a more theoretical foundation while the latter an empirical basis. While there is admittedly a large grey area in between, the conceptual distinction (just like the macro/micro distinction), is a helpful one in dealing with the labyrinth of existing and evolving models.

Macro and micro models are developed for different purposes. While macro models recognize explicitly the various economic components of the study area—from primary activities such as manufacturing to secondary activities such as the service sector—the treatment is more on an aggregate level as mentioned. For example, in both the Lowry and EMPIRIC models the three sectors—residential, blue and white-collar industries—are modelled, but only in terms of their interactions. To understand each of the sectoral components on an in-depth level (which are required in a variety of applications), micro models have to be employed.

There is another motivation for performing micro analyses. For example, the macro models outlined above often require various inputs that have to be obtained from sources outside the model itself. No where is it more evident than in the Lowry or EMPIRIC models, where basic employment or total employment and population for the forecast year have to be found exogenously. Micro models—whether they be demographic, housing, industrial and service—provide such exogenous inputs. An example of demographic forecast is found in the "Economics Methods" chapter of Chan (2000).

Similarly, theoretical vs. empirical models are developed through the years for different functional and operational considerations. Theoretical models, for example, are obviously advocated by professionals in a different school of thought than the empirical camp. While it is true that empirical models have a more pronounced track record of actual application than theoretical ones (as evidenced by the comparatively prolific application of EMPIRIC), perhaps the `theoretical' models simply produce more plausible forecasts (because of their underlying structure). They are in that sense more serviceable than empirical ones. Empirical models have one advantage, however, and that is their ability to integrate with other micro models to perform a comprehensive study—again proven by the EMPIRIC experience.

It is seen, therefore, that there are definite roles played by macro, micro, theoretical and empirical models. Aside from the richness provided by pluralism, it is important to remember the interdependencies between these models in the planning process. Viewed in this light, no single model is superior to others; several different classes of models may need to be used together in solving a specific problem. Their interdependency raises one question that comes up again and again over the years. Considering the interface problems that typically plague the practitioner in coordinating macro/micro models, is it not more sensible to come up with some guidelines for integrating models that will bypass some of the trial-and-error difficulties? We will answer this question through a case study.

9.11 The Kansai International Airport experience

In 1987, the construction of the Kansai International Airport was started on a reclaimed island located off the southern coast of Osaka Bay, Japan. The environmental and socioeconomic effects of construction and the airport's eventual operation on the surrounding region are of great concern. A comprehensive model was built to assess the impact of the new facility (Suzuki et al. 1989, Pak et al. 1989). The integrated model has a hierarchical structure composed of three tiers, as illustrated in **Figure 9-10**. We offer an overview of the model below, while technical details are documented as an exercise in the "Problems and Exercises" web page.

REGIONAL MACRO ECONOMIC MODEL: The first tier is an econometric model for the whole region. This model has personal income and various consumption expenditures as endogenous variables and public investment as asignificant exogenous variable, which allows the economic effects of the construction of the new airport to be estimated. The econometric model assumes a simple structure consisting of four exogenous variables and 13 endogenous variables. As mentioned, gross domestic product (GDP), and fixed capital formation by central and local governments are adopted as exogenous variables. This model makes it possible to and in public investment on the regional economy. Notice this model is aspatial in nature and is used for the region as a whole, rather than on a zone-by-zone basis.

REGIONAL INPUT-OUTPUT MODEL: The second tier of the integrated model is a regional input-output model, which is constructed to estimate the gross output and value added by industrial sectors of the total region. By using consumption expenditures and capital formations forecasted in the econometric model, the gross output and value added by industrial sectors are calculated based on the industrial structure expressed in the coefficients of the regional input-output table. Employment by sectors and the residential population are also estimated by making use of the per-capita value-added productivity and per-capita employment, respectively.

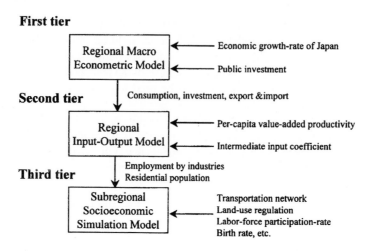

Figure 9-10   Integration scheme of the Kansai Airport model

In this model, the industry is divided into 12 sectors. The gross output is obtained; so is the total employment. The input coefficients are time-varying parameters. They take into account the change of industrial structure during the long-term projection. A large number of employees are hired for the operation of estimate the effects caused by the difference in the national economic growth rate the new airport. The multiplier effect of this input into the economy is explicitly modelled. Once the number of total employment is determined, the total population of the region can be calculated by using per-capita employment, which is also an exogenous variable of the model as mentioned. Again, this is an aspatial model for the region as a whole, rather than a subareal-level model.

SUBREGIONAL SOCIOECONOMIC IMPACT MODEL: The third tier is a multi-regional socioeconomic 'simulation' model. In this model, the Kansai region is divided into 120 zones and the industry is divided into 46 sectors. Population is treated by gender and five-year age-groups and land use is divided into seven use-categories. The economic estimates from the first two-tier models are used as the exogenous variables of the simulation model, which in turn estimates the effects of construction and operation of the new airport on each of the 120 zones.

This large scale model consists of the population, employment, land-consumption, and travel-time submodels. In the employment submodel, 46 industries are grouped into primary, secondary, and tertiary levels. The model allocates to each zone the number of employees of these 46 industries by using location functions, given the total employment of each industry forecasted by the second-tier input-output model. Similar to other components of this socioeconomic

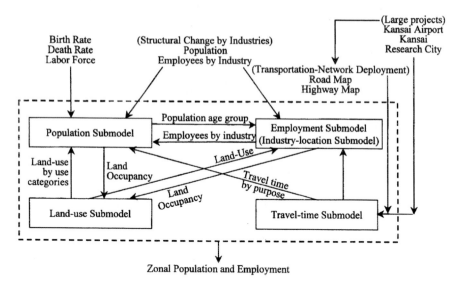

Figure 9-11 Schematic diagram of Subregional Simulation Model

'simulation' model, location functions are determined through extensive regression analyses.

In the population submodel, the population by gender and by five-year age-groups is forecasted for each zone by using demographic and various socioeconomic relationships that have been obtained by statistical analyses. A unique feature of the model is that the migration pattern in each zone is taken as an important endogenous variable that varies according to the development of various socioeconomic activities in each zone.

In the land-consumption submodel, changes of land-use for each purpose (residential, industrial, commercial, public, road and railways, agricultural and others) are computed based on the population and employment figures supplied. Competitive relations among land-uses are embedded in the model.

The travel-time submodel estimates spatial separation between zones using minimum-path algorithms. Development potentials are computed from these travel times, based on which activities are allocated. This and the other three submodels are intertwined, as illustrated in **Figure 9-11**. The final output consists of future zonal employment and population after satisfying land-use constraints.

9.11.1 Insights from this study. The Kansai Airport case study provides some valuable lessons regarding how EMPIRIC-type models should be properly integrated with submodels or micro models to perform the best study. In the Kansai model, the socioeconomic-impact and input-output models can be classified as micro models while the subregional 'simulation' model the macro model. The integration scheme used in the Kansai study is really quite straightforward. The output of one model becomes the input to the next model, as illustrated in **Figure**

9-10. No "feedback loop" is used. Similar interfaces are built into the Subregional Simulation Model among its own micro models, although the interaction there is a bit more complex (see **Figure 9-11**). Of particular interest is the employment and total population figures that are output from the input-output model. They are in turn input to the subregional socioeconomic-impact model —similar to the exogenous control totals that are fed into EMPIRIC. The three-tier model structure is quite laborious, as is typical of a comprehensive study of this sorts.

If the 1960's through the 1980's have taught us any lessons, there are many drawbacks with a large scale model of this kind. (See chapter on "Activity Allocation and Derivation"). It is speculated that a more simple structure can be built to accomplish the goals of this study (see "Spatial Equilibrium and Disequilibrium" chapter). For example, the first- and third-tier models can possibly be combined to form an EMPIRIC-type model, with information fed from a regional input-output model. The integration should be relatively straightforward since both are built upon regression models. The first step of integration consists of the inclusion of public investment as an exogenous variable, very similar to the accessibility terms in the classic EMPIRIC equations. This exogenous variable will be merged with the rest of the independent variables, including the classification of population into age groups, as with the existing Kansai model. The set of variables is then subject to factor analysis (FANAL), coming out with factor scores(FSCORE), which roughly represent the 'quantity' of each variable that a particular subarea has. This identifies the important explanatory variables for inclusion into the final model. A schematic of this procedure is shown in **Figure 9-12**. Here, y represent dependent (endogenous) variables, x's represent new independent (exogenous) variables such as public investments, and z are the existing independent variables including base-year population and employment.

Even before our integration discussion here, the reader has no doubt realized the similarity between the Subregional Simulation Model as outlined in **Figure 9-11** and the EMPIRIC simultaneous equations. Subregional population and employment are modelled as dependent variables while travel time is modelled as the accessibility independent variable. Population migration among zones can beimplicitly modelled within this framework through their allocation among subregions—if the population is stratified into the correct number of groups. Now the reduced model structure will consist of the Regional Input-Output Model (RIOM) and the Subregional Simulation Model (SSM), with feedback loops between them. Employment and population totals are fed into SSM from RIOM, while zone-specific consumption, investment, export and import are fed into RIOM from SSM. The iterations between the two models will end upon reaching an equilibrium.

The size of the expanded SSM appears to bear a linear relationship to the number of activity variables that is included. There is one equation for each dependent variable. There are linear increases in the number of independent variables as well. At least one additional independent variable, for example, is introduced corresponding to public investment into the airport. In general, the linear expansion in dimensionality appears reasonable, although one must not

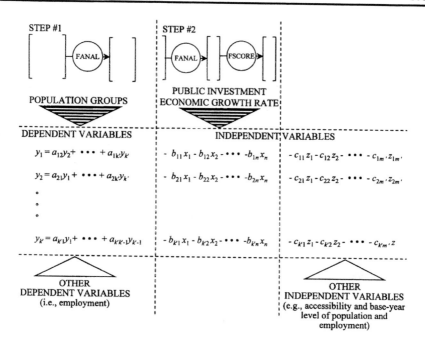

Figure 9-12 Incorporation of public-investment variable into the simultaneous-equations system

overlook the additional difficulty often associated with calibrating and solving a large set of simultaneous equations. The number of subregions that can be handled within this reduced model structure is likely to be much less than the 120 zones currently modelled by the Kansai model. Past EMPIRIC applications point toward dozens of subareas as the maximum number such models can comfortably handle.

Our comments above should in no way be construed as a critique against the builders of the Kansai model, who are extremely well qualified individuals. But they appear to be not as familiar with the detailed land-use literature in the author's opinion. There is also a tendency to apply what they are familiar with, rather than the most appropriate operational model. The literature reviewed in this chapter would have been helpful to the Kansai undertaking; so would chapters on the "Spatial Equilibrium" and "Activity Allocation Models." The point is that the state-of-the-art in land-use modelling has progressed significantly in recent years. Some of the intricacies are only familiar to people that follow the recent research intimately.

9.11.2 Implications for the future. The exercise above is an attempt to gain some insights into the integration of the macro and micro models in land-use analysis. The Kansai exercise has shown that there are more satisfactory integration methods than the commonly practiced approach, where the output of one model is fed in as input to another. In this experiment, one of the more operational macro models, EMPIRIC, is integrated with one of the more promising submodels, input-output analysis. The integrated model appears to bridge the gap between macro and micro models, allowing the needed information to be supplied in a symbioses basis. However, to fully assess its effectiveness and operationality, more detailed studies must be performed, which will refine several aspects of the outline presented. The reason is that the demonstration of the integrated model's feasibility from a methodological standpoint does not imply its receptivity in professional practice, where a good deal of 'soft' factors (to be distinguished from technical consider-ations) are involved. Short of conducting an actual field test of such a model, the author sees no fool-proof way of predicting the viability of such an integrated model. From the integration exercise, however, several pertinent remarks can be made about the appropriate level-of-aggregation and the appropriate level-of-complexity in urban planning model design—if an integrated model is to be developed (Meda-Dooley and Chan 1976). They are documented below for general reference.

9.11.2.1 FUNCTIONAL CONSIDERATIONS. When an integrated model is warranted, the first step toward the design of such a model is to assess the compatibility between a macro and micro model. Such a task is simplified if the candidate models are classified into the two-way classification scheme according to their level-of-aggregation and model-structure, since the taxonomy allows for a fast screening of the commonality (and the lack of it) between candidate models. In **Figure 9-13** is shown three possible integration strategies, ranging from the "most recommended" to the lesser ones. The author feels—partly from the insights gained in this experiment and partly from his other experiences—that empirical models are more amenable to integration than theoretical ones. This is so because empirical models are usually built upon more flexible underlying structure, making them less of a closed system and thus allowing for integration to be performed. Upon the rare occasion where both macro and micro models are founded upon a common theory, integration between theoretical models can be performed after the necessary adaptation work. An example is to integrate a macro and micro model where both are based on input-output models—as discussed in the "Spatial Equilibrium" chapter when an aspatial input-output model is extended to a spatial version.

It is possible to integrate a theoretically-based macro model and an empirically-based micro model. This is the case when integration is interpreted broadly to mean the operation of two separate models where the output of one serves as an input to another (which is predominantly the current practice as mentioned). For example, the Lowry-based models—classified as theoretical in our typology—receive as input an exogenous projection of basic employment that is usually obtained from an empirical model.

9.11.2.2 OPERATIONAL CONSIDERATIONS. After functional considerations, the next concern is the actual operational details that may decide the ultimate success or failure of the integration. The first of such details is the level-of-

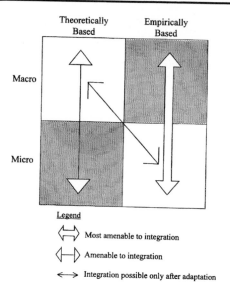

Figure 9-13 Compatibility between macro and micro models

aggregation for the data. As evident from the experiment, the geographic unit of operation may be entirely different for the macro vs. the micro model. EMPIRIC, for example, operates on a larger geographic subarea than zonal gravity models. Only after they are brought to a common level can integration be performed. **Table 9-8** provides a checklist for the variables that need to be examined for aggregation levels. They include activity variables such as population and employment, land-use data such as residential vs. industrial acreage and transportation variables such as interzonal travel time.

A second checklist can be compiled to ensure that both candidate models operate on a common set of solution methods. **Table 9-9**, for example, shows several widely used calibration methodologies and forecasting procedures. Included in the calibration methodologies are ordinary least squares and two-stage least squares—which are commonly employed in EMPIRIC. The forecast technique, whether time-series or cross-sectional, is another crucial integration consideration. Still another common distinction among the forecasting methodologies is whether the maximum developable-capacity of a subarea is modelled, and if so, the precise technique of monitoring the re-allocation of excess activities beyond capacity. The two checklists shown here are not meant to be comprehensive. They do, however, serve as a starting point for a more in-depth analysis.

9.12 Concluding comments

Spatial-econometric techniques offer an alternate way of modelling land-use and activity allocation. Experiences with this modelling approach allow these

	Macro Model		Micro Model	
	Required?	Available?	Required?	Available?
ACTIVITY VARIABLES: *Subareal* (district, zonal ☐ census, tract, or household level?) ☐ Population ☐ Employment *Regional* ☐ Population ☐ Employment				
LAND-USE VARIABLES: *Subareal* (district, zonal census tract, or household level?) ☐ Acreage by Activity Type ☐ Permissible Development Densities ☐ Capacity Limitations ☐ Locational Restrictions				
TRANSPORTATION VARIABLES: ☐ Travel Time & Cost ☐ Trip Frequency				
OTHER VARIABLES: ☐ Public Utilities ☐ Socioeconomic Attributes				

Table 9-8   Level of data-aggregation checklist

observations to be made (Griffith and Can 1996. Cressie 1991, Foot 1981): All data from remote-sensing, land-use and activity allocation have a more-or-less precise spatial and temporal label associated with them. These attributes include multichannel grey-values and spatial contiguity in remote-sensing images, and the tabular subregional population, employment and accessibility information in EMPIRIC, which make them amenable to spatial-econometric modelling. Compared with such models as input-output analysis, a pure spatial model usually has no causative component in it. Such models are useful when a space-time process has reached temporal equilibrium—equilibrium that can often be approximated over 'short' time-increments. They are also useful when short-term causal effects are aggregated over a fixed time-period. This is seen in the five-year forecast increments typically found in the EMPIRIC Model, representing the elapsed time between census updates.

	Macro Models	Micro Models

Calibration Techniques:
☐ Manual Procedures
☐ Ordinary Least
   Squares
☐ Indirect Least
   Squares
☐ Two-Stage Least
   Squares
☐ Maximum
   Likelihood		

Forecasting Procedure:
☐ Capacitated/
   Uncapacitated
☐ Cross Sectional/
   Time Series		

Table 9-9   Solution method checklist

Data that are close together in space are often more alike than those that are far apart—an observation that is sometimes called Tobler's "first law." A spatial model incorporates this spatial variation, termed spatial autocorrelation, into the modelling mechanism, in contrast with a non-spatial model. One may choose to model the spatial variation through the (non-stochastic) mean structure, as demonstrated in the mixed-regressive-spatial-autoregressive model of EMPIRIC. Alternatively, one can employ the stochastic-dependence structure by way of a spatial-autoregressive-disturbance equation-set. The choice depends on the underlying scientific problem, and can sometimes be simply a trade-off between model fit and parsimony of model description. What is one person's covariance structure may be another person's mean structure. In the "Spatial Time Series" chapter that follows, we will pick up this discussion in earnest.

In model construction, explanatory variables should be included in the mean structure first, and great care should be taken to find all of them. A missed variable that is itself varying spatially will contribute to the spatial dependence, as can a mis-specification of the functional relationship between the independent (exogenous) and explanatory (endogenous) variables. Consequently, a model that includes a spatial-dependence component, such as a spatially-lag, pays a low-cost premium that insures against mis-specification of the mean structure. Having allowed for explanatory variables, models with spatial dependence typically have a more parsimonious description than classical trend-surface models. They also have more stable spatial extrapolation properties and yield more efficient estimators of explanatory-variable effects.

We have included two examples in this chapter to drive this point home: the EMPIRIC and Kansai econometric models. Both assume a spatial form of the autoregressive model, serving as an introduction to the "Spatial Time Series" chapter. We have chosen to go into quite a bit of detail of EMPIRIC, including its successes in field implementation. Among other reasons, its success can be attributable to a model structure that has both a time and spatial correlation explicitly built in, allowing the model to analyze and forecast subareal economic activities with reasonable short-term accuracy. The Kansai study, aside from confirming similar observations, is chosen specifically to bridge the gap between a more causal model such as the input–output format—of which the Lowry–Garin model is a special case—and correlative models such as the autoregressive form. The Kansai study also bridges the gap between discrete facility-location and continuous activity-deviation-and-allocation—a central theme of this book—in that the spatial economic impact of a new airport is carefully traced through time.

Having promoted its virtues, we owe it to the reader some of the details that warrant special attention in operationalizing spatial econometric techniques:

a. Linear regression assumes linear relationships between variables. Nonlinear models have been made piecewise linear through forecasts in increments, as witnessed in the EMPIRIC experience. Approximations such as this can be used to convert non-linearities into linear functions. But there are definite limitations in a linearity assumption, particularly in systems that exhibit precipitous behavior such as bifurcation or catastrophe.

b. Regression is a statistical technique that assumes certain properties of its data. In dealing with spatial data there are continual problems of multicollinearity where the independent or exogenous variables are correlated among themselves. Residuals are often correlated as well, giving rise to another level of autocorrelation that needs to be modelled. The latter phenomenon may give rise to heterogeneity, making up yet another complication. A linear model performs well if the allocation of activities between subareas in a region conforms to a fairly general pattern—i.e., when data are homogeneous. Data clustering around selected subareas can pose great problems in calibration, particularly with respect to the sensitivity of regression coefficients to these selected subareas. Data may not be normally distributed either, and are usually spatially clustered.

c. The models allocate activities to subareas, but do not model interaction between subareas explicitly as in the gravity model. The closest way that it comes to model such interaction is exogenously using spatial weights or accessibility measures. As a result, there is no satisfactory way to deal with holding capacity of a subarea except through ad hoc means. One can say that models such as EMPIRIC are unconstrained, as to be contrasted with a doubly-constrained gravity models.

d. Since most econometric models deal with incremental changes over time, accurate data in a consistent format are required for at least two points in time. To the extent that consistent sets of data over time is often hard to obtain, data quality is greatly compromised. This affects the proper functioning of econometric models. This problem may be partially mitigated by remotely-sensed raster images that have uniform, digitized formats.

However, there are a number of practical advantages of using spatial-econometric models, as alluded to already.

a. Regression is a highly developed statistical technique and has been widely applied in many fields of study. There is considerable literature that can be consulted on ordinary regression, two-stage least squares, and other simultaneous regression techniques. Similarly, econometric techniques have long been used by quantitative economists to model the economy. Much of that experience can be cross fertilized to spatial economics. No specialized computer codes need to be developed since regression packages are available on major and personal computer systems. Calibration of econometric land-use models generally pose no great computational requirement.

b. There is tremendous flexibility in model building using regression, from the selection of variables to the structural form of the equations. This contrasts sharply with the relatively rigid structure of a gravity model for example. We have seen evidence of this not only in EMPIRIC, but also the Kansai Airport study. But both econometric models require as input control totals on population and employment that are often generated from causal models such as input-output analysis.

c. In spite of problems with spurious correlation, short-term forecasts using regression have been satisfactory if there exists a stable general pattern in the study area. This can be gleaned from the population-forecast equation in six EMPIRIC applications in the United States.

d. The execution time on a computer is short for many regression runs, including for a large study area—generally in the order of seconds. Data assembly and determination of an appropriate structural equation set, rather than computer time, is the main problem in operationalizing econometric models.

Through an experiment, the author concludes that empirical macro/micro models, by virtue of their more flexible structure, are more subject to integration than theoretical ones. The empirical integration approach recommended in previous sections can be amended and improved in the future when more is known about the unifying theory behind land-use and facility-location models. Due to the present lack of a universally accepted theory that could explain the complex mechanisms that shape land-use patterns, however, the empirical approach of spatial econometrics has to be regarded as a valuable tool in the interim. This statement holds true in spite of rapid progress in causal spatial modelling reported in the "Spatial Equilibrium" chapter.

A major task in the use of econometric models is calibration. Spatial autoregressive models pose extra demands on calibration above and beyond identification and model specification traditionally associated with non-spatial econometric models. The inclusion of spatial correlation beyond temporal correlation requires multi-step calibration procedures where subsets of the parameters are calibrated in sequence, rather than simultaneously. The sequential approach estimates a set of parameters in the upper-tier models that are in turn input to lower-tier models. Such an approach is absolutely required in real-world applications where there simply is not feasible nor are there enough data to simultaneously calibrate all the parameters at once. This does not mean that more integration efforts need not be tried to address the inherent ecological fallacies associated with a sequential calibration approach. It simply underlines a key feature of a spatial-econometric model that the users need to keep in mind—as illustrated in the three tiers of the Kansai model.

The steps taken in this chapter to integrate the three tiers of models used in the Kansai region outside Osaka, Japan appear promising. As a first attempt and as a framework, the integrated formulation proposed might be grounded on more solid state-of-the-art knowledge than the one used in the original study area, given a "20-20 hindsight." More important, the Kansai experiment allows us to formulate some guidelines for future integration efforts. Among the guidelines are functional and operational checklists for integrating a macro and micro model. Obviously, these checklists can be refined by further experimentation.

For all at last returns to the sea – to Oceanus,
the ocean river, like the everflowing stream of time,
the beginning and the end.
Rachel Louise Carson

CHAPTER TEN

SPATIAL TIME SERIES

One of the newest ways to forecast land-use development and related activities is by using time series in a spatial dimension. For example, one may wish to model the interaction between zonal population and employment over five years. Traditional time series rely heavily on historical data, which forms a 'pattern' and 'trend' for future projection. Recent development in spatial time-series combines historical trend with 'causal' modelling, capturing the spirit of both approaches. To the extent that population often follows employment, one wishes to analyze the cause and effect besides a trend line. This chapter reviews the classic methods used in forecasting, introduces the relatively new concepts of spatial-temporal modelling and ends with a controlled experiment to assess the performance of these various techniques in land-use forecasting. In the experiment, a data base is generated from a traditional (causal) activity generation-and-allocation-model, which allows us to implicitly compare these two sets of forecasting techniques. Also included in the chapter is how to deal with precipitous changes such as the 'boom' caused by a new plant or the 'bust' by a plant closure, both of which have very different effects compared with gradual changes. In other words, we wish to cope with a departure from the normal pattern and trend.

10.1 Univariate time-series analysis

The fundamental basic building block in analyzing spatial times-series is the univariate case, in which a single variable such as population is tracked. Aggregate population for the entire city is of prime interest, rather than subareal (zonal) population, let alone its relationship to employment. Autoregressive moving average (ARMA) models are integrated approaches to forecasting, usually attributed to Box and Jenkins (1976), who synthesized the up-to-then fragmented information into a coherent framework. Makridakis et al. (1983) represents a more readable primer in this field and much of the aspatial discussions in this chapter are borrowed from that text. ARMA combines the correlative and smoothing components of forecasting into a single framework using three parameters, of which we will deal with only two: p and q, leaving out the 'differencing' parameter. To fix ideas, a time series is *differenced* when the growth rate and cyclic fluctuation are taken out for analytic convenience. In other words, we assume that the data have

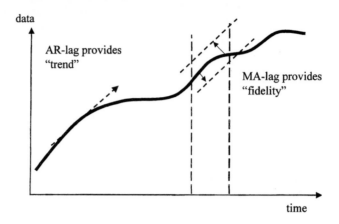

data

AR-lag provides
"trend"

MA-lag provides
"fidelity"

time

Figure 10-1 Graphical illustration of ARMA

been prepared in the 'correct' format through differencing. This is like the preparation of raw data for regular linear regression through log transformation. In this case, it is through taking the difference between two variables. An example of such 'corrected' data is given in the "Statistical Tools" Appendix in Chan (2000), where the *corrected* sum-of-squares is used instead of the raw data in ordinary-least-squares (OLS) regression. The conventional notation is to write ARMA (p,q). This notation captures the essence of a time series, which is to explain how the data correlate over time through lead-lag relationships. An example of p, the lag of 'trend' data, and q, the lag of error term, is illustrated in **Figure 10-1**, where the lag between two data-points over time is used to form a 'trend' and the lag between two error-terms is used to measure the 'fidelity' of the time series. p is associated with the autoregressive component and q the moving average component of the time series. In many ways we can relate time series with OLS regression, where the dependent variable is Z_t and the independent variables are the one- and two-period lagged-data Z_{t-1}, Z_{t-2} and so on. While regression mainly entertains the relationship between the dependent and independent variables, ARMA models take into explicit account the relationship between the dependent variable and the error terms as well. It is through the analysis of the error term that provides the *fidelity* of a forecast.

It is possible to have a Seasonal Autoregressive Moving-Average (SARMA) model. Seasonality is characterized by a periodic (cyclic) movement in the data that usually repeats itself. For example, the increase in travel during summer months occurs every year without exception. The season here is defined as the length of time over which the periodic movement occurs (in this case one year). The common designation of a SARMA model is $\text{SARMA}(p,q)_{s_p, s_q}$ where s_p is the autoregressive season-length and s_q is the moving average-season-length. Special cases of ARMA models are the pure autoregressive model AR(p) and the pure moving average

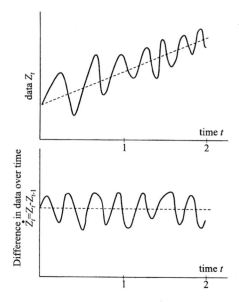

Figure 10-2 Constant growth pattern

model MA(q). In the seasonal versions, they will be $\text{SAR}(p)_{s_p}$ and $\text{SMA}(q)_{s_q}$ respectively. As will be shown, differencing of two data points separated by a season length may reduce a SARMA series to an ARMA series.

10.1.1 Stationarity. Stationarity of data is very important in forecasting. By way of a definition, a stationary series means that the data is in equilibrium around a constant value (the underlying mean) and that the variance around the mean remains constant over time. One can think of stationarity as the equivalence of the *homoscedasticity* property in regression, where the errors are spread uniformly around the regression line. More formally, a series is stationary if its statistical properties are independent of the particular time-period during which it is observed. From this definition, it is easy to see the analogy to the stationarity concept in stochastic process.[1] If a time series is generated by a constant process subject to random error, then the mean is a useful statistic and can be used as a forecast for the next period(s). A time series either involves a trend, a seasonal effect, or both. ARMA is a useful method to examine such patterns from stationary data.

Differencing is a common method to induce stationarity. **Figure 10-2** shows a pattern of constant growth. Taking the difference, which can be thought of as the first derivative over time, will yield a constant over the entire period. The data are now said to be stationary. The same can be applied to a parabolic growth pattern,

---

[1] See the "Markovian Process" Appendix in Chan (2000) for an introduction to stochastic process, including the definition of stationarity.

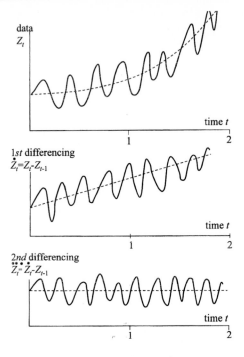

Figure 10-3 Parabolic growth pattern

such as the one shown in **Figure 10-3**. In this case, second-order differencing, again equivalent to taking the second derivative with respect to time, will yield stationarity. Obviously, in both cases the data are subject to random fluctuations around the "trend line."

Example

Given the time series in the first column, please compute the first and second-order differences by filling in the second and third columns of the following **Table 10-1** *(Makridakis et al. 1983).*

Plot the data on graph paper. By this graphical means alone, do you think the data in column 1 is stationary. How about column 2? Why

The differenced data are shown in columns 2 and 3. It can be seen from **Figure 10-4** that the first difference is not stationary, but the second difference is—the data 'hover' around the horizontal axis evenly. □

Seasonality is often present in a time series. If the data contains non-stationary seasonality, it is reasonable to assume that the data may be de-seasonalized by taking a difference equal to the number of periods in the season (or the season length). For example, if the non-stationary season consists of 12 periods (as in 12

Period t	Time series Z_t	First difference $\dot{Z}_t = Z_t - Z_{t-1}$	Second-order difference $\ddot{Z}_t = \dot{Z}_t - \dot{Z}_{t-1}$
1	2.44		
2	5.30	(2.83)	
3	8.97	(3.67)	(0.81)
4	13.88	(4.91)	(1.24)
5	19.58	(5.70)	(0.79)
6	26.99	(7.41)	(1.71)
7	35.95	(8.96)	(1.55)
8	45.86	(9.91)	(0.95)
9	55.70	(9.84)	(-0.07)
10	67.36	(11.66)	(1.82)
11	79.63	(12.27)	(0.61)
12	92.13	(12.50)	(0.23)

Table 10-1 Differencing example

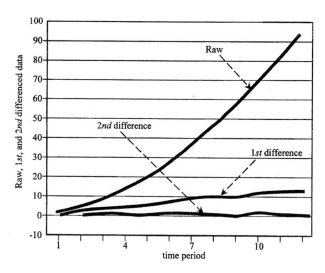

Figure 10-4 Parabolic-growth pattern

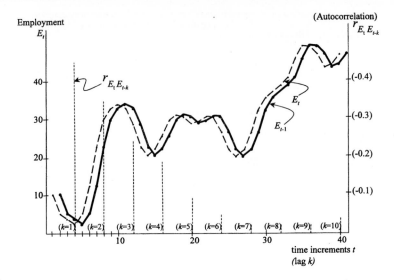

Figure 10-5 Employment over time and autocorrelation

months of a year), the data can be de-seasonalized using the following equation: z_t $= Z_t - Z_{t-12}$. This is effectively differencing the data by 12 periods (instead the normal situation of differencing by one period). It should be noted that in so doing, we have lost 12 degree-of-freedom, similar to losing one degree of freedom in the one-period differencing in **Table 10-1**.

For this discussion, it is best to work through an example to illustrate an ARMA(p,q) model. **Figure 10-5** contains a set of data from Makridakis et al. (1983), which is interpreted here as an employment growth-profile E_t with apparent random-fluctuations. The first step in the modelling procedure is to identify the model to be used, including the value of the parameters p and q.

10.1.2 Analysis of data. The first parameter we calculate is *autocorrelation.* Autocorrelation is the correlation of the time series with itself, lagged by $k = 1, 2$, or more periods. Consider the data given in **Figure 10-5**, the following time-series can be written as E_t = constant + $\varphi_1 E_{t-1} + \varphi_2 E_{t-2} + A_t$. This is nothing more than a regression of the time-series data upon itself, with a shift of one and two time-periods, complete with an intercept constant, calibration coefficients φ_1, φ_2 and an error-term A_t. Thus the three streams of data will show a complete column for E_t from $t = 1$ through 40; an identical stream in E_{t-1} column starting out at $t = 2$, with the first entry in the column missing and the last entry removed; and the same stream for E_{t-2} starting at $t = 3$. The correlation coefficient between E_t and E_{t-1}, for example, is now named an *autocorrelation coefficient*:

$$r_{E_t E_{t-1}} = \frac{Cov\ (E_t, E_{t-1})}{(std\ dev\ E_t)(std\ dev\ E_{t-1})} \tag{10-1}$$

where the covariance between E_t and E_{t-1} measures the interdependence between the two streams of data

$$\text{cov}(E_t, E_{t-1}) = \frac{\sum_{t=2}^{n}(E_t - \bar{E}_t)(E_{t-1} - \bar{E}_{t-1})}{(n-2)}.$$ (10-2)

In general the degree-of-freedom in computing covariance is $(n-k-1)$. Obviously, the autocorrelation coefficient, often abbreviated as r_k, will range between 0 and 1 in absolute value. For the 1-lag coefficient

$$r_{E_t E_{t-1}} = \frac{\sum_{t=2}^{n}(E_t - \bar{E}_t)(E_{t-1} - \bar{E}_{t-1})}{\sqrt{\sum_{t=1}^{n}(E_t - \bar{E}_t)^2}\sqrt{\sum_{t=2}^{n}(E_{t-1} - \bar{E}_{t-1})^2}}.$$ (10-3)

Example

Given the following data set in **Table 10-2**, *please compute the autocorrelation coefficient for 1-lag and 2-lag* (Makridakis et al. 1983).

For this problem, the sum and mean were calculated for the E_{t-1} and E_{t-2} columns. Beyond these calculations, a sample standard-deviation was calculated for all three columns. To calculate the autocorrelation coefficient for lag-1 and for lag-2, it was necessary to use equations such as **(10-1)** and **(10-2)**. The sample standard deviations, calculated for E_t, E_{t-1}, and E_{t-2} as part of the process, are shown in **Table 10-2**. The calculation of the covariance for the one time-lag resulted in a value of -3.458 and a value of -4.250 for the two time-lag. When these values are substituted into Equation **(10-2)** and its analogue for two time-lags, the resultant values for the autocorrelation coefficient were $r_{E_t E_{t-1}} \approx -0.207$ and $r_{E_t E_{t-2}} \approx -0.257$ for the one- and two-lag respectively. Notice with a small series such as this, round off rules—particularly regarding the degree-of-freedom—have substantial effects on the numerical results. In general, the difference will be small for a large series.

Alternate solution: Assuming the series is lengthy enough and stationary, $\bar{E}_t = \bar{E}_{t-1} = \bar{E}$, and the two standard deviations can be estimated just once using all known data for E_t, then

$$r_{E_t E_{t-1}} = \frac{\sum_{t=2}^{n}(E_t - \bar{E}_1)(E_{t-1} - \bar{E})}{\sum_{t=1}^{n}(E_t - \bar{E})^2}.$$ (10-4)

This is the same as

$$r_{E_t E_{t-1}} = \frac{\sum_{t=1}^{n-1}(E_t - \bar{E})(E_{t+1} - \bar{E})}{\sum_{t=1}^{n}(E_t - \bar{E})^2}.$$ (10-5)

In general

Time t	Original series E_t	One-time-lag series E_{t-1}	Two-time-lag series E_{t-2}
1	13		
2	8	13	
3	15	8	13
4	4	15	8
5	4	4	15
6	12	4	4
7	11	12	4
8	7	11	12
9	14	7	11
10	12	14	7
sum	100	88	74
mean	10	9.78	9.25
covariance		−3.458	−0.425
autocorrelation		−0.207	−0.257

Table 10-2 Autocorrelation example

$$r_{E_t E_{t-k}} = \frac{\sum_{t=1}^{n-k}(E_t - \bar{E})(E_{t+k} - \bar{E})}{\sum_{t=1}^{n}(E_t - \bar{E})^2}. \tag{10-6}$$

Using this simplifying formula

$$r_1 = \frac{(13-10)(8-10)+(8-10)(15-10)+(15-10)(4-10)+\ldots+(14-10)(12-10)}{(13-10)^2+(8-10)^2+(15-10)^2+(4-10)^2+\ldots+(14-10)^2+(12-10)^2} \approx -0.188 \tag{10-7}$$

$$r_2 = \frac{(13-10)(15-10)+(8-10)(4-10)+(15-10)(4-10)+\ldots+(7-10)(12-10)}{(13-10)^2+(8-10)^2+(15-10)^2+\ldots+(14-10)^2+(12-10)^2} \approx -0.201. \tag{10-8}$$

While these differ from the previous values of −0.207 and −0.257, such differences will disappear for a lengthy time-series (Makridakis et al. 1983). □

   Autocorrelations can be used to find out whether there is any pattern in a set of data. It is useful to plot the autocorrelation coefficients as in **Figure 10-5** to see if any of them are significantly different from zero. The autocorrelations of stationary data drop to zero (in absolute value) after the second or third time lag,

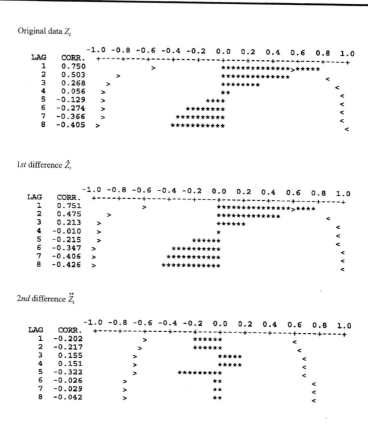

Figure 10-6 Autocorrelations for original and differenced data
(Makridakis et al. 1983)

while for a nonstationary series they are significantly different from zero for several time-periods.

The existence of a trend in the data means that successive values of the series will be positively correlated with each other. The autocorrelation for one time-lag, r_1, will be relatively large and positive. The autocorrelation for two time-lags also will be relatively large and positive, but not as large as r_1, because the random-error component has entered the picture twice, 'smoothing' out the correlation. Similarly, r_k for non-stationary data will be relatively large and positive, until k gets big enough so that the random error components begin to dominate the autocorrelation. This is illustrated in **Figure 10-5**, which contrasts sharply with the negative autocorrelations computed from **Table 10-2** for differenced data.

Stationarity example (continued)

Let us revisit the example contained in **Table 10-1**. While we have shown previously how to induce stationarity graphically, it is useful to confirm it analytically using autocorrelations. For the original data Z_t and the first difference $\dot{Z}_t$, the autocorrelations start out as high positive values and drop to zero and then become significantly negative. These are not what we would expect to see for stationary data. For $\ddot{Z}_t$, however, the autocorrelations are close to zero and by our best judgement seem to imply acceptable stationarity. The three sets of autocorrelations are presented side-by-side in **Figure 10-6** for comparison. □

A periodic pattern on the autocorrelations will suggest a seasonal pattern. In this case, it may be discerned that no autocorrelation coefficients except the seasonal ones are significantly different from zero. The cycle length can readily be picked out from such an autocorrelation plot. An example is shown in **Figure 10-7**, which shows a cycle length of 4 time periods. Similar analysis can be performed in the residual (error) terms A_t to see if they are random. The presence of seasonality, for example, will result in a periodic pattern in A_t. Standard procedures such as *smoothing* (a form of moving average) exist for removing seasonality—a point that we will come back to in sequel.

At this point, it is useful to introduce the *partial autocorrelation coefficient*. This is again analogous to the partial-correlation coefficient in regular regression. If E_t is regressed against E_{t-1} and E_{t-2}, the partial of E_{t-1} simply measures the explanatory power of E_{t-1} if the effects of E_{t-2} are "partialled out" first. An analogy is again drawn with stepwise regression, where an additional variable is added one at a time to the regression. Thus E_{t-1} is introduced first, E_{t-2} second, and so on. The coefficients $\hat{\varphi}_1$, $\hat{\varphi}_2$ etc. will be the partial-autocorrelation coefficients:

$$E_t = \hat{\varphi}_1 E_{t-1} + A_t, \tag{10-9}$$

$$E_t = \varphi_1 E_{t-1} + \hat{\varphi}_2 E_{t-2} + A_t, \tag{10-10}$$

$$E_t = \varphi_1 E_{t-1} + \varphi_2 E_{t-2} + \hat{\varphi}_3 E_{t-3} + A_t. \tag{10-11}$$

Equation **(10-11)** shows, for example, that if an additional variable E_{t-3} is added to an existing equation consisting of explanatory variables E_{t-1} and E_{t-2}, the coefficient of E_{t-3} is its partial-autocorrelation coefficient. Such partials help to determine the appropriate ARMA model. If the underlying process generating a given series is an AR(1) model, it should be understood that only $\hat{\varphi}_1$ will be significantly different from zero. If the true generating process is AR(2), then only $\hat{\varphi}_1$ and $\hat{\varphi}_2$ will be significant, and so on. While the number of significant partials indicates the order of an AR process, the presence of a single first-partial alone suggests non-stationary data. The calculations for the analogous partial correlation coefficients for linear regression are explained in the "Statistical Tools" appendix (Chan 2000), if we treat E_{t-1}, E_{t-2} as explanatory variables and E_t as a dependent

```
                    AUTOCORRELATION ANALYSIS


              -1 -.8 -.6 -.4 -.2   0  .2  .4  .6  .8   1
   LAG  VALUE  I...!...!...!...!...I...!...!...!...!...I
    1  -0.344                    *******
    2  -0.203                     *****
    3  -0.356                    ********
    4   0.764                       ***************
    5  -0.246                    ******
    6  -0.105                      ***
    7  -0.289                    *******
    8   0.572                       ***********
    9  -0.212                    *****
   10  -0.071                      **
   11  -0.234                    ******
   12   0.417                       *********
```

Figure 10-7 Periodic autocorrelation example (Makridakis et al. 1983)

variable. If the generating process is MA rather than AR, the partial autocorrelations will not suggest the order of the MA process, since they are constructed to fit an AR process. If the partial-correlation coefficients decline to zero exponentially, the series can best be explained by an MA process.

Another test that is useful in identifying an appropriate ARMA model is the *line spectra*. This is useful as a confirmation tool as well as in analyzing the periodicity of the data, and for that reason, it is also called the *periodogram*. It is well known that a seasonal time-series can be represented by a set of sine waves of various frequencies (wavelength), amplitudes and phases.[2] For example, sine waves of frequencies 1, 2, 3 ··· $\lceil n/2 \rceil - 1$ can be fitted to a time-series by simple least-squares techniques. (Here, $\lceil \cdot \rceil$ is the rounded-up integer of the argument $\cdot$.) For a discrete time-series where there is no phase, frequency or wavelength is simply the number of time units. The set of amplitudes for all fitted frequencies then forms the line spectrum.[3] Such a fit allows us to measure:

a. Randomness in the data series ('noise' or residual series);
b. Seasonality in a time series;
c. Predominance of positive or negative autocorrelations.

---

[2]In the case of discrete data, phase does not come into the picture.

[3]A more formal discussion is presented in the "Remote Sensing and Geographic Information System" chapter (Chan 2000) under Fourier transformation.

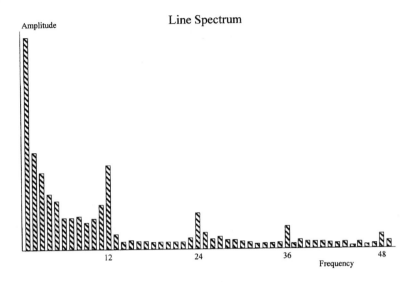

Figure 10-8 Example of seasonality in a periodogram (Makridakis et al. 1983)

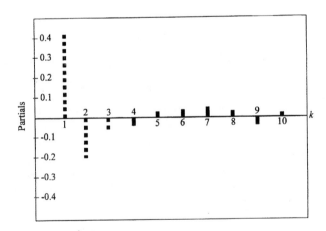

Figure 10-9 Partial correlation coefficients in the original data

In examining the positive or negative autocorrelation, one associates the dominance of low-frequency amplitudes to the former and high-frequency amplitude to the latter.

For the case where the autocorrelations and partial autocorrelations are not significantly different from zero, and the line spectrum is roughly uniform, the time-series can be modelled by an ARMA(0,0) model. As another example, a line spectrum peaking at the low frequency suggests a requirement for differencing. A

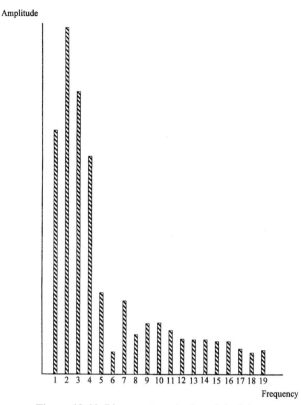

Figure 10-10 Line spectrum in the original data

stationary series, on the other hand, would have a dominant frequency that is neither high nor low. While seasonality can be detected by the autocorrelations, the observations can be confirmed by line spectrum as well. For example, if the amplitudes have a recurring pattern corresponding to x cycles over y data points, then a cycle length of y/x can be inferred. An example periodogram (line spectrum) is shown in **Figure 10-8**, where a seasonal pattern is discerned, with a frequency or wavelength of 12 months.

10.1.3 Example of an autoregressive time series. Analysis of the employment time-series data in **Figure 10-5** using autocorrelation shows nonstationarity. The autocorrelation indicates a gentle declining pattern over time, which does not go hand-in-hand with the hypothesis that the employment data E_t are stationary (see overlay in **Figure 10-5**.) The first partial in **Figure 10-9** is dominant, again suggesting nonstationarity. Finally, the line spectrum in **Figure 10-10** shows dominant slow waves, further confirming nonstationarity.

Taking the first difference of the data, $\Delta E_t = \dot{E}_t = E_t - E_{t-1}$, appears to be a good (and common) approach to induce stationarity. The differenced data, representing

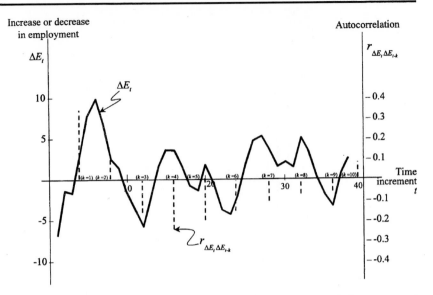

Figure 10-11 Differenced employment data and autocorrelation

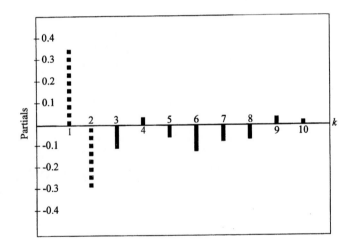

Figure 10-12 Partial autocorrelation coefficients of differenced data

the increase or decrease in employment over two consecutive time increments, is shown in **Figure 10-11**. The plot shows increases to be above the time axis and decreases below. (For data-processing convenience, it is advisable to lower the time axis to make every observation a positive number.) In the same plot is shown the autocorrelation of the differenced data, $r_{\Delta E_t \Delta E_{t-2}}$.

Autocorrelation of the differenced data shows a sine-wave pattern, and there are two significant partials. This points toward an autoregressive (AR) process of

Amplitude

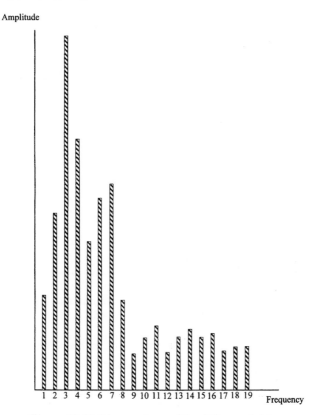

Frequency

Figure 10-13 Line spectrum of the differenced data

lag-2 observations, or model AR(2). The line spectrum in **Figure 10-12** and **Figure 10-13** has the frequencies covered more equitably. Analysis of the moving-ave rage (MA) pattern of the differenced data indicates 0-lag error terms, or the model is MA(0). The combined result is an ARMA(2,0) model using first-order differencing. This means a model of the following form:

$$E_t = constant + \varphi_1 E_{t-1} + \varphi_2 E_{t-2} + A_t. \tag{10-12}$$

When the data have been differenced to a stationary series, we denote such a series $\hat{z}_t$, rather than the original data Z_t (which in this case is the same as E_t).

10.1.4 Example of a moving-average time series. **Figure 10-14** shows a one-lag example of a general moving-average time series $\hat{z}_t$ = constant + $a_t - \theta_1 a_{t-1} - \theta_2 a_{t-2} - ... - \theta_q a_{t-q}$ (Makridakis et al. 1983). Note the line spectrum shows higher frequencies to be dominant. The autocorrelations show one large negative-value (the first one) and others that are not quite trivial. The partials show two large negative-values—but do not show a clear exponential decay. A series bearing the

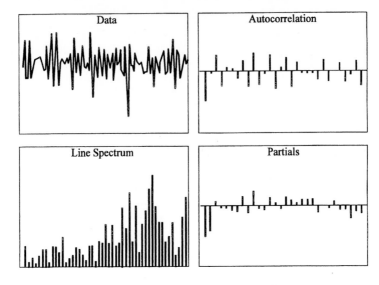

Figure 10-14 First illustration of a moving-average time series
(Makridakis et al. 1983)

equation $\hat{z}_t = 100 + a_t - 0.6a_{t-1}$ was fitted to the data. In theory, an ARMA(0,1) model
can be characterized by the following table where these characteristics of the autocorrelations, partials and spectrum are featured:

	$0 < \theta_1 < 1$	$-1 < \theta_1 < 0$
Autocorrelation	One (negative)	One (positive)
Partials	Exponential decay (all negative)	Exponential decay (alternating)
Spectrum	High frequencies dominate	Low frequencies dominate

This table shows that for an MA(1) model, one expects the coefficient θ to be
either positive or negative, but less than unity in absolute value. One significant
autocorrelation coefficient is expected, and the partials are supposed to decline
exponentially over the lags $k = 1, 2, 3, \ldots$ The partials will be alternating in the case
of a negative θ_1, suggesting that in a model such as $\hat{z}_t = 100 + a_t + 0.7a_{t-1}$, (where $\theta_1 =$
-0.7), the data series is "out-of-phase" with each other when shifted by one time
period. It turns out that for both first-order autoregressive and moving-average
processes to be stationary, similar conditions can be specified. For an AR(1)
process, $-1 < \varphi_1 < 1$; and for an MA(1) process, $-1 < \theta_1 < 1$.

Following this line of logic, we present another example of a MA(2) model.
Figure 10-15 shows an idealized second-order moving-average time-series. There

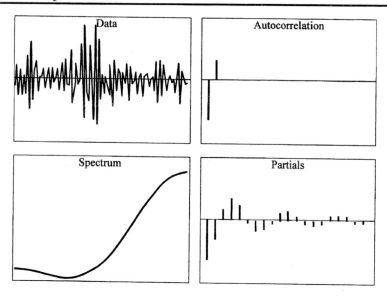

Figure 10-15 Idealized illustration of a moving-average lag-2 time-series
(Makridakis et al. 1983)

are two parameters, θ_1 and θ_2 that must obey the restrictions shown by the formulas
that govern either an AR(2) or MA(2) process:

$$\theta_1 + \theta_2 < 2 \qquad -\theta_1 + \theta_2 < 2$$
$$\varphi_1 + \varphi_2 < 2 \qquad -\varphi_1 + \varphi_2 < 2. \tag{10-13}$$

In addition, an AR(2) model must have $-2 < \varphi_1 < +2$ and $-1 < \varphi_2 < +1$. Likewise,
an MA(2) model must have $-2 < \theta_1 < +2$ and $-1 < \theta_2 < +1$. These conditions simply
show that for a two-lag series, the data must be scaled such that the dependent
variable $\hat{z}_t$ is estimated within bounds. Furthermore, the additional contribution from
a 2-lag term should be smaller than the one-lag term—roughly in the order of half as
much.

The plots show exactly two nonzero autocorrelations. The partials decay in a
damped sine wave, and that the spectrum—the theoretical equivalent of the line
spectrum in a continuous curve—shows the dominance of high frequencies. The
fitted model looks like $\hat{z}_t = a_t - a_{t-1} + 0.8 a_{t-2}$, where the mean of the series is zero. In
contrast, a pure AR(2) process, as illustrated in the previous employment growth
example, has exactly two nonzero partials, and the autocorrelations die out in a
damped sine wave. The spectrum stresses lower frequencies, but not the very lowest.

10.1.5 A mixed ARMA process. A synthesized ARMA(1,1) process is shown
in **Figure 10-16**. The autocorrelations and partials both decay exponentially, with
the partials alternating in sign. The model bears this form $\hat{z}_t = 0.3 z_{t-1} + a_t + 0.7 a_{t-1}$.

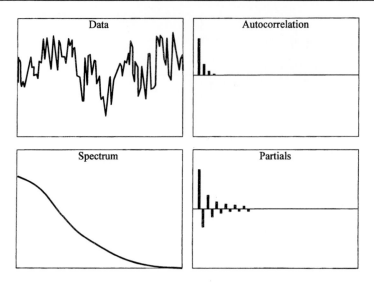

Figure 10-16 A mixed autoregressive and moving-average model
(Makridakis et al. 1983)

It is quite apparent that except for the idealized cases, identification of an ARMA model is a challenging task in general. In a sense, it is comforting to know that the family of models with $p \le 2$ and $q \le 2$ explains most of the processes. Consequently, the task is a combinatorially defined one, although the amount of trial-and-error can be substantial before arriving at a satisfactory model. Hoff (1983) provided some excellent guidelines for model identification. A summary of the theoretical autocorrelation and partial autocorrelation patterns associated with AR, MA, and ARMA models is given in **Table 10-3**.

It can be seen that identifying a proper ARMA model is an art. But a set of guidelines has been spelled out to make such a model-building exercise more systematic:

a. Plot the raw data Z_t, since a comprehensive examination of the data before analysis is the only way to provide the necessary insights for model building.

b. Compute the autocorrelations and partials of the original series. If they drop to or near zero quickly (say after the second or third periods), it suggests that the data are stationary. If they do not, nonstationarity is implied. Use the line spectrum to reinforce these findings. The amplitudes of the lowest frequencies will usually dominate in a nonstationary series.

c. Take the difference of the time series until stationarity is obtained. For most practical purposes a maximum of two differences will transform the data into a stationary series $\hat{z}_t$.

d. When stationarity has been achieved, examine the autocorrelations to see if anypattern remains: If autocorrelations for time lags of every cycle are large

Model	Parameter(s)	Auto-correlation	Partial auto-correlation
MA	q parameters of order $1, 2, \cdots, q$	Spikes at lags 1 through q; zeros elsewhere	Spikes decreasing exponentially, beginning at lag 1; for q grater than 1, damped sine waves are superimposed on the pattern
AR	p parameters of order $1, 2, \cdots, p$	Spikes decreasing exponentially, beginning at lag 1; for p greater than 1, damped sine waves are superimposed on the pattern	Spikes at lags 1 through p; zeros elsewhere
ARMA	q MA and p AR parameters of orders $1, \ldots, q$ and $1, \ldots, p$ respectively	Irregular pattern of spikes at lags 1 through q; remaining pattern as in AR autocorrelations	Irregular pattern of spikes at lags 1 through p; remaining pattern as in MA partial autocorrelations

Table 10-3 Identification of AR, MA and ARMA models (Hoff 1983)

and significantly different from zero, and amplitudes for appropriate sine waves are large, as shown in the line spectrum, seasonality is discerned. The combined patterns of autocorrelation, partials and the line spectrum will suggest the p and q values of the ARMA model.

Starting with stationary non-seasonal data, we estimate p and q by the following rules (Saaty and Vargas 1990):

a. For autoregressive AR (p) models, if the autocorrelation coefficients decrease in a steady fashion and the partial autocorrelations are different from zero at lags $k = 1, 2, \ldots, p^*$ and zero for $k > p^*$, the series can be described by an auto-regressive model AR(p^*).

b. For moving-average MA(q) models, if the autocorrelation coefficients corresponding to lags $k = 1, 2, \ldots, q^*$ are different from zero, and are zero for all $k > q^*$, and the partial-autocorrelation coefficients decrease in a steady fashion, then the series can be described by a model MA(q^*).

c. For ARMA(p, q) models, both the autocorrelations and the partial auto-correlations decrease exponentially to zero.

Should seasonality be detected, let s be the seasonality length. As mentioned, these models are formally denoted by SARMA$(p, q)_{s_p, s_q}$, where s_p and s_q denote the autoregressive season-length and moving-average season-length respectively. Use the behavior of the autocorrelations and partial autocorrelations to find the order of the seasonal SARMA(P,Q) model. The parameters P and Q here are estimated for the seasonal data using, as with non-seasonal data, the autocorrelations and partial autocorrelations of stationary data.

a. If the partial autocorrelations are significantly different from zero at the lags ks_p, $k = 1, 2, \ldots, P$; they are zero for $k > P$; and the autocorrelations tend to zero; then the model to use is a seasonal autoregressive model of order P.
b. If the autocorrelations are significantly different from zero at the lags ks_q, $k = 1, 2, \ldots, Q$; they are zero for $k > Q$; and the partial autocorrelations tend to zero; then use a seasonal moving-average model of order Q.
c. If the partial autocorrelations are significantly different from zero for lags of order less than or equal to P, and autocorrelations are significantly different from zero for lags of order less than or equal to Q, then use either a seasonal autoregressive or a seasonal moving-average model. If $P \ll Q$, use an AR(P), and if $Q \ll P$, then use a MA(Q). If P is almost the same as Q, then use either one, but not both.
d. If the autocorrelations and the partial autocorrelations are almost zero, eliminate the seasonal component by differencing.
e. If both the autocorrelations and the partial autocorrelations tend to zero quickly at the seasonal level, then use a seasonal SARMA(P,Q).

10.1.6 Estimating the parameters. Having made tentative model identification, the AR and MA parameters φ and θ have to be determined. Let us refer to Equations **(10-9)**, **(10-10)** and **(10-11)**, and tackle the φ first. The φ can theoretically be estimated incrementally from these equations, following the analogy from stepwise regression as discussed in the "Statistical Tools" appendix in Chan (2000). However, the process will be laborious at best. Here we will introduce a more efficient, recursive procedure. We still start with Equations **(10-9)**, **(10-10)** and **(10-11)** except now we will dispense with the 'hats' notation for the φ since we are no longer interested in the partial-correlation coefficients. Rather, we are interested in the values of the φ coefficients for a lag-1, lag-2 and lag-3 models etc.:

$$\hat{z}_t = \varphi_1 \hat{z}_{t-1} + a_t \tag{10-14}$$

$$\hat{z}_t = \varphi_1 \hat{z}_{t-1} + \varphi_2 \hat{z}_{t-2} + a_t \tag{10-15}$$

$$\hat{z}_t = \varphi_1 \hat{z}_{t-1} + \varphi_2 \hat{z}_{t-2} + \varphi_3 \hat{z}_{t-3} + a_t. \tag{10-16}$$

10.1.6.1 AUTOREGRESSIVE PARAMETERS. If both sides of Equation **(10-14)** are multiplied by $\hat{z}_{t-1}$, the result is $\hat{z}_{t-1}\hat{z}_t = \varphi_1\hat{z}_{t-1}\hat{z}_{t-1} + \hat{z}_{t-1}a_t$. Taking the expected value of this expression yields

$$E(\hat{z}_{t-1}\hat{z}_t) = \varphi_1 E(\hat{z}_{t-1}\hat{z}_{t-1}) + E(\hat{z}_{t-1}a_t). \tag{10-17}$$

Let the autocovariance of order-0 be $\mathrm{cov}(\hat{z}_t, \hat{z}_t)$ and the autocovariance of order-1 be $\mathrm{cov}(\hat{z}_t, \hat{z}_{t-1})$ and the autocovariance of order-k be $\mathrm{cov}(\hat{z}_t, \hat{z}_{t-k})$ in general. Here $\mathrm{cov}(\hat{z}_t, \hat{z}_t) = \mathrm{var}(\hat{z}_t)$ or the covariance of order-0 is simply the variance of the time series. Suppose the sample estimator approximates the population estimator, the autocorrelation coefficient of lag-1 is then $r_1 = \mathrm{cov}(\hat{z}_t, \hat{z}_{t-1}) / \mathrm{var}(\hat{z}_t)$ and the autocorrelation coefficient of lag-k is then $r_k = \mathrm{cov}(\hat{z}_t, \hat{z}_{t-k}) / \mathrm{var}(\hat{z}_t)$. Now since

$$E(\hat{z}_{t-1}\hat{z}_t) = \mathrm{cov}(\hat{z}_{t-1}\hat{z}_t) + E(\hat{z}_{t-1})E(\hat{z}_t) = \mathrm{cov}(\hat{z}_t, \hat{z}_{t-1}) + E(\hat{z}_t)E(\hat{z}_t),$$
$$E(\hat{z}_{t-1}\hat{z}_{t-1}) = \mathrm{cov}(\hat{z}_{t-1}\hat{z}_{t-1}) + E(\hat{z}_{t-1})E(\hat{z}_{t-1}) = \mathrm{var}(\hat{z}_t) + E(\hat{z}_t\hat{z}_t), \text{ and}$$
$$E(\hat{z}_{t-1}a_t) = \mathrm{cov}(\hat{z}_{t-1}a_t) + E(\hat{z}_t)E(a_t) = 0,$$

the above expected-value expression **(10-17)** can be rewritten as $\mathrm{cov}(\hat{z}_t, \hat{z}_{t-1}) = \varphi_1 \mathrm{var}(\hat{z}_t)$. If both sides are divided by $\mathrm{var}(\hat{z}_t)$, we obtain $r_1 = \varphi_1$. This agrees with our intuition that $r_1 = \mathrm{cov}(\hat{z}_t, \hat{z}_{t-1}) / \mathrm{var}(\hat{z}_t)$. It also agrees with our suggestion in Equation **(10-9)** that $\varphi_1 = \hat{\varphi}_1 = r_1$, or the first partial correlation is the same as the lag-1 autocorrelation, and it is the calibration coefficient φ_1 itself.

In general, since $r_k = \mathrm{cov}(\hat{z}_t, \hat{z}_{t-k}) / \mathrm{var}(\hat{z}_t)$, the above procedure can be extended. Multiplying both sides of Equation **(10-15)** by $\hat{z}_{t-1}$, take expected values, and dividing by $\mathrm{var}(\hat{z}_t)$, yields the following equation for lag-1 ($k = 1$):

$$r_1 = \varphi_1 \frac{\mathrm{cov}(\hat{z}_{t-1}\hat{z}_{t-1})}{\mathrm{var}(\hat{z}_t)} + \varphi_2 \frac{\mathrm{cov}(\hat{z}_{t-1}\hat{z}_{t-2})}{\mathrm{var}(\hat{z}_t)} + \frac{\mathrm{cov}(\hat{z}_{t-1}a_t)}{\mathrm{var}(\hat{z}_t)} \tag{10-18}$$

The first term above simplifies to φ_1 as before, and the third term is again zero. Thus $r_1 = \varphi_1 + \varphi_2 r_1$. Another equation can be generated from Equation **(10-15)** by multiplying by $\hat{z}_{t-2}$, take expected values, and dividing by $\mathrm{var}(\hat{z}_t)$: $r_2 = \varphi_1 r_1 + \varphi_2$. Together, they form a simultaneous equation set

$$r_1 = \varphi_1 + \varphi_2 r_1$$
$$r_2 = \varphi_1 r_1 + \varphi_2 \tag{10-19}$$

from which the parameters φ_1, φ_2 can be solved in terms of the autocorrelations r_1 and r_2:

$$\varphi_1 = \frac{r_1(1 - r_2)}{(1 - r_1^2)} \qquad \varphi_2 = \frac{r_2 - r_1^2}{1 - r_1^2}. \tag{10-20}$$

The procedure can be generalized to include the third equation **(10-16)**, resulting in

$$r_1 = \varphi_1 + \varphi_2 r_1 + \varphi_3 r_2$$
$$r_2 = \varphi_1 r_1 + \varphi_2 + \varphi_2 r_1 \tag{10-21}$$
$$r_3 = \varphi_1 r_2 + \varphi_2 r_1 + \varphi_3.$$

Generalization of this procedure leads to the *Yule-Walker equations* (Makridakis et al. 1983, Hoff 1983). The analogy to stepwise regression is quite transparent if one reviews the "Statistical Tools" appendix in Chan (2000) .

10.1.6.2 MOVING-AVERAGE PARAMETERS. For simplicity of presentation, consider a moving-average model

$$z_t = a_t - \theta_1 a_{t-1} - \theta_2 a_{t-2} - \theta_3 a_{t-3} - \ldots - \theta_q a_{t-q} \tag{10-22}$$

in which the mean has been taken out from the stationary series $\{\hat{z}_t\}$, thus transforming it into the 'sanitized' series $\{z_t\}$ (Makridakis et al. 1983, Box and Jenkins 1976). The autocovariance function $cov(z_t z_{t-k})$ of the series is then

$$E[(a_{t-1} - \theta_1 a_{t-1} - \ldots - \theta_q a_{t-q})(a_{t-k} - \theta_1 a_{t-k-1} - \ldots - \theta_q a_{t-k-q})]. \tag{10-23}$$

The variance of the process, $var(z_t)$, is correspondingly

$$(1 + \theta_1^2 + \theta_2^2 + \ldots + \theta_q^2)\sigma_1^2. \tag{10-24}$$

$cov(z_t z_{t-k})$ can be re-written as

$$(-\theta_k + \theta_1\theta_{k+1} + \theta_2\theta_{k+2} + \ldots + \theta_{q-k}\theta_q)\sigma_a^2; \qquad k = 1, 2, \ldots, q, \tag{10-25}$$

and assumes the value of zero for $k > q$. When sample estimates can be used to approximate population estimates, the autocorrelation function r_k can be expressed in terms of $cov(z_t z_{t-k}) / var(z_t)$, or for $k \le q$

$$r_k = \frac{-\theta_k + \theta_1\theta_{k+1} + \ldots + \theta_{q-k}\theta_q}{1 + \theta_1^2 + \ldots + \theta_q^2}. \tag{10-26}$$

For an MA(1) process where $q = 1$, the above equation reduces to

$$r_1 = \begin{cases} \dfrac{-\theta_1}{1 + \theta_1^2} & k = 1 \\[2mm] 0 & k \ge 2. \end{cases} \tag{10-27}$$

Solving for θ in terms of r_1, we have a quadratic equation $\theta_1^2 + \theta_1/r_1 + 1 = 0$. From this equation we will pick the appropriate root that lies between -1 and $+1$, the expected calibration value for θ. For an MA(2) process, or $q = 2$, Equation **(10-26)** reduces to

$$r_1 = \frac{-\theta_1(1 - \theta_2)}{1 + \theta_1^2 + \theta_2^2}$$

$$r_2 = \frac{-\theta_2}{1 + \theta_1^2 + \theta_2^2} \tag{10-28}$$

$$r_k = 0 \qquad k \ge 3.$$

While these constitute two equations and two unknowns, the θ cannot be solved in a straightforward manner. Numerical procedures have to be employed to arrive at the final calibration values for θ. For convenience, Box and Jenkins (1976) provide charts for their graphical determination.

10.1.6.3 MIXED ARMA MODELS. To estimate the parameters for a fixed ARMA model, we combine the autoregressive series

$$z_t = \varphi_1 z_{t-1} + \varphi_2 z_{t-2} + \varphi_3 z_{t-3} + \ldots + a_t \qquad \text{(10-29)}$$

with the moving-average process (10-22) (Makridakis et al. 1983). As before, we post-multiply by z_{t-k} and take the expected value. Covariance of lag-k can now be written as

$$\begin{aligned}
\mathrm{cov}(z_t z_{t-k}) = &\ \varphi_1 E(z_t z_{t-k}) + \ldots + \varphi_p E(z_{t-p} z_{t-k}) + E(a_t z_{t-k}) \\
&\ - \theta_1 E(a_{t-1} z_{t-k}) - \ldots - \theta_q E(a_{t-q} z_{t-q}).
\end{aligned} \qquad \text{(10-30)}$$

If $k > q$, the terms $E(a_t z_{t-k}) = 0$, which leaves

$$\mathrm{cov}(z_t z_{t-k}) = \varphi_1 \, \mathrm{cov}(z_t z_{t-k+1}) + \ldots + \varphi_p \, \mathrm{cov}(z_t z_{t-k+p}). \qquad \text{(10-31)}$$

When $k < q$, the past errors and the z_{t-k} will be correlated and autocovariances will be affected by the moving-average part of the process, requiring that it be included. Take an ARMA(1,1) precess, for $k = 0$, Equation (10-30) becomes

$$\begin{aligned}
\mathrm{cov}(z_t z_t) = &\ \varphi_1 \mathrm{cov}(z_t z_{t-1}) + \\
&\ E[(\varphi_1 z_{t-1} + a_t - \theta_1 a_{t-1}) a_t] - \theta_1 E[(\varphi_1 z_{t-1} + a_t - \theta_1 a_{t-1}) a_{t-1}]
\end{aligned} \qquad \text{(10-32)}$$

since $z_t = \varphi_1 z_{t-1} + a_t - \theta_1 a_{t-1}$. Now

$$\mathrm{cov}(z_t z_t) = \mathrm{var}(z_t) = \varphi_1 \mathrm{cov}(z_t z_{t-1}) + \sigma_a^2 - \theta_1(\varphi_1 - \theta_1)\sigma_a^2. \qquad \text{(10-33)}$$

Similarly, if $k = 1$

$$\mathrm{cov}(z_t z_{t-1}) = \varphi_1 \mathrm{cov}(z_t z_t) - \theta_1 \sigma_a^2. \qquad \text{(10-34)}$$

Solving Equations (10-33) and (10-34) for $\mathrm{cov}(z_t z_t)$ and $\mathrm{cov}(z_t z_{t-1})$ yields

$$\mathrm{cov}(z_t z_t) = \mathrm{var}(z_t) = \frac{1 + \theta_1^2 - 2\varphi_1 \theta_1}{1 - \varphi_1^2} \qquad \text{(10-35)}$$

$$\mathrm{cov}(z_t z_{t-1}) = \frac{(1 - \varphi_1 \theta_1)(\varphi_1 - \theta_1)}{1 + \theta_1^2 - 2\varphi_1 \theta_1}. \qquad \text{(10-36)}$$

Dividing Equation (10-36) by Equation (10-35),

$$r_1 = \frac{(1 - \varphi_1 \theta_1)(\varphi_1 - \theta_1)}{1 + \theta_1^2 - 2\varphi_1 \theta_1}. \qquad \text{(10-37)}$$

Finally, if $k = 2$, the autocovariance function from Equation **(10-30)** when divided by the var(z_t) yields the correlation equation $r_2 = \varphi_1 r_1$ or

$$\varphi_1 = \frac{r_2}{r_1}. \tag{10-38}$$

From Equations **(10-37)** and **(10-38)**, parameters φ and θ can be estimated from the autocorrelations r_1 and r_2 numerically through an iterative procedure. Unfortunately, the procedure can be quite laborious in practice.

Example 1

Suppose the autocorrelations from an ARMA(1,1) series are $r_1 = 0.77$ and $r_2 = 0.368$. Then according to Equation **(10-38)** $\varphi_1 = 0.368/0.77 = 0.478$. Estimating θ_1 must be done iteratively by starting with an initial θ_1 value, checking if it satisfies Equation **(10-37)**. If not, a subsequent value must be derived and tried. A converging numerical scheme here finally arrives at -1.09, which satisfies Equation **(10-37)**:

$$0.77 = \frac{(1 - (0.478)(-1.09))(0.478 - (-1.09))}{1 + (-1.09)^2 - 2(0.478)(-1.09)}. \tag{10-39}$$

The time-series now assumes the calibrated form $z_t = 0.478 z_{t-1} + a_t + 1.09 a_{t-1}$ (Makridakis et al. 1983). □

Example 2

To illustrate these guidelines, let us revisit our favorite example from **Table 10-1**. Remember two differencing operations were needed to induce stationarity in the original data, resulting in the stationary time-series $\ddot{z}_t$. Now we shall illustrate an intuitive way to calibrate an ARMA model. We start out with a model of this form: $\hat{z}_t = \text{constant} + \varphi_1 \hat{z}_{t-1} + \varphi_2 \hat{z}_{t-2} + a_t$. Fitting an AR model with one and two lags, we find that the coefficients φ are not significantly different from zero. The best-fit model is $\hat{z}_t = \text{constant} + a_t$, where the constant term is 0.964, representing the average of the stationary series here. Thus an AR(0) model is proposed. This is confirmed by the partial-autocorrelation coefficients. One single significant value at $k = 1$ for both Z_t and $\dot{Z}_t$ were obtained, which re-confirm nonstationarity; but the $\ddot{Z}_t$ or $\hat{z}_t$ series yields no significant partials, suggesting the AR(0) model again.

Now we compute the residuals from $a_t = \hat{z}_t - \bar{z}_t$, where $\bar{z}_t = 0.964$ is the mean of the stationary series as mentioned. We then regress a model of this form $\hat{z}_t = \bar{z}_t + a_t - \theta_1 a_{t-1} - \theta_2 a_{t-2}$. To perform this conveniently, we rewrite the subject equation in the following form: $\hat{z}_t - 0.964 = a_t - \theta_1 a_{t-1} - \theta_2 a_{t-2}$. The coefficients θ turn out to be insignificant. The final model is therefore of the MA(0) form, assuming once again the equation $\hat{z}_t = 0.964 + a_t$.

Overall, we can decide unequivocally upon an ARMA(0,0) model. This makes tremendous sense from the plot of the data $\ddot{Z}_t$ in Figure 4. Also on reflection, both auto- and partial autocorrelations are not significantly different from zero for $\hat{z}_t$. The line spectrum (where computed) is shown to be uniform. We conclude that there is no explanatory power from this time series, other than an inherent parabolic growth trend. This is observed when we 'recover' the original series Z_t from the intercept constant 0.964. □

10.1.7 Formal notations. Suppose stationary data z_t have been obtained 'hovering' symmetrically around the time axis. A model such as the one shown in Equation **(10-12)** can be represented more formally as

$$(1 - \varphi_1 B - \varphi_2 B^2) z_t = a_t \qquad (10\text{-}40)$$

where B is the backward-shift operator $Bz_t = z_{t-1}$. In other words, B, operating on time series z_t, shifts the data back one period. Two applications of B to z_t shifts the data back 2 periods: $B(Bz_t) = B^2 z_t = z_{t-2}$. φ_j is the jth autoregressive coefficient or parameter. Thus a lag-1 model can be represented by the operator $(1 - \varphi_1 B)$ and a lag-2 model $(1 - \varphi_1 B - \varphi_2 B^2)$. z_t is the difference between two numbers. The first number is an appropriately transformed (to take care of nonstationary variance) and differenced (to take care of nonstationary means) datum. The second number is the mean of that series. For example, the employment data would be transformed and differenced into $\dot{E}_t$ first, and then z_t computed according to

$$z_t = \dot{E}_t - c \qquad (10\text{-}41)$$

where c is the average of $\dot{E}_t$. a_t is the error term at time t. Since there is no time shift necessary on the error term, it appears by itself without a moving-average parameter θ. Should there be a shift, the operator $\theta(B)$ is used to indicate this. For example, moving-average process of order 1, MA(1), can be written as $(1 - \theta_1 B)\alpha_1$. A combined ARMA(2,1) model will look like

$$(1 - \varphi_1 B - \varphi_2 B^2) z_t = (1 - \theta_1 B) a_t. \qquad (10\text{-}42)$$

The notation using backshift operator is convenient, but it is more functional to write it out similar to a regression equation for forecasting purposes. For example, the ARMA(2,1) model will look like $z_t = \varphi_1 z_{t-1} + \varphi_2 z_{t-2} + a_t - \theta_1 a_{t-1}$. In general, we represent the ARMA(p,q) model as

$$\varphi(B) z_t = \theta(B) a_t \qquad (10\text{-}43)$$

where z and a are the 'sanitized' numbers. It can be shown that these backshift operators can be manipulated algebraically:

$$z_t = \frac{\theta(B)}{\varphi(B)} a_t. \qquad (10\text{-}44)$$

Take the ARMA(2,0) model, $\theta(B) = 1$ and $\varphi(B) = 1 - \varphi_1 B - \varphi_2 B^2$, which results in Equation **(10-40)** as expected. As another example, $\theta(B) = (1 - \theta_1 B) a_t$ and $\varphi(B) = 1 - \varphi_1 B - \varphi_2 B^2$ yield Equation **(10-42)**.

10.1.8 Forecasting. It is possible, but not very easy, to calculate a simultaneous "prediction region" for a set of future observations for a time series. Here we show how to predict z_{t+l} as a linear combination of past observations $z_t, z_{t-1}, \dots$. The forecast is made one period at a time through l periods ahead. Then we will look at a "one-period-ahead" way to forecasting in order to operationalize the results so derived. The discussions below parallel that of a *prediction interval* for a least-squares-ordinary regression, as documented in the "Statistical Tools" appendix in Chan (2000). To forecast, an ARMA model incorporates both the autoregressive series $z_t(l) = \varphi_0 z_t + \varphi_1 z_{t-1} + \varphi_2 z_{t-2} + \dots$ and the moving-average series $z_t(l) = \theta_0 a_t + \theta_1 a_{t-1} + \theta_2 a_{t-2} + \dots$. Similar to least-squares regression, the coefficients φ and θ are usually estimated by minimizing the mean square error $E(z_{t+l} - z_t(l))^2$. In other words, the difference between estimated and predicted values is to be minimized.

Using the ψ weights in $\psi(B) = \theta(B)/\varphi(B)$ to express the ARMA model, estimated values can be obtained from Equation **(10-44)**:

$$z_{t+l} = a_{t+l} + \psi_1 a_{t+l-1} + \dots + \psi_{l-1} a_{t+1} + \psi_l a_t + \psi_{l+1} a_{t-1} + \dots . \qquad \textbf{(10-45)}$$

To clarify the subscripts of this equation, it is best to write out several forecast series:

$$z_{t+1} = a_{t+1} + \psi_1 a_t$$
$$z_{t+2} = a_{t+2} + \psi_1 a_{t+1} + \psi_2 a_t$$
$$z_{t+3} = a_{t+3} + \psi_1 a_{t+2} + \psi_2 a_{t+1} + \psi_3 a_t.$$

The mean square error can now be expressed as

$$E[z_{t+l} - z_t(l)]^2 =$$
$$E[a_{t+l} + \psi_1 a_{t+l-1} + \dots + \psi_{l-1} a_{t+1} + (\psi_l - \theta_0) a_t + (\psi_{l+1} - \theta_1) a_{t-1} + \dots]^2. \qquad \textbf{(10-46)}$$

As a function of θ_j, the error series above is minimized if $\theta_j = \psi_{l+j}$ $(j = 0,1,2,\dots)$. Thus the minimum-error forecast is $z_t(l) = \psi_l a_t + \psi_{l+1} a_{t-1} + \dots$. The forecast error is the difference between estimated and theoretical forecast values:

$$\varepsilon_t(l) = z_{t+l} - z_t(l) = a_{t+l} + \psi_1 a_{t+l-1} + \dots + \psi_{l-1} a_{t+1} \qquad \textbf{(10-47)}$$

with a variance of $\sigma_a^2(1 + \psi_1^2 + \psi_2^2 + \dots + \psi_{l-1}^2)$.

In other words, the forecast error is a linear combination of the unobservable future shocks or noise $a_{t+1}, \dots, a_{t+l}$ entering the system after time t. These shock terms can be estimated one at a time through a recursive difference equation as we forecast one period ahead at a time, as soon as z_{t+1} becomes available: $a_{t+1} = z_{t+1} - z_t(1)$. The t-origin forecast of z_{t+l+1} can then be updated to become the $t+1$ origin forecast of the same z_{t+l+1}. This is done by adding a constant multiple of the one-step-ahead forecast error a_{t+1}, with multiplier ψ_l:

$$z_{t+1}(l) = z_t(l+1) + \psi_l a_{t+1} \qquad l = 1, 2, \dots, L-1. \qquad \textbf{(10-48)}$$

The above difference equation is obtained by subtracting the series $z_t(l+1)$ from the series $z_{t+1}(l)$.

Similar to a regression equation, the $100(1-\alpha)\%$ confidence limit for $z_{t+l}(\pm)$ is thus given by

$$z_t(l) \pm \eta_{\alpha/2} \left[1 + \sum_{j=1}^{l-1} \psi_j^2 \right]^{1/2} \sigma_a \qquad (10\text{-}49)$$

Here $\eta_{\alpha/2}$ is the $100(1-\alpha/2)$ percentile of the standard normal-distribution, σ_a^2 is an estimate of the random shock variance to be included as a part of the input, and ψ is the parameter of the random shock form of the difference equation. It can be verified that the confidence limit/interval becomes bigger as l increases. In other words, there is more uncertainty as one forecasts further into the future.

Example 1

Let us say that σ_a can be approximated by $s_a = 0.134$ in a large data-series, a 95-percent confidence-limit (interval) is sought, meaning that $\eta_{0.05/2} = 1.96$. Suppose we wish to forecast two periods ahead ($l = 2$), the corresponding term in Equation (10-49) is $\sum_{j=1}^{2-1} \psi_j^2 = \psi_1^2 = 1.8^2$. The 95% confidence-limit (or prediction interval) is $z_t(2) \pm 1.96 (1+1.8^2)^{1/2} (0.134) = z_t(2) \pm 0.55$ (Box and Jenkins 1976). □

Example 2

A calibration output for the univariate ARMA case-study will be presented later in this chapter. In the figure that contains the output (**Figure 10-19**), column 5 is the confidence limit for each period of forecast into the future. Here, $s_a = 6.987$. The $100(1-\alpha)$ confidence interval for a forecast three periods ahead ($l = 3$) is $(1.96) [1 + \sum_{j=1}^{3-1} \psi_j^2 (= \psi_1^2 + \psi_2^2)]^{1/2} (6.987) = (1.96) [1+0.24^2+0]^{1/2} (6.987) \approx 14.09$. □

10.2 Multivariate time-series analysis

The discussion thus far concentrates on univariate time-series, multivariate autoregressive moving-average (MARMA) models deal with more than one time series. It will be visited in this section as a stop-over point to spatial time-series. An example of MARMA may be the bivariate case, where the 'input' time series is employment in a study area and the 'output' time series is the population. Besides investigating the 'trend' and pattern in each individual time-series, the relationship between the two time-series is also of interest. One can think of this type of model as a combination of time-series with causal analysis.

10.2.1 Transfer function. Again, the best way to introduce this subject is via an example (Makridakis et al. 1983). Let E_t be the employment data and N_t be the population data, a transfer function maps E to N in the presence of noise. **Table 10-4** gives the employment in each time-period and how the population responds in each of the time periods. For example, the 50 jobs introduced in period 1 resulted in the

pe-r-iod	em-ploy*	1	2	3	4	5	6	7	8	9	10	11	12	13	14	15	16	17	18	19	20	pop*
1	5																					0
2	3	5																				5
3	9	25	3																			28
4	6	10	15	9																		34
5	5	5	6	45	6																	62
6	7	5	3	18	30	5																61
7	3		3	9	12	25	7															56
8	4			9	6	10	35	3														63
9	8				6	5	14	15	4													44
10	7					5	7	6	20	8												46
11	8						7	3	8	40	7											65
12	1							3	4	16	35	8										66
13	3								4	8	14	40	1									67
14	3									8	7	16	5	3								39
15	7										7	8	2	15	3							35
16	3											8	1	6	15	7						37
17	6												1	3	6	35	3					48
18	4													3	3	14	15	6				41
19	1														3	7	6	30	4			50
20	3															7	3	12	20	1		43
	emp	50	30	90	60	50	70	30	40	80	70	80	10	30	30	70	–	–	–	–	–	

*Figures are shown in units of 10's. For example, 50 jobs (instead of 5 jobs) are introduced to the study area in period 1.

Table 10-4 Data base describing the relationship between employment and population over time

migration of population over the next several time-periods. The data in **Table 10-4** shows a population of 50 generated in period 2, a population of 250 in period 3, 100 in period 4, 50 in period 5 and finally 50 in 6.

An examination of the population distribution over the 6 periods following the introduction of employment, including the period in which employment was introduced, results in the following distribution: 0% on period 1, 10% on period 2, 50% on period 3, 20 on 4, 10 on 5, and the remaining 10 on 6. This distribution is repeated for all the employment injected into the study area over the 20 periods. A distribution function can therefore be introduced to model this, which is often called *transfer-function* weights. The transfer function can be written as

$$N_t = d_0 E_t + d_1 E_{t-1} + \ldots + d_5 E_{t-5}$$
$$= (d_0 + d_1 B + d_2 B^2 + \ldots + d_5 B^5) E_t \qquad (10\text{-}50)$$
$$= d(B) E_t$$

where $d_0 = 0$, $d_1 = 0.1$, $d_2 = 0.5$, $d_3 = 0.2$, $d_4 = 0.1$ and $d_5 = 0.1$. In general, the $d(B)$ transfer-function is of order k, instead of 5:

$$d(B) = d_0 + d_1 B + \ldots + d_k B^k = N_t / E_t. \qquad (10\text{-}51)$$

This bivariate transfer-function model can be represented in general form as

$$N_t = d(B) E_t + v_t, \qquad (10\text{-}52)$$

where v_t is the combined effects of all other factors influencing N_t (called the *noise*). These include the individual-time-series AR-coefficients φ and MA-coefficients θ with their backward-shift operators B of order p_v and q_v respectively. To effect simple model-representation, we consider the properly transformed stationary-series for N_t and E_t denoted by n_t and e_t. The v_t term can now be written as v_t or $[\theta(B)/\varphi(B)]a_t$ similar to Equation (10-45): $v_t = [\theta(B)/\varphi(B)]a_t$. Equation (10-52) shows then the population time-series is explained by a combination of the employment series through the transfer function as well as the combined effects of an ARMA(p_v, q_v) model.

Similar to v_t or a regular time-series, the transfer-function $d(B)$ in Equation (10-51) can also be represented in a more parsimonious form $[\omega(B)/\delta(B)]$. Here $\omega(B)\omega_0 - \omega_1 B - \omega_2 B^2 - \ldots - \omega B^s$ and $\delta(B) = 1 - \delta_1 B - \delta_2 B^2 - \ldots - \delta_r B^r$. The ratio $\omega(B)/\delta(B)$ can then be expanded into a series of k terms, as illustrated by the following numerical example (Makridakis et al. 1983).

$$\frac{\omega(B)}{\delta(B)} = \frac{(1.2 - 0.5B)}{(1 - 0.8B)} = 1.2 + 0.46B + 0.368B^2 + 0.294B^3 + \ldots . \qquad (10\text{-}53)$$

With the appropriate coefficients for $\omega(B)$ and $\delta(B)$, the weights of the employment-population transfer-function $0 + 0.1B + 0.5B^2 + 0.2B^3 + 0.1B^4 + 0.1B^5$ can be reproduced. The combined model will look like

$$n_t = \frac{\omega(B)}{\delta(B)} e_t + \frac{\theta(B)}{\varphi(B)} a_t. \qquad (10\text{-}54)$$

10.2.2 Prewhitening. Notice that the orders r and s are usually much smaller than k, hence achieving parsimony. The constants (r,s,b), similar to (p,q), represent the parameterization of the transfer-function model. Remember that b stands for a delay of b periods before individual employment e begins to influence population n. Here the input and output series are normally transformed and differenced similar to z_t in Equation **(10-41)** and the ARMA discussions above. To identify a transfer function, it will be useful to also *prewhiten* the input series in the same way that we induce stationarity in ARMA. By removing the fluctuating employment injections as "white noise," or by eliminating all known patterns, from the residuals the transfer function can be discerned a lot more easily. In the example used above, this means streamlining the employment input to be 'constant'.

To do this, consider the input series z_t of ARMA(p,q):

$$\varphi_z(B)\,z_t = \theta_z(B)\,\alpha_t, \qquad\qquad\qquad\qquad (10\text{-}55)$$

where $\varphi_z(B)$ is the autoregressive operator and $\theta_z(B)$ is the moving-average operator for the input series, following the notation of Equation **(10-43)**. α_t is a random-shock term or white-noise term, and is introduced for this purpose:

$$\alpha_t = \frac{\varphi_z(B)}{\theta_z(B)}\,z_t. \qquad\qquad\qquad\qquad (10\text{-}56)$$

The end result is a re-write of Equation **(10-52)**

$$\beta_t = d(B)\alpha_t + \frac{\varphi_z(B)}{\theta_z(B)}\,\upsilon_t = d(B)\alpha_t + \varepsilon_t, \qquad\qquad\qquad (10\text{-}57)$$

where β is the estimated population from the prewhitened employment-series and can be equivalenced to $\left[\varphi_z(B)/\theta_z(B)\right]n_t$. ε_t is the transformed noise-series, which is presumed to be unrelated to the α_t series. Equation **(10-57)** can be thought of as the equivalent operation on the transfer function as differencing and transforming the input and output series in ARMA. Analogous to forecasting from 'sanitized' ARMA models, the same 'whitening' procedure needs to be applied toward the output series. This will preserve the integrity of the functional relationship between input and output. The procedure will not necessarily convert the output series to white noise, however, since the whitening procedure described above is geared specifically toward the input series.

10.2.3 Transfer function weights. In bivariate MARMA, cross correlation is a key feature of the model, with autocorrelation playing a secondary role. Similar to regular correlation, the cross-correlation is usually computed for one series N_{t+k} *lagged* on the other E_t by k periods and vice versa. The cross-covariances are computed according to the following formula:

$$C_{EN}(k) = \frac{1}{n}\sum_{t=1}^{n-k}(E_t - \bar{E})(N_{t+k} - \bar{N}). \qquad\qquad\qquad (10\text{-}58)$$

Depending on the form of the data given, the equation above can be rewritten. We can refer to the 'standardized' data z_t instead of the parenthesized capitalized data terms E_t and N_t, where $z_t = (Z - \bar{Z})/s_Z$. The same goes with the prewhitened data. The cross-covariances can be converted to cross-correlations by dividing by two standard deviations:

$$r_{\alpha\beta}(k) = \hat{\rho}_{EN}(k) = \frac{C_{EN}(k)}{\sqrt{C_{EE}(0)\, C_{NN}(0)}} = \frac{C_{EN}(k)}{s_E s_N} \tag{10-59}$$

where α and β are prewhitened input and output series, r is the sample cross-correlation and $\hat{\rho}$ is the population cross-correlation. We are now ready to estimate the impulse response weights for Equation **(10-52)**:

$$d_k = \frac{r_{\beta\alpha}(k)\, s_N}{s_E}. \tag{10-60}$$

This equation can be shown to be mathematically equivalent to the weight in Equation **(10-52)**.

To show this, multiply both sides of Equation **(10-57)** by α_{t-k} and take the expected values:

$$E(\alpha_{t-k}\beta) = \{d_0 E(\alpha_{t-k}\alpha_t) + \dots + d_k E(\alpha_{t-k}^2) + \dots\} + E(\alpha_{t-k}\varepsilon_t).$$

Since $E(\alpha_{t-k}\alpha_{t-j}) = 0$ for $j \neq k$ and $E(\alpha_{t-k}\varepsilon_t) = 0$ (as α_t and ε_t are uncorrelated), this yields $C_{\alpha\beta}(k) = d_k \sigma_\alpha^2$. Thus

$$d_k = \frac{C_{\alpha\beta}(k)}{\sigma_\alpha^2} = \frac{\sigma_\beta}{\sigma_\alpha} \frac{C_{\alpha\beta}(k)}{\sigma_\alpha \sigma_\beta} = \frac{\sigma_\beta}{\sigma_\alpha} \hat{\rho}_{\alpha\beta}(k); \qquad k = 0, 1, 2, \dots, \tag{10-61}$$

where σ_β^2 is the variance of the β process. The coefficient d_k is therefore proportional to the cross-correlation $\hat{\rho}_{\alpha\beta}(k)$ between α_t and β_t at lag k, which is the same as $\hat{\rho}_{EN}(k)$. Hence Equation **(10-60)** results when one is willing to approximate population standard-deviation σ with sample standard-deviation s and population cross-correlation $\hat{\rho}$ with sample cross-correlation r.

10.2.4 Summary. Now that we have discussed the various components of a MARMA model, it will be useful to summarize the entire model-building procedure in terms of several steps. Notice these steps are nothing more than a simple generalization of the univariate ARMA-modelling procedure:

a. *Identification* of a model from the data-base, such as through autocorrelation and backshift operator on transformed, differenced and whitened data;

b. *Estimation*, or the fitting of the tentative model, calibrates the values of the parameters θ, φ, ω, δ in the integrated MARMA model above;

c. *Diagnostic checking*, or the verification part of modelling, tests the adequacy of the model for possible improvement; this includes the ability of the model to replicate observed data. A numerical example of the transfer-function model is included in Section 10.5.4 of this chapter.

10.3 Space–time modelling

In general, multivariate time-series consisting of quite a few more variables than two are the norms, rather than the exception. In the same example cited above, it may be useful—for the purpose of this book—to examine space–time, autoregressive moving-average (STARMA) model. Thus for an urban four-zone study area, it may be of interest to examine 8 time-series, corresponding to the population and employment in each of the 4 zones. The STARMA model examines the relationship between two items: the current population or employment as a linear combination of past observations, and observations at neighboring sites. In other words, the cross correlation of employment or population time series is examined between zones as well as within the same zone.

10.3.1 Spatial weights. Pfeifer and Deutsch (1980) gave an excellent illustration of this type of modelling. **Figure 10-17** depicts a hierarchical ordering of spatial neighbors. Represented in a spatial grid, first order neighbors are "closest" to the site of interest. Second-order neighbors are next, then third-order and so forth. In a four-zone example, we are only dealing with first- and second-order relationships (see **Figure 10-17**).

Suppose an autoregressive model holds between *every* site and its neighbors in both time and space, then "spatial stationarity" is established. A spatial-lag operator is a simple extension of the concept established previously for the backshift operator in the time domain:

$$L^{(l)}z_i(t) = \sum_{j=1}^{n'} w_{ij}^{(l)} z_j(t),$$ (10-62)

where the weights sum to unity:

$$\sum_{j=1}^{n'} w_{ij}^{(l)} = 1 \qquad i=1,\dots,n'; \qquad l=1,\dots,K.$$ (10-63)

Here $\mathbf{W}^{(l)} = \left[w_{ij}^{(l)}\right]$, with $\left[w_{ij}^0\right] = \mathbf{I}$. The weights are defined only if sites i and j are lth-order neighbors. Obviously, a $0th$-lag operator will reproduce the time series itself: $L^{(0)}z_i(t) = z_i(t)$.

The weights work very similar to the transfer function defined above for MARMA, with only one major conceptual exception. Instead of being a purely statistical statement, the weights here may be exogenously decided, reflecting physical properties of the system. An obvious example in the spatial context is the relative accessibility between site i and j. The similarity between the matrix of weights here and the derivation-allocation matrices of $\mathbf{F}$ and $\mathbf{A}$ in the Lowry–Garin

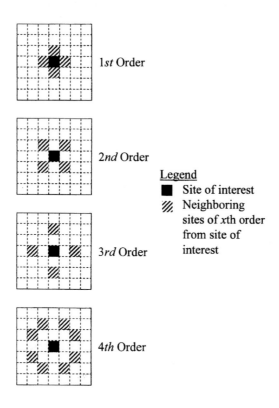

Figure 10-17 Spatial order in a grid

Model[4] is striking; the row sums (or column sum) of the embedded allocation matrices are both unity and they are of the same $n' \times n'$ dimension. The functions played by both are the same. Both express the observation at time t for site i, $z_i(t)$ is a linear combination of past observation at i and a non-temporal observation (in this case the observation at neighboring sites). Finally, the order with which sites interact corresponds to the distance-dependent function between the subject site and the neighboring sites, often implying weaker interaction with distant neighbors.

In a STARMA model, a regular univariate time-series is extended to include lth-order neighbors:

$$z_i(t) = \sum_{k=1}^{p} \sum_{l=0}^{l_k} \varphi_{kl} L^{(l)} z_i(t-k) - \sum_{k=1}^{q} \sum_{l=0}^{m_k} \theta_{kl} L^{(l)} a_i(t-k) + a_i(t) \qquad \textbf{(10-64)}$$

---

[4]See the chapter on "Chaos, Catastrophe, Bifurcation and Disaggregation in Locational Models."

where l_k is the spatial order of the kth autoregressive term, and m_k is the spatial order of the kth moving-average term. The model is usually referred to by the designation STARMA$(p_{l_1,l_2,\ldots,l_p}, q_{m_1,m_2,\ldots,m_q})$. In equivalent matrix notation, Equation (10-64) can be written as

$$z_i(t) = \sum_{k=1}^{p} \sum_{l=0}^{l_k} \varphi_{kl} \, (\mathbf{w}^{(l)})^T \mathbf{z}(t-k) - \sum_{k=1}^{q} \sum_{l=0}^{m_k} \theta_{kl} \, (\mathbf{w}^{(l)})^T \mathbf{a}(t-k) + a_i(t) \qquad \textbf{(10-65)}$$

where $\mathbf{w}^{(l)} = (w_{i1}^{(l)}, w_{i2}^{(l)}, \ldots, w_{in}^{(l)})^T$. In the STARMA model described thus far, only one type of observation—either population or employment—is considered. To the extent that both population and employment are available in each zone, and there is correlation between the population time-series $N_i(t)$ and the employment time-series $E_i(t)$, a model should be constructed to consider this relationship. This is in addition to the spatial relationship. Deutsch and Wang (1983) call this type of model multivariate STARMA model:

$$\mathbf{z}^h(t) = \sum_{g=1}^{\zeta} \sum_{k=1}^{p^{hg}} \sum_{l=0}^{l_k^{hg}} \varphi_{kl}^{hg} L^{(l)} \mathbf{z}^g(t-k) - \sum_{g=1}^{\zeta} \sum_{k=1}^{q^{hg}} \sum_{l=0}^{m_k^{hg}} \theta_{kl}^{hg} L^{(l)} \mathbf{a}^g(t-k) + \mathbf{a}^h(t). \qquad \textbf{(10-66)}$$

We refer to this as the MULSTARMA(ζ,p,q,l,m) model. Clearly in such a model, the current h-category observations are expressed as the weighted sum of previous observations, previous errors of all categories and the current error of h category. This inter-category dependency can be removed by setting $p^{hg} = 0$, $q^{hg} = 0$ where $h \neq g$, reducing the model to ζ independent STARMA models.

10.3.2 Extension of Box–Jenkins modelling procedure.

The Box–Jenkins modelling procedure for univariate ARMA—consisting of identification, estimation, diagnostic checking and forecasting—can be extended to the spatial-temporal context. Besides inducing stationarity, one very important first step in data preparation may be to form a single time series out of the original data. This consists of merging the time series of the target region and the lth time-series of weighted neighbors. In the notation below, $z_0(t)$ is the target-zone time-series where $(i = 0)$. We show below how the single time-series is formed when n observations in each time series are combined:

$$\{z_0(t=1), z_1(t=1), z_2(t=1) \ldots z_K(t=1); \ z_0(t=2), z_1(t=2), z_2(t=2) \ldots z_K(t=2);$$
$$\cdots;$$
$$z_0(t=n), z_1(t=n), z_2(t=n) \ldots z_K(t=n)\}$$

This forms the data-base toward which the Box–Jenkins procedures are applied (Greene 1992).

10.3.2.1 IDENTIFICATION Similar to the univariate case, the purpose of the identification stage is to employ statistical procedures to specify tentatively which STARMA classes are appropriate for the data. The two statistics used to identify potential STARMA classes are the spatial-temporal autocorrelations and the spatial-temporal partial-autocorrelations at temporal-lag k and spatial-lag l at time t. The autocorrelations measure the relationship, or how much spatial-temporal interdependence exists, between data points of the n' time series. Spatial-temporal autocorrelation between the lth and mth neighbors of site i is estimated by

$$\hat{r}_{lm}(k) = \frac{\sum_{i=1}^{n'} \sum_{t=1}^{n-k} L^{(l)} z_i(t)\, L^{(m)} z_i(t+k)}{\left[\sum_{i=1}^{n'} \sum_{t=1}^{n} \left(L^{(l)} z_i(t)\right)^2 \sum_{i=1}^{n'} \sum_{t=1}^{n} \left(L^{(m)} z_i(t)\right)^2\right]^{1/2}}. \qquad (10\text{-}67)$$

The space–time partial-autocorrelation function could be estimated by successively fitting spatial-temporal autoregressive STAR $(p_{l_1,l_2,\dots,l_p})$ models for $l_p = 0,1,\dots,K$ for each p, $(p = 1,2,\dots,n)$ and picking out the estimates $\hat{\varphi}_{kl}$ of the last coefficient from one of these vector models. This is similar to the univariate ARMA Equations **(10-9)**, **(10-10)**, **(10-11)** and the stepwise regression as discussed in the "Statistical" Appendix in Chan (2000). This results in the following model:

$$\mathbf{z}(t) = \sum_{j=1}^{p} \sum_{l=0}^{l_j} \varphi_{jl} \mathbf{W}^{(l)} \mathbf{z}(t-j) + \mathbf{a}(t) \qquad (10\text{-}68)$$

where $\mathbf{W}^{(l)} = [w_{ij}^{(l)}]$. The modeler uses these autocorrelations to decide if the data is stationary. A non-stationary time-series can be reduced to a stationary time-series through differencing if it is a *homogeneous* time-series (Montgomery 1990).

The modeler identifies the subclass by examining the autocorrelations and the partial autocorrelations. A spatial-temporal autoregressive (STAR) process exhibits space–time partial-autocorrelations that go to zero after p lags in time and l_p lags in space. The space–time autocorrelations of a spatial-temporal moving-average (STMA) process will go to zero after q lags in time and m_q in space (Pfeifer and Deutsch 1980). The modeler would further investigate a STARMA model class if both the autocorrelations and the partial autocorrelations exponentially go to zero.

Once the modeler identifies a model subclass, the temporal autoregressive-order p, the temporal moving-average-order q, the spatial autoregressive-order l_p, and the spatial moving-average-order m_k are identified. The identification of a candidate model and its orders is never easy because the autocorrelations and the partial autocorrelations are only estimates. As always, model building is both an art and a science and will require the judgment of the modeler to choose the best candidate model and its orders.

10.3.2.2 ESTIMATION After choosing the best potential candidate model and its orders, the modeler estimates the autoregressive parameters φ_{kl} and the moving-average parameters θ_{kl}. The efficient estimates of both parameters are maximum-likelihood estimators (Tiao and Box 1981). However, calculating the maximum-likelihood estimators of the parameters is not an easy task. The modeller usually approximates the parameters using a conditional likelihood-function that minimizes the conditional sum-of-squares (Pfeifer and Deutsch 1980).[5] Estimating the STAR parameters then is easy since the conditional-likelihood function is also a least-squares estimate; the model can simply estimate the STAR model parameters using linear regression (Pfeifer and Deutsch 1980). As an example, consider the STAR(2_{10}) vector model:

$$\mathbf{z}(t) = \varphi_{10}\mathbf{z}(t-1) + \varphi_{11}\mathbf{W}^{(1)}\mathbf{z}(t-1) + \varphi_{20}\mathbf{z}(t-2) + \mathbf{a}(t) \qquad t = 1, 2, \ldots, n. \tag{10-69}$$

In general linear-model form, this model for $t = 1, 2, \ldots, n$ can be written as

$$
\begin{bmatrix} \mathbf{z}(1) \\ \mathbf{z}(2) \\ . \\ . \\ . \\ \mathbf{z}(n) \end{bmatrix} = \begin{bmatrix} 0 & 0 & 0 \\ \mathbf{z}(1) & \mathbf{W}^{(1)}\mathbf{z}(1) & 0 \\ \mathbf{z}(2) & \mathbf{W}^{(1)}\mathbf{z}(2) & \mathbf{z}(1) \\ . & . & . \\ . & . & . \\ . & . & . \\ \mathbf{z}(n-1) & \mathbf{W}^{(1)}\mathbf{z}(n-1) & \mathbf{z}(n-2) \end{bmatrix} \begin{bmatrix} \varphi_{10} \\ \varphi_{11} \\ \varphi_{20} \end{bmatrix} + \begin{bmatrix} \mathbf{a}(1) \\ \mathbf{a}(2) \\ . \\ . \\ . \\ \mathbf{a}(n) \end{bmatrix} \tag{10-70}
$$

Here, zero vectors have been substituted for the unobserved $\mathbf{z}$ vectors, for those times before the system was under observation. The estimated values for φ are the conditional least-squares estimates of the parameters of the STAR(2_{10}) model. Similar equations have been presented as Equations **(10-14)**, **(10-15)** and **(10-16)** when we discussed univariate time-series.

On the other hand, estimating the STMA and STARMA model-parameters is not so effortless due to non-linearity. The modeler can employ various non-linear optimization techniques, such as gradient methods or linearization, to calculate the STMA and STARMA model parameters (Pfeifer and Deutsch 1980). The sum-of-squares surface $S(\boldsymbol{\Phi},\boldsymbol{\Theta})$ can be approximated by expanding about the least-squares estimates as follows $S(\delta) \approx S(\hat{\delta}) + (\delta - \hat{\delta})^T \mathbf{Q}'(\delta - \hat{\delta})$ where $\boldsymbol{\delta}^T = [\boldsymbol{\Phi}^T, \boldsymbol{\Theta}^T]$ and $\hat{\delta}$ is its least-squares estimate. $\mathbf{Q}' = \mathbf{X}^T\mathbf{X}$ where

---

[5]For a definition of conditional likelihood and conditional sum-of-squares, see the "Nonlinear regression" section of the "Statistical Tools" Appendix (Chan 2000).

$$X = \begin{bmatrix} \dfrac{\partial \mathbf{a}(1)}{\partial \delta_1}\bigg|_{\hat{\delta}} & \dfrac{\partial \mathbf{a}(1)}{\partial \delta_2}\bigg|_{\hat{\delta}} & \cdots & \dfrac{\partial \mathbf{a}(1)}{\partial \delta_k}\bigg|_{\hat{\delta}} \\[2ex] \dfrac{\partial \mathbf{a}(2)}{\partial \delta_1}\bigg|_{\hat{\delta}} & \cdot & \cdots & \dfrac{\partial \mathbf{a}(2)}{\partial \delta_k}\bigg|_{\hat{\delta}} \\[2ex] \cdot & \cdot & \cdot \\ \cdot & \cdot & \cdot \\ \cdot & \cdot & \cdot \\[1ex] \dfrac{\partial \mathbf{a}(n)}{\partial \delta_1}\bigg|_{\hat{\delta}} & \dfrac{\partial \mathbf{a}(n)}{\partial \delta_2}\bigg|_{\hat{\delta}} & \cdots & \dfrac{\partial \mathbf{a}(n)}{\partial \delta_k}\bigg|_{\hat{\delta}} \end{bmatrix}. \qquad (10\text{-}71)$$

Matrix $\mathbf{Q}'$ must be numerically estimated. Introduction to such nonlinear regressions is documented in the "Statistical Tools" appendix to Chan (2000). Calibration examples for univariate time-series have been worked out in Section 10.1.6 of this chapter.

10.3.2.3 DIAGNOSTIC CHECKING The objective of diagnostic checking is to verify that the selected model is adequate. The first test examines the statistical significance of the model parameters. The model parameter of the highest order for p, q, l_k and m_k must be significantly different from zero. If any parameter of its respective highest order is insignificant, then the modeler will return to the identification stage. The second test verifies that the model residuals are white noise. This ensures that the model does not violate the *sphericity* assumption (Pfeifer and Deutsch 1980)—an extension of the homoscedastic property in linear regression to spatial-temporal dimensions. If the space–time autocorrelations and the space–time partial-autocorrelations of the residuals are all close to zero, then the model residuals are random. If a scatter plot of the residuals shows no patterns, then it can be concluded that the model residuals are from a random process. It has been shown (Pfeifer 1979) that if the underlying process is pure white-noise, $var\left(\hat{r}_{l0}(k)\right) \approx [n'(n-k)]^{-1}$. In other words, if the residuals are approximately white noise, the sample space–time autocorrelation-functions should all be effectively zero, irrespective of the lag-k and the spatial-order l under consideration. The analysis of the space–time autocorrelations guards against model mis-specification and searches for directions of improvement. If the selected model fails this second test, the modeler will return to the identification stage to represent the residuals as a separate STARMA model and combine it with the original model.

10.3.2.4 FORECASTING The first step in the forecasting stage is to select the best and most parsimonious model. As indicated in the "Statistical Tools" appendix in Chan (2000), there are at least two criteria used to select the best and most parsimonious model.[6] The first one is the fraction of the variance explained by the model, adjusted for the degrees-of-freedom—the adjusted-R^2. The second criterion is the sum of the squared residuals, SSR. The best and most parsimonious model is that model that uses the smallest number of parameters necessary to adequately describe and represent the n' time-series, such that the adjusted-R^2 and the SSR are minimized. It is best to place more weight on the adjusted-R^2 over the SSR value, because the former accounts for the number of parameters in the model by adjusting for the degrees-of-freedom. The adjusted-R^2 can be worse sometimes as the number of parameters is increased, whereas the SSR value can only get better. Once the best and most parsimonious model is selected based on the above and other criteria, the model can be used to predict the n' time-series. The forecasting procedure follows that described in Section 10.1.8. An example will be offered in Section 10.5 to illustrate this spatial extension of the Box–Jenkins procedure. Meanwhile, we offer an illustration to bring the procedure to focus.

> **10.3.3 Assault-arrest example.** Arrest data from the 14 areas of northeast Boston were collected from 1969 to 1974 (Pfeifer and Deutsch 1980). These assault arrests were observed for $t = 1, \dots, n$ time-periods. **Table 10-5** gives the spatial neighbors of each of the 14 districts. Weights as captured in $\mathbf{W}^{(l)}$ (Equations **(10-62)** and **(10-63)**) are given below

$$\begin{bmatrix}
0 & 1/2 & 1/2 & 0 & 0 & 0 & 0 & 0 & 0 & 0 & 0 & 0 & 0 & 0 \\
1/2 & 0 & 0 & 1/2 & 0 & 0 & 0 & 0 & 0 & 0 & 0 & 0 & 0 & 0 \\
1/3 & 0 & 0 & 1/3 & 0 & 1/3 & 0 & 0 & 0 & 0 & 0 & 0 & 0 & 0 \\
0 & 1/4 & 1/4 & 0 & 1/4 & 1/4 & 0 & 0 & 0 & 0 & 0 & 0 & 0 & 0 \\
0 & 0 & 0 & 1/4 & 0 & 1/4 & 1/4 & 1/4 & 0 & 0 & 0 & 0 & 0 & 0 \\
0 & 0 & 1/4 & 1/4 & 1/4 & 0 & 1/4 & 0 & 0 & 0 & 0 & 0 & 0 & 0 \\
0 & 0 & 0 & 0 & 1/4 & 1/4 & 0 & 1/4 & 0 & 1/4 & 0 & 0 & 0 & 0 \\
0 & 0 & 0 & 0 & 1/3 & 0 & 1/3 & 0 & 1/3 & 0 & 0 & 0 & 0 & 0 \\
0 & 0 & 0 & 0 & 0 & 0 & 0 & 1/3 & 0 & 1/3 & 0 & 0 & 0 & 1/3 \\
0 & 0 & 0 & 0 & 0 & 0 & 1/4 & 0 & 1/4 & 0 & 1/4 & 0 & 1/4 & 0 \\
0 & 0 & 0 & 0 & 0 & 0 & 0 & 0 & 0 & 1/2 & 0 & 1/2 & 0 & 0 \\
0 & 0 & 0 & 0 & 0 & 0 & 0 & 0 & 0 & 0 & 1/2 & 0 & 1/2 & 0 \\
0 & 0 & 0 & 0 & 0 & 0 & 0 & 0 & 0 & 1/3 & 0 & 1/3 & 0 & 1/3 \\
0 & 0 & 0 & 0 & 0 & 0 & 0 & 0 & 1/2 & 0 & 0 & 0 & 1/2 & 0
\end{bmatrix} \tag{10-72}$$

Initial identification of the assault data suggested that the system was non-stationary and required a first difference. The sampled space–time autocorrelation (Equation **(10-67)**) and partial-autocorrelation function of the differenced series is presented in **Table 10-6**. The spatial-time partials tail off and the space–time autocorrelation-

---

[6]A discussion of an analogous process, stepwise regression, can be found in the "Statistical Tools" Appendix of Chan (2000).

Site/Order	First	Second	Third
1	2,3	4	6
2	1,4	3	5,6
3	1,4,6	2	5,7
4	2,3,5,6	1	7,8
5	4,6,7,8	-	3.9,10,2
6	3,4,5,7	-	1,2,8,10
7	5,6,8,10	11	3,4,9,13
8	5,7,9	10	4,6,14
9	8,10,14	13	4,6,14
10	7,9,11,13	8,12,14	5,6
11	10,12	7,13	9
12	11,13	10	14
13	10,12,14	9,11	7
14	9,13	10	8,12

Table 10-5 Neighbors of each site for each spatial order
(Pfeifer and Deutsch 1980)

function seems to cut off both spatially and temporally after one lag; the differenced series was tentatively identified as a $STMA(1_0)$ process.

Estimating this model gave $\theta_{10} = 0.803$, resulting in the model $z(t) - z(t-1) = -0.803\,a(t-1) + a(t)$. The residual sum-of-squares was 6504.679. The estimated variance of the parameter estimate was 0.000368, and thus an approximate 95% confidence-interval for θ_{10} is (0.764, 0.841). **Table 10-7** reveals relatively large values for $\hat{r}_{10}(1)$ (compared to an approximate variance of $1/n'(n-k) = [(14)(17)]^{-1}$ and $\hat{\phi}_{11}$, suggesting the residuals exhibit some spatial correlation. Consequently, a new model $STARMA(0,1_1)$ was entertained. The extra 'spatial' parameter θ_{11} in this model will hopefully describe the spatial structure of the residuals.

Estimation of the $STARMA(0,1_1)$ model yielded $\theta_{10} = 0.812$, $\theta_{11} = -0.092$, resulting in the residuals from the model $z(t)-z(t-1) = 0.812a(t-1) +0.092\,W^{(1)}a(t-1) + a(t)$, and a residual sum-of-squares of 6457.266. The sample correlative properties of the residuals from this model are given in **Table 10-8**. We note that the first-spatial-order autocorrelation and partial-autocorrelation have both decreased, and in general there seems to be a lack of structure in these residuals. The lone exception occurs at time-lag 6 and spatial-lag 1, pointing to a possibility that further investigation into seasonal forms of STARMA might prove useful. Testing the significance of the θ_{11} parameter is done via the *F*-statistic by observing

Space–time autocorrelations $\hat{r}_{lm}(k)$				
Spatial-lag(l) / time-lag(k)	0	1	2	3
1	−0.484	0.007	0.041	0.013
2	0.023	−0.0	0.012	−0.027
3	−0.017	0.004	−0.045	0.049
4	0.026	0.013	0.019	−0.038
5	−0.056	0.039	0.005	0.017
6	0.043	−0.0	0.045	−0.021
7	−0.003	0.015	−0.091	−0.018
8	−0.032	0.015	0.053	0.037

Space–time partial autocorrelations $\hat{\varphi}_{kl}$				
Spatial-lag(l) / time-lag(k)	0	1	2	3
1	−0.484	0.068	0.007	0.020
2	−0.281	0.015	0.034	−0.011
3	−0.196	0.068	−0.026	0.070
4	−0.111	0.015	−0.025	0.028
5	−0.140	0.004	−0.021	0.030
6	−0.092	0.025	0.071	−0.007
7	−0.048	0.138	−0.036	−0.055
8	−0.073	0.008	−0.011	0.008

Table 10-6 Sample Space-time correlation functions of the differenced series
(Pfeifer andDeutsch 1980)

that $n' = 14$, $n = 71$, $m_k = 2$, $S(\hat{\delta}^*) = 6504.679$ and $S(\hat{\delta}) = 6457.266$. Combining these gives an approximate $F_{1,992} = 7.3$. The theoretical critical F-value at $\alpha = 0.01$ with 1 and 992 degrees-of-freedom is approximately 6.7. Thus θ_{11} is significant with 99% confidence.

The sample space–time autocorrelations and partials show a lack of structural characteristic of a white-noise sequence (with the noted exception). Since all parameters have proven statistically significant, the STARMA$(0,1_1)$ model $\mathbf{z}(t) - \mathbf{z}(t-1) = 0.812\mathbf{a}(t-1) + 0.092\mathbf{W}^{(1)}\mathbf{a}(t-1) + \mathbf{a}(t)$ passes the diagnostic-checking portion of the extended Box–Jenkins modelling procedure. This model is now ready to be employed in forecasting the number of assault arrests in northeast Boston.

Space–time autocorrelations $\hat{r}_{lm}(k)$			
Spatial-lag(l) / time-lag(k)	0	1	2	3
---	---	---	---	---
1	0.024	0.084	0.051	0.058
2	0.031	0.016	0.023	0.027
3	−0.007	0.027	−0.026	0.046
4	−0.002	0.038	0.021	0.016
5	−0.056	0.024	0.029	−0.014
6	0.000	−0.061	0.026	−0.041
7	−0.031	−0.009	−0.062	−0.026
8	−0.046	0.011	0.030	0.021

Space–time partial autocorrelations $\hat{\varphi}_{kl}$			
Spatial-lag(l) / time-lag(k)	0	1	2	3
---	---	---	---	---
1	0.024	0.133	0.042	0.055
2	0.019	−0.007	0.016	0.010
3	−0.013	0.028	−0.042	0.045
4	−0.009	0.058	0.015	−0.044
5	−0.060	0.047	0.031	−0.035
6	−0.003	−0.094	0.039	−0.050
7	−0.022	0.003	−0.064	−0.021
8	−0.043	0.064	0.039	0.042

Table 10-7 Sample space–time autocorrelation function of residuals
(Pfeifer and Deutsch 1980)

10.4 Econometric Models

Econometric models for time-series data and multiple-time-series models both attempt to describe the dynamic relationship between the variables under consideration. Although they were developed historically under very different circumstances, several features are common between the two. Granger and Watson(1984) suggested a linkage between the two models. The multivariate time-series is represented by

$$\varphi(B)\,\mathbf{z}(t) = \theta(B)\,\mathbf{a}(t). \tag{10-73}$$

Space-time autocorrelations $\hat{r}_{lm}(k)$				
Spatial-lag(l)/ time-lag(k)	0	1	2	3
1	0.023	0.045	0.036	0.046
2	0.023	0.062	0.014	0.008
3	−0.010	−0.014	−0.043	0.041
4	−0.004	0.029	0.015	−0.046
5	−0.053	0.020	0.034	−0.036
6	0.007	−0.120	0.042	−0.050
7	−0.023	−0.023	−0.062	−0.023
8	−0.044	0.036	0.041	0.040
Space-time partial autocorrelations $\hat{\phi}_{kl}$				
Spatial-lag(l)/ time-lag(k)	0	1	2	3
1	0.023	0.030	0.036	0.042
2	0.028	−0.026	0.010	0.011
3	−0.008	−0.009	−0.036	0.030
4	−0.001	0.016	0.014	−0.029
5	−0.055	0.003	0.027	−0.028
6	−0.003	−0.074	0.026	−0.051
7	−0.032	−0.022	−0.058	−0.034
8	−0.046	0.004	0.034	0.015

Table 10-8 Sample space–time autocorrelation functions
(Pfeifer and Deutsch 1980)

This is a rewrite of Equation **(10-43)**, where the matrix lag-operator $\phi(B)$ is applied to $z(t)$. It results in autocorrelation of the multivariate time-series that can be represented by the multivariate moving-average model $\theta(B)a(t)$. As alluded to above, time-series identification for MARMA models is a difficult task, often with a fair amount of trial-and-error involved.

10.4.1 Econometric equivalence of time series. In traditional econometrics, the above equations can be written in terms of the following example

$$\begin{vmatrix} \varphi_{11} & \varphi_{12} \\ \varphi_{21} & \varphi_{22} \end{vmatrix} \begin{vmatrix} y(t) \\ x(t) \end{vmatrix} = \begin{vmatrix} \theta_{11} & \theta_{12} \\ \theta_{21} & \theta_{22} \end{vmatrix} \begin{vmatrix} a_1(t) \\ a_2(t) \end{vmatrix}. \tag{10-74}$$

Without loss of generality, the lag operator has not been shown for notational clarity. If $\varphi_{21} = \theta_{12} = \theta_{21} = 0$, the equation becomes

$$\varphi_{11}(B)\, y(t) + \varphi_{12}(B)\, x(t) = \theta_{11}(B)\, a_1(t) \tag{10-75}$$

and

$$\varphi_{22}(B)\, x(t) = \theta_{22}(B)\, a_2(t). \tag{10-76}$$

Should there be no contemporaneous correlation between the components of th e white-noise vector $a_1(t)$ and the white-noise vector $a_2(t)$, $z(t)$ is decomposed into $x(t)$ and $y(t)$, where $x(t)$ is called *exogenous*.

These two equations provide the link between time-series and econometric models. The first is a dynamic simultaneous equation model and the second represents the evolution of the exogenous variable. Notice such evolution is often not considered under traditional econometric models, where the exogenous variables are simply *given*. This contrasts with time-series analysts, who identify and estimate both equations. The econometricians are interested in interpretation of the coefficients estimated in the first equation, particularly their meaning regarding the economic system under study. Time-series analysts, on the other hand, are primarily interested in forecasts, using both equations in performing their tasks.

The structural equations **(10-75)** and **(10-76)** can be expressed in another form that is of equal interest. Define $\dot{\varphi}_{11}(B) = \varphi_{11}(B) - \varphi_{11}(0)$, then Equation **(10-75)** can be rewritten as

$$y(t) = -\frac{\dot{\varphi}_{11}(B)}{\varphi_{11}(0)} y(t) + \frac{\varphi_{12}(B)}{\varphi_{11}(0)} x(t) + \frac{\theta_{11}(B)}{\varphi_{11}(0)} a_1(t), \tag{10-77}$$

which is known as the *reduced form*.[7] The reader may wish to consult an example of this in the "Descriptive Tools" chapter (Chan 2000) under the "Calibration" section. Alternatively, Equation **(10-75)** can be simply written as

$$y(t) = -\frac{\varphi_{12}(B)}{\varphi_{11}(B)} x(t) + \frac{\theta_{11}(B)}{\varphi_{11}(B)} a_1(t), \tag{10-78}$$

which is a unidirectional *transfer function*. In the reduced form, endogenous variables ($y(t)$) are explained by exogenous ($x(t)$) and *lagged* endogenous variables

---

[7]For an introduction to simultaneous econometric-equation-sets, please consult the"Descriptive Tools for Analysis" chapter in Chan (2000).

(now we write $y(t = j)$ instead of $y(t)$ with $j \geq 0$). In the transfer function, $y(t)$ appears to be explained only by the exogenous variables $x(t)$. After the parameters have been estimated, both the reduced form and the transfer function can be used to forecast. The reduced form uses the past observations $x(n-j)$ and $y(n-j)$ ($j \geq 0$), plus forecasts of exogenous variables. The transfer function, on the other hand, uses simply $x(n-j)$ ($j \geq 0$) plus exogenous variables. Such forecasts, however, produce errors that are not white noise (as suggested in Subsection 10.2.2 of this chapter). This can be remedied by modelling the residuals. But this requires earlier values of the residuals, which in turn requires earlier values of $y(t)$. Effectively, both the reduced form and the transfer function use the same set of data to forecast.

Example

Consider the following bivariate example due to Granger and Newbold (1977) in which Equation **(10-77)** is written with $y_1(t)$ on the right-hand-side (the lagged endogenous variable) written as $y_1(t-1)$:

$$\begin{bmatrix} y_1(t) \\ y_2(t) \end{bmatrix} = \begin{bmatrix} 0.7 & 0.3 \\ 0.4 & 0.4 \end{bmatrix} \begin{bmatrix} y_1(t-1) \\ y_2(t-1) \end{bmatrix} + \begin{bmatrix} 1 & 0 \\ 0 & 1 \end{bmatrix} \begin{bmatrix} a_1(t) \\ a_2(t) \end{bmatrix} + \begin{bmatrix} 0 & 0.5 \\ 0 & 0.6 \end{bmatrix} \begin{bmatrix} a_1(t-1) \\ a_2(t-1) \end{bmatrix}. \tag{10-79}$$

Here $x(t)$ is written as $y_2(t-1)$ for simplicity. Also Equation **(10-76)** is now embedded in Equation **(10-77)** as an additional noise term $\theta_2(B)a_2(t-1)$, where autocovariance matrix $\text{var}(\mathbf{a}) = \begin{bmatrix} 1.09 & 0.7 \\ 0.7 & 1.16 \end{bmatrix}$. Multiplying by a nonsingular matrix $\mathbf{P'} = \begin{bmatrix} 1 & 0 \\ -0.6427 & 1 \end{bmatrix}$ will not change the correlative structure of $\mathbf{y}(t)$ and will therefore result in an equivalent representation. With such a triangular matrix $\mathbf{P'}$, it can be shown that $\mathbf{P'}^T\text{var}(\mathbf{a})\mathbf{P'}$ is a diagonal matrix as desired. As suggested in Equation **(10-56)**, define $\mathbf{w}_t = \mathbf{P}\mathbf{a}_t$, so that $\mathbf{P'}^T\text{var}(\mathbf{a})\mathbf{P'} = \text{var}(\mathbf{w}_t)$, where $\text{var}(\mathbf{w}_t)$ $= \begin{bmatrix} 1.09 & 0 \\ 0 & 0.71 \end{bmatrix}$. Multiplying the model by $\mathbf{P'}$ yields

$$\begin{bmatrix} 1 & 0 \\ -0.642 & 1 \end{bmatrix} \begin{bmatrix} y_1(t) \\ y_2(t) \end{bmatrix} = \begin{bmatrix} 0.7 & 0.3 \\ -0.049 & 0.207 \end{bmatrix} \begin{bmatrix} y_1(t-1) \\ y_2(t-1) \end{bmatrix} +$$
$$\begin{bmatrix} 1 & 0 \\ 0 & 1 \end{bmatrix} \begin{bmatrix} w_1(t) \\ w_2(t) \end{bmatrix} + \begin{bmatrix} 0.321 & 0.5 \\ 0.179 & 0.279 \end{bmatrix} \begin{bmatrix} w_1(t-1) \\ w_2(t-1) \end{bmatrix}. \tag{10-80}$$

Here $\varphi_{12} = \theta_{12} = \theta_{21} = 0$ as suggested in Equation **(10-74)**. The model is also in recursive form with a diagonal covariance-matrix, so that the errors $w_1(t)$ and $w_2(t)$ are mutually stochastically uncorrelated (Mills 1990). □

    Time-series models often involve just a few series, yet concerned with various lag-structures. Econometric models have a large number of series, but only very few lags. Econometricians pose theoretical constructs on model specification as a way to compensate for the lack of data. Time-series analysts, on the other hand, are often armed with a history of observations and "let data speak for themselves." STARMA can be thought of a bridge between the two modelling approaches, where features from both schools are amalgamated into one. For example, residuals from an econometric model are analyzed to check for model mis-specification, and in fact can be included in the model as an STMA component if required. At the same time, theoretical construct is used in specifying the weights that govern spatial interaction between a subject site and neighboring sites, often based on spatial economics.

10.4.2 Dynamic simultaneous equations. Equation **(10-78)** can also be thought of as the *final form* of a simultaneous equation set (Mills 1990). In this form, the infinite matrix polynomial

$$\Omega(B) = \sum_{i=0}^{\infty} \Omega_i B^i = -\frac{\varphi_{12}(B)}{\varphi_{11}(B)} \tag{10-81}$$

contains the *dynamic multipliers*, describing the response of vector $\mathbf{y}(t)$ to unit shocks in the exogenous variables $\mathbf{x}(t)$. $\Omega(1) = \sum_{i=0}^{\infty} \Omega_i$ represents the total response of $\mathbf{y}(t)$ to these unit shocks in $\mathbf{x}(t)$. It is therefore called the matrix of *total multipliers*, while the partial sums $\Omega_j(1) = \sum_{i=0}^{j} \Omega_i$ are referred to as the matrices of the *j*th *interim multipliers*. The matrix Ω_0 contains the immediate, or contemporaneous, effects of unit shock in $\mathbf{x}(t)$ and is called the matrix of *impact multipliers*. In the Lowry–Garin model discussed in the "Chaos, Catastrophe and Bifurcation" chapter, for example, Ω_i is the derivatiion-allocation matrix **FA** and

$$\Omega(1) = \sum_{i=0}^{\infty} \Omega_i = \Omega^0 + \Omega^1 + \Omega^2 + \ldots + \Omega^{\infty}. \tag{10-82}$$

Here, the impact multipliers Ω_0 are contained in the identity matrix $\Omega^0 = \mathbf{I}$, or the immediate effect would be the status quo. The Lowry–Garin model, interpreted in its dynamic form, can also be written as

$$\Omega(B) = \Omega^0 + \Omega^1 + \Omega^2 + \ldots + \Omega^{\infty} \tag{10-83}$$

where $B^i\Omega_i = \Omega^i$ $(i=0,1,2,\ldots,\infty)$.

Pre-multiplying the final form (Equation **(10-78)**) by the determinant $|\varphi_{11}(B)|$ and denoting the adjoint matrix[8] associated with $\varphi_{11}(B)$ as $\varphi_{11}^*(B)$, yields the set of *fundamental dynamic equations*

---

[8] If the row and column containing an element a_{ij} in an $n \times n$ square matrix **A** are deleted, the determinant of the remaining square array is called the *minor* of a_{ij}, and is denoted by M_{ij}. The *cofactor* of a_{ij}, denoted by A_{ij}, is then defined by $(-1)^{i+j}M_{ij}$. The *adjoint* of $\begin{vmatrix} A_{11} & A_{12} & A_{13} \\ A_{21} & A_{22} & A_{23} \\ A_{31} & A_{32} & A_{33} \end{vmatrix}$, for example, is then $\begin{vmatrix} A_{11} & A_{21} & A_{31} \\ A_{12} & A_{22} & A_{32} \\ A_{13} & A_{23} & A_{33} \end{vmatrix}$. It is obtained by replacing each element by its cofactor and then interchanging rows and columns. Also note the inverse $\mathbf{A}^{-1}$ can be expressed as $\mathbf{A}^*/|\mathbf{A}|$.

$$\left| \varphi_{11}^{*}(B) \right| \mathbf{y}(t) = - \varphi_{11}^{*}(B)\, \varphi_{12}(B)\, \mathbf{x}(t) + \varphi_{11}^{*}(B)\, \theta_{11}(B)\, \mathbf{a}(t). \tag{10-84}$$

Here each endogenous variable depends only on its own lagged values and on the exogenous variable, with or without lags. Each endogenous variable will have autoregressive factors having identical order and parameters.

10.4.3 Remarks. After pointing out the relationship between time-series and econometric models, particularly the various forms under which the relationship can manifest itself, let us now consider the possible uses of these different equation systems. We will also point out the prerequisites for their deployment. The multivariate time-series as expressed in Equation **(10-73)** can forecast future values of some or all of the variables in $\mathbf{y}(t)$, given that ARMA process has been fitted to them. These time-series equations cannot be used for structural analysis since they are not equations of the fundamental relationship. Neither can they be used for control purposes, since no exogenous variables are identified in the present form.

On the other hand, the reduced-form equations **(10-77)**, final-form equations **(10-78)** and fundamental dynamic equations **(10-84)** can all be used for both prediction and control, since the exogenous variables $\mathbf{x}(t)$ appear as inputs in them. However, because the lag polynomials appearing in the equations are functions of the structural parameters, none of these equation systems can be employed for structural analysis. The exception is when the structural equations are themselves in either reduced form ($\varphi_{11}(0) = \mathbf{I}$) or final form ($\varphi_{11}(B) = \mathbf{I}$). In other words, when φ_{11} disappears in Equations **(10-77)** or **(10-78)**. Furthermore, use of these equation systems implies that we have enough prior information to distinguish endogenous and exogenous variables. If data on some endogenous variables are unavailable, it may still be possible to use the final-form (or fundamental dynamic) equations relating to those endogenous variables for which data are available. This is different for the reduced-form equations that need all the endogenous variables. Absence of any endogenous variables will render these equations unusable, as alluded to earlier when we thought of the reduced form as a transfer function.

10.5 An extended numerical example

In order to show the theoretical linkages established above on multivariate time-series, an extended numerical example was performed in which the different approaches are used to forecast. The results of these forecasts are then compared, providing an insight into the strengths and weaknesses of these varied approaches. To start out, the disaggregate version of the Lowry–Garin model was used to generate a set of data for analysis.[9] Except for the calibration performed between the base and forecast year, no zoning constraint was imposed to induce bifurcation

---

[9]For a description of the Lowry–Garin model, see the "Activity Allocation" chapter.

Per-iod	Pop. at zone 1	Emp. at zone 1	Pop. at zone 2	Emp. at zone 2	Pop. at zone 3	Emp. at zone 3	Pop. at zone 4	Emp. at zone 4
1	117.5	134.5	381.4	162.0	134.0	88.7	605.5	413.4
2	142.3	127.6	463.0	159.3	165.9	79.3	671.4	458.5
3	147.6	126.0	480.1	158.6	172.5	77.1	689.3	468.9
4	148.8	125.6	454.1	158.5	174.0	76.5	693.3	471.3
5	149.1	125.5	485.0	158.5	174.4	76.4	694.3	471.9
6	149.1	125.5	485.2	158.5	174.4	76.4	694.5	472.0
7	149.2	125.5	485.2	158.5	174.5	76.4	694.5	472.1
8	149.2	125.5	485.2	158.5	174.5	76.4	694.6	472.1
9	149.2	125.5	485.3	158.5	174.5	76.4	694.6	472.1
10	149.2	125.5	485.3	158.5	174.5	76.4	694.6	472.1

Table 10-9 Time series data generated from Lowry–Garin model

.This allowed for a continuous, well-behaved (rather than truncated) set of data for analysis. For the set of data documented in the chapter on "Chaos, Bifurcation and Disaggregation," Forgues and Chan (1991) generated the set of data for zonal population and employment in **Table 10-9**.

The Lowry–Garin derivative models use Equation **(10-82)** or its equivalent Equation **(10-83)** to generate the total employment-vector **E** and the population-vector **N**:

$$\mathbf{E} = \mathbf{E}^B\left[\mathbf{I} + \mathbf{FA} + (\mathbf{FA})^2 + \ldots + (\mathbf{FA})^j\right] \qquad j \to \infty$$
$$\mathbf{N} = \mathbf{E}^B\left[\mathbf{I} + \mathbf{FA} + (\mathbf{FA})^2 + \ldots + (\mathbf{FA})^j\right]\mathbf{F} \qquad j \to \infty \tag{10-85}$$

where the basic employment for the four zones is contained in the vector $\mathbf{E}^B = (180, 150, 130)^T$, the derivation-allocation matrices for population and employment are respectively

$$\mathbf{F} = \begin{bmatrix} 0.140 & 0.572 & 0.211 & 0.886 \\ 0.184 & 0.494 & 0.199 & 0.932 \\ 0.184 & 0.572 & 0.152 & 0.900 \\ 0.187 & 0.617 & 0.222 & 0.777 \end{bmatrix} \text{ and } \mathbf{A} = \begin{bmatrix} -0.022 & -0.010 & -0.035 & 0.191 \\ -0.030 & -0.011 & -0.046 & 0.213 \\ -0.029 & -0.012 & -0.029 & 0.196 \\ -0.045 & -0.018 & -0.059 & 0.254 \end{bmatrix}.$$

Notice this is an unconstrained application of the model. No adjustment is made in the population or employment derivation and allocation to comply with zoning ordinance or minimum threshold for population-service establishment, respectively. This accounts for the negative values in the **A** matrix. A constrained application of the model would have produced bifurcation in the generation process, or precipitous zonal development or decline would have taken place. This in turn necessitates "intervention analysis" to control development density or to ensure that only viable service businesses get underway—a subject yet to be discussed in this chapter. The current application of the Lowry–Garin model would simply assume development will take its place without external interventions.

To perform univariate time-series analysis, the population at zone 1 over the 10 time-periods was examined. Data for an additional time-period 11 is generated to obtain 10 data points after differencing. The data point for zone-1 population reads 149.2, which is the same as the previous period. Please notice no noise is explicitly introduced into the data (but the analyst who is studying the data do not know this fact).

Looking at the data with our privileged knowledge that it came from the Lowry–Garin model, recall that the data was generated from a geometric series. Specifically, the data series assumes the form of Equation **(10-85)**, which asymptotically reach a limit as long as Ω is less than $\mathbf{I}$:[10]

$$\mathbf{Z} = \mathbf{Z}^0\left[\mathbf{I} + \Omega + \Omega^2 + \ldots + \Omega^i + \ldots + \Omega^\infty\right] = \mathbf{Z}^0/(\mathbf{I} - \Omega). \tag{10-86}$$

Here $\mathbf{Z}$ is a vector of zonal population or employment, and $\mathbf{Z}^0$ is the forecast-year basic-employment $\mathbf{E}^B$ or population $\mathbf{E}^B\mathbf{F}$. $\Omega = \mathbf{FA}$ represents the spatial interaction between zonal employment and population. It can be seen that such a spatial growth-pattern, while driven by a geometric series in general, is 'perturbed' by interaction between zones. The perturbation works simultaneously with the general geometric-series to effect a spatial-temporal distribution of employment and population over the four-zone study area.

According to the Malthusian growth-equation $\dot{Z} = Z^0 Z$, differentiation of Equation **(10-86)** with respect to 'time', or more precisely the discrete time-increments serialized by exponent i in Equation **(10-86)** yields $\Delta Z/\Delta i = Z^0 Z^i$ $(i=1,2,\ldots,\infty)$. Here $\mathbf{Z}^i$ is the 'current' activity levels. They become steady and infinitesimally small as i becomes large. Assuming the process is reaching steady state, the slope of the time-series might very well become zero upon the first differencing. In other words, a first differencing of the 'time'-series might readily induce stationarity. To the extent that the rate-of-change $\mathbf{Z}^0\mathbf{Z}^i$ $(i = 1, 2, \ldots, \infty)$ is a function of the previous time-period $(i-1)$, an autocorrelation exists between increments of the time series. The rate-of-change $\mathbf{Z}^0\mathbf{Z}^i$, while small, represents a perturbation term that re-allocates population and employment between zones from

---

[10]For further explanation of this convergence, please refer to the Chapter on "Chaos, Catastrophe, Bifurcation and Disaggregation."

one time-period to another—as alluded to earlier. This perturbation will follow a definite pattern should Ω be constant (as is the current case when there is no bifurcation). But observing from zone 1 (or any other individual zone), this perturbation looks random, or more precisely, it appears as noise that has an underlying pattern. The latter statement suggests a moving-average model may also make sense to explain the patterned noise, on top of an autoregressive model—from a statistical modelling standpoint.

When the time-series $f(t)$ is defined only for $t \geq 0$, the Fourier transform of the series is the same as Laplace transform $\int_0^\infty e^{-ut} f(t)\, dt$. For convenience, let us approximate the zonal time-series as an exponential series $e^{-\omega t}$ and replace the index i with the time-axis t.[11] The transform of the time series now becomes $\int_0^\infty e^{-ut} e^{-\omega t}\, dt =$ $1/(u - \omega)$ for $u > \omega$. Simple as this analysis may be, a fair amount of insight can be gained from examining the transform—or from the frequency domain. First, periodicity is absent since the spectrum is monotonically decreasing (similar to the spectrum shown in **Figure 10-16**). Second, dominance of low-frequency amplitude suggests positive autocorrelation. Third, the same low-frequency dominance suggests a requirement for differencing to achieve stationarity. These properties will come in handy in the identification and calibration of the proper ARMA model, as shown below.

10.5.1 Univariate ARMA model. These steps were carried out in the International Mathematical and Statistical Library (IMSL) software for zone-1 population: (1) The time series was differenced; (2) Autocorrelation was computed; (3) The parameters p and q were estimated; (4) Coefficients φ and θ were estimated; and (5) ARMA forecast was performed. Obviously, if unsatisfactory forecasts were obtained, the process was repeated, by either taking higher-order differences or using different p or q values or both. Judging from the computer runs (and confirmed by our privileged information on how the data were generated), only one difference of the time series was necessary to achieve stationarity. This is shown in **Figure 10-18**—the differenced data and the associated autocorrelation.

There is basically one real measure-of-merit in selecting the best model, that is its ability to forecast accurately. This accuracy is manifested in the size of the confidence interval and replication of the original time-series data. Experimentally, several combinations of p and q values were attempted, for p and q between 0 and 2, and p+q $\geq$ 1. While it is an art rather than a science, the most parsimonious model was judged to be ARMA (1,1)—assisted again by our a priori knowledge about the generated data. The corresponding output is shown as **Figure 10-19**, where the parameters φ (labelled as PAR), θ (PMA) and constant (CONST) are estimated. Forecast is performed for 12 time-periods, with lags of 1 to 4, as tabulated in columns 1 through 4 in the forecast table. Column 5 of the forecast table shows the 95-percent confidence-interval. Column 6 shows the psi weight ψ, the parameter of

---

[11]An analogous approach would be to take the geometric transform for the discrete function ω^i, for $i = 0, 1, \dots, \infty$.

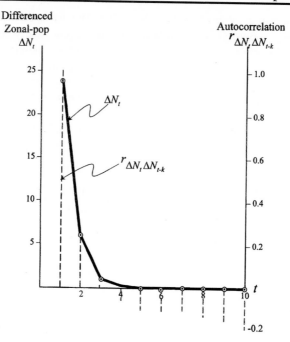

Figure 10-18 Differenced zone-1 population data and autocorrelation

the random-shock form of the difference equation as discussed in Section 10.1.8 (Box and Jenkins 1976). WMEAN is the constant used to center the time-series, and AVAR is the estimate of the random-shock variance. In summary, the fitted model looks like $\hat{z}_t = 3.056 + 0.01415\hat{z}_{t-1} + a_t - 0.2281a_{t-1}$. To read the computer output shown in **Figure 10-19**, one can see that if we are at time-period 1, forecasting one, two, three and four periods ahead will yield 3.06, 3.75, 3.91 and 3.95. Similarly, starting at period 2, the forecasts are 3.10 to 3.11, 3.11 and 3.11. Notice this is the same as the more compact notation using the backshift operator $(1-0.0142B)z_t = (1-0.228B)a_t$ except for an intercept. The discrepancy can be traceable to the fact that the former equation was calibrated on 'raw' differenced data $\hat{z}_t$ and a_t, while the latter was transformed to z_t and a_t by subtracting the mean of $\hat{z}_t$ from the 'raw' data, as indicated in Equation **(10-41)**. Notice the positive sign of the coefficient should be expected considering an a priori positive autocorrelation.

The above represents a privileged understanding of the Lowry–Garin model, which helps us to hypothesize the correct model structure. From there on, more rigorous statistical tests need to be performed to substantiate our intuition. The typical first step is to look at the autocorrelations of the raw data, the first difference, and the second difference. Since the data time-series reaches a stable level very rapidly, it appears that the original, un-differenced data—similar to the differenced data—can pass the test for stationarity. This is in consideration that growth or decline takes place very quickly only in the initial time-periods.

```
WMEAN = 3.10000
CONST = 3.05612
AVAR  = 48.8238

PAR
0.01415

PMA
-0.2281

ALPHA = 0.0500
```

	TEST					
	1	2	3	4	5	6
1	3.06	3.75	3.91	3.95	13.70	0.24
2	3.10	3.11	3.11	3.11	14.09	0.00
3	3.10	3.10	3.10	3.10	14.09	0.00
4	3.10	3.10	3.10	3.10	14.09	0.00
5	3.10	3.10	3.10	3.10	14.09	0.00
6	3.10	3.10	3.10	3.10	14.09	0.00
7	3.10	3.10	3.10	3.10	14.09	0.00
8	3.10	3.10	3.10	3.10	14.09	0.00
9	3.10	3.10	3.10	3.10	14.09	0.00
10	3.10	3.10	3.10	3.10	14.09	0.00
11	3.10	3.10	3.10	3.10	14.09	0.00
12	3.10	3.10	3.10	3.10	14.09	0.00

Figure 10-19 Result of an ARMA (1,1) model

Statistically speaking, one can either treat the first differenced data or the original undifferenced data as stationary. From there on, the ARMA modelling-process starts. If the autoregressive model is the explanatory model, the p value is 1 without too much dispute, as judged from the partial correlation-coefficient. However, the moving-average model is harder to determine.

The typical way of detecting periodicity is through autocorrelation and residuals, where seasonality will show up in the plots. It is not clear which values are 'significantly' different from zero, as one can verify from the autocorrelation plot of differenced data **Figure 10-19**. In **Figure 10-20**, the ARMA(0,1) model for differenced data is shown. From the forecasts, residuals can be computed. All these examinations, combined with the spectral analysis on the generated geometric series, suggest seasonality need not be considered in this analysis.

An ultimate test is the size of the 95-percent confidence-interval and the ability of the model to reproduce the original data. In this regard, all three models—ARMA(1,1), ARMA(0,1) and ARMA(1,0)—for both the differenced and undifferenced cases perform quite well on the surface. The ARMA(1,1) on the differenced data shows a relatively smaller 95-percent confidence interval compared to other models, with the ARMA(0,1) model for the differenced data coming next. For the undifferenced data, the ARMA(1,0) appears to be slightly better among the other models constructed on the same set of data.

Getting to the set of the goodness-of-fit statistics, one can construct a table showing the R^2, t-statistics and F-statistics, among others. **Table 10-10** summarizes all the models we hypothesized, as obtained by Greene and Chan (1991). An examination of this table rules out the ARMA(1,1) models for both the differenced and un-differenced data based on the t-value for one of the two coefficients.

```
WMEAN = 3.10000
CONST  = 3.05612
AVAR   = 48.8035
(PAR is not printed since NPAR = 0.)

  PMA
 -0.2432

ALPHA = 0.0500
```

FCST

	1	2	3	4	5	6
1	3.10	3.85	4.04	4.08	13.69	0.24
2	3.10	3.10	3.10	3.10	14.09	0.00
3	3.10	3.10	3.10	3.10	14.09	0.00
4	3.10	3.10	3.10	3.10	14.09	0.00
5	3.10	3.10	3.10	3.10	14.09	0.00
6	3.10	3.10	3.10	3.10	14.09	0.00
7	3.10	3.10	3.10	3.10	14.09	0.00
8	3.10	3.10	3.10	3.10	14.09	0.00
9	3.10	3.10	3.10	3.10	14.09	0.00
10	3.10	3.10	3.10	3.10	14.09	0.00
11	3.10	3.10	3.10	3.10	14.09	0.00
12	3.10	3.10	3.10	3.10	14.09	-3.10

Figure 10-20 Detecting periodicity through error terms in ARMA(0,1) model

ARMA	p	q	Adj. R^2	F-value	t-value for φ_1	t-value for θ_1
	1	1	0.991	482.	30.9	0.38
undifferenced	1	0	0.992	1077.	32.8	-
	0	1	0.958	229.	-	12.5
	1	1	0.941	64.5	11.4	-0.57
differenced	1	0	0.947	142.	4.8	-
	0	1	0.952	179.	-	11.1

Table 10-10 Goodness-of-fit statistics for hypothesized ARMA models

However, it leaves the rest of the models equally competitive in terms of statistical fit. In other words, based on the number of 'valid' data points, having two parameters may be overfitting.where the constant term, c, is 3.1 (mean of error term). This model has fewer parameters to deal with, has a very slightly lower 95-percent confidence interval and yields pretty much the same forecast values as the competing models. The next best-fit model after this would probably be the ARMA(1,0) model on the un-differenced data, as alluded to above. The reason is that there really is not much difference between the differenced and un-differenced series as far as statistical stationarity is concerned, judging from the autocorrelation and partial autocorrelation coefficients. When we put these facts together, one can even argue for an ARMA(0,0) model on the un-differenced data set. This is based on the additional observation that the data really are two scanty to afford a pattern to be detected.

Overall, we may wish to consider the ARMA(0,1) model for the differenced data, based on its goodness-of-fit and confidence interval in forecasting: $z_t = (1 - 0.2432B)a_t$, where the constant term, c, is 3.1 (mean of error term). This model has fewer parameters to deal with, has a very slightly lower 95-percent confidence interval and yields pretty much the same forecast values as the competing models. The next best-fit model after this would probably be the ARMA(1,0) model on the un-differenced data, as alluded to above. The reason is that there really is not much difference between the differenced and un-differenced series as far as statistical stationarity is concerned, judging from the autocorrelation and partial auto-correlation coefficients. When we put these facts together, one can even argue for an ARMA(0,0) model on the un-differenced data set. This is based on the additional observation that the data really are two scanty to afford a pattern to be detected.

Notice we ruled out the undifferenced model based on our privileged understanding of the problem—the fact that the Lowry–Garin model is a geometric series. Perhaps the point we get out of all these computer runs is th is.Depending on the pattern of data, statistical fit alone may not be able to discriminate the fidelity between competing models. Having an understanding of the underlying process helps a great deal in posing an explanatory structure on top of what seems to be a rather evasive set of data. An ARMA(0,1) model for the *differenced* data suggests that once the 'trend' is removed from the data, the only pattern remaining is the interaction among zones (or the 'noise').

10.5.2 Univariate STARMA model. To harness the richness of the *spatial* data-variation, we proceed to examine the relationship between the population at the four zones over time. A STARMA model was calibrated based on the undifferenced population time-series for the four zones, as documented in **Table 10-9**. Drawing upon the Lowry–Garin model, we extract the first row of the "relative-accessibility-to-residence" $[t_{ij}]$ matrix from the chapter on "Chaos, Bifurcation and Disaggregation," corresponding to the fact that zone 1 is the site of interest. In other words, we take the first row of this matrix

$$\begin{bmatrix} 0.079 & 0.321 & 0.118 & 0.482 \\ 0.104 & 0.278 & 0.112 & 0.506 \\ 0.104 & 0.322 & 0.085 & 0.489 \\ 0.106 & 0.347 & 0.125 & 0.423 \end{bmatrix}.$$

These become the basis for the weights in Equation **(10-63)**: $(w_{1j})^T = (0.079, 0.321, 0.118, 0.482)$. This is essentially the normalized 'trip-distribution' curve, showing the

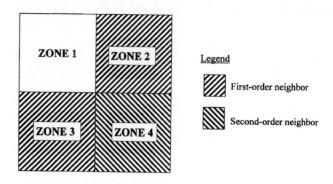

Figure 10-21 Four-zone example

Time period	0th order $L^{(0)}Z_1$	1st order $L^{(1)}Z_1$	2nd order $L^{(2)}Z_1$
1	118	315	605
2	142	383	671
3	148	397	689
4	149	401	693
5	149	401	694
6	149	401	694
7	149	401	695
8	149	401	695
9	149	401	695
10	149	401	695
11	149	401	695

Table 10-11 Spatial-lag data

percentage breakdown of trips commuting to zones 1, 2, 3 and 4 respectively. We define the *0*th-spatial-order as the site-of-interest itself, which is in this case zone 1. The *1*st-order neighbors include zones 2 and 3, while the *2nd*-order neighbor is zone 4. (Refer to **Figure 10-21**.) Correspondingly, $w_{12}^1 = w_{12}/(w_{12}+w_{13}) = 0.321/(0.321+0.118) = 0.731$. Similar calculations yield $w_{13}^1 = 0.269$ The sum $w_{12}^1 + w_{13}^1$ is unity as expected.

According to Equation **(10-62)**, the spatial-lag operator $L^{(l)}$ transformed the original data in zones 1 through 4 into the *0*th-order neighbor, the *1st*-order neighbor

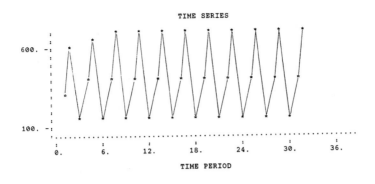

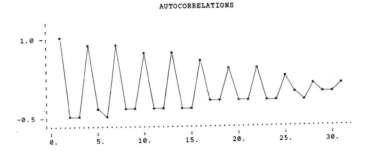

Figure 10-22 Undifferenced series and autocorrelation

and the *2nd*-order neighbor. This is shown in **Table 10-11**. Notice that the *0th* and *2nd* order columns of this Table simply repeat the zonal-population figures in the original data, since there is only one zone in each of these spatial orders. The *1st*-order column is the only one where the weights are applied to transform the original data into another observation used in the STARMA model. For example, the first time-period data-point is calculated as $w_{12}^1 Z_2 + w_{13}^1 Z_3 = (0.731)(381) + (0.269)(134) =$ 315, where the weights w_{12}^1 and w_{13}^1 are normalized among two of the original 4 weights in Equation **(10-86)**.

As suggested in Section 10.3.2 the STARMA model operates on the table, taking the data points row-wise starting from the first row to the second and through the last. As a result, the original 11 data-points are expanded into 33. The regular ARMA procedure is then applied toward the data-set, starting with differencing to obtain stationarity. We examine the original data, the *1st* difference and *2nd* difference in association with the autocorrelations. No clear decision can be made regarding the three series. They all produced a damped sinusoidal autocorrelation plot. (See **Figure 10-22**.) The fact that the autocorrelation function decays to zero and oscillates points toward a negative first autoregressive-

PAR

1	2	3
- 0.4717	- 0.4762	- 0.4759

PMA
- 0.3845

ALPHA = 0.0500

FCST

	1	2	3	4	5	6
1	150.9	389.1	664.9	139.3	160.7	- 0.1
2	387.5	675.0	165.1	393.0	161.3	- 0.4
3	674.9	166.0	395.1	677.8	175.8	0.7
4	167.8	384.2	650.2	158.3	210.6	- 0.2
5	382.7	659.7	182.5	388.4	212.5	- 0.5
6	659.6	183.6	391.1	662.8	225.5	0.6
7	185.3	380.9	636.8	176.5	248.4	- 0.2
8	379.4	645.6	199.0	384.7	249.8	- 0.5
9	645.4	200.3	387.8	648.7	260.2	0.6
10	201.8	378.1	624.1	193.6	277.5	- 0.1
11	376.8	632.3	214.5	381.7	278.5	- 0.4
12	632.1	215.8	385.0	635.2	287.1	0.6

Figure 10-23 Calibration results and forecast based on best STARMA model

parameter φ_1. Furthermore, a φ_1 value of less than unity in absolute value confirms a stationary series (see Section 10.1.4).

ARMA(p,q) models were performed on all three series, with l_p, m_q ranging from 0 through 3. The various calibrated models were evaluated in terms of (a) their ability to reproduce the cyclical input-series (column 1 of **Figure 10-23**), and (b) the magnitude of the 95% confidence-interval (column 5 of **Figure 10-23**). Based on these criteria, the run with $p = 3$ and $q = 1$ on the original (un-differenced) data produced the most parsimonious model *statistically speaking*, resulting in a STARMA$(1_3,1_1)$ model:

$$z_1(t) = -0.4717L^{(0)}z_1(t-1) - 0.4762L^{(1)}z_1(t-1)$$
$$+ 0.4759L^{(2)}z_1(t-1) + 0.3845L^{(0)}a_1(t-1) + a_1(t) \tag{10-87}$$

with mean value $c = 406.79$ and a constant term of 598.79. The series can also be written in a more compact form using the backshift operation B: $(1 + 0.4717BL^{(0)} + 0.4762BL^{(1)} - 0.4759BL^{(2)})z_1(t) = (1 + 0.3845BL^{(0)})a_1(t)$.

To perform diagnostic checking on the various models that were specified and estimated, data from Forgues and Chan (1991) is reproduced here as **Table 10-12**. In the Table, the various (l_p,m_q) combinations are evaluated. The successful ones, as measured by confidence interval and replication of time-series data, are marked with a √ and the less than successful ones with a ×. Of all 15 runs, only the (l_p,m_q) pairs (0,1), (1,0), (1,1), (2,0), (3,0), (3,1) and (3,2) are feasible according to

l_p	m_q	Undifferenced	1st difference	2nd difference
0	*1*	✓	✓	✓
0	2			
0	3			
1	*0*	✓	✓	✓
1	*1*	✓	✓	
1	2			
1	3			
2	0	✓	✓	✓
2	1		✓	✓
2	2		✓	✓
2	3		✓	✓
3	0	✓	✓	✓
3	1	✓	✓	✓
3	2	✓	✓	✓
3	3		✓	✓

Note: Italicized rows are the feasible models according to Equation **(10-13)**.

Table 10-12 Diagnostic check on various (l_p, m_q) combinations

Equation **(10-13)**. Of these, it appears that the models with (l_p, m_q) combinations of (3,0) (3,1) and (3,2) give the most consistent results. Selection between these three contenders is a difficult one based on the confidence interval and replication alone. The (3,1) combination on the original, un-differenced data appears to yield the best result.

Again at this point, it will be useful to reflect upon the process through which the data were generated. It has been pointed out that the first differencing is a logical way to obtain stationarity. The univariate ARMA model was fitted on the first-differenced data with an ARMA(1,1) model by Forgues and Chan (1991) based on confidence interval and replication. While this is a valid model specification, estimation and diagnostic checking later revealed that it might have been an overspecified model. Nonetheless, we build upon the ARMA(1,1) model and estimated a univariate STARMA model based on that foundation—following the identification procedures described in Section 10.1.5. While statistical tests are to be examined, it will not be the only criterion for model selection .Understanding of the underlying Lowry–Garin model—privileged information as it may be—is indispensable. A second examination of **Table 10-12** shows that the first-difference data-set does yield consistently quality models as well. The first-difference column has 11 checkmarks (of which 7 are associated with feasible models), compared with the un-differenced column of 7 checkmarks (all of which belong to feasible models), and the second-difference column of 10 (of which 6 are feasible). By feasibility-violation is meant that IMSL's NSPE subroutine for estimating preliminary values of the autoregressive and moving-average parameters could not converge to a solution. This may reflect in part the violation of Equation **(10-13)** and associated inequalities prescribed for the calibration coefficients. For our purpose, this routine can be thought of as a non-spatial implementation of the procedure outlined in our STARMA extension of the Box–Jenkins' iterative estimation procedure. For details please refer to description of the N-Sample Preliminary Estimation (NSPE) routine in IMSL (1990), which parallels Box-and-Jenkins' (1976) Univariate Stochastic Model Preliminary Estimation (USPE) program.

Based on this set of information, the first-difference data-set is the one adopted here. Putting all the above together, we have a STARMA$(1_3,1_1)$ model. The second best—based on confidence interval, replication and parsimony—has to be STARMA$(1_3,1_0)$. To see this graphically, we summarize the decision criteria for selecting the 'best' model under (a) magnitude of the 95-percent confidence- interval, (b) number of parameters, and (c) the model's ability to reproduce 'cyclic' trend in the merged data-set—concentrating on the (l_p,m_q) pairs (3,0), (3,1) and (3,2) as finalists. In this case all three candidates could reproduce the 'cyclic' trend and therefore not discriminatory; only two criteria remain:

confidence interval for first forecast	*No. of parameters l_p, m_q*
166.0	$3 + 0 = 3$
160.7	$3 + 1 = 4$
153.0	$3 + 2 = 5$

Parsimony suggests STARMA$(1_3,1_1)$ and STARMA$(1_3,1_0)$, precisely in that order of preference. The STARMA$(1_3,1_1)$ model looks like

$$z_1(t) = -0.8809L^{(0)}z_1(t-1) - 0.8581L^{(1)}z_1(t-1) + 0.0821L^{(2)}z_1(t-1) - 0.3555L^{(0)}a_1(t-1) + a_1(t).$$

What this STARMA$(1_3,1_0)$ model means is that, considering spatial interaction, the development trends among neighboring time-series *do* affect one another—as far away as two neighbors over (or all the neighboring zones in the current case). This is reflected by the spatial-lag operation $L^{(2)}$. While spatial interaction also introduces noise, the effect is only local—that the noise does not propagate to neighboring zones, as evidenced by the spatial-lag operator $L^{(0)}$.

10.5.3   Multichannel time-series and econometric model.   Instead of explaining the population at each of the four zones via STARMA, an alternate way is to model them as four separate 'channels' of population time-series. A regression model was constructed to explain the base time-series in terms of its *individual* 'trend' as well as *other* trends. The un-differenced time-series is the data-base, and the data was differenced once. A site of interest, here called the base channel, is specified. The number of regression parameters for each channel in the *differenced* form of the model—meaning the inclusion of t and $(t-1)$ terms—was set to one. The amount of time that each channel is to lag the explanatory base-series was also one. This means the additional inclusion of $(t-2)$ terms.

As it turned out, the following differenced predictive form for zone 1 emerged from the IMSL computer runs, based on least-squares estimation of parameters:

$$z_1(t) = -1.618 z_1(t-1) + 0.618 z_1(t-2) - 0.0493 z_2(t-1) + 0.0493 z_2(t-2)$$
$$+ 0.761 z_3(t-1) - 0.761 z_3(t-2) + 0.0180 z_4(t-1) - 0.0180 z_4(t-2). \tag{10-88}$$

As expected, the forecast for the base series is a function of itself, and of the other three series. Because of the symmetry of the data among the four zones and the stability of these series over time, non-base channels play almost no role in this model. Except for their 1*st* and 2*nd* lags, the other regression parameters "cancel each other out." The model degenerates into an autoregressive model.

One can recognize that Equation **(10-88)** is a special case of the reduced-form Equation **(10-77)** where the noise term is absent. There is one significant difference, however, and that is no simultaneous equation set is constructed though the context warrants it. This little exercise points toward the danger of mis-specification. The above represents what *not* to do in constructing a multivariate time-series model. Instead of including population trends in other zones as the only explanatory variables, perhaps employment figures should also be used. This will transform the model to a more valid econometric formulation.[12]

10.5.4 Transfer-function model. An alternative to econometric modelling is the transfer-function model. It calls for input of two *stationary* input-output time series; or a bivariate data-base is used, where the two variables involved are population and employment. Zone-1 employment is the input series and zone-1 population is the output series, both of which have been differenced to achieve stationarity (**Table 10-13**). Having read in the population/employment data, IMSL estimates the p number of autoregressive parameters and the q number of moving-average parameters. The best model for the input series is ARMA(0,1), while the output series was modelled as ARMA(1,1)—following the procedure as described previously in Section 10.5.1. Many texts explain transfer models through excellent numerical examples, including Makridakis et al. (1983), and Mills (1990). Instead of repeating these examples, here we illustrate transfer functions through a lesson learned on the computer software IMSL. It follows the same philosophy as we have espoused for the previous models (Forgues and Chan 1991). We wish to avoid future mistakes in constructing a bivariate model such as explaining population g r o w t h i n t e r m s o f e m p l o y m e n t .

---

[12]For an example of a much more credible econometric formulation, see the EMPIRIC model in "Spatial Econometric Models."

Employment		Population		Prewhitened series	
E_t	$\dot{E}_t$	N_t	$\dot{N}_t$	α_t	β_t
135	−7	118	24	−7.00	24.00
128	−2	142	6	−0.317	0.230
126	0	148	1	0.0762	0.945
125	−1	149	1	−1.02	−0.227
125	0	149	0	0.245	0.0546
125	0	149	0	−0.059	−0.013
125	0	149	0	0.0142	0.0032
125	0	149	0	−0.0034	−0.0008
125	0	149	0	0.0008	0.0002
125	0	149	0	−0.0002	−0.000

Table 10-13 Input and output series

IMSL is now used to compute estimates of the impulse-response weights and the noise series. The number of terms in the transfer function can be any number k as large as the number of observations. Counting the degree-of-freedom however, we like k to be smaller than the theoretical upper bound, so that the noise series can have more than one term. In other words $k +$ (no. of noise terms) $= n$. A transfer function with too many weights has its drawback in an application context, since it is less parsimonious than fewer number of terms. In the case of Forgues and Chan (1991), their experimental run has 10 weights. Meanwhile, the first-differenced data-series has only 10 data-points (from a total of 11 data-points in the original series.) As the differenced series was used as input, there are some questions regarding the consistency in this modelling approach. The calibrated transfer function looks like:

$$d(B) = -3.38 + 0.02B - 0.04B^2 + 0.17B^3 + 0.18B^4 + 0.23B^5 + 0.29B^6 + 0.28B^7 + 0.33B^8 + 0.35B^9.$$

This says that population trails employment over a period of 9 years, with the commensurate number of impulse-response weights.

Now by the default value of 1, Forgues and Chan (1991) suggested only one constant term for the noise series: 3.4189. This completes the model as stated in model specification Equation **(10-52)**. Prewhitened input and output series are shown as part of **Table 10-13**. We can see the smoothing effect of the prewhitening process by comparing these series with the original input series, albeit small in magnitude. Notice that the number of p and q parameters for the input series must

Differenced input series		Prewhitened input series	
`observed' population	estimated population	`observed' population	estimated population
24	27.059	24.00	27.059
6	15.622	0.230	4.358
1	3.682	0.945	3.456
1	5.688	−0.227	5.031
0	1.818	0.055	1.275
0	1.500	−0.013	2.015
0	0.777	0.003	1.115
0	0.736	−0.001	1.029
0	0.354	0	0.866
0	0	0	0.614

Table 10-14 Validation of transfer function

be specified before the execution of the transfer-function program in IMSL. (Here the model for the input series is ARMA(0,1). This contrasts with the ARMA(1,1) model of the output series.) It is interesting that the ARMA(0,1) model was used to prewhiten the input series (which consists of the differenced data). This is in concert with an earlier finding that the most parsimonious model for the first- difference data was exactly this specification (albeit for the population rather than employment series). An arbitrary judgment was made to use the simplest model for the 'exogenous' employment input series, as it is given and little can be explained about it. A more 'complex' model, however, is warranted for the 'derived' population series in both its trend and noise terms.

Overall, calibration for the transfer-function model is not necessarily a satisfactory one. First, the noise time series is specified to be a single constant term. Actually, the noise is a time series and should be modelled by an ARMA(p_v, q_v) and represented by multiple terms. To prove this point, the population time-series was calculated via Equation (**10-52**) using the transfer function as calibrated (McGee and Chan 1991). The population series so estimated do not necessarily agree with $\dot{N}$ the differenced population-series in **Table 10-13**. If the prewhitened input-series α_t is input to the transfer function, the prewhitened output time-series β_t is again not reproduced accurately. The results, while acceptable possibly from a statistical standpoint, do not correlate well with the 'observed' data—as shown in **Table 10-14**.

In short, the transfer function can have fewer terms. This will allow the noise series to play a more prominent part in the model. Undoubtedly, this will involve more effort than assuming a constant noise term. But it may pay off in the accuracy with which the 'observed' values can be reproduced. Stepping back, the transfer function is not an appropriate model for modelling population and employment in zone 1 generally speaking. As the original data in **Table 10-13** shows, population is increasing while employment is decreasing over time. This does not appear to be reasonable, until one observes that the "study area" really consists of four zones. Employment increases in other zones are the reason for the population increase in zone 1. The local-employment decreases in zone 1 will affect population in the study area, but not necessarily those in zone 1. This speaks for a *spatial*-temporal model where the interaction between zones is as important as, if not more important than, the relationship between *local* population and employment.

10.5.5 Summary. The Box–Jenkins ARMA model is a very versatile forecasting tool. But it is not necessarily suitable for a large number of distinct time-series or when these series interact with one another. As hundreds (if not thousands) of time-series often need to be modelled in land-use and remote-sensing applications, an alternate model other than multichannel time-series may be more suitable. Because of the inherent spatial interaction among geographic subunits and pixels, a spatial-temporal model may in fact be the answer to land-use forecast and remote-sensing applications. But it requires more of the modeler to pose the underlying 'causal' relationship among spatially distributed activities. Thus an intimate knowledge of the geographic area under study is required. In this section, a univariate STARMA model has been calibrated using IMSL on a set of synthesized population/employment data with partial success. The calibration was greatly assisted by an intimate knowledge of how the data were generated from a causal model.

On the other hand, a bivariate transfer function has been estimated somewhat unsuccessfully in the above example, illustrating again the requirement placed on the modeler. Multivariate transfer-function models require a considerable amount of computation. The process goes through not only identification, estimation, diagnostic checking and forecasting, but also prewhitening. While computer programs are becoming more sophisticated and powerful, the ultimate success or failure of these methodologies rests with the modeler, who has to match the problem against the commensurate level of analysis. Perhaps an allied econometric model is a more convenient way for modelling multivariate data since it gives an economic structure beyond extrapolation. Still, there is a limitation on how many variables can be modelled on a practical level. Aside from the familiar statistical assumptions of linearity and homoscedasticity etc., a thorough knowledge of the area under study is again indispensable in model specification, as amply illustrated in the above exercise.

If nothing else, this extended numerical example has served to illustrate the family of forecasting models: ARMA, transfer function, econometric and STARMA. It can be seen that the same set of data can be modelled different ways, with varying degrees of success of course. In the process of running different statistical models, we not only familiarize ourselves with model specification and calibration techniques, but also gain valuable understanding of the relationship between them. These techniques range from purely a correlative structure such as ARMA to the hybrid between causal and statistical model such as STARMA (with transfer function and econometric models somewhere in between).

10.6 Adapting a time series to change

An ARMA or STARMA model as described above assumes a recursive pattern among the data that is held constant for the time horizon for which forecasting is sought. Should there be change in the underlying pattern, techniques need to be devised to model this change. Such change is often assumed to be gradual rather than precipitous, as we will point out below. In the following section, we will deal with abrupt changes, which requires a whole host of new techniques (Greene 1992).

10.6.1 Simple exponential-smoothing. Simple exponential-smoothing is a popular forecasting technique to respond to underlying changes in pattern. A forecasting system using simple exponential-smoothing involves re-designating the model parameters each period in order to incorporate the most recent period's observation. The following represents the fundamental univariate equation in exponential smoothing

$$z_{t+1} = \lambda' x_t + (1 - \lambda') z_t. \tag{10-89}$$

Here z_t is the forecast at time t, x_t is the observation at time t, and λ' is called the *smoothing constant* $(0 \le \lambda' \le 1)$. The model simply says that the next period's forecast is obtained from a weighted combination of the current period's actual and forecast values, where actual observations may represent recent patternal changes in the time series. Notice the data are assumed to be stationary. The model says that an appropriate way to obtain the new estimate is to modify the old estimate by some fraction of the forecast error. Such error results from using the old estimate to forecast in the current period. To see this, Equation **(10-89)** can be re-written as

$$z_{t+1} = z_t + \lambda'(x_t - z_t) = z_t + \lambda' a_t. \tag{10-90}$$

In this light, the simple exponential smoothing model can be thought of as a special case of ARMA. Equation **(10-90)** suggests an ARMA(1,1) model.

A relatively small λ' is used when lots of noise is present in the data. A small λ' (0.01–0.10) will force the forecasting system to react very slowly to changes in the data. As a result, the system will smooth over the noise. Typical values for λ' lie between 0.01 and 0.30. The λ' value is selected by minimizing the mean-squared-error. A relatively large λ' (0.20–0.30) is used when it is necessary to react quickly to changes in the data. A problem arises when an intervention occurs that abruptly affects the underlying process. Values of λ' between 0.01 and 0.30 may take a long

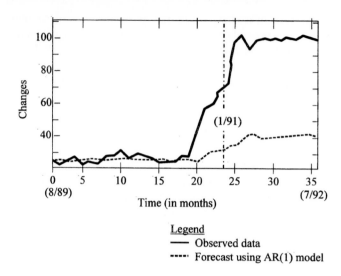

Figure 10-24 Subject-of-interest time series (Greene 1992)

time to adjust to the new level; biased forecasts will occur and will continue for some time before it catches up with the change.

Take the example of an intervention such as the outbreak of the Gulf War in 1991 as shown in the time series of **Figure 10-24**. Such an event affects the time series abruptly. A relatively high value of λ' is expected to model the effect of such an intervention, and often may not be sufficient to respond quickly enough to the change. To show this, overlaid on top of the observed data is a fitted AR(1) forecasting model, which corresponds to a exponential smoothing model with $\lambda' = 0$. It can be seen that the model predictions do not respond readily to the abrupt changes starting in January 1991. There is a definite, but slow response to the stepped-up activities and the model falls far short of responding to the events through 1991 and the first half of 1992. Clearly, it can be seen an AR(1) model is not adequate to capture the dynamics of rapid changes.

Instead of an AR(1) model, let us assume that the analyst is interested in exponentially smoothing the monthly number-of-changes in the subject-of-interest. Given the data shown in **Figure 10-24**, the λ' values and the associated mean squared errors, rounded to the nearest one-thousandth, are computed below.

λ'	mean square error
0.10	597.472
0.20	265.864
0.30	155.023
0.40	104.789
0.50	77.952
0.60	62.227
0.70	52.490

0.80	46.280
0.90	42.311
1.00	39.889

It can be seen that the error decreases as more weight is placed on the current observation (or keen adjustment is made accounting for the discrepancy between what actually happens and what is expected to happen), as one would expect. A λ' = 1 would minimize the error, since such a model simply uses the most recent observation as the prediction one time-period into the future. This way, the forecast will follow the observation x closely (within one time-period). But the mean-square-error is only one out of several goodness-of-fit parameters. Others include the percentage-of-noise smoothed out of the data. The two are certainly not the same.

There is a tradeoff between how to track a time series, which can be accomplished either by modelling abrupt changes in the data due to an 'intervention' or by smoothing noise. A multi-criteria decision-making approach of *goal-setting* and *compromise programming* can produce a desirable λ' value. By way of a definition, goal-setting is the procedure of identifying a satisficing set such that, whenever the decision outcome is an element of the set, the decision-maker will be happy and satisfied. Compromise programming attempts to minimize the gap between the satisficing set and the solution achievable (see 'Multi-criteria Decision-making' chapter in Chan (2000)). Suppose only one decision maker (DM) is involved in choosing the value of λ'. S/he defines the satisficing set as $\{y_1', y_2' \mid y_1' \geq 0.90, y_2' \geq 0.90\}$, where y_1' is the percentage of noise that is smoothed and y_2' is the percentage of the interventions that is modelled. In other words, the DM is interested in smoothing 90 percent of the noise and also capturing 90 percent of the precipitous changes as well.

There is a direct relationship between the half-life of an intervention and the best λ' value. The half life of an intervention is considered that point in time where the change in the mean observation has reached its half-way point. For example, if the number-of-observations changes from 25 in January to 100 in August 1991, or a total change of 75 over the life span of the war, the half-life is that month when the number of observations changes was approximately 37.5. In the Gulf-War example, this corresponds to the month of April. This was three months after the intervention occurred. If the change in the mean occurs very abruptly, a relatively high value of λ' is necessary to stay abreast of the intervention. If the change in the mean observation occurs slowly, a relatively small value of λ' will fit the intervention. There is also a direct relationship between the effect of the intervention and the λ' value. If the percent change in the mean observation is relatively large, a relatively large λ' value is necessary to accommodate the large change. If the percent change in the mean is relatively small, a relatively small λ' value will suffice.

The amount of noise in the data can be determined from a plot of the residual-autocorrelations of the time series and the *Q-statistic*[13] (MicroTSP 1990). An autocorrelation statistically different from zero signifies non-randomness among the residuals. Higher values of the Q-statistic therefore suggest more correlated noise in the data. The capacity of modelling inertia of a system decreases as the Q-static increases. Thus the Q-statistic represents the percentage of noise smoothed and the percentage of the intervention(s) modelled. In other words, a large Q means the model is diligently tracking the 'real' data.

Let us define the decision space (X-space) as three-dimensional with decision variables x_1, x_2 and x_3, where x_1 is the half-life of the intervention, x_2 is the percent-change-in-the-mean, and x_3 the Q-statistic. The following relationships are assumed to exist, where the percentage-of-intervention modelled is: $y_1' = f_1(x_1, x_2, x_3) = (0.50x_1 + 0.35x_2 + 0.15x_3) / 100$. The percentage-of-noise smoothed is: $y_2' = f_2(x_1, x_2, x_3) = 0.60x_3 / 100$. Compromise is the deviation from the goal of smoothing-90-percent-of-the-noise and tracking-90-percent-of-the-changes: $f(y_1', y_2') = |0.90 - y_1'| + |0.90 - y_2'|$. The compromise is to be minimized, as one maximizes the percentage-of-intervention modelled and the percentage-of-noise smoothed. It is also assumed that the above relationships exist such that the decision, outcome and preference-structure space (X-, Y'-, and Z'-spaces) are convex and continuous. An example X-space in the compromise program confines the degree with which one can control the pace of and character of the War: $x_1/100 + x_2 \leq 1.20$ and $x_3 \leq 43$. The above illustrates a compromise program using an l_1-norm for the compromise, from which the decision variables can be characterized, and from which λ' can be determined using a more systematic procedure than an *a priori* means. With a judicious choice of λ', the time series can be closely monitored. Obviously the above is simply a contrived example to illustrate the idea of 'controlling' a time series to track ongoing events, paving the way for the next two topics.

10.6.2 Adaptive-response-rate exponential-smoothing. Simple exponential-smoothing assumes a stationary process (or a constant mean and variance). If a time series contains a varying mean or variance, the forecasts produced by exponential smoothing are not statistically meaningful. If there is a change in the mean, adaptive-response-rate exponential-smoothing can account for the change by continuously updating its λ' parameter. Updating the parameters accounts for changes in the pattern and can deal with changes in trend as well, as one would expect.

---

[13]The Q-statistic is an equivalence of a chi-square statistic, measuring the residual autocorrelations. It can be used to test whether a set of residual autocorrelations, *as a whole*, is statistically significant. It is computed as $Q = j\sum_{k=1}^{m} r_k^2$, where m is the maximum-lag considered, $j = n - d'$ (or the difference between the-number-of-observations and the-amount-of-differencing-to-induce-stationarity), and r_k is the residual-autocorrelation for lag-k. As alluded to above, Q is distributed approximately as a chi-square statistic with ($m-p-q$) degree-of-freedom. A test is typically performed by comparing the Q-statistic with a critical test value (the chi-square value obtainable from typical statistical tables). If the Q-statistic is larger than the critical test value, then one can conclude (usually with a 95-percent confidence level) that the residual autocorrelations are significant. The reverse indicates otherwise.

The most obvious way to react automatically when fitted values go out of control is to increase the value of λ' so as to give more weight on recent data in order to model the change. It is important to lower the λ' value once the changes have occurred in order to reduce the amount of noise modelled. The equations of adaptive-response-rate exponential-smoothing is $\hat{z}_{t+1} = \lambda'(t)x_t + (1-\lambda'(t))\hat{z}_t$. Here

$$\lambda'(t+1) = |E(t)/M(t)| \tag{10-91}$$

in which $E(t) = [1-\lambda'(t)]a_t + \lambda'(t)E(t-1)$ and $M(t) = [1-\lambda'(t)]|a_t| + \lambda'(t)M(t-1)$. In other words, the smoothing constant is determined by the ratio of relative to absolute smoothed-errors, as indicated in Equation **(10-91)**. Both relative and absolute smoothed-errors are set to zero initially: $E(1) = M(1) = 0$. If the model does not need change, the ratio will fluctuate around the same λ' value.

For demonstration purposes, an adaptive-response-rate exponential-smoothing forecast was computed for the Gulf War example. The mean-squared-error using this approach was 54.515. A comparable mean-squared-error of 52.490 occurred when λ' reached 0.70 for simple exponential-smoothing. To the extent that heavy weighting needs to be applied in simple exponential-smoothing (at the expense of more noise) to capture the intervention, an adaptive λ' is deemed more appropriate for this set of data than a fixed value.

Example

Suppose we initialize the process by setting $z_2 = x_1$, $\lambda'(2) = \lambda'(3) = \lambda'(4) = 0.2$, and $E(1) = M(1) = 0$. Forecasts are tabulated in **Table 10-15**. The forecast for period 10, for example, is projected from period 9 by $z_{10} = \lambda'(9)x_9 + (1-\lambda'(9))z_9 = 0.438(220) + 0.562(187.3) = 201.6$. Once the actual value for period 10, x_{10}, becomes known, λ' can be updated and used for the next period's calculations. This entails computing the error, relative smoothed-error and absolute smoothed-error following Equation **(10-91)**: $a_{10} = 277.5-201.6 = +75.9$, $E(10) = 0.2(75.9) + 0.8(-10.3) = +7.0$, $M(10) = 0.2|75.9| + 0.8(45) = 51.1$, and $\lambda'(11) = |7/51.1| = 0.136$. Similarly, the forecast for period 11 can be estimated from period 10 and so on. Note that the λ' values fluctuate quite significantly ranging from 0.040 to 0.438. If a different initializing procedure had been adopted, a different series of λ' values would have been generated (Makridakis et al. 1983). Being a numerical approximation scheme, there are variants to adaptive-response-rate exponential-smoothing algorithm to perform similar functions [aside from Makridakis (1983) and Abraham and Ledolter (1983)]. □

10.6.3 Kalman filter.

Adaptive-response-rate exponential-smoothing can account for change in the mean of the data. Unfortunately, it cannot account for changes in the variance. Kalman filter is the most general model thus far in accounting for an intervention because it can account for variable models, variable parameters, and variable variances simultaneously. All forecasting methods can be cast as special cases of the Kalman filter (Mills 1990, Lutkepohl 1991). The Kalman filter combines two independent estimates of the time series to form a weighed estimate. One estimate is a prior prediction based on prior knowledge and the other is based on new information. The purpose is to combine these two

Period (t)	Observed value (x_t)	Forecast ($\hat{z}_{t+1}$)	Error (a_t)	Relative smoothed error ($E(t)$)	Absolute smoothed error ($M(t)$)	λ' value
1	200.0					
2	135.0	200.0	-65.0	-13.0	13.0	0.200
3	195.0	187.0	+8.0	-8.8	12.0	0.200
4	197.5	188.6	+8.9	-5.3	11.4	0.200
5	310.0	190.4	+119.6	+19.7	33.0	0.462
6	175.0	245.7	-70.7	+1.6	40.6	0.597
7	155.0	203.5	-48.5	-8.4	42.1	0.040
8	130.0	201.5	-71.5	-21.0	48.0	0.199
9	220.0	187.3	+32.7	-10.3	45.0	0.438
10	277.5	201.6	+75.9	+7.0	51.1	0.228
11	235.0	218.9	+16.1	+8.8	44.1	0.136
12		221.1				0.199

Table 10-15 Forecasts using adaptive-response-rate single exponential smoothing (Makridakis et al. 1983)

estimates to get an even better estimate. The Kalman filter is very similar to a Bayesian approach[14] in that prior and sampling information are used to form *a posteriori* distribution. The following is the Kalman-filter equation for a univariate time-series:

$$z_{t+1} = wx_t + (1-w)z_t$$ (10-92)

where

$$w = \sigma_Z^2 / \left(\sigma_Z^2 + \sigma_X^2 \right).$$ (10-93)

As a result

$$z_{t+1} = \left(\sigma_Z^2 x_t + \sigma_X^2 z_t \right) / \left(\sigma_Z^2 + \sigma_X^2 \right).$$ (10-94)

Now comes the philosophical aspect of modelling change. It is fine if observations x are historical. Should X represent anticipated future-events, however, the variance associated with X can be large. As the uncertainty of the future increases so too will the value of the variance of X, σ_X^2, relative to the variance of Z, σ_Z^2. This forces the denominator for the equation for w to increase and w to decrease, putting more weight on z_t relative to x_t according to Equation (10-94). Now w is a

---

[14]For an explanation of the Bayesian approach, see the "Descriptive Tools" chapter in Chan (2000).

variable that changes over time. The parameter w(t) is similar to $\lambda'(t)$ in adaptive-response-rate exponential-smoothing, but $\lambda'(t)$ is calculated using past data and $w(t)$ from *variances*. In operational terms, both the estimates and their *variances* can be computed recursively. But there will be some difficulties—in comparison to Equation **(10-91)**—in obtaining estimates of the variance of x_t for Equation **(10-93)**. The reason is that the former can be computed "as you go" while the latter requires σ_X^2 as an input a priori.

The univariate model can be easily extended to multivariate form by replacing scalars with matrices. Thus Equation **(10-92)** can be rewritten as $\mathbf{z}_{t+1} = \mathbf{W}\mathbf{x}_t + (\mathbf{I} - \mathbf{W}_t)$ Here $\mathbf{z}$ and $\mathbf{x}$ are vectors while $\mathbf{W}$ and $\mathbf{I}$ are weight and identity matrices respectively. Besides updating the weights, a variance–covariance matrix between the $\mathbf{x}_t$ estimates must be obtained (instead of the scalar σ_X^2 in the univariate case), where

$$
\mathrm{var}(\mathbf{x}) = \mathrm{var}\begin{bmatrix} x_1 \\ x_2 \\ \cdot \\ \cdot \\ \cdot \\ x_n \end{bmatrix} = \begin{bmatrix} \mathrm{var}(x_1) & \mathrm{cov}(x_1,x_2) & \cdots & \mathrm{cov}(x_1,x_n) \\ \mathrm{cov}(x_1,x_2) & \mathrm{var}(x_2) & \cdots & \mathrm{cov}(x_2,x_n) \\ \cdot & \cdot & \cdots & \cdot \\ \cdot & \cdot & \cdots & \cdot \\ \cdot & \cdot & \cdots & \cdot \\ \mathrm{cov}(x_1,x_n) & \mathrm{cov}(x_2,x_n) & \cdots & \mathrm{var}(x_n) \end{bmatrix}.
\tag{10-95}
$$

This generalization allows the variance of $\mathbf{x}$ to be computed "as you go."

The Kalman filter examined so far provides for varying parameters and variances, but still assumes a fixed model. By introducing a transition matrix $\mathbf{\Omega}_{t,t+1}$ for the model parameters from period to period, a flexible model is obtained. To be commensurate with this flexibility, let us define a forecasting-error vector $\mathbf{a}_t = \mathbf{x}_t - \mathbf{z}_t$. Similarly, a *measurement* error can be defined

$$
\mathbf{a}'_t = \mathbf{x}_t - \mathbf{\Gamma}_t \mathbf{x}'_t.
\tag{10-96}
$$

Here $\mathbf{\Gamma}_t$ is the observation matrix that specifies what is actually observable and $\mathbf{x}'_t$ is the actual pattern. The evolution of the system is represented by

$$
\mathbf{x}'_{t+1} = \mathbf{\Omega}_{t,t+1}\mathbf{x}'_t + \mathbf{w}'_t
\tag{10-97}
$$

where the transition from period to period is subject to white noise $\mathbf{w}'_t$. The above two equations combined describe the Kalman-filter system in its most general form. It is also referred to as the state-space representation of a time series and $\mathbf{x}_t$ is called the state vector.[15] They can be solved recursively using the following procedure.

---

[15]For a discussion of state-space, see the "Markovian Process" and "Control, Dynamics and System Stability" appendices in Chan (2000).

Given some initial estimates $\hat{x}'_0$, an estimate $\hat{x}'_t$ of x'_t can be made starting with $t = 1$:

$$\hat{x}'_t = \Omega_{t-1,t}\hat{x}'_{t-1} + W_t a'_t \qquad (t=1,2,\ldots,n).$$ (10-98)

Here W_t is the *gain* matrix representing the net percentage of measurement-error or noise that is left after *filtering* by the transition matrix Ω (hence the term Kalman filter). The noise term can be estimated by Equation **(10-96)** and the best estimate of Equation **(10-97)**, and the equation now looks like

$$\hat{x}'_t = \Omega_{t-1,t}\hat{x}'_{t-1} + W_t(x_t - \Gamma_t \Omega_{t-1,t}\hat{x}'_{t-1}),$$ (10-99)

where $\hat{x}'_t = \Omega_{t-1,t}\hat{x}'_{t-1}$. The gain decides how much weight should be given to the most recent forecast-errors.

Let $P_{t-1,t}$ be the covariance matrix for the estimation error $(\hat{x}'_{t-1} - x'_{t-1})$. The matrix generalization of w, or Equation **(10-93)**, can be written

$$W_t = \frac{P_{t-1,t}\Gamma_t^T}{\Gamma_t P_{t-1,t}\Gamma_t^T + R_t}.$$ (10-100)

Here the R_t matrix is given as the variance of the error-vector a'_t in Equation **(10-98)**. It can be shown that

$$P_{t-1,t} = \Omega_{t-1}P_{t-1}\Omega_{t-1,t}^T + Q_{t-1},$$ (10-101)

where Q_{t-1} is a given covariance-matrix of the white noise w_t in Equation **(10-97)**. Beyond the variance–covariance matrix, the transition matrix Ω must be estimated. In summary, we need to exogenously determine $P_{t-1,t}, \Omega_{t-1}, \Gamma_t, R_t, Q_{t-1}$ and x_t before the recursion can begin.

Estimating-a-gain-matrix example

A model is constructed for the sediment of the Des Moines River at Boone, Iowa (Fuller 1996). Here the first equation, an example of Equation **(10-97)**, is the calibrated time-series, representing the actual average-suspended-sediment in parts-per-million (expressed in logarithms). The second equation, representing Equation **(10-96)**, describes the recorded sediment as a function of the actual values:

$$Z_t - 5.28 = 0.81(Z'_{t-1} - 5.28) + \varepsilon_t$$
$$Z_t = Z'_t + a'_t.$$ (10-102)

Here $\Gamma_t = 1$, suggesting that we measure the sediments with 100 percent precision. ε_t are independently-distributed error-terms with mean 0 and variance 0.172. a'_t is the measurement errors with 0 mean and 0.053 variance. The sediment time-series has a nonzero mean,

t	Z'_t	$\hat{Z}_t$	$\sigma^2_{a'_t}$	$\sigma^2_{w'_t}$
1	5.44	5.42467	0.04792	0.50000
2	5.38	5.38355	0.04205	0.20344
3	5.43	5.41613	0.04188	0.19959
4	5.22	5.25574	0.04187	0.19948
5	5.28	5.27588	0.04187	0.19947
6	5.21	5.22399	0.04187	0.19947

Table 10-16 Estimates of sediment constructed with Kalman filter

so the difference $Z_t - 5.28$ plays the role of x_t in Equation **(10-97)**. The transition 'matrix' Ω, or the Kalman-filter, is given as 0.81.

We begin with the first observation of the field data $\{Z_t\}$ as shown in column 2 of **Table 10-16** and the entire data-set is fully documented here: {5.44, 5.38, 5.43, 5.22, 5.28, 5.21, 5.23, 5.33, 5.58, 6.18, 6.16, 6.07, 6.56, 5.93, 5.70, 5.36, 5.17, 5.35, 5.51, 5.80, 5.29, 5.28, 5.27, 5.17}. We call this first entry Z'_1. Suppose we assume Z_t to be a stationary process. We can use the population mean as an estimator of Z_0, the variance of the estimation error $P_{t-1,t}$ is 0.500, which is the variance of the Z_t process. The Kalman-filter equation system becomes $Z_1 - 5.28 = 0.16 = (Z'_1 - 5.28) + \varepsilon_1$ and $0.81\,\hat{Z}_0 = 0.81(0) = 0 = (Z_1 - 5.28) + w'_1$, where $w'_1 = \Omega a'_{0}$. The variance–covariance matrix for the measurement and estimation errors is $\mathrm{var}(\alpha'_1, w'_1) = \begin{bmatrix} 0.053 & 0 \\ 0 & 0.500 \end{bmatrix}$. Therefore, by Equation **(10-99)**. $\hat{Z}_1 = 5.28 + (18.868 + 2.000)^{-1}\,[18.868\,(Z'_1 - 5.28) + 0] = 5.28 + 0.145 = 5.425$, where the "variance of the estimation error" is 18.868, the "variance of the measurement vector" is 2.000 and $\hat{Z}_0 = 0$. The reader may verify that the coefficient for $Z'_1 - 5.28$ is the covariance between Z_1 and Z'_1 divided by the variance of Z_1. The variance of error in the estimator of Z_1 is $\sigma^2_{a_1} = 0.0479$. The estimate of Z_2 is $\hat{Z}_2 = 5.28 + [(0.053)^{-1} + (0.2034)^{-1}]^{-1}\,[(0.053)^{-1}\,(Z'_1 - 5.28) + (0.2034)^{-1}\,(0.81)\,(\hat{Z}_1 - 5.28)] = 0.2075 + 0.7933\,Z'_2 + 0.1674\,\hat{Z}_1 = 5.3836$, where the estimation error $\sigma^2_{w'_2} = \sigma^2_{\varepsilon} + \Omega^2$, $\sigma^2_{a'_1} = 0.2034$ according to Equation **(10-101)**. The variance of $\hat{Z}_2 - Z_2$ is $\sigma^2_{a'_2} = 0.2034 - (0.2034)^2\,(0.053 + 0.2034)^{-1} = 0.0420$.

The estimates and variances for the remaining observations are given in **Table 10-16**. Note that the variance of the measurement error is approaching 0.04187. This limiting variance is denoted by $\sigma^{2'}_a$. The variance of the estimation error w'_t stabilizes at $\sigma^2_{w'} = \sigma^2_{\varepsilon} + \Omega^2\sigma^2_a = 0.172 + (0.81)^2(0.0419) = 0.19947$. It follows that Equation **(10-99)** stabilizes at $\hat{Z}_t = 5.28 + 0.81\,(\hat{Z}_{t-1} - 5.28) + 0.7901\,[Z'_t - 5.28 - 0.81\,(\hat{Z}_{t-1} - 5.28)]$, where $w_t = \sigma^2_{w'}(\sigma^2_{a'} + \sigma^2_{w'})^{-1} = 0.7901$ as shown in Equation **(10-100)** and the analogy in Equation **(10-93)** and $\Omega = 0.81$ as given. □

Per-iod (t/k)	x_t	$\hat{x}_t$	cov($\hat{x}$)	# observations (n)	Sum of squares ($n\sigma^2$)	$\sum_t \ln$ [det $\mathbf{P}_t$] or $\ln(\Pi_k q_k^t)$	Error ($a_t = x_t - \hat{x}_k'$)	cov(a_t') $\overline{\sigma^2}$
1/1	4.4	4.37	0.94	1	0.00	2.833	0.400	17.00
2/1	4.0	4.37	4.94	1	0.00	2.833	0.400	17.00
2/2	4.0	4.06	0.83	2	0.03	4.615	-0.376	5.941
3/2	3.5	4.06	4.83	2	0.03	4.615	-0.376	5.941
3/3	3.5	3.59	0.82	3	0.08	6.378	-0.563	5.832
4/3	4.6	3.59	4.82	3	0.08	6.378	-0.563	5.832
4/4	4.6	4.42	0.82	4	0.26	8.141	1.003	5.829
5/4	–	4.42	4.82	4	0.26	8.141	1.003	5.829

Table 10-17 Filtered and one-step-ahead estimates

Recursion example

We will compute the filtered estimates and one-step-ahead estimates for a *scalar* discussed by Harvey (1981). The observation equations **(10-96)** through the state equation **(10-98)** are rewritten as follows for simplicity.

$$x_t = \hat{x}_t + a_t$$
$$\hat{x}_{t+1} = \hat{x}_t + w'_{t+1} \qquad (t=1,2,3,4) \qquad \textbf{(10-103)}$$

Here a_t are *identically* and independently distributed normal with mean 0 and variance σ^2, and the w_t are *identically* and independently distributed normal with mean 0 and variance $4\sigma^2$, and $\hat{x}_t$ is distributed normal with mean 4 and variance $16\sigma^2$. Notice that $\Gamma_t = 1$ and $\Omega_{t-1,t} = 1$ in this recursion example. The recursions from applying Kalman filtering are tabulated in **Table 10-17**, where cov($\hat{x}$) is an $n \times n$ matrix (as shown in Equation **(10-95)**) such that cov($\hat{x}$)σ^2 is the mean-square-error matrix for $\hat{x}$, and n is the number of observations included in the algorithm. Notice the entries in these matrices are identical. $\sum_t \ln[\det \mathbf{P}_t]$ is the natural logarithm of the determinant of var($\mathbf{x}$)$/\sigma^2$, where σ^2 is estimated by the sum of squares $a_t^T \mathbf{P}_t^- a_t /$ Here $m = \sum_t n_t$, where n_t is the size of the a' vector. If σ^2 is known, the $\mathbf{R}_t$ and $\mathbf{Q}_t$ can be input as the variance–covariance matrices exactly. The earlier discussion is then simplified by letting $\sigma^2 = 1$, as shown previously in all the equations. If $\mathbf{P}_t$ is singular, $\mathbf{P}_t^{-1}$ becomes the generalized inverse, n is now $\sum_t \text{rank}(\mathbf{P}_t)$. $\sum_t \ln[\det \mathbf{P}_t]$ is replaced by $\ln(\Pi_k q_k^t)$, where $\ln(\Pi_k q_k^t)$ is the natural log of the product of the nonzero eigenvalues of var($\mathbf{x}$)$/\sigma^2$. This natural log paves the way for maximum likelihood estimation of the transition matrix Ω (the Kalman filter). The computations in **Table 10-17** converge toward a stable variance for the estimator error (5.83), from Equation **(10-101)** (IMSL 1990).

Notice in the table that the algorithm is executed twice in each time-period t, once for the filtered estimate (k) and another for the one-step-ahead estimate (t). Another noticeable feature is that the observations x_t are included one at a time into the algorithm (instead of being computed a priori exogenously as a scalar σ_X^2). n, the number of observations included, becomes the current rank of the variance–covariance matrix for $\mathbf{x}$. The general features of this algorithm have been shown through Equation **(10-92)** to Equation **(10-98)**, where the variance of the error vector $\mathbf{R}_t = \sigma^2$ and variance of the white noise $\mathbf{Q}_{t-1} = 4\sigma^2$. Computational details are implemented in the KALMN computer routine of IMSL(1990). The example as presented here shows adequately the elaborate computational nature of the Kalman-filter procedure—even for only four time-periods. It is clear from **Table 10-17** that several intricate computations are involved. These include covariances, sum of squares, and eigenvalues. Worse still, these have to be repeated for each time-period (twice). For this reason, the computer (rather than hand calculation) is required even for a simple illustration. Hopefully, this example introduces the readers to one such computer package, IMSL. □

Varying-Trend example

In this example, we will work with a randomly varying trend with added noise (Brockwell and Davis 1991). First, the 'intercept' coefficient b is constant in time series $\{Z_t, t = 1, 2, \dots \}$. Let $\{w_t'\}$ consists of white noise normally distributed with 0 mean and variance σ^2. Z_1 is a random variable uncorrelated with the noise series $\{w_t', t = 1, 2, \dots \}$, then the process $\{Z_t, t = 1, 2, \dots \}$ defined by

$$Z_{t+1} = Z_t + b + w_t' = Z_1 + b + w_1' + w_2' + \dots + w_t' \qquad (t=1,2,\dots) \qquad \textbf{(10-104)}$$

has approximately linear sample-paths if σ is small (perfectly linear if $\sigma = 0$). In other words, $Z_2 = Z_1 + b + w_1'$, $Z_3 = Z_2 + b = Z_1 + b + w_1' + w_2'$, $Z_4 = Z_3 + b = Z_1 + b + w_1' + w_2' + w_3'$, etc. The sequence $\{w_t'\}$ introduces random variation into the slope of the sample-paths. To construct a *state-space* representation for $\{z_t\}$, we introduce the observation vector $\mathbf{x}_t' = (Z_t, b)^T$. Then the series as represented by Equation **(10-104)** can be written in the equivalent form

$$\hat{\mathbf{x}}_{t+1} = \begin{bmatrix} 1 & 1 \\ 0 & 1 \end{bmatrix} \hat{\mathbf{x}}_t + \mathbf{w}_{t+1}' \qquad (t=1,2,\dots) \qquad \textbf{(10-105)}$$

where $\mathbf{w}_t' = (w_t', 0)^T$. The process $\{Z_t\}$ is then determined by the observation equation $Z_t = [1, 0]\hat{\mathbf{x}}_t$. An additional random-noise-component can be added to Z_t, giving rise to the sequence **(10-96)**

$$x_t = [1 \ \ 0]\hat{\mathbf{x}}_t + a_t; \qquad (t=1,2,\dots), \qquad \textbf{(10-106)}$$

where $\{a_t\}$ again consists of white noise and is normally distributed with zero mean and variance s^2. Here $\mathbf{a} = (a_t)$. If $\{\hat{\mathbf{x}}_t, w_1', a_1, w_2', a_2, \dots\}$ is an orthogonal sequence,[16] the Equations **(10-105)** and **242** constitute a state-space representation of the process $\{x_t\}$, following

---

[16]The random vectors $\mathbf{X}$ and $\mathbf{Y}$ are said to be *orthogonal* if the matrix $(\sum_{i=1}^{n} \sum_{j=1}^{n} x_i y_j)/n$ is zero. In other words, $\mathbf{X}$ and $\mathbf{Y}$ are not correlated.

Equation **(10-103)** above. Specifically this is a model for data with *randomly varying trend* and added noise. For this model we have the transition matrix $\Omega = \begin{bmatrix} 1 & 1 \\ 0 & 1 \end{bmatrix}$, the observation matrix $\Gamma = [1, 0]$ in this case shows that the data are perfectly observable. The covariance matrix of white noise $\mathbf{w}'$, $\mathbf{Q} = E[\mathbf{w}'_t(\mathbf{w}'_t)^T] = \begin{bmatrix} \sigma^2 & 0 \\ 0 & 0 \end{bmatrix}$, the variance-covariance matrix of noise $\mathbf{a}$ reduces to $\mathbf{R} = E[\mathbf{a}_t(\mathbf{a}_t)^T] = s^2$ and the two noise-series $\mathbf{w}'$ and $\mathbf{a}$ are orthogonal as mentioned. $E[\mathbf{w}'_t(\mathbf{a}_t)^T] = 0$.

Hopefully, this example helps to further illustrate the various matrix notations used in Equations **(10-96)** through **(10-101)**. Simple as it may be, it also serves to review again some basic concepts about stationary vs. non-stationary time-series, particularly how a model can possibly change over time through the transition matrix Ω. Here, Ω is readily identified due to the model structure and properties imposed a priori. This is not so in general; elaborate procedures such as the maximum-likelihood estimator has been employed to approximate Ω (and other matrices such as $\mathbf{Q}$ and $\mathbf{R}$.) An example is the bifurcation case of the Lowry–Garin model, when the derivation-allocation matrix Ω changes because of zoning ordinance or policy on the threshold-retail-activity. In this case the 'transition' matrix Ω is more appropriately written in its complete form as $\Omega_{t,t+1}$, which is revised from period t to $(t+1)$ as described in the "Chaos, Catastrophe and Bifurcation" chapter. Instead of the maximum-likelihood procedure, Fibonacci search is used, as explained in the "Prescriptive Tools" chapter in Chan (2000).

Now one possible extension to this example is to go through the recursions involving the set of observation and state equations. This is similar to what we have done in the first example. This may be more computationally involved than previous examples since a more elaborate set of equations are used. Obviously, data have to be generated first (possibly through simulation) to form the observation series $\{Z_t\}$ before this exercise can begin (Koopman 1997). □

10.6.4 Concluding comments. The above discussion for all three methods—simple exponential smoothing, adaptive-response-rate exponential smoothing, and Kalman filter—started with univariate time series. Extension into multivariate time-series is relatively straightforward in theory, as shown in the most general Kalman-filter representation. But it is much more requiring in practice. Operationally speaking, all three approaches update their estimates using new and old information that exist. The weights combining new and old information for the exponential-smoothing approaches are functions of past data. The weights for the Kalman filter are functions of variances from past data. Once initial estimates are made, subsequent estimates for all three approaches can be computed recursively. But the recursions can take on rather elaborate computations each time, as illustrated in the Kalman-filter example.

There are arguments for and against using simple exponential smoothing and adaptive-response-rate smoothing. When using simple exponential smoothing, the analyst is given the freedom to use any λ' value. These values can be based upon the characteristics of the time series and the expectations on the intervention's effect. The λ' value for an adaptive-response-rate exponential-smoothing system, however, is based largely upon the data. In other words, the analyst has comparatively little input into the value of λ'. Calculating the best λ' value from the simple exponential-smoothing model requires much judgment, while the adaptive-response-rate method removes the dilemma in determining the optimal λ' value. When there is some insight into the effect of intervention, we recommend simple exponential smoothing using perhaps a compromise-programming approach to determine the best λ' value. However, the adaptive method may be more practical in forecasting multivariate time-series with many data points. The simple exponential-smoothing approach does not account for changes in the mean or the variance of the time series explicitly, but it is actually carried out implicitly by the λ' value. The adaptive-response-rate exponential-smoothing approach updates the mean of the series, but does not account for changes in the variance. The Kalman filter, being the most comprehensive of all three approaches, accounts for both changes in the mean and the variance simultaneously.

As with any analytical tool, there are shortcomings among all three approaches. Classical statistical estimation, such as the simple exponential approach, attempts to minimize the mean squared-error. Minimizing the mean squared-error is an appropriate criterion for past observations but may not be for future predictions. At the same time, there are some major problems involving the Kalman filter that raise many technical questions not being answered satisfactorily. Initial estimates for the parameters and variances are difficult to calculate simultaneously in the simple univariate-case. The problem is compounded by the need of initial covariances and the transition matrix in the pay-as-you-go case (Makridakis et al. 1983). Each update of the estimates using the Kalman filter requires the calculation of the covariances and the matrices that govern transition, which is not easily obtainable computationally.

A spatial analogy can be drawn with the Iterative Conditional Mode (ICM) algorithm discussed in the "Remote Sensing and Geographic Information System" chapter (Chan 2000). There is a "chicken and egg" paradox in the application of both the Kalman filter and ICM. In the ICM, the accurate estimate of β, the smoothing parameter among pixels, is an art. Oftentimes, it amounts to a trial-and-error process. We can say the same for estimating the Kalman filter Ω. Similar comments can be made about estimating such parameters as μ_i, the grey-value means in each land-cover feature (whether it be a lake or forest or corn field etc.) Just like the error variances in Kalman filter, the 'true' values of these parameters are in theory unobtainable. We are happy to settle with 'guestimates'. On top of all these intricacies, Kalman filtering addresses the question of observability, adding yet an additional level of complexity to an already involved process.

In summary, there are already technical problems with constructing a regular STARMA model, as pointed out in the first part of this chapter. Including a new pattern into existing time series is an even more challenging task. This is made more so considering most challenges encountered with univariate and multivariate applications would be compounding manifoldly with their spatial-temporal counterparts. Last, but not at all the least, is that in the case of external interventions, they often need to be interpreted and anticipated rather than simply gleaned from a change in historical information. The ability to quantify future intervention poses the single most challenge in this type of modelling.

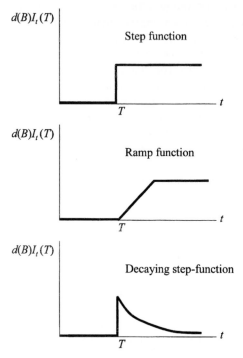

Figure 10-25 Interventions involving shifting of the mean

10.7   Intervention analysis

In the last section, we discussed the introduction of a new pattern into an existing time-series through three conventional methods. The precondition, however, for these methods to work is that the new pattern evolves gradually and not precipitously. Here in this section, we discuss abrupt changes in pattern, an analogy to bifurcation in activity-allocation models discussed in the middle part of this book. The term used in this context is *intervention analysis* (Greene 1992).

A time series of observations can sometimes be affected by external events, commonly called *interventions*. An example of intervention is a doctrinal, policy or operational change. Most interventions result in either a shift in the mean of the time series or a shift in the trend of the time series. Since the two are equivalent through one differencing of the data, a shift in the mean of the time series will be our focus for purposes of this discussion. This shift in the mean can occur in various forms. Common forms of this shift include step functions where the mean changes abruptly, ramp functions where the mean changes gradually in a linear fashion, and exponential functions where the mean changes in a step function that decays over time. The form of the shift may even be a combination of different functions. Graphic illustrations of these sample interventions are shown in **Figure 10-25**.

To model such interventions, classical statistics require the use of a control group that is not affected by the intervention. But such a control group is not often available to determine a change in mean value. It is common practice to use a Student's *t*-distribution to test for a change in mean value instead.[17] Unfortunately, the Student's *t*-distribution assumes independent observations and that the change in the mean value can be represented by a step function. There is usually a correlation, or dependence, among successive data observations in a time series, as when a ramp function intervention is involved. As a result, the Student's *t*-test is not applicable for intervention analysis in time-series data and no quantitative measure of effect of an intervention can be easily obtained.

Intervention analysis is the classical method of dealing with external events that affect a time series or a process. There are instances when other methods are preferred. These techniques can be applied to account for patternal changes: simple exponential-smoothing, adaptive-response-rate exponential-smoothing, and Kalman filtering. We have already reviewed them one-by-one and pointed out their assets and liabilities. With that background, we will pose intervention analysis as an alternative to these three techniques.

10.7.1 Intervention analysis. When using intervention analysis, a transfer function is generally used to model the effect of an intervention. Box and Tiao (1975) use a transfer-function-noise model that can describe the effects of an intervention on a time series in the following manner:

$$Y_t = d(B) I(t) + z_t \qquad\qquad (10\text{-}107)$$

where Y_t and z_t are the same time series before intervention, representable by an ARIMA model, $E(Y_t) = z_t$ precisely. $d(B)$ is a transfer function that allows for the effect of the intervention, often found empirically. $I(t)$ is the 0–1 indicator sequence reflecting the absence and presence of the intervention. Equation **(10-107)** assumes that the time- series model parameters are the same before and after the intervention. The model also assumes that the intervention can be represented as an additive effect of the dichotomous variable $I(t)$ and the original noise-series.

Mills (1990) gave an excellent discussion of how to incorporate such interventions in a time series. It is one thing to "let the data speak for themselves," and another thing to believe that such interventions are often exogenous and are best modelled independent of the time-series data. We take the view here that an independent model can be constructed. Several difficult problems of such versions of intervention analysis are

---

[17]In essence, we wish to test whether the means of two separate samples $\{x_i\}$ and $\{y_i\}$—each consisting of n independent observations—are statistically different. We take the difference between each pair of these observations, $\delta_i = x_i - y_i$. The *t*-statistic $\bar{\delta}/s_{\bar{\delta}}$ is formed. The null-hypothesis regarding this statistic is tested with the aid of a *t*-statistic table. If the null hypothesis is accepted, we say that the two means are not significantly different. If the hypothesis is rejected, we conclude that they are different. One can find a numerical example of such *paired-t test* in any introduction statistics text.

a. determining what external events are interventions to the time series;
b. deciding if and when an intervention occurs; and
c. finding exogenously the $d(B)$ function.

In view of these difficulties, it is often required to follow an opinion survey among experts in order to answer these three questions, following a multi-criteria-decision-making (MCDM) philosophy.

10.7.2 An example. To illustrate this approach, we will work out the Persian-Gulf-War example through each step (a), (b), and (c). Factor analysis can be used in the first step to decide what external events are likely to intervene on the time series. We are interested in determining what type of intervention is likely to occur that will drastically change the mean value of the time series. A survey among experts will be used to produce such information. Let us suppose there were nine U.S. Department of Defense experts who responded to the survey. Such survey asks which factors are likely to cause an abrupt increase in activities—henceforth called the subject-of-interest—at the outbreak of the Gulf War. All nine experts said that the subject-of-interest being modelled never changes due to event A, B or C. On the other hand, there was at least one expert who responded that there is a change in the subject-of-interest based upon a change in the remaining events D, E, F, G or H.

What external events are interventions?

A design matrix $\mathbf{X}$ represents the 0–1 entries of x_{ij}, where i is the expert respondent ($i = 1, 2, ..., n$) and j is the potential intervention ($j = 1, ..., m$). Thus $x_{ij} = 1$ if expert i thinks that a change in j will result in a change in the subject-of-interest being modelled. Let us say the design matrix[18] looks like

$$\mathbf{X}^T = \begin{bmatrix} 1 & 1 & 1 & 1 & 1 & 0 & 0 & 1 & 1 \\ 1 & 0 & 0 & 1 & 1 & 0 & 1 & 0 & 0 \\ 1 & 1 & 1 & 1 & 0 & 0 & 1 & 0 & 1 \\ 0 & 0 & 0 & 0 & 0 & 0 & 1 & 1 & 1 \\ 0 & 0 & 1 & 1 & 0 & 0 & 1 & 0 & 1 \end{bmatrix}. \tag{10-108}$$

The dependent-variable vector $\mathbf{Y}$ has nine elements. $Y_i = 1$ if expert i thinks that changes in the subject-of-interest occur. *Factor analysis* answers the following question: "Given an external event i has occurred, what is the probability that the external event will intervene and result in a change in the observations?"

To calculate the probability that intervention i ($i = 1, ..., 5$) can cause a change in the subject-of-interest, multiple regression is performed. The serious problem of

---

[18]Matrix representation of linear regression is explained in the "Statistical Tools" appendix (Chan 2000).

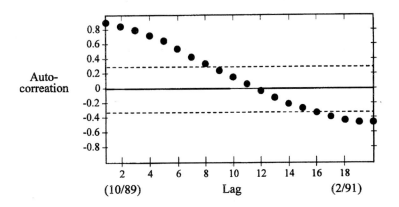

Figure 10-26 Autocorrelations of residual from AR(1) model of Gulf-War data
(Greene 1992)

near singularity will arise if the **Y** vector contains only values of 1 or only values of 0. For the current example, let us say all responses indicate that a change has occurred, except expert 6, who stated that changes in the subject-of-interest never occurred. This is represented by $y_6 = 0$ and the corresponding row vector $\mathbf{X}_6 = (0, 0, 0, 0, 0)$. The overall **Y** vector now looks like $\mathbf{Y}^T = (1, 1, 1, 1, 1, 0, 1, 1, 1)$, with the remaining entries of unity (except the sixth).

A multiple-regression model of the following form can be obtained by running a model with all five explanatory variables. From the regression coefficients (called *factor loadings*) and their t-values, one can discern the significant and insignificant variables or factors. The only factor that appears to be significant (at a 0.05 confidence level) out of such a master equation is X_1. It has a large factor-loading and a significant *t*-statistic as shown right below the loading in parenthesis:

$$Y = \quad 0.074 + 0.606X_1 + 0.170X_2 + 0.234X_3 + 0.394X_4 + 0.053X_5$$
$$\qquad\qquad (4.367) \quad (1.579) \quad (1.502) \quad (2.835) \quad (0.373)$$

Here X_i ($i = 1, \dots, 5$) is a 0–1 dichotomous variable showing whether events D, E, F, G and H have occurred. A factor loading can be interpreted as the probability for the subject-of-interest changing given that the corresponding event has occurred. The result of the factor analysis suggests that given a change has occurred in event D, there is 60.6 percent that the subject-of-interest will change.

Expert	Response	Rank	Weighted response
1	90	3	270
2	120	7	840
3	110	2	220
4	100	6	600
5	85	8	680
6	130	1	130
7	70	4	280
8	100	9	900
9	120	5	600
Sum	100 (weighted)	45	4500

Table 10-18 Experts' responses to maximum-monthly-number-of-changes

Did an intervention occur?

From the AR(1) model fitted out of the Gulf-War data in **Figure 10-24**, it can be concluded that the Gulf War is an intervention on the changes in the subject-of-interest. The following facts led to this conclusion:

a. The plot of the time series.
b. The differences in coefficients of the AR(1) model that were found from the data dating back before the Gulf War and the AR(1) model determined from all of the data.
c. The plot of the residuals of the time series fitted with the AR(1) model from the data before the Gulf War.
d. The plots of the residual autocorrelations as displayed in **Figure 10-26**. (It is obvious from the plot that after January 1991, all of the residuals are significantly different from zero, as suggested by the band drawn around the time axis). It is also concluded that the event occurs so precipitously that regular time-series model such as AR(1) is ill equipped to model the situation (see **Figure 10-24**.)

What is the shape of the transfer function?

The next step of intervention analysis is to estimate the $d(B)$ function of all interventions. The $d(B)$ function must be calibrated exogenously from the data. A way to calibrate the $d(B)$ functions is to apply the fractile method, which involves questioning a group of experts on the form and shape of the $d(B)$ function. The

Expert	Response	Rank	Weighted response
1	July (6)	3	18
2	February (1)	7	7
3	May (4)	2	8
4	April (3)	6	18
5	May (4)	8	32
6	June (5)	1	5
7	April 15 (3.5)	4	14
8	May 15 (4.5)	9	40.5
9	August 15 (7.5)	5	37.5
Total	May (4) [weighted]	45	180

Table 10-19 Experts' responses to when-50-changes-will-occur

fractile method is commonly used to measure utility functions in multi-criteria decision-making. (Please reference the chapter in Chan (2000) on this subject). The first step in calibrating the $d(B)$ function is to determine what form of the function is most likely to occur starting in January of 1991. This is done by examining a plot of the time series of interest. From the data in **Figure 10-24**, the analyst believes that the form of $d(B)$ is either an exponentially-increasing function or a ramp function.

On the average, 25 changes in the subject-of-interest were made before January 1991. The analyst asks the experts what are the maximum-monthly-number-of-changes in the observations that will occur after the Gulf War started. The same nine experts that performed factorial analysis are used, but this time, it is assumed that the experts differ in their background and knowledge of the subject-of-interest. A simple way to apply the experience ranking is to

a. Multiply an expert's response by his/her ranking,
b. Add up all the weighted responses, and
c. Divide the weighted responses by the sum of the ranks ($i = 1, 2, ... , 9$).

Table 10-18 shows the experts' responses for the maximum-number-of-changes in the subject-of-interest that will occur after the Gulf War started. Included in the Table is the weighted-sum-of-responses, which is 100.

The next step is to ask the experts in what month will the total number of changes in the subject-of-interest increase to some intermediate value between 25 and 100, let us say 50. **Table 10-19** lists the responses of the nine experts, where the number in parentheses is the number of months that response is from January.

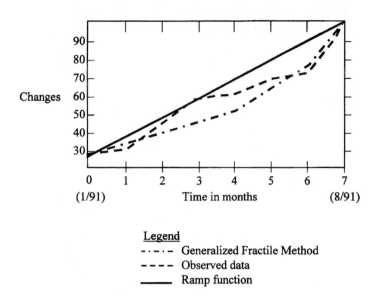

Figure 10-27 Transfer function from the Generalized Fractile Method

For example, April is 3 months from January and so on. Included in the table is the weighted-sum-of-responses, May, or 4 months from the start of the War.

Following the same line of questioning, we asked the experts in what month will the total number of changes increase to another intermediate value, such as 75. A similar table of responses can be compiled. In this case, it resulted in a weighted response of July, or more precisely 5.867 months from the start of the War. The analyst may continue this line of questioning for intermediate values. The final question for the experts is when will the number of monthly changes increase to the maximum number, already found to be 100. The data shows a weighted response of August, or more precisely a round down from 7.3 months from the start of the War.

After the questioning is complete, the analyst will have a piecewise linear approximation to the transfer function $d(B)$. The analyst initially thought that the $d(B)$ function would be either an exponentially increasing function or a ramp function. For comparison purposes, **Figure 10-27** is a plot of the actual monthly-number-of-changes in observations. It is a fit to the data using a constant slope of 10.714 from January 1991 to July 1991, and a linear approximation to the $d(B)$ function.

If the actual data is fitted using a constant slope of 10.714, the mean-squared-error from January 1991 to July 1991 is 571.0. If the actual data is fitted using the piecewise linear approximation to the transfer function $d(B)$, the mean-squared-error from January 1991 to July 1991 is 337.0. The piecewise linear approximation resembles an exponentially increasing function more than a ramp function and the mean-squared-error is considerably lower for the piecewise linear approximation. Thus it can be concluded that the ramp function is not appropriate for the calibration of the transfer function in this case. An exponential function, as obtained from the fractile method, is deemed more appropriate.

10.7.3 *Comparison of intervention analysis and Kalman filter.* Both our MCDM version of intervention analysis and the Kalman filter are ways of accounting for an external event's effect on a time series. Intervention analysis has some advantages and some disadvantages over the Kalman filter. It becomes necessary to use a Kalman filter as opposed to intervention analysis when intervention cannot be identified *ex post*. Such a situation can arise in several ways: (a) when there are no experts in the field; (b) the experts do not know the answers to the analyst's questions; (c) the experts are not willing to tell the analyst the answers, or (d) the analyst does not have time or the resources to conduct an interview . Under these circumstances, one replaces exogenously-determined interventions with endogenously-determined filters. Experts-in-the-field are among the most important aspect of intervention analysis. Without expert judgments, the intervention analysis procedure as described above cannot be conducted.

MCDM intervention-analysis, through subjective judgment, identifies when the intervention occurred or would occur. The Kalman filter, through interpreting the time series, tells approximately when an 'intervention', or more accurately a change of pattern in the data, might have occurred. The analyst would want to use MCDM intervention-analysis if the nature of the intervention, and/or the time an intervention started or would start is important. On the other hand, a Kalman filter can account for a change in pattern without recognizing the particular incident that causes the change. MCDM intervention analysis tells what the intervention is, while Kalman filtering does not. MCDM intervention analysis can be described as more causal since a determination is made regarding what is causing the precipitous change in the time series. With the Kalman filter, all one knows is that there is a change.

A Kalman filter explains changes through the time-series data, specifically in terms of the mean and the variance .There is no need to recognize when the incident occurs or how. MCDM interventions-analysis determines switches in pattern exogenously. These switches can be precipitous events, sometimes referred to as outliers or bifurcations. A Kalman filter continuously groups the data. For example, a Kalman filter may have truncated the number of subjects-of-interest changes into, say, five groups within each is assumed a uniform pattern. Pattern changes occur from group to group. MCDM intervention-analysis simply truncates the data into two groups: one accrued before January 1991 and the other occurred after January 1991 in our example. A Kalman filter uses past and present information contained in the data to calculate its parametric estimates. MCDM intervention-analysis uses exogenous information along with past and present observations to calibrate its estimates. As a result, the Kalman filter has to ferret out the information that it needs to compute its estimates and quite often, this information is difficult to discern. Because MCDM intervention-analysis determines the effect of the interventions exogenously, there is little chance of modelling noise. There is a built-in capacity, on the other hand, for the Kalman filter to pick up noise.

In short, there are two philosophies in modelling time series. One is to let-the-data-speak-for-themselves. The other is to have judgment merged with data in the modelling process. Kalman filter exemplifies the first school of thought while MCDM intervention-analysis the second. As such, Kalman filter is more adept in explaining a change in pattern, while MCDM intervention-analysis renders the additional capability of determining and anticipating the nature of the change, even when changes are abrupt. A middle ground modelling approach was taken by Wright (1995), in which he analyzed ground-water pollution combining flow equations with empirical measurements. He relied on the equations to filter out intervention by well pumping. To the extent he did not have full information on pollution diffusion, he calibrated a transfer function to let "data speak for themselves." In this way, he captured the best of both worlds in analyzing the effect of pumping in controlling the spread of pollution beyond the limits of the clean-up site. Wright's analysis is captured in a homework problem in the "Exercises and Problems" web site.

10.8   A univariate STARMA case study

The United States Department of Defense (DOD) employs a worldwide-sensor system to detect certain 'events' of interest. A tasking model was designed to allocate the scarce sensor-resources so as to optimize the detection of these events (Greene 1992). The success of the tasking algorithm depends upon good estimates for the joint probabilities P_{ijk}, aggregated monthly, of event type i ($i = 1, 2, 3$) occurring in area j ($j = 1, 2, \ldots, 22$) at time k given that an event-type i occurred at time k ($k = 1, 2, \ldots, 12$). The time-of-day k is divided into two-hour time-blocks. For example, $k = 1$ may represent the two-hour time-block from 6:00 a.m. to 8:00 a.m., $k = 2$ may represent the two-hour time-block from 8:00 a.m. to 10:00 a.m., etc. These joint probabilities P_{ijk} are obtained from the product $p_{ijk}q_{ik}$. Here p_{ijk} is the conditional probability that event-type i occurs at geographical region j at time of day k, and q_{ik} is the probability that an event type i occurs at time k. p_{ijk} is a relative probability because the p_{ijk} over a given region j sum to 1.0. It is assumed that $q_{ik} = 1.0$, or the data are from actual events that took place, and thus, the present concern is for accurately estimating the p_{ijk}, which in this case is the same as P_{ijk}.

10.8.1 The problem. Given that an event occurs, where it occurs is controlled by some physical law of detection that is known and/or some exogenous plans or doctrines that are unknown. For this known exogenous intervention, a hypothesized model exists. This model generates analytical predictions $\breve{P}_{ijk}$ of the desired probability p_{ijk} aggregated monthly. The accuracy of these predictions depends on the strength of the hypothesized intervention-model. A historical data-base exists that consists of the relative frequencies p_{ijk} of occurrences observed by the

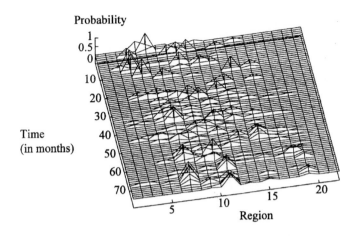

Probability

Time
(in months)

Region

Figure 10-28 Historical data p_{2j1} (Greene 1992)

worldwide sensor from January of 1985 through July of 1991. Each relative frequency is aggregated over each month, resulting in a total of 79 months of relative aggregated frequencies p_{ijk} for each event type i, each geographical region j, and each time-block-of-the-day k. The historical data-base consists of a total of 62,568 ($3 \times 22 \times 12 \times 79$) relative frequencies p_{ijk}. A combination of the historical data p_{ijk} and the corresponding prediction-data $\breve{P}_{ijk}$ should result in a prediction $\hat{p}_{ijk}$ more accurate than that of the intervention model alone. The accuracy of the predictions $\hat{p}_{ijk}$ depends upon the adequacy of the historical data-base p_{ijk} and the intervention-model estimates $\breve{P}_{ijk}$ 1, as well as the correlation of future regional-event-occurrences to past regional-event-occurrences.

The main objective of this case study is to establish a methodology for forecasting the relative monthly probabilities $\hat{p}_{ijk}$ one month into the future. This is to configure a worldwide-sensor system based on historical data p_{ijk} with estimates $\breve{P}_{ijk}$ from an existing prediction-model. To scale the problem down, this case study will be limited to event-type 2 of the three event types and to time-block 1 of the twelve time-blocks. STARMA forecasts will be made using historical monthly relative frequencies and analytical monthly predictions for event-type 2 and time-block 1 for all 22 geographical regions.

10.8.2 The original data. The development of a STARMA model is an iterative process. Step 1 calls for the analysis of historical data. Three dimensional plots using several perspectives of the historical data with time on the x-axis, geographical region on the y-axis and the p_{2j1} on the z-axis were examined. **Figure 10-28** shows time from January 1985 (notch 1), February 1985 (notch 2) and so on, until notch 79, which corresponds to July 1991. The y-axis is set up such that the first notch corresponds to geographical region 1, the second to geographical region

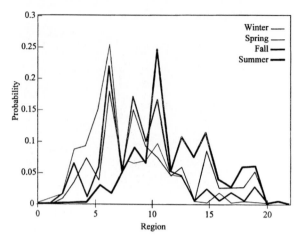

Figure 10-29 Average relative probabilities in the four seasons (Greene 1992)

2 and so on until finally, we have notch 22 for region 22. It is obvious from the historical data that there are numerous observations that have a p_{2j1} value of 0.0. Because the observations are relative frequencies, the observation values are limited to a range between 0.0 and 1.0. Geographical regions 7, 9, and 11 appear to have the most nonzero p_{2j1} values. According to the DOD experts on the historical data-base, the odd-numbered regions are more likely to have non-zero values than the even-numbered regions. Few non-zero observations occur in geographical regions 1, 2, 3, 4, 5, 6, 12, 13, 14, 15, 16, 17, 18, 19, 20, 21, 22, and the year 1986 had few non-zero p_{2j1} values.

Two-dimensional plots of the average-relative-frequency of the historical data-base for each year were examined. Average-relative-probabilities in Spring, Summer and Fall are overlaid on top of each other in **Figure 10-29**. These plots represent the average over all observations through July of 1991. Notice again the x-axis is set up such that the first notch corresponds to region 1, second region 2 and so on. The values on the y-axis are the average relative historical frequencies for the corresponding region. There is a change in the average relative frequencies for each season and thus, it can be concluded that there is seasonality in the historical data-base.

10.8.3 Intervention analysis. Turning back to the original data-base in **Figure 10-28**. The many zero p_{2j1} values, combined with the numerical range of the relative frequencies between 0 and 1, make it very difficult to model the historical data-base. The analytical estimates $\breve{P}_{ijk}$, hypothesized to be an intervention, were used as a simple filter on the historical data-base. This filtering is accomplished simply by subtracting the analytic estimates from the corresponding relative-historical-frequencies. This difference between the historical-relative-frequencies and the analytical-model-prediction can be interpreted as the difference between what was observed to occur and what was anticipated to occur.

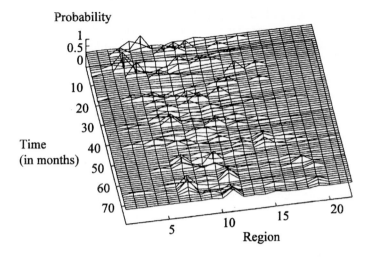

Probability

Time
(in months)

Region

Figure 10-30 Normalized transformed data (Greene 1992)

Before calculating the difference, the analytical probabilities were energy normalized, as opposed to statistically normalized, so that the probabilities became relative probabilities. Statistical normalization changes the highest-observation-value to 1.0 and the lowest-observation-value to 0.0 and adjusts for the other observations in relationship to the highest-observation-value and the lowest-observation-value. Energy normalization transformed the analytical probabilities such that the sum of the probabilities over the 22 geographical regions added up to one for every time-period using the following equation

$$\check{p}_{2j1} = \frac{\check{P}_{2j1}}{\sum_{r=1}^{22} \check{P}_{2r1}} \qquad \forall j \text{ over all 79 months.} \tag{10-109}$$

This was done because the historical frequencies are relative frequencies that sum to one for every time-period over the 22 geographical regions. This transformation, in effect, changed the intervention probabilities to relative intervention probabilities. The differences were then calculated by the following equation:

$$\tilde{p}_{2j1} = p_{2j1} - \check{p}_{2j1} \qquad \forall j \text{ over all 79 months.} \tag{10-110}$$

This is in effect an application of Equation **(10-109)** when the transfer function is $\check{p}_{zj1}$ and $I(t)$ is a negative step-function ($I(t) = -1$ for $t \geq 0$).

The difference can be interpreted as information of what was observed that was not adequately represented or explained by the normalized prediction model. For a given month, the difference is positive in sign when the observed-relative-frequency for j is greater than the predicted-intervention-relative-probability for j. This occurs when the intervention model under-estimated the relative-frequency-of-occurrence.

For a given month, the difference is negative in sign when the observed-relative-frequency is less than the normalized-analytical-estimate for j. This occurs when the intervention model over-estimated the relative-frequency-of-occurrence.

10.8.4 Deseasonalization. Three-dimensional plots of $\tilde{p}_{2j1}$ were examined. An example of such a plot is shown in **Figure 10-30**. The $\tilde{p}_{2j1}$ values do not appear to be stationary because a 12-month season is present. For example, region 7 has relatively high values during the winter months (December, January and February) and relatively low values during the summer months (June, July, and August). The winter months that exhibit the relatively high values for regions 7 can be seen when the time dimensions are 36, 48, 60 and 72. The summer months that exhibit relatively low values for region 7 can be seen when the time dimensions are 42, 43, 44, 54, 55, 56, 66, 67, 68, 78, and 79.

In examining **Figure 10-30**, several regions become 'prominent'. Regions 7 and 11 appear to be very 'mountainous' over time, having many relatively large values of $\tilde{p}_{2j1}$. This implies that the intervention model does not adequately predict regions 7 and 11. To the extent that most of the values of $\tilde{p}_{2j1}$ are positive, the analytical model greatly underestimates the p_{2j1} values for regions 7 and 11. An autocorrelation analysis was conducted on the values of $\tilde{p}_{2j1}$. We summarize the number of regions that had an autocorrelation value significantly different from zero at the kth lag:

Lag	1	2	3	4	5	6	7	8	9	10	11	12	13	14	15	16	17	18	19	20
# of Regions	13	5	2	0	1	2	2	2	2	3	6	6	3	3	1	0	1	0	1	0

Six of the regions had an autocorrelation value at the 11*th* and the 12*th* lags that are significantly different from zero. This implies that a 12-month season may be present in the data. To remove the 12-month seasonality, the values of $\tilde{p}_{2j1}$ were differenced using a 12-month lag:

$$\grave{p}_{2j1} = \tilde{p}_{2j1}(t) - \tilde{p}_{2j1}(t-12) \qquad \forall j \ over \ 79 \ time \ periods. \tag{10-111}$$

The data-set lost 12 degrees-of-freedom and thus the new data-set now contains 67 (79−12) time periods and a total of 1,474 (67×22) observations.

10.8.5 Deciding on a data set for analysis. Three-dimensional plots of the de-seasonalized data $\grave{p}_{2j1}$ from several perspectives were again examined. An example is shown in **Figure 10-31**. Differencing the data by 12 lags appears to have 'smoothed' the data in the spatial dimension. In the temporal dimension, the 12-month season appears to still be present. The seasonality is confirmed by autocorrelation analysis, which shows 13 regions have an autocorrelation value

Probability

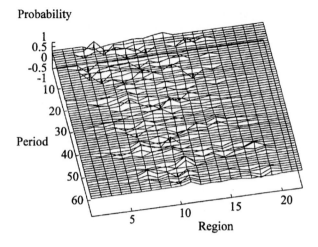

Figure 10-31 Normalized and deseasonalized data (Greene 1992)

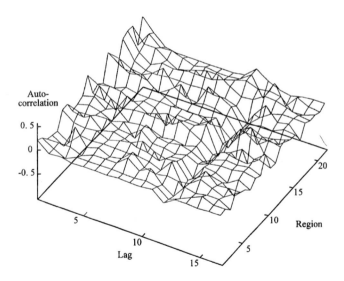

Figure 10-32 Temporal autocorrelations of transformed data (Greene 1992)

significantly different from zero at lag 12 (see **Figure 10-32**). This set of data has seven more regions with an autocorrelation value significantly different from zero at lag 12 than $\tilde{p}_{2j1}$. Differencing did not remove the 12-month seasonality, resulting in the conclusion that the seasonality may be stationary. Again, regions 7 and 11 appear to have relatively high numerical values, as seen in **Figure 10-31**.

At this point, a decision had to be made on which data-set was the best to continue analysis on. Between the current data set $\hat{p}_{2j1}$ and the previous data-set $\tilde{p}_{2j1}$, which is better? The current data are more uniformly regular over the spatial dimension, as one would detect if s/he examines carefully the difference between **Figure 10-31** and **Figure 10-30**. **Figure 10-30** appears to have a 'funnel' effect in that the values in the distant past are more spread out over the regions than the more recent months. This 'funneling' does not appear to be present in **Figure 10-31**. Uniformity in the spatial dimension is important in STARMA modelling. If the spatial dimension is not uniform over time, it becomes necessary to model this spatial non-uniformity. Thus the model must account for not only the non-uniformity in the temporal dimension represented by the 12-month season, but also   the funneling effect in the spatial dimension. To keep the model as simple as possible, the decision is to use the $\hat{p}_{2j1}$ data.

Further analysis is required on the autocorrelations of the transformed data-set as presently adopted. Region 15 is the only region that did not pass the first stationarity test based on autocorrelation. In fact, it did not pass the second stationarity test based on partial autocorrelations either. However, this should pose no problems since region 15 is a first-order neighbor in the spatial dimension with respect to the region-of-interest rather than the region-of-interest itself. More will be said about this later.

In fitting a STARMA model over this set of data, a seasonal moving-average and/or seasonal autoregressive term is expected due to the 12-month season. This 12-month season is evident in both the three-dimensional plots of $\hat{p}_{2j1}$ data and in the three-dimensional autocorrelation plots. This seasonality is further confirmed by the number of regions that had a partial autocorrelation value significantly different from zero at each lag:

Lag	1	2	3	4	5	6	7	8	9	10	11	12	13	14	15	16	17
# of Regions	9	3	1	0	1	1	2	1	0	0	1	17	2	1	0	1	0

10.8.6   Spatial weights.   DOD personnel suggested that region 11 is of particular interest and should be the target region for the model. Under their guidance, first-order neighbors and second-order neighbors are defined as shown in **Table 10-20** and **Table 10-21** respectively. It should be noted that all of the first-order neighbors are odd-numbered regions. The set of spatial weights is usually selected to represent some physical property among the geographical regions. In the $\hat{p}_{2j1}$ data, a strong seasonality still remains. **Figure 10-29** clearly shows four distinct average-relative-frequencies corresponding to each season of the year. Four sets of spatial weights were selected to represent the spatial weighting among the 22 geographical regions corresponding to each of the four seasons. The four average-

Region	Winter	Spring	Summer	Fall
7	0.68	0.33	0.04	0.44
9	0.17	0.12	0.01	0.01
13	0.12	0.10	0.26	0.08
15	0.01	0.15	0.28	0.05
17	0.01	0.05	0.06	0.04
19	0.01	0.09	0.15	0.05

Table 10-20 Weights between region 11 and its first-order neighbors

Region	Winter	Spring	Summer	Fall
1	0.00	0.00	0.00	0.00
2	0.02	0.00	0.00	0.00
3	0.03	0.02	0.00	0.04
4	0.15	0.09	0.00	0.18
5	0.16	0.18	0.01	0.03
6	0.26	0.09	0.08	0.15
8	0.13	0.16	0.14	0.13
10	0.12	0.22	0.18	0.28
12	0.08	0.11	0.13	0.15
14	0.01	0.01	0.20	0.01
16	0.03	0.06	0.10	0.02
18	0.01	0.06	0.10	0.02
20	0.00	0.00	0.00	0.00
21	0.00	0.00	0.01	0.00
22	0.00	0.00	0.00	0.00

Table 10-21 Weights between region 11 and its second-order neighbors

relative-frequencies associated with each season were selected to be the spatial weights. The average-relative-frequencies of the first-order neighbors of region 11 for each season were energy normalized so that the first-order weights for each season sum to 1.0. **Table 10-20** lists the weights for region 11 and its first-order

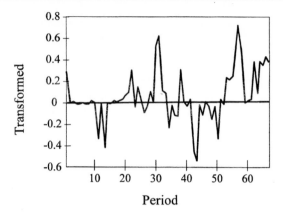

Figure 10-33 Region-11 transformed data (Greene 1992)

neighbors for all four seasons. We do the same for the second-order neighbors and list the results in **Table 10-21**.

10.8.7 Identification of univariate series. Because nine regions have an autocorrelation value and a partial autocorrelation value significantly different from zero at the first lag, it is highly probable that the STARMA model will have $p = 1$ and/or $q = 1$. Since four regions have an autocorrelation value significantly different from zero at the second lag and three regions have a partial autocorrelation value significantly different from zero at the second lag, it is also probable that the STARMA model will have $p = 2$ and/or $q = 2$. Because 13 regions have an autocorrelation value significantly different from zero at the 12*th* lag and 17 regions have a partial autocorrelation value significantly different from zero at the 12*th* lag, it is possible that the STARMA model will have an autoregressive season-length of 12 and/or moving-average season-length of 12. All other possibilities appear unlikely.

Identification of a STARMA model is usually accomplished by examination of the autocorrelations and the partial autocorrelations. The objective is to see which of the two tends to exponentially decrease and which tends to go to zero fast. Neither the autocorrelations nor the partial autocorrelations exhibit a tendency to exponentially decrease to zero or go to zero fast. Thus no conclusive comments can be made regarding the identification of a STAR, STMA or STARMA model. We turn to a two-dimensional plot of the univariate data at geographical region 11, $\dot{p}_{2111}$, as shown in **Figure 10-33**. From analyzing the plot, there appears to be no trend in the data and the target-region data is stationary. A 12-month season is not apparent from the plot of geographical-region 11. It is noted that thethird block of 12 months (time periods 25 through 36) appears similar to the fifth block of 12 months (time periods 49 through 60).

Figure 10-34 is a two-dimensional plot of autocorrelations of $\dot{p}_{2111}$. Going hand-in-hand with this is the partial-correlation plot of **Figure 10-35**. The

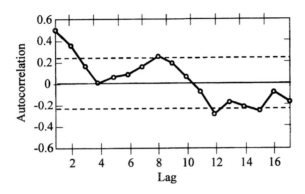

Figure 10-34 Autocorrelations of target-region 11 data

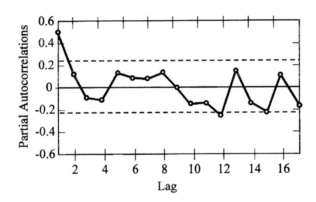

Figure 10-35 Partial autocorrelations of target-region 11 data

autocorrelations are significantly different from zero at lags 1, 2, 12 and 15. The partial autocorrelations are significantly different from zero at lags 1 and 12. Besides confirming stationarity, both the autocorrelations and the partial autocorrelations are significant at the lag 12, suggesting a 12-month season. Neither the autocorrelations nor the partial autocorrelations appear to go to zero fast or exponentially decrease to zero. Putting all the information together, we hypothesize that $p \leq 2$ and $q \leq 2$, with a seasonality of 12 for both autoregressive and moving-average models.

10.8.8 Diagnostic checking of univariate series. **Table 10-22** summarizes the six ARMA models—out of the 22 models estimated—with significant coefficients. An AR or MA model is considered significant if the coefficient of the

Model	Adjusted-R^2	Sum of squared residuals
MA(1)	0.164	2.990
AR(1)	0.249	2.654
SMA(1)$_{12}$	0.339	2.329
SAR(1)$_{12}$	0.353	1.969
MA(2)	0.239	2.682
SMA(2)$_{12}$	0.407	2.058

Table 10-22 Significant ARMA models for target-region 11

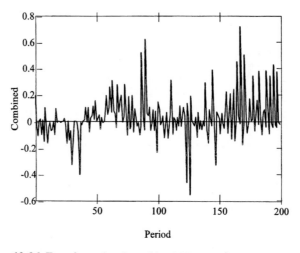

Figure 10-36 Transformed and combined STARMA series (Greene 1992)

highest order is significantly different from zero at the 90 percent confidence level. Similarly, for a seasonalized model it means that the coefficient of the SAR(12) or SMA(12) term is significant. Shown in **Table 10-22** are the goodness-of-fit parameters for each of the six most significant models. To select an appropriate model for $\dot{p}_{2111}$, the adjusted R^2 and the sum of the squared residuals or exponentially decrease to zero. Putting all the information together, we hypothesize that $p \leq 2$ and q ≤ 2, with a seasonality of 12 for both autoregressive (SSR) were examined. The criterion is to maximize the adjusted-R^2 while minimizing the SSR. The SMA(2)$_{12}$ has the superior adjusted-R^2 value and has an SSR value that is less

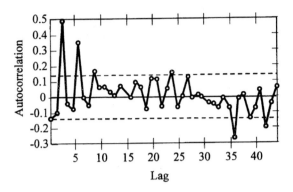

Figure 10-37 Autocorrelations of the combined series

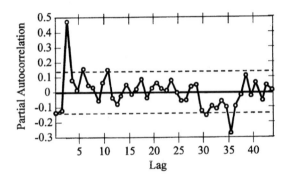

Figure 10-38 Partial autocorrelations of the combined series

than 5 percent different from the $AR(1)_{12}$ model. The best model to forecast target-region 11 is judged to be the $SMA(2)_{12}$ model.

10.8.9 The combined STARMA series. The target series, the first-order series, and the second-order series were then combined into one as shown in **Figure 10-36**. First we examine the autocorrelations and the partial autocorrelations. Because there are 201 total observations in the combined series, it is recommended the autocorrelations be computed up to lag 51 (considering there are four sets of spatial weights for the four seasons, or $201/4 = 50.25$ sets of temporally-distinct data). Example plots are presented in **Figure 10-37** and **Figure 10-38**.

The autocorrelations are significantly different from zero at lags 3, 6, 9, 24, 36 and 42. The partial autocorrelations are significantly different from zero at lags 3, 6, 24, and 36. All of the lags that exhibit either an autocorrelation or a partial autocorrelation significantly different from zero occur at multiples of 3. This is

Parameter	Coefficient	Standard error	t-statistics	2-tail significance
constant	0.027	0.009	3.031	0.002
$\theta_{1,0}$	0.424	0.064	6.594	0.000
$\theta_{1,1}$	-0.160	0.064	-2.498	0.012
$\theta_{2,0}$	0.396	0.064	6.190	0.000
$\theta_{2,1}$	-0.131	0.064	-2.037	0.042
$\theta_{Sq,0}$	-0.586	0.078	-7.486	0.000

Table 10-23 The final univariate STARMA model for region 11

expected since the series was combined from three separate series. The significant autocorrelation and partial autocorrelation at lag 36 corresponds to the significant autocorrelation at lag 12 for target-region 11.

Two candidate models were identified: MA(1) and MA(2). The coefficient for the MA(2) term in the MA(2) model was not significantly different from zero at the 90-percent confidence-level. On the other hand, the two-tailed significance values show that the coefficients for the terms in the MA(1) model are significant at the 90-percent confidence-level. The MA(1) model is selected over the MA(2) model as a result.

10.8.10 The STARMA model. The purpose of building a univariate ARMA model on the target region was to identify p, q, s_p, s_q for the STARMA model. The best ARMA model for target-region 11 was SMA$(2)_{12}$ with $p = 0$, $q = 2$, $s_p = 0$ and $s_q = 12$. The purpose of building another ARMA model on the combined series was to determine the spatial relationships between the target-region and its neighbors. The MA(1) model was judged the best and most parsimonious model of the combined time-series. The q value of the combined time-series corresponds to the m_k value identified for the STARMA model. Here, $m_k = 1$ and the identified STARMA model is the Seasonal Spatial-Temporal Moving-Average SSTMA$(2_{1,1})_{12}$ model. **Table 10-23** lists the coefficients of the terms in the SSTMA$(2_{1,1})_{12}$ model along with the standard-error of the coefficients, the t-statistic of the coefficients, and the two-tailed significance of the coefficients. The two-tailed significance shows that all of the coefficients for the model are significantly different from zero at the 90-percent-confidence-level. The SSTMA$(2_{1,1})_{12}$ model has an adjusted-R^2 value of 0.382 and a sum of squared residuals (SSR) value of 3.115. It can be written in long hand as

$$\hat{P}_{2,11,1}(t) = -\sum_{k=1}^{2} \sum_{l=0}^{1} \theta_{kl} \mathbf{w}^{(l)} \mathbf{a}(t-k) - \theta_{s_{q},0} \mathbf{w}^{(0)} \mathbf{a}(t-12) + a(t)$$

$$= -\theta_{1,0} \mathbf{w}^{(0)} \mathbf{a}(t-1) - \theta_{1,1} \mathbf{w}^{(1)} \mathbf{a}(t-1) - \theta_{2,0} \mathbf{w}^{(0)} \mathbf{a}(t-2) - \theta_{2,1} \mathbf{w}^{(1)} \mathbf{a}(t-2) - \theta_{s_{q},0} \mathbf{w}^{(0)} \mathbf{a}(t-12) + a(t) \quad \textbf{(10-112)}$$

$$= -0.424\,a(t-1) + 0.160\,\mathbf{w}^{(1)} \mathbf{a}(t-1) - 0.396\,a(t-2) + 0.131\,\mathbf{w}^{(1)} \mathbf{a}(t-2) + 0.586\,a(t-12) + a(t)$$

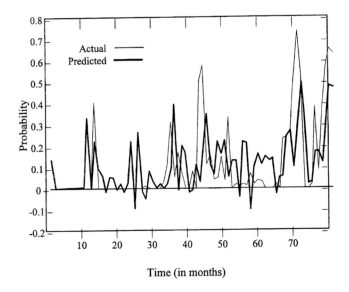

Time (in months)

Figure 10-39 Observed P_{2j1} and predicted of region 11 (Greene 1992)

recalling $\mathbf{W}^{(0)} = \mathbf{I}$ and $\mathbf{w}^{(1)}$ is the 11*th* row of weight matrix $\mathbf{W}^{(1)}$. The weight matrices $\mathbf{W}^{(1)}$ and $\mathbf{W}^{(2)}$ are given in **Table 10-20** and **Table 10-21** respectively. The calibration constant, or 'intercept', of this regression equation is 0.027.

Many transformations are required on the fitted and forecasted $\dot{p}_{2j1}$ values to return to the original historical data. The first transformation converts the fitted $\dot{p}_{2j1}$ values to fitted $\tilde{p}_{2j1}$ values and accounts for the 12-month lag that was taken to remove seasonality. The second transformation adds back the corresponding relative intervention-probabilities $\breve{p}_{2j1}$ to the fitted $\tilde{p}_{2j1}$. This results in $\hat{p}_{2j1}$. Notice it is not possible to forecast $\hat{p}_{2j1}$ without knowing the corresponding intervention $\breve{p}_{2j1}$ value. **Figure 10-39** is a two-dimensional plot of the observed and predicted relative frequencies of geographical-region 11 from January 1985 through July 1991. This is after all of the transformations have been done to return the $\hat{p}_{2j1}$ values back to forecasted values for the historical relative frequencies. The SSR value for the SSTMA$(2_{1,1})_{12}$ predictions is 2.37. This compares with the equivalent normalized intervention prediction SSR of 3.06. An attempt was made to use this identified model SSTMA$(2_{1,1})_{12}$ to forecast for another target-of-interest, region 7. The exercise was not particularly successful, thus limiting this model to be a univariate STARMA model rather than a multivariate STARMA.

10.8.11 Remarks. We chose to include this case study here for several reasons. First and foremost, it serves to illustrate "intervention analysis." To account for a known phenomenon, the data-base was screened by an exogenously-determined 'filter'. The filter represents the telecommunication capabilities as dictated by the laws-of-physics—a fact we can quantify precisely. In this context, the range (or maximum distance) of telecommunication imparts a discontinuity (or intervention) on the data—distinguishing between regions reachable from those not reachable for telecommunication. This distinction can best be filtered out. This approach substantiates our thesis that when the phenomenon being studied is known, it is more simple. In this case, it is advantageous to account for the known phenomenon, rather than relying exclusively on statistical, correlative analysis. In the case study, we factored out what can be explained causally from the rest.

Second, along the same line-of-thought, the weights used for spatial-interaction was also found exogenously. However, it was derived from historical data rather than causality. The only reason is that we do not know anything abut the underlying process, given the actual information was classified by the DOD. it was either not made available to the modelers, not known to DOD, or both. At any rate, the information is unavailable. To the extent that the interaction weights seemed to follow a regular pattern among the geographic regions (as shown in **Figure 10-29**), we felt comfortable using these weights, lacking any better information.

Third, the case study was a success, in that the STARMA model was able to forecast event-occurrence probabilities better than any prior models assembled by DOD. We felt that such an experience is worth sharing with our readers to show the potentialities of spatial-temporal modelling. Unlike other reported studies, this has a distinctly different context, falling into the telecommunication arena and defense-intelligence arena. Its inclusion can only enrich the literature in this field in our opinion.

Perhaps the least exciting of the study is the mechanics of ARMA-modelling, which has been covered in previous examples already, including the extended numerical example. But it serves to further accentuated the role of causal vs. correlative modelling. It does illustrate the proper place for correlative analysis of ARMA when the time evolution of these probabilities-of-occurrences cannot be explained causally. We show that there is a proper place for STARMA as well, emphasizing again that most STARMA procedures can be thought of as a middle ground between correlative and causal modelling. Remember that STARMA relies heavily upon exogenously-supplied spatial-weights, which are often derived from an understanding of the phenomenon being studied.

10.9   Computational considerations

It is safe to say that the analysis described in this chapter will not be possible without the arrival of the computer. In spite of the development in hardware in recent years, the software available to operationalize MARMA and STARMA remains limited. The IMSL Stat/Library (1990) and recent revisions represent one of the few software that directly addresses these types of models. While there exists a number of software packages that readily perform transformations, differencing, and autocorrelation, IMSL offers features beyond such functions. It contains routines to calculate sample cross-correlation function of two stationary time-series up to a multivariate case. Furthermore, it computes estimates of the impulse-response weights and noise series of transfer-function models, including between two mutually stationary time-series. Multichannel, or multivariate, time-series are analyzed using several routines .An example is a program to compute the least-squares estimates of a linear-regression model for a multichannel time-series with a specified "base channel"—the site-of-interest in our application.

In general, IMSL has tools to handle the comprehensive model-building process, including model identification, parameter estimation, diagnostic checking and forecasting. An extended numerical example has been used to illustrate the versatility of such a software package in building the models discussed above. Irrespective of the software employed, we will illustrate below the required capabilities of a computer program to support this type of analysis. First, a basic interactive ARMA model is a necessary tool for model identification, usually by way of plots of the time-series and autocorrelation. The user determines and initiates the differencing process, and then specifies the number of autoregressive and moving-average parameters to be used in the model. Besides parameter estimation, the program should provide the Box–Jenkins forecasts for examination. Depending on its ability to replicate the base data, the number of parameters can be accepted or altered, corresponding to the need for further experimentation. To operationalize a univariate STARMA model, we need to employ a two-phase procedure. First, the base time-series for the site of interest needs to be modelled. Then it is necessary to merge 0*th*-, 1*st*- and 2nd-order neighboring time-series into one time series. The ARMA procedure is then reapplied to the merged series for model identification and parameter estimation. The bivariate ARMA model can be analyzed through a transfer function, in which the required number of autoregressive and moving-average whitening parameters are determined using the ARMA program. The whitened input/output series, together with the noise series, then allows the impulse-response-weights to be estimated. A minimum requirement for multivariate or multichannel time-series-analysis should consist of an econometric-type model that explains the base time-series in an autoregressive fashion, as well as other associated time-series. The input consists of very specific differencing and lag specifications, which assumes certain knowledge by the modeler regarding the system under study.

According to Mills (1990), computer software for vector ARMA model building is somewhat limited. Many write their own specialized programs, particularly for spatial applications (for example, Robinson (1987)). Of the major statistical packages, SAS/ETS's (1985) PROC STATESPACE contains most of the identification procedures discussed above. PROC SYSNLIN can also be adapted to estimate vector ARMA models, but this requires rather sophisticated knowledge of the SAS language. PROC ARIMA, on the other hand, straightforwardly identifies and estimates multiple input, including intervention, and transfer functions. Mills (1990) suggests that possibly the most complete packages for vector ARMA-- modelling are the SCA system (Liu and Hudak (1983)) and the WMTS-1 program (Tiao et al. (1979)). Certainly, there are others that are around—including spatially-oriented packages—but have not been widely publicized and there are others being developed day-by-day.

Newbold et al. (1994) found considerable inconsistencies among the performances of time-series software. Their analysis results illustrate how the user may be confronted with substantially different estimates and forecasts, depending on which package is used, and consequently the difficulty in replicating published results. After a fair amount of experimentation, they suggest that a full maximum-likelihood (ML) procedure is more reliable (than say a conditional least-squares procedure). While ML is more computationally intensive, the practical significance of this factor is rapidly becoming negligible with advances in software and hardware. Software-performance inconsistencies would discourage many from using the techniques advanced in this chapter for doing spatial-temporal analysis, in spite of their potential merits. Most critically, this confusion is not likely to disappear in the near future.

10.10   Concluding comments

This chapter introduces an alternate way of land-use forecasting. A correlative approach is presented, as contrasted with causal approaches discussed previously. This is based on the autoregressive moving-average (ARMA) models. We start out with reviewing the basic concepts of ARMA, but quickly extend the ideas to multivariate and spatial applications, ending up with the spatial-temporal ARMA (or STARMA) model. Correlation in time is now extended to include correlation in space. To tie this approach to previous approaches, econometric simultaneous-equations are shown to be a specialization of ARMA. Through this econometric reference, one can show further that the Lowry–Garin model, a cornerstone of land-use models of the generation-allocation variety, is again a specialization of the econometric model. Hence, one can think of the correlative procedures of STARMA as the most general formulation of land-use models and it deserves a prominent place in any serous discussion of this field.

Through numerical example and a case study, this chapter shows the use of time-series as a way to forecast spatial activities. The approach is parallel to those discussed earlier, including econometric models, as shown by the EMPIRIC model. As part of this chapter, we review how to adept a time-series to change, particularly through the state-space formulation of Kalman filter. Also discussed is the use of intervention analysis to handle precipitous happenings, similar to bifurcation in activity generation-allocation models. Besides drawing parallels, the specific features of the time-series approach are also outlined. These include its general lack of causality, which leads toward the mandatory emphasis on careful experimental design.

Our experience with STARMA modelling points toward its taxing computational requirement and inconsistent results. Also our experience suggests its general inability to go beyond a univariate spatial-temporal framework, mainly because of the quality of the data and the often nonhomogeneous nature of the data. Thus the general flexibility of STARMA as a modelling tool has to be balanced against its operational requirements. We will further elaborate on this point as we launch the next chapter that further compares the time-series models with econometric and generation-allocation models. There the comparison will be made more formally and put into the context of additional case studies. Also of interest is the role spatial time-series play in image processing.

The world is either the effect of cause or of chance.
If the latter, it is a world for all that,
that is to say, it is a regular and beautiful structure.
Marcus Aurelius Antonius

CHAPTER ELEVEN

SPATIAL-TEMPORAL INFORMATION: STATISTICAL AND CAUSAL-MODEL DEVELOPMENT

A major function of facility-location, transport and land-use models is *analysis*. Among other uses, sensitivity analysis is performed to answer "what-if" type questions regarding investment and logistics. Another function of spatial modelling is *forecasting*, which is often carried out by time-series. As explained in previous chapters, analysis and forecasting in a spatial dimension have introduced several complications to traditional statistical and time-series analysis. First and foremost is the greatly increased dimensionality in what sometimes has been referred to as multivariate and vector-time-series models. Concomitant with this is the taxing specification and identification problem in model construction. Not only is it difficult to discern from data inspection a meaningful structural model, the procedure is further hampered by the inconsistent data-sets collected simultaneously and over time among disparate and changing geographic subdivisions. The availability of remote-sensing devices such as satellites has allowed the collection of consistent sets of raster data on a continuous basis. This opened new opportunities for this type of model development and calibration.

Coupled with the advances in theory and computation, *causal* models—to be contrasted with correlative models—have been employed to bypass specification, identification and calibration bottlenecks. Such models also relax the often restrictive homogeneity-assumptions in statistical spatial-temporal models, in which the model is assumed to hold for all locations i. A comparison will be made here in this chapter among three models (commonly found in location, transport and land-use analyses), representing auto-regressive-moving-average (ARMA), simultaneous econometric-equations, and Lowry–Garin matrix-series techniques. Case studies using these models point toward the importance of model transparency and explicit consideration of spatial interaction among geographic subdivisions.

11.1 Causal vs empirical modelling

As in many other disciplines, professions in spatial-temporal analysis have been divided roughly into two groups—those that deal with data analysis and those that deal with theoretical development. Oftentimes, philosophies and practices are distinctly different among these two camps, with only minimal crossover or synergism. For example, those in empirical analyses have been concentrating on refining the tools of data collection—from relatively crude surveying instruments (*a la* Lewis and Clark days) to today's geosynchronized global positioning satellites. Those in the theoretical, causal modelling world, on the other hand, have been proposing numerous hypotheses about the way spatial-temporal systems can be explained. Many of these are elegant and sophisticated, employing the latest advances in the mathematical sciences and computation. Both camps have been pursuing these disparate goals quite happily. It is only recently that an increasingly large degree of frustration has been expressed in both quarters regarding the limited number of problems that can be solved with only data or theory alone. Thus as bits and bytes of remote-sensing imagery get stored in reams and reams of computer tapes or disks, attempts are being focused upon making real use of the rich data-base. This is to be contrasted making use of only the face value of these data. Pattern recognition and statistical analyses are typical tools that have been proposed to examine the data, establishing relationships not obvious to the bare eyes. As a proliferous number of models are built, professionals in the theoretical camp are increasingly confused about the relative merits of these models. Until validation is performed, no one can pass judgment on the usefulness of these models. Increasingly, theorists are looking for data sets that can help in the validation task. However, they are often not aware of what is available. By the time the data is finally obtained, they typically find out that the sophisticated nature of their models cannot be calibrated or verified by the nature of the data and the format in which they are collected.

Just to give an example, remote-sensing devices have now been able to collect images of extremely fine resolution. But the modelers have rarely participated in the decision-making process regarding the data precision required of certain types of models. Likewise, it is not at all clear whether the 90-minute lag-time between revolutions of a satellite around the earth (as imposed by physical and technological limitations) is acceptable. In other words, when will it be necessary to go to geosynchronized satellites so that continuous glimpses of the target can be obtained? Conversely, empiricists have employed pattern recognition and statistical analysis with only limited success. For example, the analyses yield better results with *homogeneous* spatial-temporal processes than nonhomogeneous ones in that they work well for processes that are stationary under translation.[1] In general, the techniques have been more successful in identifying the appearance or images rather than the processes underneath the imagery. As a result, empiricists often have to settle for the phenomena rather than the structure behind the observed.

It is somewhat ironic that the advances in data collection and analysis cannot be carefully married with model development. Combining the significant advances on both sides will afford a synergistic breakthrough in this field, not only in terms

---

[1]For a review of this concept, refer to the "Spatial Time-Series" chapter, particularly the "Univariate STARMA case study".

of our current problem-solving capability, but also our future research-directions. Thus empiricists will eventually collect the pertinent data and theorists will also develop the relevant models that can be readily calibrated and validated. In this chapter, we survey the latest advances in both camps and attempt to point out places where synergism is most promising. If nothing else, it will inform both sides what is available "on the other side of the fence" so that a meaningful dialogue can be initiated. The plan of this chapter is as follows. First, we review the latest advances in data collection via satellites. Second, we use advances in statistical (and "data mining") analyses to perform model specification and identification on the spatial-temporal data so collected. Third, we extend the analysis to time-series and econometric-model formulations. Finally, a causal model is proposed, representing in our opinion, the most structured approach to analyze a general spatial-temporal process. The three analysis techniques—in increasing amount of casual relationship—are then contrasted in terms of their merits and demerits, in view of their varying data requirements and information content. A unifying theme is then proposed to integrate these approaches from recent advances in spatial statistics. To provide this integration, we will first review recent advances in spatial statistics.

11.2 Spatial stochastic process

In order to apply spatial statistics to facility-location and land-use, one needs to be acquainted with *spatial* random-process or stochastic-process (Anselin 1988, Upton and Fingleton 1985). A stochastic process, as the reader can recall, is a formalism used to designate collections or sequences of random variables that are organized in some regular fashion.[2] More specifically, let a random variable, say the ith one, stands for a location in general n-dimensional space, then the set of variables $\{Z_i, i \in I\}$ is a spatial stochastic-process. The range I, which is the collection of all possible values for the index parameter, can be any general dimensional space, discrete or continuous. In other words, a *spatial* stochastic-process is the special case of a general stochastic-process where the index pertains to a location in space.

11.2.1 Definitions. Three spatial stochastic-processes can be defined. The first two yield variables that comply with the requirements of

$$P(Z_1 < Z_1^0, \ldots, Z_{|I|} < Z_{|I|}^0) = \prod_{i=1}^{|I|} P(Z_i < Z_i^0), \tag{11-1}$$

where Z are quantitative variables. In the case where the variable is categorical (discrete) rather than continuous

$$P(Z_i = a, Z_j = b) = P(Z_i = a) P(Z_j = b), \tag{11-2}$$

---

[2]The basic concepts of stochastic process is reviewed in the "Markovian Process" appendix in Chan (2000).

with *a* and *b* being two possible categories of the variable and $i \neq j$. Put it differently, independence among the variables is assumed in equations **(11-1)** and **(11-2)**. Specifically, these are the three spatial stochastic-processes of interest:

a. Continuous variables $\{Z_j\}$ are assigned to each of $|I|$ locations by drawing $|I|$ times from a $N(0, \sigma^2)$ or Gaussian distribution.

b. Discrete (dichotomous) values are assigned to each of $|I|$ locations based on $|I|$ tosses of an unbiased coin.

c. The value Y_{ij} assigned to a location with coordinates (i,j) is to some extent influenced by values assigned to neighboring locations; for example,

$$Y_{ij} = \theta(Y_{i-1,j} + Y_{i+1,j} + Y_{i,j-1} + Y_{i,j+1}) + a_{ij} \qquad \text{(11-3)}$$

where $|\theta| < 1/4$. Here $\{a_{ij}\}$ are un-correlated error terms from a distribution with zero mean and constant variance, and where the locations are restricted, for notational and computational ease, to the intersection points in a lattice.

While processes **a.** and **b.** will normally produce chaotic maps, process **c.** will create some semblance of order (due in part to the 'smoothing' that characterizes the process.) Examples of Equation **(11-3)** can be found in image processing and spatial autoregressive series, as explained in the "Remote Sensing" (Chan 2000), "Spatial Econometric Models," and "Spatial Timer-Series" chapters previously.

One difficulty with Equations **(11-1)** and **(11-2)** is that probabilities $P(Z_i < Z^0_i)$ and $P(Z_i = a)$ require estimation. This in turn requires rather extensive knowledge of the distribution form of Z. It is insufficient to assume that, for example, this distribution is $N(0, \sigma^2)$ from flimsy evidence—although this may simplify detecting the presence of spatial auto-correlation. An alternative to assigning the form of the underlying probability distributions for the mapped variables is to adopt the so-called *randomization* approach. As an example, two maps contain the same $|I|$ locations and the same Z^0_i values. The difference between them is simply a different matching of values to locations. In general $|I|!$ different matchings could be produced by random permutations of the Z_i-values. As we have seen in a non-parametric statistic such as *entropy*, each map can be summarized by at least one coefficient like entropy, and thus there are potentially $|I|!$ values for a coefficient (though some values may be repeated).[3] Also, randomization has profound implications in treating auto-correlated residuals when the normality assumption is inappropriate.

In spite of the appearance of Equations **(11-1)** and **(11-2)**, the notion of a spatial stochastic-process is very different from the more traditional random-sampling approach taken in empirical work. Most spatial processes have dependency relationship between locations in the $|I|$ entities in space (e.g., $|I|$ pixels in a satellite image). Indeed, the dependence of the values observed at different points in space is assumed to be generated by an unknown underlying

---

[3] *Entropy* has been defined in the "Descriptive Tools" chapter in Chan (2000).

process. The specific form of this process is decided a priori, presumably based on substantive theoretical grounds. In contrast, in the random sampling case, any dependence between the observational units is an unwanted feature. The independence is needed to infer information about a population from a subset of its members. This represents a distinguishing feature of a spatial stochastic-process, which make it specifically appropriate in modelling facility-location and land-use. While many of these concepts have been alluded to earlier in this book, we will recap and expand on them in this chapter. In this sense, the current chapter serves as a "melting pot" for these ideas. Many of these concepts would come to life during this synthesis, particularly when they are discussed with respect to case studies, rather than in abstract.

11.2.2 Stationarity and isotropy. Modelling with spatial stochastic-process is a highly complex task. The stochastic-process approach to spatial data means that essentially only one observation is variable at each instance, i.e., the process of allocating values to the random variables in space or space-time is performed one at a time. This gives rise to a computationally imposing situation, and there are some operational difficulties. Since this is not particularly practicable, some restrictions need to be imposed on the degree of dependence and heterogeneity that can be allowed. Only in this way can one have a handle on the spatial stochastic-process. Essentially, in order to infer certain characteristics of the underlying process, a degree of stability needs to be assumed. This is usually achieved by imposing the requirements of *stationarity, ergodicity* and *isotropy*—terms that we shall explain below.

A stationary stochastic-process has reached a state of statistical equilibrium so that the generating mechanism can be assumed to work uniformly over space and time. In other words, the underlying joint-distribution is taken to be the same for any subset of observations. This implies a number of restrictions on the moments, such as a constant mean and a constant finite-variance.[4] In more formal terms, strict stationarity can be defined as the state where any finite subset of variables $\{Z_i, Z_j, ..., Z_n\}$ from the stochastic process $\{Z_i, i \in I\}$ has the same joint distribution as the subset $\{Z_{i+k}, Z_{j+k}, ..., Z_{n+k}\}$ for any k, where k represents a uniform shift in time, space or space-time. We refer to the data-set as *homogeneous* then. In a time-series context, the notion of a shift k is straightforward. Stationarity implies that the distribution of any Z_i is the same at any point in time, that mean and variance are constant and finite, and that the covariance between Z_i and Z_j does not depend on the particular time-period, but only on the shift in time (time-lag k). A weaker notion of stationarity, and the one used in most applications is covariance stationarity, where the time independence is required only for the moments up to order two. These ideas have been discussed as part of the "Spatial Time-Series" chapter.

---

[4]Examples of stationarity and homogeneity can be found in the "Spatial Time-Series" chapter.

The stationarity requirement in time-series analysis leads to the use of auto-covariance and auto-correlation functions indexed by the time shift. This helps in the identification and estimation of the models, as in the well-known approach pioneered by Box and Jenkins (1976). In addition, for the underlying stochastic process to be identifiable, and to ensure the existence of desirable asymptotic properties, some restrictions have to be imposed on the context of the dependence. For example, one often-used constraint consists of the requirement of ergodicity, which ensures that, on the average, two events will be independent in the limit. Now recall that by definition a Markov chain is ergodic if all states in the state-transition chain are recurrent, aperiodic, and communicate with each other. In the current context, an ergodic assumption allows for consistent estimation of the joint-probability law of various subsets $Z_1, \dots , Z_p$ of the time series. Hence, from observations $Z_{n-p+1}, \dots , Z_n$, prediction of Z_{n+1} from *estimated* joint-covariance laws can proceed in a straightforward manner. Moreover, mean-squared-errors of prediction can be estimated consistently under an ergodic assumption.

Let us give an example or two to illustrate ergodicity. According to Cressie (1992), the most obvious ergodic time-series occurs when $Z_1, Z_2, \dots$ are independent and identically distributed. But notice that if a new series Z_t' is defined as $Z_t' = \mu t + Z_t$, then Z_t' is not ergodic, and it is not stationary. Here in the new series Z_t', where $\mu = E(Z)$, stationary time-series that are m-dependent (i.e., Z_j and Z_{j+m+1} are independent) are ergodic. A stationary Gaussian-time-series (i.e., series with normality assumptions) is characterized by its mean μ and its covariance $\mathrm{cov}(Z_{t+k}, Z_t)$. A sufficient condition for the series to be ergodic is $\lim_{k \to \infty} \mathrm{cov}(Z_{t+k}, Z_t) = 0$. An example of a non-ergodic but stationary-time-series (that is not necessarily Gaussian) is $Z_t'' = A$ ($t = 1, \dots$) where A is a random variable. A single realization of this time-series gives no information on the probability law of A, and hence nothing can be gleaned about, say, $P(Z_1'' \le Z^0)$, from a time average based on the observations $Z''^{0}_1, \dots , Z''^{0}_{|1|}$.

In spatial analysis, the notion of stationarity is more complex than regular time-series, as already pointed out in the discussion of spatial weight-matrices in the "Spatial Econometrics" chapter. In a strict sense, spatial process is stationary when any joint distribution of the random variables over a subset of points depends only upon the relative position of the different locations, as determined by their relative orientation (angle) and respective distance. Since the orientation between points in two (or more) dimensions still leave many different situations (potentially over a 360-degree rotation), the stricter notion of isotropy is usually imposed as well. For an isotropic process, the joint distribution depends on the inter-location distance only, and orientation is irrelevant. Again, weak or covariance spatial-stationarity and -isotropy can be defined by confining these requirements to first- and second-order moments. Consequently, for a weak spatial-stationary-process, the covariance between the random variables at two different locations depends only on the distance between the locations and the relative orientation in a coordinate system.

For an isotropic process, these moments depend only on the distance between two locations. In analogy to the time-series situation, the analysis of spatial autocovariance and auto-correlation functions, indexed by the spatial shift, could form the basis of model identification and estimation. We have illustrated this idea with numerical examples in the "Spatial Time-Series" chapter. As illustrated, this analogy is only appropriate in a limited class of spatial processes. Without the restrictive assumption on isotropy, no concept of spatial lag is available that provides a rigorous interpretation for stationarity and the associated use of spatial auto-correlation functions. Consequently, a meaningful treatment of spatial data (and space-time data) analogous to the Box–Jenkins approach in time-series analysis is limited to these special processes. Not only is the structure of spatial processes that obtain well-defined auto-correlation properties highly restrictive, the nature of spatial stationarity and isotropy itself seems rather unrealistic for the type of data and spatial patterns encountered in applied regional science. This problem is not limited to spatial processes, but applies to the time domain as well. We will illustrate these statements with case studies in the sections to come.

Example

An example of an ergodic point-field is the homogeneous Poisson-field (Okabe et al. 1992). Relaxing homogeneity, we define the point process by

$$P(X(R)=x) = \frac{\lambda(R)|R| \ \exp(-\lambda(R)|R|x)}{x!} \qquad x = 0, 1, 2, \ldots \tag{11-4}$$

where $\lambda(R) = \int_R \lambda(\mathbf{x})d\mathbf{x}$. We call this point process the general Poisson-point-process with the intensity-function $\lambda(\mathbf{x})$. An example is shown in **Figure 11-1**, where $\lambda(\mathbf{x}) = c_1$ exp $[-c_2(x_1^2+x_2^2)^{\frac{1}{2}}, \ c_1, c_2 > 0]$. The general Poisson-point-process reduces to the regular Poisson-point-process if $\lambda(\mathbf{x}) = \lambda = $ constant. In a spatial context, λ indicates the density of points. $\lambda(\mathbf{x})$ implies that the density depends upon where R is placed. Thus the density of points may vary from location to location in a general Poisson-point process. A general Poisson-point-process is therefore non-ergotic. This contrasts with the case $\lambda = $ constant, which is a *homogeneous* Poison-point-process. □

11.2.3 Relevance of statistical approaches. The models typically considered in spatial econometrics, in the sense that has been defined in this book, are framed in regression analysis. The two main purposes are inference and point-process. A general Poisson-point-process is therefore non-ergodic. This forecasting. The focus for the former is on finding the extent to which a theoretical relationship can be substantiated by empirical evidence, i.e., on estimating parameters and testing hypotheses. In forecasting, the interest lies in using empirically established relationships to predict values for unobserved units, i.e., forecasting future values, or interpolation in space. Clearly, the imperfect state of theoretical knowledge and the imprecision in measuring the variables of interest are solid arguments for the inclusion of a stochastic error term. Furthermore, the stochastic process approach exemplifies interests in the analysis of a sample to distinguish characteristics of an underlying unknown population. Based on this, a spatial econometric methodology should be considered relevant for the analysis of spatial data. Successful applica

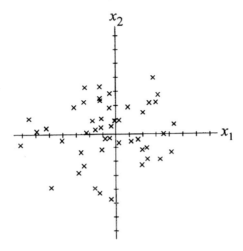

Figure 11-1 A general Poisson-point-process (Okabe et al. 1992)

tions of this approach have been documented in both the "Spatial Econometrics" and "Spatial Time-Series" chapters.

A contrasting view posits that cross-sectional data sets do not constitute a sample, only the whole population does. This means that only a descriptive approach has merit, which is operationalized by way of a 100-percent sample or by way of computer simulation when the underlying process is understood. Although this argument may seem to have intuitive appeal, it actually confuses several issues. This confusion arises in part because of the limitations imposed by spatial dependence and spatial heterogeneity. In particular, when heterogeneity is such that each spatial unit has its own unique characteristics, a cross-sectional analysis will not provide sufficient information to extract these. But this does not imply that an econometric methodology is invalid, but rather that more data (e.g., space–time data) may be needed for it to be meaningfully applied. Similarly, spatial dependence does not invalidate estimation and hypothesis testing, although it renders traditional random sampling inappropriate. Instead, the focus should be on the extent to which an infinitely large sample can be applied to the observational context at hand. The essence of the problem is whether the information in the data-set is compatible with the complexity of the model under consideration. Clearly, if this is not true, it simply means more data (or the relevant data) are needed, rather than a departure from an econometric or statistical methodology.

Homogeneity-test example

In general, only certain aspects of homogeneity can be tested (Stoyan and Stoyan 1994). Differences in the point density can be tested as follows. Let R_1 and R_2 be two subregions of the window R. They have to be chosen a priori. Under the assumption of a homogeneous Poisson field, the quantity $F = \left[A(R_1)(2n_2+1)\right]/\left[A(R_2)(2n_1+1)\right]$ has an approximate F-

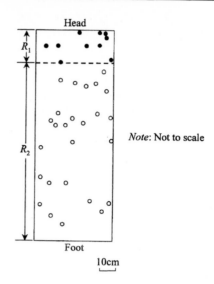

Head

R_1

R_2

Note: Not to scale

Foot

10cm

Figure 11-2 Defect pores on a side of a steel block (Stoyan and Stoyan 1994)

distribution (Sachs 1984). Here $A(R_i)$ is the area of R_i ($i = 1,2$). The degrees-of-freedom are $2n_1+1$ and $2n_2+1$, where n_i is the number of points in R_i. Here the ratio is defined such that $F > 1$. The homogeneity assumption is rejected if $F > F(2n_1+1, 2n_2+1, \alpha/2)$, where α is the probability of a type-1 error (or the probability of falsely rejecting the null hypothesis due to chance.)

Consider the point pattern as shown in **Figure 11-2**. R_1 and R_2 are the upper and lower halfs of the observed rectangle, where $n_1 = 10$, $A(R_1) = 1485$, $n_2 = 27$, $A(R_2) = 1485$. Thus $F = 2.619$. For $\alpha = 0.05$ the critical F-value is $F(21,55,0.025) = 2.20$. The difference in point density is therefore considered significant. □

Let us now turn to the idea of *asymptotics*. In classical problems and in time series it is totally clear how to embed asymptotics in as a series or problems to perform the necessary asymptotic calculations. Perhaps the best example is the pervasive central-limit theorem or what is commonly called the "law of large numbers," whereby a normal distribution becomes a good asymptotic approximation for any distribution when enough data have been collected. This is not at all the case in spatial problems. Consider an $m \times n$ portion of an image. Clearly either m or n should tend to infinity to include the entire population, but very little can be concluded about the ratio m/n. The reader could be forgiven for assuming that this might be immaterial, and in some problems it is. However, there are problems in which this asymptotic distribution has a mean depending on the asymptotic value of m/n (Ripley 1988). Similarly, when data are used for a cross section of irregular spatial units, such as counties, states or provinces, the meaning of asymptoticity is not always clear. This is in contrast to the situation for a regular infinite lattice structure. In essence, the spatial units of observation should typify a larger population, and the number of units should potentially be able to approach infinity in a regular fashion. Clearly, this is not always immediately obvious for the type of data used in applied empirical work in regional science.

For example, consider the case where a multi-regional econometric model is implemented for a given set of regions, such as all the counties in a state. The regions included in these data presumably have characteristics that set them apart from a larger population. This empirical situation can easily lead to considerable heterogeneity in the sample. In the extreme case, where each region has it own characteristics (own variables, own parameter values or own functional forms), there is clearly a need to structure this variety before meaningful statistical inference can be carried out. The same applies to potential spatial dependence between the regions in the sample. There are two ways in which this can be approached. On the one hand, information over many time-periods may be introduced. Under the assumption that the patterns of spatial heterogeneity and dependence remain constant over time, this allows for identification and estimation. Such is the usual approach taken for multi-regional econometric models. Consequently, the formal properties of the estimators and tests are based on the time-dimension in the data. Also, the precise nature of dependence and heterogeneity can be left unspecified. Alternatively, the heterogeneity itself can be structured as a specific function of space (i.e., re-parameterized in function of spatial variables), for which the data-set under consideration can be taken as a representative sample. This would be the situation where variables such as county area or county population would capture the heterogeneity between counties. The statistical properties could then be based solely on the cross-section dimension, if the sample could potentially be extended to include infinitely many *similar* counties.

When spatial dependence is present, the situation is more complex. In essence, the pattern of dependence in the sample should be assumed to represent a pattern in a hypothetical, infinitely large set of contiguous spatial units. This necessitates that the spatial nature of the dependence is general enough to be applicable in this infinitely large data-set. When the dependence is defined in terms of a well-behaved distance metric, as with the EMPIRIC model in the "Spatial Econometrics" chapter, this would not be a problem. In contrast, the extension of an ad hoc notion of first-order contiguity to an infinitely large irregular spatial-configuration is not necessarily meaningful. The general scope of the spatial dependence also implies that observations for regions outside the sample, but close enough to influence the regions that are included, may affect estimation and testing. This leads to the boundary-value problem in spatial analysis. Example of this can be found in classifying pixels in satellite images, as described in the "Remote Sensing and Geographic Information Systems" chapter of Chan (2000).

To conclude this section on the formal properties of spatial processes, it is important to stress the need for caution before a spatial econometric analysis is carried out for irregular spatial data-sets. It is highly advisable to first assess whether the complexity of the proposed models is compatible with the limitations of the data. For many interesting models, there will simply not be enough information in a cross-sectional data set to allow for meaningful analysis, and the use of space-time data is mandated. On the other hand, if the data-set is of the kind that fits with the formal requirements discussed in this section, the econometric methodology will then be based on a rigorous probabilistic foundation.

11.3 Spatial heterogeneity

The reader should appreciate by now that the assumption of stationarity and homogeneity is a limited concept at best (Anselin 1988). Heterogeneity is more of the norm than an exception. In general terms, there are two distinct aspects to spatial heterogeneity. One is structural instability as expressed by changing functional forms or varying parameters. The other aspect is heteroscedasticity, which follows from missing variables or other forms of model mis-specification that leads to error terms with non-constant variance. Ignoring either aspect has well-known consequences for the statistical validity of the estimated model: biased parameter-estimates, misleading significance-levels, and suboptimal forecasts. The extent of spatial heterogeneity that can be formally incorporated in a model is limited by the *incidental parameter problem*, i.e., the situation where the number of parameters increases directly with the number of observations. To avoid this, heterogeneity needs to be expressed in terms of a few distinct categories or parameters. For models with varying coefficients, this implies that the variation should either be determined systematically, in functions of a few additional variables, or stochastically in terms of an a priori distribution. In the case of structural instability of the functional form, the number of different regimes that can be efficiently estimated is limited by degrees-of-freedom considerations. We will illustrate how this is handled in real life in sequel.

A complicating factor in spatial analysis is that the mis-specifications and measurement errors that may lead to heteroscedasticity, such as the problems with the choice of spatial unit of observation, are also likely to cause spatial auto-correlation. It is therefore important to consider the effect of one type of mis-specification on the tests for and estimation of the other. Testing for spatial auto-correlation in the error terms, when heteroscedasticity is present, was alluded to in the previous chapter on "Spatial Time-Series." In this chapter the other combina-tion is considered, i.e., the effect of spatial auto-correlation on tests for heteroscedasticity and structural stability. In many situations, specifying a particular form of heterogeneity in spatial models can be derived from remote-sensing or regional-science theory. In particular, image-processing filters or theories of regional structure and urban form can provide insight into the spatial data-sets that are likely to cause heterogeneity, as well as provide important variables that determine its form. Again, we will illustrate.

Spatial-Dependence Example

A 9-row by 8-column set of data was extracted from a remote-sensing image to test for heterogeneity (Fortmann and Chan 1994). They consist of a SPOT channel-1 image of the Washington, D.C. Mall, as illustrated in the "Remote Sensing" chapter in Chan (2000). The

First-order regression $Z_1^1 = f(Z_2^1, Z_4^1, Z_3^1, Z_5^1)$			Second-order regression $Z_1^2 = f(Z_2^2, Z_3^2, Z_4^2, Z_5^2)$			2×2 mask of first- and second-order neighbors $Z_1 = f(Z_2, Z_3, Z_4)$		
	Estimate	Std. error		Estimate	Std. error		Estimate	Std error
Constant	8.60	47.60	Constant	54.00	62.50	Constant	123.20	29.80
Z_2^1	0.43	0.13	Z_2^2	0.33	0.15	Z_2	0.06	0.14
Z_3^1	0.12	0.14	Z_3^2	0.07	0.13	Z_3	−0.28	0.13
Z_4^1	0.38	0.12	Z_4^2	0.33	0.14	Z_4	0.49	0.13
Z_5^1	0.02	0.14	Z_5^2	−0.04	0.17			
R^2	0.40		R^2	0.21		R^2	0.30	

Table 11-1 Spatial-dependence tests (Fortmann and Chan 1994)

sample data came from the northern central part of the Mall as delineated in the "SPOT sub-image grey values" figure in that chapter. The grey values of this image are highlighted below for easy reference:

139	174	210	168	139	157	139	186
157	180	204	180	168	157	211	214
157	151	213	151	151	192	204	192
133	174	210	115	145	192	174	139
151	180	192	192	192	168	145	163
157	186	192	215	168	115	151	192
163	180	192	210	115	151	174	210
168	192	210	218	174	145	192	204
174	213	215	216	211	163	127	127

Spatial dependance was evaluated by running a least-squares linear-regression on the data-set. Besides an overall R^2, parameter estimates were compared to their standard errors in each run to evaluate whether the coefficient is significant enough to show spatial dependence in a particular direction. These tests are summarized in ?, where regressions are performed on the first-order neighbor, second-order neighbor and a 2×2 mask including first- and second-order neighbors. Consider a 'blurring' filter-mask is being used in image processing. In **Table 11-1**, Z_1, Z_2, Z_3 and reference to the subject pixel Z_1^1 in the center. Finally, Z_2^2, Z_3^2, Z_4^2, Z_5^2 are second-order neighbors clockwise from the upper-left pixel, again referenced against subject pixel Z_1^2. The result of the regression suggests that spatial dependence is not shown in general judging from the low overall R^2 and specifically the coefficients associated with the first-order neighbors and second-order neighbors. However, correlations appear to be stronger along the north-south northwest-southeast directions. □

To perform the following analyses, the spatial data were prepared in accordance with the procedure outlined in the "Spatial Time-Series" chapter under "An extended numerical example" section. These image-processing filters were used

first-order neighbor: $Z_1^1 = \frac{1}{4}(Z_2^1 + Z_3^1 + Z_4^1 + Z_5^1)$;
second-order neighbor: $Z_1^2 = \frac{1}{4}(Z_2^2 + Z_3^2 + Z_4^2 + Z_5^2)$.

These are a priori assignment of weights and are subject to validation.

Homogeneity Example

Having "fixed ideas" with a sub-image, we will now investigate in more detail spatial dependence using the full Channel-1 image of the Washington, D.C. Mall. Recall that this full image was shown in the "Remote Sensing" chapter of Chan (2000). Here, we will run through the entire set of analysis steps from the start to finish, rather than a simple regression analysis as shown in the "Spatial dependence" example above (Ference and Chan 1996). Similar to regular time-series, the first step is to induce homogeneity (the equivalance of stationarity) in the data. To do this, the data was differenced *spatially*. Differencing between t and $t-1$ time-periods now translates to differencing between a pixel and the average of its first-order neighbors. Instead of considering the first time-lag, second time-lag autocorrelation, etc., the correlations between a subject pixel and the average of its first, second, and third neighbors are considered. For convenience, here we will refer to the first-, second-, third-order neighbors as the first, second, third "onion-rings."

Following this scheme, the first, second, and third-order differencing was first performed among the data to induce homogeneity. Plot of the third-order difference against grey values suggest possible homogeneity, since the data now "hover" around the mean (almost). Autocorrelations are now computed on this third-order differenced-dataset. Specifically, autocorrelations are computed between the average grey-values of the first-onion-ring and a subject pixel, and similarly between the second-one-ring and the subject pixel, third-onion-ring and the subject. The result is shown below:

	Autocorrelation coefficient
first spatial-lag	−0.77469
second spatial-lag	0.407173
third spatial-lag	−0.016707
fourth spatial-lag	−0.13729

Again drawing the analogy with regular time-series, we try to infer possible homogeneity from these autocorrelation coefficients. Strictly speaking, the pattern among these coefficients does not suggest homogeneity. With the amount of landscaping in the D.C. Mall, this result is not surprising. □

Isotropy Example

To the extent that homogeneity cannot be induced, we retun to the original dataset for isotropy analysis (Ference and Chan 1996). For the full Channel-1 image, correlation coefficients are computed between a subject pixel and its first-order neighbors. Also, correlations are computed between a subject and its second-order neighbors. The results are summarized below:

Correlation coefficients	
North neighbors	0.6432
South neighbors	0.6542
East neighbors	0.6215
West neighbors	0.6409
Northeast neighbor	0.4440
Southeast neighbor	0.4067
Southwest neighbor	0.5026
Northwest neighbor	0.4156

These results show that the strengths of relationship between subject grey-values and their first-order neighbors are very similar. This is also true for second-order neighbors, although to a lesser degree. All signs point toward isotropy for this dataset. While there is limited data in the sub-image to draw definitive inference, the entire image is much bigger—9×8 versus 50×20. Any manmade objects, such as walkways, are only a minute part of the full image. Whatever directionality present in, say walkways, is therefore buried under the majority of the remaining data that have little orientation patterns. This explains the analystical results as reported. □

Our treatment here in this chapter is likely to be illustrative rather than theoretical. Our objective is to discuss these problems via case studies and point out their applications. We have pointed out that simple extentions of time-series statistics are often not sufficient for spatial analysis. In spite of the last numerical example above, isotropy is clearly a broader concept than "four points of the compass." For the interested reader, s/he is referred to Anselin (1988) for a more axiomatic exposition. Three cases are selected for discussion below: Case 1 deals with a spatial time-series to model deforestation in east Texas. Case 2 revisits the EMPIRIC model as applied to Minneapolis-St Paul land-use forecasting. Finally, case 3 reexamines the disaggregate/bifurcation model of Yi and Chan (1988). These studies represent a progressive degree of heterogeneity in model construction. Another way of saying this is that we progress from purely correlative relationship to models with increasing amount of embedded spatial structure.

11.4 Identification of spatial-temporal models for multivariate processes[5]

Many remote-sensing targets change dynamically, such as the growth of a city and the deforestation of tropical wilderness. If time-series models can be identified for data collected from these targets, non-characteristic changes in their behavior can

---

[5]The readers not familiar with multivariate time-series can review the basic concepts in the "Statistical" appendix (Chan 2000) and the "Spatial Time-Series" chapter in the current volume.

be detected and their future behavior forecasted. The difficulty in identifying models for this new category of multivariate time-series data is often due to the number of pixels time-series involve. Consider n' distinct pixel positions defined in a plane, the vector variable $\mathbf{Z}_t$ then consists of n' components $Z_i(t)$ referenced to position i in the plane at time t ($t = 1, 2, \ldots, n$). This amounts to a dimension of $n' \times n$, which is often a huge number.

11.4.1 Spatial time-series.

The spatial-temporal process for the target region can be represented by the vector or multivariate time-series autoregressive moving-average (ARMA) model[6] where the $\mathbf{Z}$ have been differenced and appropriately transformed to $\mathbf{z}$ with the purpose of inducing stationarity[7] (Box and Jenkins 1976, Makridakis and Wheelwright 1978):

$$\mathbf{z}(t) - \mathbf{\Phi}_1 \mathbf{z}(t-1) - \mathbf{\Phi}_2 \mathbf{z}(t-2) + \ldots + \mathbf{\Phi}_p \mathbf{z}(t-p)$$
$$= \mathbf{a}(t) - \mathbf{\Theta}_1 \mathbf{a}(t-1) - \mathbf{\Theta}_2 \mathbf{a}(t-2) - \ldots - \mathbf{\Theta}_q \mathbf{a}(t-q) \qquad \textbf{(11-5)}$$

Here $\mathbf{\Phi}$ are coefficients for the autoregressive (AR) process, $\mathbf{\Theta}$ are for the moving-average (MA) process, and $\mathbf{a}$ are the error disturbance or random terms in the ARMA (p,q) model with p-lag observation-terms and q-lag error-terms.

The basic multivariate time-series between variables $\mathbf{z}_t$ and error-terms $\mathbf{a}_t$ can be expressed here as $\mathbf{\Phi}(B)\mathbf{z}(t) = \mathbf{\Theta}(B)\mathbf{a}(t)$ where $\mathbf{z}(t) = (\mathbf{z}'(t) - \mathbf{c})$. $\mathbf{Z}'_t$ is an appropriately transformed and differenced vector-series, and $\mathbf{c}$ is the vector mean of that series (Robinson 1987). B is the backward shift operator, defined by $B\mathbf{z}_t = \mathbf{z}_{t-1}$. $\mathbf{\Phi}(B)$ and $\mathbf{\Theta}(B)$ are matrices of autoregressive operators and moving-average operators respectively with components. We write

$$\varphi_{ij}(B) = \varphi_{ij1}B - \varphi_{ij2}B^2 - \varphi_{ij3}B^3 - \ldots - \varphi_{ijp_{ij}}B^{p_{ij}} \qquad (i \neq j) \qquad \textbf{(11-6)}$$

$$\theta_{ij}(B) = \theta_{ij1}B - \theta_{ij2}B^2 - \theta_{ij3}B^3 - \ldots - \theta_{ijq_{ij}}B^{q_{ij}} \qquad (i \neq j) \qquad \textbf{(11-7)}$$

$$\varphi_{ii}(B) = 1 - \varphi_{ii1}B - \varphi_{ii2}B^2 - \varphi_{ii3}B^3 - \ldots - \varphi_{iip_{ii}}B^{p_{ii}} \qquad \textbf{(11-8)}$$

Thus for a 2×1 image, $\mathbf{z}(t) = (z_1(t), z_2(t)]^T$

$$\theta_{ii}(B) = 1 - \theta_{ii1}B - \theta_{ii2}B^2 - \theta_{ii3}B^3 - \ldots - \theta_{iiq_{ii}}B^{q_{ii}}. \qquad \textbf{(11-9)}$$

---

[6]Sometimes referred to as a special case of the space-time autoregressive moving-average (STARMA) model. See the chapter on "Spatial Time-Series" for a review of these basic concepts.

[7]For a review of the idea of stationarity, see the discussion in the last two sections and also the "Spatial Time-Series" chapter (particularly under the "Univariate time-series" section).

$$\Phi_1 = \begin{bmatrix} \varphi_{111} & \varphi_{121} \\ \varphi_{211} & \varphi_{221} \end{bmatrix} \qquad \Phi_2 = \begin{bmatrix} \varphi_{112} & \varphi_{122} \\ \varphi_{212} & \varphi_{222} \end{bmatrix} \qquad\qquad \textbf{(11-10)}$$

$$\Theta_1 = \begin{bmatrix} \theta_{111} & \theta_{121} \\ \theta_{211} & \theta_{221} \end{bmatrix} \qquad \Theta_2 = \begin{bmatrix} \theta_{112} & \theta_{122} \\ \theta_{212} & \theta_{222} \end{bmatrix}. \qquad\qquad \textbf{(11-11)}$$

a vector-ARMA(2,2) time-series will look like:

$$\begin{bmatrix} z_1(t) \\ z_2(t) \\ a_1(t) \\ a_2(t) \end{bmatrix} - \begin{bmatrix} \varphi_{111} & \varphi_{121} \\ \varphi_{211} & \varphi_{221} \\ \theta_{111} & \theta_{121} \\ \theta_{211} & \theta_{221} \end{bmatrix} \begin{bmatrix} z_1(t-1) \\ z_2(t-1) \\ a_1(t-1) \\ a_2(t-1) \end{bmatrix} - \begin{bmatrix} \varphi_{112} & \varphi_{122} \\ \varphi_{212} & \varphi_{222} \\ \theta_{112} & \theta_{122} \\ \theta_{212} & \theta_{222} \end{bmatrix} \begin{bmatrix} z_1(t-2) \\ z_2(t-2) \\ a_1(t-2) \\ a_2(t-2) \end{bmatrix} = \qquad \textbf{(11-12)}$$

A homogeneous process will postulate the same calibration coefficients for $z_1(t)$ as for $z_2(t)$, i.e., the same estimation equation is written for the first pixel as the second pixel or the subscript 1 and 2 can be replaced by i.

The main objective in identification is to determine the overall limits to the multivariate process, p and q, and to determine the specific limits, p_{ij} and q_{ij}, for each of the components within the $\Phi(B)$ and $\Theta(B)$ matrices. The most widely-used identification-methods are extensions to univariate Box-Jenkins methodology. Sample cross-correlation (r_k, $k = 1, 2, \dots , m > \max(p,q)$) and partial-correlation coefficient (r'_k) are two principal identification-tools which are used to identify parsimonious models in commercial software packages. Each univariate component series of the multivariate process is examined to determine the level of differencing and the transformation necessary to make $z_i(t)$ a stationary series ($i = 1, 2, \dots , n'$). The components of r_k are standard cross-correlations at lag-k. The sample estimate of r'_k is retained from the highest-order matrix computed in a multivariate autoregressive-model of order k.[8]

One way to view this space-time, autoregressive moving-average (STARMA) model is as an extension of the univariate ARMA model into the spatial dimension. For each of the n' time series, the STARMA model represents the current observation as a linear combination of past observations and innovations as well as observations and innovations at neighboring sites. It is these spatial terms that tie the n' series together and differentiate the STARMA model from a sequence of n' separate-identical univariate ARMA-models.

---

[8]For a review of partial-correlation coefficients, see "Statistical" appendix ibn Chan (2000) and the "Spatial Time-Series" chapter under the "Univariate Time-Series" section.

11.4.2 Spatial-temporal canonical-analysis. The basic procedure for constructing a STARMA model has been discussed and illustrated with a complete example in the "Spatial Time-Series" chapter. It was based on an extension of the Box–Jenkins four-step procedure consisting of identification, parameter estimation, diagnostic checking, and forecasting. Among the four steps, the most important and challenging is the identification step. Here we will introduce an improved procedure for identification called "spatial-temporal canonical analysis." The technique is used to identify spatial-temporal limits in the underlying process at the vicinity of many target-points from a satellite image (Robinson 1987). Instead of an a priori assumption about spatial correlation, each individual-target pixel-series with an image-series can theoretically be related to the surrounding past space-time neighborhood. This association can be evaluated using *canonical correlation analysis* (Box and Draper 1987)—a procedure encoded in such statistical software as SAS. Canonical analysis rewrites a fitted equation in a more easily understood form. This is achieved by a rotation of axes, which removes all cross-product terms. If desired, a change of origin further removes first-order terms. Suppose the following fitted second-order equation is obtained:

$$z = 78.8988 + 2.2272z_1 + 3.496z_2 - 2.08z_1^2 - 2.92z_2^2 - 2.88z_1z_2 \qquad \textbf{(11-13)}$$

Rotation of axes results in transforming this equation into:

$$z = 78.8988 + 4.16U_1 + 0.28U_2 - 4U_1^2 - U_2^2 \qquad \textbf{(11-14)}$$

where $U_1 = 0.6z_1 + 0.8z_2$, $U_2 = -0.8z_1 + 0.6z_2$. By a translation of axis, the equation can be simplified further to look like $z = 80 - 4V_1^2 - V_2^2$ where $V_1 = 0.6(z_1 - 0.2) + 0.8(z_2 - 0.5)$, $V_2 = -0.8(z_1 - 0.2) + 0.6(z_2 - 0.5)$. Specifically, canonical correlation-analysis (CCA), when applied in our present context, provides an identification of the approximate spatial-temporal lags.

   11.4.2.1 COMPUTING CANONICAL CORRELATIONS. In general, multiple criteria (dependent) variables $\mathbf{z}^{(1)} = (z_1^{(1)}, z_2^{(1)}, \dots, z_{n_1}^{(1)})^T$ can be regressed against multiple predictor (independent) variables $\mathbf{z}^{(2)} = (z_1^{(2)}, z_2^{(2)}, \dots, z_{n_2}^{(2)})^T$. Consider the following definitions:

$$\begin{aligned} U_1 &= \alpha_{11}z_1^{(1)} + \alpha_{12}z_2^{(1)} + \dots + \alpha_{1n_1}z_{n_1}^{(1)} \\ V_1 &= \beta_{11}z_1^{(2)} + \beta_{12}z_2^{(2)} + \dots + \beta_{1n_2}z_{n_2}^{(2)}. \end{aligned} \qquad \textbf{(11-15)}$$

These two equations define the new variable U_1 and V_2 in terms of the old variables $\mathbf{z}^{(1)}$ and $\mathbf{z}^{(2)}$ respectively. Let r_1 be the correlation between U_1 and V_1. The objective of canonical correlation is to estimate $\alpha_1 = (\alpha_{11}, \alpha_{12}, \dots, \alpha_{1n_1})$ and $\beta_1 = (\beta_{11}, \beta_{12}, \dots, \beta_{1n_2})$ such that r_1 is maximum. Equation **(11-15)** defines the pair of *canonical equations*. U_1 and V_1 are the *canonical variates*, and r_1 is the *canonical correlation* (Sharma 1996).

Once U_1 and V_1 have been estimated, the next step is to identify another set of canonical variates

$$U_2 = \alpha_{21}z_1^{(1)} + \alpha_{22}z_2^{(1)} + \ldots + \alpha_{2n_1}z_{n_1}^{(1)}$$
$$V_2 = \beta_{21}z_1^{(2)} + \beta_{22}z_2^{(2)} + \ldots + \beta_{2n_2}z_{n_2}^{(2)}$$

(11-16)

such that the correlation r_2 between them is also maximum, and U_1 and V_1 are uncorrelated with U_2 and V_2. That is, the two sets of canonical variates are uncorrelated. This procedure is continued until the mth set of canonical variates is identified such that r_m is maximum.

$$U_m = \alpha_{m1}z_1^{(1)} + \alpha_{m2}z_2^{(1)} + \ldots + \alpha_{mn_1}z_{n_1}^{(1)}$$
$$V_m = \beta_{m1}z_1^{(2)} + \beta_{m2}z_2^{(2)} + \ldots + \beta_{mn_2}z_{n_2}^{(2)}$$

(11-17)

The objective of canonical correlations to identify the m sets of canonical variates (U_1,V_1), (U_2,V_2), ... , (U_m,V_m); such that the corresponding canonical-correlations r_1, r_2, ... , r_m are maximum, and the correlations $r(V_i,V_j) = r(U_i,U_j) = r(U_i,V_j) = 0$ for all $i \neq j$. This is clearly a maximization problem subject to certain constraints, the details of which are given in Sharma (1996), or can be left to software packages that typically implement CCA.

Consider the n-component-vector-random-variable **z** partitioned into two sub-vectors of n_1 and n_2 lengths such that $\mathbf{z} = (\mathbf{z}^{(1)}, \mathbf{z}^{(2)})^T$. Here $n = n_1 + n_2$, $n_1 \leq n_2$, where $E(\mathbf{z}) = \mathbf{0}$ and the covariance matrix is non-singular and positive definite. CCA finds n_1 mutually-orthogonal pairs of linear combinations of the members of the two partitioning-sets $U_i = \alpha_i^T \mathbf{z}^{(1)}$ and $V_i = \beta_i^T \mathbf{z}^{(2)}$ $(i = 1, 2, \ldots, n_1)$ such that the correlations between the two new canonical variates formed in each pair, $r(U_i,V_i)$, is maximized.[9] (Here α and β are vectors of scaling constants determined as part of the CCA procedure.) These correlations are ordered from largest to smallest (Reinsel 1993). Thus $r_1 = r(U_1,V_1) \geq 0$ may be the largest correlation between any linear combinations of $\mathbf{z}^{(1)}$ and $\mathbf{z}^{(2)}$, that is, U_1 and V_1 are determined by the property that they are the linear combinations of $\mathbf{z}^{(1)}$ and $\mathbf{z}^{(2)}$ which possess the maximum possible correlation among all such linear combinations. In addition, U_2 and V_2 have correlation $r_2 = r(U_2,V_2) \geq 0$ and are characterized as those linear combinations with maximum correlation among all linear combinations that are uncorrelated with U_1 and V_1. Proceeding in this way, then, U_i and V_i have the properties that $r_i = r(U_i,V_i) \geq 0$, with $r(U_i,U_j) = 0$, $r(V_i,V_j) = 0$ and $r(U_i,V_j) = 0$ for $i \neq j$, $i,j = 1,$. ... n_1; and U_i,V_i have maximum correlation among all linear combinations which are uncorrelated with U_j,V_j; $j = 1, \ldots ,i-1$. The variables U_i, V_i are called the canonical variables, and $r_1 \geq r_2 \geq \ldots \geq r_{n_1} \geq 0$ are the canonical correlations between $\mathbf{z}^{(1)}$ and $\mathbf{z}^{(2)}$.

---

[9] The term 'canonical' suggests that the two random-variables U_i and V_i have unit variances, i.e., corr$(U_i, U_j) = 1$ if $i = j$ and $= 0$ for $i \neq j$; and similarly for the V_i. Orthogonality means corr $(U_i, V_j) = 0$ for $i \neq j$.

11.4.2.2 SPATIAL-TEMPORAL CORRELATION ANALYSIS. In the multivariate-
or vector-ARMA time series, we consider vectors such as the $n'(m+1)$-dimensional
vector

$$\mathbf{z}_m(t) = [\mathbf{z}(t), \mathbf{z}(t-1), \dots, \mathbf{z}(t-m)]. \tag{11-18}$$

We examine the canonical-correlation structure between the variables $\mathbf{z}_m(t)$ and
$\mathbf{z}_k(t-j-1) = [\mathbf{z}(t-j-1), \mathbf{z}(t-j-2), \dots, \mathbf{z}(t-j-k)]$ for various combinations of $m = 0, 1,$
$\dots$, and $j = 0, 1, \dots$, and $k \geq m$. Recall that for an ARMA(p,q) process, $\mathbf{z}(t) -$
$\sum_{j=1}^{p}\mathbf{\Phi}_j\mathbf{z}(t-j) = \mathbf{a}(t) - \sum_{j=1}^{q}\mathbf{\Theta}_j\mathbf{a}(t-j)$, the variables $\mathbf{z}(t-s)$ are uncorrelated with the
$\mathbf{a}(t-j), j = 0, 1, \dots, q$; for $s > q$. Thus if $m \geq p$, then there are (at least) s linear-
combinations of $\mathbf{z}_m(t)$,

$$\mathbf{z}(t) - \sum_{j=1}^{p}\mathbf{\Phi}_j\mathbf{z}(t-j) = \left[\mathbf{I}, -\mathbf{\Phi}_1, \dots, -\mathbf{\Phi}_p, 0, \dots, 0\right]\mathbf{z}_m(t) = \mathbf{a}(t) - \sum_{j=1}^{q}\mathbf{\Theta}_j\mathbf{a}(t-j) \tag{11-19}$$

which are uncorrelated with $\mathbf{z}_k(t-j-1)$ if $j \geq q$. In particular, for $m = k = p$ and $j =$
q there are (at least) s zero canonical correlations between $\mathbf{z}_p(t)$ and $\mathbf{z}_p(t-q-1)$, as
well as between $\mathbf{z}_p(t)$ and $\mathbf{z}_p(t-j-1)$; $j > q$, and between $\mathbf{z}_m(t)$ and $\mathbf{z}_m$(t-q-1), $m > p$.
Systematic examination of the number of zero canonical-correlations between $\mathbf{z}_m(t)$
and $\mathbf{z}_k(t-q-1)$ for combinations of m and j can reveal the detailed structure of the
nature of the ARMA(p,q) model. In particular, the smallest integer values of m and
j of which (at least) k zero canonical-correlations exist will identify the overall
orders p and q of the model. In the framework of spatial-temporal canonical-
correlation analysis, we perform a multi-variate regression (canonical correlation)
of $\mathbf{z}^{(1)}$ on $\mathbf{z}^{(2)}$ where $\mathbf{z}^{(1)}$ represents a future space. The future space is restricted to the
space generated by the present and future series at a target site. This series is
defined as

$$\mathbf{z}^{(1)} = [z_0(t), \ z_0(t+1), \ z_0(t+2), \dots, z_0(t+\infty)]^T \tag{11-20}$$

where the "0" subscript designates the position of the target-series in space, at the
central site in the neighborhood.

Space-time shells at equal Euclidean space-time distance from a given anchor
point in space-time can be constructed. Let x_1 axis be the east-west or right-left
dimension, x_2 axis be the north–south or up–down dimension, and x_3 axis be the
future–past dimension. The ordered $\mathbf{z}^{(2)}$-vector is now defined as $z^{(2)'} = [z_{(1)}, z_{(2)}, \dots, z_{(\infty)}]$.
The bracketed subscript of $\mathbf{z}_{(i)}$ specifies a 'shell' number, indicating that each
sub-vector is an ordered vector of the space–time lagged-variables resident at the
ith ordered space–time distance from the anchor point. This representation is a
version of a nearest-neighbor ring structure modified to fit a space–time requirement, as illustrated in **Figure 11-3**. The distance from the anchor point in this
coordinate system depends both on the temporal-sampling interval and the relations
of x_1-distance to x_2-distance in the data. Notice the spatial data is also a function of
the spatial sampling-scheme. Thus the distance is expressed as $(w_1x_1^2 + w_2x_2^2 + w_3x_3^2)^{1/2}$
where w_i are weighting coefficients that allow customizing the shell spaces to the

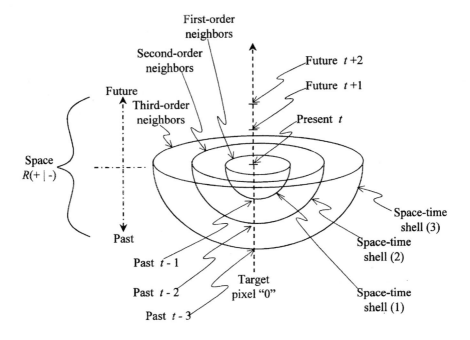

Figure 11-3 Conceptualizaton of Spatial-Temporal Canonical-Analysis
(Robinson 1987)

sampling intervals and the strength-of-relationships observed. Values of the shell distance coefficients w_i are determined iteratively. For any arbitrary spatial-temporal sample, the shell configuration that is optimal depends on information available and sampling intervals in space and time. The objective of the spatial-temporal canonical-analysis identification technique is to evaluate the dimension and quality of the space $R(+|-)$ in which the shell configuration is accommodated. Two dimensions must be considered in deciding a minimal interface:

a. The dimension that represents the number of non-zero autoregressive parameters in the eventual forecast function pertaining to the target.
b. The limiting shell-order that contains the maximum amount of information about the future subject to the minimality constraint or the art of parsimony.

It is possible to find an approximation to these two optimal dimensions using a two-step technique under an assumption of finite predictor and prediction spaces, $R(-|+)$ and $R(+|-)$. The future dimension must be selected using a sufficiently large space–time past-shell dimension in which it is assumed the minimal shell-dimension is included. The minimal order of the past projected on the future is determined. The space–time past-shell dimensions are then analyzed by projecting the minimal future-dimension on the past space and selecting the minimum space. The limit of the space–time influence-region is used as a limiting boundary in the identification of the process using other techniques. The process is iterative. If the interface between the two spaces is finite, the technique will high light a minimal definition of that interface. Ideally, the space–time influence should be as small as possible, but still contain a balanced descriptive capability.

11.4.2.3 STARMA MODELLING. Thus according to Reinsel (1993), one may first obtain estimates of the innovations-series $\mathbf{a}(t)$ in a potential vector-ARMA(p,q) model. This is accomplished by approximating the model by a (sufficiently high order) AR-model of order m^*. The order m^* of the approximating AR-model might be chosen by use of a parsimonious selection-criterion.[10] From the selected AR(m^*) model, one obtains residuals $\tilde{a}(t) = z(t) - \sum_{j=1}^{m*} \hat{\Phi}_{jm*} z(t-j)$. In the second stage of the procedure, one regresses $\mathbf{z}(t)$ on $\mathbf{z}(t-1), \dots, \mathbf{z}(t-p)$ and $\tilde{a}(t-1)$, $\dots, \tilde{a}(t-q)$, for various values of p and q. That is, one estimates (approximates) models of the form

$$\mathbf{z}(t) = \sum_{j=1}^{p} \mathbf{\Phi}_j \mathbf{z}(t-j) - \sum_{j=1}^{q} \mathbf{\Theta}_j \tilde{\mathbf{a}}(t-j) + \mathbf{a}(t) \tag{11-21}$$

by linear least-squares regression; and by another parsimonious selection criterion, the order (p,q) of the ARMA model is chosen. Use of this procedure may lead to one or two ARMA models that seem highly promising, and these models can be finally estimated by maximum-likelihood procedures. The models can subsequently be evaluated by residual-analysis checks on model adequacy. The appeal of the procedure is that computation of maximum-likelihood estimates, which are much more computationally expensive, over a wide range of possible ARMA-models is avoided. Furthermore, the parameter estimates obtained by this procedure are generally fairly efficient compared with maximum likelihood and provide excellent starting values for the maximum-likelihood iterations.

In general, the STARMA model is expressed as:

$$z_i(t) = \sum_{k=1}^{p} \sum_{l=0}^{\lambda_k} \varphi_{kl} z_{i-l}(t-k) - \sum_{k=1}^{q} \sum_{l=0}^{m_k} \theta_{kl} a_{i-l}(t-k) + a_i(t) \qquad \forall i \tag{11-22}$$

where λ_k is the spatial order of the kth autoregressive-term and m_k is the spatial order of the kth moving-average-term. Here space–time autocorrelations and partials are two-dimensional analogues of the usual autocorrelations and partials used to identify univariate ARMA-models as explained in the "Spatial Time-Series" chapter. The *sample* space–time autocorrelation at spatial-lag l and time-lag k is calculated via

---

[10] An example of such a criterion is the Akaike Information Criterion (AIC), except that the estimated error-covariances may replace the maximum-likelihood estimator if a least-squares calibration-procedure is used. Details of such selection criteria will be discussed in the following section in connection with the East-Texas case-study.

Model	Space–time autocorrelation	Space–time partial autocorrelation
$STAR(p_{\lambda_1,\ldots,\lambda_p})$	Tails off	Cuts off after p time-lags, λ_p spatial-lags
$STMA(q_{m_1,\ldots,m_q})$	Cuts off after q time-lags, m_q spatial-lags	Tails off
$STARMA(p,q)$	Tails off	Tails off

Table 11-2 Identification of STARMA models (Pfeifer and Bodily 1990)

$$r_j(k) = \frac{n}{(n-k)} \frac{\sum_{t=1}^{n-k} z_i^T(t)\, z(t+k)}{\left[\sum_t z_i^T(t)\, z_j(t)\, z^T(t)\, z(t)\right]^{1/2}}. \tag{11-23}$$

Space–time partial-autocorrelations are calculated using the space–time analogue of the Yule–Walker equations as explained in the "Spatial Time-Series" chapter (Makridakis and Wheelwright 1978, Pfeifer and Deutsch 1980). Analogous to univariate case, **Table 11-2** summarizes the cut-off and tail-off properties of these autocorrelations, which greatly facilitates the STARMA identification-process. Notice here in our discussion the data z has been appropriately prepared via canonical analysis. In fact, canonical analysis distinguishes this calibration procedure from previous ones described in "Spatial Time-Series" chapter, Although similar autocorrelations are computed in both cases. In previous attempts, spatial correlation is calibrated separate from temporal correlation. In other words, instead of a simultaneous calibration-procedure employed through canonical analysis, spatial weights are estimated outside the model.

11.4.3 Case Study 1 - Vegetation in East Texas. By spatial-temporal modelling of vegetation images collected from satellites, it is possible to better understand and predict bio-environmental change. Robinson (1987) applied identification techniques to a segment of image series collected over East Texas during a three-year period (144 weeks). The remote-sensing unit Advanced Very High Resolution Radiometer (AVHRR) aboard National-Oceanic-and-Atmospheric-Administration-n (NOAA-n) series meteorological satellites has a frequency of coverage in the order of 2-day-and-2-night passes per day. Considering the difference between Channel 2 (near infrared: 0.725–1.10 µm) and Channel 1 (visible: 0.58–0.68 µm), the Gray–McCrary Index (GMI) is a useful measure for biomass evaluation. A normalized vegetation-index

$$NVI = \frac{(CH2 - CH1)}{(CH2 + CH1)} \tag{11-24}$$

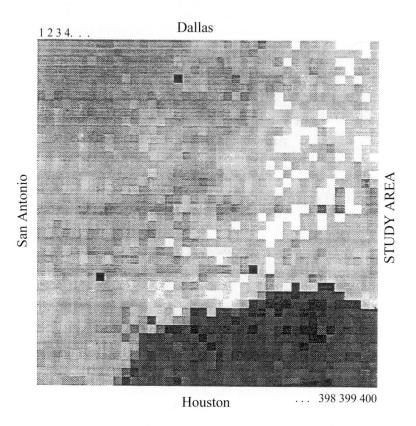

Figure 11-4 Pixel map of Texas Gulf Coast

partially compensated for the slope of the surface, viewing aspect, and illumination conditions of a target, serve as a useful composite of two channels. (See "Remote Sensing" chapter in Chan (2000).)

The raw data in this case study consists of 8-bit Channel-1 (CH1) and Channel-2 (CH2) value-pairs converted to an NVI at each pixel over time. A 20×20 pixel area over east Texas and western Louisiana is centered on the Piney Woods, where the ground cover consists primarily of young forest and grassland interspersed with pine-hardwood forest. The 20×20 grid represents approximately a 240×240 mile (384×384 km) region (See **Figure 11-4**). The 20×20×144 sample is readily manipulated with image-processing software. A five-point velocity and acceleration transformation is also applied in these images on a pixel-by-pixel basis to highlight changes—the equivalent of viewing the data as a 'movie'. By sectioning the transformed data, the seasonality, homogeneity and speed-of-change are displayed visually.

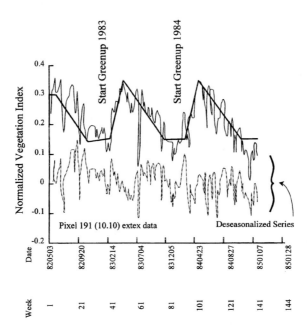

Figure 11-5 Single pixel NVI forest series with triangular
deseasonalization function (Robinson 1987)

11.4.3.1 IDENTIFICATION. The 20×20 pixel region is really not homogeneous, in that the data are not stationary under translation. It has a highly seasonal aspect and an apparently high degree of multi-collinearity. There is some consistency of behavior in the center of the image and in pockets along the edges. The univariate autocorrelation function (ACF) and partial autocorrelation-function (PACF) for 25 time-lags are computed for the image. Pixels in the 20×20 east-Texas study-area are ordered from 1 to 400 from top to bottom where each group of twenty 20×25 sections in the image represents a row in the original study-region. It is clear that there are differences in the basic processes that reflect the degree-of-homogeneity in the image. These changes in homogeneity are most noticeable between the north and south in the region.

Two major options were considered to remove the non-stationarity in the process: differencing and extraction of deterministic trends. Because of the obvious deterministic-seasonal-nature of the process, an initial trend was removed to achieve stationarity. If the ideal deterministic-trend for each individual pixel were removed, deviations from that trend would suggest the impact of climatological and sensor noise in the process. The de-seasonalization algorithm appears to work particularly satisfactorily for the central 10×10 site grid (See **Figure 11-5**). Examination of display indicates a low-order autoregressive or mixed process in most of de-seasonlized data. Models were identified and parameters estimated for limited subsets of the component series using a standard time-series package. Three separate areas were examined: a central 3×3 area, a 3×3 region in the vicinity of

Site	Nearly significant		Nearly significant residual autocorrelation function
	φ_1	θ_1	
169	0.8683	0.6151	LAG 7
189	0.8071	0.4913	
209	0.9210	0.7520	LAG 9
170	0.9113	0.6736	
190	0.8606	0.6365	
210	0.8058	0.5532	LAG 9
171	0.7328	0.3563	
191	0.9435	0.7741	
211	0.7614	0.5219	LAG 7

Table 11-3 Univariate identification and estimation
results for 9 central forest sites (Robinson 1987)

grassland incursion into the forest, and a 3×3 area about pixel 121 along the left edge of the east-Texas area. The last region is clearly non-stationary, for which estimation was not performed. Estimations for the other two sets of nine sites are exemplified in **Table 11-3**. The residuals column suggests possible deficiencies in the models where high values in the autocorrelation function (ACF) and partial autocorrelation functions (PACF) of the residuals occurred. A number of sites exhibited high ACFs at lags 7 and 9 that were borderline. The weak lag-9 influence is clearly observable in the univariate identification-images of de-seasonalized data. This influence was initially considered an outlier effect; however, its significance is increased in view of its repetition in the spatial-temporal series as a whole. This example highlights an interesting aspect of the image-construction technique—the ability to use color to visually detect patterns of significant values representing correlations. Repeated high values of the ACF at lag-9 that might be ignored in separate univariate-analysis have obvious significance when viewed in the image simultaneously. The physical meaning of this peculiar lag-9 influence is not understood at this time. Overall, the univariate influence structure is typically ARMA(1,1), indicating that the time depth for the spatial-temporal process is essentially lag-1.

The 7×7 cross-correlation target-windows around each site in the de-seasonalized 10×10 east-Texas-target-scene show a distinctly non-homogeneous process. The images primarily highlight the ecological diversity in the region, showing no consistent, identifiable diffusion gradient(s). Based on the graphical results and preliminary analysis of images, the time-biased shell-configuration appears to be appropriate. A preliminary analytical-evaluation was performed for a 16-site grid.

A standard hypothesis test can be used to test whether a canonical correlation, or multiple canonical-correlations, beyond some limit, say k^*, is equal to zero (Robinson 1987). To test the null hypothesis H_0: $r_{k^*+1} = r_{k^*+2} = ... = r_{n_1} = 0$, the following statistic is to be minimized:

$$-\left(n' - \frac{n+3}{2}\right) \prod_{i=k^*+1}^{n_1} \log(1-r_i^2) - \chi^2\left[(n_1-k^*),(n_2-k^*)\right]. \tag{11-25}$$

From this statistic, the statistical significance of the additional canonical-correlation coefficient can be assessed. This is done each time a new candidate variable is considered for the state-vector $\mathbf{z}$ by computing the appropriate percentage point for the $\chi 2$-distribution. The method is appealing because an approximate tradeoff between degrees-of-freedom determined by the number-of-parameters in the model (first term) and statistical significance (second term) is implicit in the test.

The assessment of a canonical-correlation coefficient represents an evaluation of the marginal utility of increased explanatory-power versus the cost of over-specification. Remember the cost includes loss of degrees-of-freedom if an extra dimension is added to the prediction space. When the model is too flexible, the dimension of the predictor space is assumed too large, which results in unreliable estimates of covariance structure. If the information is too inflexible, with the dimension of the predictor space too small, the true structure of the process is lost, and the model approximation of that structure is faulty. Using an information-theoretic approach, Akaike (1976) proposed the following alternate statistic to discriminate between a true model and the assumed model, called the Akaike Information Criterion (AIC):

$$(-2)\log(\text{max likelihood}) + 2(\text{no. of adjusted parameters}).$$

This statistic is an approximation of the average information available to discriminate a true model from a proposed model. The first term operates as a penalty for poor fit of the proposed model from the true model. The second provides a balance by penalizing excessive number of explanatory variables. By choosing the model with the minimum statistic, an optimum model in terms of closeness to the true model is found. In practice, however, the difficulty associated with calculation of the maximum-likelihood component precludes examining all possible models.

In practice, an approximation to this measure is used for the Markovian class of models. In this study, the significance levels are matched against the Information Criterion (IC) decision rule:

$$IC = \chi^2 - 2(df(k)). \tag{11-26}$$

Use of an IC and a χ2-hypothesis test is very similar. Let us assume that k^* components are valid members of the state vector. It is possible to check additional candidate-variables by computing the IC for variables k^*+1 up to an arbitrary limit $n1$. For univariate processes, the point at which the IC achieves a minimum designates the dimension of the process. For a multivariate process, the IC is examined for each variable as a candidate for entry one at a time. If the IC decreases from the previous value, it is assumed that the canonical-correlation coefficient is significant, and the kth variable is added. If it increases, the variable and its future projections are removed from consideration.

In this study, the IC performs fairly well; however, the IC often appears to support a higher-ordered model than the χ^2 statistic. All sites, except for site 2 (target-pixel 109) and site 15 (target-site 292), have at least one significant canonical-correlation coefficient. It is interesting that a decrease in the average significance of all canonical correlations taken together is observed as the shell space is increased. However, this is paralleled by an apparent increase in average significance of the second and third canonical-correlations. This may be interpreted as hitting a limit to the overall model-predictability at low order. Increases in the subsequent canonical-correlations may be more related to over modelling than they are related to valid gains in information. Canonical analysis provides an indicator of the relevant dimension of the predictor space for individual sites, but appears unreliable for firmly determining the optimum spatial-temporal lag structure to use. Where large increases in significance are gained by adding to the shell dimension, the additional information may be justified. On the evidence of the significance values, extending the dimension beyond shell 5 is not justified for the small gains in significance (See **Table 11-4**).[11]

11.4.3.2 RESULT OF THE STUDY. A future space plus two positive lags was conservatively selected as an example to show how statistical methods perform in selecting an appropriate shell space. Three canonical-correlations coefficients can be computed and tested for significance given this future space. It is difficult to summarize the behavior for all sixteen sites; however, the mean significance-level for the 16 sites is of interest as well as the minimum and maximum significance as an indicator of variability. These values for the 16-site grid are displayed in **Table 11-4**. For each shell configuration, three significances were computed: the significance of all canonical correlations, the significance of the second and third canonical correlations, and the significance of the third individually. It is interesting to note a decrease in the average significance of all canonical correlations taken together as the shell space is increased; however, this is accompanied by an apparent increase in average significance of the second and third, and the third canonical correlations. As mentioned, this may be interpreted

---

[11] A univariate analysis (Grant 1989) of three pixels, Dallas, Houston and San Antonio, shows that first-order time-series work remarkably well in forecasting applications.

Average	Min	Max	Shell	Canonical Correlation
0.998	0.981	1.000		1–3
0.345	0.005	0.947	3	2–3
0.197	0.001	0.666		3
0.989	0.905	1.000		1–3
0.412	0.000	0.993	4	2–3
0.239	0.000	0.719		3
0.979	0.805	1.000		1–3
0.436	0.002	0.999	5	2–3
0.213	0.005	0.971		3
0.974	0.757	1.000		1–3
0.459	0.011	0.999	6	2–3
0.253	0.012	0.961		3

*Using a time weight coefficient of 0.5.

Table 11-4 Summary of the significance of the canonical correlations for shell configurations (Robinson 1987)

as reaching a limit to the overall model-predictability at a low order. Increases in the subsequent canonical correlations may be more related to over modelling than they are related to valid gains in information. Analysis of average significance values can be valuable as an indicator of where to place the most effort in themodelling exercise. If the system was truly homogeneous, the averaging procedure would be particularly beneficial. For this spatial-temporal sample, the results suggest a low-ordered system overall.

Canonical correlation analysis provides the capability to evaluate both the dimension and the quality of the information passed from the space-time past to the future of any pixel series. This application extends canonical-identification techniques to non-homogeneous spatial-temporal processes. It was demonstrated in this study that a combination of graphical and statistical techniques is most appropriate. They both help in defining the future and past spaces in the vicinity of a target, $R(+)$ and $R(-)$ respectively. The value of this definition is the ability to define the limits of the process and the identification windows to use.

During this study, there surfaced a need for more automated-information techniques that supplement the visual-identification methods. For example, hard copy was difficult to obtain and photographic reproduction of the identification images often does not match the cathode ray tube (CRT) image. The reproduction is time consuming and expensive. This emphasizes the value of readily available interactive-processing where modelling decisions can be made while observing the CRT image. Another difficulty in the use of images is that various output devices display identical colors differently. These difficulties are compounded by the varied

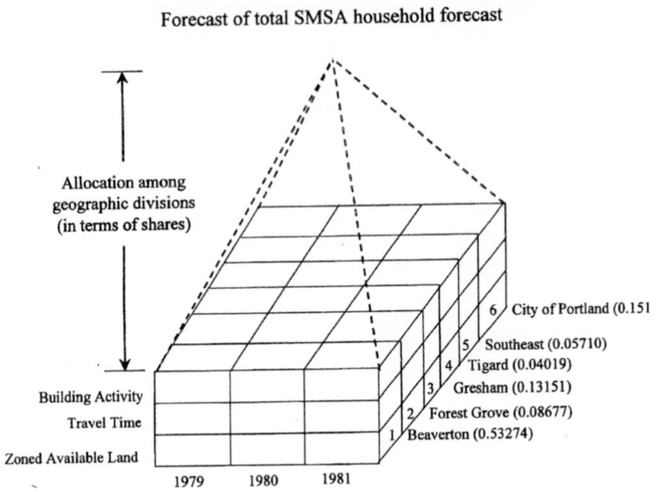

Forecast of total SMSA household forecast

Figure 11-6 Forecast adjustment using Analytic Hierarchy Process

perception of individuals. What may be a strong highlighting color for one person may not be for another.

Overall, the methods developed in this study are effective in preliminary identification of models of image-series processes; however, further refinement of the techniques is clearly required. When combined with standard multi-variate methods, it is possible to identify spatial-temporal limits in an image-series process accurately enough to estimate a model. This ability is clear with simple homogeneous models and more difficult as the degree-of-homogeneity decreases and model complexity increases. Unfortunately, most real-world data-bases tend to be non-homogeneous, as clearly shown by this study. This poses additional requirement on the modelling procedure. One way of addressing this problem is through more complex models, as we will show.

11.4.3.3 DECISION-MAKING AND FORECASTING. Multicriteria-decisionmaking techniques such as the analytic hierarchy process (AHP) can account for the impacts of these complicating factors in a forecasting model, including non-homogeneity. More often than not, there is information to include these factors—information such as supplemental cross-sectional data that cannot be explicitly captured in a time-series approach (Marshall and Oliver 1995). As explained in the "Multi-criteria Decision-making" chapter (Chan 2000), AHP evaluates these additional data together with subjective judgment. Recognizing this potential, Cook et al. (1984) developed a forecast-allocation hierarchy for the Portland standard-metropolitan-statistical-area (SMSA) divisions. This follows a regular forecasting model consisting of a univariate time-series for each division. The purpose of this hierarchy is to compensate for spatial-interaction, which is left out in the divisional univariate time-series. This is presented in **Figure 11-6**, where the hierarchy includes the year in which the supplemental data is collected, urban data and SMSA divisions. The Portland SMSA-divisions consist of a central city and five suburban-groups. Urban data such as building activity, travel time and residential zoned-vacant-lands are selected to provide indicators of household-population growth. Overall criteria determination and scaled judgments are focused on the importance of a criterion and a division. This is obtained from the interaction between forecasters and economists at General Telephone of the Northwest. The experience and analytic knowledge of those intimately involved in monitoring local activity is a valued input to the process. The process involves constructing survey matrices as shown in the example worked out in the "Multi-criteria Decision-making" chapter. For example, in the building-activity category for 1979, Beaverton compares to Gresham at a ratio scale of 6 to 1, representing strong dominance of Beaverton over Gresham as a residential location. In contrast, Beaverton compared to Southeast at a reduced ratio of 3 to 1, suggesting lesser dominance. It was also found, for example, that building-activity dominates zoned-available-land at a ratio of 7 to 1.

After the survey matrices have been developed, the criteria categories and geographic divisions are assigned a set of numerical weights according to the regular steps in AHP. The final product, representing the overall relative strengths of each division, is given as the numbers in **Figure 11-6**. In forecast adjustments, the composite weight vector provides a condensed index of contributions across multiple urban criteria. For 1979, the weights of the composite vector help to adjust population forecasts from the univariate time-series models, and likewise for 1980 and 1981. This results in the re-allocation of household-population shares according to the preferences expressed in these composite weights. Specifically, the adjustment process is listed below:

$$\max\ 0.15168z_1+0.05710z_2+0.04019z_3+0.13151z_4+0.08677z_5+0.53274z_6$$

where z_i are adjusted household-population shares. Two constraints are imposed. First, the sum of the division shares must equal unity:

$$z_1+z_2+z_3+z_4+z_5+z_6=1.$$

The second constraint allows the division share to vary only within two standard errors of the univariate forecast:

$$0.10065 < z_1 < 0.10393 \qquad 0.06606 < z_4 < 0.06814$$
$$0.07289 < z_2 < 0.07637 \qquad 0.12300 < z_5 < 0.12608$$
$$0.09126 < z_3 < 0.09526 \qquad 0.53369 < z_6 < 0.54129.$$

The optimal solution to the above problem is $z_1 = 0.10312$, $z_2 = 0.07370$, $z_3 = 0.09126$, $z_4 = 0.06763$, $z_5 = 0.12300$, $z_6 = 0.54129$. These are the final ratios that distribute the total SMSA forecast among the divisions. The adjusted forecast proved to be significantly more credible, not only in statistical measures such as mean absolute-percentage-error, but also in its acceptance among the local communities, which now felt that they have a direct input to the process. Here we show how a prescriptive model can be used alongside a descriptive model to address the problem of non-homogeneity and spatial interaction.

11.5 Econometric analysis: the EMPIRIC model

In spite of the difficulties encountered in remote sensing, the multivariate processes described above have been applied in a varied form toward urban land- and economic-development with considerable success. Perhaps the best known example, although somewhat historic in nature by now, is the EMPIRIC model. It employs a first-order autoregressive-model based on econometric principles rather than purely statistical correlation as in most time-series models. Econometric model is a special case of vector-time-series models when casualty is introduced and independence is assumed among error or noise terms. A complete numerical-example of EMPIRIC has also been worked out in a previous chapter entitled "Spatial Econometric Models."

11.5.1 First-order autoregressive-model. This type of spatial-temporal system, owing to its unique history and developmental context, has cultivated the following practice and modelling procedure (Putman 1979): (a) The region is analyzed in terms of a number of small areas called zones, which are typically irregular in shapes and varying in sizes (although this does not preclude a grid overlay similar to the east-Texas example above). (b) Forecast activities such as future population and employment distribution among the urban area are called *located (endogenous) variables*: $z_t = [z_1(t), z_2(t), ..., z_n(t)]^T$ covering all n' zones through the forecast periods t. (c) The factors inducing population and employment distributional-changes are called *locator (exogenous) variables*. For example, activity l's accessibility to zone h, X_{lh}, is a factor in the development of zone h, where, for example:

$$X_{lh}^{'} = \sum_{g=1}^{n'} z_{lg} \exp(-\alpha d_{gh}) = \sum_{g=1}^{n'} z_{lg} f(d_{gh}) \tag{11-27}$$

and $f(d_{gh})$ is the trip-distribution curve or trip-frequency function of distance-separation between zones g and h. α is the calibration constant. Instead of the generalized-cost C_{ij} combining distance and time, we will use d_{ij} to highlight the spatial dimension in this chapter. Equation (11-27) effectively replaces the "space shell" concept as discussed in Subsection 11.4.2. Isotropy is also assumed in this equation, inasmuch as it applies to any zone-pair irrespective of orientation.

Item (c) above shows the way EMPIRIC tackles Tobler's "first law" (Tobler 1970), which assumes influence of neighbors generally decreases with increasing space-time distance. Remember that Equation (11-27) above is couched within the time-series econometric-equations below. Instead of an error term as in the STARMA model such as Equation (11-22), a_i can now be thought of as being replaced by some kind of distance-measure between geographic subdivisions. The result is simply an autoregressive model. The basic structure of the EMPIRIC model is

$$z_{ih}(\Delta) = \sum_{j=1, j\neq i}^{k'} a'_{ij} z_{jh}(\Delta) + \sum_{l=1}^{n} \varphi_{il} z_{lh}(t-1) + \sum_{l=n+1}^{m'} \theta'_{il} X_{lh}(\Delta) = 0 \qquad \forall\, i,h \qquad \textbf{(11-28)}$$

Here h is a zone in the study-region taking on the values 1, 2, ... , n'; i is an endogenous (located) activity to be forecasted for 1, 2, ... , k'; l is an exogenous (locator) variable whose location and intensity is related to development patterns of the forecast activity i ($l = 1, 2, ... , m'$); $z_{jh}(\Delta)$ represents change in endogenous (located) variable j in zone h within a forecast interval, usually expressed in regional shares

$$\frac{z_{jh}(t) - z_{jh}(t-1)}{\sum_{h=1}^{n'} z_{jh}(t-1)}. \qquad \textbf{(11-29)}$$

$X_{lh}(\Delta)$ represents change-in-accessibility to activity-variable l in zone h from the beginning to the end of a forecast-interval, usually expressed in regional-shares

$$\frac{X'_{lh}(t) - X'_{lh}(t-1)}{\sum_{h=1}^{n'} X'_{lh}(t-1)}. \qquad \textbf{(11-30)}$$

$z_{lh}(t-1)$ is the value of locator (exogenous)-variable l in zone h at the beginning of a forecast-interval, $t-1$; and φ_{il}, θ'_{il}, a'_{ij} are coefficients to be calibrated.

This equation relates the growth of a single endogenous (located) variable i in zone h to: the growth of the other endogenous variables j in zone h; the change of the accessibility to activity i in zone h; and the amount of the exogenous (locator) variables l in zone h. Implicitly, the system is assumed to be homogeneous within each equation i, although *ad hoc* procedures have been imposed to handle non-homogeneous situations via the simultaneous equation set. Thus k equations are written, one for each endogenous activity, instead of a single equation such as Equation **(11-22)**. As shown in the "Spatial Econometric Models" chapter under the "Spatial linear regression models for cross-section data" subsection, the resulting model can be represented as a mixed-regressive-spatial-autogressive model with a spatial-autoregressive disturbance: $\mathbf{z} = [a'_{ij}]\mathbf{z} + \mathbf{z}(t-1)\mathbf{\Phi} + [\theta_{ij}]\mathbf{x}$. The coefficients a'_{ij} express the influence of the growths of endogenous-variables j ($j \neq i$) on the location of endogenous-variable i in zone h. The coefficients φ_{il} express the influence of exogenous-variables l on the location of endogenous-variable i in zone h. The realism of Equation **(11-28)** and therefore of the entire model depends on whether it is possible to determine values of a'_{ij}, φ_{il} and θ'_{il} which describe effectively the interrelationships between all economic activities and policy variables such as accessibility.

11.5.2 Two-stage least-squares calculation. The coefficients a'_{ij}, φ_{il} and θ'_{il} are estimated by means of 'simultaneous' multiple-linear-regression analysis carried out on data from two points in the recent past. Two-stage least-squares is the most commonly used calibration technique. There are $k \times k$ values of the a'_{ij} coefficients,

$k' \times n''$ values of the φ_{il} coefficients, and $k' \times m'$ of θ_{il}' to be estimated; or a total of $k'(k'+n''+m')$ coefficients. All a_{ij}' having $i = j$ are set equal to unity and their corresponding endogenous-variables are moved to the left-hand side of the equation. For proper identification of the equation system, at least $(k'-1)$ of the a_{ij}' and φ_{il} in each equation must be constrained (usually set equal to zero).

Once the model has been calibrated it is operated recursively for forecasting purposes—for example, from $(t-1)$ to t and then $(t+1)$. Thus information over many time-periods is introduced irrespective of the presence of heterogeneity or interregional dependence. Ergodicity is assumed here to make this possible. There is one equation for each endogenous variable in each zone, and the system of equations is solved separately for each zone during forecasting. At full utilization, therefore, the model comprises of k' equations per zone, whose simultaneous solution for a given forecasting interval will provide growths of zonal activities during this interval. This is illustrated in the "Hypothetical study" section of the "Spatial Econometric Models" chapter. Instead of an NVI reading for each pixel (Equation **(11-24)** in the Robinson model, there are now several 'readings' for each zone, representing the various activities $i(i = 1, 2, ... , k')$. But only one lag is modelled in the EMPIRIC model. On the balance, much experience and under-standing of the problem at hand goes into the calibration of EMPIRIC. This is due to the complex relationship that can be traced between the k' economic-activities over time.

While the consulting firm supporting EMPIRIC had considerable success in model use, the applied nature of their studies precluded detailed statistical experiments to be performed. Masser et al. (1971) applied EMPIRIC to Northwest England and obtained some valuable insights in calibrating such a model. The following results are reported through Putman (1979). The application of ordinary-least-squares (OLS) in estimating a simultaneous-equation model leads to biased parameter-estimates even for infinitely large samples .Nevertheless, it is more convenient to use for preliminary significance-tests before moving to an appropriate estimation procedure such as two-stage least squares (2SLS). Note that strictly speaking, the ratios of estimated slope-parameters to their estimated standard-errors do not have the t-distribution for a simultaneous-equation model. They are approximately normal, however, and large values may therefore be used as an indication of statistical significance, especially when a large sample is available. Similar conclusions, of course, apply to the use of the F-statistic. The determinant of the matrix of estimated first-order (pairwise) correlation-coefficients $\det[r_{ij}]$ for each equation is also given as an indication of the presence of multi-collinearity. (The determinant has a value of one if the estimated correlation between each pair of explanatory variables is zero).

If the variables that appear not to be significant at the 1-percent confidence-level are dropped, the resulting *reduced-form* model is shown in **Table 11-5**.[12]

---

[12]Reduced-form models are defined in the "Descriptive Tools" chapter (Chan 2000) under the "Ordinary least-squares" subsection.

Primary endogenous variable	Other endogenous variables	Lagged variables	Accessibility variables	Constant	Goodness-of-fit statistics
Δpop	0.2692 Δman (6.17) 0.3205 $\Delta serv$ (3.92)	$-0.1354\,pop$ (11.1) 0.06210 $serv$ (3.97) 0.02347 $area$ (4.91)		0.0673 (5.29)	$F = 94.1$ $\bar{R}^2 = 0.864$ $\text{var}(\varepsilon) = 0.00483$ $\det[r_{ij}] = 0.0583$
$\Delta serv$	0.3814 Δpop (3.51)	0.1569 pop (13.0) $-0.1715\,serv$ (21.3)		0.0198 (1.02)	$F = 169$ $\bar{R}^2 = 0.874$ $\text{var}(\varepsilon) = -0.00893$ $\det[r_{ij}] = 0.150$
Δman	0.7340 Δpop (3.96)	0.1238 pop (4.38) $-0.07956\,man$ (2.82)	0.7881 $\Delta accemp$ (3.10)	-0.0598 (1.95)	$F = 19.9$ $\bar{R}^2 = 0.509$ $\text{var}(\varepsilon) = 0.0205$ $\det[r_{ij}] = 0.0394$

Table 11-5 Model calibrated by ordinary least-squares

There, the variable definitions should be self-evident except for 'area', which stands for the zonal-share of regional land-area. The number of variables in the model is reduced to eight as a results of the 1-percent rule. The equation for Δpop is therefore just-identified[13] while the other two equations are over-identified. While the quality of the equations can be gleaned from the goodness-of-fit statistics and their inherent reasonableness, some comments on multi-collinearity is in order. Estimating the first and third equations is complicated by the presence of multi-collinearity among the variables. This may be suspected from the determinants of the matrices of first-order correlation-coefficients, and from the matrix of estimated first-order correlation-coefficients among the variables. Close correlation between a pair of variables will mean that their separate influences may not be estimated with confidence. This is manifested in the instability of a regression coefficient when a closely correlated variable is dropped, and in some extreme cases, by high variance of parameter estimates and, therefore, low values for t-ratios.

Some of these effects are evident in the first equation with respect to the variables $\Delta serv$ and $serv$. Although the variables as a pair are significant at the 1-percent level, neither is so significant when the other is omitted from the equation.

---

[13]For a definition of "just-identified" and "over-identified," please refer to the "Spatial Econometric Models" chapter under the subsection "Identification defined."

Their respective regression-coefficients are subject to considerable change when the other variable is omitted. It seems, therefore, that they should be included, or omitted, as a pair. If they are left out the result is:

Here $F = 126$, $\bar{R}^2 = 0.837$, estimated variance of the disturbance-term var(ε) $= 0.00582$, $\det[r_{ij}] = 0.972$, and the Student-t values are in parentheses directly underneath the appropriate calibration coefficient. The resulting equation is thus almost entirely free of multi-collinearity, although there is some shift in the estimated parameter for *pop* because of the high correlation between *pop* and *serv*. If it is considered that the place of $\Delta accemp$ in the Δman equation has no a priori foundation (as postulated by the Lowry model for example), then the maximum number of variables that may appear in any equation (if it has to be identified) is reduced to five. In this case $\Delta serv$ and *serv* may not be both present in the Δpop equation. The variable $\Delta accemp$ may thus be regarded as the identifying variable for the pair of variables $\Delta serv$ and *serv* in the *pop* equation. One alternate formulation to the model in the form of **Table 11-5** would be the above equation for Δpop, the second equation in **Table 11-5** for $\Delta serv$ and

$$\Delta man = \underset{(5.88)}{1.012}\ \Delta pop\ -\ \underset{(4.77)}{0.1235}\ man\ +\ \underset{(6.98)}{0.1731}\ pop\ -\ \underset{(2.07)}{0.0670}. \qquad \textbf{(11-32)}$$

Here $F = 20.8$, $\bar{R}^2 = 0.448$, var(ε) $= 0.0230$, and $\det[r_{ij}] = 0.0685$. It is also possible to have a "halfway house" between these two specifications of the model, as in Equation **(11-31)** and the last two equations in **Table 11-5**.

Somewhat better 2SLS-results are obtained for the model as specified in **Table 11-5**, which are summarized in **Table 11-6**. For the 'halfway-house' specification

$$\Delta pop = \underset{(5.88)}{0.2579}\ \Delta man\ -\ \underset{(18.1)}{0.09083}\ pop\ +\ \underset{(4.94)}{0.02562}\ area\ +\ \underset{(6.96)}{0.0881} \qquad \textbf{(11-31)}$$

mentioned earlier, the reduced-form equations of course remain the same as for the above specification, and the estimation results for the last two equations are similar. The results for Equation **(11-31)** are now

$$\Delta pop = \underset{(3.93)}{0.2881}\ \Delta man\ -\ \underset{(16.3)}{0.0907}\ pop\ +\ \underset{(4.39)}{0.02531}\ area\ +\ \underset{(6.31)}{0.0883}. \qquad \textbf{(11-33)}$$

Here $F = 98.6$, var(ε) $= 0.00711$, and $\det[r_{ij}] = 0.961$. The main effect of multi-collinearity for using the equations in **Table 11-6** or its "halfway-house" specification is in the last equation for Δman. Here only $\Delta accemp$ remains significant at the 1-percent level. Generally, however, the results from this 2SLS model are close to those obtained earlier for OLS estimates. The results obtained here again show that the change variables are positively related with one another. At the same time, there is a negative correlation between the change variables and the corresponding variables for initial share. Nevertheless, the estimated parameters need to be regarded with considerable caution. For instance, the parameter estimate for *pop*, which is highly correlated with *serv*, changes from -0.1635 in the first equation in

Primary endo-genous variable	Other endo-genous variables	Lagged variables	Accessi-bility variables	Con-stant	Goodness-of-fit statistics
Δpop	0.2932 Δman (3.18) 0.5372 $\Delta serv$ (2.34)	$-0.1635\,pop$ (5.66) 0.1016 $serv$ (2.56) 0.02166 $area$ (3.74)		0.0544 (2.98)	$F = 64.9$ var(ε) = 0.00663 det$[r_{ij}]$ = 0.00862
$\Delta serv$	0.4836 Δpop (2.75)	0.1649 pop (10.1) $-0.1702\,serv$ (20.0)		0.00721 (0.28)	$F = 158$ var(ε) = 0.00948 det$[r_{ij}]$ = 0.0734
Δman	0.2830 Δpop (0.68)	0.07759 pop (1.60) $-0.06701\,man$ (2.05)	1.086 $\Delta accemp$ (2.95)	-0.0143 (0.29)	$F = 13.2$ var(ε) = 0.0250 det$[r_{ij}]$ = 0.0116

Table 11-6 Model calibrated by two-stage-least-squares

Table 11-6 to -0.0907 in Equation **(11-33)**. The principal value of the estimated model must therefore be viewed mainly in terms of predicting the three jointly-dependent change-variables, rather than inferences of causality.

In short, two alternate formulations of the reduced-form equations of the model may be used: (a) the reduced-form equations as estimated by OLS in the first stage of the 2SLS procedure, or (b) reduced-form equations obtained form solutionof the 2SLS estimated-equations above. In the first case any unexplained variation is minimized, but the restrictions on the structural-form equations for *Δserv* and *Δman* are ignored. If the OLS estimated reduced-form equations are found to give better forecasts, then this is some indication that the restrictions are unnecessary, and 2SLS estimation may not be required. Calibration of this type of model is therefore highly judgmental. The model is used mainly for short-term projection purposes and little claim is made regarding causality, as mentioned.

11.5.3 Case Study - Development in Twin City, Minnesota. EMPIRIC was applied to at least seven major metropolitan areas in the U.S. The endogenous and accessibility variables were always expressed in terms of "change in regional share" (Equations **(11-29)** and **(11-30)**). Employment was broken down into a few basic and a few nonbasic sectors. Population was always divided into four groups by income, approximating quartiles. Employment was broken down into aggregate categories approximating Standard Industrial Classification (SIC) Codes. In each application, there were typically four or five population-sectors and five or six employment-sectors being forecasted. The 'locator' variables are of four types. First, there are lagged, or initial-period values of the located variables and second, there are the other located variables. The third type of locator variable is the accessibility and/or land-use 'policy' variables of which there are usually several, including developability or zoning considerations. Finally, there are the public-utility variables such as sewer and water availabilities .

Of these data requirements, remote sensing can supply surrogates, rather than the traditional census and other survey data. For example, the number of occupied dwelling-units (through long-wavelength infrared radiation) can be a surrogate for population, the same with the occupied offices and factories for employment. Disaggregation of population by income class can be identified by the type of homes that are occupied; similarly for employment in terms of the type of work place. Transportation, utility and other land-use information can also be extracted from the satellite images directly. With urban development models of this type, the once-a-day, or even once-a-week frequency of surveillance is more than sufficient; and, in fact, luxurious compared with the 5-year interval between censuses that most urban planners are accustomed to. Perhaps the only data-set that needs to be supplemented is the trip-distribution information, which shows the frequency plot of trip lengths in the study area. But this is easy to obtain considering the relatively constant shape of these curves over time. With enough resolution in satellite imagery, one can even count traffic on streets and highways (McCord et al. 1996) and synthesize a trip-distribution curve using origin–destination estimates (Chan et al. 1986, Xu and Chan 1993, Chan and Rahi 1993).

The measure of goodness-of-fit used in the EMPIRIC applications was the coefficient-of-multiple-determination R^2. For the Twin Cities of Minneapolis and St. Paul, for example, three calibrations were performed and the results shown in **Table 11-7**. The differences between the ordinary-least-squares and two-stage-least-squares calibration-runs (performed at the University of Pennsylvania) were minor. This is also true for all but one of the comparisons between the Kates, Peat, Marwick & Co. calibration and these University of Pennsylvania runs. These results are reasonably good. Estimation accuracy is also encouraging for dependent variables, which are explained in terms of a sector's zonal employment or zonal population by income over a 10-year period. As for the calibrated-model coefficients, there are a few peculiarities worth discussing. For example, why is change in a zone's share of low-income-quartile population positively related to change in local-government and educational employment and negatively related to change in the product of highway-accessibility-to-employment and used-land-area? Why is change in a zone's share-of-population in the upper-middle-income-quartile not related to any employment or access variable? Without an explicit theory or an attempt at identifying structural-equations, there can be few expectations regarding signs and magnitudes of coefficients. Suffice to say, the parameters of EMPIRIC model can be calibrated to yield relatively close fits to the data. The only consistency in the parameters from one application to the next appears in the population-group-to-population-group relationships. (See chapter on "Spatial Econometric Models.") The parameters for other variables and other equations are ad hoc in

Endogenous variable (all in "change-in-share")	Kates, Peat, Marwick & Co.[1]	University of Pennsylvania	
		Two-stage least-squares	Ordinary least-squares
Households in lowest-income quartile	0.702	0.703	0.706
Households in middle-income quartile	0.708	0.714	0.720
Households in upper-middle-income quartile	0.812	0.816	0.824
Households in highest-income quartile	0.715	0.715	0.724
Construction and other employment	0.750	0.746	0.761
Manufacturing and wholesale employment	0.718	0.708	0.714
Transportation, communication and utility employment	0.504	0.464	0.464
Retail employment	0.790	0.790	0.793
Service, finance, insurance, real-estate employment	0.755	0.754	0.758
Local government and education employment	0.545	0.545	0.546

1. Formerly Peat, Marwich, Mitchell & Co.

Table 11-7 Coefficients-of-multiple-determination (R^2) in Twin Cities (Putman 1979)

nature, and raise questions as to the similarity woven into the fabric of the model. For these reasons, the model is recommended for only short-term forecasts.

Many of these difficulties can be traceable to the well-known problems of specification and identification in econometric models, as explained in the "Spatial Econometric Models" chapter (Makridakis and Wheelwright 1978). But the most serious problem has to be associated with the estimation process. These problems, together with the quality of the data, are the biggest challenge to the operational use of econometric models such as EMPIRIC. Models of this kind will benefit from the consistency and frequent availability of remote-sensing data, which allows a more convenient update of the model. Thus the model can be constantly fine-tuned to the latest urban development trend (from time period $t-1$ to t). This in turn lends greater credence and utility to EMPIRIC's short-term forecasts (to $t+1$). Unlike the East-Texas vegetation study, more structure is posed on top of the data-set (but fall far short of offering an explanation of the underlying processes). In comparison, we remember that much of the statistical analyses on East Texas are of the "let data speak for themselves" type, with hardly any postulation of causal relationship. EMPIRIC represents a move toward the causal-modelling direction.

11.6   The Lowry–Garin derivative models

A spatial-temporal model of the econometric variety, similar to the East-Texas study, still relies heavily on 'eyeballing' in the specification and identification process. Robinson (1987) pointed out that in m''-dimensional time-series models, the requirement on judgment based on graphic display can be overwhelming even for $m'' = 3$. With $m'' = 3$ or the space-time shell configuration of **Figure 11-3**, it is necessary to visualize x_1-lag (+ and −), x_2-lag (+ and −), time-lag, and spatial-temporal autocorrelation functions (STACF) or spatial-temporal partial-auto-correlation function (STPACF) values simultaneously to support identification. Even when the estimation process is successfully executed, a good fit of the data does not necessarily guarantee a good forecast. The reason is that there may be little explanatory relationship behind the models as can be seen in the Twin-City application of EMPIRIC. In light of some difficulties encountered in the east-Texas study and EMPIRIC, perhaps the only alternative is to discern a spatially non-homogeneous model *a priori*. Then we either estimate coefficients in the model as a non-homogeneous process or as a set of homogeneous processes that interact.

   11.6.1   Spatial input–output model.   Posing an explanatory structure on spatial-temporal relationship, a whole family of Lowry–Garin derivative models has been developed to explain regional development (Putman 1983). These models explicitly depart from correlation relationship and move squarely into causal modelling. Lowry–Garin based models combine economic-base theory and spatial-interaction theory in a set of explanatory and forecasting equations, as explained in the "Activity-Allocation and -Derivation" chapter. In this chapter and the subsequent chapter on "Chaos, Catastrophe, Bifurcation and Disaggregation," we have illustrated the basic concepts with more than one numerical example. Activities such as employment and population are represented as vectors, each entry of which is a zonal variable:

$$\mathbf{Z}_j = (\mathbf{Z}_{j1}, \mathbf{Z}_{j2}, \mathbf{Z}_{j3}, \ldots, \mathbf{Z}_{jn'}) \qquad\qquad j = 1, 2. \tag{11-34}$$

These vectors are in turn forecasted over time

$$\mathbf{Z}_{jh} = \left[Z_{jh}(t), Z_{jh}(t+1), \ldots, Z_{jh}(t+\infty) \right]^T; \quad h = 1, 2, \ldots, n'. \tag{11-35}$$

We can now write

$$\mathbf{Z}_j(t+\infty) = \mathbf{Z}_j(t)\left[\mathbf{I} + \mathbf{W}' + \mathbf{W}'^2 + \mathbf{W}'^3 + \ldots + \mathbf{W}'^\infty \right]\mathbf{W}^{j'}, \tag{11-36}$$

where $\mathbf{W}^{j'} = \mathbf{I}$ if $j = j'$ and $\mathbf{W}'$ is a $n' \times n'$ matrix. Here the entries of $\mathbf{W}'$ matrix, W_{gh}, normally have the property that

$$\sum_{h=1}^{n'} W_{gh} < 1 \qquad \forall g. \tag{11-37}$$

This allows the Equation series **(11-36)** to be summed as

$$\mathbf{Z}_j(t+\infty) = \mathbf{Z}_j(t)\left[\mathbf{I} - \mathbf{W}'\right]^{-1}\mathbf{W}^{j'}. \tag{11-38}$$

Equation **(11-36)** is nothing more than a geometric series. As such, stationarity of the data is a non-issue, as explained in the "Extended numerical example" subsection of the "Spatial Time-Series" chapter. In fact there is a predetermined structure for the system being modelled. Once this structure has been proved to be acceptable, the properties of the system can be clearly delineated. For example, bifurcation occurs when

$$\sum_{h=1}^{n'} W_{gh} \geq 1; \qquad g = 1, 2, \dots, n' \tag{11-39}$$

in the Equation series **(11-36)**. It is in this situation that a system as modelled by EMPIRIC or the equations above becomes highly complex. Yi and Chan (1988) outline the detailed conditions under which bifurcation will occur. It boils down to the magnitudes of the economic-base multipliers disaggregated by zone h, ρ_h^j, and accessibility functions between zones g and h, X_{gh}^j, the product of which yield W_{gh}^j. In other words

$$\mathbf{W}' = \prod_{j=1}^{2}\left[W_{gh}^j\right] = \prod_{j=1}^{2}\mathbf{W}^j \tag{11-40}$$

where $\mathbf{W}^j$ is the product of an $n' \times n'$ matrix and a diagonal matrix:

$$\mathbf{W}^j = \left[X_{gh}^j\right]\left[\rho_h^j\right] \qquad j = 1, 2. \tag{11-41}$$

The denominator of the accessibility function

$$X_{gh}^j = \frac{Z_h^j \, f(d_{gh})}{\sum_{h=1}^{n'} Z_h^j \, f(d_{gh})} \tag{11-42}$$

is similar to Equation **(11-27)**, with $f(d_{ih})$ defined and calibrated the same way, except $X_{gh}^j \neq X_{pq}^j$ gives rise to a nonhomogeneous system. One can readily see the relationship to Equation **(11-38)** when one recognizes that, in the notations of spatial linear regression model, $\mathbf{W}_1 = [X_{gh}^1]$, $\mathbf{W}_2 = [X_{gh}^2]$, $\mathbf{\Phi} = [\rho_h^1]$, $\mathbf{\gamma} = [\rho_h^2]$ and $\mathbf{z}$ or $\mathbf{a}$ can be equivalenced to $\mathbf{Z}$.

As explained in the "Spatial Equilibrium" chapter, the economic multipliers ρ_h^j are similar to those in input–output tables, if one considers economic-base theory as a special case of input-output analysis. Thus industries and services produce to satisfy consumption by other industries and services, consumption by households, and exogenous consumption. This generates flows of goods and services for various purposes. Households supply labor to satisfy consumption by industries and services and by other households, and the spatial interaction between industries, services and households fuels upon themselves—sometimes to a frenzy. Viewed in this light, multipliers are certainly the determinants of the rate of economic-growth. Disaggregate multipliers ρ_h^j, where $\rho_g^j \neq \rho_h^j$, define a nonhomogeneous system. Likewise, the spatial-interaction relationship, as determined by X_{gh}^j, is an equally important factor in explaining discontinuities in an otherwise homogeneous development pattern. Literally, there are infinite combinations of economic multipliers and accessibility indices that can give rise to bifurcation. In this case, it appears unlikely that bifurcation relationships can be explained from spatial time-series used in remote-sensing previously (Priestley 1988), and an analytical causal model such as the Yi–Chan variety will be a much more convenient way to locate these discontinuities on a spatial-temporal dimension. It appears that the only correlative techniques that come anywhere close to tackle this type of complexity are Kalman filters and transfer functions. As explained in the "Spatial Time-Series" chapter, Kalman filters can help track a series with varying means and variances. Transfer functions can incorporate intervention analysis, where an intervention includes a number of precipitous events such as a step- or ramp-function. In general, bifurcation and catastrophes often prove too complicated for these correlative techniques to handle, particularly in spatial systems.

When applied in a causal context, bifurcation theory can be used to explain highly complex non-homogeneity in spatial-temporal systems. In regional development, there are invariable 'patchiness' due to geographic subdivisions and associated regulatory constraints such as zoning ordinances that allow or disallow location of a particular activity. This can be viewed conveniently as a special case of our general bifurcation condition. Instead of simply precipitous growth or decline over time, we place upper or lower limit on the amount of activities that can be located in a zone. As explained in the "Chaos, Catastrophe, Bifurcation and Disaggregation" chapter, structural instability of the functional form is involved here. A way needs to be found to estimate the number of different regimes within which bifurcation occurs. Input-output analysis as outlined in the "Spatial Equilibrium" chapter provides valuable insights into the spatial data-sets that are likely to cause complex heterogeneity. Yi and Chan (1988) outlined a simple and 'elegant' way to take care of developmental constraints such as maximum-residential-density and minimum-threshold for retail-activity to be located. Through modified accessibility functions, $X_{gh}^j(m)$, such constraints are included in the forecasting sequence of Equation (11-36) through travel-cost elasticities b^j ($j = 1,2$). Here

$$X_{gh}^j(m) = \frac{Z_h(m) \left[f(d_{gh}) \right]^{b^j}}{\sum_{h=1}^{n'} Z_h(m) \left[f(d_{gh}) \right]^{b^j}} \tag{11-43}$$

where m denotes the mth revision to the accessibility. A large b^j (in absolute value) will show smaller 'dampening' effects for exponential growth. In other words, all else being equal, X_{gh}^j tends to increase more rapidly (thus reaching the condition for bifurcation) for larger values of b^j than for smaller ones. Details of handling this "constraint catastrophe" are documented in the "Chaos, Catastrophe, Bifurcation and Disaggregation" chapter. In passing, the approach taken here is generally called compartmental modelling, in reference to the state-transitions exemplified in Equation **(11-41)**.[14] More will be said about this later.

11.6.2 Calibration of a non-homogenous process.

The Fibonacci-search technique (Bazarraa and Shetty 1979) is used to calibrate the elasticities b^j endogenously.[15] The procedure seeks to minimize the deviation between each revised accessibility-function $X_{ih}^j(m)$ and the original, $X_{ih}^j(0)$.

$$\min \left[X_{ih}^j(m) - X_{ih}^j(0) \right]^2 \qquad \forall\, i, h;\ j=1,2. \tag{11-44}$$

A numerical example has been worked out in the "Chaos, Catastrophe, and Bifurcation" chapter to illustrate the calibration procedure. Assume now the accessibility-functions X_{ih} have been calibrated as we have done in the EMPIRIC model using trip-distribution information. From Equations **(11-38)** and **(11-40)**, one can show that for the case of two activities $\mathbf{Z}_j$, $(j = 1,2)$, employment $\mathbf{Z}_1$ and population $\mathbf{Z}_2$, the following equations can be derived (Yi and Chan 1988). First, we rewrite Equation **(11-38)** as

$$\mathbf{Z}_1(\mathbf{I} - \mathbf{W}') = \mathbf{Z}_1(t) \tag{11-45}$$

$$\mathbf{Z}_2\left[W^1\right]^{-1} = \mathbf{Z}_1(t)(\mathbf{I} - \mathbf{W})^{-1} \tag{11-46}$$

or

$$\mathbf{Z}_1 - \mathbf{Z}_1(t) = \mathbf{Z}_1\mathbf{W} \tag{11-47}$$

$$\mathbf{Z}_2[\mathbf{W}^1]^{-1} - \mathbf{Z}_1(t) = \mathbf{Z}_2\mathbf{W}^2 \tag{11-48}$$

where

$$\mathbf{W}^1 = \left[X_{gh}^1\right]\left[\rho_h^1\right] = \mathbf{X}^1\boldsymbol{\rho}^1 \tag{11-49}$$

---

[14]A compartmental model describes the temporal evolution of the state-of-the-system defined as the number of units in the different compartments. Compartment models can be viewed as a generalization of a Markovian process. Please refer to the "Control, Dynamics" appendix in Chan (2000) and the "Activity Allocation/Derivation" chapter in the current volume.

[15]For an explanation of the Fibonacci search, see the "Prescriptive Tools" chapter in Chan (2000).

Combining the last four equations, we have

$$\mathbf{Z}_1\mathbf{X}^1\boldsymbol{\rho}^1 = \left[\mathbf{Z}_1 - \mathbf{Z}_1(t)\right](\boldsymbol{\rho}^2)^{-1}(\mathbf{X}^2)^{-1} \tag{11-51}$$

$$\mathbf{W}^2 = \left[X_{gh}^2\right]\left[\rho_h^2\right] = \mathbf{X}^2\boldsymbol{\rho}^2. \tag{11-50}$$

$$\mathbf{Z}_2\mathbf{X}^2\boldsymbol{\rho}^2 = \mathbf{Z}_2(\boldsymbol{\rho}^1)(\mathbf{X}^1)^{-1} - \mathbf{Z}_1(t). \tag{11-52}$$

Once again, $\mathbf{X}^1$ is the accessibility matrix for work trips, $\mathbf{X}^2$ is the accessibility matrix for nonwork trips, $\boldsymbol{\rho}^1$ is the multiplier matrix for population, $\boldsymbol{\rho}^2$ is the multiplier matrix for employment, and $\mathbf{Z}_1(t)$ is the seed employment vector (equivalent to $\mathbf{Z}_j(t)$ in Equation (11-36) when one considers $\mathbf{Z}_2(t) = \mathbf{Z}_1(t)\mathbf{X}^1\boldsymbol{\rho}^1$.

Equations (11-51) and (11-52) above form a simultaneous equation set of $2n'$ equations and $2n'$ unknowns, which enable us to solve for $\boldsymbol{\rho}^1$ and $\boldsymbol{\rho}^2$ explicitly. Experimentation with numerical solution techniques, however, reveals that owing to the non-linearity of the equation sets, questions of solution uniqueness and algorithmic convergence arise. In view of this difficulty, and the fact that in the equation sets thus far we have yet to recognize zonal constraints (patchiness) explicitly via the calibration data, the base-year data ($\mathbf{Z}_1(t)$ and $\mathbf{Z}_2(t)$) is to be supplemented with a calibration-year ($t-1$) data set. In this way, the effects of zoning constraints are explicitly taken into consideration—via the two sets of real-world data-sets *over time* (from $t-1$ to t). Also, the extra data may help settle the problem of solution uniqueness. This supplemental calibration procedure has the distinct advantage of establishing a "trend line" over two data-sets from two different time-periods, $t-1$ to t. It ensures temporal (on top of spatial) replication and thus paves the way for forecasting applications of the model. The results of the calibration between base-year (t) and calibration-year ($t-1$) data become the input for the forecast from base-year to "target-year" ($t+1$). Conceptually, one can think of this as solving Equations (11-51) and (11-52) with each variable being replaced by its increment from calibration year to base year (excepting $\boldsymbol{\rho}^j$ and $\mathbf{X}^j$—the multipliers to be solved and exogenous variables respectively). Remote sensing can make the information available so much more frequently and so much more consistently over time for calibration and re-calibration of this set of disaggregate parameters.

While the above model was developed for two activity types ($j = 1, 2$), it can readily be generalized to more than two activities. Let us assume we have three activities: residential ($j = 1$), white-collar employment ($j = 2$), and blue-collar employment ($j = 3$). The population sector supplies each of the two employment sectors with workers. In addition, white-collar industries also serve the population. But blue-collar industries do not, and no interaction takes place between white and blue collar industries. Again, we seed the development process with an exogenously determined amount of blue-collar employment at the target year. All the equations can be directly extended by a running index j that goes to 3 (instead of 2). Equation (11-40) would remain the same, since blue-collar industries do not supply households. $\mathbf{W}^3$ would simply be $[X_{gh}^j]$ without any multipliers associated with it. The net result is that white-collar and blue-collar industries compete for labor from the same household sector. Additional white-collar dependent-population chooses a place to live not only in relation to white-collar industries, but blue-collar industries as well. In this fashion, the people who work for blue-collar employment and their residential location may change from a two-sector model. Notice the aggregate total of employees remain the same for each zone between the 2-sector and 3-sector models.

Notice the calibration procedure used here in this model is much more well-defined than the previous two models, STARMA and EMPIRIC, as one would expect. Calibration of the multipliers ρ^j and the elasticities $\mathbf{b}$ is handled endogenously within the forecasting sequence. The only calibration performed exogenously to the model is that of the accessibility function $\mathbf{X}^j$. It is widely recognized that calibration of a nonlinear forecasting model is not at all a trivial task (see "Statistical" appendix to Chan (2000) and Seber and Wild [1989]). This is particularly true for compartmental models, which is the general class of models within which the Lowry–Garin model can be placed. Simultaneous calibration of all parameters within a stochastic or deterministic compartmental-model is highly complex, if not practically impossible. In spite of the linkage to spatial input–output model in the "Spatial Equilibrium" chapter, simultaneous calibration of the disaggregate multipliers ρ^j and the accessibilities X_{gh}^j is currently infeasible. Consequently, we have resorted to separate calibration procedures. The same applies to the elasticities b^j used to take care of constraint catastrophes. They are again estimated by a distinct procedure. We can relate this to our experience in calibrating the STARMA and spatial econometric-models, where the spatial weights are estimated exogenous to STARMA and the accessibility functions are calibrated outside the simultaneous-econometric equations. In spite of the risk of ecological fallacies, this appears to be the only prudent course of action in real-world problem solving.

11.6.3 Case Study 3 - Development in York, PA. A case study was performed on York, Pennsylvania (PA)—a medium-size city of about a quarter million people. It was divided into 42 zones in our study. In the study we compare the aggregate calibration approach with the disaggregate. By aggregate approach we mean an average ρ^1 and ρ^2 were calibrated, as distinguished from disaggregate parameters ρ_h^1 and ρ_h^2 (see calibrated values in **Table 11-8**). We also compared the constrained model formulation against the unconstrained, and forecasts with transportation-policy implementation against those without. All these cases were studied with an eye toward convergence and bifurcation behavior, comparing the case of a homogeneous system with a heterogeneous system. Data-sets and computer programs for this case study are stored in the posted computer software, as alluded to in the "Chaos, Catastrophe, Bifurcation and Disaggregation" chapter.

Instead of the exponential form shown in Equation **(11-27)**, the trip-distribution curves were calibrated for work and nonwork trips as

Zone	Aggregate			Disaggregate	
	ρ^1	ρ^2		ρ^1	ρ^2
1	↑	↑		5.242	0.156
2				3.667	0.175
3				3.214	0.146
4				7.500	0.100
5				5.011	0.132
6				2.371	0.169
7				5.298	0.080
8				3.703	0.164
9				2.677	0.136
10				2.494	0.187
11				1.727	0.188
12				2.102	0.118
13	2.145	0.130		1.707	0.188
14				2.021	0.174
15				1.742	0.185
16				1.801	0.184
17				2.029	0.187
18				1.944	0.177
19				1.765	0.162
20				3.308	0.133
21				1.821	0.180
22				1.801	0.183
23				2.124	0.163
24				2.009	0.168
25				2.222	0.184
26				1.725	0.187
27				2.141	0.177
28				1.783	0.187
29				2.221	0.181
30				2.276	0.162
31				2.249	0.172
32				1.901	0.186
33				1.718	0.185
34				3.155	0.175
35				2.306	0.174
36				2.079	0.188
37				1.779	0.185
38				1.707	0.187
39				1.780	0.184
40				1.741	0.177
41	↓	↓		2.209	0.176

Table 11-8 Calibration results for multipliers

$$f^1_{gh}(d_{gh}) = 33174 + 28838\, d_{gh}^{0.8} - 6346\, d_{gh}^{1.5} + 183\, d_{gh}^{2.5} \qquad \textbf{(11-53)}$$

$$f^2_{gh}(d_{gh}) = 4296 - 300\, d_{gh} + 5.2\, d_{gh}^{2.0} \qquad \textbf{(11-54)}$$

Iterations		1	2	3	4	5	6	7	8
Do-nothing alternative	b^1	−3.03	−3.04	−3.04					
	b^2	−1.86	−1.77	−1.77					
PRT alternative	b^1	−3.03	−3.04	−3.01	−2.99	−2.98	−2.98	−2.98	−2.98
	b^2	−1.86	−1.77	−1.66	−1.28	−1.21	−1.09	−1.00	−1.00

Table 11-9 Calibration results for elasticities in disaggregate model

respectively, where d_{gh} is the travel time between zones g and h. These two curves were the result of applying multiple regression on the base-year data, which consists both of work trips and of nonwork trips. Based on these curves, accessibility functions similar to Equation **(11-27)** were obtained, which in turn were normalized into regional shares as in the EMPIRIC model. Then the disaggregate multipliers ρ^1 and ρ^2 were obtained via Equations **(11-51)** and **(11-52)** using base and calibration-year data. The results are tabulated in **Table 11-8**.

Forecasting was performed on the "do-nothing" alternative and the "personal-rapid-transit (PRT)" alternative. There was no significant change in the "do-nothing" disaggregate-forecast from the base year to the forecast year except for normal growth. But significant growth and spatial re-allocation occur with an aggregate forecast. In comparison, the disaggregate approach was found to yield a model that can better replicate the existing nonhomogeneous pattern. Perhaps the most interesting observation is that in the disaggregate case the system converges in spite of some multiplier products ($\rho_h^1\rho_h^2$) being bigger than unity. This is possibly due to the strong dampening effects of the spatial-allocation process and the equalizing effects of zoning constraints. Convergence was obtained within three iterations, with b^1 unchanged over the iterations and b^2 decreasing in absolute value (see **Table 11-9**). Here is a clear example of when structural instability can be taken care of by varying the parameters during forecasting.

For the PRT alternative, disaggregate forecast results in bifurcation. Some zones experience exponential growth in spite of "dampening effects" of the allocation process and the zoning constraints. Notice that in this computer run, we are using the same ρ_j^1 and ρ_j^2 as the disaggregate forecast of "do-nothing" alternative. This illustrates that the conditions laid out by Equation **(11-37)** are now violated by a different set of accessibilities X_{gh}^j, due to implementation of the PRT system. The accessibilities, as suggested by the row sums in Equations **(11-37)** and **(11-39)**, also become quite uniform. In other words, in the PRT-alternative disaggregate-forecast, parameters b^1 and b^2 both steadily decrease in absolute value over eight iterations. the uniform 25 miles-per-hour (40 km/h) PRT-system is built to provide ubiquitous accessibility to every part of the town. This has the effect of equalizing development among various parts of York, resulting finally in the 'overflow' of activities from zones that have reached their capacities to those that have not—so much so that no more land is available for development. To explain this in another way, the uniform 25 miles-per-hour PRT-system homogenizes and generally stimulates development in all parts of the town, which historically was zoned for specialized land-use. This leads to the result that significant developments occur in the least expected places, and these developments are no longer compatible with an outdated zoning code. The result is a semi-chaotic situation.

In our opinion, the Lowry–Garin derivative model presented here represents one of the most advanced approaches to spatial-temporal modelling. Instead of borrowing general time-series or econometric models, specialized approaches were developed from regional-science theory specifically geared toward spatial interaction and forecasting. A noteworthy feature of this approach is the explicit inclusion of developmental constraints in the classic Lowry–Garin model. This is accomplished with the introduction of simply two endogenous parameters b^1 and b^2. The parameters carry with them the physical interpretation of travel-cost elasticity. Such parameters are policy-sensitive in that the development implication of accessibility improvement is shown analytically by the magnitude of these parameters. Modelled as a matrix time-series, general conditions for bifurcation were laid down. Expressed in terms of accessibilities and economic multipliers, the various conditions for convergence and divergence of the spatial system are delineated. These conditions include the influence of travel costs, multipliers and zonal-development constraints. They encompass and illustrate the myriad of combinations under which bifurcation can take place. Of particular interest is the discussion on the 'dampening' effects of spatial allocation and the 'explosive' growth at selected zones. This disaggregate analysis allows a better understanding of bifurcation in terms of the economic-growth over time and the accompanying spatial-interaction. It is also one of the few available means to model a nonhomogeneous system.

The Yi–Chan model belongs to a family of compartmental models, which has been applied toward activity allocation and deviation as documented starting with the chapter bearing this name.[16] A general compartmental model is a nonlinear, stochastic, dynamical-system. It is solvable only by simulation. As a *quasi-deterministic* case of a nonlinear-stochastic-dynamical system, analytical tractability is afforded for the Yi–Chan formulation. Being a discrete-time model, it is shown that the matrix series converges under normal circumstances. In general, the stability of the Yi–Chan model can be analyzed in terms of the *spectral radius* of the matrix **W**. Specifically, the model converges when the spectral radius is less than unity. When the model becomes unstable, most other analysis resorts to a qualitative one—analyzing the system behavior in its idealized, canonical form rather than the actual model constructed for the situation. It is here that we again claim a slight contribution. Unlike canonical formulations, the Yi–Chan model is an extension of the Lowry heritage—a long established means to model land-use.

---

[16]The "Control, Dynamics & System Stability" appendix (Chan 2000) provides the general background for the current discussion.

Going well beyond the classic Garin matrix-series and input–output formulations, it provides computational procedures and analytic statements on bifurcation and catastrophe, if only for a quasi-deterministic formulation. Moreover, a disaggregate, rather than aggregate model is offered.

Unlike other algorithms, the calibration and solution procedure presented here converged very fast. The disaggregate calibration procedure merges within the forecasting sequence. The calibration result from the calibration-year $(t-1)$ to the base-year (t) serves directly as input to the forecast from the base-year (t) to the target- year $(t+1)$. The fidelity and consistency of remote-sensing data could facilitate such disaggregate modelling and calibration, which can only be discussed in theory before this time. Now a non-homogeneous system is modelled explicitly in terms of bifurcation theory. Remote sensing and geographic information system allow up-to-date calibration from $t-1$ to t, which in turn permits accurate forecasts of a nonhomogeneous spatial-temporal system to period $t+1$. Most important, this paves the ground for even more causal-theory in spatial-temporal modelling.

11.7 Statistical vs. causal modelling

There are relationships between the above three approaches to spatial-temporal modelling: spatial time-series, econometric model, and Lowry–Garin matrix-series. They can be viewed as generalization and specialization of the equation set shown in the "Spatial linear regression models" subsection of the "Spatial Econometric Models" chapter:

$$z = \Phi W_1 z + Z\beta + a$$
$$a = \gamma W_2 a + \varepsilon. \tag{11-55}$$

Here β is a $K \times 1$ vector of parameters associated with exogenous variables Z, which is represented in a $n \times K$ matrix. $W_1 z$ is the spatially-lagged dependent variable, Φ is the coefficient matrix of the spatially-lagged dependent variable, and γ is the coefficient matrix in a spatial autoregressive structure for the disturbance a. The error-vector ε is normally distributed with a general diagonal-covariance-matrix $\sum$. The diagonal elements allow for hecteroscedasticity as a function of $m'+1$ exogenous variables y, including a constant term: $\Sigma_{ii} = h_i(a^T y)$. The two $n \times n$ matrices W_1 and W_2 are spatial weight matrices.

11.7.1 General comparison of three approaches. For $a = 0$ or fixing m' parameters, the mixed-regressive-spatial-autoregressive model with a spatial autoregressive-disturbance is obtained: $z = \Phi W_1 z + Z\beta + (I - \gamma W_2)^{-1}\varepsilon$. Introducing the time-dimension into this cross-sectional model generalizes it to the spatial autoregressive moving-average model: $z_t = \Phi W_1 z_{t-1} + Z_{t-1}\beta + (I - \gamma W_2)^{-1}\varepsilon_t$.

Spatial autoregressive moving-acreage (STARMA) models have been proposed to account for correlation in space much like time-series models account for correlation in time. Pfeifer and Bodily (1990) extended the STARMA model to demand-related data from eight hotels of the same chain in a large U.S. city. The driving distances between the hotels are modelled as a simple weighting matrix $\mathbf{W}^{(l)}$ in the by-now familiar spatial time-series:

$$\mathbf{z}(t) = \sum_{k=1}^{p} \sum_{l=0}^{\lambda_k} \varphi_{kl} \mathbf{W}^{(l)} \mathbf{z}(t-k) - \sum_{k=1}^{q} \sum_{l=0}^{m_k} \theta_{kl} \mathbf{W}^{(l)} \mathbf{a}(t-k) + \mathbf{a}(t). \qquad (11\text{-}56)$$

$\mathbf{W}^{(l)}$ is the $n' \times n'$ matrix for spatial order l ($\mathbf{W}^{(0)} = \mathbf{I}$). Matrix $\mathbf{W}^{(l)}$ has nonzero elements only for those pairs of sites that have been defined to be lth-order neighbors. First-order neighbors are understood to be closer than second-order ones, which are closer than third-order neighbors, etc. The modeler specifies exogenously the order and magnitude of a nonzero entry for a particular pair of sites. STARMA was shown in this study to be a better model than ARMA in that the spatial interaction is explained in terms of the exogenously determined $\mathbf{W}^{(l)}$. We include this model as an exercise in the book appendix.

As previously mentioned, the model specified in Equation **(11-56)** is labelled STARMA $(p_{\lambda_1, \lambda_2, \ldots, \lambda_p}, q_{m_1, m_2, \ldots, m_q})$ model. It can be seen that it is equivalent to the Robinson STARMA (p,q) when the $L^{(l)}$ matrix is subsumed by spatial-temporal canonical-analysis.

When $q_{m_1, \ldots, m_q}$ is set to zero, only autoregressive terms remain in Equation **(11-56)** and the model starts to resemble EMPIRIC (Equation **(11-28)** where p and λ_k are set to unity. This can also be accomplished by setting $\gamma = 0$, $\alpha = 0$ in Equation **(11-55)**, which results in the elimination of m'+1 parameters. This specialization yields the mixed-regressive-spatial-autoregressive model, which includes common-factor specifications as a special case: $\mathbf{z} = \mathbf{\Phi W}_1 \mathbf{z} + \mathbf{Z\beta} + \mathbf{a} = \mathbf{\Phi W}_1 \mathbf{z} + \mathbf{Z\beta} + \boldsymbol{\varepsilon}$. The EMPIRIC model can therefore be thought of as a spatial lag-1 model of contemporaneous general-joint-spatial-model as shown in the comparison of **Table 11-10**. The relationship between time-series and econometric models has been discussed in detail in the "Spatial Time-Series" chapter. In the section "Econometric models," equivalencing between the two is further explained when the econometric model is in its reduced form. Equivalencing through transfer function is also outlined. In the reduced form, for example, endogenous variables are explained by exogenous and lagged endogenous variables, while in the transfer-function format, endogenous variables are explained only by the exogenous variables.

Historically speaking, the EMPIRIC modelling effort was preceded by the POLYMETRIC model. Instead of simultaneous linear-difference-equations, POLYMETRIC was in simultaneous non-linear difference-equations. POLYMETRIC was very much more demanding of analysis and computer time. After comparing the results from the alternate models, development of POLYMETRIC was abandoned in favor of continued work with EMPIRIC. This parallels the experience with ARMA modelling. First-order difference did very well in comparison with higher-order differences in the case of univariate applications in three cities in Texas (Grant 1989). This is consonant with practice where only weak stationarity is required, i.e., time independence for moments up to order two. Parsimony is the key to statistical modelling and a first-order autoregressive model

Metho-dology	Robinson Metho-Vector time-series	EMPIRIC Simultaneous-equation set	Yi-Chan Spectral matrix-series
Endogenous Variables	$\mathbf{z}_t = [z_1(t), z_2(t), \ldots, z_{n'}(t)]^T$ where there are n' pixels covering $t = 1, 2, \ldots n$ periods; these 'endogenous' variables are part of the autoregressive (AR) process of ARMA.	$\mathbf{z}_i(\Delta) = [z_{i1}(\Delta), z_{i2}(\Delta), \ldots, z_{in'}(\Delta)]^T$ where there are k' activities covering each of the n' zones over one Δt period; these exogenous variables are part of the lag-1 autoregressive (AR) model.	$\mathbf{Z}(t) =$ $\begin{bmatrix} Z_{11}(t) & Z_{12}(t) & \ldots & Z_{1k'}(t) \\ Z_{21}(t) & Z_{22}(t) & \ldots & Z_{2k'}(t) \\ \cdot & \cdot & & \cdot \\ \cdot & \cdot & & \cdot \\ Z_{n'1}(t) & Z_{n'2}(t) & \ldots & Z_{n'k'}(t) \end{bmatrix}$ where there are k' activities and n' zones over one time period t.
Exogenous Variables	$\mathbf{z}_t$ and $\mathbf{a}_t = [a_1(t), a_2(t), \ldots, a_{n'}(t)]^T$ where there are n' pixels covering $t = k+1, k+2, \ldots, n$ periods; these 'exogenous' variables are the time series itself in the autoregressive (AR) process and the corresponding error/noise terms in the moving-average (MA) process.	$\mathbf{z}_i(t-1)$ and $\mathbf{X}_i(\Delta) = [X_{i1}(\Delta), X_{i2}(\Delta), \ldots, X_{in'}(\Delta)]^T$ where there are exogenous activities and k' accessibilities covering each of the n' zones over one Δt period; these exogenous variables are part of the lag-1 autoregressive (AR) model.	$\mathbf{W}' = \begin{bmatrix} W_{11} & W_{12} & \ldots & W_{1n'} \\ W_{21} & W_{22} & \ldots & W_{2n'} \\ \cdot & \cdot & & \cdot \\ \cdot & \cdot & & \cdot \\ W_{n'1} & W_{n'2} & \ldots & W_{n'n'} \end{bmatrix}$ where the same W_{ij} is calibrated between zones i and j over all time periods t to forecast $k = 1, 2, \ldots, k'$ different activities (except for bifurcation).
Operator on Endogenous Variables	N/A	$[a_{ij}]\ i,j = 1, 2, \ldots, k'$ where a_{ij} = coefficient between activity i and zone j covering each zone over Δt.	$(\mathbf{I} - \mathbf{W}')$ where the same W_{ij} is calibrated between zones i and j over all time periods t to forecast different activities ($k = 1, 2, \ldots, k'$) where $\mathbf{w} = \prod_{k=1}^{k'} \mathbf{w}^k$ (except for bifurcation).
Operator on Exogenous Variables	$\Phi(B)$ where $\Phi(B) =$ autoregressive operator between pixel i and j covering $t = 1, 2, \ldots, n$ periods. $\Theta(B)$ where $\Theta(B) =$ moving-average operator between pixel i and j covering $t = 1, 2, \ldots, n$ periods.	$\Phi = [\varphi_{ij}]\ i = 1, 2, \ldots, k';\ j = 1, 2, \ldots, n$ where φ_{ij} = autoregressive coefficient between activity i and exogenous variable covering each of the zones over one Δt period. $\Theta = [\theta_{ij}]\ i = 1, 2, \ldots, k';\ j = n+1, \ldots, m'$ where θ_{ij} = autoregressive coefficient between activity j and accessibility variable covering each of the zones over one Δt period.	Solving simultaneous equations between t and $t-1$ for multiplier diagonal matrix ρ^k given $\mathbf{Z}(t)\ \mathbf{Z}(t-1)$. Calibrate accessibility-matrix $\mathbf{X}^k$ outside the model. Also calibration of elasticities b^k ($k = 1, \ldots, k'$) to account for bifurcation and constraint catastrophe.

Table 11-10 A comparison of three spatial-temporary models

such as EMPIRIC has the lure of simplicity compared with a full STARMA model. Note once again that the spatial relationship is modelled by exogenous terms such as the accessibility-variable X_{ih}, rather than implicitly as done in STARMA through $L^{(l)}$.

As discussed, econometric model can be expressed in a transfer-function format. The transfer function can again be viewed as the final form of the simultaneous-equation set. In this form, the infinite matrix-polynomial for Lowry–Garin model—Equation **(11-36)** when $\mathbf{W}^{j'} = \mathbf{I}$ for $j' = j$—can be derived. Such a series can also be interpreted in its dynamic form: $\mathbf{W}(B) = \mathbf{W}^0 + \mathbf{W}^1 + \mathbf{W}^2 + \dots + \mathbf{W}^\infty$, in which each endogenous variable depends only on its own lagged values and on the exogenous variable (without lags). One can also think of Equation **(11-55)** with β and $\mathbf{a}$ set to zero in the first equation and ε set to zero for the second. Thus we have $\mathbf{TU} = \Phi \mathbf{W}_1 \gamma \mathbf{W}_2 \mathbf{TU} + \mathbf{K}$. Here $\mathbf{T}$ and $\mathbf{U}$ replace $\mathbf{z}$ and $\mathbf{a}$ respectively and are now diagonal matrices of zonal activities such as population and employment and $\mathbf{K}$ is a constant diagonal-matrix. Transposition of terms and simplification of this yields $\mathbf{TU} = \mathbf{K}[\mathbf{I} - \Phi \mathbf{W}_1 \gamma \mathbf{W}_2]^{-1} = \mathbf{K}[\mathbf{I} - \mathbf{W}^j]^{-1}$, which is the same form as Equation **(11-38)** above. Instead of a correlative relationship made up of spatial autoregressive and spatial autoregressive-disturbance terms, a causal relationship is postulated here.

As shown by the calibration results, spatial-temporal models of the correlative type, while simple in its concepts, post demanding requirements on model specification and identification. This often results in poorly calibrated models. A natural alternative is to employ causal models as a substitute, wherever the physical process can be meaningfully described. Forty years of urban spatial-modelling has pointed toward the simplicity and flexibility of EMPIRIC. But the relative calibration ease and attractiveness of a more causal Lowry-derivative model are becoming increasingly attractive. The Yi–Chan experiences with the Lowry–Garin model give encouragement to the use of matrix series to model nonhomogeneous spatial-temporal processes. Similar to EMPIRIC, the calibration breaks down to the equivalence of a one-lag autoregressive format, rather than the full STARMA (p,q) model. (See the last row of **Table 11-10** labeled "Operator on exogenous variables.") The Yi–Chan model also appears promising in handling nonhomogeneous processes using disaggregate (and perhaps elegant) forecasting and calibration techniques. Particularly notable is its ability to handle bifurcation and catastrophe in both the temporal and spatial dimensions.

11.7.2 Comparison among spatial terms. While we made a distinction between correlative vs. causal modelling, the distinction is not necessarily clear. Getis (1991) suggested that spatial-interaction models such as the gravity model is a special case of a general model of spatial autocorrelation. The common elements of the various spatial autocorrelation models are: (a) a matrix of values representing the association between locations, and (b) values representing a vector of the attributes of the various locations. A general form of the association between these elements is contained in the following *cross-product statistic*

$$\sum_i \sum_j w_{ij} Y_{ij}. \qquad \qquad \textbf{(11-57)}$$

a	*b*	*c*
d	*e*	*f*
g	*h*	*i*

Figure 11-7 Sample nine-quadrant locations

Here w_{ij} are elements of a matrix of spatial proximity of locations i to locations j, and Y_{ij} is a measure of the association of i and j on another dimension.[17] In this and in all subsequent formulations where we use summation signs, i does not equal j (i.e., no self-association or self-interaction), unless otherwise indicated. In addition, stationarity and isotropy are assumed where required. A common choice of Y_{ij} is $(Z^o_i - Z^o_j)^2$ where Z^o are the values observed for variate Z_i.

11.7.2.1 MORAN'S *I*. A survey of the models of spatial-autocorrelation shows that nearly all of the models are simply another specification of a cross-product statistic. Perhaps the most common form of the spatial-autocorrelation models is the *Moran's* form.

$$\frac{n' \sum_i \sum_j w_{ij}(Z_i - \bar{Z})(Z_j - \bar{Z})}{w \sum_i (Z_i - \bar{Z})^2} \tag{11-58}$$

where $w = \sum_{i=1}^{n'} \sum_{j=1}^{n'} w_{ij}$. In general, $w = 2(2\bar{R}C' - \bar{R} - C')$ in the rook's definition of contiguity (first-order neighbors), $w = 4(\bar{R}-1)(C'-1)$ in the bishop's definition (second-order neighbors), and $w = 2(4\bar{R}C' - 3\bar{R} - 3C' + 2)$ in the queen's definition (first- and second-order neighbors combined). Here $\bar{R}$ is the number of rows and C' the number of columns in the pixel image, grid, or lattice. In a square $C' \times C'$ lattice, these formulas become $4C'(C'-1), 4(C'-1)$ and $4(2C'-1)(C'-1)$ respectively. These three definitions of contiguity are illustrated in **Figure 11-7**, with their associated **W** matrices shown in **Table 11-11**, in which an entry of 1 indicates contiguity between a pair of lattices.

---

[17]It is interesting to note that when Y_{ij} is simply the interaction between i and j in terms of activity-allocations, the cross-product statistic becomes the familiar min-sum $\sum_i \sum_j c_{ij} x_{ij}$ commonly used in locating median-facilities, where $w_{ij} = c_{ij}$ and $Y_{ij} = x_{ij}$.

Rook	a	b	c	d	e	f	g	h	i
a	0	1	0	1	0	0	0	0	0
b	1	0	0	0	1	0	0	0	0
c	0	1	0	0	0	1	0	0	0
d	1	0	0	0	1	0	0	0	0
e	0	1	0	1	0	1	0	1	0
f	0	0	1	0	1	0	0	0	1
g	0	0	0	1	0	0	0	1	0
h	0	0	0	0	1	0	1	0	1
i	0	0	0	0	0	1	0	1	0

Bishop	a	b	c	d	e	f	g	h	i
a	0	0	0	0	1	0	0	0	0
b	0	0	0	1	0	1	0	0	0
c	0	0	0	0	1	0	0	0	0
d	0	1	0	0	0	0	0	1	0
e	1	0	1	0	0	0	1	0	1
f	0	1	0	0	0	0	0	1	0
g	0	0	0	0	1	0	0	0	0
h	0	0	0	1	0	1	0	0	0
i	0	0	0	0	1	0	0	0	0

Queen	a	b	c	d	e	f	g	h	i
a	0	1	0	1	1	0	0	0	0
b	1	0	1	1	1	1	0	0	0
c	0	1	0	0	1	1	0	0	0
d	1	1	0	0	1	0	1	1	0
e	1	1	1	1	0	1	1	1	1
f	0	1	1	0	1	0	0	1	1
g	0	0	0	1	1	0	0	1	0
h	0	0	0	1	1	1	1	0	1
i	0	0	0	0	1	1	0	1	0

Table 11-11 A **W** matrix pertaining to the bird species
(Upton and Fingleton 1985)

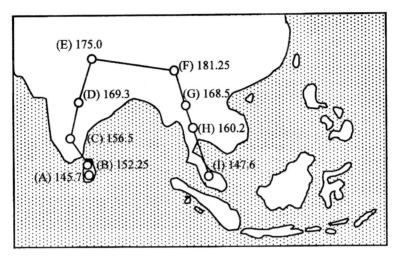

Figure 11-8 Clines in wing lengths of bird species (Upton and Fingleton 1985)

The theoretical base for these models is interval-scale observations. These are essentially Pearson product-moment correlation-coefficient models—as defined in the "Statistics." appendix as $r = \left[\Sigma_{ij}(Z_i - \bar{Z}_i)(Z_j - \bar{Z}_j) \right] / s(Z_i) s(Z_j)$—altered to consider the effect of a spatial weight matrix. The cross-product Y_{ij} is the covariance, $(Z_i - \bar{Z})(Z_j - \bar{Z})$, and the weight-matrix **W** has no restrictions. As in the Pearson statistic, Moran's measurement includes a scaling factor. While Pearson correlation employs the product of standard deviations $s(Z_i) s(Z_j) = (Z_i - \bar{Z}_i)(Z_j - \bar{Z}_j)/n$, Moran employs $\Sigma_i (Z_i - Z)^2 / n'$. Perhaps a better analogy can be made to the space-time autocorrelation of Equation **(11-23)** since both are spatial statistics.

Example

Moran's *I* is used to measure a location-specific value for the spatial gradients in analyzing an Asiatic bird (Upton and Fingleton 1985). Specifically we wish to study the existence of a wing-length *cline*—the gradual change in wing length of adjacent bird populations. The lines in the map of **Figure 11-8** connect the localities for which measurements are available and show the probable routes of bird-movement and gene-flow. These connecting links form the proximity values for the associated **W** matrix according to rook's contiguity rule. **Table 11-12** displays the **W** = $[w_{ij}]$ matrix of the species for which $w = 16$. The mean of the observations is $Z = 161.81$. Substituting these values together with the Z_i values indicated on the map for this species into Equation **(11-58)**, we obtain $I = 9 \Sigma_{i=1}^{9} \Sigma_{j=1}^{9}$ $w_{ij}(Z_i - 161.81)(Z_j - 161.81) / 16 \Sigma_{i=1}^{9}(Z_i - \bar{Z})^2 = 0.603$.

When laid side-by-side the space-time autocorrelation-function **(11-23)**, notice the asymptotic normal-distribution of the model as the number of locations n' increases, which explains much of its popularity. To show this, we recognize that the normality property hinges on the number of places under consideration, and on the extent and manner in which they are interconnected in the **W** matrix. With this latter reservation we are probably

Location	A	B	C	D	E	F	G	H	I
A	0	1	0	0	0	0	0	0	0
B	1	0	1	0	0	0	0	0	0
C	0	1	0	1	0	0	0	0	0
D	0	0	1	0	1	0	0	0	0
E	0	0	0	1	0	1	0	0	0
F	0	0	0	0	1	0	1	0	0
G	0	0	0	0	0	1	0	1	0
H	0	0	0	0	0	0	1	0	1
I	0	0	0	0	0	0	0	1	0

Table 11-12 A W matrix pertaining to the bird species (Upton and Fingleton 1985)

justified in considering 20 places sufficient to assume normality. Building upon the "Asiatic bird" example cited above, if there had been more locations than just the nine, we can safely use a normal approximation. However, $n = 9$ and the **W** matrix has few non-zero entries suggest that normal approximation is open to criticism.

Instead, we should assess the significance of the observed value of I by comparison with its randomization distribution. In 99 random-permutations of the Y_{ij} in Equation **(11-58)** above, an approximately normal distribution of I is obtained:

Cumulative distribution of I for data given in **Figure 11-8**

Value of I	-1.0	-.08	-0.6	-0.4	-0.2	0.0	0.2	0.0	0.6	0.8
Frequency I	99	98	94	78	59	40	19	7	1	0

It can be seen that on only one occasion is the observed value equalled or exceeded. We can conclude that there is significant evidence of a wing-length cline for the Asiatic bird species. The expected value of I is

$$E(I) = -1/(n'-1).$$ **(11-59)**

The variance of I under the assumption of normally-distributed data is

$$\text{Var}(I) = \left[n'\tilde{w}_1 - n'\tilde{w}_2 + 3w\right] / \left[w(n'-1)\right] - E(I)^2$$ **(11-60)**

where $\tilde{w}_1 = 1/2\sum_{i=1}^{n'}\sum_{j=1}^{n'}(w_{ij} + w_{ji})^2$ and $\tilde{w}_2 = \sum_{i=1}^{n'}\left(\sum_{i=1}^{n'}w_{ij} + \sum_{j=1}^{n'}w_{ji}\right)^2$. If randomization is assumed, then the variance is

$$\text{Var}(I) = \frac{n'\left[(n'^2 - 3n'+3)\tilde{w}_1 - n'\tilde{w}_2 + 3w^2\right] - k[n'(n'-1)\tilde{w}_1 - 2n'\tilde{w}_2 + 6w^2]}{(n'-1)^3 w^2} - E(I)^2$$ **(11-61)**

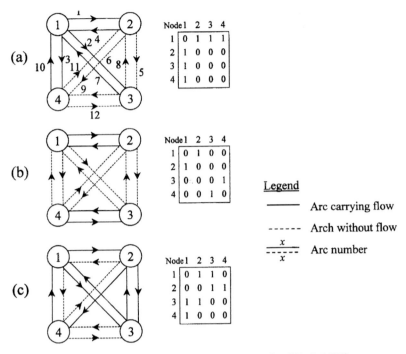

Figure 11-9 Network autocorrelation examples (Black 1992)

where $k = m_4/m_2^2$ and $m_r = [\Sigma_i(z_i - \bar{z})]^r$. $\square$

Besides a lattice pattern, Moran's measure of spatial correlation can be applied to discrete networks (Black 1992). Network autocorrelation-analysis assesses the degree to which the value of a variable on each network arc co-varies with other values of that variable. These other values are found on the network arcs to which each arc is connected. Spatial proximity is less important here than network connectivity. An advantage of network autocorrelation-analysis is its explicit recognition of co-variation for values on neighboring arcs. Here spatial separation explicitly refers to the 'routing' between two points, consisting of sequences of arcs that connect these two points.[18] Three networks are shown in **Figure 11-9**. These networks have only unit or zero flows on their arcs. The first, marked (a), has unit flows centered on node 1; they appear to be dependent on that node. The second, marked (b), has a pattern of unit flows on arcs at the top and bottom of the network. But flows do not exist in between. The third, marked (c), consists of apparently independent unit and zero flows without discernible patterns of clustering or dispersion. The unit-flow matrix is shown next to each of the networks for convenience. (The readers may notice the similarity between this matrix and the

---

[18]For a discussion of this concept in depth, see the chapter on "Measuring Spatial Separation".

Arc	1	2	3	4	5	6	7	8	9	10	11	12
1	0	1	1	1	1	1	1	1	0	1	1	0
2	1	0	1	1	1	0	1	1	1	1	0	1
3	1	1	0	1	0	1	1	0	1	1	1	1
4	1	1	1	0	1	1	1	1	0	1	1	0
5	1	1	0	1	0	1	1	1	1	0	1	1
6	1	0	1	1	1	0	0	1	1	1	1	1
7	1	1	1	1	1	0	0	1	1	1	0	1
8	1	1	0	1	1	1	1	0	1	0	1	1
9	0	1	1	0	1	1	1	1	0	1	1	1
10	1	1	1	1	0	1	1	0	1	0	1	1
11	1	0	1	1	1	1	0	1	1	1	0	1
12	0	1	1	0	1	1	1	1	1	1	1	0

Table 11-13 Arc-connection matrix for the four-node network (Black 1992)

contiguity matrix used for generating routes in the "Simultaneous Location-Routing" chapter.)

Using Equations **(11-59)**, **(11-60)**, and **(11-61)**, and calculating standard normal-deviates (some kind of standardized 't-value') for each of the networks yields the expected-value-of-Moran's-I, I, variance-under-the-normality-assumption, and variance-under-randomization. Specifically, the following results are obtained:

Network	*E(I)*	*I*	*Var(I):normal*	*Var(I):random*
(a)	−0.09091	0.11111	3.469	3.162
(b)	−0.09091	−0.33333	−4.163	−3.904
(c)	−0.09091	−0.10476	−0.238	−0.218.

These numbers illustrate positive autocorrelation ($I = 0.11111$), negative auto-correlation ($I = -0.33333$), and an absence of autocorrelation ($I = -0.10476$) in **Figure 11-9** (a), (b), and (c) respectively. Notice the unit-flow matrices of the figures do not define the autocorrelation structure for the arcs. This comes from the arc-connection relationship depicted by **Table 11-13**, which forms the w_{ij} used here. The unit-flow matrices provide the Z values in the computation of Moran's I.

Subtracting E(I), the expected-value-of-I, from each I and dividing by the standard-error derived by taking the square-root-of-the-variances from Equations **(11-60)** and **(11-61)** yields the standard-normal-deviates of *Var(I):normal* and

Var(I):random. Notice these variances differ depending on whether we assume simple random-sampling-from-a-normal-distribution or whether we assume the data represent a single-random-sample from all the different samples that could be drawn. In the former case, the variance is calculated from Equation **(11-60)**. In the randomization case, we calculate the variance using Equation **(11-61)**. The readers will notice that the levels-of-variance between these two measures are quite similar in these examples. As sample size increases, the distribution of an attribute becomes asymptotically normal. Since sample sizes in most network problems are very large, there is a tendency to prefer the use of the more straightforward *Var(I):normal* since the assumption of normality is easily met.

The Var(*I*) scores obtained—both *Var(I):normal* and *Var(I):random*—do agree with our expectations based on visual inspection. The first pattern indicates dependence influenced by node 1, which is statistically significant at the 0.01 level. This is manifested in a clustering of flow arcs around that node, that is, positive network autocorrelation. Motor-vehicle-accident-rates for road segments (arcs) in a local area under extreme-weather-conditions might display such autocorrelation, since one would expect clustering of accidents in this bad-weather zone. The second pattern has a significant level of negative dependence (at the 0.01 level-of-statistical-significance) leading to dispersion, that is, negative network auto-correlation. Motor-vehicle-accidents (or accident rates) occurring at uniformly-spaced points of access and egress along a toll road could yield such a pattern. In this case, the top and bottom arc of **Figure 11-9** (b) represent access and egress ramps of a highway. The third pattern has a Var(*I*) score near the mean of that distribution (that is, zero) suggesting we cannot reject the null-hypothesis for independence-of-the-flows. In other words, accidents occur randomly among all road segments (arcs) of the highway network. The consistency of these result with expectations is sufficient to suggest the use of these methods in a more realistic context (Black 1993). For the interested readers, another example of network autocorrelation can be found in Wright (1995) and the "Exercises and Problems" web site, where the technique is used in monitoring groundwater pollution through readings from wells.

11.7.2.2 FIRST-ORDER AUTOREGRESSIVE MODEL. **Table 11-14** compares Moran's model with other spatial models. A first-order autoregressive model is given by

$$Y_i = \theta + \varphi \sum_j w_{ij} Y_j + a_i, \tag{11-62}$$

which exemplifies a variation on the form specified in **(11-57)**. For a spatial-autoregressive interpretation, φ is the spatial autocorrelation-coefficient, w_{ij} is an element of the spatial weight-matrix, and *a* is the uncorrelated, normally distributed, non-spatially autocorrelated, homoscedastic error-term. The space-time autocorrelation function is shown in Equation **(11-23)**. The $w_{ij}Y_j$ is a spatial variable that we construct from the dependent variable itself (see Equation **(11-28)** as an example). The system is stationary. Thus the model represents the spatial-

Spatial statistic	Model	Spatial weight	Spatial relation Y_{ij}	Restrictions on Y_{ij}	Scale
Space–time auto-correlation $r_t(k)$	$\dfrac{n}{(n-k)} \dfrac{\sum_{t=1}^{(n-k)} \mathbf{z}_t^T(t)\mathbf{z}(t+k)}{\left(\sum_t \mathbf{z}_t^T(t)\mathbf{z}_t(t)\,\mathbf{z}^T(t)\mathbf{z}(t)\right)^{1/2}}$	none	$\mathbf{z}_t^T(t)\mathbf{z}(t+k)$	none	$\dfrac{n}{(n-k)\left(\sum_t \mathbf{z}_t^T(t)\mathbf{z}_t(t)\,\mathbf{z}^T(t)\mathbf{z}(t)\right)^{1/2}}$
Spatial- auto-correlation model (Moran's I)	$\dfrac{n'\sum_i\sum_j w_{ij}(Z_i-\bar Z)(Z_j-\bar Z)}{w\sum_i (Z_i-\bar Z)^2}$	w_{ij}	$(Z_i-\bar Z)(Z_j-\bar Z)$	none	$\dfrac{n'}{w\sum_i (Z_i-\bar Z)^2}$
Spatial-interaction model (Gravity model)	$V_{ij}=KZ_i^a Z_j^{a'} w_{ij}^{-b}$	w_{ij}^{-b}	$Z_i^a Z_j^{a'}$	positive	K
General location-i spatial-model (General spatial-statistic G_i)	$\left(\sum_j z_i z_j w_{ij}^{-b}\right)\left(\sum_j z_i z_j\right)^{-1}$	w_{ij}^{-b}	$z_i z_j$	positive	$\left(\sum_j z_i z_j\right)^{-1}$
General location-pair spatial-model (General spatial statistic G_{ij})	$\left(z_i z_j w_{ij}^{-b}\right)\left(z_i z_j\right)^{-1}$	w_{ij}^{-b}	$z_i z_j$	positive	$(z_i z_j)^{-1}$
Min-sum facility-location model	$\sum_i\sum_j w_{ij}Z_{ij}$	w_{ij}	z_{ij}	positive	$\sum_i\sum_j Z_{ij}$

Table 11-14 A comparison of various spatial models (Getis 1991)

dependence structure of Y. This is not a model of spatial autocorrelation per se, but a model of the effect of spatial autocorrelation on an endogenous variable. The coefficient φ is a parameter that relates the spatial dependence form of Y with itself, whereas Moran's form is strictly a value representing the spatial-autocorrelation characteristic of variable Y. The numerators of both the Moran statistic and φ are the covariance.

11.7.2.3 SPATIAL-INTERACTION MODEL. The Lowry–Garin model, similar to the accessibility term of EMPIRIC, requires the calibration of a gravity model. This can be seen from Equations **(11-42)** and **(11-27)** respectively. The form of the gravity model assumes the common form

$$V_{ij}=KZ_i Z_j f(d_{ij}).\tag{11-63}$$

This is the general unconstrained gravity-model where the Z_i and Z_j represent the magnitude of the variable under study at i and j respectively. Sometimes exponents are used in these two variables to differentiate the effect of the origin from that of the destination. A common form of the trip-frequency function $f(d_{ij})$ is d_{ij}^{-b} or

$\exp(-bd_{ij})$. K is a scalar or constant proportion, usually assuming the form $\left(\sum_{j=1}^{n'} Z_j f(d_{ij})\right)^{-1}$.

The above gravity model is generalized in terms of the "spatial interaction model" as shown in **Table 11-14**, where exponents are introduced to the activity-variables Z_i and Z_j. The characteristics-of-interaction measures that help differentiate them from autocorrelation measures can be enumerated. They consists of (a) a focus on a single i–j relationship, (b) the use of exponents to adjust variables, (c) constraints to draw attention to one or more of the variables. In terms of the *cross-product statistic* (Equation **(11-57)**, there are significant similarities between them. In **Table 11-14**, Equation **(11-63)** is generalized and rewritten to conform to the nomenclature of the cross-product statistic. Note that no summation sign is used in Equation **(11-63)**, nor its generalization in **Table 11-14**. The focus in interaction modelling is on a single association, although the derivation of the parameters usually depends on the empirical data of all associations. The point, however, is that the form of the measure is similar to the measure of spatial autocorrelation.

The expression for V_{ij} in Equation **(11-63)** or its generalization is simply one value that could be used in the development of a spatial-autocorrelation statistic. The elements of a w_{ij} matrix contain the values of d_{ij}^{-b} or its more general form $f(d_{ij})$. The Y_{ij} are simply the association values between locations i and j. As in the spatial-autocorrelation statistics, the Y_{ij} are defined in any of several ways. The various constraints placed on the Y_{ij} values in **Table 11-14** can easily be accommodated in a cross-product statistic. Thus the exponents that are used in interaction models represent more advanced development than autocorrelation models. But there is nothing standing in the way of the use of exponents to enhance spatial-auto-correlation measures in general.

11.7.3 A general spatial statistic. The statistic developed below contains the elements of the cross-product statistic but instead of it being a summary measure over an entire set of data it focuses on a single point as in Equation **(11-62)** (Getis 1991):

$$\left[\sum_j w_{ij}(d)\, z_i\right]\left(\sum_j z_j\right)^{-1}. \tag{11-64}$$

Here w_{ij} is an entry in a 0–1 spatial weight-matrix. These entries assume values of unity for all 'links' i–j within distance d of given location i, and zero otherwise. Notice the variable z_i has a natural origin such as the subject-location-of-interest and is positive. As such, the statistic is suitable for analyzing nonhomogeneous data that has spatial properties varying from point to point. Now the difficulty with the statistic shown above is its dependence on a 0–1 weight or distance matrix. Further development of the statistic would allow i to equal j in the substitution of d_{ij}^{-b}, $\exp(-bd_{ij})$, $f(b,d_{ij})$, or $f(b,d_{ij})$ in general for w_{ij}. For example, the following formulation would replace Equation **(11-64)**:

$$G_i = \sum_j z_i z_j f(b, d_{ij}) \left(\sum_j z_i z_j\right)^{-1} \qquad b > 0 \tag{11-65}$$

where i–i is allowed. In the above equation, there is an obvious correspondence between both the cross-product statistic and the general form of the interaction model. The following expected value would be based on the assumption that all z values were similar. Thus

$$E(G_i) = \frac{1}{n'} \sum_j f(b, d_{ij}) \qquad b > 0. \tag{11-66}$$

As with Equation **(11-64)**, the new statistic G_i will be bounded between 0 and 1. Tests based on the statistic would answer the fundamental question: "Are the association and the interaction between i and all j greater than chance would have it?" A variation on the above two equations would focus on the relationship between a single i and a single j. These equations are essentially Equation **(11-65)** with the summation signs removed:

$$G_{ij} = \frac{z_i z_j f(b, d_{ij})}{z_i z_j}; \qquad b > 0, \tag{11-67}$$

where z_i and z_j are random variables. Expected value of G_{ij} yields the familiar spatial-interaction term: $E(G_{ij}) = f(b, d_{ij})$.

To use the general spatial-statistic $G_i(d)$ (Equation **(11-65)** to test hypothesis, consider the following properties of this statistic. Setting all z_j to one, the pattern of z_j represents a condition of no spatial autocorrelation. In this case, the null hypothesis is: there is no difference (and thus no spatial autocorrelation) among the z_j within distance d of i. By substituting a one for each z_j in Equation **(11-65)**, we find

$$E[G_i(d)] = w/(n' - 1) \tag{11-68}$$

and

$$E[\mathrm{Var}\, G_i(d)] = \frac{(n' - 1 - w)^2}{(n' - 1)^2 (n' - 2)}. \tag{11-69}$$

Here $G_i(d)$ is an expanded version of Equation **(11-64)** using the observed z_j values and replacing $f(b, d_{ij})$ with $w_{ij}(d)$:

$$G_i(d) = \left[\sum_j w_{ij}(d)\, z_i z_j \right]\left[\sum_j z_i z_j \right]^{-1}. \tag{11-70}$$

If the norm deviate

$$R_s(d) = \frac{G_i(d) - E[G_i(d)]}{\left\{ E[\mathrm{Var}\, G_i(d)] \right\}^{1/2}} \tag{11-71}$$

is positively or negatively greater than some specified level of significance, then positive or negative spatial-autocorrelation are obtained. A large positive R_s implies that large values above the mean z_j are spatially associated. A large negative R_s

means that small z_j are spatially associated with one another. A similar statistic can be defined for the Moran's I: $R_S(I) = [I - E(I)]/[E(\text{Var }I)]^{1/2}$. By now, the reader can see the parallel to the two-tailed t-statistic in which the t-statistic can take on to both positive and negative values in hypothesis testing. Here the null hypothesis is to suggest that the regression coefficient be equal to a particular value. The alternate hypothesis is that it is not.

A null hypothesis based on this statistic can be formed. It might call for interaction no greater (or less) than one might expect when all z_j are equal. The expectations are computed in Equations **(11-68)** and **(11-69)**. Rejection of the null hypothesis would suggest that there is greater (or less) interaction than expected. Remember the G-statistic measures the degree of association that results from the concentration of a weighted point and all other weighted points included within a radius of distance d from the original point i. $G_i(d)$ measures the concentration or lack-of-concentration of the sum-of-values associated with variable z in the region under study. $G_i(d)$ is a proportion of the sum of all z_j values that are within d of i. If, for example, high value z_j are within d of point i, then $G_i(d)$ is high. Whether the $G_i(d)$ value is statistically significant depends on the statistic's distribution.

A special feature of this statistic is that the pattern of data points is neutralized when all z values are the same. This is illustrated for the case when data-point densities are high near point i, and d is just large enough to contain the clustered points. Theoretical $G_i(d)$ values are high because $w_i = \Sigma_j w_{ij}(d)$ is high. However, only if the observed z_j values in the vicinity of point i differ systematically from the mean is there the opportunity to identify significant spatial-concentration of the sum of z_j. That is, as data points become more clustered near point i, the expectation of $G_i(d)$ rises, neutralizing the density of j values. Besides the above meaning, the value of d can be interpreted as a distance that incorporates specified cells in a lattice. It is expected that neighboring G_i will be correlated if d includes neighbors. To examine this issue, consider a regular lattice. When n' is large, the denominator of each G_i is almost constant. So it follows that $r(G_i, G_j)$ stands for the proportion of neighbors that i and j have in common. Let us give an example.

Example

Consider an example involving the first-order neighbors (rook's case) [Getis and Ord 1992]. Cell i in the following pattern has no common neighbors with its four immediate neighbors, but has two with its second-order neighbors (bishop's case). The number of common neighbors—0 and 2—are coded in the illustration below where the neighbor-of-concern is. For example, the 'cells' that are marked 0 suggest that there exist no common neighbor between cell i and its immediate neighbor to the north, south, east and west. The cells that are marked 2 suggest that there are two common first-order neighbors between i and these neighbors. The cells marked 1 indicate that as far as the rook's-neighbors-once-removed, cell i and these neighbors share one common first-order neighbor. All the other cells have no common neighbors with i and therefore are coded with 0 (or not shown).

$$
\begin{array}{ccccc}
 & 0 & 1 & 0 & \\
0 & 2 & 0 & 2 & 0 \\
1 & 0 & i & 0 & 1 \\
0 & 2 & 0 & 2 & 0 \\
 & 0 & 1 & 0 &
\end{array}
$$

The G-statistic was computed for all points in the lattice above. It was found that the G-indices for the four diagonal-neighbors j—i.e., those marked with a 2—have correlations of about 0.5 with G_i. Thus considering first-order neighbors, $r[G_i(1)G_j(1)]$ = proportion of neighbors in common = $2/4 = 0.5$. Four others—the four marked with a 1—have correlations of about 0.25. Thus for the new locations of j, $r(G_iG_j) = 1/4 = 0.25$. The rest are virtually uncorrelated. For more highly connected lattices, such as the queen's case, the array of non-zero correlations stretches further. But the maximum correlation between any pair of G-indices remains about 0.5. □

11.7.4 Comparison between the G(d)-statistic and Moran's-I. The $G(d)$-statistic measures overall concentration or lack-of-concentration of all pairs of (z_iz_j) such that i and j are within d of each other. Here d is defined by the rule of contiguity such as rook's case (first-order neighbors) or bishop's case (second-order neighbors). Following this equation,

$$
G(d) = \frac{\sum_{i=1}^{n} \sum_{j=1}^{n} w_{ij}(d) z_i z_j}{\sum_{i=1}^{n} \sum_{j=1}^{n} z_i z_j} \qquad j \neq i.
\tag{11-72}
$$

one finds $G(d)$ by taking the sum-of-the-multiples of each z_i with all z_j within d of i. The statistic is, therefore, based on the degree-of-covariance within d of all z_i. Consider k_1, k_2 as constants invariant under random permutations. Using summation shorthand we have $G(d) = k_1\Sigma_i\Sigma_jw_{ij}z_iz_j$ and $I(d) = k_2\Sigma_i\Sigma_jw_{ij}(Z_i - Z)(Z_j - Z)$ = $(k_2/k_1)G(d) - k_2Z\Sigma_i(w_i + w_{\cdot i})Z_i + k_2\bar{Z}^2w$ where $w_{\cdot i} = \Sigma_jw_{ji}$. In these expressions, remember we use z to denote a positive-random-variable-with-a-natural-origin and Z to denote a regular-random-variable.

Since both $G(d)$ and $I(d)$ can measure the association among the same set of weighted points, they may be compared. They will differ when the weighted-sums $\Sigma_iw_iZ_i$ and $\Sigma_iw_{\cdot i}Z_i$ differ from wZ, that is, when the patterns of weights are unequal. The basic hypothesis is a random pattern in each case. We may compare the performance of the two measures by using their equivalent deviates of the approximate-normal-distribution, R_S (Getis and Ord 1992).

Example

Refer to the following lattice:

$$
\begin{array}{cccccccccc}
m & m & m & m & m & m & m & m & m & m \\
m & A & A & A & m & m & B & B & B & m \\
m & A & A & A & m & m & B & B & B & m \\
m & A & A & A & m & m & B & B & B & m \\
m & m & m & m & m & m & m & m & m & m
\end{array}
$$

Set A and B both to $2m$ for the values of z_i, which means all z_i are positive ($A \geq 0$; $B \geq 0$) and have a natural origin. This results in the mean value $\bar{z} = m$ where the number of locations $n' = 50$. For sensitivity analysis, put $A = m(1+c)$, $B = m(1-c)$, and $0 \leq c \leq 1$. In addition, put $a = A - m$; $B = 2m - A = m - a$; $B - m = -a$; $m \geq a$; and j not equal to i. For the rook's case, $w = \Sigma_i\Sigma_j w_{ij} = 2(\bar{R}C' - \bar{R} - C') = 2[2(5)(10) - (5) - (10)] = 170$. With an expanded version of **Table 11-11** for $\mathbf{W} = [w_{ij}]$,

$$I = \frac{n'\sum_i \sum_j w_{ij}(z_i - \bar{z})(z_j - \bar{z})}{w\sum_i (z_i - \bar{z})^2} = \frac{(50)\sum_{i=1}^{50} \sum_{j=1}^{50} w_{ij}(z_i - m)(z_j - m)}{(170)\sum_{i=1}^{50}(z_i - m)^2} =$$

$$\frac{(50)(24a^2)(2)}{(170)(18a^2)} = 0.784$$

(11-73)

for all choices of a and m. Again, referring to the expanded version of **Table 11-11** for $[w_{ij}]$, $\tilde{w}_1$

$= \frac{1}{2}\sum_{i=1}^{50} \sum_{j=1}^{50}(w_{ij} + w_{ji})^2$, $\tilde{w}_2 = \sum_{i=1}^{50}(\sum_{i=1}^{50} w_{ij} + \sum_{i=1}^{50} w_{ji})^2$, $\text{Var}(I) = \left[(50)^2\tilde{w}_1 - (50)\tilde{w}_2 + 3(170)\right]$ /

$\left[(170)(50^2 - 1)\right] - (50 - 1)^{-2} = 0.010897$, and $R_S(I) = \left[I - E(I)\right]\left[E(\text{Var } I)\right]^{-1/2} = 7.7088$ whenever $A >$ B.

$$G = \frac{\sum_i \sum_j w_{ij} z_i z_j}{\sum_i \sum_j z_i z_j} = \frac{24A^2 + 24B^2 + 24Am + 24Bm + 74m^2}{2500m^2 - 9A^2 - 9B^2 - 32m^2} =$$

$$\frac{170 + 48c^2}{2450 - 18c^2}.$$

(11-74)

When $c = 0$, $A = B = m$, G is a minimum, where $G_{min} = 170/2450 = 0.0694$ and $\text{Var}(G_{min}) = 0.000$. When $c = 1$, $A = 2m$, $B = 0$, G is a maximum, where $G_{max} = 218/2432 = 0.0896$, $\text{Var}(G_{max}) = 0.0000118555$ and $R_S(G_{max}) = 5.87$ for any m. It can be seen that G depends on the relative absolute magnitudes of the sample values. Note that I is positive for any A and B, while G-values approach a maximum when the ratio of A-to-B or B-to-A becomes large. □

From a number of experiments besides the one cited above, Getis and Ord (1992) recommended that any test for spatial association should use both Moran's I and the general-spatial-statistic G. Sums-of-products (G) and covariances (I) are two different aspects of a pattern. Both reflect the spatial-dependence structure. The $I(d)$-statistic has its peculiar weakness in not being able to discriminate between patterns that have exclusively high or low values dominant within d. Both statistics have difficulty discerning a random pattern from one in which there is little deviation from the mean. If a study requires that $I(d)$ or $G(d)$ values be traced over time, there are advantages to using both statistics to explain changes in spatial association. If data values increase or decrease at the same rate, that is, if they increase or decrease in proportion to their existing size, Moran's-I changes while $G(d)$ remains the same. On the other hand, if all z values increase or decrease by the same amount, $G(d)$ changes but $I(d)$ remains the same. Using both statistics will give a complete picture of what is happening.[19] Remember that G(d) is based on a variable that is positive and has a natural origin. Thus, for example, it is inappropriate to use $G(d)$ to study residuals from regression. Also, for both $I(d)$ and $G(d)$ one must recognize that transformations of the variable Z or z result in different values for the test statistic. As mentioned, conditions may arise when d is so small or large that tests based on the normal approximation are inappropriate.

As summarized by Getis and Ord (1992), the G statistics provide researchers with a straightforward way to assess the degree of spatial association. This can be done at various levels of spatial refinement—in an entire sample or in relation to a single observation. When used with Moran's I or another measure of spatial autocorrelation, they enable us to deepen our understanding of spatial series. One of G statistics' useful features, that of neutralizing the spatial distribution of the data points, allows for the development of hypotheses where the pattern of data points will not bias results. When G statistics are contrasted with Moran's I, it becomes clear that the two statistics measure different things (as mentioned above). Fortunately, both statistics are evaluated using normal theory so that a set of standard normal-variates taken from tests using each type of statistic are easily compared and evaluated.

11.7.5 Concluding Remarks. In the previous sections 11.4, 11.5 and 11.6, we used an autoregressive model, if not an autoregressive-moving-average model, to analyze and predict future land-use patterns. Also, a hybrid model combining statistical and matrix-series was advanced to consider nonlinear behavior—a feature above and beyond the capability of traditional autoregressive or autoregressive-moving-average models. These efforts represent an extension of distance-statistics-measures to both linear and nonlinear random-spatial-processes. In this section of the chapter, a cross-product statistic is used to show that spatial-interaction-models are a special case of a general model of spatial-autocorrelation. A series of traditional measures of spatial-autocorrelation, including Moran's-I, is shown to have a cross-product form. Several interaction models, including the gravity model, are shown to have a similar form as well. Finally, a general-spatial-statistic is developed which suggests that the relationship between the two types of models—spatial interaction and autocorrelation models—is particularly strong when the focus is on measurements from a single point i. The key words differentiating the two types of models are *interaction* and *association*. The interaction-implied-in-gravity-models refers to the possible *movement* of elements at i to or from places j. In a spatial-autocorrelation model, the *linkage* between i and j is a correlation indicating if these places have common or different characteristics. As the development of the spatial-autocorrelation model has a statistical origin, one usually considers association as having positive or negative statistical significance. For interaction modelers, interest is in the flow between places, whether the flows are greater or less than those predicted by a normal random-variable model.

---

[19]In this situation the space–time autocorrelation will obviously be an informative statistic.

We were able to show that the cross-product statistic allows for a unification of the two types of models (Getis 1991). This was accomplished by means of the development of spatial-autocorrelation-statistic that serves as a measure-of-spatial-interaction as well. An advantage to this approach is that the way is now paved for the development of statistical tests on interaction theory. These spatial statistics are constructed on the assumption that observations are from a *homogeneous* distribution over space and have an equal distributional-probability at every spatial-unit over space. In many sample-survey studies, the spatial units are usually defined arbitrarily instead of being defined in regular size. If so, problems of *heterogeneity* in spatial distribution may exist. The census tract is an example of spatial observations defined over irregular polygons since a tract is defined according to the size of population instead of the areal size. The statistical properties of the G-statistics and Moran's-I will not hold when the assumption of homogeneous distribution is violated. Bao and Henry (1996) proposed a revised G or I statistic for identifying local spatial-patterns. They offered examples to illustrate that the more general techniques can make identifying spatial association with irregular spatial-units more accurate. They showed that the contribution of each observation to the spatial measurements can be adjusted by using an appropriate weight such as the areal share.

It is interesting to note in **Table 11-14** that the min-sum facility-location model is also a manifestation, albeit a deterministic and prescriptive manifestation, of the cross-product statistic. The min-sum objective of many discrete facility-location models follows the form $\sum_i \sum_j w_{ij} Z_{ij}$ with the flows Z_{ij} representing the interaction (allocation) between two points in space. A normalization scale $\sum_i \sum_j Z_{ij}$, which represents the total demand in the study area is defined here. Suppose we are given a set of w_{ij} scaled between 0 and 1. The objective function normalized by the total demand measures the overall concentration (or lack-of-concentration) of all pairs of demands (z_i, z_j) in the study area. A median (or p medians with $p > 1$) in this case reflects the concentration-point(s) or centroid(s) of the demands in the area. This location or locations can be compared to the 'randomized' situation where all z_j are set to unity, which result in little concentration. The difference(s) or displacement between the median(s) of actual demands and the centroid(s) for randomized demands measures the amount of spatial bias or the 'autocorrelation' of the given demand pattern.

11.8 Spatial uncertainty and competitive facility location

Since we are concentrating on spatial interaction and association in both facility-location and land-use, it appears that another important variable affecting locational decisions is the level-of-uncertainty of spatial decision-makers (Pooler 1992). As an example, consider travellers choosing among a set of potential destinations, migrants choosing among cities, or shoppers choosing among stores. Assume that the destinations exert varying degrees of spatial dominance or influence because of their relative attractivities and locations. The variety or mix of spatial dominances at any point varies with respect to location, and it is hypothesized that the degree-of-uncertainty faced by decision-makers varies accordingly. It is the amount of uncertainty that ultimately determines his/her destination choice. Here we construct a slightly different paradigm for viewing spatial competition based on dominance and uncertainty—one that has a distinct computational advantage over previous spatial-interaction constructs.

11.8.1 *Entropy*. A simple index of spatial uncertainty is the entropy statistic. As introduced in the "Prescriptive Tools" and "Descriptive Tools" chapters in Chan (2000), entropy is defined here specifically as the measure of frequency with which an event occurs. We wish to define an index that is a theoretical spatial-measure a location's uncertainty with respect to potential destinations. Then we test empirically the relevance of the index to the prediction of spatial-interaction, particularly the competitiveness among destinations. The equation for the entropy of a distribution is

$$-\sum_{i=1}^{n'} P_i \ln P_i \qquad\qquad\qquad\qquad (11\text{-}75)$$

where $P_i = Z_i / \sum_{i=1}^{n'} Z_i$. Naturally, the sum of the P_i equals to 100 percent according to this definition. In this formulation, Z_i is the value of a variable (such as population or employment) at n' points, or zones.

The entropy is a well-known non-parametric measure of the diversity, variety, or uncertainty in a distribution. Unlike variance measures of deviation around the mean, it requires no assumptions about the form of the distribution. We have seen the use of entropy on several occasions throughout this book and Chan (2000), particularly in the derivation of spatial-interaction relationships. Of particular relevance is the treatment under the "Spatial interaction" section of the "Descriptive Tools" chapter in Chan (2000). While the concept is similar here, its use will be distinctly different as we will see.

11.8.2 *Spatial dominance uncertainty and deterrence-function*. Consider a geographical area with a set of n' spatially and randomly distributed potential destinations (of the same type) competing for the attention of would-be travelers. Assume that potential travelers are distributed randomly in the area of interest. This creates a geography where potential travelers in different locations (denoted by index i) are at varying distances, d_{ij}, from destinations j. The destinations are considered to exhibit varying degrees of attractiveness, W, for potential travelers. It is assumed that the impact of dominance of a destination is some decreasing function of distance, and some increasing function of the attractivity of the destination. These ideas are familiar to those dealing with gravitational interac-tions, but the concepts take on different meaning here. Now given the assumptions above, the theoretical spatial-dominance, Θ_{ij}, at a location $d_1,\cdots,d_{|J|}$ from $|J|$ potential destinations, is assumed to be of the form $\Theta_{ij} = W_j f(d_{ij}) / \sum_{j=1}^{|J|} W_j f(d_{ij})$ where $\sum_j \Theta_{ij}$ = 1 by definition. There are as many Θ_{ij} values at each location as there are destinations considered to be influencing it, and thus $|J|$ may be less than n'. *Spatial dominance*, as defined, is a measure of the way in which the total dominance at any point, or origin, is split among the potential destinations. Dominance is distinguished from spatial interaction in the sense that dominance refers to the presence and availability of information, whereas spatial interaction refers to the decisions and actual movement of people or products in response to that information. Again, this conceptual distinction has a significant bearing upon the computational requirements, as we will see shortly.

The theoretical spatial uncertainty at a point is defined as

$$-\sum_{j=1}^{|J|} \Theta_{ij} \ln \Theta_{ij}. \qquad (11\text{-}76)$$

This index measures the degree of diversity in the mix of dominancy at any location, i. The nature of the mix depends on the relative attractivities and locations of potential destinations with respect to that point. The lower the diversity in the dominancy, the higher the index and the higher the level of *uncertainty*. For example, if the mix of dominancy at a location is tending toward uniformity, it is expected that other *non-spatial* variables are more crucial in predicting accurately the movement from that location. Conversely, if one or a few destinations dominate a particular location, traditional gravity-model variables are likely to be more significant. It is again in the interpretation that Equation **(11-76)** differs from the regular definition of entropy as shown in Equation **(11-75)**. Specifically, the frequency measures P are replaced by the spatial-dominance measures, where the former is observed data while the latter is a priori information.

The "market share" of an individual firm can be defined as

$$\Theta_{ij} = W_j \exp(-b\,d_{ij}) \Big/ \sum_{j=1}^{|J|} W_j \exp(-b\,d_{ij}) \qquad (11\text{-}77)$$

where b denotes the deterring effect of distance—a familiar concept in gravitational interaction. With the specification of the spatial deterrence-function for $f(d_{ij})$, it is possible to speculate patterns of competition and uncertainty. For a regular triangular-lattice of points of equal size and with no boundary effect, the spatial uncertainty is at a minimum at the points themselves and at a maximum at interstitial points, or the intervening space.[20] When a boundary is introduced, the effect is to decrease the uncertainty of origin points near the boundary relative to those toward the center of the distribution. Therefore, one generally expects to find some correspondence between accessibility, or centrality, and spatial uncertainty. Unlike discrete trade, catchment, or market areas, spatial uncertainties (Equation **(11-76)**) are mappable as continuous *isopleth* fields—isoquant lines drawn similar to a contour map. Because no observed movement data are required in the calculation of the dominancy at a point, the theoretical spatial uncertainty is

---

[20]Reference the airport location example in "Prescriptive Tools" chapter (Chan 2000) and the Weber problem in "Facility Location" chapter.

calculable for any point on a surface. Compare this with mapping the entropies of the observed flows from origins and one can readily see the difference in data requirements.

11.8.3 The Tobler spatial-interaction model. Now we are well equipped to compare the new paradigm with the prevailing interaction model: the gravity model. (Tobler 1988). In this step, one simply compare destination choices computed from spatial uncertainty with that from the gravity model. The empirical test is hypothesized to be independent of the particular type of interaction model being used, and therefore the choice of model form should not be crucial. Tobler suggested this spatial-interaction model $\hat{V}_{ij} = (k_i + l_j)Z_iZ_j/d_{ij}$ where $\hat{V}_{ij}$ is the predicted interaction, and Z_i and Z_j are the sizes of the origins and destinations respectively. The k_i and l_j terms are proportionality constants that have the same normalizing effects as the Lagrangians in entropy-maximizing models. Likewise, they are interpretable as pushes and pulls, or emissivities and attractivities. They are given by $k_i = \left(2m_i' - \Sigma_j Z_j l_j/d_{ij}\right)/\Sigma_j Z_j/d_{ij}$ and $l_j = \left(2m_j'' - \Sigma_i Z_i k_i/d_{ji}\right)/\Sigma_i Z_i/d_{ji}$. The k are calculated from observed in and out movement rates (m_i' and m_j'' respectively), and from the population potential or accessibility of origins and destinations.[21] The results presented by Tobler indicate that the model preforms slightly better than either a doubly-constrained entropy-model with simple linear distance, or a totally constrained negative exponential entropy-model. The origin-specific interaction-model percentage-error-level, A_i, calculated with respect to the difference between the observed trips V_{ij} and the predicted trips leaving each origin, is

$$A_i = \frac{1}{2V_i}\sum_{j=1}^{|J|} |\hat{V}_{ij} - V_{ij}| \tag{11-78}$$

where $V_i = \Sigma_j V_{ij}$. This is the measure of compliance between trips predicted by spatial uncertainty and that by gravity model. As an empirical test of the relationship between spatial uncertainty and spatial interaction, an analysis is undertaken of a set of movement data. The expectation is that the error level is correlated positively with the uncertainty level.

11.8.4 Recreational travel to parks: a case study. The reader will recall we studied the recreational-travel pattern in Long Island New York as part of the "Competition and distribution in location-allocation" chapter. We will perform a similar case study here, except the techniques employed will be *spatial dominance* and *uncertainty*. As defined above, spatial dominance refers to the preeminence of a destination from the eyes of the traveller, and spatial uncertainty refers to the state of ignorance regarding his/her knowledge of a destination. The recreational-traveldata employed in this section represent the movement of people from ten countiesin northeastern Pennsylvania who visited one of five parks during a single

---

[21]Calibration procedure for a doubly-constrained model is illustrated via a numerical example in the "Descriptive Tools" chapter (Chan 2000) under the "Gravity model revisited" section.

On County	From Park				
	Big Pocono	Golds-boro	Hickory Run	Promised Land	Toby-hanna
Berks	0.080	0.056	0.621	0.121	0.119
Carbon	0.021	0.011	0.918	0.016	0.034
Lackawanna	0.012	0.765	0.008	0.064	0.152
Lehigh	0.126	0.162	0.494	0.098	0.121
Lucerne	0.017	0.268	0.656	0.017	0.043
Monroe	0.273	0.149	0.070	0.071	0.437
Northhampton	0.123	0.262	0.259	0.109	0.247
Pike	0.101	0.039	0.095	0.727	0.037
Schuylkill	0.025	0.036	0.908	0.016	0.016
Wayne	0.022	0.077	0.024	0.807	0.070

Table 11-15 Spatial dominances of parks on population of counties
(Pooler 1992)

day. These data represent 33,461 recreational day-trippers travelling up to a maximum of 115 miles (184 km) one way. Spatial dominancy is calculated from the following close cousin of Equation **(11-77)**: $\Theta_{ij} = z_j d_{ij}^{-b} / \sum_{j=1}^{|J|} z_j d_{ij}^{-b}$ with $b = 1$, and with the total observed inflows to parks as the attractivity term W_j (for lack of a better measure). The dominancy is calculated for the impacts of the five parks on the ten counties (**Table 11-15**). The spatial uncertainty—the entropy of the dominancy—is calculated with Equation **(11-76)** and the results are given in **Table 11-16**.

11.8.4.1 THEORETICAL SPATIAL UNCERTAINTY. There appears to be a pattern in the theoretical spatial-uncertainties. The four counties with spatial uncertainty greater than 1.0 are clustered on the central and southeast side of the region and are contiguous. The remaining counties, with spatial uncertainty less than unity, are on the north and northwest periphery of the region but are also adjoining. The widely-used Pearson product-moment correlation, r, as defined in the "Statistical Tools" appendix (Chan 2000) for interval-scaled variables is computed. The independent variable X is defined as the distance from Northampton County, a centrally-located county that is the peak of the "uncertainty surface." With the dependent variable defined as entropy, r is computed to be -0.61. When the related statistic $(r-0)[(1-r^2)/(n-2)]^{-1/2}$ is computed, we have an approximate t-distribution, for which a significant level of 0.10 is inferred for a two-tail test. This suggests that the further potential travellers are from Northampton County, the

County name	Total outflow	Theoretical spatial uncertainty	Percentage error	Observed spatial uncertainly
Berks	567	1.17	0.32	1.23
Carbon	1,915	0.37	0.41	0.55
Lackawanna	10,235	0.76	0.22	0.94
Lehigh	2,987	1.39	0.19	1.39
Lucerne	9,105	0.90	0.09	1.11
Monroe	1,940	1.37	0.36	1.44
Northhampton	4,249	1.55	0.11	1.52
Pike	143	0.94	0.52	1.15
Schuylkill	2,138	0.43	0.47	0.52
Wayne	182	0.73	0.50	1.05

Table 11-16 Spatial uncertainties from recreational travel data (Pooler 1992)

lower their uncertainty about choosing a park. At peripheral locations it would appear that the choices are clear-cut. For example, in Schuylkill County for which Hickory Run is the dominant destination, 88 percent of travellers go there. On the other hand, residents of Northampton, Lehigh, and Monroe Counties, having higher spatial uncertainty values, are faced with a more difficult spatial choice set, given the more 'even' mix of dominancy at their locations.

In spite of the apparent spatial-pattern discussed above, Moran's spatial-autocorrelation-statistic for this map, under an assumption of randomization, gives −0.10. This results in a significant level of 0.46 (one-tailed), suggesting that the pattern of spatial uncertainty is not positively autocorrelated and is not significantly different from random. The percentage-error level (as shown in **Table 11-16**) is calculated with Equation **(11-78)**. With these data the initial correlation between the spatial uncertainty levels and the observed percentage error is not significant and, contrary to expectations, is negative at −0.53. However there appears to be a size effect at work with the data—the percentage error is a function of how large the county is, and so does spatial uncertainty.

The correlation between the observed error level at the origins and the size of the outflows from the origin counties is an inverse one, as expected, with $r = -0.73$ (significant at 0.02 for a two-tail test). This strong inverse-relationship between size and error is consistent with studies of spatial aggregation effects, which often show that the interaction-model error decreases as the aggregation-level increases. Such results show, as do these data, that the larger the number of travellers, the easier it is to predict their aggregate interaction. At first glace, the correlation between the

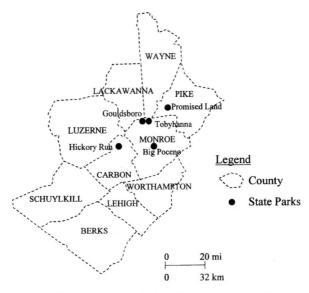

Figure 11-10 Study area consisting of ten counties and five parks

entropy uncertainty and the size of the outflows is close to zero at −0.02. However, inspection of a plot of outflow size against entropy (**Figure 11-11**) reveals that except for the two largest counties (which far exceeds others in size), there is a positive, though curvilinear relationship. When the two extreme points are omitted from the data set, a nonlinear regression of outflow on the second-degree-polynomial of theoretical-spatial-uncertainty gives R^2 of 95.7 percent. This provides evidence of a positive empirical relationship between theoretical-spatial-uncertainty and outflows.

A relationship between theoretical-spatial-uncertainty and outflow may seem unexpected. Why should one expect to find a correlation between outflows from the origins and a measure of the distribution-of-destinations-facing-individuals at the origin? A likely answer, with these particular data, is that there is a *map-pattern effect* at work. Of the eight counties employed in the regression above, those with largest outflows are clustered, for the most part, at the center of the region **Figure 11-10**, and those with the smallest outflows tend to be at the periphery of the map. In addition, the centralized counties on the map have higher spatial uncertainty simply because of their centrality. Therefore, in these data, there is a positive empirical relationship between outflow size and entropy. To eliminate size effects, both the error level and the spatial uncertainty are divided by the outflow size to make them per-capita quantities at the origin. The resulting correlation between per-capita entropy and per-capita error increases to a suspiciously high level of 0.98. The problem is that dividing each variable by a common denominator creates a correlation that is partly spurious. The amount of spurious correlation contained in a correlation-coefficient r can be estimated by

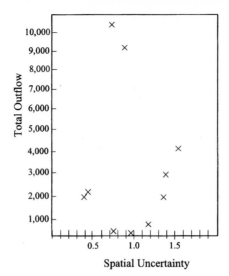

Figure 11-11 Relationship between outflows and
theoretical spatial uncertainty (Pooler 1992)

$$\frac{C_c^2}{\left[(C_a^2+C_c^2)(C_b^2+C_c^2)\right]^{1/2}}. \tag{11-79}$$

Here C is the coefficient-of-variation of variables a, b and c, defined as the standard deviation divided by the mean for each variable.[22] a is the error level, b is the spatial uncertainty, and c is the outflow size. Equation **(11-79)** applies to variables that are ratios or indices. With Equation **(11-79)**, the amount of spurious correlation in the observed covariance between per-capita error and per-capita spatial uncertainty is estimated to be 0.31. The portion of the correlation that is non-spurious is obtained by subtraction, such that $r = 0.98-0.31 = 0.67$ (significant at 0.05 for a two-tail test). This suggests that there is a positive correlation in these data between per-capita error-level and per-capita theoretical-spatial-uncertainty.

11.8.4.2   OBSERVED SPATIAL UNCERTAINTY.   Besides the *theoretical* uncertainty, it is natural to suggest an *observed* spatial-uncertainty, defined as the entropy of the actual outflows as shown in **Figure 11-11**. This index is also to be compared with the error levels. As with the theoretical case above, it is necessary first to consider size effects. For the observed data the correlation between outflow size and error level is significant and inverse (−0.73). The correlation between size and observed uncertainty is close to zero at −0.01. Again, a plot of the size against

---

[22]For a discussion of the coefficient-of-variation, see the "Statistics" appendix to Chan (2000).

observed uncertainty shows a definite curvilinear trend in all except the two largest points. When these points are omitted, a nonlinear regression of outflow on a second-degree polynomial of the observed spatial-uncertainty gives an R^2 of 88.9 percent. This suggests a positive relationship between outflow size and observed entropy. Given these size effects, the division of each variable by total outflow is warranted. The resulting correlation between per-capita observed uncertainty and per-capita error level, after elimination of the spurious portion, is 0.60 (significant at 0.05 for a two-tail test).

In many real-world applications beyond recreational travel (such as retail), a very large flow-matrix (thousands of columns and rows) corresponding to the traffic between providers and customers is involved. While gravity models used to describe these flows have been around for a long time, parameter estimation under these circumstances poses considerable computational problems (Bailey and Munford 1994). This leads to a search for efficient maximum-likelihood-estimation of parameters in such a constrained gravity model. These algorithms exploit certain theoretical properties that arise from assumption of Poisson-distributed flows. Many applications require the formulation of a model that would predict both competitor- and client-account flows at any shopping center, including those where the client institution is not represented. But detailed flow data were only available for the client institution and therefore obviously for only a subset of centers. Hence it is necessary to devise a way of extrapolating the estimated attractiveness of centers where the client is represented to all such centers. In addition, we wish to develop competitor adjustments to these. This is to be accomplished without any competitor flow-data except market shares at a regional level.

To address this, Bailey and Munford recommended the use of synthesized competitor-attractiveness-coefficients in a partially-doubly-constrained model. In the case of singly-constrained gravity model, we have discussed previously in the "Generation, Competition and Distribution" chapter that new centers can be handled easily. This is by way of the "independence of irrelevant alternatives" property. In fact, we have exploited such properties in full during the New York state-parks-study. There, we have extended this result to an unconstrained model, in which the activities are generated at the demand origin while the destination-facility end is constrained. This extension is sometimes referred to as the "independence of irrelevant actors" property (Sen and Smith 1995). Being able to analyze the patronage at new destination-parks is desirable. At the same time, there are undesirable implications of this property, including insensitivity of existing park-patronage to the introduction of new parks, as discussed in the aforementioned chapter.

11.8.5 Discussion. The empirical analysis above provides mixed results as illustrated in **Figure 11-11**. While the expected relationship between theoretical-spatial-uncertainty and the spatial-interaction-model error was found, the covariance may be attributable largely to a map-pattern effect. The result must therefore be accepted with caution. Testing the hypothesis put forward above on competitive location with a small data-set leaves inconclusive results . These pertain to the connection between the accuracy of spatial-interaction-model predictions for an origin, and the theoretical level-of-spatial-uncertainty of potential travellers at that origin. In the calculations of the observed spatial-uncertainty there is some correspondence between the entropy of the observed outflows and the origin-specific error. This provides indirect support for the idea that there is a relation between spatial uncertainty and the interaction-model error. Those origins having the more uniform observed outflow are those that also have higher error-levels on predicted flows. Conversely, those origins having the more diverse observed outflows tend to have lower error-levels. In spite of the negative results obtained with Moran's spatial-autocorrelation analysis, it is encouraging to see that the spatial uncertainties do appear to exhibit non-random spatial-patterns with respect to distance. The correlation between the uncertainty levels and distance is significant. This is an interesting result. If the general hypothesis holds true, then an isopleth map of the spatial uncertainties is also a map of the expected performance-level of a spatial-interaction model.

The data-set, being small, is inadequate with respect to the assumptions of correlation analysis. Rank correlations would have been more appropriate, but the method for eliminating spurious components is uncertain under rank correlation. Even with the adjustment method using Pearson correlation, the correlation between per-capita error and the unadjusted entropy at the origins is insignificant at -0.223. This again raises doubts about the strength of the empirical results. Notice that only one form of spatial-interaction model was employed, although Tobler (1983) reports that the model works slightly better than some well-known constrained entropy-maximizing models. Some evidence of a correlation between origin size and theoretical spatial-uncertainty was found, but was attributable to a map-pattern effect. In the case of observed spatial-uncertainty, it is expected that it will normally vary directly with size. An explanation is that smaller origins do not interact with more distant destinations, but large ones do. The observed entropy, if left unstandardized, is normally larger for larger origins simply because they send their larger numbers of travelers to more destinations. This suggests the definition of theoretical-measures-of-uncertainty where, beyond some distance (or below some specified θ_{ij} level), destinations are eliminated from the choice set. This allows the number of destinations and the *pattern of the map* to have a greater effect on the spatial uncertainty at a point. Such is a more realistic perspective than the one implicit here where observed movements, and dominance effects occur between every origin-destination pair. The amount of theoretical-spatial-uncertainty is sensitive to the effect of distance. The uncertainty reflects not only the manner in which distance is measured, but also the deterrence function used, as well as the method of centroid selection. The correlation between the interaction-model-error and spatial-uncertainty could be improved significantly with the use of alternative distance-response functions and/or calibration methods intended to maximize the fit by adjustment of the b parameter in the calculation of spatial dominancy.

The clearest finding to emerge from the analysis is that very strong relationship exists between the interaction-model-error level and origin size. In these data, the error level at each origin is so dominated by the size of the origin as to make it difficult to separate out and identify other sources of the origin-specific model-error. Any attempt to analyze the location-specific-interaction-model error must confront this important problem. It would be useful to see additional empirical evidence, especially with a larger data-set, of the relationship between theoretical-spatial-uncertainty and the difficulty of predicting spatial interaction. This is more important when the hypothesis appears valid from an intuitive perspective. It seems reasonable to assume that it is more difficult to make predictions about peoples' behavior when they are confronted with greater uncertainty or when there is less to choose among a set of alternatives. The idea tested here is implicit in the well-known concept of the zone-of-spatial-uncertainty. The greater the level-of-spatial-uncertainty at allocation, the less well a spatial-interaction model should perform at that location. It represents an empirical approach to a widely-publicized theory of destination choice. Coincidently, the result is also consistent with spatial-autocorrelation-statistics when an equal set of destination weights z_j neutralize the travel pattern.

As a parallel school of thought, Cromley et al. (1993) presented a prescriptive, rather than the above descriptive, framework for analyzing service markets. Traditional approaches use constrained spatial-interaction models to calibrate friction-of-distance coefficients based on actual transaction-flows. Guided by the spatial-st.ructure of the market, they simulated results for a range of friction-of-distance coefficients.[23] In a traditional facility-location or siting situation, any calibrated friction-of-distance b is biased because it is implicitly based on the existing spatial arrangement of opportunities. In actuality, when a new facility is added, the siting arrangement is altered and actual transaction-flows will change. The b values Cromley et al. used in their approach were selected through simulation to provide a representative value. The value is estimated within a range for which the top-ranked markets were invariant. Outside this range, there is considerable divergence in market rankings between the extreme value of $b = \infty$ where opportunity is limited to the local market, and $b = 0$ where distance is no barrier because all markets are homogeneous.[24] The simulation of b reduces the need for disaggregate data regarding actual transaction flows, thereby overcoming an information constraint that may limit strategic analysis by the new entrant. This is corroborated by some of the difficulties encountered in the descriptive approach discussed previously. A similar, prescriptive approach was taken by Allen et al. (1993). They used a theoretical construct in computing the relative accessibilities of 60 largest U.S. metropolitan areas, and validated them against empirical measurements.

---

[23]For an example of this prescriptive procedure, refer to the "Cost function of distance" Figure in the "Facility Location" chapter, and subsequent discussions on the implications of a linear, concave or convex function respectively.

[24]The interpretation of the b value is explained in the first part of the "Generation, Competition and Distribution" chapter.

Sen and Smith (1995) presented a comprehensive mathematical and statistical framework within which the behavioral properties of gravity models can be characterized explicitly. The text provides a rigorous theoretical treatment of spatial interaction. Its basic components are discerned. Based on these components, the gravity model can be applied for practical estimation and forecasting. As a result, these applications can then be anchored solidly against a set of explicit assumptions and rigorous methods for parameter estimation. A particular strength of the book is the way the past research in probabilistic choice theory can be presented in an axiomatic exposition.

11.9 Spatial tessellation: order of nature?

Perhaps the dichotomy between empiricism and model development painted in the earlier part of the book and the beginning of this chapter is only a generalization. The schism between empirical and theoretical work may not be as large as implied. Let us introduce a paradigm that may begin to bridge the gap: *spatial tessellation*. We contend that the background that gives rise to the theoretical and empirical camps in spatial analysis over time has to do with the way we represent spatial data. Historically, data such as population and employment are collected on each subarea called zone or region, while spatial-separation between these units is measured in terms of travel-time, cost or accessibility-function. To some quarters, the description of spatial pattern in terms of the number of individuals in a given area and in terms of the distribution-function for distances between individuals would be misplaced. It misses entirely the fundamental spatial nature of the pattern. They argue that intuitively one expects the need to characterize a pattern by describing the positions of each individual relative to the *position* of its immediate neighbors. We alluded to this point previously in the "Spatial Econometric Models" chapter. Again earlier this chapter we talked about the "natural origin" in the general spatial-statistic G_i.

11.9.1 Voronoi diagrams. If we consider identifying 'neighbors' by joining one to another with a line, then we have commenced subdividing the entire region into subregions or *'tiles'*, with each tile being linked to one or more points. The mosaic of tiles that results is called a *tessellation* (Upton and Fingleton 1985, Okabe et al. 1992). **Figure 11-12** illustrates the Dirichlet tessellation, which has its place in history. Each individual is contained within a polygon, which has the property that every point within that polygon is nearer to that individual than it is to the points in another polygon. It is natural to regard each polygon as the territory or service area for its own individuals. In the figure, point A has five neighbors (the points B, C, D and F). Similar to the idea of first- and second-order neighbors[25], one can define *direct* vs. *indirect* neighbors. Thus in **Figure 11-12**, points B, C, D and

---

[25]The terms first- and second-order neighbors are defined in the "Geographic Information Systems" chapter in Chan (2000) and "Spatial Econometric Models" chapter in the current volume.

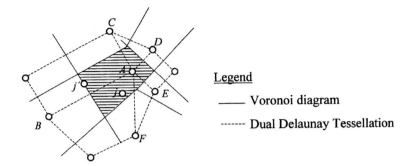

Figure 11-12 The Dirichlet tesselation

E are direct neighbors while point *F* is an indirect neighbor. Notice the tile boundaries are all segments of the perpendicular bisectors of the lines joining A to its neighbors. A neighbor is labelled as indirect if the line joining it to point A does not cross the tile boundary at right angles.

There are three obvious properties of Dirichlet tiles that might serve as the basis for a test of significance. These are the number of sides (n_s), the length of the perimeters ($\tilde{L}$), and the area (A). They have zero, one, and two dimensions respectively. Unfortunately, the underlying mathematics for these tiles is horrendous and until recently little is known of their properties. For a poisson forest (or a randomly-populated forest), $E(n_s) = 6$, $E(A) = 1/\rho''$, and $E(L) = 4/\sqrt{\rho''}$, where ρ'' is the intensity of the plants. It can be shown further that $E(A^2) = 1.28/\rho''^2$. These represent some theoretical results that have been obtained about the Dirichlet tessellation. The result $E(n_s) = 6$ can easily be shown to be true for *any* point process, as follows. Connect together all neighboring points to form a tessellation of triangles (the *Delaunay triangles*). The average interior angle of a triangle is obviously $\pi/3$ since a triangle always contains three angles that sum to π. Suppose that each point has an average of n^* neighbors. There will therefore be an average of n^* triangles meeting at each point. Now these triangles will use up the entire 2π angle available at that point and consequently the average angle per triangle will be $2\pi/n^*$. Since this must equal $\pi/3$, we have the result $2\pi/n^* = \pi/3$, which give us $n^* = 6$. One can think of this spatial pattern as a "natural order," since it applies to all point processes.

The result of the above is a partitioning of the space into a set of regions. We will call the generic diagram such as **Figure 11-12** *Voronoi diagram*. The process that allocates individual locations in Euclidean or higher-dimensional space or a bounded region to a set of *n* points $J = \{1,2,\cdots,n\}$ at positions $\mathbf{x}_1$, $\mathbf{x}_2$, $\cdots$, $\mathbf{x}_n$, respectively is called *Voronoi Assignment*. If the following assumptions are made, the resulting set of regions is equivalent to the *ordinary* Voronoi diagram:

a. Each *generator point* $i = 1, 2, \cdots, n$ is located simultaneously before any assignment procedure begins.

b.   Each point i remains fixed at $\mathbf{x}_i$ throughout the assignment process.

c.   Each point i is of equal importance.

d.   Each location that is closer in terms of Euclidean distance to i than to any other member of J is assigned to i; locations that are equi-distant from two or more members of J are assigned to the boundaries of the regions of those points. The boundaries of the polygons are defined by a set of *vertices*.

Notice here a Euclidean space and Euclidean distances are assumed. Collectively, assumptions 1 through 4 define the *Voronoi-Assignment Model*. When the Euclidean assumption is relaxed, we have the generalized Voronoi diagrams.

Social scientists studying various types of territorial structures associated with humans have made extensive use of Voronoi-Assignment Models. Unlike many of their physical counterparts, human spatial-patterns rarely show a high degree of regularity although one celebrated instance occurs in geography called *Central Place Theory*.[26] This is concerned with explaining the size and spacing of nucleated settlements (central places) which provide goods and services to rural customers distributed over a geographical region. One aspect of this theory relates to the definition of the trade areas (areas of influence) associated with the central places for a specified good or service. In classical central-place theory, the locations of the central places are assumed to form a triangular lattice as shown in **Figure 11-13**: the Delaunay triangles. Rural population is assumed to be uniformly distributed over the polygon region.

In addition, it is assumed that

a.   Every central place provides the good or service under consideration;

b.   Consumers travel to a central place to find the good or service;

c.   The market price of the good or service (excluding transportation cost) is the same at all central places;

d.   The cost of acquiring the good or service is equal to the market price plus the transportation cost to the central place;

e.   Cost of transportation is calculated by unit cost times distance;

f.   Customers minimize the cost of purchasing the good or service.

Taken collectively, assumptions (a) through (f) imply that consumers will patronize the nearest central place. Thus the trade areas are equivalent to the Voronoi polygons of the central places. In classical central-place theory they will be regular hexagons with the 'dual' Delaunay triangles shown in **Figure 11-13**. These triangles represent the most direct route from one central place to the next.

As alluded to previously, such considerations are also relevant in modelling the service areas of facilities providing public services. Suppose that we consider a set of facilities at fixed locations in a region that provides a service to a set of individual users distributed over the same region. The service may either be dispensed to the users at their locations or offered at the facility to which the user

---

[26]We introduced Central Place Theory initially in the "Geographic Information Systems" chapter (Chan 2000) as a 'natural' way to define subareas.

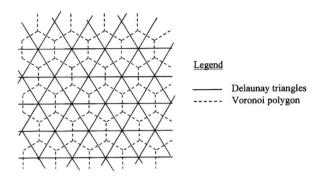

Figure 11-13 Voronoi polygons and and Delaunay triangles in Central Place Theory

must travel. Obvious examples of such facilities are those providing municipal services such as heath and education. We assume that the facilities are identical, and that there are no constraints on the number of people who can use a facility (i.e., the facility is un-capacitated) and that, to minimize cost of finding the service, users are served by the nearest facility. Under these circumstances, the service areas of the facilities will be equivalent to their Voronoi polygons. Note that the last of these assumptions means that the aggregate distance travelled between user and facility is minimized. Thus the Voronoi diagram is often used as a prescriptive structure providing a *min-sum* system optimum (as distinguished from a user optimum that, say, minimizes the maximum distance to any customer).[27]

Given n generator points in J, it is possible to generate the Voronoi regions associated with the points. Okabe et al. (1992) present several algorithms for accomplishing this. They include the *Incremental, Divide-and-Conquer*, and *Plane Sweep* Methods. Based on computational geometry, these are computer codes for producing consistent and numerically-stable multidimensional Voronoi-diagrams. The computational complexity of the quarternary version of the Incremental method is $O(n^2)$ in the worst case and $O(n)$ on the average. The complexity of the Divide-and-Conquer method is $O(n \log n)$ in the worst and on the average. Invariably, they locate a nearest neighbor node from the generator point in J and decide which Voronoi-region boundaries are involved in computing a vertex location. For details of these algorithms, the readers are referred to Okabe et al. (1992) or subsequent refinements (Sugihara 1997). The latest algorithms can also be downloaded from the Internet (Schewchuk 1996). Here we outline the basic ideas behind these algorithms in seven steps.

a. Decide on the locations of the generator points in J.
b. Identify the neighboring points for each generator.

---

[27]For a discussion on the difference between these two locational criteria, see the "Facility Location" chapter. Introductory concepts are discussed in the "Economic Methods" chapter of Chan (2000) also.

c. Write the equation for the line segment joining a generator a and each of its neighbors b (i.e., edge (a,b) of the Delaunay triangle).

d. Find the midpoints of the line segments found in step 3 by the formula using $[\mathbf{x}_b-(\mathbf{x}_a-\mathbf{x}_b)/2]$.

e. Using the negative-inverse of the slope of the line-segment found in step 3 and the respective midpoint found in step 4, determine the equations for the perpendicular bisectors.

f. Identify the intersections of the bisector lines that define the Voronoi vertices.
    ■

Example

A light-rail system is to be constructed in Dayton, Ohio. These potential station-location coordinates are identified

Station	1	2	3	4	5	6
Location	(2,20)	(12,22)	(18,13)	(11,8)	(7,13)	(7,5)

Using the above method, the equations for the line segments in step 3, the midpoints in step 4, and the perpendicular bisectors in step 5 are determined according to the following table.

Delaunay edge	Equation	Midpoint	Bisector
1 & 2	$x_2=x_1/5+98/5$	(7,21)	$x_2=-5x_1+56$
1 & 5	$x_2=-7x_1/5+114/5$	(4.5,16.5)	$x_2=5x_1/7+93/7$
2 & 3	$x_2=-3x_1/2+40$	(15,17.5)	$x_2=2x_1/3+15/2$
2 & 5	$x_2=9x_1/5+2/5$	(9.5,17.5)	$x_2=-5x_1/9+205/9$
3 & 4	$x_2=5x_1/7+1/7$	(14.5,10.5)	$x_2=-7x_1/5+154/5$
3 & 5	$x_2=13$	(12.5,13)	$x_1=12.5$
4 & 5	$x_2=-5x_1/4+87/4$	(9,10.5)	$x_2=4x_1/5+33/10$
4 & 6	$x_2=3x_1/4-1/4$	(9,6.5)	$x_2=-4x_1/3+37/2$
5 & 6	$x_1=7$	(7,9)	$x_2=9$

Correspondingly, the Voronoi vertices for each station's region were calculated and are summarized in the following table (Smetek and Chan 1997, Beeker 1998):

Station/ region	Voronoi vertices
1	(0,25),(6.2,25),(7.48,18.625),(0,13.29)
2	(6.2,25),(25,25),(25,24.17),(12.5,15.83),(7.48,18.625)
3	(25,24.17),(25,0),(0,22),(12.5,13.3),(12.5,15.83)
4	(7.13,9),(12.5,13.3),(22,0),(0.1388,0)
5	(0,13.29),(7.48,18.625),(12.5,15.83),(12.5,13.3),(7.13,9),(0,9)
6	(0,9),(7.13,9),(0,13.88)(0,0)

Once the Voronoi diagram is in place, the Delaunay triangles are also known. The rail-like network can possibly be configured by constricting the minimum spanning tree[28] out of the Delaunay triangles. □

11.9.2 *Central-place application.*

Cox and Agnew (1976) use the Voronoi diagram of the existing county towns in Ireland to create a theoretical partition, against which the existing set of counties is compared. Let $I = \{i_1, i_2, \ldots, i_n\}$ be the set of 32 counties ($n = 32$), $R = \{R_1, R_2, \ldots, R_n\}$ be the set of Voronoi polygons and R_T be the total region covered by I or R (in this case Ireland). Furthermore, let $J = \{j_1, j_2, \ldots, j_{n'}\}$ be the set of n' locations on a dense triangular-lattice (see **Figure 11-13**) imposed on R_T. Define $f_{ik'}$ as the number of members of J falling in $R_i \cap i_k$ so that $\sum_{i=1}^{n} \sum_{k=1}^{n} f_{ik} = n'$. Convert $f_{ik'}$ into an empirical probability, $p_{ik} = f_{ik}/n'$, and define $p_i.$ and $p_{\cdot k}$ as the probability of a member of J being in the ith region of R and the kth location of I respectively. If we have no information other than $p_i.$ and $p_{\cdot k}$ and we consider that $p_i.$ and $p_{\cdot k}$ are independent (as would occur if there was no association between R and I), then our estimate $\hat{p}_{ik}$ of p_{ik} would be the joint probability $\hat{p}_{ik} = p_i. \cdot p_{\cdot k}$.

If stochastic independence actually occurs, $p_{ik} = \hat{p}_{ik}$ and $\log_2 (p_{ik}/\hat{p}_{ik}) = 0$, implying that knowledge of p_{ik} provides us with no information additional to that provided by $p_i.$ and $p_{\cdot k}$ In general the total expected-mutual-information for the two patterns is given by

$$I_{RI} = \sum_{i=1}^{n} \sum_{k=1}^{n} p_{ik} \log_2(p_{ik}/\hat{p}_{ik}). \tag{11-80}$$

If there is a perfect correspondence between R and I, $p_{ik} > 0$ in Equation **(11-80)** only if $i = k$ and for each $p_{ii} > 0$, $p_{ii} = p_i. = p_{\cdot i}$. In this case the expected mutual-information I_{RI}^* for the two patterns is given by

$$I_{RI}^* = \sum_{i=1}^{n} \sum_{k=1}^{n} p_{ik} \log_2(1/p_{ik}) = \sum_{k=1}^{n} p_{\cdot k} \log_2(1/p_{\cdot k}) = \sum_{i=1}^{n} p_i. \log_2(1/p_i.), \tag{11-81}$$

which is equal to the entropy of I (i.e., no prior information).[29] Thus a percentage measure of the spatial correspondence between R and I is provided by $(I_{RI}/I_{RI}^*)100$.

Correspondence between individual members of R and I can be measured by a parallel procedure. For each member i_k of I, the expected mutual information is given by $I_{Ri_k} = \sum_{i=1}^{n} p_{ik} \log_2(p_{ik}/\hat{p}_{ik})$. The expected mutual-information under the condition of perfect correspondence of i_k to one member of R is given by $I_{Ri_k}^* = p_{\cdot k} \log_2(1/p_{\cdot k})$. The result is that a measure of correspondence of the kth member of

---

[28]The algorithm for constructing a minimum spanning tree was discussed in the "Measuring Spatial Separation" chapter under the Subsection "Double spanning tree heuristic".

[29]Entropy is defined in the "Descriptive Tools" and "Prescriptive Tools" chapters (Chan 2000) and also discussed earlier in this chapter. In this context it measures how probable a pattern is consistent with what is known about the pattern.

Theoretical Partition   Actual Counties

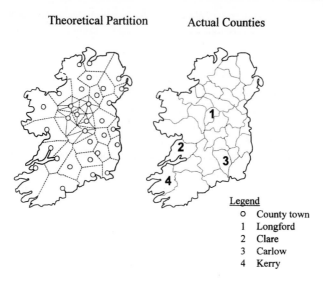

Legend

  o   County town
  1   Longford
  2   Clare
  3   Carlow
  4   Kerry

Figure 11-14 Defining Irish counties using Voronoi diagrams (Okabe et al. 1992)

I is given by $(I_{Ri_k}/I_{Ri_k}^*)100$. Applying these measures to the study of Irish counties, Cox and Agnew found that there is an overall spatial correspondence of 72 percent between the two structures. However, the measures of agreement between individual counties are quite variable, being highest for large coastal counties such as Clare (92%) and Kerry (90%) and lowest for small inland counties such as Carlow (40%) and Longford (44%). In spite of these variations, the overall correspondence and visual inspection of **Figure 11-14** suggest that there exist some correlation between the Voronoi partitions and the actual county jurisdictions. Could this example again suggest that spatial tessellations reflect the "laws of nature?"

11.9.3 Nearest neighbor's operations. A Geographic Information System (GIS) usually has a collection of nearest neighbor's operations, such as searching for the nearest facility from a given location. These tools serve a useful function for spatial analysis (Okabe et al. 1994). As an example, let us consider a region R_T in which there are n hospitals located at $j_1, j_2, \dots, j_n$. GIS users often make these queries:

a. Locate the regions in which the nearest hospital is within 1/4 mile (400 m).
b. Find the nearest hospital from a given residence.
c. Show that the region in which the nearest hospital is j_i.
d. Locate the nearest hospital from a given hospital.
e. Find the home from which the distance to the nearest hospital is the longest in a region R_T.
f. Calculate the average distance to the nearest hospital in the region.

Such queries are quite common in geographical searches and can be restated in more general and mathematical terms:

Q1. Locate the region i that the distance to the nearest point in J is less than or equal to a given distance.

Q2. Find the nearest point in J from a given location.

Q3. Show that the region in which the nearest point in J is j_i.

Q4. Find the nearest point in $J_{-i} = \{j_1, \dots, j_{i-1}, j_{i+1}, \dots, j_n\}$ from point j_i, where J_{-i} is the set of all points in J except j_i.

Q5. Find the location in a region R_T from which the distance to the nearest point in J is the longest (i.e., the maximum problem), or equivalently, find the largest circle whose center is in R_T and does not contain any point in J in its interior.

Q6. Calculate the average distance to the nearest point in J over a finite region R_T.

While the first and third queries can be handled readily within GIS, the rest are somewhat problematic, in that a *direct* determination is impossible. Voronoi diagrams can respond to these queries with dispatch. Here, we adopt the vector-based data-structure because topological relations in a Voronoi diagram can be more explicitly treated this way than in the raster-based format. Let us define a Voronoi diagram with the example discussed above, which is depicted in **Figure 11-12**. Given a set of hospitals at $j_1, j_2, \dots, j_n$, we assign every location in the region to the closest hospital (for example, demand location j is assigned to A.) If a location happens to be equally close to two or more hospitals, we assign the location to each of those hospitals. For example, location j' is assigned to both A and B. As a result, the set of locations assigned to each hospital j_i forms its own region. For example, the locations assigned to A form the shaded region, in which the nearest hospital is the one at A. The resulting regions are collectively exhaustive and mutually exclusive except for the boundaries. Most GIS packages can generate Voronoi diagrams, although efficiency and robustness vary considerably. The third query can also be answered directly once such *ordinary* Voronoi diagrams have been constructed. For example, in the same figure, the region in which the nearest hospital is the one at A is given by the shaded region.

The answer to query 2 (find the nearest facility in J for a given demand location) is obtained if we know which Voronoi polygon contains the location. Generally, for a given tessellation consisting of polygons, the problem of finding the polygon that contains a probe point is known as the *point location problem*. In **Figure 11-12**, since location j is included in the shaded region, the nearest hospital from j is the one at A.

The nearest point in J from point j_i exists in the generator points of the Voronoi polygons next to the polygon R_A. For example, the nearest point in J from A exists in the generator points B, C, D, E, and F. In this case, it is between D and E. Since the topological structure of a tessellation is managed in an ordinary GIS, we can readily obtain the adjacent Voronoi polygons. Let us say that the nearest point in J from j_i is given by the point j_i^*. The location j_i^* is the answer to query 4.

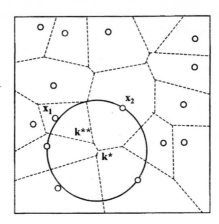

Figure 11-15 The largest empty circle and searches for local maxima

Since a Voronoi region R_{j_i} is a convex polygon, the farthest location in R_{j_i} from j_i exists in the vertices of the polygon $R(j_i)$. In **Figure 11-15**, the farthest location from j_1 is k^{**}. Thus the farthest location is obtained by searching for the longest distance among all such distances. In **Figure 11-15**, the farthest location is k^*. If we draw the circle centered at k^*, this circle is the largest circle whose center is in R_T and does not contain any point in J in its interior. We call this the *largest empty circle*. Such a circle will pass through exactly three generations. This circle helps to answer query 5, solving the maximum location-problem (or the anti-center problem).

We can obtain exactly the average value or higher moments of the distance to the nearest point in J (query 6) from the coordinates of the points $j_1, j_2, \ldots, j_n$ and those of the Voronoi vertices. First, we triangulate each Voronoi polygon by line segments combining all its generator point and vertices as shown in the center counties of **Figure 11-14**. For each triangle, we calculate the average value or moments of the distance from a vertex of a triangle to all points in the triangle. From those values, we obtain the average value or higher moments of the distance to the nearest point over the whole region, and this answers query 6. This also solves the min-sum location-problem (or the median problem).

In this illustration with the ordinary Voronoi diagram, we have shown that it is a powerful tool for spatial analysis. Since efficient computational methods for constructing these diagrams exist, we can use the above results for our purpose. Thus the proposed tool-box of nearest neighbor's operations is a feasible tool in GIS, made possible by Voronoi diagrams.

11.9.4 Spatial competition models. In spatial economics, many studies have been carried out to examine market-area stability (Okabe et al. 1992).[30] Firms are located to compete against each other in order to maximize their profits. Suppose there are n firms at $\mathbf{x}_1, \dots, \mathbf{x}_n$ in a region R_T, and these firms are selling the same products with the same mill price—i.e., price not including a delivery cost. We assume that the delivery cost from a firm at $\mathbf{x}_i$ to a consumer at $\mathbf{x}$ is proportional to the Euclidean distance $\|x_i-x\|$, and that consumers buy the products from the firm that quotes the lowest delivered price (the mill price plus the delivery cost). Under these assumptions the pattern of n market areas is represented by the ordinary Voronoi diagram $R(J) = \{R(\mathbf{x}_1), \dots, R(\mathbf{x}_n)\}$, and the market area of firm i is represented by the Voronoi polygon $R(\mathbf{x}_i)$. We further assume that the demand for the products is uniformly distributed over region R_T; the marginal cost of the products is the same for all firms; and relocation cost is negligibly small. Then the profit of firm i is proportional to the area $R(\mathbf{x}_i)$. The firms compete in terms of their locations to maximize their profits. As a result we observe a spatial-competition process of n firms over time. We specify this process by the following behavioral assumptions:

Relocation timing: Each firm considers relocation once in every period, and time periods are equally spaced apart. ∎

Zero conjectural variation: In considering possible relocation, every firm conjectures that all other firms will not change their locations (i.e., the Cournot assumption).[31] ∎

Let $X_{-i} = \{\mathbf{x}_1, \dots, \mathbf{x}_{i-1}, \mathbf{x}_{i+1}, \dots, \mathbf{x}_n\}$ be the set of locations of all firms except firm i, and $\mathbf{x}_i^*$ be the location of firm i that yields the maximum profit if the locations of the other firms are fixed at preset locations; i.e.,

$$|R(\mathbf{x}_i^*)| = \max_{\mathbf{x} \in R_T}\{|R(\mathbf{x})|\,|\,[X_{-i} \ is \ fixed, \ \mathbf{x} \neq \mathbf{x}_j, \ j \in (I-i)]\}. \tag{11-82}$$

Here $|R(\mathbf{x})|$ indicates the area of $R(\mathbf{x})$ or a proxy for profit level. As is suggested by the condition in Equation **(11-82)**, firm i is not allowed to locate at the same location as any other firms. To all these, we add one more behavioral assumption below.

---

[30]For an introduction to this subject, see the "Generation, Competition and Distribution" and "Chaos, Catastrophe, Bifurcation" chapters.

[31]The economist Cournot constructed a model of two competing firms where an initiating firm is assumed to have its price fixed when the reacting firm formulates its response. An equilibrium the two firms reach is called the Cournot-Nash equilibrium. The idea was introduced in the "Economics" chapter (Chan 2000). The reader may wish to consult the "Multi-criteria Decision-making" (Chan 2000) and the "Generation, Competition and Distribution" chapters for illustrative examples.

Relocation decision: Firm i moves to location $\mathbf{x}^*_i$ (given by Equation **(11-82)**) if location $\mathbf{x}^*_i$ yields a greater profit than the present profit, i.e., $|R(\mathbf{x}^*_i)| > |R(\mathbf{x}_i)|$. Otherwise, firm i remains at the present location $\mathbf{x}_i$. ∎

The readers will recognize by now that the above describes the *Hotelling process*, as introduced in the "Chaos, Catastrophe, Bifurcation and Disaggregation"

$$|R(\mathbf{x}^*_i)| \ge \max_{\mathbf{x}_i \in R_T} \left\{ |R(\mathbf{x}_i)| \,|\, [X_{-i} = X^*_{-i}, \ \mathbf{x}_i \ne \mathbf{x}^*_j, \ j \in (I - i)] \right\} \qquad (11\text{-}83)$$

chapter. Stated abstractly, the Hotelling process is the process in which each *generator* (firm) independently moves to maximize the area of its Voronoi polygon. Through the Hotelling process, the Voronoi diagram may change over time, or it may eventually reach an equilibrium state in which all firms have no incentive to relocate (a Nash equilibrium). To be precise, the configuration of the n firms is in *global equilibrium* if and only if no firm can find a more profitable location than the present location. In other words, the Voronoi diagram $R^* = \{R(\mathbf{x}^*_1), \dots, R(\mathbf{x}^*_n)\}$ is in global equilibrium if and only if

for any $i \in I$, where $X^*_{-i} = \{\mathbf{x}^*_1, \dots, \mathbf{x}^*_{i-1}, \mathbf{x}^*_{i+1}, \dots, \mathbf{x}^*_n\}$. Equilibrium for the Hotelling process is relatively straightforward for a bounded one-dimensional space and has been reviewed in the "Chaos, Catastrophe, Bifurcation and Disaggregation" chapter as mentioned. Generalization to a two dimensional case, however, is not as straightforward. To discuss this case, we introduce (without proof) several properties of the Voronoi diagram—some of which correspond to the one-dimensional case as the reader will recognize.

Property 1. The Voronoi diagram of regular triangular lattice points and that of square lattice points are in global equilibrium. The Voronoi diagram of regular hexagonal points is not in local equilibrium; consequently, it is not in global equilibrium. ∎

Property 2. When the region R_T is a disk, two firms paired at the center of the disk are in global equilibrium; there is no local equilibrium for three firms. ∎

The analytical difficulty encountered in a bounded two-dimension space with $n \ge 4$ firms results from the fact that the optimization problem of Equation **(11-82)** is a non-linear, non-convex programming problem. As shown in the "Prescriptive Tools" chapter (Chan 2000), it is almost impossible to obtain the global maximum by an analytical method; only a local maximum is obtainable by the numerical method. This intractability prevents analytical examination of the Hotelling process over time, not only in a bounded two-dimensional space but also in an unbounded two-dimensional space. At present it is difficult to say if firms on an unbounded two-dimensional space reach the global equilibrium configurations stated in Property 1 for any initial configuration. Giving up on analytical examination, Eaton and Lipsey (1975) employed a numerical method to examine the Hotelling process for 17 firms in a disk. They conjectured from their numerical results that the global equilibrium configuration would not be achieved. Alternatively, Okabe and Suzuki

(1987) relaxed the assumption stated under "relocation decision" and examined the Hotelling process over time with numerical simulations. Their relaxed assumption is stated as follows.

Relaxed assumption about the relocation decision: Define the *largest empty circle* as a circle whose center is inside the convex hull of a set of distinct points J and no point in J is contained within the circle. The center can be any vertex of the bounded Voronoi diagram, including the intersection points between Voronoi edges and the convex-hull boundary of J. Such a circle is illustrated in **Figure 11-15**. Let k^* be the center of the largest empty circle in the Voronoi diagram $R = \{R(\mathbf{x}_1), \dots , R(\mathbf{x}_n)\}$. With this extended set of locations, these more relaxed assumptions can be stated. (i) If $|R(\mathbf{x}_i)| < |R(k^*)|$, firm i (such as $i = 1$ in **Figure 11-15**) search for a local-maximum location $\mathbf{x}_i^*$ from k^* (for example k^* in **Figure 11-15**). If $|R(\mathbf{x}_i^*)| > |R(k^*)|$, firm i moves to $\mathbf{x}_i^*$; otherwise, firm i moves to k^*. (ii) If $|R(\mathbf{x}_i)| \geq |R(k^*)|$, firm i searches a local-maximum location x_i^{**} (such as $\mathbf{x}_2$ in **Figure 11-15**) from the present location k^*. If $|R(x_i^{**})| > |R(\mathbf{x}_i)|$, firm i moves to x_i^{**}; otherwise firm i remains at the present location $\mathbf{x}_i$. ∎

Under the stated assumptions on "relocation timing," "zero conjectural variation" and the above "relaxed relocation decision," Okabe and Suzuki (1987) carried out the numerical simulation for 256 firms in a square over 100 periods. The steepest-descent method is used,[32] and one period is the time during which every firm makes the location decision once. From these numerical results, they noticed that although the configuration does not achieve the global equilibrium by the *100*th period, the configuration of inner firms is stable after a certain time. From the simulation of $n = 17$ in Eaton and Lipsey and that of $n = 256$ in Okabe and Suzuki, we notice the stability of the Hotelling process depends upon the number of firms. Moreover, the shape of the region R_T crucially affects the stability. These effects are reconfirmed by Aoyagi and Okabe (1992).

11.9.5 Continuous p-median problem. The classic Weber problem was discussed in the "Facility Location" chapter under the "Planar location problem" section. Here we offer a Voronoi-diagram solution-algorithm to a variation of this multi-median problem (Okabe et al. 1992). Consider the problem of minimizing the average travel-cost to the nearest n generators without constraints (i.e., there is no limitation on the size of the region, facility capacity, etc.):

$$\min_{\mathbf{x}_1,\dots,\mathbf{x}_n} \left[F(\mathbf{x}_1, \dots, \mathbf{x}_n) = \sum_{i=1}^{n} \int_{R_i} f(\|\mathbf{x}-\mathbf{x}_i\|^2)\, \varphi(\mathbf{x})\, d\mathbf{x} = \sum_{i=1}^{n} |R_i(\mathbf{x}_i)| \right]. \tag{11-84}$$

Here we have written the demands as a continuous-function $\varphi(\mathbf{x})$ (instead of discrete-functions f_i). The distance function has also been generalized to a function

---

[32]For an introduction to gradient search, of which steepest descent is a special case, see the "Prescriptive Tools" chapter in Chan (2000).

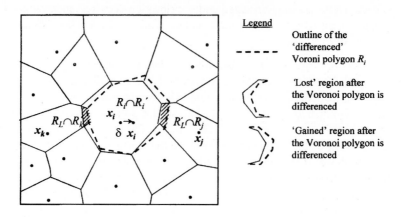

Figure 11-16 Variation in a Voronoi polygon (Okabe et al. 1992)

of the Euclidean separation between the demands and the generator, $f(\|\mathbf{x}-\mathbf{x}_i\|^2)$. The right-most term of the expression suggests solving the objective function by Voronoi diagrams, where $|R_i(\mathbf{x}_i)|$ is the total access-cost to a facility at the generator point. The function F is apparently nonlinear and is definitely non-convex. The minimization problem is non-convex based on the fact the F is a non-convex function. To see this, let us examine these facts. If $(\mathbf{x}_1^*, \dots, \mathbf{x}_p^*, \dots, \mathbf{x}_{\hat{p}}^*, \dots, \mathbf{x}_n^*)$ is a local minimum of F, then $(\mathbf{x}_1^*, \dots, \mathbf{x}_{\hat{p}}^*, \dots, \mathbf{x}_p^*, \dots, \mathbf{x}_n^*)$ is also a local minimum, because the equation $(\mathbf{x}_1^*, \dots, \mathbf{x}_{\hat{p}}^*, \dots, \mathbf{x}_p^*, \dots, \mathbf{x}_n^*) = (\mathbf{x}_1^*, \dots, \mathbf{x}_p^*, \dots, \mathbf{x}_{\hat{p}}^*, \dots, \mathbf{x}_n^*)$ holds. This implies we have multiple local-minima unless $\mathbf{x}_1^* = \dots = \mathbf{x}_n^*$. The solution $\mathbf{x}_1^* = \dots = \mathbf{x}_n^*$ means that all facilities are co-located at the same place. Obviously, having one single 'subregion' does not give a minimum. It follows that the function F has multiple local-minima, implying that the function F is non-convex.

Because of this non-convx property, we have to use a numerical, rather than analytical, solution procedure—the steepest-descent method in this case.[33] Let R_T be the Voronoi diagram generated by $\{\mathbf{x}_1, \dots, \mathbf{x}_i, \dots, \mathbf{x}_n\}$ and R_T' be that generated by $\{\mathbf{x}_1, \dots, \mathbf{x}_i+\delta\mathbf{x}_i \dots, \mathbf{x}_n\}$; R_i be the Voronoi polygon of the ith generator in R_T and R_i' be that in R_T'; and $J(i)$ be the set of Voronoi polygons adjacent to R_i. Owing to the slight move, $\delta\mathbf{x}_i$, the Voronoi diagram slightly changes as shown in **Figure 11-16**. As shown by the broken lines in the same figure, however, this change occurs only in R_i and at adjacent Voronoi polygons, i.e., R_j, $j\in J(i)$. Now consider the Voronoi edges between adjacent Voronoi-polygons R_j and R_k before, $j,k\in J(i)$. Also consider the Voronoi edges between adjacent Voronoi-polygons R_j' and R_k' after, $j,k\in J(i)$. They are on the same line. In sum, the following relationships are observed.

---

[33]The analogous "Steepest ascent" method was introduced in the "Prescriptive Tools" chapter (Chan 2000) under the "Solution of a nonlinear program" section.

a. The original Voronoi polygon is made up of the unchanged region and the 'lost' region: $R_i = (R_i \cap R_i') \cup (R_i \backslash R_i \cap R_i')$.

b. The 'differenced' Voronoi polygon is made up of the unchanged portion and the 'gained' portion: $R_i' = (R_i \cap R_i') \cup (R_i' \backslash R_i \cap R_i')$.

c. The adjacent Voronoi polygon j is made up originally of the unchanged portion and the 'lost' portion: $R_j = (R_j \cap R_j') \cup (R_i' \cap R_j)$.

d. After differencing, the adjacent Voronoi polygon k is made up of the un- changed portion and the 'gained' portion: $R_k' = (R_k \cap R_k') \cup (R_i \cap R_k')$.

Observing these relationships allows Okabe et al. (1992) to compute the partial derivatives of F and the Hessian[34] of F, which allows the method of steepest descent to be implemented. In the final stage of the algorithm, the derivatives are very close to zero, and each facility is near the point at which $|R_i(\mathbf{x}_i)|$ is minimized. Note that for R_i, the point at which $|R_i(\mathbf{x}_i)|$ is minimized is called the *continuous Weber point*. Thus each facility moves to the Weber point of its region at the end of the computational procedure.

We now state the above procedure as an algorithm (Suzuki and Okabe 1995):

Step 1. Set an initial value $\mathbf{x}_i^{(0)}$, $i = 1, \dots, n$ and counter $k = 0$.

Step 2. Construct the Voronoi diagram generated by a set of generator points $\mathbf{x}_i^{(k)}$.

Step 3. Compute the value of F, the partial derivatives of F and the auxiliary quantities (such as the Hessian) by numerical means.

Step 4. Determine the descent direction $\mathbf{d}^k$ based on the quantities computed in step 3.

Step 5. Replace $\mathbf{x}_i^{(k)}$ with $\mathbf{x}_i^{(k+1)}$ using the formula $\mathbf{x}_i^{(k+1)} = \mathbf{x}_i^{(k)} + t^k \mathbf{d}^k$, where t^k is the step size found in a line search.

Step 6. If the difference between $\mathbf{x}_i^{(k)}$ and $\mathbf{x}_i^{(k+1)}$ is small enough, stop. Otherwise, increment k by 1 and go to step 2. ∎

While this is nothing but an application of the regular steepest-descent procedure, the distinguishing feature is the repeated computation of the Voronoi-diagram algorithm. Unlike the regular procedure, the Hessian is required and represents an extensive effort, prompting the use of only diagonal elements as an approximation.

11.9.6 Network location-routing. Facilities are not always fixed at the same locations. Some facilities are mobile and stop at several points in a region to provide service. An example is a bookmobile that offers, on a rotational basis, a library service at some fixed points in a region for a certain length of time. Library users go to their nearest service points when a bookmobile stops at those points. If a bookmobile could stop at every house, the user's convenience would be maxi- mized. Because of a time constraint, however, the total amount of time for library service and travel is limited. The locational optimization-problem is therefore to

---

[34] The role the Hessian $[\partial F / \partial x_i \partial x_j]$ in the optimization of a function F is discussed in the "Control, dynamics" appendix in Chan (2000).

minimize the average distance of users to their nearest service points if the number of service points and the total travel-distance of the bookmobile are given (Okabe et al. 1992). Other examples of this generic problem have been discussed in the "Simultaneous Location-routing Models" chapter. If we assume the total travel-distance is the sum of the Euclidean distance between successive service points, the objective function of this problem is

$$
\min_{\mathbf{x}_1,\dots,\mathbf{x}_n} \left[\sum_{i=1}^{n} \int_{R_i} f(\|\mathbf{x}-\mathbf{x}_i\|^2)\,\varphi(\mathbf{x})\,d\mathbf{x} = \sum_{i=1}^{n} |R_i(\mathbf{x}_i)| \right]
\tag{11-85}
$$

Here the travel cost on each segment $f(\|\mathbf{x}-\mathbf{x}_i\|^2)$ is a function of the squared distance[35], and the density function $\varphi(\mathbf{x})$ reflects the demand on Voronoi polygons along the route. $|R_i(\mathbf{x}_i)|$ is the total access cost to the nearest bookmobile from all users within this ith Voronoi polygon. Compared to integration, $|R_i(\mathbf{x}_i)|$ can be computed much more readily using Voronoi diagrams. As an example, the simplest travel-cost function may be $|\mathbf{x}-\mathbf{x}_i|$ and the simplest demand-density may be $1/R_T$. A constraint is written on the maximum allowable amount of travel "on the road":

$$
\sum_{i=1}^{n-1} d(\mathbf{x}_i,\mathbf{x}_{i+1}) \leq U.
\tag{11-86}
$$

We may also add the constraint that the starting point m and finishing point n are fixed and distinct. Alternatively those points can be the same and (as in the travelling-salesman problem), i.e.,

$$
\mathbf{x}_m=\mathbf{x}_s,\ \ \mathbf{x}_n=\mathbf{x}_t;\qquad \mathbf{x}_m=\mathbf{x}_n=\mathbf{x}_t
\tag{11-87}
$$

respectively. In this model, variables specifying the intermediate mobile-locations in Equation **(11-86)** are $\mathbf{x}_2, \dots, \mathbf{x}_{n-1}$. Variants of this model have been solved in the "Measuring Spatial Separation" and "Case Study in Location-routing Models" chapters. Instead of repeating them here, we show a new solution using spatial tessellations. Takeda (1985) solved a problem consisting of 20 service points, the maximum time "on the road" is specified to be 0.1 unit. A tour is constructed whereby the starting position $\mathbf{x}_1$ and the finishing point $\mathbf{x}_{20}$ are both at $(0.15,0.15)^T$. The resulting tour is shown in **Figure 11-17**(a). When the starting point is $(0.1,0.1)$ and distinct from the finishing point $(0.9,0.9)$, the resulting route is shown in **Figure 11-17**(b).

In most network optimization problems the location of nodes and links, or at least the location of links, is fixed. The locational optimization of both nodes and links was rarely studied, except in the "Simultaneous location-routing model" chapter and related references. Recent progress in Voronoi diagrams, however,

---

[35]Note that the squared Euclidean distance is used only for analytic convenience; it could very well be the absolute Euclidean distance by itself (without the term being squared). However, the absolute Euclidean-distance is harder to handle analytically. For example, it is not differentiable at all points.

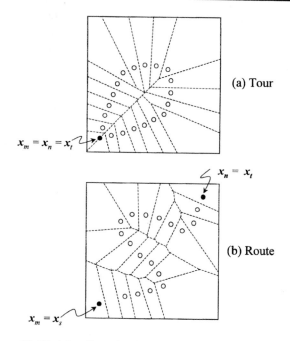

$$x_m = x_n = x_t$$

$$x_n = x_t$$

(a) Tour

(b) Route

$$x_m = x_s$$

Figure 11-17 A locally optimal railway and stations (Okabe et al. 1992)

provides a clue to this difficult problem. Here we show how spatial tessellation can simultaneously locate stations and configure rail line in a network. Let us consider a railway network in region R_T in which nodes are stations and links are railway right-of-ways connecting the stations. We assume that the speed of a train on this railway network is controlled as follows: first, the speed is accelerated from a station with constant acceleration a' until it reaches a top speed v_{max}; the speed is then kept constant at v_{max}; and finally, the train is decelerated at a constant rate $-a'$ until it stops at the next station. To avoid unnecessarily complicated equations, we assume that the distance between two adjacent stations is greater than v_{max}^2/a', implying that a train achieves the maximum speed between stations.

When a traveller makes a trip from origin x_s^0 to destination x_p^0, we assume that the traveller first walks to the nearest station $x^*(x_s^0)$ from his/her point of origin x_s^0 at walking speed v_w, arriving at the station just when a train pulls up. The traveller then takes the train from station $x^*(x_s^0)$ to the station $x^*(x_p^0)$, which is the nearest station from his/her destination x_p^0; finally, s/he disembarks and walks from $x^*(x_p^0)$ to his/her destination x_p^0 at walking speed v_w. To obtain the total-travel time for this trip, let $n[x^*(x_s^0),x^*(x_p^0)]$ be the number of stops between the boarding station $x^*(x_s^0)$ and disembarking station $x^*(x_p^0)$; $D[x^*(x_s^0),x^*(x_p^0)]$ be the railway-route distance between station $x^*(x_s^0)$ and station $x^*(x_p^0)$; and t_0 be the constant amount of dwell-time at each station. A train runs a distance of v_{max}^2/a' while it is accelerating and decelerating, and the train covers a distance $D[x^*(x_s^0),x^*(x_p^0)]$ $-$ $n[x^*(x_s^0),x^*(x_p^0)]v_{max}^2/a'$ while it keeps a constant speed v_{max}. Since it takes $2v_{max}/a'$

amount of time for acceleration and deceleration, the total travel time from origin $\mathbf{x}_s^0$ to destination $\mathbf{x}_t^0$ is given by

$$
\begin{aligned}
\tau(\mathbf{x}_s^0, \mathbf{x}_t^0) = &\left[\left\|\mathbf{x}_s^0 - \mathbf{x}^*(\mathbf{x}_s^0)\right\| + \left\|\mathbf{x}_t^0 - \mathbf{x}^*(\mathbf{x}_t^0)\right\|\right]/v_w \\
&+ \left(v_{max}/a' + t_0\right)\left\{n\left[\mathbf{x}^*(\mathbf{x}_s^0), \mathbf{x}^*(\mathbf{x}_t^0)\right] + 1\right\} \\
&+ D\left[\mathbf{x}^*(\mathbf{x}_s^0), \mathbf{x}^*(\mathbf{x}_t^0)\right]/v_{max}.
\end{aligned}
\tag{11-88}
$$

Now we consider the traffic volume between two points in R_T. Suppose that a traveller in a small region around $\mathbf{x}_s^0$ makes trips to see his/her friends who are distributed over the region R_T. The number of friends in a small region $d\mathbf{x}$ around $\mathbf{x}_t^0$ is assumed to be proportional to the number of inhabitants there. In terms of the density function of inhabitants, $\varphi(\mathbf{x}_t^0)$, this number is given by $c_1\varphi(\mathbf{x}_t^0)d\mathbf{x}$, where c_1 is a constant. We assume that the traveller makes $c_1\varphi(\mathbf{x}_t^0)d\mathbf{x}$ trips to a small region around $\mathbf{x}_t^0$ during a period of time. We also assume that the number of trip-makers in a small region around $\mathbf{x}_s^0$ is proportional to the number of inhabitants there, i.e., $c_2\varphi(\mathbf{x}_s^0)d\mathbf{x}$, where c_2 is also a constant. Thus the number of trips from a small region round $\mathbf{x}_s^0$ to a small region around $\mathbf{x}_t^0$ is given by $c_1 c_2 \varphi(\mathbf{x}_s^0)\varphi(\mathbf{x}_t^0)d\mathbf{x}_s^0 d\mathbf{x}_t^0 = \varphi(\mathbf{x}_s^0)\varphi(\mathbf{x}_t^0)d\mathbf{x}_s^0 d\mathbf{x}_t^0$, where we fix $c_1 c_2 = 1$ without loss of generality. In region R_T, the total two-way trip-time of all travellers in a unit-period-of-time is therefore given by

$$
F(\mathbf{x}_1, \ldots, \mathbf{x}_n) = 2\int_{R_T} \tau(\mathbf{x}_s^0, \mathbf{x}_t^0)\,\varphi(\mathbf{x}_s^0)\,\varphi(\mathbf{x}_t^0)\,d\mathbf{x}_s^0 d\mathbf{x}_t^0.
\tag{11-89}
$$

Here $\tau(a,b)$ is the travel time from a to b. The nearest stations, $\mathbf{x}^*(\mathbf{x}_s^0)$ and $\mathbf{x}^*(\mathbf{x}_t^0)$, are readily obtained from the Voronoi diagram $\{R_1, \ldots, R_n\}$ generated by a set of n stations at $\mathbf{x}_1, \ldots, \mathbf{x}_n$. Obviously, if $\mathbf{x}_s^0 \in R_i$, then $\mathbf{x}^*(\mathbf{x}_s^0) = \mathbf{x}_i$; if $\mathbf{x}_t \in R_j$, then $\mathbf{x}^*(\mathbf{x}_t^0) = \mathbf{x}_j$. We can simply write $n[\mathbf{x}^*(\mathbf{x}_s^0), \mathbf{x}^*(\mathbf{x}_t^0)]$ as $n[\mathbf{x}_i, \mathbf{x}_j]$ for $\mathbf{x}_s^0 \in R_i$, $\mathbf{x}_t^0 \in R_j$, and Equation (11-85) is now written as

$$
\begin{aligned}
F(\mathbf{x}_1, \ldots, \mathbf{x}_n) = &2\sum_{i=1}^{n} \int_{R_i} \frac{1}{v_w}\left\|\mathbf{x}_s^0 - \mathbf{x}_i\right\|\varphi(\mathbf{x}_s^0)d\mathbf{x}_s^0 + 2\sum_{j=1}^{n} \int_{R_j} \frac{1}{v_w}\left\|\mathbf{x}_t^0 - \mathbf{x}_j\right\|\varphi(\mathbf{x}_t^0)d\mathbf{x}_t^0 \\
&+ 2\sum_{i=1}^{n}\sum_{\substack{j=1 \\ j\neq i}}^{n}\left\{\left(\frac{v_{max}}{a'} + t_0\right)(n[\mathbf{x}_i, \mathbf{x}_j] + 1) + \frac{1}{v_{max}}D(\mathbf{x}_i, \mathbf{x}_j)\right\}\int_{R_i}\varphi(\mathbf{x}_s^0)d\mathbf{x}_s^0\int_{R_j}\varphi(\mathbf{x}_t^0)d\mathbf{x}_t^0.
\end{aligned}
\tag{11-90}
$$

The first and second terms are similar and represent the round-trip access- and egress-time from origins in region R_i to station $\mathbf{x}_i$, and from station $\mathbf{x}_j$ to destinations in region R_j respectively. The last term is the two-way trip-time from station $\mathbf{x}_i$ to $\mathbf{x}_j$.

To be concise, the network locational-optimization problem can be formulated as minimizing a simplified form of Equation (11-90), in which the first two terms can be written as $4\sum_{i=1}^{n}\int_{R_i}(1/v_w)\|\mathbf{x} - \mathbf{x}_i\|\varphi(\mathbf{x})d\mathbf{x}$, while the summation signs for the third term can be modified to read $\sum_{i=1}^{n}\sum_{j=i+1}^{n}$, with the figure 2 dropped. The figure 4 in the first term can also be subsequently dropped in the final objective function: $\min_{x_1,\ldots,x_n} F(\mathbf{x}_1, \ldots, \mathbf{x}_n)$. We may add the constraint Equation **(11-87)** to designate the starting and end points of the railway line. The result is a nonlinear non-convex programming problem. Again, one can solve it with the steepest-descent method. The first derivative is obtained from

$$
\frac{\partial}{\partial x_{ik}} F(\mathbf{x}_1, \ldots, \mathbf{x}_n) = \int_{R_i} \frac{1}{v_w} \frac{\partial}{\partial x_{ik}} \|\mathbf{x} - \mathbf{x}_i\| \varphi(\mathbf{x}) d\mathbf{x}
$$

$$
+ \sum_{i=1}^{n} \sum_{j=i+1}^{n} \frac{1}{v_{max}} \left\{ \frac{\partial}{\partial x_{ik}} D(\mathbf{x}_i, \mathbf{x}_j) \right\} \int_{R_i} \varphi(\mathbf{x}) d\mathbf{x} \int_{R_j} \varphi(\mathbf{x}) d\mathbf{x}
$$

$$
+ \sum_{i=1}^{n} \sum_{j=i+1}^{n} \left\{ \left(\frac{v_{max}}{a'} + t_0 \right) (n[\mathbf{x}_i, \mathbf{x}_j] + 1) + \frac{1}{v_{max}} D(\mathbf{x}_i, \mathbf{x}_j) \right\}
$$ **(11-91)**

$$
\times \left\{ \int_{R_j} \varphi(\mathbf{x}) d\mathbf{x} \frac{\partial}{\partial x_{ik}} \int_{R_i} \varphi(\mathbf{x}) d\mathbf{x} + \int_{R_i} \varphi(\mathbf{x}) d\mathbf{x} \frac{\partial}{\partial x_{ik}} \int_{R_j} \varphi(\mathbf{x}) d\mathbf{x} \right\} \qquad (k=1,2)
$$

The first term is computed similar to a regular derivative in two-dimensional Euclidean space. The second term shows the change in the railway distance due to the slight move of station $\mathbf{x}_i$. If the railway is assumed to be a chain of lines connecting stations with straight-line segments, the value of this term is obtained from

$$
\frac{\partial}{\partial x_{ik}} D(\mathbf{x}_i, \mathbf{x}_j) = \frac{\partial}{\partial x_{ik}} \left[\|\mathbf{x}_{i-1} - \mathbf{x}_i\| + \|\mathbf{x}_i - \mathbf{x}_{i+1}\| \right] \qquad k=1,2
$$ **(11-92)**

The last term in Equation **(11-91)** treats the corresponding change in the area of Voronoi polygons. Using the steepest-descent method with this derivative, we can solve this network locational-optimization problem. **Figure 11-18** shows a result when there are sixteen stations ($n = 16$). The study area consists of a unit square centered on $(0,0)$ (or $R_T = \{(x_1, x_2) | -0.5 \le x_1, x_2 \le 0.5\}$). The starting point of the railway line is specified to be $\mathbf{x}_1 = (-0.9, -0.9)^T$, and similarly for the end point $\mathbf{x}_{16} = (0.9, 0.9)^T$. The maximum speed is $v_{max} = 0.01$, an acceleration/deceleration rate of $a' = 0.002$ is specified, and so is the walking speed $v_w = 0.001$. Finally, a demand-density function is given as

$$
\varphi(\mathbf{x}) = \begin{cases} \exp(-25\|\mathbf{x}\|) & \mathbf{x} \in R_T \\ 0 & \mathbf{x} \notin R_T \end{cases}.
$$ **(11-93)**

As described, these steepest-descent algorithms are based on repeated construction of Voronoi Diagrams. To facilitate this, several computer codes have been developed for constructing Voronoi diagrams. VORONOI2 (Sugihara and Iri 1994) configures the 2-dimensional Voronoi diagram. SEGVOR (Imai 1994) constructs the line Voronoi diagram used in the network location-routing problem described above. L1VOR (Suzuki 1989) constructs the Voronoi diagram for the l_1-

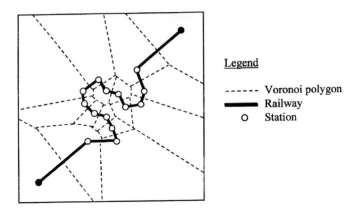

Legend

- - - - -   Voronoi polygon
━━━━━   Railway
o   Station

Figure 11-18 Example of locational optimization on a plane

metric.[36] Apparently, these computer programs are available for dissemination. Inquiries should be addressed directly to the authors. To the best of my knowledge, these computer codes have been refined to the stage where thousands of generators can be solved practically, even in higher dimensions.

11.9.7 *Continuous vs. discrete modelling*. The fundamental idea of spatial tessellation is really quite simple. Given a finite set of distinct, isolated points in a *continuous* space, we associate all *n discrete* locations in that space with the closest member of the point set. The result is a partitioning of the space into a set of regions. This gives rise to a Voronoi diagram. A second diagram can be constructed from the Voronoi diagram in *m*-dimensional space by joining those points whose regions share an (*m*−1)-dimensional face (see the "dashed lines" in **Figure 11-12**). We refer to this dual diagram as the Delaunay tessellation. While not discussed in detail here, a number of theoretical constructs stem from this primal-dual paradigm. In fact, Voronoi diagrams and Delaunay triangles are two of a few truly interdisciplinary concepts. It has a *natural* application in *discrete* location and *continuous* land-use patterns. While we have cited several examples above, other applications abound. Beasley and Goffinet (1994) outlined a triangulation-based heuristic for the *Euclidean Steiner problem*[37]—an important problem in facility location-and-

---

[36] A rectilinear-metric example is included in the "Exercises and Problems" web site. Admittedly, it is based on a simplified way of constructing the Voronoi diagram, as we began to illustrate in the numerical example of Section 11.9.1

[37] If *I* is a set of demand nodes in a plane, the Euclidean Steiner problem is to connect together the nodes in *I* so as to minimize the total length of the edges used, where an edge is the Euclidean distance between two nodes. The solution to this problem is a minimum spanning tree on *I*∪*J*, where *J* is a set of additional nodes introduced into the Euclidean plane in order to achieve a minimal solution. Nodes in *J* are called *Steiner vertices* and the nodes *I* are customer or demand nodes. This problem can be interpreted as defining a tree servicing *I* customer nodes with visits up to *J* optional nodes in order to minimize total travel-distance.

routing. Also, in location-routing problems, there is often a need to partition a region into several service zones, each of which is serviced by a facility. In seeking an effective relation among the different levels in a hierarchy of service facilities, all the zones assigned to a particular facility at one hierarchical level should belong to the same district in the next level of the hierarchy (La Figuera and Revelle 1993)[38]. Rather than a discrete demand node, there is often a need to represent demand as continuous density distribution and locate hubs based on this representation (Campbell 1993). All these point toward the potential usefulness—both in terms of model representation and computation—of Voronoi diagrams and Delaunay triangles.

While we have shown some very typical applications, there are much more to be discussed (Okabe et al. 1992, Suzuki and Okabe 1995). The field is moving rapidly, and additional results will no doubt be emerging in the near future. The discussion here is therefore nothing more than an introduction. Specifically, it paves the way to another general body of knowledge, which for lack of a better name, is referred to here in this book as *spatial statistics*. It is our intention to show that spatial-tessellation statistics use a different representation when compared with conventional spatial-autocorrelation analysis. Instead of assigning a weight z_i to location i, and a distance d from i, space is divided into tiles most proximal to discrete point i, or Voronoi diagrams. A demand density is then prescribed on the entire map. Thus geographic subdivisions or service regions are naturally modelled in this representation, complete with border lines between them. As demonstrated in this section, the new paradigm provides tremendous flexibility in modelling discrete and continuous location-allocation problems. It also allows conventional statistical and optimization tools to be applied in a most convenient manner. It truly bridges the gap between causal and correlative modelling—the two camps of thought we refer to from the very early discussions in this book. To show this point, let us discuss the notion of *spatial prediction* (Cressie 1991), which is the prediction of unobserved values from observed data. Specifically, suppose that measurements are denoted by $\{z(\mathbf{x}): \mathbf{x} \in R\}$ where $\mathbf{x}$ is a spatial location in two-dimensional space. R gives the extent of the region-of-interest and $\mathbf{x}$ varies continuously over it. Suppose that data $\mathbf{z} = [z(\mathbf{x}_1), \ldots, z(\mathbf{x}_{n'})]^T$ are observed at known sites $(\mathbf{x}_1, \ldots, \mathbf{x}_{n'})$. We are concerned with the prediction of $z(\mathbf{x}_0)$, an unknown value at a known location $\mathbf{x}_0$. We will present various suggestions for the predictor $f(\mathbf{z}; \mathbf{x}_0)$, and briefly critique them.

For Delaunay triangulation, we assume that the domain R can be triangulated according to the locations of the data $\{\mathbf{x}_1, \ldots, \mathbf{x}_{n'}\}$, and on any triangle, the predictor is a planar interpolant of the surrounding three data values. The Delaunay triangulation can be used as a predictor because the greatest distances over which

---

[38] A hierarchy of service facilities is found in health-care delivery, large regional-hospitals supply to the population all the health-care services offered by smaller local hospitals. This is an example of a successively-inclusive hierarchy which can be represented as a tree structure, with the larger hospitals at a higher level and the smaller ones at a lower level. A service zone of a smaller hospital in this case is covered by the service district of the larger hospital.

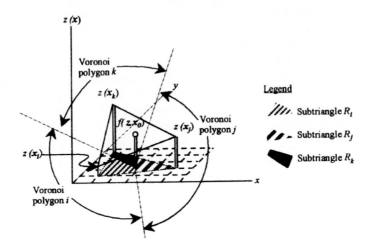

Figure 11-19 Spatial prediction using Delaunay triangulation

interpolations must be carried out are smaller than for any other triangulation. Consider the points i that are closer to $\mathbf{x}_i$ than any other data-location. The collection of all such points partitions R into Voronoi polygons, the ith polygon referring to the data-location $\mathbf{x}_i$. If the jth polygon shares a common boundary with the ith polygon, join $\mathbf{x}_i$ and $\mathbf{x}_j$ with a straight line. The set of all such joins defines the Delaunay triangulation as discussed above in Subsection 11.9.1. Then consider

$$f(\mathbf{z};\mathbf{x}_0) = F[z(\mathbf{x}_i), z(\mathbf{x}_j), z(\mathbf{x}_k); \mathbf{x}_0, \mathbf{x}_i, \mathbf{x}_j, \mathbf{x}_k] \qquad (11\text{-}94)$$

where $\mathbf{x}_0$ is contained in the Delaunay triangle defined by nodes $\mathbf{x}_i$, $\mathbf{x}_j$, and $\mathbf{x}_k$, and the right-hand-side of Equation (11-94) is the planar interpolant throughout the coordinates $(\mathbf{x}_i, z(\mathbf{x}_i))$, $(\mathbf{x}_j, z(\mathbf{x}_j))$, and $(\mathbf{x}_k, z(\mathbf{x}_k))$ at $\mathbf{x}_0$. By joining $\mathbf{x}_0$ to $\mathbf{x}_i$, $\mathbf{x}_j$ and $\mathbf{x}_k$, three subtriangles are formed. Let R_i be the area of the subtriangle opposite node i, and so on. Then $\mathbf{x}_0 = (x_0, y_0)^T$ has the property that $\mathbf{x}_0 = (|R_i|\mathbf{x}_i + |R_j|\mathbf{x}_j + |R_k|\mathbf{x}_k) / (|R_i| + |R_j| + |R_k|)$, and the predictor is

$$f(\mathbf{z};\mathbf{x}_0) = (|R_i|z(\mathbf{x}_i) + |R_j|z(\mathbf{x}_j) + |R_k|z(\mathbf{x}_k))/(|R_i| + |R_j| + |R_k|). \qquad (11\text{-}95)$$

The procedure is illustrated in **Figure 11-19**. This predictor has several interesting properties. First, The predictor is unique and requires no ad-hoc-parameter choices. Second, it is computationally much slower than the similar predictors such as *median polish*[39] or *inverse-distance-squared weighted-average.*[40] Third, the

---

[39] A median-polish filter is used in removing noise in an image. Please see a numerical example worked out in the "Remote-sensing" chapter (Chan 2000) under the "Digital image processing" section.

[40] This predictor assumes the gravity-model form $f(\mathbf{z},\mathbf{x}_0) = \sum_{i=1}^{n'} d_{0,i}^{-2} z(x_i) / \sum_{i=1}^{n'} d_{0,i}^{-2}$.

predictor is *resistant* in the sense that any one outlier affects it only locally. Finally, *f* is an exact interpolator.

11.10 Conclusion

In this chapter, we, have summarized in one place some of the latest in examining spatial information. First, spatial information is re-examined as a stochastic process, wherein the issues of stationarity and isotropy are discussed. Unlike the single time-dimension of classic stochastic-processes, the concept of stationarity has to be ascertained along all points of the compass. This greatly increases the complexity of the process. Consequently, much work remains to be done in the field in spite of dramatic advances documented in a diversity of recent literature (de Oliveira et al. 1997). Aside from data recognition and the elaborate process of specification and identification, the analyses presented here in this chapter suggest that we have a rich array of techniques to model the underlying processes, depending on what is known about the process. Such an approach will afford a much better use of the collected data, yielding a rich array of information previously judged inconceivable by researchers in this field. Rather than an end by itself, advances in remote-sensing, pattern-recognition and statistical analysis should play the additional role of influencing the direction of causal-model development. When data and model are used skillfully together, one could explain complicated processes (Aykroyd and Zimeras 1999, Pierce and Kinateder 1999). Thus disaggregate models are no longer a theoretical construct, and bifurcation theory can now be used meaningfully to explain non-homogeneous processes. This in turn will permit accurate calibration of the models for useful prediction of the process.

Conversely, causal modelling identifies the data that are required for an in-depth understanding of the underlying process. As an illustration, disaggregate bifurcation models require consistent set of detailed geographic information over two specified time-periods to perform accurate forecast of spatial change and to validate the model. It was also identified that certain information cannot be collected directly from existing remote-sensing devices, including spatial-interaction relationships such as the trip-distribution information. However, this information gap can potentially be bridged either by advances in analytical models such as spatial statistics and origin-destination estimation. If nothing else, this chapter opens a dialogue between the two disparate groups of our colleagues working in spatial-temporal research: those concerned with empiricism and those concerned with analytic-model construction. Likewise it illustrates the spectrum of modelling approaches, advanced over the past several decades, ranging from "model specification and identification through data examination" to "causal modelling." Included are such techniques as spatial autocorrelation, spatial interaction, entropy, spatial dominance and uncertainty. Instead of representing spatial data in terms of lattices and distances, spatial tessellation further character-izes geographic subunits by way of boundaries and density distribution. It represents a 'natural' way of treating spatial information, including both discrete and continuous characterizations. During the discussions in this chapter, we try to point out the linkages between these diverse sets of techniques, and refer the reader to details in previous chapters where appropriate. For example, facility location is analyzed in terms of Voronoi diagrams here instead of optimization using node-arc contiguity representation, as done in previous chapters (Sik and Woo 1999). Rather than gravitational attractions, spatial competition is analyzed in terms of the prescriptive framework spatial dominance-and-uncertainty as perceived by the trip maker. In a new paradigm, facility location and gravitational interaction is viewed as a special case of a general spatial-autocorrelation statistic. The synthesis is made in this chapter to drive home one of the objectives of this book—to relate the diverse schools of thought to one another. Hopefully, it will effect synergism between these diverse camps in their future work.

An equally important illustration is to establish the commonality among the three representative spatial-temporal models: spatial time-series, the EMPIRIC econometric-model, and the Yi–Chan Lowry–Garin model. It was found that EMPIRIC is a one-lag specialization of the spatial-temporal autoregressive moving-average (STARMA) model with exogenous spatial variables. The Yi–Chan model, on the other hand, is a generalization of STARMA. It poses a causal-modelling structure that eases the burden of calibration by breaking it into a series of 1-lag autoregressive calibrations. A side benefit is the ability to model a non-homogeneous process that includes bifurcation. A major difference between statistical and causal spatial models can be summarized in terms of two key words: *interaction* and *association*. According to Getis (1991), interaction is implied in the gravity models (which is embedded in the Yi–Chan formulation). Interaction refers to the possible movement of people or commodities at location i to or from locations j. In spatial-autocorrelation models, including STARMA and EMPIRIC, the link between i and j is statistical correlation between locational characteristics at two points. The cross-product statistic allows for a unification of the two types of models. This was accomplished by means of the development of a spatial-autocorrelation statistic that serves as a measure of spatial interaction as well. An advantage to the approach taken by Getis is that the way is now paved for the more formal development of statistical tests on interaction theory. On a general level, spatial-temporal information is represented as a random or Markovian spatial-process. Such a process is usually described on a specific lattice, where no value is specified between the lattice points. These spatial models on lattices are analogues to time-series models, where for example, February immediately follows January and an immediate measurement does not exist in between. An objective of this chapter is to provide a couple of simple methods that use the discrete locations of sites to model the probability distribution of the spatial process. Examples include the Yi–Chan model, EMPIRIC and STARMA.

The ordinary Voronoi diagram of a set of points is a tessellation. It consists of a set of regions that do not overlap, except possibly at their boundaries, and completely covers the space under study. Since there are many empirical structures that involve tessellations, a most obvious application of Voronoi concepts is in the modelling of such structures and the processes that generate them. An example is the spatial-assignment process, of which facility-location is a special case. We have seen that spatial assignment can be generalized to spatial-temporal processes, as witnessed by the classic Hotelling spatial-competition example, which was modelled by Voronoi diagrams in this chapter. Spatial tessellation is a promising paradigm to explain facility location and land use, bridging the gap between discrete and continuous models. It also integrates the prescriptive approach with the descriptive approach under one roof, allowing optimal location-routing decisions to be made with the same ease as spatial prediction. The duality between Voronoi diagrams and Delaunay triangles appears to be an exciting area for future research, so is the capacity of Voronoi diagrams in articulating multicriteria obnoxious-facility-location (Melannondis and Xanthopulos 1994). What is described here barely "scratch the surface" of this powerful technique. The interested reader may wish to refer to Okabe et al. (1992), Suzuki and Okabe (1995) and Allard and Fraley. (1997) for details.

Powerful as the techniques described in this chapter are, there are always improvements that can be made to them. Also, their deployment is inevitably colored by the perception and judgment of the analyst. The analysis results are only as credible as the interpretive skill of the analyst and how much the client of the study would choose to accept them. In this regard, there really is no truly rigorous and objective analysis in spite of our honest attempt to provide one. Subjective judgment is part and parcel to any analysis, either during the conduct of the analysis itself or in the interpretation of the final results. Multi-criteria analysis, in spite of its inevitable imperfections, is an important way to include such subjectivity. In this chapter, we have shown via an example how a population forecast in Portland, Oregon can incorporate the judgment of the analyst and insights of the local community into the process. Such a pluralistic involvement results in very favorable outcomes, both in terms of the rigor of the analysis itself and the credibility of the final forecast among its end users. I recommend this approach to the consideration of our readers (Marshall and Oliver 1995).

CHAPTER TWELVE

RETROSPECT AND PROSPECTS

In this volume, we have surveyed a wide spectrum of methodologies for analyzing facility location and land use. We identified the common denominator of these methodologies as the multicriteria analysis of spatial-temporal information. It is on this commonality that we have laid out the past 16 chapters. In coming to a closure, we wish to summarize here some of our current thoughts now that these pages have been written. There are causes for celebration and there are reasons to be contemplative as well. To be rejoiced are some significant achievements reported here, representing progress in the state-of-the-art in analyzing spatial-temporal information over the last few decades. To be contemplative are the limitations of these techniques and the many frontiers yet to be conquered before one can reach truly sound spatial decisions. We will devote equal time in this chapter discussing both accomplishments and shortcomings, with an eye out for an action plan for practitioners and researchers alike.

12.1 Facility location

Traditionally, the facility-location problem involves finding the number, scale and sites of building, installation, plants and depots--whether private or public--in a given region. Industrial plants will tend to locate in positions which take into account material, labor sources and demand markets. Public facilities will be located in positions that primarily focus on serving demands. At one extreme are emergency services for which the promptness of response is often the main concern. In locating obnoxious facilities, factors such as safety, water and air pollution temper demand and cost considerations with natural and manmade laws that govern human exposure to risk and/or pollutants. In the case of satellite-tracking stations target observability is balanced against economy-of-scale in public expenditure.

There is a remarkable difference between locational decisions made in the private sector and those made in the public sector. In the private sector, the most important concern is likely to be acceptable cost, profit or market shares. Normally decision-making is centralized among relatively few stakeholders. In the public sector, there is an articulate decision-structure involving many parties with vested interests in the various stages of the implementation process. It is not always possible to define precise indicators that measure the merits of the various alternatives. Furthermore, these alternatives are often conflicting. Perhaps another taxonomy is to view private vs. public facility-location as centralized vs. decentralized decision-making procedures. Even then we are only capturing part of the picture in describing corporations with social responsibilities.

Usually decision-makers consider a *discrete* location to be realistic and accurate portrayal of the problem at hand. Thus one chooses among the well-defined candidate locations to build an airport, a landfill, a factory, a warehouse and the like. *Continuous* spatial formulations are both uncommon and difficult to solve. This is in spite of the integration between the discrete and continuous versions put forth here in this book and recent advances in the field. There simply do not appear to be well-documented, operational algorithms for several broad classes of simple, continuous planar location problems. It contrasts dramatically with their discrete counterparts, although the theoretical linkages between the two models have been delineated through the various measures of proximity.

In recent years, there is a convergence between *prescriptive* and *descriptive* modelling. The former selects a few 'best' locations while the latter affords a better understanding of the problem by studying a replica of the system. Recent advances in interactive search-techniques allow not only the merging of these two modelling approaches, but also an explicit recognition of multicriteria decision-making. This new paradigm permits—albeit in an elementary way—the interaction between the decision-maker, the analyst, and the computer, reflecting the interfaces between location decisions and other strategic decisions. This represents a shift of focus from building more 'realistic' models to building models that better *assist* in making 'sound' decisions.

As important as the steps of defining, analyzing, generating, evaluating, selecting, and sizing facilities are, the real payoff comes from implementation. Benefits accrue if and only if the plan is implemented properly. Implementation is a tedious, often frustrating activity that must be pursued with vigor. It is that aspect of facility location that ultimately tests the usefulness of the discussions presented in this volume. The 'permitting' process, for example, can be a most frustrating task facing the analyst and planner. It often taxes our analytical and emotional capacities to the limit. The information age has added further complications to the process. Virtual workplaces, telecommuting, home-shopping (Borsuk 1996) are but a few examples of the new facility-location decisions that are more complex than ever. One thing is clear; these activities are expanding rapidly. Except for occasional studies, however, little has been done explicitly on analyzing virtual workplaces and telecommuting, not to say home shopping. For example, Gray (1995) poses these intriguing questions: "Do the benefits outweigh the costs (and there are costs) for the employer? For the worker? How far can we push agency theory?" [The theory views the relations between an employer and employee as a contract. It focuses on developing the most efficient contract between them.] Although the techniques presented in this volume have been based on traditional location studies, the spatial information revealed have a direct bearing upon this type of analysis. This is shown in the telecommuting analysis performed in the "Economics" chapter of Chan (2000).

In a recent survey (*OR/MS Today* 1995), the top-ten operations opportunities in industry were identified. A panel of six experts in the field of warehousing, warehouse-management systems, logistics, manufacturing, organization excellence and maintenance independently ranked opportunities in their respective areas. Second in ranking is "facility/space utilization." "The capacity of facilities and the traffic flow through these facilities—be they production operation, maintenance storeroom, raw-material storage, work-in-process storage, finished-good storage or distribution—are all viewed as offering tremendous opportunities for improvement. Facility consolidation, contraction and expansion are all mentioned as having significant potentials depending on the circumstances." These point toward the relevance of the combined location-routing models proposed in this book under the chapter with the same name.

In the public sector, no where is the importance of facility-location shows up more than in emergency-unit deployment, such as police, fire, medical and utility-repair services (Zografos et al. 1994). Here, the usefulness of probabilistic location, districting and dispatching models becomes crystal clear as human lives are often "on the line." As shown in the "Measuring Spatial Separation" chapter, many observations in deterministic facility-location have an analogue in the probabilistic world. For example, deterministic nodal-optimality conditions carry over when a stochastic-queuing model has a demand arrival-rate sufficiently small. In the case of random travel-times, there exist at least one set of optimal locations that fully reside at the nodes of the network. As operational problems are solved, the challenge of the future, according to Swersey (1994), will be to address broader, more strategic issues. An example is the tradeoff between two disparate factors: smoke detectors and fire companies.

12.2 Land use

Facility-location and *activity-allocation* are interrelated processes. Once a facility is sited, the demands for service (or the activities) are assigned toward these facilities, often—although not exclusively—based on proximity. Thus the population that requires services is catered for by a close-by hospital, library, recreational, and shopping facility. In turn, employees from businesses and public agencies are provided with comfortable housing and convenient support-services so that they can conduct their day-to-day dealings. Activity-distribution, alternatively considered in land-use, then becomes a natural consequence of site development. In its aggregate, facility building necessarily redistributes growth within the study area (Boarnet 1995), as it changes relative accessibilities for the population (Cervero and Landis 1995). Closely related as they are, it is surprising that there were so few attempts to view facility-location and land-use under a common light—as we are trying to do here in this volume. One possible explanation is that the job is non-trivial, to say the very least.

Besides methodological integration, facility-location and land-use planning need a coherent framework in which policy issues, objectives, criteria and impacts are systematically brought together. The emergence of new policy issues is not always predictable. A given policy issue will never have the same relevance and get the same attention over time. This implies that policy measures are to be taken in an uncertain environment (Giuliano 1995). Many external factors may change during policy implementation, so that the question of resilience of plans is in order. The same is true for their flexibility and the need for adaptation and learning by all participating parties. DeNeufville (1990), for example, advocated a flexible-planning approach that explicitly takes risks into account.

Spatial-impact analysis differs from other types of impact analysis in at least a couple of ways. It relates to a multiplicity of policies, such as transportation, housing, public services, industry and so on. Most important, it also deals with various levels of jurisdictional entities, from local, state, federal to international authorities. *Macro* land-use models are often used to analyze these more 'global' effects. As shown in the chapter on "Econometric Models," however, macro models are not entirely separate from *micro* models. While micro models deal with individual facility location, macro models are often aggregated from micro models. Ulanowicz (1987), for example, pointed out that there is inseparable linkage between micro and macro models. He proposed a thermodynamic description of growth and development that reinforces the natural exchange between macroscopic and microscopic events. In particular, he maintained that the theory of "optimal ascendancy," based on *variational principles*, as a way of embracing micro behavior within a macro, system-optimal framework.

Having said that, the use of the term macro models is far from standardized. Various macro land-use analyses can be distinguished (Nijkamp et al. 1990):

(a) *Statistical analyses without an explicit model*—These exercises may be based on differential growth indicators or on *shift-and-share* analysis. This is described in the "Economic Methods" chapter in Chan (2000) and subsequently in the "Econometric Models" chapter here in this volume. By definition, these are descriptive rather than prescriptive models.

(b) *Partial models with policy instruments*—Such models can assess the direct impacts of policy measures on spatial economic and non-economic objectives. Spatial-time-series, often called Spatial-Temporal-Autogressive-Moving-Average (STARMA) models, are good examples of these techniques. A limitation of these techniques is that only the direct effects of policy instruments on one policy objective (such as *growth*) can be estimated at a time.

(c) *Integrated models*—These simultaneous models provide a coherent picture of direct and indirect effects of policy measures on spatial economic and non-economic objectives. This class of model includes both spatial input–output models and integrated regional-economic models such as generalized multi-regional growth equilibria. One may also use simultaneous equations such as the EMPIRIC model to perform similar functions, following a statistical, rather than causal model, approach.

Contrary to the name, there is no single model that can be 'comprehensive' enough to address all issues of concern, nor do we necessarily wish such a model to be built. This is too "tall an order to fill" for the state-of-the-art. It is evident if we examine the specifications for such a model (Nijkamp et al. 1990), and the fact that a model is a decision aid rather than the decision-maker. Here is a list of the challenges facing one who wishes to build a comprehensive model:

Consistency The statistical data and the analytical relationships in a land-use analysis should provide a set of coherent and non-contradictory results on the spatial effects of public and private policies. An example is the expected signs of coefficients in an econometric model like EMPIRIC, where the expected effect of a policy is shown by the sign of a coefficient. The data-base here should also support the model specification.

Completeness An impact analysis should present all relevant (intended and unintended, direct and indirect) effects of public decisions upon the spatial system. Completeness does not necessarily mean detailed, since a comprehensive model can address all the relevant issues at an aggregate level, depending on the purpose of the study. If a qualitative insight is desired, for example, a model based on the canonical forms of catastrophe theory can be very useful, although it lacks the details required for field implementation.

Relevance The various impacts described should be meaningful from both an analytical and a policy viewpoint .The level of detail should be in agreement with policy objectives and with the analysis framework used. This echoes the point made above regarding the difference between completeness and details, except that relevance refers to a specific set of stakeholders involved in the process. The analyst is ultimately held responsible to *the other* stakeholders in expressing their policies, inasmuch as these stakeholders are held responsible in communicating their policies to the analyst.

Pluralism The choice of the various impacts serves to represent the variety and multidimensionality of the spatial system at hand. These impacts should be the consequence of related courses of action for all the interest groups and citizens. To the extent that a model is to provide the information base for all stakeholders, this is a mandatory requirement. Toward this, multicriteria decision-making procedures play an important role in trading off between impacts and among stakeholders.

Comparability Spatial effects should be measured so that a comparison among the corresponding effects at different places or in different times is possible. For example, the level of data-aggregation should be the same in each of the time periods during a forecast. This means a consistent data-base collected in a compatible format for the time periods used for calibration. Advances in remote sensing and geographic information systems (GIS) greatly ease this function. Satellites allow for almost continuous monitoring of spatial activities while GIS helps to organize the data.

Flexibility The impact analysis concerned should be adjustable to changes in policy and to new circumstances, i.e., it is resilient to change. Another name is robustness, in that the model can analyze a variety of scenarios and be valuable in the long run. This echoes previous discussions on adaptive analysis. Recent developments in control, dynamics, and system stability add to the pool of time-dependent or time-independent techniques that make adaptive decisions possible.

We have seen examples ranging from facility siting over time to handling policy changes in spatial time-series.

Data availability An impact analysis should as much as possible be based on existing information systems and analytical models or techniques. A model that cannot be calibrated or operable is of little use as a practical or research tool. Data-retrieval technologies such as GIS helps to streamline this process. Ultimately, it is the analyst's responsibility to match data to models and models to data. No where is it more critical than real-time decisions, which depends critically on how fast one can translate data into a form useful for decision-making.

Coherence The successive steps of an impact analysis should provide an integrated, instead of a segmented, picture of the spatial distribution and interaction effect of a policy plan. This is not to be equated with large-scale models necessarily. Simple models can provide such an insight if tailored correctly toward the problem at hand. In fact simple models have fewer problems in integration than complex models. For example, a spatial input–output model is a more coherent version of the original Lowry model. It is conceptually more simple and analytically more transparent than the Lowry model, while at the same time more sophisticated.

Testability The empirical values of the various impacts have to be gauged against *a priori* policy objectives, so that the effectiveness of policy measures can be tested. To see whether a model is doing its job in providing policy-sensitive output, the "job description" has to be specified. It was found, for example, that the fidelity of activity-allocation models is often reliable only on aggregate geographic subdivisions, rather than detailed units. Here, one should not expect accurate results in testing policies on a city-block level. When such a fine scale is desired, facility-location models or spatial tessellation models may be more appropriate.

Feasibility Any impact analysis has to be incorporated in the prevailing policy and its related analytical framework, in order to stimulate effective global, national, and regional management-plans. A model sitting on a book shelf is only for research use at best, it is certainly not for implementation. Feasibility in this case has to be defined in terms of the intended use of the model, whether it be implementation or research.

Transparency The results of an impact analysis have to be presented in such a way that they can be readily understood by policy makers, experts and interested citizens. If the results of an analysis cannot be communicated to the clients it serves, its use is greatly discounted. The best model is a simple model since the implications of a decision can be traced and can be easily understood.

Democratic content A spatial-impact analysis should provide the information from a broad spectrum, so that diverging views of different agencies or groups can be taken into consideration. This means that conflicting points of view need to be represented. While this point is particularly important for the industrialized nations, it remains relevant to developing countries as long as dissenting viewpoints exist. Group-decision theory is particularly useful in this regard, in spite of its embryonic stage of development.

Phasing A useful impact-analysis has to show the effects over time. Also, the steps of tracing, measuring and interpreting these effects should be transparent. This is particularly important when it comes to implementation, which is often carried out in sequential phases. Airport expansion schemes, for example, often start out with runway additions, which eventually lead up to building a new airport. In other words, capacity expansion is often the intermediate phase to a brand new facility (Min et al. 1997).

It can be seen therefore that the demands on the analyst for building comprehensive models are exorbitant. It may not be too modest to start with spatial models that can answer only one question at a time, and progressively work our way toward the more holistic models. Land-use models have gone through evolutions in the past decades. They have shifted from their comprehensive orientation in the sixties to simple forms in the seventies, and analytical developments in the eighties. The premise in this book has always been a modest one--that no model can be all inclusive. While we should keep the list above for reference, sometimes the most effective and practical models are the ones that can answer a specific question, and answer it well. Fortunately, after one has mastered the basic building blocks, transitions can be made with ease between models—whether they be land-use or facility-location models—in order to respond to a whole host of questions. This book purports to make a modest step forward in providing our readers with these basic building blocks.

While we have reiterated that both facility-location and land-use models are methodologically equivalent, the latter are usually more comprehensive in orientation than the former. This is the outcome of historical development where the former is used in a much smaller physical-scale than the latter. Whereas facility-location models analyze one facility at a time, land-use models deals with the entire urban and regional landscape, with more stakeholders involved. After saying that, however, we turn to an example that shows that facility-location models can be taxed on occasions to answer some very broad, often conflicting, questions also.

12.3 The example of obnoxious facilities

Day after day, the news headlines are replete with examples of obnoxious-facility-location problems. These headlines relate to locating landfills, waste-disposal war, siting of energy facilities such as nuclear power plants, and U. S. Defense Department's hazardous-materials-disposal sites. The Not-In-My-Back-Yard (NIMBY) syndrome is prevalent among all segments of society. At the same time, people wish to enjoy the comfort and convenience of garbage collection, cheap energy, and national security.

The NIMBY attitude is not unfounded. Numerous reasons for this attitude are founded in news accounts: the evacuation of entire communities, the nuisance to people living at the end of an airport runway, the contamination of a community's water supply by a local landfill, and the smell of air pollutants by a belching smoke stack. Almost every community has had its share of such incidences. Examples where public outcry may not be justifiable can also be enumerated. Hysteria and perception often are at the root of such episodes, while the real problem may or may not exist. Controversy is usually created by humankind, not nature. Therefore controversies exist as long as humankind exists. The recourse is planning and analysis, which—when carried out properly—help to mitigate some of these controversies (Kleindorfer and Kunreuther 1994).

Several common factors are involved in the site-selection process: tangible and intangible costs, distance separation between the facility and the population, adverse human-relations between the proponents and opponents of a project, political risks in taking a position--just to name a few. Complicated as it is, we have made progress in several fronts here in this book. In the "Prescriptive Tools" chapter in Chan (2000) and "Measuring Spatial Separation" chapter in this volume, for example, the traditional *anti-center* problem typically used in locating obnoxious facilities is expanded into a *Data Envelopment Analysis* (DEA) framework. Thus not only spatial measures such as the max-min metric is used, it is fully supplemented with a whole host of other cost-and-benefit indicator. These other cost-and-benefit measures include impacts on real-estate values, stimulation on local economic-development, and in the case of a transfer facility, the user fee for the transfer of waste to another community for final disposal. Obviously, this is only the first, albeit important, step in a long journey toward a truly-responsive obnoxious-facility location-model.

The site-selection process is correspondingly characterized by the following attributes (Keeney 1980):

High stakes The difference between site A vs. site B may involve hundreds of millions of U.S. dollars or severe local environmental-damage. At the worst case, human lives may be involved.

Complicated structure The large number of alternate sites, the many criteria for evaluation, the diverse interest-groups, the intangibles, the degree-and-uncertainty of long-range impacts, and the time delay in licensing and construction, all show the complexity inherent in most siting problems. This makes it very difficult to appraise each spatial alternative in any detail.

No overall expert The complexity of the problem requires a team approach to analysis, involving planners, economists, engineers, environmental scientists, and many others. No single individual could take on the comprehensive roles of these diverse disciplinary-experts.

Need to justify decisions To obtain authorization to build, stringent requirements are placed on accountability and in public trust. Documents such as *environmental impact statements* in the U.S. need to be carefully prepared.

Clearly analysis in this arena is by no means simple. Perhaps that is why we—among others—have devoted so many pages to help fine tune the art of spatial analysis here in this book. The intent of an analysis is not necessarily to *solve* the problem. It may *help* the stakeholders to understand the many facets of the problem and to reach responsible decisions. "It is precisely the complexity which implies that formal analysis should be done" (Keeney 1980). There is no such thing as objective, value-free analysis. Someone's personal feelings and judgment are involved in any site identification problem, not to mention identifying alternate sites. In assessing site impacts, the *correct* data have to be collected, and the *pertinent* questions to be asked--anything but objective tasks. It is precisely the opinions and the judgment of the experts that we seek in such a study. Of course the formal analysis-procedures suggested in this volume strive to be objective and value free. But the users of these models and procedure are not. They are the ones who will decide what sensitivity analyses to perform, and they are the ones that finally interpret the results of the analyses. What we strive for is a logical, systematic analysis that unveils the implicit professional and value judgments. Multicriteria decision-making process of this kind has to be responsive to the needs of the participants in the process to be of any use. The American Society of Civil Engineers (1996), in its Policy Statement 297, recommends the formation of an autonomous hazardous-waste facility-siting board in each State. The job of this board is to coalesce technical and non-technical inputs from all parties to reach binding decisions on a timely basis. The example of obnoxious facility-location is characterized by as long a list of evaluation criteria as those prescribed for land-use in Section 12.2

12.4 Multicriteria decision-making procedures

In a facility-location/land-use model, values are used to specify the criteria for evaluation of alternatives. Values are necessary to build a quantitative criterion, which provides the representation of values for evaluating alternatives (Keeney 1994). Most decision problems can be broken into two parts. The first part relates the various alternatives to the corresponding consequences. These consequences describe the degree to which criteria are adhered to. The second part evaluates the relative desirability of the consequences. The evaluation of these consequences is done with the value model. One then combines information from the two parts to derive the relative desirability of alternatives. This is obtained from the relative desirability of the consequences and the likelihood that the various consequences result from the different alternatives. This is clear from a decision-tree representation where alternatives represent one set of branches from each of which outcome branches sprout.[1] The likelihood of the consequences is quantified using judgment about available facts and the desirability of the consequences is quantified using judgment of values. This distinction is important because the techniques used in the two parts of the overall decision model are different. The individuals whose judgments are appropriate for the parts may be different.

12.4.1 Some common fallacies. Many complex decision problems have significant consequences including hundreds-of-millions-of-U.S. dollars or potential fatalities or possible large-scale environmental degradation. The only reason for an interest in such problems is that some consequences may be much

---

[1]For a review of decision tree, see the "Descriptive Tools" chapter in Chan (2000).

better than others, and so some alternatives may be much better than others. However, the amount of time used to articulate the appropriate values of the problem is minuscule compared with the time used to address other aspects of the decision problem. The "criterion function" might be chosen in an hour with very little thought. Yet several person-years of effort and millions-of-dollars can be used to model the relationships between the alternatives and the consequences. This is similar for gather inginformation about those relationships, including spatial relationships. Since the entire reason for caring about the problem is based on values, it would not seem unreasonable to use a fraction of those resources to understand and appropriately structure and quantify the relevant values. Such an effort should be used to build a value model.

According to Keeney (1994), there is one mistake that is very commonly made in constructing value models, and this mistake is sometimes ground for poor decision-making. It can be illustrated in an air-pollution problem where one is concerned about the cost of pollution emissions and pollution concentrations. Administrators, regulators, and members of the public are often asked the following questions: "In this air-pollution problem, which is more important, monetary costs or pollutant concentrations, and how much more?" One might answer that pollutant concentrations are three times as important as costs. While the sentiment of this statement may make sense, it is completely useless for building a model of values. Would such a statement mean that lowering of one-part-per-billion of air-pollutant concentrations would worth the cost of two billion dollars in a metropolitan area? The answer is "probably not" even to the person who just answered the question in less than two minutes. When asked to clarify the apparent discrepancy, s/he would naturally state that the decrease in air pollution was very small, only one part in a billion, and the cost was a very large two billion dollars. The point should now be clear. It is necessary to know how much the *relative* decrease in air-pollution will be and how much the increase in regulation costs will be. Only then can one logically discuss and quantify the concept of relative importance. In the "Multi-criteria Decision-making" chapter of Chan (2000), we emphasize the importance of 'ranging'—a process to normalize an attribute between say a 0–1 range. Current interest in interactive multicriteria decision-making also places the decision-maker in a much more specific situation, under which his/her responses can be realistic and mindful of existing constraints.

This error gets us into problems for two reasons. First, it does not afford the in-depth appraisal of values that should be done on important decision problems. If we are concerned about the health of the public due to pollution and expenditure of billions-of-dollars, more than two minutes of thought should be given to the matter before one states that pollutant concentrations are three times as important as cost. Second, decision-makers such as legislators use these indications of relative importance inappropriately. Indeed, sometimes legislators pass legislation such as the Clean Air Act in the United States which essentially says the health of the public is of paramount importance and the cost in reducing air-pollution should not be considered in setting standards for pollution levels. This is not practical, possible, or desirable in the real world. We could spend hundreds-of-billions-of-dollars and still have the opportunity to improve our air quality. If the value tradeoffs are done properly and address the question of how much of one specific attribute is worth how much of another attribute, the insights from the analysis are greatly increased and the likelihood of misuse of those judgments is greatly decreased.

12.4.2 Making values explicit. Strategic thinkers have long recognized the need of clarifying values as a key step. Peters and Waterman (1982) refer to their "one all-purpose bit of advice for management" as "figure[ing] out our value system." Values are clarified with an explicitly stated criterion—sometimes called a "mission statement." However, identifying and structuring criteria is a difficult task: ends are often confused with means, criteria are often confused with targets or constraints or even alternatives, the relationships between different criteria are unclear, and priorities among criteria are easily misconstrued. Clear criteria are very useful, but how should they be developed? The process requires significant creativity in discussing with decision-makers and the individuals concerned. Simply listing criteria is shallow; there is a need for greater depth, clear structure, and a sound conceptual base in developing criteria.

A criterion is a statement of something that one wants to strive toward. It is characterized by three features: a decision context, an object, and a direction of preference. For example, one objective of cleaning up defense-related hazardous-facilities is to "minimize environmental impacts." With this objective, the decision context is restoring the natural surroundings, the object is environmental impact, and less impact is preferred to more. We should also distinguish between *fundamental objectives* and *means objectives*. Fundamental objectives concern the ends that decision-makers value; means objectives are methods to achieve ends. Fundamental objectives refer to the broadest class of decisions; they are defined as strategic objectives. These strategic objectives provide common guidance for all decisions and form the basis of more fundamental objectives appropriate for specific decisions.

Consider the decision situation involving the transportation of nuclear wastes. One objective may be to minimize the distance the material is transported by trucks. The question should be asked: "Why is this criterion important?" The answer may be that shorter distances would reduce both the chances of accidents and the costs of transportation. However, it may turn out that shorter transportation routes go through major cities, exposing more people to the nuclear waste, and this may be recognized as undesirable. Again, for each objective concerning traffic accidents, costs, and exposure, the question should be asked: "Why is that important?" For accidents, the response may be that with fewer accidents would be fewer highway fatalities and less accidental exposure of the public to nuclear waste. And the answer to why it is important to minimize exposure may be to minimize the health impacts due to nuclear waste. To the question "Why is it important to minimize health impacts?", the response may be that it is simply important. This suggests that the objective concerning impacts on public heath is a fundamental, rather than means, objective in the decision context.

12.4.3 Value-focused thinking. According to Keeney (1994), traditional decision-making procedures usually focus on alternatives. Decision problems are thrust upon us by the actions of others: competitors, customers, governments, and stakeholders; or by circumstances such as recessions and natural disasters. Faced with a decision problem, the so called solving begins. Typically, the decision-maker defines the decision problem as a choice among identified alternatives, and only afterwards considers criteria to evaluate the alternatives. This standard problem-solving approach is referred to here as *alternative-focused* thinking. Focusing on alternatives is a limited way to think through decision situations, no matter how well these alternatives consider spatial attributes. It is reactive, not proactive. If you wish to be the master of your decision-making, it makes sense to have more control over the decision situations you face. You do not control decision situations that you approach through alternative-focused thinking. This standard mode of thinking is backwards, because it puts the cart of identifying alternatives before the horse of articulating values. It is values that are fundamentally important in any decision situation. Alternatives are relevant only because they are means to achieve values. Thus, although it is useful to iterate between articulating values and creating spatial alternatives, the principle should be "value first." This manner of thinking, which is called *value-focused* thinking, is a way to channel a critical resource—hard thinking—to lead to better decisions.

Value-focused thinking helps uncover hidden objectives and lead to more productive collection of information. It can improve communication among parties concerned about a decision, facilitate involvement of multiple stakeholders, and enhance the coordination of interconnected decisions. These all contribute to a more insightful evaluation of alternatives. But the greatest benefits of value-focused thinking result from the guidance provided for creating better spatial alternatives for any decision problems that one faces. It is also useful for identifying decision situations more appealing than the current decision problems. These better-decision situations, which one creates, should be thought of as decision *opportunities*, rather than as decision *problems*.

Progress has been made toward value-focus thinking here. Take the "Prescriptive Tools" chapter in Chan (2000) and the "Measuring Spatial Separation" chapter in the current volume, for example. There, we have combined DEA with traditional facility-location models. Thus we start with the valuation of costs and benefits and then generate the siting alternatives correspondingly. It is a step in the right direction in our judgment. Traditional DEA lists costs and benefits a priori without tracing the relationships between these measures and the interaction among siting decisions. The *combined* DEA/location model allows the best of both worlds to be addressed simultaneously. Through the convenience of modern-day linear-programming codes, various cost-benefit measures can be tried repeatedly and the corresponding alternatives evaluated—all in a matter of minutes. Thus different valuations and definitions of costs and benefits can be entertained and their consequences fully assessed to satisfy each stakeholder's decision context. It represents a way to operationalize Keeney's proposals.

12.4.4 Using values to guide analysis. Values should guide the problems that analysts work on and how they address those problems. Sometimes the values will be the only means to achieve the objectives of one's employer or client. As an example, one might work to do a better job on a problem assigned by one's superior. But usually, even in such cases, there is considerable flexibility in defining and pursuing the problem posed. Keeney (1994) cited one piece of his work to illustrate this point. The Nuclear Waste Policy Act (NWPA) of 1982 outlined the U.S. Government's plan to dispose of this material in a safe and environmentally acceptable manner. The U.S. Department of Energy (DOE) has major responsibility for the plan. This plan involved several steps to eventually identify a site for nuclear repository, which is a deep underground mine for permanently storing the nuclear waste. One step involved the identification of three states for simultaneous characterization. This would entail numerous activities including the construction of an exploratory shaft into the proposed repository area, subsurface excavation and tunneling, and testing the host rock near the shaft.

Based on draft environmental assessments in December 1984, DOE proposed five sites as suitable for characterization. These sites (and their host rock-types) were in Mississippi (a salt dome), Texas (bedded salt), Utah (bedded salt), Nevada (tuff), and Washington (basalt). With strong prodding from the Board of Radioactive Waste Management of the National Academy of Sciences, DOE chose to conduct a rigorous evaluation of the five sites using multi-attribute utility analysis. This analysis (Merkhofer and Keeney 1986) suggested the more desirab lesites to be Nevada, Mississippi, Texas, Utah, and Washington. Using insights from this analysis and other information at their disposal, DOE selected the Nevada, Texas, and Washington sites for characterization. In making this recommendation, DOE necessarily balanced concerns for the diversity of rock type with the evaluations of individual sites. This balancing was done without the aid of formal analysis; hence, it had to be based on intuitive professional judgement. The quality of this judgment was called into question by the public and the U S Congress. In 1982, the cost of characterizing a site was estimated to be less than 100 million dollars. By 1986, estimates for characterization were at least one billion per site. This raised the question of whether the information to be gathered during characterization of the three sites was worth three billion.

Keeney conducted an analysis to examine two related issues: (a) Were the three sites selected for characterization by DOE the best available sites; and (b) Was *sequential*, rather than *simultaneous*, characterization a better way to investigate potential repository sites? Notice these questions reflect judgment and value by Keeney. Kleindorfer and Kunreuther (1994) suggested this premise: If there are no fixed costs associated with obtaining detailed data on the site and interacting with other relevant stakeholders, then it will always be optimal to consider all potential sites simultaneously and find the one that maximizes expected net-benefits. On the other hand, if there are substantial fixed costs of characterizing each potential site, then it may be more desirable to sequentially examine the sites. Now one needs to define net benefits of a site before any information is collected on the details of the site.

When the NWPA was written, it was recognized that selecting three sites for characterization was a portfolio problem. Because of the dependencies between sites (e.g., due to common rock type), the best three individual sites did not necessarily form the best portfolio. But it was not understood how to appropriately evaluate the diversity of the portfolio. It was observed that the portfolio chosen and characterized was *means* to identify a single final repository site where characterization was just the gathering of sample information. Characterization would mainly provide information about the economic costs of proposed repositories. It yields little information about the other relevant *consequences* concerning heath and safety, environmental, and socioeconomic impacts. Thus, the essential features of the decision could be examined as a sequential decision problem that resolved uncertainties about economic costs only.

Decision trees were used to examine strategies for both simultaneous and sequential characterization strategies. The resulting analyses showed that selecting the Mississippi site to replace the Texas site in simultaneous characterization would make an improvement equivalent to saving 100 to 400 million dollars. More significant is that sequential characterization strategies were 1.7 to 2 billion dollars less expensive than the selected DOE simultaneous-characterization. The best strategy was to characterize the Nevada site first. We select it for the repository unless its revised equivalent cost was more than one billion dollars over the expected cost of the Mississippi site. If this occurred, which had a 20 percent chance in the analysis, the Mississippi site should then be characterized. The better of the two sites is then chosen as the repository. The analysis was available in late 1986 when Congress began discussions that led to the Nuclear Waste Policy Amendments Act of 1987. A draft of the amendments stated that, among others, its purpose was "to provide for the *sequential* characterization of repository sites [to] result in significant Federal budget savings." The final Amendment Act ended selecting the Nevada site for the repository.

Such a case study reflects how an analyst uses values to guide his/her work. Other cases cited in this book illustrate further the importance of the analyst in the conduct of a study. An example is the study of Long-Island state-parks in New York (see the "Generation, Competition and Distribution" chapter). In that particular study, the question of user-fee increase was posed initially by the client. Our analysis led toward other ways to "balance the books," including park closures and park improvements. Well-being to the public was also considered besides cash flow. Such an analysis yielded insights into park usage and the value of state parks to the Long-Island residents. Of particular interest is the intrinsic value of different park*locations*, distinguishing parks not just by the services and facilities they offer, but their accessibility to the population. This revealed the desirability of selective user-charge increases as a way to generate maximum revenue while minimizing erosion of public welfare. It was found that such a strategy is well worth considering, rather than an across-the-board fee-increase as initially envisioned. There really is no such thing as value free, or totally objective analysis! The more informed and more experienced the analyst is, the more insightful the study tends to be! The analyst can assist in the *progressive articulation of preferences* iteratively. This is explained in the MCDM chapter of Chan (2000) and discussed in Bana e Costa et al. (1995). Toward this end, we hope that this book contains the necessary information on both decision analysis and spatial modelling to serve the analyst as a text and a reference.

12.5 The example of risk assessment

There is no better example of MCDM than risk assessment. Risk assessments should help to find appropriate remedies to the timely problem of environmental pollution of hazardous facilities. Debates are continuing regarding this common practice: using worst-case assumptions about contaminant-exposure durations and concentrations. Controversies evolve around the costs in erring on the conservative side. The U.S. Environmental Protection Agency estimated the incremental cost to cleaning up to a 10^{-7} risk level from a 10^{-6} risk level. The cost ranges from 700,000 U S dollars to 10.4 million dollars per site. Doty and Travis (1989) presented their analysis of 50 Superfund-cleanup records, in which they characterize the role of risk assessment in the remedy decision-process. They showed that, in most cases, exposure-point concentrations are typically estimated conservatively. Human risks are extrapolated even when complete pathways in risk analysis do not exist. Ecological risks are rarely assessed. Also,the degree of risk-reduction offered by remedial alternatives is not considered. They concluded that the rationale for most remedy selection is not clear; most decisions to remediate superfund-sites are because contamination *per se* exists rather than on actual degree-of-risk to public health.

Proponents of the current practice, however, point out that the frequency of catastrophic events, rare as they may be, are of the greatest consequences. To the extent that it is difficult to assess the value of human lives and environmental damage, it is prudent to be acting on the safe side. In locating or remediating obnoxious facilities, including solid-waste facilities, many negative *perceptions* are held toward such facilities—e.g., accident hazard and low quality-of-life. Perceptions do not have to be fact-based. Whether correctly or incorrectly placed, the impact of public perceptions upon facility siting, however, cannot be discounted. As long as sensitivities are heightened, extra precaution and careful analysis are the only ways to calm fears.

As in other decisions, the risk-assessment process is a pluralistic multi-dimensional procedure. The outcome is likely to be a participatory consensus. Tusa (1992) offers some suggestions to the process participants for involvement:

a. Become involved in the risk-management process as early as possible.
b. Coordinate with the risk analyst(s) to focus on collecting the specific data necessary to evaluate existing risks scientifically.
c. Perform tests on the baseline risk-analyses to evaluate which input parameters have the most impact on the risk conclusions.
d. Evaluate the potential for risk reduction associated with each remedial option using realistic assumptions for source, pathway, and/or receptor terms.

e.   Consider other relevant criteria, such as ease of deletion, operability and reliability, to select a desirable remedial alternative.

It is observed also that public perceptions of risk gradually decay over distance away from the facility (Rahman et al. 1992). But equity in risk sharing by way of decentralizing obnoxious facilities among various parts of the community is not an unequivocally accepted solution. Clearly, many other factors should be considered in siting decisions besides risk, including the many commonly accepted measures of costs and benefits. Risk is singled out here in the context of hazardous facilities. This is due to its overriding influence in many NIMBY syndromes. Risk and uncertainly prevail in other facility-location and land-use decisions. Consider the following heated exchange in a public hearing about the risk of placing a mental institution amid a residential neighborhood:

Mayor: "If there are any problems with the mental patients roaming the neighborhood causing trouble, please call me at home and I will see to it personally that the nuisance is taken care of."

An angry mother: "Do you mean to let you know after my daughter has been raped?!!!"

In this volume, we try hard to incorporate uncertainty in our analysis procedures. A prime example is the probabilistic multiple-travelling-salesmen facility-location problem (PMTSFLP), considered in the "Measuring Spatial-Separation" chapter. In that study, we showed that in spite of uncertainties, a fair amount can be accomplished by the appropriate analysis procedures. In a Defense-Courier-Service example (the counterpart of commercial small-package-deliveries in the military), depot location and route configuration can be determined a priori. This should come as no surprise, since commercial small-package-couriers such as Federal Express and United Parcel Service do very well in servicing customers on an on-call basis. The significant result is that such location-routing decisions can be approximated asymptotically by deterministic analysis. In this sense, we can apply the arsenal of classic location-routing tools toward this class of probabilistic problems.

12.6 Gaming

The above Mother vice Mayor dialogue brings us to the relationship between different decision-makers in spatial organization. Of particular interest are conflicts that can rise within groups and between groups. The question is whether such interaction results in stable equilibria. Of particular interest to this book is equilibria in space over time. Where markets do not converge rapidly, strategy becomes all-important. Take spatial economical development for example. All production that concentrates on space must have increasing return over the relevant range of output and further analysis suggests that discontinuities may also be important. The corresponding dynamic and disequilibrium formulations are accommodated within the spatial input–output models. In special cases, the multipliers can in fact be endogenously calibrated in the forecasting sequence, as shown in the Yi–Chan model in the "Chaos, Catastrophe" chapter. In the case of non-convergence, the question is raised where the failure is due to mis-adjustments spatially or a more fundamental instability of the system. In the "spatial equilibrium" chapter, we have illustrated classical cases of pure competition, oligopoly and monopoly, all in the contest of gaming.

In general, gaming helps those that deal with facility-location and land-use to learn about the many dimensions of a land-development process. One can have a better appreciation for a number of political and institutional factors through gaming. A 'game', as defined by Webster's dictionary, is a "situation involving opposing interests given specific information and allowed a choice of moves with the objective of maximizing their wins and minimizing their losses." The analytical procedures discussed in this book—including the "Spatial Equilibrium" chapter—provide one with an information base to examine options. Such information is usually obtained for idealized situations, however. It is safe to say that analytical model of spatial games have not gone beyond the developmental stage. Competitive facility-location results are often interesting, but analytical solutions to these models are extremely limited. Numerical and heuristic algorithms are inevitable should one wish to apply these models to solve current problems. In this regard, the techniques advanced in the spatial input–output models are more general, since the more advanced versions have the temporal dimension incorporated, in addition to spatial. Chaos can also be observed as part of the spatial-interaction process. Numerical simulation of this kind appears, therefore, to be the practical solution procedure.

The decision-making process on land developments is often characterized by conflicting interests seeking social, political and financial gains. Such a complex decision-making process goes well beyond the facts and figures provided by the analytical and numerical procedures, including multicriteria models (Tajima 1995). Short of actual experiences (which sometimes may turn out to be costly) the use of *games* is a best way to highlight the issues. Players in a spatial game assume roles as public officials, administrators, developers and citizens. They act out public and private decisions, developing land, improving public works, and even hold elections. A number of games have been developed over the years, including SimCity (Wilson 1990), the Community Land Use Game—CLUG (Feldt 1972), City Games (House and Patterson 1969) and Metropolis Games (Meier and Duke 1966). There are at least five general areas of potential use for games (Haack and Peterson 1971): teaching and training, solving problems, evaluating alternate strategies, extending or proving theories (or both), and evaluating prospective managers. These areas are not necessarily mutually exclusive.

While its role in teaching is clear, the role of gaming in other applications needs further elaboration. Gaming has been proposed for solving complex problems because complex problems cannot be solved directly but can be approximated with a game. A game leads to different results each time the game is played. By playing the game many times, the most likely result can be identified. Games need not be played out exhaustively, however, to gain helpful insights. Despite the difficulty in determining optimum strategies, there is benefit in simply analyzing available alternatives. The insights gained from games are more important than identifying a particular alternative. Games can also be used to prove or disprove theories. For large and complex systems, it is usually necessary to construct models and then to test proposed principles by observing the model rather than the original system. In this regard, the game becomes the 'breadboard' for experiments. Guetskow (1962), for example, suggests that gaming will lead to a tighter body of social theory. During a course of games, players could be evaluated on their ability to assimilate the information presented, pick out the critical issues, and make appropriate decisions. Thus leadership skills and managerial talents can be discerned out of the gaming exercise. Gaming has long been recognized by military commanders as essential training to attain leadership positions. Extensive war-gaming models have been developed, assisted by state-of-the-art computing technology. Ironic as it may seem, these models generally do not have the *spatial* dimension included. One would expect that it is paramount to monitor troop and weapon movements on a two- or three-dimensional *map*.

One more use, the most important, of gaming encompasses all the above: improving communication (Lee 1977), particularly between professional planners and analysts (who are the bulk of our readers) and the public. The need for improved communications is evident in public hearings, where an adversarial relationship is often observed between professional planners and the public. Professional planners often feel that politicians and citizens are unwilling not only to accept the technical recommendations of the experts, but also to do their homework well enough to understand how the recommendations were reached. To them, the political side exhibits little tolerance for new ideas. It cannot digest much in the way of complexity. It only takes an interest when an immediate self-interest is threatened. It asks questions that are either simplistic or rhetorical or designed to reinforce an already-determined position. Finally, it makes decisions largely on political grounds rather than merit. On the other hand, politicians and citizens perceive that technical professionals are secretive and devious. They never answer questions directly. They couch their answers in meaningless technical jargon. They rigidly refuse to consider another viewpoint or incorporate factors not already accepted. They are obtuse either by preference or by training. They fail to comprehend political realities and take them into account. Also, they do all their plans in private before venturing to discuss anything with the public. Requests by politicians for more information from agencies supposedly working for them are met with responses that either duck the question asked, take too long to be of any use, or simply recite dogma already presented.

Gaming can assist in addressing this communication by:

a. Helping the players to consider the impacts of alternate policies on various interest groups and constituencies, and anticipate their concerns.
b. Promoting the bridging or crossing of organizational lines to facilitate agreement on issues.

c. It allows players to change roles occasionally, which can easily be done during a game.

d. Helping professionals to recognize that they are part of a political process, and that the line between political and technical is precise only in the abstract.

Hopefully, our discussion of CLUG in the "Descriptive Tools for Analysis" chapter (Chan 2000) and elsewhere has illustrated some of these points.

In a case study examining the controversial Hong-Kong Port-and-Airport-Development Strategy (PADS), Ng (1993) argued that the PADS study, which sites the proposed new airport, is restricted in nature. She stated that the planning process adopted in the study is outdated and ineffective. For instance, the assumptions and parameters were not disclosed to the public and were not subject to open comments and critiques. Since China was to take over governance of Hong Kong in 1997, the lack of consultation with China and the general public contributed to the major obscurities and inadequacies of the plan. Ng further suggested that planning in Hong Kong has been done by professionals behind closed doors and their findings have been usually presented as "scientific discoveries," rather than tentative courses of actions subject to public scrutiny, political debates, and decisions. May be a gaming approach would have been an ideal supplement to the PADS study. This way, all stakeholders have an input to the process, instead of just the 'experts'.

12.7 Perspectives and research directions

The use of formal mathematical concepts and expressions, and the application of statistics, optimization and other quantitative methods forms the most common characteristic of modelling in facility-location, transport and land-use over the past decades. These approaches have gained significant grounds in our understanding of the field (Ghosh and Rushton 1987, Anselin 1990). Here we review some of our progress and lay out plans for the future.

12.7.1 From facility-location to land-use. The underlying techniques that are employed in location, transport and land-use are similar, even though these fields might have started from very different contexts. Traditional facility-location models assume a one-to-one pairing between facility j and demands i, where i is served by j exclusively. Recently, there is a fair amount of interest in problems where service to i is rendered by more than one facility. Thus overlapping service is provided to demands i from facility $j_1, j_2, ..., j_{|J|}$. While such an extension of classical facility-location models is fully justified by real-world applications, the problem becomes a good deal more complex to solve analytically. Now consider another extension, in which only the appropriate facility of type k can be used to serve demand of type k. This will no doubt further increase the dimensionality of the problem, as the running index i–j becomes i–j–k. Imagine a facility can change from type k_1 to k_2 to cater for prevalent demand types, and several facilities of each type can be co-located in *j*. Our job now becomes one of not only pairing *i* to *j* to *k*, but also in allocating facilities of type $k_1, k_2, \ldots, k_K$ among sites $j_1, j_2, \ldots, j_U$.

As if life is not complicated enough, we further require that facilities $k_1, k_2, \ldots, k_K$ need to be assigned in bundles of m_j to site *j*. In other words, we assign either zero or m_j facilities to *j*; any number between 0 and m_j is not acceptable. This constitutes a combinatorial optimization problem that entertains active research today (Underwood 1996). Finally, we specify that the demands be generated over time, rather than fixed, and they are responsive dynamically to the services rendered. Recall that a dividing line between facility-location and land-use models is the assumptions made regarding the demand function. Traditional facility-location models assume fixed, perfectly inelastic demands, while land-use models typically use a fully-specified demand-function. Thus a facility-location model simply allocates demands among the servicing facilities, while land-use models both generate and allocate demands. As mentioned, this distinction has been gradually eroding as researchers start to merge the two distinct historical developments. Thus in the last decade one has witnessed the emergence of spatial-interaction-based location-allocation models. In these models, a full gravity-model is built on top of traditional facility-location model to more accurately represent the spatial-cost function. Most the effort, however, still fall short of generating activities based on such factors as socioeconomic variables. Here in this book, we make a genuine effort to merge the *generation* part with the *location–allocation* part, as we explained beginning with the "Including Generation into Location-Allocation" chapter. This, together with a more realistic spatial-cost function (such as the gravity-model representation), bridge the gap between facility-location and land-use models. Now we have made the transition from a traditional facility-location model to a land-use model. An objective in this book is to highlight the relationship between these two types of models, and to outline the quantitative techniques that help to analyze both.

Most applications of location-allocation models in the classic phase used a single, system-wide objective to evaluate alternate location patterns. During recent decades, many researchers have focused on methods for solving multiple-objective location-allocation problems. This brings us squarely to multicriteria decision-making where defining an objective function properly is so important. It also mandates an efficient solution procedure to take into account more than one criterion function. Most critically, interesting theoretical relationships can be identified for the nondominated solutions as alluded to in the "Facility Location" chapter. Distinction is made between a system-optimization objective and a user-optimizing objective. A classic example is the user-optimal route-choice problem by Braess, who formulated route-choice in a network using an implicit game-theoretic model. Here each shipper or traveller makes his/her decision based on the route choices of other shippers. The perfectly-inelastic demand-function assumed in the model has been subsequently extended to include a full demand-function. The route-choice decision is then analyzed in a monopoly, oligopoly and perfect-competition market respectively. This allows a better understanding of the spatial equilibria of shipping decisions and traffic flow under these various market conditions. The model has been subsequently extended to a dynamic formulation by Ran et al., where the departure-time decision is modelled besides route choice (1996).

In all location–allocation problems, demand data are usually aggregated according to some arbitrary spatial unit. It was found out that optimal locations are stable over a wide range of spatial aggregation of data; but the measures of the objective function are increasingly biased with higher degrees of spatial aggregation[2]. The finding was predicated on having a systematic bias in the spatial variation of levels-of-aggregation. Disaggregation to a common analysis unit, whether it be a household or a pixel, would be useful in addressing this problem. Geographic information system (GIS) is one way to reach such a common geographic unit. Conversion from existing data-bases to such a common format, however, needs to be handled with care (Kutz 1995). When executed in the correct way, remote-sensing and GIS have the potential to describe land-use not only in terms of two-dimensional geographic and socioeconomic information, but also to include elevation (or time) as a third dimension as well (Corbley 1995, Langran 1993).

GIS also provides a convenient environment for selecting and analyzing potential sites (Sandhu 1995). Organized around a GIS, the general location-allocation problem boils down to a processing of these data files:

data-set/coverage—the name of the data-set that will be used in the location/ allocation process, consisting of points, polygons, line (street network) with a specified coordinate system and ancillary attributes;
demand locations—a set of locations that need to be serviced, each of which may have a weight associate with them (such as the population);
candidate locations—a set of potential locations where the site(s) could be located;
distances—a method to compute the distances or travel costs, whether Euclidean or shortest-path etc.;
models—the method in which the locations will be chosen, such as min-sum, max-coverage etc.;
sites—the number of sites that need to be located.

Discrete applications of location–allocation modelling assume knowledge of time, distance, or cost of interactions between all places. In reality, such inter-place distance or travel time must often be estimated, giving rise to another potential source of error in the data. Combined location-routing models have been proposed and surveyed in this book to address this problem. They form a focus of this volume. One germane area of investigation is the use of special geographic encoding procedures such as TIGER files to compute route distances. Because of improvement in GIS, many location–allocation analyses are beginning to represent the environment far more realistically than in the past.

---

[2] See chapters on "Lowly-based Models" and "Chaos, Catastrophe, Bifurcation, and Disaggregation in Locational Models".

Of particular theoretical interest is the fact that spatial separation, as represented in the appropriate GIS data-structure, starts to ameliorate the distinction between various types of facility-location models, and between location-and-routing models. Thus through a transformation of the distance measure, one can show that the p-median, set-covering and maximal-coverage models are related. More important, the solution procedures for a discrete median-problem can be applied to a center problem and vice versa.[3] When proximity relationships are properly specified among the nodes, the quadratic-assignment problem becomes mathematically equivalent to the travelling-salesman problem.[4] In planar-location problems, the role of the spatial-separation cost-function—whether convex, concave or linear—is also confirmed. We have outlined the conditions for interior and extreme-point locations in the convex hull of the Weber problem in the "Facility Location" chapter. Seen in this light, a properly designed GIS can therefore greatly facilitate the efficient solution of location-routing problems through data-structure organization. Here, one of the GIS functions is computing distances.

In land-use models, the measure of spatial-separation is often called *accessibility*. Such a measure is at the heart of many transport-based activity derivation-and-allocation mechanisms. There is a long history behind such a measure (Pooler 1995), emanating from studies in geography through spatial statistics. In parallel, an economic interpretation of such a measure is offered in terms of consumers' surplus—a tripmaking benefit that can be attributed to the pertinent origin and destination activities (Martinez C. 1995). We have shown in the "Economics" chapter that through this interpretation, one can derive the gravity model—the engine that drives most land-use models. The measure of accessibility is important in that it constitutes the spatial-price system that organizes activities such as population and employment in a study area (Mackett 1993). We have shown in this book, both in facility-location models and land-use models, that spatial-separation is the magic parameter that explains the relationship between some very disparate spatial phenomena. In land-use models, accessibility determines whether equilibrium, bifurcation or chaos is obtained in land development, as explained in the "Chaos, Bifurcation" and "Spatial Equilibrium" chapters. The major difference between the spatial price used in facility-location models and land-use models is this. In the former case, the spatial-cost metric is not weighed directly by activities while in the latter case it is. The activity weights simply reflect the activity-generation mechanism embedded within land-use models, which is distinctly absent in a facility-location model, as mentioned previously.

Earlier in this section, we have already established that land-use models can be viewed as continuous generalization of discrete facility-location models. Subareal shares tend to be a continuous variable rather than a binary variable in a land-use model. To the extent that land-use models are means to a desirable

---

[3]For further discussions, please refer to the "Facility Location" chapter.

[4]Please review the "Spatial Separation" chapter for further explanation.

development, critiques have been levied against the traditional location–allocation
/land-use modelling. These models tend to be demand-driven, or that spatial pattern
is purely the result of consumer preference. In contrast, the "functional integration
approach" (Hansen 1992) induces linkage effects that will have much greater
beneficial, if not readily predictable, social and economic consequences. The
functional integration approach has been enhanced recently by more rigorous
economic-linkage effects, including the process by which direct investment in one
sector induces development in other sectors. Current developments have taken an
economic equilibrium and disequilibrium approach, representing an extension of
spatial input–output analysis[5]. Such efforts have great promise in a more satisfying
analysis of land-use, although its computational procedures still need a fair amount
of improvement.

12.7.2   From land-use to spatial-temporal modelling.   In many early
approaches to spatial-temporal modelling, there was an underlying assumption that
a model had to be based on a solid and consistent system of regional economics.
Recent emphasis has shifted to a more encompassing social-accounting matrix as
an extension of input–output analysis. This approach allows for a detailed
description of the interaction between sectors of the economy, including firms,
factors of production, institutions. It often has particularly detailed focus on the role
of different income classes. Such an approach also allows for some interesting
decomposition of various types of multipliers. These multipliers capture intra-
regional, interregional interactions, as well as between production sectors.

As explained, accessibility of vertices in a transportation network plays an
essential role in analyzing regional space-economy. A definition of topological
accessibility of transport network is proposed by Mackiewicz and Ratjczak (1996).
They represent a transport network in terms of an undirected graph composed of
vertices and edges. In other words, a network is represented by a symmetric
adjacency matrix whose rows and columns are the nodes in the network. This
square matrix is composed of zeros and ones, where a one is placed between row
node i and column node j when they are connected and zero otherwise. When the
adjacency matrix is raised to its powers, the separation between any pair of vertices
(i,j) in the network is defined as the length of the shortest chain composed of edges
linking i and j. It is clear how this procedure can generate the routing between
origin–destination pairs i and j, as we have shown in the "Location-Routing"
chapter. When this concept is generalized, it can also serve to model the interactions among not only geographic locations i and j, but also between production
sectors as well. Correspondingly, the entries of zero–one's are now modified to
assume any value between zero and one, and is referred to by a more general name,
the *weight* matrix. When the weight matrix is embedded in the multipliers, it plays
a critical role in spatial input–output models (and beyond, as we will see).

---

[5]See the chapter on "Spatial Equilibrium and Disequilibrium" for a more detailed discussion.

The use of social-accounting methods such as the above is not without its problems. The method is extremely data-hungry and there are problems with sectoral classification. In a multi-regional model, there is difficulty with incorporating transfer costs and factor mobility in a consistent fashion. In this book, we have tried to marry earlier development in the field with this recent effort, which results in discussions in such chapters as "Spatial Equilibrium." There, it was pointed out that the marriage is typically based on very restrictive assumptions of fixed proportional relationships. Such relationship enables the transition from a purely accounting procedure to a more causal model. More important, the formal tie between input–output models and bifurcation was established based on this assumption.

Another intricacy with input–output models is the calibration procedure. Available information suggests that the calibration of technical coefficients is no trivial task, particularly in the spatial version when the gravity model is embedded within these coefficients. If we assume that all inter-industry flows are zero, and the nonbasic industry sells to the household sector while the export sector does not, the spatial input–output model degenerates into the Lowry–Garin model. But the gravity model becomes too aggregate now for a zone-specific calibration of disaggregate labor-force participation rate and population-serving ratio. Here is when additional compromises have to be made regarding the rigor of the model formulation and the ease of calibration. Although much of the formal theory about spatial economics is abandoned, the flexible structure of econometric modelling is one way to overcome the calibration problem.

The econometric modelling of a regional economy is an example where the application of quantitative methods in regional science has found widespread acceptance. For the most part, however, regional econometric models are the poor cousins of their national counterparts. This is due to data limitations and a lack of distinct regional-growth theory that can be implemented empirically. Overall, the theory of regional econometric model seems to be stagnating. The interest in model error and model validation has not led to a more critical approach to model specification nor to a more careful assessment of predictive performance. Also, the method used in these models tends to lag behind theoretical development in statistical estimation and forecasting. Most models are still of the standard structural form and are estimated by means of ordinary least squares, in spite of well-known problems with this approach. Few alterative techniques from such approaches as vector autoregressive moving-average (VARMA) modelling or its specialization into spatial-temporal applications (STARMA) have been introduced at the regional scale.

Spatial econometric models, when viewed in a broader context, can be described as special cases of spatial time-series. Perhaps the most common form is the spatial autoregressive-model. We have established this observation through a case study of the land-use model EMPIRIC. This simultaneous-linear-equation model has been shown to be both a mixed-regressive-spatial-autoregressive model as well as a one-lag specialization of the spatial-temporal autoregressive moving-average model. The linkage lies in the way the 'spatial' weight matrix is defined.

Here, the weight matrix quantifies the contiguity relationship between economic sectors. This contrasts with the spatial relationships between network nodes or subareas as in the quadratic-assignment to travelling-salesman problem conversion. It can be seen therefore that the concept of spatial econometrics is rather far reaching. It is best viewed as a prominent branch of statistics, often called spatial statistics, as we will detail immediately below.

12.7.3   From spatial-temporal models to spatial statistics. A great majority of empirical work in regional science deals with spatially referenced data, i.e., information associated with points, lines or areas in space. This organization often leads to spatial dependence and spatial heterogeneity, which invalidate many independence properties of standard statistical and econometric methods. What is it that makes spatial data special? Anselin and Getis (1992) explain that spatial effects complicate any straightforward understanding of spatial data. "Spatial effects" has two interrelated meanings. The first is that embodies in Tobler's First Law of Geography, where "everything is related to everything else, but near things are more related than distant things." This simply implies that one should expect stronger relationships within and among variables that are sampled at places that are spatially near to one another rather than far from one another. The more troublesome second meaning, however, is that because of the size and configuration of spatial units we find relationships within or among variables that are due as much to the nature of the spatial units as to the nature of the variables being studied. The first type of spatial effect can be handled, for the most part, with conventional data-analytical procedures, but not the second. Since all spatial data are subject to the second effect, one must take it into account when devising systems for analysis (Arbia et al. 1996).

For those familiar with time-series analysis, they should recognize that—while there are similarities—spatial modelling is essentially different because it is not causal (Guyon 1995). That is, the chronology of time does not exist in spatial analysis. It is not reasonable to model the growth of a corn plant based only on that of the plant on its left. But structural difference between temporal and spatial models are more than causality and non-causality: They involve the role played by boundary conditions, such as that exist at the edges of a satellite image. They are also characterized by the absence of factorization techniques of several complex-variable polynomials that are commonly used in spectral analysis.[6] These differences led to the definition of new and specific ideas, models, results, techniques and algorithms. For example, Bayesian methodology gave a strong impulse to pattern recognition and image analysis: object reconstruction, detection of movement, and segmentation for cartography or for data reduction. Markovian techniques, with a coherent mathematical setup, can also be readily applied. They are usually efficient algorithms.

---

[6]Spectral analysis was discussed in the "Remote Sensing" chapter, and the "Control, Systems Dynamics" appendix in Chan (2000), as well as the "Spatial Time-Series" chapter in the current volume. . See also the "Glossary" appendix in Chan (2000) for a definition of the technical term if necessary.

In recent years, research in these fields has started to deal with some of the complexities associated with realistic spatial data: new estimators, techniques to deal with limited dependent, categorical and latent variables, not to say a general interest in specification testing and model validation. However, this flurry of activities on methods and techniques has not been accompanied by a wide dissemination into practice. To a large extent this is due to the lack of readily available software that incorporates the spatial tests and estimators. Currently, none of the popular statistical or econometric packages for either mainframe, workstation or personal computers include any techniques for spatial analysis. The same holds to a large extent for the commercial GISs, which have seen an explosive growth in the last few years. It should be worrisome to regional scientists that the implementations of this new technology, which according to some has the potential to evolve the ideal models of spatial information, lack features to carry out all but the most rudimentary forms of spatial analysis. Hopefully, this may change over time, as major vendors such as Intergraph and Oracle recognize that up to 85 percent of all data have spatial components, and there is a market for such routines.

One difficulty associated with the implementation of basic GIS routines in spatial analysis is the associated computational burden. In spite of the speed of today's computational machinery, the data resolution has become increasingly more detailed for many applications, and the analytical models also become more complex. In the words of Hodgson et al. (1995), we continue to create the need for faster processors to handle larger amounts of model-generated data as well as more voluminous raw-data collections. An integrated approach combining efficient algorithms (including heuristics) with the power of parallel-processing machines can be a way to overcome this problem. This harnesses the potentiality of both advances, providing us with a way to catch up with the increasing computational burden. While distributed processing is not necessarily a focus of this book, we have taken the care to present various algorithms for spatial analysis throughout. A newest approach is the use of Voronoi diagrams. We have shown that many spatial routines can be easily implemented in Voronoi diagrams, while it is much less straightforward in a regular GIS data-structure.

Thus two opposite approaches are evident in spatial-data analysis. One is model-driven and lets spatial theory determine which specifications need to be empirically validated, as in the case of spatial econometrics. The other is data driven and is geared toward elucidating theory from the data, as seen in recent developments in spatial statistics. Similar techniques have been developed in both approaches, with very little cross-reference between the two. In addition to the loss in efficiency from parallel developments, it should be clear that the two approaches are not mutually exclusive and competitive, but complementary. We witness a recent gain in power and popularity of GIS, and a growing acceptance of the viewpoint of "letting the data speak for themselves." Now it becomes feasible and in factcrucial to keep an important role for spatial theory in spatial-data analysis.

At the same time, most data used with quantitative methods in facility-location and land-use are notoriously bad: the scope of available information is limited, it is often not collected in a consistent fashion, and is only loosely connected to the concepts that underlie spatial theory. Fortunately, recent advances in remote sensing have begun to address some of these problems. Ideally, a GIS with spatial data available at a disaggregate level will allow the determination of the geographic level-of-specificity to become endogenous to the analysis itself (Choi and Jang 1999). This will hopefully mitigate the bias due to the size and configuration of spatial units.

The data model implicit in a GIS is the 'discretization' of geographical reality necessitated by the nature of computing devices. Commercial GIS can be classified as following either a raster- or vector-data format. The raster or vector structure defines the spatial-unit-of-observation that can be used in spatial analysis. In the former, the unit is the grid and all points within the grid are assumed to take on the same value. This is an implicit form of spatial sampling. Clearly, if the grid does not exactly correspond to the spatial arrangement of values in the underlying process there will be an inherent tendency of spatial dependence. Similarly, if the scale of the grid cell has an imperfect match with the scale of the process studied, various type of mis-specification may result. This is often called ecological fallacy or the modifiable-areal-unit problem.

When a vector structure is used, the choice of the points, lines and polygons that will be presented, their spatial resolution and spatial arrangement are also an implicit form of spatial sampling. Similar to the raster approach, homogeneity is assumed within the point, line or areal-unit of observation. For the latter in particular, this may only be a crude approximation; spatial dependence as well as scale problems are likely to be present. Raster format may be in its ascendancy due to the advance in remote-sensing and digital-computing technologies. Renewed interest in Voronoi diagrams, however, provides an equally invigorated interest in vector-based data-storage.

The data-sampling process structures the data-base and precedes any sampling the analyst may want to carry out. It is often dictated by administrative or policy concerns that may or may not be founded on 'accepted' theoretical concepts of the time. Examples include the delineation of administrative regions that pre-determine the collection of many socio-economic data. In a sense then, although spatial analysis may be exploratory, the data that are available and the way in which they are collected and arranged are often constrained by the accepted theoretical knowledge of the time and its implications for spatial resolution. Digitization into a raster format may help in data standardization. Concomitant with it, however, is the expense of conversion and storage. It may be a simplistic way of addressing this problem. Meanwhile, extensive efforts have been carried out recently on developing a spatial-data infrastructure (National Research Council 1994) and spatial data-base transfer-standards (Mollering 1997). The huge amount of work involved testifies to the complexity and importance of this subject (Dailey et al. 1999).

The way data is organized, either in relational format or otherwise, also has profound implications on the efficiency of location-routing or activity-allocation algorithms. To the extent that the model often comes after data collection and organization, this means that a number of spatial-data post-processing routines are necessary to transform the stored information into the appropriate format for model solution. For example, efficient location–allocation algorithms require pre-processing inter-point distance as special demand and candidate strings. Similarly, routing models using space-filling curves require the pre-processing of spatial and demand data in terms of the more conventional geo-reference layers. This practice is akin to the current consensus in discrete optimization modelling where model formulation has a uniquely significant bearing upon the efficiency of solution algorithms.

Once the spatial data are organized in the proper way, the fundamentals of spatial statistics allow some very basic processing-routines to be implemented. One such routine is the computations associated with the spatial-weight matrix, which records the relationship between first-, second-, third-order neighbors and beyond. As illustrated by the spatial econometric model, such a weight matrix needs not reflect just physical contiguity, it could equally represent functional relationships between variables, as is typical in a model-driven analysis. The important point is that once this weight matrix is in place, we can extend a rich body of statistical knowledge, including the full power of the autoregressive moving-average models (ARMA). For example, the typical Box–Jenkins ARMA steps of identification, estimation, diagnostic testing, and forecasting can now be applied toward a spatial time-series, of which econometric models form a special case. It is worth commenting at this juncture that the gravity model and its derivatives can also be implemented in a weight matrix, reflecting the relative degree-of-interaction between close and not-so-close neighbors. Once the gravity model is captured, we have the basic building block for a majority of the modelling efforts recorded in this book.

12.7.4 *Random or Poisson fields.*

12.7.4 Random or Poisson fields. Facility-location and land-use have attributes which, when viewed on an appropriate scale, exhibit complex patterns of variation in space and time (Vanmarcke 1983, Guyon 1995). Many physical and socioeconomic systems can be analyzed using a *random (Poisson)-field* paradigm. All these phenomena are examples of random fields:

a. Image processing, as illustrated in the numerous examples in the "Remote Sensing and GIS" chapter in Chan (2000) and the TS-IP software that comes with this book;

b. "The worldwide sensor monitoring" problem under the "Spatial Time-Series" chapter

c. Pollutant concentration in an aquifer (such as illustrated in "Arsenic contamination" homework problem under the same chapter);

d. The Yi–Chan land-use model as documented in the "Chaos, Catastrophe" chapter; and

e. The wing-length cline example in the "Spatial-Temporal Information" chapter.

Each phenomenon is characterized by a *distributed disordered system*. Its attribute displays a complete pattern of variation in space—in two or three dimensions—as well as variation in time. All these examples are modelled in 'probabilistic' terms, including the Yi–Chan model that has a transition of population and employment from one time-period to another. Whether *formal* treatment of uncertainty is warranted in a particular situation depends on such factors as the quality and quantity of information, the importance of the problem, and the resources at hand.

Such a *space–time process* is characterized by active and inherent uncertainty: properties at different points in space change randomly with time. Measurement may be taken at selected points in time and space, or continually (in time) at specific locations. Examples include the arsenic sampling at selected pumps to monitor ground-water pollution. The measurements may be used for forecasting the future values of the process at given locations and times, the future values of system-wide averages, or measures of performance. This is the case with the U.S. Defense Department's worldwide forecast of events-of-interest. Many other problems in such fields as energy, water, environmental quality, and natural-hazard protection fall into the random-field model. In these processes, extreme deviation from expected values will occasionally occur following laws of probability. The most dramatic illustration of this is the bifurcation forecast in the land-use development in York, Pennsylvania by the Yi–Chan model. In decision situations, it offers an opportunity for careful risk-assessment. It considers the tradeoff between benefits and risks under these extreme conditions. The decisions may involve siting, design, or maintenance of manmade facilities, regulations to enhance environmental quality, action intended to stimulate economic activity, or the design of a network of instruments to monitor or acquire basic data about a distributed disordered-system.

According to Vanmarcke (1983), a pragmatic view point of the question of scale in random-field modelling is that there invariably exist a time- and distance-scale below which micro-scale variation of the attribute are

a.   Not observed or observable, and/or
b.   Of no practical interest to the problem at hand, and/or
c.   Characterized by a deterministic micro-structure.

This is certainly the case with a remote-sensing image, where the resolution is driven by the problem at hand. Thus a 4 km (2.5 mi) resolution would be quite sufficient for a weather satellite, while a traffic-monitoring satellite may require a 1 meter (1.11 yd.) resolution (McCord et al. 1996). The Yi–Chan land-use model is really a deterministic model in its zonal population/employment attributes. It is only conceptually treated as a random field in that inter-zonal population/employment movements over time can be viewed as probabilistic, as we have done in the "Spatial Time-Series" chapter.

Often the same argument can be made with respect to very slow variations occurring on a macro-scale. The gradual change in wing length of adjacent Asiatic-bird populations certainly falls into this category. It takes generations to observe such a spatial gradient in a map of Southeast Asia. Thus it can be seen that random-field theory is applicable to a wide spectrum of distributed disordered-systems, when viewed with an appropriate range on the time or distance scale. It provides the methodology for description, analysis, and sometimes, for prediction and control of random systems. The only reason we are monitoring the ground water at the Fernald plant outside Cincinnati, Ohio—for example—is to prevent any toxicity to spread to the general population. Thus pumping is used at selected wells to divert the flow of contaminated ground-water from the population.

The statistics of the partial derivatives of a random field z_t contain important information about fluctuation level, excursion, and extremes. They govern the ability for us to model the process as a random field. In a single time- or distance-dimension, a *mean-squared derivative* is defined as $\dot{z}_t = (z_{t+\Delta t} - z_t)/\Delta t$. Being a stationery series, the mean of $\dot{z}_t$ is zero. Its variance is

$$\sigma_{\dot{z}}^2 = \frac{1}{(\Delta t)^2}\left[2\,E(z_t^2) - 2\,E(z_t z_{t+\Delta t})\right]$$
$$= \frac{2}{(\Delta t)^2}\left[\sigma_z^2 - B_z(\Delta t)\right] \qquad\qquad (12\text{-}1)$$

If the interval $\Delta t \to 0$, the mean-squared derivative will exist if and only if $B_z(\Delta t)$ has the following limiting form as $\Delta t \to 0$:

$$B_z(\Delta t) = \sigma_z^2 - \sigma_{\dot{z}}(\Delta t)^2/2. \qquad\qquad (12\text{-}2)$$

This further implies that the first derivative of $B_z(\tau)$ must be zero at $\tau = 0$: $\dot{B}_z(0) = \left[dB_z(\tau)/d\tau\right]_{\tau=0}$ $= 0$. Hence, for random processes to possess a mean-squared derivative, $\dot{B}_z(0)$ must be zero, and the covariance function must have the form of Equation **(12-2)** near the lag-origin.

The existence of a mean-squared derivative is a rather restrictive assumption. The approach based on local averages resolves a long-standing issue of how to deal with nonexistence of mean-squared derivative of stationary random-functions. This problem is solved, for all practical purposes, by permitting a small amount of local averaging, sufficient to smooth the micro-scale fluctuations—fluctuations that are either natural or manmade. The local-average processes possess stable measures of fluctuation patterns such as the mean-squared slope and the apparent frequency of oscillation. These quantities depend directly on the variance function that fully captures second-order information about the process. Here second-order characteristics describe variation and correlation in point fields. The theory based on local averages applies to random processes that are continuous in state and parameter . It is also applicable to discrete-state, discrete-parameter random-fields, and random impulse-processes. If the correlation parameters are finite, the derived local-average field becomes a continuous-parameter Gaussian-field under sufficient local-averaging.

Throughout this book, we have seen examples of local-averaging using spatial weights. As the reader may recall, spatial dependence is usually expressed in terms of weights w_{ij} between units i and j, where a larger weight connotes a heavier dependence. These weights are typically applied through the lth-order weight-matrix $\mathbf{W}^{(l)} = [w_{ij}^{(l)}]$, with the normalized spatial weights sum to unity $\sum_j w_{ij}^{(l)} = 1$. Here $\mathbf{w}^{(l)} = (\leftarrow w_{ij}^{(l)} \rightarrow)^T$ is the vector of spatial weights associated with the lth contiguity-class. An example is the weights associated with such lth-order neighbors as the rook, bishop or queen contiguity etc. Notice this is equivalent to the spatial lag-operator $L^{(l)}(\cdot)$. For example, one may pre-process image-data **y** by removing the subject ith entry. We replace it with a value resulting from 'filtering' with a spatial 'mask' $\mathbf{W}^{(l)}$ of order l.

The interaction between centroids or generators can in turn be described by different models. Among them is the venerable gravity model. Here, the weight matrix assumes the form $\mathbf{W}^j$—an activity derivation-allocation, transition-matrix for the jth activity in such a model as the Yi–Chan model. The gravity model suggests that closer-by and more-intense activities interact more than distant or low-intensity activities. The spatial mask simply summarizes these inter-centroid interactions in a matrix form. Viewing the gravity model as a generalization of 0–1 location-allocation models, min-sum facility location models are nothing other than special cases of inter-centroid spatial masks.

Thus far, w_{ij} assumes a continuous value between zero and one above ($0 \leq w_{ij} \leq 1$), as shown in the "Spatial-Temporal Information" chapter. A general spatial statistic can be defined using a more general weight $w_{ij}(d)$. Here, $w_{ij}(d)$ can take on binary valuations, assuming unity when an activity at j is within a generic-distance d from i. We have already alluded to the discrete facility-location example in the last paragraph. Instead of a summary of a set of data, the general spatial-statistic $G_i[w_{ij}(d)]$ has a natural origin such as the subject unit-of-interest. As such, the statistic is suitable for analyzing nonhomogeneous data that has spatial properties varying from point to point. We have seen examples of such a spatial weight, as illustrated in spatial econometric models. In these spatial-econometric models, we are explaining spatial dependence between the variables defined at various 'locations'. To calibrate a homogeneous model, we properly define spatially-lagged variables with predefined weights.

Local averaging in space has an analogue in the time dimension. A linear filter, for example, takes a weighted sum of a time-series to transform it into another time-series. An ARMA time-series can be thought of a 'smoothing' operation on a time-sequenced set of data over specified lags. In multivariate time-series (including spatial time-series), linear combinations of these lagged series are then formed by putting different weights on each series. This again is some sort of averaging across these individual time-series. A useful property now is *ergodicity* —a concept borrowed from memoryless random process called Markov chain. Ergodicity ensures that, on the average, two events will be independent in the limit. Now recall that by definition a Markov chain is ergodic if all states in the state-transition chain are recurrent, aperiodic, and communicate with each other. An ergodic assumption allows for consistent estimation of the joint-probability of various variables in a spatial time-series. Through proper local-averaging, such properties can often be obtained. While these concepts carry over from the time domain to spatial data in general, the process is much more complex in space, as previously mentioned. The important result is that the local-average process and the associated variance will be mean-squared differentiable, even if the mean-squared derivative of the original random field does not exist.

12.8 Decisions based on real-time information

In the discussions above, we have introduced remote-sensing as a viable source of real-time information (particularly after processing). To the extent that aerial photography is one kind of remote sensing, the idea is not new. The large number of channels available in today's satellites, however, makes available information that the bare eye cannot see, affording infrared and heat-sensing signals that are essential in transportation, reconnaissance, and target-location applications. Geosynchronized satellite constellations also provide real-time raster images to a fine level of resolution, not to say the Global Positioning System that is used extensively for navigation (U.S. Department of Transportation 1995a and 1995b, Parkinson et al. 1995). Geographic information systems (GIS) allow the merging of data from diverse sources—from remote-sensing to survey and interview data. Modern data-processing capabilities such as relational data-base and object-oriented programming do not only facilitate data fusion, but also greatly streamline modelling applications, including spatial analysis. To the extent that spatial relationship is the basic building block for transportation and locational modelling, GIS becomes an integral part of today's analysis toolkit (Chan 1996). It has been shown in this book that a very desirable focus of GIS is problem solving. With the convenience of electronic data-transfer, GIS is also a *Global* information system, affording truly distributed decision-making to take place (Ran et al. 1998, Peng and Beimborn 1998). This is facilitated by international spatial data-base standardization efforts such as the one led by Mollering (1997). It is my contention that through remote-sensing and through a very careful planning of the data structure, location, transport and land-use analyses can be readily performed. Here, we use models based on a spatial-oriented set of data-processing procedures, including spatial statistics. Aside from such descriptive tools as spatial statistics, prescriptive tools can also be easily incorporated into a GIS. For example, optimization procedures based on the "generalized algebraic modelling language" builds heavily upon arrays that are organized in certain ways, representing vectors and matrices in a mathematical-programming model. These vectors and matrices can possibly be extracted directly from a GIS through relational-data organization. The same arguments hold for recent emphasis on spreadsheet-based management-tools ranging from optimization to simulation. Voronoi diagrams also provide an exciting approach to spatial-data organization (Baccelli and Zuyev 1999). The developers of GIS should keep this in mind in their future endeavors.

We have shown in this book that there are some very basic principles involved in spatial-temporal analysis—a term that encompasses facility-location and land-use. Instead of calling upon a variety of analysis tools, data-oriented computation tools can be easily embedded into the data-base. The first example is the efficient space-filling-curve location-routing heuristic, which builds directly on the latitude-longitude coding of a location and an additional ancillary data-base[7]. By pre--

---

[7]For details, see the discussion in the "Spatial Separation and Routing" chapter.

processing the inter-point distance data as both candidate and demand strings, an $O(n)$ algorithm is found to locate facilities for a study area that can have up to 3,000 nodes.[8] A third example is a simple "look-ahead" capability in a spatial data-base that will allow real-time diversion of vehicles in case of unexpected demands (Regan et al. 1995). In reconnaissance, spatial pattern-recognition builds directly upon the concept of contiguity, which is easily implemented on top of a spatial data-base through a weight matrix. It turns out that such a weight matrix and the associated spatial-cost function constitute the common vehicle to effect gaming and competition.[9] This is above the classic function of placing facilities, population and other activities such as employment on a plane or network. Among other procedures, we have shown how the spatial-temporal-autoregressive-moving-average (STARMA) model or its specialized econometric form can be used to allocate economic activities over time. The allocation of these spatial resources is further assisted by the time-honored analysis tool called Voronoi diagrams—a technique based heavily on a spatial data-base. These examples provide convincing arguments for a brand new, simple and robust way of performing spatial-temporal analysis based directly on a data-base. Most important, these analysis tools can be truly executed on a real-time basis once the basic data-base (to be contrasted with the derived data-base) is in place.

Example

Floating cars are traditionally used to collect travel-time and delay information (U.S. Department of Transportation 1997a). Instead of manual labor, Griesenbeck et al. (1998) and Quiroga/Bullock (1998) used GPS receivers to log time and position points of a floating car. Quiroga/Bullock's spatial model uses highway segments of 0.2 miles to record time, local coordinates, and speed of probe vehicles every second. This segment length was necessary to detect changes in speed due to physical discontinuities like signalized intersections, ramps, and interchanges. This means that traditional link-based segments, which are typically longer than 0.5 miles, are not sufficient to characterize localized effects properly. The sampling rate was based on the effect of different time-intervals between consecutive GPS data points on segment GPS data coverage and segment speed variability. The sampling period between consecutive GPS points has to be less than half the shortest segment travel-time to achieve a 100-percent segment coverage. Quiroga/Bullock (1998) compared harmonic mean speeds and median speeds as estimators of central tendency of segment speed. Since traffic flow is dynamic both in space and time, segments shorter or longer than 0.2 miles maybe more appropriate in a particular site. The authors used error propagation theory to compare errors in GPS position fixes with errors in GPS speeds. They concluded that using GPS speeds was preferable for computing segment speeds. In a parallel study,

Griesenbeck et al. (1998) reported that the GPS receiver and laptop configurations required an average of 30 bytes per data point; so a typical one-hour run would require over

---

[8]For details, refer to the "Remote-sensing GIS" chapter in Chan (2000).

[9]This is so since the weight matrix is a summary of the gravity model or entropic models. For details, see the "Spatial Temporal Information" chapter.

100 kilo-bytes of memory. Once collected, the GPS data-files were corrected for selective-signal availability and/or ionospheric distortion using publicly available base-station correction files. Base-station files used for correcting the GPS points were based on five-second intervals, while the data points were taken at one-to-three second intervals. The goal was one-second intervals, but low signal-strength or quality eliminated many points, and as a result, the data file included some two- and three-second intervals. Once collected, the data points were linked spatially to a previously prepared map-layer of numbered roadway segments and buffers. This step added a segment identifier to each data-point in the corrected GPS data-file. The edited data-point files were converted to a data-based format readable by a standard statistical software-package, which was subsequently used for reduction and data processing. It was found that GPS and GIS can be used to collect travel-time and delay data. Due to the computational procedure, however, travel-time data is more reliable than the delay data. The total effort was accomplished at half the previous cost using manual labor. The GIS graphic displays of travel-time and delay data are also a highly desirable feature. □

12.9 Epilogue

In replicating real-world policies, facility-location and land-use models produce structures in terms of spatial organization, despite such assumptions as spatial homogeneity. In planar location models, for example, transportation costs are assumed isotropic or even constant over space. At the outset, accessibility and production opportunities never favor any spatial locations. However, the combination of accessibility and spatial setting eventually discriminates between center and periphery, resulting in spatial structure of the most interesting kind.

Correspondingly, facility location and land development require careful analysis procedures. Besides data analyses, a number of multicriteria decision problems need to be solved. They involve *competition* of individual criterion (Zelany 1991), rather than a predetermined set of priorities as often represented in utility theory. The weights of importance would reflect relative power or strength of competing 'individuals'. Of equal importance is the organizational embedding of decisions. Organizations that derive their structure, strategy and motivation from the singleness of purpose cannot be conducive to the notions of multiple criteria. This contrasts with pluralistic societies that we often associate with the free-enterprise system (Zelany 1992). Oftentimes, an interactive decision-making procedure is necessary for modelling the competition between criteria and the competition between interest parties.

According to Wyman and Kuby (1993), location modelling can be used not just for technology assessment but for technology specification. The technology need not exist, only that the technology's predicted benefit be greater than the predicted development cost. The engineer will gladly come with any process or design—within current capabilities—demanded by management. What is lacking from management is awareness of how the entire spatial, social, and economic picture might be changed for the better by alternative technology. The gap between engineering possibility and industrial or governmental investment would be no more than a lack of spatial-temporal information in costs, risk, and equity terms.

Such information facilitates the dialogue between the technical community and the managerial community. Currently, industries and governments are much less accustomed to requesting this type of information. For example, what should a technology's performance be to meet locational constraints over time? To be non-inferior? To dominate on two or more criteria? To what size must a facility or facilities be down-scaled or up-scaled to improve benefits and costs? These are questions that can be answered, at least in part, by the analysis techniques proposed in this volume. This way, it allows the analyst to get out in front of technological change—to show what could be—rather than simply reacting to change. In brief, let spatial-temporal information technology guide the development of other infrastructure technologies.

We started this book with a disclaimer on providing a panacea for all ailments. Neither did we subscribe to the notion of holistic planning in which all factors or purposes are embedded in the analysis framework. In spite of the advances reported in this book, we are still much more comfortable with answering one question at a time, or at most a subset of related questions. We also explicitly point out the absence of value-free analysis. The state-of-the-art have progressed by leaps-and-bounds over the past decades in facility-location and land-use, and this volume tries at length to point out the relationship between these advances. But it is only the beginning of our efforts, and the journey is a long, windy and endless one toward the debatable goal of a 'comprehensive' model. We take pleasure in traversing a path that leads toward what lies ahead—a path that may never be obliterated nor ended by virtue of future developments in the field. We have pointed out the commonalities and the differences between facility-location decisions and land-development decisions. We have highlighted the strength of these methodologies—particularly in their capacity to bring the obscure to our awareness. More important, in these pages we have grown accustomed to their limitations as well—not the least of which is their frailty to model the many profound human dimensions.

REFERENCES & BIBLIOGRAPHY

Chapter 1 – Introduction

American Society of Civil Engineers (1987). *Urban Planning Guide*. New York, New York.

Antoine, J; Fischer, G; Makowski, M (1997). "Multiple criteria land use analysis." *Applied Mathematics and Computation*, Vol. 83, pp. 195–215.

Brennan, M. W. (1999). "How e‑commerce will transform industrial RE." *National Real Estate Investor Magazine*, October 1, p. 88.

Brewer, W. E.; Alter, C. P. (1988). *The Complete Manual of Land Planning and Development*. Prentice Hall, Englewood Cliffs, New Jersey.

Chan, Y. (2000). *Locaton Theory and Decision Analysis*, South-Western, Cincinnati, Ohio.

Chapin, F. S.; Kaiser, E. J. (1979). *Urban Land Use Planning*, Third Edition, University of Illinois Press, Urbana, Illinois.

Civil Engineering (1998). "'Growth' in land use criteria defined." May, p. 29.

Churchill, C. J.; Baetz, B. W. (1998). "Development of a decision support system for sustainable community design." Working Paper, Department of Civil Engineering, McMaster University, Hamilton, Ontario, Canada.

Davis, K. P. (1976). *Land Use*. McGraw Hill, New York, New York.

Divis, D. A. (2000). "Remote regs, SRTM, and financing NSDI." *Geo Info Systems*, pp. 18–20. Engelhardt, J. (2000). "What's E‑commerce have to do with GIS?" *Geo Info Systems*, January, p. 58. *Geo Info Systems* (2000). "GIS goes mobile." January, p. 14.

Gould, J; Golob, T F (1998). "Will electronic home shopping reduce travel?" *Access*, No. 12, Spring, pp. 26–31 (University of California–Berkely).

Jha, M. K.; Schonfeld, P. (2000). "Integrating genetic algorithms and GIS to optimize highway alignments." Pre‑print 00–1060, 79th Transportation Research Board Annual Meeting, January, Washington, D.C.

Karkazis, J.; Thanassoulis, E (1998). "Assessing the effectiveness of regional development policies in Northern Greece using data envelopment analysis" *Socio‑Economic Planning Sciences*, Vol. 32, No. 2, pp. 123–137.

Kikenny, M.; Thisse, J.‑F. (1999). "Economics of location: A selective survey." *Computers & Operations Research*, Vol. 26, pp. 1369–1394.

Love, R. F.; Morris, J. G.; Wesolowsky, G. O. (1988). *Facility location: Models and Methods*. North–Holland, New York, New York.

Louis Berger & Associates (1998). *Guidance for Estimating the Indirect Effects of Proposed Transportation Projects*. Transportation Research Board, National Cooperative Highway Research Program Report 403, National Academy Press, Washington D.C.

Massam, B. H. (1988). "Multi‑criteria Decision Making Techniques in Planning." *Progress in Planning*, Vol. 30, Part I, Pergamon Press.

McNerney, M. T. (2000). "The state‑of‑the‑art in airport infrastructure management using geographic information systems." Pre‑print, 79th Transportation Research Board Annual Meeting, January, Washington, D.C.

Quiroga, C. A.; Bullock, D. (1999). "Travel time information using global positioning system and dynamic segmentation techniques." *Transportation Research Record*, No. 1660, pp. 48–57.

ReVelle, C (1997). "A perspective on location science" *Location Science*, Vol. 5, No. 1, pp. 3–13.

Rodrigue, J.-P. (1997). "Parallel modelling and neural networks: an overview for transportation/land use systems." *Transportation Research C*, Vol. 5, No. 5, pp. 259–271.

Sen, A.; Smith, T. E. (1995). *Gravity Models of Spatial Interaction Behavior.* Springer–Verlag, Berlin, Germany.

Thrall, G. I.; McClanahan, M.; Elshaw–Thrall, S. (1995). "Ninety years of urban growth as described with GIS: a historic geography." *Geo Info Systems*, April, pp 20–45.

Shefer, D; Amir, S; Frenkel, A; Law–Yone, H (1997). "Generating and evaluating alternative regional development plans" *Environment and Planning B: Planning and Design*, Vol. 15, pp. 7–22.

Solomon, I.; Mokhtarian, P. L. (1998). "What happens when mobility-inclined market segments face accessibility-enhancing policies?" *Transportation Research D*, Vol. 3, No. 3, pp. 129–140.

Tobler, W. R. (1969). "Geographical filters and their inverses." *Geographical Analysis*, Vol. 1, pp. 234–253.

Todtling, F. (1992). "Technological change at the regional level: the role of location, firm structure, and strategy." *Environment and Planning A*, Vol. 24, pp. 1565–1584.

Transportation Research Board (1997). *Information Needs to Support State and Local Transportation Decision Making into the 21st Century.* Conference Proceedings 14, National Academy Press, Washington, D.C.

Yun, D.-S.; Kelly, M. E. (1997). "Modeling the day-of-the-week shopping activity and travel patterns." *Socio-Economic Planning Sciences*, Vol. 31, No. 4, pp. 307–319.

Chapter 2 – Facility-Location Models

The field of facility location is one of the most active areas of research. The survey in this chapter represents an introduction to this subject. Readers who are interested in more in-depth study of facility location will find the following reference list helpful. The reference list contains literature up to the time of writing, and are organized into these categories for the reader's convenience: (a) general, (b) plant location, (c) planar location, (d) p-median, (e) maximal coverage and set covering, (f) center location and obnoxious facility, (g) multi-criteria facility location, and (h) others.

General

Ahituv, N.; Berman, O. (1988). *Operations Management of Distributed Service Networks: A Practical Quantitative Approach*, Plenum Press, New York.

Aykin, T. C. (1988). "On the location of hub facilities." *Transportation Sciences*, Vol. 22, pp. 155–157.

Batta, R.; Palekar, U. (1988). "Mixed planar/network facility location problems." *Computers and Operations Research*, Vol. 15, pp. 61–67.

Beasley, J. E. (1990). "Lagrangian heuristics for location problems." working paper, The Management School, Imperial College, University of London, London, England.

Berlin, G. N.; ReVelle, C. S.; Elzinga, D. J. (1976). "Determining ambulance–hospital locations for on-scene and hospital services." *Environment and Planning A*, Vol. 8, pp. 553–561.

Berman, O.; Simchi-Levi, D.; Tamur, A. (1988). "The minimax multistop location problem on a tree." *Networks*, Vol. 18, pp. 39–40.

Brandeau, M. L.; Chiu, S. S. (1991). "Parametric analysis of optimal facility locations." Networks, Vol. 21, pp. 223–243.

Chan, Y. (2000). *Location Theory and Decision Analysis*, ITP/South-Western/Cincinnati, Ohio.

Chen, R. (1988). "Conditional minimum and minimax location–allocation problems in Euclidean space." *Transportation Science*, Vol. 22, pp. 157–160.

Chhajed, D; Lowe, T. J. (1992). "m-median and m-center problems with mutual communication: solvable special cases." *Operations Research*, Vol. 40, pp. S56–S66.

Church, R.; ReVelle, C. S. (1976). "Theoretical and computational links between the p-median, location set covering and the maximal covering location problem." *Geographical Analysis*, Vol. 8, pp. 406–415.

Cohon, J. L.; Eagles, T. W.; Margalies, T. S.; ReVelle, C. S. (1982). "Population/cost tradeoffs for nuclear reactor siting policies." Operations Research Group Report Series Paper #81-04, Dept. of Geography and Environmental Engineering, The Johns Hopkins University, Baltimore, Maryland.

Colorni, A. (1987). "Optimization techniques in locational modelling." in *Urban Systems Contemporary Approaches to Modelling* (edited by O.S. Bertuglia et al.), Croom–Helm, Kent, Great Britain.

da Conceicao Cunha, M.; Pais Antunes, A. (1997) "On the efficient location of pumping facilities in an aquifer system." *International Transactions in Operational Research*, Vol. 4, No. 3, pp. 175–184.

Daskin, M. (1995). *Network and Discrete Location: Models, Algorithms and Applications*, Wiley-InterScience, New York, New York.

de Matta, R.; Hsu, V. N.; Love, T. J. (1999) "The selection allocation problem." *Naval Research Logistics*, Vol. 46, pp. 707–725.

Drezner, Z.; Gravish, B. (1985). "ε-approximation for multidimensional weighted location problems." *Operations Research*, Vol. 33, pp. 772–782.

Drezner, Z. – editor (1995). *Facility Location: A Survey of Applications and Methods*, Springer–Verlag, New York, New York.

Eiselt, H. A. (1992). "Location modeling in practice." *American Journal of Mathematical and Management Sciences*, Vol. 12, No. 1, pp. 3–8.

Engberg, D.; Linke, E.; Cohon, J.; ReVelle, C. S. (1982). "Siting an offshore natural gas pipeline using a mathematical model." *National Development*, June/July, pp. 93–100.

Erkut, E.; Francis, R.; Lowe, T.; Tamir, A. (1987). "Equivalent mathematical programming formulations of monotonic tree network location problem." working paper, Dept. of Finance and Management Science, University of Alberta, Edmonton, Canada.

Evans, J.; Minieka, E. (1992). *Optimization Algorithms for Networks and Graphs*, Dekker.

Francis, R. L.; Lowe, T. J. (1990). "On the worst-case aggregation analysis for network location problems." Research Report No. 90–13, Industrial and Systems Engineering Department, University of Florida, Gainesville, Florida.

Francis, R. L.; Lowe, T. J.; Tamir, A. (2001). "Worst-case incremental analysis for a class of p-facility location models." Working Paper, Department of Industrial and Systems Engineering, University of Florida, Gainesville, Florida.

Francis, R. L.; McGinnis, L. F.; White, J. A. (1999). *Facility layout and location: An analytical approach*, 3rd Edition, Prentice Hall, Upper Saddle River, New Jersey.

Francis, R.; McGinnis, L.; and White, J. (1983). "Locational analysis." *European Journal of Operations Research*, Vol. 12, pp. 220–252.

Galvao, R. D. (1993). "The use of Lagrangian relaxation in the solution of uncapacitated facility location problems." *Location Science*, Vol. 1, pp. 57–79.

Ghosh, A.; Harche, F. (1993). "Location–allocation models in the private sector: progress, problems, and prospects." *Location Science*, Vol. 1, pp. 81–106.

Goldman, A. J. (1972). "Approximate localization theorems for optimal facility placement." *Transportation Science*, Vol. 6, pp. 195–201.

Hakimi, S L; Labbe, M; Schmeichel, E F (1994). "Locations on time-varying networks." Discussion paper 94/05, Service de Mathematiques de la Gestion, Universite Libre de Bruxelles, Brussels, Belgium.

Hakimi, S. L. (1964). "Optimal location of switching centers and the absolute center and the medians of a graph." Operations Research, Vol. 12, pp. 450–459.

Hakimi, S. L.; Kuo, C-C (1988). "On a general network location–production–allocation problem." working paper, Dept. of Electrical Engineering and Computer Science, University of California at Davis, Davis, California.

Handler, G. Y.; Mirchandani, P. B. (1979). *Location on Networks*: Theory and Algorithms, MIT Press, Cambridge, Massachusetts.

Hansen, P.; Thisse, J-F; Wendall, R. (1986). "Equivalence of solutions to network location problems." *Mathematics of Operations* Research, Vol. 11, pp. 672–678.

Hansen, P.; Peeters, D.; Richard, D.; Thisse, J. F. (1985). "The minisum and minimax location problems revisited." *Operations Research*, Vol. 33, no. 6, pp. 1251–1265.

Hansen, P.; Labbe, M.; Thisse, J-F. (1990). "From the median to the generalized center." working paper, GERAD, Ecole des Hautes Etudes Commerciales, Montreal, Canada.

Hockbaum, D. (1984). "When are NP-hard location problems easy?" *Annals of Operations Research*, Vol. 1, pp. 201–214.

Hooker, J. N. (1989). "Solving nonlinear multiple-facility network location problems." Networks, Vol. 19, pp. 117–133.

Hurter, A. P.; Martinich, J. S. (1989). *Facility location and the theory of production*, Kluwer Academic Publishers.

Kara, B. Y. (1999). "Modeling and Analysis of Issues in Hub Location Problem." Doctoral Dissertation, Department of Industrial Engineering, Balkent University, Ankara, Turkey.

Keeney, R. L. (1980). *Siting Energy Facilities*, Academic Press.

Krarup, J.; Pruzan, P. M. (1981). "Reducibility of minimax to minisum 0–1 programming problems." *European Journal of Operational Research*, Vol. 6, pp. 125–132.

Krarup, J.; Pruzan, P. M. (1990). "Ingredients of locational analysis." in *Discrete Location Theory* (edited by P. B. Mirchandani and R. L. Francis), Wiley-Interscience, New York, New York, pp. 1–54.

Levin, Y.; Ben-Israel, A. (2001). "A heuristic method for large-scale multifacilty location problems." Working Paper, Rutcor, Rutgers University, New Brunswick, New Jersey.

Love, R. F.; Morris, J. G.; Wesolowsky, G. O. (1988). *Facility Location: Models and Methods*. North Holland, New York, New York.

Marks, D. H.; ReVelle, C.; Liebman, J. (1970). "Mathematical models of location: a review." *Journal of the ASCE Urban Planning and Development Division*, Vol. 96, No. UP1, pp. 81–93.

Marsh, M. T.; Schilling, D. A. (1994). "Equity measurement in facility location analysis: a review and framework." *European Journal of Operational Research*, Vol. 74, pp. 1–17.

Minieka, E. (1985). "The optimal location of a path or tree in a tree network." *Networks*, Vol. 15, pp. 309–321.

Mirchandani, P. (1987). "Generalized hierarchical facility locations." *Transportation Science*, Vol. 21, pp. 123–125.

Mirchandani, P. B.; Francis, R. L. – Editors (1990). *Discrete Location Theory*, Wiley-Interscience, New York, New York.

Moon, I. D.; Chaudhry, S. S. (1984). "An analysis of network location problems with distance constraints." *Management Science*, Vol. 30, pp. 290–307.

Moore, K.; Chan, Y. (1990). "Integrated location- and- routing models: Model examples." Working paper, Dept. of Operational Sciences, Air Force Institute of Technology, Wright–Patterson AFB, Ohio.

Mouchahoir, G. E. (1983). "Regional cargo transportation center: definition of concept and optimal network." *Transportation Quarterly*, Vol. 37, No. 3, pp. 355–377.

ReVelle, C.; Church, R.; Schilling, D. (1975). "A note on the location model of Holmes, Williams and Brown." *Papers of the Regional Science Association*, Vol. 7, No. 4.

ReVelle, C.; Marks, D.; Liebman, J. C. (1970). "An analysis of private and public sector location models." *Management Science*, Vol. 15, No. 11, pp. 692–707.

ReVelle, C. (1991). "Siting ambulances and fire companies: new tools for planners." *Journal of the American Planning Association*, Vol. 57, No. 4, pp. 471–484.

Rojeski, P.; ReVelle, C. (1972). "Central facilities location under an investment constraint." *Geographical Analysis*, Vol. 2, pp. 343–360.

Rushton, G. (1989). "Applications of location models." *Annals of Operations Research*, Vol. 18, pp. 25–42.

Schilling, D. A.; Jayaraman, V.; Barkhi, R. (1993). "A review of covering problems in facility location." *Location Science*, Vol. 1, pp. 25–55.

Scott, A. J. (1971). *Combinatorial programming, spatial analysis and planning*, Methuen, London.

"S-Distance: A Spatial Decision Support System."(www.prd.uth.gr/res_labs/spatial_analysis/software) Laboratory of Spatial Analysis, GIS and Thematic Mapping, Department of Planning and Regional Development, University of Thessaly, Greece.

Sule, D. R. (1988). *Manufacturing Facilities: Location, Planning and Design*, Second Edition, ITP/PWS, Boston, Massachusetts.

Swain, R. W. (1974). "A parametric decomposition approach for the solution of uncapacitated location problems." *Management Science*, Vol. 21, No. 2, pp. 189–198.

Swink, M.; Robinson, E. P. (1990). "Reason based solutions and the complexity of distribution network design problems." Working paper, Operations Management Department, School of Business, Indiana University, Bloomington, Indiana.

Tamir, A. (1992). "On the complexity of some classes of location problems." *Transportation Science*, Vol. 26, No. 4, pp. 352–354.

Tamir, A.; Perez–Brito, D.; Moreno–Perez, J. A. (1998) "A polynomial algorithm for the p-centdian problem on a tree." *Networks*, Vol. 32, pp. 255–262.

Taniguchi, E.; Noritake, M.; Yamada, T.; Izumitani, T. (1999) "Optimal size and location planning of public logistics terminals." *Transportation Research–E*, Vol. 35, pp. 207–222.

Tansel, B.; Francis, R.; Lowe, T. (1983). "Location on networks: A survey—Parts I and II." *Management Science*, Vol. 29, No. 4, pp. 482–511.

Ting, S. S. (1984). "A linear- time algorithm for maxisum facility location on tree networks." *Transportation Science*, Vol. 18, pp. 76–84.

Toregas, C. (1971). "Location under maximal travel time constraints." Ph.D. thesis, Cornell University, Ithaca, New York.

Toregas, C.; ReVelle, C. (1972). "Optimal location under time or distance constraints." *Regional Science Association Papers*, Vol. 28, pp. 133–143.

Toregas, C.; ReVelle, C. (1973). "Binary logic solution to a class of location problem." *Geographical Analysis*, Vol. 5, pp. 145–155.

Toregas, C.; Swain, R.; ReVelle, C. S.; Bergman, L. (1971). "The location of emergency service facilities." *Operations Research*, Vol. 19, No. 5, pp. 1363–1373.

Transportation Science, Vol. 28, No. 2 (1994). Focus issue on "Location Models and Algorithms".

Verter, V.; Dincer, M. C. (1994). "Facility location and capacity acquisition: an integrated approach." Research Report No. 94–2, Department of Finance and Management Science, University of Alberta, Edmonton, Alberta.

Xu, N.; Lowe, T. J. (1993). "On the equivalence of dual methods for two location problems." *Transportation Science*, Vol. 27, pp. 194–199.

Plant location

Bartezzaghi, E.; Colorni, A.; Palermo, P. C. (1981). "A search tree algorithm for plant location problems." *European Journal for Operations Research*, Vol. 7, pp. 371–379.

Cabot, A. V.; Erenguc, S. S. (1984). "Some branch–and–bound procedures for fixed cost transportation problems." *Naval Research Logistics*, Vol. 31, pp. 145–154

Dearing, P. M.; Hammer, P. L.; Simeone, B. (1992). "Boolean and graph theoretic formulations of the simple plant location problem." *Transportation Science*, Vol. 26, pp. 138–148.

Erlenkotter, D. (1978). "A dual–based procedure for uncapacitated facility location." *Operations Research*, Vol. 26, No. 6, pp. 992–1009.

Gao, L-L.; Robinson, E. P. (1994). "Uncapacitated facility location: general solution procedure and computational experience." *European Journal of Operational Research*, Vol. 76, pp. 410–427.

Hanjoul, P.; Hansen, P.; Peeters, D.; Thisse, J.F. (1990). "Uncapacitated plant location under alternative spatial price policies." *Management Science*, Vol. 36, No. 1, pp. 41–57.

Holmberg, K.; Jornsten, K. (1996). "Dual search procedures for the exact formulation of the simple plant location problem with spatial interaction." *Location Science*, Vol. 4, pp. 83–100.

Kennington, J.; Unger, E. (1976). "A new branch and bound algorithm for the fixed–charge transportation problem." *Management Science*, Vol. 22, pp. 1116–1126.

Khumawala, B. M. (1972). "An efficient branch and bound algorithm for the warehouse location problem." *Management Science*, Vol. 18, No. 12, pp. 718–731.

Klincewicz, J. G. (1990). "Solving a freight transport problem using facility location techniques." *Operations Research*, Vol. 38, No. 1, pp. 99–109.

Klincewicz, J.; Luss, H. (1987). "A dual–based algorithm for multiproduct uncapacitated facility location." *Transportation Science*, Vol. 21, No. 3, pp. 198–206.

Louveaux, F. V.; Peeters, D. (1992). "A dual–based procedure for stochastic facility location." *Operations Research*, Vol. 40, No. 3, pp. 564–573.

Monticone, L. C.; Funk, G. (1994). "Application of the facility location problem to the problem of locating concentrators on an FAA microwave system." *Annals of Operations Research*, Vol. 50, pp. 437–454.

Piersma, N. (1992). "Almost sure convergence of the capacitated facility location problem." working paper, Econometric Institute, Erasma University, Rotterdam, Netherlands.

ReVelle, C. S.; Laporte, G. (1996). "The plant location problem: new models and research prospects." *Operations Research*, Vol. 44, pp. 864–874.

Sankaran, J. K.; Raghavan, N. R. S. (1996). "Locating and sizing plants to bottle propane in South India." Working paper, Department of Management Science an Information Systems, University of Auckland, Auckland, New Zealand, 19 pp.

Sherali, H. D.; Ramchandran, S.; Kim, S-I. (1991). "A localization and reformulation discrete programming approach for the rectilinear distance location–allocation problem." working paper, Department of Industrial and Systems Engineering, Virginia Polytechnic Institute and State University, Blacksburg, Virginia.

Sherali, H. D.; Tuncbilek, C. H. (1992). "A squared–Euclidean distance location–allocation problem." *Naval Research Logistics*, Vol. 39, pp. 447–469.

Verter, V.; Dincer, M. C. (1992). "An integrated evaluation of facility location, capacity acquisition, and technology selection for designing global manufacturing strategies." *European Journal of Operational Research*, Vol. 60, pp. 1–18.

Zhu, Z. P.; ReVelle, C.; Rosing, K. (1989). "Adaptation of the plant location model for regional environmental facilities and cost allocation strategy." *Annals of Operations Research*, Vol. 18, pp. 345–366.

Planar location

Allen, W. R. (1995). "An improved bound for the multifacility location model." *Operations Research Letters*, Vol. 17, pp. 175–180.

Brimberg, J.; Love, R. F. (1993). "Global convergence of a generalized iterative procedure for the minisum location problem with l_p distances." *Operations Research*, Vol. 41, pp. 1153–1163.

Brimberg, J.; Chen, R.; Chen, D. (1998) "Accelerating convergence in the Fermat–Weber location problem." *Operations Research Letters*, Vol. 22, pp. 151–157.

Chen, P-C.; Hansen, P.; Jaumard, B.; Tuy, H. (1998). "Solution of the multi-facility Weber and conditional Weber problems by d.-c. programming." *Operations Research*, Vol. 46, No. 4, pp. 548–562.

Chen, P-C; Hansen, P.; Jaumard, B.; Tuy, H. (1992). "Weber's problem with attraction and repulsion." RUTCOR research report 13–92, State University of New Jersey, New Brunswick, New Jersey.

Drezner, Z.; Goldman, A. J. (1991). "On the set of optimal points to the Weber problem." *Transportation Science*, Vol. 25, No. 1, pp. 3–8.

Drezner, Z.; Wesolowsky, G. O. (1983). "Location of an obnoxious facility with rectangular distances." *Journal of Regional Science*, Vol. 23, pp. 241–248.

Gamal, M. D. H.; Salhi, S. (2003). "A cellular heuristic for the mutisource Weber problem." *Computers & Operations Research*, Vol. 30, Issue 11, pp. 1609–1624.

Hansen, P.; Mladenovic, N; Taillard, E. (1998) "Heuristic solution of the multisource Weber problem as a p-median problem." *Operations Research Letters*, Vol. 22, pp. 55–62.

Huang, S.; Batta, R.; Klamroth, K.; Nagi, R. (2002). "K-connection location problem in a plane." Working Paper, Department of Industrial Engineering, University of Buffalo, Buffalo, New York.

Hurry, D.; Farmer, R.; Chan, Y. (1995). "Airport–Location–Problem: Comparison of algorithms." Working paper, Dept. of Operational Sciences, Air Force Institute of Technology, Wright–Patterson AFB, Ohio.

Juel, H.; Love, R. F. (1986). "A geometrical interpretation of the existing facility solution condition for the Weber problem." *Journal of the Operational Research Society*, Vol. 37, No. 12, pp. 1129–1131.

Klamroth, K.; Wiecek, M. M. (1997) "A multiple objective planar location problem with a line barrier." Working Paper, University of Kaiserslautern, Kaiserslautern, Germany.

Mehrez, A.; Sinuany–Stern, Z.; Stulman, A. (1986). "An enhancement of the Drezner–Wesolowsky algorithm for single-facility location with maximin of rectilinear distance." *Journal of the Operational Research Society*, Vol. 37, pp. 971–977.

Melachrinoudis, E.; Xanthopulos, Z. (2001). "The Euclidean distance single facility location problem with the minisum and maximin objectives." Working Paper, Department of Mechanical, Industrial and Manufacturing Engineering, Northeastern University, Boston, Massachusetts.

Ohsawa, Y.; Imai, A. (1997). "Degree of locational freedom in a single facility Euclidean minimax location model." *Location Science*, Vol. 5, No. 1, pp. 29–45.

Rado, F. (1988). "The Euclidean multifacility location problem." *Operations Research*, Vol. 36, pp. 485–492.

Rosen, J. B.; Xue, G. L. (1993). "On the convergence of a hyperboloid approximation procedure for the perturbed Euclidean multifacility location problem." *Operations Research*, Vol. 41, pp. 1164–1171.

Sherali, H. D.; Tuncbilek, C. H. (1992). "A squared-Euclidean distance location–allocation problem." *Naval Research Logistics*, Vol. 39, pp. 447–469.

Wang, C.; Gao, C.; Shi, Z. (1997) "An algorithm for continuous type optimal location problem." *Computational Optimization and Applications*, Vol. 7, pp. 239–253.

Wesolowsky, G. O. (1993). "The Weber problem: history and perspectives." *Location Science*, Vol. 1, pp. 5–23.

p-median

Batta, R.; Mannur, N. R. (1990). "Covering-location models for emergency situations that require multiple response units." *Management Science*, Vol. 36, No. 1, pp. 16–23.

Berman, O.; Simchi–Levi, D. (1990). "Conditional location problems on networks." *Transportation Science*, Vol. 24, No. 1, pp. 77–78.

Butt, S. E.; Cavalier, T. M. (1997). "Facility location in the presence of congested regions with the rectilinear distance metric." *Socio-Economic Planning Sciences*, Vol. 31, No. 2, pp. 103–113.

Campbell, J. F. (1996). "Hub location and the p-hub median problem." *Operations Research*, Vol. 44, pp. 923–935.

Chaudhry, S. S.; Choi, I. C. (1995). "Facility location with and without maximum distance constraints through the *p*-median problem." *International Journal of Operations and Production Management*, Vol. 15, No. 10.

Chhajed, D.; Lowe, T. J. (1992). "M-median and M-center problems with mutual communication: solvable special cases." *Operations Research*, Vol. 40, pp. 56–66.

Chhajed, D.; Lowe, T. J. (1998). "Solving a selected class of location problems by exploiting problem structure: a decomposition approach." *Naval Research Logistics*, Vol. 45, pp. 791–815.

Choi, I-C.; Chaudhry, S. S. (1992). "The *p*-median problem with maximum distance constraints: a direct approach." working paper, Department of Industrial Engineering, Wichita State University, Wichita, Kansas.

Choi, I-C.; Chaudhry, S. S. (1993). "The *p*-median problem with maximum distance constraints: a direct approach." *Location Science*, Vol. 1, pp. 235–243.

Drezner, Z. (1995). "On the conditional *p*-median problem." *Computers and Operations Research*, Vol. 22, pp. 525–530.

Francis, R. L.; Lowe, T. J.; Rayco, M. B. (1993). "Row-column aggregation of rectilinear distance *p*-median problems." Research Report 93–5, Department of Industrial and Systems Engineering, University of Florida, Gainesville, Florida.

Hansen, P.; Labbe, M. (1989). "The continuous *p*-median of a network." *Networks*, Vol. 19, pp. 595–606.

Hribar, M.; Daskin, M. S. (forthcoming). "A dynamic programming heuristic of the *p*-median problem." *European Journal of Operational Research*.

Hsu, V. N.; Lowe, T. J.; Tamir, A. (1997) "Structured *p*-facility location problems on the line solvable in polynomial time." *Operations Research Letters*, Vol. 21, pp. 159–164.

Lefebvre, O.; Michelot, C.; Plastria, F. (1991). "Sufficient conditions for coincidence in minisum multifacility location problems with a general metric." *Operations Research*, Vol. 39, No. 3, p. 437–442.

Louveaux, F. (1986). "Discrete stochastic location models." *Annals of Operations Research*, Vol. 6, Baltzer, pp. 23–34.

Mirchandani, P. B. (1990). "The *p*-median problem and generalizations." in *Discrete Location Theory* (edited by P. B. Mirchandani and R. L. Francis), Wiley, New York, New York, pp. 55–118.

Mirchandani, P. B.; Odoni, A. R. (1979). "Location of medians on stochastic networks." *Transportation Science*, Vol. 13, pp. 85–97.

Sheraldi, H.; and Adams, W. (1984). "A decomposition algorithm for a discrete location–allocation problem." *Operations Research*, Vol. 32, No. 4, pp. 878–900.

Syam, S. S. (1997). "A model for the capacitated *p*-facility location problem in global environments." *Computers & Operations Research*, Vol. 11, No. 11, pp. 1005–1016.

Watson–Grandy, C. (1985). "The solution of distance constrained mini-sum location problems." *Operations Research*, Vol. 33, No. 4, pp. 784–802.

Xue, G. L.; Rosen, J. B.; Pardalos, P. M. (1996) "A polynomial time dual algorithm for the Euclidean multifacility location problem." *Operations Research Letters*, Vol. 18, pp. 201–204.

Maximal coverage and set covering

Aytug, H.; Saydam, C. (2002). "Solving large-scale maximum expected covering location problems by genetic algorithms: A comparative study."*European Journal of Operational Research*, Vol. 141, pp. 480–494.

Ball, M. O.; Lin, F. L. (1995). "Reliability, covering and balanced matrices." *Operations Research Letters*, Vol. 17, pp. 1–7.

Batta, R.; Dolan, J. M.; Krishnamurthy, N. N. (1989). "The maximal expected covering location problem: revisited." *Transportation Science*, Vol. 23, pp. 277–287.

Batta, R; Mannur, N. R. (1990). "Covering-Location Models for emergency situations that require multiple response units." *Management Science*, Vol. 36. pp. 16–23.

Berman, O. (1992). "The *p* maximal cover–*p* partial center problem on networks." working paper, Dept. of Management at Scarborough." University of Toronto, Toronto, Ontario, Canada.

Berman, O.; Drezner, Z.; Wesolowsky, G. O. (2003). "Locating service facilities whose reliability is distance dependant." *Computers & Operations Research*, Vol. 30, Issue 11, pp. 1683–1695.

Chaudhry, S. S. (1993). "New heuristics for the conditional covering problem." *Opsearch*, Vol. 30, pp. 42–47 (Operational Research Society of India).

Chaudhry, S. S.; Moon, I. D.; McCormick, S. T. (1987). "Conditional covering: Greedy heuristics and computational results." *Computers and Operations Research*, Vol. 14, pp. 1–18.

Chaudhry, S. S.; McCormick, S. T.; Moon, I. D. (1987). "Locating independent facilities with maximum weight: Greedy heuristics." *OMEGA*, Vol. 14, pp. 383–389.

Church, R.; ReVelle, C. S. (1974). "The maximal covering location problem." *Papers of the Regional Science Association*, Vol. 32, pp. 101–117.

Daskin, M.; Hogan, K.; ReVelle, C. (1988). "Integration of multiple, excess, backup and expected covering models." *Environment and Planning B.*, Vol. 15, pp. 15–35.

Daskin, M. (1983). "A maximum expected covering location model: formulation, properties and heuristic solution." *Transportation Science*, Vol. 17, pp. 48–70.

Gentili, M.; Mirchandani, P. B. (2004). "Locating active sensors on traffic networks." Working Paper, Computing Science Department, University of Salerno, Baronissi, Italy.

Hogan, K.; ReVelle, C. (1986). "Concepts and applications of backup coverage." *Management Science*, Vol. 32, pp. 1434–1444.

Hutson, V. A.; ReVelle, C. S. (1989). "Maximal direct covering tree problems." *Transportation Science*, Vol. 23, pp. 288–299.

Lin, J-J.; Feng, C-C. (2002). "A location design model for land uses, network and facilities in urban planning." Working Paper, Graduate Institute of Urban Planning, National Taipei University, Taipei, Taiwan.

Marianov, V.; ReVelle, C. (1994). "The queuing probabilistic location set covering problem and some extensions." *Socio-Economic Planning Sciences*, Vol. 28, pp. 167–178.

Martin, P. E. (1999) "A Multi-Service Location–Allocation Model for Military Recruiting." Master's Thesis, Code: SM/GK, Naval Postgraduate School, Monterey, California.

Mehrez, A.; Eben–Chaime, M.; Brimberg, J. (1996). "Locational analyses of military intelligence ground facilities." Working paper, Department of Industrial Engineering and Management, Ben-Gurion University of the Negev, Beer Sheva, Israel.

Moore, G.; ReVelle, C. (1982). "The hierarchical service location problem." *Management Science*, Vol. 28, pp. 775–780.

ReVelle, C.; Hogan, K. (1989). "The maximum availability location problem." *Transportation Science*, Vol. 23, pp. 192–199.

ReVelle, C.; Moore, G. C. (1982). "Hierarchical service location problem." *Management Science*, Vol. 28, No. 7, pp. 775–780.

Saccomanno, F. F.; Allen, B. (1988). "Locating emergency response capability for dangerous goods incidents on a road network." *Transportation Research Record*, No. 1193, National Academy of Sciences, Washington, D. C., pp. 1–9.

Saydam, C.; Aytug, H. (2003). "Accurate estimation of expected coverage: revisited." *Socio-Economic Planning Sciences*, Vol. 37, pp. 69–80.

Schick, W. G. (1992). "Locating an imaging radar in Canada for identifying space-borne objects." master's thesis, AFIT/GSO/ENS/92D–13, Department of Operational Sciences, Air Force Institute of Technology, Wright–Patterson AFB, Ohio.

Sheraldi, H. D.; Park, T. (2000). "Discrete equal-capacity p-median problem." *Naval Research Logistics*, Vol. 47, pp. 166–183.

Center location and obnoxious facility

Appa, G. M. (1993). "k-integrality, an extension of total unimodularity." *Operations Research Letters*, Vol. 13, pp. 159–163.

Averbakh, I.; Berman, O. (1997). "Minimax regret p-center location on a network with demand uncertainty." *Location Science*, Vol. 5, No. 4, pp. 247–254.

Berman, O.; Ingco, D. I.; Odoni, A. (1994). "Improving the location of minimax facilities through network modification." *Networks*, Vol. 24, pp. 31–41.

Berman, O. (1992). "The p maximal cover $-p$ partial center problem on networks." working paper, Dept. of Management at Scarborough, University of Toronto, Toronto, Ontario, Canada.

Berman, O.; Drezner, Z.; Wesolowsky, G. O. (1996). "Minimum covering criterion for obnoxious facility location on a network." *Networks*, Vol. 28, pp. 1–5.

Berman, O.; Simchi–Levi, D. (1990). "Conditional location problems on networks." *Transportation Science*, Vol. 24, No. 1, pp. 77–78.

Brimberg, J.; Juel, H. (1998) "On locating a semi- desirable facility on the continuous plane." *International Transactions in Operational Research*, Vol. 5, No. 1, pp. 59–66.

Chaudry, S. S.; Moon, I. G. (1987). "Analytical models for locating obnoxious facilities." in *Hazardous Material Disposal: Siting and Management*, Edited by M Chatterji, Gower Publishing Co.

Chhajed, D..; Lowe, T. J. (1990). "M- median and M- center problems with mutual communication: solvable special case." Faculty working paper no. 90–1654, Bureau of Economic and Business Administration, University of Illinois, Urbana–Champaign, Illinois.

Chen, M-L; Francis, R.; Lowe, T. (1988). "The 1-center problem: Exploiting block structure." *Transportation Science*, Vol. 22, p. 259–269.

Church, R. L.; Garfinkel, R. S. (1978). "Locating an obnoxious facility on a network," *Transportation Science*, Vol. 12, pp. 107–118.

Daskin, M. (1995). *Network and Discrete Location: Models, Algorithms and Applications.* Wiley-Interscience, New York, New York.

Dasrathy, B.; White, L. J. (1980). "A maxmin location problem" *Operations Research*, Vol. 28, pp. 1385–1401

Drezner, Z. (1989). "Conditional p- center problem." *Transportation Science*, Vol. 23, pp. 51–53.

Erkut, E.; Neuman, S. (1989). "Analytical models for locating undesirable facilities." *European Journal of Operational Research*, Vol. 40, pp. 275–291.

Francis, R. L.; Rayco, M. B. (1995). "Asymptotically Optimal Aggregation for Some Unweighted p- Center Problems with Rectilinear Distances." Research Report 95–4, Department of Industrial & Systems Engineering, University of Florida, Gainesville, Florida.

Jaeger, M.; Goldberg, J. (1997). "Polynomial alogrithms for center location on spheres." *Naval Research Logistics*, Vol. 44, pp. 341–352.

Keysor, G.; Totten, C.; Chan, Y. (1994). "Multi- criteria decision- making in locating p centers." Working paper, Department of Operational Sciences, Air Force Institute of Technology, Wright–Patterson AFB, Ohio.

Labbe, M. (1990). "Location of an obnoxious facility on a network: a voting approach." *Networks*, Vol. 20, pp. 197–207.

Melachrihoudis, E. (1988). "An efficient computational procedure for the recti- linear maximum location problem." *Transportation Science*, Vol. 22, pp. 217–223.

Rayco, M. B.; Francis, R. L.; Lowe, T. J. (1995). "Error- Bound Driven Demand Point Aggregation for the Rectilinear Distance p- Center Model." Research Report 95–9, Department of Industrial & Systems Engineering, University of Florida, Gainesville, Florida.

Rhee, W-S. T.; Talagrand, M. (1989). "On the k- center problem with many centers." *Operations Research Letters*, Vol. 8, pp. 309–314.

Shier, D. R. (1977). "A max- min theorem for p- center problems on a tree" *Transportation Science*, Vol. 11, pp. 243–252.

Sung, C. S.; Joo, C. M. (1994). "Locating an obnoxious facility on a Euclidean network to minimize neighborhood damage." *Networks*, Vol. 24, pp. 1–9.

Tansel, B. C., et al. (1982). "Duality and distance constraints for the non- linear p- center problem and covering problem on a tree network" *Operations Research*, Vol. 30, pp. 725–744.

Thomas, P.; Chan, Y. (1995). "Obnoxious facility location and the Data Envelopment Analysis." Master's Thesis. Department of Operational Sciences, Air Force Institute of Technology, Wright–Patterson AFB, Ohio.

Tsay, H-S.; Lin, L-T. (1989). "Efficient algorithm for locating a new transportation facility in a network." *Transportation Research Record*, No. 1251, pp. 35–44.

Woeginger, G. J. (1998). "A comment on a minimax location problem." *Operations Research Letters*, Vol. 23, pp. 41–43.

Multi-criteria facility location

Badri, M. A.; Mortagy, A. K.; Alsayed, A. (1998) "A multi- objective model for locating fire stations." *European Journal of Operational Research*, Vol. 110, pp. 243–260.

Berman, O. (1990). "Mean–variance location problems." *Transportation Science*, Vol. 24, No. 4, pp. 287–293.

Bogardi, I.; Haden, R. J.; Zhang, L. (1999). "Facility location planning of street maintenance under conflicting objectives." Pre- print of the Annual Transportation Research Board Meeting, National Research Council, Washington, D.C.

Carrizosa, E.; Conde, E.; Fernandez, F. R.; Puerto, J. (1993). "Efficiency in Euclidean constrained location problems." *Operations Research Letters*, Vol. 14, pp. 291–295.

Carrizosa, E.; Conde, E.; Munoz–Marquez; Puerto, J. (1995). "Planar point–objective location problems with no convex constraints: a geometrical construction." *Journal of Global Optimization*, Vol. 6, pp 77–86.

Carrizosa, E. J.; Puerto, J. (1995). "A discretizing algorithm for location problems." *European Journal of Operational Research*, Vol. 80, pp. 166–174.

Cohon, J. L.; ReVelle, C. S.; Current, J.; Eagles, T.; Eberhart, R.; Church, R. (1980). "Applications of a multiobjective facilities location model to power plant siting in a six–state region of the U.S." *Computer and Operations Research*, Vol. 7, pp. 107–123.

Current, J.; Min, H.; Schilling, D. (1990). "Multiobjective analysis of facility location decisions." *European Journal of Operational Research*, Vol. 49, pp. 295–307.

Forgues, P. J. (1991). "The Optimal Location of GEODSS Sensors in Canada." Master's Thesis (AFIT/GSO/ENS/91M–1), Department of Operational Sciences, Air Force Institute of Technology, Wright–Patterson AFB

Forgues, P.J.; Chan, Y. ; Kellso, T. S. (1998). "A mean–variance location problem: a case study of siting satellite–tracking stations." Working paper. Department of Operational Sciences, Air Force Institute of Technology, Wright–Patterson AFB, Ohio.

Hamacher, H. W.; Nickel, S. (1993). "Multi-criterial planar location problems." Working paper, University of Kaiserslautern, Germany.

Hamacher, H. W.; Nickel, S. (1994). "Combinatorial algorithm for some 1–facility median problems in the plane." *European Journal of Operational Research*, Vol. 79, pp. 340–351.

Hooker, J. N.; Garfinkel, R. S.; Chen, C. K. (1991). "Finite dominating sets for network location problems." *Operations Research*, Vol. 39, No. 1, pp. 100–118.

Hutson, V. A.; ReVelle, C. S. (1989). "Maximal direct covering tree problems." *Transportation Science*, Vol. 23, No. 4, pp. 288–299.

Koksalan, M. M.; Sagely, P. N. S. (1995). "Interactive approaches for discrete alternative multiple criteria decision making with monotone utility functions." *Management Science*, Vol. 41, pp. 1158–1171.

Kouvelis, P.; Carlson, R. C. (1992). "Total unimodularity application in bi–objective discrete optimization." *Operations Research* Letters, Vol. 11, pp. 61–65.

Malczewski, J.; Ogryczak, W. (1995). "The multiple criteria location problem: 1. A generalized network models and the set of efficient solutions." *Environment and Planning A*, Vol. 27, pp. 1931–1960.

Malczewski, J.; Ogryczak, W. (1996). "The multiple criteria location problem: 2. Preference–based techniques and interactive decison support." *Environment and Planning A*, Vol. 28, pp. 69–98.

Melachrinoudis, E.; Xanthopulos, Z. (1994). "A Euclidean distance single facility location problem with the minisum and maximin objectives." Working paper, Department of Industrial Engineering and Information Systems, Northeastern University, Boston, Massachusetts.

ReVelle, C.; Cohon, J.; Shobrys, D. (1981). "Multiple objectives in facility location: a review." Operations Research Group Report Series Paper #81–01, The John Hopkins University, Baltimore, Maryland.

Ye, M-H.; Yezer, A. M. (1992). "Location and spatial pricing for public facilities." *Journal of Regional Science*, Vol. 32, No. 2, pp. 143–154.

Others

Ahuja, R. K.; Jha, K. C.; Orlin, J. B.; Sharma, D. (2002). "Very large-scale neighborhood search for the quadratic assignment problem." Working Paper, Department of Industrial and Systems Engineering, University of Florida, Gainesville, Florida.

Batta, R.; Ghose, A.; Palekar, U. (1989). "Location facilities on the Manhattan metric with arbitrarily shaped barriers and convex forbidden regions." *Transportation Science*, Vol. 23, pp. 26–36.

Bazaraa, M. S.; Jarvis, J. J.; Sherali, H. D. (1990). *Linear Programming and Network Flows*, Second Edition, Wiley, New York, New York.

Berg, W.; Fair, T.; Chan, Y. (1996). "A k–medoid method approach to delineating line of sight radar degradation due to topography." Working paper. Department of Operational Sciences, Air Force Institute of Technology, Wright–Patterson AFB, Ohio.

Berman, O.; Simchi–Levi, D. (1990). "Conditional location problems on networks." *Transportation Science*, Vol. 24, pp. 77–78.

Berman, O.; Kaplan, E. H. (1990). "Equity maximizing facility location schemes." *Transportation Science*, Vol. 24, No. 2, pp. 137–144.

Brimberg, J.; Wesolowsky, G. O. (2000). "Note: Facility location with closest rectangular distances." *Naval Research Logistics*, Vol. 47, pp. 77–84.

Carreras, M.; Serra, D. (1999) "On optimal location with threshold requirements." *Socio–Economic Planning Sciences*, Vol. 33, pp. 91–103.

Chang, M. D.; Cheng, M-W.; Chen, C-H J. (1989). "Implementation of new labeling procedures for generalized networks." Working paper, Dept. of Computer Science and Operations Research, North Dakota State University, Fargo, North Dakota.

Chardaire, P.; Sutter, A. (1996). "Solving the dynamic facility location problem." *Networks*, Vol. 26, pp. 117–124.

Chepoi, V.; Dragan, F. F. (1996). "Condorcet and median points of simple rectilinear polygons." *Location Science*, Vol. 4, pp. 21–35.

Chhajed, D.; Lowe, T. J. (1990). "Locating facilities which interact: some solvable cases." Faculty working paper no. 90–1686, Bureau of Economic and Business Research, University of Illinois, Urbana–Champaign, Illinois.

Current, J.; Weber, C. (1994). "Application of facility location modeling constructs to vendor selection problems." *European Journal of Operational Research*, Vol. 76, pp. 387–392.

Daskin, M. S.; Hopp, W. J.; Medina, B. (1992). "Forecast horizons and dynamic facility location planning." *Annals of Operations Research*, Vol. 40, pp. 123–151.

Drezner, A.: Mehrez, A.; Weslowsky, G. O. (1991). "The facility location problem with limited distance." *Transportation Science*, Vol. 25, No. 3, pp. 183–187.

Eiselt, H. A.; Gendreau, M.; Laporte, G. (1992). "Location of facilities on a network subject to a single–edge failure" *Networks*, Vol. 22, pp. 231–246.

Goldberg, A. V.; Tardos, E.; Tarjan, R. E. (1989). "Network Flow Algorithms." Dept. of Computer Science, Stanford University, Stanford, California.

Goldberg, J.; Paz, L. (1990). "Locating emergency vehicle bases when service time depends on call location." working paper, Dept. of Systems and Industrial Engineering, University of Arizona, Tucson, Arizona.

Hanjoul, P.; Hansen, P.; Peeters, D. (1990). "Uncapacitated plant location under alternative spatial price policies." *Management Science*, Vol. 16, pp. 41–57.

Holmberg, K. (1994). "Solving the staircase cost facility location problem with decomposition and piecewise linearization." *European Journal of Operational Research*, Vol. 75, pp. 41–61.

Huang, G. H.; Baetz, B. W.; Patry G. (1993). "Grey Dynamic programming for waste management planning under uncertainty." Working paper, Department of Civil Engineering, McMaster University, Hamilton, Ontario, Canada.

Jacobsen, S. K. (1990). "Multiperiod capacitated location models." in *Discrete Facility Location* (Edited by P Mirchandani and R Francis), Wiley, New York, New York.

Jensen, P. (1985). *MICROSOLVE Network Flow Programming, IBM–PC Student Version Software and User Manual*, Holden–Day, Oakland, California.

Jones, P. C.; Lowe, T. J.; Muller, G.; Xu, N.; Ye Y.; Zydiak, J. L. (1995). "Specially structured uncapacitated facility location problems." *Operations Research*, Vol. 43, pp. 661–669.

Kara, B. Y.; Tansel, B. C. (2001). "The latest arrival hub location problem." *Management Science*, Vol. 47, No. 10, pp. 1408–1420.

Kaufman, L.; Rousseuw, P. J. (1990). *Finding Groups in Data: An Introduction to Cluster Analysis*, Wiley, New York, New York.

Kelso, T. S. (1990). Satellite pass scheduling software, Department of Operational Sciences, Air Force Institute of Technology, Wright–Patterson AFB, Ohio.

Klincewicz, J. G. (1990). "Solving a freight transport problem using facility location techniques." *Operations Research*, Vol. 38, pp. 99–109.

Labbe, M.; Schmeichel, E. F.; Hakimi, S. L. (1994). "Approximation algorithm for the capacitated plant allocation problem." *Operations Research Letters*, Vol. 15, pp. 115–126.

Larson, R. C.; and Sadiq, G. (1983). "Facility locations with the Manhattan metric in the presence of barriers to travel." *Operations Research*, Vol. 31, No. 4, pp. 652–669.

Lin, G-H.; Xue, G. (1998) "K- center and K- median problems in graded distances." *Theoretical Computer Science*, Vol. 207, pp. 181–192.

Loerch, A. G.; Boland, N.; Johnson, E. L.; Nemhauser, G. L. (1995). "Finding an optimal stationing policy for the US Army in Europe after the force drawdown." Working paper, US Army Concepts Analysis Agency, Bethesda, Maryland.

Louwers, D.; Kip, B. J.; Peters, E.; Souren, F.; Flapper, S. D. P. (1999). "A facility location–allocation model for reusing arpet materials." *Computers & Industrial Enineering*, Vol. 36, pp. 855–869.

Lucena, A.; Beasley, J. E. (1998). "A branch and cut algorithm for the Steiner problem in graphs." *Networks*, Vol. 31, pp. 39–59.

Magnanti, T. L.; Wong, R. T. (1990). "Decomposition methods for facility location problems." in *Discrete Facility Location*, edited by P. B. Mirchandani and R. L. Francis, Wiley Interscience, New York, New York.

Marsh, M. T.; Schilling, D. A. (1994). "Equity measurement in facility location analysis: a review and framework." *European Journal of Operational Research*, Vol. 74, pp. 1–17.

Mavridou, T. D.; Pardalos, P. M. (1997) "Simulated annealing and genetic algorithms for the facility layout problem: a survey." *Computational Optimization and Applications*, Vol. 7, pp. 111–126.

McAleer, W. E.; Naqvi, I. A. (1994). "The relocation of ambulance stations: a successful case study." *European Journal of Operational Research*, Vol. 75, pp. 582–588.

Min, H.; Melachrinoudis, E. (1997). "Dynamic expansion and location of an airport: a multiple objective approach." *Transportation Research–A*, Vol. 31, No. 5, pp. 403–417.

Nel, L. D.; Colbourn, C. J. (1990). "Locating a broadcast facility in an unreliable network." *INFOR*, Vol. 28, No. 4, pp. 363–379.

Nozick, L. K.; Turnquist, M. A. (1998) "Integrating inventory impacts into a fixed–charge models for locating distribution centers

Nuworsoo, C. (1999). "A model to minimize non–revenue costs in bus transit operations." Pre–print of the Annual Transportation Research Board Meeting, National Research Council, Washington, D. C.

Odoni, A. (1987). "Stochastic facility location problems." in *Stochastics in combinatorial optimization* (G. Andreatta and F. Mason, editors), World Scientific, N.J.

Punnen, A. P. (1994). "On a combined minmax–minsum optimization." *Computers and Operations Research*, Vol. 21, pp. 707–716.

Sasaki, M.; Suzuki, A.; Drezner, Z. (1994). "On the selection of relay points in a logistics system." Working paper, Center for Management Studies, Nagoya, Japan.

Schilling, D.; Elzinga, D. J.; Cohon, J.; Church, R.; ReVelle, C. S. (1979). "The team/fleet models for simultaneous facility and equipment siting." *Transportation Science*, Vol. 13, May.

Serra de la Figuera, D.; ReVelle, C. (1993). "The pq–median problem: location and distribution of hierarchical facilities." *Location Science*, Vol. 1, pp. 299–312.

Serra de la Figuera, D.; ReVelle, C. (1994). "The pq–median problem: location and distribution of hierarchical facilities–II: Heuristic solution methods." *Location Science*, Vol. 2, pp. 63–83.

Shulman, A. (1991). "An algorithm for solving dynamic capacitated plant location problems with discrete expansion sizes." *Operations Research*, Vol. 39, No. 3, pp. 423–436.

Spoonamore, J. H. (1996). "Solving a facility location problem using new optimization methods." Working paper, U S Army Construction Engineering Research Laboratory, P O Box 9005, Champaign, Illinois.

Steppe, J.; Chan, Y.; Chrissis, J.; Marsh, F.; Drake, D. (1996). "A hierarchical maximal coverage location–allocation model: Case study of generalized search–and–rescue." Working paper. Department of Operational Sciences, Air Force Institute of Technology, Wright–Patterson AFB, Ohio.

Steuer, R. E. (1986). *Multicriteria optimization: Theory computation and application*, Wiley, New York, New York.

Skorin–Kapov, D.; Skorin–Kapov, J.; O'Kelly, M. (1994). "Tight linear programming relaxations of uncapacitated p–hub median problems." Working paper, Harriman School for Management and Policy, State university of New York, Stony Brook, New York.

Xu, N.; Chhajed, D.; Lowe, T. J. (1994). "Tool design problems in a punch press flexible manufacturing system." Working paper, Office of Research, College of Commerce and Business Administration, University of Illinois, Urbana–Champaign, Illinois.

Wright, S.; Chan, Y. (1994). "A network with side constraints formulation of the k–medoid method for optimal plant location applied to image classification." Working paper. Department of Operational Sciences, Air Force Institute of Technology, Wright–Patterson AFB, Ohio.

Chapter 3 – Measuring Spatial Separation: Distance, Time, Routing, and Accessibility

Because of the rapid advances in this area, we have included with this chapter a bibliography. Some of the entries are referenced in the text, but many are not. We hope that this bibliography serves as a quick reference of the literature up to the time of writing. Due to the large amount of literature in recent years, we have broken down the bibliographic entries here into several categories: (a) shortest paths, l_p–norm, accessibility and other distance functions, (b) travelling-salesman, (c) vehicle routing, (d) stochastic facility location, and (e) general.

Shortest paths, l_p-norm, accessibility and other distance functions

Bach, L. (1981). "The problem of aggregation and distance for analysis of accessibility and access opportunity in location–allocation models." *Environment and Planning A*, Vol. 13, pp. 955–978.

Bartholdi III, J. J.; Platzman, L. K. (1988). "Heuristics based on spacefilling curves for combinatorial problems in Euclidean space." *Management Science*, Vol. 34, No. 3, pp. 291–305.

Bertsekas, D.; Tsitsiklis, J. (1988). "An analysis of stochastic shortest path problems." Working paper, Laboratory for Information and Decision Systems, Massachusetts Institute of Technology." Cambridge, Massachusetts.

Brimberg, J.; Love, R. F. (1992). "A new distance function for modeling travel distances in a transportation network." *Transportation Science*, Vol. 26, No. 2.

Brimberg, J.; Wesolowsky, G. O. (1992) "Probabilistic l_p distances in location models" *Annals of Operations Research*, Vol. 40, pp. 67–75.

Campbell, J. F. (1992). "Continuous and discrete demand hub location problems." Working paper CBIS 92-06-01, Center for Business and Industrial Studies, School of Business Administration, University of Missouri, St. Louis, Missouri.

Chan, Y. (1979). "A graph-theoretic method to quantify the airline route authority." *Transportation*, Vol. 8, pp. 275–291.

Carrizosa, E.; Romero–Morales, D. (2001). "Combining minsum and minmax: A goal programming approach." *Operations Research*, Vol. 49. No. 1. pp. 169–174.

Charnes, A.; Cooper, W. W.; Rhodes, E. (1978) "Measuring the efficiency of decision making unit" *European Journal of Operational Research*, Vol. 2, pp. 429–444.

Dalvi, M. Q.; Martin, K. M. (1976). "The measurement of accessibility: some preliminary results." *Transportation*, Vol. 5, pp. 17–42.

Deo, N.; Pang, C. (1984). "Shortest path algorithms: taxonomy and annotation." *Networks*, Vol. 14, pp. 275–323.

Dreyfus, S. (1969). "An appraisal of some shortest path algorithms." *Operations Research*, Vol. 17, pp. 395–412.

Fukuyama, H.; Weber, W. L. (2002). "Estimating output gains by means of Luenberger efficiency measure." Working Paper, Faculty of Commerce, Fukuoka University, Fukuoka, Japan.

Grosskopf, S.; Hayes, K. (1993) "Local public sector bureaucrats and their input choices" *Journal of Urban Economics*, Vol. 33, pp. 151–166.

Grosskopf, S.; Hayes, K.; Porter–Hudak, S. (1993) "A computationally efficient method for estimating distance functions" Working paper, Department of Economics, Southern Illinois University, Carbondale, Illinois.

Grosskopf, S.; Magaritis, D.; Valdmanis, V. (1995) "Estimating output substitutability of hospital services: a distance function approach" *European Journal of Operational Research*, Vol. 80, pp. 575–587.

Hall, R. W. (1991). "Characteristics of multi-stop multi-terminal delivery routes, with backhauls and unique items." *Transportation Research B*, Vol. 25B, No. 6, pp. 391–403.

Hsu, C-I; Hsieh, Y-P (1993) "The development of individual accessibility measure models" *Transportation Planning Journal*, Vol. 22, pp. 203–230, Ministry of Transportation and Communication, Taiwan (in Chinese).

Jin, H.; Batta, R.; Karwan, M. H. (1996) "On the analysis of two new models for transporting hazardous materials" *Operations Research*, Vol. 44, pp. 710–723.

Kim, S-I; Choi, I–C (1994) "A simple variational problem for a moving vehicle" *Operations Research Letters*, Vol. 16, pp. 231–239.

Love, R. F.; Morris, J. G.; Wesolowsky, G. O. (1988). "Mathematical models of travel distances." Chapter 10 of *Facilities Location: Models and Methods*, North–Holland, New York, New York.

Love, R. F; Walker, J.H.; Tiku, M. L. (1995) "Confidence intervals for $l_{k,p,\theta}$ distances" *Transportation Science*, Vol. 29, pp. 93–100.

Martins, E. (1984). "On a multicriteria shortest path problem." *European Journal of Operational Research*, Vol. 16, pp. 236–245.

Miller–Hooks, E.; Mahmassani, H. S. (1997). "Optimal routing of hazardous materials in stochastic, time-varying transportation networks." Working paper, Department of Civil and Environmental Engineering, Duke University, Durham, North Carolina.

Robuste, F. (1901). "Centralized hub-terminal geometric concepts I: Walking distance." *Journal of Transportation Engineering,* Vol. 117, No. 2, pp. 143–177.

Sathisan, S. K.; Srinivasan, N. (1997). "A framework for evaluating accessibility of urban transportation networks." Working paper, Transportation Research Center, University of Nevada, Las Vegas, Nevada.

Stone, R. E. (1991). "Some average distance results." Technical notes, *Transportation Science,* Vol. 25, No. 1, pp. 83–90.

Tassiulas, L. (1996) "Adaptive routing on the plane" *Operations Research,* Vol. 44, pp. 823–832.

Thomas, P. C. (1995) "Using Locational and Data Envelopment Analysis Models to Site Municipal Solid Waste Facilities" Master's Thesis, Department of Environmental Engineering and Management, Report AFIT/GEE/ENS/950-09, Air Force Institute of Technology, Wright–Patterson AFB, Ohio.

Xue, G.; Ye, Y. (1997). "An efficient algorithm for minimizing a sum of Euclidean norms with applications." *SIAM Journal of Optimization,* Vol. 7, No. 4, pp. 1017–1036.

Travelling salesman

Anily, S.; Mosheiov, G. (1994) "The traveling salesman problem with delivery and backhauls" *Operations Research Letters,* Vol. 16, pp. 11–18.

Baker, S. F. (1991). "Location and Routing of the Defense Courier Service Aerial Network." Master's Thesis, AFIT/GOR/ENS/91M-1, Department of Operational Sciences, Air Force Institute of Technology, Wright–Patterson AFB, Ohio.

Baker, E. (1983). "An exact algorithm for the time-constrained traveling salesman problem." *Operations Research,* Vol. 31, No. 5, pp. 938–945.

Butt, S. E.; Ryan, D. M. (1999). "An optimal solution procedure for the multiple tour maximum collection problem using column generation." *Computers & Operations Research,* Vol. 26, pp. 427–441.

Balas, E.; Fischetti, M. (1989). "A lifting procedure for the asymmetric traveling salesman polytope and a large new class of facets." Working paper 88-89-83, Graduate School of Industrial Administration, Carnegie–Mellon University, Pittsburgh, Pennsylvania.

Berman, O.; Simchi–Levi, D. (1988). "A heuristic algorithm for the travelling salesman location problem on networks." *Operations Research,* Vol. 36, pp. 478–484.

Berman, O.; Simchi–Levi, D. (1988a). "Finding the optimal *a priori* tour and location of a travelling salesman with non-homogeneous customers." *Transportation Science,* Vol. 22, pp. 148–154.

Bertsimas, D. (1989). "Travelling salesman facility location problems." *Transportation Science,* Vol. 23, pp. 184–191.

Bertsimas, D. J; Simchi–Levi, D. (1996) "A new generation of vehicle routing research: robust algorithms addressing uncertainty" *Operations Research,* Vol. 44, pp. 286–304.

Bienstock, D.; Goemans, M. X.; Simchi–Levi, D.; Williamson, D. (1990). "A note on the prize collecting travelling salesman problem." Working paper, Bellcore, Morristown, New Jersey.

Burkhard, R. (1990). "Locations with spatial interactions: The quadratic assignment problem." *Discrete Location Theory* (edited by P. Mirchandani and R. Francis), pp. 387–438.

Cao, B.; Glover, F. (9194) "Tabu search and ejection chains—application to a node weighted version of the cardinality-constrained TSP" Working paper, University of the Federal Armed Forces at Munich, Neubiberg, Germany.

Chan, Y.; Merrill, D. L. (1997). "The probabilistic multiple-travelling-salesmen-facility-location problem: space-filling curve and asymptotic Euclidean analyses." *Military Operations Research,* Vol. 3, No. 2, pp. 37–53.

Current, J.; Schilling, D. (1989). "The covering salesman problem." *Transportation Science,* Vol. 23, pp. 208–213.

Desrosiers, J.; Sauve, M.; Soumis, F. (1988). "Lagrangian relaxation methods for solving the minimum fleet size multiple traveling salesman problem with time windows." *Management Science,* Vol. 34, pp. 1005–1022.

Dumas, Y.; Desrosiers, J.; Gelinas, E. (1995) "An optimal algorithm for the traveling salesman problem with time windows" *Operations Research,* Vol. 43, pp. 367–371.

Eiselt, H. A.; Gendreau, M.; Laporte, G. (1995) "Arc routing problems, Part I: the Chinese Postman Problem" *Operations Research,* Vol. 43, pp. 231–242.

Eiselt, H. A.; Gendreau, M.; Laporte, G. (1995) "Arc routing problems, Part II: the Rural Postman Problem" *Operations Research,* Vol. 43, pp. 399–414.

Fischetti, M.; Gonzalez, J. J. S., Toth, P. (1994) "A branch-and-cut algorithm for the symmetric generalized travelling salesman problem" Working paper, Department of Electronic Information and Systems, University of Bologna, Italy.

Fischetti, M.; Gonzalez, J. J. S., Toth, P. (1994) "The symmetric generalized travelling salesman polytope" Working paper, Department of Electronic Information and Systems, University of Bologna, Italy.

Franca, P.; Gendreau, M.; Laporte, G.; Muller, F. M. (1995) "The m-traveling salesman problem with minmax objective" *Transportation Science*, Vol. 29, 267–275.

Gendreau, M.; Hertz, M.; Laporte, G. (1996) "The traveling salesman problem with backhauls" *Computers and Operations Research*, Vol. 23, pp. 501–508.

Gendreau, M.; Hertz, A.; Laporte, G. (1992). "New insertion and post-optimization procedures for the traveling salesman problem." *Operations Research*, Vol. 40, No. 6, pp. 1086–1094.

Gendreau, M.; Laporte, G.; Potvin, J-Y. (1994) "Heuristics for the clustered traveling salesman problem" Publication CRT-94–54, Centre de Recherche sur les Transports, Universite de Montreal, Montreal, Canada.

Gendreau, M.; Laporte, G.; Semet, F. (1997). "The covering tour problem." *Operations Research*, Vol. 45, No. 4, pp. 568-576.

Gendreau, M.; Laporte, G.; Semet, F. (1998). "A branch-and-cut algorithm for the undirected select traveling salesman problem." *Networks*, Vol. 32, pp. 262–273.

Gendreau, M.; Laporte, G.; Simchi–Levi, D. (1991). "A degree relaxation algorithm for the asymmetric generalized travelling salesman problem." Working paper CRT-800, Centre de Researche sur les Transports, Universite de Montreal, Montreal, Canada.

Gillett, B.; Miller, L. (1974). "A heuristic algorithm for the vehicle dispatch problem." *Operations Research*, Vol. 22, pp. 340–349.

Hoffman, A.; Wolfe, P. (1985). "History." in the *Traveling Salesman Problem*, edited by E. Lawler et al., Chichester, G. B., Wiley, New York, New York.

Jaillet, P. (1988). "A prior solution of a traveling salesman problem in which a random subset of the customers are visited." *Operations Research*, Vol. 36, p. 929.

Jonker, R.; Volgenaut, T. (1989). "Nonoptimal edges for the symmetric traveling salesman problem." *Operations Research*, Vol. 32, pp. 837–846.

Junger, M.; Kaibel, V. (1996) "A basic study of the QAP-polytope" Working paper, Institut fur Informatik, Universitat zu Koln, Denmark.

Kikuchi, S.; Rhee, J-H. (1989). "Scheduling method for demand-responsive transportation system." *Journal of Transportation Engineering*, Vol. 115, pp. 630–645.

Laporte, G. (1997). "Modeling and solving several classes of arc routing problems as traveling salesman problems." *Computers & Operations Research*, Vol. 24, No. 11, pp. 1057–1061.

Laporte, G.; Louveaux, F. V.; Mercure, H. (1994) "A priori optimization of the probabilistic traveling salesman problem" *Operations Research*, Vol. 42, pp. 543–549.

Lawler, E. L.; Lenstra, J. K.; Kan, A. H. G.; Shmoys, D. B. (1985) *The Traveling Salesman Problem*, Wiley-Interscience, New York, New York.

Merrill, D. (1989). "Facility Location and Routing to Minimize the en route Distance of Flight Inspection Missions." Master's thesis submitted to the Department of Operational Sciences, Air Force Institute of Technology, Wright–Patterson AFB, Ohio.

Minieka, E. (1989) "The delivery man problem on a tree network" *Annals of Operations Research*, Vol. 18, pp. 261–266.

Moore, R. S.; Pushek, P.; Chan, Y. (1991). "A travelling salesman/facility location problem." Working paper, Department of Operational Sciences, Air Force Institute of Technology, Wright–Patterson AFB, Ohio.

Mullen, D. (1989). "A Dynamic Programming approach to the Daily Routing of Aeromedical Evacuation System Missions." Master's thesis (AFIT/GST/ENS/89J–5), Dept. of Operational Sciences, Air Force Institute of Technology, Wright–Patterson AFB, Ohio.

Noon, C.; Bean, J. (1989). "An efficient transformation of the generalized traveling salesman problem." Working paper, Dept. of Management Science, University of Tennessee, Knoxville, Tennessee.

Padberg, M.; Rinaldi, G. (1991). "A branch-and-cut algorithm for the resolution of large-scale symmetric traveling salesman problems." *SIAM Review*, Vol. 33, No. 1, pp. 60–100.

Padberg, M.; Rinaldi, G. (1989). "A branch-and-cut approach to a traveling salesman problem with side constraints." *Management Science*, Vol. 35, pp. 1393–1412.

Renaud, J.; Boctor, F. F. (1996) "An efficient composite heuristic for the symmetric Generalized Traveling Salesman Problem" Working paper, Tele-universite, Saint Foy, Quebec, Canada.

Reynolds, J.; Baker, S.; Chan, Y. (1990). "Multiple travelling salesman problem: integer programming, subtour breaking and time constraints." Working paper, Department of Operational Sciences, Air Force Institute of Technology, Wright–Patterson AFB, Ohio.

Rosenkrantz, D. J.; Stearns, R. E.; Lewis, P. M. (1974). "Approximate algorithms for the traveling salesman problem." *Proc. 15th Ann IEEE Symp on Switching and Automata Theory*, pp. 33–42.

Sexton, T.; Choi, Y-M. (1986). "Pickup and delivery of partial loads with "soft" time windows." *American Journal of Mathematical and Management Sciences*, Vol. 6, pp. 369–398.

Shirley, M.; Sosebee, B.; Chan, Y. (1992). "Facility location study." Working paper, Department of Operational Sciences, Air Force Institute of Technology, Wright–Patterson AFB, Ohio.

Stephens, A.; Habash, N.; Chan, Y. (1991). "Alternate depot: Altus Air Force Base." Working paper, Department of Operational Sciences, Air Force Institute of Technology, Wright–Patterson AFB, Ohio.

Tillman, F. (1969). "The multiple terminal delivery problem with stochastic demands." *Transportation Science*, Vol. 3, pp. 192–204.

Tsitsiklis, J. N. (1992). "Special cases of traveling salesman and repairman problems with time windows." *Networks*, Vol. 22, pp. 263–282.

Tzeng, G-H.; Wang, J-C.; Hwang, M-J. (1996) "Using genetic algorithms and the template path concept to solve the traveling salesman problem" *Transportation Planning Journal*, Vol. 25, pp. 493–516. Institute of Transportation, Taiwan. (In Chinese).

Vander Wiel, R. J.; Sahinidis, N. V. (1995) "Heuristic bounds and test problem generation for the time-dependent traveling salesman problem" *Transportation Science*, Vol. 29, pp. 167–184.

Warburton, A. R. (1993) "Worst-case analysis of some convex hull heuristics for the Euclidean travelling salesman problem" *Operations Research Letters*, Vol. 13, pp. 37–42.

Yang, J.; Jaillet, P.; Mahmassani, H. S. (1998). "On-line algorithms for truck-fleet assignment and scheduling under real-time information." Working paper, Department of Management Science and Information Systems, University of Texas at Austin, Austin, Texas.

Zografos, K.; Davis, C. (1989). "Multi-objective programming approach for routing hazardous materials." *Journal of Transportation Engineering*, Vol. 115, pp. 661–673.

Zweig, G. (1995) "An effective tour construction and improvement procedure for the traveling salesman problem" *Operations Research*, Vol. 43, pp. 1049–1057.

Vehicle routing

Achutan, N. R.; Caccetta, L. (1991). Integer linear programming formulation for a vehicle-routing problem." *European Journal of Operational Research*, Vol. 52, pp. 86–89.

Agarwal, Y.; Mathur, K.; Salkin, H. (1989). "A set-partitioning-based exact algorithm for the vehicle routing problem." *Networks*, Vol. 19, pp. 731–749.

Altman, S. M.; Bhagat, N.; Bodin, L. D. (1971). "Algorithm for routing garbage trucks over multiple planning periods." Report of the Urban Science and Engineering Program, State University of New York, Stony Brook, New York.

Anily, S.; Bramel, J. (1999). "Approximation algorithms for the capacitated traveling salesman problem with pickups and deliveries." *Naval Research Logistics*, Vol. 46, pp. 654–670.

Assad, A. (1988). "Modeling and implementation issues in vehicle routing." in *Vehicle Routing Methods and Studies*, (B. Golden and A. Assad, editors), Elsevier Science Publishers.

Baker, S. F.; Chan, Y. (1996). "The multiple depot multiple travelling salesmen problem: vehicle range and servicing frequency implementations." Working paper, Department of Operational Sciences, Air Force Institute of Technology, Wright–Patterson AFB, Ohio.

Bauer, M.; Blythe, R.; Chan, Y. (1992). "Multiple vehicle travelling salesman problem" Working paper, Dept. of Operational Sciences, Air Force Institute of Technology, Wright–Patterson AFB, Ohio.

Baugh Jr, J. W.; Kakivaya, G. K. R.; Stone, J. R. (1996) "Intractability of the dial-a-ride problem and a multiobjective solution using simulated annealing" Working paper, Department of Civil Engineering, North Carolina State University, Raleigh, North Carolina.

Beltrami, E.; Bhagat, N.; Bodin, L. (1971). "A randomized routing algorithm with application to barge dispatching in New York City." USE Technical Report #71-15, November 1971, The State University of New York, Stony Brook, New York.

Beltrami, E.; Bhagat, N.; Bodin, L. (1971). "Refuse disposal in New York City; an analysis of barge dispatching." USE Technical Report #71-10, July 1971, The State University of New York, Stony Brook, New York.

Benavent, E.; Campos, V.; Corberan, A.; Mota, E. (1992) "The capacitated arc routing problem: lower bounds" *Networks*, Vol. 22, pp. 669–690.

Bergevin, R.; Pflieger, C.; Chan, Y. (1992). "The medical evacuation problem: using traveling salesman and vehicle routing formulations." Working paper, Department of Operational Sciences, Air Force Institute of Technology, Wright–Patterson AFB, Ohio.

Beroggi, G. E. G. (1994) "A real time routing model for hazardous materials" *European Journal of Operational Research*, Vol. 75, pp. 505–520.

Bertsimas, D. J. (1992). "A vehicle routing problem with stochastic demand." *Operations Research*, Vol. 40, No. 3, pp. 574–585.

Bodin, L.; Godin, B. (1981). "Classification in vehicle routing and scheduling." *Networks*, Vol. 11, pp. 77–108.

Bowerman, R. L; Calamai, P. H. (1994) "The space filling curve with optimal partitioning heuristic for the vehicle routing problem" *European Journal of Operational Research*, Vol. 76, pp. 128–142.

Bramel, J.; Coffman Jr., E. G.; Shor, P. W.; Simchi–Levi, D. (1992). "Probabilistic analysis of the capacitated vehicle routing problem with unsplit demands." *Operations Research*, Vol. 40, No. 6, pp. 1095–1106.

Bramel, J.; Simchi–Levi (1994) "On the effectiveness of set partitioning formulations for the vehicle routing problem" Working paper, Graduate School of Business, Columbia University, New York, New York.

Bramel, J.; Simchi–Levi (forthcoming) "Probabilistic analyses and practical algorithms for the vehicle routing problem with time windows" *Operations Research*.

Brodie, G. R.; Waters, C. D. (1988). "Integer linear programming formulation for vehicle routing problems." *European Journal of Operational Research*, Vol. 34, pp. 403–404.

Burnes, M. D. (1990). "Application of Vehicle Routing Heuristics to an Aeromedical Airlift Problem." Master's thesis, AFIT/GST/ENS/90M–3, Department of Operational Sciences, Air Force Institute of Technology, Wright–Patterson AFB, Ohio.

Burnes, M.; Chan, Y. (1989). "Solution of a classic vehicle routing problem." Working paper, Department of Operational Sciences, Air Force Institute of Technology, Wright–Patterson AFB, Ohio.Carter, W. B. (1990). "Allocation and Routing of CRAF MD80 Aircraft." Master's thesis, AFIT/GST/ENS/90M–4, Department of Operational Sciences, Air Force Institute of Technology, Wright–Patterson AFB, Ohio.

Clarke, G.; Wright, J. W. (1964). "Scheduling of vehicles from a central depot to a number of delivery points." *Operations Research*, Vol. 12, pp. 568–581.

Daganzo, C. F.; Hall, R. W. (1993) "A routing model for pickups and deliveries: no capacity restrictions on the secondary items" *Transportation Science*, Vol. 27, pp. 315–340.

Desorchers, M.; Desrosiers, J.; Solomon, M. (1992). "A new optimization algorithm for the vehicle routing problem with time windows." *Operations Research*, Vol. 40, No. 2, pp. 342–354.

Desrosiers, J.; Dumas, Y.; Soumis, F. (1986). "A dynamic programming solution of the large-scale single-vehicle dial-a-ride problem with time windows." *American Journal of Mathematical and Management Sciences*, Vol. 6, pp. 301–325.

Desrosiers, J.; Laporte, G. (1991). "Improvements and extensions to the Miller–Tucker–Zemlin subtour elimination constraints." *Operations Research Letters*, Vol. 10, pp. 27–36.

Dror, M.; Laporte, G.; Trudeau, P. (1989). "Vehicle routing with stochastic demands: properties and solution frameworks." *Transportation Science*, Vol. 23, No. 3, pp. 166–176.

Duhamel, C.; Potvin, J.Y.; Rousseau, J.M. (1997). "A tabu search heuristic for the vehicle routing problem with backhauls and time windows." *Transportation Science*, Vol. 31, No. 1, pp. 49–59.

Fischetti, M.; Gonzalez, J. J. S.; Toth, P. (1997). "A branch-and-cut algorithm for the symmetric generalized traveling salesman problem." *Operations Research*, Vol. 45, No. 3, pp. 378–394.

Fischetti, M.; Toth, P.; Vigo, D. (1994) "A branch-and-bound algorithm for the capacitated vehicle routing problem on directed graphs" *Operations Research*, Vol. 42, pp. 846–859.

Fisher, M.; Jaikumar, R. (1981). "A generalized assignment heuristic for vehicle routing." *Networks*, Vol. 11, pp. 109–124.

Fisher, M. L.; Jornsten, K. O.; Madsen, O. B. G. (1997). "Vehicle routing with time windows: two optimization algorithms." *Operations Research*, Vol. 45, No. 3, pp. 488–492.

Gendreau, M.; Laporte, G.; Seguin, R. (1995) "An exact algorithm for the vehicle routing problem with stochastic demands and customers" *Transportation Science*, Vol. 29, pp. 143–155.

Gendreau, M.; Hertz, A.; Laporte, G. (1991). "A tabu search heuristic for the vehicle routing problem." Working paper CRT-777, Center de Researche sur les Transports, Universite de Montreal, Montreal, Canada.

Golden, B. L. (1984). "Introduction to and recent advances in vehicle routing methods." in *Transportation Planning Models* (M. Florian, Editor), Elsevier Science publishers, pp. 383–418.

Golden, B.; Assad, A. (1986). "Perspective on vehicle routing: exciting new developments." *Operations Research*, Vol. 34, No. 5, pp. 803–810.

Haimovich, M.; Rinnooy Kan, A. H. G.; Storigie, L. (1988). "Analysis of heuristics for vehicle routing problems." in *Vehicle Routing: Methods and Studies* (B. Golden and A. Assad, editors).

Hall, R. W.; Du Y.; Lin, J. (1994) "Use of continuous approximations within discrete algorithms for routing vehicles: Experimental results and interpretation" *Networks*, Vol. 24, pp. 43–56.

Haouair, M.; Dejax, P.; Desrochers, M. (1990). "Modelling and solving complex vehicle routing problems using column generation." GERAD, 5255 Ave Decelles, Montreal, Quebec, Canada.

Harche, F.; Raghavan, P. (1991). "A generalized exchange heuristic for the capacitated vehicle routing problem." Working paper, Department of Statistics and Operations Research, Stern School of Business, New York University, New York, New York.

Hooker, J. N.; Natraj, N. R. (1995) "Solving a general routing and scheduling problem by chain decomposition and tabu search," *Transportation Science*, Vol. 29, pp. 30–44.

Jacobs–Blecha, C.; Goetschalckx, M.; Desrochers, M.; Gelinas, S. (1992). "The vehicle routing problem with backhauls: properties and solution algorithms." Working paper, Georgia Tech Research Corporation, Atlanta, Georgia.

Jacob–Blecha, C.; Goetschalckx, M. (1992) "The vehicle routing problem with backhauls: properties and solution algorithms" Working paper, Georgia Tech Research Corporation, Atlanta, Georgia.

Jansen, K. (1993) "Bounds for the general capacitated routing problem" *Networks*, Vol. 23, pp. 165–173.

Kohl, N.; Desrosiers, J.; Madsen, O. B. G.; Solomon, M. M.; Soumis, F. (1999). "2-path cuts for the vehicle routing problem with time windows." *Transportation Science*, Vol. 33, No. 1, pp. 101–116.

Kohl, N.; Madsen, O. B. G. (1997). "An optimization algorithm for the vehicle routing problem with time windows based on Lagrangian relaxation." *Operations Research*, Vol. 45, No. 3, pp. 395–406.

Koskosidis, Y. A.; Powell, W. B.; Solomon, M. M. (1992). "An optimization-based heuristic for vehicle routing and scheduling with soft time window constraints." *Transportation Science*, Vol. 26, No. 2, pp. 69–85.

Kulkarni, R.; Bhave, P. (1985). "Integer programming formulations of vehicle routing problems." *European Journal of Operational Research*, Vol. 20, pp. 58–67.

Laporte, G.; Nobert, Y.; Desrochers, M. (1985). "Optimal routing under capacity and distance restriction." *Operations Research*, Vol. 32, No. 5, pp. 1050–1065.

Li, C-L.; Simchi–Levi, D.; Desrochers, M. (1992). "On the distance constrained vehicle routing problems." *Operations Research*, Vol. 40, No. 4, pp. 790–799.

Li, C-L.; Simchi–Levi, D. (1990). "Worst-case analysis of heuristics for multidepot capacitated vehicle routing problem." *ORSA Journal on Computing*, Vol. 2, No. 1, pp. 64–74.

Li, C-L.; Simchi–Levi, D. (1989). "On the distance constrained vehicle routing problem." Dept. of Industrial Engineering and Operation Research, Columbia University, New York, New York.

Mingozzi, A.; Giorgi, S.; Baldacci, R. (1999). "An exact method for the vehicle routing problem with backhauls." *Transportation Science*, Vol. 33, No. 3, pp. 315–329.

Mosheiov, G. (1994) "The travelling salesman problem with pick-up and delivery" *European Journal of Operational Research*, Vol. 79, pp. 299–310.

Orloff, C. (1974). "A fundamental problem in vehicle routing." *Networks*, Vol. 4, pp. 35–64.

Rana, K.; Vickson, R. (1988). "A model and solution algorithm for optimal routing of a time-constrained containership." *Transportation Science*, Vol. 22, pp. 83–95.

Ribeiro, C. C.; Soumis, F. (1994) "A column generation approach to the multiple-depot vehicle scheduling problem" *Operations Research*, Vol. 42, pp. 41–52.

Rodriguez, P.; Nussbaum, M.; Baeza, R.; Leon, G., Sepulveda, M., Cobian, A. (1998). "Using global search heuristics for the capacity vehicle routing problem." *Computers & Operations Research*, Vol. 25, No. 5, pp. 407–417.

Russell, R. A. (1995) "Hybrid heuristics for the vehicle routing problem with time windows" *Transportation Science*, Vol. 29, pp. 156–166.

Savelsbergh, M. W. P.; Sol, M. (1995) "The general pickup and delivery problem" *Transportation Science*, Vol. 29, pp. 17–29.

Savelsbergh, M. W. P.; Sol, M. (1998). "DRIVE: dynamic routng of independent vehicles." *Operations Research*, Vol. 46, No. 4, pp. 474–490.

Sexton, T. (1979). "The single vehicle many to many routing and scheduling problem with desired delivery times." Ph.D. thesis, Applied Mathematics and Statistics Dept., State University of New York at Stony Brook.

Sexton, T.; Bodin, L. (1985). "Optimizing single vehicle many–to–many operations with desired delivery times: Part I—scheduling, and Part II—routing." *Transportation Science*, Vol. 19, No. 4.

Simchi–Levi, D.; Bramel, J. (1990). "On the optimal solution value of the capacitated vehicle routing problem with unsplit demands." Working paper, Department of Industrial Engineering and Operations Research, Columbia University, New York, New York.

Simchi–Levi, D.; Bramel, J. (1990). "Probabilistic analysis of heuristics for the capacitated vehicle routing problem with unsplit demands." Working paper, Department of Industrial Engineering and Operations Research, Columbia University, New York, New York.

Solomon, M. M.; Desrosiers, J. (1988). "Time window constrained routing and scheduling problems." *Transportation Science*, Vol. 22, No. 1, pp. 1–13.

Taillard, E.; Bade1au, P.; Gendreau, M.; Guertin, F.; Potvin, J.Y. (1996) "A tabu search heuristic for the vehicle routing problem with soft time windows" Publication CRT-95-66, Centre de Recherche sur les Transports, Universite de Montreal, Montreal, Canada.

Teodorovic, D.; Kikuchi, S. (1989). "Application of fuzzy sets theory to the savings based vehicle routing algorithm." Working paper, Civil Engineering Department, University of Delaware, Newark, Delaware.

Thangiah, S. R.; Potvin, J.-Y.; Sun, T. (1996). "Heuristic approaches to vehicle routing with backhauls and time windows." *Computers & Operations Research*, Vol. 23, No. 11, pp. 1043–1057.

Toth, P.; Vigo, D. (1995) "A heuristic algorithm for the vehicle routing problem with backhauls" Working paper, Dipartmento di Electtronica Informatica e Sistemistica, Universita degli Studi di Bologna, Italy.

Wei, H.; Li, Q.; Kurt, C. E. (1998). "Heuristic–optimization models for service–request vehicle routing with time windows in a GIS environment." Working paper, Transportation Center, The University of Kansas, Lawrence, Kansas.

Wiley, V. D.; Chan, Y. (1995) "A look at the vehicle routing problem with multiple depots and the windows." Working paper, Department of Operational Sciences, Air Force Institute of Technology, Wright–Patterson AFB, Ohio.

Yan, S.; Yang, Hwei–Fwa (1995) "Multiple fleet routing and flight scheduling" *Transportation Planning Journal*, Vol. 24, pp. 195–220 Institute of Transportation, Taiwan. (in Chinese).

Stochastic facility-location

Ahituv, N.; Berman, O. (1988). *Operations Management of Distributed Service Networks: A practical quantitative approach*, Plenum Press.

Batta, R.; Berman, O. (1989). "Allocation model for a facility operating as an M/G/k queue." *Networks*, Vol. 19, pp. 717–728.

Batta, R.; Larson, R.; Odoni, A. (1988). "A single–server priority queuing–location model." *Networks*, Vol. 8, pp. 87–103.

Berman, O.; Chiu, S. S.; Larson, R. C.; Odoni, A. R.; Batta, R. (1990). "Location of mobile units in a stochastic environment." in *Discrete Location Theory* (Edited by P. Mirchandani and R. Francis), Wiley-Interscience, New York, New York.

Berman, O.; Krass, D. (2001). "Facility location problems with stochastic demands and congestion." Working Paper, Rotman School of Management, University of Toronto, Toronto, Canada.

Berman, O.; Larson, R.; Parkan, C. (1987). "The stochastic queue p-median problem." *Transportation Science*, Vol. 21, pp. 207–216.

Berman, O.; Larson, R. (1985). "Optimal 2–facility network districting in the presence of queuing." *Transportation Science*, Vol. 19, pp. 261–277.

Berman, O.; Larson, R.; Chiu, S. (1985). "Optimal server location on a network operating as an M/G/1 queue." *Operations Research*, Vol. 33, pp. 746–771.

Berman, O.; Rahnama, M. (1985). "Optimal location–relocation decisions on stochastic *Networks*." *Transportation Science*, Vol. 19, pp. 203–221.

Berman, O.; LeBlanc, B. (1984). "Location–relocation of mobile facilities on a stochastic network." *Transportation Science*, Vol. 18, pp. 315–330.

Berman, O.; Odoni, A. (1982). "Locating mobile servers on a network with Markovian properties." *Networks*, Vol. 12, pp. 73–86.

Brandeau, M. L.; Chiu, S. S. (1990). "A unified family of single-server queuing location models." *Operations Research*, Vol. 38, No. 6, pp. 1034–1044.

Brandeau, M. L.; Chiu, S. S. (1990). "Trajectory analysis of the stochastic queue median in a plane with rectilinear distances." *Transportation Science*, Vol. 24, No. 3, pp. 230–243.

Brimberg, J.; Mehrez, A. (1993) "Location/allocation of queuing facilities in continuous space using a minisum criterion and arbitrary distance function" Working paper, Department of Engineering Management, Royal Military College of Canada, Kingston, Ontario, Canada.

Brotcore, L.; Laporte, G.; Semet, F. (2003). "Ambulance location and relocation models." *European Journal of Operational Research*, Vol. 147, pp. 451–463.

Chiu, S.; Berman, O.; Larson, R. (1985). "Locating a mobile server queuing facility on a tree network." *Transportation Science*, Vol. 31, pp. 764–773.

Chiu, S.; Larson, R. (1985). "Locating an *n*-server facility in a stochastic environment." *Computers and Operations Research*, Vol. 12, pp. 509–516.

Gendreau, M.; Laporte, G.; Semet, F. (1997). "Solving an ambulance location model by tabu search." *Location Science*, Vol. 5, No. 2, pp. 75–88.

Horton, D.; Chan, Y. (1993) "Service facility location in a stochastic demand service model" Working paper, Department of Operational Sciences, Air Force Institute of Technology, Wright–Patterson AFB, Ohio.

Irish, T.; May, T.; Chan, Y. (1995) "A stochastic facility relocation problem" Working paper, Department of Operational Sciences, Air Force Institute of Technology, Wright–Patterson AFB, Ohio.

Jamil, M.; Baveja, A.; Batta, R. (1999). "The stochastic queue center problem." Computers & Operations Research, Vol. 26, pp. 1423–1436.

Louveaux, F. (1986). "Discrete stochastic location models." *Annals of Operations Research*, Vol. 6, pp. 23–34.

Mirchandani, P. B. (1975). "Analysis of stochastic *Networks* in emergency service systems." Technical Report IRP-TR-15-75, MIT *Operations Research* Center, Cambridge, Massachusetts.

Mirchandani, P. B.; Odoni, A. R. (1979). "Location of medians on stochastic networks." *Transportation Science*, Vol. 13, pp. 85–97.

Mohan, R.; Chan, Y. (1993) "Combat rescue forward operating locations: a stochastic facility location problem" Working paper, Department of Operational Sciences, Air Force Institute of Technology, Wright–Patterson AFB, Ohio.

Odoni, A. (1987). "Stochastic facility location problems." in *Stochastics in combinatorial optimization* (G. Andreatta and F. Mason, editors), World Scientific, New Jersey.

Sosebee, B.; Chan, Y. (1993) "Air defense alert facility location in a theater of operations" Working paper, Department of Operational Sciences, Air Force Institute of Technology, Wright–Patterson AFB, Ohio.

Weaver, J.; Church, R. (1983). "Computational procedures for location problems on stochastic networks." *Transportation Science*, Vol. 17, No. 2, pp. 168–180.

General

Anstreicher, K. M. (1999). "Eigenvalue bounds versus semidefinite relaxations for the quadratic assignment problem." Working paper, Department of Management Sciences, University of Iowa, Iowa City, Iowa.

Balakrishnan, J.; Cheng, C. H. (1998). "Dynamic layout algorithms: a state-of-the-art survey." *Omega*, Vol. 26, No. 4, pp. 507–521.

Bartholdi, J.; Platzman, L. (1988). "Heuristics based on spacefilling curves for combinatorial problems in Euclidean space." *Management Science*, Vol. 34, pp. 291–305.

Bazaraa, M. S.; Jarvis, J. J.; Sherali, H. D. (1990). *Linear Programming and Network Flows*, Second Edition, Wiley, New York, New York.

Beckmann, M. J. (1987) "The economic activity equilibrium approach in location theory" in *Urban Systems: Contemporary Approaches to Modelling* (Edited by C. S. Bertuglia, G. Leonardi, S. Occelli, G. A. Rabino, R. Tadei, and A. G. Wilson) Croom Helm, United Kingdom.

Berlin, G. N.; Revelle, C. S.; Elizinga, D. J. (1976). "Determining ambulance-hospital locations for on-scene & hospital services." *Environment & Planning*, Vol. 8, pp. 553–561.

Bertsimas, D. J. (1988). "Probabilistic combinatorial optimization problems." Ph.D. thesis, Massachusetts Institute of Technology, Cambridge, Massachusetts.

Bodin, L. D. (1990). "Twenty years of routing and scheduling." OR Forum, *Operations Research*, Vol. 38, No. 4, pp. 571–579.

Bodin, L. D.; Golden, B. L.; Assad, A. A.; Ball, M. O. (1983). "Routing and scheduling of vehicles and crews: the state of the art" Special Issue, *Computers and Operations Research*, Vol. 10, No. 2.

Cela, E. (1998). *The Quadratic Assignment Problem: Theory and Algorithms*, Kluwer, Norwell, Massachusetts.

Charnes, A.; Cooper, W. W.; Rhodes, E. (1978) "Measuring the efficiency of decision making units" *European Journal of Operational Research*, Vol. 2, pp. 429–444.

Combs, T.; Cartlin, A.; Chan, Y. (1996) "The Quadratic Assignment Problem." Working paper, Department of Operational Sciences, Air Force Institute of Technology, Wright–Patterson AFB, Ohio.

Desrosiers, J.; Dumas, Y.; Solomon, M. M.; Soumis, F. (1992) "Time constrained routing and scheduling" forthcoming in *Handbooks in Operations Research and Management Science: Networks*, North–Holland, Amsterdam, Netherlands.

Evans, J. R.; Minieka, E. (1992). *Optimization Algorithms for Networks and Graphs*, Second Edition, Dekker.

Fuller, W. (1987) *Measurement Error Models*, Wiley, New York, New York.

Hahn, P.; Grant, T. (1998). "Lower bounds for the quadratic assignment problem based upon a dual formulation." *Operations Research*, Vol. 46, No. 6, pp. 912–922.

Hargrave, W.; Nemhauser, G. (1962). "On the relation between the travelling salesmen problem and the longest path problem." *Operations Research*, Vol. 10, pp. 647–657.

Jaillet, P. (1991). "Rates of convergence for quasi-additive smooth Euclidean functionals and application to combinatorial optimization problems." to appear in *Mathematics of Operations Research*.

Jaillet, P. (1990). "Cube versus Torus models for combinatorial optimization problems and the Euclidean minimum spanning tree constant." Working paper, Department of Management Science and Information Systems, University of Texas at Austin, Austin, Texas.

Love, R. F.; Morris, J. G.; Wesolowsky, G. O. (1988). *Facilities Location: Models and Methods*, North–Holland, New York, New York.

Mathaisel, D. F. X. (1996). "Decision support for airline system operations control and irregular operations." *Computers & Operations Research*, Vol. 23, No. 11, pp. 1083–1098.

Mavridou, T. D.; Pardalos, P. M. (1997). "Simulated annealing and genetic algorithms for the faility layout problem: A survey." *Computational Optimization and Applications*, Vol. 7, pp. 111–126.

Meller, R. D.; Narayanan, V.; Vance, P. H. (1999). "Optimal facility layout design." *Operations Research Letters*, Vol. 23, pp. 117–127.

Moore, K. R.; Chan, Y. (1990). "Integrated location-and-routing models—part I: model examples." Working paper, Department of Operational Sciences, Air Force Institute of Technology, Wright–Patterson AFB, Ohio.

Parker, R. G.; Rardin, R. L. (1988). *Discrete Optimization*, Academic Press.

Phillips, K. J.; Beltrami, E. J.; Carroll, T. O.; Kellog, S. R. (1982). "Optimization of areawide wastewater management." *Journal WPCF*, Vol. 54, No. 1, pp. 87–93.

Teodorovic, D.; Kikuchi, S.; Hohlacov, D. (1990). "A routing and scheduling method in considering trade-off between user's and operator's objectives." Working paper, Faculty of Transport and Traffic Engineering, University of Belgrade, Vojvode stepe 305, 11000 Belgrade, Yugoslavia.

Toregas, C.; Swain, R.; Revelle, C.; Bergman, L. (1972). "The location of emergency service facilities." *Operations Research*, Vol. 20, pp. 1363–1373.

Yaman, R.; Balibek, E. (1999). "Decision making for facility layout problem solutions." *Computers & Industrial Engineering*, Vol. 37, pp. 319–322.

Zhou, J.; Liu, B. (2003). "New stochastic models for capacitated location–allocation problem." *Computers & Industrial Engineering*, Vol. 45, pp. 111–125.

Chapter 4 – Simultaneous Location-and-routing Models

Literature in this subject has literally exploded over the last couple of decades. For this reason, we are presenting the bibliography in four categories. The first category contains general methodological references. The remaining categories are application areas. While the three case-studies documented here in this chapter are illustrative, many other location-routing studies have been performed. We list in the following bibliography the various documents available. They are grouped under three categories corresponding to the three generic problems in simultaneous location-routing: (a) Obnoxious facility location, (b) Depot location and commodity distribution, and (c) Transportation terminal location.

Methodology

Anily, S. (1996). "The vehicle-routing problem with delivery and back-haul options." *Naval Research Logistics*, Vol. 43, pp. 415–434.

Arkic, U.; Srikanth, K. (1992). "Optimal routing and process scheduling for a mobile service facility." *Networks*, Vol. 22, pp. 163–183.

Aykin, T. (1995). "The hub location and routing problem." *European Journal of Operational Research*, Vol. 83, pp. 200–219.

Baker, S. F.; Chan, Y. (1991). "The multiple depot multiple travelling salesmen problem: vehicle range and servicing frequency implementations." Working paper, Department of Operational Sciences, Air Force Institute of Technology, Wright–Patterson AFB, Ohio 45433.

Baker, S. F.; Chan, Y. (1990). "The travelling salesman problem: Extensions in demand frequency and linear programming relaxation," working paper, Department of Operational Sciences, Air Force Institute of Technology, Wright–Patterson AFB, Ohio.

Balakrishnan, A.; Ward, J.; Wong, R. (1987). "Integrated facility location and vehicle routing models: recent work and future prospects," *American Journal of Mathematical and Management Science*, Vol. 7 Nos. 1 and 2, pp. 35–61.

Barnhart, C.; Johnson, E. L.; Nemhauser, G. L.; Sigismondi, G.; Vance, P. (1993). "Formulating a mixed integer programming problem to improve solvability." *Operations Research, Vol 41, No. 6*, pp. 1013–1019.

Bazaraa, M. S.; Jarvis, J. J.; Sherali, H. D. (1990). *Linear Programming and Network Flows*. Second Edition, Wiley.

Bender, T.; Hennes, H.; Kalcsics, J.; Melo, T.; Nickel, S. (2001). "Location software and interface with GIS and supply chain management." in *Location Analysis: Applications and Methods*, Z. Drezner and H. W. Hamacher (Editors), Springer-Verlag, New York, New York.

Berger, R. T. (1997). "Location-Routing Models for Distribution System Design." Doctoral dissertation, Industrial Engineering and *Management Sciences*, Northwestern University, Evanston, Illinois.

Berman, O.; Simchi–Levi, D. (1989). "The travelling salesman location problem on stochastic *Networks*." *Transportation Science*, Vol. 23, pp. 54–57.

Berman, O.; Simchi–Levi, D. (1988). "Finding the optimal a priori tour and location of a traveling salesman with nonhomogeneous customers." *Transportation Science*, Vol. 22, pp. 148–154.

Berman, O.; Simchi–Levi, D. (1986). "Minisum location of a traveling salesman." *Networks*, Vol. 16, pp. 239–254.

Bertazzi, L.; Speranza, M. G.; Ukovich, W. (1996). "Exact and heuristic solutions for a shipment problem with given frequencies." Dipartimento Metodi Quantitativi, Universita degli Studi di Brescia, Brescia, Italy.

Bertsimas, D. J. (1989). "Traveling salesman facility location problems." *Transportation Science*, Vol. 23, No. 3, pp. 184–191.

Bertsimas, D. (1988). "Some Combinatorial Optimization Problems." Doctoral dissertation, Operations Research Center, MIT, Cambridge, Massachusetts.

Bhadury, J.; Chandrasekharan, R.; Gewali, L. (1998). "On the complexity of designing a transportation network subject to a single facility location." Working paper, Faculty of Administration, University of New Brunswick, Fredericton, New Brunswick, Canada.

Bramel, J.; Simchi–Levi, D. (1995). "A location based heuristic for general routing problems." Operations Research, Vol. 43, pp. 649–660.

Bramel, J.; Simchi–Levi, D. (1996). "Probabilistic analyses and practical algorithms for the vehicle routing problem with time windows." *Operations Research*, Vol. 44, pp. 501–509.

Campbell, J. F. (1993). "One-to-many distribution with transshipments: an analytical model. "*Transportation Science*, Vol. 27, No. 4, pp. 330–340.

Campbell, A.; Clarke, L.; Kleywegt, A.; Savelsbergh, M. (1997). "The inventory routing problem." Working paper, School of Industrial and Systems Engineering, Georgia Institute of Technology, Atlanta, Georgia.

Carpaneto, G.; Del'Amico, M.; Fischetti, M.; Toth, P. (1989). "A branch-and-bound algorithm for the multiple depot vehicle scheduling problem." *Networks*, Vol. 19, pp. 531–548.

Chan, L. M. A.; Federgruen, A.; Simchi–Levi, D. (1998). "Probabilistic analyses and practical algorithms for inventory-routing models." *Operations Research*, Vol. 46, No. 1, pp. 96–106.

Chan, L. M. A.; Muriel, A.; Simchi–Levi, D. (1997). "Supply-chain management: integrating inventory and transportation." Working paper, Philips Laboratories, Briarcliff Manor, New York.

Chan, L. M. A.; Simchi–Levi, D. (1994). "Probabilistic analyses and practical algorithms for multi-echelon distribution systems." Working paper, Department of Industrial Engineering and Operations Research, Columbia University, New York, New York.

Chan, Y. (2000). *Location Theory and Decision Analysis*. Thomson/South-Western, Cincinnati, Ohio.

Chan, Y. (1997). "A general class of location-routing problems: formulation and solution." Working paper, Department of Operational Sciences, Air Force Institute of Technology, Wright–Patterson AFB, Ohio; also presented at the 1988 Spring Joint Meeting of the Institute of Management Science and Operations Research Society of America, Washington, D. C.

Chan, Y. (1979). "A Graph Theoretic Method to Quantify the Airline Route Authority." *Transportation*, Vol. 19, No. 1, pp. 309–320.

Chan, Y. (1974). "Configuring a Transportation Route Network via the Method of Successive Approximations." *Computers and Operations Research*, Vol. 1, pp. 385–420.

Chan, Y. (1972). "Route Network Improvement in Air Transportation Schedule Planning," Flight Transportation Laboratory, Department of Aeronautics and Astronautics, MIT, Cambridge, Massachusetts.

Chan, Y.; Merrill, D. (1997). "A probabilistic multiple-travelling-salesman-facility-location problem: asymptotic analysis using space-filling curve heuristic," *Military Operations Research*, Vol. 3, No. 2, pp. 37–53.

Coutinho–Rhodrigues, J.; Current, J.; Climaco, J.; Ratick, S. (1997). "Interactive spatial decision-support system for multiobjective hazardous materials location-routing problems." *Transportation Research Record*, No. 1602, pp. 101–109.

Current, J. (1988). "The design of a hierarchical transportation network with transportation facilities." *Transportation Science*, Vol. 22, pp. 270–277.

Current, J. R.; ReVelle, C. S.; Cohon, J. L. (1987). "The median shortest path problem: a multiobjective approach to analyze cost vs. accessibility in the design of transportation networks." *Transportation Science*, Vol. 21, No. 3, pp. 188–197.

Current, J.; Pirkul, H.; Holland, E. (1994). "Efficient algorithms for solving the shortest covering path problem." *Transportation Science*, Vol. 28, No. 4, pp. 317–327.

Current, J. R.; Cohon, J. L.; ReVelle, C. S. (1983). "The maximum covering/shortest path problem: a multi-objective network design and routing formulation." *European Journal of Operations Research*, Vol. 21, pp. 189–199.

Day, R. (1973). "Recursive programming models: a brief introduction." in G.G. Judge and T. Takayama (editors), *Studies in Economic Planning Over Space and Time*, North Holland, Amsterdam.

Del Rosario, M.; Chan, Y. (1992). "Determining routes and route frequencies of an airline." Working paper, Department of Operational Sciences, Air Force Institute of Technology, Wright–Patterson AFB, Ohio.

Dror, M.; Ball, M.; Golden, B. (1985/6). "A computational comparison of algorithms for the inventory routing problem." *Annals of Operations Research*, pp. 3–23.

Erkut, E.; Glickman, T. (1997). "Minimax population exposure in routing highway shipments of hazardous materials." *Transportation Research Record*, No. 1602, pp. 93–100.

Erkut, E.; Verter, V. (1995). "Hazardous Materials Logistics: A Review." Chapter in *Facility Location: A Survey of Applications and Methods*, Z. Drezner, editor. Springer–Verlag, New York, New York.

Federgruen, A.; Zipkin, P. (1984). "A combined vehicle routing and inventory allocation problem." *Operations Research*, Vol. 32, No. 5, pp. 1019–1037.

Fourer, R.; Gay, D. M.; Kernighan, B. W. (1993). *AMPL: A Modeling Language for Mathematical Programming*. Scientific Press/Boyd & Fraser, Fencroft Village, Massachusetts.

Graham, R. P.; Chan, Y. (1994). "Facility location and routing: multi-facility multi-tour example." Working paper, Department of Operational Sciences, Air Force Institute of Technology, Wright–Patterson AFB, Ohio.

Green, D. J.; Chan, Y. (1993). "Possible improvement to the Laporte et al. algorithm for symmetric location-routing." Working paper, Department of Operational Sciences, Air Force Institute of Technology, Wright–Patterson AFB, Ohio.

Hachicha, M.; Hodgson, M. J.; Laporte, G.; Semet, F. (1999). "Heuristics for the multi-vehicle covering tour problem." *Computers & Operations Research*, Vol. 27, pp. 29–42.

Helander, M. E.; Melachrinoudis, E. (1997). "Facility location and reliable route planning in hazardous material transportation." *Transportation Science*, Vol. 31, No. 3, pp. 216–226.

Herer, Y.; Roundy, R. (1997). "Heuristic for one–warehouse multi-retailer distribution problem with performance bounds." *Operations Research*, Vol 45, pp. 102–115.

Hooker, J. N. (1996). "Inference duality as a basis for sensitivity analysis." Working paper, Graduate School of Industrial Administration, Carnegie Mellon University, Pittsburgh, Pennsylvania.

Huang, G. H.; Baetz, B. W.; Patry, G. G. (1994). "Grey dynamic programming for waste–management planning under uncertainty" *Journal of Urban Planning and Development*, Vol 120, No. 3, pp. 132–156.

Haughton, M. A.; Rao, K. (1995). "The stochastic vehicle routing problem: extensions to consider impact of demand variability and route failure risk on logistic performance" presented at *The 24th Western Decision Sciences Institute Meeting.*" San Francisco.

Irwin, C. L.; Wang, C. (1982). "Iteration and sensitivity for a spatial equilibrium problem with linear supply and demand functions," *Operations Research*, Vol. 30, No. 2, pp. 319–335.

Jacobs, T. L.; Warmerdam, J. M. (1994). "Simultaneous routing and siting for hazardous–waste operations." *Journal of Urban Planning and Development*, Vol 120, No. 3, pp. 115–131.

Johnson, K. E.; Chan, Y. (1990). "Integrated location and routing models solution methodologies." Working paper, Department of Operational Sciences, Air Force Institute of Technology, Wright–Patterson AFB, Ohio.

Lam, W. H. K.; Gao, Z. Y.; Chan, K. S.; Yang, H. (1999). "A stochastic user equilibrium assignment model for congested transit *Networks.*" *Transportation Research B*, Vol. 33B, pp. 351–368.

Laporte, G. (1988). "Location–routing problems." in *Vehicle Routing: Methods and Studies*, B. Golden and A. Assad, editors. Elsevier, New York, New York, and Amsterdam, Netherlands, pp. 163–197.

Laporte, G.; Nobert, Y.; Taillefer, S. (1988). "Solving a family of multi–depot vehicle routing and location–routing problems." *Transportation Science*, Vol. 22, pp. 161–172.

Laporte, G.; Nobert, Y.; Arpin, D. (1986). "An exact algorithm for solving a combined location–routing problem." *Annals of Operations Research*, Vol. 6, Baltzer, pp. 293–310.

Loftus, M.; Habash, N.; Chan, Y. (1992). "Enumerative approach to an inventory replenishment and delivery problem." Working paper, Department of Operational Sciences, Air Force Institute of Technology, Wright–Patterson AFB, Ohio.

Loftus, M.; Chan, Y. (1992). "The multiple–travelling–salesmen facility–location problem." Working paper, Department of Operational Sciences, Air Force Institute of Technology, Wright–Patterson AFB, Ohio.

Madsen, O. B. G. (1983). "Methods for solving combined two–level location–routing problems of realistic dimensions." *European Journal of Operational Research*, Vol. 12, pp. 295–301.

Mahaba, N.; Gunes, E.; Chan, Y. (1987). "Terminal location and routing." Working paper, Department of Operational Sciences, Air Force Institute of Technology, Wright–Patterson AFB, Ohio.

Melkote, S.; Daskin, M. S. (1997). "Polynomially solvable cases of combined facility location–network design problems." Working paper, Department of Industrial Engineering and Management Sciences, Northwestern University, Evanston, Illinois.

Mentzer, J. T.; Schuster, A. D. (1982). "Computer modeling in logistics, existing models and future outlook." *Journal of Business Logistics*, Vol. 3, No. 2, pp. 1–55.

Merrill, D. L. (1989). "Facility Location and Routing to Minimize the en route Distance of Flight Inspection Mission." Master's thesis, Department of Operational Sciences, Air Force Institute of Technology, Wright–Patterson AFB, Ohio.

Moore, K. R.; Chan, Y. (1990). "Integrated location–and–routing models — Part I: Model examples," Working paper, Department of Operational Sciences, Air Force Institute of Technology, Wright–Patterson AFB, Ohio.

Peeters, D.; Thisse, J.F.; Thomas, I. (1998). "Transportation networks and the location of human activities." *Geographical Analysis*, Vol. 30, No. 4, pp. 355–371.

Perl, J. (1987). "The multi–depot routing allocation problem." *American Journal of Mathematical and Management Science*, Vol. 7, nos. 1 and 2, pp. 7–34.

Perl, J.; Daskin, M. S. (1985). "A warehouse location–routing problem," *Transportation Research B*, Vol. 19B, pp. 117–136.

ProLogis: The Global Distribution Solution. (www.prologis.com)

Sheraldi, H.; Carter, T. B.; Hobeika, A. G. (1991). "A location–allocation model and algorithm for evacuation planning under hurricane/flood conditions." *Transportation Research B*, Vol. 25B, No. 6, pp. 439–452.

Shirley, E.; Chan, Y. (1993). "Single run facility location study." Working paper, Department of Operational Sciences, Air Force Institute of Technology, Wright–Patterson AFB, Ohio.

Simchi–Levi, D. (1989). "The capacitated travelling salesman location problem." Working paper, Dept. of Industrial Engineering and *Operations Research*, Columbia University, New York, New York.

Simchi–Levi, D.; Berman, O. (1988). "A heuristic algorithm for the traveling salesman location problem on networks." *Operations Research*, Vol. 36, pp. 478–484.

Simchi–Levi, D.; Berman, O. (1987). "Heuristics and bounds for the travelling salesman location problem on the plane." *Operations Research Letters*, Vol. 6, pp. 243–248.

Soumis, F.; Sauve, M.; LeBeau, L. (1988). "Simultaneous origin–destination assignment and vehicle routing problem." Working paper, Ecole PolyTechnique, Montreal, Quebec, Canada.

Srivastava, R.; Benton, W. C. (1990). "The location–routing problem: considerations in physical distribution system design," *Computers and Operations Research*, Vol. 17, No. 5, pp. 427–435.

Trudeau, P.; Dror, M. (1992). "Stochastic inventory routing: route design with stockouts and route failures." *Transportation Science*, Vol. 26, No. 3, pp. 171–184.

Tsay, H.-S. (1985). "The Combined Facility Location and Vehicle Routing Problem: Formulation and Solution." Doctoral Dissertation, Purdue University, W. Lafayette, Indiana.

Vickerman, R. W. (1980). *Spatial Economic Behavior: The Microeconomic Foundations of Urban and Transport Economics*, St. Martin Press, New York, New York.

Webb, I. R. (1992). "Lower bound for the strategic inventory routing problem using a nonlinear assignment formulation." Working paper WP92–032, School of Business Administration, University of Connecticut, Storrs, Connecticut.

Webb, I. R.; Larson, R. C. (1995). "Period and phase of customer replenishment: a new approach to the strategic inventory/routing problem" *European Journal of Operational Research*, Vol 85, pp. 132–148.

Weir, J.; Stewart, J.; Chan, Y. (1994) "Single–facility multiple–tour allocation problem." Working paper, Department of Operational Sciences, Air Force Institute of Technology, Wright–Patterson AFB, Ohio.

Zograftos, K. G.; Douligeris, C.; Chaoxi, L. (1992). "Model for optimum deployment of emergency repair trucks: Application in electric utility industry." *Transportation Research Record*, No. 1359, pp. 88–94.

Obnoxious facility location

Britt, R.; Paulsen, R.; Chan, Y. (1987). "Nuclear power network optimization for safety and reliability." Working paper, Dept. of Operational Sciences, Air Force Institute of Technology, Wright–Patterson AFB, Ohio 45433.

Campbell, K. W.; Eguchi, R. T.; Duke, C. M. (1979). "Reliability in Lifeline Earthquake Engineering." *Journal of the Technical Councils*, American Society of Civil Engineers, Vol. 105, No. TC2, pp. 259–270.

Chan, Y. (1992). "Facility location and routing analysis: parts I and II," Working paper, Department of Operational Sciences, Air Force Institute of Technology, Wright-Patterson AFB, Ohio 45433.

Chan, Y. (1990a). "Seismic–reliability and environmental–cost tradeoffs in nuclear–power distribution systems: A model of obnoxious facility location." *Proceedings of the International Symposium on Reliability and Maintainability*, Tokyo, Japan, pp. 487–491.

Christian, J. T.; Borjeson, R. W.; Tringale, P. T. (1978). "Probabilistic Evaluation of OBE for a Nuclear Power Plant." *Journal of the Geotechnical Division*, American Society of Civil Engineers, Vol. 104, GT7, pp. 907–919.

Cohen, J. L.; Revelle, C. S.; Current, J.; Eagles, T.; Eberhart, R.; Church, R. (1980). "Application of a multiobjective facility location model to power plant siting in a six-state region in the U.S.," *Computers and Operations Research*, Vol. 7, pp. 107-123.

Doan, J. T. (1985). "A Model for the Location of a Nuclear Power Plant in areas of Seismic Risk," Master of Science thesis, Civil and Environmental Engineering, Washington State University, Pullman, Washington.

Donovan, N. C. (1973). "A statistical evaluation of strong motion data including the February 9, 1971, San Fernando earthquake," *Proceedings of the Fifth World Conference on Earthquake Engineering (Rome)*, Vol. 1, pp. 1252–1261.

Estera, L.; Rosenblueth, E. (1964). "Espectros de temblores a Distancias Moderadas y Grandes." *Sociedad Mexicana de Ingenieria Seismica*, Vol. 2, No. 1.

Giannikos, I. (1998). "A multiobjective programming model for locating treatment sites and routing hazardous wastes." *European Journal of Operational Research*, Vol. 104, pp. 333–342.

Hobbs, B. F. (1980). "A comparison of weighting methods in power plant siting." *Decision Sciences*, Vol. II, No. 4, pp. 725-737.

Kilmer, K. A.; Anandalingam, G.; Malcolm, S. A. (1998). "Siting noxious facilities under uncertainty." Working paper 98–10, Department of Systems Engineering, University of Pennsylvania, Philadelphia, Pennsylvania.

Krinitzsky, E. L.; Marcason, W. F. III (1983). "Principles for selecting earthquake motions in engineering design." *Bulletin of the Association of Engineering Geologists*, Vol. 20, No. 3, pp. 253-265.

Kubo, K.; Katayama, T.; Masamitsu, O. (1979). "Lifeline earthquake engineering in Japan." *Journal of the Technical Councils*, American Society of Civil Engineers, Vol. 105, No. TC1, pp. 221-238.

List, G.; Mirchandani, P. (1991). "An integrated network/planar multi-objective model for routing and siting for hazardous materials and wastes." *Transportation Science*, Vol. 25, No. 2, pp. 146–156.

List, G.; Mirchandani, P.; Turnquist, M. (1990). "Logistics for Hazardous Materials Transportation: Scheduling, Routing and Siting." DOT–T–92–09, U. S. Department of Transportation, Washington D.C.

Matsuda, T. (1982). "Earthquake Scars." in *Earthquake Prediction Techniques: Their Application in Japan*, T. Asada, editor. University of Tokyo Press, Tokyo, Japan.

Revelle, C.; Cohen, J.; Shrobrys, D. (1991). "Simultaneous siting and routing in the disposal of hazardous wastes," *Transportation Science*, Vol. 25, No. 2, pp. 138–145.

Richter, C. F. (1958). *Elementary Seismology*. W. H. Freemont and Co., San Francisco, California.

Schiff, A. J. (1981). "Earthquakes and Power Systems." in *Earthquakes and Earthquake Engineering: The Eastern United States — Vol. 1*, J. E. Beavers, editor. Ann Arbor Science Publishers, Ann Arbor, Michigan, pp. 493-513.

Schiff, A. J.; Newcom, D. E. (1979). "Fragility of electrical power equipment," *Journal of the Technical Councils*, American Society of Civil Engineers, Vol. 105, No. TC2, pp. 451-465.

Schiff, A. J. (1978). "Seismic emergency operation of power systems." *Journal of the Technical Councils*, American Society of Civil Engineers, Vol. 14, No. TC1, pp. 1-11.

Steinhardt, O. W. (1979). "Protecting a power lifeline against earthquakes." *Journal of the Technical Councils*, American Society of Civil Engineers, Vol. 104, No. TC1, pp. 49-57.

Steppe, J.; Chan, Y. (1990). "Nuclear power network optimization for safety and reliability." Working paper, Department of Operational Sciences, Air Force Institute of Technology, Wright-Patterson AFB, Ohio.

Steuer, R. E. (1986). *Multiple Criteria Optimization: Theory, Computation and Application*, Wiley, New York, New York.

Verter, V.; Erkut, E. (1994). "Hazardous Material Logistics: An Annotated Bibliography." Research Report No. 94–1, Department of Finance and *Management Science*, Faculty of Business, University of Alberta, Edmonton, Alberta, Canada.

Wesson, R. L.; Helley, E. J.; Jajoie, K. R.; Wentworth, C. W. (1975). "Faults and future earthquakes," in *Studies for Seismic Zonation of the San Francisco Bay Region*, R. D. Becherdt, editor. USGS Professional Paper 941-A, U.S. Geological Survey, Reston, Virginia.

Depot location and commodity distribution

Achutan, N. R.; Caccetta, L. (1991). "Integer linear programming formulation for a vehicle routing problem." *European Journal of Operational Research*, Vol 52, pp. 86–89.

Ackley, M. (1990). Telephone interview. Military Airlift Command, Analysis Group, Scott AFB, Illinois, 9 July.

Anily, S.; Federgruen, A. (1991). "Structured partitioning problems." *Operations Research*, Vol. 39, pp. 130–149.

Araque, J. R. (1989). "Contributions to the Polyhedral Approach to the Vehicle Routing Problem." Doctoral dissertation, State University of New York, Stony Brook, New York.

Baker, S. F. (1991). "Location and Routing of the Defense Courier Service Aerial Network," Master's thesis, Department of Operational Sciences, Air Force Institute of Technology, Wright–Patterson AFB, Ohio.

Beltrami, E. J.; Bodin, L. D. (1976). "Networks and Vehicle Routing for Municipal Waste Collection," in *Large-scale Networks: Theory and Design*, F. T. Boesch, editor. IEEE Press, New York, New York.

Brodie, G. R.; Waters, C. D. (1988). "Integer linear programming formulation for vehicle routing problems." *European Journal of Operational Research*, Vol. 34, pp. 403–404.

Burns, M. D. (1990). "Application of Vehicle Routing Heuristics to an Aeromedical Airlift Problem" Master's thesis, Department of Operational Sciences, Air Force Institute of Technology, Wright–Patterson AFB, Ohio.

Clarke, G.; Wright, J. W. (1964). "Scheduling of vehicles from a central depot to a number of delivery points." *Operations Research*, Vol. 12, pp. 568–581.

Dror, M.; Trudeau, P. (1989). "Savings by split delivery routing," *Transportation Science*, Vol. 23, pp. 141–145.

Fisher, M. L. (1994). "Optimal solution of vehicle routing problems using minimum K-trees" *Operations Research*, Vol 42, pp. 626–642.

Fisher, M. L. (1994a). "A polynomial algorithm for the degree-constrained minimum K-tree problem." *Operations Research*, Vol. 42, pp. 775–779.

Fisher, M. L.; Jaikumar, R. (1981). "A generalized assignment heuristic for vehicle routing." *Networks*, Vol. 11, pp. 109–124.

Gillett, B.; Miller, L. (1974). "A heuristic algorithm for the vehicle dispatch problem." *Operations Research*, Vol. 22, pp. 340–349.

Glover, F.; Klingman, D. (1984). "Note on admissible exchanges in spanning trees." *Advances in Management Studies*, Vol 3.2, pp. 101–104.

Kulkarni, R. V.; Bhave, P. R. (1985). "Integer programming formulations of vehicle routing problems." *European Journal of Operational Research*, Vol. 20, pp. 58–67.

Muyldermans, L.; Cattrysse, D.; Van Oudheusden, D.; Lotan, T. (2002). "Districting for salt spreading operations." *European Journal of Operational Research*, Vol. 139, pp. 521–532.

Naddef, D. (1994). "A remark on 'integer linear programming formulation for a vehicle routing problem' by N. R. Achutan and L. Caccetta, or how to use the Clarke and Wright savings to write such integer linear programming formulations." *European Journal of Operational Research*, Vol. 75, pp. 238–241.

Nemhauser, G. L.; Wolsey, L. A. (1988). *Integer and Combinatorial Optimization*, Wiley-Interscience, New York, New York.

Rosenkrantz, D. J.; Stearns, R. E.; Lewis, P. M. (1974). "Approximate algorithms for the traveling salesman problem." *Proc. 15th Ann. IEEE Symp. on Switching and Automata Theory*, pp. 33–42.

Syslo, M. M., et al. (1983). *Discrete optimization algorithms with PASCAL programming*, Prentice Hall, Englewood Cliffs, New Jersey.

Thouvenot, T. (1990). Telephone interview, Headquarters Military Aircraft Command, Contract Requirements Division (TRCAS), Scott AFB, IL, 30 July.

Ware, K. (1990). Personal interview at the Military Airlift Command headquarters, Scott AFB, Illinois.

Transportation terminal location

Chan, Y. (1992). "Integrated location-and-routing models — Part I: a review of model formulations." Working paper, Department of Operational Sciences, Air Force Institute of Technology, Wright-Patterson AFB, Ohio.

Chan, Y. (1992). "Integrated location-and-routing models — Part II: a review of solution algorithms." Working paper, Department of Operational Sciences, Air Force Institute of Technology, Wright-Patterson AFB, Ohio.

Chan, Y. (1986). "Transportation network-investment problem: a synthesis of tree-search algorithms." *Transportation Research Record*, No. 1074, pp. 32-40.

Golden, B. (1984). "Introduction to and Recent Advances in Vehicle Routings Methods," in *Transportation Planning Models*, Elsevier, New York, New York and North Holland, Amsterdam, Netherlands.

Haghani, A.; Chen, M.-C. (1998). "Optimizing gate assignments at airport terminals." *Transportation Research A*, Vol. 32A, pp. 437–454.

Hooker, J. N. (1996). "Inference duality as a basis for sensitivity analysis." Working paper, Graduate School of Industrial Administration, Carnegie–Mellon University, Pittsburgh, Pennsylvania.

Nero, G.; Black, J. A. (1998). "Hub-and-spoke networks and the inclusion of environmental costs on airport pricing." *Transportation Research D*, Vol. 3D, No. 5, pp. 275–296.

Puscher, M.; Seitz, J.; Chan, Y. (1988). "Development of a Military Airlift Command aircraft parts delivery system." Working paper, Department of Operational Sciences, Air Force Institute of Technology, Wright-Patterson AFB, Ohio.

Tang, Z. (1987). "Facility Location and Routing in Airline Management," Master's thesis, Department of Civil Engineering, Washington State University, Pullman, Washington.

Van Nes, R.; Hamerslag, R.; Immers, L. (1988). "The design of public transport networks." Paper No. 870536, 67*th* Annual Meeting of the Transportation Research Board, Washington, D.C.

Yu, P. L. (1986). *Multicriteria Decision Making*, Plenum, New York, New York.

Chapter 5 – Including Generation, Competition and Distribution in Location Allocation

Agnihothri, S.; Karmarkar, U. S.; Kubat, P. (1982). "Stochastic allocation rules" *Operations Research,* Vol. 30, pp. 545–555.

Anderson, S. P.; Never, D. J. (1991). "Cournot competition yields spatial agglomeration." *International Economic Review,* Vol. 32, pp. 793–808.

Arasan, V. T.; Rajesh, R. (1996). "Land–use allocation for minimum travel intensity in Indian cities." Working paper, Transportation Engineering Division, Department of Civil Engineering, Indian Institute of Technology, Madras, India.

Baumal, W. J. (1965). *Economic Theory and Operations Analysis,* Second Edition. Prentice–Hall.

Beaumont, J. R. (1987). "Location–allocation models and central place theory." in *Spatial Analysis and Location–Allocation Models* (A. Ghosh and G. Rushton; editors). Van Nostrand, pp. 21–54.

Ben-Akiva, M.; Bowman, J. L. (1996). "Integration of an activity based model system and a residential location model." Paper presented at the Lincoln Institute of Land Policy, TRED Conference on Land use and Transportation, Cambridge, Massachusetts, October.

Berman, O.; Krass, D. (1997). "Flow intercepting spatial interaction model: a new approach to optimal location of competitive facilities." Working paper, School of Management, University of Toronto, Toronto, Canada.

Bhat, C. R. (1997) "Covariance heterogeneity in nested logit models: econometric structure and application to intercity travel." *Transportation Research B*, Vol. 31, pp. 11–21.

Borgers, A.; Hofman, F.; Ponje, M.; Timmermans, H. (1998). "Toward a conjoint–based context–dependent model of task allocation in activity settings: some numerical experiments." Pre–print 98–0325, 77th Annual Transportation Research Board Meeting, Washington, D.C.

Brandeau, M. L.; Chiu, S. S. (1991). "Location of competing facilities in a user–optimizing environment with market externalities." Working paper, Depts. of Industrial Engineering and Engineering–Economic Systems, Stanford University, Stanford, California.

Brog, W. (1982). "Subjective perception of car costs." *Transportation Research Record*, No. 858, Transportation Research Board, Washington, D.C.

Brotchie, J. F.; Sharpe R. (1975). "A general land use allocation model: applications to Australian cities." *Urban Development Models*, (R. Baxter, M. Echenique and J. Owers; Editors). Construction Press, Lancaster, United Kingdom.

Brotchie, J. F.; Dickey, J. W.; Sharpe R. (1976). "TOPAZ: Planning Technique and Applications." Draft Technical Report, Division of Building Research, CSIRO, Melbourne, Australia.

Burton, M. R.; Chan, Y. (1993). "Facility location under spatial competition." Working paper, Department of Operational Sciences, Air Force Institute of Technology, Wright–Patterson AFB, Ohio.

Carroll, O. T.; Chan, Y.; Sexton, T.; Silkman, R. (1982). "An Assessment of Regional and State–wide Economic Impacts of New York State Parks." Institute for Urban Science Research, State University of New York at Stony Brook, Stony Brook, New York.

Cesario, F. J. (1975). "A combined trip generation and distribution model." *Transportation Science*, Vol. 9, pp. 211–223.

Cesario, F. J.; Knetsch, J. L. (1976). "A recreation site demand and benefit estimation model." *Regional Studies*, Vol. 10, pp. 97–104.

Chan, Y.; Carroll, T. O. (1985). "Estimating recreational travel and economic values of state parks." *Journal of Urban Planning and Development*, Vol. 111, pp. 65–79.

Clawson, M.; Knetsch, J. L. (1966). *Economics of Outdoor Recreation.* Johns Hopkins University Press, Baltimore, Maryland.

Cochrane, R. A. (1975). "A possible economic base for the gravity model." *Journal of Transport Economics and Policy*, Vol. 9, pp. 34–49.

de la Barra, T. (1989). *Integrated Land Use and Transport Modelling,* Cambridge University Press, Cambridge, United Kingdom.

dePalma, A.; Grinsburgh, V.; Labbe, M.; Thisse, J-F. (1989). "Competitive location with random utilities." *Transportation Science*, Vol. 23, No. 4, pp. 244–252.

Dickey, J. W.; Leone, P. A.; Scharte, A. R. (1973). "Use of TOPAZ for generating alternate land use schemes" *Highway Research Record*, No. 422, Highway Research Board, Washington, D.C.

Dobson, G.; Karmarkar, U. (1987). "Competitive location on a network." *Operations Research*, Vol. 35, pp. 565–574.

Drezner, A.; Wesolowsky, G. O. (1999). "Allocation of demand when cost is demand-dependent." *Computers & Operations Research*, Vol. 26, pp. 1–15.

Fang, S. C.; Tsao, H.-S. J. (1995). "Linearly-constrained entropy maximization problem with quadratic cost and its applications to transportation planning problems" *Transportation Science*, Vol. 29, pp. 358–365.

Foot, D. (1981). *Operational Urban Models*. Methuen, London, United Kingdom.

Forsythe, S.; Chan, Y. (1993). "Facility location and price equilibrium." Working paper, Department of Operational Sciences, Air Force Institute of Technology, Wright–Patterson AFB, Ohio.

Fotheringham, A. S. (1983). "A new set of spatial-interaction models: the theory of competing destinations." *Environment and Planning A*, Vol. 15, pp. 15–36.

Fujii, S.; Kitamura, R.; Monma, T. (1997). "A micro-simulation model system that produces individuals' activity-travel patterns based on random utily theory." Working paper, Department of Transportation Engineering Systems, Kyoto University, Kyoto, Japan.

Fujita, M.; Ogawa, H.; Thisse, J-F. (1988). "A spatial competition approach to central place theory." *Journal of Regional Science*, Vol. 28, pp. 477–94.

Gensch, D. H.; Ghose, S. (1997). "Differences in independence of irrelevant alternatives at individual vs aggregate levels, and at single pair vs full choice set." *Omega*, Vol. 25, pp. 201–214.

Goodchild, M. F.; Booth, P. J. (1976). "Modelling human spatial behavior in urban recreation facility site location." DP–007, Department of Economics, University of Western Ontario, London, Ontario, Canada.

Haag, G. (1989). *Dynamic Decision Theory: Applications to Urban and Regional Topics*, Kluwer.

Hakimi, S. L. (1990). "Locations with spatial interactions: competitive locations and games." in *Discrete Facility Location*, (P. Mirchandani and R. Francis; editors). Wiley Interscience, New York, New York.

Hansen, P.; Thisse J-F.; Wendell, R. E. (1990). "Equilibrium analysis for voting and competitive location problems" in *Discrete Facility Location*, (P. Mirchandani and R. Francis; editors). Wiley Interscience, New York, New York.

Hansen, P.; Labbe, M. (1988). "Algorithms for voting and competitive location on a network." *Transportation Science*, Vol. 22, pp. 278–288.

Healy III, R. W.; Jackson, R. W. (2001). "Competition ad complementarity local economic development: A nonlinear dynamic approach." *Studies in Regional & Urban Planning*, Issue 8, December.

Hicks, J. E.; Abdel-Al, M. M. (1998). "Maximum likelihood estimation for combined travel choice model parameters." Pre-print, 77th Annual Transportation Research Bard Meeting, Washington, D.C.

Hodgson, M. J. (1978). "Toward more realistic allocation in location–allocation models: an interaction approach." *Environment and Planning A*, Vol. 10, pp. 1273–1285.

Hughes, J. M., Lloyd R. D., editors (1977). "Outdoor Recreation—Advances in Applications of Economics." Technical Report WO–2, U. S. Forest Service, Washington, D.C.

Hunt, J. D.; Teply, S. (1993). "A nested logit model of parking location choice." *Transportation Research B*, Vol. 27B, pp. 253–265.

Hunt, J. D.; McMillan, J. D. P.; Abraham, J. E. (1994). "A stated preference investigation of influences on the attractiveness of residential locations." Pre-print, Transportation Research Board Annual Meeting, Washington D.C.

Ingene, C. A.; Ghosh, A. (1990). "Consumer and producer behavior in a multi-purpose shopping environment." *Geographical Analysis*, Vol. 22, pp. 70–93.

Johnston, R. A.; Rodier, C. J. (1999). "Synergism among land use, transit, and travel pricing policies." Pre-print, 78th Annual Transportation Research Board Meeting, Washington, D.C.

Koppelman, F.; Wen, C.-H. (1998). "Different nested logit models: which are you using?" Pre-print, 77th Annual Transportation Research Board Meeting, Washington, D.C.

Koppelman, F.; Wen, C.-H. (1998a). "Alternative nested logit models: Structure, properties and estimation." *Transportation Research B*, Vol. 32, No. 5, pp. 289–298.

Labbe, M.; Hakimi, S. L. (1991). "Market and locational equilibrium for two competitors." *Operations Research*, Vol. 39, No. 5, pp. 749–756.

Lee, C-K. (1987). "Implementation and evaluation of network equilibrium models of urban residential location and travel choices." Doctoral Dissertation, Department of Civil Engineering, University of Illinois at Urbana–Champaign, Illinois.

Leonardi, G. (1983). "The use of random-utility theory in building location–allocation models." in *Locational Analysis of Public Facilities* (J-F. Thisse and H.G. Zoller, editors). North–Holland, Amsterdam, pp. 357–383.

Li, S.; Tirupati, G. (1994). "Dynamic capacity expansion problem with multiple products: technology selection and timing of capacity additions" *Operations Research, Vol. 42*, pp. 959–976.

Matsushima, N. (2001). "Cournot competition and spatial agglomeration revisited." *Economic Letters*, Vol. 73, Issue 2, pp. 175–177.

Murray, A. T.; Gerrard, R. A. (1997). "Capacitated service and regional constraints in location–allocation modeling." *Location Science*, Vol. 5, No. 2, pp. 103–118.

Neuberger, H. L. T. (1971). "User benefit in the evaluation of transport and land use plans." *Journal of Transport Economics and Policy.*" Vol. 5, pp. 52–75.

O'Kelly, M. E. (1987). "Spatial interaction based location–allocation models" in *Spatial Analysis and Location–Allocation Models* (A Ghosh and G Rushton, editors). pp. 302–326, Van Nostrand, New York, New York.

O'Kelly, M. E.; Bryan, D. L. (1998). "Hub location with flow economies of scale." *Transportation Research B*, Vol. 32, No. 8, pp. 605–616.

Oppenheim, N. (1995). *Urban Travel Demand Modeling—from Individual Choices to General Equilibrium.* Wiley-Interscience, New York, New York.

Pappis, C. P.; Karacapilidis, N. I. (1994). "Applying the service level criterion in a location–allocation problem" *Decision Support Systems,* Vol. 11, pp. 77–81.

Pastor, J. T. (1994). "Bicriterion programs and managerial location decisions: application to the banking sector." *Journal of the Operational Research Society*, Vol. 45, No. 12, pp. 1351–1362.

Puu, T. (1997). *Mathematical Location and Land Use Theory: An Introduction.* Springer–Verlag, Berlin, Germany.

Sherali, H. D.; Carter, T. B.; Hobeika, A. G. (1991). "A location–allocation model and algorithm for evacuation planning under hurricane/flood conditions" *Transportation Research B*, Vol. 25B, No. 6, pp. 439-452.

Smith, K.; Krishnamoorthly, M.; Palaniswami, M. (1996). "Neural versus traditional approaches to the location of interacting hub facilities." *Location Science*, Vol. 3, pp. 155–171.

Stahl, K. (1982). "Location and spatial pricing theory with non–convex transportation cost schedules." *Bell Journal of Economics*, Vol. 12, pp. 575–82.

Stopher, P. R.; Ergun, G. (1982). "The effect of location on demand for urban recreation trips." *Transportation Research A*, Vol. 16A, No. 1, pp. 25–34.

Thill, J-C. (1992). "Spatial competition and market interdependence." *Papers in Regional Science*, Regional Science Association, Vol. 71, No. 3, pp. 259–75.

Thill, J-C.; Wheeler, A. (2000). "Tree induction of spatial choice behavior." *Transportation Research Record*, No. 1719, pp. 250–258.

van Lierop, W. (1986). *Spatial Interaction Modelling and Residential Choice Analysis*, Gower.

Vickerman, R. W. (1974). "The evaluation of benefits from recreational projects." *Urban Studies*, Vol. 11, pp. 277–288.

Wang, F.; Guldmann (1996). "Simulating urban population density with a gravity-based model" *Socio-Economic Planning Sciences*, Vol. 30, pp. 245–256.

Webber, M. J. (1978). "Spatial interaction and the form of the city" in *Spatial Interaction Theory and Planning Models* (A. Karlquist, L. Lundquist, F. Snickar, and J. W. Weibull; editors). North-Holland, Amsterdam, pp. 203–226.

Webber, M. J.; O'Kelly, M. E.; Hall, P. D. (1979). "Empirical tests on an information minimizing model of consumer characteristics and facility location." *Ontario Geography, Vol. 13*, pp. 61–80.

Williams, H. C. W. L. (1976). "Travel demand models, duality relations and user benefit analysis." *Journal of Regional Science*, Vol. 16, No. 2, pp. 147–166.

Wu, T.-H.; Lin, J.-N. (2003). "Solving the competitive discretionary service facility location problem." *European Journal of Operational Research*, Vol. 144, pp. 366–378.

Chapter 6 – Activity Allocation and Derivation: Evolution of the Lowry-Based Land-Use Models

Allen, P. M.; Sanglier, M.; Boon, F.; Denneubourg, J. L.; Depalma, A. (1981). "Models of Urban Settlement and Structure as Dynamic Self-organizing Systems," for the U. S. Department of Transportation by Universite Libre de Bruxelles.

Batty, M. (1976). *Urban Modelling*, Cambridge University Press, Cambridge, United Kingdom.

Beaumont, J. R. (1982). "Towards a conceptualization of evolution in environment systems." *Int J Man–Machine Studies*, Volume 16, pp. 113–145.

Bertuglia, C. S.; Leonardi, G.; Occelli, S.; Rabinao, G. A.; Tadei, R.; Wilson, A. G. — Editors (1987). *Urban Systems: Contemporary Approaches to Modelling*, Croon Helm, Beckenham, United Kingdom.

Casetti, E. (1981). "A catastrophe model of regional dynamics." *Annals of the Association of American Geographers*, Vol. 71, No. 4, pp. 572–579.

Cesario, F. J. (1975). "A primer in entropy modelling." *Journal of the American Institute of Planners*, January, pp. 40–48.

Chan, Y. (2000). *Location Theory and Decision Analysis*, Thomson/South-Western, Cincinnati, Ohio.

Chan, Y.; Fitzroy, S. S. (1985). "Urban spatial modelling: some experimental observations." *Transportation Planning Journal*, Vol. 14, No. 2, Taiwan, Republic of China, pp. 287–310.

Cochrane, R. M. (1975). "A possible economic basis for the gravity model." *Journal of Transport Economic and Policy*, January, pp. 34–39.

Cordey–Hays, M.; Broadbent, T. A.; Massey, B. (1970). "Towards operational urban development models." in *Regional Forecasting: proceedings of the twenty-second symposium of the Colston Research Society* (M. Chisholm, A. E. Frey and P. Haggett — editors), University of Bristol, April 6–10, 1970, Colston Papers, No. 22, Hamden, Conn., Archon Books, Shoestring Press, pp. 221–254.

Dendrinos, D. S.; Sonis, M. (1990). *Chaos and Socio-Spatial Dynamics*, Spring–Verlag, New York, New York.

Fitzroy, S. S. (1978). "A Comparative Analysis of the Land Use Allocation Models on a Common Data Base," Master's Thesis, Urban and Regional Planning Program, Pennsylvania State University, University Park, Pennsylvania.

Foot, D. H. S. (1982). *Operational Urban Models*, Methuen, London, United Kingdom.

Foot, D. (1978). "Urban Models I: A Computer Program for the Garin–Lowry Model." *Geographical Papers No. 65*, Department of Geography, University of Reading, Reading, United Kingdom. (Printed by George Over Ltd. London, United Kingdom).

Frankhauser, P. (1987). "Entkopplung der stationaren Losung des Haag–Wilson Modells durch einen neuen Ansatz fur die Nutzen–Funktion." Arbeitspapier, 2, Institute fur Theoretisch physik, Germany.

Gabriel, S. A.; Faria, J. A.; Moglen, G. E. (2002). "A multiobjective optimization approach to smart growth in land development." Working Paper, Department of Civil and Environmental Engineering, University of Maryland, College Park, Maryland.

Garin, R. A. (1966). "A matrix formulation of the Lowry model for intra-metropolitan activity location" *Journal of the America Institute of Planners*, Vol. 32, pp. 361–364.

Golan, A.; Vogel, S. J. (1998). "Estimation of non-stationary social accouting matrix coefficints with supply-side information." Working paper, Deparment of Economics, American University, Washingon, D.C.

Goldner, W. (1968). "Projective Land Use Model." Bay Area Transportation Study Commission, Technical Report 219, September, Berkeley, California.

Goldner, W.; Rosenthal, S. R.; Meredith, J. R. (1972). "Projective Land Use Model: Vol. II — Theory and Application." Prepared for the U. S. Department of Transportation, Contract FH-11-7132, Washington, D.C.

Greenhut, M. L. (1956). "Contrasts in site-selection: factors which influence the location of a wood-metal pattern shop and a narrow fabric mill." *Land Economics*, pp. 152–166.

Haag, G. (1989). *Dynamic Decision Theory: Applications to Urban and Regional Topics*, Kluwer Academic Publishers, Norwell, Massachusetts; and Dordrecht, Netherlands.

Hansen, W. G. (1959). "How accessibility shapes land use." *Journal of the American Institute of Planners*, pp. 152–166.

Huang, Y. (1997) "Identifying the link between transportation and land use density with accessibility." Pre-print 97–0293, 76th Annual Transportation Research Board Meeting, Washington, D.C.

Kain, J. F. (1962). "The journey-to-work as a determinant of residential location." *Regional Science Association Paper*, Vol. 9, pp. 137–160.

Kockelman, K. M. (1997) "Travel behavior as a function of accessibility, land use mixing, and land use balance: evidence from the San Francisco Bay area." Pre-print 97-0048, 76th Annual Transportation Research Board Meeting, Washington, D. C.

Krishnamurthy, S.; Kockelman, K. M. (2002). "Propagation of uncertainty in transportation-land use models: An investigation of DRAM-EMPAL and UTPP prediction in Austin, Texas." Presentation at the 82nd Annual Meeting of the Transportation Research Board, Washington, D.C.

Lane, T. (1966). "The urban base multiplier: an evaluation of the state of the art." *Land Economics*, pp. 339–347.

Lowry, I. S. (1964). "A Model of Metropolis." Report RM–4035 RC, Rand Corporation, Santa Monica, California.

Lowry, I. S. (1992). Personal correspondence.

Mihram, G. A. (1972). "Some practical aspects of the verification and validation of simulation models." *Operational Research Quarterly*, Vol. 23, No. 1, pp. 17–29.

Moody, H. T.; Puffer, F. W. (1970). "The empirical verification of the urban base multiplier: traditional and adjustment process models." *Land Economics*, pp. 91–98.

Nicolas, G.; Prigogine, I. (1977). *Self-Organization in Non-equilibrium Systems*. Wiley, New York, New York.

Oppenheim, N. (1995). *Urban Travel Demand Modeling: From Individual Choices to General Equilibrium*, Wiley-Interscience.

Putman, S. H. (1972). A "Workbook" for Educational and Limited Research use of the Lowry Model,Department of City and Regional Planning, University of Pennsylvania, Philadelphia, Pennsylvania.

Putman, S. (1983). *Integrated Urban Models: Policy Analysis of Transportation and Land Use*. Pion, London, United Kingdom.

Putman, S.; Ahmad, H. Z. Z.; Choi, K.-M.; McCarthy, W. P.; Yan, Y. (2000). "Integrated transportation and land use policy analysis for Sacramento." Pre-print, 79th Annual Meeting of the Transportation Research Board, Washington D.C.

Puu, T. (1997). *Mathematical Location and Land Use Theory*, Springer–Verlag, Berlin, Germany.

Reif, B. (1973). *Models in Urban and Regional Planning*, Intext Educational Publishers, New York.

Rosenthal, S. R.; Meredith, J. R.; Goldner, W. (1972). *Projective Land Use Model: Vol. I — Plan Making with a Computer Model*, prepared for the U. S. Department of Transportation, Contract FH-11-7132, Washington, D.C.

Sen, A.; Smith, T. E. (1995). *Gravity Models of Spatial Interaction Behavior*, Springer–Verlag, Berlin, Germany.

Shen, P. (1997). "Mathematical programming formulation of a combined transportation and land use model" Pre-print 97-0635, 76th Annual Transportation Research Board Meeting, Washington, D.C.

Sirkin, G. (1959). "The theory of the regional economic base." *Review of Economic Statistics*, Vol. 41, pp. 426–424.

Tiebout, C. M. (1956). "A pure theory of local expenditure." *Journal of Political Economy*, Vol. 65, pp. 223–245.

Ulanowitz, R. (1980). "An hypothesis on the development of natural communities." *Journal of Theoretical Biology*, Vol. 85, pp. 223–245.

U. S. Department of Transportation (1996). "Land Use Modeling Conference Proceedings — February 19–21, 1995." Report DOT-T-96-09, Washington, D.C.

Vanden Bosch, P.; Dietz, D. C.; Pohl, E. A. (1999). "Choosing the best approach to matix exponentiation." *Computers & Operations Research*, Vol. 26, pp. 871–882.

Varaiya, P.; Wiseman, M. (1981). "Bifurcation models of urban development: a survey." Working paper WP81-2, Institute of Business and Economic Research, University of California, Berkeley, California.

Webber, M. (1984). *Prediction, Explanation and Planning: the Lowry Model*, Pion, London, United Kingdom.

Webber, H. J. (1984). *Explanation, prediction and planning: the Lowry Model*, Pion, London, United Kingdom.

Williams, G.; Tempio, J. L. (1974). "PLUM: Projective Land Use Model." Working paper, Urban and Regional Planning Program, Pennsylvania State University, University Park, Pennsylvania.

Wilson, A. G. (1974). *Urban and Regional Models in Geography and Planning*, Wiley, Chichester, United Kingdom.

Wilson, A. G. (1981). *Catastrophe Theory and Bifurcation: Applications to Urban and Regional* Systems. Croon Helm, Beckenham, United Kingdom; and University of California Press, Berkeley, California.

Yen, Y.-M.; Fricker, J. D. (1997). "An integrated transportation land use modeling system." Pre-print 97-0659, 76th Annual Transportation Research Board Meeting, Washington, D.C.

Yi, P. (1986). "Infrastructure Management: A Bifurcation Model in Urban/Regional Planning." Master's Thesis, Department of Civil and Environmental Engineering, Washington State University, Pullman, Washington.

Chapter 7 - Chaos, Catastrophe, Bifurcation and Disaggregation in Locational Models

Allen, P. M.; Sanglier, M. (1979). "A dynamic model of growth in a central place system." *Geographical Analysis, Vol 11*, pp. 226–272.

Andersson, A. E.; Kuenne, R. E. (1986). "Regional economic dynamics." in *Handbook of Regional and Urban Economics*, Vol. I, (P. Nijkamp – Editor), Elsevier Science Publishers B.V., Amsterdam, Netherlands.

Bazarraa, M. S.; Shetty, C. M. (1979). *Nonlinear Programming: Theory and Algorithms*, John Wiley, New York, New York.

Belenky, A. S. (1998). *Operations Research in Transportation Systems*, Kluwer Acadmic Publishers, Norwell, Massachusetts.

Bertuglia, C. S.; Leonardi, G.; Occelli, S.; Rabino, G. A.; Tadei, R.; Wilson, A. G. – Editors (1987). *Urban Systems: Contemporary Approaches to Modelling*, Croom Helm, Beckenham, United Kingdom.

Brown, S. (1992). "The wheel of retail gravitation." *Environment and Planning A*, Vol. 24, pp. 1409–1429.

Chan, Y. (2000). *Location Theory and Decision Analysis*, Thomson/South-Western, Cincinnati, Ohio.

Chan, Y.; Yi, P. (1987). "Bifurcation and disaggregation in urban/regional spatial modelling: a technical note." *European Journal of Operational Research*, Vol. 30, pp. 321–326.

Clarke, M.; Wilson, A. G. (1983). "The dynamics of urban spatial structure: problems and progress." *Journal of Regional Science*, Vol. 23, pp. 1–18.

Dendrinos, D. S.; Sonis, M. (1990). *Chaos and Socio-Spatial Dynamics*, Springer–Verlag, New York, New York.

Dendrinos, D. S.; Mulally, H. (1985). *Urban Evolution: Studies in the Mathematical Ecology of Cities*, Oxford University Press, Oxford, United Kingdom.

Eaton, B. C.; Lipsey, R. G. (1975). "The non-uniqueness of equilibrium in Loschan location model." *The American Economic Review, Vol. 66*, pp. 77–93.

Feichtinger, G. (1996). "Chaos theory in Operations Research." *International Transactions in Operations Research*, Vol 3, pp. 23–36.

Fischer, M. M.; Nijkamp, P.; Papageorgiou, Y. Y. – Editors (1990). *Spatial Choices and Processes*, North Holland/Elsevier Science Publishers B.V., Amsterdam, Netherlands.

Golan, A.; Vogel, S. J. (1998). "Estimation of non-stationary social accouting matrix coefficints with supply-side information." Working paper, Deparment of Economics, American University, Washingon, D.C.

Hakimi, S. L. (1990). "Locations with spatial interactions: competitive locations and games." in *Discrete Location Theory* (P. B. Mirchandani and R. L. Francis – Editors), Wiley-Interscience, New York, New York.

Hof, J.; Bevers, M. (1998). *Spatial Optimization for Managed Ecosystems*, Columbia University Press, New York, New York.

Iooss, G.; Joseph, D. D. (1990). *Elementary Stability and Bifurcation Theory*, Second Edition, Springer–Verlag, New York, New York.

Lorenz, E. N. (1983). "Deterministic nonperiodic flow." *Journal of the Atmospheric Sciences*, Vol. 82, No. 2, pp. 130–141.

Lorenz, H.W. (1993). *Nonlinear Dynamical Economics and Chaotic Motion, Second Edition*, Springer–Verlag, Berlin, Germany.

Manheim, M. (1984). *Fundamentals of Transportation Systems Analysis – Volume 1: Basic Concepts*, MIT Press, Cambridge, Massachusetts.

May, R. (1976). "Simple mathematical models with very complicated dynamics." *Nature, Vol 261*, pp. 459–467.

Miyao, T. (1987). "Dynamic urban models." chapter 22 in *Handbook of Regional and Urban Economics, Vol II* (E S. Mills – Editor), Elsevier Science Publishers B.V., Amsterdam, Netherlands.

Moon, F. C. (1992). *Chaotic and Fractal Dynamics*, Wiley-Interscience, New York, New York.

Nijkamp, P.; Reggiani, A. (1992). *Interaction, Evolution and Chaos in Space*, Springer–Verlag, Berlin, Germany.

Nijkamp, P.; Reggiani, P. (1993). "Stability and complexity in spatial *Networks*." Discussion paper TI93–249, Tinbergen Institute, Amsterdam–Rotterdam, the Netherlands.

Noble, B. (1969). *Applied Linear Algebra*, Prentice Hall, Englewood, New Jersey.

Puu, T. (1997). *Mathematical Location and Land Use Theory: An Introduction*, Springer–Verlag, Berlin, Germany.

Sen, A.; Smith, T. (1995). *Gravity Model of Spatial Interaction Behavior*, Springer–Verlag, Berlin, Germany.

Silberberg, E. (1990). *The Structure of Economics: A Mathematical Analysis*, Second Edition, McGraw–Hill, New York, New York.

Smith, B. A. (1996). Private communication.

Upton, G. J. G.; Fingleton, B. (1989). *Spatial Data Analysis by Example: Categorical and Directional Data - Volume 2*, Wiley, Chichester, United Kingdom.

Webber, M. (1984). *Prediction, Explanation, and Planning: the Lowry Model*, Research in Planning and Design, Volume 11, Pion, London, United Kingdom.

Wilson, A. G. (1981). *Catastrophe theory and bifurcation: applications to urban and regional systems*, Croon Helm, Beckenham, United Kingdom; and University of California Press, Berkeley, California.

Wong, C. K.; Wong, S. C.; Tong, C. O. (1998). "A new methodology for calibrating the Lowry Model." *Journal of Urban Planning and Development*, Vol. 124, No. 2, pp. 72–91.

Yi, P.; Chan, Y. (1988). "Bifurcation and disaggregation in Lowry–Garin derivative models: theory, calibration, and case study." *Environment and Planning A*, Vol. 20, pp. 1253–1267.

Yi, P. (1986). "Infrastructure Management: a Bifurcation Model in Urban Regional Planning." Master's thesis, Department of Civil and Environmental Engineering, Washington State University, Pullman, Washington.

Chapter 8 - Spatial Equilibrium And Disequilibrium: Locational Conflict and the Input-Output Model

Ansari, A.; Economides, N.; Ghosh, A. (1992). "Competitive positioning in markets with non-uniform preferences." Working paper, Stern School of Business, New York University, New York, New York.

Aoyagi, M.; Okabe, A. (1993). "Spatial competition of firms in a two-dimensional bounded market." *Regional Science and Urban Economics*, Vol. 23, pp. 259–289.

Apte, U. M. (1999). "Decision model for planning of regional industrial programs." *IIE Transactions*, Vol. 31, pp. 881–898.

Andersson, A. E.; Kuenne, R. E. (1986). "Regional Economic Dynamics." in *Handbook of Regional and Urban Economics*, Vol. I (P. Nijkamp – Editor), Elsevier Science Publishers B.V., Amsterdam, Netherlands.

Barondes, C.; Chan, Y. (1991). "User vs system optimization of network flow: Multicriteria decision-making." Working paper, Department of Operational Sciences, Air Force Institute of Technology, Wright–Patterson AFB, Ohio.

Beckmann, M. J. (1987). "The economic activity equilibrium approach in location theory." in *Urban Systems: Contemporary Approaches to Modelling*, (C. S. Bertuglia, G. Leonardi, S. Occelli, G. A. Rabino, R. Tadei, and A. G. Wilson – editors), Croom Helm, Beckenham, United Kingdom.

Bhadury, J. (1993). "Competitive location under uncertainty." Working paper 92–029, Faculty of Administration, University of New Brunswick, Fredericton, New Brunswick, Canada.

Bhadury, J. (1995). "Competitive location and entry deterrence: a variant of Hotelling's duopoly model." Working paper, Faculty of Administration, University of New Brunswick, Fredericton, New Brunswick, Canada.

Bhadury, J.; Eiselt, H. A. (1995). "Stability of Nash equilibria in locational games." *RAIRO/Recherche Operationelle*, Vol. 29, No. 1, pp. 19-33.

Billups, S. C.; Murty, K. G. (2000). "Complementarity problems." *Journal of Computational and Applied Mathematics*, Vol. 124, pp. 303–318.

Boyce, D. E.; Lundquist, L. (1987). "Network equilibrium models of urban location and travel choices: alternative formulations for the Stockholm region." *Papers of the Regional Science Association*, Vol. 61, pp. 93–104.

Boyce, D.; Mattsson, L.-G. (1999). "Modeling residential location choice in relation to housing location and road tolls on congested urban highway *Networks*." *Transportation Research B*, Vol. 33, pp. 581–591.

Brandeau, M. L.; Chiu, S. S. (1994). "Location of competing facilities in a user-optimizing environment with market externalities." *Transportation Science, Vol. 28*, pp. 125–140.

Catoni, S.; Pallotino, S. (1991). "Traffic equilibrium paradoxes." *Transportation Science*, vol 25, No 3.

Chan, Y. (2000). *Location Theory and Decision Analysis* Thomson/South-Western, Cincinnati, Ohio.

Chang, L.; Ziliaskopoulos, A.; Boyce, D.; Waller, S. T. (2001). "Solution algorithm for combined interregional commodity flow and transportation network model with link capacity constraints." *Transportation Research Record*, No. 1771, pp. 114–123.

Chiang, Y. S.; Roberts, P. O.; Ben-Akiva, M. (1980). "Development of a Policy Sensitive Model for Forecasting Freight Demand." Center for Transportation Studies, Report CTS81-1, M.I.T., Cambridge, Massachusetts.

Chu, Y.-L. (1999). "Network equilibrium model of employment location and travel choices." *Transportation Research Record*, No. 1667, pp. 60–66.

de Palma, A.; Ginsburgh, V.; Labbe, M.; Thisse, J.-F. (1989). "Competitive location with random utilities." *Transportation Science*, Vol. 23, pp. 244–252.

Dembo, R. (1983). *NLPNET User's Manual*, Yale School of Organization and Management, New Haven, Connecticut.

Dennis, S. M. (1999). "Using spatial equilibrium models to analyze transportation rates: an application to steam coal in the United states." *Transportation Research E*, Vol. 35, pp. 145–154.

Eaton, B. H.; Lipsey, R. G. (1977). "The introduction of space into the neoclassical model of value theory." in *Studies in Modern Economics* (M. J. Artis and A. R. Nobay – editors), Basil Blackwell, Oxford, United Kingdom.

Eaton, B. H.; Lipsey, R. G. (1978). "Freedom of entry and the existence of pure profit." *Economic Journal*, Vol. 88, pp. 455–469.

Eiselt, H. A. (1995). "Perception and information in a competitive location model." Working paper, Faculty of Administration, University of New Brunswick, Fredericton, New Brunswick, Canada.

Evans, S. (1976). "Derivation and analysis of some models for combining trip distribution and assignment." *Transportation Research*, Vol. 10, pp. 37–57.

Feng, C-M.; Lin J-J. (1992). "Impact analysis of major construction plans on regional development—a case study of Taiwan northern region." *Transportation Planning Journal*, Vol. 21, pp. 367–400, Institute of Transportation, Ministry of Transportation and Communication, Taiwan (in Chinese).

Florian, M.; Nguyen, S.; Ferland, J. (1975). "On the combined distribution-assignment of traffic." *Transportation Science*, Vol. 9, pp. 43–53.

Friesz, T. L.; Bernstein, D.; Mehta, N. J.; Tobin, R. L.; Ganjalizadeh, S. (1994). "Day-to-day dynamic network disequilibria and idealized traveler information systems." *Operations Research*, Vol. 42, pp. 1120–1136.

Friesz, T. L.; Harker, P. T. (1983). "Multicriteria spatial price equilibrium network design: theory and computational results." *Transportation Research, Vol. 17B*, pp. 411–426.

Friesz, T. L.; Suo, Z.-G.; Bernstein, D. (1998). "A dynamic disequilibrium interregional commodity flow model." *Transportation Research B*, Vol. 32, No. 7, pp. 467–483.

Gabszewicz, J. J.; Thisse, J.F. (1986). "Spatial competition and the location of the firms." in *Location Theory* (J. J. Gabszewicz, J..F. Thisse, M. Fujita, and U. Schweizer – editors), Harwood Academic Publishers, Chur, Switzerland.

Guder, F.; Morris, J. G.; Yoon, S. H. (1992). "Parallel computation of inter-temporal multicommodity spatial price equilibria in the presence of quotas." Working paper, Loyola University of Chicago, Chicago, Illinois.

Harker, P. T. (1986). "Alternative models of spatial competition." *Operations Research*, Vol. 34, No 3, pp. 410–425.

Hof, J. G.; Bevers, M. (1998). *Spatial Equilibrium for Managed Ecosystems*, Columbia University Press, New York, New York.

Infante–Macias, R.; Munoz–Perez, J. (1995). "Competitive location with rectilinear distances." *European Journal of Operational Research, Vol. 80*, pp. 77–85.

Jin, Y.-X.; Leigh, C. M.; Wilson, A. G. (1991). "Construction of an input-output table for Yorkshire and Humberside." Working paper 544, School of Geography, University of Leeds, Leeds, United Kingdom.

Jones, P. C.; Theise, E. S. (1989). "On the equivalence of competitive transportation markets an congestion in spatial price equilibrium models." *Transportation Science, Vol. 23*, pp. 112–117.

Kanafani, A. (1983). *Transportation Demand Analysis*, McGraw–Hill, New York, New York.

Ke, Y-S. (1997). "Alternative Implementations of Expanding Algorithms for Multi-commodity Spatial Price Equilibrium." Master's thesis, Air Force Institute of Technology, Wright–Patterson AFB, Ohio.

Kim, T. J. (1989). *Integrated Urban Systems Modeling: Theory and Applications, Studies in Operational Regional Science*, Kluwer Academic Publishers, Norwell, Massachusetts.

Knoblauch, V. (1991). "Generalizing location games to a graph." *The Journal of Industrial Economics*, Vol. 19, pp. 683–688.

Knoblauch, V. (1996). "A pure strategy Nash equilibrium for a 3-firm location game on a sphere." *Location Sciences*, Vol. 4, No. 4, pp. 247–250.

Miller, T. C.; Tobin, R. L.; Friesz, T. L. (1991). "Stackelberg games on a network with Cournot–Nash oligopolistic competitors." *Journal of Regional Science*, Vol. 31, pp. 435–454.

Miller, T.; Tobin, R. L.; Friesz, T. L. (1992). "Network facility-location models in Stackelberg–Nash–Cournot spatial competition." *Papers in Regional Science*, Vol. 71, pp. 277–291.

Miller, T. C.; Friesz, T. L.; Tobin, R. L. (1996). *Equilibrium Facility Location on Networks*, Springer–Verlag, Berlin, Germany.

Nagurney, A.; Aronson, J. (1989). "A general dynamic spatial price network equilibrium model with gains and losses." *Networks*, Vol. 19, pp. 751–769.

Nagurney, A.; Zhao, L. (1991). "A network equilibrium formulation of market disequilibrium and variational inequalities." *Networks*, Vol. 21, pp. 109–132.

Nagurney, A. (1993). *Network Economics: A Variational Inequality Approach*, Kluwer Academic Publishers, Norwell, Massachusetts.

Nagurney, A. (2000). "Congested urban transportation *Networks* and emission paradoxes." *Transportation Research D*, Vol. 5, pp. 145–151.

Nagurney, A.; Eydeland, A. (1992). "A splitting equilibration algorithm for the computation of large-scale constrained matrix problems: theoretical analysis and applications." *Computational Economics and Econometrics, vol. 22* (H. Amman, D. Belsley and L. Pau – editors), Kluwer Academic Publishers, Norwell, Massachusetts, pp. 65-105.

Nagurney, A.; Thore, S.; Pan, J. (1996). "Spatial market policy modeling with goal targets." *Operations Research*, Vol. 44, pp. 393–406.

Oppenheim, N. (1980). *Applied Models in Urban and Regional Analysis*, Prentice–Hall, Englewood Cliffs, New Jersey.

Oppenheim, N. (1995). *Urban Travel Demand Modeling: From Individual Choices to General Equilibrium*, Wiley-Interscience, New York, New York.

Penchina, C. M. (1997). "Braess paradox: maximal penalty in a minimal critical network." *Transportation Research A*, Vol. 31, No. 5, pp. 379–388.

Portugal, L.; Judice, J. (1996). "A hybrid algorithm for the solution of a single commodity spatial equilibrium model." *Computers and Operations Research*, Vol. 23, pp. 623–639.

Putman, S. H. (1991). *Integrated Urban Models 2: New Research and Applications of Optimization and Dynamics*, Pion, London, United Kingdom.

Ran, B.; Boyce, D. (1994). *Dynamic Urban Transportation Models*, Springer–Verlag, New York, New York.

Schneider, M. H.; Zenios, S. A. (1990). "A comparative study of algorithms for matrix balancing," *Operations Research*, Vol. 38, pp. 439–455.

Sheffi, Y. (1985). *Urban Transportation Networks: Equilibrium Analysis with Mathematical Programming Methods*, Prentice Hall, Englewood Cliffs, New Jersey.

Soumis, F.; Nagurney, A. (1990). "A stochastic, dynamic airline network equilibrium model." Working paper, GERAD, Polytechnic School, Montreal, Canada.

Suzuki, Y.; Pak, P. S.; Kim, G. (1988). "Impact analysis of construction of the Kansai International Airport." Working paper, Faculty of Engineering, Osaka University, Osaka, Japan.

Theise, E. S.; Jones, P. C. (1988). "Alternative implementations of a diagonalization algorithm for multiple commodity spatial price equilibria." Technical Report 8–06, Department of Industrial Engineering and Management Sciences, Northwestern University, Evanston, Illinois.

Theise, E. S.; Jones, P. C. (1988). "A microcomputer-based implementation of the expanding equilibrium algorithm for linear, single commodity spatial price equilibrium problems." *Operations Research Letters, Vol. 7*, pp. 241–245.

Theise, E. S.; Jones, P. C. (1988). "Nonlinear, single commodity spatial price equilibria and the expanding equilibrium algorithm: two strategies for implementation." Technical report 88–16, Department of Industrial Engineering and *Management Science*, Northwestern University, Evanston, Illinois.

Tobin, R. L.; Miller, T.; Friesz, T. L. (1996). "Incorporating competitors' reactions in facility location decisions: a market equilibrium approach." *Location Science*, Vol. 3, pp. 239–253.

Toyomane, N. (1988). *Multiregional Input-Output Model in Long-run Simulation*, Kluwer Academic Publishers, Norwell, Massachusetts.

Vickerman, R. W. (1980). *Spatial Economic Behavior*, St. Martin's, New York, New York.

Wardrop, J. G. (1952). "Some theoretical aspects of road traffic research." *Proceedings of the Institute of Civil Engineers*, Part 2, United Kingdom, pp. 325–378.

Webber, M. J. (1984). *Explanation, Prediction, and Planning: the Lowry Model*, Research in planning and design series, Pion, London, United Kingdom.

Wilson, A. G.; Coelho, J. D.; Macgill, S. M.; Williams, H. C. W. L. (1981). *Optimization in Locational and Transport Analysis*, Wiley, Chichester, United Kingdom.

Wu, Y. J.; Fuller, J. D. (1996). "An algorithm for the multiperiod market equilibrium model with geometric distributed lag demand." *Operations Research*, Vol. 43, pp. 1002–1012.

Zhang, D.; Buongiorno, J.; Ince, P. J. (1993). "PELPS III—A Microcomputer Price Endogenous Linear Programming System for Economic Modeling." Version 1.0, Research paper FPL-RP-526, Forest Products Laboratory, Forest Service, U. S. Department of Agriculture, Washington, D.C.

Chapter 9 – Spatial Econometric Models: The Empirical Experience

Allen, W. B.; Liu, D.; Singer, S. (1993). "Accessibility measure of U. S. metropolitan areas." *Transportation Research B*, Vol. 27B, No 6, pp. 439–449.

Alterkawi, M. (1991). "Land economic impact of fixed guideway rapid transit systems on urban development in selected metropolitan areas." Doctoral dissertation, Texas A&M University, College Station, Texas.

Anselin, L. (1988). *"Spatial Econometrics: Methods and Models."* Kluwer Academic Publishers, Norwell, Massachusetts.

Armstrong Jr, R. J. (1994). "Impact of commuter rail service as reflected in single-family residential property values." Working paper, EG&G Dynatrend, Volpe National Transportation Systems Center, Cambridge, Massachusetts, Pre-print, 73rd Annual Transportation Research Board Meeting, National Research Council, Washington D.C.

Apogee Research and Greenhorne & O'Mara (1998). "Research on the Relationship between Economic Development and Transportation Investment." National Cooperative Highway Research Program, Report 418, National Academy Press, Washington, D.C.

Bach, L. (1981). "The problem of aggregation and distance for analysis of accessibility and access opportunity in location–allocation models." *Environment and Planning A*, Vol. 13A, pp. 955–978.

Bennett, R. J.; Horduk, L. (1986). "Regional Econometric and Dynamic Models." in *Handbook of Regional and Urban Economics* (P. Nijkamp, Editor), Elsevier Science Publishers B.V., Amsterdam, Netherlands.

Chan, Y. – Editor (1973). "A Review of Operational Urban Transportation Planning Models." Report DOT–TSC–496, performed by Peat Marwick Mitchell & Co. for the U. S. Department of Transportation, Washington, D.C.

Chan, Y. (1991). "Route-specific and time-of-day demand elasticities." *Transportation Research Record*. No. 1328. pp. 43–48.

Chan, Y. (2000). *Location Theory and Decision Analysis.* Thomson/South-Western, Cincinnati, Ohio.

Cliff, A. D.; Ord, J. K. (1981). *Spatial Processes: Models and Applications*, Pion, London, United Kingdom.

Cressie, N. (1991). *Statistics for Spatial Data.* Wiley-Interscience, New York, New York.

Foot, D. (1981). *Operational Urban Models*, Methuen, London, United Kingdom.

Forgionne, G. A. (1996). "Forecasting army housing supply with a DSS-delivered econometric model." *Omega*, Vol 24, pp. 561–576.

Getis, A. (1991). "Spatial interaction and spatial correlation: a cross-product approach." *Environment and Planning* A. Vol. 23A, pp. 1269–1277.

Griffith, D. A.; Can, A. (1996). "Spatial, Statistical/Econometric Versions of Simple Urban Population Density Models." *Practical Handbook of Spatial Statistics* (S. L. Arlinghaus, Editor-in-Chief), CRC Press, Boca Raton, Florida.

Hamed, M. M.; Mannering, F. L. (1993). "Modeling travelers' postwork activity involvement: Toward a new methodology." *Transportation Science*, Vol 27, No. 4, pp. 381–394.

Kanafani, A. (1983). *Transportation Demand Analysis.* McGraw–Hill, New York, New York.

Kane, E. J. (1968). *Economic Statistics and Econometrics: An Introduction to Quantitative Economics.* Harper & Row, New York, New York.

Kates, Peat, Marwick & Co. (1971). "EMPIRIC Activity Allocation Model Summary." April, Washington, D.C.

Kawamura, K. (2001). "Empirical examination of relationship between firm location and transportation facilities." *Transportation Research Record*, No. 1747, pp. 97–103.

Kitamura, R.; Akiyama, T.; Yamamoto, T.; Golob, T. (2001). "Accessability in a metropolis: Toward a better understanding of land use and travel." *Transportation Research Record,* No.1780, pp. 64–75.

Kitamura, R.; Kermanshah, M. (1983). "Identifying time and history dependencies of activity choice." *Transportation Research Record,* No. 944, pp. 22–30.

Kitamura, R.; Kermanshah, M. (1984). "A sequential model of interdependent activity and destination choices." *Transportation Research Record,* No. 987, pp. 81–89.

Meda–Dooley, R. H.; Chan, Y. (1978). "Models of the urban spatial structure: a framework for the integration of the macro and micro approaches." presentation at the Annual American Planning Association Meeting at New Orleans, August.

Miller, E. J.; Ibrahim, A. (1998). "Urban form and vehicular tsravel: Some empirical findings." Pre-print, 77th Annual Meeting of the Transportation Research Board, National Research Council, Washington, D.C.

Pak, P. S.; Kim, G.; Suzuki, Y. (1989). "Multizonal model of industrial activities based on journey time model." Working paper, Faculty of Engineering, Osaka University, Yamada–oka 2–1, Suita, Osaka 565, Japan.

Putman, S. H. (1979). *Urban Residential Location Models,* Martinus Nijhoff Publishing, Hingham, Massachusetts.

Ripley B. D. (1988). *Statistical Inference for Spatial Process: An Essay awarded the Adams Price of the University of Cambridge,* Cambridge University Press. Cambridge, United Kingdom.

SRI International (1977). "The SRI–WEFA Soviet Econometric Model: Phase Three Documentation – Volumes I and II." Performed for the Defense Advanced Research Projects Agency, Report AD A 043287, Defense Technical Information Center, Alexandria, Virginia.

Suzuki, Y.; Pak, P. S.; Kim, G. (1989). "Impact analyses of construction of Kansai International Airport." *Journal of Urban Planning and Development,* Vol. 115, No. 1, pp. 33–49.

Thompson, H. T. (1972). "The EMPIRIC model." Working Paper, Dept. of Civil Engineering, Pennsylvania State University, University Park, Pennsylvania.

Vadali, S. R.; Sohn, C. (2001). "Using a geographic information system to track changes in spatially segregated location premium." *Transportation Research Record,* No. 1766, pp. 180–191.

Chapter 10 - Spatial Time Series

Abraham, B.; Ledolter, J. (1983). *Statistical Methods for Forecasting,* Wiley, New York, New York.

Ahuja, R. K.; Orlin, J. B, (1998). "Solving the convex ordered set problem with applications to isotoneregression." Working paper, Sloan School of Management, Massachusetts Institute of Technology, Cambridge, Massachusetts.

Arinze, B.; Kim, S-L.; Anandarajan, M. (1997). "Combining and selecting forecasting models using rule based induction." *Computers and Operations Research,* Vol. 24, pp. 423–433.

Bennett, R. J. (1975). "The representation and identification of spatio-temporal systems: an example of population diffusion in North–West England." *Transactions – Institute of British Geographers,* Vol. 66, pp. 73–94.

Bhattacharjee, D.; Sinha, K. C.; Krogmeier, J. V. (1997). "Modeling the effects of ITS on route switchingbehavior in a freeway and its effect on the performance of O–D prediction algorithm." Working paper, Department of Civil Engineering, Purdue University, West Lafayette, Indiana.

Box, G. E. P.; Jenkins, G. M. (1976). *Time Series Analysis: Forecasting and Control,* Holden–Day, Oakland, California.

Box, G. E. P.; Tao, G. C. (1975). "Intervention analysis with applications to economic and environmental applications." *Journal of the American Statistical Association,* Vol. 70, pp. 70–79.

Chan, Y. (2000). *Location Theory and Decision Analysis.* Thomson/South-Western, Cincinnati, Ohio.

der Voort, M.; Dougherty, M.; Watson, S. (1996). "Combining Kohonen maps with ARIMA time series models to forecast traffic flow." *Transportation Research C,* Vol. 4, pp. 307–318.

Deutsch, S. J.; Wang, C. C. (1983). "Multivariate consideration in space-time modeling." Final report for grant #81-IJ-CX-0020 from the National Institute of Justice, School of Industrial and Systems Engineering, Georgia Institute of Technology, Atlanta, Georgia.

Forgues, P. J.; Chan, Y. (1991). "Forecasting models applied to remote sensing data," working paper, Department of Operational Sciences, Air Force Institute of Technology, Wright–Patterson AFB, Ohio.

Fuller, W. A. (1996). *Introduction to Statistical Time Series,* Wiley, New York, New York.

Garrido, R. A. (2000). "Spatial interaction between the truck flow through the Mexico–Texas border." *Transportation Research A,* Vol.. 34, pp. 23–33.

Garrido, R. A.; Mahmasani, H. S. (1998). "Forecasting short term freight transportation demand: the Poison STARMA model." Pre-print, 77th Annual Meeting of the Transportation Research Board, National Research Council, Washington, D.C.

Granger, C. W. J.; Newbold, P. (1977). *Forecasting Economic Time Series*, Academic Press, New York, New York, .

Granger, C. W. J.; Watson, M. W. (1984). "Time series and spectral methods in econometrics," in *Handbook of Econometrics*, Vol. 11 (Edited by Z. Griliches and M. D. Intriligator), ElsevierScience Publishers B.V., Amsterdam, Netherlands.

Greene, K. A. (1992). "Causal Univariate Spatial–Temporal Autoregressive Moving Average Modelling of Target Information to Generate Tasking of a World-wide Sensor System." Master's Thesis, Department of Operational Sciences, Air Force Institute of Technology, Wright–Patterson AFB, Ohio.

Greene, K. A.; Chan, Y. (1991). "Analysis and modelling of spatial-temporal information." Working paper, Dept. of Operational Sciences, Air Force Institute of Technology, Wright–Patterson AFB, Ohio.

Griffith, D. A. (1996). "Some Guidelines for Specifying the Geogrphic Weights Matrix cotnianed in Spatial Staistical Models." *Practical Handbook of Spatial Statistics* (S. L. Arlinghaus, Editor-in-Chief), CRC Press, Boca Raton, Florida.

Harvey, A. C. (1981). *Time Series Models*, Wiley, New York, New York.

Harvey, A. C. (1989). *Forecasting, Structural Time Series and the Kalman Filter*, Cambridge University Press, Cambridge, United Kingdom.

Hoff, J. C. (1983). *A Practical Guide to Box–Jenkins Forecasting*, Lifetime Learning Publications, Belmont, California.

IMSL (1990). *Statistical Library User's Manual – FORTRAN subroutines for statistical analysis*, IMSL Inc., Houston, Texas.

Kamarianakis, Y. (2003). "Spatial-time series modeling: A review of the proposed methodologies." Discussion Paper REAL 03-T-19, Regional Economics Applications Laboratory, University of Illinois, Urbana, Illinois.

Koopman, S. J. (1997). "Exact initial Kalman filtering and smoothing for nonstationary time series models." *Journal of the American Statistical Association*, Vol. 92, No. 440, pp. 1630–1638.

Liu, L-M.; Hudak, G. B. (1983). "An Integrated Time Series Analysis Computer Program – The SCA Statistical System." *Time Series Analysis: Theory and Practice* (O. D. Anderson, Editor), Vol. 4, pp. 291–310, North Holland, Amsterdam, Netherlands.

Lutkepohl, H. (1991). *Introduction to Multiple Time Series Analysis*, Springer–Verlag, Berlin, Germany.

Makridakis, S.; Wheelwright, S. C.; McGee, V. E. (1983). *Forecasting methods and applications*, Second Edition, Wiley, New York, New York.

Martin, R. L.; Oeppen, J. E. (1975). "The identification of regional forecasting models using space-time correlation functions." *Transactions – Institute of British Geographers*, Vol. 66, pp. 95–118.

McGee, W.; Chan, Y. (1991). "Analysis and modelling of spatial-temporal information II." working paper, Dept. of Operational Sciences, Air Force Institute of Technology, Wright–Patterson AFB, Ohio.

MicroTSP (1990). Quantitative Software, Irvine, California.

Mills, T. C. (1990). *Time Series Techniques for Economists*, Cambridge University Press, Cambridge, United Kingdom.

Montgomery, D. C., et al. (1990). *Forecasting and Time Series Analysis*, Second Edition, McGraw–Hill, New York, New York.

Newbold, P.; Agiakloglou, C.; Miller, J. (1994). "Adventures with ARIMA software." *International Journal of Forecasting*, Vol. 19, pp. 573–581

Pfeifer, P. E. (1979). "Spatial-Dynamic Modeling." Unpublished PhD dissertation, Georgia Institute of Technology, Atlanta, Georgia.

Pfeifer, P. E.; Deutsch, S. J. (1981). "Seasonal space-time ARMA modeling." *Geographical Analysis*, Vol. 13, pp. 117–133.

Pfeifer, P. E., Deutsch, S. J. (1980). "A three-stage iterative procedure for space-time modelling." *Technometrics*, Vol. 22, No. 1, pp. 35–47.

Pfeifer, P. E.; Bodily (1990). "A test of space-time ARMA modelling and forecasting of hotel data." *Journal of Forecasting*, Vol. 9, pp. 255–272.

Saaty, T. L.; Vargas, L. G. (1990). *Prediction, Projection and Forecasting*. Kluwer Academic Publishers, Norwell, Massachusetts.

SAS (1985). *SAS/ETS User's Guide*, Version 5 Edition, SAS Institute, Cary, North Carolina.

Tiao, G. C.; Box, G. E. P. (1981). "Modeling multiple time series with applications." *Journal of the American Statistical Association Theory and Methods*, Vol. 76, pp. 802–816.

Tiao, G. C.; Box, G. E. P.; Grupe, M. R.; Hudak, G. B.; Bell, W. R.; Chang, I. (1979). "The WisconsinMultiple Time Series (WMTS–1) Program: A Preliminary Guide." Department of Statistics, University of Wisconsin, Madison, Wisconsin.

Vasiliev, I. R. (1996). "Visualization of Spatial Dependence: An Elementary View of Spatial Autocorrelation." *Practical Handbook of Spatial Statistics* (S. L. Arlinghaus, Editor–in–Chief), CRC Press, Boca Raton, Florida.

Wright, S. A. (1995). "Spatial time–series–pollution pattern recognition under irregular interventions." Master thesis, AFIT/GOR/ENS/95M–19, Department of Operational Sciences, Air Force Institute of Technology, Wright–Patterson AFB, Ohio.

Yi, P., and Chan, Y. (1988). "Bifurcation Disaggregation in Garin–Lowry Models: Theory, Calibration and Case Study," *Environment and Planning*, pp. 1253–1267.

Chapter 11 - Spatial Temporal Information: Statistical and Causal-Model Development

Akaike, H. (1976). "Canonical correlation analysis of time series and the use of an information theoretic criterion." in *System Identification: Advances and Case Studies* (R. K. Mehra and D. G. Lainiotis, Editors), Academic Press, New York, New York.

Alam, S. B.; Goulias, K. G. (1999). "Dynamic emergency evacuation management system using GIS and spatio–temporal models of behavior." Pre-print, 78*th* Annual Meeting of the Transportation Research Board, National Research Council, Washington, D.C.

Allard, D.; Fraley, C. (1997). "Nonparmetric maximum likelihood estimation of features in spatial point processes using Voronoi tessellation." *Journal of the American Statistical Association*, Vol. 92, No. 440, pp. 1485–1493.

Anselin, L. (1988). *Spatial Econometrics: Methods and Models*. Kluwer Academic Publishers, Norwell, Massachusetts.

Arbia, G.; Benedetti, R.; Espa, G. (1996). "Effects of the Modifiable–Areal–Unit–Program on image classification." *Geographical Systems*, Vol. 3, pp. 123–141.

Armstrong Jr., R. J. (1994). "Impact of commuter rail service as reflected in single–family residential property values." Preprint No. 940774, 73*rd* Annual Meeting, Transportation Research Board, Washington, D.C.

Aykroyd, R. G.; Zimeras, S. (1999). "Inhomogeneous prior models for image reconstruction." *Journal of the American Statistical Association*, Vol. 94, No. 447, pp. 934–946.

Bagley, M. N.; Mokhtarian, P. L.; Kitamura, R. (2000). "A methodology for the disaggregate, multimmensioanl measurement of residential neighborhood type." Working Paper, Institute of Transportation Studies, University of California, Davis, California.

Bazarraa, M. S.; Shetty, C. M. (1979). *Nonlinear Programming: Theory and Algorithms*. Wiley, New York, New York.

Bailey, T. C.; Munford, A. G. (1994). "Modelling a large, sparse spatial interaction matrix using data relating to a subset of possible flows." *European Journal of Operational Research*, Vol. 79, pp. 489–500.

Bao, S.; Henry, M. (1996). "Heterogeneity issues in local measurements of spatial association." *Geographical Systems*, Vol 3, pp. 1–13.

Beasley, J. E.; Goffinet, F. (1994). "A Delaunay triangulation–based heuristic for the Euclidean Steiner problem." *Networks*, Vol. 24, pp. 215–224.

Beeker, E. (1998). "An efficient algorithm for calculating shortest paths while avoiding obstacles." Abstracts, Working Group 31, 66*th* Military Operations Research Society Symposium, Military Operations Research Society, Alexandria, Virginia.

Besag, J. (1974). "Spatial interaction and the statistical analysis of lattice systems." *Journal of the Royal Statistical Society*, Vol. 36, pp. 192–236.

Besag, J. (1986). "On the statistical analysis of dirty pictures." *Journal of the Royal Statistical Society*, Vol. 48, pp. 259–302.

Black, J. A.; Paez, A.; Suthanaya, P. A. (2002). "Sustainable urban transportation: Performance indicators and some analytical approaches." *Journal of Urban Planning and Development*, Vol. 128, No. 4, pp. 184–209.

Black, W. R. (1992). "Network autocorrelation in transport network and flow systems." *Geographical Analysis*, Vol. 24, No. 3, pp. 207–222.

Black, W. R. (1993). "Network autocorrelation in the analysis of flows on transportation *Networks.*" presented at the Operations Research Society of America National Meeting in Chicago, May 16–19.

Box, G. E. P.; Draper, N. R. (1987). *Empirical Model Building and Response Surface*, Wiley, New York, New York.

Box, G. E. P.; Jenkins, G. M. (1976). *Time Series Analysis – Forecasting and Control*, Holden–Day, Oakland, California.

Campbell, J. F. (1993). "Continuous and discrete demand hub location problems." *Transportation Research B*, Vol. 27B, pp. 473–482.

Chan, Y. (2000). *Location Theory and Decision Analysis.* Thomson/South-Western, Cincinnati, Ohio.

Chan, Y.; Regan, E.; Pan, W-M. (1986). "Inferring an origin-destination matrix directly from network-flow sampling." *Transportation Planning and Technology*, Vol. 11, pp. 27–46.

Chan, Y.; Rahi, M. Y. (1993). "An l_p-norm origin-destination estimation method that minimizes site-specific data requirements." *Transportation Research Record*, No. 1413, pp. 130–140.

Cook, T.; Falch, P.; Marino, R. (1984). "An urban allocation model combining time series and analytic hierarchial methods." *Management Science*, Vol. 30, pp. 198–208.

Cox, K. R.; Agnew, J. A. (1976). "The spatial correspondence of theoretical partitions." Department of Geography, Discussion paper series, No. 24, Syracuse University, New York.

Cromley, R. O.; Hempel, D. J.; Hillyer, C. L. (1993). "Dimensions of market attractiveness: competitively interactive spatial models." *Decision Sciences, Vol. 24*, pp. 713–738.

Dahl, C. M. Gonzales–Rivera, G. (2003). "Testing for neglected nonlinearity in regression models based on the theory of random fields." *Journal of Econometrics*, Vol. 114, pp. 141–164.

Dasci, A.; Laporte, G. (2002). "Location and pricing decisions of a multi-store monopoly in a spatial market." Working Paper, Ecole des Hautes Etudes Commerciales, Montreal, Canada.

de la Figuera, D. S.; Ravelle, C. (1993). "The *pq*-median problem: location and districting of hierarchical facilities." *Location Science*, Vol. 1, pp. 299–312.

de Oliveira, V.; Kadem, B.; Short, D. A. (1997). "Bayesian prediction of transformed Gaussian random fields." *Journal of the American Statistical Association*, Vol. 92, No. 440, pp. 1422–1433.

Eaton, B. C.; Lipsey R. G. (1975). "The non-uniqueness of equilibrium in the Löschan location model." *The American Economic Review*, Vol. 66, pp. 77–93.

Feng, C-M.; Chang, C-Y. (1993). "A transit-oriented land use model." *Transportation Planning Journal*, Vol. 22, pp. 429–444, Ministry of Transportation and Communication, Republic of China (in Chinese).

Ference, A.; Chan, Y. (1996). "Spatial dependence, homogeneity, and isotropy of the Washington DC Mall image." Working paper, Department of Operational Sciences, Air Force Institute of Technology, Wright–Patterson AFB, Ohio.

Fernandez, E.; Garfinkel, R.; Arbiol, R. (1998). "Mosaicking of aerial photographic maps via seams defined by bottleneck shortest paths." *Operations Research*, Vol. 46, No. 3, pp. 293–304.

Fortmann, K.; Chan, Y. (1994). "Analysis of spatial-temporal information: testing for spatial heterogeneity." Working paper, Department of Operational Sciences, Air Force Institute of Technology, Wright–Patterson AFB, Ohio.

Fotheringham, A. S.; O'Kelley, M. E. (1989). *Spatial Interaction Models and Applications*, Kluwer Academic Press.

Frost, R. B. (1992). "Directed graphs and misconceptions and more efficient methods." *Engineering Optimization*, Vol. 20, pp. 225–239.

Getis, A. (1991). "Spatial interaction and spatial autocorrelation: a cross-product approach." *Environment and Planning A*, Vol. 23, pp. 1269–1277.

Getis, A.; Ord, J. K. (1992). "The analysis of spatial association by use of distance statistics." *Geographical Analysis*, Vol. 24, No. 3, pp. 189–206.

Grant, M.; Robinson, J. N. (1989). "Time Series Forecasting." Working paper, Department of Operational Sciences, Air Force Institute of Technology, Wright–Patterson AFB, Ohio.

Guy, C. M. (1991). "Spatial interaction in retail planning practice: the need for robust statistical methods." *Environment and Planning B*: Vol. 18, pp. 191–203.

Hunt, J. D.; Simmonds, D. C. (1993). "Theory and applications of an integrated land-use and transport modeling framework." *Environment and Planning B*, Vol. 20, pp. 221-244.

Imai, T. (1994). "Manual of SEGVOR: a FORTRAN program for constructing the Voronoi Diagram of Line Segments." Department of Mathematical Engineering and Information Physics, University of Tokyo, Tokyo, Japan.

Ioannides, Y. M.; Overman, H. G. (2003). "Zipf's law for cities: an empirical examination." *Regional Science & Urban Economics*, Vol. 33, pp. 127–137.

Kelejian, H. H.; Prucha, I. R. (2001). "On the asymptotic distribution of the Moran *I* test statistic with applications." *Journal of Econometrics*, Vol. 104, Issue 2, pp. 219–257.

Kim, D. S.; Chung, H. W. (2000). "Spatial location–allocation models for multiple center villages." Working Paper, College of Agriculture a& Life Sciences, Seoul National University, Suwon, Republic of Korea.

Ko, C- W.; Lee, J.; Queyranne, M. (1995). "An exact algorithm for maximum entropy sampling." *Operations Research*, Vol. 43, pp. 684–691.

Lautso, K.; Toivanen, S. (1997). "The SPARTCUS system for analysing urban sustainability." Working paper, LT Consultants Ltd., Helsinki, Finland.

Mackett, R. (1993). "Structure of linkages between transport and land use." *Transportation Research B*, Vol. 27B, pp. 189–206.

Makridakis, S.; Wheelwright, S. C. (1978). *Forecasting Methods and Applications*, Wiley, New York, New York.

Marshall, K. T.; Oliver, R. M. (1995). *Decision Making and Forecasting*, McGraw–Hill, New York, New York.

Masser, I.; Coleman, A.; Wynn, R. F. (1971). "Estimation of a growth allocation model for north–west England." *Environment and Planning*, Vol. 3, pp. 451–463.

Maurkio, J. A. (1995). "Exact maximum likelihood estimation of stationary vector ARMA model." *Journal of the American Statistical Association*, Vol. 90, pp. 282–291.

McCord, M. R.; Jafar, F.; Merry, C. J. (1996). "Estimated satellite coverage for traffic data collection." Working paper, Department of Civil & Environmental Engineering and Geodetic Science, Ohio State University, Columbus, Ohio.

Melachrinondis, E.; Xanthropulos, Z. (1994)"A Euclidean distance single facility location problem with minisum and maximin objectives." Working paper, Department of Industrial Engineering and Information Ssytems," Northeastern University, Boston, Massachusetts.

Oishi, Y.; Sugihara, K. (1995). "Topology- oriented divide- and- conquer algorithm for Voronoi diagrams." *Graphical Models and Image Processing*, Vol. 57, pp. 303–314.

Okabe, A.; Boots, B.; Sugihara, K. (1994). "Nearest neighborhood operations with generalized Voronoi diagrams: a review." *International Journal of Geographic Information Systems*, Vol. 8, pp. 43–71.

Okabe, A.; Suzuki (1987). "Stability of spatial competition for a large number of firms on a bounded two-dimensional space." *Environment and Planning A*, Vol. 19, pp. 1067–1082.

Okabe, A.; Boots, B.; Sugihara, K. (1992). *Spatial Tessellations Concepts and Applications of Voronoi Diagrams*, Wiley, New York, New York.

Peruggia, M.; Santner, T. (1995). "Bayesian analysis of time evolution of earthquakes." Working paper, Department of Statistics, Ohio State University, Columbus, Ohio.

Pfeifer, P. E.; Deutsch, S. J. (1980). "A three-stage iterative procedure for space-time modelling." *Technometrics*, Vol. 22, pp. 35–47.

Pfeifer, P. E.; Bodily, S. E. (1990). "A test of space-time ARMA modelling and forecasting of hotel data." *Journal of Forecasting*, Vol. 9, pp. 255–272.

Pierce, B. K.; Kinateder, J. G. (1999). "Incorporating geographic correlation when sampling a transportation network." *Transportation Research Record*, No. 1665, pp. 13–21.

Pooler, J. (1992). "Spatial uncertainty and spatial dominance in interaction modelling: a theoretical perspective on spatial competition." *Environment and Planning A*, Vol. 24, pp. 995–1008.

Priebe, C. E.; Olson, T.; Healy, D.M. (1997). "A spatial scan statistic for stochastic scan partitions." *Journal of the American Statistical Association*, Vol. 92, No. 440, pp. 1476–1484.

Priestley, M. B. (1988). *Non -linear and Non -stationary Time Series Analysis*, Academic Press, London, United Kingdom.

Putman, S. (1979). "Urban Residential Location Models." Martinus Nijhoff Publishing, Hingham, Massachusetts.

Putman, S. (1983). "Integrated Urban Models: Policy analysis of transportation and land use." Pion, London, United Kingdom.

Reinsel, G. C. (1993). *Elements of Multivariate Time Series Analysis*, Springer- Verlag, New York, New York.

Ripley, B. D. (1988). *Statistical Inference for Spatial Processes*, Cambridge University Press, Cambridge, United Kingdom.

Robinson, J. N. (1987). "Identification of spatial-temporal models of remote-sensing data." Doctoral dissertation, University of Texas, Austin, Texas.

Sachs, L. (1984). *Applied Statistics – A Handbook of Techniques*, Springer–Verlag, New York, New York.

Saavedra, L. A. (2003). "Tests for spatial lag dependence based on method of moments estimation." *Regional Science & Urban Economics*, Vol. 33, pp. 27–58.

Schewchuk, J. (1996). "Triangle: Engineering a 2D quality mesh generator and Delaunay triangulator." Technical report, School of Computer Science, Carnegie Mellon University, Pittsburgh, Pennsylvania. (Also available at http://www.cs.cmu.edu/~quake/triangle.research.html)

Schurmann, J. (1996). *Pattern classification: a unified view of statistical and neural approaches*, Wiley-Interscience, New York, New York.

Sen, A.; Smith, T. E. (1995). *Gravity Models of Spatial Interaction Behavior*, Springer–Verlag, Berlin, Germany.

Sharma, S. (1996). *Applied Multivariate Techniques*, Wiley, New York, New York.

Sharma, S. (2003). "Persistence and stability in city growth." *Urban Economics*, Vol. 53, pp. 300–320.

Sik, K. D.; Woo, C. H. (1999). "A spatial location–allocation model for rural center villages in Republic of Korea." Working paper, Institute of Agricultural Development, College of Agriculture & Life Sciences, Seoul National University, Republic of Korea.

Smetek, T.; Chan, Y. (1997). "Solving a multi-facility location problem using Voronoi diagrams and linear programming." Working paper, Department of Operational Sciences, Air Force Institute of Technology, Wright–Patterson AFB, Ohio.

Solna, K.; Switzer, P. (1996). "Time trend estimation for a geographic region." *Journal of the American Statistical Association*, Vol. 91, pp. 577–589.

Storer, J. (1994). "Shortest path in the plane with polygonal obstacles." *Journal of the Association of Computing Machinery*, Vol. 41, No. 5, pp. 982–1012.

Stoyan, D.; Stoyan, H. (1994). *Fractals, Random Shapes and Point Fields*, Wiley, New York, New York.

Sugihara, K. (1993). "Approximation of generalized Voronoi diagrams by ordinary Voronoi diagrams." *Graphical Models and Image Processing*, Vol. 55, pp. 522–531.

Sugihara, K. (1997). "Experimental study on acceleration of an exact-arithmetic geometric algorithm." *Proceedings of the 1997 International Conference on Shape Modeling and Applications*, pp. 160–168.

Sugihara, K.; Iri, M. (1994). "A robust topology-oriented incremental algorithm for Voronoi diagrams." *International Journal of Computational Geometry & Applications*, Vol. 4, No. 2, pp. 179–228.

Suzuki, A.; Okabe, A. (1995). "Using Voronoi Diagrams." in *Facility Location – A Survey of Applications and Methods* (Z. Drezner, Editor), Springer–Verlag, New York, New York.

Suzuki, A. (1989). "On the average complexity of the incremental algorithm for the Voronoi diagrams." Working paper, Department of Electrical Engineering and Computer Science, Northwestern University, Evanston, Illinois.

Srinivasan, S. (2001). "Quantifying spatial characteristics for travel behavior models." *Transportation Research Record*, No. 1777, pp.1–15.

Takeda, S. (1985). "On geographical optimization dynamic facility location problem." Master's Thesis, Department of Mathematical Engineering and Information Physics, University of Tokyo [in Japanese].

Tobler, W. R. (1970). "A computer movie simulating urban growth in the Detroit region." *Economic Geography*, Vol. 46.

Tobler, W. (1983). "An alternative formulation for spatial-interaction modelling." *Environment and Planning A*, Vol. 15, pp. 693–703.

Tobler, W. (1988). "The quadratic transportation problem as a model of spatial interaction patterns." in *Geographical Systems and Systems of Geography: Essays in Honor of William Warntz* (W. J. Coffey, Editor), Department of Geography. University of Western Ontario, London Ontario, pp. 75–88.

Upton, G. J. G.; Fingleton, B. (1985). *Spatial Data Analysis by Example – Volume I: Point Pattern and Quantitative Data*, Wiley.

Wackernagel, H. (2003). *Multivariate Geostatistics*. Springer-Verlag, Heidelberg, Germany.

Wright, S. (1995). "Spatial Time–Series: Pollution Pattern Recognition Under Irregular Intervention."Masters Thesis, Dept. of Operational Sciences, Air Force institute of Technology, Wright–Patterson AFB, Ohio.

Xu, W.; Chan, Y. (1993a). "Estimating an origin-destination matrix with fuzzy weights – Part I: Methodology." *Transportation Planning and Technology*, Vol. 17, pp. 127–144.

Xu, W.; Chan, Y. (1993b). "Estimating an origin-destination matrix with fuzzy weights – Part II: Case studies." *Transportation Planning and Technology*, Vol. 17, pp. 145–163.

Yi, P.; Chan, Y. (1988). "Bifurcation and disaggregation in Lowry–Garin based models: theory, calibration and case study." *Environment and Planning A*, Vol. 20, pp. 1253–1267, Pion, London, United Kingdom.

Zimmerman, P. (1988). "Photos from space – why restrictions won't work." *Technology Review*, May–June, Massachusetts Institute of Technology, pp. 45–53.

Chapter 12 - Retrospects and Prospects

American Society of Civil Engineers, *1996 Policies and Priorities*, January.

Anselin, L. (1990). "Quantitative methods in regional science: perspectives on research directions," in *Regional Science*, edited by D. Boyce, P. Nijkamp, D. Shefer, Springer–Verlag.

Anselin, L.; Getis, A. (1992). "Spatial statistical analysis and geographic information systems," The *Annals of Regional Science*, Vol. 26, pp. 19–33.

Arasan, V. T.; Rajesh, R. (1997) "Land use allocation for minimum travel intensity in Indian cities" Working paper, Department of Civil Engineering, Indian Institute of Technology, Madras, India.

Arbia, G.; Benedetti, R.; Espa, G. (1996) "Effect of the Modifiable-Areal-Unit-Problem on image classification" *Geographical Systems*, Vol.3, pp. 123–141.

Aysan, M.; Demir, O.; Altan, Z.; Dokmeci, V. (1997) "Industrial decentralization in Istanbul and its impact on transport" *Journal of Urban Planning and Development*, Vol 123, pp. 40–58.

Bana e Costa, C.; Stewart, T. J.; Vansnick, J.C. (1995). "Multicriteria decision analysis: some thoughts based on tutorial and discussion sessions of the ESIGMA meetings," Working paper, Technical University of Lisbon, IST/CESUR, Lisbon, Portugal.

Batty, M. (1992) "Urban modeling in computer-graphic and geographic information system environments" *Environment and Planning B*, Vol. 19, pp. 663–688.

Bevers, M.; Hof, J.; Uresk, D. W.; Schenbeck, G. L. (1997) "Spatial optimization of Prairie dog colonies for black-footed ferret recovery" *Operations Research*, Vol 45, pp. 495–507.

Boarnet, M. G. (1995). "New highways and economic growth: rethinking the link," *Access*, No. 7, University of California Transportation Center, pp. , 11–15.

Borsuk, M. (1996) [mborsuk@ix.netcom.com] "Third wave wipeout: INFOTECH's impact on retail space demand" Presentation to the Institute of Real Estate Management Mid-Year Convention in Chicago, June 8.

Cervero, R.; Landis, J. (1995). "The transportation-land use connection still matters," *Access*, Vol. 7, University of California Transportation Center, pp. 2–10.

Cervero, R. (1995) "Creating a Linear City with a Surface Metro" Technical Report, Institute of Urban and Regional Development, University of California at Berkeley.

Chan, Y. (1996). "Real-time information and transportation decisions: analysis of spatial data," *Military Operations Research*, Vol. I, No. 4, pp. 23–48.

Civil Engineering (1997) "The future is spatial for Intergraph and Oracle" September, p. 28.

Corbley, K. P. (1995). "Making connections: using satellite-derived Digital Elevation Models and GIS to design Japanese cellular network," *Geo Info Systems*, pp. 26–32.

de Neufville, R. (1990). *Applied Systems Analysis*, McGraw-Hill.

de Neufville, R. (1990). "Successful citing of airports: Sydney example," *Transportation Engineering* journal, Vol. 116, No. 1, American Society of Civil Engineers. pp. 37–48.

Doty, C. B.; Travis, C. C. (1989). "Role of risk assessment in the remedy decision process," *Journal of the Air Pollution Control Association*.

Feldt, A. G. (1972). *CLUG: Community Land Use Game – Player's Manual*, Free Press.

Ference, A. A.; Shelley, M. L.; Pohl, E. A. (1997) "Habitat suitability mapping through integration of multicriteria evaluation techniques with a geographic information system" Working paper, Department of Environmental Engineering Management, Air Fore Institute of Technology, Wright–Patterson AFB, Ohio.

Friedrich, J. (2004). *Spatial Modeling in Natural Sciences and Engineering*. Springer-Verlag, Heidelberg, Germany.

Garvey, P. (1997) "The MOD squad: EPA brings GIS to the world wide web" *Geo Info System*, June, pp. 28–33.

Ghosh, A.; Rushton, G. (1987). *Spatial Analysis and Location–Allocation Models*, Van Nostrand Reinhold.

Giuliano, G. (1995). "The weakening transportation–land use connection," *Access*, No. 6, University of California Transportation Center, pp. 3–11.

Gray, P. (1995). "The virtual workplace," *OR/MS Today*, August, pp. 22–26.

Griesenbeck, B.; Gebert, K.; Fitch, J. (1998) "Travel time and delay data collection using GPS and GIS" Preprint 98–0422, Transportation Research Board 77th Annual Meeting, January, Washington DC.

Guetzkow, H., editor (1962). *Simulation in Social Science, Readings*, Prentice Hall.

Guyon, X. (1995). *Random Fields on a Network - Modeling, Statistics, and Applications*, Springer–Verlag.

Haack, H.; Peterson, G. L. (1971). "Games for simulating urban development process," *Journal of the Urban Planning and Development Division*, American Society of Civil Engineers.

Hamed, M. M.; Easa, S. M. (1997) "Integrated modelling of urban shopping activities" Working paper, Department of Civil Engineering, Jordan University of Science and Technology, Irbid, Jordan.

Hansen, N. (1992). "The location–allocation versus functional integration debate: a assessment in terms of linkage effects," *International Regional Science Review*, Vol. 14, No. 3, pp. 299–305.

Hatipkarasulu, Y.; Wolshon, B.; Quiroga, C. (2000). "A GPS approach for the anysis for car following behavior." Working Paper, Department of Civil and Environmental Engineering, Louisiana State University, Baton Rouge, Louisiana.

Hodgson, M. E.; Cheng, Y.; Coleman, P.; Durfee, R. (1995). "Computational GIS burdens: solution with heuristic algorithms and parallel processing," *Geo Info System*, April.

House, P.; Patterson, P. D. Jr. (1969). "An environmental gaming-simulation laboratory," *Journal of the American Institute of Planners*, Vol. 35, No. 6, pp. 383–389.

Iyer, G.; Pazgal, A. (2003). "Internet shopping agents: Virtual co-location and competition." *Marketing Science*, Vol. 22, No. 1, pp 85–106.

Keeney, R. (1980). *Siting Energy Facilities*, Academic Press.

Keeney, R. L. (1994) "Using values in operations research," Philip McCord Morse Lecture, Institute-of-Management-Science/Operations-Research-Society-of-America Joint National Meting, Boston Massachusetts, 25 April.

Kennedy, M. (1996) *The Global Positioning System and GIS – An Introduction*, Ann Arbor Press.

Kleindorfer, P. R.; Kunreuther, H. C. (1994). "Siting of Hazardous Facilities," in *Handbooks in Operations Research and Management Science*, Vol. 6 (Edited by S. M. Pollock et al.), Elsevier, pp. 403–437.

Kutz, S. A. (1995). "Chicago remaps itself," *Civil Engineering*, June, pp. 52–54.

Langran, G. (1993) *Time in Geographic Information Systems*, Taylor and Francis.

Lee, D. B. (1977). "Improving communication among researchers, professionals and policy makers in land use and transportation planning," Report DOT-OS-60113, prepared for the Office of the Secretary of Transportation, Forecasting and Evaluation Division, U.S. Department of Transportation.

Leung, Y.; Yan, J. (1997) "A note on the fluctuation of flows under the entropy principle" Transportation Research B, Vol. 31, pp. 417–423.

Mackett, R. L. (1993). "Structure of linkages between transport and land use," *Transportation Research B*, Vol. 27B, pp. 189–206.

Mackiewicz, A.; Ratajczak, W. (1996). "Toward a new definition of topological accessibility," *Transportation Research B*, Vol. 30, pp. 47–49.

Martinez C.; F. J. (1995) "Access: the transport–land use economic link," *Transportation Research B*, Vol. 29, pp. 457–470.

McCord, M. R.; Jafar, F.; Merry, C. J. (1996) "Estimated satellite coverage for traffic data collection" Working paper, Department of Civil & Environmental Engineering and Geodetic Science, The Ohio State University, Columbus, Ohio.

Meier, R. L.; Duke, R. D. (1966). "Gaming-as-simulation for urban planning," *Journal of the American Institute of Planners*, Vol. 32, No. 1, pp. 3–17.

Merkhofer, M. L.; Keeney, R. L. (1987). "A multi-attribute utility analysis of alternative sites for the disposal of nuclear waste," *Risk Analysis*, Vol. 7, pp. 173–194.

Min, H.; Melachrinoudis, E.; Wu, X. (1997) "Dynamic expansion and location of an airport: A multiple objective approach" *Transportation Research A*, Vol. 31, pp. 403–417.

Mollering, H. (Editor) [1997] *Spatial Database Transfer Standards 2: Characteristics for Assessing standard and full Descriptions of the National and international Standard in the World*, International Cartographic Association and Pergamon Press.

National Research Council (1994) *Promoting National Spatial Data Infrastructure Through Partnerships*, Commission on Geosciences, Environment, and Resources, National Academy Press.

Nierat, P. (1997) "Market area of rail-truck terminals: pertinence of the spatial theory" *Transportation Research A*, Vol 31, pp. 109–127.

Nijkamp, P.; Rietveld, P.; Voogd, H. (1990). *Multicriteria Evaluation in Physical Planning*, North Holland.

Ng, M. N. (1993) "Port and Airport Development Strategy in Hong Kong," *Planner's Casebook 7*, American Institute of Certified Planners.

OR/MS Today (1995). "Top ten operations opportunities identified," December, pp. 17–18.

Parkinson, B. W.; Spilker Jr., J. J.; Axelrad, P.; Enge, P. (1995). *Global Positioning System: Theory and Applications*, Volumes I and II, American Institute of Aeronautics and Astronautics.

Peters, T. J.; Waterman, R. H., Jr. (1982). *In Search of Excellence*, Harper and Row, New York.

Pooler, J. A. (1995). "The use of spatial separation in the measurement of transportation accessibility," *Transportation Research A*, Vol. 29A, pp. 421–427.

Puu, T. (1997) *Mathematical Location and Land Use Theory: An Introduction*, Springer–Verlag.

Rahman, M.; Radwan, A. E.; Upchurch, J.; Kuby, M. "Modelling Spatial Impacts of Siting a NIMBY Facility," presented at the 71st annual meeting of the Transportation Research Board, January 1992.

Ran, B.; Hall, R. W.; Boyce, D. E. (1996). "A link–based variational inequality model for dynamic departure time/route choice," *Transportation Research B*, Vol. 30, pp. 31–46.

Regan, A. C.; Mahmassani, H. S.; Jaillet, P. (1995). "Improving efficiency of commercial vehicle operations using real–time information – potential uses and assignment strategies," *Transportation Research Record*, No. 1993, pp. 188–198.

ReVelle, C.; McGarity, A. E. (1997) *Design and Operation of Civil and Environmental Engineering Systems*, Wiley-Interscience.

Ridgley, M. A.; Team, A. (1995) "Determining rural land–use goals: Methodological primer and an application to agroforestry in Italy" *The Environmental Professional*, Vol 17, pp. 209–225.

Russ, J. C. (1995) *The Image Processing Handbook*, 2nd Edition, CRC Press.

Sanders, N. R. (1997) "The impact of task properties feedback on the time series judgmental forecasting tasks" *Omega*, Vol. 25, pp. 135–144.

Sandhu J. (1995). "Incorporating location/allocation models within a GIS," presented in the Fall meeting of the Institute for Operations Research and Management Science, New Orleans, pp. 13.

Sweet, R. J. (1997) "An aggregate measure of travel utility" *Transportation Research B*, Vol. 31, pp. 403–416.

Swersey, A. J. (1994). "The Deployment of Police, Fire, and Emergency Medical Units," *Handbook in Operations Research and Management Science*, Vol. 6, (Edited by S. M. Pollack, et al.) Elsevier, pp. 151–200.

Tajima, M. (1995). "Modelling multiple criteria– multiple participant problems: an integrated approach to multicriteria games." Proposal for a doctoral thesis in Management Sciences, Department of Management Sciences, University of Waterloo, Waterloo, Ontario, Canada.

Thrall, G. E. (1997) "GIS and Business Geography: a Retrospective" *Geo Info Systems*, June, pp. 46–50.

Tusa, W. K. (1992). "Reassessing the risk assessment," *Civil Engineering*, March, pp. 46–48.

Ulanowicz, R. E. (1987). "Growth and development: variational principles reconsidered," University of Maryland, Chesapeake Biological Laboratory, Solomons, Maryland 20688 (presented at the Workshop on Modelling Complex Systems, Prigogine Center for Nonlinear Dynamics, University of Texas at Austin, March 1985.)

Underwood, A. (1996). "A modified Melzak procedure for computing node–weighted Steiner trees," *Networks*, Vol. 27, pp. 73–79.

U. S. Department of Transportation (1995a). "GPS Standard Positioning service Signal specifications now Available," *News*, DOT 129-95, August 1.

U. S. Department of Transportation (1995b). "Coast Guard announces expected Differential Global Positioning System Operational Startup," *News*, CG 45-95, December 22.

U. S. Department of Transportation (1997) "National Transportation Atlas Databases" (in CD ROM) Bureau of Transportation Statistics.

U. S. Department of Transportation (1997a) "Travel Time Data Collection Handbook, Part One (draft)" Report by the Texas Transportation Institute for the Office Highway information management, June.

Vanmarcke, E. (1983) *Random Fields: Analysis and Synthesis*, MIT Press.

Wilson, J. L. (1990). *The SimCity: Planning Commission Handbook*, McGraw–Hill.

Wyman, M. M.; Kuby, M. (1993). "Turning the tables: using location science to choose technology rather than allowing technology to constrain location," *Locator*, Vol. 3, No. 2, pp. 3–4.

Zelany, M. (1991). "Editorial," *Human Systems Management*, Vol. 11, pp. 237–238.

Zelany, M. (1992). "An essay into a philosophy of MCDM: a way of thinking or another algorithm?" *Computers and Operations Research*, Vol. 19, pp. 563–570.

Zografos, K. G.; Douligeris, C.; Chaoxi, L. (1994). "Simulation model for evaluating the performance of emergency response fleets," *Transportation Research Record*, No. 1452, National Research Council, National Academy Press, pp. 27–34.

AUTHOR INDEX

SUBJECT INDEX

LIST OF SYMBOLS

a	A calibration constant; for example, it is the service-employment multiplier or population-serving ratio (number of service jobs generated from one household or resident)
$\tilde{a}$	Intercept regression-coefficient as a random variable
$\tilde{a}*$	Specific value of $\tilde{a}$ corresponding to a sample of data points
a'	Acceleration of a vehicle; also a constant parameter, such as unit cost of commuting (cost per unit-of-distance travelled), or the exponent of the development opportunity W_j at destination j
a_i	Calibration parameter corresponding to the utility increase in zone i, where utility is some measure of composite accessibility to the zone; also the population-serving ratio at zone i
a_t	Estimation-error or noise term for a series of data $(t = 1, 2, \dots)$ usually in a 'normalized' time-series, or after the data have been differenced to a stationary series; the estimated error or noise in Kalman filtering; also referred to as innovations when it is white noise
a'_i	Physical area of geographic sub-unit i or the demand-generating potential of i
a'_t	Measurement error in a Kalman-filter time-series, representing the difference between observed and measured data
a_D	Error term in a demand econometric-equation
a_S	Error term in a supply econometric-equation
a^W	Weighted labor-force-participation-rate, where the weights are the percentages of regional population in each zone
a^p	The pth-sector employment-growth-rate in the entire study-area
a_{ij}	Parallel to its single-dimension analogue, a_{ij} is an error- or noise-term in the spatial context; it has a zero mean and a constant variance; also stands for the entries in the $\bar{\mathrm{A}}$ matrix
a''_i	Convex combination of the population-serving ratios, with normalized accessibilities to zone i as weights
a^p_j	Employment multiplier considering the population-serving ratio, i.e., $(1+a_j)$— segregated both by economic-sector p and by zone j here
a^{kl}	Calibration parameter in a predictor-prey equation-set showing the interaction between the kth and lth species
a'_{kj}	The kth output (benefit) measures due to decision-making-unit j considering both non-spatial and spatial attributes (see also $\bar{\mathrm{A}} = [a_{ij}]$)
$a^{pq}_{ij}(k)$	Impact of the pth-state-variable-in-zone-i-at-time-k on the qth-state-variable-in-zone-j-at-time-k+1
a''	Threshold for a high-pass noise-filter
$\mathbf{a} = (\leftarrow a_i \rightarrow)^T$	Vector of calibration coefficients in the second stage of 2-stage least-squares, consisting of q entries; also stands for the vector of the error (noise) terms in a spatial-temporal forecasting-model
$\mathbf{a}'$	Vector whose ith element is the ratio of-the-household-income to the gross-output-in-the-ith-industrial-sector

$\tilde{a}$	Interim error-vector or noise-term in a more efficient calibration-procedure for STARMA
$\mathbf{a}_{ij} = (\leftarrow a_{ij}^k \rightarrow)^T$	Each entry of the $\overline{A} = [a_{ij}]$ payoff-matrix is replaced by a vector in a linear program, mainly to facilitate a multicriteria, two-person, zero-sum, non-cooperative game; here k is the index for a criterion
α	Calibration constant, or step size in "hill-climbing" algorithms; also the tail of a distribution
α'	Angle between two criterion-functions in multicriteria linear-programming; also a calibration constant
α''	Resulting problem-type after the original problem has been polynomially reduced
α_t	Random-shock or white-noise input-time-series in a transfer-function model
α_{ji}^{qp}	Exponent in a Cobb–Douglas production-function corresponding to the factor input x_{ji}^{qp}
A	Accessibility expenditure for a household (part of locational expenditure); also the area
$A(\cdot)$	Area of $\cdot$
A_i	Weighted labor-force participation-rate, with accessibility from zone i as weights
A_j	Gross acreage of subarea j
A_j'	Useable gross-acreage of subarea j
A_t	Error term in a 'raw-data' time-series
$\overline{A}_j$	Developable acreage in subarea j
A^B	Basic land-use (A_j^B is basic land-use in zone j)
A^R	Retail land (A_j^R is retail land in zone j)
A^U	Unusable land (A_j^U is unusable land in zone j)
$\underline{A}$	Set of arcs in a network
$\hat{A}_j^k$	Net acreage in subarea j devoted to the kth land-use
$\mathbf{A} = (\leftarrow A_t \rightarrow)^T$	Vector of disturbance or error terms in econometric or spatial time-series models, consisting of n observations; in 2-stage least-squares, it consists of q entries, where q is the number of endogenous variables
$\mathbf{A}$	As a matrix (instead of a vector), $\mathbf{A}$ stands for node-arc incidence-matrix in network-flow programming
$\mathbf{A}' = [A_{ij}']$	An $n \times n$ square matrix; for a compartmental model, it is the rate-of-change matrix; and for the matrix of secondary (retail)-employment it is the distribution-rate by zone, where $n = n'$.
$\mathbf{A}_0(t)$	Vector showing rate-of-change with the "outside world" over time
$\mathbf{A}'' = [(i,j)]$	Contiguity matrix with nonzero arc-entries where i is incident upon j
$\hat{\mathbf{A}}$	An $n \times n$ matrix, which converts value-added output vector by industrial sectors to the same vector measured in labor-force base
$\mathbf{A}_j$	Vector of socioeconomic variables at location j, representing such activities as population and employment
$\mathbf{A}(j)$	Column vector in the network-simplex tableau for arc j
$\overline{\mathbf{A}} = [a_{ij}]$	Coefficient matrix of linear-programming constraints, where a_{ij} expresses the incidence relationship between row i and column j; an example is the kth output measures due to decision-making-unit j, a_{kj}, in a data-envelopment analysis.
$\mathbf{A}_B$	Basis of a linear program
$\mathbf{A}_N$	Nonbasic part of the tableau in a linear program
$\mathbf{A}^1$	The complicated set of constraints in a mixed integer-program
$\mathbf{A}^2$	The straightforward set of constraints in a mixed integer-program
b	Generally a constant parameter, denoting a growth rate, intercept or slope in a linear equation, or the positive exponent of a spatial cost-function etc.
$\tilde{b}$	'Slope' regression-coefficient as a random variable
$\tilde{b}^*$	Specific value of $\tilde{b}$ for a sample of data points
b^U	Household budget
b_j	The fixed cost of siting a depot at node j

b^j	Travel-cost elasticity for activity j
$b^k(m)$	A scale factor used to adjust the kth zonal-retail-employment from one loop of the Lowry model m to another $m+1$, where $m = 1, 2, \ldots$
b_{ki}, b_{ik}	Slack-flow capacity on slack arc (k,i) or (i,k); also the benefit variable in data-envelopment analysis, denoting the weight placed on the kth benefit of the ith alternative
b_{kji}	Benefit variable used in the combined data-envelopment-analysis-and-location model, showing the relative importance of assigning the kth benefit to the demand-facility pair i–j
$\mathbf{b} = (\,\text{-}b_{i\rightarrow})^T$	Vector of estimated parameters in ordinary least-squares regression or other calibration procedures, consisting of $k+1$ parameters (including the 'intercept'); also the right-hand-side of a linear or mixed integer program
$\mathbf{b}' = (\,\text{-}b_i'\rightarrow)^T$	A given vector of the right-hand-side of a mathematical program; also the fixed external-flows in a network-flow program
$\overline{\mathbf{b}}$	Updated right-hand-side of a linear program during a simplex procedure; also the birth rates in a cohort-survival analysis
$\mathbf{b}^1$	The portion of the right-hand-side corresponding to the complicated set of constraints in a mixed-integer-program
$\mathbf{b}^2$	The portion of the right-hand-side corresponding to the straightforward set of constraints in a mixed-integer-program
β	A calibration constant, such as the positive exponent of a spatial cost-function or the round-trip factor in stochastic facility-location. (This same constant β is also referred to as b)
β'	A calibration constant
β_i	Current level of inventory at location i
β_t	Prewhitened output time-series in a transfer-function model
B	An arbitrarily large integer; also the backshift operator in a time series
B'	Bifurcation set of control variables
B''	Blue-collar employment
B_k	Percentage reflectance in band k of a satellite sensor
B_L, B_R	Left and right boundaries of a firm's market area
B'_k	Number of times a facility is exposed to demands in period k
B^k	Bound value for distance from a vertex, used to locate the intersecting point q_k or a candidate location for a center
$B^M_{\min}, B^M_{\max}$	Lower and upper bounds for the border-line length of a subregion
$\mathbf{B} = [b_j]$	Birth matrix with nonzero diagonal-elements showing the 'birth' rate within subarea j
$\mathbf{B} = [b_{ij}]$	Arbitrary matrix in a tableau of network-with-side-constraint program, corresponding to the flow variables
$\mathbf{B}' = [\beta_{ij}]$	Calibration-coefficient matrix in the first stage of a 2-stage least-squares, which measures $q \times k$, where q is the number of endogenous variables and k the exogenous variables
$\tilde{\mathbf{B}} = \left[\tilde{b}_{ij}\right]$	Quasi-deterministic transition-matrix in a compartmental model
$\mathbf{B}_i$	Diagonal-block i of the inverse of a network node-arc incidence-matrix, expressed in terms of a spanning subgraph
$\mathbf{B}'' = [b'_{ij}]$	Fixed cyclic-permutation δ' expressed in terms of a matrix operation, where $b'_{i,\delta'(i)} = 1$ and all other elements $b'_{ij}=0$
$\overline{B}$	Initial basis for a network-with-side-constraint model
c	Cost of operation, unit-cost, or a constant in general (e.g., c_i is the unit cost at location i; c_{kl} is the "interaction cost" of moving materials between workstations k and l in an assembly line)
c'	Proportionality constant
c^k	Weight reflecting the relative importance of home-based retail-trips for purpose k
$^r\mathbf{c}^s(\mathbf{x})$	rth-stop coverage of state s by routing-variable $\mathbf{x}$

$\mathbf{c} = (\leftarrow c_j \rightarrow)$	Cost vector in the objective function of a linear program, which is also the gradient of the objective function; here c_j is the constant unit-cost
$\mathbf{c}'$	Consumption-coefficient vector, whose ith element is the ratio of the purchased-value-of-the-commodity-from-the-ith-industrial-sector to the household income
$\mathbf{c}_B$	The part of the cost-vector $\mathbf{c}$ corresponding to the basic variables
$\mathbf{c}_N$	The part of the cost-vector $\mathbf{c}$ corresponding to the nonbasic variables
$\mathbf{c}^r$	Binary vector of rth-stage coverage-requirements in the decomposed recursive-program
$\mathbf{c}^{k+r}(k)$	Binary vector of rth-stage coverage-requirements on each origin–destination pair in cycle k; $\mathbf{C}(k) = [\leftarrow \mathbf{c}^{k+r}(k) \rightarrow]$
$\mathrm{conv}(\tilde{Q}')$	Convex combination of discrete points $\tilde{Q}'$ in a feasible region of an integer program
C	Generalized cost to include both time and monetary outlay, or unit composite-cost in general (e.g., C_i is the generalized cost of operation or the inventory-carrying cost at location i, C_{ij} is the composite transportation-cost from location i to j, C_{ij}^p is the composite transportation-cost from location i to j for commodity p etc.)
C'	Number of columns in a lattice, grid or a pixel image; also household expenditure on community amenities (which is part of non-locational expenditure)
C_0	Overhead of a firm
C_o	Operating cost
C_s	Capital cost
C_j	Equity factor in districting algorithms
C_X	Coefficient-of-variation of variable X, or $s_X / \bar{X}$
C_{XY}	Cross-covariance between random variables X and Y
$C(C_{ij})$	Propensity, distribution, or accessibility function between i and j, assuming such forms as exponential function or power function of spatial-cost C_{ij}
$C[a](\mathbf{x})$	Performance of arc or path a as a function flow-vector $\mathbf{x}$
$C'(\tau)$	Accessibility to work-opportunities as a function of time τ
$C^k(\tau)$	Accessibility to the kth non-work-opportunity as a function of time τ
$C_i(\cdot)$	The cost function (including land rent), or performance function, of firm i—expressed in terms of the supply volume V_i^π or other arguments
$C_{ij}(V_{ij})$	Transportation cost between origin–destination pair i–j as a function of flow V_{ij} between them
$C^{k,l}$	Transportation cost between origin k and destination l
$C^{mn}(r)$	Connectivity requirement between origin–destination pair m–n via at most rth-stop itineraries
$\mathbf{C} = [C_{ij}]$	Arbitrary matrix in a tableau of network-with-side-constraint program, corresponding to the non-flow variables; also the covariance matrix
$\mathbf{C} = [\mathbf{c}^1, \ldots, \mathbf{c}^q]^T$	A $q \times n$ matrix of cost coefficients in a multicriteria linear-program, where each criterion j has a cost and a gradient vector $\mathbf{c}^j$
$\mathbf{C}(\cdot)$	State-connectivity function linking to past decisions and connectivity requirements in a recursive program
$\mathbf{C}'$	Diagonal matrix converting the gross-output vector to value-added vector
$\hat{\mathbf{C}}$	Matrix of estimated coefficients in stage 1 of 2-stage least-squares, measuring $q \times k$
$\bar{C}$	Number of cell columns in a grid region or in a raster image
γ	Unit price at the market, Lagrange multiplier, and a calibration constant in general
γ'	Capacity-utilization rate, bounded between zero and unity
γ_j^{pq}	Dual variable associated with the input–output coefficients in an entropy-maximization model
$\gamma' = [q_j']$	Matrix of subareal growth-rates along its diagonal
$\bar{\gamma}$	Economic-base multiplier over a time-increment Δt, combining the activity-rate f and the population-serving-ratio a; $\bar{\gamma}_{ij}$ (with the subscript) would include the locational attributes as captured in work- and nonwork-accessibilities t_{ij} and u_{ij}
$\gamma_i(p,s)$	General 'strain' or the savings from including new-demand i via a triangular-inequality-style route-replacement between points p and s

Γ	The gross economic-multiplier deriving the total employment from the initial basic-employment
Γ	Vector of economic-multipliers deriving the total employment in the study area from the initial basic-employment, including c_j, f and a
Γ_t	Observation matrix in Kalman filter; when multiplied against the observed time-series, specifies what is actually observable
$\Gamma(W,p)$	Optimization results from a facility-location model where p facilities are relocated to respond to a maximum demand of W
$\Gamma(k) = [\leftarrow\gamma_i(k)\rightarrow]$	Vector of payoff-function consisting of q entries, where $q \le \acute{\eta}\mu$
d	Distance or spatial separation; also a proxy for a particular spatial order
d'	Amount of differencing to induce stationarity in a time-series
d''	A decision in a Markovian decision-process
d_i	Distance from location i (notice this is not necessarily Euclidean distance); or deviation from a standard or ideal in dimension i; also the capacity of arc i or the weights in a transfer function
d^k	Minimum threshold of retail-employment by trade-class k; d^R is the threshold for the case when there is only one trade class
$d_j(\mathbf{x})$	Multidimensional decision-boundary in a Bayesian classifier
$d(B) =$ $d_0+d_1B+d_2B^2 + \ldots$	Transfer function in a multivariate time-series, consisting of weights d_0, d_1, d_2, etc. and backshift operators B
d_{ij}	Euclidean distance or the spatial-cost in general between locations i and j
d_{ijk}	Euclidean distance or the spatial cost between locations i and j in state k
d_{ij}^h	Distance or travel time between nodes i and j by salesman or vehicle h
d^{i}	Time a salesman or vehicle visits node i in a tour or a route
d^{ij}	Distance or time between locations i and j, starting with arrival at i and terminating at arrival at j (notice this is not necessarily the Euclidean distance)
$d(\mathbf{i},\mathbf{j})$	Planar Euclidean distance between two Cartesian coordinate points $\mathbf{i}$ and $\mathbf{j}$
$d(\mathbf{x}_i,\mathbf{x}_{i+1})$	Spatial separation between consecutive stops $\mathbf{x}_i$, $\mathbf{x}_{i+1}$
$\mathbf{d}, \mathbf{d'}$	Vector of arc capacities in network-flow programming
$\mathbf{d}^j$	Extreme direction along the jth axis in a linear program
$\mathbf{d}^k = (\leftarrow d_i^k\rightarrow)$	Direction of steepest ascent in the kth step of a hill-climbing optimization-algorithm, as characterized by n components of the vector
δ	Change in a quantity (e.g., δx is the increase or decrease in quantity x); δ_{ij} is the distance savings in directly going from i to j, instead of through an intermediate point k
$\delta(i)$	The steady-state decision whenever the state is i in a Markovian-decision-process
$\tilde{\delta}$	Policy in a Markovian decision-process
$\tilde{\delta}^{/}$	Improved stationary-policy in the policy-iteration procedure of a Markovian-decision-process
δ^*	Optimal policy in a Markovian-decision-process
δ', δ''	Fixed cyclic-permutation
δ_i	Binary decision-variable to be switched on, conditional upon another decision-variable being engaged; also a calibration constant; or a nonnegative real-number denoting the number of legs in a subtour-breaking constraint
$\delta\Omega$	Boundary of the bounded-domain Ω
$\delta(k)$	Savings by using route k
$\delta^+(i)$	Set of nodes reachable from i
$\delta^-(i)$	Set of nodes incident upon i
δ_{ij}	Route-distance savings by including demands i and j in a single, rather than separate tours, in accordance with the Clarke–Wright heuristic
$\boldsymbol{\delta}$	Vector of estimated-parameters in nonlinear regression
$\hat{\boldsymbol{\delta}}$	Least-squares estimate of $\boldsymbol{\delta}$, usually obtained as a conditional estimate
$\boldsymbol{\delta}_j = (\leftarrow\delta_{ji}\rightarrow)^T$	Orthonormal base of the transition-rate space when the system is in compartment j

D	Distance or time of specified length
D'	Data, population density, or a measure of crowding
D''	Dual polyhedron of a linear program; or a subset of nodes/vertices
D_{ab}	Shortest distance from demand or customer a to demand b along a path, or along a tour from depot a to demand b
$D(i)$	Decision set in a Markovian decision-process
$D(a,b)$	Shortest distance along a vehicle route from terminal a to terminal b
D_i	Cumulative distance (along a path) to demand i from a facility
$D_l(V_l^d)$	Demand at location l showing price against flow-quantity; in other words, price paid at demand quantity V_l^d
D_i'	Cumulative distance (along a path) to demand i from all facility candidate-sites
D^k	Total sales or service from facility k
D^H	Upper-bound distance
D^L	Lower-bound distance
D_j^H	Maximum allowable household-density in zone j
$\mathbf{D} = [d_j]$	Death matrix with non-zero diagonal elements, showing the 'death' rate within subarea j
$\mathbf{D}'$	Calibration-coefficient matrix in the first stage of 2-stage least-squares, measuring $q \times$ where q is the number of endogenous variables
$\overline{\mathbf{D}} = [D_{ab}] \; |I| \times |I|$	matrix of shortest cumulative-distances along a path from vertex a to vertex b
$\overline{\mathbf{D}}' = [D_{qk}]$	$|I| \times m$ matrix of distances from vertex q to arc k
Δ_i^j	The difference between two utility measures i and j
$\nabla f(\mathbf{x}) = (-\partial f / \partial x_i \cdots)^T = (G_x, G_y, G_z, \ldots)^T$	Gradient of a function over n variables
e	The exponent value of 2.7183; also a calibration constant
e'	Number of exogenous variables left in the econometric model after estimation
e''	Number of endogenous variables left in the econometric model after estimation
e_i	Index to denote the ith type of industrial employment; also the ith arc in a network
e_{j_i}	Arc j associated with node/vertex i
$\mathbf{e}^{j(i)}$	Unitary column-vector for arc j with unitary entry in the ith row
ε	A very small number or a random perturbation
ε_k	Efficiency-measurement error-term associated with the kth input–output pair in empirically curve-fitting a distance function
$\boldsymbol{\varepsilon}$	Normally-distributed error-vector with zero mean; when it has a constant variance, it could be a vector of random perturbations in the forecast using a transfer function, due to white noise in the inputs
E	Total employment
E'	Number of exogenous variables
E''	Number of endogenous variables
E^B	Basic employment (E_j^B is basic employment in zone j)
E^R	Service employment
E^k	Retail employment by trade-class k (E_j^k is retail employment by trade class k in zone j)
$E(t)$	Relative smoothed-errors in adaptive-response-rate exponential-smoothing
$\tilde{E}_j$	Employment in the jth zone as projected from an areawide growth-rate for each sector
E_{ijk}	Expected number of demands i in period k at location j
$E(i_1, i_2, h_1, h_2)$	Net change in travel-distance from an exchange of demands i_1 and i_2 between tours h_1 and h_2
$E'(i_1, i_2, h_1, h_2)$	Modified generalized-savings-measure from an exchange of demands i_1 and i_2 between tours h_1 and h_2
$\mathbf{E}$	Row vector of employment-levels, made up of individual zonal employment E_i
f	Average household-size in terms of the number of employed residents per household, or reciprocal of the labor-force participation-rate (also called the activity rate)
$f(\cdot)$	Regular function of the argument (e.g., the criterion function in dynamic programming)

$f(\mathbf{x}_q, \mathbf{x}-\mathbf{x}_q)$	A functional for which the directional derivative is being considered, approaching point $\mathbf{x}_q$ from point $\mathbf{x}$
f'	Functional-attribute score, including spatial separation
$f'(t)$	Cumulative demand at time-period t
f_i	Demand-for-service frequency at location i; also the natural growth-rate of population in subarea i (the activity rate)
f^w	Weighted activity-rate, where the weights are the percentages of regional population at each zone
f_{ik}	Demand-for-service frequency at location i in state k
f'_{ik}	Number of demands k serviced by facility i
f'_i	Convex combination of activity-rate f_i, where the weights are the normalized accessibilities into zone i
$f_j^{(l)}(\cdot)$	Speed-of-adjustment function for the jth zone and lth activity
f_r^{mn}	rth-stop demand between origin–destination m–n
$\dot{f}(x)= df/dx$	Derivative of function f over variable x
$\mathbf{f}$	Partial-flow pattern in the decomposed RISE algorithm
F	Set of candidate or new facilities to be sited, or an objective functional
$F(f(x)) = F(u')$	Fourier transform of function $f(x)$ in frequency u'
$F'(\mathbf{z})$	Production function with input rates $\mathbf{z} = (\leftarrow z\rightarrow)^T$
$F'(\cdot)$	Regional-growth-rate function
F_k	Fibonacci numbers; also the weighted activity-rate, with work-accessibilities from zone k as the weights
F_X	Derivative of function F with respect to variable X
$\dot{F}=\nabla F$	Gradient of the function F being maximized
F'_i	Unsatisfied demand or remaining service-capacity at each demand-node i to entertain additional vehicle-deliveries
F_{ij}	Accessibility factor between locations i and j, expressed as an inverse function of travel cost
F_{ik}	Probability that a demand from i is of type k
$\mathbf{F} = [F_{ij}]$	Square matrix of population-distribution rate by zone, measuring $n'\times n'$
$\mathbf{F}'(\mathbf{x})=(\leftarrow F_i(\mathbf{x})\rightarrow)^T$	A vector of functions whose interactions $\partial F'_i(\mathbf{x})\,/\,\partial x_j \neq \partial F'_j(\mathbf{x})\,/\,\partial x_i$ are asymmetric, where $\mathbf{x} = (\leftarrow x_i\rightarrow)^T$ for $i = 1,\dots,n$
g	A scale factor; when serialized against argument m for example, $g(m)$, it is used to adjust zonal population from one loop of the Lowry model m to another $m+1$, where $m = 1, 2, \dots$
$g(\cdot)$	A special function of $\cdot$, such as the state equation; the relocation-cost function in stochastic facility-location; or the expected-master-travelling-salesman-tour length in probabilistic travelling-salesman-problem
g_k	Generalized unit-cost at facility k or for vehicle k
g'_i	Load to be picked up at node/vertex i
g''_i	Spatial 'drift' of activities toward location i, in accordance with a profit/benefit motive or some gravitational potential-function
g_{ij}	Short-hand notation for nonwork accessibility between i and j
$\mathbf{g}$	Vector of coefficients associated with the discrete-variables $\mathbf{y}$; when used as a function, it is the subgradient
$\mathbf{g}(j) = (\leftarrow g_{h(j)}\rightarrow)^T$	Vector of input measures for a decision-making unit j
G	Number of salespersons in a travelling-salesman problem, or the number of vehicle-tours out of a depot
G'	Maximum fleet-size available at a depot; or share of the population which are immigrants
$G(\cdot)$	Multiple-travelling-salesmen expected-tour-length-function involving k salespersons
$G(\xi)$	Generating function for the probability distribution $P_0, P_1, P_2, \dots, P_n$ where ξ takes on values of $0, 1, 2, \dots, n$
$G(\xi, t)$	Generating function for the probability distribution $P(\mathbf{X}_0^*, \mathbf{X}^*, t)$; where $\mathbf{X}_0^*$ is the initial-condition vector, $\mathbf{X}^* = [X_1^*(t), X_2^*(t), \dots, X_n^*(t)]^T$, and where the n-dimensional-vector ξ takes on values of $\xi^{\mathbf{X}^*} \equiv (\xi_1^{x_1^*}, \xi_2^{x_2^*}, \dots, \xi_n^{x_n^*})^T$, for $|\xi_j|<1$. Thus for the stationary,

	irreducible Markov-process, it assumes the form $P(X_0^\bullet) + \xi_1^{\gamma_1^\bullet} P[X_1^\bullet] + \xi_2^{\gamma_2^\bullet} P[X_2^\bullet] + \ldots + \xi_n^{\gamma_n^\bullet} P[X_n^\bullet]$
G_i	Class or group i; also a generalized spatial-statistic for point i
$G_i^\bullet(p,s)$	Generalized savings-measure from including demand node i between demand points p and s in a location–routing heuristic
$G_i'(p,s)$	Modified generalized-savings-measure from including node i between points p and s, after considering different depot-based tours
$G_i^\bullet(h'')$	Net change in cost from displacing demand i from tour h to h''
$G_i^{\leftrightarrow}(h'')$	Net change in cost from displacing demand i from tour h to h'' considering different fleets
G^{ij}	Transaction of goods and services between the ith and jth industrial sectors
G_{ij}	General location-pair spatial-statistic
G_{ij}^{pq}	Monetary transaction between the qth industrial sector in zone j and the pth economic-sector in zone i in an input–output model; with shorthand notation being G_{ij}^{pq} for consumption and G_{ij}^q for production respectively, considering only the nonzero elements
$\mathbf{G} = [G_{ij}]$	The growth matrix showing the growth springing off from group/location i to group/location j (within a period of time); also a basic-feasible-solution to a simplex-on-a-graph
$\mathbf{G}(\cdot)$	Vector return-function in a recursive program
$\mathbf{G}' = [g_{hj}]$	Input matrix containing the hth input for decision-making-unit j
$\zeta_j^{(l)}(\cdot)$	Economic surplus- or deficit-function at zone j of the lth type
h	Index for a variable; generally to show a fleet type, a category of inputs (costs) in data-envelopment analysis, or the iteration number in a recursive program
h'	Minimum fleet size
$h'(\cdot)$	State-transition function in dynamic programming
h''	Calibration parameter in a dynamic version of a spatial-location model; an example is the time-scale parameter to convert activity to a rate-of-change
h_k	Height of a subregion k
h_{ij}	A rate- or calibration-constant in a deterministic compartmental-model; for example, the interaction between regions i and j in a multiple-region predictor–prey equation-set, or a short-hand notation for work-accessibility
$\mathbf{h}(j) = (\leftarrow h_{k(j)} \rightarrow)^T$	Vector of output-measures for target decision-making-unit j
H	Housing expenditure for a household (part of locational expenditure)
$H(\cdot)$	The Hamiltonian function in terms of the state equation, the costate or adjoint variable, and the figure-of-merit at the present; it also stands for a general function
H'	An upper limit of discrete index h
$H'(\cdot)$	Regional growth-rate function
H''	Set of vehicles in a fleet, or the set of vehicle types in the fleet
$|H''|$	Cardinality of set H'', or the number of members in the set; here it is the fleet size
H^i	Transaction of goods and services to the ith household-sector
H_i	Set of potential tours in which demand i can be included
H_p	Cost of one dispatch on route p
H_r^G	Imports to region r
H_i'	Hazard a node i is exposed to
H_{ij}	Hazard a link (i,j) is exposed to
$H_{ij}'(\cdot)$	Flow-rate function from compartment i to compartment j
H_{ij}^p	Monetary transaction between the household sector in zone i and the pth economic-sector in zone j in an input–output model
η	Elasticity of demand
$\eta_{\alpha/2}$	$100(1-\alpha/2)$ percentile of the standard normal-distribution
θ	A parameter in general; for example, it can show decline in demand per unit-of-spatial-separation; θ_i is the rate-of-decline (or diffusion rate) of inflows into i
θ_t	Coefficient of the tth term in a moving-average time-series
$\theta(B)$	The backshift operation of a moving-average model
θ_{ij}	Proportion of activities (or trips) from origin-location i that end up in destination-location j based strictly on accessibility alone

Θ_{ij}	A short-hand notation for the spatial-interaction term, indicating the proportion of activities (or trips) from origin-location i that end up in destination j—based on both accessibility and the attractiveness at the destination; i.e., the normalized accessibility-function between i and j
$\mathbf{\Theta}_k = [\theta_{ijk}]$	A kth-order spatial-matrix of moving-average coefficients
$\mathbf{\Theta}(B) = [\theta_{ij}(B)]$	A spatial matrix of moving-average operators
i,j	Indices for nodes/vertices; i normally stands for a demand node and j a facility node; or they can just be any counter
$i(k)$	Beginning node of arc k
$j(k)$	Terminating node of arc k
$\mathbf{i}$	Cartesian coordinates of a demand i
I	Set of nodes/vertices in a network
$I(d)$	The spatial-statistic Moran's-I for a particular spatial-order as defined by the distance-parameter d
$|I|$	Cardinality of set I, or the number of members in the set
I_k	Profit or income for facility k
I_N	Set of unlabelled nodes
I_D	Dual objective-function in recursive program
I'	Household or aggregate income
I'_t	Aggregate income at time t
I'_h	Set of potential demands for exchange, with an existing demand on the tour h
I''	Subset of potential demand nodes within the set I, where demands are non-zero
I_{p_k}	Any subset of nodes in the kth-stop route p_k
$I(i)$	Set of nodes/vertices which are input markets
$I(0)$	Set of nodes/vertices which are output markets
$I(t)$	0–1 indicator-sequence reflecting the absence and presence of an intervention, overlaying the transfer-function on top of the time-series
$I_{i\kappa}$	A binary variable assuming unity if the combination of facilities κ provides a satisfactory service to demand i
I_{Rx}	Total expected-mutual-information between the facility pattern in the region R and the demand spatial-pattern (when $x=I$), or between the facility pattern and an individual demand (when $x=i_k$); i.e., how probable the facility pattern is consistent with what is known about the demand pattern I or individual demand i_k
$I[\mathbf{X}(k), \Gamma(k)]$	kth-stage payoff or objective-function of a recursive program, defined in terms of decision-variables $\mathbf{X}$ and constraint parameters Γ
$I(\mathbf{P};\mathbf{Q})$	Information that allows updating a prior probability-distribution $\mathbf{Q}$ to probability $\mathbf{P}$
$^rI^s(\cdot)$	Net-benefit function in a decomposed recursive-program
$\mathbf{I}$	Identity matrix
$j^*(k)$	Optimal facility location in state k
$\mathbf{j}$	Cartesian coordinates of a facility j
J	Subset of nodes/vertices in a network, generally the candidate sites for facility location
J_q	Set of candidate production sites
$|J|$	Cardinality of set J, or the number of members in the set
J'	A particular control-point in the bifurcation set
J''	The double values that the state variable assumes, corresponding to the control variable J' in the bifurcation set
$J(i)$	Set of Voronoi polygons adjacent to the ith polygon
$\mathbf{J}_k$	Basis k of a multicriteria linear-program
k	Index to show category k (e.g., Z^k is the kth activity); it marks a node, the commodity, the tree in a forest, or just serves as a counter
$k(\cdot)$	Equation for the control variable over time, expressed in terms of the state, the costate or adjoint variables
k_i	Calibration or scaling constant for zone i in a doubly-constrained gravity model; the Moran's-I or General Spatial statistic; alternatively, it is the propensity to save (invest)

$\mathbf{k}$	row vector consisting of 0, $+1$, -1 entries marking an orthonormal base of the transition-rate space
K	A discrete or continuous constant, or the upper limit of running index k
$K(t)$	Capital-stock investment over time
K_i, K_i'	Trip-production and -attraction rate at zone i respectively
$\bar{K}_j^p$	A scaling constant; it ensures that the inter-sectorial and inter-zonal flows sum up to the non-labor input to the input–output table for sector-p and zone-j
$\dot{K}_r$	Instantaneous rate-of-capital-accumulation in region r
κ	Combination of three or more facilities that perform a certain function
κ'	The complement of the set κ
κ^h	Cost of operating vehicle h
κ_i^h	Marginal cost of serving demand-node i
K	Combination of three or more facilities
$l(T)$	Total cost of spanning-tree T, which is sum of the arc costs
l'	Discount rate (e.g., on the number of commuting trips, or traditionally in the time stream of cost or benefits)
l^i	Lower bound of a specified time window for a salesman or vehicle to visit node/vertex i
l_j	Calibration constant for zone j in a doubly-constrained gravity model
l_k	Spatial order of the kth autoregressive-term in a spatial time-series
$l^{h''}$	Ordered set of neighboring points (p,s) representing candidate tour h''
$l_{h''/i'}$	Ordered set of neighboring points (p,s) in tour h'' after removing demand i'
$l^{mn}(r)$	Length of an r-stop route originating in m and terminating in n
$^r\mathbf{l}^s(\mathbf{x})$	Route-length vector at stage r and in state s of a decomposed recursive-program, expressed as a function of the decision variable $\mathbf{x}$
L	Nonempty subset of demand nodes/vertices, where a demand instance may be characterized by having actual demands realized in a node subset L of the network nodes/vertices I; the symbol also denotes twice the boundary length of a district
$|L|$	Cardinality of set L, or the number of members in the set
$\tilde{L}$	Length of the perimeter of a subarea
$\bar{L}$	The length of a queue, including the entity being served
$L(\cdot)$	Lagrangian or maximum-likelihood function
L'	Probability that the location visited is the termination point for the trip
L''	A calibration constant in a bivariate predictor–prey difference-equation-set
L_q	Queue length (excluding the one being served)
L_r	Regional labor-input-factor
$L^{(l)}x_i$	Spatial-lag operator on the value of spatial unit i, where l refers to the lth contiguity-class such as the lth-order neighbors; alternatively, we can write $L^{(l)}x_i$ as a matrix operation to compute the weighted sum of the neighboring values of i contained in vector $\mathbf{x}$, or $(\mathbf{w}^{(l)})^T\mathbf{x}$. In general, $L^{(l)}(\cdot)$ stands for spatial-lag operator of the lth-order, with the 0th-order operator reproducing the observation itself, or $L^{(0)}(\cdot) = \cdot$
$L_T(\cdot)$	Length of a master travelling-salesman-tour, constructed out of the set of nodes/vertices $\cdot$
L_{ij}	Error (in terms of a "loss measure") when a Bayesian classifier mis-assigns a multi-attribute observation $\mathbf{x}=(x_1, x_2, \dots)^T$ to group j when it actually belongs to group i; usually $L_{ij}=0$ if there is no error and $L_{ij}=1$ if there is a misclassification
$L_j(\mathbf{x})$	Average misclassification error (in terms of a "loss measure") when assigning multi-attribute observation $\mathbf{x}=(x_1, x_2, \dots)^T$ to group j; a couple of computational transformations of this measure are $L_j'(\mathbf{x})$ and $L_j''(\mathbf{x})$
$\mathbf{L} = \left(\mathbf{x}_L(q_1'), \mathbf{x}_L(q_2'), \dots\right)^T$	Matrix containing the left eigenvectors $\mathbf{x}_L$
λ	Dual variable or Lagrange multiplier, with a specific (not necessarily feasible) solution $\bar{\lambda}$ and the optimal solution λ^*
λ_i'	A normalized weight, where $\Sigma_i\lambda_i'=1$ unless noted otherwise
λ''	Arrival rate for a queuing process
$\lambda^k = (-\lambda_i^k \rightarrow)^T$	The kth solution-vector in a Lagrange-relaxation procedure

λ'^*	Dual optimal-solution to the linear-program subproblem at the last iteration within Benders' decomposition
$\Lambda(\mathbf{J}_k)$	The weight cone for multicriteria linear-program, showing the λ'-weight combinations that characterize a particular solution $\mathbf{J}_k$ among the nondominated set of solutions
m, n	Indices for dimension or for a node/vertex
m'	A calibration constant in a bivariate predictor–prey difference-equation-set
m^*	A critical bifurcation-value in a bivariate predictor–prey difference-equation-set
m^1	A collection of entities of characteristic 1; e.g., the number of complicated constraints in a Lagrangian-relaxation problem
m^j	A collection of entities of characteristic j; e.g., the number of high-frequency direction finders in a bundle located at station j
m_k	Spatial-order of the kth moving-average term in a spatial time-series
m_r	Vehicle-fleet requirement at depot r, or the number of deployed vehicles at depot r
m'_i, m''_i	In- and out-movement rate to and from subarea i
$m(k)$	Median for a median-filter using a $k \times k$ mask
$m_1, m_2, \ldots, m_{k'}$	Groups of demand nodes to be served by route $1, 2, \ldots, k'$, with $m_1 + m_2 + \ldots + m_{k'} \leq |I|$
$m'(q)$	Maximum shortest-distance from point q
m'_{ji}	Binary variable that is "switched on" when demand i is allocated to facility j in a combined data-envelopment-analysis/location model; also the benefit valuation for such i–j pair
M	Area specification for a districting model
M_i	Maximum inventory carried at node i
$M_{\max}$	Maximum number of nodes in a vehicle route
$\tilde{M}, \tilde{M}'$	A couple of matchings in a spanning-tree/perfect-matching heuristic for the travelling-salesman-problem
$M(t)$	Absolute smoothed-error (used in conjunction with relative smoothed-error) for adaptive-response-rate exponential-smoothing over time
$M(\Xi)$	Maximum of the weighted distances from the center candidates to each of the demands in the candidate facility-locations Ξ
M'	Non-locational expenditure such as food, clothing, education, savings etc.
M''	A very large number or weight
M_{ij}	Minor of a square matrix
$M(W,p)$	Simulation results of a facility-location model where p facilities are relocated to respond to a maximum load of W
$\mathbf{M} = [m_{ij}]$	Migration matrix showing the migration rate between locations i and j
μ	Mean of a probability distribution
μ'	Service rate of a queuing process; also the number of intermediate stops in the longest vehicle-route
μ_j	Positive weights placed upon an extreme direction $\mathbf{d}^j$ in a linear program
$\mu_i, \boldsymbol{\mu}_i$	Mean of observations in group i in both scalar and vector form
$\mu^{(l)}$	Scaling constant of the error ε associated the value v being measured, resulting in $v^{(l)} + \mu^{(l)}\varepsilon^{(l)}$
v	A collection of integer numbers
v_i	Route shape parameter (serialized by i) used in location-routing heuristics, assuming values such as 1 or 2
v_t	Noise series in a transfer-unction multivariate time-series
μ^p	Dual variable associated with the control total of areawide-transportation-cost constraint in an entropy-maximization model
Ξ	Collection of candidate facility-locations
$\Xi(X)$	Collection of all candidate facility-locations in the decision space X
$\Xi(\mathbf{y})$	Collection of candidate facility-locations which are open (i.e., for those locations where $y_j=1$)
$\Xi(z)$	Collection of candidate facility-locations in the Z space, whose distance bounds are within z units
ξ	As used in the Minkowski's distance-function, it is the proportion by which factor inputs have to be reduced to reach the efficient point on the production frontier

n'	The number of units in a spatial entity (e.g., the number of zones in a region, the number of subareas in a study area, or the total number of pixels in an image)
n_s	Number of sides in a subareal polygon (e.g., in a Dirichlet tesselation)
$n(a,b)$	Number of stops between origin-terminal a and destination-terminal b
N	Population or number of households (e.g., N_i is the population at location i)
N_j	Number of pattern vectors from class G_j, or the number of nodes or pixel vectors belonging to class j
N'(large)	A large number
$\bar{N}$	Total working population in the study area
N^p	Population working in economic-sector p
N_j^c	Capacity for residential development in zone j
N_i'	Set of spatial units (including facilities) within a distance S from demand i
N_{ij}	Binary decision-variables in a districting model, serving as a 'pointer' across a district boundary separating a geographic sub-unit i and one that is not j; it is unitarily value if subunit j is acquired and i is not
N	Row vector of zonal population N_i
N(k)	The nonbasic column associated with variable k in a linear-programming tableau
o_i	Export share of region i
$O(l^k)$	Worst-case kth-polynomial computational-complexity for input-data-length l
O_i	Export from the ith region
$O'(P) = \{\leftarrow O_i'(P) \rightarrow\}$	Orientation sequence of a path P, consisting of $+1$ and -1 entries, depending on the orientation of the arc in the path sequence
O^i	Export from the ith industrial sector, measured in dollars
O_j^i	Export from the ith industrial sector in subarea j, measured in dollars
$\mathbf{O} = (0 \leftarrow O^i \rightarrow)^T$	Export vector in an aspatial input–output model, showing the convention that the first sector (the household sector) has no exports
$O = (0 \leftarrow O_j^i \rightarrow)^T$	Export vector in a spatial input–output model, where i is the economic sector and j is the subarea
p	An integer number for the number of facilities, the number of services provided, the index for the pth vehicle route, the parameter in the l_p-metric, or the differencing parameter in a time-series
p'	Number of facilities in a subset of the p facilities (i.e., $p' \leq p$)
p_f	Price of fuel
p_g	Price of the good
p_k	Price of a commodity k, with **p** standing for a vector of commodity prices
p_i'	Probability of adopting strategy i in a two-person game
$p^{(j)}(\cdot)$	Probability function of choosing alternative j, $j = 1, ..., n$
p_{ik}	Empirical probability that demand k patronizes facility i; or the probability of transitioning from state i to state k
$\hat{p}_{ik}$	Estimated value of p_{ik}
$p_{i\cdot}$	Empirical probability that a demand patronizes facility i
$p_{\cdot k}$	Empirical probability that a demand k is being served
$p_j'^q$	qth factor-of-production input-prices at subarea j
p_k'	Number of facilities of the kth type (as used in a multi-product facility-location formulation)
$\bar{p}(t)$	Capacity expansion at time t
p''	Price of composite consumption-good
P_{ijk}	*Conditional* probability that event-type i occurs at geographic-region j at time-of-day k
$\hat{p}_{ijk}$	Prediction of p_{ijk} based both on the hypothesized intervention model and historical data
$\breve{p}_{ijk}$	Analytical prediction of the *relative* probabilities p_{ijk}, for field implementation as a transfer function
$\tilde{p}_{ijk}$	Relative probabilities after intervention probabilities have been implemented, using the transfer function $\breve{p}_{ijk}$

$\dot{p}_{ijk}$	Deseasonalized relative-probabilities after intervention probabilities have been implemented
$\mathbf{p} = (\leftarrow p^{(l)} \rightarrow)$	Perron vector whose components are positive and sum to unity
$\mathbf{p}_i(t) = (\leftarrow p_{ij}(t) \rightarrow)^T$	Vector of transitioning probabilities from state i to state j (where $j = 1, \dots, n$)
$\dot{\mathbf{p}}_i(t) = (\leftarrow \dot{p}_{ij}(t) \rightarrow)^T$	Time-derivative vector of probabilities transitioning transitioning from state i to state j (where $j = 1, \dots, n$)
P	A path; also a set of vehicle routes generated for a network
P'	Potential surface for destination choice, whose derivative dP'/dC_{ij} is often operationalized by the trip-distribution function
P_D	Dual space of the linear-programming relaxation problem
$P(p)$	Probability that p servers are occupied (busy)
$P(\cdot)$	Probability of an event ·
P_i	Nearest location for demand or customer i; also the probability that the system is in state i
$P_i(t)$	Probability that the system is in state i at time t
P_k', $P_{(k)}$	Steady-state probability of being in state k
$P_{id''}$	Steady-state probability that decision d'' is reached while in state i
P_{ij}	Binary decision-variables in a districting model, serving as a 'pointer' across a district boundary separating a geographic sub-unit i from one that is not j; it is unitarilly value if i is acquired and j is not
P_{ijk}	*Joint* probability of event-type i occurring in area j at time k, given that an event-type i occurred at time k
$\tilde{P}_{ijk}$	Analytical predictions of p_{ijk} aggregated monthly, based on the hypothesized intervention-model
P_k^{mn}	Set of vehicle routes covering origin–destination pair m–n via k-stop itineraries
P_c^{mn}	Set of vehicle routes covering origin–destination pair m–n via connect itineraries
$\bar{P}$	Scale of a facility as represented by its capacity, capital outlay etc.
$\tilde{P}_l, \bar{P}_l$	Lower and upper bound of the supply at location l
$\bar{P}'$	Aggregate production-function with capital as input
$\mathbb{P}(\cdot)$	Logical predicate over the argument ·
$P_j(p)$	Steady-state saturation-probability of all p service-units (in stochastic facility-location)
$\mathbf{P} = (\leftarrow P_i \rightarrow)$ or $(\leftarrow V_{ij} \rightarrow)$	Updated probability-distribution for each of the n' subareas or $|I|$ nodes written in a vector form; also be the updated travel-vector between i and j, $V_{11}, V_{12}, \dots, V_{ij}, \dots, V_{|I||J|}$, measuring $|I| \cdot |J|$ long
$\mathbf{P}(t) = (\leftarrow P_i(t) \rightarrow)$	Vector of the state probabilities $P_i(t)$; also the square matrix of transition probabilities over time
$\dot{\mathbf{P}}(t) = (\leftarrow \dot{P}_i(t) \rightarrow)$	Time-derivative vector of state probabilities $P_i(t)$
$\mathbf{P}' = [\mathbf{x}_1, \dots, \mathbf{x}_n]$	Matrix containing independent eigenvectors $\mathbf{x}(q_j')$, $j = 1, \dots, n$.
$\mathbf{P}_{t-1, t}$	Variance–covariance matrix for the difference between the observed and estimated Kalman-filter time-series-vector (or the estimation-error vector)
π_i	Dual variable in a network; such as the shadow price at node i, or a real number showing the amount of load carried on board a vehicle at node/vertex i
$\pi^{(j)}$	Probability that an individual reviews his/her choice of the jth compartment in a compartmental model
$\pi_{ij}(\cdot)$	Probability a given individual moves from compartment i to compartment j—as a function of, say, the state variable and time
π_i^j	Dual variable associated with the ith column of the spanning-tree ($j=1$) or non-spanning-tree ($j=2$) part of the basis (in a network-with-side-constraint tableau)
$\pi(\cdot)$	Permutation operator on the argument ·
$\pi(j \mid i, d'')$	The probability of transitioning from state i to state j during one period of the Markov process, given a decision d'' has been made
$\varPi^n$	n-dimensional transition-rate space

$^r\boldsymbol{\pi}^s(\mathbf{x}, \mathbf{y})$	Vector gross-return-function of decisions $\mathbf{x}$ and $\mathbf{y}$ (in a decomposition implementation of recursive-program)
$\boldsymbol{\Pi}(\cdot)$	Vector of gross return-functions of decisions in a recursive program
$\boldsymbol{\Pi}_0(t) = (\leftarrow \pi_{i0}(t) \rightarrow)^T$	Vector of transition rates with the "outside world" over time
$\boldsymbol{\Pi} = [\pi_{kl}]$	Transition-probability matrix in a Markov chain or compartmental model, with each entry denoting the given probability of transitioning from state k to state l; also the matrix of transition rates from state k to state l
$\tilde{\boldsymbol{\Pi}}$	Matrix of transition rates from state k to state l, considering both arrival and service in a queue
q	Index to show a node number, center number, median number, number of substations, or the number of attributes, criteria, endogenous variables, eigenvalues, or differencing parameter in a time series
q_k	Candidate location for a center k
q_{ik}	Probability that an event-type i occurs at time k
q'	Eigenvalue, with $q'_{\max}$ as the principal eigenvalue; also the growth rate of an area (with q'_j being the subareal growth-rate)
q'_i	Probability that strategy i is followed (in a two-person game); also the ith eigenvalue
$q_i(\cdot)$	Inventory-cost functions at demand-node i; or simply the unit cost-of-time (a constant) from demand-origin i
$\bar{q}_j$	Mean queuing delay
Q	Total economic-activity in the study area, such as consumption in dollars or number of trips executed
Q_i	Ratio of two accessibility definitions from location i
$\tilde{Q}_l, \bar{Q}_l$	Lower and upper bounds for the demand at location l
Q'	Total number of servers, or number of suppliers
$\tilde{Q}'$	Set of discrete points in the feasible region of an integer program
Q''	Cost per rejected demand in a loss-system location-model
$\mathbf{Q} = [\bar{\gamma}_{ij}]$	A matrix of economic-base multiplier over a time-increment Δt
$\mathbf{Q} = (\leftarrow Q_i \rightarrow)$ or $[Q_{ij}]$	Prior-probability distribution for locating in each of the n' subareas (written in a vector form); or the vector of prior-travel between i and j, Q_{ij}
$\mathbf{Q}_{t-1}$	Variance–covariance matrix of the white-noise vector $\boldsymbol{\alpha}_t$
$\mathbf{Q}'$	The $\mathbf{X}^T\mathbf{X}$ data-matrix in the nonlinear regression of a STARMA model; where $\mathbf{X}$ is not explicitly given, and has to be numerically estimated
$\mathbf{Q}'' = [q_j]$	Matrix with eigenvalues $q_1, q_2, \ldots$ along its diagonal
r	Rent or mortgage, as part of locational expenditure (e.g., r^i is the rent for a unit of land i at a distance d_i from market, and $\mathbf{r}$ is the vector of rents among these land units)
r_0	Pearson correlation-coefficient
r_k	Satisficing-level of criterion k; also the autocorrelation of lag-k in a time-series
r'^k	Land-consumption rate per retail-employee of trade-class k
r'	An l_p-metric deviational-measure from a standard or an ideal
$r'(\mathbf{y}', \mathbf{x})$	Generalized-Leontief distance-measure, as a function of inputs $\mathbf{x}$ and outputs $\mathbf{y}'$
$r(\cdot)$	Spatial-separation or response-time function of argument $\cdot$; or the return function in dynamic programming
r'_0	Partial correlation coefficient
r'_k	Partial-correlation-coefficient of lag-k in a time-series
$\bar{r}_j$	The expected response-time of service-unit j, consisting of mean queuing-delay and mean-travel-time to the demand
r_{ij}	Direct user-charge at facility j for user from origin i
$r(i, d'')$	Reward expected at state i by making decision d'' (in a Markovian-decision-process)
$r_{XY}, r(X, Y)$	Sample (cross) correlation-coefficient between random-variables X and Y
$r_{Y|X_j \ldots}(X_k)$	Partial-correlation-coefficient between Y and X_k, given $X_j, X_p, \ldots$ are in the equation already
$r'_i(\cdot)$	Euclidean distance between demand i and a facility
$\hat{r}_{lm}(k)$	kth-order spatial-temporal-autocorrelation between the lth and mth neighbors of the subject site

R	A closed region in Euclidean 2-space; the set of n subregions $\{R_1, R_2, \ldots, R_n\}$; or the multiple correlation-coefficient
$R(J)$	The set of n subregions, each identified by its service-facility location $\mathbf{x}_i$: $\{R(\mathbf{x}_1), \ldots, R(\mathbf{x}_n)\}$
R_T	Total physical region made up of subregions $R_1, R_2, \ldots, R_n$; these regions can be of higher dimensions than the Euclidean 2-space
R_+^n	Domain of continuous non-negative variables in Euclidean n-space
$\lvert R(k^*)\rvert$	The area of the largest empty-circle with center at k^*, located at any vertex of the bounded Voronoi-diagram
$\lvert R(\mathbf{x})\rvert$	The area of subregion $R(\mathbf{x})$; $\lvert R(\mathbf{x}_i^*)\rvert$ is the area of the optimal ith Voronoi polygon, with its facility at $\mathbf{x}_i^*$
R'	In stochastic facility-location models, R' is the required time in dispatching a special reserve-service-unit from a neighboring jurisdiction
R^2	Coefficient-of-multiple-determination in regression
$\bar{R}^2$	Coefficient-of-multiple-determination after adjusted for the degree-of-freedom
$R^2_{Y\mid X_1, X_2, \ldots, X_k}$	Coefficient-of-multiple-determination between Y and $X_1, X_2, \ldots, X_k$
$R'(\mathbf{y}')$	Set of input requirements $\mathbf{x}$ to produce $\mathbf{y}'$ in a production function
R''	The entire image or entire region
$R(+\mid-), R(-\mid+)$	Finite predictor/prediction-space used in spatial-temporal canonical-analysis
R_i	Subregion i within the entire region R''; also the production in subregion i
R_i'	Normalizing constant in a spatial-interaction function, or the denominator of the function Θ_{ij}
R_i^p	Production output of the pth industry in zone i
$\bar{R}$	Number of row cells in a grid region, a raster image, or a lattice
R^i	Monetary output from the ith industrial-sector
R_j^i	Monetary output from the ith industrial-sector located in subarea j
$R_s(d)$	Norm deviate of the generalized spatial-statistic (analogous to the two-tailed t-statistic)
$\bar{R}_j^p$	The observed value of non-labor input to the input–output table for sector-p and zone-j
$\mathbf{R} = (yz^e \leftarrow R^i \rightarrow)$	Output vector in an aspatial input–output model, showing the production in each economic-sector, starting with output from the household (or labor) sector (measured in wages) and followed by the first, second, … industrial sectors i
$\mathbf{R} = (yz_j^e \leftarrow R_j^i \rightarrow)$	Output vector in a spatial input–output model, showing the subareal production in each economic-sector i, starting with the subareal output from the household (or labor) sector (measured in wages) and followed by the first, second, … industrial-sectors by subarea $j\mathbf{R}' =$
$[\mathbf{x}_R(q_1'), \mathbf{x}_R(q_2'), \ldots]$	Matrix containing the right eigenvectors $\mathbf{x}_R$
$\mathbf{R}''$	Commodity-value-added output-vector
$\mathbf{R}_t$	Variance–covariance matrix of the measurement error (or noise) in a Kalman-filter time-series
ρ	Parameter or dual variable to account for the delivery-vehicle capacity
$\breve{\rho}(\tilde{\mathbf{B}})$	Spectral radius of matrix $\tilde{\mathbf{B}}$
$\rho' = \lambda''/\mu'$	Utilization rate of a server in a queuing system, or ratio of the arrival rate λ'' and service rate μ'
ρ''	Intensity of activity in a subarea
ρ_j	Utilization-rate of a service-unit j in stochastic facility-location; also the import rate of region j
ρ^p	Productivity-in-the-pth-economic-sector per unit-of-labor
ρ_{ij}	Trade coefficient between regions i and j
ρ^{pq}	Technical coefficients showing the transactions between the pth and qth economic-sectors in an input–output model
ρ_j^{pq}	Technical coefficient at the receiving-sector zone-j
ρ_{ij}^{pq}	Technical coefficients showing the transactions between the pth economic-sector in zone i and the qth economic-sector in zone j in an input–output model

ρ	Matrix of technical or input–output coefficients $[\rho^{pq}]$, trade coefficients $[\rho_{ij}]$, or combined spatial-technical coefficients $[\rho_{ij}^{pq}]$
$\hat{\rho}$	Diagonal matrix of trade coefficients, $[\rho_{ii}]$
$\rho^j = [\rho_h^j]$	A matrix of economic-multipliers for the jth economic-sector, disaggregated by each zone-h
ρ_S, ρ_T	The consumption and production multi-sectorial components of the input/output-coefficient-matrix ρ, derived from row- and column-sum normalization of transaction flows respectively, with $\rho_S\rho_T = \rho$; the spatial, multi-subareal version assumes $G_j^{pq}/G_j^{\cdot q} = \rho_j^{pq}$ and $G_{ij}^q/G_{\cdot j}^q = \rho_{ij}^q$
$\hat{\rho}_{XY}$	Population cross-correlation between random-variables X and Y
$\hat{\rho}^2$	Relative size of the variance; $(1-\hat{\rho}^2)$ is the variance reduction
s	Source of a network
s_p	Autoregressive season-length in a seasonal time-series
s_q	Moving-average season-length in a seasonal time-series
$\underline{s}$	Prescribed frequency-of-visit at a node/vertex
s_X	Standard deviation of the random-variable X
$s(j)$	Sum of vertex(node)–arc(link) distances for facility j (the smallest sum identifies the general median)
$s'(j)$	Sum of point–arc distances for facility j (the smallest sum identifies the general absolute median)
s^2	Sample variance, with s being the standard deviation
s_{ij}	Length of the border separating geographic sub-unit i from sub-unit j; also surviving ratio of cohort-group j from cohort-group i
s'	Average size of a site; or the ratio between the demand potentials at sites i and j
s''	Slack node/vertex in a network
S	A set of alternatives (e.g., the set of solutions that satisfies a predetermined goal or standard, the branch-and-bound search-space in a linear-programming relaxation etc.)
$S(\cdot)$	Sum-of-squares surface constructed out of the parameters $\cdot$ in nonlinear regression
S'	Consumers' surplus (or net benefit) to a tripmaker in making a trip; alternatively it refers to a predetermined maximum-service-distance in discrete facility-location
S''	Another set of alternatives (for example, the set after introducing a new alternative)
S_k	Set of demand vertices or nodes that would be covered by a center at q_k
$S_i(p',q')$	The increase (or savings) via a triangular-inequality-style inclusion (or exclusion) of demand i between the adjacent points (p',q')
$S_i(l_{h''}/i')$	Increase in travel-distance from serving demand i via tour h'' (after the former-demand i' has been removed)
S^i	Marginal-cost function for path i
$S_l(V_l^S)$	Supply function showing price against flow quantity, in other words price charged at supply-quantity V_l^S
S'_{kj}	Unit benefit of assigning the kth activity (or activity from zone k) to zone j
S_{jk}	The kth site-specific attribute of the jth facility (such as the acreage of a state park)
$S^{k,l}$	Marginal-cost function between origin k and destination l
$\mathbf{S}_{ij}$	Vector of level-of-service variables between locations i and j, including such variables as travel time and travel cost
σ	Standard deviation of a probability distribution
σ^2	Variance of a probability distribution (see also the sample-variance s^2)
σ'	Vendor score or simply a constant in a model
σ_i	Real number showing the 'odometer' reading of a vehicle at node/vertex i
$\sigma_{\hat{Y}}^2$	'Tilting' effect, as measured in terms of the variance, on the regression line (due to the randomness of the regression coefficients)
σ_M^2	'Tilting' effect, as measured in terms of the variance, on the regression line—when an additional data-point x' is added to the regression
σ_Y^2	Total regression-based prediction- or estimation-error, as expressed in terms of the variance of the predicted- or estimated-values Y

σ_T^2	Total regression-based prediction-error, as expressed in terms of the variance of the predicted values Y'
$\sigma_{M^\bullet}^2$	Variance of a normally-distributed set of residuals, around the sample regression-line at $X = x^*$
σ_{ij}^{pq}	Calibration coefficient such as the subareal investment-coefficient or the marginal capital-output-ratio, quantifying the multiplier effect of investment among economic sectors and between subareas
σ_j^2	Variance (or second moment) of service-time at service-unit j
$\sigma^h = (\leftarrow \sigma_i^h \rightarrow)$	Vector of dual-variables corresponding to the ith constraints defining the hth travelling-salesman-polytope
$\sigma = [a_j]$	Zonal population-serving-ratios along the diagonal of an $n' \times n'$ matrix
$\Sigma = [\mathrm{cov}(\varepsilon_i \varepsilon_j)]$	Error covariance-matrix
t	Time dimension or simply a counter for a series of data (e.g., $N(t)$ is the population at time t, Δt is a time increment)
t'	Subareal share of transportation-accessibility-to-employment
t_b	Student-t statistic for calibration-parameter b
$t_{\alpha/2,\, n-2}$	t-statistic at $100(1-\alpha)\%$ confidence-level and $n-2$ degrees-of-freedom
t_N	Sink node/vertex of a network
t''	Technical-attribute score
t^k	Step size in iteration-k of a hill-climbing optimization-algorithm
t_0	Dwell time at a terminal
t_j^h	Delivery- or dwell-time at node j by salesman or vehicle h
t_{ij}	Normalized work-accessibility-function between i and j
$^r t \, ^s(\mathbf{x}, \mathbf{\Phi}, \mathbf{V})$	Cost of providing service at state s and stage r of a recursive-program
$\tilde{t}$	Random service-time on-scene $\tilde{t}_i$ or off-scene $\tilde{t}_j$
$\bar{t}$	Expected value of on-scene service-times to all demands i
$\bar{t}'$	Ratio between intra-nodal distances at i and j
$\bar{t}_j$	Average service-time for a service-unit stationed at depot j, consisting of on-scene service-time at the demand t_j^1 and the off-scene service-time at the depot t_j^2
$\mathbf{t} = [t_{ij}]$	Matrix of normalized work-accessibilities, measuring $n' \times n'$
$\mathbf{t}^k = [\tau_{ij}^k]$	Matrix of travel-times between i and j
τ	Time duration (e.g., τ_{ij} or $\tau(i,j)$ is the travel time from location i to j)
τ'	Calibration constant in a dynamicized input–output model
τ_k	A user-defined scalar in the subgradient optimization routine ranged (say) between 0 and 2
$\tilde{\tau}$	Random variable for service-time in a queuing process; $\tilde{\tau}_{j|i}$ is the random service-time for demand i from depot j
$\bar{\tau}_j$	Expected one-way travel-time to a random demand from depot j
$\bar{\tau}_j'(k)$	Expected travel-time from j to all demands in state k
T	Transportation cost as part of locational expenditure; also quantifies other technological factors
$T.$ or $T(\cdot)$	A-priori travelling-salesman-tour as a function of $\cdot$
T'	Minimum spanning-tree of a graph
T''	Multi-graph, derived from the minimum spanning-tree by duplicating every arc of the graph; also an instance of the travelling-salesman problem
T_N	Alternate sink-node/vertex in a network for excess flows
T_j	Number-of-neighbors surrounding geographic sub-unit j
T_i'	Proportion of sales from subject location to demand at i
T_i''	Electrical-flow capacity of a substation i
T_{ij}	Number of ith-group neighbors for a jth-group geographic sub-unit
$\hat{T}_{ij}$	Current estimate on random-variable T_{ij}
$\mathbf{T}$	Diagonal matrix of zonal activities such as population
$\mathbf{T}(\cdot)$	Vector of cost-functions in a recursive-program

T_B	Basis for a simplex-on-a-graph, represented graphically as a tree
u	Accessibility-to-population, or a calibration parameter in general; for example, u_{ij} is the normalized nonwork-accessibility between i and j
$u(t)$	The set of infinite control-paths between the initial point $t=a$ and end point $t=b$
u'	Frequency of a signal
u''	Ratio of the maximum travel-distances between nodes i and j
$u_t(t)$	Dual variables in a recursive-program for $t = 1, 2, \ldots$
u^i	Upper bound of a specified time-window when a salesman or a vehicle visits node/vertex i
$\overline{u}_{ij}$	Capacity on arc (i,j) in a network
${}^r u^s(\mathbf{x}, \mathbf{y})$	Inference dual-variable to show the value (or shadow price) of relaxing an rth connectivity-requirement at state s
$\mathbf{u}$	Surplus variables in a linear program; also a subset of control-variables $\mathbf{U}$
$\mathbf{u}' = [u_{ij}]$	Matrix of nonwork accessibilities, measuring $n' \times n'$
U	Utilities (e.g., U^* is the maximum amount of utility from a given income or budget)
$U(h)$	Route length or the range of a vehicle tour for vehicle type h ($h = 1, 2, \ldots$)
U'	Maximum route-length or range among a fleet of vehicles, $U' = \max_h [U(h)]$
$U(t)$	Control variables over time t
$\mathbf{U} = (\leftarrow U_j \rightarrow)$	Vector of control-variables in control theory (slow variables), usually expressed as a function of t; U_j also stands for just the jth canonical-variate
$\mathbf{U}$	Diagonal matrix of zonal activities such as employment
$\mathbf{U}(k) = [\leftarrow \mathbf{u}^{k+r} \rightarrow]$	Matrix of inference dual-variables in a binary recursive-program
v	Value or utility function, or simply the metric resulting from such a measurement
v_{ij}	The composite travel-cost, or the "utility function," between zones i and j, combining time, cost and other travel impedances into a single metric
$v(k)$	Average filter using the kth-order neighbors
v'	A given parameter (such as housing subsidy per household)
v''	Velocity of a service vehicle in stochastic facility-location
v_i	Dual variable associated with node/vertex i
v_w	Walking speed
$v_{\max}$	Maximum velocity of a vehicle
$v^{()}(\cdot)$ or $v^j(\cdot)$	Deterministic value-function for alternative j
$\overline{v}_{ij}$	The reduced-cost for arc (i,j) in network-flow programming
$\mathbf{v}_i = (\leftarrow v_i^j \rightarrow)$	An eigenvector consisting of as many entries as the number of alternatives; this is equivalent to $\mathbf{x}_i = \mathbf{x}(q_i')$
$\mathbf{v}$	Surplus variables in a linear program
V	The amount of economic activities, traffic flow or patronage (e.g., V_i is the amount of activities or trips originating or terminating at location i, and V_{ij} is the exchange of economic activities or traffic movement between locations i and j); $\hat{V}$ is the estimated value and V^* is the observed value.
$V(h)$	Capacity of vehicle-type h, where $h = 1, 2, \ldots$
$V'(h)$	Capacity remaining on each vehicle h
$V'(\cdot)$	Normalized vehicle-capacity
V^d	Inverse demand-function, or the price schedule expressed as a function of a firm's (firms') total output; V_i^d is the excess demand at subarea i
V'	Set of vertices or nodes in a graph or network
V_i	The ith canonical-variate
V_{ij}	Flow between origin–destination pair i–j; $\tilde{V}_{ij}$ is the lower bound and $\overline{V}_{ij}$ is the upper bound
V_{ijk}	Probability that a demand i of type k is received by facility j
V_{ij}^k	Trips of type k from i to j
$\hat{V}_{ij}$	Predicted interactions between subareas i and j
$\tilde{V}_{iq}^d$	Amount supplied by all the firms other than q to demand-location i
V_i^s	Output of firm i; also standing for the excess supply of a firm located in subarea i

φ	Calibration constant representing such parameters as the trip-generation rate or response rate of the system
φ^h	Polytope (feasible region) defined by the hth travelling-salesman-problem
φ'	Probability distribution (e.g., probability that the surplus resulting from the trip to j has a value in the neighborhood of S')
Φ	Cumulative distribution (e.g., $\Phi(v) = [F(v)]^n$ is the cumulative distribution-function of the largest-utility v among n independent samples; $\Phi_{ij}(S')$ is the cumulative-distribution-function of the surplus accruing from the preferred (optimal) trip between location i and j)
φ_k	Coefficient of the kth-lag term in an autoregressive-time-series
$\varphi(B)$	The backshift operation of an autoregressive model
$\hat{\varphi}_k$	Partial-autocorrelation-coefficient for the kth-lag term in an autoregressive-time-series
$\hat{\varphi}_{kl}$	Partial-autocorrelation-coefficient at temporal-lag k and spatial-lag l in an autoregressive spatial-time-series
$\varphi(\cdot)$	Flow-vector function at stage s of a decomposed recursive-program
$\varphi(\mathbf{x})$	Demand density-function on Voronoi polygons
$\boldsymbol{\varphi}$	Vector of pertinent flows at stage r and state s of a decomposed recursive-program; these flows can be expressed in terms of the pertinent demand-vector $\mathbf{f}$
$\boldsymbol{\varphi}^T = (\leftarrow \varphi_i \rightarrow)$	Vector of autoregressive coefficients in a conditional spatial-econometric model
$\boldsymbol{\Phi}_k = [\varphi_{ijk}]$	A spatial autoregressive-coefficient matrix of order-k
$\boldsymbol{\Phi}(B) = [\varphi_{ij}(B)]$	A spatial autoregressive-operator matrix
$\boldsymbol{\Phi}^k(\cdot) = [\leftarrow \boldsymbol{\Phi}^{k+r} \rightarrow]$	Matrix of flow-vectors $\boldsymbol{\Phi}^{k+r}[\leftarrow \mathbf{x}^{k+r}(k) \rightarrow]$
$\boldsymbol{\Phi}^{k+r}[\leftarrow \mathbf{x}^{k+r}(k) \rightarrow]$	Flow-vector at the kth cycle and rth stage, showing origin-destination-connectivity as a function of the iterative multi-stop routing-decisions
χ^2	Chi-square statistic
φ	Expected cost between stockout and storage in a newsboy problem
$\boldsymbol{\varphi} = [f_j]$	Zonal activity-rates along the diagonal of the $n' \times n'$ matrix
ψ	Value of a given function; e.g., Sierpinski's-curve value
ψ_j	Weights used in time-series forecasting
$\Psi^n_{k=1}(l_k)$	Dynamic-program recursion-function for computing the shortest-route-length l
Ω	Dual variable corresponding to the terminal capacity constraint—a parameter to account for the given warehouse capacity; also regular vector space
$\overline{\Omega}$	A bounded domain including the boundary $\delta\Omega$
$\Omega_q = \{\mathbf{x}_q\}$	A feasible region within the vector space Ω; e.g., a set of constraints in a spatial-equilibrium model, expressed in terms of the flow decision-variables $\mathbf{x}_q$ for each of the suppliers q
Ω_{ij}	Percentage-change-of-patronage at facility j from the demands that originate at i
$\Omega(B) = \sum_i \Omega_i B^i$	Backshift operator containing the dynamic multipliers Ω_i in a set of dynamic simultaneous-equations
$\boldsymbol{\Omega}_{t-1,t}$	Transition matrix in a Kalman filter
$\acute{\eta}_k$	Connectivity requirement on the origin–destination pairs during the kth cycle
$\acute{\eta}_k(r)$	Connectivity requirement on a subset of the origin–destination pairs during the rth stage in the kth cycle; i.e., the number of constraint functions defining the local flow-pattern in a recursive program for the RISE algorithm
w	A constant, or an aggregate weight-parameter, placed on a variable or an estimator-measure (such as Moran's-I, and its variance, plus the mean and expected variance of the general spatial-statistic)
w_k	A constant or a weight placed on entity or attribute k; when these weights are normalized and summed to unity, we write $\sum_i \lambda'_i = 1$
w^k	Weight reflecting the relative importance of workplace-based retail-trips for purpose k
$\tilde{w}_1, \tilde{w}_2$	Weight-parameters used in the formulas for the variance of Moran's-I

w'_k	Width of a subregion k
w'_t	A white-noise series, consisting of a sequence of uncorrelated random-variables, each with zero mean and finite variance; engineers consider them as independent 'shocks' that are transformed by a "transfer function" to another time-series whose successive values are highly dependent.
w''_p	Frequency on route p
w_{ij}	Weight placed on the demand-facility pair $i-j$ or the weight placed on arc flow (i,j), otherwise referred to as cost coefficients in the equivalent linear-program; also denotes the weight entry in a spatial-weight-matrix $\mathbf{W}$, with $0 \le w_{ij} \le 1$
$w_{ij}(d)$	Binary valuations of w_{ij} when an activity at j is within a distance d from i
w_{ijp}	Frequency on the (i,j) segment of route p
w^j_i	Weight contribution toward criterion i by alternative j
$^r\mathbf{w}^s(\varphi,\mathbf{V})$	Vector of route-frequencies at stage r and state s of a decomposed recursive-program
$\mathbf{w}=(\leftarrow w_i\rightarrow)^T$	Eigenvector consisting of q entries—this is equivalent to $\mathbf{v}_i$ and $\mathbf{x}_i$; also the cost vector in a network-flow program
$\mathbf{w}^{(l)}=(\leftarrow w^{(l)}_{ij}\rightarrow)^T$	The vector of spatial-weights associated with the lth contiguity-class; an example is the weights associated with the lth-order neighbors—notice this is equivalent to the spatial operator $L^{(l)}(\cdot)$
W	White-collar employment; also work load or demand placed on a service-unit
W_i	Size of demand or activity at i, which is proxy for development opportunity at the zone; $\mathbf{W}'$ is the vector of development-opportunities among all zones
W_q	Delay time in queue
W_T	Total time in system, including delay time in queue and the time being served
$W(t)$	Rate of investment in new capacity over time
W'_i	Revised size of demand or activity at i
W_{ij}	Service-effectiveness weight expressed as a function of the separation between demand i and facility j; i.e., the further apart i and j are, the less effective it is for service to be rendered
$\bar{W}^p_i$	Observed value of attractiveness or the opportunity of zone-i as a location for industry-p
W^h_i	Observed zonal-residence attractiveness or opportunity
W^p_i	Observed zonal-shopping attractiveness or opportunity
$\mathbf{W}=[w_{ij}]$	A $q \times q$ pairwise-comparison weight-matrix used in the analytic hierarchy process; also denotes the weight matrix in spatial econometric-models, measuring $n \times n$
$\mathbf{W}'=[W_{gh}]$	An $n' \times n'$ activity derivation-and-allocation matrix of Lowry–Garin model, with each entry denoting a zone pair $g-h$
$\mathbf{W}''=[w_j]$	The diagonal matrix consisting of per-capita value-added productivity (wage rate)
$\mathbf{W}^j$	Activity derivation–allocation, transition or spatial-weight matrix for the jth activity in a Lowry–Garin model
$\mathbf{W}^{(l)}=[w^{(l)}_{ij}]$	Spatial weight-matrix for the lth-contiguity class; with the normalized spatial-weights sum to unity $\sum_j w^{(l)}_{ij}=1$, and $\mathbf{W}^{(0)}=[w^{(0)}_{ij}]=\mathbf{I}$, or the 0th-order neighbors being the subject entry itself.
$\mathbf{W}_t$	Gain matrix in Kalman filter, representing the net percentage of measurement-error or noise that is left after filtering
$(\mathbf{W}^{(l)}\mathbf{y})_{-y_i}$	Preprocessing of data $\mathbf{y}$ by removing the subject ith-entry, and then replace it with a value resulting from 'filtering' with a spatial-'mask' $\mathbf{W}^{(l)}$ of order l
x^*	Sample observation or the optimal value of x
x'	A particular observation for the random-variable X
x'_t	Actual, accurate data in a Kalman-filter time-series (to be differentiated from what is observable)
$x_0, x'_0, x''_0, \dots$	Decision boundary between pattern groups 1 and 2, 2 and 3, 3 and 4, etc.
x_{ij}	Allocation of demand i to facility j; or flow from i to j
x^i	Flow on path i in a network
x^p_i	Equilibrium economic-activity at each subarea i and sector p
$\tilde{x}^p_i$	Projected sales of product p in subarea i
x_{ijk}	Allocation of demand i to facility j in state k
x^{mn}	Lost calls between origin–destination pair $m-n$

$x^{m,n}(C^{m,n})$	Demand-for-transportation between origin–destination pair m–n as a function of the transportation cost between them
x_p^{mn}	Binary link-allocation of demand between origin–destination pair m–n to non-stop route or itinerary p
x_{ji}^{qp}	Input of commodity-q from subarea-j in the production of commodity-p in subarea-i
x_{mip}^{mn}	Binary allocation of demand m–n on route p as indicated by the usage of segment (m,i) in the itinerary
$\mathbf{x} = (\leftarrow x_j \rightarrow)^T$	Vector of decision-variables, or empirical readings (such as change-in-accessibility for all the activities j)
$\mathbf{x}_q$	An interior point in the feasible-region Ω_q
$\mathbf{x}_t = (x_1, x_2, \dots)^T$	Observed readings in a time-series
$\mathbf{x}_t' = (x_1', x_2', \dots)^T$	Actual readings over time in a Kalman-filter time-series
$\mathbf{x}_L$	The left eigenvector of a square matrix
$\mathbf{x}_R$	The right eigenvector of a square matrix
$\mathbf{x}^i$	The ith discrete-point proposal in a branch-and-bound tree, corresponding to a constraint in the Lagrangian-dual linear-program
$\mathbf{x}^k$	The kth basic-solution in a linear-program, or the kth set of decision-variables (e.g., solution alternative in a branch-and-bound tree, the location of the kth-facility, or the routing decision-variables for the kth-vehicle)
$\mathbf{x}_s^0$	Coordinates for the origin of a trip
$\mathbf{x}_t^0$	Coordinates for the destination of a trip
$\mathbf{x}^{i_k}$	The ith discrete-point proposal in a branch-and-bound tree during the kth-step of the subgradient-optimization procedure of Lagrangian relaxation
$\mathbf{x}^*(\mathbf{x}_s^0)$	Nearest public-transportation terminal for a trip starting at origin $\mathbf{x}_s^0$
$\mathbf{x}^*(\mathbf{x}_t^0)$	Nearest public-transportation terminal for a trip terminating at destination $\mathbf{x}_t^0$
$\hat{\mathbf{x}}_t' = (\hat{x}_1, \hat{x}_2, \dots)^T$	Estimated-values of the observations in a Kalman-filter time-series
$\mathbf{x}^{k+r}(k)$	The kth iterative multi-stop-routing decision-variables for the rth-stage
$\overline{\mathbf{x}}^{k+r}(k)$	Realized values for the kth iterative multi-stop routing-decision-variables in the rth-stage
X	The decision-variable X; or the decision or alternative space in multi-criteria decision-making
X'	The state-space in Markovian-decision-processes
$\overline{X}$	Average of the independent random-variable X in a regression model
X^p	Control-total of areawide-transportation-cost for commodity p
$X(t)$	Random-variable for the state at time t
X_k	Random-variable for the state at stage or time k
$X_i(\cdot)$	Accessibility from origin i to all destinations as a function of such parameter as travel cost
X_{lj}^t	Activity-l's accessibility to zone-j
X_{ij}	Observed patronage of facility j by demand from location-i
X_i^k	Amount of activity k in zone i
$\mathbf{X} = \begin{bmatrix} \leftarrow \mathbf{x}_1 \rightarrow \\ \leftarrow \mathbf{x}_2 \rightarrow \\ \vdots \end{bmatrix} = [X_{ij}]$	Exogenous- or independent-variable $n \times (k+1)$ matrix in ordinary-least-squares regression, corresponding to n observations and $(k+1)$ calibration-parameters
$\mathbf{X}(t) = (\leftarrow X_i(t) \rightarrow)^T$	Vector of state-variables in control theory (fast variables), expressed as a function of t in terms of the individual state variables $X_i(t)$ for states $i = 1, \dots, n$
$\mathbf{X}(0) = \mathbf{X}_0 = (\leftarrow X_i(0) \rightarrow)^T$	Initial condition of the state-vector at time 0 for states $i = 1, \dots, n$
$\mathbf{X}_{max}^*(t)$	The most-likely state
$\mathbf{X}(k) = [\leftarrow \mathbf{x}^{k+r}(k) \rightarrow]$	Matrix of binary-decision-variables in a recursive-program during the kth-cycle and the rth-stage
$\mathbf{X}_l(\Delta) = (\leftarrow X_{lj}(\Delta) \rightarrow)$	Activity-l's accessibility to individual-zone-j expressed as change in the regional-share-in-accessibility
$\mathbf{X}'' = (\leftarrow X_i'' \rightarrow)$	Stationary states in system of interacting differential-equations

$\mathbf{X}^j = [X^j_{gh}]$	A matrix of accessibilities between zones g and h for activity j
y	Wage rate for a household or total wages across the labor-force
y^*	Sample observation or the optimal-value of y
y'	Regression-based prediction corresponding to a given x'
y_p	The pth-component of the $\mathbf{y}'$-vector in a network-tableau
y_t	Ordinate of an observed-data-point in the series $t = 1, 2, \ldots$
$\hat{y}_t$	Estimated ordinate of an observed-data-point in the series $t = 1, 2, \ldots$
y^q	Household-wage expenditure on the qth industrial-sector
y_{jk}	Binary-decision-variable to assign facility to node-j in state-k
y_{ijk}	Binary-decision-variable to indicate that node/vertex-i is served by facility-j in state-k
y_k^{mn}	Binary indicator that there is kth-stop service between origin–destination pair $m-n$
$y_{u(k),v(l)}$	Binary decision-variable to indicate moving a facility from node/vertex u to v as the state transitions from k to l
$\mathbf{y} = (\leftarrow y_j \rightarrow)^T$	Vector of integer-variables in a mathematical-program, or simply a point within the regular vector-space
$\mathbf{y}_q$	A point other than $\mathbf{x}_q$ within the feasible-region Ω_q
$\mathbf{y}' = (\leftarrow y_i' \rightarrow)^T$	A vector of criterion-measures consisting of attributes i; also the updated or 'refreshed' column in a network-flow-tableau during the simplex-iterations
$\mathbf{y}''$	Interim solution in Benders' decomposition
$\mathbf{y}(k)$	The updated (or 'refreshed') kth column in a network-tableau
$\mathbf{y}^j = (\leftarrow y_i^j \rightarrow)^T$	A vector of criterion-measures for alternative j, or the jth group of y_i-variables (e.g., the delivery commitment of vehicle j toward demand i)
Y	The decision-variable Y, or random-variable notation of the explanatory or dependent variable in ordinary-least-squares regression; also the regional income
$\bar{Y}$	Mean of the random-variable Y
Y'	Outcome or criterion space of multi-criteria decision-making; also the prediction random-variable in regression
Y''	The combinatorial space of the discrete-variables y_i
Y_{ij}	A spatial-variable defined by the coordinates i and j—a variable that is related to its neighbors in both axes of this coordinate system; this cross-product is the covariance between the observations at i and j
$\mathbf{Y} = (\leftarrow y_i \rightarrow)^T$	Explanatory- or dependent-variable vector in ordinary-least-squares regression, consisting of n observations; $\hat{Y}$ denotes the estimated-values of random-variable Y
$\mathbf{Y}^{ij}(k) = [\leftarrow y_j^{ij} \rightarrow]$	Binary parameters of each constraint-function in recursive-programming (p' in total), where i is the state-index and j the stage-index; $\mathbf{Y}(k) = \begin{bmatrix} \uparrow \\ \mathbf{Y}^{s,k-r} \\ \downarrow \end{bmatrix}$
$\mathbf{Y}(\cdot)$	State-connectivity linkage-function of past decisions and available vehicle-capacity in a recursive-program
$\mathbf{Y}'$	Labor-force-value-added output-vector
z	Objective-function of an optimization-problem; also used to denote the activity-generation rate
z'	A bound on z
$z(j)$	Objective-function value of the jth alternative
z_c	Largest demand-facility assignment-distance
z_i	Amount of product or services sold at demand-point i; or a transformed observation from the raw-data Z_i
z_t	Stationary time-series with zero mean
z_{IP}	An integer-programming objective-function that is to be estimated by Lagrangian-relaxation
$\hat{z}_t$	Stationary time-series with non-zero mean; also the estimated-value in an adaptive time-series
z_j'	Binary variable to denote the location of a facility at j; z_j is used after y_j when there is more than one type of facility to be located
z_0^j	Amount-of-output produced at supply-facility or plant j
z_{0i}^j	Amount-of-output produced at plant j and sent to output-market i

z_j'	The optimal benefit of opening facility-j in a generalized p-median-problem (as defined in a subproblem of Lagrangian-relaxation solution)
z_{ij}	'Trunk' traffic from supply-source i to distribution-center j
z_i^j	Amount of input-i used by plant-j
z^{e_i}	Employment by the e_ith-type industry
z_i^e	Number-of-households in zone-i employed by industry
$z_{ij}^{e_i}$	Supply-of-labor by household in zone-i to zone-j for employment by the e_ith-type industry
z_L^i	Lower-bound of objective-function-value at iteration-i
z_U^i	Upper-bound of objective-function-value at iteration-i
z'	Lower or upper bound of objective-function-value
z_{ij}	Binary indicator-variable to show whether a multiattribute observation $\mathbf{x} = (x_1, x_2, \ldots)^T$ for a pixel of color j has been properly classified into group i; $z_{ij}=1$ when it is properly classified into group i (or $i=j$) and $z_{ij}=0$ when it is improperly classified ($j̸$). In vector notation for two groups i and j, we write $\mathbf{z}_i = (z_{ii}, z_{ij})^T = (1,0)^T$; and the random variable corresponding to $\mathbf{z}_i = (z_{ii}, z_{ij})^T$ is $\tilde{\mathbf{z}}_i = (\tilde{z}_{ii}, \tilde{z}_{ij})^T$.
z_{ij}'	Impedance between zones i and j
z_{LD}	Objective-function-value of a Lagrangian-dual
z_{LP}	Objective-function-value for a linear-program relaxation
z_{LR}	Objective-function-value for a Lagrangian-relaxation problem
$\dot{z}_i$	Goods in storage at location-i
$\mathbf{z}$	vector of $\mathbf{Z}$ values induced for stationary and with mean set to zero; also stands for endogenous variables in an econometric model
$\mathbf{z}_j$	Vector-of-pixels $\mathbf{z}$ for group j in a Bayesian classifier
Z	Activity level (where the activity can be population, employment, grey values, or any economic or non-economic activity)
$Z(i)$	Expected-value of the decision made at state-i
$Z'(i)$	Expected-value of the improved-decision made at state-i according to Howard's policy-iteration
Z_j	Objective-function value or activity level at location-j
Z_t	Raw-data time-series before inducing stationarity
Z_t'	Actual, accurate daa in a Kalman-filter time series (to be differentiated from what is observable)
$\dot{Z}_t, \ddot{Z}_t$	First and second differencing of time-series Z_t
Z'	Preference structure in multi-criteria decision-making
Z''	Deviation-measures from the efficient-contour of unity in the Minkowski distance-function
Z_{ij}	Value of spatial-data at grid-point $i - j$; often simplified to read Z_j to stand for the spatial-data value at location-j
Z_j^l	Value of the jth spatial-data at spatial-lag l
Z_+^n	n-dimensional Euclidean-space of positive discrete-values
$\mathbf{Z} = (\leftarrow Z_i \rightarrow)$	Vector of exogenous-variables Z_i of such activities as population and employment in each zone-i; $\mathbf{Z}_0$ is the initial-values of $\mathbf{Z}$
$\mathbf{Z}(t)$	Density or relative-frequency of the state-vector $\mathbf{X}(t)$; in other words, the normalized state-vector
$\mathbf{Z}_j = (\leftarrow Z_{ji} \rightarrow)$	The jth-activity assigned to zone-i
$\mathbf{Z}^i$	Vector of the *total*-population/employment activity-levels at time-period (iteration) i, with $\mathbf{Z}^0$ as the given final-period *basic-activities* (from which other activities are generated)